Photography by J. Gerard Smith

WAYNE GISSLEN

PROFESSIONAL EIGHTH EDITION
COOKING

FOR CANADIAN CHEFS

WILEY

This book is dedicated to
the many Chef-Instructors
preparing a new generation of culinary professionals.

SENIOR ACQUISITIONS EDITOR	Mary J. Cassells
ACQUISITIONS EDITOR	Darren Lalonde
SENIOR DEVELOPMENTAL EDITOR	Julie R. Kerr
PROJECT EDITOR	Andrea Brescia
EDITORIAL ASSISTANT	Jenni Lee
EDITORIAL ASSISTANT	Helen Seachrist
SENIOR PRODUCTION EDITOR	William A. Murray
CONTENT MANAGER	Micheline Frederick
MEDIA SPECIALIST	James Metzger
PRODUCT DESIGNERS	Beth Tripmacher and Lydia Cheng
EDITORIAL OPERATIONS MANAGER	Melissa Edwards
MEDIA CONSULTANT	Lynne Marsala
MARKETING MANAGER	Suzanne Bochet
MARKETING MANAGER	Aida Krneta
COVER AND INTERIOR DESIGNER	Maureen Eide
COVER AND INTERIOR PHOTOGRAPHY	J. Gerard Smith

This book was set in 9/12 MyriadPro by cMPreparé and Manufactured in the United States by RR Donnelley.

This book is printed on acid free paper. ∞
Copyright © 2015, 2011, 2007 by John Wiley & Sons, Inc. All rights reserved.
Published by John Wiley & Sons, Inc., Hoboken, New Jersey.
Published simultaneously in Canada.

For general information on our other products and services, or technical support, please contact our Customer Care Department within the United States at 800-762-2974, outside the United States at 317-572-3993 or fax 317-572-4002.

Wiley also publishes its books in a variety of electronic formats. Some content that appears in print may not be available in electronic books. For more information about Wiley products, visit our web site at www.wiley.com.

Evaluation copies are provided to qualified academics and professionals for review purposes only, for use in their courses during the next academic year. These copies are licensed and may not be sold or transferred to a third party. Upon completion of the review period, please return the evaluation copy to Wiley. Return instructions and a free of charge shipping label are available at www.wiley.com/go/returnlabel. Outside of the United States, please contact your local representative.

ISBN-13   978-1-118-63660-2

Printed in the United States of America

10 9 8 7 6 5 4

CONTENTS

PREFACE

The **Eighth Edition** of **Professional Cooking for Canadian Chefs** reflects the changing nature of our understanding of cooking and related fields such as food safety, nutrition, and dietary practices, as well as new thinking about how best to teach this material. What has not changed is the core material that focuses on the essentials—the comprehensive understanding of ingredients and basic cooking techniques that are the foundation of success in the kitchen, and the development of manual skills to apply this knowledge.

WHAT'S NEW

The **Eighth Edition** of **Professional Cooking for Canadian Chefs** contains a great deal of new and revised material. Among the most important changes are:

- 125 new images detailing cooking procedures and modern plating styles.

- A new chapter, Chapter 14, *combines common cooking methods for meats, poultry, and fish* to sharpen the focus on basic techniques and procedures. It also provides background, guidelines, and standards of quality for finished dishes. Specific methods unique to these products are detailed in later chapters.

- Further reorganization repositions the chapters on vegetable cookery *ahead* of meats and fish—giving the text a smoother, more logical flow.

- Chapter 29 on food presentation and plating is rewritten and newly illustrated for designing modern platings, appropriate for all types of food-service operations.

- Yields are reduced in most large-quantity recipes to no more than 8–12 portions.

- Food science topics get expanded treatment, with more detail on caramelization, gelatinization, dextrinization, and emulsification.

- Sous vide methods are enhanced by a larger recipe collection.

- Techniques of modernist cuisine, or molecular gastronomy, are recognized as increasingly important with an expanded recipe library. Hydrocolloids are incorporated in Chapter 8 along with conventional thickeners for sauces.

- Nutritional information is updated with *Eating Well with Canada's Food Guide*.

- Developing vegetarian menus receives more attention and updates.

- Increasing use of combi ovens in today's restaurants is supported with new combi recipe variations.

- Basic cooking methods are supplemented with detailed guidelines for evaluating the quality of finished products.

THE RECIPES

The recipes are planned and organized to reinforce the basic skills being taught. In each case, specific recipes follow theories, guidelines, and general procedures applicable to a defined category of foods and/or cooking methods. Recipe variations encourage students to apply these procedures to other ingredients, and to see the similarities and differences among preparations.

Attention to the basics is the hallmark of this text. Because the purpose is to teach fundamental cooking techniques, it is important to illustrate them—and allow the student to experience them—with fundamental, straightforward recipes that reveal the connection between general theory and specific application. Many new recipes provide updates and alternatives that will achieve these goals.

Core recipes, usually those that directly follow a procedure, are chosen as clear and direct applications of a fundamental technique. These recipes help students learn the technique by applying it in the most straightforward way, without the distraction of unusual techniques or unfamiliar ingredients. Of the more than 1,400 **core recipes** recipes in the book (and additional recipes on **CulinarE-Companion**), about 70 are designated as core recipes and are indicated by this icon.

The book builds on these primary techniques to more advanced styles of preparation. More challenging recipes, including many new to this edition, enable students to refine their techniques and prepare dishes of increasing sophistication.

International and regional cuisines play important roles in the evolution of cooking in North America, and the text reflects that importance with approximately 130 international recipes. These recipes **international** are indicated by a globe. Sidebars throughout **recipes** give background information on these recipes and the cuisines and cultures they come from.

Even more than with previous editions, important information devoted to vegetarianism and cooking for the various types of vegetarian diets is included. Approximately 475 recipes suitable for a vegetarian **vegetarian** ian diet are indicated by the symbol shown here. **recipes**

What makes a dish feel modern is as much a matter of presentation as of ingredients or recipe instructions. How an item, along with garnish and sauce, is plated can make it look rustic or elegant, traditional or modern. Photographs accompanying the recipes illustrate a variety of preparations and plating styles.

Readers are urged to study Chapter 4, "Menus, Recipes, and Cost Management," before proceeding with the recipes. This will ensure they know how to use the recipes in this book as well as understand the structure and limitations of the many recipes they will use in their careers.

While every culinary program has different requirements, the recipes in this book are adaptable to any purpose. New to this *Eighth Edition* is that most major recipes are now written for *12 or fewer portions,* quantities that can be converted easily to higher or lower yields, either by hand or with the accompanying *CulinarE-Companion™* Recipe Management Software. Those recipes requiring more costly ingredients, made to order, or that are particularly complex have smaller yields. In addition, variations often indicate ingredient substitutions so the recipes will fit different budgetary requirements and local or regional tastes.

Nutritional Information

Cooks and chefs are increasingly aware of the importance of preparing healthful foods. To support this, nutritional analyses are included for each main recipe. These analyses were done using the software program Genesis R&D 8.4.0, which calculates nutrients based on ingredients. It is important to realize that the actual nutrients in a prepared dish will vary depending on many factors. The following should be taken into account when reading the nutritional analyses:

- Where a portion size is indicated, the analysis is per portion.

- Where there is no portion size, as for stock and sauce recipes and most recipes in the baking chapters, the analysis is usually per ounce (28.35 g) or per fluid ounce (29.57 mL); for most hors d'oeuvre recipes, analysis is per piece.

- Not included in the analyses are ingredients listed: "to taste" or "as needed"; ingredients in sachets and bouquets-garnis; optional ingredients; garnishes such as parsley sprigs.

- Stocks are adjusted for removal of bones, mirepoix, and other strained out ingredients.

- Mirepoix ingredients are not included, except for a small amount of sodium.

- If a quantity range is given for an ingredient, the smaller number was used.

- Adjustments are made for recipes in which food is degreased or fat is skimmed off. The amount of fat remaining will vary depending on how thoroughly the item is degreased.

- Fat was calculated for pan-fried and deep-fried foods based on a percentage of the total weight. The amount of fat actually absorbed will vary depending on the temperature of the fat, the cooking time, and the surface area of the food.

- For marinated foods, 10 percent of the marinade is included in the analysis, unless the marinade is used to make a sauce, in which case all the marinade is included.

- The amount of fat used for sautéing was estimated for the analysis.

- The numbers for each nutrient are rounded according to FDA rounding rules for food labeling.

- The "(% cal.)" information following the fat content in each analysis refers to percentage of calories from fat, and is required to determine whether a recipe can be labeled as low in fat. It can't be used to determine percentage of fat in the total diet.

For more awareness of the fat content of prepared foods, dishes especially low in fat are designated by the heart symbol. *Low in fat* means, according to FDA labeling laws, that the food contains 3 grams of fat or less per reference amount (or serving size indicated in the analysis) if the reference amount is greater than 30 grams (about 1 ounce). This is to prevent making foods sound low in fat just by making the portion size smaller. Main-dish items and meals (weighing at least 6 ounces per serving and containing two or more from the four food groups—bread,

low in fat recipes

cereal, rice and pasta; fruits and vegetables; milk, yogurt and cheese; or meat, poultry, fish, dry beans, eggs, and nuts) must contain 3 grams of fat or less per 100 grams and not more than 30 percent of calories from fat.

Goals and Organization of This Book

This book has a dual goal: *understanding*—that is, of cooking theory, of how to cook—and *performing*—that is, mastery of a set of manual skills and the ability to apply them to a wide range of cooking styles and products.

This *Eighth Edition* retains the book's basic approach, but sharpens the focus on the fundamentals with its new organization. The basic cooking methods (dry-heat methods, moist-heat methods, and so on) are introduced early. Then, within the main cooking chapters, the material is arranged by cooking method. In addition, the new Chapter 14, "Cooking Methods for Meat, Poultry, and Fish", focuses directly on the common cooking methods for these products, allowing the student to master fundamental techniques, before exploring variations on these procedures unique to individual products in later chapters.

Professional Cooking for Canadian Chefs focuses on the development of flexible skills, which are essential for a successful cooking career. The graduate who understands the workings of foods and the interplay of ingredients, cooking methods, cost factors, and other elements can function successfully in any type of food-service operation.

The Role of the Chef-Instructor

No book can substitute for practical kitchen experience. Nor can a book replace an experienced chef-instructor. Although this book presents methods and recipes that are widely used and accepted, many instructors will prefer procedures that differ and may wish to supplement the recipes with their own.

FEATURES

Pronunciation Guides and Glossaries

Phonetic guides are included for difficult words, giving the approximate pronunciation using English sounds. Definitions of terms introduced in the text are summarized in the Glossary.

Illustrations

Hundreds of full-color photographs, including 125 new to this edition, illustrate basic manual techniques *shown from the point of view of the person performing them.* Additional photographs illustrate ingredients and finished dishes. Numerous line drawings also enhance the text, illustrating hundreds of pieces of equipment you'll encounter in the professional kitchen.

Format

The very readable format emphasizes and highlights key points in bold type, italics, and numbered sequences, so key information can be located and reviewed at a glance.

Realistic Procedures

Though supported by cooking theory, procedures here are based on actual industry practice. Attention is given to quantity production, but also the special problem of cooking to order. Presentation and service of finished product are considered in detail, as is pre-preparation, or mise en place—so essential to the organization of a working restaurant. At the same time, the major emphasis is on quality, too often neglected in the quest for convenience. Finally, although much of what we talk about is strongly influenced by the cooking of other nations, the practices discussed are primarily those of North American food service.

DIGITAL RESOURCES AND SUPPLEMENTS

CulinarE-Companion™ Recipe Management Software

CulinarE-Companion™ is a web-based database of recipes from *Professional Cooking for Canadian Chefs.* You can set up an account and have instant access to the software, *viewable from any device's browser, whether a laptop, desktop, tablet, or mobile device.*

In addition to the recipes from the book and additional bonus recipes, the software includes a range of useful features. The registration code included with each copy of *Professional Cooking, Eighth Edition,* allows you to access this valuable asset at no additional cost—and your account does not expire so it can be used throughout your professional career.

Feature Highlights

- **Enhanced Recipe Management Tools:** Edit, scale, view nutritional information, convert from U.S. to metric measures and vice versa; print and share recipes. Users can also add their own recipes and create and revise shopping lists.

- **Search recipes** by main ingredient, primary cooking method, and cuisine type.

- **Calculate nutritional analyses** and update if an ingredient is changed.

- **My Files:** Organize your recipes, your images, and your videos in one location.

- **Audio Pronunciations:** Within the extensive glossary, *CulinarE-Companion™* has over 1,000 terms with audio pronunciations to make learning a snap.

- **Food Costing:** Calculate food costs based on each ingredient's individual cost.

- **Unit Conversions:** Scale recipes and units of measures are converted to the next logical unit.

- **Adding New Ingredients:** Add new ingredients that do not exist in *CulinarE-Companion™* and they are automatically added into the ingredient database.

- **Nutritional Analysis:** Add ingredients to a recipe that do not have nutritional information and select from an existing list of ingredients with possible USDA matches so nutritional analysis is complete.

WILEYPLUS

WileyPLUS is an online teaching and learning environment used by educational programs. It integrates the entire digital textbook with the most effective instructor and student resources, fitting every learning style. For instructors, help your students become kitchen ready...or flip your classroom for more engaged learning.

With **WileyPLUS**:

- Students achieve concept mastery in a rich, structured environment available 24/7.

- Instructors manage their course more effectively with assessment, assignments, grade tracking, and more.

For Students

Different learning styles, different levels of proficiency, different levels of preparation—each student is unique. **WileyPLUS** empowers them to take advantage of individual strengths with:

- Timely access to resources that address demonstrated needs, with immediate feedback and remediation.

- Integrated multimedia resources—including Math Tutor, audio pronunciations, technique videos, visual learning activities, flashcards with audio pronunciations and quizzes, and much more—for multiple study paths to fit each student's learning preferences and encourage more active learning.

- Instant feedback and context-sensitive help. Students take control of their learning and practice until they master the material.

For Instructors

WileyPLUS supports you with tools and resources for even more effective teaching. You can:

- Customize presentations with a wealth of resources and functionality. You can even add your own materials to your **WileyPLUS** course.

- Identify students who are falling behind and intervene accordingly, without waiting for them to come to office hours.
- Take advantage of the program's simplification and automation of such tasks as making assignments, scoring student work, managing your gradebook, and more.

TECHNIQUE VIDEOS

Nearly 200 technique videos that clearly demonstrate essential kitchen skills are available as part of your **WileyPLUS** course. They can be used for study prior to class or review afterward, or as a step-by-step demonstration before lab.

MATH TUTOR

Integrated within **WileyPLUS, Math Tutor** white-board type tutorials demonstrate common math used in the kitchen by walking through examples and calculating results. In addition, practice exercises are included for students to apply and reinforce these skills. Math Tutor is available in U.S. and metric versions.

ADDITIONAL STUDENT AND INSTRUCTOR RESOURCES

The following student and instructor supplements are also available:

Student Study Guide (978-1-118-63655-8) contains review materials, practice problems, and exercises. (Answers are found in the *Instructor's Manual*.)

Online *Instructor's Manual with Study Guide Solutions* includes teaching suggestions and test bank questions and is available to qualified adopters from this book's web site at www.wiley.com/college/gisslen. Instructors who adopt *Professional Cooking for Canadian Chefs* can download the Respondus test bank free of charge.

PowerPoint slides and an image gallery are also available to provide additional support in delivering course material.

ACKNOWLEDGMENTS

During a long and productive session of planning, photography, and recipe testing for this edition, I was fortunate to have the expert assistance of Chef Tim Bucci of Joliet Junior College. Tim is a master of modern culinary technique and the creation of artistic food presentations. Many of the plates that enhance the new photos in this edition are his creations, and many of the new recipes are also his, including those employing sous vide techniques and avant-garde ingredients. I am grateful for Tim's participation, as well as his friendship.

Chef James Feustel of Kingsborough Community College was generous with his time and talents to develop combi oven variations for a number of the book's recipes. I thank him and his students for this work. Additionally, James created new Math Tutor tutorials for WileyPLUS, which provide additional kitchen math practice to help students master this critical skill.

Our photography sessions once again benefited from the contributions of time and talent by Chef Rick Forpahl of Minneapolis Community & Technical College, as well as Eric Ervasti and Choden Bhutia, students of Chef David Eisenreich of Hennepin Technical College. These chefs have been valuable collaborators

on several editions, and have been significant in helping shape each revision. In addition, Rick Elsenpeter of Lund's market was untiring in response to innumerable special requests. My wife, Meg, assisted our kitchen crew and helped with my research. I am grateful to one and all for their help. They were a pleasure to work with.

Photographer Jim Smith has been my partner in these texts for nearly 30 years. His hundreds of photographs are an indispensable part of *Professional Cooking for Canadian Chefs* and valuable teaching tools. I can never thank him enough. Thanks also to Michael Haight for his work in Jim's studio and on the set in my kitchen.

The technique videos in **WileyPLUS** could not have been accomplished so successfully without the on-air talent of Chef Ambarush Lulay, Chef Klaus Tenbergen, Chef Melina Kelson, and most especially, Chef Lisa Brefere and Chef Andy Chlebana. Both Lisa and Andy played an incalculable role in scripting, planning, executing, and ensuring each video meets professional kitchen standards. Many thanks to Kendall College and the College of DuPage for the gracious use of their kitchens in the filming of many of the technique videos.

Christin Loudon has again contributed her expertise to provide analyses for the new recipes in this edition, for which I thank her most warmly. I would also like to thank Drew Appleby, whose expertly written test questions form an important part of the support materials.

The list of culinary and hospitality professionals who have provided support, guidance, advice, and constructive criticism for all eight editions of this book has grown so long that I can only hope I have not omitted many in the list of reviewers that follows.

The updated and enhanced *CulinarE-Companion™* that accompanies this *Eighth Edition* is the result of a coordinated team effort. Thank you to Lydia Cheng, Beth Tripmacher, and Lynne Marsala for their consultation in reviewing, conceptualizing, and coordinating the development of the web-based recipe management software. Thanks also to Chef Jean Vendeville of Savannah Technical College for his review of the audio pronunciations.

Thank you to Danielle Gleason, Chef, Sullivan University, for creating the PowerPoints for the *Eighth Edition*.

Finally, I would like to thank everyone at John Wiley & Sons who worked so hard on this project: Jenni Lee, Helen Seachrist, James Metzger, Andrea Brescia, Jeff Rucker, Suzanne Bochet, Bill Murray, Micheline Frederick, Maureen Eide, and Harry Nolan. Thanks to Cheryl Ferguson, who copyedited the manuscript. Special thanks to Julie Kerr, who has worked with me for so long and with such dedication, and to my editor and friend, Mary Cassells.

CONTRIBUTORS

I would like to thank the following contributors and reviewers for their valuable input and feedback for the *Eighth Edition* of *Professional Cooking for Canadian Chefs*:

Micheal Mcfadden	**Richard Hendy**
Humber College	*St. Lawrence College*
Mario Ramsey	**Leslie Leacy**
Algonquin College	*St. Lawrence College*
Claude AuCoin	**Scott Warrick**
Nova Scotia Community College	*Algonquin College*
Jason Inniss	**David Fairbanks**
George Brown College	*Algonquin College*

Allen Clampitt
Lethbridge Community College

Gilbert Noussitou
Camosun College

Derek Lawday
Confederation College

Peter Storm
Niagara College

Steve Price
Algonquin College

David Keindel
Algonquin College

Scot Woods
George Brown College

REVIEWERS

I would like to acknowledge the following instructors who have contributed to this book over eight editions by suggesting revisions and additions and by answering survey questions:

George Akau
Clark College

George Allen Akmon
Sullivan University

Erik Anderson
Camosun College

Angela M. Anderson
Miami Dade Community College

Robert Anderson
Des Moines Area Community College

Tim Appleton
Red River College

Alan Argulski
Genesee Community College

Valerie Arthurs
Cambrian College of Applied Arts and Technology

Allen Asch
University of Nevada–Las Vegas

Kirk Bachmann
Le Cordon Bleu Schools North America

Moses Ball
Atlantic Vocational-Technical Center

Charles Becker
Pueblo Community College

Carl A. Behnke
Purdue University

Joseph L. Belvedere Jr.
Paul Smith's College

Perry Bentley
Okanagan University College

Claire A. Berg
Brookdale Community College

Nancy Berkoff
Los Angeles Trade Technical College

Leslie Bilderback
California School of Culinary Arts

George Bissonette
Pikes Peak Community College

Peter Blakeman
George Brown College

LeRoy Blanchard
Los Angeles Trade Technical College

Roy Blundell
College Lasalle

Pete Bordi
Penn State University

Kenneth P. Bourgoin
Valencia Community College

Jeff Bunting
Ivy Tech Community College

Roy Butterworth
New Brunswick Community College

Alfredo Cabaccingan
Kapiolani Community College

Joanne Canada-Somers
Assiniboine Community College

Harska Chacko
University of New Orleans

Daniel Charna
Ashland College

Richard Chiasson
New Brunswick Community College

Albert D. Cipryk
Niagara College

Scott Clapp
Central New Mexico Community College

Rob Cleland
Liaison College

Jesse Clemons
State Technical Institute at Memphis

Alec O. Cline
American Culinary Federation, Inc.

JoAnne Cloughly
State University of New York at Cobleskill

Michael M. Collins
San Jacinto College

Randall Colman
Sullivan County Community College

Paulette Cormier-MacBurnie
Mount Saint Vincent University

Steven Cornelius
Sinclair Community College

Michael Cortese

Mike Costello
St. Cloud Area Vocational-Technical Institute

Tim Curnow
College of the Rockies

Chris Cutler
Georgian College

Gerardo Jesi D'Amore
St. Pius X Culinary Institute

William J. Daly
State University of New York at Cobleskill

Juanita M. Decker
Waukesha County Technical Institute

Bob Demers
Niagara College

Larry DeVries
Brandon School District

Frank Diamond
Academy Canada

Elizabeth Dickson
Bossier Parish Community College

Marian Dobbins

Brendan Dolan
Atlantic Culinary Academy

Rodney Donne
George Brown College

Robert Dowden
Nova Scotia Community College

James Drake
Onondaga Community College

Dominique and Cindy Duby
DC Duby School

Dale Dunham
Oklahoma State University

Michael Durrer
Algonquin College

William Easter
Des Moines Area Community College

Louis Eguaras
Los Angeles Mission College

Thomas Elia
St. Lawrence College

Michael A. Elliott
Liaison College

Thom England
Ivy Tech Community College

Evan Enowitz
Grossmont College

Daniel Esposito
Canadore College

Guy Ethier
Canadian Federation of Chefs and Cooks

Brenda Excell
Liaison College

Mark Facklam
The Cooking and Hospitality Institute of Chicago

Fred T. Faria
Johnson and Wales University

Alan Fleming
Algonquin College

Robyn L. Flipse
Brookdale Community College

Sandra Flowerday
Charlotte, MI

Sandra Foley
Gogebic Community College

Keith E. Gardiner
Guilford Technical Community College

Willard L. Geach
Long Beach City College

Joan W. Geerken
State University of New York at Cobleskill

Fred German
Saskatchewan Institute of Applied Science and Technology

Dave Gibson
Niagara College

Jeff Gill
Red River College

Gilles Godin
Nova Scotia Community College

James Goering
El Centro College

Meyer Goldhar
Liaison College

Mario Gozzi
George Brown College

Ian Grady
George Brown College

Jeff Graves
Purdue University

Michael Graves
Atlantic Culinary

Nancy Graves
University of Houston

Bill Gregorash
Confederation College

Ava R. P. Gritzuk
Central Piedmont Community College

Juliet Groux
West Liberty State College

James Hallett
Wake Technical Community College

Denise Perry Hanson
Lincoln Land Community College

Gregory Hare
Baltimore International College

Robert Harrington
Nichols State University

George Michael Harris
Mohave Community College

David Hawey
Fleming College

Robert S. Haynes
Lambton College

John D. Hedley
Los Angeles Trade-Technical College

Rosemary L. Hedlund
Des Moines Area Community College

Iris Helveston
State of Florida Department of Education

Maynard G. Hemmah
Moorhead Area Vocational-Technical Institute

Jean Hertzman
University of Nevada–Las Vegas

Andrew Hewson
Southern Alberta Institute of Technology

Margaret A. Howard
Sheridan Vocational-Technical Center

Russ Hudson
North Island College

Lynn Huffman
Texas Tech University

Anne L. Jachim
Moraine Valley Community College

Daniel K. Jeatran
Milwaukee Area Technical College

Frank F. Johnson Jr.
University of South Florida

Roosevelt Johnson
Sidney North Colver Vocational-Technical Institute

Todd Jones
Mattatuck Community College

Mike Jung
Hennepin Technical Center

Sandy Kapoor
Cal Poly Pomona

David Keindel
Algonquin College

Tom King
Cabrillo College

Darren Kreps
CSC Frontenac Institution

Jackson Lamb
*Metropolitan State College of
Denver*

Heinz Lauer
Le Cordon Bleu, Las Vegas

Sunshine Layton
Winnipeg Technical College

Suzanne Little, M.S., R.D.
San Jacinto College

Shirley Lotze
*Western Wisconsin Technical
Institute*

Robert M. Lyna
*Southern Maine Vocational-
Technical Institute*

Deborah Lynch
Middlesex Community College

George Macht
College of DuPage

Merle Maerz
Selkirk College

Sylvia Marple
University of New Hampshire

Valeria S. Mason
State Department of Education

John McDonald
St. Augustine Technical Center

Terence F. McDonough
Erie Community College

Marcia W. McDowell
Monroe Community College

Linda McDuffie
North Seattle Community College

Robert McLean
The Fay School

Deborah Midkiff
Indian River State College

Ann Miglio
*Williamsport Area Community
College*

Harold O. Mishoe
*Southern Maine Vocational-
Technical Institute*

Ken W. Myers
University of Minnesota–Crookston

Kathy Niemann
Portland State University

William T. Norvell
William Rainey Harper College

Therese O'Connor
Cornell University

Michael Olson
Niagara College

Andrew Ormiston
Brandon, Manitoba

Gary Page
Grand Valley State College

Michael Palmer
Western Culinary Institute

Philip Panzarino
New York City Technical College

Jayne Pearson
Manchester Community College

Richard Petrello
*Withlacoochee Vocational-Technical
Center*

William F. Petsch
*Pinellas Vocational Technical
Institute*

Michael PiccZinino
Shasta College

Steven Popp
Liaison College

Steve Price
Algonquin College

Tony Rechsteiner
Northern Lights College

Deborah Reid
George Brown College

John Reimers
Red River College

Larry Richardson
*Hinds Junior College–Jackson
Branch*

Neil Rittenaur
*North Dakota State College of
Science*

Hubert E. Robert
Holyoke Community College

Richard Roberts
Wake Technical Community College

Jean Guy Robichaud
Cambrian College

Michael Rohatynsky
*Saskatchewan Institute of Applied
Science and Technology*

Vance Roux
Delgado Community College

Ricardo G. Saenz
*Renton Vocational-Technical
Institute*

Jeff Santicola
*Le Cordon Bleu Schools North
America*

Robert R. Santos
Maui Community College

Philippe Saraiva
Conestoga College

Frank Schellings
Diablo Valley College

Matthew Schenk
Bradford School, Columbus

David Schneider
Macomb Community College

Charlotte Schwyn
Delta College

Elaine Seniuk
Mount Saint Vincent University

Marshall J. Shafkowitz
Le Cordon Bleu, Chicago

Roy John Sharp
Portland Community College

Chris D. Shields
*The State University of New York–
Delhi*

Settimio Sicoli
Vancouver Community College

Clarence Steadman
Hocking Technical College

Norman C. Smith
Sir Sanford Fleming College

Reuel J. Smith
*Truckee Meadows Community
College*

Jennifer Sohonie
Le Cordon Bleu, Seattle

William Sprowl
Mt. San Jacinto College

Ben Stanford
*Saskatchewan Institute of Applied
Science and Technology*

Nancy S. Steryous
*St. Petersburg Vocational-Technical
Institute*

Siegfried Stober
Joliet Junior College

Peter Storm
Niagara College

Peter Sugameli
Wayne Community College

Julia Sullivan
Copiah-Lincoln Junior College

Christopher Allen Tanner
*Montgomery County Community
College*

Christopher Thielman
College of DuPage

William Thornton
St. Phillips College

Sheila A. Tillman
*Asheville-Buncombe Technical
Community College*

S.W. (Stan) Townsend
*Northern Alberta Institute of
Technology*

Robin W. Turner
State University of New York at Delhi

Vinod Varshney
*Northern Alberta Institute of
Technology*

Ken Veneruz
Confederation College

Jerry L. Vincent
Johnson County Community College

Barbara Vredeveld
Iowa Western Community College

Don Waddell
Institute of Technology

Christine Walker
George Brown College

D. Michael Wallace
North Island College

R.G. Werth
*Asheville-Buncombe Technical
Community College*

Eberhard Werthman
St. Paul Technical College

J. William White
Pinellas County School System

Edward Whitfield
Prince George's Community College

Diane Wilkerson
Cape Fear Community College

Doris Wilkes
The Florida State University

Shelly Wilkinson-Adair
Fleming College

Wes J. Wilkinson
Lambton College

Ronald S. Wolf
Florida Community College

Ron Wong
*Northern Alberta Institute of
Technology*

Tony Wood
*Hospitality Industry Education
Advisory Committee*

Ronald Zabkiewicz
South Technical Education Center

Mike Zema
Elgin Community College

Charles Ziccardi
Drexel University

Ronald Zwerger
Elgin Community College

Culinary Media Library Reviewers

Marco Adornetto
Zane State College

Charlton Alvares
George Brown College

Belinda Brooks
Kendall College

Alan Brown
George Brown College

Deane Cobler
Columbus State Community College

Patricia Deal
Sinclair Community College

Rodney Donne
George Brown College

Collen Engle
Miami Culinary Institute

Albert I. M. Imming
Joliet Junior College

Joanne Jacus
New York City College of Technology

William Jolly
Clover Park Technical College

John Kapusta
Indiana University of Pennsylvania

Amede Lamarche
George Brown College

Jesus Lugo
El Paso Community College

Robynne Maii
Kingsborough Community College

Elaina Ravo
Liaison College

Sherry Stolfo
The Chef's Academy

Christopher Thielman
College of DuPage

Jean Yves Vendeville
Savannah Technical College

Christine Walker
George Brown College

WILEY CULINARE-COMPANION™ RECIPE MANAGEMENT SOFTWARE

Supporting chefs and foodservice managers throughout their careers, **CulinarE-Companion™** includes all recipes from **Professional Cooking for Canadian Chefs, Eighth Edition,** plus hundreds of bonus recipes, color photographs, audio pronunciations, and illustrated procedures. Create shopping lists, resize recipes, perform metric conversions, and analyze nutritional content of ingredients and recipes with the software. You can also add your own recipes and photos, link to external videos, and create your own cookbooks.

The software is now Web-based! Your personal registration code and instructions are included with your purchased copy of **Professional Cooking for Canadian Chefs, Eighth Edition**. Go to cec.wiley.com to access the software. Once you create a user name and password, you can log onto **CulinarE-Companion™** from your computer, tablet, or mobile device.

THE HOMEPAGE

- Link directly to any of the cookbooks, including cookbooks you created.

- View recipes that have been recently seen or all recipes by selecting the RECIPES tab. Perform either a basic or an advanced search based on specified criteria, such as recipe name or even part of a name, cookbook, ingredients, and cooking method.

- View recipes and procedures organized by kitchen skill when you click on the SKILLS tab.

- Select the GLOSSARY tab to access definitions from **Professional Cooking's for Canadian Chefs, Eighth Edition** glossary, as well as hundreds of additional defined terms and audio pronunciations.

RECIPE LIST

- Scroll through an alphabetical list of all recipes in the application.

- Refine the recipe listing by category, course, cuisine, main ingredient, primary cooking method, or dietary considerations.

- Add recipes to your shopping list, as well as export and print recipes.

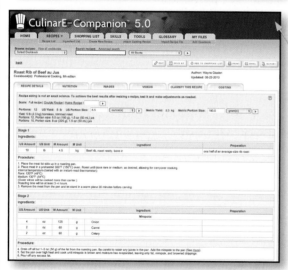

RECIPE SCREEN

- Resize recipes, perform metric conversions, show recipe notes, variations, and more!

- View referenced procedures by simply clicking on the relevant highlighted term.

- Click the IMAGES tab to see photos of plated dishes or to add your own photos and links to external videos.

COSTING INFORMATION

- Calculate food costs for a total recipe cost or a cost per portion of a recipe by selecting the COSTING tab.

- Add or edit existing cost data for individual ingredients from a shopping list or a recipe.

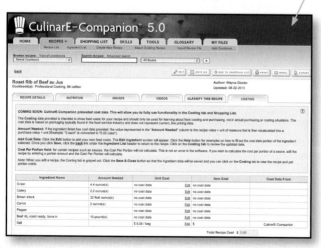

NUTRITIONAL INFORMATION

- View nutritional information for ingredients and recipes.

WileyPLUS

WileyPLUS is a research-based online environment for effective teaching and learning.

WileyPLUS builds students' confidence because it takes the guesswork out of studying by providing students with a clear roadmap:

- what to do
- how to do it
- if they did it right

It offers interactive resources along with a complete digital textbook that help students learn more. With *WileyPLUS*, students take more initiative so you'll have greater impact on their achievement in the classroom and beyond.

© Dan Lipow.

THE FOOD-SERVICE INDUSTRY

This is an exciting time to begin a career in food service. Interest in dining and curiosity about new foods are greater than ever. More new restaurants open every year. Many restaurants are busy every night, and restaurant chains number among the nation's largest corporations. The chef, once considered a domestic servant, is now respected as an artist and skilled craftsperson.

The growth of the food-service industry creates a demand for thousands of skilled people every year. Many people are attracted by a career that is challenging and exciting and, above all, provides the chance to find real satisfaction in doing a job well.

Unfortunately, many people see only the glamorous side of food service and fail to understand that this is a tiny part of the picture. The public does not often see the years of training, the long hours, and the tremendous pressures that lie behind every success.

Before you start your practical studies, covered in later chapters, it is good to know a little about the profession you are entering. This chapter gives you a brief overview of modern food service, including how it got to where it is today and where it is headed.

AFTER READING THIS CHAPTER, YOU SHOULD BE ABLE TO

1. Name and describe four major developments that significantly changed the food-service industry in the twentieth century.

2. Identify seven major stations in a classical kitchen.

3. Explain how the size and type of an operation influence the organization of the modern kitchen.

4. Identify and describe three skill levels of food production personnel.

5. Identify eight behavioral characteristics food-service workers should develop and maintain to achieve the highest standards of professionalism.

A HISTORY OF MODERN FOOD SERVICE

The value of history is that it helps us understand the present and the future. In food service, knowledge of our professional heritage helps us see why we do things as we do, how our cooking techniques have been developed and refined, and how we can continue to develop and innovate in the years ahead.

An important lesson of history is that the way we cook now is the result of the work done by countless chefs over hundreds of years. Cooking is as much science as it is art. Cooking techniques are not based on arbitrary rules some chefs made up long ago. Rather, they are based on an understanding of how different foods react when heated in various ways, when combined in various proportions, and so on. The chefs who have come before us have already done much of this work so we don't have to.

This doesn't mean there is no room for innovation and experimentation or that we should never challenge old ideas. But it does mean a lot of knowledge has been collected over the years, and we would be smart to take advantage of what has already been learned. Furthermore, how can we challenge old ideas unless we know what those old ideas are? Knowledge is the best starting point for innovation.

THE ORIGINS OF CLASSICAL AND MODERN CUISINE

Quantity cookery has existed for thousands of years, as long as there have been large groups of people to feed, such as armies. But modern food service is said to have begun shortly after the middle of the eighteenth century. At this time, food production in France was controlled by guilds. Caterers, pastry makers, roasters, and pork butchers held licenses to prepare specific items. An innkeeper, in order to serve a meal to guests, had to buy the various menu items from those operations licensed to provide them. Guests had little or no choice and simply ate what was available for that meal.

In 1765, a Parisian named Boulanger began advertising on his shop sign that he served soups, which he called **restaurants** or **restoratives**. (Literally, the word means "fortifying.") According to one version of the story, one of the dishes he served was sheep's feet in a cream sauce. The guild of stew makers challenged him in court, but Boulanger won by claiming he didn't stew the feet in the sauce but served them with the sauce. In challenging the rules of the guilds, Boulanger unwittingly changed the course of food-service history.

The new developments in food service received a great stimulus as a result of the French Revolution, beginning in 1789. Before this time, the great chefs were employed in the houses of the French nobility. With the revolution and the end of the monarchy, many chefs, suddenly out of work, opened restaurants in and around Paris to support themselves. Furthermore, the revolutionary government abolished the guilds. Restaurants and inns could serve dinners reflecting the talent and creativity of their own chefs rather than being forced to rely on licensed caterers to supply their food. At the start of the French Revolution, there were about 50 restaurants in Paris. Ten years later, there were about 500.

Another important invention that changed the organization of kitchens in the eighteenth century was the stove, or **potager**, which gave cooks a more practical and controllable heat source than an open fire. Soon commercial kitchens became divided into three departments: the rotisserie, under the control of the meat chef, or **rôtisseur**; the oven, under the control of the pastry chef, or pâtissier; and the stove, run by the cook, or **cuisinier**. The meat chef and pastry chef reported to the cuisinier, who was also known as **chef de cuisine**, which means "head of the kitchen."

CARÊME

All the changes that took place in the world of cooking during the 1700s led to, for the first time, a difference between home cooking and professional cooking. One way we can try to understand this difference is to look at the work of the greatest chef of the period following the French Revolution, **Marie-Antoine Carême** (1784–1833). As a young man, Carême learned all the branches of cooking quickly, and he dedicated his career to refining and

Marie-Antoine Carême. *L'Art de la Cuisine Française au Dix-Neuvième Siècle.* Paris: L'auteur, 1833–1844. Courtesy of the Rare Manuscript Collections, Cornell University Library.

organizing culinary techniques. His many books contain the first systematic account of cooking principles, recipes, and menu making.

At a time when the interesting advances in cooking were happening in restaurants, Carême worked as a chef to wealthy patrons, kings, and heads of state. He was perhaps the first real celebrity chef, and he became famous as the creator of elaborate, elegant display pieces and pastries, the ancestors of our modern wedding cakes, sugar sculptures, and ice and tallow carvings. But it was Carême's practical and theoretical work as an author and an inventor of recipes that was responsible, to a large extent, for bringing cooking out of the Middle Ages and into the modern period.

Carême emphasized procedure and order. His goal was to create more lightness and simplicity. The complex cuisine of the aristocracy—called *Grande Cuisine*—was still not much different from that of the Middle Ages and was anything but simple and light. Carême's efforts were a great step toward modern simplicity. The methods explained in his books were complex, but his aim was pure results. He added seasonings and other ingredients not so much to add new flavors but to highlight the flavors of the main ingredients. His sauces were designed to enhance, not cover up, the food being sauced. Carême was a thoughtful chef, and, whenever he changed a classic recipe, he was careful to explain his reasons for doing so.

Beginning with Carême, a style of cooking developed that can truly be called international, because the same principles are still used by professional cooks around the world. Older styles of cooking, as well as much of today's home cooking, are based on tradition. In other words, a cook makes a dish a certain way because that is how it always has been done. In Carême's *Grande Cuisine*, and in professional cooking ever since, a cook makes a dish a certain way because the principles and methods of cooking show it is the best way to get the desired results. For example, for hundreds of years, cooks boiled meats before roasting them on a rotisserie in front of the fire. But when chefs began thinking and experimenting rather than just accepting the tradition of boiling meat before roasting, they realized either braising the meat or roasting it from the raw state were better options.

ESCOFFIER

Georges-Auguste Escoffier (1847–1935), the greatest chef of his time, is still revered by chefs and gourmets as the father of twentieth-century cookery. His two main contributions were (1) the simplification of classical cuisine and the classical menu, and (2) the reorganization of the kitchen.

Escoffier rejected what he called the "general confusion" of the old menus, in which sheer quantity seemed to be the most important factor. Instead, he called for order and diversity and emphasized the careful selection of one or two dishes per course, dishes that followed one another harmoniously and delighted the taste with their delicacy and simplicity.

Escoffier's books and recipes are still important reference works for professional chefs. The basic cooking methods and preparations we study today are based on Escoffier's work. His book *Le Guide Culinaire*, which is still widely used, arranges recipes in a simple system based on main ingredient and cooking method, greatly simplifying the more complex system handed down from Carême. Learning classical cooking, according to Escoffier, begins with learning a relatively few basic procedures and understanding basic ingredients.

Escoffier's second major achievement, the reorganization of the kitchen, resulted in a streamlined workplace better suited to turning out the simplified dishes and menus he instituted. The system of organization he established is still in use, especially in large hotels and full-service restaurants, as we discuss later in this chapter.

Georges-Auguste Escoffier. Courtesy of Adjointe à la Conservation du Musée Escoffier de l'Art Culinaire.

MODERN TECHNOLOGY

Today's kitchens look much different from those of Escoffier's day, even though our basic cooking principles are the same. Also, the dishes we eat have gradually changed due to the innovations and creativity of modern chefs. The process of simplification and refinement, to which Carême and Escoffier made monumental contributions, is ongoing, adapting classical cooking to modern conditions and tastes.

TWO IMPORTANT COOKBOOKS

In the Middle Ages, cooking consisted mostly of roasting meats on spits in front of a fire and suspending pots from hooks over the fire. Ovens, which were used in ancient Rome, had disappeared, so there was no baking. Roasted meats and poultry were usually boiled before being placed on the spit, and most foods were heavily spiced. It wasn't until the thirteenth century that ovens were used again and that stews and sauces started to appear on the dining table.

Perhaps the first important cookbook to appear at the end of the Middle Ages was *Le Viandier* ("The Cook"), by Guillaume Tirel, usually known as Taillevent, born about 1310.

Taillevent invented many dishes, especially sauces and soups. He refined old recipes to depend less on heavy use of spices and more on the flavors of the foods themselves. He wrote his book before the invention of the printing press, and handwritten copies of it remained in use for more than a century, until 1490, when it became perhaps the first cookbook ever printed.

By the seventeenth century, cooking practices still had not advanced much beyond Taillevent's day. Perhaps the next most important cookbook after Taillevent's was *Le Cuisinier François* ("The French Chef"), by François-Pierre de La Varenne (1615–1678). This book, published in 1651, was a summary of the cooking practices in households of the aristocracy. It was one of the first books to present recipes and cooking techniques in an orderly fashion rather than as an unsystematic collection. *Le Cuisinier François* was one of the main reference works for cooks for more than 150 years.

These two chefs are memorialized in the names of two important culinary institutions. Taillevent is the name of a Paris restaurant that has long been one of the finest in France, and La Varenne is the name of a distinguished cooking school.

Before we discuss the changes in cooking styles that took place in the twentieth century, let's look at some of the developments in technology that affected cooking.

Development of New Equipment

We take for granted such basic equipment as gas and electric ranges and ovens and electric refrigerators. But even these essential tools did not exist until fairly recently. The easily controlled heat of modern cooking equipment, as well as motorized food cutters, mixers, and other processing equipment, has greatly simplified food production.

Research and technology continue to produce sophisticated tools for the kitchen. Some of these products, such as tilting skillets and steam-jacketed kettles, can do many jobs and are popular in many kitchens. Others can perform specialized tasks rapidly and efficiently, but their usefulness depends on volume because they are designed to do only a few jobs.

Modern equipment has enabled many food-service operations to change their production methods. With sophisticated cooling, freezing, and heating equipment, it is possible to prepare some foods further in advance and in larger quantities. Some large multiunit operations prepare food for all their units in a central commissary. The food is prepared in quantity, packaged, chilled or frozen, and then heated or cooked to order in the individual units.

Development and Availability of New Food Products

Modern refrigeration and rapid transportation caused revolutionary changes in eating habits. For the first time, fresh foods of all kinds—meats, fish, vegetables, and fruits—became available throughout the year. Exotic delicacies can now be shipped from anywhere in the world and arrive fresh and in peak condition.

The development of preservation techniques—not just refrigeration but also freezing, canning, freeze-drying, vacuum-packing, and irradiation—increased the availability of most foods and made affordable some that were once rare and expensive.

Techniques of food preservation have had another effect. It is now possible to do some or most of the preparation and processing of foods before shipping rather than in the food-service operation itself. Thus, convenience foods have come into being. Convenience foods continue to account for an increasing share of the total food market.

Some developments in food science and agriculture are controversial. Irradiation, mentioned above, caused much controversy when it was introduced because it exposes foods to radioactivity to rid them of organisms that cause spoilage and disease. Scientists say, however, that no traces of radioactivity remain in the foods, and the procedure is now used more widely.

A more controversial technique is genetic engineering, which involves artificially changing the gene structure of a food to give it some desirable trait, such as resistance to disease, drought, or insect damage.

Food Safety and Nutritional Awareness

The development of the sciences of microbiology and nutrition had a great impact on food service. One hundred years ago, there was little understanding of the causes of food poisoning and food spoilage. Food-handling practices have come a long way since Escoffier's day.

Also, little knowledge of nutritional principles was available until fairly recently. Today, nutrition is an important part of a cook's training. Customers are also more knowledgeable and therefore more likely to demand healthful, well-balanced menus. Unfortunately, nutrition science is constantly shifting. Diets considered healthful one year become eating patterns to be avoided a few years later. Fad diets come and go, and chefs often struggle to keep their menus current. It is more important than ever for cooks to keep up to date with the latest nutritional understanding.

Complicating the work of food-service professionals is a growing awareness of food allergies and intolerances. Not only are chefs called upon to provide nutritious, low-fat, low-calorie meals, they must also adapt to the needs of customers who must eliminate certain foods from their diets, such as gluten, soy, dairy, or eggs.

COOKING IN THE TWENTIETH AND TWENTY-FIRST CENTURIES

All these developments have helped change cooking styles, menus, and eating habits. The evolution of cuisine that has been going on for hundreds of years continues. Changes occur not only because of technological developments, such as those just described, but also because of our reactions to culinary traditions.

Two opposing forces can be seen at work throughout the history of cooking. One is the urge to simplify, to eliminate complexity and ornamentation, and instead to emphasize the plain, natural tastes of basic, fresh ingredients. The other is the urge to invent, to highlight the creativity of the chef, with an accent on fancier, more complicated presentations and procedures. Both these forces are valid and healthy; they continually refresh and renew the art of cooking.

A generation after Escoffier, the most influential chef in the middle of the twentieth century was Fernand Point (1897–1955). Working quietly and steadily in his restaurant, La Pyramide, in Vienne, France, Point simplified and lightened classical cuisine. He was a perfectionist who sometimes worked on a dish for years before he felt it was good enough to put on his menu. "I am not hard to please," he said. "I'm satisfied with the very best." Point insisted every meal should be "a little marvel."

Point's influence extended well beyond his own life. Many of his apprentices, including Paul Bocuse, Jean and Pierre Troisgros, and Alain Chapel, later became some of the greatest stars of modern cooking. They, along with other chefs in their generation, became best known in the 1960s and early 1970s for a style of cooking called **nouvelle cuisine**. Reacting to what they saw as a heavy, stodgy, overly complicated classical cuisine, these chefs took Point's lighter approach even further. They rejected many traditional principles, such as the use of flour to thicken sauces, and instead urged simpler, more natural flavors and preparations, with lighter sauces and seasonings and shorter cooking times. In traditional classical cuisine, many dishes were plated in the dining room by waiters. Nouvelle cuisine, however, placed a great deal of emphasis on artful plating presentations done by the chef in the kitchen.

Very quickly, however, this "simpler" style became extravagant and complicated, famous for strange combinations of foods and fussy, ornate arrangements and designs. By the 1980s, nouvelle cuisine was the subject of jokes. Still, the best achievements of nouvelle cuisine have taken a permanent place in the classical tradition. Meanwhile, many of its excesses have been forgotten. It is probably fair to say that most of the best new ideas and the longest-lasting accomplishments are those of classically trained chefs with a solid grounding in the basics.

New Emphasis on Ingredients

Advances in agriculture and food preservation have had disadvantages as well as advantages. Everyone is familiar with hard, tasteless fruits and vegetables developed to ship well and last long, without regard for eating quality. Many people, including chefs, began to question not only the flavor but also the health value and the environmental effects of genetically engineered foods, of produce raised with chemical pesticides and fertilizers, and of animals raised with antibiotics and other drugs and hormones.

A prominent organization dedicated to improving food quality is Slow Food, begun in Italy in 1986 in reaction to the spread of fast-food restaurants. Slow Food has since become a global movement, with chapters in cities around the world. It emphasizes fostering locally grown food, using organic and sustainable farming practices, preserving heirloom varieties of plants and animals, and educating consumers about the food they eat.

A landmark event in the history of modern North American cooking was the opening of Alice Waters's restaurant, Chez Panisse, in Berkeley, California, in 1971. Waters's philosophy is that good food depends on good ingredients, so she set about finding dependable sources of the best-quality

SLOW FOOD TODAY

As with any movement, the growth of Slow Food has not been without controversy. For more than 20 years after its founding, Slow Food had little impact in North America, finding greater popularity in Europe. As recently as 2008, the organization had only 16,000 members in the United States out of more than 100,000 in all.

In its earlier years, the movement was sometimes criticized for elitism and snobbishness, for focusing primarily on pleasure, and for being against technology and globalization. Detractors said opposition to global food trade and rejection of industrial agricultural practices are unrealistic in today's world.

In recent years, however, Slow Food has expanded its focus and has addressed issues of race, poverty, and hunger as well as its more traditional concerns—the disappearance of local food traditions and people's dwindling interest in the food they eat, where it comes from, and how it tastes. A surge of interest in Slow Food has come at the same time as growing concerns about the environment and climate change. Members encourage sustainable, ecologically sound agriculture and stewardship of the land as part of their efforts to educate people about their food and their eating habits.

vegetables, fruits, and meats, and preparing them in the simplest ways. Over the next decades, many chefs and restaurateurs followed her lead, seeking out the best seasonal, locally grown, organically raised food products. A few years after Chez Panisse opened, Larry Forgione picked up the banner of local ingredients and local cuisine in his New York City restaurant, An American Place. Other chefs quickly followed suit, and soon chefs across the continent made names for themselves and their restaurants at least in part by emphasizing good-quality local ingredients. Half a century ago, nearly all the most respected chefs working in the United States and Canada were European-born. Today, the movement begun by the pioneering, quality-oriented chefs of the 1970s and 1980s has fostered a great number of creative North American–born chefs who are among the most respected in the world.

Concern for quality of ingredients has led many chefs to support and to purchase from farmers who practice **sustainable agriculture**. This term refers to methods of raising healthful food in a way that is profitable to farms and farming communities and that provides living wages and benefits to workers while at the same time preserving and enhancing the soil, water, and air. Sustainable farming treats workers justly and raises animals in humane conditions. Farmers continually work to increase the fertility and conservation of soil and avoid the use of synthetic pesticides and herbicides as much as possible. The goal is to manage farmlands so that they not only will be profitable but will continue to be productive indefinitely.

Chefs can carry the concept of sustainability into their own operations by using renewable power sources, installing energy-efficient equipment, and recycling as many waste materials as possible.

The public has benefited greatly from these efforts. Today, in supermarkets as well as in restaurants, a much greater variety of high-quality foods is available than there was 40 or 50 years ago. Many chefs have modified their cooking styles to highlight the natural flavors and textures of their ingredients, and their menus are often simpler now for this reason.

International Influences

After the middle of the twentieth century, as travel became easier and as new waves of immigrants arrived in Europe and North America from around the world, awareness of and taste for regional dishes grew. Chefs became more knowledgeable not only about the traditional cuisines of other parts of Europe but about those of Asia, Latin America, and elsewhere. Many of the most creative chefs have been inspired by these cuisines and use some of their techniques and ingredients. For example, many North American and French chefs, looking for ways to make their cooking lighter and more elegant, have found ideas in the cuisine of

CATERINA DE MEDICI

The Medicis were a powerful Italian family that ruled Florence from the fourteenth to the sixteenth century and provided, in addition to the rulers of Florence, three popes and two queens of France.

Until recently, the accepted and often-told story is that when Caterina de Medici went to France in 1533 to marry the future King Henry II, she brought with her a staff of cooks as part of her household. This introduction of Italian cooking practices into France supposedly changed and modernized the cooking not only of France but of all of Western Europe. According to this story, Caterina and her Italian cooks should be credited with fostering modern cuisine.

When cookbooks and other culinary writings of the period are examined, however, it appears that French cooking didn't begin to modernize until at least a century later. During the hundred years after Caterina's arrival in France, no new, important cookbooks were written. There is no sign of a revolution in cooking. In fact, banquet menus that survive from the period are not much different from menus of the Middle Ages.

Banquets during the Middle Ages were like huge, sit-down buffets. For each course, the table was loaded with large quantities of meats, poultry, and fish dishes, usually heavily spiced, and an assortment of side dishes and sweets. Diners generally ate only what they could reach. The course was then removed and another course, also meats and side dishes, was loaded onto the table. Again, each person ate only a fraction of the dishes present, depending on what was within reach.

The modern idea of a menu in which everyone at the table eats the same dishes in the same order did not appear until the 1700s.

So on the one hand, it is not historically accurate to give the Italian princess Caterina credit for modernizing French cuisine. On the other hand, it is fair to say she and her offspring brought more refined manners and elegance to European dining rooms. Italian innovations included the use of the fork as well as greater cleanliness in general. An additional Italian contribution was the invention of sophisticated pastries and desserts.

Japan. In the southwestern United States, a number of chefs have transformed Mexican influences into an elegant and original cooking style. Throughout North America, traditional dishes and regional specialties combine the cooking traditions of immigrant settlers and the indigenous ingredients of a bountiful land. For many years, critics often argued that menus in most North American restaurants offered the same monotonous, mediocre food. In recent decades, however, American and Canadian cooks have rediscovered traditional North American dishes.

The use of ingredients and techniques from more than one regional, or international, cuisine in a single dish is known as **fusion cuisine**. Early attempts to prepare fusion cuisine often produced poor results because the dishes were not true to any one culture and were too mixed up. This was especially true in the 1980s, when the idea of fusion cuisine was new. Cooks often combined ingredients and techniques without a good feeling for how they would work together. The result was sometimes a jumbled mess. But chefs who have taken the time to study in depth the cuisines and cultures they borrow from have brought new excitement to cooking and to restaurant menus.

Today chefs make good use of all the ingredients and techniques available to them. It is almost second nature to give extra depth to the braising liquid for a beef pot roast by adding Mexican ancho chiles, for example, or to include Thai basil and lemongrass in a seafood salad. In the recipe sections of this book, classic dishes from many regions of the world are included among more familiar recipes from home. To help you understand these recipes and the cuisines they come from, background information accompanies many of them. The international recipes are identified in the Recipe Contents.

New Technologies

As described on page 4, new technologies, from transportation to food processing, had a profound effect on cooking in the twentieth century. Such changes continue today, with scientific developments that are only beginning to have an effect on how cooks think about food and menus.

One of these technologies is the practice of cooking **sous vide** (soo veed, French for "under vacuum"). Sous vide began simply as a method for packaging and storing foods in vacuum-sealed plastic bags. Modern chefs, however, are exploring ways to use this technology to control cooking temperatures and times with extreme precision. As a result, familiar foods have emerged with new textures and flavors. (Sous vide cooking is discussed further in Chapter 6.)

Another approach to cooking precision was pioneered by the Spanish chef Ferran Adrià in his acclaimed restaurant, El Bulli. Adrià explores new possibilities in gels, foams, powders, infusions, extracts, and other unexpected ways of presenting flavors, textures, and aromas. This approach to cooking is called **molecular gastronomy**, a name coined by the French chemist Hervé This, who has done much of the research in the field. More recently, chefs and other food experts have looked for more approachable terms to describe these techniques. The term *modernist cuisine* has been popularized by Nathan Myhrvold in his massive set of books of the same name.

Molecular gastronomy, or modernist cuisine, has been taken up by Heston Blumenthal in England, Wylie Dufresne, Grant Achatz, and Homaro Cantu in North America, and other chefs who continue to experiment and to explore what science and technology can contribute to food and food presentation. Many of the techniques make use of unfamiliar ingredients, such as natural gums, and put familiar ingredients, such as gelatin and pectin, to unfamiliar uses. Although this approach to cooking may be best known for its unusual ingredients and techniques, its finest chefs are focused on the food, treating the techniques primarily as new tools in the chef's repertoire.

Cooking and cooking styles continue to change. Men and women are needed who can adapt to these changes and respond to new challenges. Although automation and convenience foods will no doubt grow in importance, imaginative chefs who can create new dishes and develop new techniques and styles will always be needed, as will skilled cooks who can apply both old and new techniques to produce high-quality foods in all kinds of facilities, from restaurants and hotels to schools and hospitals.

KEY POINTS TO REVIEW

- How have the following developments changed the food-service industry: development of new equipment; availability of new food products; greater understanding of food safety and nutrition?

- How have international cuisines influenced and changed cooking in North America?

THE ORGANIZATION OF MODERN KITCHENS

THE BASIS OF KITCHEN ORGANIZATION

The purpose of kitchen organization is to assign or allocate tasks so they can be done efficiently and properly and so all workers know what their responsibilities are.

The way a kitchen is organized depends on several factors.

1. The menu.

   The kinds of dishes to be produced obviously determine the jobs that must be done. The menu is, in fact, the basis of the entire operation. Because of its importance, we devote a major section of Chapter 4 to a study of the menu.

2. The type of establishment.

   The major types of food-service establishments are as follows:

   - Hotels
   - Institutional kitchens

     Schools

     Hospitals, nursing homes, and other health care institutions

     Retirement community and assisted living facilities

     Employee lunchrooms and executive dining rooms

     Airline catering

     Military food service

     Correctional institutions

   - Private clubs
   - Catering and banquet services
   - Fast-food restaurants
   - Carry-out or take-out food facilities, including supermarkets
   - Full-service restaurants
   - Private homes (personal chefs)

3. The size of the operation (the number of customers and the volume of food served).

4. The physical facilities, including the equipment in use.

THE CLASSICAL BRIGADE

As you learned earlier in this chapter, one of Escoffier's important achievements was the reorganization of the kitchen. This reorganization divided the kitchen into departments, or stations, based on the kinds of foods produced. A station chef was placed in charge of each department. In a small operation, the station chef might be the only worker in the department. But in a large kitchen, each station chef might have several assistants.

This system, with many variations, is still in use, especially in large hotels with traditional kinds of food service. The major positions are as follows:

1. The **chef** is the person in charge of the kitchen. In large establishments, this person has the title of **executive chef**. The executive chef is a manager who is responsible for all aspects of food production, including menu planning, purchasing, costing, planning work schedules, hiring, and training.

2. If a food-service operation is large, with many departments (for example, a formal dining room, a casual dining room, and a catering department), or if it has several units in different locations, each kitchen may have a **chef de cuisine**. The chef de cuisine reports to the executive chef.

3. The **sous chef** (*soo* shef) is directly in charge of production and works as the assistant to the executive chef or chef de cuisine. (The word *sous* is French for "under.") Because the

executive chef's responsibilities may require a great deal of time in the office, the sous chef often takes command of the actual production and the minute-by-minute supervision of the staff.

4. The station chefs, or *chefs de partie*, are in charge of particular areas of production. The following are the most important station chefs.

- The *sauce chef*, or **saucier** (so-see-*ay*), prepares sauces, stews, and hot hors d'oeuvres, and sautés foods to order. This is usually the highest position of all the stations.

- The *fish cook*, or **poissonier** (pwah-so-*nyay*), prepares fish dishes. In some kitchens, this station is handled by the saucier.

- The *vegetable cook*, or **entremetier** (awn-truh-met-*yay*), prepares vegetables, soups, starches, and eggs. Large kitchens may divide these duties among the vegetable cook, the fry cook, and the soup cook.

- The *roast cook*, or **rôtisseur** (ro-tee-*sur*), prepares roasted and braised meats and their gravies and broils meats and other items to order. A large kitchen may have a separate *broiler cook*, or **grillardin** (gree-ar-*dan*), to handle the broiled items. The broiler cook may also prepare deep-fried meats and fish.

- The *pantry chef*, or **garde manger** (gard mawn-*zhay*), is responsible for cold foods, including salads and dressings, pâtés, cold hors d'oeuvres, and buffet items.

- The *pastry chef*, or **pâtissier** (pa-tees-*syay*), prepares pastries and desserts.

- The *relief cook*, *swing cook*, or **tournant** (toor-*nawn*), replaces other station heads.

- The *expediter*, or **aboyeur** (ah-bwa-yer), accepts orders from waiters and passes them on to the cooks on the line. The expediter also calls for orders to be finished and plated at the proper time and inspects each plate before passing it to the dining room staff. In many restaurants, this position is taken by the head chef or the sous chef.

5. *Cooks* and *assistants* in each station or department help with the duties assigned to them. For example, the assistant vegetable cook may wash, peel, and trim vegetables. With experience, assistants may be promoted to station cooks and then to station chefs.

MODERN KITCHEN ORGANIZATION

As you can see, only a large establishment needs a staff like the classical brigade just described. In fact, some large hotels have even larger staffs, with other positions such as separate day and night sous chefs, assistant chef, banquet chef, butcher, baker, and so on.

Most modern operations, though, are smaller than this. The size of the classical brigade may be reduced simply by combining two or more positions where the workload allows it. For example, the *second cook* may combine the duties of the sauce cook, fish cook, soup cook, and vegetable cook.

A typical medium-size operation may employ a chef, a second cook, a broiler cook, a pantry cook, and a few cooks' helpers.

A **working chef** is in charge of operations not large enough to have an executive chef. In addition to being in charge of the kitchen, the working chef also handles one of the production stations. For example, he or she may handle the sauté station, plate foods during service, and help on other stations when needed.

Small kitchens may have only a chef, one or two cooks, and perhaps one or two assistants to handle simple jobs such as washing and peeling vegetables. Cooks who prepare or finish hot à la carte items during service in a restaurant may be known as **line cooks**. Line cooks are said to be on the hot line, or simply on the line.

In many small operations, the **short-order cook** is the backbone of the kitchen during service time. This cook may handle the broiler, deep fryer, griddle, sandwich production, and even some sautéed items. In other words, the short-order cook's responsibility is the preparation of foods that are quickly prepared to order.

One special type of short-order cook is the **breakfast cook**. This worker is skilled at quickly and efficiently turning out egg dishes and other breakfast items to order.

By contrast, establishments such as school cafeterias may do no cooking to order at all. Stations and assignments are based on the requirements of quantity preparation rather than cooking to order.

SKILL LEVELS

The preceding discussion is necessarily general because there are so many kinds of kitchen organizations. Titles vary also. The responsibilities of the worker called the *second cook*, for example, are not necessarily the same in every establishment. Escoffier's standardized system has evolved in many directions.

One title that is often misunderstood and much abused is *chef*. The general public tends to refer to anyone with a white hat as a chef, and people who like to cook for guests in their homes refer to themselves as amateur chefs.

Strictly speaking, the term *chef* is reserved for one who is *in charge of a kitchen* or a part of a kitchen. The word *chef* is French for "chief" or "head." Studying this book will not make you a chef. The title must be earned by experience not only in preparing food but also in managing a staff and in planning production. New cooks who want to advance in their careers know they must always use the word *chef* with respect.

Skills required of food production personnel vary not only with the job level but also with the establishment and the kind of food prepared. The director of a hospital kitchen and the head chef in a luxury restaurant need different skills. The skills needed by a short-order cook in a coffee shop are not exactly the same as those needed by a production worker in a school cafeteria. Nevertheless, we can group skills into three general categories:

1. *Supervisory.* The head of a food-service kitchen, whether called *executive chef*, *head chef*, *working chef*, or *dietary director*, must have management and supervisory skills as well as a thorough knowledge of food production. Leadership positions require an individual who understands organizing and motivating people, planning menus and production procedures, controlling costs and managing budgets, and purchasing food supplies and equipment. Even if he or she does no cooking at all, the chef must be an experienced cook in order to schedule production, instruct workers, and control quality. Above all, the chef must be able to work well with people, even under extreme pressure.

2. *Skilled and technical.* While the chef is the head of an establishment, the cooks are the backbone. These workers carry out the actual food production. Thus, they must have knowledge of and experience in cooking techniques, at least for the dishes made in their own department. In addition, they must be able to function well with their fellow workers and to coordinate with other departments. Food production is a team activity.

3. *Entry level.* Entry-level jobs in food service usually require no particular skills or experience. Workers in these jobs are assigned such work as washing vegetables and preparing salad greens. As their knowledge and experience increase, they may be given more complex tasks and eventually become skilled cooks. Many executive chefs began their careers as pot washers who got a chance to peel potatoes when the pot sink was empty.

Beginning in an entry-level position and working one's way up with experience is the traditional method of advancing in a food-service career. Today, however, many cooks are graduates of culinary schools and programs. But even with such an education, many new graduates begin at entry-level positions. This is as it should be and certainly should not be seen as discouragement. Schools teach general cooking knowledge, while every food-service establishment requires specific skills according to its own menu and its own procedures. Experience as well as theoretical knowledge is needed to be able to adapt to real-life working situations. However, students who have studied and learned well should be able to work their way up more rapidly than beginners with no knowledge at all.

OTHER PROFESSIONAL OPPORTUNITIES

Not all those who train to be professional culinarians end up in restaurant careers. Professional cooking expertise is valuable in many callings. The following are just a few of the employment opportunities available in addition to standard cooking positions. Most of these require advanced training in other fields in addition to food production:

- Hospitality management in hotels, restaurants, large catering companies, and other organizations with a food-service component
- Product development and research for food manufacturers
- Product sales representatives for food and beverage distributors
- Product sales representatives for equipment companies
- Restaurant design and consulting
- Food styling for photography in books, magazines, and other publications, as well as for food packaging and marketing materials
- Food writing for newspapers, magazines, food industry journals, and other publications—not only restaurant criticism but analysis and reporting on food-related topics such as nutrition and health, agriculture, and food supply
- Training the next generation of chefs in culinary schools and in large hospitality companies with in-house training programs

STANDARDS OF PROFESSIONALISM

What does it take to be a good food-service worker?

The emphasis of a food-service education is on learning a set of skills. But in many ways, *attitudes* are more important than skills because a good attitude will help you not only learn skills but also persevere and overcome the many difficulties you will face.

The successful food-service worker follows an unwritten code of behavior and set of attitudes we call **professionalism**. Let's look at some of the qualities a professional must have.

POSITIVE ATTITUDE TOWARD THE JOB

To be a good professional cook, you have to like cooking and want to do it well. Being serious about your work doesn't mean you can't enjoy it. But the enjoyment comes from the satisfaction of doing your job well and making everything run smoothly.

Every experienced chef knows the stimulation of the rush. When it's the busiest time of the evening, the orders are coming in so fast you can hardly keep track of them, and every split second counts—then, when everyone digs in and works together and everything clicks, there's real excitement in the air. But this excitement comes only when you work for it.

A cook with a positive attitude works quickly, efficiently, neatly, and safely. Professionals have pride in their work and want to make sure it is something to be proud of.

Pride in your work and in your profession is important, but humility is important too, especially when you are starting out. Sometimes new culinary school graduates arrive on the job thinking they know everything. Remember that learning to cook and learning to manage a kitchen is a lifelong process and that you are not yet qualified to be executive chef.

The importance of a professional attitude begins even before you start your first job. The standard advice for a successful job interview applies to cooks as well as to office professionals: Dress and behave not for the group you belong to but for the group you want to join. Arrive neat, clean, appropriately dressed, and on time. Get noticed for the right reasons. Carry this attitude through every day on the job.

STAYING POWER

Food service requires physical and mental stamina, good health, and a willingness to work hard. It is hard work. The pressure can be intense and the hours long and grueling. You may be working evenings and weekends when everyone else is playing. And the work can be monotonous. You might think it's drudgery to hand-shape two or three dozen dinner rolls for your baking class, but wait until you get that great job in the big hotel and are told to make 3,000 canapés for a party.

Overcoming these difficulties requires a sense of responsibility and a dedication to your profession, to your coworkers, and to your customers or clients. Dedication also means staying with a job and not hopping from kitchen to kitchen every few months. Sticking with a job at least a year or two shows prospective employers you are serious about your work and can be relied on.

ABILITY TO WORK WITH PEOPLE

Few of you will work in an establishment so small you are the only person on the staff. Food-service work is teamwork, and it's essential to be able to work well on a team and to cooperate with your fellow workers. You can't afford to let ego problems, petty jealousy, departmental rivalries, or feelings about other people get in the way of doing your job well. Today's kitchens hold people of many races, nationalities, and origins, some of whom speak languages different from yours. You have to be able to work on the same team as everyone. In the old days, many chefs were famous for their temper tantrums. Fortunately, self-control is more valued today.

EAGERNESS TO LEARN

There is more to learn about cooking than you will learn in a lifetime. The greatest chefs in the world are the first to admit they have more to learn, and they keep working, experimenting, and studying. The food-service industry is changing rapidly, so it is vital to be open to new ideas. No matter how good your techniques are, you might learn an even better way.

Continue to study and read. Seek extra work that gives you the opportunity to learn from people with more experience. For example, if you are working on the hot line in a restaurant, ask the pastry chef if you could come in early, on your own time, to help out and, in the process, gain new knowledge and experience.

Many culinary schools and programs have continuing education programs that can help you develop new skills. Professional associations such as the American Culinary Federation (ACF), the Canadian Culinary Federation – Fédération Culinaire Canadienne (CCFCC), and the International Association of Culinary Professionals (IACP) provide opportunities for learning as well as for making contacts with other professionals. The ACF, as well as other professional organizations such as the Retail Bakers of America (RBA) and the International Food Service Executives Association (IFSEA), sponsors certification programs that document a professional's skill level and encourage ongoing study.

A FULL RANGE OF SKILLS

Most people who become professional cooks do so because they like to cook. This is an important motivation, but it is also important to develop and maintain other skills necessary for the profession. To be successful, a cook must understand and manage food costs and other financial matters, manage and maintain proper inventories, deal with purveyors, and understand personnel management.

EXPERIENCE

One of our most respected chefs said, "You don't really know how to cook a dish until you have done it a thousand times."

There is no substitute for years of experience. Studying cooking principles in books and in schools can get your career off to a running start. You may learn more about basic cooking theories from your chef instructors than you could in several years of working your way up from washing vegetables. But if you want to become an accomplished cook, you need practice, practice, and more practice. A diploma does not make you a chef.

DEDICATION TO QUALITY

Many people think only a special category of food can be called *gourmet food*. It's hard to say exactly what that is. Apparently, the only thing so-called gourmet foods have in common is high price.

The only distinction worth making is between well-prepared food and poorly prepared food. There is good roast duckling à l'orange and there is bad roast duckling à l'orange. There are good hamburgers and French fries, and there are bad hamburgers and French fries.

Whether you work in a top restaurant, a fast-food restaurant, a college cafeteria, or a catering house, you can do your job well, or not. The choice is yours.

High quality doesn't necessarily mean high price. It costs no more to cook green beans properly than to overcook them. But in order to produce high-quality food, you must want to. It is not enough to simply know how.

GOOD UNDERSTANDING OF THE BASICS

Experimentation and innovation in cooking are the order of the day. Brilliant chefs are breaking old boundaries and inventing dishes that would have been unthinkable years ago. There is apparently no limit to what can be tried.

However, the chefs who seem to be most revolutionary are the first to insist on the importance of solid grounding in basic techniques and in the classic methods practiced since Escoffier's day. In order to innovate, you have to know where to begin.

As a beginner, knowing the basics will help you take better advantage of your experience. When you watch a practiced cook at work, you will understand better what you are seeing and will know what questions to ask. In order to play great music on the piano, you first must learn to play scales and exercises.

That's what this book is about. It's not a course in French cooking or American cooking or gourmet cooking or coffee shop cooking. It's a course in the basics. When you finish the book, you will not know everything. But you should be ready to take good advantage of the many rewarding years of food-service experience ahead of you.

KEY POINTS TO REVIEW

- What are the major stations in a classical kitchen? What are their responsibilities?

- How do the size and type of a food-service operation affect how the kitchen is organized?

- What are the three basic skill levels of modern kitchen personnel?

- What are eight personal characteristics that are important to the success of a food-service professional?

TERMS FOR REVIEW

Marie-Antoine Carême	**chef de cuisine**	**pâtissier**
Georges-Auguste Escoffier	**sous chef**	**tournant**
nouvelle cuisine	**station chef**	**expediter**
sustainable agriculture	**saucier**	**aboyeur**
fusion cuisine	**poissonier**	**working chef**
sous vide	**entremetier**	**line cook**
molecular gastronomy	**rôtisseur**	**short-order cook**
chef	**grillardin**	**breakfast cook**
executive chef	**garde manger**	**professionalism**

QUESTIONS FOR DISCUSSION

1. Escoffier is sometimes called the father of modern food service. What were his most important accomplishments?

2. Discuss several ways in which modern technology has changed the food-service industry.

3. Discuss how an emphasis on high-quality ingredients beginning in the late twentieth century has influenced cooks and cooking styles.

4. What is fusion cuisine? Discuss how successful chefs make use of international influences.

5. What is the purpose of kitchen organization? Is the classical system of organization developed by Escoffier the best for all types of kitchens? Why or why not?

6. True or false: A cook in charge of the sauce and sauté station in a large hotel must have supervisory skills as well as cooking skills. Explain your answer.

7. True or false: If a culinary arts student in a professional school studies hard, works diligently, gets top grades, and shows real dedication, he or she will be qualified to be a chef upon graduation. Explain your answer.

SANITATION AND SAFETY

In the last chapter, we talked about professionalism in food service. Professionalism is an attitude that reflects pride in the quality of your work. One of the most important ways of demonstrating professional pride is in the area of sanitation and safety. Pride in quality is reflected in your appearance and work habits. Poor hygiene, poor grooming and personal care, and sloppy work habits are nothing to be proud of.

Even more important, poor sanitation and safety can cost a lot of money. Poor food-handling procedures and unclean kitchens cause illness, unhappy customers, and even fines, summonses, and lawsuits. Food spoilage raises food costs. Poor kitchen safety results in injuries, medical bills, and workdays lost. Finally, poor sanitation and safety habits show lack of respect for your customers, for your fellow workers, and for yourself.

In this chapter, you will study the causes of food-borne diseases and kitchen injuries, and you will learn ways of preventing them. Prevention, of course, is the most important thing to learn. It is not as important to be able to recite the names of disease-causing bacteria as it is to be able to prevent their growth in food.

AFTER READING THIS CHAPTER, YOU SHOULD BE ABLE TO

1. Describe steps to prevent food poisoning and food-borne diseases in the following areas: personal hygiene, food handling and storage techniques, cleaning and sanitizing techniques, and pest control.

2. Demonstrate safe workplace habits that prevent injuries from the following: cuts, burns, operation of machinery and equipment, and lifting.

3. Identify safe workplace habits that minimize the likelihood of fires and falls.

SANITATION

Rules of personal hygiene and sanitary food handling were not invented just to make your life difficult. There are good reasons for all of them. Instead of starting this chapter with lists of rules, we first talk about the causes of food-borne diseases. Then, when we get to the rules, you will understand why they are important. This will make them easier to remember and to practice.

The rules presented in this chapter are basic guidelines only. Local health departments have more detailed regulations. *All food-service operators are responsible for knowing the health department regulations in their own city and province.* Because food safety guidelines and regulations change frequently, foodservice workers should keep up to date by checking information published by agencies such as Health Canada (www.hc-sc.gc.ca).

The information presented here is practical as well as theoretical. It should not merely be learned but also put to use systematically. One effective system food-service establishments can use to ensure food safety is the Hazard Analysis Critical Control Point (HACCP) system. This practical program identifies possible danger points and sets up procedures for corrective action. HACCP is introduced later in this chapter.

FOOD HAZARDS

Preventing food-borne illness is one of the most important challenges facing every food service worker. In order to prevent illness, a food worker must understand the sources of food-borne disease.

Most food-borne illness is the result of eating food that has been **contaminated**. To say a food is contaminated means it contains harmful substances not originally present in it. In other words, contaminated food is food that is not pure. In this section, we first discuss the various substances that can contaminate food and cause illness. Afterward, we consider how these substances get into food to contaminate it and how food workers can prevent contamination and avoid serving contaminated food.

Any substance in food that can cause illness or injury is called a **hazard**. Food hazards are of four types:

1. Biological hazards
2. Chemical hazards
3. Physical hazards
4. Allergens

Notice it was said *most* food-borne illness is caused by eating food contaminated with foreign substances. Some illness is caused not by contaminants but by substances that occur naturally in foods. These include plant toxins (toxin means "poison"), such as the chemicals in poisonous mushrooms, and certain natural food components to which some people are allergic. This section considers all these kinds of food hazards.

PATHOGENS

The most important kind of biological hazards to consider are microorganisms. A **microorganism** is a tiny, usually single-celled organism that can be seen only with a microscope. A microorganism that can cause disease is called a **pathogen**. Although these organisms sometimes occur in clusters large enough to be seen with the naked eye, they are not usually visible. This is one reason why they can be so dangerous. Just because food looks good doesn't mean it is safe.

Four kinds of microorganisms can contaminate food and cause illness:

1. Bacteria
2. Viruses
3. Fungi
4. Parasites

Most food-borne diseases are caused by bacteria, so most of our attention in this chapter is focused on them, but the other types can be dangerous as well. Many of the measures we take to protect food from bacteria also help prevent the other three kinds of microorganisms.

Bacteria

Bacteria are everywhere—in the air, in the water, in the ground, on our food, on our skin, inside our bodies. Scientists have various ways of classifying and describing these bacteria. As food workers, we are interested in a way of classifying them that may be less scientific but is more practical to our work.

1. *Harmless bacteria.* Most bacteria fall into this category. They are neither helpful nor harmful to us. We are not concerned with them in food sanitation.

2. *Beneficial bacteria.* These bacteria are helpful to us. For example, many live in the intestinal tract, where they fight harmful bacteria, aid the digestion of food, and produce certain nutrients. In food production, bacteria make possible the manufacture of many foods, including cheese, yogurt, and sauerkraut.

3. *Undesirable bacteria.* These are the bacteria that are responsible for food spoilage. They cause souring, putrefying, and decomposition. These bacteria may or may not cause disease, but they offer a built-in safety factor: They announce their presence by means of sour odors, sticky or slimy surfaces, and discoloration. As long as we use common sense and follow the rule that says, "When in doubt, throw it out," we are relatively safe from these bacteria.

    We are concerned with these bacteria for two reasons:

    - Food spoilage costs money.
    - Food spoilage is a sign of improper food handling and storage. This means the next kind of bacteria is probably present.

4. *Disease-causing bacteria, or pathogens.* These are the bacteria that cause most food-borne illness, the bacteria we are most concerned with.

    Pathogens do not necessarily leave detectable odors or tastes in food. In other words, you can't tell if food is contaminated by smelling, tasting, or looking at it. The only way to protect food against pathogenic bacteria is to use proper hygiene and sanitary food-handling and storage techniques.

    Each kind of bacterial pathogen causes disease in one of three ways:

1. **Intoxications** are caused by poisons (toxins) the bacteria produce while they are growing in the food, before it is eaten. It is these poisons, not the bacteria themselves, that cause the diseases.

2. **Infections** are caused by bacteria (or other organisms) that get into the intestinal system and attack the body. Disease is caused by the bacteria themselves as they multiply in the body.

3. **Toxin-mediated infections** are also caused by bacteria that get into the body and grow. Disease is caused by poisons the bacteria produce as they grow and multiply in the body. Most food-borne diseases are toxin-mediated infections.

Bacterial Growth

Bacteria multiply by splitting in half. Under ideal conditions for growth, they can double in number every 15–30 minutes. This means that one single bacterium could multiply to one million in less than six hours!

Conditions for Growth

1. *Food.* Bacteria require food in order to grow. They like many of the foods we do. Foods with sufficient amounts of proteins are best for bacterial growth. These include meats, poultry, fish, dairy products, and eggs, as well as some grains and vegetables.

2. *Moisture.* Bacteria require water to absorb food. Dry foods do not support bacterial growth. Foods with a very high salt or sugar content are also relatively safe, because these ingredients make the bacteria unable to use the moisture present.

    The availability of water to bacteria is indicated by a measure called **water activity**, abbreviated a_w. The scale runs from 0 (meaning no water available) to 1.0. Most pathogens grow best in an environment from 0.85 to 1.0 a_w.

3. *Temperature.* Bacteria grow best at warm temperatures. *Temperatures between 40°F and 140°F (4°C and 60°C) promote the growth of disease-causing bacteria.* This temperature range is called the **Food Danger Zone**. (By comparison, the United States has changed its danger zone standards to 41°–135°F or 5°–57°C.)

BACTERIA AND pH

In general, food-borne pathogens grow best in an environment with a pH of 4.6–10. Every type of bacteria is different, however, and some grow when there is a higher or lower pH than this range. Salmonella bacteria, for example, can grow when there is a pH of 4.1–9.0. In general, however, acidity is an enemy of bacterial growth.

BACTERIA AND TEMPERATURE

The world is full of bacteria, and many kinds do not fit the food safety guidelines outlined here. Some bacteria, for example, need cool or cold temperatures to grow. These are called **psychrophiles**. Others thrive at high temperatures. These are called **thermophiles**. Some extreme thermophiles even grow at temperatures above the boiling point of water (212°F or 100°C). Nevertheless, most food-borne pathogens are **mesophiles**, bacteria that grow fastest at moderate temperatures (77°–113°F or 25°–45°C).

4. *Acidity or alkalinity.* In general, disease-producing bacteria like a neutral environment, neither too acidic nor too alkaline (see sidebar top left). The acidity or alkalinity of a substance is indicated by a measurement called *pH*. The scale ranges from 0 (strongly acidic) to 14 (strongly alkaline). A pH of 7 is neutral. Pure water has a pH of 7.

5. *Oxygen.* Some bacteria require oxygen to grow. These are called **aerobic**. Some bacteria are **anaerobic**, which means they can grow only if there is no air present, such as in metal cans. Botulism, one of the most dangerous forms of food poisoning, is caused by anaerobic bacteria. A third category of bacteria can grow either with oxygen or without it. These bacteria are called **facultative**. Most bacteria in food that cause disease are facultative.

6. *Time.* When bacteria are introduced to a new environment, they need time to adjust to their surroundings before they start growing. This time is called the **lag phase**. If other conditions are good, the lag phase may last one hour, or somewhat longer.

   If it weren't for the lag phase, there would be much more food-borne disease than there is. This delay makes it possible to have foods at room temperature *for very short periods* in order to work on them.

Potentially Hazardous Foods or TCS Foods

Foods that provide a good environment for the growth of disease-causing microorganisms are called **potentially hazardous foods**. Looking back at our list of conditions for growth of bacteria, we can see that protein foods with sufficient moisture and neutral pH are the most likely to host bacteria that cause disease. Of the conditions in the list, the one over which we have most control is *temperature.*

These foods are also called *TCS foods.* The abbreviation stands for time/temperature control for safety. In other words, our guidelines for keeping foods out of the Food Danger Zone temperatures, except for limited times, must be followed to keep these foods safe.

Potentially hazardous foods fall into two general categories, plus four specific items that do not fit into these categories. All these foods, plus any foods prepared with any of them, are potentially hazardous:

1. Any food derived from animals, or any food containing animal products, including meat, poultry, fish, shellfish, eggs, and dairy products.

2. Any food derived from plants that has been cooked, partially cooked, or otherwise heat-treated. This category includes not only cooked vegetables but also such items as cooked pasta, cooked rice, and tofu (soybean curd).

3. Raw seed sprouts.

4. Sliced melons (because the edible flesh can be contaminated by organisms on the rind's exterior, which was in contact with soil).

5. Cut tomatoes (for the same reason as sliced melons).

6. Garlic and oil mixtures that haven't been specifically treated to prevent growth of pathogens (because the oil seals the garlic from the air, fostering the growth of anaerobic bacteria, as explained above).

Foods that are not potentially hazardous include dried or dehydrated foods, foods that are strongly acidic, and commercially processed foods that are still in their original unopened, sealed containers.

Locomotion

Bacteria can move from place to place in only one way: They must be carried. They can't move on their own.

Foods can become contaminated by any of the following means:

Hands	Air
Coughs and sneezes	Water
Other foods	Insects
Equipment and utensils	Rats and mice

Protection against Bacteria

Because we know how and why bacteria grow, we should be able to keep them from growing. Because we know how bacteria get from place to place, we should be able to keep them from getting into our food.

There are three basic principles of food protection against bacteria. These principles are the reasons behind nearly all the sanitation techniques we discuss in the rest of this chapter.

1. *Keep bacteria from spreading.* Don't let food touch anything that may contain disease-producing bacteria, and protect food from bacteria in the air.

2. *Stop bacteria from growing.* Take away the conditions that encourage bacteria to grow. In the kitchen, our best weapon is temperature. *The most effective way to prevent bacterial growth is to keep foods below 40°F (4°C) or above 140°F (60°C).* These temperatures won't necessarily kill bacteria; they'll just slow their growth greatly.

3. *Kill bacteria.* Most disease-causing bacteria are killed if they are subjected to a temperature of 170°F (77°C) for 30 seconds, or higher temperatures for shorter times. This enables us to make food safe by cooking and to sanitize dishes and equipment with heat. The term **sanitize** means to kill disease-causing bacteria.

   Certain chemicals also kill bacteria. These may be used for sanitizing equipment.

Bacterial Diseases

For the most common bacterial diseases it is important to pay particular attention to the way it is spread, the foods involved, and the means of prevention. To see a table of the most common bacterial diseases, visit www.wiley.com/college/gisslen. General practices and procedures for prevention of food-borne diseases are discussed in a later section.

Viruses

Viruses are even smaller than bacteria. They consist of genetic material surrounded by a protein layer. Unlike bacteria, they can't reproduce or multiply unless they are inside a living cell, but they can be carried on almost any surface and can survive for days or even months. Viruses are inactive or dormant until they enter a living cell. Then they use that cell to make more viruses and release them into the organism. The new viruses can then enter new cells and continue to multiply.

Because viruses do not multiply in food like bacteria, food-borne viral diseases are usually caused by contamination from people, food contact surfaces, or, in the case of seafood, contaminated water.

Visit www.wiley.com/college/gisslen for a table that identifies the most important food-borne viral diseases.

Parasites

Parasites are organisms that can survive only by living on or inside another organism. The organism a parasite lives in and takes nourishment from is called the *host*. Parasites may pass from one host organism to another and complete a different stage of their life cycle in each organism. Human parasites are generally transmitted to them from animal hosts.

Human parasites are usually very small, and although they may be microscopic, they are larger than bacteria. They can usually be killed by proper cooking or by freezing.

The most important diseases caused by human parasites transmitted by food are found at www.wiley.com/college/gisslen.

Fungi

Molds and yeasts are examples of fungi. These organisms are associated primarily with food spoilage rather than food-borne disease. Most molds and yeasts, even those that cause spoilage, are not dangerous to most human beings. Some, in fact, are beneficial— for example, those responsible for the veining in blue cheese and the fermentation of bread dough.

Some molds, however, produce toxins that can cause allergic reactions and severe disease in those people who are susceptible. For example, certain molds produce a toxin called *aflatoxin* in such foods as peanuts and other nuts, corn, cottonseed, and milk. This toxin can cause serious liver disease in some people.

Other Biological Hazards

In addition to the biological hazards associated with bacteria and other organisms, some hazards occur naturally in foods and are not the result of contamination. These hazards include plant toxins, seafood toxins, and allergens.

Plant Toxins

Put simply, some plants are naturally poisonous to human beings. The only way to avoid plant toxins is to avoid the plants in which they occur, as well as products made with those plants. In some cases, the toxins can be transferred in milk from cows that have eaten the plant (such as jimsonweed and snakeroot) or in honey from bees that have gathered nectar from the plants (such as mountain laurel).

The best-known plant toxins are those found in certain wild mushrooms. There are many kinds of poisonous mushrooms, and eating them causes symptoms that range from mild intestinal discomfort to painful death. Some mushroom toxins attack the nervous system, some attack and destroy the digestive system, and some attack other internal organs.

Other toxic plants to avoid are rhubarb leaves, water hemlock, apricot kernels, and nightshade.

Seafood Toxins

Some toxins occur in fish or shellfish that have eaten a kind of algae that contains the toxins. Because cooking does not destroy these toxins, the only method of protection against them is to purchase fish and shellfish from approved suppliers who can certify that the seafood comes from safe water.

Some fish naturally contain toxins. The best-known fish toxin is the one present in pufferfish, known in Japanese as *fugu*. Raw fugu is considered a delicacy in Japan, but it must be prepared only by certified chefs who have been trained to remove the glands that produce the toxin without breaking them so they don't contaminate the flesh of the fish. This toxin attacks the nervous system and can be fatal.

Some other species of fish, such as moray eels, contain natural toxins and should be avoided.

CHEMICAL AND PHYSICAL HAZARDS

Some kinds of chemical poisoning are caused by the use of defective or improper equipment or equipment that has been handled improperly. The following toxins (except lead) create symptoms that show themselves very quickly, usually within 30 minutes of eating poisoned food. By contrast, symptoms of lead poisoning can take years to appear. To prevent these diseases, do not use the materials that cause them.

1. *Antimony.* Caused by storing or cooking acid foods in chipped gray enamelware.
2. *Cadmium.* Caused by cadmium-plated ice cube trays or containers.
3. *Cyanide.* Caused by silver polish containing cyanide.
4. *Lead.* Caused by lead water pipes, solder containing lead, or utensils containing lead.
5. *Copper.* Caused by unclean or corroded copper utensils, acid foods cooked in unlined copper utensils, or carbonated beverages in contact with copper tubing.
6. *Zinc.* Caused by cooking foods in zinc-plated (galvanized) utensils.

Other chemical contamination can result from exposure of foods to chemicals used in commercial food-service establishments. Examples include cleaning compounds, polishing compounds, and insecticides. Prevent contamination by keeping these items physically separated from foods. Do not use them around food. Label all containers properly. Rinse cleaned equipment thoroughly.

Physical contamination is contamination of food with objects that may not be toxic but may cause injury or discomfort. Examples include pieces of glass from a broken container, metal shavings from an improperly opened can, stones from poorly sorted dried beans, soil from poorly washed vegetables, insects or insect parts, and hair. Proper food handling is necessary to avoid physical contamination.

ALLERGENS

An **allergen** is a substance that causes an allergic reaction. Allergens affect only some people, and these people are said to be *allergic* to that specific substance. Not all allergens are biological hazards, but the most important ones are, so we discuss them together in this section.

Allergic reactions to food may occur as soon as the food is eaten or, in some cases, merely touched, or they may not occur until hours after the food is eaten. Common symptoms of allergic reaction to foods include itching, rash or hives, shortness of breath, tightness in the throat, and swelling of the eyes and face. In severe cases, allergic reactions may lead to unconsciousness or death.

Foods to which some people are allergic include wheat products, soy products, peanuts and other nuts, eggs, milk and dairy products, fish, and shellfish. Nonbiological allergens include food additives such as nitrites, used in cured meats, and monosodium glutamate (MSG), often used in Asian foods.

Health Canada lists 10 priority food allergens that must be clearly identified: peanuts, eggs, milk, mustard, tree nuts, wheat, soy, sesame, seafood (fish, crustaceans, shellfish), and sulphites (a food additive).

Because these products are common and are perfectly safe for most people, it is difficult to avoid serving them. For the sake of people who are sensitive to these foods, food-service personnel, especially all dining room staff, must be well informed of the ingredients in all menu items and be able to inform customers as needed. If any staff member does not know, when asked by a customer, if a food might contain an allergen, that employee should tell the customer so and then find someone who does know or else urge the customer to order a different item.

PERSONAL HYGIENE

Earlier in this chapter, we said most food-borne disease is caused by bacteria. Now we expand that statement slightly to say that *most food-borne disease is caused by bacteria spread by food workers*.

At the beginning of this chapter, we defined *contamination* as harmful substances not present originally in the food. Some contamination occurs before we receive the food, which means proper purchasing and receiving procedures are important parts of a sanitation program. But most food contamination occurs as a result of **cross-contamination**, defined as the transference of hazardous substances, mainly microorganisms, to a food from another food or another surface, such as equipment, worktables, or hands. Examples of situations in which cross-contamination can occur include the following:

- Mixing contaminated leftovers with a freshly cooked batch of food.
- Handling ready-to-eat foods with unclean hands.
- Handling several types of food without washing hands in between.
- Cutting raw chicken, then using the same cutting board, unsanitized, to cut vegetables.
- Placing ready-to-eat foods on a lower refrigerator shelf and allowing juices from raw fish or meat to drip onto them from an upper shelf.
- Wiping down work surfaces with a soiled cloth.

For the food worker, the first step in preventing food-borne disease is good personal hygiene. Even when we are healthy, we have bacteria all over our skin and in our nose and mouth. Some of these bacteria, if given the chance to grow in food, will make people ill. The following are steps to practicing good personal hygiene:

1. Do not work with food if you have any communicable disease or infection.
2. Bathe or shower daily.
3. Wear clean uniforms and aprons.
4. Keep hair neat and clean. Always wear a hat or hairnet. Hair longer than shoulder length must first be tied back and then secured under a net or hat.
5. Keep mustaches and beards trimmed and clean. Better yet, be clean-shaven.
6. Remove all jewelry: rings, low-hanging earrings, watches, bracelets. Avoid facial piercings; if you have them, don't touch them.

ALLERGIES AND INTOLERANCES

Health professionals make a distinction between food allergies and intolerances. Allergies are reactions by the body's immune system. The body sees a food substance as a foreign invader and attacks it, harming the body in the process. A food intolerance, on the other hand, is the inability of the body to process the food properly. For example, some people can't drink milk because of lactose intolerance. This means they can't digest milk sugar, or lactose. By contrast, a milk allergy is the reaction by the immune system to milk proteins.

KEY POINTS TO REVIEW

- What six conditions are necessary for the growth of bacteria?
- What are potentially hazardous foods?
- What are the three ways to protect against bacteria?
- Besides bacteria, what other hazards can make food unsafe?

7. Wash hands and exposed parts of arms before work and as often as necessary during work, including:
   - After eating, drinking, or smoking.
   - After using the toilet.
   - After touching or handling anything that may be contaminated with bacteria.

8. Cover coughs and sneezes, then wash your hands.

9. Keep your hands away from your face, eyes, hair, and arms.

10. Keep fingernails clean and short. Do not wear nail polish.

11. Do not smoke or chew gum while on duty.

12. Cover cuts or sores with clean bandages. If the sore is on the hands, you must wear gloves.

13. Do not sit on worktables.

PROCEDURE for Washing Hands

1. Wet your hands with hot running water. Use water as hot as you can comfortably stand, but at least 100°F (38°C).

2. Apply enough soap to make a good lather.

3. Rub hands together thoroughly for 20 seconds or longer, washing not only the hands but also the wrists and the lower part of the forearms.

4. Using a nail brush, clean beneath the fingernails and between the fingers.

5. Rinse hands well under hot running water. If possible, use a clean paper towel to turn off the water to avoid contaminating the hands by contact with soiled faucets.

6. Dry hands with clean single-use paper towels or a warm-air hand dryer.

USE OF GLOVES

If used correctly, gloves can help protect foods against cross-contamination. If used incorrectly, however, they can spread contamination just as easily as bare hands. Health departments may require the use of some kind of barrier between hands and any foods that are ready to eat—that is, foods that will be served without further cooking. Gloves, tongs, and other serving implements, and bakery or deli tissue can serve as barriers. To be sure gloves are used correctly, observe the following guidelines.

GUIDELINES for Using Disposable Gloves

1. Wash hands before putting on gloves or when changing to another pair. Gloves are not a substitute for proper handwashing.

2. Remove and discard gloves, wash hands, and change to a clean pair of gloves after handling one food item and before starting work on another. In particular, never to fail to change gloves after handling raw meat, poultry, or seafood. Gloves are for single use only. Remember that the purpose of using gloves is to avoid cross-contamination.

3. Change to a clean pair of gloves whenever gloves become torn, soiled, or contaminated by contact with an unsanitary surface.

FOOD STORAGE

The following rules of safe food storage have two purposes:

1. To prevent contamination of foods.

2. To prevent growth of bacteria that may already be in foods.

Temperature control is an important part of food storage. Perishable foods must be kept out of the Food Danger Zone—40° to 140°F (4° to 60°C)—as much as possible, because these temperatures support bacterial growth. See Figure 2.1 for a chart of important temperatures.

RECEIVING

1. Safe food handling begins the moment food is unloaded from the delivery truck. In fact, it begins even earlier than this, with the selection of good, reputable suppliers. Keep the receiving area clean and well lit.

2. Inspect all deliveries. Try to schedule deliveries during off-peak hours to allow proper time to inspect the items. For the same reason, try to schedule deliveries so they arrive one at a time.

3. Reject shipments or parts of shipments that are damaged or not at the proper temperature. Frozen foods should show no signs of having been thawed and refrozen.

4. Label all items with the delivery date or a use-by date.

5. Transfer items immediately to proper storage.

DRY FOOD STORAGE

Dry food storage pertains to those foods not likely to support bacterial growth in their normal state. These foods include

> Flour
>
> Sugar and salt
>
> Cereals, rice, and other grains
>
> Dried beans and peas
>
> Ready-prepared cereals
>
> Breads and crackers
>
> Oils and shortenings
>
> Canned and bottled foods (unopened)

1. Store dry foods in a cool, dry place, off the floor, away from the wall, and not under a sewer line.

2. Keep all containers tightly closed to protect from insects, rodents, and dust. Dry foods can be contaminated, even if they don't need refrigeration.

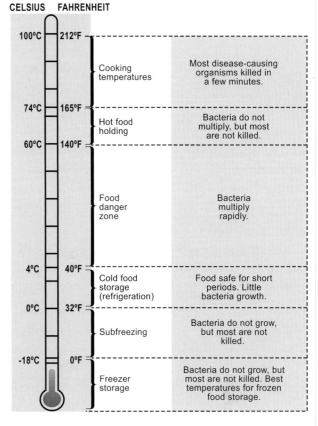

FIGURE 2.1 Important temperatures in sanitation and food protection.

FREEZER STORAGE AND PROPER THAWING

1. Keep frozen foods at 0°F (–18°C) or lower.

2. Keep all frozen foods tightly wrapped or packaged to prevent freezer burn.

3. Label and date all items.

4. Thaw frozen foods properly. Do not thaw at room temperature, because the surface temperature will go above 40°F (4°C) before the inside is thawed, resulting in bacterial growth. These methods may be used:

   - In a refrigerator, so that the food maintains a temperature of 40°F (4°C) or less.
   - Under cold running water, as long as the water temperature is at a temperature of 70°F (21°C) or lower. For ready-to-eat foods, no part of the food may go above 40°F (4°C). For raw meat, poultry, and fish, the temperature of the outside of the item may rise above 40°F (4°C) while thawing, but it must not remain at this temperature for more than 4 hours in total until it is cooked.
   - In a microwave oven, but only if the item is to be cooked or served immediately.

REFRIGERATOR STORAGE

1. Keep all perishable foods properly refrigerated. Note the lower limit of the Food Danger Zone (40°F/4°C) is only the upper limit for refrigerator storage. Most foods keep even better at lower temperatures. The major exception is fresh fruits and vegetables, which are not considered potentially hazardous foods. See Table 2.1 for preferred storage temperatures for various foods.

TABLE 2.1 Food Storage Temperatures		
Raw vegetables and fruits (see note)	40°–45°F	4°–7°C
Eggs	38°–40°F	3°–4°C
Milk and cream	36°–40°F	2°–4°C
Poultry and meat	32°–36°F	0°–2°C
Fish and seafood	30°–34°F	−1°–1°C

Note: Potatoes, onions, and winter squash are best held at cool temperatures (50°–65°F or 10°–18°C).

2. Do not crowd refrigerators. Leave space between items so cold air can circulate.

3. Keep refrigerator doors shut except when removing or putting in foods.

4. Keep shelves and interiors of refrigerators clean.

5. Store raw and cooked items separately, if possible.

6. If raw and cooked foods must be kept in the same refrigerator, keep cooked foods above raw foods. If cooked foods are kept below raw foods, they can become contaminated by drips and spills. Then, if they are not to be cooked again before serving, they may be hazardous.

7. Keep refrigerated foods wrapped or covered and in sanitary containers.

8. Do not let any unsanitary surface, such as the bottoms of other containers, touch any food.

9. Chill foods as quickly as possible over ice or in a cold-water bath before placing in the refrigerator. A gallon of stock placed in a refrigerator hot off the stove may take 10 hours to go below 40°F (4°C), giving bacteria plenty of time to grow.

10. When holding foods such as protein salads in a cold bain-marie or refrigerated table for service, do not heap the food above the level of the container. The food above this level will not stay cold enough.

HOT FOOD HOLDING

1. To keep foods hot for service, use steam tables or other equipment that will keep all parts of all foods above 140°F (60°C) at all times.

2. Keep foods covered.

3. Bring foods to holding temperature as quickly as possible by using ovens, steamers, rangetop pots and pans, or other cooking equipment. Do not warm cold foods by placing them directly in the steam table. They will take too long to heat, and bacteria will have time to grow.

4. Do not let ready-to-eat foods come in contact with any contaminated surface.

KEY POINTS TO REVIEW

- What is cross-contamination?

- What are the important rules of personal hygiene? List as many as you can.

- What is the Food Danger Zone?

FOOD HANDLING AND PREPARATION

We face two major sanitation problems when handling and preparing food. The first is *cross-contamination*, defined on page 21.

The second problem is that, while we are working on it, food is usually at a temperature of 40°–140°F (4°–60°C), or in the Food Danger Zone. The lag phase of bacteria growth (p. 18) helps us a little but, to be safe, we must keep foods out of the danger zone whenever possible:

1. Start with clean, wholesome foods from reputable purveyors. Whenever applicable, buy government-inspected meats, poultry, fish, dairy, and egg products.

2. Handle foods as little as possible. Use clean tongs, spatulas, or other utensils instead of hands when practical.

3. Use clean, sanitized equipment and worktables.

4. Clean and sanitize cutting surfaces and equipment after handling raw poultry, meat, fish, or eggs and before working on another food.

5. Place only food items and sanitary knives or other tools on cutting boards. Do not set food containers, toolboxes, or recipe books, for example, on cutting boards, as the bottoms of these items are likely to be unsanitary.

6. Clean as you go. Don't wait until the end of the workday. Keep clean cloths and sanitizing solution handy at your workstation and use them often.

7. Wash raw fruits and vegetables thoroughly.

8. When bringing foods out of refrigeration, do not bring out more than you can process in 1 hour.

9. Keep foods covered unless in immediate use.

10. Limit the time that foods spend in the Food Danger Zone.

11. Cook foods to minimum internal cooking temperatures (see next section).

12. Taste foods properly. With a ladle or other serving implement, transfer a small amount of the food to a small dish. Then taste this sample using a clean spoon. After tasting, do not use either the dish or the spoon again. Send them to the warewashing station or, if using disposables, discard them.

13. Boil leftover gravies, sauces, soups, and vegetables before serving.

14. Don't mix leftovers with freshly prepared foods.

15. Chill all ingredients for protein salads and potato salads before combining.

16. Cool and chill foods quickly and correctly, as explained in the following section. Chill custards, cream fillings, and other hazardous foods as quickly as possible by pouring them into shallow, sanitized pans, covering them, and refrigerating. Do not stack the pans.

MINIMUM INTERNAL COOKING TEMPERATURES

The **minimum internal cooking temperature** is the internal temperature for a given food product at which microorganisms are killed. The product must be held at that temperature for a specified period for the food to be considered safe. See Table 2.2.

Be sure to measure internal temperatures in at least two or three places, always inserting the thermometer into the thickest part of the food. Use sanitary thermometers that are accurate to within 2°F or 1°C.

COOLING PROCEDURES

If cooked foods are not to be served immediately or kept hot for service, they must be cooled quickly so they do not spend too much time in the Food Danger Zone. The rate at which foods cool depends on their total volume in relation to how much surface area they have to transfer heat away. In other words, a large batch of food cools more slowly because it has less surface area per unit of volume. One of the hazards of cooking foods in large volumes is cooling them so slowly they spend too much time in the Food Danger Zone.

To help gauge the time you may safely take to cool large volumes of food, use either the **two-stage cooling method** or the **one-stage cooling method**.

For the two-stage cooling method, cool foods from 140°F (60°C) to 70°F (21°C) in no more than 2 hours, and then from 70°F (21°C) to below 40°F (4°C) within an additional 4 hours, for a total cooling time of no more than 6 hours. The temperature range between 70°F (21°C) and 125°F (52°C) is the most dangerous part of the Food Danger Zone. This method ensures the food spends a minimum of time in that temperature range. If food has not cooled to 70°F (21°C) within 2 hours, it must be reheated to 165°F (74°C) and held at that temperature at least 15 seconds and then cooled again.

TABLE 2.2 Minimum Internal Cooking Temperatures

Beef, veal, and lamb (pieces and whole cuts)—medium-rare	63°C	145°F
Beef, veal, and lamb (pieces and whole cuts)—medium	71°C	160°F
Beef, veal, and lamb (pieces and whole cuts)—well done	77°C	170°F
Pork (pieces and whole cuts)	71°C	160°F
Poultry (chicken, turkey, duck)—pieces	74°C	165°F
Poultry—whole	85°C	185°F
Ground meat and meat mixtures (burgers, sausages, meatballs, meatloaf, casseroles)—beef, veal, lamb, pork	71°C	160°F
Ground meat and meat mixtures—poultry	74°C	165°F
Fish	70°C	158°F
Shellfish	74°C	165°F
Egg dishes	74°C	165°F
Others (hot dogs, stuffing, leftovers)	74°C	165°F
Game birds and animals, except for game birds cooked whole	74°C	165°F
Game birds cooked whole	82°C	180°F

For the one-stage cooling method, cool foods to below 40°F (4°C) in no more than 4 hours. If the food does not reach this temperature in 4 hours, it must be reheated to 165°F (74°C) and held at that temperature at least 15 seconds and then cooled again. The one-stage method should be used if the item was made from potentially hazardous foods that were at room temperature when preparation was begun.

REHEATING COOKED AND COOLED FOODS

Cooled foods must be reheated in such a way that all parts of the food reach at least 165°F (74°C) in as short a time as possible and held at that temperature for a minimum of 15 seconds. They should then be transferred to holding equipment, such as steam tables, and held at a temperature above 140°F (60°C) until served.

Foods reheated in a microwave oven must be reheated to 165°F (74°C), then stirred, covered, and allowed to stand for 2 minutes after reheating.

GUIDELINES for Cooling Foods

1. Never put hot foods directly into the cooler. Not only will they cool too slowly but also they will raise the temperature of other foods in the cooler.
2. If they are available, use quick-chill units or blast chillers to cool foods quickly before transferring them to cold storage.
3. Use ice-water baths to bring down the temperature of hot foods quickly.
4. Stir foods as they are cooling to redistribute the heat and help them cool more quickly.
5. Divide large batches into smaller batches. This increases the amount of surface area for the volume of food and helps it cool more quickly. Pouring foods into flat, shallow pans also increases surface area and cooling speed.

CLEANING AND SANITIZING EQUIPMENT

Cleaning means removing visible soil. *Sanitizing* means killing disease-causing bacteria. Two ways of killing bacteria are by *heat* and by *chemicals*.

MANUAL DISHWASHING

Figure 2.2 shows the setup of a three-compartment sink for washing dishes, glassware, and eating utensils by hand.

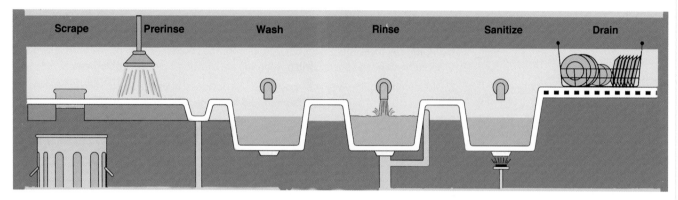

FIGURE 2.2 Setup of a three-compartment sink for manual dishwashing.

PROCEDURE **for Manual Dishwashing**

1. *Scrape and rinse.* The purpose of this step is to keep the wash water cleaner longer.
2. *Wash.* Use warm water at 110°–120°F (43°–49°C) and a good detergent. Scrub well with a brush to remove all traces of soil and grease.
3. *Rinse.* Use clean, warm water to rinse off detergent. Change the water frequently, or use running water with an overflow, as in Figure 2.2.
4. *Sanitize.* Place utensils in a rack and immerse in hot water at 171°F (77°C) for 30 seconds. (A gas or electric heating element is needed to hold water at this temperature.)
5. *Drain and air dry.* Do not towel dry. This may recontaminate utensils. Do not touch food contact surfaces of sanitized dishes, glasses, and silverware.

MECHANICAL DISHWASHING

The steps in washing dishes by machine are the same as in the hand method, except the machine does the washing, rinsing, and sanitizing.

PROCEDURE **for Mechanical Dishwashing**

1. Scrape and rinse.
2. Rack dishes so the dishwasher spray will strike all surfaces.
3. Run the dishwasher for a full cycle.
4. Critical temperatures:

    For machines that sanitize by heat, final rinse must be at least 180°F (82°C), although some stationary rack machines operate correctly at 165°F (74°C). Follow the instructions for your model.

    For machines that sanitize by chemical disinfectant, washing temperature should be above 120°F (49°C), and sanitizing temperatures should be 68°– 120°F (20°–49°C). Sanitizing chemicals are not as effective above 120°F (49°C).

5. Air dry and inspect dishes. Do not touch food contact surfaces.

CHEMICAL SANITIZING

The three most commonly used chemical sanitizers or disinfectants are chlorine, iodine, and quaternary ammonium, usually known as quats. Each of these chemicals has advantages and disadvantages. Read package instructions for each, and follow the appropriate guidelines for their use. All chemicals must be used with care.

Chlorine is widely used and inexpensive. It kills a broad range of pathogens and is effective in hard water, but it is less effective above 115°F (46°C).

Iodine is more expensive than chlorine and can stain surfaces with its brown color. However, it is not as quickly inactivated by dirt as chlorine is.

Quats works well in a wide range of temperatures, and it does not corrode metals, while chlorine and iodine can both be corrosive. However, it does not kill certain organisms.

WASHING KITCHEN UTENSILS AND EQUIPMENT

1. Use the same three-compartment sink setup and procedure as for manual dishwashing.

2. Do not use scouring powder or steel wool. These may make scratches where bacteria can hide. Also, pieces of steel wool break off and can remain in the pan and thus get into food.

3. Utensils with baked-on foods should be scraped and rinsed, soaked in the first compartment to loosen the baked-on food, and then scraped and rinsed again.

4. Kitchen equipment may be sanitized with *chemical disinfectants* (see sidebar) instead of heat. Use an approved disinfectant, and follow the instructions on the label.

CLEANING AND SANITIZING STATIONARY EQUIPMENT AND WORK SURFACES

1. Unplug electrical equipment before cleaning. You could seriously injure yourself if you accidentally hit the power switch while you are cleaning a piece of equipment.

2. Disassemble equipment when possible. (This obviously doesn't apply to such equipment as worktables.) All immersible parts should be cleaned and sanitized like kitchen utensils.

3. Wash all food contact surfaces, using a detergent solution and clean cloths.

4. Sanitize all surfaces with a double-strength sanitizing solution and with clean cloths used only for this purpose.

5. Allow to air dry.

6. Reassemble equipment.

RODENT AND INSECT CONTROL

Rats, mice, flies, and cockroaches can spread disease by contaminating food and food contact surfaces. Any sign of rodent or insect infestation is usually considered a serious violation of health codes.

There are four basic methods of pest control. We start with the most important and most effective.

BUILD THEM OUT

1. Block all possible rodent entrances, including structural defects in the building.

2. Put screens on all windows and doors.

3. Make sure all doors are self-closing, or install fly fans or air curtains.

4. Inspect incoming supplies for signs of insect infestation.

ELIMINATE HARBORAGE AND BREEDING PLACES

1. Repair holes and all other structural defects in walls and floors.

2. Eliminate narrow spaces between and behind equipment, counters, and other fixtures, and hollow spaces made by false bottoms in counters, cabinets, and so on.

3. Store food and supplies off the floor.

4. Seal all cracks and crevices. Repair loose tiles, wall coverings, and so on.

5. Remove all fly-breeding places inside and out: garbage, manure, and general filth.

ELIMINATE FOOD SUPPLIES

1. Keep all foods tightly covered or wrapped.

2. Keep garbage containers tightly covered, and use metal (ratproof) garbage cans.

3. Clean up all spilled food.

4. General sanitation: Keep floors, walls, and equipment clean.

EXTERMINATE

Hire a qualified, licensed exterminator who knows how to use poisons, insecticides, and traps. Most poisons should not be used in a food production operation, so it's better not to do the job yourself.

Extermination is a temporary solution only. For permanent freedom from rodents and insects, you must rely on the other methods of control.

KEY POINTS TO REVIEW

- What does *minimum internal cooking temperature* mean?
- How should cooked foods be cooled?
- What are the steps in the procedure for manual dishwashing? for mechanical dishwashing?
- What are the four ways to protect against rodents and insects? Which of these is the most effective?

SETTING UP A SYSTEM FOR FOOD SAFETY

Once you have learned the information in the first part of this chapter, you must apply it in the kitchen.

Many food-service operations have designed food safety systems that enable food workers to keep a close check on food items whenever there is a risk of contamination or of the growth of pathogens. In the most effective systems, nothing is left to chance. At each stage of food production and storage, workers refer to written guidelines that explain what to look for and what action to take if the standards are not met. Having written guidelines helps everyone avoid costly mistakes.

THE HACCP SYSTEM

One effective food safety system is called the *Hazard Analysis Critical Control Point* system, or **HACCP** (pronounced HASS-up). Versions of this system have been widely adopted throughout the food-service industry.

The following discussion is a brief introduction to the basic concepts of HACCP. For a more detailed explanation, you may refer to other published material listed in the Bibliography (pp. 1031–1032). The following discussion is based on information presented in those books.

The Steps of the HACCP System

The purpose of HACCP is to identify, monitor, and control dangers of food contamination. It is a system of seven steps:

1. Assess hazards.
2. Identify critical control points (CCPs).
3. Set up standards or limits for CCPs.
4. Set up procedures for monitoring CCPs.
5. Establish corrective actions.
6. Set up a recordkeeping system.
7. Verify the system is working.

These steps are the basis of the following discussion.

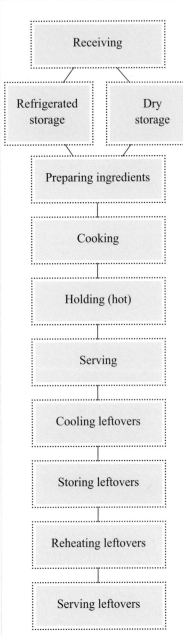

FIGURE 2.3 Flow of food.

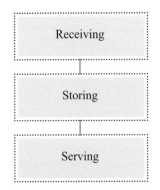

FIGURE 2.4 Basic flow of food.

The Flow of Food

HACCP begins with a concept called the **flow of food**. This term refers to the movement of food through a food-service operation, from receiving through storage, preparation, and service, until it gets to the final consumer.

The flow of food is different for each item being prepared. Some menu items involve many steps. For example, a luncheon dish of creamed chicken and vegetables over rice might have the steps shown in Figure 2.3.

Even the simplest items undergo several steps. For example, a cake bought already prepared from a commercial baker and served as dessert will go through at least the three steps in Figure 2.4 on its way to the customer.

Assessing Hazards

At each step of the flow of foods through the operation, risks can lead to dangerous conditions, or *hazards*. Assessing hazards is the process of identifying which of these dangerous conditions can occur at every step of the process.

These hazards can be divided into three categories:

1. *Contamination*, such as cross-contamination from a soiled cutting surface, torn packaging that permits insect infestation, working on food without washing hands, and spilling cleaning chemicals on food.

2. *Growth of bacteria and other pathogens* due to such conditions as inadequate refrigeration or storage, and holding hot foods below 140°F (60°C).

3. *Survival of pathogens or the continued presence of toxins*, usually because of inadequate cooking or heating or inadequate sanitizing of equipment and surfaces.

Note these hazards correspond to the sanitation techniques discussed on page 19 (keep bacteria from spreading, stop bacteria from growing, kill bacteria). The important difference is that the hazards addressed by HACCP include chemical and other hazards in addition to disease-causing organisms. Naturally, however, most of the hazards we are concerned with are those that affect potentially hazardous foods (see page 18).

Identifying Critical Control Points

Once the potential hazards are identified, the next step is to decide at what stages a worker can control the hazards. These points are called *control points*. For any given hazard, there may be several control points, or several chances to control the hazard. The last control point at which a worker can control a particular hazard is especially important to determine because this is the last chance to prevent a possible danger. These control points are called **critical control points (CCPs)**. Identifying CCPs is the second step in a HACCP program.

In simple language, setting up a HACCP system starts with reviewing the flow of food to figure out where something might go wrong, and then deciding what can be done about it. In the language of HACCP, these steps are called *assessing the hazards* and *identifying critical control points*.

Setting Standards or Limits for CCPs

The next step in designing a HACCP food safety system is setting up procedures for CCPs. At each such point, food workers need to know what standards must be met, what procedures to follow to meet the standards, and what to do if they aren't met. To reduce the chances for making mistakes, these standards and procedures are written out. Whenever possible, they should be included in the operation's recipes. In Chapter 4, you will see how CCPs are incorporated into a standardized recipe.

Some procedures are general and include the sanitation rules discussed earlier in this chapter. For example: Wash hands before handling food and after handling raw foods; hold foods above 140°F (60°C) or below 40°F (4°C). Others apply to specific items. For example: Cook a beef roast to an internal temperature of at least 145°F (63°C) and ensure that it stays at that temperature for at least 4 minutes. The minimum internal cooking temperatures discussed on page 26 are an important part of the standards of a HACCP system.

Setting Up Monitoring Procedures

Careful observation is needed to know when standards are met. This often involves measuring. The only way to know, for example, that a roast has reached the required internal temperature is to measure it, using a clean, sanitized thermometer.

Managers must ensure that all employees are trained to follow procedures and have the equipment needed to do the job.

Establishing monitoring procedures includes determining how a CCP is to be monitored or measured, when it is to be monitored, who is responsible for doing the measuring, and what equipment is needed to do the monitoring.

Taking Corrective Action

A **corrective action** is a procedure that must be followed whenever a critical limit is not met. Corrective actions should be identified in written procedures that clearly tell the worker what must be done in each situation.

For example, a monitoring procedure might show the internal temperature of a pan of chicken breast just out of the oven is 155°F (68°C). But the critical limit for chicken breasts is 165°F (74°C). The corrective action might be to return the chicken breast to the oven until the temperature reaches the critical limit.

Other corrective actions might be more complicated, but the written procedure should describe clearly what steps must be taken and who must take them.

Setting Up a Recordkeeping System

Keeping records of all the procedures described previously is important if a HACCP system is to succeed. Time and temperature logs, records of corrective actions taken, and documentation of when and how measuring devices were calibrated are examples of the kinds of records that enable an establishment to ensure food safety. Each establishment should develop clear, easy-to-use forms for entering all needed information.

Verifying the System Works

Accurate records enable you to make sure a HACCP system is working as intended. Review records regularly to check that all CCPs are being correctly monitored and that corrective actions are taken according to the proper procedures and adequate to control hazards. Revise procedures as necessary.

Accurate records also demonstrate to health inspectors that your operation is following correct safe procedures. In addition, records help you determine what went wrong if a foodborne illness does occur.

In addition, whenever purchasing specifications are changed, new items are added to the menu, or new equipment is put into use, review procedures and change them if needed. For example, if an operation starts buying larger beef steamship rounds for roasting, the internal temperature of the roasts will not meet critical limits unless the roasting time allowed for the beef is increased.

As this brief introduction to HACCP implies, establishing such a system to control all aspects of food production requires more information than this chapter has space for. Refer to the Bibliography for more detailed information.

LEARNING MORE ABOUT FOOD SAFETY

It is important for you to understand that food safety and sanitation is a large and complex topic. The first half of this chapter is only an introduction to the study of food safety. To advance in a food-service career, you must demonstrate a detailed knowledge of the subject well beyond what can be presented in such a short space. Entire textbooks are devoted to kitchen sanitation and safety. Many organizations, including local and regional health departments, sponsor training programs leading to certificates of competency in food safety. In Canada, many provinces have their own safety regulations, and food-service operators should be familiar with these as well as with federal regulations. Food-service employees in supervisory positions in the United States may be required to hold such a certificate by state or local law. The health and safety of your clientele depend on your diligent study.

SAFETY

Kitchen work is usually considered a relatively safe occupation, at least in comparison with many industrial jobs. Nevertheless, the kitchen presents many hazards. Minor injuries from cuts and burns are common, and more serious injuries are all too possible. The quantity of hot equipment and powerful machinery, combined with the busy, sometimes frantic pace, make it important for everyone to work carefully and with constant attention to the rules of safety.

In Canada, information on legislation, both national and provincial, governing workplace safety rules is provided by the Canadian Centre for Occupational Health and Safety (CCOHS). In the United States, the Occupational Safety and Health Administration (OSHA) established comparable sets of rules. Employers are required to follow these rules and guidelines.

THE SAFE WORKPLACE

Most of this section is concerned with ways workers can prevent certain kinds of accidents, such as cuts, burns, and falls. However, it is much easier to develop and practice habits that prevent accidents if safety is built into the workplace.

The management of a food-service operation must see to it that the structure and equipment have these 10 safety features:

1. Structure, equipment, and electrical wiring in good repair.
2. Adequate lighting on work surfaces and in corridors.
3. Nonslip floors.
4. Clearly marked exits.
5. Equipment supplied with necessary safety devices.
6. Heat-activated fire extinguishers over cooking equipment, especially deep fryers.
7. Conveniently located emergency equipment, such as fire extinguishers, fire blankets, and first-aid kits.
8. Clearly posted emergency telephone numbers.
9. Clearly posted emergency procedures, including the Heimlich maneuver for victims of choking. One or more employees should have received formal training in this procedure. In addition, it is a good idea to train one or more employees in cardiopulmonary resuscitation (CPR).
10. Smooth traffic patterns to avoid collisions between workers.

PREVENTING CUTS

1. Keep knives sharp. A sharp knife is safer than a dull one because it requires less pressure and is less likely to slip.
2. Use a cutting board. Do not cut against a metal surface. Place a damp towel under the board to keep it from slipping.
3. Pay attention to your work when using a knife or cutting equipment. Have only one knife at a time on the cutting board. Knives not in use should be on the worktable near but not on the cutting board.
4. Cut away from yourself and other workers.
5. Use knives only for cutting, not for such jobs as opening bottles.
6. Don't try to catch a falling knife. Step back and let it fall.
7. Don't put knives in a sink, under water, or in any other place they can't be seen.
8. Clean knives carefully, with the sharp edge away from you.
9. Store knives in a safe place, such as a rack, when not in use.

10. Carry knives properly. Hold the knife beside you, point down, with the sharp edge back and away from you. Don't swing your arm. Whenever possible, carry knives in a sheath. Warn people when you are walking past them with a knife in hand.

11. Keep breakable items, such as dishes and glassware, out of the food production area.

12. Don't put breakable items in the pot sink.

13. Sweep up—don't pick up—broken glass.

14. Discard chipped or cracked dishes and glasses.

15. Use special containers for broken dishes and glasses. Don't throw them in with other garbage.

16. If there is broken glass in the sink, drain the sink before trying to take out the glass.

17. Remove all nails and staples when opening crates and cartons, and dispose of them.

PREVENTING BURNS

1. Always assume a pot handle is hot. Don't just grab it with your bare hand.

2. Use dry pads or towels to handle hot pans. Wet ones will create steam, which can burn you.

3. Keep pan handles out of the aisle so people won't bump into them. Also, keep handles away from the open flames of gas burners.

4. Don't fill pans so full they are likely to spill hot foods.

5. Get help when moving heavy containers of hot food.

6. Open lids away from you to let steam escape safely.

7. Use care when opening compartment steamers.

8. Make sure gas is well vented before trying to light ovens or pilot lights. Strike matches before turning on the gas. Also, strike matches away from your body.

9. Wear long sleeves and a double-breasted jacket to protect yourself from spilled or spattered hot foods or fat. Also, wear sturdy leather shoes with closed toes.

10. Dry foods before putting them in frying fat, or hot fat may splatter on you.

11. When placing foods in hot fat, let them fall away from you so fat will not splash on you.

12. Keep liquids away from the deep fryer. If a liquid were spilled into the fryer, the sudden rush of steam could spray hot fat on anyone nearby.

13. Always warn people when you are walking behind them with hot pans or when you are walking behind someone who is working with hot items.

14. Warn service staff about hot plates.

PREVENTING AND DEALING WITH FIRES

1. Know where fire extinguishers are located and how to use them.

2. Use the right kind of fire extinguisher. There are four classes of fire, and fire extinguishers should be labeled according to the kind of fire for which they can be used.

   • **Class A**: wood, paper, cloth, ordinary combustibles

   • **Class B**: burning liquids, such as grease, oil, gasoline, solvents

   • **Class C**: switches, motors, electrical equipment, and so forth

   • **Class K**: cooking appliances involving combustible cooking products such as vegetable or animal oils and fats

   Never use water or a Class A fire extinguisher on a grease fire or electrical fire. You will only spread the fire.

3. Keep a supply of salt or baking soda handy to put out fires on rangetops.

4. Keep hoods and other equipment free from grease buildup.

5. Don't leave hot fat unattended on the range.

6. Smoke only in designated areas. Do not leave burning cigarettes unattended.

7. If a fire alarm sounds and if you have time, turn off all gas and electrical appliances before leaving the building.

8. Keep fire doors closed.

9. Keep exits free from obstacles.

10. Establish and post a plan for emergency evacuation, clearly identifying routes and exits. The plan should include procedures for evacuating customers and other nonemployees.

PREVENTING INJURIES FROM MACHINES AND EQUIPMENT

1. Do not use any equipment unless you understand its operation.

2. Use all guards and safety devices on equipment. Set slicing machines at zero (blade closed) when not in use.

3. Don't touch or remove food from any kind of equipment while it is running, not even with a spoon or spatula.

4. Unplug electrical equipment before disassembling or cleaning.

5. Make sure the switch is off before plugging in equipment.

6. Do not touch or handle electrical equipment, including switches, if your hands are wet or if you are standing in water.

7. Wear properly fitting clothing. Tuck in apron strings to avoid getting them caught in machinery.

8. Use equipment only for its intended purpose.

9. Stack pots and other equipment properly on pot racks so they are stable and not likely to fall.

FIGURE 2.5 Proper lifting technique.

(a) Squat on one knee, and then lift with the leg muscles.

(b) Do not bend over and lift with the back.

PREVENTING FALLS

1. Clean up spills immediately.

2. Throw salt on a slippery spot to make it less slippery while a mop is being fetched.

3. Keep aisles and stairs clear and unobstructed.

4. Don't carry objects too big to see over.

5. Walk, don't run.

6. Stand on a safe ladder, not a chair or piles of boxes, to reach high shelves or to clean high equipment.

PREVENTING STRAINS AND INJURIES FROM LIFTING

1. Lift with the leg muscles, not the back. Figure 2.5 shows proper lifting technique.

2. Don't turn or twist your back while lifting. Make sure your footing is secure.

3. Use a cart to move heavy objects long distances, or get help.

KEY POINTS TO REVIEW

- What does the term *flow of food* mean?

- What does the term *critical control point* mean?

- What are the seven steps of the HACCP system?

- What workplace habits can help you prevent injuries from cuts and burns and from kitchen equipment?

- How can you prevent fires in the kitchen?

TERMS FOR REVIEW

contaminated

hazard

microorganism

pathogen

bacteria

intoxication

infection

toxin-mediated infection

water activity (a_w)

Food Danger Zone

aerobic

anaerobic

facultative

lag phase

potentially hazardous food

sanitize

parasite

physical contamination

allergen

cross-contamination

minimum internal cooking temperature

two-stage cooling method

one-stage cooling method

HACCP

flow of food

critical control point (CCP)

corrective action

Class A, B, C, and K fires

QUESTIONS FOR DISCUSSION

1. True or false: Holding food in a steam table above 140°F (60°C) kills disease-causing bacteria and eliminates the problem of food poisoning. Explain your answer.

2. True or false: Canning foods eliminates air so disease-causing bacteria can't grow. Explain your answer.

3. Which of the following foods can become contaminated by disease-causing organisms?

Chocolate éclairs	Dinner rolls
Potato salad	Shrimp cocktail
Roast beef	After-dinner mints
Lettuce	Saltine crackers
Turkey sandwich	Rice pudding

4. How often should you wash your hands when working on food?

5. Why is temperature control one of the most effective weapons against bacterial growth? What are some important temperatures to remember?

6. What is the importance of cleaning and sanitizing equipment and cutting boards immediately after working on raw poultry?

7. You are making egg salad, and you have just cooked the eggs. What step do you take before chopping the eggs and mixing them with the other ingredients? Why?

8. Is it possible for a dish to be clean but not sanitized? sanitized but not clean?

9. Explain the concepts of hazards and critical control points. Give at least three examples of each.

10. What are the three general categories of potentially hazardous foods? Give examples of each category. Give examples of foods that are not potentially hazardous.

11. True or false: The lower limit of the Food Danger Zone is the proper refrigeration temperature for perishable foods. Discuss.

3

TOOLS AND EQUIPMENT

Thorough knowledge of equipment is essential for success in the kitchen. Few food-service operations depend on nothing more than a range and an oven, an assortment of pots and pans, and knives and other hand tools. Modern technology continues to develop more and more specialized and technically advanced tools to reduce kitchen labor.

Much of this equipment is so complex or so sophisticated that only first-hand instruction and practice will teach you how to operate it effectively and safely. Other items, especially hand tools, are simple and need no explanation but require much practice to develop good manual skills.

A vast array of specialized equipment is available for today's kitchens. It would take a large book, not just a short chapter, to describe all of the many items you will encounter in your career—items such as pasta machines, crêpe machines, burger formers, breading machines, cookie droppers, beverage machines, Greek gyro broilers, doughnut glazers, conveyor fryers, and so on. In this technological age, nearly every year brings new tools to simplify various tasks.

This chapter introduces you to the most commonly used equipment in food-service kitchens. It cannot, in this short space, serve as an operating manual for every model of every machine you will use. It cannot take the place of demonstration by your instructor and of actual experience.

AFTER READING THIS CHAPTER, YOU SHOULD BE ABLE TO

- Identify the dos and don'ts associated with the safe and efficient use of standard kitchen equipment; processing equipment; holding and storage equipment; measuring devices; and knives, hand tools, and small equipment.

INTRODUCTION TO QUANTITY FOOD EQUIPMENT

Before we look at specific items, we must first consider points relating to the use of equipment in general.

FOOD EQUIPMENT CAN BE DANGEROUS

Modern cooking and food processing equipment has an extraordinary capacity to burn, cut, smash, mangle, and amputate parts of the tender human body. This may sound like a harsh way to begin a chapter, but the intent is not to intimidate you or scare you but to inspire a healthy respect for the importance of proper safety and operating procedures.

Never use a piece of equipment until you are thoroughly familiar with its operation and all its features. You must also learn how to know when a machine is not operating correctly. When this happens, shut it down immediately and report the malfunction to a supervisor.

NOT ALL MODELS ARE ALIKE

Each manufacturer introduces slight variations on the basic equipment. While all convection ovens operate on the same basic principle, each model is slightly different, if only in the location of the switches. It is important to study the operating manual supplied with each item or to be taught by someone who already knows that item well and has operated it.

CLEANING IS PART OF THE OPERATING PROCEDURE

Thorough, regular cleaning of all equipment is essential. Most large equipment can be partially disassembled for cleaning. Again, every model is slightly different. Operating manuals should describe these procedures in detail. If a manual is not available, you must get the information from someone who knows the equipment.

When purchasing equipment, look for models that have been tested and certified by recognized agencies that certify products and write standards for food, water, air, and consumer goods. Three prominent agencies are NSF International (formerly the National Sanitation Foundation), CSA International (formerly Canadian Standards Association), and Underwriters Laboratory. These three agencies are recognized internationally. Products meeting their testing requirements are labeled or marked accordingly (Figure 3.1). Criteria govern such factors as design and construction (for example, sealed joints and seams), materials used (for example, nontoxic materials, smooth and easily cleanable surfaces), and performance testing.

CONSERVE ENERGY

At one time, it was standard procedure for the chef to turn on the ovens and ranges first thing in the morning and keep them on all day. Today, high energy costs have made this practice expensive. Fortunately, modern equipment takes less time to heat.

Know the preheating time for all your cooking equipment so you don't need to turn it on before necessary. Plan production so equipment that requires a lot of energy is not on for long periods when not in use.

YOUR HANDS ARE YOUR BEST TOOLS

Machines are intended to be laborsaving devices. However, the usefulness of specialized processing equipment often depends on the volume of food it handles. It takes less time for a cook to slice a few pounds of onions by hand than to set up a slicing attachment, pass the onions through it, and break down and clean the equipment. This is why it is important to develop good manual skills.

FIGURE 3.1

(a) NSF International certification mark.
Courtesy of NSF International.

(b) The CSA International logo
Courtesy of the Canadian Standard Association.

(c) The Underwriters Laboratory logo
Reproduced with permission of Underwriters Laboratory, Inc.

COOKING EQUIPMENT

RANGETOPS

The range is still the most important piece of cooking equipment in the kitchen, even though many of its functions have been taken over by other tools such as steamers, steam kettles, tilting skillets, and ovens.

Types of Cooktops

1. **Open elements** (burners), either electric coils or gas flames. These tops are the fastest to heat and can be turned off after short use. However, cooktop space is limited to one pot per burner.

2. **Flattop** or **hot top** (lightweight). Burners covered with steel plate. More cook space is available. Top supports moderately heavy weights.

3. **Heavy-duty flattop.** Burners covered with heavy cast steel. The top supports many heavy pots. A thick top requires longer preheating. Set burners for different levels, and adjust cooking heat by moving pots to different spots on the top. A **ring-top range** is a type of flattop that has removable rings, allowing access to even more intense heat from the flames below.

4. **Induction cooktops.** The top of an induction unit does not become hot. Rather, it works by magnetically agitating the molecules in steel or iron cookware so the cookware becomes hot. As a result, much less energy is used and the kitchen stays cooler, because only the pots and pans and their contents become hot. There are no hot surfaces or open flames. Also, no warm-up is required. The top can be turned instantly on or off. Small, easily portable induction burners are available. These are useful for off-premise catering operations, for buffet service, and even for tableside heating and cooking. The disadvantage of this cooktop is that only iron or steel pots can be used. Traditional aluminum or copper cookware will not work. Some manufacturers of cookware have responded to the new demand by producing pots and pans made of aluminum sandwiched between layers of stainless steel. In this way, the good heat-conducting qualities of aluminum are preserved as well as adapted to this new technology.

Open-burner gas range with griddle
Courtesy of Vulcan Hart Company.

Flattop range
Courtesy of Vulcan Hart Company.

Ring-top range
Courtesy of Manitowoc Foodservice.

Portable induction cooktop
Courtesy of Fagor America, Inc.

Dos and Don'ts

1. Make sure gas pilots are lit before turning on burners. If burners do not light, turn off gas and allow the gas to ventilate before trying again to light pilots or burners.

2. For maximum heat, adjust air intake so gas flames are blue with a white tip.

3. Do not keep flattop ranges on high heat unless items are being cooked over them. Damage to tops could result.

OVENS

The oven and the rangetop are the two workhorses of the traditional kitchen, which is why they are so often found in the same unit. Ovens are enclosed spaces in which food is heated, usually by hot air or, in some newer kinds of ovens, by microwaves or infrared radiation.

In addition to roasting and baking, ovens can do many of the jobs normally done on the rangetop. Many foods can be simmered, stewed, braised, or poached in the oven, freeing the rangetop and the chef's attention for other tasks.

There are many kinds of ovens beyond those discussed here, but they are often for specialty or high-volume uses. These include **conveyor ovens**, which carry foods through the oven on a steel conveyor belt; **holding ovens** or warmers, which are designed to hold many types of foods at serving temperatures for extended periods without drying out or overcooking (this category includes ovens that also cook the food, then automatically switch to holding temperature); and high-volume **roll-in ovens**, with large doors into which one can roll carts loaded with trays of food.

CONVENTIONAL OVENS

Conventional ovens operate simply by heating air in an enclosed space. The most common ovens are part of the range unit, although separate oven units or ovens as part of a broiler unit are also available. **Stack ovens** are units that consist of individual shelves or decks arranged one above the other. Pans are placed directly on the oven deck rather than on wire shelves. Temperatures are adjustable for each deck.

Stack or deck ovens
Copyright Blodgett Corporation.

Dos and Don'ts

Many of these points apply to other types of ovens as well.

1. Preheat ovens thoroughly, but no longer than necessary, to avoid excess energy use.
2. To avoid high energy loss and interruption of cooking, do not open the door more often than necessary.
3. Space items well to allow for heat circulation.
4. Be sure the pilot light is on before turning on gas ovens.

CONVECTION OVENS

Convection ovens contain fans that circulate the air and distribute the heat rapidly throughout the interior. Because of the forced air, foods cook more quickly at lower temperatures. Also, shelves can be placed closer together than in conventional ovens without blocking the heat flow.

Dos and Don'ts

1. For most products, set the temperature 25°–50°F (15°–30°C) lower than you would a conventional oven. Check the manufacturer's recommendations.
2. Watch cooking times closely. The forced heat cooks foods more quickly and tends to dry out some foods if they are overcooked. Roasts shrink more than they do in conventional ovens.
3. Many convection oven models should not be operated with the blower switch off, as the motor may burn out.
4. The forced air of a convection oven may deform soft items. Cake batters, for example, develop ripples. Check the manufacturer's recommendations.

Convection oven
Courtesy of Vulcan Hart Company.

Revolving Ovens

Revolving ovens, also called **reel ovens**, are large chambers containing many shelves or trays on an attachment like a Ferris wheel. This oven eliminates the problem of hot spots, or uneven baking, because the mechanism rotates the foods throughout the oven.

Revolving ovens are used in bakeshops and in high-volume operations.

Slow-Cook-and-Hold Ovens

The traditional oven is nothing more than a heated box equipped with a thermostat. Some modern ovens have more sophisticated features, such as computerized electronic controls and special probes that sense when a roast is done and tell the oven to switch from cooking temperature to holding temperature.

Many of these ovens are designed to be especially useful for low-temperature roasting (see p. 415). The sensitive controls make it possible to cook at steady, reliable temperatures of 200°F (95°C) or lower and to hold foods at 140°F (60°C) for long periods. Large cuts of meat take many hours to roast at a low temperature like 200°F (95°C). By setting the controls in advance, the operator can even let meats roast overnight, unattended.

These ovens are available as convection ovens and as regular stationary-air ovens.

Combination Steamer Ovens

The **combination steamer oven**, also called a **combi oven**, can be operated in three modes: as a convection oven, as a convection steamer (see p. 45), and, with both functions on at once, as a high-humidity oven. Injecting moisture into an oven while roasting meats can help reduce shrinkage and drying.

Modern combi ovens are equipped with computerized controls than enable them to be programmed for multiple changes of temperature and humidity during the cooking cycle, customizing the program for each product to be cooked. In addition, built-in temperature probes to be inserted in the product being cooked can be set to turn the oven off or change it to holding mode when the item reaches the desired internal temperature. Some special units are also equipped to inject smoke into the oven.

Because combi ovens have precise temperature control and can be set for very low cooking temperatures, they can also be used in some sous vide cooking (see page 118).

Combination steamer oven
Courtesy of Vulcan Hart Company.

Barbecue Ovens or Smoke Ovens

Barbecue ovens are like conventional ovens, but with one important difference: They produce wood smoke, which surrounds the food and adds flavor while it bakes or roasts. Special woods such as hickory, mesquite, or fruitwoods such as apple or cherry must be added to the smoke-producing part of the oven according to the manufacturer's instructions. This device is usually nothing more complicated than an electric heating element that heats small blocks or chips of the wood so they are hot enough to smoke but not hot enough to burst into flame.

Depending on the model, various cooking features are available. Thus, ovens may have smokeless roast/bake cycles, cold-smoke cycles (with the smoke element on but the oven off), holding cycles, and broiling capabilities.

A barbecue oven that uses wood smoke should not be confused with a *smoker*, used for making hot-smoked and cold-smoked foods, as discussed in Chapter 27. The interior of a smoker is shown on page 841.

Smoke oven
Courtesy of Cookshack, Inc.

Infrared or Reconstituting Ovens

Infrared units contain quartz tubes or plates that generate intense infrared heat. These ovens are used primarily for reconstituting frozen foods. They bring large quantities of foods to serving temperature in a short time. The heat is even and controllable.

Wood-burning Ovens

Ancient ovens were made of heavy masonry, brick, or clay and heated by building a wood fire inside them. In this type of oven, the brick absorbs the intense heat of the fire and cooks foods long after the fire has gone out and the ashes removed. Items such as breads and pizzas are baked directly on the floor of the oven, just as in modern deck ovens. Wood-burning ovens have once again come into fairly wide use, mostly in specialty restaurants that feature roasted meats, pizzas, and similar items. The foods absorb some of the appealing wood-smoke flavors and aromas.

Because a traditional wood-burning oven lacks a temperature control knob, some experience is needed to produce baked goods of consistent quality. During the firing of the oven, a wood fire is built directly on the hearth inside the oven. The flue and the door are kept open to allow the smoke to escape. After the fire is stopped, the ashes and embers are swept out.

Modern wood-burning oven
Courtesy of Earthstone Ovens.

Microwave oven
Courtesy of Vulcan Hart Company.

The door and flue are closed to allow the heat to equalize inside the dome of the oven. Initial temperatures inside the oven at this point may be as high as 900°F (480°C). Pizza can be baked at this temperature. For meats and breads, the oven is left to stand until the temperature gradually drops to the desired range. Alternatively, a fire can be kept going in the back of the oven while foods are roasted toward the front, with the flue kept open.

Combination ovens, also available, are more controllable. These can be fired by gas alone, by wood alone, or by gas and wood together.

Microwave Ovens

In these ovens, special tubes generate microwave radiation, which creates heat inside the food. Microwave cooking is discussed in detail in Chapter 6.

BROILERS AND SALAMANDERS

Broilers are sometimes called **overhead broilers** to avoid confusing them with grills. Overhead broilers generate heat from above, and food items are placed on a grate beneath the heat source. Broiling is a favorite way of preparing steaks, chops, chicken, and many other items.

Heavy-duty broilers produce very high heat and consume vast quantities of energy. Some broilers are said to go as high as 2,000°F (1,100°C) at the burner.

Heavy-duty broiler
Courtesy of Vulcan Hart Company.

Salamander (above range)
Courtesy of Vulcan Hart Company.

Gas grill
Courtesy of Vulcan Hart Company.

Foods must be watched closely to avoid burning. Cooking temperature is adjusted by raising or lowering the grate that holds the food.

Salamanders are small broilers used primarily for browning or glazing the tops of some items. They may also be used for broiling small quantities during off-peak hours. Salamanders are usually mounted above the range, as illustrated in the photo. In addition, the photo of an open-burner gas range on page 39 shows a salamander under the griddle.

GRILLS

Grills are used for the same cooking operations as broilers, except the heat source is below the grid that holds the food rather than above it. Many people like grilled foods because of their charcoal taste, which is created by smoke from meat fats that drip into the heat source.

Although smoke from meat fats creates the taste people associate with grilled foods, actual wood-smoke flavors such as hickory or mesquite can be added to foods if those woods are burned in the grill under the food. In order to do this, you must use a grill designed to burn such fuels.

Types

Many grill models are in use. The major differences in operation among them are due to the difference in heat source—gas, electricity, or charcoal.

To operate, set areas of the grill to different temperatures and place foods in the areas with the appropriate cooking temperature. Keep grills clean, as the high temperatures can easily start grease fires.

GRIDDLES

Griddles are flat, smooth, heated surfaces on which food is cooked directly. Pancakes, French toast, hamburgers and other meats, eggs, and potato items are the foods most frequently cooked on a griddle. Griddles are available as separate units or as part of a rangetop (there is a griddle on the right side of the range pictured on page 39).

Clean griddle surfaces after every use so they will cook at peak efficiency. Polish with a griddle stone or griddle cloth until the surface shines. Follow the grain of the metal to avoid scratching.

Condition griddles after each cleaning or before each use to create a nonstick surface and to prevent rusting. Procedure: Spread a thin film of oil over the surface and heat to 400°F (200°C). Wipe clean and repeat until griddle has a smooth, nonstick finish.

ROTISSERIES

Rotisserie broilers cook meats and other foods by turning them slowly in front of electric- or gas-powered heating elements. Even though classical cooking theory categorizes spit-cooking as roasting, these cookers are more closely related to broilers in that the foods are cooked by the infrared heat of the elements.

Although they are especially suitable for chicken and other poultry, rotisseries can be used to cook any meat or other food that can be held on a spit or in any of various attachments or accessories.

Both enclosed (ovenlike) rotisseries and open or unclosed units are available. Small units hold about 8 chickens, and sizes range all the way to very large models that can hold as many as 70 chickens.

Because the heating elements are on the side (or sometimes above), the fats and juices don't drip into the flames as they do with grills. Drip pans catch juices, which can be used for basting or gravy making.

DEEP FRYERS

A deep fryer has only one use: to cook foods in hot fat. Yet because of the popularity of fried foods, this function is an important one.

Standard deep fryers are powered by either gas or electricity and have thermostatic controls that maintain fat at preset temperatures.

Automatic fryers remove food from the fat automatically after a preset time.

Pressure fryers are covered fry kettles that fry foods under pressure. Foods cook faster, even at a lower fat temperature.

Dos and Don'ts

Frying procedures and the care of frying fat are discussed in detail in Chapter 6. The following points relate to the operation of the equipment:

1. When filling kettles with solid fats, set the thermostat at 250°F (120°C) until the fat has melted enough to cover the heating elements.

2. Keep the kettles filled to the fill line.

3. Make sure the drain valve is shut before adding fat to the empty kettle.

4. Check the accuracy of the thermostat regularly by reading the fat temperature with a thermometer.

Deep fryers
Courtesy of Vulcan Hart Company.

Cleaning

Cleaning procedures differ greatly depending on the model. Here is a general procedure:

1. Shut off the power.
2. Drain the fat through a filter into a dry container (unless you are discarding it). Before you start, be sure the container is large enough to hold all the fat.
3. Flush food particles from the sides and bottom of the kettle with some of the hot fat.
4. Wash the kettle with a mild detergent solution. If the kettle is not removable, turn on the fryer and bring the detergent solution almost to a boil (beware of foaming over). Scrub with a stiff brush.
5. Drain and rinse thoroughly with clean water.
6. Dry the kettle, heating elements, and baskets thoroughly.
7. Refill with strained or fresh fat. (If the fat is drained by a spigot or valve, be sure it is closed before you refill the fryer.)

TILTING SKILLET

The tilting skillet, also known as the **tilting brazier** or **tilting fry pan**, is a versatile and efficient piece of equipment. It can be used as a griddle, fry pan, brazier, stewpot, stockpot, steamer, and bain-marie or steam table.

The tilting skillet is a large, shallow, flat-bottomed pot. To look at it another way, it is a griddle with sides 6 inches (24 cm) high, plus a cover. It has a tilting mechanism that enables liquids to be poured out of it. Power may be gas or electric.

Clean the skillet immediately after each use, before food has time to dry on. Add water, turn on the skillet to heat it, and scrub thoroughly.

Tilting skillet
Courtesy of Vulcan Hart Company.

STEAM-JACKETED KETTLES

Steam-jacketed kettles, or **steam kettles**, are sometimes thought of as stockpots heated not just on the bottom but on the sides as well. This comparison is only partly accurate. Steam kettles heat much more quickly and have more uniform and controllable heat than pots on the range.

Types

Steam kettles range in capacity from 2 gallons (7.5 L) to over 100 gallons (378.5 L). Some large institutional kettles hold 4,000 gallons (15,142 L). **Tilt** or **trunnion kettles** can be tilted for emptying, either by turning a wheel or by pulling a lever. **Nontilt kettles** are emptied by a spigot and drain on the bottom. Heat is controlled by regulating the steam flow or by adjusting the thermostat. Steam may be from an outside source or self-generated. *Exercise caution when operating all steam equipment. Steam can cause serious burns.*

Clean immediately after use to avoid food drying on surfaces. Disassemble the spigot and drain, and clean with a bottle brush.

Small tilt (trunnion) kettle
Courtesy of Vulcan Hart Company.

Large floor-model steam kettle
Courtesy of Vulcan Hart Company.

STEAM COOKERS

Steam cookers are ideal for cooking vegetables and many other foods rapidly and with minimum loss of nutrients and flavor. For this reason, they are becoming more common in both large and small kitchens.

Types

Pressure steamers cook foods under a pressure of 15 pounds per square inch (1.05 kg/cm) in high-pressure steamers or 4–6 pounds per square inch (0.28–0.42 kg/cm) in low-pressure steamers. They are operated by a timer, which shuts the equipment off after a preset time. The door cannot be opened until the pressure returns to zero.

Pressureless or **convection steamers** do not operate under pressure. Jets of steam are directed at the food to speed the heat transfer, just as the fan in a convection oven speeds cooking. The door can be opened any time during cooking.

All steamers hold standard-size counter pans (12 × 20 inches or 325 × 530 mm) or fractions thereof. Their capacity varies from one to many pans.

Steamer operation varies greatly depending on the model. Check the operating manual and be sure you understand a particular model well before attempting to operate it.

Caution is important with all steam equipment because of the danger of severe burns.

Convection steamer
Courtesy of Vulcan Hart Company.

Pressure steamer
Courtesy of Vulcan Hart Company.

KEY POINTS TO REVIEW

- What are the four basic types of cooktops?

- What is the difference between a standard oven and a convection oven? List four guidelines for operating each type.

- What is the difference between a broiler, a salamander, a grill, and a griddle?

- What are four guidelines for operating a standard deep fryer? Describe the general procedure for cleaning a deep fryer.

PROCESSING EQUIPMENT

MIXERS

Vertical mixers are important and versatile tools for many kinds of food mixing and processing jobs, both in the bakeshop and in the kitchen. (The kind of mixer shown here is called the *planetary mixer,* because the beater attachment revolves on its own axis as it also rotates around the bowl, like the motion of planets around the sun. The baking industry also uses a number of specialized mixers not discussed here.)

Types

Bench-model mixers range in capacity from 5 to 20 quarts (5 to 20 L). Floor models are available as large as 140 quarts (133 L). Adaptor rings enable several bowl sizes to be used on one machine. Most mixers have three operating speeds.

Small table-model mixer
Courtesy of Hobart Corporation.

Large floor-model mixer
Courtesy of Hobart Corporation.

Agitator Attachments

There are three main mixing attachments, plus some specialized ones. The **paddle** is a flat blade used for general mixing. The **wire whip** is used for such tasks as beating cream and eggs and making mayonnaise. The **dough arm** or **hook** is used for mixing and kneading yeast doughs.

Mixer attachments: (left) paddle, (center) whip, (right) dough arm
Courtesy of Hobart Corporation.

Dos and Don'ts

1. Make sure the bowl and the mixing attachment are firmly in place before turning on the machine.
2. Make sure you are using the right size attachment for the bowl. Using a 40-quart paddle with a 30-quart bowl, for example, could cause serious damage. Sizes in quarts are marked on the sides of large bowls and on the tops of attachments.
3. Turn off the machine before scraping down the bowl or inserting a spoon, scraper, or hand into the bowl. Mixer motors are powerful and can cause serious injury.
4. Turn off the machine before changing speeds.

FOOD CUTTER

Rotation chopper
Courtesy of Hobart Corporation.

The food cutter or rotation chopper, familiarly known as the **buffalo chopper**, is a common piece of equipment used for general food chopping. A variety of attachments (described in the next section) makes it a versatile tool.

General Operation

Food is placed in a rotating bowl, which carries the food to a pair of knives spinning rapidly under a cover. The fineness of the cut depends on how long the food is left in the machine.

Dos and Don'ts

1. Always make sure the machine is completely assembled before use.
2. Close the cover lock knob, or the machine will not turn on.
3. Never reach under the bowl cover while the machine is running.
4. For uniform chopping, place the food in the bowl all at one time.
5. Keep the knives sharp. Dull knives bruise food rather than cut it cleanly.

Attachments for Mixers and Food Choppers

Grinder attachment (on separate motor)
Courtesy of Hobart Corporation.

The following are the most common of the many attachments designed to fit both the food chopper and the vertical mixer:

1. The **food grinder** is used mostly for grinding meats, although other moist foods may be ground also. Food is forced through a feed tube into a screw, which pushes the food through holes in a plate, at which point it is cut by a rotating blade. The size of the holes regulates the fineness of the grind.

Make sure the rotating blade is attached properly, cutting edge out, when assembling the grinder.

2. The **slicer/shredder** consists of a hopper and a lever that feeds the food onto a rotating disk or plate. The plate cuts or shreds the food and drops it into a receiving container. The slicing plate may be adjusted to cut various thicknesses.

3. The **dicer** attachment forces foods through a grid-type blade that cuts them into perfect dice. Blades of different sizes may be used.

Slicer/shredder attachment
Courtesy of Hobart Corporation.

SLICER

The **slicer** is a valuable machine because it slices foods more evenly and uniformly than can be done by hand. This makes it valuable for portion control and for reducing cutting loss.

Types

Most modern slicers have blades set at an angle. Slices fall away from these blades with less breaking and folding than from vertical blades.

With manual machines, the operator must move the carriage back and forth to slice the food. Automatic machines move the carriage with an electric motor.

Dos and Don'ts

1. Be sure the machine is properly assembled before using.

2. Always use the end weight to press the food against the blade. This protects the hand from serious cuts and provides a more even pressure on the food, resulting in more uniform slices.

3. Set the thickness control knob to zero when the machine is not in use or is being cleaned.

4. Always unplug the machine before dismantling and cleaning.

5. Keep the blade sharp with the sharpening stones provided with the slicer.

Slicer
Courtesy of Hobart Corporation.

VERTICAL CUTTER/MIXER

The vertical cutter/mixer (VCM) is like a large, powerful, high-speed blender. It is used to chop and mix large quantities of foods rapidly. It can also be used for puréeing (soups, for example) and for mixing liquids.

Types

VCMs range in size from 15 to 80 quarts (14 to 75 liters). The small models have a hand-operated mixing baffle, which moves the foods into the blades. Larger machines have automatic baffles.

Dos and Don'ts

1. Watch processing times closely. Chopping times are so short that an extra second can make cabbage soup out of coleslaw.

2. Make sure the machine is properly assembled before use.

3. After turning off the machine, allow the blades to come to a full stop before opening the cover.

4. Keep the blades sharp. Dull blades bruise food.

FOOD PROCESSOR

Food processors were used in commercial kitchens long before home models were introduced. Professional models are two to four times larger than the largest home models. They consist of a motor in a heavy base topped by a cylindrical work bowl containing an **S**-shaped

blade. Processors are used to chop or purée foods, including raw or cooked meats, and to mix or emulsify such items as sauces and flavored butters. With special disk attachments in place of the standard blade, they can also slice, shred, and julienne solid foods such as vegetables.

In basic design, a food processor is similar to a vertical cutter/mixer. The same dos and don'ts should be observed.

BLENDER

Like the VCM and the food processor, a blender consists of a motor in a base, topped by a container with a spinning blade. However, because the blender's container is tall and narrow, it is more suited for mixing and puréeing liquids than for chopping solid foods. In the commercial kitchen, the blender is used to mix, purée, and emulsify liquids such as soups, sauces, and batters. It is also used in bars and coffeehouses to prepare certain drinks.

Blender motors may have from 2 to 10 speeds, or even more. The containers are made of stainless steel, glass, or plastic. The blade assembly at the base of the container can be disassembled for thorough cleaning.

Food processor
Courtesy of Robot Coupe USA.

Professional blender
Courtesy of Waring Products.

IMMERSION BLENDER

An immersion blender, also called a **stick blender** or **burr mixer**, consists of the blade of a blender, protected by a guard, at the bottom end of a long wand or shaft with a motor at the top. Using an immersion blender, the cook can purée or mix hot or cold foods in any container without transferring them to a blender jar. This makes it possible to blend hot foods in their cooking pots. The largest stick blenders are long enough to purée foods in large steam kettles.

SOUS VIDE EQUIPMENT

Two types of equipment are essential for operations that use sous vide (see pages 7 and 117) cooking techniques: vacuum-packaging equipment and cooking equipment.

The most commonly used packaging equipment in food service is a *chamber vacuum packer*. Food is placed in a specially designed plastic bag and inserted in the chamber. The chamber is then closed and the machine pulls the air from the bag and seals the bag. Vacuum pressure may be varied from low, for delicate, easily crushed foods, to high, for firmer foods.

Precise control of cooking temperature is at the heart of sous vide cooking, and for this, *immersion circulators* are used to heat the water in a hot-water bath. Immersion circulators have a heating element, a pump that constantly circulates the water, and a temperature control that can keep the water at a steady temperature to within a fraction of a degree.

Combi ovens (see page 41), with their precise temperature control and low-heat cooking capabilities, are also used for sous vide cooking.

Rapid cooling is also important in sous vide cooking. Larger operations may want to invest in a blast cooler. For smaller kitchens, a regular ice bath may be sufficient.

Chamber vacuum packer
Courtesy of PolyScience.

Immersion circulator
Courtesy of PolyScience.

HOLDING AND STORAGE EQUIPMENT

HOT FOOD HOLDING EQUIPMENT

Several types of equipment are used to keep food hot for service. This equipment is designed to hold foods above 135°F (57°C) in order to prevent the growth of bacteria that can cause disease. Because food continues to cook at these temperatures, it should be held for as short a time as possible.

1. **Steam tables** are standard holding equipment for serving lines. Standard-size counter pans or hotel pans are used as inserts to hold the foods. Flat or domed covers may be used to cover the foods.

    Check water levels in steam tables periodically to make sure they don't go dry. Electrically heated counters that operate dry—without steam—are also available.

2. A **bain-marie** is a hot-water bath. Containers of foods are set on a rack in a shallow container of water, which is heated by electricity, gas, or steam. The bain-marie is used more in the production area, while the steam table is used in the service area.

3. **Overhead infrared lamps** are used in service areas to keep plated food warm before it is picked up by the service staff. They are also used for keeping large roasts warm.

    Foods dry out quickly under holding lamps. This is a disadvantage for almost all foods except French fries and other deep-fried foods, which lose their crispness if they are kept moist.

Steam table
Courtesy of Hobart Corporation.

COLD FOOD STORAGE EQUIPMENT

The quality of the food you serve depends to a great degree on refrigeration equipment. By keeping foods cold, usually below 41°F (5°C), the refrigerator (known in the trade as the *cooler* or the *box*) guards against spoilage and bacterial growth.

Several types of refrigerator are used in food service. The **walk-in** is a room-size refrigerator with built-in shelves on the walls. Walk-ins can be customized to fit nearly any available space.

The **reach-in** is a standard upright refrigerator similar in shape to a large home refrigerator, but without the freezer unit. It may have shelves or simply brackets for holding sheet pans.

Small reach-ins that fit under counters, as well as refrigerated drawers, are used in prep and service areas of the kitchen. Finally, refrigerated display cases are used in retail and the dining areas of delis, bakeries, diners, coffee shops, and some restaurants.

Freezers are used to hold foods for longer times, or to store foods purchased in frozen form. Like refrigerators, freezers are available as walk-ins, reach-ins, and smaller units.

To enable refrigerators and freezers to work at top efficiency, observe the following rules:

1. Place items far enough apart and away from the inside walls of refrigerators so cold air can circulate. Freezers, however, work most efficiently when they are full.

2. Keep the door closed as much as possible. When storing or removing an item, do it quickly and shut the door.

3. Keep stored foods well wrapped or covered to prevent drying and transfer of odors. Meats are an exception to this rule (see p. 482).

4. Keep refrigerators spotlessly clean.

KEY POINTS TO REVIEW

- What are four guidelines for the safe operation of a vertical mixer? What three types of beater attachment are used most often on a mixer, and what are they used for?

- What are five guidelines for the correct operation of a rotation chopper (buffalo chopper)?

- What are five guidelines for the safe operation of a slicing machine?

- What are five kinds of equipment used to keep foods out of the Food Danger Zone?

POTS, PANS, AND CONTAINERS

METALS AND CONDUCTIVITY

A good cooking utensil distributes heat evenly and uniformly. A poor cooking utensil develops hot spots that are likely to burn or scorch the food being cooked. Two factors affect a pan's ability to cook evenly:

1. *Thickness of the metal*. A heavy-gauge pot cooks more evenly than one made of thin metal. Thickness is most important on the bottom.

2. *Kind of metal*. Different metals have different conductivity, or the speed at which they transfer or disperse heat. The following materials are used for cooking equipment:

   - *Aluminum* is used for most cooking utensils in food-service kitchens. It is a good conductor, and its light weight makes pots and pans easy to handle. Because it is a relatively soft metal, it should not be banged around or abused.

     Do not use aluminum for storage or for long cooking of strong acids because it reacts chemically with many foods. Also, it tends to discolor light-colored foods such as sauces, especially if they are stirred or beaten with a metal spoon or whip.

     Pans made of *anodized aluminum*, sold under such brand names as Calphalon, have surfaces that are harder and more corrosion-resistant than regular aluminum pans. Although this is not, strictly speaking, a nonstick finish, it is less porous than untreated aluminum, so foods are less likely to stick. Also, it is more resistant to acids than regular aluminum, and it will not discolor light-colored foods. Its disadvantages are that it is more expensive than and not quite as durable as standard aluminum.

     Aluminum, of course, cannot be used on induction cooktops, which work only with steel or iron (see p. 39).

   - *Copper*, the best heat conductor of all, was once widely used for cooking utensils. However, it is extremely expensive and requires a great deal of care. In addition, it is heavy. Today it is used mostly for show, although a few high-end restaurants use it for cooking as well.

     Copper reacts chemically with many foods to create poisonous compounds, so copper pans must be lined with another metal, such as tin or stainless steel.

   - *Stainless steel* is a poor heat conductor. Cooking pots and pans made of it tend to scorch foods easily because the heat does not disperse throughout the pan quickly and evenly. Stainless steel is ideal for storage containers because it does not react with foods as aluminum does. It is also used for low-temperature cooking or holding equipment, such as steamer pans and counter pans, where scorching or hot spots are not a problem.

     Stainless-steel pots and pans are available with a heavy layer of copper or aluminum bonded to the bottom. Heavy aluminum pans may also be lined with stainless steel on the inside, or on both the inside and outside. This feature gives the advantages of stainless steel (hardness, durability, nonreactivity with acid foods, and nondiscoloration of light sauces) with the heat-conducting qualities of copper or aluminum. These pans are usually expensive.

   - *Cast iron* is a favorite material with many chefs because of its ability to distribute heat evenly and to maintain high temperatures for long periods. It is used in griddles and heavy skillets. Cast iron cracks easily if dropped. It rusts quickly unless kept properly conditioned (see p. 773) and dry.

   - *Porcelain enamel-lined pans* should not be used. In fact, they are forbidden by some health departments. They scratch and chip easily, providing good hiding places for bacteria. Also, certain kinds of gray enamel can cause food poisoning if chipped.

   - *Nonstick plastic-type coatings*, known by brand names including Teflon and Silverstone, provide a slippery finish, but one that requires a lot of care because it is easily scratched. Do not use metal spoons or spatulas with this equipment. Instead, use tools made of plastic, silicone, or wood. Do not use abrasive materials to clean the nonstick surface.

Nonstick pans are best reserved for eggs and other items that are likely to be damaged if they stick. Many chefs keep a set of nonstick egg pans and use them for no other purpose. In addition, these pans are useful for dietary cooking because they enable cooks to sauté foods with little or no added fat.

Nonstick coatings should not be used for sautéing and braising procedures that involve deglazing to make a sauce (see p. 177). Foods do not brown as well in nonstick pans as in traditional metal pans, and they do not form a fond (the flavorful browned bits that stick to the pan) that can be deglazed to make a sauce or braising liquid.

- *Glass* and *earthenware* have limited use in commercial kitchens because they break easily. They are poor conductors of heat but are resistant to corrosion and food acids.

POTS AND PANS AND THEIR USES

1. Stockpot.

A large, deep, straight-sided pot for preparing stocks and simmering large quantities of liquids. Stockpots with spigots allow liquid to be drained off without disturbing the solid contents or lifting the pot. Sizes: 8–200 quarts (liters).

2. Saucepot.

A round pot of medium depth. Similar to a stockpot but shallower, making stirring or mixing easier. Used for soups, sauces, and other liquids. Sizes: 6–60 quarts (liters).

Stockpot   **Stockpot with spigot**

3. Brazier.

A round, broad, shallow, heavy-duty pot with straight sides. Also called a *rondeau*. Used for browning, braising, and stewing meats. Sizes: 11–30 quarts (liters).

Saucepot

Brazier

4. Saucepan.

Saucepan

Similar to a small, shallow, light saucepot, but with one long handle instead of two loop handles. May have straight or slanted sides. Used for general rangetop cooking. Sizes: 1 1/2–15 quarts (liters).

5. Sauté pan, straight-sided.

Also called a *sautoir*. Similar to a shallow, straight-sided saucepan, but heavier. Used for browning, sautéing, and frying. Because of its broad surface area, the sauté pan is used for cooking sauces and other liquids when rapid reduction is required. Sizes: 2 1/2–5 inches (65–130 mm) deep; 6–16 inches (160–400 mm) in diameter.

Straight-sided sauté pan

6. Sauté pan, slope-sided.

Also called a *sauteuse*. Used for general sautéing and frying of meats, fish, vegetables, and eggs. The sloping sides allow the cook to flip and toss items without using a spatula, and they make it easier to get at the food when a spatula is used. Sizes: 6–14 inches (160–360 mm) top diameter.

Slope-sided sauté pan

7. Cast-iron skillet.

Very heavy, thick-bottomed fry pan. Used for pan-frying when steady, even heat is desired.

8. Double boiler.

A pot with two sections. The lower section, similar to a

Cast-iron skillet

stockpot, holds boiling water. The upper section holds foods that must be cooked at low temperatures and cannot be cooked over direct heat. Size of top section: 4–36 quarts (liters).

Double boiler

Sheet pan

Fish poacher
Courtesy of RSVP International, Inc.

Wok

Hotel pan

9. **Sheet pan or bun pan.**
A shallow, rectangular pan (1 inch/25 mm deep) for baking cakes, rolls, and cookies, and for baking or broiling certain meats and fish. Sizes: full pan, 18 × 26 inches (46 × 66 cm); half-pan, 18 × 13 inches (46 × 33 cm).

10. **Bake pan.**
A rectangular pan about 2 inches (50 mm) deep. Used for general baking. Available in a variety of sizes.

Bake pan

11. **Roasting pan.**
A large rectangular pan, deeper and heavier than a bake pan. Used for roasting meats and poultry.

12. **Fish poacher.**
A long, narrow, straight-sided pan with a removable rack insert. Used for poaching whole fish.

Roasting pan

13. **Wok.**
A round-bottomed steel pan with two loop handles. Used for stir-frying, especially in Chinese cuisine. Woks are best used with special burner units that have a high heat output and a broad ring-shaped support that holds the wok steady during cooking.

14. **Hotel pan, also called counter pan, steam table pan, or service pan.**
A rectangular pan, usually made of stainless steel. Designed to hold foods in service counters. Also used for baking, steaming, and subsequent serving. Also used for storage. Standard size: 12 × 20 inches. Fractions of this size (1/2, 1/3, etc.) are also available. Standard depth: 2 1/2 inches (65 mm). Deeper sizes are also available. (Standard metric pan is 325 × 530 mm.)

15. **Bain-marie insert, usually called simply bain-marie.**
A tall, cylindrical stainless-steel container. Used for storage and for holding foods in a bain-marie (water bath). Sizes: 1–36 quarts (liters).

Bain-marie inserts

16. **Stainless-steel bowl.**
A round-bottomed bowl. Used for mixing, whipping, and producing hollandaise, mayonnaise, whipped cream, and egg white foams. Round construction enables whip to reach all areas. Available in many sizes.

MEASURING DEVICES

The following equipment is discussed in terms of U.S. measurements. Comparable items in metric units are also available.

1. **Scales.** Most recipe ingredients are measured by weight, so accurate scales are important. **Portion scales** are used for measuring ingredients as well as for portioning products for service. Traditional portion scales are spring-operated and usually have a dial to indicate weight. More accurate **digital scales** are electrically operated and provide a digital readout. The **baker's balance scale** is discussed in Chapter 30.

2. **Volume measures** used for liquids have lips for easy pouring. Sizes are pints, quarts, half-gallons, and gallons. Each size is marked off into fourths by ridges on the sides.

3. **Measuring cups** are available in 1-, 1/2-, 1/3-, and 1/4-cup sizes. They can be used for both liquid and dry measures.

4. **Measuring spoons** are used for measuring very small volumes: 1 tablespoon, 1 teaspoon, 1/2 teaspoon, and 1/4 teaspoon. They are used most often for spices and seasonings.

Portion scale

Digital scale

Liquid volume measure

TABLE 3.1 Scoop Sizes

| Scoop Number | U.S. Measure | | Metric Measure | |
	Volume	Approximate Weight	Volume	Approximate Weight
6	²/₃ cup	5 oz	160 mL	140 g
8	¹/₂ cup	4 oz	120 mL	110 g
10	3 fl oz	3–3¹/₂ oz	90 mL	85–100 g
12	¹/₃ cup	2¹/₂–3 oz	80 mL	70–85 g
16	¹/₄ cup	2–2¹/₂ oz	60 mL	60–70 g
20	1¹/₂ fl oz	1³/₄ oz	45 mL	50 g
24	1¹/₃ fl oz	1¹/₃ oz	40 mL	40 g
30	1 fl oz	1 oz	30 mL	30 g
40	0.8 fl oz	0.8 oz	24 mL	23 g
60	¹/₂ fl oz	¹/₂ oz	15 mL	15 g

Note: Weights vary greatly with different foods, depending on how compact they are. Best practice is to weigh a scoopful of an item before proceeding with portioning.

5. **Ladles** are used for measuring and portioning liquids. The size, in ounces, is stamped on the handle.

6. **Scoops** come in standard sizes and have a lever for mechanical release. They are used for portioning soft solid foods. Scoop sizes are listed in Table 3.1. The number of the scoop indicates the number of level scoopfuls per quart. In actual use, a rounded scoopful is often more practical than a level scoopful, so exact weights will vary.

Scoop

Ladles

7. **Thermometers** measure temperature. There are many kinds for many purposes.

- A **meat thermometer** indicates internal temperature of meats. It is inserted before cooking and left in the product during cooking.

Meat thermometer

- An **instant-read thermometer** gives readings within a few seconds of being inserted in a food product. It reads from 0°F to 220°F. Many chefs carry these in their jacket pocket like a pen, ready whenever needed. Instant-read thermometers must not be left in meats during roasting, or they will be damaged.

- **Fat thermometers** and **candy thermometers** test temperatures of frying fats and sugar syrups. They read up to 400°F.

- Special thermometers are used to test the accuracy of oven, refrigerator, and freezer thermostats.

Instant-read thermometer with holder

KNIVES, HAND TOOLS, AND SMALL EQUIPMENT

KNIFE MATERIALS

The metal a knife blade is made of is an important consideration, as the metal must be able to take and hold a very fine edge:

1. **Carbon steel** was for many years the traditional favorite because it can be honed to an extremely sharp edge. Its disadvantages are that it corrodes and discolors easily, especially when used with acid foods and onions. Also, it discolors some foods (such as

hard-cooked eggs) and may leave a metallic taste. Because of these disadvantages, it has given way to high-carbon stainless steel (described below), which is now the preferred material for the best knives.

2. **Traditional stainless-steel alloys** will not rust or corrode, but they are much harder to sharpen than carbon steel. Stainless steel is used mostly for low-cost, lightweight knives.

3. **High-carbon stainless steel** is a relatively new alloy that combines the best aspects of carbon steel and stainless steel. It takes an edge almost as well as carbon steel, and it will not rust, corrode, or discolor. Knives made of this material are highly prized and relatively expensive. Most high-quality knives today are made of high-carbon stainless steel.

KNIFE PARTS

Chef's knives and other knives have a number of parts, and you should be familiar with their names. These parts are illustrated in the diagram.

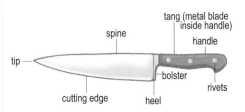

The parts of a chef's knife

The **spine** is the back of the blade. It is the edge opposite the **cutting edge**. The tip is the pointed end of the blade, while the **heel** is the back end of the blade closest to the handle. On some knives, the blade has a raised part called a bolster at the heel end. The bolster is a sort of guard that helps protect the hand from slips and also helps balance the weight of the knife.

The **tang** is the portion of the metal blade inside the handle. The highest-quality, most durable knives have a **full tang**, which means the tang runs the full length of the handle. On knives with traditional wood handles, **rivets** hold the **handle** to the tang. The rivets should be perfectly smooth and flush with the handle. Composite molded handles are bonded to the tang without rivets.

KNIVES AND THEIR USES

1. **French knife or chef's knife.**
   Most frequently used knife in the kitchen, for general-purpose chopping, slicing, dicing, and so on. The blade is wide at the heel and tapers to a point. Blade length of 10 inches (260 mm) is most popular for general work. Larger knives are for heavy cutting and chopping. Smaller blades are for more delicate work.

   This is your most important tool, so you must learn to handle it and care for it well. Chapter 7 explains its use in detail.

French knife or chef's knife

2. **Santoku knife or Japanese cook's knife.**
   A wide-bladed knife that is often used as a substitute for the traditional chef's knife. Blades are usually 5 inches (13 cm) or 7 inches (18 cm) long.

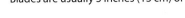

Santoku knife or Japanese cook's knife

3. **Utility knife or salad knife.**
   A narrow, pointed knife 6–8 inches (160–200 mm) long. Used mostly for pantry work, cutting and preparing lettuce, fruits, and so on. Also useful for carving roast chicken and duck.

Utility knife

4. **Paring knife.**
   A small, pointed blade 2–4 inches (50–100 mm) long. Used for trimming and paring vegetables and fruits.

Paring knife

5. **Boning knife.**
   A thin, pointed blade about 6 inches (160 mm) long. Used for boning raw meats and poultry. Stiff blades are used for heavier work. Flexible blades are used for lighter work and for filleting fish.

Boning knife

6. **Slicer.**
   A long, slender, flexible blade up to 14 inches (360 mm) long. Used for carving and slicing cooked meats.

Slicer

7. Serrated slicer.

Like a slicer, but with a serrated edge. Used for cutting breads, cakes, and similar items.

Serrated slicer

8. Butcher knife.

A heavy, broad, slightly curved blade. Used for cutting, sectioning, and trimming raw meats in the butcher shop.

Butcher knife

9. Scimitar or steak knife.

A curved, pointed blade. Used for accurate cutting of steaks.

Scimitar

10. Cleaver.

A heavy, broad blade. Used for cutting through bones. Do not confuse a cleaver with a similarly shaped Chinese cook's knife, which is lighter in weight.

Cleaver

Chinese cook's knife

11. Oyster knife.

A short, rigid, blunt knife with a dull edge. Used for opening oysters.

Oyster knife

12. Clam knife.

A short, rigid, broad-bladed knife with a slight edge. Used for opening clams.

Clam knife

13. Vegetable peeler.

A short tool with a slotted, swiveling blade. Used for peeling vegetables and fruits.

Vegetable peeler

Swiss-style vegetable peeler

14. Steel.

Not a knife, but an essential part of the knife kit. Used for truing and maintaining knife edges (not for sharpening them—see Chapter 7).

15. Cutting board.

An important partner to the knife. Hardwood boards are favored by many chefs. Hard rubber or plastic boards are thought to be more sanitary, but there is some evidence that bacteria actually survive longer on plastic and rubber than on wood. Cutting boards must be kept very clean, and they must be sanitized regularly. Color-coded composite boards are designed to help reduce cross-contamination, as each color is used for a different category of food (for example, green for vegetables, red for meats).

> *Note:* In some communities, wooden boards are prohibited by health regulations.

Steel

HAND TOOLS AND SMALL EQUIPMENT

1. Ball cutter, melon ball scoop, or parisienne knife.

The blade is a small, cup-shaped half-sphere. Used for cutting fruits and vegetables into small balls.

Parisienne melon baller

2. Cook's fork.

A heavy, two-pronged fork with a long handle. Used for lifting and turning meats and other items. Must be strong enough to hold heavy loads.

Cook's fork

3. Straight spatula or palette knife.

A long, flexible blade with a rounded end. Used mostly for spreading icing on cakes and for mixing and bowl scraping.

Straight spatula

4. Sandwich spreader.

A short, stubby spatula. Used for spreading fillings and spreads on sandwiches.

Sandwich spreader

5. Offset spatula.

A broad blade, bent to keep the hand off hot surfaces. Used for turning and lifting eggs, pancakes, and meats on griddles, grills, sheet pans, and so on. Also used as a scraper to clean benches and griddles.

Offset spatula

Rubber spatula

6. Rubber spatula or scraper.

A broad, flexible rubber or plastic tip on a long handle. Used to scrape bowls and pans. Also used for folding in egg foams and whipped cream. Heat-resistant spatulas can be used for stirring foods while cooking, but regular spatulas will melt at high heat.

7. Pie server.

A wedge-shaped offset spatula. Used for lifting pie wedges from pan.

Pie server

Bench scraper

8. Bench scraper or dough knife.

A broad, stiff piece of metal with a wooden handle on one edge. Used to cut pieces of dough and to scrape workbenches.

9. Pastry wheel or wheel knife.

A round, rotating blade on a handle. Used for cutting rolled-out doughs and pastry and baked pizza.

Spoons: slotted, perforated, solid

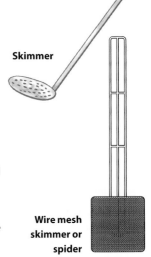

Pastry wheel

10. Spoons: slotted, perforated, and solid.

Large stainless-steel spoons that hold about 3 ounces (90 mL). Used for stirring, mixing, and serving. Slotted and perforated spoons are used when liquid must be drained from solids.

Skimmer

11. Skimmer.

A perforated disk, slightly cupped, on a long handle. Used for skimming froth from liquids and for removing solid pieces from soups, stocks, and other liquids. Wire mesh versions are sometimes called *spiders*.

Tongs

12. Tongs.

Spring-type or scissors-type tools used to pick up and handle foods.

13. Wire whip.

Loops of stainless-steel wire fastened to a handle. There are two kinds of whips (also called *whisks*):

Wire mesh skimmer or spider

- Heavy whips are straight, stiff, and have relatively few wires. Used for general mixing, stirring, and beating, especially heavy liquids.

- Balloon whips, or piano-wire whips, have many flexible wires. Used for whipping eggs, cream, and hollandaise, and for mixing thinner liquids.

Wire whip

14. China cap.

A cone-shaped strainer. Used for straining stocks, soups, sauces, and other liquids. Pointed shape allows the cook to drain liquids through a relatively small opening.

China cap

15. Fine china cap or chinois (shee-nwah).

A china cap with very fine mesh. Used when great clarity or smoothness is required in a liquid.

Chinois

16. Strainer.

A round-bottomed, cup-shaped tool made of screen-type mesh or perforated metal. Used for straining pasta, vegetables, and so on.

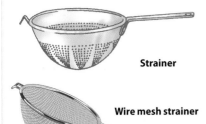

Strainer

Wire mesh strainer

17. Drum sieve or tamis.

A screen-type mesh supported in a round metal frame. Used for sifting flour and other dry ingredients and for puréeing soft foods.

Sieve

18. Colander.

A large, perforated bowl made of stainless steel or aluminum. Used to drain washed or cooked vegetables, salad greens, pasta, and other foods.

19. Food mill.

A tool with a hand-turned blade that forces foods through a perforated disk. Interchangeable disks produce varying degrees of coarseness or fineness. Used for puréeing foods.

Colander

Food mill

Grater

20. Grater.

A four-sided metal box with grids of varying sizes. Used for shredding and grating vegetables, cheese, citrus rinds, and other foods.

21. Plane grater.

Usually known by the brand name Microplane. These graters shave off thin shreds of the item being grated, the way a carpenter's plane shaves wood. Available in varying degrees of fineness or coarseness.

Microplane

22. Zester.

A small hand tool used for removing the colored part of citrus peels in thin strips.

Zester

23. Channel knife.

A small hand tool used mostly in decorative work.

Channel knife

24. Mandoline.

A manual slicing implement consisting of blades fitted in a flat metal or wood framework. Folding legs position the mandoline on the worktable at a 45-degree angle for use. Levers allow the blades to be adjusted to control the thickness of the slices. A traditional mandoline has a flat blade and a serrated blade. Additional blades can be used in combination with the flat blade to cut julienne and bâtonnet. The serrated blade is used to cut gaufrette or waffle slices.

For safest use, a detachable guard is also supplied. The guard holds the food and allows it to be sliced without getting the fingers near the blades.

Mandoline

25. Pastry bag and tubes.

Cone-shaped cloths or plastic bags with an open end that can be fitted with metal tubes or tips of various shapes and sizes. Used for shaping and decorating with items such as cake icing, whipped cream, duchesse potatoes, and soft dough. Disposable plastic pastry bags are the most sanitary.

Pastry bag and tubes

26. Pastry brush.

Used to brush items with egg wash, glaze, etc.

Pastrybrush

27. Can opener. Heavy-duty can openers are mounted on the edge of the workbench. They must be carefully cleaned and sanitized every day to prevent contamination of foods. Replace worn blades, which can leave metal shavings in the food.

Can opener

4

MENUS, RECIPES, AND COST MANAGEMENT

A menu is a list of dishes served or available to be served at a meal. But a menu is more than this. It is an important management tool. Nearly every aspect of the operation of a food-service business depends on the menu. In fact, it is fair to say the menu is the single most important document in the business. Purchasing, production, sales, cost accounting, labor management, even the kitchen layout and equipment selection of a new facility—all are based on the menu.

Recipes can be said to be the building blocks of the menu. Each item on the menu can be represented by the recipe or the procedure for preparing it. Therefore, recipes, like menus, are important management tools. They indicate ingredients to be purchased and stored, and they give measuring and preparation instructions to the kitchen staff.

In addition, recipes are important tools for the cook because they are a means of recording and passing along essential information. Learning to cook without being able to consult recipes would be like learning to play the piano without using written music.

This chapter discusses menus and recipes as they are used in commercial kitchens. How do we construct a menu that builds sales by offering the best choices to the customer and that also promotes efficiency and productivity? How do we read and understand recipes? How do we measure ingredients and portions, convert recipes to different yields, and calculate food cost with the aid of written recipes?

AFTER READING THIS CHAPTER, YOU SHOULD BE ABLE TO

1. Explain how the makeup of a menu depends on the type of meal and on the institution using it.

2. Describe the differences between static and cycle menus, and between à la carte and table d'hôte menus.

3. List in order of their usual service the courses that might appear on modern menus.

4. Devise balanced menus that contain an adequate variety of foods and that can be efficiently and economically prepared.

5. Describe the problems and limitations of written recipes and the importance of using judgment when cooking.

6. Discuss the structure and functions of standardized recipes.

7. Use and understand the recipes in this book to practice basic cooking techniques.

8. Measure ingredients and portions.

9. Use metric measurements.

10. Convert recipes to higher or lower yields.

11. Perform yield-cost analysis.

12. Calculate raw food costs.

MENU FORMS AND FUNCTIONS

Menus must be planned for the people eating the food. This sounds like a simple rule, but it is frequently forgotten. You must never forget the customer is the main reason for being in business.

This rule means that, in most operations, the taste and preferences of the cooks or chefs are of less importance when planning the menu. True, some of the most famous restaurants exist primarily as showcases for the chef's own artistry, but these are a small percentage of all food-service establishments. Instead, the taste and preferences of the clientele must be given top priority if the business is to succeed. The kind of clientele the business serves influences the form the menu takes.

THE CLIENTELE

Type of Institution

Each kind of operation has a different menu because each serves the needs of a different clientele.

Hotels must provide a variety of services for their guests, from budget-minded tourists to businesspeople on expense accounts. Thus, their offerings may range from quick breakfast and sandwich counters to elegant dining rooms and banquet halls.

Hospitals must satisfy the dietary needs of the patients.

Schools must consider the ages of the students and their tastes and nutritional needs.

Employee food services need menus that offer substantial but quickly served and reasonably priced food for working customers.

Catering and banquet operations depend on menus that are easily prepared for large numbers but that are lavish enough for parties and special occasions.

Fast-food and take-out quick-service operations require limited menus featuring inexpensive, easily prepared, easily served foods for people in a hurry.

Full-service restaurants range from simple neighborhood diners to expensive, elegant restaurants. Menus, of course, must be planned according to the customers' needs. A menu of high-priced, luxurious foods in a café situated in a working-class neighborhood will probably not succeed.

Customer Preferences

Even facilities with captive audiences, such as school cafeterias and hospital kitchens, must produce food that is appealing to their customers and in sufficient variety to keep those customers from getting bored with the same old things. Grumbling about the food is a favorite sport among students, but at least it can be kept to a minimum.

Restaurants have an even harder job because their customers don't just grumble if they don't like the selections. They don't come back. People are becoming more and more interested in trying unfamiliar foods, especially ethnic foods. Nevertheless, tastes vary by region, by neighborhood, by age group, and by social and ethnic background. Foods enjoyed by some people are completely rejected by others.

Prices must be kept in line with the customers' ability and willingness to pay. Prices, of course, place limits on what foods can be offered.

KIND OF MEAL

Menus vary not only by kind of operation but by meal as well.

Breakfast

Breakfast menus are fairly standard within any one country. In North America, for example, a restaurant has to offer the usual selection of fruits, juices, eggs, cereals, breads, pancakes, waffles, breakfast meats, and regional specialties because this is what customers want and expect.

In addition, featuring one or two unusual items on the menu—such as an English muffin topped with creamed crabmeat and a poached egg, a special kind of country ham, or an assortment of freshly made fruit sauces or syrups for the pancakes and waffles—often attracts additional customers. Breakfast menus must feature foods that can be prepared quickly and eaten in a hurry.

Lunch

The following factors are important to consider when planning lunch menus.

1. **Speed.**
   Like breakfast customers, luncheon diners are usually in a hurry. They are generally working people who have limited time to eat. Foods must be prepared quickly and be easy to serve and eat. Sandwiches, soups, and salads are important items on many lunch menus.

2. **Simplicity.**
   Menu selections are fewer, and fewer courses are served. In many cases, customers select only one course. Luncheon specials—combinations of two or three items, such as soup and a sandwich or omelet and salad, offered at a single price—satisfy the need for simplicity and speed.

3. **Variety.**
   In spite of the shortness of the menu and the simplicity of the selections, luncheon menus must have variety. This is because many customers eat at the same restaurant several times a week or even every day. In order to keep the menu short, many operations offer several luncheon specials every day, so there is always something new on the menu.

Dinner

Dinner is usually the main meal and is eaten in a more leisurely fashion than either breakfast or lunch. Of course, some people are in a hurry in the evening, too, but, in general, people come to a restaurant to relax over a substantial meal. Dinner menus offer more selections and more courses. Not surprisingly, prices and check averages are also higher than at lunch.

TYPES OF MENUS

Static and Cycle Menus

A **static menu** is one that offers the same dishes every day. These menus are used in restaurants and other establishments where the clientele changes daily or where enough items are listed on the menu to offer sufficient variety. A static menu may be in place indefinitely, or it may change at regular intervals, such as every season, every month, or even every week.

Some restaurants use a menu that is part static and part variable. This means they have a basic menu of foods prepared every day, plus daily specials to offer variety without putting too much strain on the kitchen. The daily specials may take advantage of seasonal produce and other occasionally available foods the chef or purchaser finds in the wholesale market.

A **cycle menu** is one that changes every day for a certain period; after this period, the daily menus repeat in the same order. For example, a seven-day cycle menu has a different menu every day for a week and repeats each week. This kind of menu is used in such operations as schools and hospitals, where the number of choices must be kept small. The cycle menu is a way of offering variety.

À la Carte and Table d'Hôte

An **à la carte** menu (Figure 4.1) is one in which each individual item is listed separately, with its own price. The customer makes selections from the various courses and side dishes to make up a meal. (**Note:** The term *à la carte* is also used to refer to cooking to order, as opposed to cooking ahead in large batches.)

Table d'hôte (tobbluh dote) originally meant a fixed menu with no choices—like a meal you would be served if you were invited to someone's home for dinner. Banquet menus are familiar examples of this kind of menu. The term has also come to mean a menu that offers a selection of complete meals at set prices. In other words, a customer may choose from among several selections, each of which includes an entrée and side dishes plus other courses, such as appetizer, salad, and dessert. Each full meal selection has a single package price.

Appetizers

SOUP OF THE DAY
MARKET

PIADINA SERVED WITH ROASTED FRUIT, SAGE HONEY AND ARTISAN BLUE CHEESE
$10.50

BISTRO HOUSE SALAD WITH BABY GREENS, BALSAMIC VINAIGRETTE AND MANCHEGO
$7.00

STAR PRAIRIE SMOKED TROUT SALAD WITH BIBB LETTUCE AND
WARM HORSERADISH/CREME FRAICHE DRESSING
$10.50

CLASSIC CAESAR SALAD WITH BABY ROMAINE, PARMESAN AND SEMOLINA CROUTONS
$8.50
WITH OIL-PACKED SPANISH ANCHOVIES ADD $2

SASHIMI OF TUNA SERVED ON BABY ARUGULA, PICKLED PINEAPPLE AND FRESH MINT
$12.50

BISTRO CALAMARI WITH THAI DIPPING SAUCE
$8.00

JP'S RUSTIC PIZZA WITH APPLEWOOD SMOKED BACON,
CASHEW/CURRANT PESTO AND PECORINO-ROMANO
$11.00

PIZZA WITH SAN MARZANO TOMATO SAUCE, BASIL AND HOUSE-MADE MOZZARELLA
$10.50

PIZZA OF THE DAY
MARKET

SEARED MINNESOTA FOIE GRAS SERVED WITH CARAMELIZED SHALLOTS AND QUINCE CONFIT
$15.00

SIDES

YUKON GOLD POTATO PURÉE $4.50 OR STEAMED JASMINE RICE $3.50

POMMES FRITES WITH SMOKED PAPRIKA OR VEGETABLE OF THE DAY
$4.50

BOWL OF OLIVES
$3.50

SPLIT PLATE CHARGE $1.00

Entrées

HOUSE-MADE FETTUCINI TOSSED WITH CHICKEN CONFIT,
ORGANIC BROCCOLI, HARISSA AND PARMESAN
$18.50

POTATO GNOCCHI WITH ORGANIC GREEN BEANS,
TOASTED HAZELNUT CREME FRAICHE, AND CHIVE OIL
$17.50

CARNAROLI RISOTTO WITH WILD MUSHROOMS, SWEET ONIONS AND GREMOLATA
$22.50

PAN-SEARED GROUPER SERVED ON YUKON GOLD POTATO PUREE WITH
BROWN BUTTER SPINACH AND BLOOD ORANGE RELISH
$27.50

WILD ACRES DUCK CONFIT RAVIOLI TOSSED WITH AN APPLE CIDER
DEMI-GLACE, SHAVED APPLE AND FRIED SAGE
$19.50

SEARED RIB-EYE RUBBED WITH GREEN THAI CURRY AND SERVED WITH
HOUSE-PICKLED PINEAPPLE, JICAMA AND RED ONION SALAD
$28.50

PAN-ROASTED VENISON STEAK ON YUKON GOLD POTATO PURÉE WITH
ROASTED BRUSSEL SPROUTS AND A CRANBERRY RED WINE REDUCTION
$32.50

BONE-IN PORK TENDERLOIN BRAISED WITH APPLE, ONION AND SWEET CURRY,
SERVED WITH GOLDEN RAISIN-RED BELL PEPPER RELISH
$24.50

SLOW-ROASTED DUCK BREAST AND CONFIT OF DUCK LEG WITH
SPINACH/MANGO SALAD AND ORANGE-MISO VINAIGRETTE
$28.50

ENTRÉE OF THE DAY
MARKET

FIGURE 4.1 An example of an à la carte menu from a fine-dining restaurant.
Courtesy of JP American Bistro, Minneapolis, Minnesota.

Many restaurants use a combination of à la carte and table d'hôte selections. For example, a steak house may include salad, potato, vegetable, and beverage with the entrée choice, while additional dishes like appetizers and desserts may be offered at extra cost.

Closely related to the table d'hôte menu is the **prix fixe** (preefeex), meaning "fixed price," menu. On a pure prix fixe menu, only one price is given. Each guest may choose one selection from each course offered, and the total meal costs the single price indicated. Often, on such menus, a few items featuring costly ingredients carry an extra charge, called a **supplement**. The supplement is usually indicated in parentheses after the listing. It is best to limit the number of supplements as much as possible. Too many extra charges on a prix fixe menu can leave customers frustrated and angry.

A special variety of the prix fixe menu sometimes used in fine restaurants is the **tasting menu**, also known by its French name, *menu dégustation*. A tasting menu (Figure 4.2) is offered in addition to the regular menu and gives patrons a chance to try a larger number of the chef's creations. The menu may feature 5 or 6 or even as many as 10 or 12 individual courses served in small portions. Because of the complexity of service, a restaurant may require that the tasting menu can be served only if everyone at the table orders it. Tasting menus may change daily, depending on the chef's choices and the availability of ingredients.

BUILDING THE MENU

A **course** is a food or group of foods served at one time or intended to be eaten at the same time. In a restaurant, the courses are normally served in sequence, allowing enough time for each to be eaten before the next is served. In a cafeteria, the customers may select all their courses at once—appetizer, salad, main dish and vegetables, and dessert, for example—but eat them in a particular order.

In the following pages, we discuss the principles that apply to planning the courses that make up a menu. The main purpose of these principles is to lend variety and interest to a meal. They are not arbitrary rules you must follow for no reason.

VEGETABLE TASTING MENU

ARTICHOKE BARIGOULE WITH ROASTED PEPPER TAPENADE
AND HERB SALAD

WHITE BEAN PURÉE WITH ESCAROLE, ROASTED TOMATO
AND LEMON CONFIT

ROASTED ROOT VEGETABLES WITH SOUBISE
AND TRUFFLE VINAIGRETTE

POLENTA CAKE WITH FIGS AND WILTED GREENS

MUSHROOM TART TATIN

DESSERT AMUSE

TASTING DESSERT

PETITS FOURS

75

TASTING MENUS REQUIRE THE PARTICIPATION OF THE ENTIRE TABLE

AUTUMN TASTING MENU

TORCHON OF FOIE GRAS WITH FIGS, PISTACHIO
AND MINT

SEA SCALLOP WITH SALSIFY, SAUTERNES AND VANILLA

TURBOT WITH LENTILS, SAVOY CABBAGE,
ROOT VEGETABLES AND RED WINE

BRAISED FRESH BACON WITH MOSTARDA
AND CHESTNUT-HONEY GLAZED TURNIPS

RACK OF LAMB WITH ARTICHOKES, TOMATO
AND OLIVE GRATIN

DESSERT AMUSE

TASTING DESSERT

PETITS FOURS

90

TASTING MENUS REQUIRE THE PARTICIPATION OF THE ENTIRE TABLE

FIGURE 4.2 An example of two tasting menus from a fine-dining restaurant. Courtesy of Gramercy Tavern, New York, New York.

THE CLASSICAL MENU

Today's menus are descendants of elaborate banquet menus served in the nineteenth and early twentieth centuries. These menus had 12 or more courses, and the sequence in which they were served was well established by tradition.

The following sequence of courses is typical of one that may have been served at a great banquet early in the twentieth century.

1. **Cold hors d'oeuvres.**
   Small, savory appetizers.
2. **Soup.**
   Clear soup, thick soup, or broth.
3. **Hot hors d'oeuvres.**
   Small, hot appetizers.
4. **Fish.**
   Any seafood item.
5. **Main course, or pièce de resistance.**
   A large cut of roasted or braised meat, usually beef, lamb, or venison, with elaborate vegetable garnishes.
6. **Hot entrée.**
   Individual portions of meat or poultry, broiled, braised, or pan-fried, etc.
7. **Cold entrée.**
   Cold meats, poultry, fish, pâté, and so on.
8. **Sorbet.**
   A light ice or sherbet, sometimes made of wine, to refresh the appetite before the next course.
9. **Roast.**
   Usually roasted poultry, accompanied by or followed by a salad.
10. **Vegetable.**
    Usually a special vegetable preparation, such as artichokes or asparagus, or a more unusual vegetable such as cardoons.
11. **Sweet.**
    What we call *dessert*—cakes and tarts, pudding, soufflés, etc.
12. **Dessert.**
    Fruit and cheese and, sometimes, small cookies or petits fours.

MODERN MENUS: COURSES AND ARRANGEMENT

Such extensive classical menus are rarely served today. Even grand, elegant banquets comprising many courses are usually shorter than the menu we just described. However, if you study that menu, you will be able to see the basic pattern of modern menus hiding amid all those courses.

The main dish is the centerpiece of the modern meal. If the meal consists of only one dish, it is considered the main course, even if it is a salad or a bowl of soup. There is usually only one main course, although large banquets may still have more than one, such as a poultry dish followed by a meat dish.

One or more dishes may be served before the main dish. These are usually light in character so the customer is not satiated before the main course.

Study the following outline of the modern menu and compare it to the classical menu. The notes that follow explain several aspects that may be puzzling. Then, in the next sections, we discuss how to select specific dishes for each course to arrive at a balanced menu.

The Modern Menu

First Courses	Appetizer
	Soup
	(Fish)
	Salad
Main Dish	Meat, poultry, or fish
	Vegetable accompaniment
Dessert Dishes	Salad
	Fruits and cheeses
	Sweets

Notes

- Appetizer, soup, and salad are the three courses usually served before the main course. One, two, or all three of them may be served, and they are usually served in this order. Thus, meals may have the following courses:

Appetizer Main dish	Soup Main dish	Salad Main dish
Appetizer Soup Main dish	Soup Salad Main dish	Appetizer Salad Main dish
	Appetizer Soup Salad Main dish	

- A fish course is sometimes included in more formal dinners, after the appetizer and soup courses. It should be a relatively small portion, and the main dish should not also be fish.

- Salads may be served either before or after the main course (but not both). In more traditional meals, they are served after the main course to refresh the appetite before the cheese and sweet courses. Serving the salad before the main course is a comparatively recent idea.

- Sometimes, one or more of the first courses are served at the same time as the main dish, possibly on the same plate. This is especially popular on luncheon menus, where quick service is desired. Thus, you will find soup and sandwich combinations, salad and omelet combinations, and so on.

- If both cheese and sweets are served for dessert, they may come in either order. English menus have cheese after the sweets, while French menus generally place the sweets last.

VARIETY AND BALANCE

Balancing a menu means providing enough variety and contrast for the meal to hold interest from the first course to the last. To balance a menu, you must develop a feeling for which foods complement each other or provide pleasing contrasts. And you must avoid repeating flavors and textures as much as possible.

These principles apply whether you are planning a banquet menu, where the diners have no choices; a school cafeteria menu, where students have only a few choices; or a large à la carte menu, where customers have many choices. Of course, with an à la carte menu, the customers' own choices determine how balanced their meals are. There's nothing wrong with listing a creamed dish among the appetizers and another creamed dish among the main dishes. But you should offer enough choices so customers can easily select balanced meals if they desire. In other words, if half the appetizers and half the entrée selections are served in a cream sauce, you're not offering enough variety.

The following factors must be considered in balancing a menu.

1. **Flavors.** Don't repeat foods with the same or similar tastes. This applies to any predominant flavor, whether of the main ingredient, of the spices, of the sauce, and so on. For example:

   - Don't serve broiled tomato halves with the main dish if the appetizer has a tomato sauce.

   - Don't serve both a spicy, garlicky appetizer and a spicy, garlicky main dish. However, don't make everything too bland.

   - Unless you operate a specialty restaurant like a steak house or a seafood restaurant, balance the menu among meats (beef, pork, lamb, veal), poultry, and fish.

   - Acid or tart foods are often served as accompaniments to fatty foods because they help cut the fatty taste. This is why applesauce and pork, mint sauce and lamb, and orange sauce and duckling are such classic combinations.

2. **Textures.** Texture refers to the softness or firmness of foods, their feel in the mouth, whether or not they are served with sauces, and so on. Don't repeat foods with the same or similar texture. For example:

   - Serve a clear soup instead of a thick soup if the main course is served with a cream sauce. By contrast, a cream soup goes well before a simple sautéed or broiled item.

   - Don't serve too many mashed or puréed foods.

   - Don't serve too many heavy, starchy items.

3. **Appearance.** Serve foods with a variety of colors and shapes. Colorful vegetables are especially valuable for enlivening the appearance of meats, poultry, fish, and starches, which tend to be mostly white or brown. (Creating attractive food is discussed in Chapter 29).

4. **Nutrients.** The importance of a nutritionally balanced menu is obvious in the case of menus for hospitals and nursing homes, for example. But even à la carte menus in restaurants should provide enough nutritional variety to allow customers to select nutritionally balanced meals. Dietary health and nutrition are considered in detail in Chapter 5.

5. **Cooking method.** Cooking methods play an important role in determining the flavors, textures, and appearance of food. For the typical full-service restaurant, it is a good idea to offer a variety of roasted, braised, grilled, sautéed, and simmered foods. Obviously, this guideline doesn't apply to specialty operations such as a barbecue restaurant or a fish-and-chips shop. Using a variety of cooking methods also eases the pressure on each department in the kitchen, as discussed in the next section.

The possible combinations of foods are so many that it is impossible to give rules that cover all of them. Besides, creative chefs are continually experimenting with new combinations, breaking old rules, and coming up with exciting menus. Years of experience, however, are required to develop this kind of creativity and a feel for what makes certain combinations work. In the meantime, pay close attention to the principles discussed.

KITCHEN CAPABILITIES AND AVAILABILITY OF FOODS

Physical conditions place limitations on your menu. Depending on your equipment, your labor force, and the foods available to you, certain items will be inconvenient, difficult, or even impossible to serve.

Equipment Limitations

Know the capacities of your equipment and plan menus accordingly. If your broiler capacity is 200 steaks an hour and you plan a banquet menu for 400 people that features broiled shrimp as an appetizer and broiled steaks as a main course, you're in big trouble.

Spread the workload evenly among your equipment. If you have ovens, a broiler, and a fryer, balance the roasted and braised items, the broiled items, and the fried items. Don't let the broiler stand idle while orders are backed up at the deep fryer. Also, using a variety of cooking methods adds variety of taste and texture to the menu.

Personnel Limitations

Spread the workload evenly among the workers. As with equipment, you don't want the fry cook to have more than he or she can handle, while the broiler cook has little to do.

Spread the workload throughout the day. Balance the cooked-to-order items against the cooked-ahead items so you don't have to do everything at the last minute.

Offer items the cooks are able to prepare. Don't put items on the menu that are above the skill level of the staff.

Availability of Foods

Use foods in season. Foods out of season are expensive and often low in quality, and their supply is undependable. Don't put asparagus on the menu if you can't get good asparagus.

Use locally available foods. Fresh seafood is an obvious example of a food that is hard to get in some regions unless you—and your customers—are willing to pay premium prices.

MENUS AND COST CONTROL

Food costs are a major part of the expenses of any food-service operation. You can't afford to throw away food any more than you can afford to throw away money. Total utilization of foods must be planned into menus. Whether or not this is done can make or break an operation.

Cost control requires not only planning but also careful arithmetic. See pages 81–88 for a discussion of kitchen mathematics as well as other factors in cost control, including the role of the menu.

*MENU TERMINOLOGY AND ACCURACY

After you have selected the items you want to include on your menu, you face the problem of what to call them. Decades ago, when the classical cuisine of Escoffier was the normal offering in fine restaurants, a standard terminology existed. Everyone knew what was meant by Tournedos Chasseur, Suprême de Volaille Orly, and Sole Nantua, and these terms meant the same dish on any menu. Today, however, there is little standardization. Chefs feel obliged to give as much information on their menus as possible in order to describe their dishes adequately to their customers. As a result, one often sees menu descriptions that list almost every ingredient in a dish, including the farms the ingredients came from.

It is probably better to give too much information than too little. The important thing is to provide enough information so the customer will understand the basic character of the dish and not have any unpleasant surprises. An alternative to writing everything out on the menu is to educate the serving staff so they know the dishes well enough to fill in any missing details for the customer.

The menu is a sales tool, so it is understandable to try to make every dish sound as appealing as possible. Accurate and truthful descriptions, however, are required. Giving misleading names to menu items is not only dishonest and unfair to the customer, it is actually illegal in some localities that have adopted truth-in-menu laws, and you can be prosecuted for fraud for doing so. Furthermore, customers who feel confused or cheated may not come back.

Calling something chicken salad if it is made with turkey, veal cutlet if it is made with pork, or whipped cream if it is actually artificial whipped topping is such obvious mislabeling that it can hardly be accidental. However, some kinds of menu inaccuracies result not from intentional deception but from simple misunderstanding. In particular, look out for these types of labeling problems:

1. **Point of origin.**

   If your menu lists "Maine lobsters" or "New Brunswick lobsters," they must be from Maine or New Brunswick. Roquefort dressing must be made with Roquefort cheese from Roquefort, France. On the other hand, generally accepted names or names that indicate type rather than origin can be used. For example: Swiss cheese, French bread, Swedish meatballs.

2. **Grade or quality.**

   U.S. Choice and Canada A are names of grades, and you'd better be using those grades if you say you are. Incidentally, the word prime in "prime rib" indicates a cut, not a grade.

3. **Cooking method.**

   A menu item described as "grilled" or "roasted" should be cooked by the method indicated. Billing a pan-fried item as "roasted" because it sounds better on the menu misrepresents the item and risks disappointing the customer.

4. **"Fresh."**

   If you call something **fresh**, it must be fresh, not frozen, canned, or dried. There is no such thing as "fresh frozen."

5. **"Imported."**

   An item labeled **imported** must come from outside the country. Better and more specific than using the term *imported* is to indicate the country of origin.

6. **"Homemade."**

   The word **homemade** means the item was made on the premises. Adding a few fresh carrots to canned vegetable soup does not make it homemade.

7. **"Organic."**

   For a food to be labeled **organic**, it must be raised without the use of hormones, antibiotics, synthetic pesticides, irradiated components, genetically modified organisms, or reprocessed sewage. In some countries, including the United States, the use of the word *organic* on labels is defined by law (p. 275).

8. **Size or portion.**

   If you indicate a portion size on the menu, be sure you serve that size (within allowable tolerances). A "10-ounce steak" must weigh at least 10 ounces before cooking. "Jumbo shrimp" are not just big shrimp. They are a specific size.

9. **Appearance.**

   Family-style restaurants that use illustrated menus should serve food that looks like the illustrations. If an illustration of a shrimp plate shows six shrimp, it would be a mistake to serve only five.

Here are other examples of common violations:

> Listing "maple syrup" and serving maple-flavored syrup.
>
> Listing a product, such as a carbonated beverage, by brand name and serving another brand.
>
> Listing "butter" and serving margarine.
>
> Listing coffee or breakfast cereal "with cream" and serving milk or artificial cream.
>
> Listing "ground round" and serving other ground beef.

Finally, please use a dictionary. Unfortunately, it is common to see menus full of misspellings. These errors reflect poorly on the restaurant. Customers may think that if you don't care enough even to spell words on the menu correctly, you may not care enough to cook the food correctly, either. If you use terms from French, Italian, or another language, make sure the terms are spelled and used correctly.

KEY POINTS TO REVIEW

- How do the contents and design of a menu depend on the type of meal served and the type of food-service institution using it?

- What is a static menu? a cycle menu? an à la carte menu? a table d'hôte menu?

- What courses are usually listed on a modern dinner menu?

- What techniques can you use to create a menu that offers a good balance of choices for the customer?

- Why should the design and limitations of the kitchen and the skill level of the staff be considered when designing a menu?

THE WRITTEN RECIPE

A **recipe** is a set of instructions for producing a certain dish. In order to duplicate a desired preparation, it is necessary to have a precise record of the ingredients, their amounts, and the way in which they are combined and cooked. This is the purpose of a recipe.

In spite of their importance, written recipes have many limitations. No matter how detailed a recipe may be, it assumes you already have certain knowledge—that you understand the terminology it uses, for example, and that you know how to measure ingredients.

THE USES AND LIMITATIONS OF RECIPES

Many people believe learning to cook means simply learning recipes. Knowledgeable cooks, in contrast, are able to prepare food without written recipes, if they have to, because they have a good understanding of basic principles and techniques. A recipe is a way of applying basic techniques to specific ingredients.

If you have already read Chapter 6, or have even casually leafed through this book, you know it is not just a book of recipes. Although it contains hundreds of recipes, they take up a relatively small part of it. Your main concern is learning techniques and procedures you can apply to any recipe.

The main purpose of learning basic cooking principles is not to be able to cook without recipes, however, but to understand the recipes you use. As we said in the beginning of this section, every recipe assumes you have certain knowledge that enables you to understand the instructions and follow them correctly.

Some recipes supply very little information, and some supply a great deal. But no matter how detailed it is, a written recipe can't tell you everything, and some judgment by the cook is always required. There are several reasons for this:

1. **Food products are not uniform.**
   Food ingredients are natural products, so they are not uniform like machine bolts, ballpoint pens, and printer paper. One tomato may be riper than another, one carrot more tender or sweeter than another, one oyster saltier than another. Such variations may affect how the ingredients are handled, how long they are cooked, what proportions are needed, and how much seasoning is required.

2. **Kitchens do not have the same equipment.**
   Different pans distribute heat at different rates. Different broilers heat to different temperatures. Liquid evaporates from wide pots faster than from tall, narrow ones, and so on.

3. **It is impossible to give exact instructions for many processes.**
   How do you set the burner if the instructions say "Cook over medium heat"? How thick is a "thick" sauce? How long do you broil a rare steak? Training and experience will help you learn to make accurate judgments about such questions.

The difference between an experienced cook and a beginning cook is the ability to make judgments about these variables.

STANDARDIZED RECIPES

1. **Definition.**
   A **standardized recipe** is a set of instructions describing the way a particular establishment prepares a particular dish. In other words, it is a customized recipe developed by an operation for the use of its own cooks, using its own equipment, to be served to its own patrons.

2. **The structure of a standardized recipe.**
   Recipe formats differ from operation to operation, but nearly all of them try to include as much precise information as possible. The following details may be listed:

   • Name of the recipe.

   • Yield, expressed as total yield (volume or weight), or number of portions and portion size.

   • Ingredients and exact amounts, listed in order of use. Exact description of ingredients is included.

- Directions for preparing or fabricating each ingredient, such as trimming and cutting instructions.

- Expected trim yields for any produce or other ingredients that must be fabricated.

- Equipment needed, including measuring equipment, pan sizes, portioning equipment, and so on.

- Directions for preparing the dish. Directions are kept as simple as possible.

- Preparation and cooking temperatures and times.

- Directions for portioning, plating, and garnishing.

- All necessary food safety instructions, including critical storage, holding, and cooking temperatures and cleaning and sanitation procedures. See discussion of Critical Control Points below and on pages 29–30.

- Directions for breaking down the station, cleaning up, and storing leftovers.

3. **The function of standardized recipes.**

An operation's own recipes are used to control production. They do this in two ways:

- *They control quality.* Standardized recipes are detailed and specific. This is to ensure the product is the same every time it is made and served, no matter who cooks it.

- *They control quantity.* First, they indicate precise quantities for every ingredient and how they are to be measured. Second, they indicate exact yields and portion sizes, and how the portions are to be measured and served.

By controlling quality and quantity, recipes are a key tool in controlling costs.

4. **The limitations of standardized recipes.**

Standardized recipes have the same problems as all recipes—the problems we discussed earlier regarding variations in foods and equipment and vagueness of instructions. These problems can be reduced by writing the recipe carefully, but they cannot be eliminated. Even if an operation uses good standardized recipes, a new employee making a dish for the first time usually requires supervision to make sure he or she interprets the instructions the same way as the rest of the staff. These limitations don't invalidate standardized recipes. If anything, they make exact directions even more important. But they do mean that experience and knowledge are still very important.

Table 4.1 gives an example of a standardized recipe based on a recipe in this book. Compare and note the differences between this recipe and the instructional recipes in this book (explained in the following section). In particular, note the following differences:

- There are no metric units. Because this recipe is designed for a single kitchen that always uses US meaures, only one set of measurements is needed.

- The procedure appears below the ingredients rather than in a column to the right. An operation can choose any recipe format, but the operation using this recipe wants to emphasize collecting and measuring all ingredients before beginning to cook.

- The recipe includes Critical Control Points. The operation using this recipe has established a HACCP system. Food safety instructions are included as part of the standardized recipe. (Read or review the section on HACCP, pp. 29–31, if necessary.)

INSTRUCTIONAL RECIPES

Recipes published in cookbooks and textbooks are different from standardized recipes, because they are not designed for use in a single specific kitchen. The recipes in this book are used in a great variety of situations and in different ways.

The purpose of a standardized recipe is to direct and control the production of a particular food item. Directions must be as complete and exact as possible.

The purpose of the instructional recipes in this book is to teach basic cooking techniques. They provide an opportunity for you to practice, with specific ingredients, the general procedures you have learned.

TABLE 4.1 Chicken Breasts Parmesan

Portion size: 1 chicken breast, 4 oz Total yield: 12 portions

Quantity	Ingredients	Equipment
4 oz	Flour	2 half-size hotel pans
1¹/₄ tsp	Salt	one 2-qt stainless-steel bowl
¹/₂ tsp	Ground white pepper	1 wire whip
5	Whole eggs, size large	1 meat mallet
3¹/₂ oz	Grated parmesan cheese	four 12-in. sauté pans
1¹/₂ oz	Whole milk	1-oz ladle
12	Boneless, skinless chicken breasts, 4 oz each	tongs
4 oz	Clarified butter	plastic wrap
		instant-read thermometer, sanitized

For Service

12 slices, ¹/₈ inch thick	Fresh lemon (washed before slicing)	full-size hotel pan

PROCEDURE

ADVANCE PREP

CCP   **1.** Collect and measure all ingredients. *Refrigerate eggs, cheese, milk, and chicken at 40°F or lower until needed.*

     2. Collect all equipment.

     3. Place the flour in the hotel pan. Season with the salt and white pepper.

     4. Break the eggs into the stainless-steel bowl and discard the shells. Beat with the wire whip until foamy. Add the grated cheese and milk. Mix in with the whip.

CCP   **5.** *Cover the bowl with plastic wrap and refrigerate at below 40°F until needed.*

     6. Flatten the chicken breasts lightly with the meat mallet until ¹/₂ in. thick. Place the breasts in a hotel pan.

CCP   *Cover with plastic wrap. Refrigerate at below 40°F until ready to cook.*

CCP   **7.** *Clean and sanitize the mallet and the work surface. Wash hands thoroughly.*

COOKING

     8. Place one of the sauté pans over moderate heat. Allow to heat 2 minutes.

     9. Measure 1 oz clarified butter into the pan.

   10. One at a time, dip 3 chicken breasts in the seasoned flour until completely coated on both sides. Shake off
CCP   excess. Dip in the egg mixture. Coat both sides completely. *Return remaining chicken and egg mixture to refrigerator.*

CCP **11.** Place the 3 breasts in the sauté pan. *Wash hands after handling the raw chicken and before handling cooked food.*

   12. Cook the chicken over moderate heat until golden brown on the bottom. Using the tongs, turn over and con-
CCP   tinue to cook *until chicken reaches an internal temperature of 165°–170°F. Test internal temperature with sanitized instant-read thermometer.*

CCP **13.** Repeat with the remaining chicken breasts, using clean sauté pans. *If your work is interrupted before completion, cover and refrigerate chicken and egg mixture.*

   14. Arrange in single layer in hotel pan. Top each chicken breast with 1 slice of lemon, placed in center of breast.

CCP **15.** Send pan of chicken breasts immediately to steam table in hot buffet line. *Ensure that steam table is turned on and food maintains an internal temperature of 145°F.*

CCP **16.** *Discard leftover egg mixture and seasoned flour. Do not use for any other products. Clean and sanitize all equipment.*

If you glance at any of the recipes in this book, you will see they do not contain all the features of a standardized recipe, as described in the previous section. In particular, you will see the following differences:

1. Instructions for preparation.

In most cases, recipes in this book follow a discussion of a basic procedure. The recipes are examples of the general procedure, and they give you experience in applying what you have learned. The information you are given in the recipe instructions is intended primarily to encourage you to think and to learn a technique, not just to turn out a product. You should consult your instructor when you have a question about a procedure.

2. Variations and optional ingredients.

Many recipes are followed by variations. These are actually whole recipes given in abbreviated terms. It is possible to write them out as separate, full-length recipes. (You are encouraged to do this before preparing a variation, as a learning experience.)

Giving recipes as variations rather than as separate recipes encourages you to see the patterns behind each. Again, you are learning techniques, not just recipes. You develop a lot more understanding of what you are doing if you see Spanish rice and Turkish pilaf, for example, or coconut cream pie and chocolate pudding as variations of the same basic techniques rather than as separate, unrelated recipes.

Your instructors may have their own variations, or they may wish to make changes in the basic recipes in order to teach you certain points. Unlike standardized recipes, instructional recipes are not engraved in stone.

✳ COOKING WITH JUDGMENT

When you make a recipe for the first time, you should apply your knowledge and think about the recipe in relation to the skills you have. The first step in preparing a recipe is to read it carefully. Then read it again. Make sure you understand every ingredient and every instruction in the procedure. A common fault is to prepare a recipe according to what you think it says rather than what it actually says. You are especially likely to make this error if the procedure in the recipe is different from the way you usually prepare a similar dish.

After you have reread the recipe carefully, you should determine the following points:

1. What are the basic cooking methods?

When you read the recipe for Sauerbraten (p. 519), you will quickly figure out the cooking method used is braising (even if the word *braise* is never used in the recipe). Then you should review in your mind everything you know about basic braising procedures.

2. What are the characteristics of the ingredients?

If the sauerbraten recipe calls for bottom round of beef, for example, you should ask yourself, "What do I know about bottom round? Is it lean or fatty? tough or tender? How do these traits affect cooking?"

3. What are the functions of the ingredients?

What does the vinegar do in the sauerbraten recipe? What about the vegetables? the gingersnaps?

When you have gained more experience, you will be able to easily answer these questions. You will know what ingredients contribute to flavor, to texture, or to body, and how they do it.

4. What are the cooking times?

Most of the recipes in this book do not give cooking times, except as general guidelines to help you plan production. This is because cooking times are too variable to be stated exactly.

Instead, you will learn how to test for doneness by observing changes in the product. You must be able to judge when a product has reached the right temperature, the proper texture or consistency, or the desired taste.

When you learn to cook with judgment, you will be able to prepare foods with most recipes, even poorly written ones. You will be able to see what might be wrong with a new recipe before you try it and to make adjustments in it. You will know how to substitute ingredients or use different equipment. You will even be able to create new recipes.

Remember we said that some recipes supply very little information and depend largely on the cook's knowledge. With enough experience, you will even be able to cook from recipes like the following, a complete recipe for Filets of Sole Bercy, quoted in its entirety from *Le Répertoire de la Cuisine*, a favorite book used by chefs in classical French cooking: "Poached with shallots and chopped parsley, white wine and fish stock. Reduce the stock, add butter, and coat the fish, glaze."

KEY POINTS TO REVIEW

- Is this statement true? "If you have a good recipe, you don't need to know how to cook, because the recipe tells you what to do." Explain.

- What are standardized recipes? How are they used?

MEASUREMENT

Many restaurants budget a net profit of 10 percent or less. This means a sandwich selling for $5.00 makes a profit of only 50 cents. If the cook happens to put a half-ounce too much meat in the sandwich, the operation may be losing money on it. No wonder so many restaurants go out of business. (See the sidebar on page 81 for an explanation of net profit and gross profit.)

Careful measurement is one of the most important parts of food production. It is important for consistent quality each time a recipe is prepared and served. And it is important for cost controls.

There are two important kinds of measurement in the kitchen:

1. Ingredient measurement

2. Portion measurement, or portion control

INGREDIENT MEASUREMENT

Weight

Weighing measures how heavy an item is. Weighing is the most accurate method of measuring ingredients. It is the method used for most solid ingredients.

Accurate scales are necessary for weighing. Small portion scales are often used in the kitchen because of their convenience. Balance scales are used in the bakeshop (see p. 895 for procedure).

To be able to weigh ingredients, you must observe the difference between AP (as purchased) weight and EP (edible portion) weight.

AP weight is the weight of the item as purchased, before any trimming is done. Also known as *APQ* (as purchased quantity).

EP weight is the weight after all inedible or nonservable parts are trimmed off. Also known as *EPQ* (edible portion quantity).

Recipes sometimes specify which weight they are referring to. When they don't, you must judge from the instructions.

1. If a recipe calls for "2 lb potatoes" and the first instruction is "scrub, peel, and eye the potatoes," then you know AP weight is called for.

2. If the recipe calls for "2 lb peeled, diced potatoes," then you know EP weight is called for. You will need more than 2 lb AP.

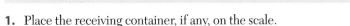

PROCEDURE for Weighing Ingredients on a Portion Scale

1. Place the receiving container, if any, on the scale.

2. Set the scale so it reads zero.

3. Add the item being weighed to the container (or place directly on the scale, if no container is used) until the scale reads the desired weight.

Volume

Volume measures how much space an item fills. Volume measures are used for liquids. Measuring a liquid by volume is usually faster than weighing it, and accuracy is good.

Solid ingredients are usually not measured by volume because they cannot usually be measured accurately by this method. One pint of chopped onions will vary considerably in weight, depending on how large or small the onions are cut and whether the pint measure is filled loosely or packed.

Dry ingredients such as flour or sugar are usually weighed in the bakeshop. However, they are sometimes measured by volume in the kitchen, when speed is more important than accuracy. To measure dry ingredients by volume, fill a dry-volume measure until the ingredient is mounded over the top. Then level it off with a spatula or other straightedge.

Very small quantities, such as ¼ teaspoon salt, may be measured by volume when the amount is too small to weigh.

Count

Measuring ingredients by count is done in these circumstances:

1. When units are in fairly standard sizes. *Examples:* 6 large eggs for a pancake batter; 8 parsley stems for a stock.
2. When serving portions are determined by numbers of units. *Examples:* 1 baked apple per portion; 6 fried shrimp per portion.

PORTION CONTROL

Portion control is the measurement of portions to ensure the correct amount of an item is served. In order for portion control to be carried out, cooks and service personnel must be aware of proper portion sizes. These are usually indicated on the house recipes and on the working menu used in the kitchen and service areas.

Portion Control in Preparation

Portion control actually begins with the measuring of ingredients. If this is not done correctly, then the yield of the recipe will be thrown off.

When portions are determined by count—1 hamburger patty, 2 tomato slices, 1 wedge of pie—then the units must be measured or cut according to instructions: 4 ounces meat per patty; 1/4-inch slices of "5 × 6" tomatoes; 8 equal wedges per pie.

Portion Control in Plating and Service

Portioning for service may be done by the cook, as in a short-order restaurant, or by the service personnel, as in a cafeteria. The following tools and techniques are used.

1. **Count.**
   Examples: 1 slice of ham per order; 5 shrimp per order. This is accurate if cutting and other prep work have been done correctly.

2. **Weight.**
   Example: 4 ounces of sliced ham per order. A portion scale must be at the serving station for this method of portion control.

3. **Volume.**
   Ladles, scoops, and kitchen spoons come in standard volume sizes and are used for portioning. The exact size of the ladle or scoop needed for a portion must be determined in advance and indicated on service instructions.

   Kitchen spoons, either solid or perforated, are not as accurate for portioning but are often used for convenience and speed. You must be able to judge by eye how full to fill the spoon (rounded, heaped, etc.). Check a spoonful on a portion scale from time to time to make sure you are being consistent.

4. **Even division.**
   Examples: Cut a pie into 8 equal wedges; cut a pan of lasagne 4 × 6 to make 24 equal portions.

5. **Standard fill.**
   Standard-size dishes, cups, or glasses are filled to a given level, as judged by eye. Example: a glass of orange juice. This is actually a form of volume measure.

UNITS OF MEASURE

The system of measurement used in the United States is complicated. Even when people have used the system all their lives, they still sometimes have trouble remembering things like how many fluid ounces are in a quart or how many feet are in a mile.

Table 4.2 lists abbreviations used in this book. Table 4.3 lists equivalents among the units of measure used in the kitchen. You should memorize these thoroughly so you don't have to lose time making simple calculations.

THE METRIC SYSTEM

The United States is the only major country that uses almost exclusively the complex system of measurement we have just described. In Canada, this system is also used, at least part of

TABLE 4.2 Abbreviations of U.S. Units in This Book

pound	lb
ounce	oz
gallon	gal
quart	qt
pint	pt
cup	cup (abbreviation not used)
fluid ounce	fl oz
tablespoon	tbsp
teaspoon	tsp
inch	in.

TABLE 4.3 Units of Measure—U.S. System

Weight		
1 pound	=	16 ounces
Volume		
1 gallon	=	4 quarts
1 quart	=	2 pints or 4 cups or 32 fluid ounces
1 pint	=	2 cups or 16 fluid ounces
1 cup	=	8 fluid ounces
1 fluid ounce	=	2 tablespoons
1 tablespoon	=	3 teaspoons
Length		
1 foot	=	12 inches

Note: One fluid ounce (usually called simply ounce) of water weighs 1 ounce. One pint of water weighs 1 pound.

the time, although the metric system has been introduced. Other countries use a much simpler system called the **metric system**. It is possible that someday the metric system may be used in U.S. kitchens. In fact, many kitchens practicing avant-garde cuisine (p. 7) use metric measures exclusively, because of their ease of use and precision. In any case, it is useful, in this age of international influences on cooking, to be able to read and use recipes from around the world. So it is a good idea to become familiar with the metric system.

Basic Units

In the metric system, there is one basic unit for each type of measurement:

The **gram** is the basic unit of weight.

The **liter** is the basic unit of volume.

The **meter** is the basic unit of length.

The **degree Celsius** is the basic unit of temperature.

Larger or smaller units are made simply by multiplying or dividing by 10, 100, 1000, and so on. These divisions are expressed by *prefixes*. The ones you will need to know are:

kilo- (kill-o) = 1,000

deci- (dess-i) = 1/10

centi- (sent-i) = 1/100

milli- (mill-i) = 1/1,000

Once you know these basic units, there is no longer any need for complicated tables like Table 4.3. Table 4.4 summarizes the metric units you will need to know in the kitchen. (*Note:* The prefix deci- is rarely used in Canada, although the deciliter is a regularly used unit of volume in Europe.)

Converting to Metric

Most people think the metric system is much harder to learn than it really is. This is because they think about metric units in terms of U.S. units. They read there are 28.35 grams in an ounce, and they are immediately convinced they will never be able to learn metrics.

Do not worry about being able to convert between U.S. and metric units. This is a very important point to remember, especially if you think the metric system might be hard to learn.

The reason for this is simple. You will usually be working in either one system or the other. You will rarely, if ever, have to convert from one to the other. (An exception might be if you have equipment based on one system but want to use a recipe written in the other.) If U.S. kitchens change to the metric system, everyone will use scales that measure in grams and kilograms, volume measures that measure in liters and deciliters, and thermometers that indicate degrees Celsius. And everyone will use recipes that indicate these units. No one will have to worry about how many grams are in an ounce. All one will have to remember is the information in Table 4.4.

To become accustomed to working in metric units, it is helpful to have a feel for how large the units are. The following equivalents may be used to help you visualize metric units. They are not exact conversion factors. (When you need exact conversion factors, see Appendix 1.)

A *kilogram* is slightly more than 2 pounds.

A *gram* is about 1/30 ounce. (1/2 teaspoon flour weighs a little less than 1 gram.)

A *liter* is slightly more than 1 quart.

A *meter* is slightly more than 3 feet.

A *centimeter* is about 3/8 inch.

0°C is the freezing point of water (32°F).

100°C is the boiling point of water (212°F).

An increase or decrease of 1 degree Celsius is equivalent to about 2 degrees Fahrenheit.

TABLE 4.4 Metric Units

Basic Units		
Quantity	**Unit**	**Abbreviation**
weight	gram	g
volume	liter	L
length	meter	m
temperature	degree Celsius	°C

Divisions and Multiples		
Prefix/Example	**Meaning**	**Abbreviation**
kilo-	1,000	k
kilogram	1,000 grams	kg
deci-	1/10	d
deciliter	0.1 liter	dL
centi-	1/100	c
centimeter	0.01 meter	cm
milli-	1/1,000	m
millimeter	0.001 meter	mm

Metric Recipes

Many recipe writers in the United States print exact metric equivalents in their recipes. As a result, you will see recipes calling for 454 grams potatoes, 28.35 grams butter, or a baking temperature of 191°C. No wonder many Americans are afraid of the metric system!

Kitchens in countries that use the metric system do not work with such impractical numbers, any more than cooks in the United States normally use figures like 1 lb 1¹/₄ oz potatoes, 2.19 oz butter, or a baking temperature of 348°F. That would defeat the purpose of the metric system, which is to be simple and practical. If you have a chance to look at a French cookbook, you will see nice, even numbers like 1 kg, 200 g, and 4 dL. (Note that the metric abbreviations used in this book are consistent with common usage in Canada. Abbreviations used in Europe are somewhat different, such as lowercase l instead of uppercase L for liter.)

The metric equivalents in the recipes in this book are rounded off. What's more, they are not always rounded off in the same way. In some places, you may see 1 pound rounded off to 500 grams, in other places to 450 grams. The object is to keep the recipe proportions and the total yield as close as possible to the original while arriving at practical measurements. Unfortunately, it is not always possible to keep the proportions exactly the same because the U.S. system is not decimal-based like the metric system. In some cases, the metric quantities may produce slightly different results due to these varying proportions, but these differences are small. If you have U.S. equipment, use the U.S. units, and if you have metric equipment, use the metric units. You should rarely have to worry about converting between the two. ***Never use U.S. measures for some ingredients and metric for other ingredients in the same recipe in this book***. Think of the U.S. and metric columns as belonging to separate recipes.

Please note that in a few recipes in this book, quantities are given in metric units only. These recipes use some of the specialty ingredients of avante-garde cuisine (p. 7) that are used in extremely small amounts. For example, dealing with a weight of 1 gram is fairly simple, while scaling its equivalent in U.S. units (approximately 0.035 oz) is comparatively difficult.

ROUNDING OFF NUMBERS

Rounding off means bumping a number up or down to the closest number that is "clean" and easiest to work with. For example, 197 is not a round number, but 200 is. In most cases, round numbers end in one or more zeros (such as 10, 300, 6,000), although sometimes we use round numbers that end in 5.

The first step in rounding is to find the "rounding digit." If you are rounding off to the nearest 10, your rounding digit is the tens' place (the second digit from the right). If you are rounding to the nearest hundred, the rounding digit is the hundreds' place.

Second, look to the place just to the right of the rounding digit. If that number is 1, 2, 3, or 4, the number is rounded down. Leave the rounding digit alone, and change all digits to the right to zero. But if the number to the right of the rounding digit is 5, 6, 7, 8, or 9, the number is rounded up. Change the rounding digit to the next higher number, and change all digits to the right to zero.

For example, 236, rounded off to the nearest ten, is 240. Thus, 240 is the closest "clean" number."

234, rounded off to the nearest ten, is 230.

3,216, rounded off to the nearest hundred, is 3,200, while 3,278 rounded off to the nearest hundred is 3,300.

KEY POINTS TO REVIEW

- How are most recipe ingredients measured?
- What tools are used for measuring in the kitchen?
- How are portion sizes measured?
- In the metric system, what are the units of measure for weight, volume, and length?

CONVERTING RECIPES

Unless you are working in an operation that uses only its own standardized recipes, you will frequently be required to increase or decrease recipes to different quantities. Each recipe is designed to make a specific amount of finishes product. This amount is called the **yield**. For example, you may have a recipe for 50 portions of Swiss steak but need only 25 portions. You need to change the yield of your recipe. This is called *converting the recipe*.

Converting recipes is an important technique. It is a skill you will probably need to use many times in this book. There is no "best" yield to write recipes for, as every operation, every school, and every individual has different needs.

CHANGING TO COMMON UNITS

Before we start converting recipe yields, we must deal with one more problem with our measuring system. As explained earlier, the U.S. system of measurement has many units that simply must be memorized: 16 ounces in a pound, 2 cups in a pint, 4 quarts in a gallon, and so on. When you are converting recipes, you often have to change back and forth between large and small units to get measurements you can work with. In most cases, the easiest way to make the calculations is to change larger units to smaller units.

- To change pounds to ounces, multiply by 16.
- To change cups to fluid ounces, multiply by 8.
- To change pints to fluid ounces, multiply by 16.
- To change quarts to fluid ounces, multiply by 32.

Note that these are the same numbers as found in Table 4.3.

After you have finished converting the recipe, you usually must convert the ounces and fluid ounces back to larger units in order to get units that are easy to work with. To do this, divide by the same numbers you used to multiply in the previous calculations. For example, to change ounces back to pounds, divide by 16.

If you divide by hand, you often get a remainder. Leave this remainder in the smaller unit. For example:

Change 60 ounces to pounds.
$60 \div 16 = 3$, with a remainder of 12
60 ounces = 3 lb 12 ounces

If you are using a calculator, you often get a decimal. In order to change the numbers to the right of the decimal point to ounces, you have to again multiply this decimal portion by 16. Either way, you end up with the same answer.

$60 \div 16 = 3.75$
$0.75 \times 16 = 12$
60 ounces = 3 lb 12 ounces

YIELDS

Recipe yields are expressed in several ways. For example:

- As a total quantity. (A soup recipe that makes 3 qt; a pot roast recipe that makes 5 lb meat.)
- As a total number of portions. (A recipe for Eggs Benedict that makes 12 portions; a recipe for grilled breast of duck that makes 4 portions.)
- As a total number of portions of a specified size. (A recipe for beef stew that makes 8 portions of 6 oz each; a recipe for individual chocolate soufflés that makes 6 soufflés, 4 oz each.)

Most recipe conversions require changing the total quantity or the total number of portions. The math for doing both of these is the same. As long as the portion size stays the same, you can use either the **total quantity** or the **total number of portions** as the yield when you make your calculations.

A little later in the discussion, after you are familiar with solving these problems, we discuss how to change both the number of portions and the portion size when converting recipes.

CONVERSION FACTORS

Nearly everyone instinctively can double a recipe or cut it in half. It seems more complicated, though, to change a recipe from 10 to 18 portions, say, or from 50 to 35. Actually, the principle is exactly the same: You find a number called a **conversion factor**, and then multiply every quantity by this number.

The conversion factor can be defined as the number used to increase or decrease each ingredient when converting a recipe to a different yield. The first step in converting a recipe is to calculate the conversion factor (see Procedures for Calculating Conversion Factors and Converting Total Yields, below and p. 77).

PROCEDURE for Calculating Conversion Factors

There is only one step in this procedure:

Divide the desired yield by the old yield stated on the recipe. This formula may be written like a mathematical calculation, as on a calculator, or as a fraction:

Mathematical Calculation: New yield ÷ Old yield = Conversion factor

$$\textbf{Fraction: } \frac{\textbf{New yield}}{\textbf{Old yield}} = \textbf{Conversion factor}$$

Example 1: You have a recipe with a yield of 8 portions, and you want to make 18 portions.

$$\textbf{18} \div \textbf{8} = \textbf{2.25}$$

Your conversion factor is 2.25. If you multiply each ingredient in your recipe by 2.25, you will prepare 18 portions, not the 8 of the original recipe.

Example 2: You have a recipe that makes 20 liters of soup, and you want to make 5 liters.

$$\textbf{5} \div \textbf{20} = \textbf{0.25}$$

Your conversion factor is 0.25. That is, if you multiply each ingredient by 0.25, you will prepare only 5 portions.

Notice in the second example the conversion factor is a number less than 1. This is because the recipe yield is decreased. You are making the recipe smaller. This is a good way to check your math. Decreasing the recipe yield will involve a conversion factor less than 1. Increasing the yield of a recipe will involve a conversion factor larger than 1.

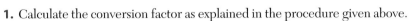

PROCEDURE for Converting Total Yield

1. Calculate the conversion factor as explained in the procedure given above.

2. Multiply each ingredient quantity by the conversion factor:

$$\textbf{Old quantity} \times \textbf{Conversion factor} = \textbf{New quantity}$$

Note: In order to do this in the U.S. system, you may have to convert all weights to ounces and all volumes to fluid ounces. (This is not necessary in the metric system.) See page 76 for more explanation.

Example 1: You have a recipe for 10 portions of Broccoli Mornay requiring 3 lb AP broccoli and $2\frac{1}{2}$ cups Mornay Sauce. Convert to 15 portions.

$$\frac{\textbf{New yield}}{\textbf{Old yield}} = \frac{15}{10} = \textbf{1.5}$$

Broccoli: 3 lb = 48 oz

$$48\ oz \times 1.5 = 72\ oz = 4\ lb\ 8\ oz$$

Sauce: $2\frac{1}{2}$ cups = 20 fl oz

$$20\ fl\ oz \times 1.5 = 30\ fl\ oz = 3\frac{3}{4}\ cups$$

Example 2: You have a recipe for 10 portions of Broccoli Mornay requiring 1,500 grams AP broccoli and 600 mL Mornay Sauce. Convert to 15 portions.

$$\frac{\textbf{New yield}}{\textbf{Old yield}} = \frac{15}{10} = \textbf{1.5}$$

$$\textbf{Broccoli: 1,500 g} \times \textbf{1.5} = \textbf{2,250 g}$$

$$\textbf{Sauce: 600 mL} \times \textbf{1.5} = \textbf{900 mL}$$

In order to make these procedures clearer, let's work through the conversion of a full recipe to give you practice with the equations. The following examples are in the U.S. system of measures. For metric examples, see Appendix 4, page 1028.

In the sample recipe that follows are the ingredients and quantities for a sautéed beef dish. As you can see, the quantities indicated are enough to make 8 portions each.

Beef Tenderloin Tips and Mushrooms à la Crème

Portions: 8   Portion size: 8 oz

Butter	2 oz	Prepared mustard	2 tsp
Onion	4 oz	Brown sauce	$1\frac{1}{2}$ pt
Flour	1 tbsp	Heavy cream	1 cup
Mushrooms	$\frac{1}{2}$ lb	Salt	to taste
Beef tenderloin	$2\frac{1}{2}$ lb	Pepper	to taste
White wine	$\frac{1}{2}$ cup		

Let's say we need 18 portions instead of 8. To find the conversion factor, we divide the new yield by the old yield:

$$\frac{\text{New yield}}{\text{Old yield}} = \frac{18}{8} = 2.25$$

To convert the recipe to 18 portions, we multiply each ingredient quantity by the conversion factor of 2.25.

First, to make this easier, we should change pounds to ounces and cups, pints, and quarts to fluid ounces, using the figures in Table 4.3. This is called *converting to common units of measure*, as explained earlier. For example, to change the measurement for beef tenderloin to ounces, multiply $2\frac{1}{2}$ (the weight in pounds) by 16 (the number of ounces in a pound) to get 40 ounces.

The equivalents we need for this recipe are as follows:

$1/2$ lb equals 8 ounces

$2^1/2$ pounds equals 40 ounces

$1/2$ cup equals 4 fluid ounces

$1^1/2$ pints equals 24 fluid ounces

In Example 1, we have substituted these equivalent quantities. Then we have multiplied all the ingredient quantities by the conversion factor to get the quantities we need for 18 portions. Check through all the calculations to make sure you follow them. The quantities for salt and pepper will still, of course, be indicated as "to taste."

EXAMPLE 1

Ingredient	Quantity	Times	Conversion Factor	Equals	New Quantity
Butter	2 oz	×	2.25	=	4.5 oz
Onion	4 oz	×	2.25	=	9 oz
Flour	1 tbsp	×	2.25	=	2.25 tbsp or 2 tbsp plus $3/4$ tsp
Mushrooms	8 oz	×	2.25	=	18 oz or 1 lb 2 oz
Beef tenderloin	40 oz	×	2.25	=	90 oz or 5 lb 10 oz
White wine	4 fl oz	×	2.25	=	9 fl oz
Prepared mustard	2 tsp	×	2.25	=	$4^1/2$ tsp or $1^1/2$ tbsp
Brown sauce	24 fl oz	×	2.25	=	54 fl oz or 3 pt plus 6 fl oz
Heavy cream	8 fl oz	×	2.25	=	18 fl oz or $2^1/4$ cups

Now let's suppose we want to find the quantities needed to give us 40 portions, 6 ounces each. Because the portion size changes, we have to add extra steps to our procedure for calculating the conversion factor. The Procedure for Calculating Conversion Factors When Portion Size Changes (p. 79) explains these steps. Then we use this new conversion factor in the same way as we did in Example 1 above. For our Beef Tenderloin Tips recipe, we first calculate our new conversion factor:

$$8 \text{ (portions)} \times 8 \text{ oz} = 64 \text{ oz}$$

Do the same calculation for the desired yield:

$$40 \text{ (portions)} \times 6 \text{ oz} = 240 \text{ oz}$$

When we divide the new yield by the old yield ($240 \div 64$), we arrive at a conversion factor of 3.75. In Example 2, we have done the conversions using the new factor of 3.75.

The new calculations, using the conversion factor of 3.75, are shown in the Example 2 table.

EXAMPLE 2

Ingredient	Quantity	Times	Conversion Factor	Equals	New Quantity
Butter	2 oz	×	3.75	=	7.5 oz
Onion	4 oz	×	3.75	=	15 oz
Flour	1 tbsp	×	3.75	=	3.75 tbsp or 3 tbsp plus $2^1/4$ tsp
Mushrooms	8 oz	×	3.75	=	30 oz or 1 lb 14 oz
Beef tenderloin	40 oz	×	3.75	=	150 oz or 9 lb 6 oz
White wine	4 fl oz	×	3.75	=	15 fl oz
Prepared mustard	2 tsp	×	3.75	=	$7^1/2$ tsp or $2^1/2$ tbsp
Brown sauce	24 fl oz	×	3.75	=	90 fl oz or 5 pt plus 10 fl oz
Heavy cream	8 fl oz	×	3.75	=	30 fl oz or $3^3/4$ cups

PROCEDURE for Converting Factors When Portion Size Changes

1. Determine the total yield in the old recipe by multiplying the number of portions by the portion size:

$$\textbf{Old recipe number of portions} \times \textbf{Portion size} = \textbf{Old yield}$$

2. Determine the total yield desired (new yield) by multiplying the number of portions desired by the portion size desired.

$$\textbf{Desired number of portions} \times \textbf{Desired portion size} = \textbf{New yield}$$

3. Divide the new yield by the old yield. This formula may be written like a mathematical calculation, as done on a calculator, or as a fraction:

$$\textbf{Mathematical Calculation: New yield} \div \textbf{Old yield} = \textbf{Conversion factor}$$

$$\textbf{Fraction:} \ \frac{\textbf{New yield}}{\textbf{Old yield}} = \textbf{Conversion factor}$$

Example: You have a recipe with a yield of 20 portions, 4 ounces each, and you want to make 30 portions, 5 ounces each.

1. Calculate the total yield of the original recipe.

$$\textbf{20 portions} \times \textbf{4 ounces} = \textbf{80 ounces}$$

2. Calculate the total yield desired.

$$\textbf{30 portions} \times \textbf{5 ounces} = \textbf{150 ounces}$$

3. Divide the new yield by the old yield.

$$\textbf{150} \div \textbf{80} = \textbf{1.875}$$

The calculation gives us a conversion factor of 1.875. We use this conversion factor the same way as explained above, using the Procedure for Converting Total Yield.

PROBLEMS IN CONVERTING RECIPES

For the most part, these conversion procedures work well. But when you make some very large conversions—from 10 to 400 portions, for example, or from 500 to 6—you may encounter problems.

For example, you may have to make major equipment changes, like from a 2-quart saucepot to a large steam kettle. Consequently, you have to adjust your techniques and, sometimes, even ingredients. Evaporation rates may be different, thickening agents may need increasing or decreasing, seasonings and spices may have to be cut back. Sometimes quantities are too large or too small to mix properly.

This is one more example of the importance of cooking with judgment. Experienced chefs develop a feel for these problems over the years. When you make such adjustments on converted recipes, be sure to make a note of them for future reference.

Although there are no fixed rules you can learn for these adjustments, it is possible to list the most common types of problems encountered so you can be on the alert for them when making recipe conversions. In general, most of the pitfalls fall into one of the following categories.

Measuring

This is most often a problem when you are expanding small recipes, such as when you want to take a consumer recipe for 4 portions and adapt it to a high-volume operation such as a large cafeteria. Many such recipes use volume measures for both solids and liquids. As we explained earlier, volume measurement of solids is inaccurate. Of course, small inaccuracies become large ones when a recipe is multiplied. Therefore, it is important to be cautious and to test carefully when you are converting a recipe that uses volume measures for solid ingredients.

Problems also occur in the opposite situation—when converting a large-quantity recipe to a much smaller yield. This is a typical problem when adapting a quantity recipe to single-portion size for à la carte service. Some quantities in the converted recipe may be so small as to be difficult to measure.

These problems usually can be avoided when all solids are measured by weight. But such items as spices and seasonings may be too small to be measured easily by weight, unless you use a scale that is accurate to a tiny fraction of an ounce. For this reason, it is usually a good idea to cut back on spices and salt in a converted recipe. You can always add more if you taste the product and decide it needs more seasoning.

Surface and Volume

If you have studied geometry, you may remember that a cube with a volume of 1 cubic foot has a top surface area of 1 square foot. But if you double the volume of the cube, the top surface area is not doubled but is in fact only about $1\frac{1}{2}$ times as large.

What in the world, you ask, does this have to do with cooking? Consider the following example.

Suppose you have a good recipe for $\frac{1}{2}$ gallon cream soup, which you normally make in a small soup pot. You want to make 16 gallons of the soup, so you multiply all ingredients by a conversion factor of 32 and make the soup in a steam kettle. To your surprise, not only do you end up with more soup than you expected but also it turns out rather thin and watery. What happened?

Your converted recipe has 32 times as much volume to start, but the amount of surface area has not increased nearly as much. Because the ratio of surface area to volume is less, there is less evaporation. This means there is less reduction and less thickening, and the flavors are not as concentrated. To correct this problem, you would have to use less stock, and preferably a stock that is more concentrated.

Suppose instead that you made the expanded recipe in a tilting skillet. In this case, there is so much surface area that the liquid would evaporate very quickly, resulting in an overly thickened and overly seasoned soup.

Differences in surface area and volume can cause other problems as well. Food-service operations must be more careful than home cooks do about food spoilage and the Food Danger Zone (see Chapter 2) because large volumes of food cool and heat much more slowly than small volumes do.

For the same reason, a home baker worries about keeping a bread dough warm so it will ferment, but a commercial baker worries about keeping a dough cool enough so it doesn't ferment too fast. This is because a large batch of dough has so much volume in comparison with its surface area that it tends to retain heat rather than cool quickly to room temperature.

Equipment

When you change the size of a recipe, you must often change the equipment, too. This change often means the recipe does not work in the same way. Cooks must be able to use their judgment to anticipate these problems and to modify their procedures to avoid them. The example just given, of cooking a large batch of soup in a steam kettle or in a tilting skillet, is among the kinds of problems that can arise when you change cooking utensils.

Other problems develop because of mixers or other processing equipment. For example, if you break down a salad dressing recipe to make only a small quantity, you might find there is so little liquid in the mixing machine that the beaters don't blend the ingredients properly.

Or you might have a recipe for a muffin batter you usually make in small quantities and mix by hand. When you increase the recipe greatly, you find you have too much to do by hand. Therefore, you use a mixer but keep the mixing time the same. Because the mixer does the job so efficiently, you overmix the batter and end up with poor-quality muffins.

Many mixing and stirring jobs can be done only by hand. This is easy with small quantities but difficult with large batches. The result is often an inferior product. In contrast, some hand-made products are better if they are done in large batches. It is hard, for example, to make a very small batch of puff pastry because the dough cannot be rolled and folded properly.

Time

Some people make the mistake of thinking that if you double a recipe, you must also double the cooking time. That this is an error can be shown by a simple example. Assume it takes 15 minutes to cook a steak in a broiler. If you put two steaks in the broiler, it still takes 15 minutes to cook them, not 30 minutes.

If all other things are equal, cooking times stay the same when a recipe is converted. Problems arise, however, because all other things are not always equal. For example, a large pot of liquid takes longer to bring to a boil than a small pot. Therefore, the total cooking time is longer.

On the other hand, a big kettle of vegetable soup you are making ahead for tomorrow's lunch takes longer to cool than a small pot. Meanwhile, the vegetables continue to cook in the retained heat during the cooling. In order to avoid overcooking, you may need to undercook the large batch slightly.

In cases where the cooking time must be increased, you might find you must increase the amount of herbs and spices. This is because the flavors are volatile (see p. 125), and more flavor is lost because of the increased cooking time. (Another answer to this problem is to add the spices later.)

Changing recipe sizes can affect not only cooking times but also mixing times. The best way to avoid this problem is to rely not on printed cooking and mixing times but on your own judgment and skills to tell you when a product is properly cooked or properly blended.

Recipe Problems

Many recipes have flaws, either in the quantities or types of ingredients or in the cooking procedures. When the item is made in small quantities, these flaws may not be noticeable, or the cook may almost unconsciously or automatically make adjustments during production. When the recipe is multiplied, however, the flaws may suddenly become apparent and the product quality lower. The only solution here is to carefully test recipes and to have a good understanding of basic cooking principles.

FOOD COST CALCULATIONS

Food-service operations are businesses. This means someone in the operation must worry about budgets, cost accounting, bills, and profits. Usually this is the job of the manager, while the cook takes care of food production.

Cooks have a great deal of responsibility for food cost controls, however. They must always be conscious of accurate measurement, portion control, and careful processing, cooking, and handling of foods to avoid excess trimming loss, shrinkage, and waste.

The manager, on the other hand, is concerned with determining budgets, calculating profits, and expenses, and so on. We cannot deal with these subjects here, as this is a book about food preparation. But you may encounter them later in your studies or in your career.

Regardless, every cook should understand three areas of cost accounting: doing yield analyses, calculating raw food cost or portion cost, and using food cost percentages.

FOOD COST PERCENTAGES

An individual operation's food cost percentage is usually determined by the budget. The chef is interested in this figure because it tells him or her whether the menu prices and the costs for each item are in line.

The **food cost percentage** of a menu item equals the raw food cost or portion cost divided by the menu price:

$$\text{Percentage} = \frac{\text{Food cost}}{\text{Menu price}}$$

You can use this figure in two ways:

1. If you know the menu price and want to see what your food cost should be in order to be within the budget, multiply the menu price by the percentage:

$$\text{Food cost} = \text{Menu price} \times \text{Percentage}$$

Example: Menu price is $6.75 and food cost percentage is 35 percent.
$$35\% = 0.35$$
$$6.75 \times 0.35 = \$2.36$$

2. If you know the food cost and want to determine what the menu price should be at a particular percentage, divide the cost by the percentage:

$$\text{Menu price} = \frac{\text{Food cost}}{\text{Percentage}}$$

Example: Food cost is $1.60 and food cost percentage is 40 percent.
$$\frac{\$1.60}{40} = \frac{\$1.60}{0.40} = \$4.00$$

Food cost

menu price | FCP

per menu item & also for the whole menu

YIELD COST ANALYSIS

In order to calculate portion costs of recipes, you must first determine the costs of your ingredients. For many ingredients, this is relatively easy. You just look at your invoices or at price lists from your purveyors.

Many recipes, however, specify trimmed weight rather than the weight you actually pay for. For example, a stew might call for 2 pounds sliced onion. Let's say you pay 24 cents a pound for onions, and to get 2 pounds sliced onions, you need $2^1/_4$ pounds untrimmed onions. In order to calculate the cost of the recipe correctly, you must figure out what you actually paid for the onions. In this case, the true cost is 54 cents ($2^1/_4$ lb $\times$ \$0.24 per lb), not 48 cents (2 lb $\times$ \$0.24 per lb).

The following are two frequently used abbreviations you must understand:

- AP stands for *as purchased*. This means the untrimmed quantity, in the same form in which it is purchased. This is the amount you pay for.
- EP stands for *edible portion*. This means the raw, uncooked quantity after all trimming is done. This is the quantity you actually cook.

In the case of fruits and vegetables, the best way to determine AP quantities for use in costing recipes is to make a note of them when you are preparing the item. Tables of vegetable and fruit trimming yields in Chapters 10 and 21 will also help you. The next section explains how to use these numbers.

Trimming Loss: Calculating Yields and Amounts Needed

The descriptions of vegetables and fruits in Chapters 10 and 21 include a number called *percentage yield*. The percentage yield of a vegetable or fruit indicates, on the average, how much of the AP weight is left after pre-prep to produce the ready-to-cook item, or EP weight. You can use this figure to perform two basic calculations.

1. **Calculating yield.**

   Example: You have 10 lb AP Brussels sprouts. Yield after trimming is 80 percent. What will your EP weight be?

   First, change the percentage to a decimal number by moving the decimal point two places to the left.

   $$80\% = 0.80$$

   Multiply the decimal by your AP weight to get EP yield.

   $$10\ lb \times 0.80 = 8\ lb$$

2. **Calculating amount needed.**

   Example: You need 10 lb EP Brussels sprouts. What amount of untrimmed vegetable do you need?

   Change the percentage to a decimal number.

   $$80\% = 0.80$$

   Divide the EP weight needed by this number to get the AP weight.

   $$\frac{10\ lb}{0.80} = 12^1/_2\ lb$$

In the case of ingredients such as meats and fish, figuring the cost can be a little more complicated. If you buy precut portion-controlled steaks or fish filets and use them just as you receive them, your AP and EP costs are the same. But if you buy whole loins of beef or whole fish and cut them yourself, you have to do a yield cost analysis in order to determine your actual costs.

The examples discussed in the remainder of this chapter use U.S. measures. For metric examples, see Appendix 4, pages 1028–1029.

Raw Yield Test

Suppose you work in a restaurant that serves veal scaloppine. The restaurant buys whole legs of veal. It is your job to bone out the veal, trim off all fat and connective tissue, separate the muscles at the seams, and cut the large pieces into scaloppine.

A typical whole leg of veal might weigh 30 pounds and cost \$5 per pound for a total cost of \$150. After finishing your trimming and cutting, you find you have 18 pounds of veal scaloppine. How do you figure the cost per pound of this meat?

The simplest example would be if you threw away all the trimmings, bones, and scrap meat. Then you would know your 18 pounds of veal cost you $150. Dividing $150 by 18 pounds gives you a cost per pound of $8.33.

But in your restaurant, you don't throw away the trimmings. You make stock with the bones, grind up the small trimmings for meatballs, use the larger trimmings for veal stew, and sell the fat to the fat collector who picks up all your waste fat once a week. Now you must do a **yield test** to figure your costs.

Table 4.5 shows a typical form you might use for a yield test. For simplification, the blanks in the form are of two types. The dotted lines are to be filled in by reading your invoices and by taking the weights from your actual yield test. The solid lines are to be filled in by doing calculations.

Note that in Table 4.6, the form has been filled in with the results of a yield test on a leg of veal. We go through the form step by step.

TABLE 4.5 Raw Yield Test Form

Item .... Test number .... Date ....
Purveyor .... Price per pound .... Total cost ....
AP weight **(1)** .... Lb price **(2)** .... Total cost **(3)** ....

Trim, salvage, and waste:

Item	Weight	Value/lb	Total Value (lb × Value)
(4)	....	....	_____
(5)	....	....	_____
(6)	....	....	_____
(7)	....	....	_____
(8)	....	....	_____
(9)	....	....	_____
(10)	....	....	_____

Total weight of trim (4 thru 10) **(11)** _____ Total value of trim (4 thru 10) **(12)** _____

Total yield of item **(13)** ....

Net cost (3 − 12) **(14)** _____

Cost per lb (14 ÷ 13) **(15)** _____

Percentage of increase (15 ÷ 2) **(16)** _____

TABLE 4.6 Completed Raw Yield Test Form

Item .... veal leg to scaloppine .... Test number .... 3 .... Date .... 09/05/13 ....
Purveyor .... ABC Meats .... Price per pound .... $5 .... Total cost .... $150 ....
AP weight **(1)** .... 30 lb .... Lb price **(2)** .... $5 .... Total cost **(3)** .... $150 ....

Trim, salvage, and waste:

Item	Weight	Value/Lb	Total Value (Lb × Value)
(4) fat	2½ lb	$0.12	$0.30
(5) bone	3 lb 5 oz	$0.38	$1.26
(6) ground veal	2 lb 2 oz	$4.89	$10.39
(7) stew meat	3 lb	$5.29	$15.87
(8) unusable trim	14 oz	0	0
(9) cutting loss	3 oz	0	0
(10)			

Total weight of trim (4 thru 10) **(11)** 12 lb   Total value of trim (4 thru 10) **(12)** $27.82

Total yield of item **(13)** 18 lb

Net cost (3 − 12) **(14)** **$122.18**

Cost per lb (14 ÷ 13) **(15)** $6.79

Percentage of increase (15 ÷ 2) **(16)** 1.36 (136%)

The executive chef in this restaurant fills out the first two lines based on the invoice, gives you the form, and requests you to do the test. You fill out the rest of the form, beginning with blank 1 on the third line. You proceed as follows:

1. Weigh the whole leg of veal and enter the weight in blank 1. Copy the price per pound and total cost from line 2 to blanks 2 and 3.

   Note that blank 3 can also be arrived at by multiplying the weight by the price per pound. However, suppose the veal were left in the cooler for several more days and dried out a bit. The weight then might be 29$\frac{1}{2}$ pounds. By multiplying 29.5 by $5.00, you would get a total cost of $147.50. But because the price you paid was actually $150, it is important to use that figure and not fill in the blank by multiplying.

2. Break down the veal into all its component parts and record the weights of the trim and waste, starting in blank 4. In this case, there are only six items: fat, bones, small meat scraps for grinding, meat for stew, unusable waste, and cutting loss.

   Record the weight of the finished scaloppine in blank 13.

   What is **cutting loss?** This is not something you can actually weigh. However, there is always some loss of weight due to particles of meat and fat sticking to the cutting board, to drying, and to other factors. So when you add up all your weights, you find they total less than 30 pounds. To determine cutting loss, add up blanks 4 through 8 and blank 13. Subtract this total from line 1.

3. Enter the values per pound of the trim, salvage, and waste on lines 4 through 10. In this case, these numbers are given to you by the executive chef from the invoices.

   - The fat collector pays 12 cents per pound for waste fat.
   - When you have to buy extra bones for your stockpot, you pay 38 cents per pound for them, so this is their value to you. This is also the figure you use when you cost out your stock recipe. If you didn't make stock and threw out the bones, you'd enter 0 in this blank.
   - Similarly, the values entered for ground veal and stew meat are the prices you'd have to pay if you bought them.
   - Unusable trim and cutting loss have no value, so you enter 0.

4. Calculate the total value of each item on lines 4 through 10 by multiplying the weight by the value per pound. Note that this particular form tells you how to do all the calculations.

5. Add the weights in lines 4 through 10 and enter the total in blank 11. Add the total values in lines 4 through 10 and enter this figure in blank 12.

6. Subtract the total value of all the trim (blank 12) from the price you paid for the veal (blank 3). This gives you the net cost of your 18 pounds of scaloppine.

7. To find the cost per pound of the scaloppine, divide the net cost (blank 14) by the weight (blank 13). This is the figure you will use in costing recipes for veal scaloppine.

8. The percentage of increase in the last line is determined by dividing the net cost per pound (blank 15) by the price per pound of the whole leg (blank 2). This figure can be used as follows:

   Suppose next week you buy another leg of veal from the same purveyor, but the price has gone up to $5.29 per pound. Instead of doing another yield test, you can simply multiply this new price by the percentage of increase ($5.29 times 1.36) to get a new cost per pound of $7.19.

Cooked Yield Test

Earlier we introduced two important abbreviations: AP (as purchased) and EP (edible portion). A third expression sometimes used is *AS*, meaning **as served**. When foods such as fruits are served raw, AS may be the same as EP. But if the food is cooked, these weights are different.

In the case of the veal scaloppine, your recipe portions, and therefore your portion costs, are based on raw weight. For example, your scaloppine recipe might call for 5$\frac{1}{2}$ ounces raw meat per portion.

In some cases, your portions may be based on cooked weight. This is most often true of roasts. For example, let's say you buy whole fresh hams, bone and trim them, and serve them as roasts, allowing 6 ounces sliced, cooked meat per portion. To arrive at your cost, you will have to do a cooked yield test, as illustrated in Tables 4.7 and 4.8. (This form may be printed on

TABLE 4.7 Cooked Yield Test Form

Item .... Test number .... Date ....
AP price per lb ....
Cooking temperature ....
Net raw weight (**1**) .... Net cost per lb (**2**) ....
.... Total net cost (**3**) ....
Weight as served (**4**) ....
Cooked cost per lb (3 ÷ 4) (**5**) ....
Shrinkage (1 − 4) (**6**) ....
Percentage of shrinkage (6 ÷ 1) (**7**) ....
Total percentage of cost increase (5 ÷ AP price per lb) (**8**)

TABLE 4.8 Completed Cooked Yield Test Form

Item ....roast fresh ham........ Test number ....2........ Date ......09/05/13......
AP price per lb ......$3.49......
Cooking temperature .....325°F.....
Net raw weight (**1**) ....12 lb....... Net cost per lb (**2**) ....$3.93........
.... Total net cost (**3**) ....$47.16........
Weight as served (**4**) ......8 lb 4 oz......
Cooked cost per lb (3 ÷ 4) (**5**) .....$5.72.....
Shrinkage (1 − 4) (**6**) ......3¾ lb......
Percentage of shrinkage (6 ÷ 1) (**7**) .....31%.....
Total percentage of cost increase (5 ÷ AP price per lb) (**8**) .....164%.....

the same sheet of paper as the raw yield test form so the operation can have a complete cost analysis on one form.)

This form has been filled in with the results of a cooked yield test done on a roast, boneless fresh ham. Let's assume this same ham has already had a raw yield test done on it.

The first half of the form, through blank 3, is filled in before the test starts. The numbers for blanks 1, 2, and 3 are taken from the raw yield test form, but you should double-check the net raw weight by weighing the item again before roasting.

Enter the total weight of cooked ham served in blank 4. You arrive at this figure by recording the total number of portions served and multiplying this number by the portion size. Let's say 22 portions are served at 6 ounces each. This gives us a total of 132 ounces (22 × 6), or 8¼ pounds.

You might be tempted to simply weigh the whole roast after cooking and trimming. Remember, though, that there will be some waste—crumbs on the slicer or cutting board, spillage of juices, and so on. It is more accurate to record the weight you actually sell.

If this had been a bone-in roast, you would have another reason to carve the meat before weighing, because you could not include the weight of the bone in your as-served figure.

The remaining blanks on the form are determined by doing the calculations, just as you would do the calculations for the raw yield test.

PORTION COSTS

Portion cost, or raw food cost, is the total cost of all the ingredients in a recipe divided by the number of portions served:

$$\text{Portion cost} = \frac{\text{Cost of ingredients}}{\text{Number of portions}}$$

Here we cost out a sample recipe to show you how the procedure works. First, note the following points and keep them in mind when you are calculating portion costs. Many errors in costing are caused by forgetting one of these points.

1. Costs must be based on AP (as purchased) amounts, even though recipes often give EP (edible portion) quantities. These terms are explained in the preceding section.

2. Include ***everything***. That means the lemon wedge and parsley garnish for the fish filet, the cream and sugar that go with the coffee, and the oil that used for pan-frying the eggplant. These are sometimes called **hidden costs**.

   Seasonings and spices are a typical example of hidden costs that are difficult to calculate. Some operations add up the cost of all seasonings used in a year and divide that by the total food cost to get a percentage. This percentage is added to each item. For example, if the cost of an item is $2.00 and the seasoning cost percentage is 5 percent, the total cost is $2.00 plus 5 percent of $2.00, or $2.10.

   Other hidden costs can be calculated in the same way. For example, you could figure out your cost percentage for frying fat and add the percentage to all deep-fried foods.

   Some restaurants take an arbitrary figure for all hidden costs, usually 8 to 12 percent, and add this to all menu items.

3. Record the number of portions ***actually served***, not just the number the recipe is intended to serve. If the roast shrank more than you expected during cooking, or if you dropped a piece of cake on the floor, those costs still have to be covered.

PROCEDURE for Calculating Portion Cost

1. List ingredients and quantities of recipe as prepared.

2. Convert the recipe quantities to AP (as purchased) quantities.

3. Determine the price of each ingredient (from invoices, price lists, etc.). The units in this step and in step 2 must be the same in order for you to do the calculation.

4. Calculate the total cost of each ingredient by multiplying the price per unit by the number of units needed.

5. Add the ingredient costs to get the total recipe cost.

6. Divide the total cost by the number of portions served to get the cost per portion.

EXAMPLE: Costing a Recipe	**Item: Baked Rice**			
Ingredient	**Amount**	**Recipe Quantity**	**Price**	**Total**
Rice, long grain	4 lb	4 lb	$0.62/lb	$2.48
Butter	12 oz	0.75 lb	1.97/lb	1.48
Onions	1 lb	1.2 lb	0.36/lb	0.43
Chicken stock	4 qt	4 qt	0.25/qt	1.00
Salt	1 oz	1/16 lb	0.15/lb	0.01
			Total cost	$5.40
			Number of portions	50
			Cost per portion	$0.11

Note: Cost of chicken stock is determined by costing out the operation's recipe for chicken stock.

CONTROLLING FOOD COSTS

Calculating food costs is a critical part of the business of selling and serving food. The preceding section explains the basic mathematics of determining these costs. But cost control is much more than calculating costs based on written recipes. Cost control begins with menu planning and encompasses every phase of the operation, from purchasing to service.

MENU PLANNING

A well-planned menu takes care to utilize ingredients efficiently and in a way that avoids waste. When writing a menu, consider the following guidelines for total utilization of foods.

1. **Use all edible trim.**
   Unless you use only portion-controlled meats, poultry, and fish and only frozen and canned vegetables, you will have edible trim. You can either throw it away and call it a loss, or you can use it and make money on it.

Plan recipes that utilize these trimmings and put them on the menu. For example:

* Use small meat scraps for soups, chopped meat, pâtés, creamed dishes, croquettes.
* Use larger meat trimmings for soups, stews, braised items.
* Use bones for stocks, soups.
* Use vegetable trimmings for purées, soups, stews, stocks, fillings for omelets and crêpes.
* Use day-old breads for stuffings, breading, French toast, croutons, meat extender.

2. **Don't add an item to the menu unless you can use the trimmings.**
   This is really the same as the preceding item, looking from the opposite angle. In other words, don't put rissolé potatoes on your menu unless you also plan to serve an item that uses the trimmings, such as whipped potatoes or croquettes.

3. **Plan production to avoid leftovers.**
   The best way to use leftovers is not to create them in the first place. Handling food twice—once as a fresh item and once as a leftover—is more expensive and time-consuming than handling it once, and it almost always results in loss of quality. Limited menus—that is, with fewer selections—decrease the likelihood of leftovers.

4. **Plan ahead for use of leftovers.**
   Careful planning of production can keep leftovers to a minimum. But some leftovers are almost inevitable, and it's better for your costs to use them than to throw them out.

   Whenever you put an item on the menu that could become a leftover, you should have a recipe ready that will use it. This is better than being surprised with leftovers you don't know what to do with.

   For example, if you served roast chicken for dinner one day, you might plan on chicken salad for a luncheon special the next day.

   Remember to handle all leftovers according to proper sanitary procedures.

5. **Avoid minimum-use perishable ingredients.**
   Minimum-use ingredients are those that are used in one or two items on your menu. For example, an operation might serve chicken breast topped with sautéed mushrooms but not use mushrooms in any other item. When the ingredient is perishable, the result is a high percentage of spoilage or waste.

   This situation can be remedied in any of three ways:

* Change the recipe to eliminate the minimum-use ingredient.
* Eliminate the item from the menu.
* Add other items to the menu using the ingredient.

   Be careful not to unbalance the menu, however, by using an ingredient in too many dishes. Try to avoid both extremes.

PURCHASING

Managing Inventory

Par stock is the inventory of goods an operation must have on hand to continue operating between deliveries. It is important to maintain a proper par stock to avoid running out of essential items. However, it is costly to maintain a larger inventory than necessary. This is partly because some of your operating cash is tied up in unused inventory. Second, excessive inventory of perishable items can lead to spoilage and thus to loss. Accurate forecasting of future business, based on careful study of past business, is key to managing inventory and establishing par stock.

Carefully write specifications for each item purchased so the item is described accurately. Compare price and quality offered by several vendors in order to get the best quality for the best price. Order from up-to-date quotes and price lists.

RECEIVING

Pay careful attention to receiving procedures:

* Schedule receipts so an employee can devote full attention to checking the shipment without being rushed. Try to schedule deliveries so they don't arrive at the same time.

If the employee receiving the goods has kitchen duties as well, schedule deliveries during off hours so he or she has adequate time to check the shipments.

- Check the delivery immediately, while the driver is there. Don't allow the driver to unload the shipment and depart, leaving the order on the loading dock.
- Compare the delivery invoice to the order forms to make sure the right goods are delivered.
- Compare the delivery invoice to the actual goods received to make sure you have received everything you are paying for. Keep a scale at the receiving entrance so items can be checked for proper weight.

Check the delivery for quality:

- Check the temperature of refrigerated goods.
- Check the temperature of frozen goods, and check for signs of thawing and refreezing.
- Check all perishables for freshness.
- Check all containers and packaging for damage.
- Immediately transfer all goods to proper storage.

STORING

Proper storage is essential to avoid spoilage and loss of food items. The basic principles of storing dry foods, refrigerated foods, and frozen foods are detailed in Chapter 2 (see pp. 22–24). In addition, storage of specific categories of foods, especially meats, poultry, seafood, produce, and dairy products, is included in the discussions of each. Consult the index for further reference.

MEASURING AND OTHER COOKING PROCEDURES

As discussed earlier in this chapter, the concept of measurement applies to two principal areas: measurement of ingredients in recipe preparation, and measurement of portions to be served, also known as portion control. The fundamentals of both types of measurement are explained earlier in this chapter (see pp. 72–74) and need not be repeated here.

Measurement is an indispensable part of good kitchen procedures. Think of measurement as part of your basic cooking procedures, not as something separate. All cooking procedures, including measuring, combine to help control food costs, because proper cooking procedures help avoid waste due to improperly prepared foods. For example, lack of attention at the grill station often results in meats returned by the customer for being overcooked. Having to discard and replace badly cooked foods greatly increases food cost.

KEY POINTS TO REVIEW

- What is the procedure for converting a recipe to a different yield?

- Why do recipes sometimes not give the same results when converted to a different quantity? What kinds of adjustments might you have to make to avoid this problem?

- What are AP quantities and EP quantities? Explain how to perform yield calculations when you know the AP quantity and the percentage yield.

- What is the procedure for calculating food costs?

- What are some good ways to keep food costs as low as possible?

SERVING

All the cook's attention to measurement and portion control goes for nothing if a server drops the plate on the way to the dining room. A well-trained serving staff is an important part of controlling food costs. Serving staff must be trained in serving techniques and in portion control for those items, such as desserts or salads, they are responsible for plating. Servers must understand the menu well so they can explain each item to customers and avoid returned dishes. The chef often may wish to sell more of some items than others as part of inventory and cost control, so serving staff must also be trained to be effective but pleasant salespeople. The demands made on the service personnel can best be met when the chef and the service manager cooperate in their training.

TERMS FOR REVIEW

static menu	standardized recipe	milli-
cycle menu	AP (as purchased) weight	yield
à la carte	EP (edible portion) weight	conversion factor
table d'hôte	portion control	food cost percentage
prix fixe	metric system	yield test
tasting menu	gram	cutting loss
course	liter	as served (AS)
fresh	meter	portion cost
imported	degree Celsius	hidden cost
homemade	kilo-	minimum-use ingredient
organic	deci-	par stock
recipe	centi-	

QUESTIONS FOR DISCUSSION

1. What role is played by the chef's favorite dishes when a menu is written?

2. What are the main differences among breakfast, lunch, and dinner menus?

3. Which of the following are most likely to have static menus?

   Fast-food restaurant   French restaurant   Army mess
   High school cafeteria   Employee lunchroom

4. The following menus are made up of dishes prepared from recipes in this book. Evaluate each for variety and balance.

Clear vegetable soup	Cream of mushroom soup
Green salad with French dressing	Waldorf salad
Chicken fricassée	Veal scaloppine à la crème
Cauliflower au gratin	Broccoli Mornay
	Rice pilaf
Scotch broth	Oxtail soup
Cucumber and tomato salad	Coleslaw
Roast rack of lamb with spring vegetables	Beef pot roast
	Braised green cabbage
	Bouillon potatoes
Oysters casino	Gazpacho
Vichyssoise	Tomato and avocado salad
Broiled steak	Chicken Pojarski
Baked potato	Baked acorn squash
Buttered green beans	Duchesse potatoes

5. What is the best solution to the problem of using up leftovers? What is the next best solution?

6. What are some reasons written recipes can't be 100 percent exact and must depend on the cook's judgment? Select two or three recipes (from this book or any other) and try to determine where they depend on the cook's judgment.

7. What is the purpose of a standardized recipe?

8. What are the three basic ways of measuring ingredients? Which method is used for most solid ingredients, and why?

9. What is the first step in portion control? List four other techniques of portion control.

10. Make the following conversions in the U.S. system of measurement:

    $3^1/_2$ pounds = _____ ounces

    6 cups = _____ pints

    $8^1/_2$ quarts = _____ fluid ounces

    $^3/_4$ cup = _____ tablespoons

    46 ounces = _____ pounds

    $2^1/_2$ gallons = _____ fluid ounces

    5 pounds 5 ounces × 2 = _____

    10 teaspoons = _____ fluid ounces

11. Make the following conversions in the metric system:

    1.4 kilograms = _____ grams

    53 deciliters = _____ liters

    15 centimeters = _____ millimeters

    2,590 grams = _____ kilograms

    4.6 liters = _____ deciliters

    220 centiliters = _____ deciliters

12. Turn to the recipe for Thai Green Curry of Pork with Vegetables on p. 505. Convert it to yield 18 portions.

13. Discuss the main types of problems you may face when converting recipe yields.

14. What is the difference between AP weight and EP weight? Explain how these terms are related to calculating costs per portion of menu items.

15. The following problems are calculations with food cost percentages, portion cost, and menu price. For each problem, two of the figures are given. Find the third.

Food Cost Percentage	Portion Cost	Menu Price
a. _____	$1.24	$4.95
b. 40%	_____	$2.50
c. 30%	$2.85	_____

5

NUTRITION

Menu planners must have a basic understanding of nutrition because the human body requires a variety of foods in order to function and be healthy.

The food-service worker's responsibility to provide nutritious food and well-balanced menus depends, in part, on the operation. School and hospital food services must, of course, plan menus carefully to meet basic nutritional needs. A qualified dietitian is usually required in such establishments.

The obligations of restaurateurs are more subtle. Because they are in business to sell food, they must offer foods that will attract customers. People who plan menus are as concerned with presenting attractive, flavorful foods as they are with serving nutritious foods. Also, if the menu is à la carte, there is no way to ensure a customer will order items that make up a nutritionally balanced meal.

But restaurateurs do have an obligation to offer a choice. That is, menus should be planned so customers can order well-balanced meals if they desire. People are becoming more concerned with fitness and health, so a nutritiously balanced menu may even help attract customers.

AFTER READING THIS CHAPTER, YOU SHOULD BE ABLE TO

1. List and describe the six categories of nutrients, explain their functions in the body, and name some food sources of each.

2. Define the term *calorie* and describe the relationship between calories and weight gain.

3. List and describe the eight guidelines for maintaining a healthful diet.

4. Describe ways that cooks can incorporate nutrition principles into their cooking and their menu construction.

NUTRIENTS

Nutrients are chemical compounds that are present in foods and that fulfill one or more of the following functions:

- Supply energy for body functions.
- Build and replace cells that make up body tissues.
- Regulate body processes.

There are six categories of nutrients:

1. Carbohydrates
2. Vitamins
3. Fats
4. Minerals
5. Proteins
6. Water

7. Phyto chemicals & anti-oxidants

plant (greek)

CALORIES

The **calorie** is a unit of measurement of energy. It is defined as the amount of heat needed to raise the temperature of 1 kilogram water by 1°C.

Remember that one of the functions of nutrients is to supply energy to the body. The calorie is used to measure how much energy certain foods supply for these functions. In our overfed society, calories have come to be viewed as something to be avoided. Nevertheless, without sufficient food energy, we could not live.

The body uses carbohydrates, proteins, and fats to supply energy:

1 gram carbohydrate supplies 4 calories.

1 gram protein supplies 4 calories.

1 gram fat supplies 9 calories.

There is a direct connection between calorie intake, physical activity, and weight gain. Simply put, if you consume more calories than you burn, you gain weight. If you consume fewer calories than you burn, you lose weight. All the diet schemes and fashions in the world—at least the ones that are medically sound—can be reduced to this. In other words, losing weight is possible only by eating fewer calories, by burning more calories through exercise, or, preferably, by both.

In order to lose weight while maintaining good nutrition and health, one should avoid **empty calories** as much as possible and eat primarily foods of high **nutrient density**. Empty-calorie foods are those that provide few nutrients per calorie. Refined sugars and starches are examples of empty calories. Foods of high nutrient density are those that provide relatively many nutrients per calorie. Fruits, vegetables, and whole grains are examples of foods with high nutrient density.

KINDS OF NUTRIENTS AND THEIR IMPORTANCE

Each of the nutrients listed has certain characteristics and functions in the body. These are discussed in general terms. For a summary of individual nutrients and the foods in which they are found, see Table 5.1.

Many fad diets emphasize either the good or the bad qualities of a single nutrient. For example, a particular weight loss diet may advise eating mostly protein foods and eliminating carbohydrates as much as possible. However, because all nutrients are essential, such diets can lead to nutritional imbalances. Thus, it is necessary to understand the function and importance of all nutrients.

The amount of each nutrient needed daily varies from person to person, depending on such factors as age, sex, general health, and level of activity.

CALORIES AND CALORIES

In scientific terminology, the amount of heat needed to raise the temperature of 1 kilogram water by 1°C is called a *kilocalorie*, sometimes written *Calorie* (with a capital C) and abbreviated *kcal*. When written with a small *c*, the term *calorie* refers to a unit of energy measurement that is only 1,000th as large—the amount of heat needed to raise the temperature of 1 *gram* of water by 1°C.

Nevertheless, in discussions of nutrition, the word *calorie* is commonly substituted for *kilocalorie*. Just remember that when you see *calorie* in connection with food, the real meaning is *Calorie* (or *kilocalorie*).

Carbohydrates

Carbohydrates are compounds consisting of carbon, hydrogen, and oxygen atoms bound together in chains of varying lengths.

Sugars are simple carbohydrates. Simple sugars, such as glucose, are small compounds containing 6 carbon atoms. Table sugar, or sucrose, is a larger sugar molecule with 12 carbon atoms. Sugars are found in sweets and, to a lesser extent, in fruits and vegetables.

Starches are complex carbohydrates consisting of long chains of simple sugars bound together. They are found in such foods as grains, bread, peas and beans, and many vegetables and fruits.

Carbohydrates are the body's most important source of food energy. Fats and proteins can also be burned for energy, but the body uses carbohydrates first. If no carbohydrates are available, the body then burns fat. However, if fats are burned with no carbohydrates present, toxic compounds called **ketone bodies** are produced. If too many ketone bodies accumulate, a condition called **ketosis** develops, and the blood becomes unable to carry oxygen properly. The result can be fatal. Thus, one of the important functions of carbohydrates is to help the body burn fat properly. About 50 to 100 grams carbohydrate are needed every day to prevent ketosis.

Most authorities believe complex carbohydrates, especially those from whole grains and unrefined foods, are better for you than simple carbohydrates. This is partly because starchy foods also have many other nutrients, while sweets have few other nutrients. Also, there is some evidence that a lot of sugar in the diet may contribute to heart and circulatory diseases. Simple sugars and refined starches are primary sources of empty calories.

Another reason carbohydrates from whole grains and unrefined foods are preferable to those from refined sugars and starches is that these unrefined foods are sources of fiber. The term **fiber** refers to a group of carbohydrates that cannot be absorbed and used by the body. Therefore, fiber supplies no food energy. However, it is important for the proper functioning of the intestinal tract and the elimination of body waste. In addition, there is evidence that sufficient dietary fiber helps prevent some kinds of cancers and helps decrease cholesterol in the blood. Fruits and vegetables, especially raw, and whole grains supply dietary fiber.

Fiber can be classified as either soluble or insoluble. **Soluble fiber** absorbs water and forms a kind of gel. It is found inside and between plant cells. **Insoluble fiber** also absorbs water, but less, and forms bulk in the intestines. It is found in cell walls and other structural parts of plants.

Fats

Fats supply energy to the body in highly concentrated form. Also, some fatty acids are necessary for regulating certain body functions. Third, fats act as carriers of fat-soluble vitamins (vitamins A, D, E, and K). Because of these important functions, it is necessary to have some fats in the diet.

Fats may be classified as **saturated**, **monounsaturated**, or **polyunsaturated**. These terms reflect chemical differences in the composition of fats. Cooks do not need to know the chemical structure of fats, but they should understand their nutritional characteristics and the foods in which they are found. Many foods contain a combination of these three types, with one type predominating.

Saturated fats are solid at room temperature. Animal products—meats, poultry, fish, eggs, dairy products—and solid shortenings are the major source of saturated fats. Tropical oils such as coconut oil and palm kernel oil are also rich in saturated fats. Health experts believe these fats contribute significantly to heart disease and other health problems.

Polyunsaturated fats and monounsaturated fats are liquid at room temperature.

Although too much of any kind of fat is unhealthy, these fats are considered more healthful than saturated fats. Polyunsaturated fats are found in vegetable oils such as corn oil, safflower oil, sunflower oil, and cottonseed oil. High levels of monounsaturated fats are found in olive oil and canola oil. Both kinds of unsaturated fats are found in other plant products as well, including whole grains, nuts, and some fruits and vegetables.

One group of saturated fats of special concern is **trans fats**. These fats occur naturally in small amounts only. Most of the trans fats in our diet are from manufactured fats subjected to a process called *hydrogenation*. *Hydrogenated fats* are fats changed from liquid to solid by adding hydrogen atoms to the fat molecules. This is the process used to make products such as solid shortening and margarine. Trans fats are of concern because they limit the body's ability to rid itself of cholesterol that builds up on the walls of arteries (see sidebar, p. 96).

ESSENTIAL FATTY ACIDS

Fats are made up of smaller compounds called *fatty acids*. Two of these are called **essential fatty acids** because they cannot be made by the body. The essential fatty acids are *linoleic acid* and *alpha-linoleic acid*. Linoleic acid is a member of a group of compounds called *omega-6 fatty acids* (the term refers to its chemical structure). It is found in vegetable oils and is usually abundant in North American diets.

Alpha-linoleic acid is a member of the group of **omega-3 fatty acids**. The body can change alpha-linoleic acid into other omega-3 fatty acids called *DHA* and *EPA*. These are all important nutrients that play vital roles in growth, in the immune system, in proper eyesight, and in cell structure. Alpha-linoleic acid is found in certain vegetable oils. DHA and EPA, in addition to being made by the body, are also found in some fatty fish, including salmon, mackerel, tuna, and sardines. Unlike omega-6 fatty acids, omega-3 acids are not usually abundant in North American diets, so it is important to be aware of foods that supply them.

Fats are members of a group of compounds called **lipids**. Another lipid found in the body is **cholesterol**, a fatty substance closely linked with heart disease because it collects on the walls of arteries and blocks the flow of blood to the heart and other vital organs. It is found only in animal products and is especially high in egg yolks, butterfat, and organ meats such as liver and brains. The human body also manufactures its own cholesterol, so not all the cholesterol in the blood is necessarily from foods. Although some cholesterol is necessary for body functions, it is not considered a nutrient because the body is able to manufacture all the cholesterol it needs. Experts generally agree it is best to keep the cholesterol in the diet as low as possible.

Recent research has suggested that monounsaturated fat may actually lower the levels of the most harmful kinds of cholesterol in the body. This may explain the relatively low incidence of heart disease in Mediterranean regions, where olive oil is the most widely used fat. This research has helped popularize the use of olive oil in other parts of the world, especially in North America.

Remember, however, that *too much fat of any kind is bad for the health*. Do not make the mistake of thinking monounsaturated fats are good for you and can be used in excess.

Proteins

Proteins are known as the building blocks of the body. They are essential for growth, for building body tissues, and for basic body functions. They can also be used for energy if the diet does not contain enough carbohydrate and fat.

Proteins consist of substances called *amino acids*. The body is able to manufacture many of them, but there are nine amino acids it cannot manufacture and must get from foods. A food protein that contains all nine essential amino acids is called a **complete protein**. Meats, poultry, fish, eggs, and dairy products contain complete proteins.

Proteins that lack one or more of these essential amino acids are called *incomplete proteins*. Foods high in incomplete proteins include nuts, grains, and dried beans and other legumes. Foods that, *if eaten together*, supply all the amino acids are called **complementary proteins**. For example, cornmeal tortillas topped with chili beans supply complete protein because the corn supplies the amino acids lacking in the beans. Beans and rice is another example of a food combination supplying complementary proteins.

Complementary proteins are especially of interest to vegetarians, especially vegans, and are discussed in more detail in Chapter 26.

The average adult needs 50 to 60 grams protein a day. For most North Americans, getting enough protein daily is not a problem; most get about twice as much as they need. Greatly excessive protein in the diet can lead to a variety of health problems, including kidney and liver damage.

Vitamins

Vitamins are present in foods in extremely small quantities, but they are essential for regulating body functions. Unlike proteins, fats, and carbohydrates, they supply no energy, but some of them must be present in order for energy to be utilized in the body. Also, lack of certain vitamins causes *deficiency diseases*.

Vitamins are classified as *water-soluble* and *fat-soluble*. The water-soluble vitamins (the B vitamins and vitamin C) are not stored in the body and must be eaten every day. Foods containing these vitamins should be handled so the vitamins are not dissolved into the cooking water and lost (as discussed in Chapter 10).

Fat-soluble vitamins (A, D, E, and K) can be stored in the body, so they do not need to be eaten every day as long as the total amount eaten over time is sufficient. Consuming too much of a fat-soluble vitamin daily, as sometimes happens when people take too many vitamin supplements, can result in toxic levels of the vitamin stored in the tissues.

More detail about individual vitamins, their functions, and their sources, can be found in Table 5.1.

TABLE 5.1 Major Nutrients

Nutrient	Major Dietary Sources	Functions in the Body
Carbohydrates	Grains (including breads and pasta) Dried beans Potatoes Corn Sugar	Major source of energy (calories) for all body functions. Necessary for proper utilization of fats. Unrefined carbohydrates supply fiber, important for proper waste elimination.
Fats	Meats, poultry, and fish Dairy products Eggs Cooking fats and shortening Salad dressings	Supply food energy (calories). Supply essential fatty acids. Carry fat-soluble vitamins.
Proteins	Meats, poultry, and fish Milk and cheese Eggs Dried beans and peas Nuts	Major building material of all body tissues. Supply food energy (calories). Help make up enzymes and hormones, which regulate body functions.
Vitamin A	Liver Butter and cream Egg yolks Green and yellow vegetables and fruits	Helps skin and mucous membranes resist infection. Promotes healthy eyes and makes night vision possible.
Thiamin (vitamin B_1)	Pork Whole grains and fortified grains Nuts Legumes Green vegetables	Needed for utilization of carbohydrates for energy. Promotes normal appetite and healthy nervous system. Prevents beriberi.
Riboflavin (vitamin B_2)	Organ meats Milk products Whole grains and fortified grains	Needed for utilization of carbohydrates and other nutrients. Promotes healthy skin and eyes.
Niacin (vitamin B_3)	Liver Meat, poultry, and fish Legumes	Needed for utilization of energy foods. Promotes healthy nervous system, skin, and digestion. Prevents pellagra.
Folic acid or folate (vitamin B_9)	Leafy vegetables Legumes Egg yolks Grain products Liver	Needed for cell functions, including cell growth and division and synthesizing and repairing DNA.
Vitamin B_{12}	Most animal and dairy products	Promotes healthy blood and nervous system.
Vitamin C (ascorbic acid)	Citrus fruits Tomatoes Potatoes Dark green leafy vegetables Peppers, cabbage, and broccoli Cantaloupe Berries	Strengthens body tissues. Promotes healing and resistance to infection. Prevents scurvy.
Vitamin D	Fortified milk products Formed in skin when exposed to sunlight	Necessary for utilization of calcium and phosphorus to promote healthy bones, teeth, and muscle tissue.
Vitamin E	Unsaturated fats (vegetable oils, nuts, whole grains, etc.)	Protects other nutrients.
Calcium	Milk products Leafy vegetables Canned fish with bones	Forms bones and teeth. Necessary for healthy muscles and nerves.
Iron	Liver and red meat Raisins and prunes Egg yolks Leafy vegetables Dried beans Whole grains	Needed for formation of red blood cells.

Lipoproteins are combinations of protein and fat that carry cholesterol and fat through the bloodstream. Two of these compounds are of concern to us. **Low-density lipoprotein (LDL)** is the most important carrier of cholesterol. Although it performs a needed function, if too much of it is present it deposits excess cholesterol inside arteries, blocking the flow of blood. **High-density lipoprotein (HDL)**, on the other hand, helps remove cholesterol from the blood and eliminate it from the body. Consequently, HDL is seen as a major preventer of heart disease.

Certain saturated fats called trans fats (see p. 93) are considered especially bad in the diet because they apparently interfere with the action of HDL and thus raise cholesterol levels in the blood.

PHYTOCHEMICALS

A phytochemical is any of several hundred compounds that are found in foods in tiny amounts and may help reduce the chance of cancer or heart disease. The list of phytochemicals includes flavonoids, isoflavonoids, phenols, lycopene, indoles, dithiolthiones, and sulforaphane. These chemicals are not yet well understood, and research is continuing. Phytochemicals are found most abundantly in whole grains, fruits, and vegetables.

Minerals

Minerals, like vitamins, are consumed in very small quantities and are essential for regulating certain body processes. Minerals that must be consumed in relatively large amounts—more than 100 milligrams daily—are called **major minerals**. These include calcium, chloride, magnesium, phosphorus, sulfur, sodium, and potassium. Minerals that must be present in smaller amounts are called **trace minerals**. These include chromium, copper, fluoride, iodine, iron, manganese, molybdenum, selenium, and zinc. Less is known about the functions of some of the trace minerals. It is important to understand, however, that although small quantities are needed by the body, too much of any of them can be harmful.

Sodium, a component of table salt, is well known as a health problem. Too much sodium is thought to contribute to high blood pressure. Health authorities try to convince people to reduce the sodium in their diets, primarily by salting foods less.

Water

The adult human body is 50 to 60 percent water by weight. Water plays a role in all the body's functions, including metabolism and other cell functions, digestion, delivery of nutrients, removal of waste, temperature regulation, and lubrication and cushioning of joints and tissues.

Water forms a large part of most of the food we eat and all the beverages we drink. The body is good at regulating its own water content and tells us when we need more by making us feel thirsty. This signal should not be ignored. Even better is to drink enough fluids to *prevent* feeling thirsty. Required daily water intake varies greatly from person to person, depending on age, level of activity, and environmental factors such as heat. The common recommendation of 8 glasses of water a day is not enough for some people, such as athletes and others who exercise strenuously, and is too much for others, such as older, sedentary adults.

KEY POINTS TO REVIEW

- What foods are good sources of proteins? What functions do proteins have in the body?

- What foods are good sources of carbohydrates? What functions do proteins have in the body?

- What foods are especially high in fats? Of the three categories of fat compounds, which kind is the most healthful, and which is the least healthful?

- What does the term *calorie* mean? What role do calories play in weight gain or loss?

THE BALANCED DIET

In order to stay healthy, we must consume a varied diet that contains all the essential nutrients. In addition, we must limit our intake of foods that can be harmful in large quantities. Although researchers still have much to learn about nutrition and our knowledge is constantly changing, there is strong evidence about what good eating patterns are. According to government health agencies, the following guidelines are suggested for maintaining a healthful diet. It should be noted these are only general recommendations for people who are already healthy and want to stay that way. They are not necessarily for those who need special diets because of disease or other abnormal conditions.

1. Get adequate nutrients within calorie needs.
The greater the variety of nutrient-dense foods and beverages within and among the basic food groups we consume, the more likely we are to get all the nutrients we need.

Choosing nutrient-dense foods and avoiding empty calories is necessary in order for us to get adequate nutrition without consuming too many calories in the process. Choose foods that limit the intake of saturated and trans fats, cholesterol, added sugars, salt, and alcohol.

2. Manage weight.

To maintain a healthy body weight, balance the calories you consume with the calories you burn. People who are greatly overweight are more likely to develop certain chronic diseases, including high blood pressure, heart disease, and stroke. People who consume more calories than they burn off will gain weight.

To prevent gradual weight gain, make small decreases in the calories you consume and increase your physical activity. Rather than depending on crash diets, it is usually better to lose weight slowly and gradually, to develop better habits of eating, and to increase physical activity. To get all the nutrients you need while cutting down on calories, decrease foods that are high in calories but low in nutrients, especially fat and fatty foods, sugar and sweets, and alcohol.

3. Engage in physical activity.

Engaging in regular physical activity promotes health, psychological well-being, and a healthy body weight. For general health and reducing the risk of chronic diseases, getting at least 30 minutes of moderately vigorous exercise every day is desirable, and more and longer vigorous exercise can be even more beneficial. In order to avoid gaining weight, adults should try to get 60 minutes of exercise most days while at the same time not consuming too many calories. People who wish to lose weight gradually should try to get 60 to 90 minutes of exercise most days, again while limiting calorie intake.

4. Select from the right food groups.

Fruits, vegetables, whole grains, and low-fat or fat-free milk and milk products are the foods with the highest nutrient density. These foods should be strongly emphasized in a healthy diet. In particular, someone who consumes 2,000 calories a day should try to eat the following daily:

- 2 cups (4 servings) fruit, selecting from a variety of fruits
- 2¹/₂ cups (5 servings) vegetables, selected from as many of the basic vegetable groups as possible: dark green vegetables, orange vegetables, legumes, starchy vegetables, and others
- 3 servings of whole grains
- 3 cups of fat-free or low-fat milk or its equivalent in other dairy products, such as yogurt and cheese

In the United States, these foods are represented graphically by the icon called MyPlate (Figure 5.1a), developed by the U.S. Department of Agriculture (USDA) as a way to help consumers make healthier food choices. The figure emphasizes the fruit, vegetable, grain, protein, and dairy food groups and represents them in proper proportions on a plate. The sidebar indicates important nutrients supplied by these food groups. Note in particular that proteins take up less than a quarter of the plate. This is in contrast with more traditional platings in which the meat, poultry, or fish item covers about half the plate or more, with vegetable and starch accounting for the rest. The average North American already consumes about twice as much protein as needed in the diet. A goal of MyPlate is to decrease emphasis on high-fat, high-calorie protein items and to add more fruits, vegetables, and grains to the diet.

Consumers are urged to seek more information on portions, portion sizes, and other healthful food options by consulting www.choosemyplate.gov. To help consumers choose the right food groups, the website offers the following tips:

- Balance calories by determining how many you need a day based on age, sex, and level of physical activity, and by increasing your activity.
- Enjoy your food, but eat less.
- Avoid oversized portions.

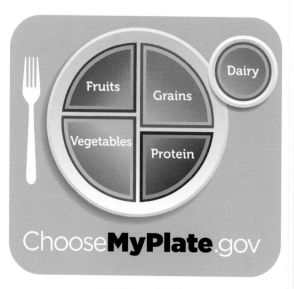

FIGURE 5.1a MyPlate
Courtesy the US Department of Agriculture.

MYPLATE NUTRIENTS

The five food groups emphasized by the MyPlate icon supply important nutrients that may be lacking in diets high in fatty foods and sweets.

Fruits are low in fat, sodium, and calories, contain no cholesterol, and are good sources of potassium, fiber, vitamin C, and folic acid.

Vegetables are also low in fat, sodium, and calories and contain no cholesterol, and supply potassium, fiber, vitamin A, vitamin C, and folic acid.

Grains are sources of fiber, B vitamins (thiamin, riboflavin, niacin, and folic acid), iron, magnesium, and selenium.

Protein foods are, of course, high in proteins. Foods from this group should be chosen carefully, as many meats and cheeses are high in solid fats, cholesterol, and calories. Egg yolks are also high in cholesterol.

Dairy products are good sources of calcium, potassium, and vitamin D.

- Make half your plate fruits and vegetables.
- Make at least half your grains whole grains.
- Switch to fat-free or low-fat (1%) milk.
- Compare sodium in foods like soup, bread, and frozen meals, and choose foods with lower numbers.
- Drink water instead of sugary drinks.
- Cut back on foods high in solid fats, added sugars, and salt.

The more traditional way to illustrate the healthful diet is with a pyramid diagram, in which the broad lower layers represent foods that should form the bulk of a healthy diet, while the smaller upper layers represent foods to be eaten less frequently or in smaller quantities. The food pyramid continues to be used to show healthful food choices in other food cultures, such as Mediterranean, Asian, and Latin American diets (Figures 5.1b, 5.1c, and 5.1d).

Health Canada, working closely with three expert advisory boards, devised *Eating Well with Canada's Food Guide* in 2007 (Figure 5.1e). The recommended number of food guide servings of each group per day, as well as information about standard portion sizes, is indicated in the chart.

5. Manage consumption of fats.

Keep total fat intake between 20 and 35 percent of calories, with most fats coming from sources of polyunsaturated and monounsaturated fatty acids, such as fish, nuts, and vegetable oils. This means for a diet of 2,000 calories daily, calories from fat should be between 400 and 700.

Why not lower than 20 percent? Remember that some fatty acids are essential nutrients, and fats also carry fat-soluble vitamins. Consuming less fat than 20 percent of daily calories could be unhealthy.

Keep consumption of saturated fats, especially trans fats, as low as possible. Consume less than 10 percent of calories from saturated fatty acids.

Consume less than 300 milligrams cholesterol per day.

When selecting and preparing meat, poultry, dry beans, and milk or milk products, make choices that are lean, low-fat, or fat-free.

Remember: High fat intake, especially of saturated fats and cholesterol, is associated with such conditions as heart disease and high blood pressure. Although other factors contribute to these diseases, such as heredity and smoking, following this dietary recommendation should increase the chances of staying healthy.

6. Manage consumption of carbohydrates.

Choose fiber-rich fruits, vegetables, and whole grains. These foods are the sources of the most healthful carbohydrates. Avoid prepared foods high in added sugars.

Reducing refined sugars and starches in the diet has the added benefit of helping reduce tooth decay.

7. Manage consumption of sodium and potassium.

Consume less than 2,300 milligrams (about 1 teaspoon or 5 milliliters salt) sodium per day. (Those with high blood pressure, diabetes, or kidney disease, and those over 50 should consume less than 1,500 milligrams daily.) Sodium, as noted earlier, appears to contribute to high blood pressure. For people who already have high blood pressure, it is especially important to reduce sodium in the diet. The best ways to do this are to decrease the use of salt in the kitchen and at the table and to limit the intake of prepared foods that are high in salt, such as potato chips, salted nuts, pretzels, pickled foods, cured meats, and salty condiments like soy sauce.

Reduce the harmful effects of sodium by eating potassium-rich foods, such as fruits and vegetables.

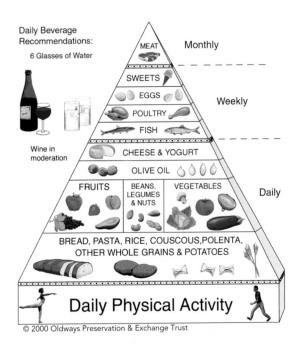

FIGURE 5.1b The traditional healthy Mediterranean diet pyramid. © 2000 Oldways Preservation & Exchange Trust.

The Traditional Healthy Asian Diet Pyramid

Daily Beverage Recommendations:
6 Glasses of Water or Tea

Sake, Wine, or Beer in moderation

MEAT — Monthly

SWEETS — Weekly
EGGS & POULTRY

FISH & SHELLFISH or DAIRY — Optional Daily

VEGETABLE OILS

FRUITS | LEGUMES, SEEDS & NUTS | VEGETABLES — Daily

RICE, NOODLES, BREADS, MILLET, CORN & OTHER WHOLE GRAINS

Daily Physical Activity

© 2000 Oldways Preservation & Exchange Trust

FIGURE 5.1c The traditional healthy Asian diet pyramid.
© 2000 Oldways Preservation & Exchange Trust.

The Traditional Healthy Latin American Diet Pyramid

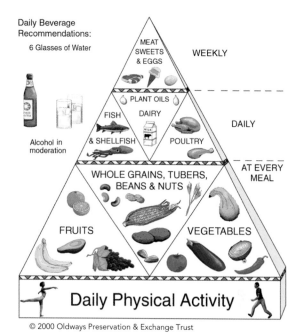

Daily Beverage Recommendations:
6 Glasses of Water

Alcohol in moderation

MEAT SWEETS & EGGS — WEEKLY

PLANT OILS

FISH & SHELLFISH | DAIRY | POULTRY — DAILY

WHOLE GRAINS, TUBERS, BEANS & NUTS — AT EVERY MEAL

FRUITS | VEGETABLES

Daily Physical Activity

© 2000 Oldways Preservation & Exchange Trust

FIGURE 5.1d The traditional healthy Latin American diet pyramid.
© 2000 Oldways Preservation & Exchange Trust.

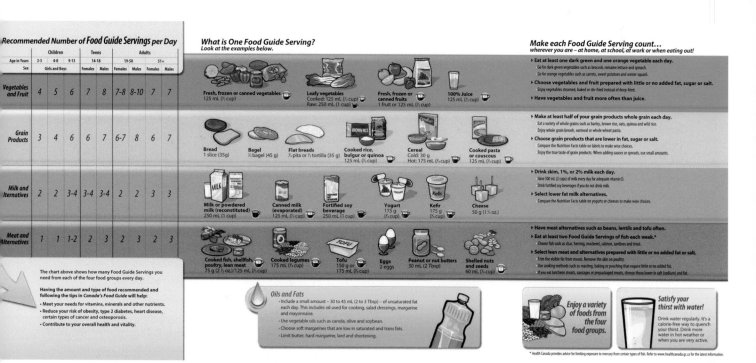

FIGURE 5.1e Eating Well with Canada's Food Guide.
Source: © All rights reserved. *Eating Well with Canada's Food Guide*. Health Canada, 2011.
Modified and reproduced with permission from the Minister of Health, 2014.

8. Manage consumption of alcoholic beverages.

People who choose to drink alcoholic beverages should do so sensibly and in moderation—defined as the consumption of up to one drink per day for women and up to two drinks per day for men.

Alcoholic beverages are high in calories but provide few nutrients. Heavy drinking may cause a variety of serious diseases. Moderate drinking—one or two drinks a day—appears to do little harm and may, in fact, be of some benefit.

Many people, including children and adolescents, pregnant and lactating women, people taking medications that interact with alcohol, and people with certain medical conditions, should avoid alcohol completely. In addition, alcoholic beverages should be avoided by people engaging in activities that require attention, skill, or coordination, such as driving or operating machinery.

WORKING WITH EXCHANGE GROUPS

Another way to help devise a healthful diet involves a concept called exchange groups. In the nutrition exchange system, foods are divided into seven groups:

Starches   Dairy
Fruits   Fats
Vegetables   Sweets
Proteins

Serving sizes for individual foods within each group are strictly defined, and various health agencies publish lists of these serving sizes. Table 5.2 gives examples of serving sizes for common foods. (Note that, because these lists are intended for consumer use, most serving sizes are indicated by volume rather than by weight, so their accuracy is limited.) Table 5.3 lists how many calories and how much carbohydrate, protein, and fat is supplied by one serving from each of these groups.

To set up a diet based on exchange groups, one must first determine how many calories you need and can consume in a day, depending on your age, sex, activity level, and weight-loss needs. Then consult various published tables to plan a daily diet consisting of fixed numbers of servings from each group. For example, a 1,500-calorie diet would consist of the following exchanges daily:

Starches	8 exchanges
Fruits	3 exchanges
Vegetables	3 exchanges
Proteins	5 exchanges
Dairy	3 exchanges
Fats	2 exchanges
Sweets	(use sparingly and substitute one sweet exchange for one starch, dairy, or fruit exchange)

Consulting Table 5.3, we can calculate that this diet would supply about 215 g carbohydrate, 90 g protein, and 20 to 30 g fat (if only low-fat dairy and protein foods are consumed). These exchanges should be divided among breakfast, lunch, and dinner, and the menu plan should stick to the servings and serving sizes allowed for each meal.

TABLE 5.2 Food Exchanges: Sample Serving Sizes

Food Exchange	U.S.		Metric
Starches			
bread	1	slice	1 slice
cooked cereal	1/2	cup	125 mL
dry cereal, unsweetened	3/4	cup	175 mL
dry flour or grain	3	tbsp	45 mL
pasta, cooked	1/2	cup	125 mL
rice, cooked	1/3	cup	80 mL
potato, mashed	1/2	cup	125 mL
Vegetables			
raw vegetables	1	cup	250 mL
cooked vegetables	1/2	cup	125 mL
vegetable juice	1/2	cup	125 mL
Fruit			
fresh fruit	1	small	1 small
canned fruit	1/2	cup	125 mL
fruit juice, unsweetened	1/2	cup	125 mL
dried fruit	1/4	cup	60 mL
Proteins			
meat, poultry, fish	1	oz	30 g
cheese	1	oz	30 g
egg	1		1
tofu	4	oz (1/2 cup)	115 g (125 mL)
Dairy			
milk	1	cup	250 mL
yogurt	1	cup	250 mL
Fat			
oil	1	tsp	5 mL
mayonnaise	1	tsp	5 mL
butter	1	tsp	5 mL
salad dressing	1	tbsp	15 mL
Sweets			
ice cream	1/2	cup	125 mL
cookies	2	small	2 small
syrup	1	tbsp	15 mL
sugar	2	tbsp	30 mL
jam or jelly	1	tbsp	15 mL

Remember that serving sizes must be followed closely for this system to work. For example, 2 ounces of turkey on your plate counts as two servings or exchanges. A sandwich consisting of 2 ounces of ham, 1 ounce of cheese, two slices of bread, and 1 tablespoon of mayonnaise contains three protein exchanges, two starch exchanges, and three fat exchanges, as indicated by Table 5.3.

TABLE 5.3 Nutrients per Exchange Group

Exchange Group	Calories	Carbohydrates	Protein	Fat
Starches	80	15 g	3 g	1 g
Fruits	60	15 g	0	0
Vegetables	25	5 g	2 g	0
Proteins	35–145	0	7 g	0–13 g
Dairy	80–150	12 g	8 g	0–8 g
Fats	45	0	0	5 g
Sweets	variable	15 g	variable	variable

COOKING HEALTHFUL MEALS

Restaurateurs and chefs are becoming more and more attentive to people's health and diet concerns. Many of them are reexamining their menus, modifying their cooking practices, and adding new, healthful items to their menus. Some have developed new menus intended to follow as closely as possible the eight recommendations just listed.

An increased health consciousness has affected the way we think about food and the way we cook. Professional cooks are making their foods more healthful in several ways:

1. **Using less fat in cooking.**

   Cooking methods that require no added fat, such as simmering, poaching, baking, steaming, and grilling, can be considered the most healthful.

   For sautéing, nonstick pans are becoming more widely used because little or no fat is needed. With regular pans, one can be careful to use as little fat as possible.

   Grilling is popular because it can be done without first coating the food with fat. If this is done, however, one must be careful not to let the food dry out.

   Using less fat in cooking also means using ingredients with less fat. Excess external fat can be trimmed from meats and poultry. Low-fat sauces, such as salsas and vegetable purées, can often be used instead of high-fat sauces. Recipes can often be modified to reduce quantities of high-fat ingredients, such as butter, cheese, and bacon.

2. **Using unsaturated fats.**

   When you do use fats, try to substitute monounsaturated fats, such as olive oil or canola oil, for saturated fats when appropriate.

3. **Emphasizing flavor.**

   Taste is the most important factor in preparing nutritious food. The most vitamin-packed dish does no one any good if it is uneaten because it doesn't taste good. Preparing flavorful foods requires knowledge of the principles of cooking. You can't rely simply on nutritional information.

   Rely more on the natural flavors of foods and less on salt and other additives that should be decreased in the diet.

4. **Using the freshest, highest-quality foods possible.**

   In order to prepare delicious foods with little or no added salt and with less reliance on high-fat, high-sodium sauces and condiments, it is important to use high-quality natural ingredients at their peak of flavor. Healthful cooking means letting the true flavors of foods dominate.

   To enhance natural flavors without added salt, cooks are using more fresh herbs, hot seasonings such as chiles, ginger, and pepper, and flavorful ingredients like garlic, browned onions, and flavored vinegars.

5. **Storing foods properly.**

Foods in storage lose nutrients as they age. The loss of nutrients can be slowed, however, by proper storage. This applies particularly to proper refrigeration. For each category of perishable food discussed in this book, pay close attention to how the foods should be stored.

6. **Modifying portion sizes.**

It is not necessary to feature huge slabs of meat to serve satisfying meals. Smaller portions of well-trimmed meat, poultry, or fish, nicely balanced on the plate with an assortment of attractive fresh vegetables and complex carbohydrates, are likely to be more healthful.

Sauces often get the blame for adding calories to a meal, but if a sauce is flavorful, you don't need much. Make a better sauce and serve less of it. Also, if a sauce isn't too thick, it won't cling as heavily to the food, and a little will go farther.

7. **Giving customers a healthful choice.**

Offer a menu with a variety of foods so customers can choose a well-balanced meal suited to their needs and desires. It's not necessary to cook only "diet food," but a menu that offers French fries as the only available starch is not well balanced.

Place more emphasis on fruits, vegetables, and whole grains. Offer a menu with choices from all the groups in the USDA MyPlate or the Canadian rainbow.

Be flexible in the kitchen. A good chef is willing to modify menu items to meet dietary requirements and to satisfy special requests from customers.

8. **Training the dining room staff.**

Some restaurants offer special "spa menus" in addition to their regular menus, or they highlight "healthy" items with a special symbol. Unfortunately, this approach may suggest to some people that the highlighted menu items are boring "health food," while the other menu items are unhealthful because they aren't flagged. Consequently, many chefs prefer to train their dining room personnel to answer customers' questions about the menu and to offer suggestions when asked.

9. **Using nutritional information.**

Study the nutritional content of foods in order to plan healthful menus. Many publications are available that list the nutritional content of common food items. Some restaurants have even hired registered dietitians to analyze their menus and give advice on how to make their food more healthful.

Hiring a dietitian is, of course, not practical for every operation. However, a basic awareness of nutrition helps every professional minimize the fat, cholesterol, and sodium in and maximize the nutritional content and balance of the foods they serve.

KEY POINTS TO REVIEW

- In the eight guidelines for maintaining a healthy diet, what are the recommended food groups, and how many servings of each should the average person consume each day?

- What two major categories of nutrients do most people eat too much of? What are the recommendations for managing these nutrients in the diet?

- Which mineral is of special health concern because of excess consumption? Why?

- What are eight ways chefs can help their customers eat a more healthful diet?

TERMS FOR REVIEW

calorie	fat	protein
empty calorie	saturated fat	complete protein
nutrient density	monounsaturated fat	complementary protein
carbohydrate	polyunsaturated fat	vitamin
ketone body	trans fat	major mineral
ketosis	lipid	trace mineral
fiber	cholesterol	low-density lipoprotein (LDL)
soluble fiber	essential fatty acid	high-density lipoprotein (HDL)
insoluble fiber	omega-3 fatty acid	

QUESTIONS FOR DISCUSSION

1. Describe the difference between foods with empty calories and foods with high nutrient density. Give examples of foods in each category.

2. Why are unrefined carbohydrates more healthful than refined starches and sugars?

3. Why is it necessary to have some fat in the diet?

4. Which vitamins are water-soluble? Which are fat-soluble? Which of the two groups is more important to include in the diet every day? Why?

5. According to the recommendations of government health agencies, which food groups should we consume more of than we now do, on average? Which foods should we consume less of?

6. Discuss and compare the healthful or unhealthful qualities of saturated fats, polyunsaturated fats, and monounsaturated fats. Give examples of each type.

7. What are some ways you, as a cook, can reduce the fat and sodium content of your menu offerings?

8. How can you ensure a nutritionally balanced menu without actually calculating the nutrient content of every item?

- found in plants, help "keep doctor away" → phytochemicals
- made up of starches, sugars & fibres → carbs
- Made up of amino acids → Protein
- our brain contains aLOT of this nutrient → FAT
- discovered in the early 20th century → Vitamins
- found in large quantities in our bones → minerals

BASIC PRINCIPLES OF COOKING AND FOOD SCIENCE

No written recipe can be 100 percent accurate. No matter how carefully a recipe is written, the judgment of the cook is still the most important factor in a preparation turning out well. A cook's judgment is based on experience, on an understanding of the raw materials available, and on knowledge of basic cooking principles and food science.

Ever since Carême began to define the principles and methods that lie behind the recipes we use (see p. 3), professional cooking has been based on procedures and methods rather than only on recipes. In the twenty-first century, chefs have improved their knowledge not just of cooking theory but also of the science behind it.

This chapter deals with basic principles. You will learn about what happens to food when it is heated, about how food is cooked by different methods, and about rules of seasoning and flavoring. It is important to understand the science of food and cooking so you can successfully use these principles in the kitchen.

AFTER READING THIS CHAPTER, YOU SHOULD BE ABLE TO

1. Name the most important components of foods and describe what happens to them when they are cooked.

2. Explain the composition of temporary and permanent emulsions.

3. Name and describe three ways in which heat is transferred to food in order to cook it.

4. Describe the two factors or changes in cooked foods that determine doneness.

5. List three factors that affect cooking times.

6. Explain the advantage of boiling or simmering in a covered pot. Describe three situations in which a pot should not be covered during simmering or boiling.

7. Explain how cooking temperature affects the doneness characteristics of a food item.

8. Define each basic cooking method used in the commercial kitchen.

9. Describe the two main steps in the process of cooking sous vide.

10. List six safety guidelines for cooking sous vide.

11. Explain the difference between a seasoning and a flavoring ingredient and give examples of each.

12. Identify appropriate times for adding seasoning ingredients to the cooking process in order to achieve optimal results.

13. List eleven guidelines for using herbs and spices in cooking.

BASIC CONCEPTS IN FOOD SCIENCE

To cook food means to heat it in order to make certain changes in it. Skillful cooks know exactly what changes they want to make and what they have to do to get them right. To learn these cooking skills, it is important for you to know why foods behave as they do when heated. For this, you have to study the theory.

Most of this section is devoted to understanding the results of applying heat to food. In addition, another basic concept of food science, that of *emulsions*, is discussed in detail. This concept is important in the production of sauces and salad dressings, as well as various bakeshop preparations, such as cake batters.

Perhaps not all of this section will make sense to you at first. But the ideas should become clearer to you after you think about them in relation to specific techniques, as demonstrated by your instructor. Later in your studies, when you are learning about cooking meats, fish, vegetables, and making sauces and salad dressings, review this section from time to time. Not only will you understand it better but also it should help you make more sense of the procedures you are learning and practicing.

WHAT IS HEAT?

Heat is a form of energy associated with the motion of atoms or molecules. When a substance absorbs heat, its molecules move faster. In liquids and gases, the molecules move more quickly from place to place and bounce off each other more frequently. In solids, the molecules stay mostly in place, but they vibrate with more energy. *Temperature* can be defined as a measure of this molecular activity. The higher the temperature, the faster the molecules are moving.

When fast-moving molecules in hot substances come in contact with slower molecules in cold substances, the fast molecules bump into the slower ones and transfer some of their energy, making the slower molecules move faster, or heat up. Thus, as heat is transferred, the hot substance loses energy and the colder substance gains energy.

The moving molecules in a liquid such as water sometimes move to the surface with enough energy to break through and escape to become a gas. This is called *evaporation*. When the molecules in the liquid move faster, more of them can escape in a shorter time. This is why hot water evaporates more quickly than cold water.

When we add enough heat to foods, the molecules may move so fast the structure of the food changes. For example, sucrose (regular sugar) may break apart and form new molecules that happen to have a brown color and the taste of caramel. Or protein molecules may break apart and reform with a different structure. Creating these molecular changes is called **cooking**.

EFFECTS OF HEAT ON FOODS

Foods are composed of proteins, fats, carbohydrates, and water, plus small amounts of other compounds such as minerals (including salt), vitamins, pigments (coloring agents), and flavor elements. It is important to understand how these components react when heated or mixed with other foods. You will then be better equipped to correct cooking faults when they occur and to anticipate the effects of changing cooking methods, cooking temperatures, or ingredient proportions.

In other words, when you know why foods behave as they do, you can understand how to get them to behave as you want them to.

The following discussion is concerned with the physical and chemical reactions that affect the components of food. The nutritional aspects of these components are discussed in Chapter 5.

CARBOHYDRATES

Starches and sugars are carbohydrates. Both compounds are present in foods in many forms. They are found in fruits, vegetables, grains, beans, and nuts. Meats and fish also contain a small amount of carbohydrate.

Caramelization

Sugars are simpler compounds than starches (see sidebar). For the cook, the most important change in sugar caused by cooking is **caramelization**, which can be defined as the browning of sugar caused by heat. The browning of sautéed vegetables is an example of caramelization. The browning of the crusts of baked goods, especially sweet doughs, is caused in part by caramelization and in part by another reaction called the Maillard reaction (discussed under Proteins, below).

Different kinds of sugars caramelize at different temperatures. Regular sucrose (table sugar) begins to melt at 320°F (160°C). When it reaches approximately 338°F (170°C) it begins to turn golden brown.

As sugar is heated, its molecules break down into many new compounds. These compounds, in combination, have the typical flavor and brown color of caramel. The more the sugar is heated, the more the compounds break down, resulting in a darker color and a bitter flavor. Heavily caramelized sugar is less sweet than lightly caramelized sugar, because more of the sugar molecules are destroyed.

Gelatinization

Starches are important to cooks because of their function as thickeners for sauces and other liquids, and they are important to bakers because they supply most of the bulk of breads and other baked goods.

As described in the sidebar, starch molecules consist of long chains. In their raw form, these molecules are packed into tiny bundles called *starch granules*. These granules change form in the presence of water and heat.

To demonstrate how starches behave when heated, first mix a little cornstarch with cold water and let it stand. You will see that the starch settles to the bottom. It does not dissolve in water as sugar does.

When mixed with cold water, starch granules absorb very little water. As the water is heated, the individual granules absorb water and swell to a much larger size (although they are still too small to be seen except with a microscope). The granules become so big that they bump into each other when the liquid is stirred or poured, and this slows down the flow. Thus, the liquid becomes thicker.

As the temperature gets close to boiling, some of the granules break open and release long starch chains. These chains connect with each other and form a network that blocks the flow of water and large starch granules. At this point, the starch has its greatest thickening effect.

This thickening process just described is called **gelatinization**.

As the thickened liquid is cooled, more reactions take place. Remember that molecules in a hot substance move faster than in a cold substance. So as the liquid cools, the water molecules slow down and are less able to resist the network created by the starch. This means that the liquid gets even thicker. If there is enough starch in the liquid, the mixture will form a solid gel and not flow at all. The formation of a solid gel is called *gelation*.

Cooks should always keep in mind that sauces get thicker as they cool. The consistency of a sauce simmering on the range should be thinner than you want it to be when served. Remember that by the time the sauce is plated and brought to the customer, it will be cooler than it was on the stove. Test the consistency of a sauce by ladling a small amount onto a plate and letting it cool for a few moments.

Acids make sauces thinner because they break apart some of the starch chains. If you will be adding an acid to a sauce, you may need to make the base sauce a little thicker than usual. For best results, don't add acids to sauces until the gelatinization or thickening is complete.

Dextrinization

In Chapter 8 you will learn how to make roux to thicken sauces. You will learn that roux can be cooked lightly to be kept white or cooked more heavily to a brown color.

When starches are heated dry (that is, with no water but with or without fat), some of the long starch chains are broken down in to simpler compounds called *dextrins,* which give the starch a golden or a brown color. This process is called *dextrinization*. Because dextrins have less thickening ability than starches, a brown roux has less thickening power than a white roux.

FRUIT AND VEGETABLE FIBER

Fiber is the name for a group of complex substances that give structure and firmness to plants. Fiber cannot be digested.

The softening of fruits and vegetables in cooking is, in part, the breaking down of fiber.

Sugar makes fiber firmer. Fruit cooked with sugar keeps its shape better than fruit cooked without sugar.

Baking soda (and other alkalis) makes fiber softer. Vegetables should not be cooked with baking soda because they become mushy and lose vitamins.

PROTEINS

Protein is a major component of meats, poultry, fish, eggs, milk, and milk products. It is present in smaller amounts in nuts, beans, and grains.

Coagulation

Proteins consist of long chains of components called amino acids. These chains normally form tight coils. As proteins are heated, the coils gradually unwind. At this point, the protein is said to be **denatured**.

For the cook, the important fact about denaturing is that, when the protein coils unwind, they become attracted to each other and form bonds. This bonding is called **coagulation**. The coagulated proteins form a solid network of bonds and become firm. As the temperature increases, the proteins shrink, become firmer, and lose more moisture. Exposure of proteins to excessive heat toughens them and makes them dry. Most proteins complete coagulation or are cooked at 160°–185°F (71°–85°C).

Maillard Reaction

Many protein foods, such as meats, contain small quantities of carbohydrate. When proteins are heated to about 310°F (154°C), the amino acids in the protein chains react with the carbohydrate molecules and undergo a complex chemical reaction. The result is that they turn brown and develop richer flavors. This reaction is called the **Maillard reaction**. It is what happens when meat browns. Because of the high temperature it requires, the Maillard reaction takes place only on the dry surface of the food. Because of its water content, the interior of the meat cannot get this hot.

Connective tissue

Connective tissues are special proteins present in meats. Meats with a great deal of connective tissue are tough, but some connective tissues are dissolved when cooked slowly with moisture. Cooking tough meats properly, therefore, makes them more tender. These techniques are explained in Chapter 14.

Acids and proteins

Acids, such as lemon juice, vinegar, and tomato products, have two effects on proteins:

- They speed coagulation.
- They help dissolve some connective tissues.

FATS

Fats are present in meats, poultry, fish, eggs, milk products, nuts, whole grains, and, to a lesser extent, vegetables and fruits. Fats are also important as cooking mediums, as for frying.

Fats can be either solid or liquid at room temperature. Liquid fats are called oils. When solid fats are heated, they melt, or change from solid to liquid. The melting point of solid fats varies.

When fats are heated, they begin to break down. When hot enough, they deteriorate rapidly and begin to smoke. The temperature at which this happens is called the **smoke point**, and it varies by type of fat. A stable fat—one with a high smoke point—is an important consideration in deep-fat frying.

Many flavor compounds dissolve in fat, so fats are important carriers of flavor. When fats melt and are lost from food, some flavors, as well as some vitamins, are lost with them.

For the functions of fats in baked goods, see page 901. For the functions of fats in emulsions, see below.

MINERALS, VITAMINS, PIGMENTS, AND FLAVOR COMPONENTS

Minerals and vitamins are important to the nutritional quality of the food. Pigments and flavor components are important to a food's appearance and taste and may determine whether the food is appetizing enough to eat. So it is important to preserve all these elements.

Some of these components are soluble in water, and others are soluble in fats. All of these components may be leached out, or dissolved away, from foods during cooking.

Vitamins and pigments may also be destroyed by heat, by long cooking, and by other elements present during cooking.

It is important, then, to select cooking methods that preserve, as much as possible, a food's nutrients, taste, and appearance. This is addressed whenever cooking techniques are explained in the remainder of this book.

Water

Nearly all foods contain water. Dried foods may contain as little as a fraction of 1 percent water, but fresh meats, fish, vegetables, and fruits consist mostly of water.

Water exists in three states: solid (ice), liquid, and gas (water vapor or steam). At sea level, pure liquid water becomes solid, or freezes, at 32°F (0°C) and turns to steam at 212°F (100°C). When water molecules turn to steam and energetically escape into the atmosphere, water is said to be **boiling**.

Water can also turn from liquid to gas at lower temperatures. When water turns to gas at any temperature, the process is called **evaporation**. Evaporation occurs more slowly the lower the temperature is. Evaporation is responsible for the drying of foods. The drying of food surfaces as they are cooked enables them to be browned.

Many minerals and other compounds dissolve in water, so water can be a carrier of flavor and of nutritional value. When water carries dissolved compounds, such as salt or sugar, its freezing point is lowered and its boiling point is raised.

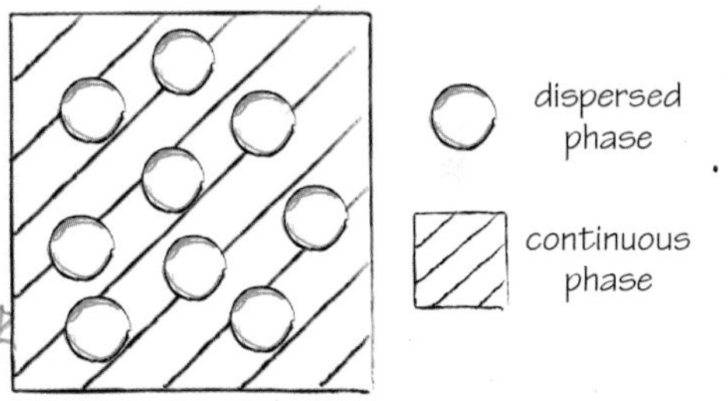

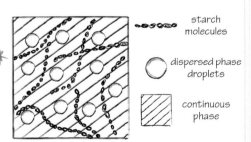

FIGURE 6.1 **In an emulsion, droplets of one substance (called the *dispersed phase*) are evenly mixed in another substance (called the *continuous phase*).**
Reprinted with permission of John Wiley & Sons, Inc.

EMULSIONS

An **emulsion** is a uniform mixture of two substances that are normally unmixable. A familiar example of an emulsion in the kitchen is mayonnaise, which is a mixture of oil and vinegar. Most of the emulsions we are concerned with in the kitchen are mixtures of fat and water. However, emulsions can be mixtures of any two substances. For example, when we cream together butter and sugar until light and fluffy, we create an emulsion of fat and air.

In an emulsion of two liquids, one of the liquids is said to be in **suspension** in the other. This means that one of the liquids is broken into tiny droplets and mixed evenly throughout the other liquid. The liquid that is broken into droplets is called the **dispersed phase,** while the liquid it is mixed in is called the **continuous phase** (see Figures 6.1 and 6.2).

FIGURE 6.2 **Particles (such as starch) in the continuous phase stabilize an emulsion by helping keep droplets of the dispersed phase from coming together and merging.**
Reprinted with permission of John Wiley & Sons, Inc.

STABILIZERS AND EMULSIFIERS

As the text explains, some ingredients in a vinaigrette help the vinegar stay suspended in the oil longer. Such ingredients are called *stabilizers*. When a stabilizer is suspended throughout a vinaigrette, its tiny particles get between the droplets of vinegar and help keep them from bumping into each other and recombining. Thus, the vinegar droplets stay suspended longer. Given enough time, however, the vinaigrette will gradually separate.

One of the most useful and effective stabilizers for vinaigrettes is mustard. Any other ingredients, such as spices and finely chopped herbs, that form small particles can also help stabilize a temporary emulsion by getting between droplets.

To make a permanent *emulsion*, a true emulsifier is needed. Lecithin, a component of egg yolks, is a powerful emulsifier that makes possible the production of mayonnaise.

In mayonnaise, the oil is broken into tiny droplets and suspended in the vinegar. Lecithin is a long molecule with an unusual characteristic. The long chain of the molecule dissolves in oil, while one end of the molecule has an electrical charge that makes it attracted to water and repulsive to oil. The long tails dissolve in the droplets of oil, while the charged ends stick out and attract the water in the vinegar and also push away other oil droplets. Thus, each droplet of oil has a coating of water around it that is kept in place by the lecithin. Because of this arrangement, the oil droplets can't bump into each other and recombine, so the emulsion is permanent.

Depending on how we mix the two liquids, we can create a water-in-oil emulsion (in which water droplets are suspended in oil), or an oil-in-water emulsion (in which oil droplets are suspended in water). A vinaigrette is an example of a water-in-oil emulsion. Mayonnaise is an example of an oil-in-water emulsion.

TEMPORARY EMULSIONS

A simple oil-and-vinegar dressing is called a **temporary emulsion** because the two liquids always separate after being shaken or beaten together. In a standard vinaigrette, the vinegar is broken into tiny droplets. These droplets are mixed evenly, or *suspended*, throughout the oil.

The harder the mixture is beaten or shaken, the longer it takes for it to separate. This is because the vinegar is broken into smaller droplets, so the droplets take longer to recombine with each other and separate from the oil. When a mixture of oil and vinegar is mixed in a blender, the resulting mixture stays in emulsion considerably longer.

Some ingredients act as stabilizers (see sidebar). For example, when mustard is added to the mixture, the vinaigrette stays emulsified longer. Other ingredients that work this way include vegetable purées and stocks with a good gelatin content. Even these mixtures, however, will separate when they stand long enough. They should be mixed again before use.

PERMANENT EMULSIONS

Mayonnaise is also a mixture of oil and vinegar—in this case, an oil-in-water emulsion—but the two liquids do not separate. This is because the formula also contains egg yolk, which is a strong emulsifier. A substance in the egg yolk called lecithin forms a layer around each of the tiny oil droplets and holds them in suspension so that they do not recombine (see sidebar for more complete explanation).

The harder the mayonnaise is beaten to break up the droplets, the more stable the emulsion becomes. All emulsions, whether permanent or temporary, form more easily at room temperature, because chilled liquid is harder to break up into small droplets.

Other stabilizers are used in some preparations. Cooked dressing uses starch in addition to eggs. Commercially made dressings may use such emulsifiers as gums, starches, and gelatin.

KEY POINTS TO REVIEW

- What is caramelization? gelatinization? dextrinization? coagulation?

- What is the Maillard reaction?

- What is an emulsion? How do stabilizers work in a temporary emulsion? How is a permanent emulsion formed?

HEAT TRANSFER

In order for food to be cooked, heat must be transferred from a heat source (such as a gas flame or an electric element) to and through the food. Understanding the ways in which heat is transferred and the speed at which it is transferred helps the cook control the cooking process.

Heat is transferred in three ways: conduction, convection, and radiation. It is important to remember that, during a cooking process, more than one of these methods of transfer may be happening at the same time. For example, food on a grill may be heated by conduction from the hot metal grill, by convection from hot air rising from the burner or charcoal, and by radiation from the glowing burner or coals.

CONDUCTION

Conduction occurs in two ways:

e.g. flat top

1. When heat moves directly from one item to something touching it—for example, from the top of the range to a soup pot placed on it, from the pot to the broth inside, and from the broth to the solid food items in it.

2. When heat moves from one part of something to an adjacent part of the same item—for example, from the exterior of a roast to the interior, or from a sauté pan to its handle.

Different materials conduct heat at different speeds. Heat moves rapidly through copper and aluminum, more slowly in stainless steel, more slowly yet in glass and porcelain. Air is a poor conductor of heat.

CONVECTION

Convection occurs when heat is spread by the movement of air, steam, or liquid (including hot fat). There are two kinds of convection:

Convection vs conventional oven

1. **Natural.**
   Hot liquids and gases rise, while cooler ones sink. Thus, in any oven, kettle of liquid, or deep-fat fryer a constant, natural circulation distributes heat.

2. **Mechanical.**
   In convection ovens and convection steamers, fans speed the circulation of heat. Thus, heat is transferred more quickly to the food, and the food cooks faster.

Stirring is a form of mechanical convection. Thick liquids cannot circulate as quickly as thin ones, so the rate of natural convection is slower. This explains, in part, why it is so easy to scorch thick soups and sauces. The heat is not carried away from the bottom of the pan quickly enough, so it stays concentrated on the bottom and scorches the food. Stirring redistributes the heat and helps prevent this. (Using heavy pots made of a material that conducts heat well also helps prevent scorching because the pot conducts the heat more quickly and evenly across the bottom and up the sides.)

Convection is the process that carries the heat from the heat source to the food. Once the carrier of the heat (air or liquid) comes in contact with the food, the heat is transferred from the carrier to the food by conduction.

RADIATION

Radiation occurs when energy is transferred by waves from a source to the food. The waves themselves are not actually heat energy but are changed into heat energy when they strike the food being cooked. (Light waves, radio waves, and X-rays are examples of radiation not used for cooking.)

Two kinds of radiation are used in the kitchen:

1. **Infrared.**
   Broiling is the most familiar example of infrared cooking. In a broiler, an electric element or a ceramic element heated by a gas flame becomes so hot it gives off infrared radiation, which cooks the food. High-intensity infrared ovens are designed to heat food rapidly.

2. Microwave.

In microwave cooking, the radiation generated by the oven penetrates partway into the food, where it agitates the molecules of water. Recall from our earlier discussion (p. 106) that heat is the motion of molecules. Thus, as the molecules move faster, the temperature of the food rises and the food is cooked.

- Because microwave radiation affects only water molecules, a completely waterless material will not heat in a microwave oven. Plates become hot only when heat is conducted to them by hot foods.
- Because most microwaves penetrate no more than about 2 inches (50 mm) into foods, heat is transferred to the center of large pieces of food by conduction, just as in roasting.

Cooking with microwaves is discussed in more detail later in this chapter.

HEAT MANAGEMENT

The final temperature to which we cook a food ranges from about 120°F (49°C) for rare meats and fish to about 400°F (200°C) for the crisp exterior of such foods as breads and seared meats. The boiling point of water, 212°F (100°C), falls within this range. Notice, however, the heat sources we use in the kitchen, from electric elements to gas flames, are much hotter than this temperature. Managing the heat to cook foods to the desired degree is an important part of cooking.

In the discussion that follows, we first consider cooking time—that is, the time it takes to heat food until it changes to a condition that we call *done*. We then look at other problems with controlling heat in cooking.

DONENESS AND COOKING TIMES

We say a food is "done" when two things have happened:

1. The interior temperature has risen to the desired degree.

Interior temperature is the most important factor when we are cooking tender meats. The difference between rare, medium, and well done (see p. 474) is a difference in temperature, and we can measure this doneness with a thermometer. Interior temperature is also important for food safety, as we learned in the discussion of minimum internal cooking temperatures on page 25.

2. The desired changes have taken place in the food.

Earlier in this chapter, we discussed the changes that take place in foods as they are heated. These changes include gelatinization of starches, coagulation of proteins, breaking down of connective tissues, caramelization of sugars, and Maillard browning.

In many foods, creating these changes is more important than simply heating the interior to a desired temperature. For example, the inside of a small piece of stew meat quickly becomes just as hot as the liquid in which it is simmering. However, we don't say it is "done" until enough connective tissue has broken down so it has a tender texture. It's not enough just to heat it to the desired degree.

Similarly, the inside of a strand of spaghetti quickly rises to the temperature of boiling water, but it is not done until enough starch has absorbed water and gelatinized, so it has the desired texture.

Standards of doneness are different for every type of food and for every cooking method. As we discuss individual foods throughout the remainder of this book, we learn more about doneness in meats, poultry, fish, vegetables, starches, and other foods.

The time it takes to achieve doneness is affected by three factors:

1. Cooking temperature.

This means the temperature of the air in the oven, the fat in the fryer, the surface of a griddle, or the liquid in which a food is cooking.

2. The speed of heat transfer.

Different cooking methods transfer heat at different rates, as shown by these examples:

Air is a poor conductor of heat, while steam is much more efficient. A jet of steam (212°F/100°C) will easily burn your hand, but you can safely reach into an oven at 500°F (260°C). This is why it takes longer to bake potatoes than to steam them.

A convection oven cooks faster than a conventional oven, even if both are set at the same temperature. The forced air movement transfers heat more rapidly.

3. **Size, temperature, and individual characteristics of the food.**
   For example:

   A small beef roast cooks faster than a large one.

   A chilled steak takes longer to broil than one at room temperature.

   Fish items generally cook more quickly than meats.

   Beef shank, which has a lot of connective tissue, takes longer to cook than beef tenderloin.

Because there are so many variables, it is difficult or even impossible to determine exact cooking times in most recipes. Individual ovens, fryers, and steamers, for example, may transfer heat more or less efficiently or have different recovery times. Roasting charts that give cooking times for various cuts of meat can be used only as guidelines, and the cook must use his or her judgment to make the final determination of doneness. Cooking times are discussed again in the next chapter.

CONTROLLING HEAT

To control cooking, we must control how heat is transferred. The kitchen contains dozens of kinds of heat sources as well as a great array of pots, pans, and other cooking tools. Controlling cooking with so many options is a skill a cook gains with experience, by performing cooking tasks over and over.

In this section, we introduce the topic of heat management with a summary of two of the most common kinds of heat control problems.

How to Boil Water

It's a common joke that boiling water is a cooking skill many noncooks have never learned. However, boiling water is a little more complex than such quips suggest. There is more to boiling water than just putting a pot on the stove.

Covering the Pot

To bring water to a boil on a cooktop, we apply heat to the bottom of a pot containing the water. The heat is transferred to the water, raising its temperature. Some of this heat energy quickly escapes from the top of the pot. If the pot is covered, much of the heat is trapped inside, and the water comes to a boil much more quickly. To raise the temperature of 1 gram of water 1 degree Celsius takes only 1 calorie of heat energy. But to turn 1 gram of boiling water to steam takes 539 calories. When the steam escapes, it takes this energy with it. A lot of energy is lost from an uncovered pot. By covering it, we save energy and shorten heating times.

After a liquid has come to a boil, keeping the pot covered can still be helpful. You have probably had the experience of removing the lid from a pot simmering over a low flame and seeing the bubbling slow down as soon as the lid is off. This is because so much heat escapes as soon as the cover is removed. By keeping the pot covered, you can maintain the desired cooking temperature using a lower burner setting.

Although covering pots is a more efficient use of energy, sometimes you must keep them uncovered:

- **When evaporation is desired.** In many cooking operations, one of the goals is to evaporate moisture to concentrate flavors or change textures. Keep the pot uncovered to speed evaporation.

- **When the contents must be visually monitored.** In some cases, you must keep an eye on the food as it simmers or boils, if only to make sure it continues to simmer at the proper rate, not too fast or too slow.

- **When green vegetables are cooked.** Plant acids that destroy green pigments must be allowed to escape, as explained on page 271.

BOILING OIL

The term *boiling oil* is commonly heard, but oil doesn't boil, at least in the way water does. The bubbling of boiling water, as the text explains, occurs when heated molecules of liquid water turn to gas—that is, steam—and rise to the surface. Oil that is free of water does not boil at normal cooking temperatures. The bubbling we see in cooking fat, such as that in deep fryers, is caused by water in submerged foods turning to steam. It's not the oil that is boiling.

Controlling the Heat

Water boils at 212°F (100°C) at sea level and at standard atmospheric pressure. When water is boiling, any additional heat is used to turn water to steam, which then carries the heat away. No matter how high you turn the heat, the water can never rise above 212°F (100°C). In other words, turning up the heat under a pot that is already boiling is a waste of energy and does not decrease cooking time. Furthermore, the increased agitation of rapidly boiling water does more damage to delicate foods. Remember, a rapid boil is no hotter than a slow boil.

Cooking to the Center

As we read earlier, heat is transferred from the outside of food to the inside by conduction. Conduction takes time, so cooking takes time.

Think of a steak cooking on a grill. Let's say we want to cook the steak to an interior temperature of 140°F (60°C), for medium doneness. When we first put the steak on to cook, the interior temperature is room temperature, or possibly refrigerator temperature. The outside, however, rises to perhaps 400°F (200°C) very soon after we place it on the grill. Gradually, this heat moves to the center. By the time the center reaches the target temperature, the outside is much hotter. If we cut the steak through the center, we see a gradation from very well done at the outside to medium done in the middle.

Often this is just what we want. This is how people are used to eating steaks, so a person might be surprised to get a steak that was a uniform medium done all the way through.

By contrast, if we cook the steak at a low temperature, there is less temperature difference between the outside and inside, so the doneness of the meat is more uniform from outside to inside.

The same is true of large roasts. Roasting at a high temperature produces a strong gradation of doneness, from well done on the outside to less done in the center. Roasting at a low temperature gives more uniform doneness throughout. The roasting temperature we use depends on the results we want.

Of course, cooking at a low temperature doesn't create the well-browned crust most diners desire. We have two options to solve this problem:

- Brown the exterior with high heat, then cook to doneness at lower heat.
- Cook to doneness at low heat, then brown the exterior with a quick blast of high heat.

KEY POINTS TO REVIEW

- What are the three ways in which heat energy is transferred to foods?

- What do we mean when we say a cooked food is "done"? What factors affect how long it takes to cook a food until it is "done"?

- When you are cooking food in a saucepan, when would you cover it, and when would you keep it uncovered?

- How does cooking temperature affect the doneness of foods such as roasts?

COOKING METHODS

Now that we understand the three basic methods of heat transfer—conduction, convection, and radiation—we can explore the basic techniques that cooks use to apply heat to food. These techniques are known as *cooking methods.*

Cooking methods are classified as moist heat or dry heat.

incl. fat!
(boil)

- **Dry-heat methods** are those in which the heat is transferred without moisture—that is, by hot air, hot metal, radiation, or hot fat. We usually divide dry-heat methods into two categories: without fat and with fat.

boil, braise

- **Moist-heat methods** are those in which the heat is transferred to the food product by water or water-based liquids such as stock and sauces, or by steam.

Different cooking methods are suited to different kinds of foods. For example, some meats are high in connective tissue and are tough unless this tissue is broken down slowly by moist heat. Other meats are low in connective tissue and naturally tender. They are at their best and juiciest when cooked with dry heat to a rare or medium-done stage.

Many other factors must be considered when choosing cooking methods for meats, fish, and vegetables, such as the flavor and appearance imparted by browning, the flavor imparted by fats, and the firmness or delicacy of the product. These factors are discussed in later chapters with respect to individual foods.

Following the definitions of the conventional cooking methods, we proceed to explanations of microwave cooking, sous vide cooking, and molecular gastronomy.

CONVENTIONAL COOKING METHODS

The basic cooking methods discussed in this section are considered the conventional or traditional cooking methods. Introduced here with brief definitions, they are discussed in greater detail in Chapter 11, as they are applied to vegetables, and again in Chapter 14, as they are applied to meats, poultry, and fish. In Chapter 14, each method is illustrated by a core recipe that gives you an opportunity to practice the technique in its most basic form before proceeding to the recipes in the chapters that follow.

DRY HEAT METHODS

Roast and Bake

To **roast** and to **bake** both mean to cook foods by surrounding them with hot, dry air, usually in an oven. Cooking on a spit in front of an open fire may also be considered roasting.

The term *roasting* usually applies to meats and poultry. The term *baking* usually applies to breads, pastries, vegetables, and fish. It is a more general term than roasting, although, in practice, there is little or no difference in actual technique, and the terms are often interchangeable (except for breads and pastries).

Cooking uncovered is essential to roasting. Covering holds in steam, changing the process from dry-heat to moist-heat cooking, such as braising or steaming.

Barbecue

To **barbecue** means to cook with dry heat created by the burning of hardwood or by the hot coals of this wood. In other words, barbecuing is a roasting technique requiring a wood fire or wood smoke.

Broil

To **broil** means to cook with radiant heat from above.

Note: The terms *broiling*, *grilling*, and *griddling* are sometimes confused. Grilling (see following) is often called *broiling*, and griddling is called *grilling*. This book uses the terms that refer to the equipment involved. Thus, broiling is done in an overhead broiler, grilling on a grill, and griddling on a griddle.

Broiling is a rapid, high-heat cooking method used mainly for tender meats, poultry, fish, and a few vegetable items.

Grill, Griddle, and Pan-Broil

Grilling, griddling, and pan-broiling are all dry-heat cooking methods that use heat from below.

Grilling is done on an open grid over a heat source, which may be charcoal, an electric element, or a gas-heated element. Cooking temperature is regulated by moving the items to hotter or cooler places on the grill. Grilled meats should be turned to achieve desired grill marks, just as in broiling.

Griddling is done on a solid cooking surface called a griddle, with or without small amounts of fat to prevent sticking. The temperature is adjustable and much lower (around 350°F/177°C) than on a grill. In addition to meats, items such as eggs and pancakes are cooked on a griddle.

Pan-broiling is like griddling except it is done in a sauté pan or skillet instead of on a griddle surface. Fat must be poured off as it accumulates, or the process becomes pan-frying. No liquid is added, and the pan is not covered, or else the item would steam.

DRY-HEAT METHODS USING FAT

Sauté

To **sauté** means to cook quickly in a small amount of fat. High heat is required, and the procedure is most often done in a broad, flat pan called a *sauté pan* (p. 51) or *sauteuse*.

Pan-Fry

To **pan-fry** means to cook in a moderate amount of fat in a pan over moderate heat. Pan-frying is similar to sautéing except more fat is used, the heat is lower, and the cooking time is longer. The method is used for larger pieces of food, such as chops and chicken pieces, and the items are not tossed by flipping the pan, as they often are in sautéing.

Deep-Fry

To *deep-fry* means to cook a food submerged in hot fat. The equipment most often used for this cooking method is the deep-fryer (p. 43), which is equipped with a thermostat to control the temperature of the fat, and with baskets to lower food into the fat and to remove and drain them when fried.

Many foods are given a coating of breading or batter before being deep-fried. These procedures are explained on pages 150 and 152.

MOIST-HEAT METHODS

Poach, Simmer, and Boil

Poaching, *simmering*, and **boiling** all involve cooking a food in water or a seasoned or flavored liquid. The temperature of the liquid determines the method.

To **boil** means to cook in a liquid that is bubbling rapidly and greatly agitated. Water boils at 212°F (100°C) at sea level. No matter how high the burner is turned, the temperature of the liquid will go no higher.

To **simmer** means to cook in a liquid that is bubbling gently at a temperature of about 185°F to 200°F (85°C to 94°C).

To **poach** means to cook in a liquid, usually a small amount, that is hot but not actually bubbling. Temperature is 160°–185°F (71°–85°C).

Steam

 en papilotte

To **steam** means to cook foods by exposing them directly to steam. In quantity cooking, this is usually done in special steam cookers. Steaming can also be done on a rack above boiling water.

Braise

To **braise** means to cook covered in a small amount of liquid, usually after preliminary browning. In almost all cases, the liquid is served with the product as a sauce.

Braising is sometimes referred to as a *combination cooking method* because the product is first browned, using dry heat, before it is cooked with a liquid. Nevertheless, in most cases, moist heat is responsible for most of the cooking process, and the browning may be thought of as a preliminary technique. The purpose of the browning step is not so much to cook the item as to develop color and flavor.

Some references describe braising and **stewing** as two different cooking methods. The term *braising* is used for large cuts of meat, and *stewing* is used for smaller items. In this book, however, we use the term *braising* for both methods because the basic procedure in both cases is the same—first browning with dry heat, then cooking with moist heat. (Note that the term *stewing* is also used for simmering in a small amount of liquid without preliminary browning.)

MICROWAVE COOKING

Microwave cooking refers to the use of a specific tool rather than to a basic dry-heat or moist-heat cooking method. The microwave oven is used mostly for heating prepared foods and for thawing raw or cooked items. However, it can be used for primary cooking as well.

Microwave oven models range in power from about 500 watts up to about 2,000 watts. The higher the wattage, the more intense the energy the oven puts out and the faster it heats foods. Most models have switches that allow you to cook at different power levels.

One of the most important advantages of the microwave oven in à la carte cooking is that it enables you to heat individual portions of many foods to order quickly and evenly. Instead of keeping such foods as stews hot in the steam table, where they gradually become overcooked, you can keep them refrigerated (either in bulk or in individual portions) and re-heat each order as needed. This is perhaps the main reason why most restaurants have one or more microwave ovens, even though they may not use them for primary cooking.

Because the microwave oven is a unique tool in food service, the cook should observe the following special points regarding its use:

1. Small items will not brown in a standard microwave. Large roasts may brown somewhat from the heat generated in the item itself. Some models have browning elements that use conventional heat.

2. Watch timing carefully. Overcooking is the most common error in microwave use. High energy levels cook small items very rapidly.

3. Large items should be turned once or twice for even cooking.

4. An on/off cycle is often used for large items to allow time for heat to be conducted to the interior.

5. If your equipment has a defrost cycle (which switches the oven to lower power), use this cycle rather than full power to thaw frozen foods. Lower power enables the item to thaw more evenly, with less danger of partially cooking it. If your oven does not have this fea-ture, use an on/off cycle.

6. Sliced, cooked meats and other items that are likely to dry out in the microwave should be protected either by wrapping them loosely in plastic or wax paper or by covering them with a sauce or gravy.

7. Because microwaves act only on water molecules, foods with high water content, such as vegetables, heat faster than denser, drier foods, such as cooked meats.

8. Foods at the edge of a dish or plate heat faster than foods in the center because they are more directly exposed to the oven's rays. Therefore:

   - Depress the center of casseroles so the food is not as thick there as at the edges. This will help it heat more evenly.

   - When you are heating several foods at once on a plate, put the moist, quick-heating items like vegetables in the center and the denser, slower-heating items at the edges.

9. Because microwaves do not penetrate metal, aluminum foil and other metals shield foods from the radiant energy. For example, a potato wrapped in foil will not cook in a microwave oven.

   With older machines, it was a general rule not to put any metal in the oven, as the radiation could bounce off the metal and damage the magnetron (the oven's generator). With newer machines, it is possible to heat foods in foil pans and to shield certain parts of the food by covering them with pieces of foil so they do not overheat. Do not place two pieces of metal close to one another, because sparks may be generated, causing a fire. Follow the procedures recommended by the manufacturer of the oven.

Because microwaves cook so rapidly, they will not break down the connective tissues of less tender meats. Slow, moist cooking is necessary for dissolving these connective tissues.

The more food placed in a microwave at once, the longer the cooking time. Thus, the primary advantage of microwave cooking—speed—is lost with large roasts and other large quantities.

COOKING SOUS VIDE

A new technology that has had a rapid growth in popularity among the world's top chefs is **sous vide** (soo veed) cooking. French for "under vacuum," the term is applied to cooking foods that have been vacuum-sealed in plastic bags.

In simplest terms, this food preparation technique is a two-step process:

1. Vacuum-pack the food item, plus any seasonings or marinades, in an appropriate plastic bag.
2. Cook the food item, while in the bag, at a constant low temperature, usually in a special water bath.

PRECISION COOKING

Although the name of the technique refers to the vacuum packing, the heart of sous vide cooking—and the reason many chefs are so excited about it—is the precise temperature control it permits.

As an example, think of roasting a boneless loin of lamb. On the one hand, we could place the meat in an oven at 400°F (200°C) and roast it until the center reaches a temperature of 140°F (60°C) for medium doneness. As we discussed on pages 113–114, however, the lamb will be medium done only in the center and more done everywhere else. In addition, we would have to monitor the cooking closely to make sure we remove it from the oven at the right time.

On the other hand, we could vacuum-pack the lamb loin in plastic and place it in a water bath heated to an exact 140°F (60°C). The temperature of the lamb would never go above that temperature, no matter how long we left it in the water bath. And it would be at exactly the same doneness from outside to center.

Because we like a browned exterior on the lamb, we could then remove it from the bag, brown it quickly in a hot sauté pan, and serve it immediately.

SOUS VIDE APPLICATIONS

Sous vide cooking is such a new science that chefs are only beginning to explore its possibilities. Techniques will surely evolve and change in years to come.

Some of the main applications of sous vide cooking as it is practiced today are detailed below. Please note temperature ranges are approximate. Immersion circulators (p. 48) are extremely accurate, and chefs may specify temperatures to within a fraction of a degree to get the precise results they want. Although not as precise as immersion circulators, combi ovens (p. 41) are also used for some sous vide cooking.

Basic sous vide procedures for meat, poultry, and seafood are explained in Chapter 14, and additional recipes are included in following chapters.

Vegetables can benefit from sous vide techniques, especially those that discolor when exposed to air, such as artichokes. See Chapter 11 for further information on sous vide techniques for vegetables.

Tender Meats and Poultry

Tender meats and poultry are usually cooked in a water bath heated to the exact doneness temperature desired. The lamb loin described above is a typical example. Cooking temperatures usually range from 140° to 149°F (60° to 65°C), although higher or lower temperatures are also used, and cooking times may range from 20 to 60 minutes, or sometimes longer.

Remember that if the cooking temperature is within the Food Danger Zone (p. 17), cooking time must be counted as part of the four-hour rule (p. 22). For this reason, tender meats and poultry are almost always finished and served immediately after cooking and not chilled and stored for later use.

Tough Meats

Tough meats can be cooked to tenderness while retaining more moisture than if they were braised or simmered. Cooking temperatures range from 149° to 158°F (65° to 70°C). These temperatures are much lower than usual braising temperatures, but they are high enough to break down connective tissue. At the higher end of this range, meats become falling-apart tender, while at the lower end they retain more of their shape and have a firmer texture while still being tender and juicy.

Because it takes time to break down connective tissue, often 12–48 hours, tough meats are usually cooked to doneness, then immediately chilled and refrigerated for later use.

Because of the long cooking times, avoid cooking temperatures below 149°F (65°C). Higher temperatures provide more safety from bacterial growth.

Fish and Seafood

Fish are naturally tender and have even more delicate connective tissue than tender meats. Cooking procedures are similar to those for tender meats, but often with even lower temperatures and/or shorter cooking times. Typical cooking temperatures may range from 122° to 140°F (50° to 60°C), or sometimes higher, with cooking times as short as 10–15 minutes. Use only the freshest, cleanest fish, and finish and serve it immediately after cooking.

Shellfish such as lobster and shrimp can be toughened by the higher heat of traditional cooking techniques, so sous vide cooking can be a benefit. Cooking temperature for these items is typically around 140°F (60°C).

Vegetables

Vegetables can benefit from sous vide techniques, especially those that discolor when exposed to air, such as artichokes. Vegetables are usually cooked at 185°F (85°C).

SAFETY FACTORS

Working with vacuum-packed foods increases the health dangers caused by anaerobic and facultative bacteria (p. 18)—that is, bacteria that can grow without oxygen. Because cooking temperatures are so low, there is danger that bacteria will not be killed if cooking is not carefully done. Salmonella, *E. coli*, listeria, and botulism bacteria are the most serious risks.

For these reasons, chefs and health officials are especially cautious about sous vide cooking and have instituted strict guidelines. The following steps summarize the most important aspects of these safety rules:

1. Know the rules for sous vide set up by your local health department, and follow them carefully. Health departments usually require that you establish a HACCP system (p. 29) for sous vide cooking. To satisfy HACCP requirements, you must keep detailed records of every step in the production process for each item, including cooking and storage times and temperatures.

2. Use only the freshest, most wholesome foods from reputable purveyors.

3. Chill all foods thoroughly before vacuum packing. If you sear a food item before packing, chill it after searing and before packing.

4. After packing, cook the food at once, or immediately refrigerate it at 38°F (3.3°C) or lower—or, even better, freeze it.

5. After cooking, serve the food immediately, or chill it as quickly as possible in an ice bath or a blast chiller.

6. Thaw cooked food frozen in its package in the refrigerator.

Sous vide cooking involves complex procedures, and the above summary is only a short introduction to its techniques and safety factors. Entire books have been written on the subject, some of which are listed in the Bibliography. Later in this text, you will find recipes illustrating some of these techniques, including a recipe for long cooking of a tough meat (p. 448), for short cooking of fish (p. 649), meat (p. 513), and poultry (p. 574), and for cooking vegetables (pp. 317 and 318). Once again, these are only examples that can help you begin to understand the nature of sous vide.

Finally, using the proper equipment, especially a chamber vacuum packager and an immersion circulator (described on p. 48), is important. Home-style vacuum packers might be useful for experimenting, but only a chamber packer can achieve the vacuum pressures chefs find they need. Furthermore, home machines are not able to pack liquids, so they can't pack meats with marinades. Most important, precise temperature control is critical when you are cooking at such low temperatures, and you need appropriate equipment for cooking sous vide.

MOLECULAR GASTRONOMY

The approach to cooking known as **molecular gastronomy**, introduced on page 7, is the latest effort by creative chefs to find new ways of preparing and presenting food. The manipulation of food ingredients in new ways by the use of technology is known as *molecular gastronomy*.

However, this description is misleading, because it suggests to many people that the technology is the most important part of this way of cooking. Even the name, molecular gastronomy, suggests scientists making artificial food in test tubes. Perhaps a better name might be *avant-garde cuisine* (see sidebar).

AVANT-GARDE TECHNIQUES AND INGREDIENTS

It is a challenge, in a short space, to describe molecular gastronomy because it consists of so many unrelated techniques. Also, every chef has his or her own style of cooking and uses a different set of favorite techniques and plating styles. Furthermore, it is important to understand that many or even most items on an avant-garde chef's menu are made with traditional techniques. The chef uses whatever cooking methods he or she feels are appropriate to the dish, whether a traditional technique or a molecular-gastronomy technique. You shouldn't think that such a menu is composed entirely of foams, bubbles, powders, and gels. In the definition in the preceding paragraph, the word *selective* is used to mean the chef selects a nonstandard technique when, and only when, it helps intensify a flavor or aroma or in another way to improve the dining experience.

The chef may use these techniques to change familiar foods into unfamiliar forms, to make unexpected combinations of foods, or to make one food look like another. Tricking the diner's expectations is another way the chef draws attention to flavors and aromas. One does not eat this food absentmindedly. Every bite is intended to be an exploration or an adventure.

In this style of cooking, using the best ingredients is necessary. Because the techniques are used to focus attention on flavors, colors, textures, and aromas, only the freshest foods have the quality to work in these dishes.

Chefs use countless individual techniques to create their versions of molecular gastronomy, and new ones are invented all the time. Just a few of the better-known and most talked-about of these techniques are described below. Following this list is a description of some of the ingredients used to achieve these effects.

- **Nontraditional thickeners.** In addition to the traditional starches, chefs have new ways to thicken sauces and to change the texture of liquids. Some of these thickening agents work without heating and are simply blended with the cold liquid. This allows the chef to create sauces and other liquids with a fresh, uncooked taste.

- **Foams, froths, and bubbles.** For many years, chefs have made foamy sauces by whipping or blending a sauce just before plating. These foams collapse quickly, however. Avant-garde chefs stabilize foams with gelatin, lecithin, and other ingredients. A well-made foam adds an additional flavor dimension to the plate without adding bulk.

- **Gels.** Turning a liquid, such as a vegetable juice, into a solid not only gives it a different texture but also enables the food to be cut into many shapes, allowing the chef to create different visual presentations. Chefs use regular gelatin as well as other jelling agents, such as agar-agar, which is derived from seaweed.

- **Drying and powdering.** Drying a food intensifies its flavor and, of course, changes its texture. Eating a cauliflower floret that was deep-fried and then dehydrated until crisp is a much different experience than eating steamed, buttered cauliflower. If the dehydrated food is powdered, it becomes yet another flavor and texture experience.

- **Spherification.** This technique creates spheres of liquid contained inside a thin gel wall. In the standard method, the liquid is mixed with a hydrocolloid (see p. 120) called sodium alginate. In a separate container is a water bath containing calcium. When the liquid is dropped into the calcium bath, the alginate and the calcium react to form a thin wall of gel surrounding a liquid center. The chef can make tiny spheres by using an eye dropper or larger ones by freezing the liquid in a mold before dropping it into the calcium bath.

  The opposite method is to dissolve the calcium in the flavorful liquid and drop it in an alginate bath. This is called *reverse spherification*.

AVANT-GARDE CUISINE

The term *avant garde* means "advance guard," and it was originally used to indicate the front ranks of an army advancing into battle. Today the term is used to describe any group that pushes the boundaries of a discipline beyond what is considered normal. The term is used most often in the arts and culture. Members of an avant-garde group experiment with and invent new techniques and new ways of applying new as well as old techniques.

For the best chefs in the field, molecular gastronomy is not so much a culinary movement as a new collection of tools for their toolkit. Their focus is still the food and the dining experience, and they use all their judgment and skills, including their training in classical techniques, to put good food on the plate. In the hands of great chefs, what could be nothing more than clever stunts with food becomes a great dining experience.

With this in mind, we can offer a second definition of *molecular gastronomy*: the selective use of technology and nonstandard ingredients to help enhance the flavors, aromas, appearance, and textures of natural foods.

To give you a sense of how some of these ingredients and techniques are used, several recipes are included in Chapter 8 (pp. 218–222), Chapter 9 (p. 233) and Chapter 11 (pp. 319 and 320). Understand, however, this material gives you only a brief introduction to molecular gastronomy. To help you get a fuller sense of the subject, further references are included in the Bibliography (page 1031).

SUMMARY OF COOKING TERMS

The following is an alphabetical list of terms that describe ways of applying heat to foods. Basic cooking methods described earlier are included, as are more specific applications of these basic methods.

bake. To cook foods by surrounding them with hot, dry air. Similar to **roast**, but the term bake usually applies to breads, pastries, vegetables, and fish.

barbecue. (1) To cook with dry heat created by the burning of hardwood or by the hot coals of this wood. (2) Loosely, to cook over hot coals, such as on a grill or spit, often with a seasoned marinade or basting sauce.

blanch. To cook an item partially and very briefly in boiling water or in hot fat. Usually a pre-preparation technique, as to loosen peels of vegetables, fruits, and nuts, to partially cook French fries or other foods before service, to prepare for freezing, or to remove undesirable flavors.

boil. To cook in water or other liquid that is bubbling rapidly, about 212°F (100°C) at sea level and at normal pressure.

braise. (1) To cook covered in a small amount of liquid, usually after preliminary browning. (2) To cook certain vegetables slowly in a small amount of liquid without preliminary browning.

broil. To cook with radiant heat from above.

deep-fry. To cook submerged in hot fat.

deglaze. To swirl a liquid in a sauté pan, roast pan, or other pan to dissolve cooked particles of food remaining on the bottom.

dry-heat cooking methods. Methods in which heat is conducted to foods without the use of moisture.

fry. To cook in hot fat.

glaze. To give shine to the surface of a food by applying a sauce, aspic, sugar, or icing, and/or by browning or melting under a broiler or salamander or in an oven.

griddle. To cook on a flat, solid cooking surface called a griddle.

grill. To cook on an open grid over a heat source.

moist-heat cooking methods. Methods in which heat is conducted to foods by water or other liquid (except fat) or by steam.

pan-broil. To cook uncovered in a skillet or sauté pan without fat.

pan-fry. To cook in a moderate amount of fat in an uncovered pan.

(en) papillote. Wrapped in paper (or sometimes foil) for cooking so the enclosed food is steamed in its own moisture.

parboil. To cook partially in a boiling or simmering liquid.

parcook. To cook partially by any method.

poach. To cook gently in water or other liquid that is hot but not actually bubbling, 160°–180°F (71°–82°C).

reduce. To cook by simmering or boiling until the quantity of liquid is decreased, often to concentrate flavors.

roast. To cook foods by surrounding them with hot, dry air in an oven or on a spit in front of an open fire.

sauté. To cook quickly in a small amount of fat, usually while mixing or tossing the foods by occasionally flipping the pan.

sear. To brown the surface of a food quickly at a high temperature.

simmer. To cook in water or other liquid that is bubbling gently, 185°–205°F (85°–96°C).

smoke-roast. To cook with dry heat in the presence of smoke, as on a rack over wood chips in a covered pan.

sous vide. Vacuum-packed. Refers to techniques for cooking foods that are packaged under vacuum in plastic bags.

steam. To cook by direct contact with steam.

stew. To simmer or braise a food or foods in a small amount of liquid, which is usually served with the food as a sauce.

stir-fry. To cook quickly in a small amount of fat by tossing cut-up foods in a wok or pan with spatulas or similar implements. Similar to sauté, except the pan is stationary.

sweat. To cook slowly in fat without browning, sometimes under a cover.

KEY POINTS TO REVIEW

- What cooking methods are classified as dry-heat methods? What methods are dry-heat methods with fat? What methods are moist-heat methods? Define each of these cooking methods.

- What are the two main steps in sous-vide cooking? List six safety guidelines to observe when cooking sous vide.

BUILDING FLAVOR

People eat because they enjoy the flavors of good food, not just because they must fill their stomachs to stay alive. Appearance, texture, and nutrition are important, too, but good taste is the first mark of good cooking. Enhancement and adjustment of flavors are among a cook's most critical tasks, requiring experience and judgment.

The most important flavors of a given preparation are those of its main ingredients. A grilled beef tenderloin steak should taste like beef, green beans should taste like green beans, tomato soup should taste primarily of tomato. Plain, unseasoned foods, however, usually taste a little bland and one-dimensional, so it is the cook's job to add interest by combining ingredients to build depth of flavor. The harmony of ingredient flavors and aromas the cook creates by combining ingredients skillfully is sometimes called a **flavor profile**.

BUILDING FLAVOR PROFILES

Foods offer complex experiences for the senses. When composing a new dish, a cook must first of all understand that more than just taste should be considered. The senses of sight, smell, taste, and touch all come into play. The fifth sense, hearing, also plays a role, as when we react to the sizzle of a steak or the crunch of a potato chip, but this sense is less of a concern for the cook than the other four. Consider how we perceive these characteristics of a dish:

- Appearance (color and color contrast, shape, shine, arrangement on the plate)
- Aroma
- Taste
- Mouthfeel (texture, moistness or dryness, softness or crispness) and temperature

UMAMI

Although European and North American tradition recognizes four basic tastes—salty, sweet, bitter, and sour—food authorities have recently identified a fifth, called **umami**, which has long been recognized by Asian cultures. Sense receptors on the tongue react to certain amino acids (see p. 94). Because amino acids are components of proteins, this taste is strong in foods high in protein. In fact, umami is often translated as "meatiness." Beef, lamb, certain cheeses, and soy sauce are especially high in umami. The food additive monosodium glutamate (MSG), used as a seasoning or flavor enhancer in some Asian cuisines, produces strong umami.

All of these factors are important to making a dish appealing to the diner. The discussion in this section, however, is concerned mostly with aroma and taste. Why not just taste? Taste buds on the tongue perceive only four basic sensations: salty, sweet, bitter, and sour (but see sidebar). What we think of as flavor is a combination of taste and aroma. When the sense of smell is lacking, such as when you have a cold, foods seem to have little flavor.

The flavors in a dish can be thought of as primary flavors and supporting, or secondary, flavors. The primary flavors are the flavors of the main ingredients. For example, in blanquette of veal (p. 508) the **primary flavor** is veal; the primary flavors in calf's liver lyonnaise (p. 505) are liver and onions; and the primary flavors in Irish lamb stew (p. 509) are lamb, onions, leeks, and potatoes. These are the flavors that predominate. When you taste each of these dishes, the first tastes you encounter are the main ingredients. Other flavors, which we can call **supporting flavors**, support and enhance the primary flavors of the main ingredients.

Examining how a flavor profile is built in a single classic and fairly simple recipe can help you begin to understand the general principles involved.

AN EXAMPLE OF FLAVOR BUILDING

Let's look more closely at the recipe for blanquette of veal. Veal, by itself, does not have a strong or pronounced flavor. Unlike meats such as beef or venison, veal has a mild, subtle flavor. In this recipe, the veal is simmered, so the flavor is even milder than it would be if the meat were browned by roasting, sautéing, or braising. For this reason, when choosing seasonings and other supporting flavors, we want to avoid strong flavors that will mask the delicate flavor of the veal. Using white veal stock as a cooking medium reinforces and strengthens the primary flavor. We could use water, but the result would be a less flavorful dish. Brown stock would be too strong for our purpose and would completely change the character of the dish. White chicken stock might be an acceptable substitute, but it wouldn't reinforce the veal flavor as well. The onion and bouquet garni are added to the stock to give it more depth and fullness of flavor.

Continuing to the finishing ingredients, we find roux, a cream-and-egg liaison, lemon juice, nutmeg, and white pepper. The roux functions as a thickener and contributes primarily to texture, although the butter in the roux also gives some richness to the flavor. The liaison is used for both texture and flavor, adding richness and creaminess. The cream and the simmered veal are a classic marriage of flavors that work well together in many dishes. However, too much richness, combined with the mildness of the veal, could make the dish cloying. The acidity of the lemon juice cuts through the richness of the cream and egg yolks, gives a more balanced flavor, and perks up the taste buds. Just enough lemon is used to balance the richness of the cream and egg, not so much as to make the dish taste lemony. Finally, the smallest amount of nutmeg and white pepper gives a pleasing complexity to the finished taste without adding identifiable flavors. If one of the first things you taste is nutmeg, then too much nutmeg has been used.

If the dish is well composed, all of these flavors, primary and supporting, combine to form a complex but unified whole we identify as the taste of veal blanquette.

GENERAL CONCEPTS IN FLAVOR BUILDING

There are no fixed rules for combining flavors, but the example just discussed suggests some general principles. When you are developing or modifying a recipe, think about the following points.

Every ingredient should have a purpose. Start with the main ingredients, and then think about what will work with them. Continue to build the flavor, using just the ingredients you need.

Ingredients can work together by harmonizing or by contrasting. In the example above, the rich taste of the liaison and the mild taste of the veal harmonize, and the tartness of the lemon contrasts with the cream.

When two ingredients contrast, be sure they balance. For example, add just enough lemon juice to the blanquette to balance the cream, not too much or too little.

Consider not only the components of the single recipe but also the other items that will be served with it on the plate. For example, think of how we use lemon to balance the richness or fattiness of the cream in the blanquette. We can use the same idea to balance the fattiness of

These are just a few of the many traditional flavoring combinations from around the world. Keep in mind that, although only one or two combinations are given for each country or region mentioned, they are not the only combinations used there. These are merely examples to stimulate your thinking.

Sour cream, paprika, caraway (Hungary)

Sour cream or mustard, dill (Scandinavia)

Caraway, onion, vinegar (Germany)

Apples, apple cider or apple brandy, cream (France—Normandy)

Shallot, garlic, parsley (France—Burgundy)

Tomato, basil, olive oil (Italy)

Olive oil, garlic, anchovy (Italy)

Lemon, oregano (Greece)

Cinnamon, nuts, honey (eastern and southern Mediterranean, Middle East)

Ginger, onion, garlic (India)

Fish sauce (nam pla), lemongrass, chiles (Thailand)

Ginger, soy sauce (Japan)

Soy sauce, sake or mirin, dried bonito (Japan)

Ginger, garlic, scallion (China)

a pork pâté or sausage by serving it with a tart mustard or chutney on the side. In other words, think of building the flavor profile of the entire plate. Plan sauces, accompaniments, and garnishes to balance, enhance, and contrast with the main item and with each other, just as the flavors in an individual recipe do.

Simplicity and Complexity

Simpler is usually better. Some cooks mistakenly think that adding more ingredients is always preferable to adding fewer. But the more flavors you combine, the harder you have to work to balance them all. Further, the more competing flavors you have, the more you have to take care that the primary flavors of the main ingredients aren't lost.

This is true whether you are planning the ingredients in a single recipe or the components on a plate. Some cooks are tempted to put too many things on a plate. When you have a meat item perched on layers of three or four vegetables and starches, with additional garnishes and two or three sauces, the result is often a confused jumble.

It would be incorrect, however, to say that simpler is *always* better. Classic dishes from many of the world's regions have complex flavor profiles. Look through any collection of recipes from India, China, or Mexico, and you will find dishes that use a large number of spices and other flavoring ingredients. The recipe for mole poblano on page 573 is an example. When these dishes work, all the ingredients blend well. In a good curry, for example, it is difficult, if not impossible, to taste each of the individual spices.

Classic Flavor Profiles

How do you know what flavors work together? Perhaps the best place to start is to study traditional recipes from around the world as well as from the classical cuisine passed down to us through Escoffier. These are dishes that have stood the test of time. We know the flavor combinations work because they have been used over and over for decades or even centuries.

We have already seen some classic flavor combinations in our discussion of veal blanquette. The combination of white meat, cream, lemon, and a hint of nutmeg is a quartet of flavors you will find repeatedly in classic and regional dishes.

International or ethnic dishes provide other examples. In northern India, many dishes are based on a mixture of onion, garlic, and fresh ginger puréed together and fried in a little oil. Studying these recipes suggests to us that these three flavors might be used together in new dishes as well. Similarly, the combination of ginger and soy sauce from Japan, paprika and cured ham from Spain, garlic, tomato, and parsley from Provence, and olives and anchovies from around the Mediterranean are all successful flavor mixtures we learn to use when we study classic dishes.

For chefs who want to create their own dishes, studying classic recipes is a good place to start.

SEASONING AND FLAVORING INGREDIENTS

The preceding discussion of flavor building concerns all ingredients that add flavor to or change the flavor of a dish. These include the primary ingredients and the supporting or secondary ingredients. The remainder of this chapter is concerned primarily with herbs and spices as well as common flavoring ingredients such as onion, garlic, and mustard.

To repeat the most important concept of flavoring, the main ingredients are the primary sources of flavor. Use good-quality main ingredients, handle all foods with care, and employ correct cooking procedures. Remember that herbs and spices play only a supporting role. Badly prepared foods can't be rescued by a last-minute addition of herbs and spices.

Although chefs do not always use the terms this way, it might be said there is a difference between seasoning and flavoring. **Seasoning** means enhancing the natural flavor of a food without significantly changing its flavor. Salt is the most important seasoning ingredient. **Flavoring** means adding a new flavor to a food, thus changing or modifying the original flavor.

The difference between seasoning and flavoring is often one of degree. For example, salt is usually used only to season, not to flavor. But in the case of potato chips or pretzels, the salt is so predominant it can be considered an added flavoring. By contrast, nutmeg is normally used for its distinctive flavor, but just a dash can perk up the flavor of a cream sauce without being detectable by most people.

SEASONING

1. The most important time for seasoning liquid foods is at the end of the cooking process.

   The last step in most recipes, whether written or not, is "adjust the seasoning." This means you have to first taste and evaluate the product. Then you must decide what should be done, if anything, to improve the taste. Often, a little salt in a stew or a dash of fresh lemon juice in a sauce is enough.

   The ability to evaluate and correct flavors takes experience, and it is one of the most important skills a cook can develop.

2. Salt and other seasonings are also added at the beginning of cooking, particularly for larger pieces of food, when seasonings added at the end would not be absorbed or blended in but just sit on the surface.

3. Adding some of the seasoning during the cooking process aids in evaluating the flavor along the way.

4. Do not add much seasoning if it will be concentrated during cooking, as when a liquid is reduced.

FLAVORING

Flavoring ingredients can be added at the beginning, middle, or end, depending on the cooking time, the cooking process, and the flavoring ingredient.

1. Only a few flavorings can be added successfully at the end of cooking. These include fresh (not dried) herbs, sherry or flamed brandy, and condiments like prepared mustard and Worcestershire sauce.

2. Most flavorings need heat to release their flavors and time for the flavors to blend. Whole spices take longest. Ground spices release flavors more quickly and thus don't require as long a cooking time.

3. Too much cooking results in loss of flavor. Most flavors, whether in spices or in main ingredients, are **volatile**, which means they evaporate when heated. That is why you can smell food cooking.

We can conclude that herbs and spices should cook with the foods long enough to release their flavors but not so long that their flavors are lost. If cooking times are short, you can generally add spices and herbs at the beginning or middle of cooking time. If cooking times are long, it is usually better to add them in the middle or toward the end of cooking time.

Note: Food safety experts recommend adding dried spices and herbs at least 30 minutes before the end of cooking so any microorganisms they might carry are destroyed.

COMMON SEASONING AND FLAVORING INGREDIENTS

Any food product can be used as a flavoring ingredient, even meat (as when crumbled bacon is added to sautéed potatoes or diced ham is included in a mirepoix). Sauces, which are complex preparations containing many flavoring ingredients, are themselves used as flavorings for meat, fish, vegetables, and desserts. The term **condiment** means any edible substance that can be added to a food to contribute flavoring. Clearly, the list of condiments is nearly limitless.

We obviously cannot treat all possible flavoring ingredients here, but we discuss some of the most important. A survey of herbs and spices is provided in Table 6.1. Ingredients used primarily in the bakeshop are discussed in Chapter 30.

TABLE 6.1 Herbs, Spices, and Spice Blends

Product	Market Forms	Description	Examples of Use
Allspice	Whole, ground	Small brown berry; flavor resembles blend of cinnamon, cloves, and nutmeg	Sausages and braised meats, poached fish, stewed fruits, pies, puddings
Anise seed	Whole, ground	Small seed; licorice flavor	Cookies, pastries, breads
Annatto (Spanish name: achiote)	Whole seed	Small, deep red, very hard seeds; very mild, slightly peppery flavor; imparts strong yellow or yellow-orange color to foods	Used primarily for coloring some Latin American dishes, including rice, soups, stews, and sauces
Asafetida	Powdered or lump	Dried sap of an Asian herb; strong aroma.	Dropped into hot oil so that it toasts and foams up slightly; the flavored oil then used in Indian and Middle Eastern dishes
Basil	Fresh and dried leaves	Aromatic leaf; member of mint family	Tomatoes and tomato dishes, pesto (Italian basil sauce), egg dishes, lamb chops, eggplant, peas, squash
Bay leaf	Whole	Stiff, dark green, oblong leaves; pungent aroma	One of the most important herbs for stocks, sauces, stews, braised meats
Caraway seed	Whole	Dark brown, curved seeds; familiar rye bread seasoning	Rye bread, cabbage, sauerkraut, pork, cheese spreads, Eastern European dishes
Cardamom	Whole pod, ground seed	Tiny brown seeds inside white or green pod; sweet and aromatic; expensive	Pickling, Danish pastries, curries
Cayenne (red pepper)	Ground	Ground form of hot red chile; looks like paprika but is extremely hot	In small amounts in many sauces, soups, meat, fish, egg, and cheese dishes (see p. 129)
Celery seed	Whole, ground, ground mixed with salt	Tiny brown seeds with strong celery flavor	Salads, coleslaw, salad dressings, tomato products
Chervil	Fresh and dried leaves	Herb with mild flavor of parsley and tarragon	Soups, salads, sauces, egg and cheese dishes
Chili powder	Ground blend	Blend of spices including cumin, chiles, oregano, garlic	Chili and other Mexican dishes, egg dishes, appetizers, ground meat
Chive	Fresh, dried, frozen	Grasslike herb with onion flavor	Salads, egg and cheese dishes, fish, soups
Cilantro (fresh coriander, Chinese parsley)	Fresh leaves	The plant that produces coriander seeds; delicate texture; assertive, herbaceous aroma and flavor; leaves resemble flat parsley	Widely used in Asian and Southwestern cooking and in dishes with various ethnic influences
Cinnamon	Sticks, ground	Aromatic bark of cinnamon or cassia tree	Pastries, breads, desserts, cooked fruits, ham, sweet potatoes, hot beverages
Clove	Whole, ground	Dried flower buds of a tropical tree; pungent, sweet flavor	Whole: marinades, stocks, sauces, braised meats, ham, pickling; Ground: cakes, pastries, fruits
Coriander	Whole, ground	Round, light brown, hollow seed, slightly sweet, musty flavor	Pickling, sausage, pork, curried dishes, gingerbread
Cumin seed	Whole, ground	Small seed resembling caraway, but lighter in color	Ingredient of curry and chili powders, sausages and meats, egg and cheese dishes
Curry powder	Ground blend	A mixture of 16–20 spices, including chile, turmeric, cumin, coriander, ginger, cloves, cinnamon, black pepper; brands vary greatly in flavor and hotness	Curried dishes, eggs, vegetables, fish, soups, rice
Dill	Fresh and dried leaves (called dill weed), whole seed	Herb and seed with familiar dill pickle flavor; seed is more pungent than the herb	Seed: pickling, sauerkraut, soups; herb: salads, cheese dishes, fish and shellfish, some vegetables
Epazote	Fresh and dried leaves	A pungent herb with coarse-textured leaves	Used in Mexican cooking; often cooked with beans

Product	Market Forms	Description	Examples of Use
Fennel	Whole seed	Greenish-brown seeds similar in flavor to anise, but larger in size	Italian sausage, tomato sauce, fish
Fenugreek	Whole and ground seed	Small, hard, light-brown seed; pungent, bittersweet aroma	Curries and other Indian dishes
Filé powder	Ground	From the dried leaf of the sassafras tree; acts as thickener as well as flavoring.	Gumbos and other creole and Cajun dishes. Texture becomes stringy if boiled, so add at end of cooking.
Five-spice powder, Chinese	Ground	Blend of star anise, cinnamon, Sichuan pepper, cloves, fennel seed	Chinese dishes
Garam masala	Blend of ground or whole toasted spices (usually ground before using)	Variable blend of spices similar to curry powder but without turmeric and often without chiles. May contain pepper, cinnamon, cloves, nutmeg, cumin, cardamom, coriander	Indian dishes: vegetables, meat, poultry, fish, sauces, soups
Garlic	Fresh: whole bulbs; dried: granulated, powder, and mixed with salt	Strong, aromatic member of onion family; fresh bulbs composed of many small cloves	Wide variety of foods
Ginger	Whole, ground (also fresh and candied or crystallized)	Light brown, knobby root of ginger plant	Baked goods and desserts, fruits, curried dishes, braised meats; fresh in Chinese and other Asian dishes
Juniper berry	Whole	Slightly soft, purple berries with piney flavor; principal flavoring of gin	Marinades, game dishes, sauerkraut
Lemongrass	Fresh stalks	A tropical grass with a slightly bulbous base and an aroma of lemon	Used in Southeast Asian dishes and in dishes influenced by Asian cuisine
Mace	Whole (blade), ground	Orange outer covering of nutmeg; similar flavor, but milder	Baked goods, desserts, fruits, sausages, pork, fish, spinach, squash, other vegetables
Marjoram	Fresh and dried leaves	Gray-green herb with pleasant aroma and slightly minty flavor, similar to oregano, but much milder	Pâtés and ground meats, braised meats, sauces, roast lamb, poultry and poultry stuffings
Mint	Fresh leaves	Aromatic herb with familiar cool flavor; two varieties: spearmint and peppermint	Lamb, fruits, tea and fruit beverages, peas, carrots, potatoes
Mustard seed	Whole, ground (also prepared mustard; see p. 129)	Very pungent seed in white or yellow and brown varieties—brown is stronger	Cheese and egg dishes, pickling, meats, sauces and gravies
Nutmeg	Whole, ground	Sweet, aromatic kernel of nutmeg fruit	Soups, cream sauces, chicken, veal, many vegetables (spinach, mushrooms, squash, potatoes), desserts, custards, breads, pastries
Oregano	Fresh and dried leaves, whole and ground	Pungent herb known as the "pizza herb"	Italian and Mexican dishes, tomato products
Paprika	Ground	Ground form of a dried, sweet red chile. Spanish variety (pimento) is often smoked; Hungarian is darker and more pungent	Spanish: used as flavoring in stewed dishes and as garnish on light-colored foods; Hungarian: goulash, braised meats and poultry, sauces
Parsley	Fresh: whole sprigs, in bunches; dried: in flakes	Most widely used herb; dark green curly or flat leaves with delicate, sweet flavor	Almost all foods
Pepper, black and white	Whole (peppercorns); ground fine, medium, or coarse	Small black or creamy white hard berry; pungent flavor and aroma	Most widely used spice (see p. 129)
Pepper, red	(see Cayenne)		

(Continued)

TABLE 6.1 Herbs, Spices, and Spice Blends *(Continued)*

Product	Market Forms	Description	Examples of Use
Peppercorn, pink	Whole	Bright pink dried seed or berry; pungent, floral taste; unrelated to black pepper	Limited uses in meat, poultry, and fish dishes; sauce garnish; used in peppercorn mixtures
Poppy seed	Whole	Tiny blue-black seeds with faint but distinctive flavor	Garnish for breads and rolls, buttered noodles; ground: in pastry fillings
Quatre épices	Ground	French name means "four spices." Blend may include white or black pepper, nutmeg, cloves, cinnamon, ginger, allspice	Soups, stews, vegetables, sausages and other charcuterie (see p. 843)
Rosemary	Fresh and dried leaves	Light green leaves resembling pine needles	Lamb, braised meats and poultry, soups, tomato and meat sauces
Saffron	Whole (thread)	Red stigma of saffron crocus; gives bright yellow color to foods; mild, distinctive flavor; very expensive	Steeped in hot liquid before use; rice dishes, poultry, seafood, bouillabaisse, baked goods
Sage	Fresh leaves; dried whole and rubbed (finer consistency than whole leaves), ground	Pungent gray-green herb with fuzzy leaves	Pork, poultry, stuffings, sausage, beans, tomatoes
Savory	Fresh and dried leaves	Fragrant herb of mint family; summer savory is preferred to winter	Many meat, poultry, fish, egg, and vegetable dishes
Sesame seed	Whole (hulled or unhulled)	Small yellowish seed with nutlike taste; familiar hamburger bun garnish; high oil content	Bread and roll garnish
Sichuan peppercorn	Whole	Brown seed pod, usually partially opened; spicy, peppery flavor, but unrelated to black peppercorns	Spicy meat and poultry dishes
Star anise	Whole or broken	Dried, star-shaped seed pod with an aniselike flavor (but unrelated to anise) but more aromatic	Braised Chinese dishes
Tarragon	Fresh and dried leaves	Delicate green herb with flavor both minty and licoricelike	Béarnaise sauce, tarragon vinegar, chicken, fish, salads and dressings, eggs
Thyme	Fresh leaves; dried leaves: whole and ground	Tiny brownish-green leaves; very aromatic	One of the most important and versatile of herbs; stocks, soups, sauces, meats, poultry, tomatoes
Turmeric	Ground	Intense yellow root of ginger family; mild but distinctive peppery flavor	A basic ingredient of curry powder; pickles, relishes, salads, eggs, rice
Wasabi	Fresh root, dried powder; paste (in tubes)	A type of Japanese horseradish; pungent horseradish flavor (dried wasabi and paste are often regular horseradish tinted green)	Basic seasoning or condiment with sushi and sashimi; Asian-style dishes. Mix powder with water to form thick paste.
Zaatar	Powdered spice blend	Blend of herbs and spices including oregano, thyme, marjoram, sesame seeds, dried sumac berries, and salt	Arabian dishes

1. *Salt* is the most important seasoning ingredient. Don't use too much. You can always add more, but you can't take it out.

   • Table salt has a fine granulation. It may contain iodine as a dietary additive. Table salt also may contain other additives to prevent caking.

   • Kosher salt is prized in the kitchen because of its purity. Unlike table salt, it contains no additives. Because of its coarse or flaky granulation, it does not dissolve as quickly as table salt, but it is easier to use when added to foods by hand, so many chefs prefer it to table salt at their cooking stations.

- Sea salts of many origins and types are available. Many of them have colors ranging from gray to green to red, from various minerals and other impurities. These impurities also add subtle flavors to the salt. In addition, their coarse granulation gives them a pleasant mouthfeel. More expensive than other salts, sea salts are used primarily as garnishes for plated foods.

2. *Pepper* comes in three forms: white, black, and green. All three are actually the same berry, but processed differently. Black pepper is picked unripe; white is ripened and the hull is removed; green peppercorns are picked unripe and preserved before their color darkens.

- Whole and crushed **black pepper** are used primarily in seasoning and flavoring stocks and sauces and, sometimes, red meats. Ground black pepper is used in the dining room by the customer.

- Ground **white pepper** is more important as a seasoning in the food-service kitchen. Its flavor is slightly different from that of black pepper, and it blends well (in small quantities) with many foods. Its white color makes it visually undetectable in light-colored foods.

- **Green peppercorns** are fairly expensive and are used in special recipes. The types packed in water, brine, or vinegar (those in water and in brine have better flavor) are soft. Wet-pack peppercorns are perishable. Water-packed peppercorns keep only a few days in the refrigerator after they are opened, while the others keep longer. Dried green peppercorns are also available.

3. *Red pepper* or *cayenne* is completely unrelated to black and white pepper. It belongs to the same family as chiles, paprika, and fresh sweet bell peppers. Used in tiny amounts, it gives a spicy hotness to sauces and soups without actually altering the flavor. In larger amounts, it gives both heat and flavor to many spicy foods, such as those of Mexico and India.

4. *Lemon juice* is an important seasoning, particularly for enlivening the flavor of sauces and soups.

5. *Fresh herbs* are almost always superior to dried herbs. They should be used whenever cost and availability permit. Not long ago, the only fresh herbs generally available in many areas of North America were parsley, chives, and sometimes mint and dill. Now, however, most herbs are available fresh. The accompanying photos illustrate the most commonly used fresh herbs as well as some unusual fresh flavoring ingredients.

6. *Onion, garlic, shallots*, and other members of the onion family, as well as carrots and celery, are used as flavorings in virtually all stations of the kitchen and even in the bakeshop. Try to avoid the use of dried onion and garlic products, except as a component of spice blends. They have less flavor, and the fresh product is always available.

7. Grated *lemon* and *orange rind* is used in sauces, meats, and poultry (as in duckling à l'orange) as well as in the bakeshop. Only the colored outer portion, called the *zest*, which contains the flavorful oils, is used. The white pith is bitter.

8. *MSG*, or *monosodium glutamate*, is a flavor enhancer widely used in Asian cooking. MSG doesn't actually change the flavor of foods, but it acts on the taste buds. It has a reputation for causing chest pains and headaches in some individuals.

9. *Wine, brandy*, and other alcoholic beverages are used to flavor sauces, soups, and many entrées. Brandy should be boiled or flamed to eliminate the high percentage of alcohol, which would be unpleasant in the finished dish. Table wines usually need some cooking or reduction (either separately or with other ingredients) to produce the desired flavors. Fortified wines like sherry and Madeira, on the other hand, may be added as flavorings at the end of cooking.

10. *Prepared mustard* is a blend of ground mustard seed, vinegar, and other spices. It is used to flavor meats, sauces, and salad dressings and as a table condiment. For most cooking purposes, European styles such as Dijon (French) or Dusseldorf (German) work best, while the bright yellow American ballpark style is more appropriate as a table condiment

Basil

Chervil

Chives

Garlic chives

Cilantro

Dill

Epazote

Regular ginger and green ginger

Lemongrass

than as a cooking ingredient. A coarse, grainy style is sometimes called for in specialty recipes.

11. *Vinegars* consist of acetic acid, water, and various flavoring compounds, depending on the source. They are made by bacterial fermentation of wine, hard apple cider, grain alcohol, and other liquids containing alcohol. Vinegars are discussed in more detail in Chapter 21.

12. *Soy sauce* is a thin, dark brown, earthy, salty liquid made by fermenting soy beans and sometimes other grains, especially wheat, with water and various mold cultures. Widely used in Asian cuisines, it is available in various strengths from light to dark. A low-salt version is also available, but it is still salty. Japanese soy sauces are generally lighter in flavor than Chinese soy sauces. *Tamari* is a special kind of dark, rich Japanese soy sauce made with little or no wheat. Other Asian soy sauces include Indonesian, Korean, Burmese, Vietnamese, and Malaysian.

13. *Fish sauce* (also called *nam pla, nuoc mam,* and *nuoc nam*) is a thin, salty liquid with a pungent aroma, made from raw or dried fish or shellfish. It is widely used in Southeast Asian cuisines.

Marjoram

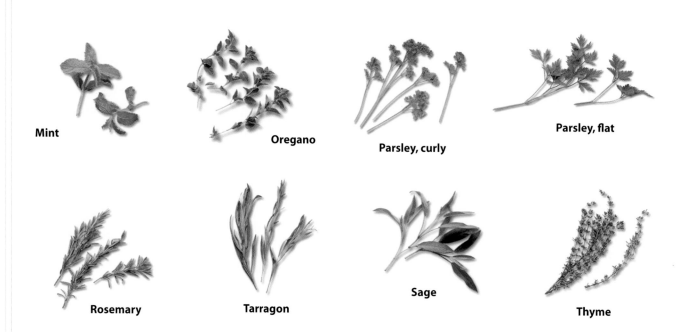

Mint **Oregano** **Parsley, curly** **Parsley, flat**

Rosemary **Tarragon** **Sage** **Thyme**

USING HERBS AND SPICES

DEFINITIONS

Herbs are the leaves of certain plants that usually grow in temperate climates.

Spices are the buds, fruits, flowers, bark, seeds, and roots of plants and trees, many of which grow in tropical climates.

The distinction is often confusing, but it is not as important to know which flavorings are spices and which are herbs as it is to use them skillfully.

Table 6.1 is not a substitute for familiarity with the actual products. Eventually, you should be able to identify any spice on your shelf by aroma, taste, and appearance without looking at the label. The accompanying photos illustrate a number of whole spices.

Top row, left to right: black peppercorns, green peppercorns, pink peppercorns. Bottom row, left to right: white peppercorns, Sichuan peppercorns

Top row, left to right: cloves, nutmeg, allspice, cinnamon sticks. Bottom row, left to right: juniper berries, cardamom, saffron, star anise

Top row, left to right: celery seed, dill seed, coriander seed, caraway seed. Bottom row, left to right: fennel seed, cumin seed, anise seed

EVALUATING HERBS AND SPICES

Fresh herbs should look fresh, not wilted, with good color and no yellowing, browning, or rotting, and no insect damage. Look for a fresh aroma characteristic of the particular herb.

Dried herbs and ground spices lose much flavor and aroma after six months. Buy properly sealed containers from reputable purveyors. Freshly opened spices and herbs should have a good aroma. Lack of aroma indicates staleness and lack of flavor. For freshest flavor, buy whole spices and grind them as needed.

GUIDELINES for Using Herbs and Spices

1. Be familiar with each spice's aroma, flavor, and effect on food. Looking at a spice chart, including the one in this book, is no substitute for familiarity with the actual product.

2. Store dried herbs and spices in a cool place, tightly covered, in opaque containers. Heat, light, and moisture cause herbs and spices to deteriorate rapidly.

3. Don't use stale spices and herbs, and don't buy more than you can use in about 6 months. Whole spices keep longer than ground, but both lose much flavor after 6 months. For maximum flavor, grind spices just before use.

4. Be cautious after you have replaced old spices. The fresher products are more potent, so the amount you used before might now be too much.

5. Use good-quality spices and herbs. It doesn't pay to economize here. The difference in cost is only a fraction of a cent per portion.

6. Whole spices take longer to release flavors than ground spices, so allow for adequate cooking time.

7. Whole herbs and spices for flavoring a liquid are tied loosely in a piece of cheesecloth (called a sachet) for easy removal.

8. When in doubt, add less than you think you need. You can always add more, but it's hard to remove what you've already added.

9. Except in dishes like curry or chili, spices should not dominate. Often, they should not even be evident. If you can taste the nutmeg in the creamed spinach, there's probably too much nutmeg.

10. Herbs and spices added to uncooked foods such as salads and dressings need several hours for flavors to be released and blended.

11. Taste foods before serving whenever possible. How else can you adjust the seasoning?

KEY POINTS TO REVIEW

- How do chefs use the idea of flavor balance to combine a variety of ingredients into a single dish?

- What is the difference between seasoning and flavoring?

- What guidelines are used for correctly adding herbs and spices to foods?

TERMS FOR REVIEW

cooking	microwave	steam
caramelization	dry-heat methods	braise
gelatinization	moist-heat methods	stewing
fiber	roast	sous vide
denature	bake	molecular gastronomy
coagulation	barbecue	flavor profile
Maillard reaction	broil	umami
connective tissues	grilling	primary flavor
smoke point	griddling	supporting flavor
evaporation	pan-broiling	seasoning
emulsion	sauté	flavoring
temporary emulsion	pan-fry	volatile
conduction	deep-fry	condiment
convection	boil	herbs
radiation	simmer	spices
infrared	poach	

QUESTIONS FOR DISCUSSION

1. Your broiler cook has just broiled a codfish fillet that turned out dry, rubbery, and shrunken. Explain what happened to it.

2. Why might adding some tomato product to a beef stew help make the meat more tender?

3. You are roasting a large quantity of ducklings and must use both your conventional ovens and your convection oven. You set all the ovens at the same temperature, but find the ducklings in the convection oven are done first. Why did this happen?

4. You are roasting two beef tenderloins of the same size, one in an oven set at 450°F (230°C), and the other in an oven at 250°F (120°C). You remove both of them from the oven when the temperature at the center is 135°F (57°C). Describe the doneness of each tenderloin from outside to inside.

5. Arrange the following cooking methods in three groups, depending on whether they are moist-heat methods, dry-heat methods without fat, or dry-heat methods with fat: braising, roasting, deep-frying, sautéing, poaching, steaming, broiling, pressure frying, grilling, simmering.

6. A cook in your restaurant is roasting several pans of chickens. He thinks they are browning too fast, and he covers the pans with foil to keep the chickens from browning much more. What is wrong with this?

7. What food safety problems are posed by the vacuum packaging and the low cooking temperatures of sous vide cooking?

8. Describe the difference between *primary flavor* and *supporting flavor*. Select a favorite recipe and explain the function of each ingredient, indicating which are primary flavors and which are secondary flavors.

9. What is meant by the phrase "adjust the seasoning"?

10. What is wrong with adding whole caraway seed to a portion of goulash just before serving?

MISE EN PLACE

To be successful in the food-service industry, cooks need more than the ability to prepare delicious, attractive, and nutritious foods. They also must have a talent for organization and efficiency. In every kitchen, a great many tasks must be completed over a given time and by a limited number of workers. No matter when these tasks are done, they all must come together at one crucial point: service time. Only if advance preparation is done thoroughly and systematically can service go smoothly.

Good chefs take pride in the thoroughness and quality of their advance preparation, or **mise en place** (meez-on-plahss). This French term, meaning "everything put in place," is an important and often-heard expression in North American kitchens because food-service professionals understand its importance to the success of the establishment.

This chapter deals with the basic concepts of mise en place as well as specific operations that are normally part of the mise en place.

AFTER READING THIS CHAPTER, YOU SHOULD BE ABLE TO

1. Define *mise en place*, and explain why care must be taken in its planning.

2. Describe five general steps used in planning mise en place.

3. Explain the difference in preparation requirements for set meal service and extended meal service.

4. List five guidelines to observe when sharpening a chef's knife.

5. Demonstrate major cutting techniques required in food preparation.

6. Describe basic precooking and marinating procedures.

7. Set up and use a standard breading station.

8. Define *convenience foods* in the context of mise en place, and list eight guidelines for their use.

PLANNING AND ORGANIZING PRODUCTION

Even on the simplest level, pre-preparation is necessary. If you prepare only one short recipe, you must first:

- Assemble your tools.
- Assemble your ingredients.
- Wash, trim, cut, prepare, and measure your raw materials.
- Prepare your equipment (preheat oven, line baking sheets, etc.).

Only then can you begin the actual preparation.

When many items are to be prepared in a commercial kitchen, the situation is much more complex. Dealing with this complexity is the basis of kitchen organization.

THE PROBLEM

Every food-service operation faces a basic conflict between two unavoidable facts:

1. There is far too much work to do in a kitchen to leave until the last minute, so some work must be done ahead.

2. Most foods are at their best quality immediately after preparation, and they deteriorate as they are held.

THE SOLUTION

To address this conflict, the chef must plan preparation carefully. Planning generally follows these steps:

1. **Break down each menu item into its stages of production.**
   Turn to any recipe in this book. Note the procedures are divided into a sequence of steps that must be done in a certain order to make a finished product.

2. **Determine which stages may be done in advance.**
   - The first step of every recipe, written or not, is always part of advance preparation: *assembling and preparing the ingredients*. This includes cleaning and cutting produce, cutting and trimming meats, and preparing breadings and batters for frying.
   - Succeeding steps of a recipe may be done in advance *if the foods can then be held without loss of quality*.
   - Final cooking should be done as close as possible to service for maximum freshness.

   Frequently, separate parts of a recipe, such as a sauce or a stuffing, are prepared in advance, and the dish is assembled at the last minute.

   In general, items cooked by dry-heat methods, such as broiled steaks, sautéed fish, and French-fried potatoes, do not hold well. Large roasts are an important exception to this rule. Items cooked by moist heat, such as braised beef, soups, and stews, are usually better suited to reheating or holding in a steam table. Delicate items should always be freshly cooked.

3. **Determine the best way to hold each item at its final stage of pre-preparation. Holding temperature is the temperature at which a product is kept for service or for storage. Holding temperatures for all potentially hazardous foods must be outside the Food Danger Zone.**
   - Sauces and soups are frequently kept hot, above 135°F (57°C), for service in steam tables or other holding equipment. Foods such as vegetables, however, should be kept hot only for short periods because they quickly become overcooked.
   - Refrigerator temperatures, below 41°F (5°C), are best for preserving the quality of most foods, especially perishable meats, fish, and vegetables, before final cooking or reheating.

4. **Determine how long it takes to prepare each stage of each recipe. Plan a production schedule beginning with the preparations that take the longest.**

   Many operations can be carried on at once because they don't all require your complete attention the full time. It may take 6 to 8 hours to make a stock, but you don't have to stand and watch it all that time.

5. **Examine recipes to see if they might be revised for better efficiency and quality as served.**

   For example:

   - Instead of preparing a full batch of green peas and holding them for service in the steam table, you might blanch and chill them, then heat portions to order in a sauté pan, steamer, or microwave oven.

   - Instead of holding a large batch of veal scaloppine in mushroom sauce in the steam table, you might prepare and hold the sauce, sauté the veal to order, combine the meat with a portion of the sauce, and serve fresh from the pan.

        Caution: Unless you are in charge of the kitchen, do not change a recipe without authorization from your supervisor.

THE GOAL

The goal of pre-preparation is to do as much work in advance as possible *without loss of quality*. Then, at service time, all energy can be used for finishing each item immediately before serving, with the utmost attention to quality and freshness.

Many preparation techniques in common use are designed for the convenience of the cooks at the expense of quality. Remember, quality should always take highest priority.

ADAPTING PREPARATION TO STYLE OF SERVICE

The way you plan production and do your mise en place depends in large part on the style of meal service. The following discussion of **set meal service** and **extended meal service** illustrates the basic differences.

Set Meal Service

- All customers eat at one time.
- Often called *quantity cooking* because large batches are prepared in advance.
- Examples: school cafeterias, banquets, employee dining rooms.

The traditional method of set meal preparation, still widely used, is to prepare the entire quantity of each item in a single large batch and to keep it hot for the duration of the meal service. This method has two major disadvantages:

- Deterioration of quality due to long holding.
- Large quantities of leftovers.

Modern high-speed equipment, such as pressure steamers, convection ovens, infrared ovens, and microwave ovens, make possible a system called *small-batch cooking*. Required quantities are divided into smaller batches, placed in pans ready for final cooking or heating, and then cooked only as needed. The advantages of this system are

- Fresher food, because it is not held as long.
- Fewer leftovers, because servings not needed are not cooked.

Small-batch cooking also accommodates items prepared in advance and frozen or chilled for storage.

Extended Meal Service

- Customers eat at different times.
- Often called *à la carte* cooking because customers usually select items from a written menu (*carte* in French).
- Examples: restaurants, short-order counters.

Individual items are cooked to order rather than cooked ahead, but pre-preparation is extensive, down to the final cooking stage.

The short-order cook, for example, must have everything ready to go: cold meats, tomatoes, and other sandwich ingredients sliced and arranged, spreads prepared and ready, hamburger patties shaped, garnishes prepared, and so on. If the cook has to stop during service to do any of these things, orders will back up and service will fall behind.

A steak that takes 10 minutes to broil may be cut and trimmed in advance, but broiling should be started 10 minutes before it is to be served.

Obviously, if the last step in a recipe is to braise the item for $1^1/_2$ hours, one cannot wait until an order comes in before beginning to braise. An experienced cook can estimate closely how many orders will be needed during the meal period and prepare a batch that, ideally, will finish braising just when service begins.

Note the differences in these two methods for Chicken Chasseur. In both cases, the final product is chicken in a brown sauce with mushrooms, shallots, white wine, and tomatoes.

1. Quantity method—Chicken Chasseur:

   Brown chicken in fat; remove.

   Sauté shallots and mushrooms in same fat.

   Add flour to make a roux.

   Add white wine, tomatoes, brown stock, seasonings; simmer until thickened.

   Add chicken; braise until done.

2. À la carte method—Chicken Chasseur:

   Prepare Sauce Chasseur in advance; refrigerate or keep hot in bain-marie.

   For each order:

   • Brown chicken in sauté pan; finish cooking in oven.

   • Deglaze pan with white wine; reduce.

   • Add one portion of sauce; add chicken and simmer briefly; serve.

MISE EN PLACE: THE REQUIRED TASKS

Up to this point, we have discussed planning the production schedule. Our planning helps us determine the tasks we must do before beginning the final cooking during the meal service period. Chefs refer to performing these preliminary tasks as "doing the mise en place." In many restaurants, especially large ones, the mise en place is extensive. It includes the preparation of stocks, sauces, breadings, and batters as well as the cutting and trimming of all the meat, poultry, fish, and vegetables the chef expects will be needed during the meal service.

A large part of a cook's workday is spent doing mise en place. This means that a large part of learning how to cook is learning how to do mise en place. In fact, a large part of this book is devoted to these tasks of preparation. There are many more such tasks than can be included in a single chapter.

The remainder of this chapter discusses the most basic and general skills required for a mise en place. The most basic of these are knife skills. Fundamentals such as how to hold the chef's knife, how to maintain a sharp edge, and how to make basic cuts are illustrated. More specific techniques required for individual food products are explained in appropriate chapters later in the book. For example, mise en place techniques can be found in the following chapters:

Stock and sauce mise en place: Chapter 8

Vegetable trimming techniques: Chapter 10

Meat fabrication: Chapter 15

Poultry fabrication: Chapter 17

Fish and seafood fabrication: Chapter 19

KEY POINTS TO REVIEW

• What does mise en place mean? What kinds of tasks are included in this term?

• What are the five steps in planning mise en place?

• How does mise en place for à la carte service differ from mise en place for set meal service?

USING THE KNIFE

Many laborsaving tools are available for cutting, chopping, and slicing fresh foods. Chapter 3 lists the basic kinds.

The chef's knife or French knife, however, is still the cook's most important and versatile cutting tool. The knife is more precise than a machine. Unless you are cutting a large quantity, the knife can even be faster. Cleaning a large machine takes time.

To get the best use out of your knife, *you must learn to keep it sharp and to handle it properly.*

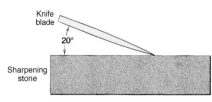

FIGURE 7.1 When sharpening a knife, hold the blade at a 20-degree angle to the stone.

KEEPING A SHARP EDGE

The Sharpening Stone

A stone is the traditional tool for sharpening a chef's knife. The best electric sharpeners do an excellent job of sharpening chef's knives, but many models wear away too much of your expensive knife without making a good edge. Modern professional knives are much harder than the old carbon steel knives, so they are more difficult to sharpen on a stone. Nevertheless, using a stone correctly is a valuable skill.

Follow these guidelines:

1. Hold the blade at a constant 20-degree angle to the stone, as shown in Figure 7.1.
2. Make light, even strokes, the same number on each side of the blade.
3. Sharpen in one direction only to get a regular, uniform edge.
4. Do not oversharpen.
5. Finish with a few strokes on the steel (see next page), and then wipe the blade clean.

Figure 7.2 illustrates one of several sharpening methods. There are other good ones, too, and your instructor may prefer a method not illustrated here.

FIGURE 7.2 Using a sharpening stone.

(a) Hold the knife firmly. Start with the tip of the knife against the stone as shown, and hold the edge against the stone at a 20-degree angle. Use the guiding hand to keep an even pressure on the blade.

FIGURE 7.3 To use a manual sharpener, draw the blade through the sharpener from the heel to the tip of the knife. Do not press down hard, but make several light strokes.

Other Knife Sharpeners

As mentioned previously, today's professional knives are made of an especially hard material that is more difficult to sharpen by traditional methods. As a result, chefs often use other sharpeners to simplify the task. Such sharpeners typically have two stones set at the correct angle, so it is necessary only to draw the knife between them. Manual and power models of these sharpeners are available. Figure 7.3 shows a manual sharpener in use.

The Steel

This tool is used not to sharpen the edge but to *true the edge* (to perfect it, or to smooth out irregularities) and to *maintain the edge* (to keep it sharp as it is used).

Observe these guidelines for using the steel:

1. Hold the blade at a constant 20-degree angle to the steel, just as when using the stone (Figure 7.2). A smaller angle will be ineffective. A larger one will dull the edge.
2. Make light strokes. Do not grind the knife against the steel.
3. Make even, regular strokes. Alternate each stroke, first on one side of the blade, then on the other.
4. Use no more than five or six strokes on each side of the blade. Too much steeling can actually dull the blade.
5. Use the steel often. Then you will rarely have to sharpen the knife on the stone.

(b) Start to draw the knife over the stone. Press very gently on the blade.

(c) Keep the motion smooth, using even, light pressure.

(d) Draw the knife across the stone all the way to the heel of the blade.

FIGURE 7.4 Using a steel.

(a) Hold the steel and the knife away from your body. With the knife in a vertical position and at a 20-degree angle to the steel, touch the steel with the heel of the blade.

(b) Pass the knife lightly along the steel, bringing the blade down in a smooth arc.

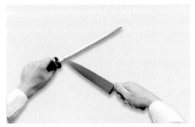

(c) Complete the movement. Do not strike the guard of the steel with the tip of the blade.

(d, e, f) Repeat the motion on the other side of the steel.

Figure 7.4 illustrates one of several steeling methods. This one is popular, but several others are equally correct. Carefully observe your instructors' demonstrations of their preferred methods.

HANDLING THE KNIFE

The Grip

A proper grip gives you maximum control over the knife, increases your cutting accuracy and speed, prevents slipping, and lessens the chance of an accident. The type of grip you use depends, in part, on the job you are doing and the size of the knife.

The grip illustrated in Figure 7.5 is one of the most frequently used for general cutting and slicing. Many chefs feel that grasping the blade with the thumb and forefinger in this manner gives them greatest control.

Holding the knife may feel awkward at first, but practice will make it seem natural. Watch your instructors demonstrate the grips they use, and then practice under their supervision.

The Guiding Hand

While one hand controls the knife, the other hand controls the product being cut. Proper positioning of the hand achieves three goals:

1. **Hold the item being cut.**

   In Figure 7.6, the item is held firmly so it will not slip.

2. **Guide the knife.**

   Note the knife blade slides against the fingers. The position of the hand controls the cut.

3. **Protect the hand from cuts.**

   Fingertips are curled under, out of the way of the blade.

BASIC CUTS AND SHAPES

Cutting food products into uniform shapes and sizes is important for two reasons:

1. It ensures even cooking.
2. It enhances the appearance of the product.

Figure 7.7 shows common shapes, with their names and dimensions. The following terms describe other cutting techniques:

Chop: to cut into irregularly shaped pieces.

Concasser (con-cass-say): to chop coarsely.

Mince: to chop into very fine pieces.

Emincer (em-man-say): to cut into very thin slices (does not mean "to mince").

Shred: to cut into thin strips, either with the coarse blade of a grater (manual or power) or with a chef's knife.

FIGURE 7.5 Grasping the blade of the knife between the thumb and forefinger gives the worker good control over the blade.

FIGURE 7.6 The position of the guiding hand, which holds the item being cut or sliced and also guides the blade, from two points of view.

CANADIAN CUT SIZES

In Canadian kitchens the following differ slightly cuts from those in the United States, and there are no cuts called Fine Brunoise or Fine Julienne:

Julienne: 1–2 mm × 1–2 mm × 2.5–4 cm

Alumette: 3 mm × 3 mm × 5–6 cm

Brunoise: 1–2 mm × 1–2 mm × 1–2 mm

Small dice: 3 mm × 3 mm × 3 mm

Medium dice: 6 mm × 6 mm × 6 mm

Large dice: 12 mm × 12 mm × 12 mm

FIGURE 7.7 Basic cuts and shapes.

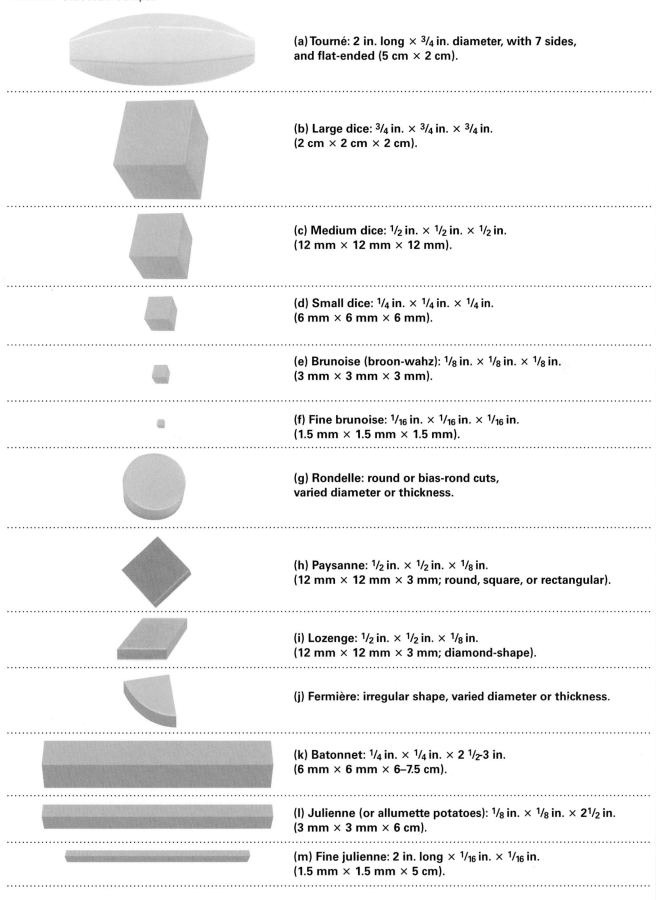

(a) Tourné: 2 in. long × ³/₄ in. diameter, with 7 sides, and flat-ended (5 cm × 2 cm).

(b) Large dice: ³/₄ in. × ³/₄ in. × ³/₄ in. (2 cm × 2 cm × 2 cm).

(c) Medium dice: ¹/₂ in. × ¹/₂ in. × ¹/₂ in. (12 mm × 12 mm × 12 mm).

(d) Small dice: ¹/₄ in. × ¹/₄ in. × ¹/₄ in. (6 mm × 6 mm × 6 mm).

(e) Brunoise (broon-wahz): ¹/₈ in. × ¹/₈ in. × ¹/₈ in. (3 mm × 3 mm × 3 mm).

(f) Fine brunoise: ¹/₁₆ in. × ¹/₁₆ in. × ¹/₁₆ in. (1.5 mm × 1.5 mm × 1.5 mm).

(g) Rondelle: round or bias-rond cuts, varied diameter or thickness.

(h) Paysanne: ¹/₂ in. × ¹/₂ in. × ¹/₈ in. (12 mm × 12 mm × 3 mm; round, square, or rectangular).

(i) Lozenge: ¹/₂ in. × ¹/₂ in. × ¹/₈ in. (12 mm × 12 mm × 3 mm; diamond-shape).

(j) Fermière: irregular shape, varied diameter or thickness.

(k) Batonnet: ¹/₄ in. × ¹/₄ in. × 2 ¹/₂-3 in. (6 mm × 6 mm × 6–7.5 cm).

(l) Julienne (or allumette potatoes): ¹/₈ in. × ¹/₈ in. × 2¹/₂ in. (3 mm × 3 mm × 6 cm).

(m) Fine julienne: 2 in. long × ¹/₁₆ in. × ¹/₁₆ in. (1.5 mm × 1.5 mm × 5 cm).

CUTTING TECHNIQUES

Different parts of the blade are appropriate for different purposes, as shown in Figure 7.8.

FIGURE 7.8 Using different parts of the knife blade.

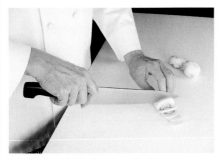

(a) The tip of the knife, where the blade is thinnest and narrowest, is used for delicate work and small items.

(b) The center of the blade is used for most general work.

(c) The heel of the knife is used for heavy or coarse work, especially when greater force is required.

1. **Slicing.**

   Two basic slicing techniques are illustrated in Figures 7.9 and 7.10. When carrots and similar items are cut into round slices as shown, the cut is called **rondelle**.

FIGURE 7.9 Slicing technique 1.

(a) Start the knife at a sharp angle, with the tip of the knife on the cutting board.

(b) Move the knife forward and down to slice through the carrot.

(c) Finish the cut with the knife against the board. For the second slice, raise the heel of the knife and pull it backward, but be sure the tip stays on the board.

Figure 7.10 Slicing technique 2.

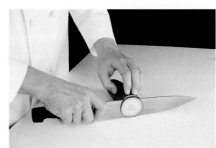

(a) Start the blade at a 45-degree angle, with the tip on the cucumber against the fingers of the guiding hand.

(b, c) Slice downward and forward through the item.

2. **Cutting dice, brunoise, bâtonnet, allumette, and julienne.**

   Figure 7.11 shows the steps in dicing a product, using a potato to illustrate. Note in Figure 7.11c that the process of cutting dice first requires you to cut stick shapes, such as bâtonnet. Thus, this illustration demonstrates the method used to cut not only **dice** and **brunoise** (broon wahz) but also **bâtonnet** (bah toh nay), **allumette** (ah lyoo met), and **julienne** (zhoo lee enn).

FIGURE 7.11 Cutting a potato.

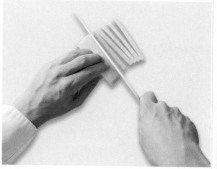

(a) Square off the peeled, eyed potato by cutting a slice from all sides. Use the trimmings for mashed potatoes or soup.

(b) Cut the potato into even slices of the desired thickness. Here we are making $1/4$-inch (6-mm) dice, so the slices are $1/4$-inch (6-mm) thick.

(c) Stack the slices and again slice across the stack in even $1/4$-inch (6-mm) slices. You now have bâtonnet potatoes, slightly smaller than regular French fries. Slices $1/8$ inch (3 mm) thick would give you allumette potatoes.

(d) Looking from this angle shows how the slices have been stacked up.

(e) Pile the bâtonnets together and cut across in slices $1/4$ inch (6 mm) apart. You now have perfect $1/4$-inch (6 mm) dice.

FIGURE 7.12 Cut the vegetable into sticks $1/2$ inch (12 mm) square. To cut the sticks into paysanne, cut them crosswise into thin slices.

3. **Cutting paysanne.**

   Paysanne are thin square, or roughly square, cuts. The procedure begins the same as for cutting medium dice. However, in the last step, cut the $1/2$-inch (12-mm) -thick sticks into thin slices rather than into dice. Figure 7.12 illustrates.

4. **Cutting lozenges.**

   A **lozenge** is a diamond-shape cut, as illustrated in Figure 7.13.

5. **Cutting fermière.**

   Fermière is an irregular slice. Shapes may vary, depending on the item, but the pieces should be of uniform size. Thickness must also be uniform, usually around $1/8$ inch (3 mm). Cut the item lengthwise into pieces of roughly uniform size and shape, and then slice as shown in Figure 7.14.

6. **Making oblique cuts.**

   Also called the *roll cut*, the **oblique cut** is for long, cylindrical vegetables such as carrots. As illustrated in Figure 7.15, hold the knife at an angle, cut, roll the vegetable one quarter-turn, and make the next cut. For tapered vegetables, change the angle as you go to keep the pieces of approximately equal size.

FIGURE 7.13 To cut lozenges, first cut the vegetable into thin slices, and then cut these slices lengthwise into strips about $1/8$ inch (3 mm) wide. Cut the strips at an angle to form diamond shapes.

FIGURE 7.14 To cut fermière, cut the item lengthwise into roughly equal pieces, and then slice uniformly.

FIGURE 7.15 To make oblique cuts, cut the vegetable at a sharp angle, roll one quarter-turn, and make another cut.

7. Dicing an onion.

Dicing an onion presents a special problem for cutting because its form is in layers, not a solid piece. This technique is illustrated in Figure 7.16.

FIGURE 7.16 Dicing an onion.

(a) Cut the peeled onion in half lengthwise, through the root end. Place one half on the cutting board, cut side down.

(b) With the root end away from you, make a series of vertical lengthwise cuts. Do not cut through the root end. The closer together you make the cuts, the smaller the dice will be.

(c) Holding the onion carefully at the top, make a few horizontal cuts toward but not through the root end, which is holding the onion together.

(d) Finally, slice across the onion to separate it into dice. Again, the closer together the cuts, the smaller the dice.

(e) Continue making slices almost to the root end. The root end may be rough cut for mirepoix, to be used for stocks, sauces, and roasts.

FIGURE 7.17 To chop mirepoix, cut onions, celery, and carrots roughly into pieces of approximately equal size. The exact size depends on what the mirepoix is to be used for.

8. Chopping mirepoix.

Mirepoix is a mixture of coarsely chopped vegetables, primarily onions, carrots, and celery, used to flavor stocks, gravies, sauces, and other items, as explained in Chapter 8. Because mirepoix is not served—rather, it is almost always strained out of the product before finishing—neatness of cut is not important. The products are cut roughly into pieces of approximately uniform size—small pieces if cooking time will be short, larger pieces for longer cooking times. Figure 7.17 illustrates mirepoix ingredients being cut.

9. Chopping herbs.

This chopping technique is used to cut a product when no specific shape is needed. Figure 7.18 illustrates chopping parsley.

In the case of chives and scallions, a more regular cut is used, similar to the slicing cut used for larger items like carrots. Figure 7.19 illustrates this procedure.

FIGURE 7.18 Chopping with a French knife. Holding the tip of the knife against the cutting board, rock the knife rapidly up and down. At the same time, gradually move the knife sideways across the product on the board so the cuts pass through all parts of the pile of food. After several cuts, redistribute the pile and begin again. Continue until the product is chopped as fine as you want.

FIGURE 7.19 Stack chives and cut crosswise into very thin slices.

10. Cutting parisienne.

Cuts made with a ball cutter are perhaps most often used for potatoes. Potatoes cut into large balls, about 1¹/₃ inch (3 cm), are called **parisienne** (pah ree zee enn). When cut into smaller balls, about ⁷/₈ inch (2.5 cm), they are called **noisette** (nwah zet). Of course, other solid vegetables, such as turnips, as well as many fruits, can be cut the same way. The procedure is illustrated in Figure 7.20.

FIGURE 7.20 Cutting parisienne potatoes.

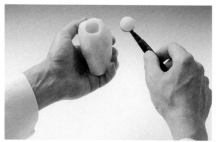

(a) Place the ball cutter against the potato as shown.

(b) With the thumb, press the cutter firmly into the potato as far as it will go.

(c) Lift the handle of the cutter outward, twist the cutter around, and remove the ball.

11. Cutting tournéed vegetables.

To **tourné** (toor nay) a vegetable is to cut it into a neat seven-sided oval shape, as illustrated in Figure 7.21. Many root vegetables, such as carrots and turnips, are cut this way. When potatoes are tournéed, they are named according to their size. **Cocotte** potatoes are about 1¹/₂ inches (4 cm) long. **Château** potatoes are about 2 inches.

FIGURE 7.21 Tournéing potatoes and other root vegetables.

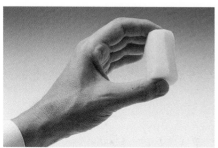

(a) Cut the potatoes roughly into pieces slightly larger than the final size desired. Cut off the top and bottom of each piece so the top and bottom are flat and parallel.

(b) Hold the potato between the thumb and forefinger. Place the paring knife against the top edge as shown and the thumb of the cutting hand firmly against the potato. Your hand should be far enough up on the blade to maintain steady control.

(c) Cut down toward your thumb with a curving movement of the blade.

(d) Turn the potato slightly (one-seventh of a full turn, to be exact) and repeat the motion.

(e) The finished product. If perfectly done, the potato has seven sides (but customers rarely count them).

12. Peeling grapefruit.

This technique, as shown in Figure 7.22, can also be used for peeling yellow turnips or other round vegetables and fruits with heavy peels.

FIGURE 7.22 Peeling a grapefruit.

(a) Cut off the ends of the grapefruit and turn it on a flat end so it is stable. Slice off a section of the peel, following the contour of the grapefruit.

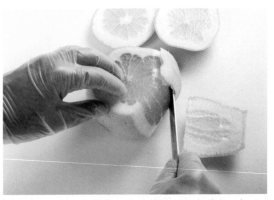

(b) Make sure the cut is deep enough to remove the peel but not so deep as to waste the product.

(c) Continue making slices around the grapefruit until all the peel is removed.

(d) Slice or section the fruit. Squeeze the remaining pulp for juice. The membrane-free citrus section cut by this method are called suprêmes.

13. Chiffonade.

This term refers to cutting leaves into fine shreds. It is applied most often to lettuce and sorrel but can also be used for fresh herbs such as parsley. To cut **chiffonade**, remove the heavy leaf ribs, roll the leaves into a tight cylinder, and then slice the cylinder cross-wise into thin shreds, as shown in Figure 7.23.

FIGURE 7.23 Cutting a chiffonade of sorrel.

(a) Roll the leaves into a cylinder.

(b) Cut crosswise into thin strips or shreds.

14. Cutting citrus zest.

With a paring knife, cut strips from the citrus peel, removing only the colored part, not the white part below it. Then, with a chef's knife, cut the zest into thin strips or julienne, as shown in Figure 7.24b. An alternative method is to use a citrus zester, as shown in Figure 7.24c.

FIGURE 7.24 Cutting citrus zest.

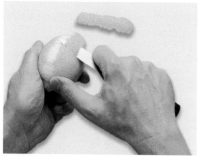

(a) Use a paring knife to cut thin strips from the peel, being careful to cut only the outer colored part, not the inner white pith.

(b) Cut the strips of peel into julienne.

(c) Alternatively, draw a zesting tool over the fruit to cut thin strips of zest.

KEY POINTS TO REVIEW

- How do you sharpen a knife?

- When performing most cutting techniques, what is the correct and safe position of the guiding hand—that is, the hand not holding the knife?

- What are the dimensions of the following cuts: large dice, medium dice, small dice, brunoise, fine brunoise, bâtonnet, julienne, fine julienne, rondelle, paysanne, lozenge, tourné?

PRELIMINARY COOKING AND FLAVORING

Advance preparation often requires precooking and flavoring of ingredients to make them ready for use in the finished recipe.

On the most obvious level, if a recipe for chicken salad calls for cooked, diced chicken, you must first cook the chicken before you can proceed with the recipe. A complete cooking procedure, in such a case, is part of the mise en place, or pre-preparation.

BLANCHING AND PARCOOKING

Partial cooking is a significant part of advance preparation. It requires a degree of culinary skill and judgment to determine when and how much cooking is necessary or desirable.

Partial cooking may be done by any moist-heat or dry-heat method. Those commonly used are simmering or boiling (parboiling), steaming, and deep-frying (especially for potatoes). The term **blanching** may mean any of these methods, but it usually implies *very brief* cooking.

There are four main reasons for blanching or parcooking:

1. To increase holding quality.

Heating helps preserve foods by:

- Destroying bacteria that cause spoilage.

- Destroying enzymes that discolor foods (as when potatoes turn brown) and cause them to deteriorate.

2. To save time.

It takes less time to finish parboiled vegetables for service than it does to finish raw vegetables. Large batches of foods may be blanched and chilled, and individual portions then finished to order.

Items, such as roast duck, that take too long to cook completely to order are often roasted half to three-quarters done and then finished as the orders are received.

3. To remove undesirable flavors.

Some variety meats and certain strong-flavored vegetables, such as rutabaga, are sometimes blanched to make them milder and more acceptable to the customer.

4. To enable the product to be processed further.

For example, vegetables and fruits such as tomatoes and peaches, as well as some nuts, are blanched to loosen the skins for peeling.

Sweetbreads are blanched so they are firm enough for slicing and breading or other kinds of handling.

MARINATING

To **marinate** means to soak a food product in a seasoned liquid in order to:

1. Flavor the product.

2. Tenderize the product.

The tenderizing effect of the acids in the marinade is relatively small. It is still essential to match the proper cut of meat with the proper cooking techniques for greatest tenderness.

The marinade can also serve as the cooking medium and become part of the sauce. Vegetable marinades, called *vinaigrettes*, are served cold with the vegetables as salads or hors d'oeuvres without further cooking or processing.

Marinades have three categories of ingredients:

1. Oil.

Oil helps preserve the meat's moisture. Sometimes it is omitted, especially for long marinations, when the oil would only float on top, out of contact with the product being marinated.

Tasteless vegetable oils are used when a neutral flavor is required. Specialty oils, such as olive oil, are used to add flavor to the item being marinated.

2. Acid from vinegar, lemon juice, wine.

Acid helps tenderize protein foods.

It carries flavors (its own and dissolved flavors from spices and herbs).

Use caution when employing strong acids, such as vinegar and lemon juice. A marinade that is too acidic will partially coagulate the protein of the meat, making it seem partially cooked. When the meat is then cooked, its texture will not be as desirable. Strong acids can be used in marinades if they are used in small quantities or if the meat is marinated only a few hours.

3. Flavorings—spices, herbs, vegetables.

A wide choice is available, depending on the purpose.

Whole spices release flavors more slowly, so they are more suitable for long marination.

Kinds of Marinade

1. Cooked.

Used when long keeping quality is important. Modern refrigeration has made cooked marinades less widely used. An advantage of cooked marinades is that spices release more flavor into the marinade when it is cooked.

2. Raw.

Most widely used for long marination under refrigeration. For example, see the recipe for Sauerbraten (p. 519). Figure 7.25 shows meat in a raw marinade.

FIGURE 7.25 Beef chuck in a raw marinade of red wine, wine vinegar, spices, and aromatic vegetables.

3. **Instant.**

The range of flavors and purposes is wide. Used for marinating a few minutes up to several hours or overnight. For example, see the recipe for London Broil (p. 499).

4. **Dry.**

A dry marinade, also called a *dry rub* or a spice rub, is a mixture of salt, spices, and herbs that is rubbed or patted onto the surface of a meat, poultry, or fish item. In some cases, a little oil or a moist ingredient, such as crushed garlic, is mixed with the spices to make a paste, called a *wet rub*. The item is then refrigerated to allow it time to absorb the flavors. The rub may be left on the item or scraped off before cooking. This technique is widely used for barbecued meats. For an example of a dry marinade, see page 545. Figure 7.26 shows a dry rub being applied to a large cut of meat.

Dry marinades are an effective way to flavor meats. Naturally, because they usually don't contain an acid, you can't expect dry marinades to produce the slight tenderizing effects of liquid marinades containing acids.

FIGURE 7.26 Applying a dry rub to a rack of spareribs.

GUIDELINES for Marinating

1. Marinate under refrigeration (unless product is to be cooked only a few minutes).
2. Remember: The thicker the product, the longer it takes for the marinade to penetrate.
3. Use an acid-resistant container, such as stainless steel, glass, crockery, or some plastics.
4. Tie spices in a cheesecloth bag (sachet) if easy removal is important.
5. Cover product completely with marinade. When marinating small items for a short time, you may use less liquid, but you must then turn the product frequently for even penetration.

Brines

A **brine** may be considered a special kind of marinade. The primary use of brines is in curing, as explained in Chapter 27 (see p. 832). However, many chefs also use them for roast poultry and pork because of their tenderizing and moisturizing effects (see sidebar, p. 150). Brines are rarely used for red meats.

PROCEDURE for Brining Meats and Poultry

1. Assemble the following ingredients:

Water	1 gal	4 L
Kosher salt	4 oz	125 g
Sugar	3 oz	90 g
Bay leaves	2	2
Dried thyme	2 tsp	10 mL
Whole cloves	4	4
Peppercorns	1 tbsp	15 mL

2. Combine the water, salt, and sugar in a stockpot. Make a sachet by tying the herbs and spices in a piece of cheesecloth. Add the sachet to the pot.
3. Bring the water to a boil, stirring to make sure the salt and sugar are dissolved.
4. Let cool, then refrigerate until completely cold. Remove the sachet.
5. Put the meat or poultry into the brine. Poultry must be weighted to keep it submerged. Refrigerate.
6. For large cuts, marinate at least 6 hours or as long as 2 days. For small pieces, such as chops and cutlets, marinate 2 to 6 hours.
7. Remove the meat from the brine, dry it, and proceed with the recipe. Treat the brined meat like fresh meat.
8. Discard the used brine.

HOW A BRINE WORKS

Brines moisturize and tenderize roast meats. At first, salt draws moisture from meat tissues. But some of the brine is also absorbed into the meat. When the salt of the brine interacts with the proteins, it dissolves some of the protein structure, and this has a tenderizing effect on the meat. Also, because the salt interacts with the proteins, the capacity of the meat to hold moisture is increased. The weight of the raw brined meat is about 10 percent greater than the weight of the meat before brining.

A brine consists primarily of salt dissolved in water. Because of the harsh taste a high salt concentration can give to meats, chefs usually add sugar to the brine as well to counteract the strong salt flavor. In addition, herbs and aromatics may be added to the brine, although these have only a mild flavoring effect.

Salt concentration in brines ranges from 3 to 6 percent. The formula given in the procedure makes a concentration of 3 percent, so you could double the salt (and sugar) if desired.

Because of the salt and sugar concentration in the meat, pan drippings may not be usable for deglazing. Also, the sugar may burn to the bottom of the pan, so you may want to put a little water in the bottom of the roasting pan.

PREPARATION FOR FRYING

Most foods to be deep-fried, with the major exception of potatoes, are first given a protective coating of breading or batter. This coating serves four purposes:

1. It helps retain moisture and flavor in the product.
2. It protects the fat against the moisture and salt in the food, which would speed the deterioration of the frying fat.
3. It protects the food from absorbing too much fat.
4. It gives crispness, flavor, and good appearance to the product.

BREADING

Breading means coating a product with bread crumbs or other crumbs or meal before deep-frying, pan-frying, or sautéing. The most widely used method for applying these coatings is called the **Standard Breading Procedure**.

The Three Stages of the Standard Breading Procedure

1. **Flour.**
   Helps the breading stick to the product.

2. **Egg wash.**
   A mixture of eggs and a liquid, usually milk or water. More eggs give greater binding power but increase the cost. A small quantity of oil is occasionally added to the egg wash.

3. **Crumbs.**
   Combine with the egg wash to create a crisp, golden coating when fried. Fine, dry bread crumbs are most often used and give good results. Also popular are Japanese-style dry bread crumbs called **panko** (Japanese for "bread crumbs"). These coarser crumbs give a pleasing texture to fried items. Other products used are fresh bread crumbs, crushed corn flakes or other cereal, cracker meal, and cornmeal.

   For small items like scallops and oysters, breading may be done with the aid of a series of wire baskets placed in the flour, egg wash, and crumbs, instead of by hand. The procedure is the same except the baskets are used to lift and shake small quantities of the product and to transfer them to the next basket.

   To keep one hand dry during breading, use your right hand (if you are right-handed; if left-handed, reverse the procedure) only for handling the flour and crumbs. Use your other hand for handling the product when it is wet. In order to keep your dry hand dry, never handle a wet product with that hand. For example, to complete the breading of an item that has been dipped in egg wash, place it in the pan of crumbs and push more crumbs over the top of the item, as shown in Figure 7.27, and then pat them down, so all sides of the item are covered in dry crumbs before you pick it up.

FIGURE 7.27 To keep your dry hand dry, push crumbs over the top of egg-washed items in the crumb pan before touching them.

PROCEDURE for Making Bread Crumbs

1. For fresh bread crumbs, use bread that is one or two days old. If the bread is fresh, its moisture content will make it difficult to process into crumbs without making gummy wads of bread. For dry bread crumbs, lightly toast the bread in a warm oven until the bread is dry but not browned. Do not use stale bread, which has an off flavor.

2. Trim off crusts to make crumbs of a uniform light color.

3. Cut or tear the bread into smaller pieces.

4. Depending on the quantity, place the pieces in a food processor or a vertical cutter/mixer. Process until the crumbs are of the desired fineness.

5. For dry bread crumbs of uniform size, pass the processed crumbs through a sieve (also called a tamis). See page 56.

PROCEDURE for Proper Breading

Figure 7.28 illustrates a station setup for the Standard Breading Procedure.

1. Dry the product to get a thin, even coating of flour.

2. Season the product—or, for greater efficiency, season the flour (step 3). Do not season the crumbs. The presence of salt in contact with the frying fat breaks down the fat and shortens its life.

3. Dip the product in flour to coat evenly. Shake off excess.

4. Dip in egg wash to coat completely. Remove. Let excess drain off so the crumb coating will be even.

5. Dip in bread crumbs. Cover with crumbs and press them gently onto product. Make sure it is coated completely. Remove. Carefully shake off excess.

6. Fry immediately, or hold for service.

7. To hold for later service, place the breaded items in a single layer on a pan or rack and refrigerate. Do not hold very moist items, such as raw clams or oysters. The breading will quickly become soggy.

8. Strain the egg wash and sift the flour and crumbs as often as necessary to remove lumps.

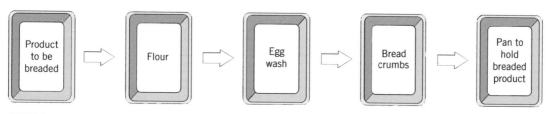

FIGURE 7.28 Setup of station for Standard Breading Procedure. Right-handed cooks work from left to right. Left-handed cooks work from right to left, with order of pans reversed, as well.

DREDGING WITH FLOUR

Purpose

The purpose of dredging is to give a thin, even coating of flour to a product.

Meats to be sautéed or pan-fried are often dredged with flour to give them an even, brown color and to prevent sticking.

Vegetables, such as sticks of zucchini, are sometimes coated only in flour before deep-frying to give them a light golden color and a very thin coating.

Procedure

Follow steps 1 to 3 of the Standard Breading Procedure above.

BATTERS

Batters are semiliquid mixtures containing flour or other starch. They are used in deep-frying to give a crisp, flavorful, golden brown coating. There are many formulas and variations for batters:

1. Many liquids are used, including milk, water, and beer.

2. Eggs may or may not be used.

3. Thicker batters make thicker coatings. Too thick a batter makes a heavy, unpalatable coating.

4. Leavenings are frequently used to give a lighter product. These may be:

- Baking powder
- Beaten egg whites
- Carbonation from beer or seltzer used in the batter

Two recipes for basic, typical batters are given in the recipe for deep-fried Onion Rings (p. 341), and the recipe for Fish and Chips (p. 433). These batters may be used on a wide variety of products. In addition, a recipe for tempura batter is on page 637.

HANDLING CONVENIENCE FOODS

Convenience foods play an increasingly prominent role in the food-service industry. Their use has become so important that no student of professional cooking can afford to be without knowledge of them.

GUIDELINES for Handling Convenience Foods

1. **Handle with the same care you give fresh, raw ingredients.**
   Most loss of quality in convenience foods comes from assuming they are damageproof and can be treated haphazardly.

2. **Examine as soon as received.**
   Particularly, check frozen foods—with a thermometer—to make sure they did not thaw in transit. Put away at once.

3. **Store properly.**
   Frozen foods must be held at 0°F (–18°C) or lower. Check your freezer with a thermometer regularly. Refrigerated foods must stay chilled, below 41°F (5°C), to slow spoilage. Shelf-stable foods (dry products, canned goods, etc.) are shelf-stable only when stored properly in a cool, dry place, tightly sealed.

4. **Know the shelf life of each product.**
   Nothing keeps forever, not even convenience foods. (Some, like peeled potatoes, are even more perishable than unprocessed ingredients.) Rotate stock according to the first in, first out principle. Don't stock more than necessary.

5. **Defrost frozen foods properly.**
   Ideally, defrost in a tempering box set at 28° to 30°F (–2° to –1°C) or, lacking that, in the refrigerator at 41°F (5°C) or lower. This takes planning and timing, because large items take several days to thaw.

   If you are short of time, the second-best way to defrost foods is under cold running water, in the original wrapper.

   Never defrost at room temperature or in warm water. The high temperatures encourage bacterial growth and spoilage.
   Do not refreeze thawed foods. Quality will greatly deteriorate.
   Certain foods, like frozen French fries and some individual-portion prepared entrées, are designed to be cooked without thawing.

6. **Know how and to what extent the product has been prepared.**
   Partially cooked foods need less heating in final preparation than do raw foods. Some cooks prepare frozen, cooked crab legs, for example, as though they were raw, but by the time the customer receives them, they are overcooked, dry, and tasteless. Frozen vegetables, for a second example, have been blanched and often need only to be heated briefly.

   Manufacturers are happy to give full directions and serving suggestions for their products. At least you should read the package directions.

7. **Use proper cooking methods.**
   Be flexible. Much modern equipment is designed especially for convenience foods. Don't restrict yourself to conventional ranges and ovens if compartment steamers, convection ovens, or microwave ovens might do a better job more efficiently.

8. **Treat convenience foods as though you, not the manufacturer, did the pre-preparation.**
   Make the most of your opportunity to use creativity and to serve the best quality you can. Your final preparation, plating, and garnish should be as careful as though you made the dish from scratch.

A **convenience food** may be defined as "any product that has been partially or completely prepared or processed by a manufacturer." In other words, when you buy a convenience product, you are having the manufacturer do some or all of your preparation for you.

Of course, you must pay for this service, as reflected in the price of the product. Although buying the convenience product will likely cost you more than buying the raw materials, you save in increased kitchen efficiency. As you remember from Chapter 4, labor costs as well as food costs must be figured into your menu prices.

Processed foods for restaurants and institutions range from partially prepared items that can be used as components in your recipes, such as frozen fish fillets, peeled potatoes, concentrated stock bases, and frozen puff pastry dough, to fully prepared items that need only be reconstituted or served as is, such as frozen prepared entrées and frozen pies and pastries. Some items, like frozen French fries, have wide acceptance, while other more fully prepared foods continue to be resisted by both customer and operator.

In general, the more completely a product is prepared by the manufacturer, the less it reflects the individuality of the food-service operator—and the less opportunity the cooks have to give it their own character and quality.

Is a stock made from scratch better than a product made from a convenience base? Most quality-conscious chefs would probably answer "Yes!" But the correct answer is, "Not if the homemade stock is poorly made." No matter what products you use, there is no substitute for quality and care. The fresh product is potentially the best, but not if it is badly stored or handled. Of course, convenience foods also require proper handling to maintain their quality.

The key to understanding and handling convenience foods is considering them as normal products with part of the pre-prep completed rather than as totally different kinds of products unlike your normal raw materials. *Convenience products are not a substitute for culinary knowledge and skill*. They should be a tool for the good cook rather than a crutch for the bad cook. It takes as much understanding of basic cooking principles to handle convenience products as it does fresh, raw ingredients, particularly if you want the convenience product to taste as much like the fresh as possible.

KEY POINTS TO REVIEW

- What is blanching? Why are foods blanched?

- What are the basic types of marinades?

- What are the three stages of the Standard Breading Procedure? Describe how to set up a breading station.

- What are convenience foods? Describe eight guidelines for handling and using convenience foods.

TERMS FOR REVIEW

mise en place	brunoise	cocotte
holding temperature	bâtonnet	château
set meal service	allumette	chiffonade
extended meal service	julienne	blanching
chop	paysanne	marinate
concasser	lozenge	brine
mince	fermière	Standard Breading Procedure
emincer	oblique cut	panko
shred	parisienne	batter
rondelle	noisette	convenience food
dice	tourné	

QUESTIONS FOR DISCUSSION

1. How does preparation differ for set meal service and extended meal service?

2. It has been said that à la carte cooking, or cooking to order, is nothing more than small-batch cooking carried to its extreme. Based on what you know about pre-preparation, what do you think this statement means?

3. Why is it important to learn to cut foods accurately and uniformly?

4. Name six basic vegetable cuts, and give their dimensions.

5. Give six examples of foods that might be blanched or par-cooked during pre-preparation, and give a reason for each.

6. Describe in detail how to set up a breading station and how to use it to bread veal cutlets.

7. The manager of the restaurant in which you are a cook has decided to try using frozen, breaded shrimp instead of having you bread shrimp by hand, but she is worried about customer acceptance and asks for your help. How will you handle the new product?

STOCKS AND SAUCES

The importance of stocks in the kitchen is indicated by the French word for stock: *fond*, meaning "foundation" or "base." In classical cuisine, the ability to prepare good stocks is the most basic of all skills because so much of the work of the entire kitchen depends on them. A good stock is the foundation of soups, sauces, and most braised foods and stews.

In modern kitchens, stocks have lost much of the importance they once had. In the first place, increased reliance on portion-controlled meats has made bones for stock a rarity in most establishments. Second, making stocks requires extra labor, which most restaurants today aren't able to provide. Finally, more food today is served without sauces, so stocks aren't seen to be quite as necessary.

Nevertheless, the finest cuisine still depends on soups and sauces based on high-quality stocks, so stock-making remains an essential skill you should learn early in your training. Stocks and sauces are almost never served by themselves but are components of many other preparations. You will need to refer to this chapter in connection with many other subjects.

AFTER READING THIS CHAPTER, YOU SHOULD BE ABLE TO

1. Prepare basic mirepoix.

2. Flavor liquids using a sachet d'épices, or spice bag.

3. Prepare white veal or beef stock, chicken stock, fish stock, and brown stock.

4. Cool and store stocks correctly.

5. Prepare meat, chicken, and fish glazes.

6. Evaluate the quality of convenience bases, and use convenience bases.

7. Explain the functions of sauces, and list five qualities a sauce adds to food.

8. Prepare white, blond, and brown roux, and use them to thicken liquids.

9. Prepare and use beurre manié.

10. Thicken liquids with cornstarch and other starches.

11. Prepare and use egg yolk and cream liaison.

12. Finish a sauce with raw butter (monter au beurre).

13. Prepare the five leading sauces: béchamel, velouté, brown sauce or espagnole, tomato, and hollandaise.

14. Prepare small sauces from leading sauces.

15. Identify and prepare five simple butter sauces.

16. Prepare compound butters and list their uses.

17. Prepare pan gravies.

18. Prepare miscellaneous hot and cold sauces.

STOCKS

The preparation of stocks has been simplified in many ways since the days of Escoffier, although this does not mean it demands less care or skill. Few chefs today bother to tie vegetables for a stock into a bundle, for example. They're going to be strained out anyway. The number and variety of ingredients is usually not as great as it once was. Nor is it common to cook stocks for as many hours as was once thought necessary. All these details are taken up one by one in this section.

A **stock** may be defined as a clear, thin—that is, unthickened—liquid flavored by soluble substances extracted from meat, poultry, and fish, and their bones, and from vegetables and seasonings. Our objective in preparing stocks is to select the proper ingredients and then to extract the flavors we want—in other words, to combine the correct ingredients with the correct procedure.

INGREDIENTS

BONES

Bones are the major ingredient of stocks (except water, of course). Most of the flavor and body of stocks are derived from the bones of beef, veal, chicken, fish, and, occasionally, lamb, pork, ham, and game. (Vegetable stocks, an exception, draw their flavor entirely from vegetables; see p. 158.)

The kinds of bones used determine the kind of stock.

Chicken stock, of course, is made from chicken bones.

White stock is made from beef or veal bones, or a combination of the two. Chicken bones or even pork bones are sometimes added in small quantity.

Brown stock is made from beef or veal bones that have been browned in an oven.

Fish stock is made from fish bones and trimmings left over after filleting. Bones from lean white fish give the best stock. Fat fish are not normally used. The term *fumet* is often used for a flavorful fish stock, especially one made with wine. See the note at the beginning of the recipe for Fish Fumet (p. 165).

Lamb, game, turkey, and other stocks have specialized uses.

In Chapter 6, we discussed a group of proteins called *connective tissue*. Remember that some of these proteins are dissolved when cooked with slow, moist heat. Chapter 15, "Understanding Meats," offers more information about these substances. You should learn and understand these two basic facts:

1. When certain connective tissues (called *collagen*) break down, they form *gelatin*. This gives body to a stock, an important feature of its quality. A well-made stock thickens or even solidifies when chilled.

2. *Cartilage* is the best source of gelatin in bones. Younger animals have lots of cartilage in their skeletons. As they become older, this hardens into solid bone, which is harder to dissolve into stocks. *Knuckle bones*, on the joints of major bones, have a lot of cartilage and are valued in stock-making. Neck bones and shankbones are also used a great deal.

Cut large bones into pieces about 3 inches (8 cm) long. This exposes more surface area and aids extraction. Also, the bones are easier to handle.

MEAT

Because of its cost, meat is rarely used in stock-making anymore. (Exception: Chicken hearts and gizzards are often used in chicken stock.)

Occasionally, a broth is produced as a result of simmering meat or poultry, as when fowl is cooked for dishes like creamed chicken. This broth can then be used like a stock. However, the chicken is considered the object of the game in this case. The broth is just a byproduct.

In this book, we use the word **broth** to mean a flavorful liquid obtained from the simmering of meats and/or vegetables.

MIREPOIX

Aromatic vegetables are the second most important contributors of flavor to stocks. (In the case of vegetable stocks, they are the most important.)

Mirepoix (meer-pwah) is a combination of onions, carrots, and celery. It is a basic flavoring preparation used in all areas of cooking—not only for flavoring stocks but also for sauces, soups, meats, poultry, fish, and vegetables. Mirepoix ingredients have a fairly neutral taste, so that they contribute important flavor to stock without making it taste like any particular vegetable. (The classical mirepoix of decades ago contained a wider variety of ingredients, sometimes including ham or bacon, leeks and other vegetables, and one or more fresh herbs. The modern version is considerably simplified.)

Learn the proportions in Table 8.1 well. Mirepoix is a basic preparation you will need throughout your career.

A *white mirepoix* is used when it is necessary to keep the stock as colorlessas possible, especially for fish stock. Carrots are not used in white mirepoix because they give too much color. Parsnips may be substituted for the carrots, or else the carrots may simply be omitted and the quantities of the other ingredients increased. Celery root, also called celeriac, can be substituted for the stalk celery for an even whiter mirepoix. Mushroom trimmings may also be added. When cost permits, it is a good idea to include leeks in the mirepoix in place of part of the onions in a white mirepoix, as indicated in Table 8.1. They give an excellent flavor. (Note: Many chefs prefer to use a standard mirepoix rather than white mirepoix for all stocks. In particular, it is common practice to use a standard mirepoix for fish fumet.)

In vegetable stocks, a variety of vegetables is used in addition to or in place of the traditional mirepoix; see page 158 for a brief discussion.

Standard mirepoix (left) and white mirepoix (right)

Cutting Mirepoix

Chop the vegetables coarsely into pieces of relatively uniform size. As mirepoix is rarely served, it is not usually necessary to cut it neatly.

The size depends on how long the mirepoix will cook. If it will cook a long time, as for beef stock, cut the vegetables into large pieces (1–2 inches [3–5 cm]). Cutting into small pieces is necessary for releasing flavors in a short time, as when the mirepoix will be used for fish stock.

TABLE 8.1 Mirepoix		
To Make:	**1 Pound**	**500 Grams**
Standard Mirepoix		
Onions	8 oz	250 g
Celery	4 oz	125 g
Carrots	4 oz	125 g
White Mirepoix (with parsnips)		
Onions	4 oz	125 g
Leeks	4 oz	125 g
Celery or celeriac	4 oz	125 g
Parsnips	4 oz	125 g
White Mirepoix (without parsnips)		
Onions	6 oz	190 g
Leeks	5 oz	155 g
Celery or celeriac	5 oz	155 g
Optional addition to white mirepoix		
Mushroom trimmings	2 oz	60 g

ACID PRODUCTS

Acids, as noted in Chapter 6 (p. 108), help dissolve connective tissues. Thus, they are sometimes used in stock-making to extract flavor and body from bones, as well as contributing their own flavor.

Tomato products contribute flavor and some acid to brown stocks. They are not used for white stocks because they would give an undesirable color. Similarly, when making brown stocks, be careful not to add too much tomato, which can make the stock cloudy.

Wine is occasionally used, especially for fish stocks. Its flavor contribution is probably more important than its acidity.

SCRAPS AND LEFTOVERS

In some kitchens, a stockpot is kept going all day, and scraps are constantly being thrown in. This may or may not be a good idea.

Scraps may be used in stocks if they are ***clean, wholesome, and appropriate to the stock being made***. If done correctly, stock-making is a good way of utilizing trimmings that would otherwise be thrown out. It is better to save trimmings and use them in a planned way than to throw them into the stock randomly.

A stockpot is not a garbage disposal. The final product is only as good as the ingredients and the care that go into it.

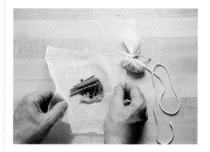

FIGURE 8.1 To make a sachet, place the spices and herbs in the center of a square of clean cheesecloth. Draw the corners together and tie with a length of twine. For making stock, use a piece of twine long enough to be tied to the handle of the stockpot for easy removal.

FIGURE 8.2 Tie the herbs and aromatic vegetables for a bouquet garni in a bundle. To tie small herbs securely, enclose them between the two halves of leek.

SEASONINGS AND SPICES

Salt is usually not added when making stocks. Stocks are never used as is but are reduced, concentrated, and combined with other ingredients. If salt were added, it might become too concentrated. Some chefs salt stocks very lightly because they feel it aids in extracting flavor.

Herbs and spices should be used only lightly. They should never dominate a stock or have a pronounced flavor.

Herbs and spices are usually tied in a cheesecloth bag called a **sachet d'épices** (sa-shay day peace; French for "spice bag"), often called simply *sachet* for short. The sachet (Figure 8.1) is tied by a string to the handle of the stockpot so it can be removed easily at any time.

A **bouquet garni** is an assortment of fresh herbs and other aromatic ingredients tied in a bundle with string. A basic bouquet garni contains pieces of leek and celery, thyme sprigs, bay leaf, and parsley stems (see Figure 8.2). The ingredients can be changed to suit different recipes. Escoffier includes only parsley, thyme, and bay leaf in the classic bouquet garni.

The following seasonings, in varying quantities, are commonly used for stocks:

Thyme	Parsley stems
Bay leaves	Cloves, whole (optional)
Peppercorns	Garlic (optional)

Onions for Flavoring

In addition to the onions in the mirepoix, an *oignon brûlé* (awn yohn broo lay; French for "burnt onion") is sometimes added to brown stock to give it color as well as flavor. To prepare, cut a large onion in half crosswise and place it, cut side down, on a flattop range or in a heavy skillet. Cook until the cut surface is dark brown. Add to the stock.

Another form of onion for flavoring is the *oignon piqué* (pee kay). This is used not so much for stocks but for soups and sauces. To prepare, stick a bay leaf to a whole, peeled onion with a whole clove. Adding the bay leaf and clove attached to the onion makes removing them easier when cooking is finished.

Oignon piqué (left) and oignon brûlé (right)

INGREDIENT PROPORTIONS

The proportions in Tables 8.2, 8.3, 8.4, and 8.5 are basic, effective, and widely used, but using them is not an ironclad rule. Nearly every chef uses some variations.

Many cooks use ratios to help them remember the basic proportions, as follows:

Bones—50 percent

Mirepoix—10 percent

Water—100 percent

INGREDIENTS FOR VEGETABLE STOCKS

Vegetable stocks, made without any animal products, play an important role in vegetarian cooking and are also used in more traditional kitchens in response to customers' requests for light, healthful dishes. The basic ingredients for vegetable stocks are vegetables, herbs and spices, water, and, sometimes, wine (see Figure 8.3).

FIGURE 8.3 Ingredients for vegetable stock.

Ingredients and proportions can vary greatly. If you want a particular flavor to predominate, use a larger quantity of that vegetable. For example, if you want a broth tasting primarily of asparagus, use a large quantity of asparagus to make it, with smaller quantities of more neutral vegetables (like onion and celery) to round out the flavor. For a more neutral, all-purpose vegetable stock, avoid strong-flavored vegetables and use more balanced proportions of ingredients.

Here are five additional guidelines for making vegetable stocks or broths:

1. Starchy vegetables, such as potatoes, sweet potatoes, and winter squash, make a stock cloudy. Use them only if clarity is not important.

2. Some vegetables, especially strong-flavored ones, are best avoided. Brussels sprouts, cauliflower, and artichokes can overwhelm a stock with a strong flavor or odor. Dark green leafy vegetables, especially spinach, develop an unpleasant flavor when cooked a long time. Beets turn a stock red.

3. Cook long enough to extract flavors but not so long that flavors are lost. Best cooking time is 30–45 minutes.

4. Sweating the vegetables in a small amount of oil before adding water gives them a mellower flavor, but this step can be omitted. Butter can be used if it is not necessary to avoid all animal products.

5. Ratios of vegetables to water may vary considerably, but the following proportions are a good starting point:

Vegetables:	4 lb	2 kg
Water:	1 gal	4 L
Sachet:	1	1

TABLE 8.2 White Stock (Chicken)

To Make:	1 Gallon	4 Liters
Bones	5 lb	2.5 kg
Mirepoix, standard or white	1 lb	500 g
Water	5–6 qt	5–6 L
Sachet	1	1

TABLE 8.3 White Stock (Beef and Veal)

To Make:	1 Gallon	4 Liters
Bones	5 lb	2.5 kg
Mirepoix, standard or white	1 lb	500 g
Water	5–6 qt	5–6 L
Sachet	1	1

TABLE 8.4 Brown Stock

To Make:	1 Gallon	4 Liters
Bones	5 lb	2.5 kg
Mirepoix	1 lb	500 g
Tomato product (see note)	8 oz	250 g
Water	5–6 qt	5–6 L
Sachet	1	1

Note: Quantity is for canned tomatoes or tomato purée. For tomato paste, use half this quantity.

TABLE 8.5 Fish Stock

To Make:	1 Gallon	4 Liters
Bones	4–6 lb	2–3 kg
Mirepoix, white	1 lb	500 g
Water	1 gal	4 L
White wine	24 fl oz	750 mL
Sachet	1	1

KEY POINTS TO REVIEW

- What is mirepoix? How is it prepared?
- What is a sachet d'épices? How is it prepared?
- What are the basic proportions of bones, mirepoix, and water to make a standard white stock or brown stock?

PROCEDURES

Making stock may seem, at first glance, a simple procedure. However, many steps are involved, each with a rather complicated set of reasons. If you are to be successful at making consistently good stocks, you must understand not only what to do but also why you are doing it.

The following outlines give procedures for making basic stocks as well as the reasons for every step. After learning these procedures and checking with your instructors for any modifications or variations they may have, you will be able to turn to the individual recipes, where the steps are given again, but without explanations.

BLANCHING BONES

In Chapter 6, we discussed proteins coagulating when heated. Many proteins dissolve in cold water but solidify into small particles or into froth or scum when heated. It is these particles that make a stock cloudy. Much of the technique of stock-making involves avoiding cloudiness to produce a clear stock.

The purpose of blanching bones is to rid them of some of the impurities that cause cloudiness. The bones of young animals, especially veal and chicken, are highest in blood and other impurities that cloud and discolor stocks.

Chefs disagree on the importance of blanching. Many feel it is needed to produce clear white stocks. Others feel blanching causes valuable flavors to be lost. Fish bones, at any rate, are not blanched because of their short cooking time.

PROCEDURE for Blanching Bones

1. **Rinse the bones in cold water.**
   This washes off blood and other impurities from the surface. It is especially important if the bones are not strictly fresh.
2. **Place the bones in a stockpot or steam-jacketed kettle and cover with cold water.**
   Impurities dissolve more readily in cold water. Hot water retards extraction.
3. **Bring the water to a boil.**
   As the water heats, impurities solidify (coagulate) and rise to the surface as scum.
4. **Drain the bones and rinse them well.**
   The bones are now ready for the stockpot.

PREPARING WHITE STOCKS

A good white stock has rich, full flavor, good body, clarity, and little or no color. Chicken stocks may have a light yellow color.

PROCEDURE **for Preparing White Stocks**

1. **Cut the bones into pieces, 3–4 inches (8–10 cm) long.**

   This exposes more surface area and helps extraction. A meat saw is used to cut heavy veal and beef bones. Fish and chicken bones don't need to be cut, but whole carcasses should be chopped for more convenient handling.

2. **Rinse the bones in cold water. (If desired, chicken, veal, or beef bones may be blanched.)**

   This removes some impurities that cloud the stock or, if the bones are old, give an off taste.

3. **Place the bones in a stockpot or steam-jacketed kettle and add cold water to cover.**

   Starting in cold water speeds extraction. Starting in hot water delays it because many proteins are soluble in cold water but not in hot.

4. **Bring water to a boil, and then reduce to a simmer. Skim the scum that comes to the surface, using a skimmer or perforated spoon.**

   Skimming is important for a clear stock because the scum (which is fat and coagulated protein) will cloud the stock if it is broken up and mixed back into the liquid.

5. **Add the chopped mirepoix and the herbs and spices.**

   Remember, the size to which you cut mirepoix depends on how long it is to be cooked.

6. **Do not let the stock boil. Keep it at a low simmer.**

   Boiling makes the stock cloudy because it breaks solids into tiny particles that get mixed into the liquid.

7. **Skim the surface as often as necessary during cooking.**

8. **Keep the water level above the bones. Add more water if the stock reduces below this level.**

   Bones cooked while exposed to air will turn dark and thus darken or discolor the stock. Also, they do not release flavor into the water if the water doesn't touch them.

9. **Simmer for the recommended length of time:**

   Beef bones—8 to 10 hours   Chicken bones—3 to 4 hours
   Veal bones—6 to 8 hours   Fish bones—30 to 45 minutes

   Most modern chefs do not simmer stocks as long as earlier generations of chefs did. It is true that longer cooking extracts more gelatin, but gelatin isn't the only factor in a good stock. Flavors begin to break down or degenerate over time. The above times are felt to be the best for obtaining full flavor while still getting a good portion of gelatin into the stock.

10. **Skim the surface and strain off the stock through a china cap lined with several layers of cheesecloth.**

    Adding a little cold water to the stock before skimming stops the cooking and brings more fat and impurities to the surface.

11. **Cool the stock as quickly as possible, as follows:**
    - Set the pot in a sink with blocks, a rack, or some other object under it. This is called **venting**. It allows cold water to flow under the pot as well as around it.
    - Run cold water into the sink, but not higher than the level of the stock, or the pot will become unsteady. An overflow pipe keeps the water level right and allows for constant circulation of cold water (see image on p. 162).
    - Stir the pot occasionally so all the stock cools evenly. Hang a ladle in the pot so you can give it a quick stir whenever you pass the sink without actually taking extra time to do it.

    Cooling stock quickly and properly is important. Improperly cooled stock can spoil in 6–8 hours because it is a good breeding ground for bacteria that cause food-borne disease and spoilage.

    Do not set the hot stock in the walk-in or, worse yet, the reach-in. All that heat and steam will overload the refrigerator and may damage other perishables as well as the equipment. Refer to food safety guidelines for cooling on page 26.

12. **Evaluate the quality of the finished stock** (see p. 166).

13. **When cool, refrigerate the stock in covered containers.** Stock will keep 2–3 days if properly refrigerated. Stock can also be frozen and will keep for several months.

FIGURE 8.4 Preparing white stock.

(a) Place the bones in a stock pot and cover with cold water.

(b) Skim the scum from the surface regularly.

(c) Add white mirepoix to the pot.

(d) Add a sachet d'épices. Tying the sachet to the pot handle enables it to be retrieved whenever necessary.

Basic White Stock (Beef or Veal) ♥

YIELD: 1 GAL (4 L)

U.S.	METRIC	INGREDIENTS
5 lb	2.5 kg	Beef or veal bones
5–6 qt	5–6 L	Water, cold
1 lb	500 g	Mirepoix, standard or white (see p. 157)
		Sachet:
1	1	Dried bay leaf
½ tsp	2 mL	Dried thyme
½ tsp	2 mL	Peppercorns
3–4	3–4	Parsley stems
1	1	Whole clove (optional)

PROCEDURE

1. Review instructions for stock preparation (pp. 160–161).

2. If beef or veal bones are whole, cut into pieces 3–4 in. (8–10 cm) long with a meat saw. Rinse bones in cold water.

3. Blanch the bones: Place in a stockpot, cover with cold water, and bring to a boil. Drain and rinse.

4. Place the bones in the stockpot and cover with cold water. Bring to a boil, reduce heat to simmer, and skim the scum carefully.

5. Add mirepoix and sachet ingredients (tied in cheesecloth).

6. Simmer for required length of time, skimming the surface as often as necessary.

    Veal: 6–8 hours

    Beef: 8–10 hours

    Add water if necessary to keep bones covered.

7. Strain through a china cap lined with several layers of cheesecloth.

8. Cool the stock, vented, in a cold-water bath (see image below), and refrigerate.

Per 1 fl oz (29.57 mL): Calories, 5; Protein, .6 g; Fat, .1 g (12% cal.); Cholesterol, 0 mg; Carbohydrates, 1 g; Fiber, 0 g; Sodium, 10 mg.

VARIATIONS

Chicken Stock

Chicken stock is usually made with a standard mirepoix, since chicken bones impart a slight yellow color of their own, so a little more color from the carrots is usually not harmful. However, white mirepoix may be used, if desired. Follow procedure in basic recipe, but reduce cooking time to 3–4 hours.

White Lamb Stock, Turkey Stock, Ham Stock

Prepare according to basic procedure, substituting appropriate bones.

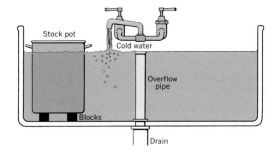

Setup for cooling stocks in a cold-water bath

PREPARING BROWN STOCKS

The difference between brown stocks and white stocks is that the bones and mirepoix are browned for the brown stock. This causes a few complications, as you will see. Otherwise, the procedure is essentially the same.

A second method for browning the mirepoix is given in the alternative procedure.

PROCEDURE for Preparing Brown Stocks

1. Cut the bones into pieces, 3–4 inches (8–10 cm) long, as for white stock. Veal and/or beef bones are used for brown stock.

2. Do not wash or blanch the bones. The moisture would hinder browning.

3. Place the bones in a roasting pan in one layer and brown in a hot oven at 375°F (190°C) or higher. The bones must be well browned to color the stock sufficiently. This takes over 1 hour. Some chefs prefer to oil the bones lightly before browning.

4. When the bones are well browned, remove them from the pan and place them in a stockpot. Cover with cold water and bring to a simmer.

5. Drain and reserve the fat from the roasting pan. Deglaze the pan by adding water and stirring over heat until all the brown drippings are dissolved or loosened. Add to the stockpot.

6. While the stock is getting started, place the mirepoix in the roasting pan with some of the reserved fat and brown the vegetables well in the oven.

7. Add the tomato product to the mirepoix. Continue to brown until the tomato product turns a rusty brown color. (See alternative procedure below.)

8. When the water in the stockpot comes to a simmer, skim and continue as for white stock.

9. Add the browned vegetables and the tomato product to the stockpot. If desired, they may be held out until 2–3 hours before the end of the cooking time.

10. Continue as for white stock.

FIGURE 8.5 Preparing brown stock.

(a) Roast the bones in a moderately hot oven until well browned.

(b) Place the bones in a stockpot and add the appropriate amount of water.

(c) While the bones are beginning to simmer, brown the mirepoix, using the same roasting pan set on top of the stove or in the oven. Add the browned mirepoix to the stockpot.

(d) Deglaze the roasting pan with water. Add the liquid to the stockpot.

(e) This stock has simmered slowly for 8 hours. Note the sachet is tied to the handle of the pot with twine for easy removal.

(f) Strain the stock through a china cap lined with cheesecloth.

ALTERNATIVE PROCEDURE

The mirepoix may be browned with the bones. When the bones are half browned, add the mirepoix to the pan and continue roasting until bones and vegetables are browned.

Some chefs use this method because it eliminates some steps. Others prefer to brown the mirepoix separately so it can be added to the stock later in the cooking time.

Tomato paste is the best choice of tomato product if it is browned with the mirepoix, as it contains less water. Tomato purée or canned tomatoes may also be used, but they are more likely to be added directly to the stockpot.

 Basic Brown Stock ♥

YIELD: 1 GAL (4 L)

U.S.	METRIC	INGREDIENTS	PROCEDURE
5 lb	2.5 kg	Bones: veal or beef	1. Review instructions for stock preparation (p. 160–161).
5–6 qt	5–6 L	Water, cold	2. If bones are whole, cut into pieces 3–4 inches (8–10 cm) long with a meat saw. Do not wash.
			3. Place bones in a roasting pan in a hot oven (400°F/200°C) and brown them well.
			4. Remove bones from pan and place in a stockpot. Cover with water and bring to a simmer. Skim and let stock continue to simmer.
			5. Drain and reserve the fat from the roasting pan. Deglaze the pan with water and add to stockpot.
1 lb	500 g	Mirepoix, standard (see p. 157)	6. Toss the mirepoix with some of the reserved fat and brown well in oven.
2 oz	60 g	Tomato paste (see Note)	7. Add tomato paste to mirepoix. Continue to brown until tomato paste turns a rusty brown color.
		Sachet:	8. Add browned mirepoix, tomato product, and sachet to the stockpot.
1	1	Bay leaf	9. Continue to simmer for required length of time, skimming surface as necessary.
1/2 tsp	2 mL	Dried thyme	Veal: 6–8 hours
1/2 tsp	2 mL	Peppercorns	Beef: 8–10 hours
3–4	3–4	Parsley stems	Add water as needed to keep bones covered.
1	1	Whole cloves	10. Strain through a china cap lined with several layers of cheesecloth.
			11. Cool the stock, vented, in a cold-water bath, and refrigerate.

Per 1 fl oz (29.57 mL): Calories, 6; Protein, 0.5 g; Fat, 0.1 g (18% cal.); Cholesterol, 0 mg; Carbohydrates, 0.5 g; Fiber, 0 g; Sodium, 15 mg.

Note: If desired, use 1 lb (500 g) tomato purée or canned tomatoes in place of the tomato paste. Add it to the browned mirepoix and continue to brown, or add it directly to the stockpot without browning.

VARIATIONS

Brown Lamb Stock, Game Stock

Prepare according to basic procedure, substituting appropriate bones.

PREPARING FISH STOCKS

A standard fish stock is made the same way as a white stock, using the proportions given in the table on page 159. This type of stock is useful for soups and similar seafood preparations. For sauces, chefs usually prefer a more flavorful fish stock called a *fumet*. A fish fumet is made by first sweating the bones and mirepoix in fat. White wine is then added to deglaze, and finally water is added and the fumet finished in the normal manner.

Recipes for a standard fish stock and a fish fumet are given below.

Fish Stock 🖤

YIELD: 1 GAL (4 L)

U.S.	METRIC	INGREDIENTS
5 lb	2.5 kg	Bones from lean fish
1 lb	500 g	Mirepoix, white, sliced thin or chopped fine (see p. 157)
		Sachet:
1	1	Bay leaf
¹/₂ tsp	2 mL	Peppercorns
3–4	3–4	Parsley stems
¹/₂ tsp	2 mL	Dried thyme
1 gal	4 L	Water, cold
24 fl oz	750 mL	White wine (optional; see Note)

PROCEDURE

1. Review instructions for stock preparation (pp. 160–161).
2. Place all ingredients in a stockpot. Bring to a simmer.
3. Simmer 45 minutes, skimming as necessary to remove scum.
4. Strain through a china cap lined with several layers of cheesecloth.
5. Cool stock, vented, in a cold-water bath, and refrigerate.

Per 1 fl oz (29.57 mL): Calories, 4; Protein, 0.8 g; Fat, 0.3 g (40% cal.); Cholesterol, 0 mg; Carbohydrates, 0.2 g; Fiber, 0 g; Sodium, 2 mg.

Note: Wine is often omitted for ordinary fish stock (although it is always used in fish fumet). Alternatively, reduce the quantity of wine as desired.

Fish Fumet 🖤

YIELD: 1 GAL (4 L)

U.S.	METRIC	INGREDIENTS
1 oz	30 g	Clarified butter
1 lb	500 g	Mirepoix, white, sliced thin or chopped fine (see p. 157)
5 lb	2.5 kg	Bones from lean fish
24 fl oz	750 mL	White wine (dry; see Note)
		Sachet:
1	1	Bay leaf
¹/₂ tsp	2 mL	Peppercorns
3–4	3–4	Parsley stems
¹/₂ tsp	2 mL	Dried thyme
1 gal	4 L	Water, cold

PROCEDURE

1. Butter the bottom of a heavy stockpot or saucepot. Place the mirepoix in bottom of pot and the bones over top of it. Cover bones loosely with a round of brown paper or parchment.
2. Set pot over low heat and cook slowly about 5 minutes, or until bones are opaque and begin to exude juices.
3. Add the wine, bring to a simmer, and simmer 10 minutes. Then add the sachet and water to cover.
4. Bring to a simmer again, skim, and let simmer 30–45 minutes.
5. Strain through a china cap lined with several layers of cheesecloth.
6. Cool, vented, in a cold-water bath, and refrigerate.

Per 1 fl oz (29.75mL): Calories, 10; Protein, 0.8 g; Fat, 0.2 g (17% cal.); Cholesterol, 1 mg; Carbohydrates, 0.3 g; Fiber, 0 g; Sodium, 3 mg

Note: If desired, use a standard mirepoix (substituting carrots for the parsnips) instead of white mirepoix. The stock will have slightly more color. Alternatively, simply omit the parsnips, if desired. Many chefs use only half this quantity of wine. Reduce quantity if desired.

FIGURE 8.6 Making fish fumet.

(a) Sweat the mirepoix and fish bones in butter.

(b) Add the white wine and bring to a simmer.

Vegetable Stock ♥ 🌸

YIELD: 1 GAL (4 L)

U.S.	METRIC	INGREDIENTS	PROCEDURE
1¹/₂ fl oz	45 mL	Oil	1. Heat oil in a stockpot over medium heat.
		Mirepoix:	2. Add mirepoix, leeks, mushrooms, turnip, fennel, and garlic. Sweat for 10 minutes. Stir as necessary so vegetables do not brown.
1 lb	500 g	Onion, chopped	
8 oz	250 g	Carrot, chopped	
8 oz	250 g	Celery, chopped	
8 oz	250 g	Leeks, chopped	
4 oz	125 g	Mushrooms or mushroom trimmings, chopped	
4 oz	125 g	Turnip, chopped	
2 oz	60 g	Fennel, chopped	
¹/₂ oz	15 g	Garlic, chopped	
4 oz	125 g	Tomatoes, chopped	3. Add tomatoes, water, and sachet.
5 qt	5 L	Water	4. Bring to a simmer, and simmer 45 minutes.
		Sachet:	5. Strain and cool in a water bath.
1	1	Bay leaf	
¹/₂ tsp	2 mL	Dried thyme	
¹/₂ tsp	2 mL	Peppercorns	
3–4	3–4	Parsley stems	
2	2	Whole cloves	

Per 1 fl oz (29.57 mL): Calories, 2; Protein, 0.1 g; Fat, 0.1 g (27% cal.); Cholesterol, 0 mg; Carbohydrates, 0.5 g; Fiber, 0 g; Sodium, 3 mg.

VARIATIONS

Other vegetables may be used as desired.
See discussion on page 158–159.

STANDARDS OF QUALITY FOR STOCKS

After you have finished a stock, evaluate its quality. Look for the following characteristics:

Clarity. When hot, a well-made beef, veal, or chicken stock is nearly clear, with only a slight amount of cloudiness. Fish and vegetable stocks may have slightly more cloudiness.

Color. When hot, white stocks made from beef, veal, and fish bones should be nearly colorless. Chicken stocks may have a slight golden color. Brown stocks should have a deep amber or rich golden brown color, although not as brown as canned beef broth. Vegetable stocks vary on color, depending on the vegetables used, although a neutral vegetable stock usually has a light golden color when carrots are used.

Body. When hot, well-made meat and poultry stocks have a rich mouth feel, due to their gelatin content. When chilled, they should be very thick or even gelled. Fish stocks should also have a rich mouth feel from gelatin, although the gelatin content may be somewhat less, depending on the fish used. Vegetable stocks have no gelatin content and therefore are much thinner.

Aroma and Flavor. All stocks should have a rich, full-bodied, and well-balanced flavor and aroma. Meat, poultry, and fish stocks should taste and smell mainly of their main ingredients, with no strong flavors from the mirepoix or the herbs and spices in the sachet. Brown stocks should have a richer, deeper flavor, but with no charred flavor from bones that were browned too heavily. Fish fumets should have a smooth, mellow flavor from the white wine content, with no harsh alcohol flavors. Flavors and aromas of neutral vegetable stock should be rich and well-balanced, with no strong flavor from any single vegetable. Vegetable stocks made for specific purposes with a high proportion of a single vegetable should have the fresh flavor and aroma of that vegetable, not a strong or unpleasant flavor of overcooked vegetables.

REMOUILLAGE AND OTHER STOCK-RELATED PREPARATIONS

Remouillage is a stock made from bones that were already used once to make stock. The literal meaning of the French term is "rewetting." Because not all possible flavor and gelatin is extracted from bones when making a stock, making a remouillage allows the chef to extract a little more value from the bones. The resulting liquid will not be as clear or flavorful as the original stock, but it does have some uses. A remouillage can be used for soups, for braised dishes, and in place of water for making stocks. It can also be reduced to a glaze and used for enriching sauces, soups, and braising liquids.

To make a remouillage, discard the mirepoix and herb sachet after draining a finished stock. Add fresh mirepoix and sachet to the bones, cover with fresh cold water, and simmer about 4 hours. Drain and cool as for regular stock.

Court bouillon is often discussed along with stocks. However, it is rarely used for making soups and sauces, as stocks are. Rather, it is used as a cooking medium for fish, and therefore is discussed in Chapter 20 (see page 638).

Basic Japanese stock, called *dashi*, is quickly and easily made from only three ingredients: water; shaved, dried bonito, called *katsuobushi* (kaht soo oh boo shee); and a type of dried seaweed or kelp called *kombu*. Dashi is used in soups and dipping sauces (page 637) as well as other dishes.

Finally, broths and jus are discussed later in this chapter (page 199).

Dashi

YIELD: 2 QT (2 L)

U.S.	METRIC	INGREDIENTS
4¹/₂ pt	2.25 L	Water, cold
2 oz	60 g	Kombu (giant kelp for stock)
1¹/₂ oz	50 g	Katsuobushi (dried bonito flakes)

PROCEDURE

1. Put the water in a pot and add kombu. Bring to a boil over moderately high heat.

2. Just as water comes to a boil, remove the kombu.

3. Remove from heat and immediately add bonito flakes. Let the flakes settle to bottom. This will take 1–2 minutes.

4. Strain through a china cap lined with cheesecloth. Use dashi within 1 day.

Per 1 fl oz (29.57 mL): Calories, 2; Protein, 0 g; Fat, 0 g (0% cal.); Cholesterol, 0 mg; Carbohydrates, 0 g; Fiber, 0 g; Sodium, 3 mg.

Note: Instant dashi is also available. Its quality is good enough for simmered dishes and miso soup, but not for good clear soup. Follow label instructions.

VARIATIONS

Vegetarian Dashi

Omit bonito flakes and use only kombu.

FIGURE 8.7 Dashi ingredients: katsuobushi and kombu.

REDUCTIONS AND GLAZES

Stocks are concentrated by boiling or simmering them to evaporate part of the water. This is called making a **reduction**, or reducing.

Reduction is an important technique in sauce-making and in many other areas of cooking because it produces a more flavorful product by concentrating it. A reduced stock also has more body because the gelatin is concentrated.

WHAT ARE GLAZES?

A **glaze**—or, in French, *glace* (glahss)—is a stock reduced until it coats the back of a spoon. It is so concentrated—reduced by three-fourths or more—that it is solid and rubbery when refrigerated.

Glazes are used as flavorings in sauce-making and in some meat, poultry, fish, and vegetable preparations. Only small amounts are needed because they are so concentrated.

Glazes diluted to original strength do not taste like the stocks they were made from. The long cooking changes the flavors somewhat.

PROCEDURE for Preparing Glazes

1. Reduce the stock over moderate heat.
2. Skim the surface frequently.
3. When reduced by half to two-thirds, strain into a smaller, heavy saucepan and continue to reduce over lower heat until the liquid is syrupy and coats a spoon.
4. Pour into containers, cool, cover, and refrigerate.
5. Glazes will keep for several weeks or longer if properly stored. They may also be frozen.

Glace de viande

KINDS OF GLAZES

1. Meat glaze, or **glace de viande** (glahss duh vee awnd)—made from brown stock.
2. Chicken glaze, or **glace de volaille** (voh lye)—made from chicken stock.
3. Fish glaze, or **glace de poisson** (pwah sohn)—made from fish stock.

CONVENIENCE BASES

The cost, both in time and materials, of making stocks in modern kitchens has led to the widespread use of concentrated convenience products known as *bases*. These are diluted with water to make flavored liquids similar to stocks.

Glazes can be considered bases and, in fact, they are the original bases, used long before today's manufacturers started producing convenience products.

JUDGING QUALITY

Bases vary greatly in quality. The best ones are composed mainly of meat extracts. These are perishable products and must be refrigerated.

Many bases are made primarily from salt, however—an expensive way to buy salt, we might add. ***Read the list of ingredients***. Avoid products that list salt first. The best way to judge the quality of a base is to dilute it and compare its flavor to that of a well-made stock.

USING BASES

Bases can be improved with little labor by simmering the diluted or made-up product for a short time with some mirepoix, a sachet, and a few bones or meat trimmings, if possible. This helps give a fresher, more natural taste to a highly processed product.

Bases are also added to stocks to supplement them when only a small quantity of stock is on hand.

Bases are sometimes added to weak stocks to give them more flavor, but this is not as good a practice as making the stock properly in the first place.

Using bases requires taste and judgment, just as other areas of cookery do. If used without care and restraint, bases can detract from the quality of your cooking. But, used carefully, they can be a valuable tool. Always taste and evaluate as you cook.

There is no substitute for a well-made stock. But it is also true that a good base may be better than a poorly made stock. It all depends on the skills you are learning now.

KEY POINTS TO REVIEW

- What are the steps in the procedure for preparing white stock?

- How does making brown stock differ from making white stock?

- What is the best way to cool a stock?

- What is glace de viande? How is it prepared?

- How do you judge the quality of a base?

 SAUCES

Like stocks, sauces have lost some of the importance they once had in commercial kitchens—except, of course, in the best restaurants serving what may be considered luxury cuisine. Some of this decline is due to changes in eating habits and to increased labor costs.

However, much of the change is due to misunderstanding. How many times have you heard someone say, "I don't go for all those sauces all over everything. I like good, simple food." No doubt this person puts ketchup—a sweetened tomato sauce—on hamburgers, gravy on mashed potatoes, and tartar sauce on fried fish.

The misunderstandings arise from poorly made sauces. No one likes thick, pasty cream sauces on vegetables or oversalted but otherwise flavorless brown sauces gumming up their meat. But just because some cooks make bad sauces is no reason to reject all sauce cookery.

In fact, many chefs believe good sauces are the pinnacle of all cooking, both in the skill they require and in the interest and excitement they can give to food. Very often, the most memorable part of a really fine meal is the sauce that enhances the meat or fish.

A sauce works like a seasoning. It enhances and accents the flavor of the food; it should not dominate or hide the food.

A good cook knows that sauces are as valuable as salt and pepper. A simple grilled steak is made even better when it has an added touch, something as simple as a slice of seasoned butter melting on it or as refined as a spoonful of béarnaise sauce.

No matter where you work, sauce-making techniques are basic skills you will need in all your cooking. Croquettes, soufflés, and mousses have sauces as their base, nearly all braised foods are served with sauces made of their cooking liquids, and basic pan gravies, favorites everywhere, are made with the same techniques as the classic sauces.

UNDERSTANDING SAUCES

THE FUNCTIONS OF SAUCES

A **sauce** may be defined as a flavorful liquid, usually thickened, used to season, flavor, and enhance other foods.

A sauce adds the following qualities to foods:

Moistness

Flavor

Richness

Appearance (color and shine)

Interest and appetite appeal

THE STRUCTURE OF SAUCES

The major sauces we consider here are made of three kinds of ingredients:

1. A liquid, the body of the sauce

2. A thickening agent

3. Additional seasoning and flavoring ingredients

To understand sauce-making, you must first learn how to prepare these components and then how to combine them into finished sauces.

Liquid

A liquid ingredient provides the body or base of most sauces. Most classic sauces are built on one of five liquids or bases. The resulting sauces are called **leading sauces** or **mother sauces**.

White stock (chicken, veal, or fish)—for velouté sauces

Brown stock—for brown sauce or espagnole (ess pahn yohl)

Milk—for béchamel

Tomato plus stock—for tomato sauce

Clarified butter—for hollandaise

The most frequently used sauces are based on stock. The quality of these sauces depends on the stock-making skills you learned in the previous section.

Thickening Agents

A sauce must be thick enough to cling lightly to the food. Otherwise, it will just run off and lie in a puddle in the plate. This doesn't mean it should be heavy and pasty. Chefs use the term **nappé** (nap pay; from the French *napper*, meaning "to top") to describe the texture of a sauce that has the right texture to coat foods.

Starches are still the most commonly used thickening agents, although they are used less often than in the past. We discuss starches and other thickening agents in detail below.

Other Flavoring Ingredients

Although the liquid that makes up the bulk of the sauce provides the basic flavor, other ingredients are added to make variations on the basic themes and to give a finished character to the sauces.

Adding specified flavoring ingredients to basic sauces is the key to the catalog of classic sauces. Most of the hundreds of sauces listed in the standard repertoires are made by adding one or more flavoring ingredients to one of the five basic sauces or leading sauces.

As in all of cooking, sauce-making is largely a matter of learning a few building blocks and then building with them.

ROUX

STARCHES AS THICKENERS

1. Starches are the most common and most useful thickeners for sauce-making. Flour is the principal starch used. Others available to the chef include cornstarch, arrowroot, waxy maize, instant or pregelatinized starch, bread crumbs, and other vegetable and grain products, like potato starch and rice flour. These are discussed later.

2. Starches thicken by *gelatinization*, which, as discussed in Chapter 6, is the process by which starch granules absorb water and swell to many times their original size, and starch molecules uncoil into long threads.

Another important point made in Chapter 6 is that acids inhibit gelatinization. Whenever possible, do not add acid ingredients to sauces until the starch has fully gelatinized.

3. Starch granules must be separated before heating in liquid to avoid lumping. If granules are not separated, lumping occurs because the starch on the outside of the lump quickly gelatinizes into a coating that prevents the liquid from reaching the starch inside.

Starch granules are separated in two ways:

- *Mixing the starch with fat.* This is the principle of the roux, which we discuss now, and of beurre manié, which is discussed in the next section.
- *Mixing the starch with a cold liquid.* This is the principle used for starches such as cornstarch. It can also be used with flour, but, as we note later, the result is an inferior sauce. A mixture of raw starch and cold liquid is called a **slurry**.

ROUX INGREDIENTS

Roux (roo) is a cooked mixture of equal parts by weight of fat and flour.

Fat

The cooking fats employed for making roux are as follows:

Clarified butter is preferred for the finest sauces because of its flavor. The butter is clarified (p. 190) because the moisture content of whole butter tends to gelatinize some of the starch and makes the roux harder to work.

Margarine is sometimes used in place of butter because of its lower cost. However, its flavor is inferior to butter, so it does not make as fine a sauce. The quality of margarine varies from brand to brand.

Animal fats, such as chicken fat, beef drippings, and lard, are used when their flavor is appropriate to the sauce. Thus, chicken fat can be used for chicken velouté, and beef drippings can be used for beef gravy. When properly used, animal fats can enhance the flavor of a sauce.

Vegetable oil and shortening can be used for roux but, because they add no flavor, they are not preferred. Solid shortening also has the disadvantage of having a high melting point, which gives it an unpleasant fuzzy feeling in the mouth. It is best reserved for the bakeshop and the fry kettle.

Today, roux-thickened sauces are often condemned for health reasons because of the fat content of the roux. It should be remembered, however, that when a roux-bound velouté or brown sauce is properly made, most of the fat is released and skimmed off before the sauce is served.

Flour

The thickening power of flour depends, in part, on its starch content. Bread flour has less starch and more protein than cake flour. Eight parts (such as ounces or grams) of cake flour has the same thickening power as 10 parts of bread flour.

Bread flour frequently is used for general cooking purposes in commercial kitchens even though it has less thickening power than cake flour or pastry flour. Most sauce recipes in this book, as well as in other books, are based on bread flour or on all-purpose flour, which has similar thickening power. The proportions of roux to liquid must be adjusted if another flour is used.

Flour is sometimes browned dry in the oven for use in brown roux. A heavily browned flour has only one-third the thickening power of unbrowned flour.

In addition to starch, wheat flour contains proteins and other components. As a roux-thickened sauce is simmered, these components rise to the surface as scum. They then can be skimmed off. Sauces are generally simmered for a time even after the starch is completely gelatinized so these "impurities" can be cooked off. This improves the texture, gloss, and clarity of a sauce. When a high-protein flour such as bread flour is used in a roux, the sauce must be cooked longer and skimmed more often to achieve good clarity.

Sauces made with wheat flour do not freeze well because some of the starch breaks down when frozen, reducing its thickening power.

Ingredient Proportions

Correct amounts of fat and flour—*equal parts by weight*—are important to a good roux. There must be enough fat to coat all the starch granules, but not too much. In fact, Escoffier called for even less fat than our standard proportions (8 parts fat to 9 parts flour).

A good roux is stiff, not runny or pourable. A roux with too much fat is called a **slack roux**. Excess fat increases the cost of the roux unnecessarily; the excess fat rises to the top of the sauce, where it either is skimmed off or makes the sauce look greasy.

PREPARING ROUX

A roux must be cooked so the finished sauce does not have the raw, starchy taste of flour. The three kinds of roux differ in how much they are cooked:

FIGURE 8.8 Cooking white roux.

White roux is cooked for just a few minutes, just enough to cook out the raw taste. Cooking is stopped as soon as the roux has a frothy, chalky, slightly gritty appearance, before it has begun to color. White roux is used for béchamel and other white sauces based on milk. In spite of its name, white roux is actually a pale yellow because it is made from butter and (usually) unbleached flour. Figure 8.8 illustrates the production of white roux.

Blond roux, or pale roux, is cooked a little longer, just until the roux begins to change to a slightly darker color. Cooking must then be stopped. Blond roux is used for veloutés, or sauces based on white stocks. The sauces have a pale ivory color.

Brown roux is cooked until it takes on a light brown color and a nutty aroma. Cooking must take place over low heat so the roux browns evenly without scorching. For a deeper brown roux, the flour may be browned in an oven before adding it to the fat. A heavily browned roux has only about one-third the thickening power of white roux, but it contributes flavor and color to brown sauces.

White, blond, and brown roux

Operations that depend on roux-based sauces and soups generally make quantities of roux in bulk and keep it available throughout the production period for thickening sauces. If you don't make batches of roux as part of the mise en place, you can also make it as part of the production process for an individual sauce. In the recipes in this section, making the roux is part of the recipe. However, they can easily be changed to use a prepared roux. For example, in the Velouté Sauce recipe (p. 182), omit the flour and use just enough butter to sweat the mirepoix. After adding the stock, beat in 8 oz (250 g) prepared blond roux.

BASIC PROCEDURE for Making All Roux

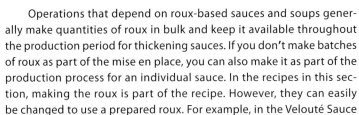

1. Melt fat.

2. Add correct amount of flour and stir until fat and flour are thoroughly mixed.

3. Cook to required degree for white, blond, or brown roux.

Cooking is done in a saucepan on top of the stove, and the roux is stirred for even cooking. Use low heat for brown roux, moderate heat for white or blond roux. Large quantities may be baked in an oven. Some restaurants make up batches large enough to last for several days.

INCORPORATING THE ROUX

Combining the roux and liquid to obtain a smooth, lump-free sauce is a skill that takes practice to master. It's a good idea to practice the various techniques with water, under the guidance of your instructor, so you understand what you are doing before you start working with valuable stocks.

General Principles

Liquid may be added to roux, or roux may be added to liquid.

The liquid may be hot or cooled, but not ice cold. A very cold liquid will solidify the fat in the roux.

The roux may be warm or cold, but not sizzling hot. Adding a hot liquid to a very hot roux causes spattering and, possibly, lumps.

Most chefs find they get the best results by combining a cold (or cool) liquid with a hot roux, or a hot liquid with a cold roux.

Within these general guidelines, there is room for a number of variations. Two of them are described here. Because successful use of roux is largely a matter of experience, you are advised to profit from your instructors' experience when they demonstrate these techniques or whichever methods they prefer.

Equipment note: Stainless-steel pans are best for white sauces. Whipping in an aluminum pan makes the sauce gray.

PROCEDURES for Incorporating Roux

METHOD 1: ADDING LIQUID TO ROUX

This method is used when a roux is made up specifically for the sauce, gravy, or soup being prepared.

1. Use a heavy saucepot to prevent scorching either the roux or the sauce.

2. When the roux is made, remove the pan from the fire for a few minutes to cool slightly.

3. Slowly pour in the liquid, all the while beating vigorously with a wire whip to prevent lumps from forming.

If the liquid is hot (such as simmering milk for béchamel sauce), you must beat especially well because the starch will gelatinize quickly.

If the liquid is cool, you can add a quantity of it, beat to dissolve the roux, and then add the remainder of the liquid, hot or cool.

4. Bring the liquid to a boil, continuing to beat well. The roux does not reach its full thickening power until near the boiling point.

5. Simmer the sauce, stirring from time to time, until all the starchy taste of the flour is cooked out. This takes at least 10 minutes, but the flavor and consistency of the sauce improve if it is cooked longer. Many chefs feel 20 minutes of simmering is a bare minimum. Others cook some sauces for an hour or longer.

6. When the sauce is finished, it may be kept hot in a bain-marie or cooled for later use. Either way, it should be covered or have a thin film of butter melted onto the top to prevent a skin from forming.

METHOD 2: ADDING THE ROUX TO THE LIQUID

Many restaurants make up large batches of roux to last all day or even all week. This method may be used in these situations.

1. Bring the liquid to a simmer in a heavy pot.

2. Add a small quantity of roux and beat vigorously with a whip to break up all lumps.

3. Continue to beat small quantities into the simmering liquid until the desired consistency is reached. Remember that roux must simmer for a time to thicken completely, so do not add roux too quickly or you risk overthickening the sauce.

4. Continue to simmer until the roux is cooked and no starchy taste remains.

5. If the sauce is to simmer a long time, underthicken it because it will thicken as it reduces.

PROPORTIONS OF ROUX TO LIQUID

Table 8.5 indicates the quantities of roux needed to thicken 1 gallon (4 L) liquid to thin, medium, and thick consistencies.

TABLE 8.5 Roux Proportions in Sauces				
Sauce	**Butter**	**Flour**	**Roux**	**Liquid**
Thin or light	6 oz/190 g	6 oz/190 g	12 oz/375 g	1 gal/4 L
Medium	8 oz/250 g	8 oz/250 g	1 lb/500 g	1 gal/4 L
Thick or heavy	12 oz/375 g	12 oz/375 g	1½ lb/750 g	1 gal/4 L

How thick is a thick sauce? Obviously, these are not precise, scientific terms that can be defined easily. Experience can be the only teacher in this case. This is another good reason to practice with roux and water—so you can, with experience, produce the exact consistency you want.

You also have available the techniques of dilution and reduction to adjust the consistency of a sauce (see pp. 176–177), and you will learn how to use beurre manié and other thickening agents.

OTHER THICKENING AGENTS

STARCHES

1. **Beurre manié** (burr mahnyay) is a mixture of equal parts soft, raw butter and flour worked together to form a smooth paste. It is used for quick thickening at the end of cooking to finish a sauce. The raw butter adds flavor and gives a sheen to the sauce when it melts.

    To use, drop very small pieces into a simmering sauce and stir with a whip until smooth. Repeat until desired consistency is reached. Simmer just a few minutes more to cook the flour, and then remove from the fire.

2. **Whitewash** is a thin mixture of flour and cold water. Sauces made with whitewash have neither as good a flavor nor as fine a texture as those made with roux. *Whitewash is not recommended for use*.

3. *Cornstarch* produces a sauce that is almost clear, with a glossy texture.

    To use, mix with cold water or other cold liquid until smooth. Stir into the hot liquid. Bring to a boil and simmer until the liquid turns clear and there is no starchy taste. Do not boil for a long period or the starch may break down and the liquid become thin. Sauces thickened with cornstarch may thin out if held on the steam table for long periods. Cornstarch is used extensively in sweet sauces to accompany certain meats as well as in desserts and dessert sauces. It has roughly twice the thickening power of flour.

4. *Arrowroot* is used like cornstarch, but it gives an even clearer sauce. Its use is limited by its high cost. Nevertheless, because of its quality, it is the preferred starch for thickening jus lié. It is less likely than cornstarch to break down when heated for a long time.

5. *Waxy maize* is used for sauces that are to be frozen. Flour and other starches break down and lose their thickening power when frozen. Waxy maize does not. It is handled like cornstarch.

6. *Pregelatinized* or *instant starches* have been cooked, or gelatinized, and then redried. Thus, they can thicken a cold liquid without heating. These starches are rarely used in sauce-making but are frequently used in the bakeshop.

7. *Bread crumbs* and other crumbs will thicken a liquid quickly because they have already been cooked, like instant starches. Bread crumbs may be used when smoothness of texture is not desired. A common example is the use of gingersnap crumbs to thicken sauerbraten gravy.

Cornstarch settles out and must be restirred before use.

8. ***Vegetable purées, ground nuts, and other solids*** can also be used. A simple tomato sauce is basically a seasoned vegetable purée. The sauce gets its texture from the thickness of the main ingredient. No additional thickener is needed.

Using this same principle, we can add body or texture to sauces by adding a smooth vegetable purée, or by puréeing mirepoix or other vegetables with the sauce. Other puréed or finely ground ingredients, such as ground nuts, add texture as well as flavor to a sauce.

EGG YOLK AND CREAM LIAISON

In classical cooking, a **liaison** is a mixture of egg yolks and cream, used to enrich and lightly thicken a sauce or other liquid. Egg yolks have the power to thicken a sauce slightly due to the coagulation of egg proteins when heated.

Caution must be used when thickening with egg yolks because of the danger of curdling. This happens when the proteins coagulate too much and separate from the liquid.

Pure egg yolks coagulate at 140°–158°F (60°–70°C). For this reason, they are beaten with heavy cream before use. This raises their curdling temperature to 180°–185°F (82°–85°C). (Note this is still well below the boiling point.) The heavy cream also adds thickness and flavor to the sauce.

Egg yolks have only slight thickening power. The liaison is used primarily to give richness of flavor and smoothness of texture to a sauce and only secondarily to give a slight thickening. Also, because of the instability of the egg yolks, it is used only as a finishing technique. Incorporating a liaison is illustrated in Figure 8.9.

PROCEDURE for Using a Liaison

1. Beat together the egg yolks and cream in a stainless-steel bowl. Normal proportions are 2–3 parts cream to 1 part egg yolks.

2. Very slowly add a little of the hot liquid to the liaison, beating constantly. This is known as tempering.

3. Off the heat, add the warmed, diluted liaison to the rest of the sauce, stirring well as you pour it in.

4. Return the sauce to low heat to warm it gently, but do not heat it higher than 180°F (82°C) or it will curdle. Under no circumstances should it boil.

5. Hold for service above 135°F (57°C) for sanitation reasons, but lower than 180°F (82°C).

EGG YOLK EMULSIFICATION

Egg yolks are used as the thickening agent for hollandaise and related sauces, but in this case the principle is entirely different. The entire procedure is discussed in detail when we get to the hollandaise family of sauces, page 193.

REDUCTION

Simmering a sauce to evaporate some of the water thickens the sauce because only the water evaporates, not the solids. As the solids become more concentrated, the sauce becomes thicker. This technique has always been important for finishing sauces (see the next section). It has become more important as a basic thickening technique as modern chefs use less starch for thickening.

Use caution when reducing stock-based sauces. If such a sauce is reduced too much, the concentration of gelatin may give it a gluey or sticky texture, and it will congeal quickly on plates. Also, the sauce may have a heavily cooked taste that is not as appealing as the fresher, livelier taste of a stock that has not been cooked as much.

FIGURE 8.9 Adding a liaison to a sauce.

(a) Slowly stir a little of the hot sauce (chicken velouté, in this picture) into the mixture of cream and egg yolks to warm it and dilute it.

(b) Stir the tempered liaison back into the remaining sauce.

FINISHING TECHNIQUES

Remember that the three basic elements of a finished sauce are a liquid, a thickening agent, and additional seasoning and flavoring ingredients. We have discussed in detail how liquids are combined with thickening agents to make the basic sauces. In the next section, we look at the way families of sauces are built on these bases by the addition of flavoring ingredients.

Sauces may be modified or added to in a great many ways. Among these methods are a number of basic techniques used over and over again for making sauces. Before we study the structure of the sauce families, it will be helpful to look at these basic finishing techniques.

REDUCTION

1. Using reduction to concentrate basic flavors.

If we simmer a sauce for a long time, some of the water evaporates. The sauce becomes more concentrated, and the resulting product is more flavorful. This is the same technique used when making glazes from stocks. Some reduction takes place in nearly all sauces, depending on how long they are simmered.

2. Using reduction to adjust textures.

Concentrating a sauce by reduction also thickens it because only the water evaporates, not the roux or other solids. A skilled sauce chef uses both reduction and dilution to give a sauce the precise texture sought. If a sauce is too thin, it may be simmered until it reaches desired thickness. Or the chef may add a large quantity of stock or other liquid to a thickened sauce to thin it out greatly, then simmer it again until it is reduced to just the right consistency. By doing this, the chef also gives more flavor to the sauce.

3. Using reduction to add new flavors.

If we can add a liquid to a sauce, then reduce it to concentrate it, why can't we reduce a liquid first and then add it to a sauce?

In fact, this is one of the most important techniques in sauce-making. We have already mentioned that glazes—reduced stocks—are used to flavor sauces. Reductions of other liquids, especially red and white wines, are used a great deal in this way.

Skip ahead to the recipe for Bordelaise Sauce (p. 187). Note how the red wine is cooked down with shallots, pepper, and herbs to one-fourth its original volume. Not only is the flavor of the wine concentrated but also the flavor from the other spices is extracted. This reduction is a powerful flavoring agent that gives bordelaise sauce its distinctive taste. Reduction allows you to add a great deal of flavor to a sauce without adding much liquid.

Terminology

To reduce by one-half means to cook away one-half of the volume so that half is left.

To reduce by three-fourths means to cook away three-fourths of the volume so that only one-fourth is left.

To reduce **au sec** (oh seck) means to reduce until dry or nearly dry.

STRAINING

If you have learned how to use a roux properly, you should be able to make a smooth, lump-free sauce. However, to bring a sauce's texture to perfection, to create the velvety smoothness that is important to a good sauce, straining is necessary. Even a slight graininess that you can't see can still be felt on your tongue.

Straining through a china cap lined with several layers of cheesecloth is effective. Very fine sieves are also available for straining sauces. Straining is usually done before final seasoning.

DEGLAZING

To **deglaze** means to swirl a liquid in a sauté pan or other pan to dissolve cooked particles of food remaining on the bottom.

This term was used in connection with the production of brown stock and will be encountered again in Chapter 14 in relation to the basic technique of sautéing. It is also an important technique for finishing sauces that accompany sautéed items.

A liquid, such as wine or stock, is used to deglaze a sauté pan and then is reduced by one-half or three-fourths. This reduction, with the added flavor of the pan drippings, is then added to the sauce served with the item.

ENRICHING WITH BUTTER AND CREAM

1. Liaison.

In addition to being a thickening agent, a liaison of egg yolks and cream is used to finish a sauce by giving it extra richness and smoothness.

2. Heavy cream.

Heavy cream has long been used to give flavor and richness to sauces. The most obvious example is adding cream to basic béchamel sauce to make cream sauce.

3. Butter.

A useful enriching technique, both in classical and in modern cooking, is called *finishing with butter*, or **monter au beurre** (mohn tay oh burr).

To finish a sauce with butter, simply add a few pieces of softened butter to the hot sauce and swirl them in until melted. The sauce should then be served immediately; if it is allowed to stand, the butter may separate.

Finishing a sauce with butter gives it a little extra shine and smoothness as well as adding to it the rich, fresh taste of raw butter.

Monter au buerre

SEASONING

Whether or not a sauce is to be given a final enrichment of liaison, cream, or butter, it must be checked carefully for seasonings before serving. Remember that the last step in any recipe, whether written or not, is "adjust the seasonings."

1. *Salt* is the most important seasoning for sauces. *Lemon juice* is also important. These two seasonings emphasize the flavors already present by stimulating the taste buds. *Cayenne* and *white pepper* are perhaps third and fourth in importance.

2. *Sherry* and *Madeira* are frequently used as final flavorings. These wines are added at the end of cooking (unlike red and white table wines, which must be cooked in a sauce) because their flavors are easily evaporated by heat.

KEY POINTS TO REVIEW

- What five qualities do sauces add to foods?

- What is roux? How is it made? How is it used?

- What is beurre manié? How is it made? How is it used?

- How do you prepare cornstarch to use in thickening liquids?

- What is a liaison? How is it made? How is it used?

- What is the meaning of the expression *monter au beurre*?

SAUCE FAMILIES

LEADING SAUCES

One more time, let's look at the three basic building blocks of sauce cookery, this time from a slightly different angle.

Liquid + Thickening agent = Leading sauce

Leading sauce + Additional flavorings = Small sauce

We have talked about five basic liquids for sauces: milk, white stock, brown stock, tomato purée (plus stock), and clarified butter. From these we get our five **leading sauces**, also known as *grand sauces* or *mother sauces*, as shown in Chart 8.1.

CHART 8.1 The Leading Sauces

Liquid	Thickening Agent	Leading Sauce
Milk	+ White roux	= Béchamel sauce
White stock (veal, chicken, fish)	+ White or blond roux	= Velouté (veal velouté, chicken velouté, fish velouté)
Brown stock	+ Brown roux	= Brown sauce or espagnole
Tomato plus stock	+ (Optional roux, see Note)	= Tomato sauce
Butter	+ Egg yolks	= Hollandaise

Note: Roux is not used in all tomato sauces, as tomato purée is naturally thick.

To these five sauces, we add one more: **fond lié** (fone lee ay), meaning "thickened stock." It is sometimes used in place of espagnole.

Brown stock + Arrowroot or cornstarch = Fond lié

You should understand that these charts are a bit oversimplified. Most of these sauces have a few other ingredients for flavoring. Yet knowing this basic structure is the key to making sauces.

SMALL SAUCES

The major leading sauces—béchamel; veal, chicken, and fish veloutés; and espagnole—are rarely used by themselves as sauces. They are more important as the bases for other sauces, called **small sauces**. Tomato sauce and hollandaise are used as they are, but they, too, are important as bases for small sauces.

Let's expand our sauce family chart one more generation to include examples of the small sauces in order to show the relationships (see Chart 8.2).

Chart 8.2 is probably a little more complicated than you expected because of the extra arrows and the extra category of secondary leading sauces. These are relatively easy to explain.

1. **Secondary leading white sauces.**

   These three sauces—allemande, suprême, and white wine—are really finished sauces, like other small sauces. But they are used so often to build other small sauces that they rate a special category.

   For example, to make suprême sauce, you add cream to chicken velouté.

   To make Albufera sauce, you can add meat glaze (glace de viande) to your suprême sauce. Or, if you don't have suprême sauce, you can make it by adding both cream and meat glaze to chicken velouté. This is why there are two sets of arrows in the chart.

   Allemande, suprême, and white wine sauces are also known as the *main small sauces*. If the concept of secondary leading white sauces seems confusing at first, you may simply think of them as small sauces. The important thing is to understand how the sauces are derived.

2. **Demi-glace.**

   - **Demi-glace** is defined as half brown sauce plus half brown stock, reduced by half. Most chefs prefer demi-glace to espagnole as a base for small sauces because of its more concentrated, more fully developed flavor.

     Note: It is possible to make small sauces directly from espagnole, but they will not be as fine.

CHART 8.2 The Small Sauces

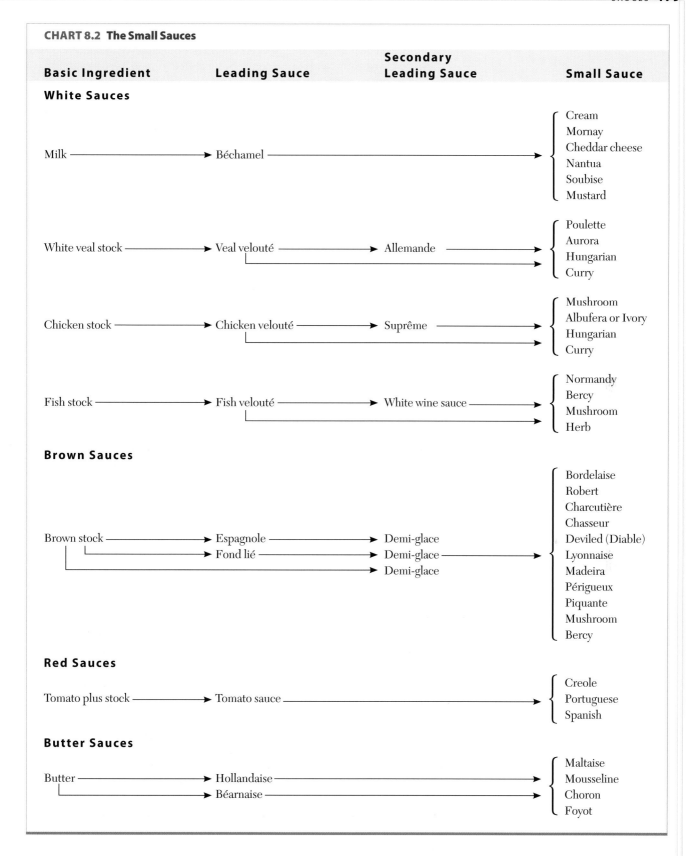

Basic Ingredient	Leading Sauce	Secondary Leading Sauce	Small Sauce
White Sauces			
Milk	Béchamel		Cream, Mornay, Cheddar cheese, Nantua, Soubise, Mustard
White veal stock	Veal velouté	Allemande	Poulette, Aurora, Hungarian, Curry
Chicken stock	Chicken velouté	Suprême	Mushroom, Albufera or Ivory, Hungarian, Curry
Fish stock	Fish velouté	White wine sauce	Normandy, Bercy, Mushroom, Herb
Brown Sauces			
Brown stock	Espagnole / Fond lié	Demi-glace / Demi-glace / Demi-glace	Bordelaise, Robert, Charcutière, Chasseur, Deviled (Diable), Lyonnaise, Madeira, Périgueux, Piquante, Mushroom, Bercy
Red Sauces			
Tomato plus stock	Tomato sauce		Creole, Portuguese, Spanish
Butter Sauces			
Butter	Hollandaise / Béarnaise		Maltaise, Mousseline, Choron, Foyot

- Some modern chefs feel espagnole is too heavy for modern tastes and that lighter sauces are required. These chefs prepare demi-glace from fond lié by reducing it with mirepoix, white wine, and seasonings, or by simply reducing by half a flavorful brown stock. In other words, demi-glace may be considered a well-flavored brown stock, reduced by half (demi means "half"), thickened with roux or other starch, or left unthickened (except by natural gelatin).

3. **Small sauces listed twice.**

   Notice, for example, that mushroom sauce is listed under both chicken velouté and fish velouté. This means you should use the stock of the product you are serving with the sauce. Mushroom sauce for chicken should be made with chicken velouté, for fish, with fish velouté. To be even more confusing, mushroom sauce is also made with brown sauce. Bercy sauce is also made as both a white and a brown sauce. These are considered unrelated sauces that happen to have the same name.

4. **Hollandaise and béarnaise.**

   These are essentially two variations of the same kind of sauce, with different flavorings. Each has its own small family of small sauces.

OTHER SAUCES

As usual, not everything fits into one package. Beyond the five major sauce families, a number of other preparations don't follow these basic patterns. We encounter these later in the chapter.

These other preparations include these groups:

Simple and compound butters, including simple browned butter as well as butter combined with flavorings.

Pan gravies, or sauces made with the pan drippings of the meat or poultry they are served with.

Miscellaneous hot sauces, which are not made like any of the five basic sauces. These include such items as raisin sauce (for ham) and sour cream sauce.

Miscellaneous cold sauces include not only sauces for meats, like Cumberland sauce and horseradish sauce, but also vinaigrettes, mayonnaise, and their variations, covered in Chapter 21.

MODERN SAUCES

Most of the emphasis in this chapter is on techniques for producing classic sauces. It is a mistake to argue that these sauces are not important and that modern sauce-making is entirely different. Modern sauces still depend on the basic classical techniques, even though the emphasis may have changed. For example, a chef in a modern kitchen may prepare a sauce for a sautéed meat item at the last minute by deglazing a sauté pan with a little wine, adding some reduced brown stock, and finishing the sauce by swirling in a little butter. As you can see, these are all techniques used in the production of classical sauces. Learning to make classical sauces is an important foundation for learning modern cooking.

While many of the recipes in this book, especially the traditional ones, incorporate sauces prepared in advance, many of the others, especially the more modern ones, incorporate sauces made at the last minute.

PRODUCTION

BÉCHAMEL

The classic version of the standard white sauce, **béchamel**, was made with lean veal and herbs and spices simmered with the sauce for an hour or with white veal stock added to the sauce and then reduced. This is rarely done today.

Nevertheless, the plain béchamel often used today—simply milk and roux—can be improved by simmering the sauce with onion and spices. These may be omitted, of course, but the sauce will have less flavor.

 Béchamel Sauce

YIELD: 1 GAL (4 L)

U.S.	METRIC	INGREDIENTS
		Roux:
8 oz	250 g	Clarified butter
8 oz	250 g	Bread flour
1 gal	4 L	Milk
1	1	Bay leaf, small
1	1	Small whole onion, peeled
1	1	Whole clove
to taste	to taste	Salt
to taste	to taste	Nutmeg
to taste	to taste	White pepper

PROCEDURE

1. Review instructions for making and incorporating roux (pp. 172–174).
2. Heat the butter in a heavy saucepot over low heat. Add the flour and make a white roux. Cool roux slightly.
3. Gradually add the milk to the roux, beating constantly.
4. Bring the sauce to a boil, stirring constantly. Reduce heat to a simmer.
5. Stick the bay leaf to the onion with clove and add to sauce. Simmer at least 15 minutes or, if possible, 30 minutes or more. Stir occasionally while cooking.
6. Adjust consistency with more hot milk, if necessary.
7. Season very lightly with salt, nutmeg, and white pepper. Spice flavors should not dominate.
8. Strain the sauce through a chinois or through a china cap lined with cheesecloth. Cover or spread melted butter on surface to prevent skin formation. Keep hot in a bain-marie, or cool in a cold-water bath for later use.
9. Evaluate the finished sauce (see p. 196).

Per 1 fl oz (29.57 mL): Calories, 40; Protein, 1 g; Fat, 3 g (63% cal.); Cholesterol, 10 mg; Carbohydrates, 3 g; Fiber 0 g; Sodium, 30 mg

VARIATIONS

Light Béchamel
Use 12 oz (375 g) roux.

Heavy Béchamel
Use 1¹/₂ lb (750 g) roux.

FIGURE 8.10 Preparing béchamel sauce.

(a) Combine butter and flour to make a roux.

(b) Cook the roux, keeping it white.

(c) Whip in the milk.

(d) Strain the finished sauce through a china cap lined with cheesecloth.

(e) Finished béchamel sauce.

Small Sauces

For each of the following sauces, add the ingredients indicated to *1 qt (1 L) béchamel sauce*. Season to taste.

Cream Sauce

4–8 fl oz (125–250 mL) heavy cream, heated or tempered

Mornay Sauce

4 oz (125 g) grated Gruyère cheese and 2 oz (60 g) Parmesan, stirred in until just melted. Finish, off heat, with 2 oz (60 g) raw butter. Thin out with a little hot milk, if necessary, or use a stock or broth appropriate for the dish being prepared.

Mornay Sauce for Glazing or Gratinéeing

Finish Mornay Sauce with liaison of 2 egg yolks and 2 fl oz (60 mL) heavy cream.

Cheddar Cheese Sauce

8 oz (250 g) cheddar cheese, ½ tsp (2 mL) dry mustard, 2 tsp (10 mL) Worcestershire sauce

Mustard Sauce

4 oz (125 g) prepared mustard

Soubise Sauce

1 lb (500 g) onions, finely diced, cooked slowly in 2 oz (60 g) butter without browning. Simmer with sauce 15 minutes and force through a fine sieve.

Tomatoed Soubise Sauce

Add 1 pt (500 mL) thick tomato purée to 1 qt (1 L) soubise sauce.

Nantua Sauce

6 oz (175 g) Shrimp Butter (p. 194), 4 fl oz (125 mL) heavy cream
(Note: Classic Nantua sauce is made with crayfish, not readily available in many regions.)

VELOUTÉ

The three **velouté** sauces are the bases of many variations. Instructions for the small sauces indicate which of the three to use. If more than one is given, the choice depends on what you are serving it with.

Note: In North America, chicken velouté is used much more often than veal velouté. Many of the sauces at one time made with veal stock are now made with chicken stock.

 # Velouté Sauce (Veal, Chicken, or Fish)

YIELD: 2 QT (2L)

U.S.	METRIC	INGREDIENTS
4 fl oz	125 mL	Clarified butter
4 oz	125 g	White mirepoix, small dice (see Note)
4 oz	125 g	Flour (see Note)
2½ qt	2.5 L	White stock (veal, beef, chicken, or fish)
		Sachet d'épices:
1	1	Bay leaf
½ tsp	2 mL	Dried thyme
½ tsp	2 mL	Peppercorns
3–4	3–4	Parsley stems
as needed	as needed	Salt (see step 7)
as needed	as needed	White pepper (see step 7)

PROCEDURE

1. Review instructions for making and incorporating roux (pp. 172–174).

2. Heat the clarified butter in a heavy saucepot over low heat. Add mirepoix and sweat the vegetables without browning them.

3. Add flour and make a blond roux. Cool roux slightly.

4. Gradually add the stock to the roux, beating constantly. Bring to a boil, stirring constantly. Reduce heat to a simmer.

5. Add the sachet.

6. Simmer the sauce very slowly for 1 hour. Stir occasionally, and skim surface when necessary. Add more stock if needed to adjust consistency.

7. If the velouté is to be used as is, season to taste with salt and white pepper. But if it is to be used as an ingredient in other preparations, do not season velouté.

8. Strain through a china cap lined with cheesecloth. Cover or spread melted butter on surface to prevent skin formation.
   Keep hot in a bain-marie, or cool in a cold-water bath for later use.

9. Evaluate the finished sauce (see p. 196).

Per 1 fl oz (29.57 mL): Calories, 30; Protein, 1 g; Fat, 2 g (53% cal.); Cholesterol, 5 mg; Carbohydrates, 3 g; Fiber, 0 g; Sodium, 10 mg.

Note: Use a standard white mirepoix, or substitute leeks for the parsnips, as desired.
If you have blond roux on hand, you can use it instead of making roux as part of this procedure.
Reduce the clarified butter to 1 fl oz (30 mL). After sweating the mirepoix, add 8 oz (250 g) blond roux.

VARIATION

For a quicker, simpler velouté, omit clarified butter, mirepoix, and sachet.

FIGURE 8.11 Preparing velouté sauce.

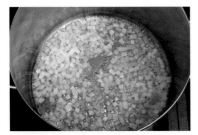

(a) Sweat the mirepoix in butter.

(b) Add the flour and make a blond roux.

(c) Whip in white stock.

(d) As the sauce simmers, skim the surface regularly.

(e) Strain the finished sauce through a china cap lined with cheesecloth.

(f) Finished velouté sauce.

Suprême Sauce

YIELD: 2 QT (2 L)

U.S.	METRIC	INGREDIENTS
2 qt	2 L	Chicken velouté
1 pt	500 mL	Heavy cream
2 oz	60 g	Butter, in pieces
to taste	to taste	Salt
to taste	to taste	White pepper
to taste	to taste	Lemon juice

PROCEDURE

1. Place velouté in a saucepan and simmer over moderate heat until reduced by about one-fourth. Stir occasionally.
2. Pour the cream into a stainless-steel bowl and temper it by slowly stirring in a little of the hot sauce. Stir this mixture slowly back into the sauce in the pan and return the sauce just to a simmer.
3. Swirl in raw butter pieces. Season to taste with salt, white pepper, and a few drops of lemon juice.
4. Strain through cheesecloth.

Per 1 fl oz (29.57 mL): Calories, 50; Protein, 1 g; Fat, 5 g (79% cal.); Cholesterol, 20 mg; Carbohydrates, 2 g; Fiber, 0 g; Sodium, 30 mg.

Allemande Sauce

YIELD: 2 QT (2 L)

U.S.	METRIC	INGREDIENTS
2 qt	2 L	Veal velouté (see Note)
		Liaison:
4	4	Egg yolks
8 fl oz	250 mL	Heavy cream
1/2 fl oz	15 mL	Lemon juice
to taste	to taste	Salt
to taste	to taste	White pepper

PROCEDURE

1. Review instructions for incorporating liaison (p. 175).
2. Place velouté in a saucepan and simmer a few minutes over moderate heat until slightly reduced.
3. Beat the egg yolks and cream together in a stainless-steel bowl.
4. Temper the liaison by slowly beating in about one-third of the hot sauce. Then slowly stir this mixture back into the sauce in the pan.
5. Reheat to just below simmering. Do not boil.
6. Add lemon juice, salt, and white pepper to taste. Strain through cheesecloth.

Per 1 fl oz (29.57 mL): Calories, 40; Protein, 1 g; Fat, 3.5 g (72% cal.); Cholesterol, 25 mg; Carbohydrates, 2 g; Fiber, 0 g; Sodium, 20 mg.

Note: Allemande sauce, strictly speaking, should be made with veal velouté. However, as chicken velouté is much more common in North America, allemande sauce and the small sauces derived from it are often made with chicken velouté.

White Wine Sauce

YIELD: 2 QT (2 L)

U.S.	METRIC	INGREDIENTS	PROCEDURE
8 fl oz	250 mL	White wine (dry)	1. Reduce the wine by half in a saucepan.
2 qt	2 L	Fish velouté	2. Add the velouté and simmer until reduced to desired consistency.
8 fl oz	250 mL	Heavy cream, hot	3. Slowly stir in the hot (or tempered) cream.
2 oz	60 g	Butter, in pieces	4. Remove from heat and swirl in raw butter pieces.
to taste	to taste	Salt	5. Season to taste with salt, white pepper, and a few drops of lemon juice.
to taste	to taste	White pepper	
to taste	to taste	Lemon juice	6. Strain through cheesecloth.

Per 1 fl oz (29.57 mL): Calories, 45; Protein, 0 g; Fat, 4 g (85% cal.); Cholesterol, 10 mg; Carbohydrates, 1 g; Fiber, 0 g; Sodium, 30 mg.

VARIATION

Instead of adding hot or tempered heavy cream, make a liaison with 5 egg yolks and 8 fl oz (250 mL) cold heavy cream. Incorporate liaison using the procedure on page 175. Then continue with step 4 in recipe.

Small Sauces

For each of the following sauces, add the listed ingredients to *1 qt (1 L) veal, chicken, or fish velouté, suprême sauce, allemande sauce, or white wine sauce* as indicated. Season sauce to taste.

Poulette

Simmer 8 oz (250 g) white mushrooms or mushroom trimmings with velouté when making allemande. Make allemande; strain. Finish with 2 tbsp (30 mL) chopped parsley and lemon juice to taste.

Aurora

Add 6 oz (175 g) tomato purée to 1 qt (1 L) veal or chicken velouté, suprême sauce, or allemande sauce.

Hungarian

Sweat 2 oz (60 g) minced onion and 1 tbsp (15 mL) paprika in 1 oz (25 g) butter until soft. Add ½ cup (100 mL) white wine and reduce by half. Add 1 qt (1 L) veal or chicken velouté, simmer 10 minutes, and strain.

Ivory or Albufera

Add 2 oz (60 g) meat glaze (glace de viande) to 1 qt (1 L) suprême sauce.

Curry

Cook 4 oz (125 g) mirepoix, cut brunoise, in 1 oz (25 g) butter until tender but not brown. Add 1 tbsp (15 mL) curry powder, 1 crushed garlic clove, pinch dried thyme, ½ bay leaf, and 2–4 parsley stems and cook another minute. Add 1 qt (1 L) veal, chicken, or fish velouté. Simmer 20 minutes, add ½ cup (125 mL) cream, strain, and season with salt and lemon juice.

Mushroom

Sauté 4 oz (125 g) sliced mushrooms in 1 oz (25 g) butter, adding 1 tbsp (15 mL) lemon juice to keep them white. Add to suprême, allemande, or white wine sauce or to appropriate velouté.

Bercy

Reduce by two-thirds 2 oz (60 g) chopped shallots and ½ cup (125 mL) white wine. Add 1 qt (1 L) fish velouté, reduce slightly, and finish with 2 oz (60 g) raw butter, 2 tbsp (30 mL) chopped parsley, and lemon juice to taste.

Herb

To white wine sauce add chopped parsley, chives, and tarragon to taste.

Normandy

To 1 qt (1 L) fish velouté add 4 oz (125 mL) mushroom cooking liquid (or 4 oz/125 g mushroom trimmings) and 4 oz (125 mL) oyster liquid or fish fumet. Reduce by one-third. Finish with liaison of 4 egg yolks and 1 cup (250 mL) cream. Strain and swirl in 3 oz (75 g) raw butter.

Anchovy

Follow instructions for Normandy sauce but, in place of the raw butter used to finish the sauce, substitute 6 oz (175 g) anchovy butter.

Shrimp

To 1 qt (1 L) white wine sauce add 4 oz (125 g) shrimp butter and a dash of cayenne. If desired, garnish with 4 oz (125 g) diced, cooked shrimp.

Venetian

Combine ½ cup (125 mL) each white wine and tarragon vinegar, ½ oz (15 g) chopped shallots, and 2 tsp (10 mL) chopped fresh chervil. Reduce by two-thirds. Add 1 qt (1 L) white wine sauce and simmer 2–3 minutes. Strain. Add fresh tarragon to taste.

Horseradish

Add 2 oz (60 g) drained horseradish, ½ cup (125 mL) heavy cream, and 2 tsp (10 mL) dry mustard dissolved in 1 fl oz (30 mL) vinegar to 1 qt (1 L) velouté made with beef or veal stock or broth from Simmered Fresh Beef (p. 436).

ESPAGNOLE OR BROWN SAUCE

As one glance at the procedure for making **espagnole** will tell you, this sauce is more complicated than béchamel or velouté. Because it is the starting point for the hearty, flavorful sauces that accompany red meats, it is necessary to give it extra flavor and richness with mirepoix. Some chefs even add more browned bones and cook the sauce as long as a stock.

Note how the roux is made in the espagnole recipe. Though mirepoix is also cooked in the fat, the basic principle is the same as when you make a simple roux in a separate pot.

Fond Lié

In its simplest form, fond lié, or jus lié, is a brown stock thickened lightly with arrowroot or cornstarch. Its quality can be improved, however, by applying the technique used for making espagnole—that is, reduce brown stock with browned mirepoix and tomato purée or tomato paste, and then thicken with a starch slurry and strain. You can use the same ratio of stock to mirepoix as for espagnole.

GASTRIQUE

A classic technique to add a balanced sweet-sour accent to a sauce is to add a **gastrique**, which is caramelized sugar dissolved in vinegar. For example, a gastrique gives necessary sweetness and acidity to sauce bigarade, the classic orange sauce served with roast duck.

A quantity of gastrique can be made in advance, stored, and used as needed. Heat 4 oz (120 g) sugar until it melts and then caramelizes to a light golden brown. Let cool briefly. Add 3 fl oz (90 mL) wine vinegar and simmer until the caramel is dissolved.

 Brown Sauce or Espagnole ♥

YIELD: 1 GAL (4 L)

U.S.	METRIC	INGREDIENTS
		Mirepoix:
1 lb	500 g	Onions, medium dice
8 oz	250 g	Carrots, medium dice
8 oz	250 g	Celery, medium dice
8 oz	250 g	Butter
8 oz	250 g	Bread flour
6 qt	6 L	Brown veal stock
8 oz	250 g	Tomato purée (see Variations)
		Sachet:
¹/₂	¹/₂	Bay leaf
¹/₄ tsp	1 mL	Thyme
6–8	6–8	Parsley stems

PROCEDURE

1. Sauté the mirepoix in butter until well browned (**Figure 8.12**).

2. Add the flour and stir to make a roux. Continue to cook until the roux is browned. (See Variations for alternative procedure.)

3. Gradually stir in brown stock and tomato purée, stirring constantly until the mixture comes to a boil.

4. Reduce heat to simmer and skim surface. Add the sachet and let simmer about 2 hours, or until the sauce is reduced to 1 gal (4 L). Skim as often as necessary.

5. Strain through a china cap lined with several layers of cheesecloth. Press on mirepoix gently to extract juices.

6. Cover or spread melted butter on surface to prevent skin formation. Keep hot in a bain-marie, or cool in a cold-water bath for later use.

7. Evaluate the finished sauce (see p. 196).

Per 1 fl oz (29.57 mL): Calories, 25; Protein, 1 g; Fat, 1.5 g (53% cal); Cholesterol, 5 mg; Carbohydrates, 2 g; Fiber, 0 g; Sodium, 20 mg.

VARIATIONS

If you have brown roux on hand, you may use it instead of making roux as part of the sauce procedure. Reduce the quantity of butter to 2 oz (60 g), or just enough to brown mirepoix. After adding stock and bringing to a simmer in steps 3 and 4, whip in roux.

2 oz (60 g) tomato paste may be used instead of tomato purée. Add paste to the browned mirepoix and continue to brown until the paste turns a rusty brown color.

Jus de Veau Lié I

Reduce the quantity of butter to 2 oz (60 g). Omit flour. In addition to mirepoix, add veal bones or trimmings if desired. Brown with the mirepoix. After straining (step 5), make a slurry of 1 oz (30 g) arrowroot or cornstarch and enough cold water to make a thin paste. Stir into sauce and simmer until clear and thickened.

Jus de Volaille Lié, Jus d'Agneau Lié, Jus de Canard Lié, or Jus de Gibier Lié

In place of veal stock, use brown chicken stock, lamb stock, duck stock, or game stock. If desired, add appropriate bones or trimmings and brown with mirepoix, as for Jus de Veau Lié.

FIGURE 8.12 Preparing brown sauce or espagnole.

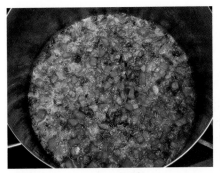

(a) Brown the mirepoix well in fat.

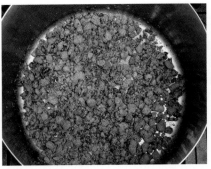

(b) Mix in flour and cook to make a brown roux.

(c) Whip in the stock. Add the sachet and simmer about 2 hours.

(d) Strain the finished sauce through a china cap lined with cheesecloth.

(e) Finished brown sauce or espagnole.

Fond Lié or Jus Lié II ♥

YIELD: 1 QT (1 L)

U.S.	METRIC	INGREDIENTS
1 qt	1 L	Brown stock
1 oz	30 g	Cornstarch or arrowroot

PRODUCTION

1. Bring the stock to a boil in a saucepan. Reduce heat to a simmer.

2. Dissolve the starch in a small amount of cold stock or water. Stir it into the simmering stock.

3. Simmer until thickened and clear.

Per 1 fl oz (29.57 mL): Calories, 5; Protein, 0 g; Fat 0 g (0% cal.); Cholesterol, 0 mg; Carbohydrates, 1 g; Fiber; 0 g; Sodium, 0 mg.

VARIATION

For added flavor, the stock can be reduced with browned mirepoix and tomato (as for espagnole) before being thickened. Browned bones may also be added.

Demi-Glace ♥

YIELD: 1 GAL (4 L)

U.S.	METRIC	INGREDIENTS	PRODUCTION
1 gal	4 L	Brown sauce (espagnole)	1. Combine the sauce and stock in a saucepan and simmer until reduced by half.
1 gal	4 L	Brown stock	

2. Strain through a chinois (fine china cap) or a regular china cap lined with cheesecloth. Cover to prevent a skin from forming. Keep hot in a bain-marie, or cool in a cold-water bath for later use.

Per 1 fl oz (29.57 mL): Calories 25; Protein, 1 g; Fat 1.5 g (53% cal.); Cholesterol, 5 mg; Carbohydrates, 2 g; Fiber, 0 g; Sodium, 20 mg.

Small Sauces

For each of the following sauces, add the listed ingredients to *1 qt (1 L) demi-glace*, as indicated.

Bordelaise

Reduce by three-fourths 1 cup (250 mL) dry red wine, 2 oz (60 g) chopped shallots, 1/4 tsp (1 mL) crushed peppercorns, a pinch of dried thyme, and 1/2 bay leaf. Add 1 qt (1 L) demi-glace, simmer 15–20 minutes, and strain. Swirl in 2 oz (60 g) raw butter, cut in pieces. Garnish with diced or sliced beef marrow poached in salted water.

Marchand de Vin (Wine Merchant)

Reduce 6 fl oz (200 mL) red wine and 2 oz (60 g) chopped shallots by three-fourths. Add 1 qt (1 L) demi-glace, simmer, and strain.

Robert

Cook 4 oz (125 g) chopped onion in butter without browning. Add 1 cup (250 mL) white wine and reduce by two-thirds. Add 1 qt (1 L) demi-glace and simmer 10 minutes. Strain and add 2 tsp (10 mL) dry mustard and a pinch of sugar dissolved in a little lemon juice.

Charcutière

Garnish Robert sauce with sour pickles, cut julienne.

Chasseur

Sauté 6 oz (175 g) sliced mushrooms and 2 oz (60 g) minced shallots in 2 oz (60 g) butter. Add 1 cup (250 mL) white wine and reduce by three-fourths. Add 1 qt (1 L) demi-glace and 8 oz (250 g) diced tomato. Simmer 5 minutes and add 2 tsp (10 mL) chopped parsley.

Diable (Deviled)

Reduce by two-thirds 8 fl oz (250 mL) white wine, 4 oz (125 g) chopped shallots, 1/2 tsp (2 mL) crushed peppercorns. Add 1 qt (1 L) demi-glace and simmer 20 minutes. Season with cayenne to taste and strain.

Madeira

Reduce 1 qt (1 L) demi-glace by about 1/2 cup (100 mL). Add 3–4 fl oz (100 mL) Madeira wine.

Périgueux

Garnish Madeira sauce with finely diced truffle.

Poîvrade

Brown 1 lb (500 g) mirepoix in butter. Add 4 fl oz (125 mL) red wine and 1 1/2 pt (750 mL) Red Wine Marinade for Game (p. 497) and reduce by one-half. Add 1 qt (1 L) demi-glace and reduce by one-third over low heat. Add 1/2 tsp (2 mL) crushed peppercorns and simmer 10 minutes. Strain.

Port Wine

Follow instructions for Madeira sauce, but use port wine instead of Madeira.

Italian Sauce

Sauté 1 lb (500 g) finely chopped mushrooms and 1/2 oz (15 g) minced shallots in 2 oz (60 g) butter until all moisture is evaporated. Add 1 cup (250 mL) white wine and reduce by half. Add 1 oz (30 g) tomato paste and 1 qt (1 L) demi-glace and simmer 10 minutes. Add 2 tbsp (30 mL) chopped parsley.

Mushroom

Sauté 8 oz (250 g) sliced mushrooms and 1 oz (30 g) minced shallots in 2 oz (60 g) butter until browned. Add 1 qt (1 L) demi-glace and simmer about 10 minutes. Add 2 oz (60 mL) sherry and a few drops of lemon juice.

Bercy

Reduce by three-fourths 1 cup (250 mL) dry white wine and 4 oz (125 g) chopped shallots. Add 1 qt (1 L) demi-glace and simmer 10 minutes.

Piquante

Reduce by two-thirds 4 oz (125 g) minced shallots, 4 fl oz (125 mL) wine vinegar, and 4 fl oz (125 mL) white wine. Add 1 qt (1 L) demi-glace and simmer until slightly reduced. Add 2 oz (60 g) capers, 2 oz (60 g) sour pickles, cut brunoise, 1 tbsp (15 mL) chopped parsley, and 1/2 tsp (2 mL) dried tarragon.

Lyonnaise

Sauté 4 oz (125 g) onions in 2 oz (60 g) butter until slightly browned. Add 1/2 cup (125 mL) white wine vinegar and reduce by half. Add 1 qt (1 L) demi-glace and simmer 10 minutes.

Bigarade

Prepare a gastrique (p. 185) from 6 oz (180 g) sugar and 4 fl oz (125 mL) wine vinegar. Add 8 fl oz (250 mL) orange juice, 3 fl oz (90 mL) lemon juice, 1 qt (1 L) demi-glace, and, if available, juices from a roast duck. Simmer and reduce to desired consistency. Garnish with blanched julienne of orange zest.

TOMATO SAUCE

Classical tomato sauce, as explained by Escoffier, is made with a roux, but this is rarely done in modern kitchens. The texture of the puréed tomatoes is sufficient to give the sauce the proper texture, even when no starch thickener is used.

This type of sauce may be referred to as a **coulis** (koo-lee). This French term means, in modern kitchens, a purée of vegetables or fruits, used as a sauce. A recipe for another coulis, of sweet peppers, is found on page 204.

Three main techniques are used to purée vegetables and other ingredients for coulis:

1. Puréeing the product in a food processor or blender

2. Passing the product through a food mill

3. Forcing the product through a fine sieve

Of these three methods, the third, forcing through a fine sieve, usually makes the smoothest purée, but it is also the most time-consuming. If you want a smooth purée but the product is difficult to force through a sieve, you can use one of the other methods first, then pass the purée through the sieve to make it smoother.

Tomato Sauce I ♥

YIELD: 2 QT (2 L)

U.S.	METRIC	INGREDIENTS	PROCEDURE
2 oz	60 g	Salt pork	1. Render the salt pork in a heavy saucepot, but do not brown it.
4 oz	126 g	Onion, medium dice	2. Add the onion and carrots and sauté until slightly softened, but do not brown.
4 oz	125 g	Carrots, medium dice	
2 qt	2 L	Tomatoes, canned or fresh, coarsely chopped	3. Add the tomatoes and their juice, tomato purée, bones, and sachet. Bring to a boil, reduce heat, and simmer over very low heat (see Note) 1^1/$_2$–2 hours, or until reduced to desired consistency.
1 qt	1 L	Tomato purée, canned	
8 oz	250 g	Ham bones or browned pork bones	
		Sachet:	
1 clove	1 clove	Garlic, crushed	4. Remove sachet and bones. Strain sauce or pass it through a food mill.
^{1}/$_2$	^{1}/$_2$	Bay leaf	
^{1}/$_8$ tsp	0.5 mL	Dried thyme	5. Adjust seasoning with salt and a little sugar.
^{1}/$_8$ tsp	0.5 mL	Dried rosemary	6. Evaluate the finished sauce (see p. 196)
^{1}/$_8$ tsp	0.5 mL	Peppercorns, crushed	
to taste	to taste	Salt	
to taste	to taste	Sugar	

Per 1 fl oz (29.57 mL): Calories, 20; Protein, 1 g; Fat, 1 g (35% cal.);
Cholesterol, 0 mg; Carbohydrates, 3 g; Fiber, 1 g; Sodium, 120 mg.

Note: Tomato sauce scorches easily, so heat must be very low. The sauce may be cooked in a slow oven (300°F/150°C), loosely covered, to reduce the danger of scorching.

VARIATION

Tomato Sauce II (Vegetarian)

Omit the salt pork. Sweat the vegetables in 1 fl oz (30 mL) olive oil. Omit bones.

Tomato Sauce III

See Italian Tomato Sauce for Pasta, page 399.

Small Sauces

For each of the following sauces, add the listed ingredients to *1 qt (1 L) tomato sauce*, as indicated.

Portugaise (Portuguese)

Sauté 4 oz (125 g) onions, cut brunoise, in 1 fl oz (30 mL) oil. Add 1 lb (500 g) tomato concassé (see p. 295) and 1 tsp (5 mL) crushed garlic. Simmer until reduced by about one-third. Add 1 qt (1 L) tomato sauce, adjust seasonings, and add 2–4 tbsp (30–60 mL) chopped parsley.

Spanish

Lightly sauté in oil without browning 6 oz (175 g) onion, small dice; 4 oz (125 g) green bell pepper, small dice; and 1 clove garlic, chopped fine. Add 4 oz (125 g) sliced mushrooms and sauté. Add 1 qt (1 L) tomato sauce, and season to taste with salt, pepper, and hot red pepper sauce.

Creole

Sauté in oil 4 oz (125 g) onion, small dice; 4 oz (125 g) celery, sliced; 2 oz (60 g) green bell pepper, small dice; 1 tsp (5 mL) chopped garlic. Add 1 qt (1 L) tomato sauce, 1 bay leaf, pinch dried thyme, and ½ tsp (2 mL) grated lemon rind. Simmer 15 minutes. Remove bay leaf and season to taste with salt, pepper, and cayenne.

Fresh Tomato Coulis with Garlic

YIELD: 1 PT (500 ML)

U.S.	METRIC	INGREDIENTS
3 oz	90 g	Shallots, chopped fine
6	6	Garlic cloves, chopped fine
2 fl oz	60 mL	Olive oil
1 lb 8 oz	750 g	Tomatoes, peeled, seeded, and chopped
to taste	to taste	Salt
to taste	to taste	White pepper

Per 1 fl oz (29.57 mL): Calories, 45; Protein 1 g; Fat, 3.5 g (66% cal.); Cholesterol, 0 mg; Carbohydrates, 3 g; Fiber,1 g; Sodium, 5 mg.

PROCEDURE

1. Sweat the shallots and garlic in olive oil until soft.
2. Add the tomatoes and cook until excess liquid has cooked out and the sauce is thick.
3. Season to taste.

Fresh Tomato Coulis

BUTTER SAUCES

The fifth leading sauce is hollandaise. Hollandaise and its cousin, béarnaise, are unlike the sauces we have been studying because their major ingredient is not stock or milk but butter.

Before tackling the complexities of hollandaise, we first look at simpler butter preparations used as sauces.

1. **Melted butter.**

   This is the simplest butter preparation of all, and one of the most widely used, especially as a dressing for vegetables.

   Unsalted or *sweet butter* has the freshest taste and is ideal for all sauce-making.

2. **Clarified butter.**

   Butter consists of butterfat, water, and milk solids. **Clarified butter** is purified butterfat, with water and milk solids removed (see Figure 8.13). It is necessary for many cooking operations. Clarified butter is used in sautéing because the milk solids of unclarified butter would burn at such high temperatures. It is used in making hollandaise because the water of unclarified butter would change the consistency of the sauce. Method 1 in the procedure below is the most widely used in North American and European kitchens. Method 2 is used for the style of clarified butter called *ghee,* used in Indian cuisine.

3. **Brown butter.**

   Known as **beurre noisette** (burr nwah zett) in French, this is whole melted butter that has been heated until it turns light brown and gives off a nutty aroma. It is usually prepared at the last minute and served over fish, white meats, eggs, and vegetables.

   Care must be taken not to burn the butter, as the heat of the pan will continue to brown it even after it is removed from the fire.

PROCEDURE for Clarifying Butter

METHOD 1

1. Melt the butter in a heavy saucepan over moderate heat.

2. Skim the froth from the surface.

3. Carefully pour off the clear melted butter into another container, leaving the milky liquid at the bottom of the saucepan.

METHOD 2

1. Melt the butter in a heavy saucepan over moderate heat.

2. Skim the froth from the surface.

3. Turn the heat to very low to avoid burning the butter. Leave the pan on the heat and continue to skim the froth from the surface at intervals. The water in the bottom will boil and gradually evaporate.

4. When the butter looks clear and no longer forms a scum on top, strain off the butter through cheesecloth into another container.

You need 1¼ lb (625 g) raw butter to make 1 lb (500 g) clarified butter; 1 lb (500 g) raw butter yields 12–13 oz (about 400 g) clarified butter.

FIGURE 8.13 Clarifying butter.

(a) Skim the foam from the top of the melted butter.

(b) Ladle off the clear, melted fat.

(c) Continue until only the milky liquid remains in the bottom of the pan.

4. Black butter.

Black butter, or ***beurre noir*** (burr nwahr), is made like brown butter but heated until it is a little darker, and it is flavored with a few drops of vinegar. Capers, chopped parsley, or both are sometimes added.

To avoid dangerous spattering of the vinegar in the hot butter, many chefs pour the butter over the food item, then deglaze the pan with the vinegar and pour that over the item.

5. Meunière butter.

This is served with fish cooked à la Meunière (see p. 628). Brown butter is seasoned with lemon juice and poured over the fish, which has been sprinkled with chopped parsley.

As in the case of black butter, dangerous spattering can result when moisture is added to hot butter. To avoid this, cooks often sprinkle the lemon juice directly on the fish before pouring on the brown butter.

6. Compound butters.

Compound butters are made by softening raw butter and mixing it with flavoring ingredients. The mixture is then rolled into a cylinder in waxed paper.

Compound butters have two main uses:

- Slices of the firm butter are placed on hot grilled items at service time. The butter melts over the item and sauces it.

- Small portions are swirled into sauces to finish them and give them a desired flavor.

Easy as they are to make, compound butters can transform a plain broiled steak into a truly special dish.

The favorite compound butter for steaks is maître d'hôtel (may truh doh tel) butter. Variations are given after the recipe (p. 192).

7. Beurre blanc.

Beurre blanc (burr blon) is a sauce made by whipping a large quantity of raw butter into a small quantity of a flavorful reduction of white wine and vinegar so the butter melts and forms an emulsion with the reduction. The technique is basically the same as monter au beurre (p. 177), except the proportion of butter to liquid is much greater.

Beurre blanc can be made quickly and easily by adding cold butter all at once and whipping vigorously over moderately high heat. The temperature of the butter keeps the sauce cool enough to prevent it from separating. Be sure to remove it from the heat before all the butter is melted, and continue whipping. It is better to remove the sauce from the heat too soon rather than too late because it can always be rewarmed slightly if necessary. Figure 8.14 illustrates this procedure.

Some chefs prefer to use low heat and add the butter a little at a time in order to reduce the chance of overheating and breaking the sauce. The process takes a little longer, but the result is the same.

FIGURE 8.14 Preparing beurre blanc.

(a) Reduce the liquids (usually wine and vinegar) with chopped shallots.

(b) Whip in the raw butter just until the butter is melted and forms a smooth sauce.

(c) Leave in the shallots, or strain them out. Strained beurre blanc has a light, smooth, creamy texture.

Beurre blanc should be held at a warm, not a hot, temperature and stirred or whipped from time to time so the fat and water do not separate. For more stable mixtures of fat and water—called *emulsions*—see the discussion of hollandaise beginning on page 193.

Butter-Enriched Sauces

As already noted, the technique for making beurre blanc is the same as monter au beurre, except the proportion of butter is much higher. The same technique can be used to finish a great variety of sauces, usually white sauces, although brown sauces can be finished the same way.

To improvise a butter-enriched version of a classic white sauce, refer to the sauce variations on page 184. In place of the 1 qt (1 L) velouté or other white sauce base, substitute 1 pt (500 mL) concentrated white stock. Combine with the flavoring ingredients indicated in the variation. Reduce to a slightly syrupy consistency. Whip in 8 oz (250 g) raw butter and strain.

Many other sauces for sautéed meat, poultry, or fish items can be improvised using the same technique. Deglaze the sauté pan with wine, stock, or other liquid, add desired flavoring ingredients, reduce, and finish by whipping in a generous quantity of raw butter. Season and strain.

Maître d'Hôtel Butter

YIELD: APPROX. 1 LB (500 G)

U.S.	METRIC	INGREDIENTS
1 lb	500 g	Butter, unsalted (see Note)
2 oz	60 g	Chopped parsley
1¹/₃ fl oz	50 mL	Lemon juice
2 tsp	10 mL	Salt (see Note)
pinch	pinch	White pepper

PROCEDURE

1. Using a mixer with the paddle attachment, beat the butter at low speed until smooth and creamy.

2. Add remaining ingredients and beat slowly until completely mixed.

3. Roll the butter into a cylinder about 1 inch (2¹/₂ cm) thick in a sheet of parchment or waxed paper. Chill until firm.

4. To serve, cut slices ¹/₄ inch (¹/₂ cm) thick and place on broiled or grilled items just before service.

Per 1 ounce (28.35g): Calories, 230; Protein 0 g; Fat, 25 g (98% cal.); Cholesterol, 70 mg; Carbohydrates, 1 g; Fiber, 0 g; Sodium, 300 mg.

Note: If using salted butter, omit the salt.

VARIATIONS

For each kind of seasoned butter, add to *1 lb (500 g) butter* the listed ingredients instead of the parsley, lemon juice, and pepper.

Anchovy Butter

2 oz (60 g) anchovy fillets, mashed to a paste

Garlic Butter

1 oz (30 g) garlic, mashed to a paste (see p. 285)

Escargot (Snail) Butter

Maître d'Hôtel Butter plus 1 oz (30 g) garlic, mashed to a paste

Shrimp Butter

¹/₂ lb (250 g) cooked shrimp and shells, ground very fine. Force shrimp butter through a fine sieve to remove pieces of shell.

Mustard Butter

3–4 oz (100 g) Dijon-style mustard

Herb Butter

Chopped fresh herbs to taste

Scallion or Shallot Butter

2 oz (60 g) minced scallions or shallots

Curry Butter

4–6 tsp (20–30 mL) curry powder heated gently with 1 oz (30 g) butter, then cooled

Beurre Blanc

YIELD: 1 PT (500 ML)

U.S.	METRIC	INGREDIENTS
8 fl oz	250 mL	Dry white wine
1¹/₂ fl oz	50 mL	White wine vinegar
1 oz	30 g	Shallots, chopped
1 lb	500 g	Cold butter
to taste	to taste	Salt

PROCEDURE

1. Combine wine, vinegar, and shallots in a saucepan. Reduce until about 1 fl oz (30 mL) liquid remains.

2. Cut the butter into small pieces.

3. Add the butter to the hot reduction. Set pan over moderately high heat and whip vigorously. When butter is nearly all melted and incorporated, remove from heat and continue to whip until smooth.

4. Season to taste. Shallots may be left in sauce or strained out.

5. Hold the sauce in a warm, not hot, place until served. Stir or whip it from time to time.

Per 1 fl oz (29.57 mL): Calories, 210; Protein, 0 g; Fat, 23 g (94% cal.); Cholesterol, 60 mg; Carbohydrates, 1 g; Fiber, 0 g; Sodium, 240 mg.

VARIATIONS

Herbed Butter Sauce

Add your choice of chopped fresh herbs to finished beurre blanc, or use an herbed compound butter instead of plain raw butter to make beurre blanc.

Beurre Rouge (Red Butter Sauce)

Use dry red wine instead of white wine to make the reduction, and increase the quantity of wine to 1 pt (500 mL). As in basic recipe, reduce the wine to about 1 fl oz. (30 ml). For good color, use a young, bright red wine.

Beurre Rouge for Fish

After reducing red wine for Beurre Rouge, add 6 fl oz (180 ml) fish stock. Reduce the liquid to 2 fl oz (60 ml).

HOLLANDAISE AND BÉARNAISE

Hollandaise is considered an egg-thickened sauce, but the egg doesn't thicken by coagulation as it does in a liaison or in a custard sauce. Instead, it works by emulsification.

An **emulsion** is a uniform mixture of two unmixable liquids. In the case of hollandaise, the two liquids are melted butter and water (including the water in the lemon juice or the vinegar reduction). The two stay mixed and thick because the butter is beaten into tiny droplets and the egg yolks hold the droplets apart. You will encounter emulsion again when you prepare mayonnaise and other salad dressings in Chapter 21.

Two recipes for hollandaise are given. The first is the classic version, flavored with lemon and a vinegar reduction (see Figure 8.15). (You may also see recipes that include shallots in the reduction, in imitation of beurre blanc, but this is not traditional, according to Escoffier.) The second recipe, flavored with just lemon juice, is slightly quicker and easier to make.

FIGURE 8.15 Making hollandaise sauce.

(a) Combine the egg yolks and reduction in a stainless-steel bowl.

(b) Whip over a hot-water bath.

(c) Continue to whip over a hot water until thick and light.

(d) Very slowly whip in the butter. (Set the bowl in a saucepan lined with a kitchen towel to hold it steady.)

(e) The finished sauce should be thick but pourable.

Holding Hollandaise Sauce

Hollandaise sauce, as well as other sauces in this family, poses a special safety problem. It must be kept warm for service, but it must be held at 145°F (63°C) so the eggs don't curdle. Unfortunately, this is only slightly above the Food Danger Zone, at which bacteria grow quickly. Therefore, extra care must be taken to avoid food-borne diseases.

The following sanitation procedures must be observed to avoid the danger of food poisoning:

1. Make sure all equipment is perfectly clean.
2. Hold sauce no longer than 2 hours. Make only enough to serve in this time, and discard any that is left over.
3. Never mix an old batch of sauce with a new batch.
4. Never hold hollandaise or béarnaise—or any other acid product—in aluminum. Use stainless-steel containers.

GUIDELINES for Preparing Hollandaise and Béarnaise

Students tend to be afraid of hollandaise because it has a reputation for being difficult to make. True, precautions are necessary to avoid overcooking the eggs and to get the right consistency. But if you follow the instructions in the recipe carefully and keep in mind these guidelines, you should have no trouble.

Many of these rules have one object in common: Don't overcook the egg yolks, or they will lose their ability to emulsify.

1. **Cool the reduction before adding the yolks, or they will overcook.**

2. **Use the freshest eggs possible for the best emulsification.**
   For safety, pasteurized eggs are recommended (see pp. 766 and 1030).

3. **Beat the yolks over hot water.**
   An experienced cook is able to beat them over direct heat, if care is taken, without making scrambled eggs. Until you have gained some confidence, it is safer to use a hot-water bain-marie.

4. **Use a round-bottomed stainless-steel bowl.**
   The whip must be able to reach all the eggs to beat them evenly. Also, stainless steel will not discolor the sauce or give it a metallic flavor.

5. **Have the butter warm but not hot, or it may overcook the eggs. If it is too cool, it might solidify.**

6. **Add the butter slowly at first.**
   The yolks can absorb only a little at a time. Add a few drops at first and beat in thoroughly before adding more. If you add butter faster than it can be absorbed, the emulsion may break.

7. **Don't add more butter than the egg yolks can hold.**
   Remember this standard proportion:
   6 egg yolks per 1 pound (450 g) clarified butter

8. **Broken or curdled hollandaise can be rescued.**
   First, try adding 1 teaspoon (5 mL) cold water and beating vigorously. If this doesn't work, start over with a couple of egg yolks and repeat the procedure from step 6 in the recipe, adding the broken sauce as you would the butter.

 Hollandaise Sauce I

YIELD: 24 FL OZ (720 ML)

U.S.	METRIC	INGREDIENTS	PROCEDURE
1 lb 4 oz	600 g	Butter	1. Review guidelines for preparing hollandaise and béarnaise (above).
			2. Clarify the butter (see p. 190). You should have about 1 pt (480 mL) clarified butter. Keep the butter warm but not hot.
¹/₈ tsp	0.5 mL	Peppercorns, crushed	3. Combine peppercorns, salt, and vinegar in a saucepan and reduce until nearly dry (au sec). Remove from heat and add the cold water.
¹/₈ tsp	0.5 mL	Salt	
3 fl oz	90 mL	White vinegar or wine vinegar	
2 fl oz	60 mL	Water, cold	4. Pass the diluted reduction through a fine strainer into a stainless-steel bowl. Use a clean rubber spatula to make sure you transfer all flavoring material to bowl.
6	6	Egg yolks (see Note after Hollandaise II)	5. Add the egg yolks to bowl and beat well.
1–2 tbsp	15–30 mL	Lemon juice	6. Hold the bowl over a hot-water bath and continue to beat the yolks until thickened and creamy.
to taste	to taste	Salt	7. Remove the bowl from the heat. Using a ladle, slowly and gradually beat in warm clarified butter, drop by drop at first. If the sauce becomes too thick to beat before all the butter is added, beat in a little lemon juice.
to taste	to taste	Cayenne	
			8. When all the butter is added, beat in lemon juice to taste and adjust seasoning with salt and cayenne. If necessary, thin the sauce with a few drops of warm water.
			9. Strain through cheesecloth if necessary and keep warm (not hot) for service. Hold no longer than 2 hours (see above).
			10. Evaluate the finished sauce (see p. 196).

Per 1 fl oz (29.57 mL): Calories, 280; Protein, 1 g; Fat, 31 g (99% cal.); Cholesterol, 155 mg; Carbohydrates, 0 g; Fiber, 0 g; Sodium, 310 mg.

Hollandaise Sauce II

YIELD: 24 FL OZ (720 ML)

U.S.	METRIC	INGREDIENTS	PROCEDURE
1 lb 4 oz	600 g	Butter	1. Review guidelines for preparing hollandaise and béarnaise (p. 194). 2. Clarify the butter (see p. 190). You should have about *1 pt (480 mL) clarified butter.* Keep the butter warm but not hot.
6	6	Egg yolks (see Note)	3. Place the egg yolks and cold water in a stainless-steel bowl and beat well. Beat in a few drops of lemon juice.
1 fl oz	30 mL	Water, cold	4. Hold the bowl over a hot-water bath and continue to beat until the yolks are thickened and creamy.
1¹/₂ fl oz	45 mL	Lemon juice	5. Remove the bowl from the heat. Using a ladle, slowly and gradually beat in the warm butter, drop by drop at first. If the sauce becomes too thick to beat before all the butter is added, beat in a little of the lemon juice.
to taste	to taste	Salt	6. When the butter is all added, beat in lemon juice to taste and adjust seasoning with salt and cayenne. If necessary, thin the sauce with a few drops of warm water.
to taste	to taste	Cayenne	7. Keep warm (not hot) for service. Hold no longer than 2 hours (see p. 193).
			8. Evaluate the finished sauce (see p. 196).

Per 1 fl oz (29.57 mL): Calories, 280; Protein, 1 g; Fat, 31 g (99% cal.); Cholesterol, 155 mg; Carbohydrates, 0 g; Fiber, 0 g; Sodium, 300 mg.

Note: For safety, pasteurized eggs are recommended.

Small Sauces

Maltaise

To 1 pt (500 mL) hollandaise add 1–2 fl oz (30–60 mL) orange juice (from blood oranges, if possible) and 1 tsp (5 mL) grated orange rind. Serve with asparagus.

Mousseline

Whip 4 fl oz (125 mL) heavy cream until stiff and fold into 1 pt (500 mL) hollandaise.

Béarnaise Sauce

YIELD: 24 FL OZ (720 ML)

U.S.	METRIC	INGREDIENTS	PROCEDURE
1 lb 4 oz	600 g	Butter	1. Review guidelines for preparing hollandaise and béarnaise (p. 194). 2. Clarify the butter (see p. 190). You should have about *1 pt (480 mL) clarified butter.* Keep the butter warm but not hot.
1 oz	30 g	Shallots, chopped	3. Combine the shallots, vinegar, tarragon, and peppercorns in a saucepan and reduce by three-fourths. Remove from heat and cool slightly.
4 fl oz	120 mL	White wine vinegar	4. To make it easier to beat with a wire whip, transfer the reduction to a stainless-steel bowl. Use a clean rubber spatula to make sure you get it all. Let the reduction cool a little.
2–3	2–3	Tarragon stems, chopped (see Note)	5. Add the egg yolks to the bowl and beat well.
¹/₂ tsp	2.5 mL	Peppercorns, crushed	6. Hold the bowl over a hot-water bath and continue to beat the yolks until thickened and creamy.
6	6	Egg yolks (see Note)	7. Remove the bowl from the heat. Using a ladle, slowly and gradually beat in the warm clarified butter, drop by drop at first. If the sauce becomes too thick to beat before all the butter is added, beat in a little lemon juice or warm water. 8. Strain the sauce through cheesecloth.
to taste	to taste	Salt	9. Season to taste with salt, cayenne, and a few drops of lemon juice. Mix in the parsley and tarragon.
to taste	to taste	Cayenne	10. Keep warm (not hot) for service. Hold no longer than 2 hours (see p. 193).
to taste	to taste	Lemon juice	11. Evaluate the finished sauce (see p. 196).
1 tbsp	15 mL	Chopped parsley	
2 tbsp	30 mL	Fresh tarragon, chopped (see Note)	

Per 1 fl oz (29.57 mL): Calories, 280; Protein, 1 g; Fat, 31 g (97% cal.); Cholesterol, 155 mg; Carbohydrates, 1 g; Fiber, 0 g; Sodium, 300 mg.

Note: For safety, pasteurized eggs are recommended.

If fresh tarragon is not available, used 1 tsp (5 mL) dried tarragon in place of the tarragon stems in the reduction, and use 1/2 tsp (2 mL) dried tarragon in place of the fresh leaves to finish the sauce.

Small Sauces

Foyot

Add 1 oz (30 g) melted meat glaze (glace de viande) to 1 pt (500 mL) béarnaise.

Choron

Add 1 oz (30 g) tomato paste to 1 pt (500 mL) béarnaise.

Foyot Sauce

STANDARDS OF QUALITY FOR CLASSICAL SAUCES

After you have finished a sauce, evaluate its quality. Look for the following characteristics:

Roux-thickened Sauces (white and brown)

1. Consistency and body.

Smooth, with no lumps, no graininess.

Not too thick or pasty, but thick enough to coat the food lightly.

2. Flavor.

Distinctive but well-balanced flavor. No scorched taste. White sauces should have a full but fairly neutral flavor. Brown sauce should have a rich, meaty flavor from roasted bones, but with no burned taste of excessively browned bones.

Proper degree of seasoning.

No starchy taste.

3. Appearance.

Smooth, with a good shine.

Good color for its type (rich, deep brown, for brown sauce, pale ivory for velouté, white—not gray—for cream sauce). Uniform color, no dark specks in brown sauce. Brown sauce should not have too reddish a color from using too much tomato product.

A brown sauce thickened by a starch slurry (jus lié) is clearer than one thickened by roux.

Tomato Sauce

1. Consistency and body.

Smooth, with no lumps. Slightly coarse; not as smooth as white and brown sauces.

Easily pourable; not too thick or too thin.

2. Flavor.

Distinctive but well-balanced and concentrated flavor of tomatoes. Not bitter, not too acidic, and not too sweet.

Proper degree of seasoning. Tomato flavor should not be overwhelmed by herbs and spices.

3. Appearance.

Opaque, with good red color.

Hollandaise and Béarnaise

1. Consistency and body.

Perfectly smooth texture, with no graininess from over-cooked egg. Light and easily pourable but thick enough to coat foods; somewhat thicker than properly made roux-based sauces.

2. Flavor.

Taste of fresh butter, balanced by a subtle acidity from the reduction and/or lemon juice. Béarnaise has, in addition to these flavors, a mild herb flavor from the tarragon and parsley. Proper amount of salt.

3. Appearance.

Lemon-yellow color. Good shine, but no oiliness.

KEY POINTS TO REVIEW

- What are the five leading sauces (also called mother sauces and grand sauces)? Describe the procedure for making each one.

- What is a small sauce?

- What is clarified butter? How is it made? What is it used for?

- What is beurre noisette?

- What is a compound butter?

- What is the procedure for making beurre blanc?

PAN GRAVIES AND OTHER INTEGRAL SAUCES

An **integral sauce** is a sauce based on the juices released during the cooking of a meat, poultry, fish, or vegetable item. Most of the sauces we have discussed so far are not integral sauces—that is, they are made separately from and independently of the items they are served with. An integral sauce, on the other hand, can't be made separately because it incorporates cooking juices from the item it is served with.

The most important technique required for integral sauces is deglazing (see pp. 177 and 180). Juices released by sautéed and roasted meats are reduced and caramelized in the bottom of the pan during cooking. Deglazing dissolves these caramelized juices and incorporates them into the desired sauce. For the simplest example, if you sauté a chicken breast and then deglaze the sauté pan with a little stock and season the resulting liquid, you end up with an integral sauce that can be served with the chicken.

The most basic and familiar integral sauces are pan gravy and jus. **Pan gravy** is a sauce made with the juices or drippings of the meat or poultry with which it is being served. Standard pan gravies are similar to brown sauces. Instead of being made with espagnole or demiglace as a base, however, they are made from pan drippings plus roux plus stock or water and, sometimes, milk or cream.

Jus (zhoo) refers to unthickened juices from a roast. When the roast is served with these clear, natural juices, it is said to be served **au jus** (oh zhoo), meaning "with juice." Stock is usually added to the pan juices to obtain enough quantity to serve.

The preparation of both pan gravy and jus are properly part of meat cookery, and recipes and detailed procedures are included in the meat and poultry chapters. Similarly, recipes for all integral sauces are included as part of the meat, fish, or vegetable in the appropriate chapters.

The principal recipes are on page 487 (Roast Beef Gravy) and page 540 (Roast Turkey with Giblet Gravy). Gravy-making is also incorporated in the recipes for Roast Stuffed Shoulder of Lamb (p. 488) and Roast Chicken with Natural Gravy (p. 418).

Now that you have studied sauce-making in detail, read the Basic Procedure for Making Pan Gravy on page 198 so you can see how similar it is to making brown sauce and how the same techniques you have just learned are applied to a different product.

MODERN SAUCES

As suggested in the introduction to this chapter, sauce-making has changed a great deal since Escoffier's day. Although our basic methods for making many of the sauces in the modern kitchen are derived from classical cuisine, details have changed. Perhaps the most important change is that chefs rely less on roux for thickening a sauce, while reduction has become more important to give sauces body (see p. 176). When starches are used, they are often purer starches, such as arrowroot.

BASIC PROCEDURE for Making Pan Gravy

Method 2 has fewer steps, but Method 1 is actually quicker for large quantities and gives greater control over final consistency.

METHOD 1

1. **Remove the roast from the roasting pan.**
   If you did not add mirepoix to the pan during roasting, you can do so now.

2. **Clarify the fat.**
   Set the roasting pan over high heat and cook until all the moisture has evaporated, leaving only the fat, mirepoix, and the brown (caramelized) drippings. Pour off and save the fat.

3. **Deglaze the pan.**
   Pour stock or other liquid into the roasting pan. Stir over heat until the caramelized drippings are dissolved.

4. **Combine with stock and simmer.**
   Pour the deglazing liquid, plus mirepoix, into a large pot with desired amount of stock. Simmer until mirepoix is well cooked. Skim the surface well to remove fat and scum.

5. **Make a roux or, alternatively, a slurry of arrowroot or cornstarch and water.**
   For roux, measure enough of the fat from step 2 to make the correct amount of roux for the volume of gravy. Make a blond or brown roux, as desired. For starch slurry, see page 171.

6. **Thicken the gravy with the roux or starch slurry.**

7. **Strain.**

8. **Adjust seasonings.**

METHOD 2

1. Remove the roast from the roasting pan.

2. Clarify the fat.

3. Add flour to the roasting pan and make a roux.

4. Add stock. Stir until thickened and the pan is deglazed.

5. Strain. Skim excess fat.

6. Adjust consistency, if necessary, with more stock or more roux.

7. Season.

Chefs have also been influenced by other cuisines, such as those of Asia and Latin America, and have borrowed ingredients and procedures from many countries and regions to give variety to their repertoire of sauces.

Because of the ongoing experimentation with and development of new sauces, it is difficult to classify and define them exactly, the way Escoffier did in the last century. We can, however, describe general groups that many of today's popular sauces fall into. The remaining recipes in this chapter include examples of these types of sauces.

A number of other popular sauces, such as barbecue sauce, that don't fit into any of the categories described in the following sections, are included in this chapter. Other sauce recipes are included elsewhere in this book, often as components of other recipes. Among the more important of these are vinaigrette and mayonnaise variations. These are traditionally used as salad dressings but are also used as sauces for meat, seafood, and vegetable items.

The following recipes from other chapters appear on the pages indicated.

Roast Beef Gravy, page 487

Jus Lié, page 487

Italian Tomato Sauce for Pasta, with variations, page 399

Pesto, page 399

Mole Poblano, page 573

Salsa Verde, Cocida, p. 206

Basic Vinaigrette and variations, page 660

American French Dressing, page 661

Oriental Vinaigrette, page 662

Mayonnaise and variations, page 665

Broths and Jus

Beginning with the introduction of nouvelle cuisine in the 1970s, chefs looked for ways to eliminate starch thickeners in sauces in order to make them lighter. The technique of reduction to concentrate a sauce has been the most important tool in this effort. Reduction hasn't been a cure-all, however. First, some of the fresher, lighter flavors of a sauce are lost when a liquid is subjected to the long cooking required for reduction. In addition, reduced sauces sometimes become so gelatinous that they solidify when they cool—not an appetizing result.

Nevertheless, we have become accustomed to sauces that do not cling thickly to the meat, poultry, or seafood. Sauces, often in smaller quantities, are served under or around the item as often as over the top of it, perhaps even more often. Some chefs have gone to the extreme of serving the item in a little broth in place of a sauce. This technique has long been popular with seafood, as in the case of Seafood à la Nage (p. 439), but it is becoming more common with meat as well. The result is often something like a garnished consommé (pp. 232–233), but with very little consommé and a full portion of meat and garnish. The recipe on page 512 is an example.

For a broth to work well as a substitute for a sauce, it should be well flavored and aromatic. Taste the broth, reduce it as necessary to concentrate the flavor, and check the seasonings carefully.

A jus is very much like a broth except it is usually more concentrated, although still unthickened. The term *jus* has two basic meanings:

1. The unthickened, natural juices resulting from a roast. This is the more traditional meaning of *jus* (see p. 197). To make a traditional jus, the drippings of a roast are deglazed with stock or other liquid, reduced slightly, seasoned, strained, and served unthickened.

   To make a meat jus without a roast, follow the procedure below.

2. An unthickened liquid carrying the concentrated flavor of a specific ingredient. This type of jus is often made from vegetables and is sometimes called an **essence**. To make a vegetable essence, the vegetable is simmered with a stock or broth until the liquid is concentrated and flavorful. The recipe for Mushroom Jus (p. 203, also called **mushroom essence**) is an example of this type of preparation.

PROCEDURE for Making a Meat Jus

1. Cut trimmings of the desired meat or poultry product into small pieces. Place them in a heavy pot over moderate heat.

2. Cook until well browned on all sides. Some liquid will be released from the meat. If the trimmings begin to simmer in these juices instead of browning, just let them continue to cook until the liquid has evaporated and browned on the bottom of the pot.

3. Deglaze with a small quantity of white wine or stock. Continue to cook until the liquid is reduced and the juices again caramelize on the bottom.

4. Add enough stock to cover the meat. Stir to dissolve the caramelized juices on the bottom of the pot. Simmer until the liquid is completely reduced and caramelized.

5. Again add enough stock to cover the meat. Stir to dissolve the caramelized juices. Simmer 10–15 minutes. Strain and degrease.

Purées

Vegetable purées have long been used as sauces. Tomato sauce is the classic example. However, nearly any vegetable can be puréed and used as a sauce, provided it is flavorful, properly seasoned, and of an appropriate consistency or thickness. A vegetable purée is sometimes called a **coulis**.

Purées of starchy vegetables, such as squash or dried beans, may need to be thinned with stock, broth, or water. Even potato purée is sometimes thinned and used as a sauce, usually enriched with a little raw butter stirred in. In addition, potato and other thick purées are used as thickeners for other sauces.

Some vegetables, such as asparagus, make a watery purée. These purées can be reduced to thicken them, but be careful not to lose the fresh vegetable taste and color. This should especially be avoided in the case of green vegetables, which quickly lose their color (see p. 271). Although thin vegetable purées may be thickened with a starch, it is more common to leave them thin or to bind them lightly by finishing them with raw butter (monter au beurre, p. 177) or by reducing them with a little cream until they have the desired consistency.

Cream Reductions

In the era of nouvelle cuisine, sauces based on reduced cream became a popular substitute for roux-thickened white sauces. When heavy cream is reduced, it thickens slightly. A common fault with cream reduction sauces is reducing the cream too much, giving it a heavy texture. If it is reduced beyond this point, it is likely to break, and the butterfat will separate. For an appealing, light texture, reduce the cream until it is about two-thirds its original volume.

A reduced cream sauce is a mixture of reduced cream and a concentrated, flavorful stock. White stock is most often used, although brown cream sauces may also be prepared using brown stock. For good results, the stock should be reduced by about three-fourths. Flavored sauces can be made by reducing the stock with flavoring ingredients, as in the recipe for Chipotle Cream Sauce on page 206.

Two methods are possible:

1. Reducing the cream to the desired consistency and then adding it to the stock reduction.

2. Adding fresh cream to the stock reduction and reducing the mixture to the desired consistency.

Many chefs feel the first method is more controllable. See the following procedure.

PROCEDURE for Making a Cream Sauce Reduction

1. Reduce white stock or brown stock by about three-fourths, or until it is concentrated and flavorful.

2. Measure the reduction. For each pint (500 mL) of reduction, measure about 1½ pints (750 mL) heavy cream.

3. Place the cream in a heavy saucepan over moderate heat and reduce until lightly thickened, or until reduced by about one-third. Stir from time to time with a whip.

4. Bring the stock reduction to a simmer in a sauce pan. Stir in the reduced cream.

5. Check the consistency. Thicken, if necessary, by reducing further, or thin with additional heavy cream.

6. Season and strain.

Salsas, Relishes, and Chutneys

It is said that, in the United States, salsa has become even more popular than ketchup. The salsa referred to is, of course, the Mexican mixture of chopped tomatoes, onions, chiles, herbs, and other ingredients. *Salsa* is actually the Spanish and the Italian word for "sauce," so the word refers to many types of preparations, both raw and cooked, not just this one Mexican relish. Nevertheless, in English-speaking countries, the word **salsa** usually refers to a mixture of raw or cooked chopped vegetables, herbs, and, occasionally, fruits.

Salsas are easily improvised. Select a suitable mixture of vegetables, fruits, or both, and chop coarsely or finely, as desired. Mix with appropriate chopped fresh herbs and season to taste. Salt draws juices out of the ingredients to provide moisture for the mixture. Add citrus juice or vinegar if the mixture is lacking in acidity. Acidity should balance any sweetness from fruits because salsas are usually intended for savory dishes, not desserts.

The words *relish* and *chutney* have no exact definitions. One meaning of *relish* is any raw or pickled vegetable used as an appetizer (see pp. 740–741). For example, a dish of celery sticks, carrot sticks, and olives is sometimes called a *relish dish*, for many years a traditional appetizer in steakhouses and other restaurants. As used in a discussion of sauces, a **relish** is a mixture of chopped vegetables (and sometimes fruits), at least one of which is pickled in vinegar or a salt solution. By this definition, a salsa may be considered a type of relish, especially if it contains an acid such as vinegar or citrus juice.

The word **chutney** originated in India, where it refers to several types of spicy condiments or relishes, including strongly spiced sweet-and-sour cooked fruit or vegetable mixtures, as well as raw or partially cooked mixtures of chopped herbs or vegetables, also spicy and often containing chiles. Almost all chutneys contain an acid ingredient. Western cooks have been especially inspired by the sweet-and-sour types of chutney, so when the word *chutney* appears on a menu, it usually refers to a cooked fruit or vegetable condiment that is sweet, spicy, and tangy.

Several examples of salsas, relishes, and chutneys are included in this chapter.

Asian Sauces

Sauces from many Asian cuisines, including Japanese, Thai, Vietnamese, and Indian, have entered the Western cook's repertoire in recent years. Asia is, of course, a huge continent, and it would take years of study to become familiar with all its varied cooking traditions. This chapter can only begin that familiarization process by providing a selection of popular recipes with sidebars containing background information on ingredients and techniques.

Incidentally, Chinese cuisines have relatively few standalone sauces. Sauces in stir-fried dishes, for example, are made as part of the cooking process by adding liquids and thickeners to the meat and vegetables as they cook. Ready-made condiments such as oyster sauce and hoisin sauce are also used.

When adopting Asian-style sauces into Western cuisine, cooks should have some familiarity with the regional cuisine they are borrowing from and how the sauces are used in that cuisine. Unless the cook is careful, mixing Asian-style sauces with Western dishes can have strange results.

Flavored Oils

Flavored oils make a light, interesting alternative to vinaigrettes and other sauces when used to dress a wide variety of dishes. They are especially suitable for simple steamed, sautéed, or grilled items, but they can be used with cold foods as well. When used as a sauce, the oil is usually drizzled around or, sometimes, over the item on the plate. A tablespoon (15 mL) or so per portion is often enough.

The simplest way to flavor an oil is simply to put some of the flavoring ingredient in the oil and let it stand until the oil has taken on enough of the flavor. For most flavorings, however, this is not the best way to extract the most flavor. The flavoring ingredient may need some kind of preparation before adding it to the oil. For example, dry spices develop more flavor if they are first heated gently with a little bit of the oil.

PROCEDURE for Making Flavored Oils

1. Prepare the flavoring ingredient in one of the following ways:

    Chop fresh roots (such as horseradish, garlic, shallots, ginger, garlic) or strong herbs (fresh rosemary, sage, thyme, oregano) by hand or in a food processor.

    Grate citrus zests.

    Blanch tender herbs (parsley, basil, tarragon, chervil, cilantro) in boiling water for 10 seconds. Drain immediately and refresh under cold water. Dry well.

    Gently heat dried, ground spices (cinnamon, cumin, curry powder, ginger, mustard, paprika) in a small amount of oil just until they start to give off an aroma.

2. Place the flavoring ingredient in a jar or other closable container. Add oil.

3. Close the jar and shake it well. Let stand 30 minutes at room temperature, then refrigerate.

4. The oil is ready to use as soon as it has taken on the desired flavor, which may be as soon as 1 hour, depending on the ingredient. After 2 days, strain the oil through a chinois lined with a paper coffee filter. Store in the refrigerator.

Refrigerating flavored oils is recommended. As you will recall from Chapter 2, botulism is caused by a kind of bacteria that grows in the absence of air. Because oil prevents air from reaching the flavoring ingredients, if any botulism bacteria are present in the flavorings (especially possible with fresh, raw roots), those bacteria could grow while covered with oil if not refrigerated.

The procedure on page 201 outlines the basic method for making flavored oils, depending on the type of ingredients. Unless otherwise indicated, use a mild or flavorless oil, such as safflower, canola, corn, or grapeseed. In some cases, as with garlic, the flavoring goes well with olive oil, but usually the goal is to have the pure taste of the flavoring ingredient unmasked by the flavor of the oil.

Molecular Gastronomy

As you read in the discussion of molecular gastronomy on page 7, chefs at the forefront of modern cuisine are exploring and inventing new techniques in food preparation and presentation. One category of new techniques is the use of nontraditional thickeners or binding agents for sauces.

Following is an alphabetical list of some of the principal thickeners (often called *hydrocolloids;* see sidebar). Many of these ingredients are used in tiny quantities. For this reason, avant-garde chefs are likely to be working in the metric system (see p. 73). It is easier to scale 2 grams of xanthan gum, for example, than 0.07 ounces. (For this reason, some ingredients in the recipes near the end of this chapter indicate only metric measures, not U.S. measures.) This need for precision carries over into every aspect of the chef's work, from scaling all ingredients and measuring temperatures to cutting ingredients and creating plate arrangements.

To give you a taste of these techniques, the end of the chapter features several unusual sauces and other condiments employing these ingredients. In addition, you will also find vegetable recipes in Chapter 11 (pp. 319–320) that employ these thickeners.

> **Agar-agar.** A jelling agent derived from seaweed. Agar-agar is a traditional ingredient in Asian cuisines and has long been used in Western cooking as a vegetarian substitute for gelatin. To use, stir into a cold liquid and bring to a boil. The liquid sets to a gel when cooled to 95°F (35°C). Once jelled, it can be reheated to serve warm because it does not melt until heated to at least 185°F (85°C). To use as a thickener, jell the liquid and then blend in a blender.
>
> **Calcium lactate** and **calcium chloride.** Calcium compounds used in the process of spherification (see p. 120).
>
> **Carageenan.** A hydrocolloid or thickener derived from seaweed. Carageenans are mixed with cold liquids and then heated to thicken.
>
> **Gellan gums** are used to make firm or solid gels. Both dissolve when the liquid they are mixed with is heated, and both form gels when the liquid is cooled. *High acyl gellan* forms soft, elastic gels which melt if heat is reapplied. *Low acyl gellan* forms hard gels that are more resistant to melting. These two gums can be used alone or in combination to produce a wide variety of textures.
>
> **Guar gum.** A hydrocolloid or thickener derived from the guar bean plant. This is a powerful stabilizer and thickener that has long been used in commercial ice creams.
>
> **Methylcellulose.** A hydrocolloid derived from plant fiber. It has long been used as a dietary fiber supplement. It is an unusual thickener and jelling agent because it thickens as it is heated and thins out or melts when cooled. This allows the chef to create unusual effects. Methylcellulose is also used to stabilize foams.
>
> **Sodium alginate.** A hydrocolloid derived from seaweed. This ingredient is used in the process of spherification. It does not have to be heated to make a gel, and the gel is *nonreversible*. This means once the gel formed, it stays solid even when heated.
>
> **Soy lecithin.** A powerful emulsifier. Lecithin is the component of egg yolks that makes mayonnaise possible (see p. 110). Also extracted from soybeans, lecithin is used to stabilize many mixtures that would separate without it.
>
> **Tapioca maltodextrin.** A modified food starch that, when mixed with fat, changes it to a powder. Because maltodextrin dissolves in water, an oil, such as olive oil, that has been powdered changes back to an oil in the mouth.

HYDROCOLLOIDS

A **colloid** is a mixture in which one substance (called the *dispersed phase*) is evenly mixed throughout another substance (called the *continuous phase*). Many colloids are familiar to us. Milk, for example, is liquid water in which milk solids and tiny butterfat globules are evenly mixed. Foams are colloids in which air bubbles are evenly distributed throughout a liquid. Smoke is a colloid of soot particles mixed in air. Emulsions such as mayonnaise (p. 663) are familiar colloids from the kitchen.

A **hydrocolloid** is a colloid in which the continuous phase is water. The presence of particles mixed throughout the water changes the water's characteristics. For example, gelatin is a hydrocolloid familiar to all chefs. In this product, strands of proteins are mixed in water. Because these strands bind to each other, even a small amount of gelatin is enough to turn water from a liquid to a soft solid, called a gel. Another familiar example of a hydrocolloid is a sauce thickened with starch. The strands of starch change the water base from a thin liquid to a thicker one.

When chefs who practice molecular gastronomy use the word *hydrocolloid* to describe unusual thickeners and jelling agents, such as sodium alginate and other gums, it makes them sound especially scientific and mysterious. But remember, chefs have been using hydrocolloids their entire careers. They just haven't called them that.

Ultra-Tex 3 and Ultra-Tex 8. Modified food starches extracted from tapioca. The name is a trademark of the National Starch Company, which makes it. Both products thicken cold liquids without heating. If the thickened liquid is poured into a thin layer, it dries to form a thin film or sheet. Ultra-Tex 8 is a newer product that works better in acidic liquids and can be used in smaller quantities.

Xanthan gum. A hydrocolloid or thickener made by fermenting sugar with a special bacterium. It has been used for years to give structure to gluten-free breads and other baked goods. Liquids thickened with xanthan gum have the same thickness whether hot or cold.

KEY POINTS TO REVIEW

- What does the term *integral sauce* mean?

- What is the basic procedure for making a pan gravy?

- What is the basic procedure for making a cream reduction sauce?

- What is the basic procedure for making a flavored oil?

Mushroom Jus ♥

YIELD: 2 QT (2 L)

U.S.	METRIC	INGREDIENTS	PROCEDURE
3 lb	1.5 kg	Mushrooms, cleaned, coarsely chopped	1. Place the mushrooms and the first quantity of water in a stockpot.
2 gal	8 L	Water	2. Bring to a boil. Reduce the heat to a rapid simmer and cook until most of the liquid has evaporated.
2 gal	8 L	Water	3. Add the second quantity of water and repeat the reduction process.
2 gal	8 L	Water	4. Add the third quantity of water. Reduce by three-quarters.
			5. Strain through a china cap lined with cheesecloth, pressing on the mushrooms to extract as much liquid as possible.
			6. To use or to finish as a sauce, see Variations.

Per 1 fl oz (29.57 mL): Calories, 5; Protein, 1 g; Fat, 0 g (0% cal.); Cholesterol, 0 mg; Carbohydrates, 1 g; Fiber, 0 g; Sodium, 0 mg.

VARIATIONS

The jus can be used as is, seasoned with salt and pepper. A small amount of arrowroot or other starch may be used to bind the sauce lightly. Alternatively, finish by enriching with cream (see p. 200 for information on cream reductions) or butter (see p. 191). The jus can also be added as a flavoring ingredient to meat or poultry broths and to demi-glace, and it can be used as a deglazing liquid.

Lamb Reduction

YIELD: APPROX. 12 FL OZ (360 ML)

U.S.	METRIC	INGREDIENTS	PROCEDURE
1½ tsp	7 mL	Vegetable oil	1. Heat the oil in a saucepan over medium heat. Add the lamb bones. Brown evenly. Remove and set aside
1 lb	480 g	Lamb bones and/or trimmings	
8 oz	240 g	Mirepoix	2. Add the mirepoix and cook until lightly browned.
1 tbsp	15 mL	Tomato paste	3. Add the tomato paste and cook until lightly caramelized.
12 fl oz	360 mL	Red wine	4. Deglaze the pan with the red wine. Reduce by three-quarters.
1½ qt	1.5 L	Brown lamb stock	
		Sachet:	5. Add the stock and sachet, and return the lamb bones to the pot. Bring to a boil, reduce heat to a simmer, and simmer until reduced by half. Skim frequently during cooking.
1	1	Bay leaf	
2 sprigs	2 sprigs	Fresh thyme	6. Strain through a chinois into a clean saucepan.
8	8	Peppercorns	7. Continue to reduce until just a little lighter than nappé consistency (see p. 170). Continue to skim while reducing.
6	6	Parsley stems	
to taste	to taste	Salt	8. Add salt and pepper to taste.
to taste	to taste	Pepper	9. Swirl in raw butter (monter au beurre).
2 oz	60 g	Butter, unsalted	10. Add the lemon juice or vinegar to cut the richness of the flavor.
½ tsp	2 mL	Lemon juice or sherry vinegar	

Per 1 fl oz (29.57 mL): Calories 80; Protein, 2 g; Fat, 4.5 g (51% cal.); Cholesterol, 10 mg; Carbohydrates, 2 g; Fiber, 0 g; Sodium, 250 mg.

Bell Pepper Coulis ♥

YIELD: 2¹/₂ PT (1.25 L)

U.S.	METRIC	INGREDIENTS	PROCEDURE
4 lb	2 kg	Red or yellow bell peppers	1. Split peppers in half lengthwise. Remove cores, seeds, and membranes. Chop coarsely.
2 fl oz	60 mL	Olive oil	2. Heat the olive oil in a saucepot over low heat.
2 oz	60 g	Shallots, chopped	3. Add the shallots and peppers. Cover and sweat over low heat until vegetables are soft, about 20 minutes.
4 fl oz	125 mL	Chicken stock, vegetable stock, or water	4. Add the stock or water. Simmer 2–3 minutes.
			5. Purée the vegetables and liquid in a blender, then pass through a strainer.
1–4 fl oz	30–125 mL	Additional stock or water	6. Adjust the texture by adding water or stock to thin it.
to taste	to taste	Salt	7. Add salt and white pepper to taste.
to taste	to taste	White pepper	

Per 1 fl oz (29.57 mL): Calories, 25; Protein, 0 g; Fat, 1.5 g (53% cal.); Cholesterol, 0 mg; Carbohydrates, 3 g; Fiber, 1 g; Sodium, 0 mg.

VARIATION

Bell Pepper and Tomato Coulis

Combine bell pepper coulis with an equal volume of tomato purée.

Root Vegetable Purée

YIELD: 2 LB (960 G)

U.S.	METRIC	INGREDIENTS	PROCEDURE
6 oz	180 g	Rutabaga (see Note)	1. Chop the rutabaga, carrots, parsnips, and turnips into small pieces (this is a good use of trimmings left from other recipes).
6 oz	180 g	Carrots	2. Combine the vegetables and stock in a saucepan. Bring to a boil, reduce to a simmer, and cook until the vegetables are very tender.
6 oz	180 g	Parsnips	
4¹/₂ oz	135 g	Turnips	3. Drain and reserve the stock.
1 qt	1 L	Chicken stock or vegetable stock	4. Place the vegetables in a blender and blend to a purée. Keep the vegetables in the blender for the remaining steps.
3 oz	90 g	Butter, unsalted	5. Add the raw butter and blend in.
to taste	to taste	Salt	6. Add enough of the reserved stock to make a soft, smooth purée.
to taste	to taste	White pepper	7. Season to taste with salt and pepper.

Per 1 fl oz (29.57 mL): Calories 40; Protein, <1 g; Fat, 2.5 g (56% cal.); Cholesterol, 5 mg; Carbohydrates, 3 g; Fiber, <1 g; Sodium, 50 mg.

Note: For all the vegetables, use trimmings left over from other preparations if available.

Carrot Purée

YIELD: APPROX. 1 LB 8 OZ (720 G)

U.S.	METRIC	INGREDIENTS	PROCEDURE
1 lb	480 g	Carrots, large dice	1. Combine carrots, juice, and stock in a saucepan.
1 pt	480 mL	Carrot juice	2. Bring to a boil, reduce to a simmer, and simmer until carrots are very tender.
8 fl oz	240 mL	Chicken stock or vegetable stock	3. Drain the carrots and reserve the cooking liquid.
2 oz	60 g	Butter	4. Purée the carrots in a blender until smooth.
to taste	to taste	Salt	5. Add the raw butter and blend it in.
to taste	to taste	Pepper	6. Adjust the consistency as necessary by adding enough of the reserved cooking liquid to make a soft purée.
			7. Season with salt and pepper.

Per 1 fl oz (29.57 mL): Calories 35; Protein, <1 g; Fat, 2 g (51% cal.); Cholesterol, 5 mg; Carbohydrates, 3 g; Fiber, <1 g; Sodium, 55 mg.

Cauliflower Purée

YIELD: APPROX. 1 LB 8 OZ (720 G)

U.S.	METRIC	INGREDIENTS	PROCEDURE
1 lb	480 g	Cauliflower, cut up	1. Combine cauliflower, cream, and stock or water in a saucepan. Bring to a boil, then reduce heat to a simmer.
8 fl oz	240 mL	Heavy cream	
8 fl oz	240 mL	Chicken stock, vegetable stock, or water	2. Simmer until the cauliflower is very tender.
			3. Drain and reserve the cooking liquid.
2 oz	60 g	Butter	4. Purée the cauliflower in a blender.
to taste	to taste	Salt	5. Add the butter and purée until smooth.
to taste	to taste	White pepper	6. Add a little of the reserved cooking liquid if necessary to adjust the consistency to a soft purée.
to taste	to taste	Lemon juice	7. Add salt, pepper, and lemon juice to taste.

Per 1 fl oz (29.57 mL): Calories 60; Protein, <1 g; Fat, 6 g (90% cal.); Cholesterol, 20 mg; Carbohydrates, 1 g; Fiber, 0 g; Sodium, 35 mg.

Cream Sauce for Fish

YIELD: 1 PT (500 ML)

U.S.	METRIC	INGREDIENTS	PROCEDURE
1 oz	30 g	Shallots, chopped	1. Sweat the shallots in butter until soft.
1/2 oz	15 g	Butter	2. Add stock, wine, parsley, and bay leaf. Over moderate or high heat, reduce the liquid by three-fourths.
1 pt	500 mL	Fish stock	
8 fl oz	250 mL	Dry white wine	
4	4	Parsley stems	
1/2	1/2	Bay leaf	
1 pt	500 mL	Heavy cream	3. Add the cream and bring to a boil. Boil 1–2 minutes, or until reduced to a light, saucelike consistency.
to taste	to taste	Lemon juice	
to taste	to taste	Salt	4. Season to taste with lemon juice, salt, and white pepper.
to taste	to taste	White pepper	

Per 1 fl oz (29.57 mL): Calories, 130; Protein, 2 g; Fat, 13 g (81% cal.); Cholesterol, 45 mg; Carbohydrates, 2 g; Fiber, 0 g; Sodium, 20 mg.

VARIATIONS

Cream Sauce for Poultry and Meat

Substitute chicken stock or white veal stock for fish stock.

Herbed Cream Sauce

Garnish the sauce with appropriate chopped fresh herbs, such as tarragon, chives, or parsley.

Chipotle Cream Sauce

YIELD: 12 FL OZ (375 ML)

U.S.	METRIC	INGREDIENTS	PROCEDURE
1 qt	1 L	Brown stock	1. Combine the stock and chiles in a saucepot.
2	2	Whole, dried chipotle chiles	2. Bring to a slow simmer. Simmer until reduced by three-fourths.
12 fl oz	375 mL	Heavy cream	3. While the stock is reducing, place the cream in a second saucepot and reduce by one-third.
to taste	to taste	Salt	4. Stir the reduced cream into the reduced stock. If necessary, adjust consistency of the sauce by reducing a little more to thicken or adding fresh cream to thin. The sauce should be about the thickness of heavy cream. The consistency depends on the gelatin content of the stock and fat content of the cream.
to taste	to taste	White pepper	
			5. Season to taste with salt and white pepper.
			6. Strain. Discard chiles.

Per 1 fl oz (29.57 mL): Calories, 110; Protein, 1 g; Fat, 11 g (93% cal.); Cholesterol, 45 mg; Carbohydrates, 0 g; Fiber, 1 g; Sodium, 20 mg.

VARIATIONS

White stock, chicken stock, or vegetable stock may be substituted for brown stock.

Herb Cream Sauce

Use white stock or chicken stock instead of brown stock. Omit chiles. Add chopped fresh herbs (such as parsley, tarragon, chervil, basil) as desired to the sauce after straining.

Salsa Verde Cocida

YIELD: 1 QT (1 L)

U.S.	METRIC	INGREDIENTS	PROCEDURE
4	4	13-oz (368-g) cans whole tomatillos (Mexican green tomatoes)	1. Drain the tomatillos.
2 oz	60 g	Onion, chopped	2. Combine tomatillos, onion, garlic, chiles, and cilantro in a blender. Blend to a smooth purée.
4	4	Garlic cloves, chopped	
2–4 oz	60–125 g	Green chiles, such as jalapeño or serrano, canned or fresh	
1 oz	30 g	Fresh cilantro leaves (optional)	
1 fl oz	30 mL	Oil	3. Heat the oil in a large saucepan. Add the purée and cook 4–5 minutes, or until slightly thickened.
to taste	to taste	Salt	4. Season to taste with salt.

Per 1 fl oz (29.57 mL): Calories, 25; Protein, 2 g; Fat, 1 g (27% cal.); Cholesterol, 0 mg; Carbohydrates, 4 g; Fiber, 1 g; Sodium, 55 mg.

VARIATIONS

Salsa Roja

Substitute 2 lb (1 kg) red, ripe tomatoes, peeled, or canned red tomatoes for the tomatillos. Onion may be included or omitted to create slightly different flavors.

Tomato Broth for Chiles Rellenos

Prepare as for Salsa Roja, using onion but omitting chiles and cilantro. After step 3, add 3 pt (1.5 L) pork stock (including the cooking liquid from making Picadillo (p. 510) for the filling for the chiles) and/or chicken stock. Also, add a sachet containing 6 whole cloves, 10 peppercorns, 2 bay leaves, and 1 small cinnamon stick. Simmer until slightly thickened to the consistency of a thick broth or thin sauce.

Salsa Verde Cocida

Salsa Cruda

YIELD: 1 QT (1 L)

U.S.	METRIC	INGREDIENTS
1 lb 4 oz	600 g	Fresh tomatoes
4 oz, or to taste	125 g, or to taste	Fresh green chiles, such as jalapeño or serrano
6 oz	175 g	Onion
1/2–1 oz	15–30 g	Fresh cilantro leaves, chopped
1 tbsp	15 mL	Lime juice or vinegar
2–4 fl oz	60–125 mL	Water or tomato juice, cold
1 1/2 tsp	7 mL	Salt

PROCEDURE

1. Chop the tomatoes fine. (You may peel them, but it is not necessary.)
2. Remove the stem ends of the chiles. Chop the chiles fine.
3. Mince the onion.
4. Mix together the tomato, chiles, onion, cilantro, and lime juice or vinegar. Dilute with water or tomato juice to make a thick, chunky sauce.
5. Add salt to taste.

Per 1 fl oz (29.57 mL): Calories, 10; Protein, 0 g; Fat, 0 g (0% cal.); Cholesterol, 0 mg; Carbohydrates, 2 g; Fiber, 0 g; Sodium, 110 mg.

Note: This sauce is used as a table condiment with many dishes, including eggs, broiled meats, tacos, tortillas, and beans. It is best if used within a few hours.

Salsa Cruda

Ancho Sauce

YIELD: APPROX. 1 1/2 PT (750 ML)

U.S.	METRIC	INGREDIENTS
8	8	Dried ancho chiles
2 oz	60 g	Onion, chopped
3	3	Garlic cloves, chopped
1 tsp	5 mL	Ground cumin
1/2 tsp	2 mL	Dried oregano
1 pt	500 mL	Water or chicken stock
1 fl oz	30 mL	Oil
to taste	to taste	Salt

PROCEDURE

1. Toast the chiles lightly in a dry skillet until softened. Split open. Remove and discard seeds and core.
2. Soak chiles about 30 minutes in enough hot water to cover. Drain.
3. Combine chiles, onion, garlic, cumin, oregano, and water or stock in a blender. Blend to a smooth purée.
4. Heat the oil in a saucepan and add the chile purée. Simmer 2–3 minutes.
5. Season to taste with salt.

Per 1 fl oz (29.57 mL): Calories, 30; Protein, 1 g; Fat, 1.5 g (46% cal.); Cholesterol, 0 mg; Carbohydrates, 3 g; Fiber, 1 g; Sodium, 5 mg.

VARIATION

Blend 8 oz (250 g) chopped tomato with the chiles in step 3.

Apple Relish

YIELD: 11 OZ (330 G)

U.S.	METRIC	INGREDIENTS	PROCEDURE
7 oz	210 g	Granny Smith apple, peeled and cut into small dice	1. Combine all ingredients except the parsley and mix until evenly combined.
1/3 cup	80 mL	Red onion, small dice, rinsed and drained	2. Just before serving, mix in the parsley.
2 tbsp	30 mL	Red bell pepper, small dice	
2 tbsp	30 mL	Cider vinegar	
1 tbsp	15 mL	Honey	
1 1/2 tsp	7 mL	Olive oil	
to taste	to taste	Salt	
to taste	to taste	Black pepper	
1 tbsp	15 mL	Italian parsley, cut chiffonade	

Per 1 fl oz (29.57 mL): Calories 20; Protein, 0 g; Fat, .5 g (23% cal.); Cholesterol, 0 mg; Carbohydrates, 4 g; Fiber, 0 g; Sodium, 0 mg.

Tomato Raisin Chutney ♥

YIELD: 1 1/4 PT (625 ML)

U.S.	METRIC	INGREDIENTS	PROCEDURE
2 lb 8 oz	1.2 kg	Tomatoes, fresh (see Note)	1. Blanch the tomatoes in boiling water for 10 seconds to loosen skins. Peel and trim.
			2. Cut tomatoes into quarters.
2 oz	60 g	Garlic, coarsely chopped	3. Combine the garlic, ginger, and first quantity of vinegar in a blender and blend until smooth.
1 oz	30 g	Fresh ginger root, peeled, coarsely chopped	
4 fl oz	120 g	Red wine vinegar	
6 fl oz	180 g	Red wine vinegar	4. Combine tomatoes, garlic mixture, second quantity of vinegar, sugar, and salt in a stainless-steel saucepan.
8 oz	240 g	Sugar	5. Bring the mixture to a boil, stirring to dissolve the sugar.
1 tsp	5 mL	Salt	6. Cook at a slow simmer, stirring occasionally, until the mixture becomes thick. There should be about 1 1/2 pints (750 mL).
2 tbsp	30 mL	Chopped hot green chiles	7. Add the chiles and raisins. Simmer another 5–10 minutes. Remove from heat and cool.
1 1/2 oz	45 g	Golden raisins	8. Taste and add more salt if necessary.

Per 1 ounce (28.35 g): Calories, 70; Protein, 1 g; Fat, 0 g (0% cal.); Cholesterol, 0 mg; Carbohydrates, 17 g; Fiber, 1 g; Sodium, 125 mg.

Note: If good-quality ripe tomatoes are not available, substitute 2 lb (900 g) canned tomatoes with their juice.

Tomato Raisin Chutney

Pear Chutney

YIELD: APPROX. 9 OZ

U.S.	METRIC	INGREDIENTS	PROCEDURE
2	2	Pears	1. Peel the pears and cut into small dice.
2 tbsp	30 mL	Honey	2. Combine the pears, honey, vinegar, juice, and spices in a small saucepan. Cook over low heat until the pears are tender and the liquid is reduced to a syrup. (Heat must be low, or the liquid will reduce before the pears are cooked.)
2 tbsp	30 mL	Cider vinegar	
2 fl oz	30 mL	Pear juice or apple juice	
1/8 tsp	0.5 mL	Cinnamon	3. Season to taste with salt.
pinch	pinch	Nutmeg	
pinch	pinch	Ground cloves	
to taste	to taste	Salt	

Per 1 fl oz (29.57 mL): Calories 40; Protein, 0 g; Fat, 0 g (0% cal.); Cholesterol, 0 mg; Carbohydrates, 11 g; Fiber, 1 g; Sodium, 0 mg.

Cucumber Raita

YIELD: 1 1/2 PT (750 ML)

U.S.	METRIC	INGREDIENTS	PROCEDURE
1/2 tsp	2 mL	Cumin seed	1. In a small, dry skillet over moderate heat, toast the cumin seeds until aromatic and a slightly darker shade of brown. Remove from the heat and grind in a spice grinder.
8 oz	250 g	Cucumber	2. Peel and grate the cucumbers coarsely.
1 pt	500 mL	Plain yogurt	3. Place the yogurt in a bowl and whip until smooth.
1 tsp	5 mL	Salt	4. Add the ground cumin, grated cucumber, salt, pepper, and cayenne. Mix well.
1/8 tsp	1 mL	Black pepper	
1/8–1/4 tsp	1–2 mL	Cayenne	

Per 1 fl oz (29.57 mL): Calories, 15; Protein, 1 g; Fat, 1 g (53% cal.); Cholesterol, 0 mg; Carbohydrates, 1 g; Fiber, 0 g; Sodium, 105 mg.

Nuoc Cham (Vietnamese Dipping Sauce) ♥

YIELD: 15 FL OZ (450 ML)

U.S.	METRIC	INGREDIENTS	PROCEDURE
2	2	Garlic cloves, chopped	1. Pound the garlic, chiles, and sugar in a mortar to make a paste.
2–4	2–4	Fresh Thai chiles, seeded and chopped	
3 tbsp	45 mL	Sugar	
4 fl oz	125 mL	Lime juice	2. Place the garlic paste in a bowl and mix in the lime juice, stirring until the sugar is dissolved.
4 fl oz	125 mL	Nuoc nam (Vietnamese fish sauce) or nam pla (Thai fish sauce)	3. Add the remaining ingredients and mix well.
4 fl oz	125 mL	Water	
1 oz	30 g	Carrot, cut fine julienne	

Per 1 fl oz (29.57 mL): Calories, 20; Protein, 1 g; Fat, 0 g (0% cal.); Cholesterol, 0 mg; Carbohydrates, 4 g; Fiber, 0 g; Sodium, 740 mg.

Nuoc Cham

Thai Red Curry Sauce

YIELD: 18 FL OZ (550 ML)

U.S.	METRIC	INGREDIENTS	PROCEDURE
1 fl oz	30 mL	Vegetable oil	1. Heat the oil over moderate heat in a saucepan.
1¹/₂ oz	45 mL	Red curry paste (see sidebar)	2. Add the curry paste and cook until aromatic.
2 fl oz	60 mL	Water or stock	3. Stir in the water or stock and bring to a simmer.
14 fl oz	410 mL	Coconut milk, canned, unsweetened	4. Add the coconut milk and nam pla. Bring to a simmer and simmer several minutes to blend flavors.
1¹/₂ oz	45 mL	Nam pla (Thai fish sauce)	

Per 1 fl oz (29.57 mL): Calories, 60; Protein, 1 g; Fat, 7 g (89% cal.); Cholesterol, 0 mg; Carbohydrates, 1 g; Fiber, 0 g; Sodium, 190 mg.

VARIATION

Thai Green Curry Sauce

Substitute green curry paste for red paste in the recipe.

THAI CURRIES

In Thai cuisine, curry sauces are typically made as integral sauces (see p. 197). With a variety of ready-made curry pastes on hand, it is simple to sauté the desired meat or seafood, prepare a sauce flavored with a curry paste, and stew the meat in the sauce with the desired vegetables.

In European and North American kitchens, on the other hand, curry sauces are often prepared separately and added to the plate at service time. The recipes included here can be used either way. To use the curry sauce recipe as an integral sauce, add the curry paste to the pan in which you have sautéed your meat or fish item. Finish the sauce as directed in the recipe, and return the meat to the sauce to finish cooking.

Thai curry pastes are available commercially. If none is available, use the recipes included here. Commercial pastes are usually of excellent quality and contain ingredients that may be hard to obtain in Western kitchens. The curry paste recipes here are useful when commercial curry pastes and unusual Asian ingredients are not available.

Red Curry Paste

YIELD: 1 LB 4 OZ (560 G)

U.S.	METRIC	INGREDIENTS	PROCEDURE
2 oz	60 g	Galangal, peeled and chopped (see *Note*)	1. Combine all ingredients in a food processor. Process until the mixture forms a smooth paste.
6 oz	180 g	Shallots, chopped	
2 oz	60 g	Garlic, chopped	2. Refrigerate and use as needed to make curry sauces.
3 oz	90 g	Cilantro roots (see Note)	
2 tbsp	30 mL	Lime zest, chopped	
5 tbsp	75 mL	Cayenne	
4	4	Lemongrass stalks, tender parts only, chopped	
3 fl oz	90 mL	Vegetable oil	

Per 1 ounce (28.35 g): Calories, 50; Protein, 4 g; Fat, 4.5g (77% cal.); Cholesterol, 0 mg; Carbohydrates, 3 g; Fiber, 1 g; Sodium, 0 mg.

Note: Galangal is a root that resembles fresh ginger in appearance but has a somewhat different taste. If it is not available, use fresh ginger. Cilantro roots are the roots attached to ordinary cilantro that are usually discarded. Save, wash thoroughly, and use for this preparation. If not available, use cilantro stems.

Green Curry Paste

YIELD: 15 OZ (475 G)

U.S.	METRIC	INGREDIENTS
15	15	Serrano chiles, seeded and chopped
2 oz	60 g	Shallots, chopped
2 oz	60 g	Garlic, chopped
4	4	Lemongrass stalks, tender parts only, chopped
2 oz	60 g	Cilantro roots, chopped (see Note to Red Curry Paste)
2 oz	60 g	Cilantro leaves, chopped
1/2 oz	15 g	Galangal, peeled and chopped (see Note to Red Curry Paste)
2 tbsp	30 mL	Lime zest, chopped
1 tsp	5 mL	Nutmeg
3 fl oz	90 mL	Vegetable oil

PROCEDURE

1. Combine all ingredients in a food processor. Process until the mixture forms a smooth paste.

2. Refrigerate and use as needed to make curry sauces.

Per 1 ounce (28.35 g): Calories, 60; Protein, 1 g; Fat, 6 g (77% cal.); Cholesterol, 0 mg; Carbohydrates, 3 g; Fiber, 1 g; Sodium, 195 mg.

Indonesian Peanut Sauce

YIELD: 1 PT (500 ML)

U.S.	METRIC	INGREDIENTS
8 oz	250 g	Peanut butter
8 fl oz	250 mL	Hot water
4–6	4–6	Garlic cloves, crushed to a paste
2–3	2–3	Serrano chiles, seeded, chopped fine
1 tbsp	15 mL	Fresh ginger root, peeled and chopped fine
1 tbsp	15 mL	Brown sugar
1 tbsp	15 mL	Lime juice
1 fl oz	30 mL	Soy sauce

PROCEDURE

1. Combine all ingredients in a food processor. Process until the mixture forms a smooth sauce.

2. Refrigerate.

Per 1 fl oz (29.57 mL): Calories, 90; Protein, 4 g; Fat, 7 g (66% cal.); Cholesterol, 0 mg; Carbohydrates, 4 g; Fiber, 1 g; Sodium, 190 mg.

Indonesian
Peanut Sauce

Asian Sweet-and-Sour Sauce

YIELD: APPROX. 14 FL OZ (420 ML)

U.S.	METRIC	INGREDIENTS	PROCEDURE
1 fl oz	30 mL	Oil	1. Heat the oil in a saucepan or sauté pan over moderate heat.
4 oz	120 g	Onion, small dice	2. Add the onion. Cook until soft, but do not brown.
3 oz	90 g	Fresh ginger, peeled and chopped	3. Add the ginger, sugar, and pineapple. Cook until sugar and fruit just start to caramelize.
4 oz	120 g	Sugar	
12 oz	360 g	Pineapple, fresh, diced	
3 fl oz	90 mL	Lime juice	4. Add the lime juice, vinegar, and cayenne. Bring to a simmer, then remove from the heat.
6 fl oz	180 mL	Red wine vinegar	5. Put all ingredients into a blender and blend to a smooth purée.
1/4 tsp	1 mL	Cayenne	6. Pass the sauce through a tamis or fine strainer, forcing through as much of the solids as possible. Alternatively, pass through a food mill fitted with the fine blade.

Per 1 fl oz (29.57 mL): Calories, 80; Protein, 0 g; Fat, 2 g (24% cal.); Cholesterol, 0 mg; Carbohydrates, 14 g; Fiber, 1 g; Sodium, 0 mg.

Sesame Miso Sauce

YIELD: 12 FL OZ (360 ML)

U.S.	METRIC	INGREDIENTS	PROCEDURE
2 oz	60 g	Sesame seeds	1. Over low heat, toast the sesame seeds in a dry sauté pan until aromatic and slightly darkened.
			2. Grind the toasted seeds in a mortar or spice grinder.
8 oz	240 g	Red miso (see p. 812)	3. Mix the ground sesame seeds with the remaining ingredients. Stir well.
2 1/2 oz	75 g	Sugar	
2 fl oz	60 mL	Mirin (sweet Japanese rice wine)	

Per 1 fl oz (29.57 mL): Calories, 110; Protein, 3 g; Fat, 3 g (26% cal.); Cholesterol, 0 mg; Carbohydrates, 15 g; Fiber, 1 g; Sodium, 680 mg.

Sesame Miso Sauce

Applesauce ♥

YIELD: APPROX. 1 QT (1 L)

U.S.	METRIC	INGREDIENTS
4 lb	2 kg	Apples (see Note)
to taste	to taste	Sugar (see step 4)
to taste	to taste	Lemon juice

PROCEDURE

1. Cut the apples into quarters and remove cores. Skins may be left on because they will be strained out later. (Red peels will color the sauce pink.) Dice the apples coarsely.

2. Place apples in a heavy saucepan with about 2 oz (60 mL) water. Cover.

3. Set the pan over low heat and cook the apples slowly until very soft. Stir occasionally.

4. Add sugar to taste. The amount needed depends on the sweetness of the apples; sauce should be tart, not too sweet.

5. Add a little lemon juice to taste.

6. Pass the sauce through a food mill.

7. If the sauce is too thin or watery, let simmer, uncovered, until thickened.

8. Serve warm or cold with roast pork or duck.

Per 1 fl oz (29.57 mL): Calories, 25; Protein, 0 g; Fat, 0 g (0% cal.); Cholesterol, 0 mg; Carbohydrates, 6 g; Fiber, 1 g; Sodium, 0 mg.

Note: Use tart cooking apples, such as Rome. Do not use Red Delicious.

Barbecue Sauce ♥

YIELD: ¹/₂ GAL (2 L)

U.S.	METRIC	INGREDIENTS
1 qt	1 L	Tomato purée
1 pt	500 mL	Water
²/₃ cup	150 mL	Worcestershire sauce
¹/₂ cup	125 mL	Cider vinegar
¹/₂ cup	125 mL	Vegetable oil
8 oz	250 g	Onion, chopped fine
4 tsp	20 mL	Finely chopped garlic
2 oz	60 g	Sugar
1 tbsp	15 mL	Dry mustard
2 tsp	10 mL	Chili powder
1 tsp	5 mL	Black pepper
to taste	to taste	Salt

PROCEDURE

1. Place all ingredients in a heavy saucepan and bring to a boil. Reduce heat and simmer about 20 minutes, or until slightly reduced and flavors are well blended. Stir occasionally during cooking so the sauce does not scorch on the bottom.

2. Adjust seasoning.

Per 1 fl oz (29.57 mL): Calories, 25; Protein, 0 g; Fat, 2 g (60% cal.); Cholesterol, 0 mg; Carbohydrates, 3 g; Fiber, 0 g; Sodium, 30 mg.

Note: This sauce is not intended to be eaten as is but to be cooked with other foods. See recipes for Barbecued Spareribs (p. 423) and for Barbecued Pork Sandwich (p. 728).

Kansas City Barbecue Sauce

YIELD: 1¼ PT (625 ML)

U.S.	METRIC	INGREDIENTS
1 pt	500 mL	Tomato purée
4 fl oz	125 mL	Cider vinegar
½ oz	15 g	Sugar
2 tbsp	30 mL	Molasses
1 tsp	5 mL	Ground cumin
1-2 tsp	5-10 mL	Hot sauce
1 tsp	5 mL	Celery seed
2 tsp	10 mL	Onion powder
1 tsp	5 mL	Garlic powder
2 tsp	10 mL	Salt
1 tsp	5 mL	Black pepper

PROCEDURE

1. Combine all ingredients in a saucepan and stir.
2. Bring to a boil, then reduce heat to a simmer. Simmer 10-15 minutes, until slightly reduced and flavors are well blended. Stir occasionally during cooking to prevent scorching.
3. Adjust seasoning with additional salt if necessary.

Per 1 fl oz (29.57 mL): Calories 20; Protein, <1 g; Fat, 0 g (0% cal.); Cholesterol, 0 mg; Carbohydrates, 5 g; Fiber, <1 g; Sodium, 340 mg.

Eastern Carolina Barbecue Sauce 🌹

YIELD: 1 QT (1 L)

U.S.	METRIC	INGREDIENTS
1 qt	1 L	Cider vinegar
1½ oz	45 g	Brown sugar
2 tbsp	30 mL	Hot pepper sauce
2 tsp (or to taste)	10 mL (or to taste)	Cayenne
2 tsp	10 mL	Salt
2 tsp	10 mL	Black pepper

PROCEDURE

Combine all ingredients and mix well.

Per 1 fl oz (29.57 mL): Calories 10; Protein, 0 g; Fat, 0 g (0% cal.); Cholesterol, 0 mg; Carbohydrates, 2 g; Fiber, 0 g; Sodium, 170 mg.

Mustard Barbecue Sauce 🌹

YIELD: 1 PT (500 ML)

U.S.	METRIC	INGREDIENTS
2 oz	60 g	Butter
8 oz	250 g	Onion, grated
8 oz	250 g	Prepared yellow mustard
8 fl oz	250 mL	Cider vinegar
5 oz	150 g	Brown sugar
1 tsp	5 mL	Cayenne
½ tsp	2 mL	Salt
½ tsp	2 mL	Black pepper

PROCEDURE

1. Melt the butter in a saucepan.
2. Add the grated onion. Cook the onion slowly without browning for about 4 minutes, until soft.
3. Stir in the remaining ingredients. Simmer about 15 minutes, until the flavors are well blended.

Per 1 fl oz (29.57 mL): Calories 80; Protein, <1 g; Fat, 3.5 g (39% cal.); Cholesterol, 10 mg; Carbohydrates, 11 g; Fiber, 1 g; Sodium, 260 mg.

Chile Barbecue Sauce

YIELD: 2¹/₂ PT (1.25 L)

U.S.	METRIC	INGREDIENTS
1¹/₂ pt	750 mL	Bottled chili sauce
8 fl oz	250 mL	Soy sauce
2 oz	60 g	Dark brown sugar
1¹/₂ fl oz	45 mL	Worcestershire sauce
1 pt	500 mL	Water
6 fl oz	175 mL	Lemon juice
1 tbsp	15 mL	Hot red pepper sauce
2	2	Whole chipotle chiles
3 tbsp	45 mL	Chili powder

PROCEDURE

1. Combine all ingredients in a heavy saucepot. Bring to a boil.
2. Simmer 15 minutes.
3. Strain.
4. Adjust seasoning with salt if necessary. (It is not likely more salt will be needed; soy sauce is salty.)

Per 1 fl oz (29.57 mL): Calories, 35; Protein, 1 g; Fat, 0 g (0% cal); Cholesterol, 0 mg; Carbohydrates, 8 g; Fiber, 1 g; Sodium, 620 mg.

Soy Barbecue Sauce

YIELD: 1 QT (1 L)

U.S.	METRIC	INGREDIENTS
12 fl oz	375 mL	Japanese-style soy sauce
4 oz	125 mL	Brown sugar
8 fl oz	250 mL	Sake (Japanese rice wine) or dry sherry
8 fl oz	250 mL	Water
6 fl oz	175 mL	Lemon juice
2 tbsp	30 mL	Chopped fresh ginger root
1 tbsp	15 mL	Chopped garlic
8 fl oz	250 mL	Vegetable oil

PROCEDURE

1. In a saucepot, combine all ingredients except oil.
2. Bring to a simmer. Reduce by one-third.
3. Strain.

4. Add the oil. Before using, mix well to emulsify; this is easily done in a blender.

Per 1 fl oz (29.57 mL): Calories, 90; Protein, 1 g; Fat, 7 g (66% cal.); Cholesterol, 0 mg; Carbohydrates, 5 g; Fiber, 0 g; Sodium, 610 mg.

Fruit Salsa

YIELD: 2 LB (1 KG)

U.S.	METRIC	INGREDIENTS
8 oz	250 g	Honeydew melon, seeded, rind removed
8 oz	250 g	Papaya, peeled and seeded
8 oz	250 g	Mango, peeled and pitted
4 oz	125 g	Red bell pepper, cored and seeded
1 oz	30 g	Jalapeno, stemmed and seeded
3 oz	90 g	Red onion, cut brunoise
4 fl oz	125 mL	Lime juice
2 tbsp	30 mL	Chopped cilantro
to taste	to taste	Salt

PROCEDURE

1. Chop the melon, papaya, mango, bell pepper, and jalapeno into fine dice. Be careful to save the juices that are released.
2. Combine the chopped fruit with their juices, peppers, onion, lime juice, and cilantro in a bowl.
3. Season to taste with salt.
4. Refrigerate until served.

Per 1 ounce (28.35 g): Calories, 15; Protein, 0 g; Fat, 0 g (0% cal.); Cholesterol, 0 mg; Carbohydrates, 3 g; Fiber, 0 g; Sodium, 0 mg.

Fruit Salsa

Shallot Oil

YIELD: 1 PT (500 ML)

U.S.	METRIC	INGREDIENTS
2–3 tbsp	30 g	Shallots, chopped
1 pt	500 mL	Flavorless oil, such as canola, corn, safflower, or grapeseed

PROCEDURE

1. Combine the chopped shallots and the oil in a jar. Shake well.
2. Let stand 30 minutes. Refrigerate.
3. The oil is ready to use as soon as it has taken on the desired flavor, which may be in 1–2 hours. After 2 days, strain the oil through a paper coffee filter. Store in refrigerator.

Per 1 fl oz **(29.57 mL):** Calories, 240; Protein, 0 g; Fat, 27 g (100% cal.); Cholesterol, 0 g; Carbohydrates, 0 g; Fiber, 0 g; Sodium, 0 mg.

VARIATIONS

Ginger Oil, Horseradish Oil, or Garlic Oil

Substitute ginger root, horseradish, or garlic for shallots in the basic recipe. For best results, chop ginger or horseradish very fine in a food processor, or grate with a fine-holed grater. For garlic oil, substitute olive oil for the flavorless oil if desired.

Lemon or Orange Oil

Substitute 3–4 tbsp (30 g) grated lemon or orange zest for shallots in the basic recipe.

Rosemary Oil, Sage Oil, Thyme Oil, or Oregano Oil

Substitute 3½ oz (100 g) chopped fresh rosemary, sage, thyme, or oregano for shallots in the basic recipe.

Cinnamon Oil, Cumin Oil, Curry Oil, Ginger Oil, or Paprika Oil

Substitute 3 tbsp (45 mL) of one of the above ground, dried spices for shallots in the basic recipe. In a small pan, combine the spice with just enough oil to make a thin paste. Heat gently just until the spice starts to give off an aroma. Be careful not to burn the spice. Paprika, especially, darkens quickly. Add the spice mixture to remaining oil. Let stand, refrigerate, and filter as in basic recipe.

Basil Oil, Parsley Oil, Chervil Oil, or Cilantro Oil

Select the desired quantity of one of the above fresh herbs. Drop into boiling water. Blanch 10 seconds. Drain and refresh under cold water. Drain again and pat dry with towels. Put the herbs in a blender and add a small amount of olive oil. Blend to make a paste. Measure the volume of the paste and add 4 times that volume of olive oil. Shake and let stand. Refrigerate and strain as in basic recipe.

Tartar Sauce

YIELD: APPROX. 1 QT (1 L)

U.S.	METRIC	INGREDIENTS
4 oz	125 g	Dill pickles or sour gherkins
2 oz	60 g	Onions
2 oz	60 g	Capers
1 qt	1 L	Mayonnaise
2 tbsp	30 mL	Chopped parsley

PROCEDURE

1. Chop pickles and onions very fine. Chop the capers if large, or leave whole if small.
2. Press the pickles and capers in a fine sieve, or squeeze out in a piece of cheesecloth so they don't make the sauce too liquid.
3. Combine all ingredients in a stainless-steel bowl and mix well.

Per 1 fl oz **(29.57 mL):** Calories, 200; Protein, 0 g; Fat, 22 g (98% cal.); Cholesterol, 15 mg; Carbohydrates, 1 g; Fiber, 0 g; Sodium, 240 mg.

VARIATIONS

Rémoulade Sauce

Add 1 tbsp (15 mL) anchovy paste or mashed anchovies to tartar sauce.

Aïoli I

YIELD: APPROX. 1 PT, 4 OZ (600 ML)

U.S.	METRIC	INGREDIENTS
10	10	Garlic cloves
⅛ tsp	0.5 mL	Salt
3	3	Egg yolks
1 pt	500 mL	Olive oil
1–2 tbsp	15–30 mL	Lemon juice

PROCEDURE

1. Crush the garlic to a fine paste with the salt.
2. Add the egg yolks and beat until thoroughly combined.
3. A few drops at a time, begin adding the olive oil, beating constantly. Do not add oil any faster than it can be absorbed.
4. After about half the oil is added, the mixture will be very stiff. Add a few drops of the lemon juice at this point.
5. Continue adding the remaining oil gradually. From time to time, add a few more drops of the lemon juice. The finished aioli should be like a stiff mayonnaise. Adjust seasoning with salt if necessary.

Per 1 fl oz **(29.57 mL):** Calories, 200; Protein, 1 g; Fat, 22 g (99% cal.); Cholesterol, 30 mg; Carbohydrates, 1 g; Fiber, 0 g; Sodium, 15 mg

Cocktail Sauce ♥

YIELD: 2 QT (2 L)

U.S.	METRIC	INGREDIENTS
1 qt	1 L	Ketchup
2¹/₂ cups	600 mL	Chili sauce
1 cup	250 mL	Prepared horseradish
4 fl oz	125 mL	Lemon juice
2 tbsp	30 mL	Worcestershire sauce
dash	dash	Hot red pepper sauce

PROCEDURE

1. Combine all ingredients and mix.
2. Chill.

Per 1 fl oz (29.57 mL): Calories, 30; Protein, 1 g; Fat, 0 g (0% cal.); Cholesterol, 0 mg; Carbohydrates, 7 g; Fiber, 0 g; Sodium, 330 mg.

Note: Serve as a dip with shrimp, crab, lobster, raw clams, or raw oysters.

Mignonette Sauce ♥ 🌹

YIELD: 1 QT (1 L)

U.S.	METRIC	INGREDIENTS
1 qt	1 L	Wine vinegar, red or white
8 oz	250 g	Shallots, cut brunoise
1 tsp	5 mL	Salt
1 tsp	5 mL	White pepper
2 tsp	5 mL	Dried tarragon

PROCEDURE

1. Combine all ingredients.
2. Chill.
3. Serve 1 fl oz (30 mL) per portion as a cocktail sauce for oysters or clams on the half-shell.

Per 1 fl oz (29.57 mL): Calories, 5; Protein, 0 g; Fat, 0 g (0% cal.) Cholesterol, 0 mg; Carbohydrates, 1 g; Fiber, 0 g; Sodium, 75 mg.

Mignonette Sauce

Fig Compote ♥ 🌹

YIELD: 1 LB, 4 OZ (625 G)

U.S.	METRIC	INGREDIENTS
1 lb	500 g	Dried figs, preferably light rather than black figs
4 fl oz	125 mL	Lemon juice
8 fl oz	250 mL	Water
2 tbsp	30 mL	Sugar
¹/₄ tsp	1 mL	Ground cumin
¹/₄ tsp	1 mL	Cinnamon
¹/₈ tsp	0.5 mL	Ground cloves
¹/₈ tsp	0.5 mL	Cayenne

PROCEDURE

1. Trim figs by cutting off hard stem ends.
2. Cut the figs into medium dice.
3. Combine all ingredients in a saucepan.
4. Bring to a boil. Reduce heat and simmer until the liquid has evaporated.
5. Serve warm or cooled.

Per 1 ounce (28.35 g): Calories, 60; Protein, 1 g; Fat, 0 g (0% cal.); Cholesterol, 0 mg; Carbohydrates, 17 g; Fiber, 2 g; Sodium, 0 mg.

Cold Snap Pea Coulis ♥

YIELD: 14 FL OZ (420 ML)

U.S.	METRIC	INGREDIENTS	PROCEDURE
1 lb	480 g	Snap peas, trimmed	1. Blanch the snap peas in boiling salted water 1 minute. 2. Drain and immediately refresh in ice water. Leave peas in the ice water until completely cold. Drain well.
10 fl oz	150 g	Water, ice cold	3. Place peas in a blender and add the water, salt, and sugar. Blend to a smooth purée. If blender is too small, purée in two batches, using about half the peas, water, salt, and sugar in each batch. 4. Strain the purée through a chinois, pressing down on the solids to extract all the liquid.
¹/₃ oz	10 g	Salt	
¹/₄ tsp	1 g	Sugar	
¹/₁₆ oz	2 g	Xanthan gum	5. Rinse out and drain blender jar. Return the pea purée to the jar and add the xanthan gum. Blend until lightly thickened. 6. Strain again through a chinois. 7. Refrigerate until needed.

Per 1 fl oz (29.57 mL): Calories, 15; Protein, 1 g; Fat, 0 g (0% cal.); Cholesterol, 0 mg; Carbohydrates, 3 g; Fiber, 1 g; Sodium, 280 mg.

Basil Sauce ♥ 🌸

YIELD: 7 FL OZ (210 ML)

U.S.	METRIC	INGREDIENTS	PROCEDURE
2 oz	60 g	Basil leaves	1. Blanch the basil leaves in boiling salted water 2 minutes. 2. Drain and immediately refresh in ice water. Drain well.
1 pt	480 mL	Water, ice cold	3. Place the basil leaves and cold water in a blender and blend until smooth. 4. Strain through a chinois.
0.2 oz	5 g	Salt	5. Place the basil liquid in a clean blender jar and add salt.
0.4 oz	10 g	Ultra-Tex 3	6. Turn the blender on low speed. With the machine running, add the Ultra-Tex and blend until the mixture is thickened, 2 minutes or more. 7. Strain through a chinois. 8. Refrigerate until needed. Use within 24 hours, or preferably the same day.

Per 1 fl oz (29.57 mL): Calories, 10; Protein, 0 g; Fat, 0 g (0% cal.); Cholesterol, 0 mg; Carbohydrates, 2 g; Fiber, 0 g; Sodium, 280 mg.

VARIATION

Parsley Sauce

Substitute parsley leaves for the basil. (Other herb sauces can be made the same way.)

Orange Fluid Gel 🌸

YIELD: 1 LB 4 OZ (600 G)

U.S.	METRIC	INGREDIENTS	PROCEDURE
³/₄ cup	180 fl oz	Sugar	1. Combine the sugar, water, lemon juice, orange juice, and agar-agar in a stainless steel saucepan. Whip with a wire whip to combine all ingredients.
8 fl oz	240 fl oz	Water	
2 fl oz	60 mL	Lemon juice	2. Bring to a boil and boil for 2 minutes.
7 fl oz	210 mL	Orange juice	3. Pour into a flat, stainless steel pan. Let sit 6 hours or overnight. The mixture will set up until very firm.
1¹/₂ tbsp	22 mL	Lemon zest, grated	
2 tbsp	30 mL	Orange zest, grated	4. Break up the gel and place in a blender. Purée until smooth, scraping down the sides as necessary.
1 tbsp	15 mL	Agar-agar	
1 oz	30 g	Butter, unsalted	5. Add the raw butter and salt and blend in.
to taste	to taste	Salt	

Per 1 oz (28.35 g): Calories 45; Protein, 0 g; Fat, 1 g (20% cal.); Cholesterol, 5 mg; Carbohydrates, 9 g; Fiber, 0 g; Sodium, 0 mg.

Corn Purée

YIELD: 24 OZ (720 G)

U.S.	METRIC	INGREDIENTS	PROCEDURE
6 ears	6 ears	Sweet corn	1. Cut the kernels from the ears of sweet corn. Scrape the cobs with the back of a knife to remove remaining pulp. Combine the pulp and the kernels. (From 6 ears you should have a yield of 1 lb, or 480 g.)
1¹/₂ pt	720 mL	Milk	
to taste	to taste	Salt	
to taste	to taste	White pepper	2. Cut each cob into a few small pieces so that they fit into a saucepan.
			3. Combine the cobs, the corn, milk, salt, and pepper in a saucepan. Simmer until the corn is tender and corn flavor has infused the milk.
			4. Remove the cobs and discard.
4 fl oz	120 mL	Heavy cream	5. In a small pan, reduce the heavy cream to half its volume.
2 tbsp	13 g	Ultra-Tex 8	6. Strain the corn and reserve the liquid.
3 oz	90 g	Butter	7. Put the corn in a blender and add enough of the reserved milk to make a soft purée. Blend until smooth.
			8. With the blender running, gradually add the Ultra-Tex through the opening in the lid to thicken.
			9. Blend in the butter and heavy cream.
			10. Check the seasonings and add additional salt and pepper if needed.

Per 1 oz (28.35 g): Calories 60; Protein, 1 g; Fat, 5 g (75% cal.); Cholesterol, 15 mg; Carbohydrates, 5 g; Fiber, 0 g; Sodium, 30 mg.

Parmesan Foam

YIELD: VARIABLE

U.S.	METRIC	INGREDIENTS	PROCEDURE
5 fl oz	150 mL	Milk	1. In a small saucepan, bring the milk to a boil.
4 oz	125 g	Grated Parmesan cheese	2. Remove from the heat, add the cheese, and stir.
4 fl oz	125 mL	Heavy cream	3. Let stand 30 minutes at room temperature.
to taste	to taste	Salt	4. Strain through a chinois. Press on the solids to extract all the liquid, but do not force the solids through strainer.
			5. Mix in cream. Add salt to taste and refrigerate.
			6. Pour the cream mixture into a foaming canister. Close canister and charge with a nitrous oxide (N_2O) cartridge. Make foam by inverting canister and pulling on lever.

Per 1 ounce (28.35 g): Calories, 70; Protein, 4 g; Fat, 6 g (73% cal.); Cholesterol, 20 mg; Carbohydrates, 1 g; Fiber, 0 g; Sodium, 140 mg.

Maple Emulsion 🌹

YIELD: APPROX. 1 PT (480 ML)

U.S.	METRIC	INGREDIENTS	PROCEDURE
1 tbsp	15 mL	Shallot, minced	1. Combine shallot, wine, and thyme in a small saucepan. Over moderate heat, reduce by half.
1/4 cup	60 mL	White wine	
1 sprig	1 sprig	Thyme, fresh	
3/4 cup	180 mL	Heavy cream	2. Add heavy cream, stock, and maple syrup. Bring to a boil.
1 cup	240 mL	Chicken stock	3. Remove from heat. Remove the sprig of thyme.
1/3 cup	80 mL	Maple syrup (genuine, not artificial)	
1 tsp	5 mL	Lecithin powder	4. Mix in the lecithin powder.
to taste	to taste	Salt	5. Season to taste with salt and white pepper.
to taste	to taste	White pepper	6. To make foam, froth with an immersion blender (**Figure 8.16**). Spoon off the froth to use as desired.

Per 1 fl oz (29.57 mL): Calories, 70; Protein, <1 g; Fat, 4.5 g (58% cal.); Cholesterol, 15 mg; Carbohydrates, 6 g; Fiber, 0 g; Sodium, 25 mg.

VARIATION

Bacon Thyme Emulsion

In step 2, add 3 strips of bacon and 2 additional sprigs of thyme to the heavy cream and stock. Increase stock to 1 1/4 cup (300 mL). After bringing the liquid to a boil, simmer 5–6 minutes to infuse flavors into stock. Strain, pressing down on the solids to extract all liquid. Continue with step 4.

FIGURE 8.16 Making a foam with an immersion blender.

(a)

(b)

Mustard Foam

YIELD: APPROX. 8 OZ (240 G)

U.S.	METRIC	INGREDIENTS	PROCEDURE
1 tbsp	15 mL	Shallot, chopped fine	1. Combine the shallot, wine, and thyme in a small saucepan. Reduce liquid to 3/4 fl oz.
3 fl oz	90 mL	White wine	
1 sprig	1 sprig	Fresh thyme	2. Add the cider and reduce to 1 fl oz.
4 fl oz	120 mL	Apple cider	3. Add the cream and bring to a simmer.
8 fl oz	240 mL	Heavy cream	4. Stir in the mustard, lemon juice, and carrageenan. Season to taste with salt and pepper. Strain to remove thyme sprig.
2 1/2 tbsp	37 mL	Dijon mustard	
1/4 tsp	1 mL	Lemon juice	5. Place the mixture into a foaming canister and charge with 2 charges of nitrous oxide.
1 g	1 g	Iota carrageenan (see Note)	
to taste	to taste	Salt	6. At service time, foam by inverting the canister and pulling on the lever.
to taste	to taste	White pepper	

Per 1 fl oz (29.57 mL): Calories, 125; Protein, <1 g; Fat, 11 g (79% cal.); Cholesterol, 40 mg; Carbohydrates, 4 g; Fiber, 0 g; Sodium, 140 mg.

Note: In general, a ratio of 1 g carrageenan to 8 fl oz (240 mL) heavy cream is a good ratio to use for most foams. Note that, because of the small quantity, the quantity in the U.S. measures column is given in grams for ease of measurement.

Raspberry Beads ♥

YIELD: VARIABLE

U.S.	METRIC	INGREDIENTS	PROCEDURE
3.3 oz	100 g	Apple juice	1. Combine the apple juice, raspberry juice, sugar, and sodium alginate in a blender. Blend until well combined.
1.3 oz	40 g	Raspberry juice	
¹/₂ oz	15 g	Sugar	2. Let the mixture stand several hours in the refrigerator.
0.03 oz	1 g	Sodium alginate	
0.08 oz	2.5 g	Calcium chloride	3. Dissolve the calcium chloride in water.
8 fl oz	250 g	Water	4. Using a syringe, drop droplets of raspberry mixture into the calcium solution. Leave in calcium bath about 1 minute (the longer they are left in, the firmer they become).
			5. Remove the beads from calcium bath and rinse in a strainer.

Per 1 ounce (28.35 g): Calories, 10; Protein, 0 g; Fat, 0 g (0% cal.); Cholesterol, 0 mg; Carbohydrates, 2 g; Fiber, 0 g; Sodium, 0 mg.

Basil "Caviar" or Pearls

YIELD: APPROX. 4 OZ (120 G)

U.S.	METRIC	INGREDIENTS	PROCEDURE
as needed (see procedure)	as needed (see procedure)	Vegetable oil	1. Pour oil into a tall container, so that the oil is at least 5 inches (22 cm) deep.
			2. Place the container in the freezer and chill the oil to about 35°F (2°C).
³/₄ oz (about ¹/₂ cup)	22 g (about 120 mL)	Fresh basil	1. Blend the basil and water in a blender until smooth. Strain through a fine strainer, pressing on the solids to extract all liquid.
4 fl oz	120 mL	Water	2. Weigh out 125 g (4.4 oz) of the liquid. If necessary, add water to bring it up to the correct weight.
1¹/₂	1¹/₂	Gelatin sheets	3. Bloom the gelatin: soak in cold water until soft, remove from water, and squeeze gently. Set aside.
1.5 g (Use metric; quantity too small for U.S. measure.)	1.5 g	Agar-agar	4. Combine the basil liquid and agar-agar in a stainless steel saucepan. Whip together to combine.
			5. Bring to a boil and boil for 2 minutes.
			6. Remove from heat and stir in the gelatin until dissolved.
			7. Place the warm liquid in a squeeze bottle with a fine tip.
			8. Remove the oil from the freezer. Drop small droplets of the basil mixture into the oil (**Figure 8.17**). The droplets will set as they sink slowly.
			9. Let sit for 3 minutes, then strain. Rinse the pearls carefully under warm water to remove excess oil. Reserve the strained oil for another use.

Per 1 ounce (28.35 g): Calories, 0; Protein, <1 g; Fat, 0 g (0% cal.); Cholesterol, 0 mg; Carbohydrates, 0 g; Fiber, 0 g; Sodium, 0 mg.

FIGURE 8.17 Basil caviar.

(a)

(b)

Balsamic Sheet

YIELD: 250 G

U.S.	METRIC	INGREDIENTS
(Use metric measures. Agar-agar quantity is too small for U.S. measures.)	4	Gelatin sheets
	250 g	Balsamic vinegar
	3 g	Agar-agar

PROCEDURE

1. Bloom the gelatin: soak in cold water until soft, remove from water, and squeeze gently. Set aside.
2. Combine the vinegar and agar in a stainless steel saucepan. Combine by whipping with a wire whip.
3. Bring to a boil and boil for 2 minutes.
4. Remove from the heat and stir in the gelatin. Let cool slightly.
5. Line the bottom of a perfectly flat pan (9 × 13 in/23 × 33 cm or comparable size) with a sheet of acetate.
6. Pour the vinegar mixture into the pan. The liquid should be $^1/_{16}$–$^1/_8$ in (approx. 2 mm) deep. Refrigerate until set (at least 1 hour).
7. Cut into desired shapes.

Per 1 ounce **(28.35 g):** Calories, 30; Protein, <1 g; Fat, 0 g (0% cal.); Cholesterol, 0 mg; Carbohydrates, 5 g; Fiber, 0 g; Sodium, 10 mg.

FIGURE 8.18 Balsamic sheet.

Olive Oil Powder

YIELD: 2$^1/_2$ OZ (75 G)

U.S.	METRIC	INGREDIENTS
1$^1/_2$ oz	45 g	Olive oil
1 oz	30 g	Tapioca maltodextrin
pinch		Salt

PROCEDURE

1. Combine all ingredients in a blender until a powder forms.

Per 1 ounce **(28.35 g):** Calories, 200; Protein, 0 g; Fat, 17 g (77% cal.); Cholesterol, 0 mg; Carbohydrates, 10 g; Fiber, 0 g; Sodium, 75 mg.

VARIATIONS

Other oils can be made into powders using the same ratio of 60 percent oil to 40 percent maltodextrin. The maltodextrin is much lighter than the oil, so 40 percent by weight requires a much larger volume than the volume of oil. (The photo illustrates the ingredients only, not the proportions needed.)

Oily ingredients containing quantities of solids, such as peanut butter, can also be made into a powder, but the proportions change because of the solids. For **Peanut Butter Powder**, use a ratio of 75 percent peanut butter to 25 percent tapioca maltodextrin by weight. For example, 3 oz (90 g) peanut butter and 1 oz (30 g) tapioca maltodextrin yield 4 oz (120 g) peanut butter powder.

Back: olive oil and tapioca maltodextrin; Front: olive oil powder (Please note that the images are for visual identification only. The quantities shown are not in proportion to those needed in the recipe. Maltodextrin is very light, and 1 oz of it is a much larger quantity than that shown.)

ADDITIONAL RECIPES

These additional recipes may be found on your CulinarE-Companion recipe management program:

Chimichurri Sauce; Cumberland Sauce; Horseradish Sauce; Pineapple Chutney; Ponzu Sauce; Port Wine Sauce; Sour Cream Sauce; Sweet Corn and Chile Purée; Vegetable Caper Relish; White Bean Purée; Yakitori Sauce.

TERMS FOR REVIEW

stock	slurry	espagnole
broth	roux	gastrique
mirepoix	white roux	coulis
sachet d'épices	blond roux	clarified butter
bouquet garni	brown roux	beurre noisette
venting	beurre manié	compound butter
remouillage	whitewash	emulsion
reduction	liaison	integral sauce
glaze	au sec	pan gravy
glace de viande	deglaze	jus
glace de volaille	monter au beurre	au jus
glace de poisson	fond lié	salsa
sauce	small sauce	relish
leading sauce	demi-glace	chutney
mother sauce	béchamel	
nappé	velouté	

QUESTIONS FOR DISCUSSION

1. Which bones make a more gelatinous stock, beef or veal?

2. The stockpot is often considered a good way to use trimmings from meats and vegetables. Do you agree? Explain.

3. How should vegetables for mirepoix be cut?

4. Explain the importance of blanching bones before making stocks.

5. Why should stock not be boiled? Should a stockpot be covered? Why or why not?

6. Explain the procedure for cooling stock. Why is it important?

7. Why is an understanding of stocks important even if you work in an establishment that uses only bases?

8. You have just prepared a suprême sauce, but your supervisor says it's too thin. It must be served in five minutes. What can you do to correct the sauce?

9. What are the two methods for preparing starches so they can be incorporated into hot liquids? Why are they necessary, and how do they work?

10. Why is it necessary to be able to thicken a sauce with a roux without making lumps if the sauce is going to be strained anyway?

11. You are preparing a gravy for a batch of Swiss steaks that are to be frozen for later use. What thickening agent will you use?

12. Name the five leading sauces and their major ingredients. List at least two small sauces made from each.

13. What precautions must be taken when finishing and holding allemande sauce?

14. What are the similarities between espagnole and pan gravy? the differences?

15. What precautions are necessary when making hollandaise to avoid overcooking the eggs or curdling the sauce?

9

SOUPS

The popularity of soups today may be due to increased nutrition consciousness, to a desire for simpler or lighter meals, or to an increased appreciation of how appetizing and satisfying soups can be. Whatever the reasons, they emphasize the importance of soup-making skills.

If you have already studied the preparation of stocks and sauces in Chapter 8, you now have at your disposal the major techniques for the preparation of soups. You know how to make stocks and how to use thickening agents such as roux and liaison.

A few more techniques are necessary for you to master before you are able to prepare all the types of soups that are popular today. As in sauce making, basic techniques are the building blocks you can use to create a wide variety of appetizing soups.

AFTER READING THIS CHAPTER, YOU SHOULD BE ABLE TO

1. Describe two basic categories of soups.
2. Identify standard appetizer and main-course portion sizes for soups.
3. State the procedures for holding soups for service and for serving soups at the proper temperature.
4. Prepare clarified consommé.
5. Prepare vegetable soups and other clear soups.
6. Prepare cream soups.
7. Prepare purée soups.
8. Prepare bisques, chowders, specialty soups, and national soups.

UNDERSTANDING SOUPS

Soup, according to the dictionary, is a liquid food derived from meat, poultry, fish, or vegetables. This definition is all right as far as it goes, but there's a lot it doesn't tell us. Is a stock, straight from the stockpot, a soup? Is beef stew liquid enough to be called a soup?

We're interested more in production techniques than in definitions. However, a few more definitions are necessary before we can go into the kitchen, so we can talk to each other in the same language. Definitions aren't rules, so don't be alarmed if you hear other books or chefs use these terms differently. What matters is that you learn the techniques and are able to adapt them to many uses.

CLASSIFICATIONS OF SOUPS

Most soups can be divided into two basic categories: clear or unthickened soups and thick soups. In addition, this chapter discusses a number of special soups that don't fit easily into these first two categories.

Most of these soups, no matter what their final ingredients may be, are based on stock. Thus, the quality of the soup depends on the stock-making skills discussed in Chapter 8.

CLEAR SOUPS

Clear soups are all based on a clear, unthickened broth or stock. They may be served plain or garnished with a variety of vegetables and meats.

1. *Broth* and **bouillon** are two terms used in many ways. In general, they both refer to simple, clear soups without solid ingredients except, sometimes, with a small amount of garnish. We have already defined broth (Chapter 8) as a flavorful liquid obtained from the simmering of meats and/or vegetables. Broths are discussed in more detail on page 228.

2. **Vegetable soup** is a clear, seasoned stock or broth with the addition of one or more vegetables and, sometimes, meat or poultry products and starches.

3. **Consommé** is a rich, flavorful stock or broth that has been clarified to make it perfectly clear and transparent. The process of clarification is a technique we study in detail.

   Far from being just a plain old cup of broth, a well-made consommé is one of the greatest of all soups. Its sparkling clarity is a delight to the eye, and its rich, full flavor, strength, and body make it a perfect starter for an elegant dinner.

THICK SOUPS

Unlike clear soups, thick soups are opaque rather than transparent. They are thickened either by adding a thickening agent, such as a roux, or by puréeing one or more of their ingredients to provide a heavier consistency.

1. **Cream soups** are soups thickened with roux, beurre manié, liaison, or other added thickening agents, plus milk and/or cream. They are similar to velouté and béchamel sauces—in fact, they may be made by diluting and flavoring either of these two leading sauces.

   Cream soups are usually named after their major ingredient, as in cream of chicken or cream of asparagus.

2. **Purée soups** are soups naturally thickened by puréeing one or more of their ingredients. They are not as smooth and creamy as cream soups.

   Purées are normally based on starchy ingredients. They may be made from dried legumes (such as split pea soup) or from fresh vegetables with a starchy ingredient, such as potatoes or rice, added. Purées may or may not contain milk or cream.

3. **Bisques** are thickened soups made from shellfish. They are usually prepared like cream soups and are almost always finished with cream.

   The term **bisque** is sometimes used on menus for a variety of vegetable soups. In these cases, it is really a marketing term rather than a technical term, so it is impossible to give a definition that covers all uses.

4. **Chowders** are hearty soups made from fish, shellfish, and/or vegetables. Although they are made in many ways, they usually contain milk and potatoes.

SPECIALTY AND NATIONAL SOUPS

This is a catch-all category for soups that don't fit well into the main categories and soups that are native to particular countries or regions.

Specialty soups are distinguished by unusual ingredients or methods, such as turtle soup, gumbo, peanut soup, and cold fruit soup.

Cold soups are sometimes considered specialty soups, and, in fact, some are. But many other popular cold soups, such as jellied consommé, cold cream of cucumber soup, and vichyssoise (vee shee swahz) are simply cold versions of basic clear and thick soups.

VEGETARIAN SOUPS AND LOW-FAT SOUPS

A great variety of vegetable-based soups are suitable for vegetarian menus. To plan vegetarian menus, review the categories of vegetarianism discussed on page 808. Vegetable soups for vegans must contain no meat or any other animal product and must be made with water or vegetable stock. To bind thick soups, use a starch slurry or a roux made with oil rather than butter. Lacto-vegetarians, on the other hand, accept soups containing butter, milk, or cream.

Because the appeal of vegetarian vegetable soups depends entirely on the freshness and the quality of the vegetables and not on the richness of meat stocks, be especially careful to use high-quality ingredients and to avoid overcooking.

Clear soups are especially suitable for people seeking low-fat foods. Consommés and clear vegetable soups are virtually fat-free, especially if the vegetables were not sweated in fat before being simmered.

Thick soups can be kept low in fat by thickening them with a slurry of starch (such as arrowroot, potato starch, or cornstarch) and cold water rather than with a roux. For cream soups, reduce or omit the cream and instead use evaporated skim milk. Purée soups are usually more adaptable than cream soups to low-fat diets because the vegetable purée adds body and richness to the soup without requiring added fat. A little yogurt or evaporated skim milk can be used to give creaminess to a purée soup. Even garnishing a serving of soup with a teaspoonful of whipped cream gives a feeling of richness while adding only a gram or two of fat.

SERVICE OF SOUPS

STANDARD PORTION SIZES

Appetizer portion: 6–8 oz (200–250 mL)

Main course portion: 10–12 oz (300–350 mL)

TEMPERATURE

Serve hot soups hot, in hot cups or bowls.

Serve cold soups cold, in chilled bowls or even nested in a larger bowl of crushed ice.

HOLDING FOR SERVICE

Strangely enough, some chefs who take the greatest care not to overcook meats or vegetables nevertheless keep a large kettle of soup on the steam table all day. You can imagine what a vegetable soup is like after four or five hours at that temperature.

1. Small-batch cooking applies to soups as well as to other foods. Heat small batches frequently to replenish the steam table with fresh soup.

2. Consommés and some other clear soups can be kept hot for longer periods if the vegetable garnish is heated separately and added at service time.

GARNISH

Soup garnishes may be divided into three groups.

1. Garnishes in the soup.

Major ingredients, such as the vegetables in clear vegetable soup, are often considered garnishes. This group of garnishes also includes meats, poultry, seafood, pasta products, and grains such as barley or rice. They are treated as part of the preparation or recipe itself, not as something added on.

Consommés are generally named after their garnish, such as consommé brunoise, which contains vegetables cut into brunoise shape ($1/8$-inch, or 3-mm, dice).

Vegetable cream soups are usually garnished with carefully cut pieces of the vegetable from which they are made.

An elegant way to serve soup with a solid garnish is to arrange the garnish attractively in the bottom of a heated soup plate. This plate is set before the diner, and then the soup is ladled from a tureen by the dining room staff.

2. Toppings.

Clear soups are generally served without toppings to let the attractiveness of the clear broth and the carefully cut vegetables speak for themselves. Occasional exceptions are toppings of chopped parsley or chives.

Thick soups, especially those that are all one color, are often decorated with a topping. Toppings should be placed on the soup just before service so they won't sink or lose their fresh appearance. Their flavors must be appropriate to the soup.

Do not overdo soup toppings. The food should be attractive in itself. Topping suggestions for thick soups include the following:

Fresh herbs (such as parsley or chives), chopped	Paprika
Fine julienne of vegetables	Flavored butters
Sliced almonds, toasted	Flavored oils
Grated cheese	Fried herbs, such as parsley, sage, chervil, celery leaves, leek julienne
Sieved egg yolk	Sour cream, crème fraîche, or whipped cream, either plain or flavored with herbs or spices
Chopped or riced egg whites	
Croutons	
Crumbled bacon	

3. Accompaniments.

American soups are traditionally served with crackers. In addition to the usual saltines, other suggestions for crisp accompaniments are:

Melba toast	Whole grain wafers
Corn chips or tortilla chips	Profiteroles (tiny unsweetened cream puff shells)
Breadsticks	
Cheese straws	

CLEAR SOUPS

BROTHS

The difference between a broth and a stock is that a broth, according to the most common definition, is made by simmering meat and vegetables, while a stock is made by simmering bones and vegetables. Because of this difference, a well-made stock is generally richer in gelatin content than a broth, because gelatin is derived from cartilage and connective tissue. A broth, on the other hand, usually has a more pronounced flavor of meat or poultry than a stock. A more neutral flavor is desired in a stock, which is used as the base for many sauces as well as soups. A broth, on the other hand, is an excellent choice as the base of a soup when a distinct meat flavor is desired.

Nevertheless, broths are not often specially made in food-service operations. The cost of the meat makes them expensive, unless the meat can be used for another purpose, or unless the restaurant has a good supply of meat trimmings that might otherwise be wasted. Instead, broth is usually a byproduct of simmering meat or poultry. The recipes for Simmered Fresh Beef Brisket (p. 436) and for "Boiled" Fowl (p. 567) produce not only the cooked meat or poultry but also flavorful broths that can be served as soups when properly seasoned and garnished.

Note that the broths resulting from both these recipes are white. To prepare a brown meat broth, follow the procedure in the recipe for Simmered Fresh Beef Brisket (p. 436), but brown the meat and mirepoix well before adding water. Flavorful cuts such as beef shank, chuck, and neck are good for making broths.

For those operations that determine that making broths especially for soups is cost-effective, a recipe for beef broth is provided below. In all broth-making operations, just as in stock making, never let the temperature rise above a simmer. Boiling makes the broth cloudy.

Broths can be served as is, with only seasoning and perhaps a light garnish added. For example, plain chicken broth is commonly served as a restorative for invalids. More often, however, broths are used in place of stocks in vegetable soups and other clear soups, as discussed in the section beginning on page 234.

Like stock, broth can be made with water. For especially rich, flavorful broths, use stock or remouillage in place of water in the broth recipe.

STANDARDS OF QUALITY FOR BROTHS

1. Appearance.

Broth should be clear, with no cloudiness caused by hard boiling. Color may range from pale gold to rich brown, depending on the ingredients and on whether or not the meat and mirepoix were browned.

The broth should be degreased, so that no more than a few droplets of fat show on the surface.

2. Flavor and Aroma.

Flavor and aroma should be full and rich, not thin, with the distinct flavor of the main ingredients (meat, poultry, or fish).

Seasoning should be well balanced, with just the right amount of salt.

3. Texture.

Broth should have good body and not feel thin and watery in the mouth. Any garnish should be cooked to the right degree of tenderness.

Beef Broth

YIELD: 3 QT (3 L)

U.S.	METRIC	INGREDIENTS	PROCEDURE
5 lb	2.5 kg	Beef: shank, chuck, oxtail, short ribs, or mixed trimmings, or a mixture of these	1. In a stockpot, brown beef in the first quantity of oil.
1 fl oz	30 mL	Vegetable oil	2. Add water or stock and bring to a simmer. Skim the surface to remove any froth or scum.
1 gal	4 L	Water, beef stock, or remouillage (p.167)	3. Continue to simmer 2 hours.
		Mirepoix:	4. Brown the mirepoix in the second quantity of oil.
8 oz	250 g	Onion, chopped	5. Add the mirepoix and tomatoes to the stockpot.
4 oz	125 g	Carrots, chopped	6. With a little water, deglaze the pan used to brown the mirepoix and add liquid to the stockpot.
4 oz	125 g	Celery, chopped	
1 tbsp	15 mL	Vegetable oil	7. Continue to simmer an additional 30 minutes.
4 oz	125 g	Tomatoes, canned or fresh	
		Sachet:	8. Add the sachet. Simmer an additional 30–45 minutes.
1	1	Bay leaf	9. Strain the broth. Discard the sachet, vegetables, and any unusable meat trimmings. Reserve any good cuts of meat for garnishing the broth or another use.
1/4 tsp	1 mL	Dried thyme	
1/4 tsp	1 mL	Peppercorns	
6–8	6–8	Parsley stems	
2	2	Whole cloves	
to taste	to taste	Salt	10. To serve as a soup, season and garnish to taste.
as desired	as desired	Meat, vegetables, or other garnish for soup	

Per serving: Calories, 130; Protein, 8 g; Fat, 11 g (3% cal.); Cholesterol, 26 mg; Carbohydrates, 1 g; Fiber, 0 g; Sodium, 55 mg.

CONSOMMÉ

When we define *consommé* as a clarified stock or broth, we are forgetting the most important part of the definition. The word *consommé* means, literally, "completed" or "concentrated." In other words, a consommé is a strong, concentrated stock or broth. In classical cuisine, this was all that was necessary for a stock to be called a consommé. In fact, two kinds were recognized: ordinary (or unclarified) consommé and clarified consommé.

Rule number one for preparing consommé is that the stock or broth must be strong, rich, and full-flavored. Clarification is second in importance to strength. A good consommé, with a mellow but full aroma and plenty of body (from the natural gelatin) you can feel in your mouth, is one of the great pleasures of fine cuisine. But clarification is an expensive and time-consuming procedure and, frankly, not worth the trouble if the soup is thin and watery.

HOW CLARIFICATION WORKS

Coagulation of proteins was an important subject in our discussion of stock making because one of our major concerns was how to keep coagulated proteins from making the stock cloudy. Strangely enough, this same process of **coagulation** enables us to clarify stocks to perfect transparency.

Remember that some proteins, especially those called *albumins*, dissolve in cold water. When the water is heated, they gradually solidify or coagulate and rise to the surface. If we control this process carefully, these proteins collect all the tiny particles that cloud a stock and carry them to the surface. The stock is then left perfectly clear.

If we are not careful, however, these proteins break up as they coagulate and cloud the liquid even more, just as they can do when we make stock.

BASIC INGREDIENTS

The mixture of ingredients we use to clarify a stock is called the **clearmeat** or the **clarification**.

1. *Lean ground meat* is one of the major sources of protein that enables the clearmeat to do its job. It also contributes flavor to the consommé. The meat must be lean because fat is undesirable in a consommé. Beef shank, also called *shin beef*, is the most desirable meat because it is high in albumin proteins as well as in flavor and gelatin, and it is very lean.

    Beef and/or chicken meat are used to clarify chicken consommé. Meat is not used, obviously, to make fish consommé. Ground lean fish may be used, but it is normal to omit flesh altogether and use only egg whites.

2. *Egg whites* are included in the clearmeat because, being mostly albumin, they greatly strengthen its clarifying power.

3. *Mirepoix* and other seasoning and flavoring ingredients are usually included because they add flavor to the finished consommé. They do not actually help in the clarification, except possibly to give solidity to the raft. The **raft** is the coagulated clearmeat, floating in a solid mass on top of the consommé.

    The mirepoix must be cut into fine pieces so it will float with the raft.

    A large amount of a particular vegetable may be added if a special flavor is desired, as in, for example, essence of celery consommé.

4. *Acid ingredients* (tomato products for beef or chicken consommé, lemon juice or white wine for fish consommé) are often added because the acidity helps coagulate the protein. They are not absolutely necessary—the heat will coagulate the protein anyway—but many chefs like to use them.

STANDARDS OF QUALITY FOR CONSOMMÉ

1. **Appearance.**
    Broth should be crystal clear and perfectly degreased. Color should be pale amber for poultry consommé and darker amber, not dark brown, for meat consommé.

    Garnish should be perfectly and uniformly cut.

2. Flavor and Aroma.

Flavor and aroma should be full, rich, well balanced, and concentrated, with the distinct flavor of the main ingredients (meat, poultry, or fish). It should be carefully seasoned, with just the right amount of salt.

3. Texture.

Broth should have good body and mouthfeel, even more so than a good broth, due to a noticeable gelatin content from the well-made stock. Garnish should be cooked to the right degree of tenderness.

PROCEDURE for Preparing Consommé

1. Start with a well-flavored, cold, strong stock or broth. If your stock is weak, reduce it until it is concentrated enough, then cool it before proceeding, or plan on simmering the consommé longer to reduce while clarifying.

2. Select a heavy stockpot or soup pot, preferably one with a spigot at the bottom. The spigot enables you to drain off the finished consommé without disturbing the raft.

3. Combine the clearmeat ingredients in the soup pot and mix them vigorously.

4. Optional step: Mix in a small amount of cold water or stock—4 to 8 oz per pound (250 to 500 mL per kg) of meat—and let stand 30–60 minutes. This allows more opportunity for the proteins that do the clarifying to dissolve out of the meat.

   Note: Chefs disagree on the importance of this step. Some let the mixture stand overnight in the refrigerator. Others skip the step. Check with your instructor.

5. Gradually add the cold, degreased stock and mix well with the clearmeat.

   The stock must be cold so it doesn't cook the proteins on contact.

   Mixing distributes the dissolved proteins throughout the stock so they can collect all the impurities more easily.

6. Set the pot over a moderately low fire and let it come to a simmer very slowly.

7. Stir the contents occasionally so the clearmeat circulates throughout the stock and doesn't burn to the bottom.

8. When the simmering point is approaching, stop stirring. The clearmeat will rise to the surface and form a raft.

9. Move the pot to lower heat so the liquid maintains a slow simmer. Do not cover. Boiling would break up the raft and cloud the consommé. The same principle operates in stockmaking.

10. Let simmer 1½ hours without disturbing the raft.

11. Strain the consommé through a china cap lined with several layers of cheesecloth.

    If you are not using a stockpot with a spigot, ladle the consommé out carefully without breaking up the raft.

    Let the liquid drain through the cheesecloth by gravity. Do not force it, or fine particles will pass through and cloud the consommé.

12. Degrease.

    Remove all traces of fat from the surface. Strips of clean brown paper passed across the surface are effective in absorbing every last speck of fat without absorbing much consommé.

13. Adjust the seasonings.

    Kosher salt is preferred to regular table salt because it has no impurities or additives that could cloud the stock.

14. Evaluate the finished consommé (see p. 230).

FIGURE 9.1 Preparing consommé.

(a) The stock is well mixed with the clarification ingredients and set on a burner to begin heating.

(b) The raft begins to rise to the top.

(c) The raft has almost completely formed. The consommé will continue to simmer for a total of 1½ hours and the soup will become completely clear. (Note: A glass beaker is used in the illustrations only so that the clarification process can be seen. Use metal pots, never glass, for consommé production.)

EMERGENCY PROCEDURES

1. Clarifying hot stock.

If you do not have time to cool the stock properly before clarifying, at least cool it as much as you can. Even 10 minutes in a cold-water bath helps. Then, mix ice cubes or crushed ice with the clearmeat. This will help keep it from coagulating when the hot stock hits it. Proceed as in the basic method.

Finally, review your production planning so you can avoid this emergency in the future.

2. Clarifying without meat.

In a pinch, you can clarify a stock with egg whites alone. Use at least 3 or 4 egg whites per gallon (4 L) stock, plus mirepoix if possible. Great care is necessary because the raft will be fragile and easily broken up.

Egg whites and mirepoix alone are often used for clarifying fish stocks.

3. Failed clarification.

If the clarification fails because you let it boil, or for some other reason, it can still be rescued, even if there is no time for another complete clarification.

Strain the consommé, cool it as much as you can, then slowly add it to a mixture of ice cubes and egg whites. Carefully return to a simmer as in the basic method and proceed with the clarification.

This should be done in emergencies only. The ice cubes dilute the consommé, and the egg white clarification is risky.

4. Poor color.

Beef or veal consommé made from brown stock should have an amber color. It is not dark brown like canned consommé. Chicken consommé is a very pale amber.

It is possible to correct a pale consommé by adding a few drops of caramel color to the finished soup, but for best results, check the color of the stock before clarification. If it is too pale, cut an onion in half and place it cut side down on a flattop range until it is black, or char it under a broiler. Add this to the clearmeat. The caramelized sugar of the onion will color the stock.

Consommé ♥

YIELD: 1 GAL (4 L)   PORTIONS: 16   PORTION SIZE: 8 FL OZ (250 ML)
                    20   6 FL OZ (250 ML)

U.S.	METRIC	INGREDIENTS	
1 lb	500 g	Lean beef, preferably shin, ground	
		Mirepoix, chopped into small pieces:	
8 oz	250 g	Onion	
4 oz	125 g	Celery	
4 oz	125 g	Carrot	
8 oz	250 g	Egg whites	
8 oz	250 g	Canned tomatoes, crushed	
6–8	6–8	Parsley stems, chopped	
pinch	pinch	Dried thyme	
1	1	Bay leaf	
2	2	Whole cloves	
¹/₂ tsp	2 mL	Peppercorns, crushed	
5 qt	5 L	Beef or veal stock, cold (brown or white)	

PROCEDURE

1. Review the information on preparing consommé, page 231.
2. Combine the beef, mirepoix, egg whites, tomatoes, herbs, and spices in a tall, heavy stockpot. Mix vigorously with a wooden paddle or a heavy whip.
3. Add about 1 pint (500 mL) cold stock and stir well. Let stand about 30 minutes. (Optional step: see p. 231 for explanation.)
4. Gradually stir in the remaining cold stock. Be sure the stock is well mixed with the other ingredients.
5. Set the pot on moderately low heat and let it come to a simmer very slowly. Stir occasionally.
6. When the simmering point is approaching, stop stirring.
7. Move the pot to lower heat and simmer very slowly about 1¹/₄ hours. Do not stir or disturb the raft that forms on top.
8. Very carefully strain the consommé through a china cap lined with several layers of cheesecloth.
9. Degrease thoroughly.
10. Season to taste.

Per serving: Calories, 30; Protein, 3 g; Fat, 1 g (31% cal.); Cholesterol, 15 mg; Carbohydrates, 2 g; Fiber, 0 g; Sodium, 75 mg.

(Variations continue on next page)

VARIATIONS *(continued)*

Double Consommé

Use twice the quantity of beef in basic recipe. Add 8 oz (250 g) leeks to mirepoix.

Chicken Consommé

Use chicken stock instead of beef or veal stock. Add to the clearmeat 8 oz (250 g) chicken trimmings (such as wing tips and necks) that have been chopped and browned in a hot oven. Omit tomato and add 1 fl oz (30 mL) lemon juice.

Cold Jellied Consommé

Unflavored gelatin must often be added to consommé to make jellied consommé. Amount needed depends on the strength of the stock and amount of jelling desired. Classically, a chilled consommé is only half jelled, more like a thick syrup. Some people, however, prefer a gelatin content high enough to solidify the consommé. In the following guidelines, use the lower quantity of gelatin for a partially jelled soup, the higher quantity for a fully jelled soup. Also, for tomatoed consommé (madrilène), increase the gelatin slightly because the acidity of the tomatoes weakens the gelatin.

1. If stock is thin when cold, add 1–2 oz (30–60 g) gelatin per gallon (4 L).
2. If stock is slightly jelled and syrupy when cold, add ½–1 oz (15–30 g) gelatin per gallon (4 L).
3. If stock is jelled when cold, no gelatin is needed. Add up to ½ oz (15 g) per gallon (4 L) if firmer texture is desired.

Gelatin may be added to clearmeat (in step 2 of recipe). This is the best method because there is no danger of clouding the consommé. It may also be added to finished consommé after softening it in cold water. See page 710 for instructions on use of gelatin.

Consommé Madrilène

Increase the tomatoes in the basic recipe to 24 oz (750 g). Use beef, veal, or chicken stock. Serve hot or jellied.

Essence of Celery Consommé

Increase the celery in the basic recipe to 1 lb (500 g).

Consommé au Porto

Flavor finished consommé with 6–8 fl oz (200–250 mL) port wine per gallon (4 L).

Consommé au Sherry

Flavor finished consommé with 6–8 fl oz (200–250 mL) sherry wine per gallon (4 L).

GARNISHED CONSOMMÉS

For the following consommés, prepare and cook the garnish separately. At service time, add 1–2 tbsp (15–30 mL) garnish to each portion. See page 140 for description of cuts. All cuts should be small enough to fit the bowl of a spoon. For example, julienne may need to be cut shorter than for other applications.

Consommé Brunoise

Onion or leek, carrot, celery, and turnip (optional), cut brunoise. Sweat lightly in butter and simmer in a little consommé until tender.

Consommé Julienne

Onion or leek, carrot, and celery, cut julienne. Prepare like brunoise garnish.

Consommé Printanière

Small dice of spring vegetables: carrot, turnip, celery, green beans. Prepare like brunoise garnish.

Consommé Paysanne

Thin slices of leeks, carrots, celery, turnip, and cabbage. Prepare like brunoise garnish.

Consommé with Pearl Tapioca

Cooked pearl tapioca.

Consommé Vermicelli

Cooked broken vermicelli (very thin spaghetti).

Consommé Fettuccine ♥

YIELD: 10 OZ (300 G)

U.S.	METRIC	INGREDIENTS	PROCEDURE
10 fl oz	300 mL	Consommé	1. Place the consommé in a saucepan and mix in the agar-agar.
0.2 oz (2 tsp)	6 g	Agar-agar	2. Bring to a boil, then remove from heat. Stir with a wire whip to be sure the agar-agar is thoroughly mixed in.
			3. Select a half-sheet pan that is perfectly level. Have ready a refrigerator shelf that is perfectly level.
			4. Pour the consommé into the sheet pan. Refrigerate until set. This will take only a few minutes.
			5. Cut the jelled consommé lengthwise into uniform strips ¼ inch (6 mm) wide. Remove from pan.
			6. To serve, mound a small portion on a plate. Garnish as desired (see Variation for suggestion).

Per serving: Calories, 5; Protein, 0 g; Fat, 0 g (0% cal.); Cholesterol, 0 mg; Carbohydrates, 1 g; Fiber, 1 g; Sodium, 10 mg.

VARIATION

Consommé Fettuccine with Parmesan and Basil

Mound 2 oz (60 g) in the center of a small plate. Using a squeeze bottle, squeeze an arc of Basil Sauce (p. 218) around one side of the mound. On the other side, spoon a mound of Parmesan Foam (p. 219). Garnish the fettuccine with a few small, whole basil leaves.

MOLECULAR SOUP

One of the ways chefs working in molecular gastronomy (see p. 120) surprise diners' expectations and thus get them to focus on the food is to transform foods into unexpected shapes and forms. The recipe for Consommé Fettuccine, transforming soup into noodles, based on an idea developed by the Spanish chef Ferran Adrià, is an example of this approach.

VEGETABLE SOUPS

Clear vegetable soups are made from a clear stock or broth, not necessarily clarified, with the addition of one or more vegetables and, sometimes, meat or poultry and/or pasta or grains. Most vegetable soups are made from meat or poultry stock or broth. Meatless or vegetarian soups are made from vegetable broth or water.

GUIDELINES for Preparing and Evaluating Vegetable Soups

Procedures for making these soups are not complicated. Most of them are made simply by simmering vegetables in stock until done. But care and attention to details are still necessary for producing a high-quality soup.

These guidelines are also useful as quality factors for evaluating soups. Each of the guidelines below is based on a desired quality in the finished soup (for example: flavorful broth, balance of vegetables, garnish neatly cut, and so on).

1. **Start with a clear, flavorful stock or broth.**
   This is one reason it's important to be able to make stocks that are clear, not cloudy. Generally, vegetable soups are a little cloudier than broths, because the vegetables are generally cooked in the soup. Nevertheless, the broth of the finished soup should be fairly clear.

2. **Select vegetables and other ingredients whose flavors go well together.**
   Don't just throw in everything you've got. Judgment, combined with experience, must be used to create a pleasing combination. Five or six vegetables are usually enough. More than that often makes a jumble.

3. **Cut vegetables uniformly.**
   Neat, careful cutting means uniform cooking and attractive appearance. Sizes of cuts are important, too. Pieces should be large enough to be identifiable but small enough to eat conveniently with a spoon.

4. **Cook vegetables slowly in a little butter before combining with liquid.**
   This process improves the flavor of the vegetables and gives the soup a mellower, richer taste.

5. **Cook starches such as grains and pasta separately and add to the soup later.**
   Cooking them in the soup makes it cloudy. Potatoes are sometimes cooked directly in the soup, but they should be rinsed of excess starch after cutting if you want to keep the soup as clear as possible.

6. **Observe differences in cooking times.**
   Add long-cooking vegetables first, short-cooking vegetables near the end. Vegetables that are already fully cooked and hot can be added to the hot soup after it is removed from the fire.

7. **Don't overcook.**
   Some cooks feel soups must be simmered a long time to extract flavors into the liquid. But you should have already done this when you made the stock. Vegetables in soup should be tender but still firm enough to hold their shapes. Vegetables should show a fresh color and not look like they were cooked too long. This is especially true of green vegetables, which turn olive brown when overcooked.

Clear Vegetable Soup

YIELD: 3 QT (3 L)   **PORTIONS: 12**   **PORTION SIZE: 8 FL OZ (250 ML)**

U.S.	METRIC	INGREDIENTS	PROCEDURE
2 oz	60 g	Butter or chicken fat	1. Heat the butter in a heavy saucepot over medium-low heat.
12 oz	375 g	Onions, small dice	2. Add the onions, carrots, celery, and turnip. Sweat vegetables in the butter over low heat until about half cooked. Do not let them brown.
8 oz	250 g	Carrots, small dice	
8 oz	250 g	Celery, small dice	
6 oz	180 g	Turnip, small dice	
3 qt	3 L	Chicken stock	3. Add the stock. Bring to a boil and skim carefully. Simmer until vegetables are just barely tender.
8 oz	250 g	Drained canned tomatoes, coarsely chopped	4. Add the tomatoes and simmer another 5 minutes.
to taste	to taste	Salt	5. Degrease the soup and season with salt and white pepper.
to taste	to taste	White pepper	
6 oz	180 g	Frozen peas, thawed	6. Just before serving, add the peas.

Per serving: Calories, 80; Protein, 3 g; Fat, 4.5 g (46% cal.); Cholesterol, 15 mg; Carbohydrates, 9 g; Fiber, 2 g; Sodium, 125 mg.

VARIATIONS

Other vegetables may be used in addition to or in place of one or more of the vegetables in basic recipe. Add with the vegetables sweated in butter:

Leeks   Green cabbage
Rutabagas   Parsnips

Add to simmering soup, timing the addition so all vegetables are done at the same time:

Potatoes   Lima beans
Green beans   Corn

Other cuts may be used for the vegetables instead of small dice, such as bâtonnet, julienne, or paysanne (see p. 141).

Vegetable Rice Soup

Add ¾–1 cup (180–250 mL) cooked rice to finished soup.

Chicken Vegetable Rice Soup

Add 6 oz (180 g) cooked, diced chicken to vegetable rice soup.

Vegetable Beef Soup

Use beef stock instead of chicken stock. Add 6 oz (180 g) cooked, diced beef when the tomatoes are added. Also, add the juice from the tomatoes.

Vegetable Beef Barley Soup

Add ¾–1 cup (180–250 mL) cooked barley to vegetable beef soup.

Clear Vegetable Soup with Cranberry Beans

Piquant Vegetable Soup with Chickpeas

YIELD: 3 QT (3 L)   PORTIONS: 12   PORTION SIZE: 8 FL OZ (250 ML)

U.S.	METRIC	INGREDIENTS
1 fl oz	30 mL	Vegetable oil
10 oz	300 g	Red onion, small dice
1½ tsp	7 mL	Garlic, chopped
8 oz	250 g	Green bell pepper, small dice
1–2 oz	30–60 g	Jalapeño or other green chile, cut brunoise
2½ qt	2.5 L	Chicken stock or vegetable stock
10 oz	300 g	Tomatoes, peeled, seeded, and chopped
10 oz	300 g	Cooked chickpeas
4 oz	125 g	Corn kernels, frozen or fresh
4 oz	125 g	Green beans, cooked until just tender and cut into ½-in. (1-cm) pieces
to taste	to taste	Salt
to taste	to taste	White pepper
to taste	to taste	Hot red pepper sauce (optional)
		Garnish:
6 oz	180 g	Grated cheddar cheese
3 tbsp	45 mL	Chopped cilantro or whole cilantro leaves

PROCEDURE

1. Heat the oil in a soup pot over moderate heat.
2. Add onion, garlic, bell pepper, and green chile. Sweat the vegetables in the oil over low heat until they are about half-cooked. Do not let them brown.
3. Add the stock. Bring to a boil and skim carefully. Simmer until the vegetables are just barely tender.
4. Add the tomatoes, chickpeas, and corn. Simmer another 5 minutes.
5. Shortly before serving, add the cooked green beans. (Beans should be cooked separately and added at the end so their color won't be destroyed by the acidity of the tomatoes.)
6. Add salt and white pepper to taste. Add hot pepper sauce, if desired.
7. To serve, ladle a portion into soup plates and sprinkle with grated cheese and a little cilantro.

Per serving: Calories, 160; Protein, 8 g; Fat, 8 g (45% cal.); Cholesterol, 20 mg; Carbohydrates, 14 g; Fiber, 3 g; Sodium, 105 mg.

VARIATION

Piquant Vegetable Soup with Roasted Garlic

Roast ½ head of garlic, as explained on page 334. Separate and peel the cloves. Add to soup in step 4.

Piquant Vegetable Soup

Mushroom Barley Soup ♥

YIELD: 3 QT (3 L)   PORTIONS: 12   PORTION SIZE: 8 FL OZ (250 ML)

U.S.	METRIC	INGREDIENTS	PROCEDURE
4 oz	125 g	Barley	1. Cook the barley in boiling water until tender. Drain.
5 oz	150 g	Onion, cut brunoise	2. In a heavy saucepot or stockpot, sweat vegetables in the fat until about half-cooked. Do not let them brown.
2¹/₂ oz	75 g	Carrot, cut brunoise	3. Add the chicken stock. Bring to a boil. Reduce heat and simmer until vegetables are just tender.
2¹/₂ oz	75 g	White turnip, cut brunoise	
1 oz	30 g	Butter or chicken fat	
2¹/₂ qt	2.55 L	Chicken stock	
1 lb	500 g	Mushrooms, diced	4. While the soup is simmering, sauté the mushrooms briefly in fat without letting them brown.
2 oz	60 g	Butter or chicken fat	5. Add mushrooms and drained, cooked barley to the soup. Simmer another 5 minutes.
to taste	to taste	Salt	6. Degrease the soup. Season to taste with salt and pepper.
to taste	to taste	White pepper	

Per serving: Calories, 60; Protein, 3 g; Fat, 1 g (15% cal.); Cholesterol, 20 mg; Carbohydrates, 10 g; Fiber, 3 g; Sodium, 75 mg.

OTHER CLEAR SOUPS

In addition to vegetable soups, many other clear or unthickened soups are known to various cuisines. They range from simple broths to elaborate concoctions of meats, vegetables, starches, and other ingredients. Although many contain vegetables, we don't classify them as vegetable soups because other ingredients are generally more prominent.

KEY POINTS TO REVIEW

- What are the two basic categories of soup? What are some examples of each?
- What are normal portion sizes for soups?
- After soups are cooked, how are they best held for service?
- What is the procedure for making consommé?
- How are clear vegetable soups made?

Chicken Noodle Soup ♥

YIELD: 3 QT (3 L)   PORTIONS: 12   PORTIONS SIZE: 8 FL OZ (250 ML)

U.S.	METRIC	INGREDIENTS	PROCEDURE
5 oz	150 g	Egg noodles	1. Cook noodles in boiling, salted water. (See Procedure for Cooking Pasta, p. 398.) Drain and rinse in cold water.
5 oz	150 g	Cooked chicken meat (see Note)	2. Cut the chicken into small dice.
3 qt	3 L	Chicken stock	3. Bring the stock to a simmer. Season to taste with salt and white pepper. If stock doesn't have enough flavor, add more stock and reduce to concentrate the flavor.
to taste	to taste	Salt	
to taste	to taste	White pepper	4. Just before service, add the chicken and noodles to the stock. Let them heat through before serving.
as desired	as desired	Chopped parsley	5. Garnish each portion with a little chopped parsley.

Per serving: Calories, 70; Protein, 6 g; Fat, 1.5 g (21% cal.); Cholesterol, 25 mg; Carbohydrates, 7 g; Fiber, 0 g; Sodium, 20 mg.

Note: See Chapter 18 (p. 567) for preparing "boiled" chicken and broth for use in soups. Other leftover cooked chicken may also be used.

VARIATIONS

Beef Noodle Soup

Prepare as in basic recipe, using beef and beef stock.

Chicken or Beef Noodle Soup with Vegetables

Before adding the chicken and noodles, simmer 5 oz (150 g) diced carrots and 2¹/₂ oz (75 g) diced celery in the stock until tender.

Oxtail Soup

YIELD: 3 QT (3 L)   PORTIONS: 12   PORTION SIZE: 8 FL OZ (250 ML)

U.S.	METRIC	INGREDIENTS	PROCEDURE
3 lb	1.35 kg	Oxtails	1. Using a heavy chef's knife, cut the oxtails into sections at the joints.
		Mirepoix:	
5 oz	150 g	Onion, medium dice	2. Place oxtails in a bake pan and brown in a 450°F (230°C) oven. When partially browned, add the mirepoix to the pan and brown it along with the oxtails.
2½ oz	75 g	Carrot, medium dice	
2½ oz	75 g	Celery, medium dice	3. Place the oxtails and mirepoix in a stockpot with the stock.
3 qt	3 L	Brown stock (see Note)	4. Pour off the fat from the pan in which the meat was browned. Deglaze the pan with a little of the stock and add this to the stockpot.
		Sachet:	
1	1	Bay leaf	
pinch	pinch	Dried thyme	5. Bring to a boil. Reduce heat to a simmer and skim well. Add the sachet.
6	6	Peppercorns	6. Simmer until the meat is tender, about 3 hours. Add a little water if necessary during cooking to keep the meat completely covered.
2	2	Whole cloves	
1	1	Garlic clove	
			7. Remove the oxtail pieces from the broth. Trim the meat from the bones and dice it. Place in a small pan with a little broth. Keep warm if the soup is to be finished immediately, or chill for later use.
10 oz	300 g	Carrots, small dice	8. Strain the broth. Degrease carefully.
10 oz	300 g	White turnips, small dice	9. Sweat the carrots, turnips, and leeks in the butter until about half cooked.
5 oz	150 g	Leeks, white part only, cut julienne	
2 oz	60 g	Butter	10. Add the broth. Simmer until vegetables are tender.
5 oz	150 g	Tomatoes (canned), drained, coarsely chopped	11. Add the tomatoes and reserved oxtail meat. Simmer another minute.
1 fl oz	30 mL	Sherry (optional)	12. Add the sherry, if desired. Season to taste with salt and pepper.
to taste	to taste	Salt	
to taste	to taste	Pepper	

Per serving: Calories, 240; Protein, 24 g; Fat, 11 g (45% cal.); Cholesterol, 90 mg; Carbohydrates, 6 g; Fiber, 2 g; Sodium, 220 mg.

Note: Water is sometime used instead of stock. If this is done, brown 4–5 lb (about 2 kg) beef or veal bones with the oxtails and simmer both bones and oxtails in the soup. Double the quantity of mirepoix.

VARIATION

Oxtail soup is often clarified. Chill broth after step 7 and clarify like consommé. See page 231 for procedure.

THICK SOUPS

CREAM SOUPS

Learning to cook professionally, as you have already heard, is not learning recipes but learning basic techniques you can apply to specific needs.

The basic techniques of saucemaking were discussed in Chapter 8. If we tell you that cream soups are simply diluted velouté or béchamel sauces, flavored with the ingredient for which they are named, you should almost be able to make a cream of celery soup without further instructions.

It's not *quite* that simple. There are some complications, but they are mostly a matter of detail. You already know the basic techniques.

THE CLASSIC CREAM SOUPS

In the great kitchens of several decades ago, cream soups were exactly as we have just described: diluted, flavored sauces. In fact, what we now call **cream soups** were divided into two groups, veloutés (made from velouté sauce and finished with a liaison) and creams (made from béchamel sauce and finished with cream).

These methods were natural to large kitchens that always had quantities of velouté and béchamel sauces on hand. Making a soup was simply a matter of finishing off a sauce.

Modern cooks view these methods as complicated and have devised other methods that seem simpler. But most of the sauce steps are involved—you still have to thicken a liquid with roux (or other starch), cook and purée the ingredients, and add the milk or cream.

The classical method is still important to learn. It will give you versatility, it makes excellent soup, and besides, it really isn't any harder or longer, in the final analysis. In addition, we explain two other methods much in use today.

But first, we consider a problem frequently encountered with cream soups.

CURDLING

Because cream soups contain milk or cream or both, curdling is a common problem. The heat of cooking and the acidity of many of the other soup ingredients are the causes of this curdling.

Fortunately, we can rely on one fact to avoid curdling: *Roux and other starch thickeners stabilize milk and cream*. Caution is still necessary because soups are relatively thin and do not contain enough starch to be completely curdle-proof.

Observe the following guidelines to help prevent curdling:

1. Do not combine milk and simmering soup stock without the presence of roux or other starch. Do one of the following:
   - Thicken the stock before adding milk.
   - Thicken the milk before adding it to the soup.

2. Do not add cold milk or cream to simmering soup. Do one of the following:
   - Heat the milk in a separate saucepan.
   - Temper the milk by gradually adding some of the hot soup to it. Then add it to the rest of the soup.

3. Do not boil soups after milk or cream is added.

STANDARDS OF QUALITY FOR CREAM SOUPS

1. **Appearance.**
   Shiny surface. Good color from main ingredient. Not too pale (from not using enough vegetable or other flavoring ingredient). Not discolored from overcooking. Garnish should be attractive and properly cut.

2. **Taste.**
   Distinct flavor of the main ingredient (asparagus in cream of asparagus, etc.). No starchy taste from uncooked roux. Pleasant creaminess, but not too much cream, which can overpower the other flavors and make the soup too rich.

3. **Texture.**
   Velvety texture. About the consistency of heavy cream. Not too thick. Smooth; no graininess or lumps (except garnish, of course).

FIGURE 9.2 Using a food mill.

(a) Purée soft foods by turning the crank of the mill to force them through small holes.

(b) Scrape the solids from the bottom of the mill after all the food is forced through the plate.

FIGURE 9.3 Using an immersion blender to purée a soup.

BASIC PROCEDURES for Making Cream Soups

The following methods apply to most cream soups. Individual ingredients may require variations.

METHOD 1

1. Prepare Velouté Sauce (p. 182) or Béchamel Sauce (p. 181), using roux.
2. Prepare the main flavoring ingredients. Cut vegetables into thin slices. Sweat them in butter about 5 minutes to develop flavor. Do not brown. Green leafy vegetables must be blanched before stewing in butter. Cut poultry and seafood into small pieces for simmering.
3. Add flavoring ingredients from step 2 to the velouté or béchamel and simmer until tender. Exception: Finished tomato purée is added for cream of tomato; further cooking is not necessary.
4. Skim any fat or scum carefully from the surface of the soup.
5. Purée the soup using a food mill (Figure 9.2) or an immersion blender (Figure 9.3), and then strain it through a fine china cap. Alternatively, just strain it through a fine china cap, pressing down hard on the solid ingredients to force out the liquid and some of the pulp. The soup should be very smooth.

   Poultry and seafood ingredients may be puréed or reserved for garnish.
6. Add hot white stock or milk to thin the soup to proper consistency.
7. Adjust seasonings.
8. At service time, finish with liaison (p. 175) or heavy cream.
9. Evaluate the finished soup (see p. 239).

METHOD 2

1. Sweat vegetable ingredients (except tomatoes) in butter (Figure 9.4a); do not let them color.
2. Add flour. Stir well to make a roux. Cook the roux a few minutes, but do not let it start to brown (b).
3. Add white stock, beating with a whip as you slowly pour it in (c).
4. Add any vegetables, other solid ingredients, or flavorings that were not sautéed in step 1.
5. Simmer until all ingredients are tender.
6. Skim any fat that rises to the surface.
7. Purée and/or strain (as in Method 1).
8. Add hot white stock or milk to thin soup to proper consistency.
9. Adjust seasonings.
10. At service time, finish with heavy cream (d) or liaison.
11. Evaluate the finished soup (see p. 239).

METHOD 3

1. Bring white stock to a boil.
2. Add vegetables and other flavoring ingredients. If desired, first slowly cook some or all of the vegetables in butter a few minutes to develop flavors.
3. Simmer until all ingredients are tender.
4. Thicken with roux, beurre manié, or other starch.
5. Simmer until no starch taste remains.
6. Skim fat from surface.
7. Purée and/or strain (as in Method 1).
8. Add hot or tempered milk and/or cream. A light cream sauce may be used, if desired, to avoid thinning the soup or curdling the milk.
9. Adjust seasonings.
10. Evaluate the finished soup (see p. 239).

FIGURE 9.4 Cream Soup Method 2.

(a) Sweat vegetables in butter.

(b) Stir in flour and cook to make roux.

(c) Whip in hot stock.

(d) Finish with hot cream.

Cream of Celery Soup (Cream Soup Method 1)

YIELD: 3 QT (3 L)   PORTIONS: 12   PORTION SIZE: 8 FL OZ (250 ML)

U.S.	METRIC	INGREDIENTS	PROCEDURE
1 lb 8 oz	750 g	Celery, small dice	1. Review cream soup guidelines and Method 1 (p. 240).
6 oz	180 g	Onion, small dice	2. Sweat the celery and onions in the butter in a heavy saucepot until
1¹/₂ oz	45 g	Butter	almost tender. Do not let them brown.
2¹/₄ qt	2 25 L	Velouté sauce, made with chicken or veal stock (see Note)	3. Add the velouté to the pot. Simmer until vegetables are very tender. 4. Skim any fat or scum from the soup. 5. Pass the soup through a food mill to purée it. 6. Pass the puréed soup through a fine china cap or cheesecloth.
1¹/₂ pt	750 mL	Milk or white stock, hot	7. Add enough hot milk or stock to bring the soup to the proper
to taste	to taste	Salt	consistency.
to taste	to taste	White pepper	8. Heat the soup again, but do not let it boil. 9. Season to taste.
12 fl oz	375 mL	Heavy cream, hot (see Note) Optional garnish:	10. At service time, add the cream. Add garnish if desired.
3 oz	90 g	Celery, cut julienne, cooked	

Per serving: Calories, 320; Protein, 5 g; Fat, 27 g (75% cal.); Cholesterol, 90 mg; Carbohydrates, 15 g; Fiber, 1 g; Sodium, 240 mg.

Note: Béchamel may be used in place of velouté if desired. This is often done for vegetarian menus. The quantity of cream may be decreased as desired to reduce dietary fat.

VARIATIONS, METHOD 1

For the following cream soups, make substitutions in the basic recipe as indicated. Frozen and canned vegetables may be used, where appropriate, in place of fresh. Also, trimmings may be used if clean and of good quality, such as bottom ends of asparagus and broccoli stalks.

Cream of Asparagus

Use 1¹/₂ lb (750 g) asparagus stalks in place of celery. Optional garnish: cooked asparagus tips.

Cream of Broccoli

Use 1¹/₂ lb (750 g) broccoli in place of celery. Optional garnish: small cooked broccoli florets.

Cream of Carrot

Use 1¹/₂ lb (750 g) carrots in place of celery. Garnish: chopped parsley.

Cream of Cauliflower

Use 1¹/₂ lb (750 g) cauliflower in place of celery. Optional garnish: tiny cooked cauliflower florets.

Cream of Corn

Use 1¹/₂ lb (750 g) whole-kernel corn (fresh, frozen, or canned) in place of celery. Do not sweat the corn with the onions. Instead, sweat the onions alone, add velouté, then add corn. Garnish: corn kernels.

Cream of Cucumber

Use 1¹/₂ lb (750 g) peeled, seeded cucumber in place of celery. Optional garnish: small, diced, cooked cucumber.

Cream of Mushroom

Use 112 oz (375 g) mushrooms in place of celery. Optional garnish: julienne, brunoise, or sliced cooked mushrooms.

Cream of Pea

Use 1¹/₂ lb (750 g) frozen green peas in place of celery. Do not sweat the peas with the onions. Add them after velouté is added.

Cream of Spinach

Use 1¹/₂ lb (750 g) fresh spinach or 1 lb (450 g) frozen spinach in place of celery. Do not sweat the spinach with the onion. Blanch it, drain well, and add to velouté in step 3.

Cream of Watercress

Use 12 oz (375 g) watercress in place of celery.

Cream of Chicken

Reduce celery to 3 oz (90 g) and add 3 oz (90 g) carrot (note that, together with the onion, this makes 12 oz [360 g] mirepoix). Use a velouté sauce made with a strong, flavorful chicken stock. After soup is strained, add 3 oz (90 g) cooked chicken meat, cut into julienne or fine dice.

Cold Cream Soups

Most cream soups are delicious cold as well as hot. For example, cold cream of cucumber soup is a special favorite in summer. Procedure:

1. Chill soup after step 9 in recipe.

2. Add cold cream after soup is well chilled.

3. Dilute with extra milk, cream, or stock if soup becomes too thick.

4. Season carefully. Cold foods require more seasonings.

Cream of Mushroom Soup (Cream Soup Method 2)

YIELD: 3 QT (3 L)   PORTIONS: 12   PORTION SIZE: 8 FL OZ (250 ML)

U.S.	METRIC	INGREDIENTS	PROCEDURE
6 oz	180 g	Butter	1. Review cream soup guidelines and Method 2 (p. 240).
6 oz	180 g	Onion, chopped fine	2. Heat the butter in a heavy saucepot over moderate heat.
12 oz	375 g	Mushrooms, chopped	3. Add the onions and mushrooms. Sweat the vegetables without letting them brown.
4¹/₂ oz	140 g	Flour	4. Add the flour and stir to make a roux. Cook the roux a few minutes, but do not let it start to brown.
2¹/₄ qt	2.25 L	White stock, chicken or veal, hot	5. Gradually beat in the stock. Bring to a boil, stirring with a whip as it thickens.
			6. Simmer until vegetables are very tender.
			7. Skim the soup carefully.
			8. Pass the soup through a food mill to purée it.
			9. Pass the puréed soup through a fine china cap or cheesecloth.
1¹/₂ pt	750 mL	Milk, hot	10. Add enough hot milk to bring the soup to the proper consistency.
to taste	to taste	Salt	11. Heat the soup again, but do not let it boil.
to taste	to taste	White pepper	12. Season to taste.
12 fl oz	375 mL	Heavy cream, hot (see Note)	13. At service time, add the cream. Add garnish, if desired.
		Optional garnish:	
3 oz	90 g	Mushrooms, cut brunoise, sautéed in butter	

Per serving: Calories, 300; Protein, 5 g; Fat 25 g (75% cal.); Cholesterol, 85 mg; Carbohydrates, 14 g; Fiber, 1 g; Sodium, 170 mg.

Note: The quantity of cream may be decreased as desired to reduce dietary fat.

VARIATIONS, METHOD 2

For each variation, replace the mushrooms with the vegetable in quantity indicated. See Note to the variations for Cream of Celery Soup, page 241.

Cream of Asparagus
1¹/₂ lb (750 g) asparagus

Cream of Broccoli
1¹/₂ lb (750 g) broccoli

Cream of Carrot
1¹/₂ lb (750 g) carrots

Cream of Cauliflower
1¹/₂ lb (750 g) cauliflower

Cream of Celery
1¹/₂ lb (750 g) celery

Cream of Corn
1¹/₂ lb (750 g) whole-kernel corn

Cream of Cucumber
1¹/₂ lb (750 g) peeled, seeded cucumber

Cream of Green Pea
1¹/₂ lb (750 g) frozen peas. Add after step 5.

Cream of Spinach
1¹/₂ lb (750 g) fresh or 1 lb (450 g) frozen spinach. Blanch, drain, and add after step 5.

Cream of Watercress
12 oz (375 g) watercress

Cream of Chicken
3 oz (90 g) celery and 3 oz (90 g) carrot. Use strong chicken stock. Add 3 oz (90 g) cooked chicken meat, cut into julienne or fine dice, to finished soup after straining.

Cream of Mushroom Soup

Cream of Broccoli Soup (Cream Soup Method 3)

YIELD: 3 QT (3 L)   PORTIONS: 12   PORTION SIZE: 8 FL OZ (250 ML)

U.S.	METRIC	INGREDIENTS	PROCEDURE
2¹/₄ qt	2.25 L	White stock, chicken or veal	1. Bring the stock to a boil in a heavy saucepot.
1 lb 8 oz	750 g	Broccoli (fresh or frozen), chopped	2. Add broccoli and onion. (Optional: Vegetables may be sweated in butter first to develop flavors.)
6 oz	180 g	Onion, chopped fine	3. Simmer until the vegetables are tender. Do not overcook, or the broccoli will lose its fresh green color.
4¹/₂ oz	140 g	Butter, clarified	4. Combine the butter and flour in a saucepan to make a roux. Cook the roux a few minutes, but do not let it color. Cool the roux slightly. (Note: Beurre manié may be used instead of roux.)
4¹/₂ oz	140 g	Flour	5. Beat the roux into the soup. Simmer until no starch taste remains.
			6. Pass the soup through a food mill, then through a fine china cap or cheesecloth.
1¹/₂ pt	750 mL	Milk, hot	7. Add enough hot milk to bring the soup to proper consistency.
to taste	to taste	Salt	8. Heat the soup again, but do not let it boil.
to taste	to taste	White pepper	9. Season to taste.
12 fl oz	375 mL	Heavy cream, hot (see Note)	10. At service time, add the heavy cream. If desired, add garnish.
		Optional garnish:	
3 oz	90 g	Small broccoli florets, cooked	

Per serving: Calories, 280; Protein, 6 g; Fat, 22 g (69% cal.); Cholesterol, 75 mg; Carbohydrates, 16 g; Fiber, 2 g; Sodium, 150 mg.

Note: The quantity of cream may be decreased as desired to reduce dietary fat.

VARIATIONS, METHOD 3

For other cream soups, replace the broccoli with 1¹/₂ lb (750 g) of any of the following:

Asparagus
Carrots
Cauliflower
Celery
Corn
Green peas
Spinach

Cream of Tomato Soup

YIELD: 3 QT (3 L)   PORTIONS: 12   PORTION SIZE: 8 FL OZ (250 ML)

U.S.	METRIC	INGREDIENTS	PROCEDURE
2 oz	60 g	Salt pork, diced	1. In a heavy saucepot, cook the salt pork over medium heat to render the fat.
2 oz	60 g	Onion, medium dice	
1 oz	30 g	Carrots, medium dice	2. Add the onion, carrots, and celery. Sweat until slightly softened.
1 oz	30 g	Celery, medium dice	
1 oz	30 g	Flour	3. Add the flour and stir to make a roux. Cook the roux a few minutes.
1½ qt	1.5 L	White stock	4. Slowly beat in the stock. Bring to a boil, stirring while the liquid thickens slightly.
1 lb	500 g	Canned tomatoes	
1 lb	500 g	Tomato purée	5. Add the tomatoes, tomato purée, and sachet. Simmer about 1 hour.
		Sachet:	
1	1	Bay leaf	6. Strain through a china cap. Press down on the solids with a ladle to force out all the juices and some of the pulp. (Alternative method: Pass through a food mill, then strain.)
pinch	pinch	Dried thyme	
1	1	Whole clove	
2	2	Peppercorns, crushed	7. If the soup is being made ahead, chill the tomato base and proceed to the next step just before service.
1 qt	1 L	Cream sauce, hot	8. Return the tomato base to the saucepot and bring back to a simmer.
to taste	to taste	Salt	9. Stir in the hot cream sauce.
to taste	to taste	White pepper	10. If the soup is too thick, thin with a little stock.
			11. Season to taste with salt and pepper.

Per serving: Calories, 210; Protein, 5 g; Fat, 15 g (64% cal.); Cholesterol, 45 mg; Carbohydrates, 14 g; Fiber, 1 g; Sodium, 230 mg.

VARIATIONS

If you study this recipe, you will see the first part (through step 6) is essentially a tomato sauce.

The recipe can be broken down as follows:

    1 part   Tomato sauce
    1 part   Stock
    1 part   Cream sauce

Using this formula, you can also make cream of tomato soup from Tomato Sauce I (p. 188) or from canned tomato sauce. You can also make it from canned tomato purée if you simmer it with extra herbs, seasonings, and mirepoix. Check all seasonings and flavors carefully when using canned, prepared products.

PURÉE SOUPS

TECHNIQUES

Purée soups are made by simmering dried or fresh vegetables, especially high-starch vegetables, in stock or water, then puréeing the soup. Thus, they are relatively easy to prepare. Purée soups are not as smooth and refined as cream soups but are heartier and coarser in texture and character.

Techniques vary greatly, depending on the ingredients and the desired result.

STANDARDS OF QUALITY FOR PURÉE SOUPS

1. **Appearance and Texture.**
   Slightly coarse texture, not as creamy as cream soups.Slightly thicker than cream soups, but still easily pourable. Good color from the main ingredients.

2. **Flavor and Aroma.**
   Robust flavor of the main ingredients should predominate, with the flavor of the stock or broth in the background. Purée soups may have a slightly rich flavor from added cream or butter, or they may have a leaner but still full-bodied flavor if no cream or butter enrichment is used.

BASIC PROCEDURE for Making Purée Soups

1. Sweat mirepoix or other fresh vegetables in fat.
2. Add liquid.
3. Add dried or starchy vegetables.
4. Simmer until vegetables are tender. Fresh vegetables should be completely cooked but not overcooked or falling apart.
5. Purée soup in a food mill or with an immersion blender.

   Variation: Some soups made from dried legumes, such as bean soup and lentil soup, are not puréed but are served as is or slightly mashed.

6. Purée soups are generally not bound with an added starch but rely on the starches present in the vegetables. Some fresh vegetable purées, however, settle out. These may be thickened with a little starch if desired.
7. Add cream if required.
8. Adjust seasonings.
9. Evaluate the finished soup (see above).

FIGURE 9.5 Preparing a purée soup.

(a) Sweat onions, mirepoix, or other fresh vegetables in fat.

(b) Add stock or other liquid.

(c) Add starchy vegetables or other remaining vegetables.

(d) Purée the soup with an immersion blender, a food processor, or food mill.

Purée of Carrot Soup (Potage Crècy)

YIELD: 3 QT (3 L)   PORTIONS: 12   PORTION SIZE: 8 FL OZ (250 ML)

U.S.	METRIC	INGREDIENTS
2 oz	60 g	Butter
2 lb	1 kg	Carrots, small dice
8 oz	250 g	Onions, small dice
2¹/₂ qt	2.5 L	Chicken stock or white veal stock
8 oz	250 g	Potatoes, small dice
to taste	to taste	Salt
to taste	to taste	White pepper

PROCEDURE

1. Heat the butter in a heavy saucepot over moderately low heat.
2. Add the carrots and onions, and sweat the vegetables until they are about half cooked. Do not let them brown.
3. Add the stock and potatoes. Bring to a boil.
4. Simmer until the vegetables are tender.
5. Purée the soup by passing it through a food mill or by using an immersion blender.
6. Bring the soup back to a simmer. If necessary, add more stock to thin the soup to the proper consistency.
7. Season to taste.

		Optional:
6–8 fl oz	180–250 mL	Cream, hot

8. If desired, finish the soup with hot cream at service time.

Per serving: Calories, 90; Protein, 2 g; Fat, 4.5 g (40% cal.); Cholesterol, 15 mg; Carbohydrates, 13 g; Fiber, 3 g; Sodium, 95 mg.

VARIATIONS

Rice may be used in place of potatoes as the binding agent in the above recipe or in any variation below except purée of potato, purée of potato and leek, and purée of watercress. Use 4 oz (125 g) raw rice in place of 8 oz (250 g) potatoes. The soup must be simmered until the rice is very soft.

Purée of Cauliflower Soup (Purée Dubarry)

Use 2 lb (1 kg) cauliflower in place of carrots.

Purée of Celery or Celery Root Soup

Use 2 lb (1 kg) celery or celery root in place of carrots.

Purée of Jerusalem Artichoke Soup

Use 2 lb (1 kg) Jerusalem artichoke in place of carrots.

Purée of Potato Soup (Potage Parmentier)

Omit carrots from basic recipe, add 5 oz (150 g) leeks to the onions, and increase the potatoes to 2¹/₂ lb (1.25 kg).

Purée of Potato and Leek Soup

Use 1 lb (500 g) leeks in place of the carrots. Increase the potatoes to 1 lb 4 oz (600 g).

Purée of Turnip Soup

Use 2 lb (1 kg) white turnips in place of carrots.

Purée of Watercress Soup

Prepare like purée of potato soup, but add 2¹/₂ bunches watercress, chopped, when the potatoes are almost tender.

Purée of Mixed Vegetable Soup

Decrease carrots to 10 oz (300 g). Add 5 oz (150 g) each celery, turnips, leeks, and cabbage.

Potage Solferino

Combine equal parts purée of potato and leek soup and cream of tomato soup.

Purée of Split Pea Soup

YIELD: 3 QT (3 L)   **PORTIONS: 12**   **PORTION SIZE: 8 FL OZ (250 ML)**

U.S.	METRIC	INGREDIENTS	PROCEDURE
3 oz	90 g	Salt pork	1. Cut the salt pork into fine dice or pass through a grinder.
		Mirepoix:	
5 oz	150 g	Onion, small dice	2. Cook the salt pork slowly in a heavy saucepot to render the fat. Do not brown the pork.
2¹/₂ oz	75 g	Celery, small dice	
2¹/₂ oz	75 g	Carrot, small dice	3. Add the mirepoix and sweat in the fat until the vegetables are slightly softened.
3 qt	3 L	Ham stock (see Note)	4. Add the ham stock and ham bone. Bring to a boil.
1	1	Ham bone or ham hock (optional)	5. Rinse the split peas under cold water. Drain in a strainer and add to the stock. Add the sachet.
1 lb 8 oz	750 g	Green split peas	
		Sachet:	6. Cover and simmer until the peas are tender, about 1 hour.
1	1	Bay leaf	7. Remove the ham bone and sachet.
2	2	Whole cloves	8. Pass the soup through a food mill.
6	6	Peppercorns	9. Bring the soup back to a simmer. If it is too thick, bring it to proper consistency with a little stock or water.
to taste	to taste	Salt	
to taste	to taste	Pepper	10. Season to taste.
			11. If a ham hock was used, trim off the meat. Dice it, and add to the soup.

Per serving: Calories, 230; Protein, 13 g; Fat, 7 g (26% cal.); Cholesterol, 5 mg; Carbohydrates, 32 g; Fiber, 12 g; Sodium, 590 mg.

Note: Water may be used if ham stock is not available. In this case, the optional ham bone or ham hock should be used to provide flavor. Simmer the water and bone together for 1 hour or more before making the soup to extract more flavor.

VARIATIONS

For a coarser, more rustic texture, do not pass the soup through a food mill but serve as is. In this case, name soup simply Split Pea Soup rather than Purée of Split Pea Soup. Other bean, pea, and lentil soups may also be served without puréeing.

Other dried vegetables are made into soups using the same procedure. Most dried beans should be soaked in cold water overnight to reduce cooking time. (Split peas may be soaked, but they cook quickly enough without soaking.)

Purée of White Bean Soup

Use 1 lb 8 oz (750 g) navy beans. Soak the beans overnight. Use chicken or veal stock in place of ham stock.

Purée of Yellow Split Pea Soup

Use yellow split peas instead of green.

Purée of Lentil Soup

Use 1 lb 8 oz (750 g) brown lentils. Soak overnight. Use either ham stock or white stock. Garnish with diced cooked bacon or ham or sliced frankfurters.

Purée of Kidney Bean Soup

Use 1 lb 8 oz (750 g) red kidney beans. Soak beans overnight. Use white stock and add 10 fl oz (300 mL) red wine to the soup when the beans are almost tender. Garnish with croutons sautéed in butter.

Purée of Black Bean Soup

Use 1 lb 8 oz (750 g) black turtle beans. Soak beans overnight. Use white stock and ham bone. Add 4 fl oz (125 mL) Madeira or sherry to the finished soup. Garnish with lemon slices and chopped hard-cooked egg.

Purée Mongole

Combine 1¹/₂ qt (1.5 L) purée of green split pea soup and 1 qt (1 L) tomato purée. Dilute to proper consistency with about 1–2 pt (0.5–1 L) white stock. Garnish with cooked peas and cooked julienne of carrots and leeks.

Nonpuréed Bean Soups

Prepare any of the above soups as directed, but purée only about one-fourth of the beans. Add this purée to the soup as a thickening agent.

Purée of Green Pea Soup with Mint Cream

YIELD: 3 QT (3 L)   PORTIONS: 12   PORTION SIZE: 8 FL OZ (250 ML)

U.S.	METRIC	INGREDIENTS	PROCEDURE
1 oz	30 g	Butter	1. Heat the butter in a heavy saucepot over moderately low heat.
3 oz	190 g	Onions, small dice	2. Add the onions and sweat them without letting them brown.
2 qt	4 L	Chicken stock	3. Add the stock and bring to a boil.
3 lb	1.5 kg	Peas, fresh or frozen	4. Add the peas. Simmer until the peas are soft, about 5 minutes.
to taste	to taste	Salt	5. Pass the soup through a food mill or purée with an immersion blender and then pass through mill to remove coarse pea husks.
to taste	to taste	White pepper	6. Return the stock to a simmer. Add additional stock if necessary to bring to a proper consistency.
			7. Adjust the seasoning.
4 fl oz	125 mL	Heavy cream	8. At service time, heat the heavy cream and stir into the soup.
		Garnish:	9. Whip the cream until it forms soft peaks. Fold in the shredded mint leaves.
6 fl oz	180 mL	Heavy cream	10. Immediately before serving, place 1 tbsp (15 mL) mint cream on top of each portion. (Within a few minutes, the whipped cream will melt into a mint-flecked foam.)
2 tbsp	30 mL	Mint leaves, finely shredded	

Per serving: Calories, 200; Protein, 7 g; Fat, 11 g (50% cal.); Cholesterol, 40 mg; Carbohydrates, 18 g; Fiber, 6 g; Sodium, 40 mg.

Purée of Green Pea Soup with Mint Cream

Purée of Spring Vegetable Soup

YIELD: 3 QT (3 L)   PORTIONS: 12   PORTION SIZE: 8 FL OZ (250 ML)

U.S.	METRIC	INGREDIENTS	PROCEDURE
1 oz	30 g	Butter	1. Heat the butter in a heavy soup pot over low heat.
1 lb	500 g	Leeks, white part only, small dice	2. Add the leeks, potatoes, and celery. Sweat over low heat until fairly tender, about 10–15 minutes.
1 lb	500 g	New potatoes, peeled, small dice	3. Add the peas, parsley, and tarragon. Sweat another 5 minutes.
8 oz	250 g	Celery, small dice	4. Add the stock. Simmer 10 minutes or more, until the vegetables are tender but not overcooked.
8 oz	250 g	Peas, fresh or frozen	5. Purée the soup with a food mill. If you use an immersion blender, pass the soup through a food mill or sieve after puréeing it in order to remove celery fibers.
2 tbsp	30 mL	Chopped parsley	
1 1/2 tsp	7 mL	Chopped fresh tarragon	
3 qt	3 L	Light vegetable stock or chicken stock	
4 fl oz	125 mL	Heavy cream (see Note)	6. Heat the cream and stir it into soup.
to taste	to taste	Salt	7. Adjust the seasoning with salt and white pepper.
to taste	to taste	White pepper	
		Garnish:	8. Whip the cream until it forms soft peaks. Fold in the chives.
4 fl oz	125 mL	Heavy cream (see Note)	9. At service time, garnish each portion of soup with a spoonful of chive cream. (Within a few minutes, the whipped cream will melt into a chive-flecked foam.)
1 tbsp	15 mL	Chopped chives	

Per serving: Calories, 160; Protein, 4 g; Fat, 10 g (53% cal.); Cholesterol, 35 mg; Carbohydrates, 16 g; Fiber, 3 g; Sodium, 65 mg.

Note: The first quantity of cream may be decreased as desired and the whipped cream garnish omitted to reduce dietary fat.

Butternut Squash Soup with Caramelized Apples

YIELD: 3 QT (3 L)   PORTIONS: 12   PORTION SIZE: 8 FL OZ (250 ML)

U.S.	METRIC	INGREDIENTS	PROCEDURE
4 oz	125 g	French bread	1. Cut the bread into slices $1/2$ in. (1 cm) thick.
1 oz	30 g	Butter	2. Fry bread in the butter until golden brown. (If desired, prepare additional croutons for garnish at the same time; see step 8.)
1 oz	30 g	Butter	3. Heat the butter in a heavy saucepot over moderately low heat.
4 oz	125 g	Onions, small dice	4. Add the onions, leeks, and carrots. Sweat them until about half cooked. Do not let them brown.
4 oz	125 g	Leeks, small dice	5. Add the squash, the stock, and the browned bread from step 2. Simmer until the vegetables are tender.
6 oz	180 g	Carrots, small dice	6. Purée the soup with a food mill or an immersion blender.
2 lb	1 kg	Butternut squash, medium dice	7. Bring the soup back to a simmer. Taste and adjust seasonings.
$2^1/2$ qt	2.5 L	Chicken stock	
$3/4$ tsp	3 mL	Salt	
$1/4$ tsp	1 mL	White pepper	
$1/4$ tsp	1 mL	Ground allspice	
$1/4$ tsp	1 mL	Ground ginger	
12	12	Croutons (see procedure)	8. Prepare croutons by browning slices of French bread in butter as in steps 1 and 2. For best appearance, use a slender loaf so croutons aren't too big.
12 oz	375 g	Tart, firm cooking apples	9. Peel and core apples. Cut into small dice.
$1/2$ oz	15 g	Butter	10. Heat the butter in a sauté pan and add the apples and sugar. Cook over moderate heat until apples are brown and caramelized.
1 oz	30 g	Brown sugar	
6 fl oz	180 mL	Heavy cream (optional)	11. At service, heat the heavy cream (if using) and add to the soup.
as needed	as needed	Additional heavy cream or crème fraîche for garnish, if desired	12. For each portion, ladle soup into a broad soup plate. Decorate the top of the soup with a swirl of cream, if desired. Heap a generous tablespoon (15 mL) of apple onto a crouton and carefully place in the soup.

Per serving: Calories, 150; Protein, 3 g; Fat, 6 g (34% cal.); Cholesterol, 20 mg; Carbohydrates, 23 g; Fiber, 4 g; Sodium, 280 mg.

Navy Bean Soup

YIELD: 3 QT (3 L)   PORTIONS: 12   PORTION SIZE: 8 FL OZ (250 ML)

U.S.	METRIC	INGREDIENTS	PROCEDURE
1 lb	500 g	Dried navy beans	1. Soak the beans overnight in cold water.
2 oz	60 g	Bacon, diced	2. Place the bacon in a heavy saucepot over medium heat. Render the fat from the bacon, but do not cook until crisp.
2 oz	60 g	Onions, small dice	3. Add the vegetables and cook over low heat until almost tender.
4 oz	125 g	Carrots, small dice	
4 oz	125 g	Celery, small dice	
2 oz	1260 g	Leeks, sliced	
2	2	Garlic cloves, chopped	
$2^1/2$ qt	2.5 L	Stock or water	4. Add the stock or water and the ham hock. Bring to a boil.
1	1	Ham hock	5. Drain the beans and add to the liquid. Also add the bay leaf, thyme, and pepper.
1	1	Bay leaf	6. Cover and simmer until the beans are tender.
$1/4$ tsp	1 mL	Dried thyme	
$1/4$ tsp	1 mL	Pepper	
1 pt	500 mL	Canned tomatoes, with juice, crushed	7. Add the tomatoes to the soup and simmer another 15 minutes. Remove the bay leaf.
			8. Remove the ham hock from the soup. Cut off and dice the meat and add it to the soup. Discard the bones.
			9. Mash the beans lightly with a paddle or pass about one-fourth of them through a food mill. Return this purée to the soup to thicken it.
			10. If the soup is too thick, thin with a little stock.
			11. Adjust the seasoning.

Per serving: Calories, 180; Protein, 10 g; Fat, 4 g (19% cal.); Cholesterol, 5 mg; Carbohydrates, 28 g; Fiber, 7 g; Sodium, 510 mg.

Chilled Leek and Potato Soup with Shrimp and Fennel Salad

YIELD: 3 QT (3 L), PLUS GARNISH   PORTIONS: 12   PORTION SIZE: 8 FL OZ (250 ML), PLUS GARNISH

U.S.	METRIC	INGREDIENTS	PROCEDURE
1¹/₂ lb	750 g	Leeks, white part only	1. Cut the leeks and potatoes into thin slices.
1¹/₂ lb	750 g	Potatoes, peeled	2. Sweat the leeks in olive oil without letting them brown.
2 fl oz	60 mL	Olive oil	3. Add the stock and potatoes and bring to a boil. Simmer until the vegetables are tender.
4¹/₂ pt	2.25 L	Chicken stock	4. Pass the soup through a food mill.
to taste	to taste	Salt	5. If the soup is too thick, add a little stock to bring it to proper consistency.
to taste	to taste	White pepper	6. Chill the soup thoroughly.
			7. Add salt and white pepper to taste.
12 oz	375 g	Tender fennel bulbs, trimmed (see Note)	8. While the soup is chilling, prepare the garnish. Cut the fennel in half lengthwise through the base, then cut lengthwise into paper-thin slices (a slicing machine may be used). Cut these slices lengthwise into rough julienne.
4 tsp	20 mL	Lemon juice	9. Toss the fennel with the lemon juice, then with half the olive oil. Add salt and pepper to taste.
1 fl oz	30 mL	Olive oil	10. Toss the shrimp with the remaining olive oil.
to taste	to taste	Salt	
to taste	to taste	Pepper	
36	36	Small cooked shrimp (see Note)	
8 fl oz	250 mL	Heavy cream	11. Stir the heavy cream and buttermilk into the chilled soup.
8 fl oz	250 mL	Buttermilk	12. Arrange the fennel salad in the center of broad soup plates. Arrange 3 shrimp on top of each mound of fennel.
			13. Place the soup plate in front of the diner, then ladle in a portion of soup from a tureen.

Per serving: Calories, 240; Protein, 8 g; Fat, 15 g (55% cal.); Cholesterol, 55 mg; Carbohydrates, 22 g; Fiber, 3 g; Sodium, 90 mg.

Note: The fennel must be young and tender, as it is eaten raw. If only large bulbs are available, use the interior part and save the outer layers for another purpose.

    Grilled shrimp are especially good for this soup.

VARIATIONS

The fennel and shrimp garnish may be omitted for a simpler presentation. Water may be substituted for stock. Leeks are flavorful enough to make a delicious soup even without stock.

Vichyssoise

Substitute butter for olive oil. Omit buttermilk and increase heavy cream to 16–20 fl oz (500–600 mL). Omit fennel and shrimp salad garnish. Garnish with chopped chives.

Chilled Leek and Potato Soup with Shrimp and Fennel Salad

BISQUES

A bisque (bisk) is a cream soup made with shellfish. At one time, bisques were thickened with rice, but today they are more frequently thickened with roux. Bisques are made basically like other cream soups, but they seem more complex because of the handling of the shellfish and the variety of flavoring ingredients often used. Expensive to prepare and rich in taste, they are considered luxury soups.

Because of the cost of shellfish bisques, it is especially important to check the quality of stocks and all other ingredients before proceeding. In particular, if stock has been in storage, bring it to a simmer and check for off odors or flavors. This helps to avoid spoiling a quantity of expensive seafood because a poor-quality ingredient ruins the entire batch.

The term *bisque* has come to be used for a great variety of soups, primarily because the word sounds nice. In this book, we reserve the term for shellfish cream soups. Nevertheless, you will also see the word *bisque* applied to many of the vegetable purée soups and cream soups discussed in earlier sections.

BISQUE

Linguists say the most likely origin of the word *bisque* is Biscay, the name of the bay off the coast of southwestern France and northwestern Spain.

It is sometimes said the word comes from *biscuit*, because the soup was once thickened by dried bread, but language experts say there is no evidence for this origin.

STANDARDS OF QUALITY FOR BISQUES

1. **Appearance and Texture.**
   Attractive creamy reddish pink color.
   Creamy texture with a slight graininess from the crustacean shells. Not too thick—about the thickness of heavy cream.

2. **Flavor and Aroma.**
   Full flavor of shrimp, lobster, or whatever shellfish is used. Wine, spices, and tomato paste should enhance the shellfish flavor but not be so strong as to overwhelm the flavor or give harshness. Rich and creamy, but not so creamy as to be cloying.

FIGURE 9.6 Bisque procedure.

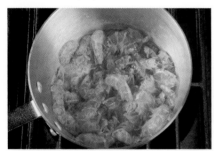

(a) Cook the shells in butter.

(b) Add fish velouté.

(c) Blend with an immersion blender.

(d) Strain through cheesecloth.

Shrimp Bisque

YIELD: 2 QT (2 L)   PORTIONS: 10   PORTION SIZE: 6 FL OZ (200 ML)

U.S.	METRIC	INGREDIENTS	PROCEDURE
1 oz	30 g	Butter	1. Heat the butter in a saucepan over medium heat.
12 oz	360 g	Shrimp shells, rinsed and drained	2. Add the shrimp shells. Cook until the shells have turned pink or red (**Figure 9.6a**). Remove and set aside.
2 oz	60 g	Onions, cut brunoise	3. Add onions and carrots to the pot. Sauté until just lightly browned.
2 oz	60 g	Carrots, cut brunoise	
small piece	small piece	Bay leaf	4. Add the bay leaf, thyme, parsley stems, and tomato paste. Cook for 2 minutes, until the tomato paste develops an aroma. Stir well.
pinch	pinch	Dried thyme	
4	4	Parsley stems	5. Return the shells to the pan.
1 oz	30 g	Tomato paste (see variations)	6. Add the brandy and wine. Simmer until reduced by half.
2 oz	60 mL	Burnt brandy (see Note)	
6 oz	200 mL	White wine	
1 qt	1 L	Fish velouté	7. Add the fish velouté (b) and stock to the saucepan. Simmer 45 minutes.
1 pt	500 mL	Fish stock	8. Blend the mixture using an immersion blender (c). This step is optional but improves the body and flavor of the bisque.
			9. Strain through a fine mesh strainer lined with cheesecloth. After most of the liquid has passed through the cheesecloth, force out all remaining liquid as shown in the illustration (d): each worker holds two corners of the cheesecloth and twists tightly in opposite directions. (If no help is available, extract liquid by pressing down well on the solids in the strainer.) Return the soup to the saucepan and bring back to a simmer.
5–10 oz	150–300 g	Raw shrimp, peeled and deveined	10. Cut the shrimp into small dice.
½ oz	15 g	Butter	11. Heat the last quantity of butter in a sauté pan. Add the shrimp and sauté just until cooked through, about 1 or 2 minutes. Add to the soup and simmer 2 minutes.
1 pt	500 mL	Heavy cream, hot	
1½ fl oz	45 mL	Dry sherry	12. Add the hot cream and the sherry. Season to taste.
to taste	to taste	Salt	
to taste	to taste	White pepper	

Per serving: Calories, 220; Protein, 8 g; Fat 17 g (70% cal.); Cholesterol, 110 mg; Carbohydrates, 6 g; Fiber, 0 g; Sodium, 180 mg.

Note: Burnt brandy is brandy that has been heated in a saucepan and flamed (carefully) to burn off the alcohol.

VARIATIONS

This recipe is based on Method 1 for making cream soups (p. 240) in that it uses velouté as a base. You can also use fish stock instead of velouté and thicken the soup in other ways:

1. Beat in beurre manié (p. 174), a little at a time, after step 8, until properly thickened.

2. Stir in a cornstarch slurry (cornstarch in cold water), a little at a time.

3. Simmer 2 oz (60 g) rice in 1 pt (500 mL) of the stock until the rice is completely cooked. Liquefy in a blender or force through a fine sieve to purée the rice, and add to the soup. (This is the classical method.)

Paprika is often used instead of tomato paste to color and flavor bisques. Substitute 1 tbsp (15 mL) Spanish paprika for the 1 oz (30 g) tomato paste.

Lobster Bisque

In place of shrimp, use raw lobster shells and lobster meat. Chop the shells into small pieces before cooking.

Shrimp Bisque

CHOWDERS

Chowders are chunky, hearty soups so full of good things they sometimes are more like stews than soups. Many types of chowder are simply cream soups or purée soups that are not puréed but left chunky. Like other specialty regional soups, chowders resist categorization. However, most of them are based on fish or shellfish or vegetables, and most contain potatoes and milk or cream.

KEY POINTS TO REVIEW

- What are the three basic procedures for making cream soups?

- What is a purée soup? Describe the basic procedure for making a purée soup.

- What is a bisque? Describe the basic procedure for making a shrimp bisque.

- What is a chowder?

Potato Chowder

YIELD: 3 QT (3 L)   **PORTIONS: 12**   **PORTION SIZE: 8 FL OZ (250 ML)**

U.S.	METRIC	INGREDIENTS	PROCEDURE
4 oz	125 g	Salt pork	1. Grind the salt pork or cut into very fine dice.
6 oz	180 g	Onions, medium dice	2. Render the pork fat in a heavy saucepot.
1½ oz	45 g	Celery, medium dice	3. Add the onions and celery. Cook in the fat over moderate heat until nearly tender. Do not brown.
2 oz	160 g	Flour	4. Add the flour. Stir into the fat to make a roux. Cook the roux slowly 4–5 minutes, but do not let it brown.
3½ pt	1.75 L	Chicken stock	5. Using a wire whip, slowly stir in the stock. Bring to a boil, stirring to make sure the liquid is smooth.
1 lb 8 oz	750 g	Potatoes, medium dice	6. Add the potatoes. Simmer until all the vegetables are tender.
1½ pt	750 mL	Milk, hot	7. Stir in the hot milk and cream.
4 fl oz	125 mL	Heavy cream, hot	8. Season to taste with salt and white pepper.
to taste	to taste	Salt	
to taste	to taste	White pepper	
as needed	as needed	Chopped parsley	9. Sprinkle each portion with a little chopped parsley for garnish.

Per serving: Calories, 210; Protein, 5g; Fat 14g (59% cal.); Cholesterol, 35 mg; Carbohydrates, 18g; Fiber, 1 g; Sodium, 180 mg.

VARIATIONS

Corn Chowder

Version 1. Prepare as in basic recipe, but reduce potatoes to 1 lb 2 oz (550 g). When vegetables are tender, add 1½ lb (750 g) frozen or drained canned whole-kernel corn. (If using canned corn, replace part of the chicken stock with corn liquid.)

Version 2. Prepare as in basic recipe, but reduce potatoes to 12 oz (375 g). Add 1½ lb (750 g) canned cream-style corn when vegetables are tender.

New England Clam Chowder

YIELD: 3 QT (3 L)   **PORTIONS:12**   **PORTIONS SIZE: 8 FL OZ (250 ML)**

U.S.	METRIC	INGREDIENTS	PROCEDURE
1 qt	1 L	Canned, minced clams, with their juice, or fresh shucked clams, with their juice (see Note)	1. Drain the clams, reserving the juice. If you are using fresh clams, chop them, being sure to save all the juice.
1¹/₂ pt	750mL	Water	2. Combine the juice and water in a saucepan. Bring to a boil.
			3. Remove from the heat and keep the liquid hot for step 7.
5 oz	150 g	Salt pork, ground or cut into fine dice	4. In a heavy saucepot or stockpot, render the salt pork over medium heat.
8 oz	250 g	Onions, small dice	5. Add the onions and cook slowly until soft, but do not brown.
2 oz	60 g	Flour	6. Add the flour and stir to make a roux. Cook the roux slowly 3–4 minutes, but do not let it brown.
1 lb	500 g	Potatoes, small dice	7. Using a wire whip, slowly stir the clam liquid and water into the roux.
			Bring to a boil, stirring constantly to make sure the liquid is smooth.
			8. Add the potatoes. Simmer until tender. (If you are using large, tough chowder clams, pass them once through a grinder and add with the potatoes.)
2¹/₂ pt	1.25 L	Milk, hot	9. Stir in the clams and hot milk and cream. Heat gently, but do not boil.
4 fl oz	125 mL	Heavy cream, hot	10. Season to taste with salt and white pepper.
to taste	to taste	Salt	
to taste	to taste	White pepper	

Per serving: Calories, 300; Protein, 16 g; Fat, 17 g (52% cal.); Cholesterol, 65 mg; Carbohydrates, 19 g; Fiber, 1 g; Sodium, 350 mg.

Note: If whole clams in the shell are used, you will need about 4–5 qt (4–5 L). Scrub them well. Combine with the 1¹/₂ pt (750mL) water in a stockpot and simmer until the shells open. Remove the clams from the shells and chop. Strain the liquid.

VARIATIONS

Manhattan Clam Chowder

Substitute 2 fl oz (60 mL) oil or 2 oz (60 g) butter for the salt pork. Add 5 oz (150 g) celery, small dice; 5 oz (150 g) carrots, small dice; and ¹/₂ tsp (2 mL) chopped garlic to the onions in step 5. Omit flour. Instead of milk, use 2¹/₂ pt (1.25L) chopped canned tomatoes and their juices. Omit cream.

New England Fish Chowder

Follow the procedure for New England Clam Chowder, but omit clams and water. Use 1¹/₂ qt (1.5 L) fish stock instead of the clam juice and water mixture in step 7. Remove all skin and bones from 10 oz (300 g) haddock fillets. Cut into ³/₄-in. (2-cm) chunks. Add to the finished soup and keep hot (do not boil) until the fish is cooked, about 5 minutes.

SPECIALTY SOUPS AND NATIONAL SOUPS

French Onion Soup Gratinée

YIELD: 3³/₄ QT (3.75L)   **PORTIONS: 12**   **PORTION SIZE: 10 FL OZ (300 ML)**

U.S.	METRIC	INGREDIENTS	PROCEDURE
2 oz	60 g	Butter	1. Heat the butter in a stockpot over moderate heat. Add the onions and cook until golden. Stir occasionally. *Note:* The onions must cook slowly and become evenly browned. This is a slow process and will take about 30 minutes. Do not brown too fast or use high heat.
2 lb 8 oz	1.25 kg	Onions, sliced thin	
3¹/₄ qt	3.25 L	Beef stock, or half beef and half chicken stock	2. Add the stock and bring to a boil. Simmer until the onions are very tender and the flavors are well blended, about 20 minutes.
to taste	to taste	Salt	3. Season to taste with salt and pepper. Add the sherry, if desired.
to taste	to taste	Pepper	
2–3 fl oz	60–90 mL	Sherry (optional)	4. Keep the soup hot for service.
as needed	as needed	French bread (see procedure)	5. Cut the bread into slices about ³/₈ in. (1 cm) thick. You need 1 or 2 slices per portion, or just enough to cover the top of the soup in its serving crock.
12 oz	375 g	Gruyère or Swiss cheese, or a mixture, coarsely grated	6. Toast bread slices in the oven or under the broiler.
			7. For each portion, fill an individual-service soup crock with hot soup. Place 1 or 2 slices of the toast on top and cover with cheese. Pass under the broiler until the cheese is bubbling and lightly browned. Serve immediately.

Per serving: Calories, 320; Protein, 15 g;
Fat, 15 g (42% cal.); Cholesterol, 50 mg; Carbohydrates, 31 g;
Fiber, 3 g; Sodium, 410 mg.

VARIATIONS

Onion soup may be served without gratinéeing and with cheese croutons prepared separately. Toast the bread as in basic recipe. Place on a sheet pan. Brush lightly with butter and sprinkle each piece with grated cheese. (Parmesan may be mixed with the other cheese.) Brown under the broiler. Garnish each portion with 1 cheese crouton. (This method is less expensive because it uses much less cheese.)

French Onion Soup Gratinée

Gulyas

YIELD: 3 QT (3 L)   PORTIONS: 12   PORTION SIZE: 8 FL OZ (250 ML)

U.S.	METRIC	INGREDIENTS
2 oz	60 g	Pork fatback or bacon
8 oz	250 g	Onions, medium dice
1 lb 8 oz	750 g	Beef chuck or shank
2	2	Garlic cloves, chopped fine
1/2 tsp	2 mL	Caraway seeds
2 tbsp	12 g	Hungarian paprika
1 tbsp	15 mL	Tomato paste
2 1/2 qt	2.5 L	Water or brown stock, hot
1 lb	500 g	Potatoes, peeled, medium dice
to taste	to taste	Salt
to taste	to taste	Pepper

PROCEDURE

1. Grind the fatback or bacon or cut into fine dice. Render the fat in a heavy pot. After the fat has rendered, remove the solids with a slotted spoon and discard or save for another use.
2. Add the onions to the fat in the pot and sweat. Do not brown.
3. Cut the beef into medium dice. Add to the onions and cook over low heat 10 minutes.
4. Add the garlic, caraway seeds, paprika, and tomato paste. Stir in well.
5. Add the liquid. Simmer until the beef is almost tender. This may take 1 hour or more, depending on the type and quality of the meat.
6. Add the potatoes and simmer until done.
7. Season to taste with salt and pepper.

Per serving: Calories, 210; Protein, 12 g; Fat, 14 g (60% cal.); Cholesterol, 40 mg; Carbohydrates, 9 g; Fiber, 1 g; Sodium, 45 mg.

Gulyas

Avgolemono

YIELD: 3 QT (3 L)   PORTIONS: 12   PORTION SIZE: 8 FL OZ (250 ML)

U.S.	METRIC	INGREDIENTS
3 qt	3 L	Chicken stock
4 oz	125 g	Rice, raw
		Liaison:
4	4	Eggs, beaten
3 fl oz	90 mL	Lemon juice
to taste	to taste	Salt
to taste	to taste	White pepper
to taste	to taste	Chopped parsley

PROCEDURE

1. Bring the stock to a boil in a large sauce pot or in a stockpot.
2. Place the rice in a strainer and rinse under cold water. Add to the stock. Simmer until the rice is cooked.
3. Remove from the heat.
4. Just before serving, beat the eggs with the lemon juice in a stainless-steel bowl.
5. Gradually beat in about 8 fl oz (250 mL) of the hot soup to temper the liaison. Stir the mixture back into the soup.
6. Return the soup to heat and warm it to below the simmering point. Do not boil, or the soup will curdle.
7. Season to taste with salt and pepper.
8. At service time, top each portion with a little chopped parsley.

Per serving: Calories, 10; Protein, 1 g; Fat, 0 g (0% cal.); Cholesterol, 10 mg; Carbohydrates, 1 g; Fiber, 0 g; Sodium, 5 mg.

Southwestern Corn and Tomato Soup

YIELD: 2 QT (2 L)   PORTIONS: 8   PORTION SIZE: 8 FL OZ (250 ML)

U.S.	METRIC	INGREDIENTS	PROCEDURE
1 fl oz	30 mL	Oil	1. Heat the oil in a soup pot and sauté the onion, green pepper, garlic, coriander, and cayenne until the onion is soft.
8 oz	250 g	Onion, small dice	
4 oz	125 g	Green pepper, small dice	
2 tsp	10 mL	Garlic, chopped	
1 tsp	5 mL	Ground coriander	
1/4 tsp	1 mL	Cayenne	
2 lb	1 kg	Tomatoes, canned or fresh, with their juice, chopped	2. Add the tomatoes and stock. Bring to a boil and simmer about 20 minutes.
1 qt	1 L	Chicken stock	3. Add the corn and simmer another few minutes.
1 lb	500 g	Corn kernels, fresh or frozen	4. Season to taste with salt and pepper.
to taste	to taste	Salt	
to taste	to taste	Pepper	
		Garnish:	5. Immediately before serving, top each corn chip with a spoonful of sour cream. Place one chip in each bowl and top with a sprinkling of cheese.
8	8	Corn chips	
4 fl oz	125 mL	Sour cream	
2 oz	60 g	Grated cheddar cheese	

Per serving: Calories, 190; Protein, 6 g; Fat, 10 g (43% cal.); Cholesterol, 20 mg; Carbohydrates, 24 g; Fiber, 5 g; Sodium, 270 mg.

VARIATIONS

Corn, Zucchini, and Tomato Soup

Add 8 oz (250 g) zucchini, in small dice, along with the corn in step 3.

Vegetarian Corn Chili Soup

Add 2 tbsp chili powder along with the coriander and cayenne. In place of chicken stock, use vegetable stock. For vegan diets, omit sour cream and cheese garnish.

Southwestern Corn and Tomato Soup

Seafood Gumbo

YIELD: 2¹/₂ QT (2.5 L)   PORTIONS: 8   PORTION SIZE: 10 FL OZ (300 ML) PLUS 1¹/₂ OZ (45 G) RICE

U.S.	METRIC	INGREDIENTS	PROCEDURE
3 tbsp	45 g	Vegetable oil	1. Heat the oil in a stockpot or heavy soup pot. Add the onion, celery, green pepper, garlic, herbs, and spices. Cook over low heat until the vegetables are soft.
3 oz	90 g	Onion, small dice	
2 oz	60 g	Celery, small dice	
2 oz	60 g	Green bell pepper, small dice	
¹/₂ oz	15 g	Garlic, chopped fine	
³/₄ tsp	3 mL	Dried thyme	
³/₄ tsp	3 mL	Dried basil	
¹/₂ tsp	2 mL	Dried oregano	
1	1	Bay leaf	
¹/₂ tsp	2 mL	Cayenne	
¹/₂ tsp	2 mL	Black pepper	
1 tsp	5 mL	Paprika	
2 qt	2 L	Fish stock, hot	2. Add the stock. Bring the mixture to a boil.
4 oz	125 g	Blond roux (p. 172)	3. Beat in the roux to thicken the soup.
5 oz	150 g	Canned tomatoes, chopped	4. Add the tomatoes. Simmer 10 minutes.
8 oz	250 g	Okra, trimmed, cut into ³/₄-in. (2-cm) slices	5. Add the okra. Simmer until the okra is tender, about 10–15 minutes more.
8 oz	250 g	Small shrimp, peeled and deveined	6. Add the shrimp and crabmeat. Simmer 10 minutes.
8 oz	250 g	Lump crabmeat	7. Sprinkle the filé powder over the soup and stir it in. Simmer another 2 minutes.
1¹/₂ tsp	7 mL	Filé powder	8. Remove from heat and season to taste with salt.
to taste	to taste	Salt	
12 oz	375 g	Cooked rice	9. To serve, spoon about 1¹/₂ oz (45 g) cooked rice into a soup bowl. Ladle 10 fl oz (300 mL) soup over the rice.

Per serving: Calories, 290; Protein, 18 g; Fat, 14 g (44% cal.); Cholesterol, 80 mg; Carbohydrates, 22 g; Fiber, 2 g; Sodium, 580 mg.

VARIATIONS

Substitute diced white fish fillets (such as grouper, snapper, or monkfish) for part of the crabmeat. Add 6 oz (180 g) shucked oysters.

Chicken and Andouille Gumbo

Substitute chicken stock for fish stock. Omit the shrimp and crabmeat. Instead, use 12 oz (375 g) diced raw chicken meat and 4 oz (125 g) sliced andouille sausage. Brown the chicken in oil, and add the chicken and sausage in step 6. Simmer until chicken is completely cooked.

Seafood Gumbo

Chinese Hot and Sour Soup

YIELD: 3 QT (3 L)   **PORTIONS: 12**   **PORTION SIZE: 8 FL OZ (250 ML)**

U.S.	METRIC	INGREDIENTS	PROCEDURE
8	8	Dried black mushrooms	1. Place the two types of dried mushroom in separate bowls. Add enough boiling water to each to cover mushrooms. Let stand until soft.
16	16	Dried cloud ear (tree ear) mushrooms	
as needed	as needed	Boiling water	2. Drain the mushrooms, squeezing them lightly. Reserve the soaking liquid.
			3. Cut off and discard the stems from the black mushrooms. Cut the caps into thin strips.
			4. Cut off any coarse, woody stems from the cloud ear mushrooms. Cut the mushrooms into $^1/_2$-inch (1-cm) pieces.
2$^1/_2$ qt	2.5 L	Chicken stock	5. Combine the stock and the mushroom liquid in a saucepot. Bring to a boil.
8 oz	250 g	Lean pork, cut julienne (see Note)	6. Add the mushrooms, pork, bamboo shoots, soy sauce, and white pepper.
6 oz	180 g	Bamboo shoots, cut julienne	7. Simmer 3 minutes.
3 fl oz	80 g	Soy sauce	
2 tsp	10 mL	White pepper	
8 tbsp	120 mL	Cornstarch	8. Mix the cornstarch with the cold water until smooth.
8 fl oz	250 mL	Cold water	9. Add the cornstarch mixture to the simmering soup, stirring constantly.
			10. Simmer until the soup is lightly thickened.
3 fl oz	90 mL	Rice vinegar or wine vinegar	11. Add the vinegar and sesame oil to soup.
2 tsp	10 mL	Oriental sesame oil	12. Slowly drizzle in the beaten egg, stirring slowly but constantly so that the egg coagulates in thin shreds.
2	2	Eggs, lightly beaten	
1	1	Scallion, sliced thin	13. At service time, top each portion with a few slices of scallion and a pinch of chopped cilantro.
2 tbsp	30 mL	Cilantro, chopped	

Per serving: Calories, 170; Protein, 11 g; Fat, 5 g (27% cal.); Carbohydrates, 20 g; Cholesterol, 50 mg; Fiber, 5 g; Sodium, 710 mg.

Note: To cut raw pork into julienne more easily, partially freeze it first.

Chinese Hot and Sour Soup

Tortilla Soup

YIELD: 3 QT (3 L)   PORTIONS: 12   PORTION SIZE: 8 FL OZ (250 ML) BROTH PLUS GARNISH

U.S.	METRIC	INGREDIENTS
12 oz	375 g	Tomatoes, whole, ripe
12 oz	375 g	Onion, chopped
1¹/₂ tbsp	22 mL	Garlic, chopped
1 fl oz	30 mL	Vegetable oil
3 qt	3 L	Chicken stock or broth
to taste	to taste	Salt
		Garnishes in the soup:
1 lb	500 g	Fresh, mild cheese, such as Mexican queso fresco or cotija, or Monterey jack, crumbled or grated
12	12	Corn tortillas, cut into strips ¹/₄ in. (6 mm) wide and deep-fried until crisp
as desired	as desired	Garnishes served separately:
		Dried pasilla chiles, fried crisp and crumbled
		Lime wedges
		Avocado, diced
		Sour cream or Mexican crema

PROCEDURE

1. Place the tomatoes on a foil-lined pan and place under a preheated broiler. Broil until softened and the skins are blistered and blackened.
2. Peel and core the tomatoes.
3. Purée the tomatoes, onion, and garlic together in a food processor.
4. Heat the oil in a heavy saucepot over moderately high heat.
5. Add the tomato mixture. Cook, stirring frequently, until the mixture is thick and several shades darker in color.
6. Add the stock or broth, stirring to make a smooth mixture. Simmer 30 minutes.
7. Season to taste with salt.
8. Place about 1¹/₂ oz (40 g) crumbled or grated cheese in the bottom of each soup bowl and top with tortilla strips.
9. Ladle the soup over the tortillas.
10. Serve immediately, with the remaining garnish on the side in small bowls.

Per serving: Calories, 250; Protein, 13 g; Fat, 10 g (37% cal.); Cholesterol, 20 mg; Carbohydrates, 26 g; Fiber, 2 g; Sodium, 440 mg.

VARIATION

For a heartier soup with more protein, add about 2 oz (60 g) cooked, shredded chicken per portion.

Tortilla Soup

Pho Bo (Vietnamese Beef and Rice Noodle Soup)

YIELD: 3 QT (3 L)   PORTIONS: 12   PORTION SIZE: 8 FL OZ (250 ML) BROTH PLUS GARNISH

U.S.	METRIC	INGREDIENTS	PROCEDURE
3 qt	3 L	Beef broth	1. Combine the broth, ginger, lemongrass, shallots, and cinnamon stick in a stockpot. Simmer 30 minutes.
12 slices	12 slices	Fresh ginger root	2. Strain. Discard the solids.
3 stalks	3 stalks	Lemongrass, coarsely chopped	3. Add the fish sauce.
6 oz	180 g	Shallots, sliced thin	4. Taste and add salt if necessary.
1	1	Cinnamon stick	
3 fl oz	90 mL	Nuoc nam (Vietnamese fish sauce) or nam pla (Thai fish sauce)	
to taste	to taste	Salt	
1 lb 8 oz	750 g	Beef tenderloin or other tender beef steak	5. Partially freeze the beef to make it easier to slice.
			6. Slice the beef into paper-thin slices.
1 lb 8 oz	750 g	Rice sticks (see p. 408)	7. Just before service, drop the rice sticks into boiling water. Return the water to a boil and drain immediately.
		Garnishes	8. For service, return the broth to a boil.
as desired	as desired	Shallots, sliced thin	9. Place each garnish in a separate bowl and arrange on the dining table.
as desired	as desired	Mung bean sprouts	10. Place 2 oz (60 g) beef slices in the bottom of each soup bowl.
as desired	as desired	Fresh mint leaves	
as desired	as desired	Fresh cilantro leaves	11. Ladle 8 fl oz (250 mL) broth over the beef slices. The boiling broth cooks the beef in moments.
as desired	as desired	Fresh red chiles, seeded and sliced thin	12. Add rice noodles to the bowl.
as desired	as desired	Lime wedges	13. Allow guests to help themselves to garnishes to flavor soup as desired.
as desired	as desired	Chile paste with garlic	

Per serving: Calories, 380; Protein, 23 g; Fat, 25 g (58% cal.);
Cholesterol, 105 mg; Carbohydrates, 17 g; Fiber, 1 g; Sodium, 800 mg.

Pho Bo

Gazpacho

YIELD: 2¹/₂ QT (2.5 L)   PORTIONS: 12   PORTION SIZE: 6 FL OZ (200 ML)

U.S.	METRIC	INGREDIENTS	PROCEDURE
2¹/₂ lb	1.2 kg	Tomatoes, peeled and chopped fine	1. If a blender is available, combine all ingredients in the blender and process until liquefied.
1 lb	500 g	Cucumbers, peeled and chopped fine	
8 oz	250 g	Onions, peeled and chopped fine	2. If a blender is not available, combine all ingredients except olive oil. Pass through a food mill. If a smoother soup is desired, then pass through a fine sieve. Rub the solids through the sieve to purée them. Place the mixture in a stainless-steel bowl. Using a wire whip, slowly beat in the olive oil.
4 oz	125 g	Green bell peppers, seeded and chopped fine	
¹/₂ tsp	2 mL	Crushed garlic	
2 oz	60 g	Fresh white bread crumbs	
1 pt	500 mL	Water or tomato juice, cold	
3 fl oz	90 mL	Red wine vinegar	
4 fl oz	125 mL	Olive oil	
to taste	to taste	Salt	3. Add salt, pepper, and cayenne or pepper sauce to taste.
to taste	to taste	Pepper	4. If necessary, adjust tartness by adding a little lemon juice or vinegar.
to taste	to taste	Cayenne or hot red pepper sauce	
to taste	to taste	Lemon juice or vinegar	5. Chill the soup thoroughly.
		Garnish:	6. Combine garnish ingredients in a small bowl or bain-marie.
2 oz	60 g	Onion, small dice	7. At service time, ladle 6 oz (200 mL) gazpacho into chilled soup cups. Top with 1–2 tbsp (15–30 g) diced vegetable garnish. If desired, gazpacho may be served with ice cubes.
2 oz	60 g	Cucumber, small dice	
2 oz	60 g	Green pepper, small dice	

Per serving: Calories, 130; Protein, 2 g; Fat, 10 g (63% cal.); Cholesterol, 0 mg; Carbohydrates, 11 g; Fiber, 2 g; Sodium, 36 mg.

Gazpacho

Minestrone

YIELD: 3 QT (3 L) PORTIONS: 12   PORTION SIZE: 8 FL OZ (250 ML)

U.S.	METRIC	INGREDIENTS	PROCEDURE
2 fl oz	60 mL	Olive oil	1. Heat the oil in a heavy pot over medium heat.
8 oz	250 g	Onions, sliced thin	2. Add the onions, celery, carrots, and garlic. Sweat them in the oil until almost tender. Do not brown.
4 oz	125 g	Celery, small dice	
4 oz	125 g	Carrots, small dice	3. Add the cabbage and zucchini. Stir to mix the vegetables. Continue to sweat another 5 minutes.
1 tsp	5 mL	Garlic, chopped	
4 oz	125 g	Green cabbage, shredded (see p. 281 for technique)	
4 oz	125 g	Zucchini, medium dice	
8 oz	250 g	Canned tomatoes, crushed	4. Add the tomatoes, stock, and basil. Bring to a boil, reduce heat, and simmer until the vegetables are almost cooked. (Do not overcook. The soup will continue to cook when the pasta is added.)
2¹/₂ qt	2.5 L	White stock	
¹/₂ tsp	2 mL	Dried basil	
3 oz	90 g	Small macaroni, such as ditalini	5. Add the pasta and continue to simmer the soup until the pasta is cooked. (Alternatively, cook pasta separately and add to the soup just before serving.)
12 oz	375 g	Drained, canned cannellini or other white beans (2 No. 2 cans)	6. Add the beans and return soup to a boil.
2 tbsp	30 mL	Chopped parsley	7. Add the parsley. Season to taste with salt and pepper.
to taste	to taste	Salt	
to taste	to taste	Pepper	
as needed	as needed	Parmesan cheese, grated	8. Just before service, top with the parmesan cheese, or serve cheese separately.

Per serving: Calories, 150; Protein, 7 g, Fat, 7 g (40% cal.);
Cholesterol, 10 mg; Carbohydrates, 17 g; Fiber, 3 g; Sodium, 200 mg.

Minestrone

Scotch Broth 🍽

YIELD: 3 QT (3 L)   PORTIONS: 12   PORTION SIZE: 8 FL OZ (250 ML)

U.S.	METRIC	INGREDIENTS	PROCEDURE
3 qt	3 L	White lamb stock	1. Bring ¹/₂ qt (500 mL) of the stock to a boil in a saucepan.
2 oz	60 g	Barley	2. Add the barley and cover the pan. Simmer until tender.
12 oz	375 g	Lean, boneless lamb shoulder or shank, cut in small dice	3. Bring the remaining 2¹/₂ qt (2.5 L) stock to a boil in another pot. 4. Add the lamb and cover the pot. Simmer until the meat is almost tender.
2 oz	60 g	Butter	5. Heat the butter in a heavy pot over medium heat. Add the vegetables and sweat until nearly tender.
6 oz	180 g	Onions, cut brunoise	6. Add the lamb and stock from step 4. Simmer until the meat and vegetables are tender.
4 oz	125 g	Carrots, cut brunoise	
4 oz	125 g	Celery, cut brunoise	7. Add the cooked barley and stock from step 2. Simmer about 5 minutes, or until the flavors are well blended.
2 oz	60 g	Leeks, cut brunoise	
2 oz	60 g	Turnips, cut brunoise	
to taste	to taste	Salt	8. Season to taste with salt and pepper.
to taste	to taste	White pepper	
as needed	as needed	Chopped parsley	9. At serving time, sprinkle each portion with a little chopped parsley.

Per serving: Calories, 120; Protein, 8 g; Fat, 7 g (51% cal.); Cholesterol, 50 mg; Carbohydrates, 7 g; Fiber, 2 g; Sodium, 80 mg.

Caldo Verde 🍽

YIELD: 2¹/₂ QT (2.4 L)   PORTIONS: 8   PORTION SIZE: 10 FL OZ (300 ML)

U.S.	METRIC	INGREDIENTS	PROCEDURE
1 fl oz	30 mL	Olive oil	1. Heat oil in a soup pot. Add the onion and garlic. Cook slowly until soft, but do not brown.
6 oz	180 g	Onion, chopped fine	
1 clove	1 clove	Garlic, chopped fine	2. Add the potatoes and water. Simmer until the potatoes are very tender.
2 lb	900 g	Potatoes, peeled and sliced	3. Purée the soup or, for a coarser texture, simply mash it in the pot.
2 qt	2 L	Water	
8 oz	225 g	Hard, spicy garlic sausage (see Note)	4. Cut the sausage into thin slices. Heat it slowly in a sauté pan to cook off some of the fat. Drain.
to taste	to taste	Salt	5. Add the sausage to the soup. Simmer 5 minutes. Season to taste.
to taste	to taste	Pepper	
1 lb	450 g	Kale	6. Remove the hard center ribs from the kale. Shred the leaves as fine as possible, about as thin as threads. 7. Add to the soup. Simmer 5 minutes. Check the seasoning. 8. This soup should be accompanied by chunks of coarse peasant bread.

Per serving: Calories, 270; Protein, 10 g; Fat, 15 g (48% cal.); Cholesterol, 25 mg; Carbohydrates, 26 g; Fiber, 3 g; Sodium, 370 mg.

Note: Because authentic Portuguese *chouriço* sausage is not widely available, you may substitute Spanish *chorizo* or Italian pepperoni.

Caldo Verde

ADDITIONAL RECIPES

These additional recipes may be found on your CulinarE-Companion recipe management program:

Borscht; Brunswick Soup; Chicken Tomato Bouillon with Pesto; Chilled Tomato Basil Soup; Chilled Tomato Soup with Moroccan Spices; Corn and Crab Chowder with Basil; Ecuadorian Quinoa and Peanut Soup; Honeydew Melon Soup; Japanese Clear Soup with Shrimp; Mediterranean Tomato and Bean Soup; Mulligatawny Soup; Red Lentil and Apricot Soup; Spicy Black Bean Soup; Summer Vegetable and Smoked Tomato Soup; Wild Rice and Mushroom Soup; Wisconsin Cheddar and Broccoli Soup; Zuppa di Ceci e Riso.

TERMS FOR REVIEW

clear soup	**purée soup**	**clearmeat**
vegetable soup	**bisque**	**clarification**
consommé	**chowder**	**raft**
cream soup	**coagulation**	

QUESTIONS FOR DISCUSSION

1. You have 3 gallons (12 L) vegetable soup in the walk-in, prepared by a cook on the morning shift. You are going to serve the soup this evening, and your dinner service lasts from 6 until 10 p.m. How should you prepare the soup for service?

2. What are the most important characteristics of a good consommé?

3. Why is it important not to boil consommé during clarification?

4. What is the function of egg whites in clearmeat? mirepoix? tomato product?

5. In what order would you add the following items to a vegetable soup during cooking?

Carrots	Shredded cabbage	Tomatoes
Barley	Diced cooked beef	

6. Using Method 1 or 2, describe how you would prepare cream of watercress soup.

10

UNDERSTANDING VEGETABLES

Vegetables were, at one time, abused and neglected, relegated to the minor role of unimportant side dishes, to be taken or left, or not even noticed on the table.

Today, however, lowly vegetables are much more appreciated, not only for their nutritional importance but for the variety, flavor, eye appeal, and even elegance and sophistication they bring to the menu. Modern cooks owe it to themselves and their customers to treat vegetables with understanding, respect, and imagination.

Because they are so perishable, vegetables require extra care from receiving to service. Freshness is their most appealing and attractive quality, and one must be especially careful to preserve it. The goals of proper vegetable cookery are to preserve and enhance fresh flavor, texture, and color, and to prepare and serve vegetables that are not just accepted but sought after.

AFTER READING THIS CHAPTER, YOU SHOULD BE ABLE TO

1. Describe the factors that influence texture, flavor, color, and nutritional changes when cooking vegetables.

2. Cook vegetables to their proper doneness.

3. Judge quality in cooked vegetables based on color, appearance, texture, flavor, seasonings, and appropriateness of combination with sauces or other vegetables.

4. Perform pre-preparation tasks for fresh vegetables.

5. Determine the quality of frozen, canned, and dried vegetables.

6. Prepare vegetables using the batch cooking method and the blanch-and-chill method.

7. Store fresh and processed vegetables.

CONTROLLING QUALITY CHANGES DURING COOKING

As a cook, you have a choice of many kinds of vegetables and many cooking methods. Not surprisingly, then, you are also faced with the necessity of learning many rules for cooking vegetables.

Many guides to vegetable cookery simply present a long list of rules to memorize. You will be able to understand the principles more easily, however, if you first learn how vegetables change as they are cooked and how to control those changes. In other words, it is suggested you not just memorize what to do but understand why you do it.

Cooking affects vegetables in four ways. It changes the following:

1. Texture
2. Flavor
3. Color
4. Nutrients

How much these four characteristics change determines whether your final product is attractive and delicious to the customer or whether it ends up in the garbage. You can control these changes if you understand how they occur.

Unfortunately, there is still legitimate controversy among chefs about proper vegetable cooking techniques. Modern technology has not yet solved all the problems experienced chefs tackle successfully every day in the kitchen.

CONTROLLING TEXTURE CHANGES

Changing texture is one of the main purposes of cooking vegetables.

FIBER

The fiber structures of vegetables (including cellulose and pectins) give them shape and firmness. Cooking softens some of these components.

The *amount of fiber* varies:

1. In different vegetables. Spinach and tomatoes have less fiber than carrots and turnips, for example.
2. In different examples of the same vegetables. Old, tough carrots have more fiber than young, fresh carrots.
3. In the same vegetable. The tender tips of asparagus and broccoli have less fiber than their tougher stalks.

Fiber is made *firmer* by acids or sugars:

1. Acids.

Lemon juice, vinegar, and tomato products, when added to cooking vegetables, extend the cooking time.

2. Sugars.

Sugar strengthens cell structure. You will use this principle primarily in fruit cookery. For firm poached apples or pears, for example, cook in a heavy syrup. For applesauce, cook apples until soft before sweetening.

Fiber is *softened* by heat or alkalis:

1. Heat.

In general, longer cooking means softer vegetables.

2. Alkalis.

Do not add baking soda to green vegetables. Not only does it destroy vitamins but it also makes the vegetables unpleasantly mushy.

STARCH

Starch is another vegetable component that affects texture:

1. *Dry starchy foods* like dried legumes (beans, peas, lentils), rice, and macaroni products must be cooked in enough water for the starch granules to absorb moisture and soften. Dried beans are usually soaked before cooking to replace lost moisture.

2. *Moist starchy vegetables* like potatoes and sweet potatoes have enough moisture of their own, but they must still be cooked until the starch granules soften.

DONENESS

A vegetable is said to be done when it reaches the desired degree of tenderness. This stage varies from vegetable to vegetable. Some, such as winter squash, eggplant, and braised celery, are considered properly cooked when they are quite soft. Most vegetables, however, are best cooked briefly, until they are crisp-tender or **al dente** (firm to the bite). At this stage of tenderness they not only have the most pleasing texture but also retain maximum flavor, color, and nutrients. Note, however, that we use the term "crisp-tender" to describe texture. The "tender" part of this term is as important as "crisp." While the vegetables retain texture, they are still cooked enough to be pleasingly edible. Vegetables such as green beans are sometimes served barely cooked, difficult to cut, and tough to chew. Avoid this fault as much as overcooking.

GUIDELINES for Achieving Proper Doneness in Vegetables

1. Don't overcook.

2. Cook as close to service as possible. Holding vegetables in a steam table continues to cook them.

3. If vegetables must be cooked in advance, slightly undercook them, cool rapidly in cold water, drain, and refrigerate, then reheat to order.

4. For uniform doneness, cut vegetables into pieces of uniform size before cooking.

5. Vegetables with both tough and tender parts need special treatment so the tender parts are not overcooked by the time the tougher parts are done. For example,

   Peel the woody stalks of asparagus.
   Peel or split broccoli stalks.
   Pierce the base of Brussels sprouts with a sharp knife.
   Remove the heavy center stalks of lettuce leaves before braising.

6. Don't mix batches of cooked vegetables. They are likely to be cooked to slightly different levels of doneness.

CONTROLLING FLAVOR CHANGES

COOKING PRODUCES FLAVOR LOSS

Many flavors are lost during cooking by dissolving into the cooking liquid and by evaporation. The longer a vegetable is cooked, the more flavor it loses.

Flavor loss can be controlled in several ways:

1. Cook for as short a time as possible.

2. Use boiling salted water. Starting vegetables in boiling water shortens cooking time. The addition of salt helps reduce flavor loss.

3. Use just enough water to cover to minimize leaching. Note that this rule contradicts rule 1 in that adding vegetables to a small quantity of water lowers the temperature more, so cooking time is extended. Save your questions on this until you have finished reading the sections on color and nutritional changes.

4. Steam vegetables whenever appropriate. Steam cooking reduces leaching out of flavor and shortens cooking time.

Strong-Flavored Vegetables

With certain strong-flavored vegetables, it is desirable to lose some of the flavor to make them more appealing to the taste. These include the onion family (onions, garlic, leeks, shallots), the cabbage family (cabbage, Brussels sprouts, cauliflower, broccoli), and some root vegetables (turnips, rutabagas).

When cooking strong-flavored vegetables, leave uncovered to allow these flavors to escape, and use larger amounts of water.

COOKING PRODUCES FLAVOR CHANGES

Cooked vegetables do not taste like raw vegetables because cooking produces certain chemical changes. As long as the vegetables are not overcooked, this change is desirable. It produces the flavors one looks for in vegetable dishes.

Overcooking produces undesirable changes in members of the cabbage family. They develop a strong, unpleasant flavor. Cabbage and its relatives, as well as strong root vegetables such as rutabaga and turnip, should be cooked quickly, uncovered.

COOKING AND SWEETNESS

Young, freshly harvested vegetables have a relatively high sugar content that makes them taste sweet. As they mature, or as they sit in storage, the sugar gradually changes to starch. This is especially noticeable in corn, peas, carrots, turnips, and beets.

To serve sweet-tasting vegetables:

1. Try to serve young, fresh vegetables that have been stored as short a time as possible.
2. For older vegetables, especially those just listed, add a small amount of sugar to the cooking water to replace lost sweetness.

CONTROLLING COLOR CHANGES

It is important to preserve as much natural color as possible when cooking vegetables. Because customers may reject or accept a vegetable on the basis of its appearance, it can be said that its visual quality is as important as its flavor or nutritional value.

Pigments are compounds that give vegetables their color. Different pigments react in different ways to heat and to acids and other elements that may be present during cooking, so it is necessary to discuss them one at a time. Table 10.1 summarizes this information.

TABLE 10.1 Vegetable Color Changes During Cooking

Color	Examples of Vegetables	Cooked with Acid	Cooked with Alkali	Overcooked
White	Potatoes, turnips, cauliflower, onions, white cabbage	White	Yellowish	Yellowish, gray
Red	Beets, red cabbage (not tomatoes, whose pigment is like that in yellow vegetables)	Red	Blue or blue-green	Greenish blue, faded
Green	Asparagus, green beans, lima beans, broccoli, Brussels sprouts, peas, spinach, green peppers, artichokes, okra	Olive green	Bright green	Olive green
Yellow (and orange)	Carrots, tomatoes, rutabagas, sweet potatoes, squash, corn	Little change	Little change	Slightly faded

WHITE VEGETABLES

Pigments called **anthoxanthins** (an tho zan thins) and **flavonoids** range from pale yellow to white. These are the primary coloring compounds in potatoes, onions, cauliflower, and white cabbage and in the white parts of such vegetables as celery, cucumbers, and zucchini.

White pigments stay white in acid and turn yellow in alkaline water. To keep vegetables such as cauliflower white, add a little lemon juice or cream of tartar to the cooking water. (Don't add too much, though, as this may toughen the vegetable.) Covering the pot also helps keep acids in.

Cooking for a short time, especially in a steamer, helps maintain color (and flavor and nutrients as well). Overcooking or holding too long in a steam table turns white vegetables dull yellow or gray.

RED VEGETABLES

Red pigments, called **anthocyanins**, are found in only a few vegetables, mainly red cabbage and beets. Blueberries also are colored by these red pigments. (The red color of tomatoes and red peppers is due to the same pigments that color carrots yellow or orange.)

Red pigments react very strongly to acids and alkalis:

Acids turn them a brighter red.

Alkalis turn them blue or blue-green (not a very appetizing color for red cabbage).

Red beets and red cabbage, therefore, have their best color when cooked with a small amount of acid. Red cabbage is often cooked with tart apples for this reason.

When a strongly acid vegetable is desired, as for Harvard beets or braised red cabbage, add just a small amount of acid at first. Acids toughen vegetables and prolong cooking time. Add the rest when the vegetables are tender.

Red pigments dissolve easily in water, so follow these guidelines:

1. Use a short cooking time. Overcooked red vegetables lose a lot of color.

2. Use only as much water as is necessary.

3. Cook beets whole and unpeeled, with root and an inch of stem attached, to protect color. Skins easily slip off cooked beets.

4. When steaming, use solid pans instead of perforated pans to retain the red juices.

5. Whenever possible, serve the cooking liquid as a sauce with the vegetable.

GREEN VEGETABLES

Green coloring, or **chlorophyll**, is present in all green plants. Green vegetables are common in the kitchen, so it is important to understand the special handling required by this pigment.

Acids are enemies of green vegetables. Both *acid* and *long cooking* turn green vegetables a drab olive green.

Protect the color of green vegetables by observing these guidelines:

1. Cook uncovered to allow plant acids to escape.

2. Cook for the shortest possible time. Properly cooked green vegetables are tender-crisp, not mushy.

3. Cook in small batches rather than holding for long periods in a steam table.

Steaming is a good method for cooking many green vegetables. Steam cooks food rapidly, lessens the dissolving out of nutrients and flavor, and does not break up delicate vegetables. Overcooking, however, can occur rapidly in steamers.

Do not use baking soda to maintain green color. Soda destroys vitamins and makes texture unpleasantly mushy and slippery.

How much water should be used when boiling? A large quantity of water helps dissolve plant acids, helps preserve colors, and speeds cooking. But some cooks feel an excessive amount of nutrients are lost. See the next section for further discussion.

YELLOW AND ORANGE VEGETABLES

Yellow and orange pigments, called **carotenoids**, are found in carrots, corn, winter squash, rutabaga, sweet potatoes, tomatoes, and red peppers. These pigments are very stable. They are little affected by acids or alkalis. Long cooking can dull the color, however. Short cooking not only prevents dulling of the color but also preserves vitamins and flavors.

CONTROLLING NUTRIENT LOSSES

Vegetables are an important part of our diet because they supply a wide variety of essential nutrients. They are our major sources of vitamins A and C and are rich in many other vitamins and minerals. Unfortunately, many of these nutrients are easily lost.

Six factors are responsible for most nutrient loss:

1. High temperature
2. Long cooking
3. Leaching (dissolving out)
4. Alkalis (baking soda, hard water)
5. Plant enzymes (which are active at warm temperatures but destroyed by high heat)
6. Oxygen

Some nutrient loss is inevitable because it is rarely possible to avoid all of these conditions at the same time. For example:

- Pressure steaming shortens cooking time, but the high temperature destroys some vitamins.
- Braising uses low heat, but the cooking time is longer.
- Baking eliminates the leaching out of vitamins and minerals, but the long cooking and high temperature cause nutrient loss.
- Boiling is faster than simmering, but the higher temperature can be harmful and the rapid activity can break up delicate vegetables and increase loss through leaching.
- Cutting vegetables into small pieces decreases cooking time, but it increases leaching by creating more exposed surfaces.
- Even steaming allows some leaching out of nutrients into the moisture that condenses on the vegetables and then drips off.

COOKING IN A LITTLE LIQUID VERSUS A LOT OF LIQUID

This is an area of controversy with good arguments on both sides.

1. Using a lot of liquid increases vitamin loss by leaching. Use just enough liquid to cover. Save the cooking liquid for reheating the vegetables or for stocks or soups.
2. Using a little liquid increases cooking time. When the vegetables are combined with the small quantity of boiling water, the temperature is lowered greatly and the vegetables must sit in warm water while it again heats up. Also, plant enzymes may destroy some vitamins before the water again becomes hot enough to destroy them.

Tests have shown that, for these reasons, no more nutrients are lost when vegetables are cooked in a lot of water than when vegetables are cooked in just enough water to cover.

When cooking green vegetables, there is an added advantage to using a lot of water. Plant acids are more quickly diluted and driven off, better preserving the color.

The best cooking methods, nutritionally, are usually those that produce the most attractive, flavorful products.

- They are more likely to be eaten. Discarded vegetables benefit no one, no matter how nutritious they are.
- Factors that destroy nutrients are often those that also destroy color, flavor, and texture.

GENERAL RULES OF VEGETABLE COOKERY

Now that you understand how vegetables change as they cook, let's summarize that information in some general rules. You should now be able to explain the reasons for each of these rules:

- Don't overcook.

- Cook as close to service time as possible, and in small quantities. Avoid holding for long periods on a steam table.

- If the vegetable must be cooked ahead, undercook slightly and chill rapidly. Reheat at service time.

- Never use baking soda with green vegetables.

- Cut vegetables uniformly for even cooking.

- Start with boiling, salted water when boiling green vegetables and other vegetables that grow above the ground. Roots and tubers are started in cold, salted water for more even cooking.

- Cook green vegetables and strong-flavored vegetables uncovered.

- To preserve color, cook red and white vegetables in a slightly acid (not strongly acid) liquid. Cook green vegetables in a neutral liquid.

- Do not mix a batch of freshly cooked vegetables with a batch of the same vegetable that was cooked earlier and kept hot in a steam table.

STANDARDS OF QUALITY IN COOKED VEGETABLES

1. **Color.**
   Bright, natural colors.
   Green vegetables, in particular, should be a fresh, bright green, not olive green.

2. **Appearance on plate.**
   Cut neatly and uniformly. Not broken up.
   Attractively arranged or mounded on plate or dish.
   Not swimming in cooking water.
   Imaginative and appropriate combinations and garnishes are always well received.

3. **Texture.**
   Cooked to the right degree of doneness.
   Most vegetables should be crisp-tender, not overcooked and mushy, but not tough or woody, either.
   Vegetables intended to be soft (potatoes, squash, sweet potatoes, tomatoes, vegetable purées) should be cooked through, with a pleasant, smooth texture.

4. **Flavor.**
   Full, natural flavor and sweetness, sometimes called *garden-fresh flavor*. Strong-flavored vegetables should be pleasantly mild, with no off flavors or bitterness.

5. **Seasonings.**
   Lightly and appropriately seasoned. Seasonings should not be too strong and should not mask the natural garden flavors.

6. **Sauces.**
   Butter and seasoned butters should be fresh and not used heavily; vegetables should not be greasy.
   Cream sauces and other sauces should not be too thick or too heavily seasoned. As with seasonings, sauces should enhance, not cover up.

7. Vegetable combinations.

Interesting combinations attract customers.

Flavors, colors, and shapes should be pleasing in combination.

Vegetables should be cooked separately and then combined to allow for different cooking times.

Acid vegetables (like tomatoes) added to green vegetables will discolor them. Combine just before service.

KEY POINTS TO REVIEW

- What factors affect changes in texture when vegetables are cooked?

- What are the guidelines for achieving proper doneness in vegetables?

- What factors affect changes in flavor when vegetables are cooked?

- What factors affect changes in color when vegetables are cooked? Describe factors specific to white vegetables, green vegetables, red vegetables, and orange and yellow vegetables.

- What steps can you take to maintain the highest nutritional values when cooking vegetables?

- What standards are used to judge the quality of cooked vegetables?

HANDLING VEGETABLES

FRESH VEGETABLES

WASHING

1. Wash all vegetables thoroughly.

2. Root vegetables that are not peeled, such as potatoes for baking, should be scrubbed very well with a stiff vegetable brush.

3. Wash green, leafy vegetables in several changes of cold water. Lift the greens from the water so the sand can sink to the bottom. Pouring off into a colander dumps the sand back onto the leaves.

4. After washing, drain well and refrigerate lightly covered. The purpose of covering is to prevent drying, but covering too tightly cuts off air circulation. This can be a problem if the product is stored more than a day because mold is more likely to grow in a damp, closed space. Use a drain insert in the storage container to allow drainage.

SOAKING

1. With a few exceptions, do not soak vegetables for long periods. Flavor and nutrients leach out.

2. Cabbage, broccoli, Brussels sprouts, and cauliflower may be soaked 30 minutes in cold salted water to eliminate insects, if necessary.

3. Limp vegetables can be soaked briefly in cold water to restore crispness.

4. Dried legumes are soaked for several hours before cooking to replace moisture lost in drying. Dried beans absorb their weight in water.

PEELING AND CUTTING

1. Peel most vegetables as thinly as possible. Many nutrients lie just under the skin.

2. Cut vegetables into uniform pieces for even cooking.

3. Peel and cut vegetables as close to cooking time as possible to prevent drying and loss of vitamins through oxidation.

4. For machine paring, sort vegetables for evenness of size to minimize waste.

5. Treat vegetables that brown easily (potatoes, eggplant, artichokes, sweet potatoes) with an acid, such as lemon juice, or an antioxidant solution, or hold under water until ready to use (some vitamins and minerals will be lost).

6. Save edible trim for soups, stocks, and vegetable purées.

CLASSIFYING VEGETABLES

Many people are bothered by the fact that tomatoes are referred to as vegetables when they are, in fact, fruits. Yes, tomatoes are fruits, and carrots are roots, and spinach is a leaf, and they are all vegetables.

To a botanist, the term *fruit* refers to a specific part of a plant, just as do the terms *stem*, *root*, and *leaf*. A fruit is defined as the ripened ovary or ovaries of a seed-bearing plant, and it contains the seeds. In other words, if it has seeds, it's a fruit. We shouldn't be misled by the fact that a few fruits are sweet. Many, if not most, fruits in nature are not sweet at all. Some examples of fruits used in the vegetable kitchen are tomatoes, eggplant, peppers, green beans, okra, cucumbers, squash, pea pods, and avocados.

There are many ways of classifying vegetables. Some are more helpful to the cook than others. Putting vegetables into groups based on their botanical origin is not always helpful. For example, okra and eggplant are both fruits, but they are handled and cooked so differently that this information doesn't really help us.

The following vegetable categories are based, in part, on how the vegetables are used in the kitchen. For example, the vegetables listed as roots and tubers come from several unrelated families, but they all have fairly solid, uniform textures and are handled in similar ways. This is not a scientific classification, and it is not the only way to group vegetables.

Note that the first three categories are all fruits or, in a few cases, seeds from fruits.

- The gourd family: cucumber, winter and summer squashes, pumpkin, chayote (Some sources put eggplant in the gourd family, but this is incorrect; see below.)

- Seeds and pods: beans, peas, corn, okra

- Other tender-fruited vegetables: avocado, eggplant, sweet and hot peppers, tomato (Eggplant, tomato, and peppers are all members of the nightshade family; the avocado is unrelated.)

- Roots and tubers: beet, carrot, celery root, parsnip, radish, turnip, rutabaga, Jerusalem artichoke, potato, sweet potato, jícama

- The cabbage family: cabbage, broccoli, cauliflower, Brussels sprouts, kohlrabi, bok choy

- The onion family: onion, scallion, leek, garlic, shallot

- Leafy greens: spinach, beet greens, lettuces, endive and chicory, Swiss chard, sorrel, watercress, collards, kale, turnip greens (The last three are also members of the cabbage family.)

- Stalks, stems, and shoots: globe artichoke, asparagus, celery, fennel, fiddlehead ferns, bamboo shoots

- Mushrooms

FRESH VEGETABLES: EVALUATING AND PREPARING

This section lists the fresh vegetables commonly used in North American kitchens, including many products that have become familiar from Asian and Latin cuisines. Tips for evaluating and trimming the products, as well as the average trimming yield, are indicated. Vegetables are listed alphabetically, rather than by family classification, to make them easier to find.

Artichokes, Globe

FIGURE 10.1 Trimming artichoke bottoms.

(a) Cut or break off the stem flush with the bottom of the artichoke, as shown.

Identification: Artichokes are the unopened or immature flowers of a type of thistle. They vary in size and coloration but are usually round to somewhat elongated, colored light to medium green, sometimes with purple tints.

Related Varieties: Baby artichokes are not actually babies but come from a different place on the plant and are at their full size. Especially young baby artichokes may be tender enough to eat whole, with little trimming except for removing the top points of the leaves.

Evaluation: Look for compact, tight leaves; heavy for size; few or no brown blemishes.

Preparation: Wash. Cut 1 inch (2–3 cm) off tops. Cut off stem and lower leaves. Scrape out choke (fuzzy center) with melon ball cutter. (Remove choke before or after cooking.) Dip in lemon juice immediately. To prepare bottoms, see Figures 10.1 and 10.2.

Percentage Yield: 80% (whole, trimmed), 30% (bottoms only)

(b) Break off the outer leaves.

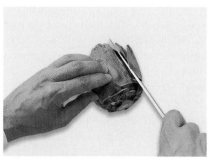

(c) Alternatively, trim the outer leaves with a knife as shown, being careful not to cut into the base of the artichoke.

(d) Cut off the remaining leaves above the base.

(e) With a paring knife, trim off the green outer peel to give the base a smooth, neat appearance.

(f) With a ball cutter or tablespoon, scrape out the fuzzy choke.

(g) A trimmed artichoke bottom on the left; a trimmed whole artichoke on the right. Note that the points of the leaves have been cut off and the center choke removed.

FIGURE 10.2 To keep the cut stem end of an artichoke from darkening during steaming or boiling, tie a slice of lemon over the cut surface.

Artichokes, Jerusalem

See Sunchoke.

Asparagus

Identification: Spear-shaped new shoot or stem that emerges from the plant's roots in the spring. The pointed spear-tip sprouts branches when the shoot is allowed to grow.

Related Varieties: *White asparagus* is the same plant as green, but soil is mounded over the shoots, protecting them from the sun so they do not turn green. In Europe, white asparagus is more common than green. The flavor is milder than that of green, although North American white asparagus is usually more bitter than European. *Purple asparagus* turns dark green when cooked. It is tender and sweet.

Evaluation: Look for tightly closed tips; firm, not withered, stalks. For white asparagus, buy only product that has been kept chilled for its entire storage time; unchilled white asparagus becomes fibrous.

Preparation: Break off woody lower ends. Remove lower scales, which may harbor sand, or peel lower part of stalk. Figure 10.3 shows an alternative method. Cut tips to uniform lengths and/or tie them in bundles for cooking. White asparagus should be peeled the entire length of the stalk. Purple asparagus needs no peeling; just trim the bottoms.

Percentage Yield: 55% (green, peeled)

FIGURE 10.3 Trimming asparagus.

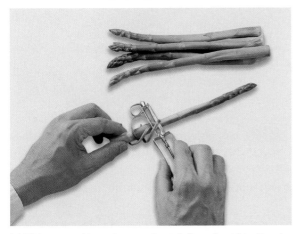

(a) With a vegetable peeler, pare the stalk from about 2 in. (5 cm) below the tip down to the base.

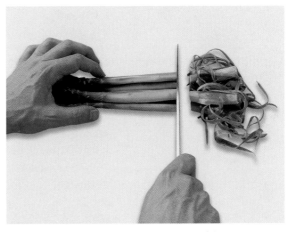

(b) Cut or break off the hard, woody bottoms of the stems.

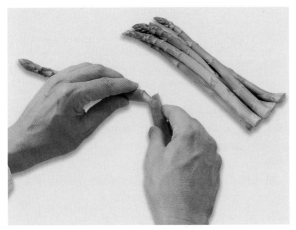

(c) Another method used by many chefs is to break off the stems first . . .

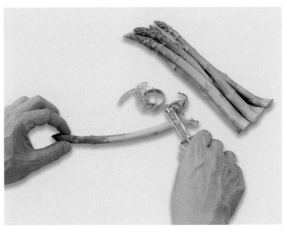

(d) . . . and then peel the stem.

Avocados

Identification: The egg-shaped fruit of a small tree, with a leathery skin, tender, pale-green flesh, and a single large seed or pit in the center.

Related Varieties: There are several varieties that fall into two main categories: (1) The Mexican or Californian avocados, mostly the Hass variety, which have rough, dark green skins that turn black when ripe. These have a rich, buttery flesh with a high oil content. (2) The West Indian or Florida type, which has smoother skin that remains green. These are juicier and have a lower oil content.

Evaluation: Look for fresh appearance; fruit heavy for size; no blemishes or bruises.

Preparation: Ripen at room temperature, 2 to 5 days. Cut in half lengthwise and remove pit (see Figure 10.4). Peel (skin pulls away easily from ripe fruit). Dip into or rub with lemon juice immediately to prevent browning.

Percentage Yield: 75%

FIGURE 10.4 Preparing avocados.

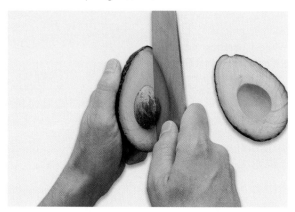

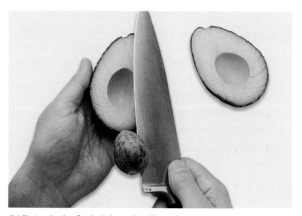

(a) To remove the pit or seed from the cut avocado, strike it sharply (but carefully) with the heel of a chef's knife.

(b) Twist the knife slightly and pull out the pit.

Bamboo Shoots

Identification: The young shoots of various species of bamboo plants, harvested as they just begin to emerge from the ground. They are roughly cone-shaped, with tough brown skins and a creamy, crisp, tender interior.

Evaluation: Look for solid, heavy shoots with no soft spots or cracks; no trace of sour smell.

Preparation: Peel down to the creamy white or pale yellowish cone-shaped core. Slice and boil in salted water until tender, then cut as desired for use in recipes.

Percentage Yield: Varies greatly, depending on size of shoots, which range from a few ounces to a pound (less than 100 grams to 500 grams) or more.

Beans, Dried

See Chapter 13.

Beans, Fava

Identification: Also called *broad bean*. Unlike most of our common beans, which originated in the Western Hemisphere, favas are Old World beans. The large pods hold four to six beans in a soft, white lining. The flat beans slightly resemble limas, but they are not as starchy. Flavor is subtle and nutlike.

Evaluation: Select small to medium pods that are fresh green in color, not overly large. Yellowing pods may be too mature. Some spots on pods is normal.

Preparation: Preparation is labor-intensive. Shell the beans, parboil, then peel off skins or husks (Figure 10.5).

Percentage Yield: 15–20%

FIGURE 10.5 Preparing fava beans.

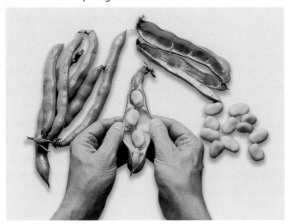

(a) Split open the pods and remove the beans.

(b) Blanch the beans for a few minutes. This cooks the beans and loosens the skins. Peel off the outer skins. Unpeeled beans are on the left, peeled beans on the right.

Beans, Fresh Shell

Identification: These are the fresh, moist versions of the many types of dried beans.

Related Varieties: Although many types of beans are grown to be shelled, most of these are dried, and, with the exception of *cranberry* or *borlotti beans*, *southern peas* or *cowpeas*, and *black-eyed peas*, few are available fresh. Soybeans and lima beans have their own entries below.

Evaluation: Look for firm, fresh, moderately filled-out pods containing firm but not hard seeds. Avoid yellow or brownish pods, which are likely to be too mature.

Preparation: Shell and rinse. Cook before serving; raw beans can be harmful.

Percentage Yield: 40%

Beans, Lima

Identification: Flat, pale-green beans in flat, slightly fuzzy pods, with three or four beans per pod. Larger limas tend to be starchier than small or baby limas.

Related Varieties: Many varieties are grown, but they are not usually distinguished from one another in the market.

Evaluation: Beans should be plump, with tender skins.

Preparation: Shell, wash, and drain.

Percentage Yield: 40%

Beans, Snap

Identification: Fresh green beans and other varieties are in the same family as shell beans, except they are grown to be picked immature for their tender, edible pods.

Related Varieties: Green beans are the most common. Some green varieties, picked when very small and tender, are known as *haricots verts* (ah ree coh vehr, French for "green beans"). Other varieties include yellow or wax beans, purple beans, and flat, Italian-style green beans.

Evaluation: Look for firm and straight beans, with few shriveled ends; even color, without blemishes. Should be tender and crisp enough to break when bent to a 45-degree angle. Enclosed seeds should be small, not large and bulging.

Preparation: Wash. Cut or snap off ends. Remove any spots. Leave whole or cut into desired lengths.

Percentage Yield: 88%

Green beans

Beans, Soy

See Soybeans.

Beets

Red beets

Identification: A tender, bulbous, somewhat sweet root, usually but not always dark red. The tops can also be cooked like other greens; handle and treat beet tops like Swiss chard (see below), which is in the beet family.

Related Varieties: Beets come in several shapes and colors in addition to the common red, including yellow, pink, orange or brick red, and white. Chioggia (kee oh ja) are striped red and white. Shapes include round, elongated oval, and cylindrical.

Evaluation: Look for firm, round, uniform size; smooth skin. Tops, if any, should be fresh or just wilted, but not yellow or deteriorated. Large, rough beets are often woody.

Preparation: Cut off tops, leaving 1 inch (2–3 cm) of stem attached to beets. Leave roots on to avoid loss of color and juice during cooking. Scrub well. Steam, boil, or bake before peeling.

Percentage Yield: 40–45% (75% if purchased without tops)

Bok Choy

Bok Choy

Identification: An Asian member of the cabbage family, related to Chinese cabbage, except the stalk is thicker and fleshier and the green portion is smoother and darker in color. Leaves and stalks are tender, crisp, and juicy.

Related Varieties: Several varieties all form compact, elongated bunches. The most common variety forms large heads or bunches with white stems. *Shanghai bok choi* is smaller, with pale green stems. *Choy sum* is allowed to become more mature, so the central stem or core is longer, and there may be small yellow flowers among the leaves.

Evaluation: Fresh, green appearance for all varieties, with no wilting or discoloration.

Preparation: Cut in half lengthwise and wash under running water to remove soil from the crooks of the stalks. Trim bottom by cutting off a thin slice, or cut out core. Cut as desired. Small varieties may be cooked whole.

Percentage Yield: 80%

Broccoli

FIGURE 10.6 Prepare tough bottoms of broccoli stalks by pulling off the fibrous peel, as shown.

Identification: A green vegetable in the cabbage family, consisting of tight clusters of tiny green flower buds on fleshy stalks.

Related Varieties: Pale green *broccoflower* is closer in character to cauliflower than to broccoli and should be handled like cauliflower. *Broccolini* is a relatively new broccoli hybrid with slender stems and small, loose florets (flower bud heads). *Broccoli rabe* (also called *broccoli raab*, *broccoletti di rape*, *rapini*, and *cima de rapa*), like broccolini, has slender stems and loose florets, but it also has tender leaves attached to the stems, and the stems are slightly ribbed rather than smooth. Broccoli rabe has a stronger, more mustardy flavor. Several varieties of *Chinese broccoli* are similar to broccoli rabe in both shape and flavor, but the stems are smoother and fleshier.

Evaluation: Look for dark green, tightly closed buds in regular broccoli and broccoflower. Broccoli rabe and Asian types should be dark green with crisp, not wilted, leaves.

Preparation: Wash well. Soak in salted water 30 minutes if necessary to remove insects. Split large stalks into smaller sizes for portioning. Split thick stalks partway for faster cooking, or cut tops from stalks. Tougher stalks may be peeled (Figure 10.6).

Percentage Yield: 65–75%

Brussels Sprouts

Identification: This cabbage relative resembles a tiny cabbage head. The clusters of leaves grow in rows along a thick stalk. Flavor is somewhat stronger than that of green cabbage.

Related Varieties: Red or purple (the color of red cabbage) Brussels sprouts exist but are not often seen, probably because their flavor and texture is less appealing to most people.

Evaluation: Look for bright green, tight heads; uniform size.

Preparation: Trim bottom ends and remove yellowed outer leaves (but don't cut off too much of the bottom or you will lose too many leaves). For more even cooking, pierce base with sharp knife point. Rinse well. Soak in cold salted water 30 minutes if necessary to remove insects.

Percentage Yield: 80%

Brussels sprouts

Cabbage: Green, Red, and Savoy

Identification: Cabbages are large leaf clusters in dense, round heads (heads may also be flat or elongated, depending on the variety). Cabbage and mustard are in the same family, a relationship that can be detected in the faintly peppery taste.

Evaluation: For both green and red cabbage, look for a firm head, heavy for size. Good color. Crisp leaves, finely ribbed. Savoy cabbage is not as heavy, with darker green, ruffled leaves.

Preparation: Remove coarse or discolored outer leaves. Remove core and rinse whole, or cut into quarters and then remove core. For wedges, core is left in, but with bottom trimmed, to hold sections together. (See Figure 10.7.)

Percentage Yield: 80%

Green cabbage

FIGURE 10.7 Cutting and shredding cabbage.

Red cabbage

(a) Cut the cabbage head into quarters. Then cut out the core as shown.

(b) With a French knife, cut into thin shreds.

Cabbage, Chinese

Identification: Heads of tender, crinkly, pale green leaves with broad, flat center stalks. The flavor is mild and slightly spicy.

Related Varieties: The two main varieties have different head shapes, although their flavor and texture are virtually the same. The head of *napa cabbage* is stout and barrel-shaped. A more elongated, slender head cabbage is often called *celery cabbage* for its shape. See also Bok Choy, a closely related cabbage.

Evaluation: Look for firm, tightly packed heads with no dry or browned tips, crisp and not limp or wilted.

Preparation: Wash. Cut as desired, discarding the center core.

Percentage Yield: 85%

Savoy cabbage

Cactus pad

Cactus Pads or Nopales

Identification: Nopales are tender pads of the prickly pear cactus. When cooked, their texture is slippery, like that of okra, and their flavor slightly resembles cooked green pepper or possibly green beans with a slight sourness.

Evaluation: Try to purchase pads with the spines removed. Look for full, firm, crisp pads, not wilted or soft ones.

Preparation: If spines are present, wear rubber gloves and brush them vigorously under running water, then with a vegetable peeler remove the eyes that held the spines. Trim the base if necessary. Cut as desired. Can be sautéed, boiled, steamed, or grilled.

Percentage Yield: 90%

Carrots

Identification: Long, pointed, orange roots. Among the most widely used vegetables. Specialty carrots are also available in other colors, including yellow, red, and purple.

Evaluation: Look for bright orange color; crisp, straight, and well shaped shaft; smooth surface. Large carrots are sometimes woody.

Preparation: Trim top and bottom ends. Pare with hand peeler.

Percentage Yield: 75–80%

Carrot
Courtesy of Grimmway Farms

Cauliflower

Cauliflower

Identification: White or off-white, tight cluster of tiny flowers (florets) branching off a central stalk. Member of the cabbage family.

Related Varieties: *Broccoflower* is a light green relative that looks like a cross between broccoli and cauliflower. Handle like regular cauliflower.

Evaluation: Look for white color, not yellow or brownish; fine-grained, tightly closed buds; fresh green, well-trimmed leaves.

Preparation: Remove leaves and trim tough part of stalk. Cut away discolored parts. Wash. Soak in salted water 30 minutes if necessary to remove insects. Separate into florets, leaving portion of center stalk attached to each one to minimize trim loss (see Figure 10.8). If cooking whole, cut out center of stalk for more even cooking.

Percentage Yield: 55%

FIGURE 10.8 Cauliflower disassembly.

(a) With a paring knife or small utility knife, cut out the central core.

(b) Separate the cauliflower into florets.

Celery

Identification: Pale green stems that form bunches or clusters attached at root end. One of the most common vegetables.

Evaluation: Look for bunches that are straight, compact, well trimmed; fresh green color.

Preparation: Cut off root end. Separate stems and scrub well. Reserve leaves and tough outer stems for stocks, soups, mirepoix. Ribbed outer side of stems may be peeled to remove strings.

Percentage Yield: 75%

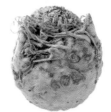
Celery

Celery Root or Celeriac

Identification: A variety of celery grown for its round, fleshy white root.

Evaluation: Select firm and heavy roots. Large ones are often soft and spongy in the center.

Preparation: Wash well, peel, and cut as desired.

Percentage Yield: 75%

Celery root

Chayote

Identification: A member of the gourd or squash family, this vegetable has a crisp texture and a mild flavor somewhere between zucchini and cucumber. It is roughly pear-shaped, although more rounded, with a thin yellow-green skin and creamy white flesh. It is also known as *mirliton* and *vegetable pear*.

Evaluation: Look for firm, solid chayotes without blemishes.

Preparation: Skin can be left on if served raw, but peel chayotes to be cooked, because the skin toughens. The soft seed in the middle can also be eaten.

Percentage Yield: 100% if unpeeled; 85–90% if peeled

Chayote

Chestnut

Identification: Chestnuts are the seed or nut of a variety of trees grown primarily in Europe. They have a thin, hard, dark brown outer shell and a bitter inner skin surrounding a soft, starchy meat.

Evaluation: Look for plump, firm nuts with no blemishes on the shells.

Preparation: The shell and skin must be removed before use. With the point of a paring knife, cut an X in the shell on the flat side of each nut. Roast at 350°F (180°C) for 15 minutes, or blanch in boiling water. Peel while still hot.

Percentage Yield: 75%

Corn

Identification: Corn is a grain used as a vegetable if immature. It grows in rows on the outside of a central woody cob and is covered with a leafy husk. Varieties grown for vegetable use are sweeter than grain varieties.

Evaluation: Look for fresh, moist husks, not dry; no worm damage; kernels well filled, tender, and milky when punctured.

Preparation: Strip off husks, remove silk, and cut off bottom stump. Cut into two or three sections as desired, if serving on cob, or cut from cob. Keep refrigerated and use as soon as possible.

Percentage Yield: 28% after husking and cutting from cob

Corn

FIGURE 10.9 Cucumber garnishes can be made more decorative by scoring the cucumber before slicing or cutting.

(a) Scoring with a fork.

(b) Scoring with a channel knife.

Cucumber

Identification: A member of the gourd family, like squash, the cucumber has a thin green skin and a crisp, mild, juicy off-white flesh.

Related Varieties: Cucumbers are of two types, slicing and pickling. The common green slicing cucumber, with dark green skin, is most often used in food service. The long, slender hothouse cucumber, also called *seedless* and *burpless*, is another common slicing cucumber. Pickling cucumbers, also called *Kirbys*, range in size from tiny gherkins to the large ones used to make dill pickles. They are somewhat drier and a little more flavorful than the slicing types. Their green skin is a little lighter in color than that of slicing cucumbers, and the surface is bumpy.

Evaluation: Look for vegetables that are firm, crisp, dark green, well shaped. Yellow color means the cucumber is overmature.

Preparation: Wash. Trim ends. Peel if skin is tough or has been waxed. Skin may be scored to make decorative slices (see Figure 10.9).

Percentage Yield: 75–95%, depending on peeling

Eggplant

Identification: Eggplants are members of the same family as tomatoes, peppers, and potatoes. They have purple, white, or pale green skin and a firm but spongy off-white flesh containing tiny edible seeds. Also called *aubergine.*

Related Varieties: Worldwide, there are dozens of varieties. In the West, the most common is the large, oblong, purple-skinned variety. Smaller, elongated varieties are sometimes called *Italian eggplant*. White-skinned varieties are round, egg-shaped (the origin of the name *eggplant*), or long and slender. *Japanese eggplants* are small, long, and slender, with a purple instead of green stem. Asian types include small round, green-skinned fruits, including a Thai eggplant no larger than a pea. The flesh of these varieties is similar in flavor, varying in texture or density from soft to almost hard.

Evaluation: Eggplants should be firm, not soft. Common purple types should be shiny, dark purple color; heavy and plump; without blemishes or soft spots.

Preparation: Wash. Trim off stem end. Peel if skin is tough. Cut just before use. Dip in lemon juice or antioxidant solution to prevent discoloration if not to be cooked within a few minutes after peeling and cutting.

Percentage Yield: 90% (75% if peeled)

Italian eggplant

White eggplant

Fennel

Identification: The clustered or bunched stems form a bulbous base. The fleshy bulb is white with a pale hint of green, while the stems are darker green. The aroma and flavor have a faint suggestion of licorice.

Related Varieties: The vegetable fennel is more properly known as *Florence fennel*, to distinguish it from the varieties grown as a spice (seed) and herb. The vegetable is sometimes incorrectly called *anise*, which is a different plant.

Evaluation: Look for bright, pale green color with few or no brown spots. Fresh green tops, not wilted or spoiled. Compact, heavy for size.

Fennel

Preparation: Trim stems and feathery leaves. Split in half through the base. If the fennel is large, cut out the core as for cabbage if it is fibrous or tough.

Percentage Yield: 80%

Fiddlehead Fern

Identification: This spring vegetable is the early shoot of a fern, usually ostrich fern, harvested just as it emerges from the ground. It consists of a tightly curled stem lined with the tiniest beginnings of leaves. It is usually handled and cooked like tender asparagus.

Evaluation: Look for jade green, firm and resilient, not wilted, fiddleheads with fresh-looking, not spoiled, cut ends and leaf tips.

Preparation: Trim the cut ends. Rinse well.

Percentage Yield: 85%

Fiddlehead ferns

Garlic

Identification: A pungent member of the onion family, garlic forms bunches of cloves attached at the root end and covered with a papery skin.

Related Varieties: There are many varieties, with white or purplish skins. Some are more pungent than others. Elephant garlic is the size of an apple, with relatively few large cloves; it is milder than other varieties.

Evaluation: Skin may be white or pink. No brown spots, soft spots, or spoilage; dry skin; no green shoots.

Preparation: Separate cloves as needed, or strike whole bulb with heel of hand to separate. To peel cloves, crush slightly with side of heavy knife. Peel and trim root end (see Figure 10.10).

Percentage Yield: 88%

FIGURE 10.10 Peeling and crushing garlic.

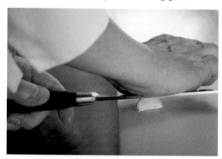

(a) Place the garlic on the worktable. Hold a broad knife blade over it as shown and strike it firmly with the palm of the hand.

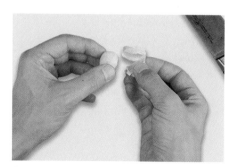

(b) You can now peel the garlic easily.

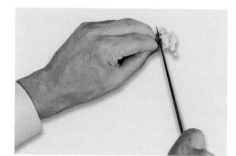

(c) Chop or mince the garlic.

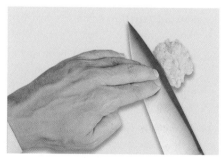

(d) To make a paste of the garlic, sprinkle it with salt and mash it firmly with the back of the knife blade.

Collard greens

Turnip greens

Kale

Greens, Cabbage Family (Collards, Turnip Greens, Kale)

Identification: These sturdy, flavorful greens are nearly always cooked because they are too tough to eat raw, unless very young. The kale varieties have curly or ruffled dark green leaves (except the ornamental varieties, which may have touches of red or other colors). Turnip greens resemble large, lobed mustard or arugula leaves. Collards are similar, generally with heavier stems and more rounded leaves.

Evaluation: Avoid browned, yellowed, or dried leaves. Look for smaller leaves that are dark green and have a moist feel.

Preparation: Strip the leaves off the stems if they are tough. Cut off any discolored parts. Wash well in several changes of water.

Percentage Yield: 80%

Jícama

Identification: Jícama (HEE kama) is a large, round tuber with a thin brown skin and crisp, mild, juicy, creamy white flesh similar to water chestnuts.

Evaluation: Look for smooth, almost shiny skin with no blemishes.

Preparation: Peel. Cut as desired.

Percentage Yield: 85%

Jícama

Kohlrabi

Identification: Kohlrabi, when trimmed, resembles a root vegetable, but it is actually a swollen stem. A member of the cabbage family, its pale white, crisp flesh resembles, in texture and flavor, the interior of a broccoli stem. The skin is tough and must be removed.

Related Varieties: Purple-skinned kohlrabi are similar to the more common green variety, except for skin color.

Evaluation: Look for uniform light green color; 2–3 inches (5–8 cm) in diameter. Crisp and firm. No woodiness.

Preparation: Peel like turnips, being sure to remove the full thickness of skin.

Percentage Yield: 55%

Kohlrabi

Leeks

Identification: A long, slender member of the onion family with distinctive flavor, milder than onions.

Related Varieties: *Ramps* are wild leeks with broad, flat leaves and bulbs that resemble those of scallions. They have a pungent, garlicky flavor.

Evaluation: Look for fresh green leaves; 2–3 inches (5–8 cm) or more of white. White part should be crisp and tender, not fibrous.

Preparation: Cut off roots and green tops. Cut deeply through white part, separate the layers slightly, and wash carefully to remove all embedded soil (see Figure 10.11).

Percentage Yield: 50%

Lettuce

Identification: See Chapter 21 for full description of salad greens.

Percentage Yield: 75%

FIGURE 10.11 Cleaning leeks.

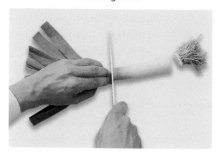

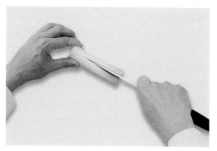

(a) Trim off the root end and as much of the green as desired.

(b) Make one or two deep cuts to within 1 in. (2.5 cm) of the root end.

(c) Spread apart the layers, as shown. Carefully wash out all embedded dirt under running water.

Mushrooms, White

Identification: Mushrooms are the fruiting body of a fungus—that is, the part of the fungus that produces spores for reproduction.

Evaluation: Select firm, white caps, closed at the stem. Stems should be relatively short. No dark spots, bruises, or mold.

Preparation: Trim bottoms of stems. Just before cooking, wash quickly in cold water; drain well. If you desire to keep the mushrooms white, add a small amount of acid (lemon juice, vinegar, ascorbic acid) to the rinse water. To flute mushrooms, see Figure 10.12.

Percentage Yield: 90%

White mushrooms

Mushrooms, Exotic and Wild Varieties

See pages 296–297.

FIGURE 10.12 Basic method for fluting mushrooms.

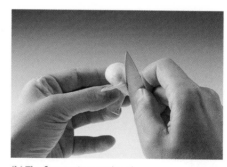

(a) Grasping the blade of a paring knife, hold the edge against the center of the mushroom cap at a sharp angle.

(b) The first cut is completed.

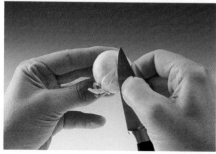

(c) Continue making cuts all around the mushroom.

(d) The fluted mushroom cap.

FIGURE 10.13 To prepare okra, cut off the stem ends.

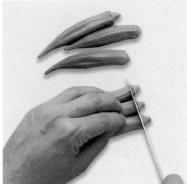

Okra

Identification: Okra is the pod or fruit of a member of the mallow family. It is long and tapered, with ridges running its length. Inside the pod are tender seeds and a thick, sticky juice. The flavor is very mild.

Related Varieties: Common okra varieties are green. Red-skinned varieties, which turn green when cooked, are uncommon but available.

Evaluation: Look for tender, full pods, not dry or shriveled. Ridges should be soft. Seeds should be soft and white. Uniform color.

Preparation: Wash. Trim ends (see Figure 10.13). Slice or leave whole.

Percentage Yield: 82%

Onions, Dry

Identification: A pungent, many-layered bulbous root that is the kitchen's most common vegetable.

Related Varieties: Onions come in almost limitless varieties, from tiny white, yellow, and red pearl onions and cipolline (chip oh lee neh, Italian pearl onions) to medium and large yellow onions, the workhorse of the kitchen, to large red and white onions. Mild varieties include Vidalia, Walla Walla, and Maui. These are often said to be sweeter, but they do not contain more sugar than regular varieties. Rather, they have less of the chemical compounds that make onions strong.

Evaluation: Select onions that are clean, hard, well shaped; no mold or black fungus; no green shoots. Skins should be very dry.

Preparation: Cut off root and stem ends. Peel. Wash. Cut or slice as needed (see Figure 7.16).

Percentage Yield: 90%

Yellow onions

Red onions

White onion

Pearl onions

Onions, Green (Scallions)

Identification: These are very young, immature bulb onions, sold with their stems. Scallions are almost always white varieties of onion, although red varieties can sometimes be found.

Evaluation: Scallions should have fresh, crisp green tops; little or no bulb formation at white part.

Preparation: Cut off roots and wilted ends of green tops. Amount of green left on varies with recipe or use.

Percentage Yield: 60–70%

Scallions

Parsley

Identification: Parsley is the most common fresh herb used in kitchens. It is used in such quantities that it is included here with vegetables.

Related Varieties: The two main varieties are curly parsley and flat or Italian parsley.

Evaluation: Select bright green, unwilted leaves with no rot.

Preparation: Wash well and drain. Remove yellow leaves and large stems (save stems for stocks). Separate into sprigs for garnish, or chop leaves.

Percentage Yield: 85%

Parsnips

Identification: The parsnip is a long, conical root vegetable that looks like a carrot, only more tapered and with a wider top. Its skin is tan and its interior is yellow-beige, with a distinctive, sweet flavor.

Evaluation: Look for firm, smooth, well-shaped parsnips, with light, uniform color. Large ones are often woody.

Preparation: Refrigerating for two weeks develops sweetness. Trim ends and peel. Rinse.

Percentage Yield: 70–75%

Parsnips

Pea Greens or Pea Shoots

Identification: These are the tender tips, with leaves, of pea vines, usually snow peas.

Evaluation: Look for fresh, medium-green leaves and tender, not woody, stems.

Preparation: Wash well; trim ends of stems, especially if tough. Cook like spinach.

Percentage Yield: 90–95%

Peas, Green

Identification: The tender, immature seed of a legume (see Chapter 13).

Evaluation: Look for firm, fresh, moderately filled-out pods. Peas lose sweetness quickly after harvest, so locally grown, just-harvested peas are best. Frozen peas are likely to be sweeter than fresh peas that have been stored for some time.

Preparation: Shell and rinse. (Peas are not often purchased by food-service operations because of the labor required for shelling.)

Percentage Yield: 40%

Pea shoots

Peas, Edible Pod

Identification: Unlike shell peas, above, whose pods are too fibrous to be eaten, these peas have tender, edible pods and so are always served in the pod.

Related Varieties: The two main varieties are the flat-podded snow pea, with tiny undeveloped seeds, and the rounder *sugar pea* or *sugar snap pea*, with larger seeds.

Evaluation: Look for fresh green color, crisp pods, no blemishes.

Preparation: Remove stem end. Pull off strings at side veins. Wash.

Percentage Yield: 90%

Snow peas   Sugar snap peas

FIGURE 10.14 Roasting peppers.

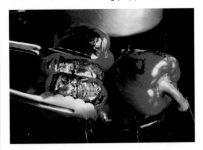

(a) Roast the peppers over an open flame until the skins blacken. For large quantities, you may do this under a broiler or in a hot oven. In this case, the skins will not darken as much but can still be peeled off.

(b) Wrap the peppers in plastic film while they are still hot. This helps loosen the skins.

(c) Peel off the loosened skin. You may do this under running water to help rinse off charred skin, but nutrients and flavor may be lost.

Peppers, Sweet

Identification: Peppers are the fruit of a plant related to tomatoes, eggplant, and potatoes. Both sweet and hot peppers are members of the *Capsicum* family. They have fleshy walls and a hollow interior, with a seedy core just below the stem. Also called *bell peppers*.

Related Varieties: Green, red, yellow, orange, and purple peppers are widely available.

Evaluation: Shiny color; well shaped; no soft spots or shriveling.

Preparation: Wash. Cut in half lengthwise and remove core, seeds, and white membranes. Peppers to be julienned or diced may have the interior ribs trimmed. Or leave whole (as for stuffed peppers) and cut out core from the end. Peppers are often roasted and peeled as preparation for use in recipes (see Figure 10.14).

Percentage Yield: 82%

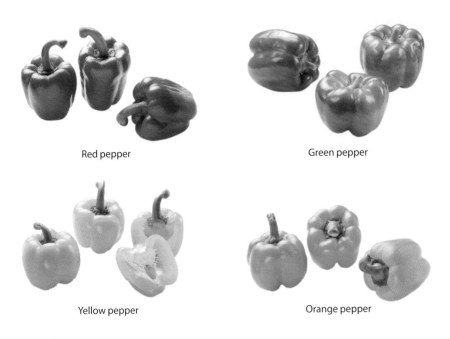

Red pepper

Green pepper

Yellow pepper

Orange pepper

Peppers, Hot, or Chiles

Identification: Relatives of sweet peppers, but containing a compound called *capsaicin*, which makes them spicy hot. (*Chile* is the original Spanish spelling; also spelled *chili* and *chilli*.)

Related Varieties: Many varieties are available worldwide. The heat of any particular pepper is determined by how much capsaicin it contains and is measured in Scoville units. A mildly hot jalapeño averages 2,500–3,000 units, while the intensely hot habanero averages around 200,000 units. Commonly used fresh chiles include the jalapeño, serrano, poblano, California, New Mexico, Thai green, and cayenne.

Evaluation: See Peppers, Sweet.

Green cayenne chiles

Italian hot peppers

Jalapeños and cherry peppers

Preparation: Larger fresh chile peppers, such as poblano, mulato, New Mexico, and Anaheim, are usually roasted and prepared like sweet peppers (see p. 290). Small peppers, such as cayenne, jalapeño, and serrano, are usually chopped or sliced and used as seasoning. Remove core, veins, and seeds carefully; wear rubber gloves if you are sensitive to the hot oils, and avoid touching the eyes or any sensitive part of the skin after working with chiles.

Percentage Yield: 80–90%

Hot pepper varieties: red habanero, caribe, orange habanero, and green chiles

Potatoes, White

Identification: See Chapter 12.

Percentage Yield: 80%

Potatoes, Sweet

Identification: Sweet potatoes are any of several types of tubers, usually bulbous, with tapered or pointed ends. They are unrelated to potatoes and to yams. (True yams are an entirely different vegetable, with starchy white flesh. They are not often seen in North America, except in Hispanic and some other specialty markets.)

Related Varieties: There are two basic groups of sweet potatoes, those with a creamy white or pale yellow flesh and those with orange flesh. The variety that has a moister, deeper orange flesh is sometimes referred to, incorrectly, as the **yam**. These varieties are interchangeable for most purposes. Skin colors range from beige to orange to red, depending on the variety.

Evaluation: Look for clean, dry surface. Firm, not shriveled or blemished. Fat, regular shapes are preferable because of less waste in trimming and portioning.

Preparation: Scrub, boil, or steam, then peel. May be peeled before cooking but must be dipped in antioxidant to prevent discoloring. Machine paring is wasteful with irregular shapes.

Percentage Yield: 80%

Sweet potato

Radishes

Identification: Crisp, juicy root vegetables in the mustard family, with a peppery taste.

Related Varieties: In North America and Western Europe, the word **radish** usually indicates the small, round or tapered, red-skinned and white-fleshed variety, or sometimes the white icicle radish. Other important radishes are the turnip-size **black radish**, eaten raw or cooked, and the very large, long Asian **daikon** (the name is Japanese for "big root"), with white skin and flesh, also served raw or cooked. Cooked radishes are somewhat like turnips in both flavor and texture.

Evaluation: Select firm, tender, crisp radishes, with good shape and color.

Preparation: Cut off root and stem ends. Wash. Decorative cuts are shown in Figure 10.15.

Percentage Yield: 90%

Radishes

Black radish

Daikon

FIGURE 10.15 Radishes can be cut into many decorative forms, including those shown here.

Rutabagas

See Turnip.

Salsify

Salsify

Identification: The name *salsify* (sal si fee) refers to either of two distantly related long, slender root vegetables with off-white flesh and a flavor, when cooked, sometimes said to resemble artichoke.

Related Varieties: True salsify, or white salsify, has brownish-white roots, often forked and branched, with many root hairs attached. Scorzonera (score tso nerra), or black salsify, is straight, long, and slender, with dark brown skin. Both have off-white flesh.

Evaluation: Select medium-size roots, as larger ones may be fibrous. They should be as smooth as possible.

Preparation: Scrub with a brush. Peel with a swivel peeler and immediately drop into water containing an acid, such as lemon juice or vinegar, to prevent discoloration. Wear gloves to prevent staining the hands.

Percentage Yield: 65–70%

Shallots

Shallots

Identification: Shallots are small, dry-skinned onions that grow in clusters attached at the root end. They have a brown, papery skin and white and purple flesh. Their flavor is milder than that of onions.

Evaluation: Same as for dry onions.

Preparation: Same as for dry onions.

Percentage Yield: Same as for dry onions.

Sorrel

Sorrel

Identification: A tender leafy green with a tart, astringent taste, used in small quantities, usually to flavor sauces. The leaf resembles spinach but is longer and not as dark green.

Evaluation: Same as for spinach.

Preparation: Pull off coarse stems. Wash in several changes of water if sandy. For most purposes, sorrel is cut into chiffonade (p. 146).

Percentage Yield: 90%

Soybean

Identification: Most soybeans are allowed to mature and dry and are harvested and processed as a grain. The use of tender, unripe soybeans, known as *edamame* (edd ah mah meh), is a popular adoption from Japanese cuisine. The small, fuzzy pods usually hold two or three light green, roundish beans.

Evaluation: Same as shell beans.

Preparation: Rinse, boil in the pod until the beans are tender, and shell.

Percentage Yield: 50%

Green soybeans

Soybeans in the pod

Spinach

Identification: A tender, dark green leaf, probably the most popular of leafy vegetables after lettuce.

Related Varieties: Two main types are curly-leafed and flat-leafed.

Evaluation: Look for fresh, crisp, dark green leaves. No rot or slime or badly bruised leaves.

Preparation: Remove stems and damaged leaves (Figure 10.16). Wash in several changes of water. Use a large quantity of water and lift spinach up and down to float off sand and dirt. Lift from water and drain well.

Percentage Yield: 50–70%

FIGURE 10.16 When trimming spinach leaves, remove the heavy center rib along with the stem.

Squash, Summer

Identification: Several types of tender-skinned members of the gourd family are called *summer squash*. All have tender seeds in a firmly filled seed cavity, unlike the hard seeds encased in a fibrous mesh in the hollow interior of winter squash.

Related varieties: Green-skinned *zucchini* is well known and the most popular. *Yellow straight* and *crookneck* squashes are similar in flavor and use. *Pattypan* squash has a squat, round shape, slightly resembling the popular idea of a flying saucer. They can be yellow, dark green, or pale green. Many other, less common types are available.

Yellow summer squash

Zucchini

Evaluation: Look for firm, heavy, and crisp squash, with tender skin, no blemishes.

Preparation: Wash or scrub well. Trim ends.

Percentage Yield: 90%

Squash, Winter, including Pumpkin

Identification: Winter squash are mature members of the gourd family, with thick skins and hard seeds in a hollow cavity. They are starchier and less moist than summer squash and, unlike summer squash, are not generally eaten raw.

Related Varieties: Many types, of all sizes and shapes, are available, including Hubbard, acorn, butternut, buttercup, kabocha, turban, delicata, sweet dumpling, and pumpkin. Spaghetti squash has a fibrous flesh that, when cooked and scraped out of the shell, resembles a tangled pile of spaghetti.

Spaghetti squash

Evaluation: Heavy and firm. Hard rind. No blemishes.

Preparation: Wash. Cut in half. Scrape out seeds and fibers. Cut into portion sizes. For puréed or mashed squash either steam or bake, then remove peel; or peel, dice, then steam.

Percentage Yield: 65–85%

Acorn squash

Butternut squash

Buttercup squash

Squash blossoms

Squash Blossoms

Identification: Young, yellow blossoms of any summer squash.

Related Varieties: Female flowers are attached to the ends of new squashes. They fall off after the squash is a day or two old so must be harvested, attached to the tiny fruit, in a timely manner. Male blossoms contain the pollen and grow on a separate part of the same plant. Both types are used, often for stuffing, or simply sliced and sautéed.

Evaluation: Squash blossoms should be fresh and bright yellow, not wilted. They are best if used within a day or two of harvest.

Preparation: Open the flower petals gently and carefully rinse in water.

Percentage Yield: 100%

Sunchokes

Sunchokes or Jerusalem Artichokes

Identification: Sunchoke is a relatively new marketing term for a Jerusalem artichoke, created because the vegetable is not an artichoke and has nothing to do with Jerusalem. Sunchokes are knobby brown tubers with crisp, creamy white flesh.

Evaluation: Look for firm sunchokes with clean brown skin, no soft or green spots, and no sprouts.

Preparation: Wash and peel off brown skin.

Percentage Yield: 80%

Red-stemmed Swiss chard

Swiss Chard

Identification: A leafy green related to the beet. Chard leaves resemble beet greens in both flavor and appearance, but the center stalks of chard are broader.

Related Varieties: Red-stemmed, white-stemmed, and yellow-stemmed chard are available.

Evaluation: Fresh, dark green color, not wilted. Crisp, not wilted or rubbery stalks.

Preparation: Wash thoroughly. Trim ends of stalks. Remove leafy greens from stalks by cutting along the margins of the stalk down the center of the leaf. Center stalks may be cooked separately like asparagus.

Percentage Yield: 85%

White-stemmed Swiss chard

Tomatoes

Identification: One of the most widely used vegetables, the fruit of a plant in the nightshade family, related to eggplant, potatoes, and peppers.

Related Varieties: In addition to the common red slicing tomato and the small cherry tomato, dozens of varieties of heirloom tomatoes are grown and marketed. These come in all shades of red, pink, yellow, and green, all with somewhat different flavors.

Evaluation: Firm but not hard, with little or no green core. Smooth, without bruises, blemishes, cracks, or discoloration. If underripe, let stand two to three days at room temperature. Do not refrigerate.

Preparation: For use with skin on: Wash, remove core. To peel: Plunge into boiling water 10–20 seconds (riper tomatoes take less time). Cool immediately in ice water. Slip skins off and remove core. (Note: Many chefs core the tomatoes and cut a shallow x in the skin

Tomato varieties

FIGURE 10.17 Preparing tomato concassé.

(a) Blanch and peel the tomato and cut it in half crosswise. Gently squeeze out the seeds, as shown.

(b) Dice the seeded tomato, or chop it coarsely.

at the bottom before blanching in boiling water. Other chefs, wishing to avoid even the slightest loss of flavor and nutrients from the exposed flesh into the water, first blanch the tomatoes and then core them and pull off the skins, which slip off just as easily.) See Figure 10.17 for further techniques.

Percentage Yield: 90% (peeled)

Tomatillos

Identification: A small green fruit vegetable resembling a green cherry tomato encased in a papery husk. It is not a tomato, although it is in the same family as tomatoes, eggplant, peppers, and potatoes. Widely used in Mexican cuisine.

Evaluation: Look for firm, dry tomatillos with tight husks.

Preparation: Remove the husks and rinse.

Percentage Yield: 95%

Turnips and Rutabagas

Identification: Root vegetables in the mustard family, related to cabbages.

Related Varieties: Turnips are white with purple skin around the stem end. The flesh is white. Rutabagas, also called *swedes*, are much larger and have yellow-orange flesh.

Evaluation: Look for roots that are firm and heavy, with good color and no blemishes. White turnips over 2^1/$_2$ inches (6–7 cm) in diameter may be woody or spongy.

Preparation: Peel heavily by hand or in machine to remove thick skin (see Figure 10.18). Rinse.

Percentage Yield: 75–80%

Tomatillos

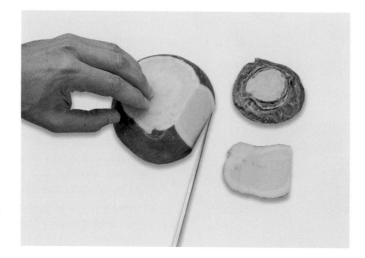

FIGURE 10.18 Peel rutabagas and turnips deeply enough to remove the full thickness of skin, as pointed out in this photograph.

Water Chestnuts

Identification: Water chestnuts are corms, or swollen underground stem bases, of plants that grow in moist, warm environments. They are small and roundish, with dark brown skin and crisp white flesh.

Evaluation: Look for solid water chestnuts with no sign of softness, bruising, rot, or shriveling. They spoil easily.

Preparation: Scrub well with a brush, as they are often muddy. Peel with a paring knife.

Percentage Yield: 75%

Watercress

Identification: See discussion of salad greens in Chapter 21.

Evaluation: Bright green, crisp, unbruised leaves.

Preparation: Wash well. Remove heavy stems and discolored leaves.

Percentage Yield: 90%

Watercress

MUSHROOMS

Because of the great interest today in exotic mushrooms, and because eating some poisonous species of wild mushroom can be fatal, it is important for the cook to be familiar with at least the most popular varieties of exotic mushrooms, both cultivated and wild.

Although hundreds of mushroom varieties are edible, until recently only the common cultivated button mushroom was used with any frequency in most commercial kitchens. Now, however, many varieties are available. Some of these mushrooms, especially the wild ones, are expensive, but the demand always seems to exceed the supply.

Strictly speaking, the term *wild* should be used only for those mushrooms that are not cultivated but rather hunted and gathered in the wild. In the kitchen and on menus, however, exotic cultivated varieties, such as shiitakes, are often referred to as "wild mushrooms" because they are seen as rare and unusual, like true wild mushrooms, and they are generally more flavorful than the button mushroom.

One important advantage of cultivated exotic mushrooms is that they are available all year, while certain wild mushrooms may be in season only a few weeks annually.

Cultivated Exotic Mushrooms

1. **Shiitake.**
   Sometimes known as **Black Forest mushroom** or **golden oak mushroom**, the shiitake is also available in dried form as **Chinese black mushroom**. The fresh mushroom is golden brown to dark brown. It has a firm, fleshy texture and a broad, dome-shaped cap with creamy white gills. The stem is rather tough, so it is trimmed off and chopped fine or used in stocks.

2. **Oyster mushroom.**
   Also called *pleurotte*, it is a light tan or cream-colored fan-shaped mushroom with a short stem at the side. Tender, with delicate flavor, it is best prepared simply so its mild flavor is not overwhelmed by stronger-tasting ingredients. (**Note**: The name *oyster* refers to the shape of the mushroom, not its taste.)

3. **Enoki mushroom.**
   Also called *enokitake* or *enokidake*, this mushroom has a tiny white cap on a long, slender stem, and it grows in clusters or bunches attached at the base. The base is trimmed off before use. The enoki mushroom has a crisp texture and a fruity, slightly acidic but sweet flavor. It is often used raw (for example, in salads or as garnish) or in clear soups. When used in cooked dishes, it should be added in the last few minutes so as not to be overcooked.

Shiitake mushrooms

Oyster mushrooms

4. Cremini mushroom.

The cremini is a variety of the common cultivated button mushroom, but it has a brown or tan skin. It may have a slightly more robust flavor than white cultivated mushrooms.

5. Portobello mushroom.

This is a mature cremini whose cap has opened and spread into a broad, flat disk. It may be 6 inches (15 cm) or more across. Portobello mushrooms are often grilled, brushed with olive oil, and served plain as a first course.

Cremini mushrooms

Wild Mushrooms

Of the many varieties of edible wild mushrooms, those described here are among the most prized as well as the most likely to be found on menus. As a rule, they are expensive and of limited availability.

Wild mushrooms should be carefully examined for spoilage and insect infestation. Cut away any damaged parts.

The four varieties described here are also available dried (see p. 300). Dried mushrooms have a high price per pound but are more economical to use than fresh wild mushrooms because they are equivalent to 7 or 8 times their weight of fresh mushrooms. In addition, they have a more intense, concentrated flavor, so a little goes a long way.

Caution: Never eat any wild mushroom that has not been identified by an expert. Many mushrooms are poisonous, and some are deadly. Many species are difficult to identify, and some poisonous varieties resemble edible ones.

Chanterelle mushrooms

1. Morel.

Several varieties exist, including black, golden, and nearly white. The morel is shaped somewhat like a conical sponge, with a pitted surface, on a smooth stem. It is completely hollow. The most prized of spring mushrooms, it is usually sautéed in butter or cooked in a sauce and is especially good with cream.

2. Bolete.

Other names for this mushroom include *cep*, *cèpe* (sepp; the French term), *porcino* (por chee no; the Italian term; the plural is *porcini* [por chee nee]), and *steinpilz* (shtine pilts; the German term). It is a brown-capped mushroom with a light-colored, bulbous stem. The interior flesh is creamy white. The underside of the cap has no gills but many tiny pores. With a meaty but smooth texture and rich, earthy flavor, it is often sautéed or braised with garlic and olive oil or butter. It is available late summer to fall.

3. Chanterelle.

Also called *girolle*, the chanterelle is yellow to orange in color and shaped like an umbrella that has turned inside out. The underside of the cone-shaped cap has ridges instead of gills. It has a rich, woodsy aroma and flavor and is best cooked simply, such as sautéed in butter, perhaps with garlic. It is available summer and fall.

4. Black trumpet.

This mushroom is closely related to the chanterelle but is black in color and has much thinner flesh. It is also called *black chanterelle*, *horn of plenty*, and *trompette de la mort* (French name, meaning "trumpet of death," so called because of its black color). In spite of this French name, it is edible and delicious.

KEY POINTS TO REVIEW

- What basic preparation techniques are common to all vegetables?

- For each vegetable commonly used in food service, what standards are used to evaluate freshness?

- For each vegetable commonly used in food service, what basic preparation techniques are used?

- Outside of standard white button mushrooms, what are the five most commonly used varieties of cultivated mushroom? Name four varieties of wild mushroom that are often purchased dried.

PROCESSED VEGETABLES

It is generally agreed that the quality of frozen and canned vegetables can never equal that of the best-quality fresh product at its peak of maturity, prepared properly, and cooked while still fresh. However, because of the high perishability of fresh produce, seasonal variations in availability and price, and the amount of labor required to handle fresh produce in commercial kitchens, food service relies, to a great extent, on processed vegetables. Therefore, it is important to know how to handle processed foods properly. Your goal should be to make them as close as possible in quality to the best fresh produce.

The quality of processed vegetables varies greatly. For example, frozen cauliflower always lacks the slightly crunchy texture of properly cooked fresh cauliflower. In fact, most frozen vegetables are a bit mushier than fresh because cell walls rupture during freezing. On the other hand, frozen peas are almost universally accepted, not just for their convenience but for their dependably high quality in comparison with the highly perishable fresh product.

In the section of Chapter 7 called "Handling Convenience Foods," we learned that convenience foods are products that are partially or completely prepared or processed by the manufacturer. This means you should treat frozen and canned vegetables as though they are partially or fully cooked fresh vegetables, which deserve the same care in handling, heating, seasoning, and presentation.

HANDLING FROZEN VEGETABLES

Checking Quality

Examine all frozen products when received to make sure they have experienced no loss of quality. Check in particular for the following:

1. **Temperature.**
   Check the temperature inside the case with a thermometer. Is it still 0°F (–18°C) or below, or have the vegetables begun to thaw during shipment?

2. **Large ice crystals.**
   A little frost is normal, but lots of ice means poor handling.

3. **Signs of leaking on the carton.**
   This is another obvious sign of thawing.

4. **Freezer burn.**
   Open a package and check the vegetables themselves. Is the color bright and natural, or is there yellowing or drying on the surface?

Cooking

Frozen vegetables are partially cooked, so final cooking time is shorter than for fresh products.

Cook from the frozen state. Most vegetables need no thawing. They can go directly into steamer pans or boiling salted water.

Exceptions: Corn on the cob and vegetables that freeze into a solid block, such as spinach and squash, should be thawed in the cooler first for more even cooking.

Seasoning: Most frozen vegetables are slightly salted during processing, so add less salt than you would to fresh products.

HANDLING CANNED VEGETABLES

Checking Quality

1. **Reject damaged cans on receipt.**
   Puffed or swollen cans indicate spoilage. Small dents may be harmless, but large dents may mean the can's protective lining has been damaged. Avoid rusted or leaking cans.

2. **Know the drained weight.**
   This varies with different grades of different vegetables and should be specified when ordering. Typical drained weights are 60 to 65 percent of total contents. You must know this drained weight in order to calculate the number of servings the can contains.

   Some canned products, such as tomato sauce and cream-style corn, have no drained weight because the entire contents are served.

3. **Check the grade.**
   Grades are determined by the packers or by federal inspectors. They are based on factors like color, absence of defect, and **sieve size** (size of individual pieces). Check to make sure you receive the grade you ordered (and paid for).

In the United States, the federal grades are

U.S. Grade A or Fancy

U.S. Grade B or Extra Standard (for vegetables) or Choice (for fruits)

U.S. Grade C or Standard

Cooking

1. Wipe the top of the can clean before opening. Use a clean can opener.

2. Drain the vegetable and place half the liquid in a cooking pot. Bring it to a boil. This shortens the heating time of the vegetable.

3. Add the vegetable and heat to serving temperature. Do not boil for a long time. Canned vegetables are fully cooked—in fact, usually overcooked. They need reheating only.

   Note: Health officials recommend holding canned vegetables at 190°F (88°C) 10 minutes or more—20 to 30 minutes for nonacid vegetables like beets, green beans, or spinach—to eliminate the danger of botulism. See Chapter 2.

4. Heat as close to serving time as possible. Do not hold in steam table for long periods.

5. Season and flavor with imagination. Canned vegetables require more creativity in preparation than fresh because they can be dreary when served plain.

6. Season the liquid while it is coming to a boil, before you add the vegetable. This will give the flavors of the herbs and spices time to blend.

7. Butter enhances the flavor of most vegetables, and it carries the flavors of the other seasonings you choose to add.

8. Dress up the vegetables with flavors and garnishes, such as beets or sauerkraut with caraway, limas or green beans with crisp crumbled bacon, corn with sautéed minced onion and green or red pepper, carrots with butter and tarragon or orange juice and brown sugar.

The combinations suggested in the table in Chapter 11 apply to canned vegetables as well as to fresh and frozen.

HANDLING DRIED VEGETABLES

There are two basic kinds of dried vegetable.

Dried Legumes

Dried beans and peas have been used as food for thousands of years, and they continue to be important foods today. In fact, with today's increased interest in healthful eating and in vegetables of all sorts, many more interesting varieties of beans are widely available now than only a few years ago.

Although legumes are dried forms of seed vegetables, they are hard and starchy and handled much like grains (which are also dried seeds). Grains and legumes are discussed in more detail in Chapter 13.

Freeze-Dried and Other Dehydrated Vegetables

Drying has always been an important method for preserving vegetables, especially before modern canning and freezing techniques were developed. Modern technology has developed new methods for drying foods, so a great variety of dried products is on the market, including dried potatoes, onions, carrots, celery, beans, peppers, tomatoes, and mushrooms.

Follow manufacturers' directions for reconstituting these products. Many must be soaked in cold or warm water for specific lengths of time. They continue to absorb water as they are simmered.

Instant dried products, especially potatoes, require only the addition of a boiling liquid and seasonings to be ready to serve. Again, manufacturers' directions vary with their brands.

Dried mushrooms, clockwise from top left: morels, porcini, shiitake, chanterelles.

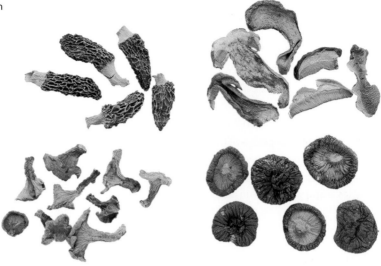

An important category of dried vegetable is dried mushrooms. Many flavorful wild mushrooms are in season for only a short time and are in limited supply at high prices. They are available year round in dried form, however. The most popular types—morels, chanterelles, and porcini—are illustrated, along with dried shiitake, a cultivated mushroom that originated in Japan and China.

Dried mushrooms should be soaked in hot water until soft, then drained and lightly squeezed before being cooked. The flavorful soaking liquid is strained and used as a flavoring for cooking liquids and sauces.

PRODUCTION AND HOLDING PROBLEMS IN QUANTITY COOKING

We have emphasized throughout this chapter that vegetables should be cooked as close as possible to serving time. They lose quality rapidly when held in a steam table.

In quantity cooking, however, it is rarely possible to cook individual vegetable portions to order. After 20–30 minutes at steam table temperatures, even carefully prepared vegetables are usually overcooked.

Two systems have been devised to help solve this problem. **Batch cooking** is especially well suited to set meal service, and the **blanch-and-chill** method is most helpful in extended meal service. Needs vary from institution to institution, and you will probably find both techniques useful in one kitchen.

BATCH COOKING

Rather than cooking all your vegetables in one batch large enough for the entire meal service, this method (described in Chapter 7, p. 137) involves dividing the food into smaller batches and cooking them one at a time, as needed.

BLANCH-AND-CHILL METHOD

It is usually impractical to cook vegetables completely to order. Too much time is required. But if the vegetables have been partially cooked, the time needed to finish them to order is short.

Partially cooking, chilling, and finish-cooking is not as good, nutritionally, as cooking completely to order, but it is almost as good. It's certainly better than holding vegetables for hours at serving temperature, and it gives the cook complete control over the degree of doneness when served.

PROCEDURE **for Batch Cooking**

1. Steamers and small tilting trunnion kettles behind the service line are the most useful kinds of equipment for vegetable batch cooking.

2. Divide each vegetable into batches small enough to be served within 20 to 30 minutes. Arrange in steamer pans ready to be placed in steamers or in containers ready for pouring into the kettles.

3. Keep the prepped vegetables in the cooler until needed.

4. Cook batches as needed. In planning, allow time for loading and unloading the equipment, for cooking, for finishing the product with the desired seasoning, sauce, or garnish, and for carrying to the serving line.

5. Undercook slightly if the vegetable must be held before serving.

6. Have all your seasonings, sauces, and garnishes ready for finishing the dish.

7. Do not mix batches. They will be cooked to different degrees, and colors and textures usually will not match.

PROCEDURE **for Blanching and Chilling**

1. Steam or simmer the vegetable until partially cooked to the desired degree. (In the case of French fries, blanch by deep-frying.)

   The amount of cooking required depends on the vegetable and on the method by which it will be reheated or finished. Frozen vegetables need less cooking than fresh. Often, they need only be thawed.

2. Chill immediately in ice water. (Needless to say, French fries are an exception.)

3. Drain and keep chilled until needed.

4. Finish to order by desired cooking method.

   For example, one or more portions can be placed in a strainer and lowered briefly into a ready pot of boiling water.

   Sautéing in butter is a popular method for finishing such items as peas, green beans, and carrots.

   Potato croquettes are an example of a more complicated application of this same method. The potatoes are boiled or steamed, puréed, seasoned, formed, and breaded in advance. They are then deep-fried to order.

STORAGE

FRESH VEGETABLES

1. Potatoes, onions, and winter squash are stored at cool temperatures (50°–65°F/10°–18°C) in a dry, dark place.

2. Other vegetables must be refrigerated. To prevent drying, they should be kept covered or wrapped, or the humidity in the cooler must be high. Allow for some air circulation to help prevent mold.

3. Peeled and cut vegetables need extra protection from drying and oxidation. Cover or wrap, and use quickly to prevent spoilage. Potatoes, eggplants, and other vegetables that brown when cut should be treated with an acid or antioxidant. As an alternative, they can be blanched to destroy the enzymes that cause browning. Raw cut potatoes are sometimes held in cold water for a short time.

4. Store all fresh vegetables as short a time as possible. They lose quality rapidly. Peas and corn lose sweetness even after a few hours in storage.

5. Keep refrigerators and storage areas clean.

FROZEN VEGETABLES

1. Store at 0°F (–18°C) or colder, in original containers, until ready for use.

2. Do not refreeze thawed vegetables. Quality will be greatly reduced.

DRIED VEGETABLES

1. Store in a cool (less than 75°F/24°C), dry, well-ventilated place.

2. Keep well sealed and off the floor.

CANNED VEGETABLES

1. Keep in a cool, dry place, away from sunlight and off the floor.

2. Discard cans that show signs of damage or spoilage (swollen, badly dented, or rusted cans). When in doubt, throw it out.

LEFTOVERS

1. The best way to store leftovers is not to create them in the first place. Careful planning and small-batch cooking reduce leftovers.

2. Don't mix batches.

3. Store leftover creamed vegetables for one day only. Then either use or discard. Before storing, cool rapidly by placing the container on ice.

KEY POINTS TO REVIEW

- How do you check frozen vegetables and canned vegetables for quality?

- What guidelines are used for handling and cooking frozen vegetables and canned vegetables?

- What are the steps in the procedure for batch-cooking vegetables? For what kind of meal service is this method usually used?

- What are the steps in the blanch-and-chill method for cooking vegetables? For what kind of meal service is this method usually used?

- What guidelines are used for storing fresh vegetables? frozen vegetables? canned vegetables? dried vegetables? vegetable leftovers?

TERMS FOR REVIEW

al dente

pigment

anthoxanthins

flavonoids

anthocyanins

chlorophyll

carotenoids

sieve size

batch cooking

blanch-and-chill

QUESTIONS FOR DISCUSSION

1. Give two reasons for not adding baking soda to the cooking water for green vegetables.

2. Besides appearance, why is proper uniform cutting of vegetables important?

3. What are some advantages of steam-cooking vegetables over boiling or simmering?

4. You are trying a recipe for blueberry muffins. When you break open a finished muffin, you see the baked dough around each berry is green. What caused this? How can you correct it? (Hint: The batter is made with buttermilk and leavened with baking soda. Even though berries aren't discussed in this chapter, the information about color changes will enable you to answer this question.)

5. Discuss the reasons for cooking green vegetables in a large quantity of water and in just enough water to cover.

11

COOKING VEGETABLES

Now that you have studied the whys and wherefores of vegetable cooking, you should be able to proceed to actual preparation with a clear understanding of what you are doing.

This chapter outlines the basic methods for cooking vegetables. Successful performance of these methods relies on your knowledge of the principles we have discussed.

The recipes given here reinforce your understanding through actual practice. The emphasis is on the method rather than on the particular vegetable used because each method applies to many vegetables. For this reason, variations are listed after basic recipes rather than presented as separate complete recipes. As in other chapters, recipes for sauces that appear in Chapter 8 and are used as components of subsequent preparations are not repeated here.

Most of the recipes are applicable to fresh, frozen, or canned vegetables, even though variations are not listed for each. You have learned how to handle these products in order to make proper substitutions. Review pages 297–300 if necessary.

Potatoes and other starchy foods, such as rice and pasta, are covered in following chapters. However, the basic cooking methods here apply to potatoes as well as to other vegetables.

AFTER READING THIS CHAPTER, YOU SHOULD BE ABLE TO

1. Identify vegetables that are well suited to the different vegetable cooking methods.
2. Cook vegetables by boiling and steaming.
3. Cook vegetables by sautéing and pan-frying.
4. Cook vegetables by braising.
5. Cook vegetables by baking.
6. Cook vegetables by broiling and grilling.
7. Cook vegetables by deep-frying.

BOILING AND STEAMING

Nearly all vegetables may be cooked by boiling or by steaming. These are the two most frequently used methods because they are easy, economical, and adaptable to a great variety of preparations.

Boiling and steaming are basic cooking methods. In most cases, additional steps are required after the basic cooking is completed in order to make the product ready for serving. These steps include adding butter, seasonings, flavorings, and sauces.

Unless they are to be served immediately, boiled or simmered vegetables are drained as soon as they are cooked and then cooled quickly under cold water or in an ice-water bath. This is called **shocking** or **refreshing**, and it prevents the vegetables from being overcooked in their retained heat. The vegetables should be drained as soon as cool. Normally, they are then reheated quickly by sautéing in butter or other fat. Seasonings and sauces can be added at this stage. We classify this cooking method as boiling even though the vegetables are finished by sautéing to reheat them. This section includes recipes for boiled vegetables finished with a variety of sauces, flavorings, and seasonings.

In other cases, the product is only partially cooked by boiling or steaming and is finished by another cooking method, such as sautéing or baking. Recipes of this sort are usually included under the final cooking method.

We speak of the cooking method as **boiling** even though, in many cases, **simmering** is a more appropriate term. Green vegetables are generally boiled so they cook quickly, preserving color and nutrients. If the vegetables are delicate, however, the agitation and high temperature of boiling is likely to break them up, and simmering is more appropriate.

Except for starchy vegetables, most vegetables are best cooked in salted water, as explained in the Salted Water sidebar.

Steaming as a method for cooking vegetables is becoming more and more widely used, especially as more varieties of advanced equipment become available. It may be the ideal method for cooking certain vegetables, such as broccoli, that easily break or turn watery or mushy when simmered.

A useful technique that combines boiling and steaming is called **pan-steaming**. In this method, vegetables are cooked in a covered pan in a small amount of water—not enough water to cover them—so they are partly cooked by the steam trapped in the pan. After the vegetables are cooked, they are removed from the pan, and the flavorful cooking liquid may be reduced to a glaze or made into a sauce. This is a quick method for cooking many vegetables and so is useful for cooking small quantities to order.

Finally, this section also contains recipes for vegetables cooked sous vide. It is essential that you read the discussion of sous vide cooking, and especially the safety precautions, on page 119 before you try these recipes.

SALTED WATER

Cooking vegetables in salted water helps their flavor and nutritional value. If vegetables are cooked in unsalted water, the water extracts salts and sugars from the vegetable cells. When salted water is used, on the other hand, more of these compounds stay inside the vegetables.

Salted water also affects the texture of vegetables. In some regions, tap water is hard, meaning that it contains many dissolved minerals, especially calcium. When vegetables are cooked in hard water, the calcium makes the vegetables firmer, because the calcium reinforces the cellulose fiber in the cell walls, making it stronger. If salt is added to the water, the sodium from the salt replaces the calcium, which weakens and partially dissolves the fiber. For this reason, it is often advised to cook starchy vegetables such as potatoes, in unsalted water (see p. 349).

PROCEDURE for Boiling Vegetables

1. Collect all equipment and food products.

2. Trim, peel, and cut vegetables as required. See pages 276–296 for prep requirements.

3. Add the required amount of water to the pot (saucepot, steam-jacketed kettle, tilting skillet, or whatever equipment you are using).

   Most vegetables are cooked in just enough water to cover, but many green vegetables and strong-flavored vegetables may be cooked in a large quantity of water (2 or 3 times their volume). See page 272 for discussion.

4. Add salt (1½–2 tbsp per gal water/6–8 g per L) and bring to a boil.

5. Place the vegetables in the pot and return the water to a boil.

6. Reduce heat to a simmer and cook the vegetables, covered or uncovered, as indicated, to required doneness.

   Green vegetables and strong-flavored vegetables are cooked uncovered. Other vegetables are cooked covered.

7. Drain the vegetables quickly to avoid overcooking.

8. Evaluate the quality of the cooked vegetable (see p. 273).

9. If the vegetables are to be served at once, complete the recipe and serve.

10. If the vegetables are not to be served at once, cool them (except potatoes and starchy vegetables) in ice-cold water, drain as soon as cool, and refrigerate until needed.

PROCEDURE **for Steaming Vegetables**

This method is used both for pressurized and nonpressurized compartment steam cookers and for simple rangetop steamers that consist of a perforated basket over a pot of boiling water.

1. Know your equipment. Read all operating instructions supplied with your equipment. Each model is a little different.
2. Collect all equipment and food products.
3. Trim and cut vegetables as required.
4. Preheat the steamer.
5. Arrange vegetables in pans or baskets for cooking. Make shallow, even layers for uniform cooking.
   - Use perforated pans for best steam circulation.
   - Use solid pans if cooking liquid must be retained.
6. Insert pans or baskets in steamer and close door or lid (Figure 11.1).
7. Steam for required period. Consult timing charts supplied with your model of steamer.
8. Remove vegetables from steamer. If it is a pressure steamer, pressure must return to zero before door is opened.
9. Evaluate the quality of the cooked vegetable (see p. 273).
10. Finish vegetables according to recipe and serve at once, or cool quickly for later use.

Vegetables that form compact layers do not steam well. They do not allow the steam to circulate, so they cook unevenly. Examples: spinach and other greens, peas, whole-kernel corn, frozen puréed squash.

FIGURE 11.1 Compartment steaming.

PROCEDURE **for Pan-Steaming Vegetables**

1. Collect all equipment and food products.
2. Trim, peel, and cut vegetables as required. See pages 276–296 for prep requirements.
3. Add the vegetables and a small amount of water to an appropriate pan, such as a sautoir, large enough to hold the vegetables in a shallow layer. The amount of water needed depends on the cooking time of the vegetable. Use less liquid for quick-cooking vegetables, more liquid for long-cooking vegetables. In most cases, the water should cover the vegetables by no more than about three-quarters.
4. Add salt, and bring the water to a boil (Figure 11.2). Other seasonings may be added to the water if desired.
5. Add any other ingredients as directed by the recipe.
6. Cover the pan and, if necessary, lower the heat to maintain a simmer. Cook until the vegetable is done. During cooking, remove the cover from time to time to check doneness and to make sure the water hasn't all evaporated. Add more water if the pan is going dry.
7. Remove the vegetable with a slotted spoon.
8. If desired, reduce the liquid to a glaze to toss with the vegetable. As an alternative, reduce the liquid as desired and finish it to make a sauce. For example: Thicken with a starch slurry or monter au beurre.
9. Evaluate the quality of the finished vegetable (see p. 273).

FIGURE 11.2 Pan-steaming vegetables.

(a) Place the vegetables in a sauté pan and add a small amount of water. Bring to a boil, and then cover.

(b) When the vegetables are cooked, remove them from the pan with a slotted spoon.

(c) Reduce the cooking liquid and finish the sauce as desired, such as with butter.

(d) Return the vegetables to the pan and toss with the sauce. Add herbs or other flavorings and seasonings as desired.

PURÉEING VEGETABLES

Vegetable purées are served as side dishes or garnishes, and they are also used as ingredients in other preparations. Purées were introduced in the discussion of soups in Chapter 9. The vegetable purées we are discussing here are similar, although they are of course much thicker than soups. Procedures for puréeing vary depending on the vegetable and how the purée is to be used, but some general guidelines can be stated.

GUIDELINES for Puréeing Vegetables

1. Vegetables to be puréed should be cooked until tender. Undercooked vegetables make a grainy purée. Be careful not to overcook green vegetables, which discolor when cooked too long.

   Exception: A few vegetables, such as avocados and tomatoes, are tender enough to be puréed raw. Whether or not they are cooked depends on the vegetable and the desired use for the purée.

2. Vegetables for puréeing can be cooked by any method, but the most commonly used methods are boiling, steaming, and baking. Baking is used most often for starchy vegetables such as sweet potatoes and white potatoes.

3. Vegetables cooked by boiling or simmering should be drained well when done. If a stiff purée is desired, it may be necessary to dry the cooked vegetable before puréeing. Place the colander of drained vegetables on a sheet pan and set in a moderate oven for a few minutes to dry.

4. Always purée cooked vegetables while they are still hot. Cold vegetables are harder to purée to a smooth texture.

5. Select appropriate equipment. Ricers, food processors, vertical cutter/mixers (VCMs), food mills, and sieves can be used for puréeing. Food mills, ricers, and sieves make lump-free purées, but they may leave a grainy texture, depending on the vegetable, although a fine drum sieve (tamis) makes a very smooth purée (Figure 11.3). Food processors and VCMs can create a smooth texture, especially with starchy vegetables, but they do not eliminate vegetable fibers and may leave a few lumps. First using a processor and then passing the purée through a food mill or tamis yields the smoothest result.

6. Season and finish the purée as desired. Butter or cream is often used to finish vegetable purées. Alternatively, use the purée as an ingredient in another recipe as directed.

FIGURE 11.3 Puréeing vegetables using a tamis.

(a) Rub the cooked vegetable through the tamis using a pestle (as shown) or a plastic scraper.

(b) Scrape the purée from the bottom of the sieve.

KEY POINTS TO REVIEW

- What are the steps in the procedure for boiling vegetables?

- What are the steps in the procedure for steaming vegetables? for pan-steaming vegetables?

- Vegetables to be puréed should be cooked to what doneness? What equipment is used to purée vegetables? List other guidelines for making puréed vegetables.

Peas, Carrots, and Pearl Onions with Tarragon Butter

PORTIONS: 15   PORTION SIZE: 3 OZ (100 G)

U.S.	METRIC	INGREDIENTS
12 oz	375 g	Pearl onions
1 lb	500 g	Carrots
1 lb 8 oz	750 g	Peas, frozen
2 oz	60 g	Butter
1½ tsp	7 mL	Dried tarragon
1 tbsp	15 mL	Chopped parsley
to taste	to taste	Salt
to taste	to taste	White pepper

Per serving: Calories, 90; Protein, 3 g; Fat, 3.5 g (34% cal.); Cholesterol, 10 mg; Carbohydrates, 12 g; Fiber, 3 g; Sodium, 90 mg.

PROCEDURE

1. Blanch the onions 20 seconds in boiling water. (Blanching makes them easier to peel.) Drain, refresh under cold water, and drain again. Peel the onions.
2. Bring salted water to a boil in a saucepan. Add the onions, return to a boil, reduce heat to a simmer, and simmer until tender. Drain, refresh under cold water, and drain again.
3. Peel the carrots and cut into bâtonnet.
4. Bring salted water to a boil in a saucepan. Add the carrots, return to a boil, reduce heat to a simmer, and simmer until tender. Drain, refresh briefly, and drain again.
5. Bring a third pan of salted water to a boil. Add the frozen peas. Return to a boil, reduce heat, and simmer until tender. This takes only a few seconds. Frozen peas have already been blanched and need very little cooking. Drain, refresh, and drain again.
6. Mix together the three vegetables.
7. Heat the butter in as many sauté pans as necessary to hold the vegetables without overcrowding.
8. Add the vegetables and the tarragon. Toss over heat until the vegetables are hot and coated with the butter. Add the parsley and toss to mix.
9. Season to taste with salt and white pepper.

VARIATIONS

Herbs may be omitted for a simpler preparation, if desired.

Quantity Preparation: Cook and drain the vegetables and combine in a steam table pan. Heat the butter with the herbs and ladle over the vegetables. Season and serve.

Buttered Vegetables

The following vegetables may be cooked by simply boiling or steaming and dressed with butter for service, as in the basic recipe:

Asparagus	Cabbage	Kohlrabi
Beans, green or yellow	Carrots	Parsnips
Beans, lima	Cauliflower (see Note)	Peas
Beets	Celery	Rutabagas
Broccoli (see Note)	Corn (on cob or	Spinach
Brussels sprouts	whole kernel)	Turnips

Note: Dress each portion of broccoli spears with butter just when served. Butter runs off broccoli quickly. Do not sauté for à la carte service. Reheat in boiling water, then add butter. Other large vegetables, such as cauliflower, may also be prepared like broccoli.

Herbed Vegetables

Season buttered vegetables with fresh chopped parsley or other appropriate fresh or dried herbs (see table on p. 343). Dried herbs should be heated with the vegetable a few minutes to release flavor.

Amandine

Especially for green beans, broccoli, celery, cauliflower. For each 2 lb (900 g) EP of vegetable, sauté 2 oz (60 g) slivered or sliced almonds in 2–3 oz (60–90 g) butter until lightly browned. (Caution: Almonds darken quickly.) Combine with cooked vegetable.

Hollandaise

Especially for broccoli, asparagus, cauliflower, Brussels sprouts, leeks, and artichoke hearts or bottoms. At service time, nap each portion of vegetable with 2 fl oz (60 mL) hollandaise sauce.

Polonaise

Especially for cauliflower, broccoli, Brussels sprouts, and, sometimes, asparagus and green beans. For each 5 lb (2.3 kg) EP of vegetable, sauté 1½ pt (750 mL) fresh bread crumbs in about 6 oz (175 g) butter until golden. Chop the whites and yolks of 2–4 hard-cooked eggs separately. Combine the crumbs, chopped egg, and 4 tbsp (60 mL) chopped parsley. Sprinkle this mixture over the cooked vegetable immediately before serving.

Peas, Carrots, and Pearl Onions
with Tarragon Butter

Creamed Spinach

PORTIONS: 12   PORTION SIZE: 3¹/₃ OZ (100 G)

U.S.	METRIC	INGREDIENTS
5 lb AP	2.25 kg AP	Spinach, fresh
1¹/₄ pt	600 mL	Cream sauce, hot
to taste	to taste	Nutmeg
to taste	to taste	Salt
to taste	to taste	White pepper

Per serving: Calories, 120; Protein, 6 g; Fat, 7 g (50% cal.); Cholesterol, 20 mg; Carbohydrates, 10 g; Fiber, 3 g; Sodium, 150 mg.

Note: For frozen chopped spinach, partially thaw 2¹/₂ packages (2¹/₂ lb/1.1 kg each). Cover with boiling salted water and break spinach apart. Cook only until hot and drain. Squeeze out excess liquid and combine with cream sauce.

PROCEDURE

1. Trim spinach and wash carefully in several changes of water. Drain. (See Note.)
2. Place 2 in. (5 cm) water in a heavy pot, cover, and bring to a boil. Add the spinach. Stir several times so it cooks evenly.
3. As soon as the spinach is thoroughly wilted, drain in a colander, pressing with the back of a kitchen spoon to squeeze out excess liquid.
4. Chop the spinach coarsely.
5. Combine with the cream sauce in a hotel pan. Season to taste with nutmeg, salt, and pepper. (The spinach must not taste strongly of nutmeg.)

VARIATIONS

Creamed Vegetables

The following vegetables, cut into small pieces if necessary, may be cooked by boiling or steaming and combined with cream sauce, as in the basic recipe. For 12 portions, use 2¹/₂–3 lb (about 1.25 kg) EP vegetables and 1¹/₄–1³/₄ pt (600–850 mL) cream sauce.

Asparagus	Cabbage	Okra
Beans, green or yellow	Carrots	Onions, small white
Beans, lima	Cauliflower	Peas
Broccoli	Celery	
Brussels sprouts	Kohlrabi	

Orange-Scented Spinach

YIELD: 12 OZ (360 G)

U.S.	METRIC	INGREDIENTS
1 lb		Spinach leaves
2 tbsp		Olive oil
2 cloves		Garlic, sliced as thin as possible
1 tbsp		Shallot, chopped fine
1 tsp		Grated orange zest

Per 1 ounce (28.35 g): Calories, 30; Protein, 1 g; Fat, 2.5 g (75% cal.); Cholesterol, 0 mg; Carbohydrates, 2 g; Fiber, <1 g; Sodium, 30 mg.

PROCEDURE

1. Blanch the spinach. Drain, refresh in cold water, and squeeze dry.
2. Heat the olive oil in a sauté pan over low to medium heat.
3. Add the garlic and shallot. Cook until soft, about 1–2 minutes.
4. Add the orange zest and spinach. Cook until the spinach is hot.
5. Add salt and pepper to taste.

Broccoli Mornay

PORTIONS: 12   PORTION SIZE: 3½ OZ (100 G) BROCCOLI, 2 FL OZ (60 ML) SAUCE

U.S.	METRIC	INGREDIENTS
3 lb 12 oz	1.7 kg	Broccoli
1½ pt	720 mL	Mornay sauce, hot

PROCEDURE

1. Trim and wash broccoli. Separate large pieces into smaller serving pieces. Split or peel stems for even cooking.
2. Arrange broccoli in hotel pan with flowers to the outside, stems in center.
3. Pour in boiling salted water to partially cover. Cover with clean, wet towels and set on rangetop.
4. Simmer until blossom parts are nearly tender. Fold back towels from edges to uncover blossoms. This releases steam and helps avoid overcooking. Leave stems covered and continue to simmer until stems feel tender but al dente when pierced with a knife. Drain well.
5. Nap each portion with 2 fl oz (60 mL) Mornay sauce at service time. Ladle the sauce across the stems without covering the blossoms.

Per serving: Calories, 160; Protein, 9 g; Fat, 10 g (53% cal.); Cholesterol, 30 mg; Carbohydrates, 11 g; Fiber, 4 g; Sodium, 180 mg.

Note: This method of cooking in a shallow pan is used to prevent damaging the blossom ends, which are easily broken. Other delicate vegetables, such as asparagus, are also sometimes cooked in shallow water in hotel pans or sauté pans.

Broccoli may be cooked in a steamer, following the basic steaming method.

VARIATIONS

Other vegetables may be served with cheese sauce, such as cauliflower and Brussels sprouts.

Broccoli with Cheddar Cheese Sauce

Prepare as in the basic recipe, but substitute cheddar cheese sauce for the Mornay sauce.

Broccoli Mornay

Puréed Butternut Squash

PORTIONS: 12   PORTION SIZE: 3 OZ (90G)

U.S.	METRIC	INGREDIENTS
3 lb 12 oz	1.75 kg	Butternut squash
3 oz	90 g	Butter
1½ oz	45 g	Brown sugar
1 tsp	5 mL	Salt
to taste	to taste	White pepper
to taste	to taste	Nutmeg or ground ginger

PROCEDURE

1. Peel the squash, cut in half, and scrape out seeds. Cut into large dice.
2. Place in perforated steamer pan. Steam until tender. (Alternative *pan-steaming method:* Place in heavy pot. Add 1 in. [3 cm] water, cover, and cook slowly until tender. Drain well.)
3. Purée the squash with a food mill. Add the butter, sugar, and seasonings. Whip until light, but do not overwhip or squash will become watery.

Per serving: Calories, 90; Protein, 1 g; Fat, 6 g (51% cal.); Cholesterol, 15 mg; Carbohydrates, 12 g; Fiber, 2 g; Sodium, 250 mg.

Note: If squash is too wet, cook out some of the moisture in a shallow pan over medium heat after step 2.

VARIATIONS

Add 1½–2 fl oz (45–60 mL) heavy cream, heated. Sugar may be reduced or omitted if the squash has a good flavor.

Mashed Rutabagas or Yellow Turnips

Prepare as in the basic recipe. If desired, add a small amount of whipped potato.

Cauliflower au Gratin

PORTIONS: 12   **PORTION SIZE: 3 OZ (90 G) CAULIFLOWER, 1¹/₂ OZ (45 G) SAUCE, PLUS TOPPING**

U.S.	METRIC	INGREDIENTS
2¹/₂ lb EP	1.15 kg EP	Cauliflower
1¹/₂ tsp	7 mL	Lemon juice (see Note)
18 fl oz	540 mL	Béchamel or Mornay sauce, hot
³/₄ oz	22 g	Dry bread crumbs
³/₄ oz	22 g	Parmesan cheese, grated
1¹/₄ oz	38 g	Butter, melted

PROCEDURE

1. Separate the cauliflower into florets.
2. Place the cauliflower and lemon juice in boiling salted water. Return to boil, lower heat, and cover. Simmer until just tender. Do not overcook, as the cauliflower will cook further in the sauce. Drain.
3. Butter the bottom of a baking pan or hotel pan and place the cauliflower in it about 2 in. (5 cm) deep. (Individual ovenproof serving dishes may be used instead.)
4. Cover with the hot sauce.
5. Mix together the bread crumbs and cheese and sprinkle evenly over the top. Drizzle melted butter over the top.
6. Bake at 350°F (175°C) about 20 minutes to heat through. Brown the top under the broiler or salamander.

Per serving: Calories, 60; Protein, 3g; Fat, 3.5 g (50% cal.); Cholesterol, 10mg; Carbohydrates, 5 g; Fiber, 2 g; Sodium, 80 mg.

Note: Adding lemon juice to cooking water helps keep white vegetables white. It may be omitted if desired.

VARIATIONS

Substitute cheddar cheese sauce for the béchamel or Mornay, and use grated cheddar cheese instead of Parmesan for topping.

Other vegetables may be prepared au gratin, such as asparagus, Belgian endive, broccoli, Brussels sprouts, celery, celery root, leeks, and turnips.

Cauliflower au Gratin

Green Beans with Sesame Dressing

PORTIONS: 16   **PORTION SIZE: 2 OZ (60 G)**

U.S.	METRIC	INGREDIENTS
1 cup	250 mL	White sesame seeds
1¹/₂ fl oz	50 mL	Soy sauce
2 tbsp	30 g	Sugar
2 fl oz	60 mL	Dashi (p. 167) or water
2 lb	1 kg	Green beans

PROCEDURE

1. Toast the sesame seeds in a dry skillet, stirring and tossing regularly, until light golden.
2. Using a spice grinder or mortar and pestle, grind the sesame seeds to a paste.
3. Mix in the soy sauce and sugar. Thin with dashi or water. Set aside.
4. Wash the green beans and trim the ends. Cut into 1-in. (2.5-cm) lengths.
5. Cook the beans in boiling salted water until crisp-tender. Drain, cool under cold running water, and drain again.
6. Mix the beans with the dressing.

Per serving: Calories, 80; Protein, 3 g; Fat, 4.5 g (48% cal.); Cholesterol, 0 mg; Carbohydrates, 8 g; Fiber, 3 g; Sodium, 150 mg.

VARIATIONS

Other mild green vegetables, such as spinach and asparagus, can be served the same way.

Sesame Miso Dressing

Reduce the sesame seeds to ¹/₂ cup (125 mL). Omit the soy sauce and sugar, and add 4 oz (125 g) white or red miso. Use mirin (sweet rice wine) instead of dashi or water to thin the dressing. Use as a dressing for vegetables in the same manner as sesame dressing.

Green Beans with Sesame Dressing

Pan-Steamed Kohlrabi with Parsley

PORTIONS: 12   PORTION SIZE: 3 OZ (90 G)

U.S.	METRIC	INGREDIENTS	PROCEDURE
2 lb 4 oz	1.1 kg	Kohlrabi, medium dice	1. Select a sautoir or saucepan large enough to hold the kohlrabi in a layer no more than 2 in. (5 cm) deep.
1/2 tsp	2 mL	Salt	2. Put about 1/2 in. (1 cm) of water in the pan and add the salt. Bring to a boil.
			3. Add the kohlrabi. Cover the pan, lower the heat to a simmer, and cook until the kohlrabi is tender but not too soft, about 5 minutes.
			4. Uncover and remove the kohlrabi with a slotted spoon.
2 oz	60 g	Butter	5. Reduce the cooking liquid over moderately high heat to about 2 fl oz (60 mL).
1 1/2 tsp	7 mL	Chopped parsley	6. Add the butter and swirl in the pan to melt.
			7. Return the kohlrabi to the pan and add the parsley. Toss over heat so the liquid reduces to a glaze and coats the vegetable.
			8. Taste and add more salt if necessary. Serve immediately.

Per serving: Calories, 60; Protein, 2 g; Fat, 4 g (53% cal.); Cholesterol, 10 mg; Carbohydrates, 6 g; Fiber, 3 g; Sodium, 125 mg.

Collards with Ham

PORTIONS: 16   PORTION SIZE: 4 OZ (125 G)

U.S.	METRIC	INGREDIENTS	PROCEDURE
2 lb	1 kg	Smoked ham hocks	1. Simmer the ham hocks in water to cover until the meat is tender, 1 1/2 to 2 hours.
			2. Remove the hocks from the liquid, and reserve the liquid in the pot.
			3. Remove the meat from the bone and cut into small dice.
6 lb	3 kg	Collard greens	4. Trim the stems from the greens. Wash the greens in several changes of cold water.
to taste	to taste	Black pepper	5. Add the greens to the ham stock. Simmer until very tender, 45 to 60 minutes.
2 tsp, or to taste	20 mL, or to taste	Hot pepper sauce	6. Drain the greens, reserving the liquid.
			7. Chop the greens coarsely.
			8. Combine the greens with the ham.
			9. Add enough of the reserved cooking liquid to moisten the greens well. Reheat as necessary.
			10. Add pepper and hot pepper sauce to taste.
as needed	as needed	Cider vinegar	11. Serve with vinegar on the side, for diners to add to taste.

Per serving: Calories, 240; Protein, 20 g; Fat, 14 g (51% cal.); Cholesterol, 60 mg; Carbohydrates, 10 g; Fiber, 6 g; Sodium, 95 mg.

Note: This is a traditional Southern preparation that disregards all the rules of cooking green vegetables and preserving vegetable color.

VARIATIONS

Turnip greens, mustard greens, or kale may be substituted for the collards.

Collards with Ham

Onion Compote

YIELD: 1 LB 8 OZ (800 G)

U.S.	METRIC	INGREDIENTS	PROCEDURE
4 lb	2 kg	Onions, sliced	1. In a nonreactive saucepan (such as stainless steel), combine the onions and the first quantity of water. Cover and simmer 5 minutes.
8 fl oz	250 mL	Water	
3 pt	1.5 L	Red wine	2. Add the red wine and the second quantity of water. Simmer, uncovered, very slowly until the onions are tender and most of the liquid has evaporated.
2 pt	1 L	Water	
8 fl oz	250 mL	Red wine vinegar	3. Add the vinegar. Mix well and continue to simmer.
			4. By the time the vinegar has reduced, the onions should be very tender. If not, add a little more water, cover, and steam gently until they are soft. Remove the cover and reduce again.
4 oz	125 g	Butter	5. Stir in the butter and sugar. Stir and heat gently until the sugar is dissolved.
2 oz	60 g	Sugar	
to taste	to taste	Salt	6. Season with salt and pepper.
to taste	to taste	Pepper	7. Serve as a condiment with meat, poultry, and fish dishes.

Per serving: Calories, 135; Protein, 1 g; Fat, 4 g (27% cal.); Cholesterol, 10 mg; Carbohydrates, 12 g; Fiber, 1 g; Sodium, 40 mg.

Onion Compote

Cipolline in Agrodolce (Sweet-Sour Onions)

PORTIONS: 16   PORTION SIZE: 3¹/₂ OZ (100 G)

U.S.	METRIC	INGREDIENTS	PROCEDURE
4¹/₂ lb	2 kg	Pearl onions	1. Blanch the onions 1 minute. Drain and peel.
1 pt	500 mL	Water	2. Put the onions in a sauté pan in a single layer. Add the water and butter and cook slowly, uncovered, about 20 minutes, or until fairly tender. Add a little water if necessary during cooking so that the pan does not become dry. Stir gently from time to time.
2 oz	60 g	Butter	
3 fl oz	90 mL	Wine vinegar	3. Add the vinegar, sugar, and salt. Cover lightly. Cook over low heat until the onions are very tender and the liquid is syrupy, about 30 minutes. If necessary, remove the cover toward the end of the cooking time to let the liquid reduce. The onions should be lightly browned by the time they are done.
1¹/₂ oz	45 g	Sugar	
1¹/₂ tsp	7 mL	Salt	

Per serving: Calories, 120; Protein, 2 g; Fat, 3 g (23% cal.); Cholesterol, 10 mg; Carbohydrates, 21 g; Fiber, 1 g; Sodium, 270 mg.

Glazed Root Vegetables ❤

PORTIONS: 10   **PORTION SIZE: 3½–4 OZ (110–120 G)**
PORTIONS: 12   **PORTION SIZE: 3 OZ (90 G)**

U.S.	METRIC	INGREDIENTS
1 lb 8 oz	700 g	Carrots
12 oz	300 g	Parsnips
1 lb 4 oz	500 g	Turnips
1½ oz	45 g	Butter
1 oz	30 g	Sugar
to taste	to taste	Salt

PROCEDURE

1. Trim and peel the carrots, parsnips, and turnips. Tournéed root vegetables (see **Figure 7.21**) are often prepared by this method.
2. Place the vegetables in a saucepan with boiling salted water to cover. Simmer until tender. Drain.
3. Heat the butter in a sauté pan. Add the vegetables and sprinkle with the sugar. Season to taste with salt. Sauté until the vegetables are well glazed.

Per serving: Calories, 90; Protein, 1 g; Fat, 3 g (30% cal.); Cholesterol, 5 mg; Carbohydrates, 15 g; Fiber, 4 g; Sodium, 85 mg.

VARIATIONS

Single vegetables as well as other combinations of vegetables may be glazed by this method, including celery root, pearl onions, rutabagas, and chestnuts, in addition to carrots, parsnips, and turnips.

Glazed Carrots

Glazed Carrots (Carrots Vichy)

PORTIONS: 12   **PORTION SIZE: 3 OZ (90 G)**

U.S.	METRIC	INGREDIENTS
3 lb 4 oz	1.5 kg	Carrots
as needed	as needed	Water (see Note)
2½ oz	75 g	Butter
1 tbsp	15 g	Sugar
1 tsp	5 mL	Salt
to taste	to taste	White pepper
as needed	as needed	Chopped parsley

PROCEDURE

1. Trim, peel, and slice the carrots.
2. Place them in a saucepan or straight-sided sauté pan. Add water to barely cover. Add the butter, sugar, and salt.
3. Bring to a boil. Lower heat and simmer until the carrots are tender and the water is nearly evaporated. If done properly, these should happen at the same time. Toss the carrots so they are well coated with the glaze left in the pan.
4. Season with pepper. Garnish with chopped parsley.

Per serving: Calories, 100; Protein, 2 g; Fat, 5 g (43% cal.); Cholesterol, 10 mg; Carbohydrates, 13 g; Fiber, 2 g; Sodium, 240 mg.

Note: Sometimes Vichy water or other bottled mineral water is used, but it is not necessary.

Fennel and Coriander Sweet Potatoes

YIELD: ABOUT 2 LB (960 G)

U.S.	METRIC	INGREDIENTS	PROCEDURE
3 lb	1.35 kg	Sweet potatoes, peeled	1. Cut the potatoes into sticks 3 in. (7–8 cm) long and ¹/₂–³/₄ in. (12–20 mm) across. 2. Cook the potatoes in simmering salted water until just tender. Do not overcook or the potatoes will be too soft and will break easily. Drain.
3 tbsp	45 mL	Butter, unsalted	3. Heat the butter in a sauté pan and sweat the shallots until soft.
1 tbsp	15 mL	Shallots, chopped fine	4. Add the coriander and fennel and cook until fragrant, about 2–3 minutes.
1 tsp	5 mL	Coriander seed, crushed	5. Toss the spice mixture carefully with the potatoes and season with salt and pepper. Be careful not to break the potatoes.
1 tsp	5 mL	Fennel seed, crushed	6. Reheat gently in the sauté pan as needed.
to taste	to taste	Salt	
to taste	to taste	Pepper	

Per 1 ounce (28.35 g): Calories, 35; Protein, 0 g; Fat, 1 g (26% cal.); Cholesterol, 5 mg; Carbohydrates, 6 g; Fiber, <1 g; Sodium, 10 mg.

Spinaci alla Romana (Roman-Style Spinach)

PORTIONS: 16   PORTION SIZE: 3 OZ (90 G)

U.S.	METRIC	INGREDIENTS	PROCEDURE
6 lb	2.7 kg	Spinach	1. Trim and wash the spinach. Cook in a small quantity of boiling water until wilted. Drain, cool under running water, and drain again. Press excess water out of the spinach, but do not squeeze too dry.
1¹/₂ fl oz	45 mL	Olive oil	2. Heat the oil in a pan. Add the fat and render it. Remove and discard the cracklings (solid pieces remaining from the fat).
1¹/₂ oz	45 g	Fat from prosciutto or pork, small dice	3. Add the spinach, pine nuts, and raisins. Sauté until hot.
1¹/₂ oz	45 g	Pine nuts	4. Season with salt and pepper.
1¹/₂ oz	45 g	Raisins	
to taste	to taste	Salt	
to taste	to taste	Pepper	

Per serving: Calories, 100; Protein, 5 g; Fat, 7 g (57% cal.); Cholesterol, 5 mg; Carbohydrates, 7 g; Fiber, 3 g; Sodium, 90 mg.

VARIATIONS

Chopped garlic may be sautéed in the fat before the spinach is added. Lean prosciutto, sliced thin, then diced, may be added.

Spinaci alla Piemontese

Omit the oil, fat, nuts, and raisins. Heat 8 chopped anchovy fillets and 2 chopped garlic cloves in 4 oz (125 g) butter, then add the boiled, drained spinach and sauté.

Spinaci alla Romana

Ragoût of Summer Vegetables

PORTIONS: 12   **PORTION SIZE: 4 OZ (120 G)**

U.S.	METRIC	INGREDIENTS	PROCEDURE
9 oz	270 g	Carrots, trimmed and peeled	1. Cut the carrots into bâtonnet.
9 oz	270 g	Yellow summer squash, trimmed	2. Cut the squash into medium dice.
9 oz	270 g	Green beans, trimmed	3. Cut the green beans into 2-in. (5-cm) lengths
9 oz	270 g	Baby turnips, peeled	4. Quarter the turnips.
6 oz	180 g	Lima beans, shelled	5. Cook the six vegetables separately in a steamer or in boiling salted water until crisp-tender. Shock each vegetable in ice water as soon as it is cooked. Drain and refrigerate.
6 oz	180 g	Pearl onions, blanched and peeled	
5 fl oz	150 mL	Vegetable stock, vegetable cooking liquid, or chicken stock	6. Bring the desired stock or cooking liquid to a simmer in a sauté pan large enough to hold the vegetables in a shallow layer. (Of course, do not choose chicken stock as the cooking liquid if the dish is for a vegetarian menu.)
3 oz	90 g	Butter, cut into small pieces	
2 tsp	10 mL	Fresh chervil, chopped	7. Add the vegetables to the pan and toss over heat until the vegetables are hot (see *Note*).
2 tsp	10 mL	Fresh chives, chopped	
2 tsp	10 mL	Parsley, chopped	8. Add the butter and the herbs. Toss just until the butter is melted and coats the vegetables.
to taste	to taste	Salt	
to taste	to taste	White pepper	9. Season to taste with salt and white pepper.

Per serving: Calories, 110; Protein, 2 g; Fat, 6 g (49% cal.); Cholesterol, 15 mg; Carbohydrates, 12 g; Fiber, 3 g; Sodium, 75 mg.

Note: As an alternative finishing method, heat the stock in a saucepan, reduce slightly, monter au beurre, and season. Reheat the vegetables separately and mix with the butter sauce.

Ragout of Summer Vegetables

Orange and Tarragon Fennel Sous Vide

YIELD: 6–8 OZ (180–240 G), DEPENDING ON SIZE OF FENNEL BULB

U.S.	METRIC	INGREDIENTS	PROCEDURE
1	1	Fennel bulb	1. Trim fennel; cut off stalks. Cut in half vertically and trim all brown parts from the base.
1	1	Fresh tarragon stem	2. Cut each half into 6 to 8 wedges, depending on the size of the bulb.
2 fl oz	60 mL	Orange juice	
1/2 tsp	2 mL	Orange zest	3. Wrap the tarragon stem in a small piece of plastic wrap and cut off the ends so the ends are open.
1 tbsp	15 mL	Olive oil	4. Place the fennel and the tarragon stem in a plastic vacuum bag.
1/4 tsp	1 mL	Salt	5. Combine the orange juice, orange zest, olive oil, salt, and pepper. Add to the bag with the fennel. Vacuum and seal the bag.
pinch	pinch	White pepper	
			6. Cook in an immersion circulator heated to 185°F (85°C) for 30–40 minutes, until the fennel feels tender when pressed through the bag.
1 tsp	5 mL	Fresh tarragon leaves (small whole leaves or large leaves cut chiffonade)	7. Remove the fennel from the bag and sprinkle with the fresh tarragon.

Per 1 ounce (28.35 g): Calories, 40; Protein, <1 g; Fat, 2 g (45% cal.); Cholesterol, 0 mg; Carbohydrates, 4 g; Fiber, 1 g; Sodium, 115 mg.

Artichokes Cooked Sous Vide, in a Warm Vinaigrette with Aromatic Vegetables

PORTIONS: 10   **PORTION SIZE: 1 ARTICHOKE**

U.S.	METRIC	INGREDIENTS
2 qt	2 L	Water, cold
3 fl oz	90 mL	Lemon juice
10	10	Artichokes, large
as needed	as needed	Lemon halves
3 fl oz	90 mL	Olive oil
1 tsp	5 mL	Salt
3 oz	90 g	Carrots, brunoise
3 oz	90 g	Shallots, brunoise
1	1	Garlic clove, minced
3 fl oz	90 mL	Olive oil
to taste	to taste	Salt
1 fl oz	30 mL	Lemon juice
1 tbsp	15 mL	Chopped parsley

PROCEDURE

1. Review the guidelines for safe sous vide cooking on page 119.
2. Prepare acidulated water: Mix the water and lemon juice in a bowl large enough to hold the artichokes.
3. Trim the artichoke bottoms as shown in **Figure 10.1**, but leave 1 in. (2.5 cm) of stem attached. Rub cut surfaces with cut lemon as you work to keep the artichokes from darkening. Drop into acidulated water as soon as each one is finished.
4. Vacuum-pack as follows: Remove each artichoke bottom from the water, dry quickly on clean towels, and dip in olive oil to coat. Sprinkle lightly with salt. Place in plastic bags and vacuum-pack.
5. Cook in a hot-water bath at 194°F (90°C) for 45 minutes, or until the artichokes are tender.
6. If the artichokes are not to be served immediately, cool in an ice-water bath. Refrigerate, unopened, until needed.
7. Over low heat, sweat the carrots, shallots, and garlic in half the olive oil just until tender. Do not let them brown.
8. Remove from the heat and add salt to taste.
9. Add the remaining olive oil and the lemon juice.
10. Reheat the artichokes, in their plastic bags, in a hot-water bath.
11. Remove the artichokes from their bags and cut each into 8 wedges.
12. Toss the artichokes with the warm mixture of olive oil, lemon juice, and aromatic vegetables. Plate at once and sprinkle with a little chopped parsley.

Per serving: Calories, 190; Protein, 2 g; Fat, 17 g (76% cal.); Cholesterol, 0 mg; Carbohydrates, 10 g; Fiber, 5 g; Sodium, 280 mg.

VARIATION

If sous vide equipment is not available, cook the artichokes in a *blanc* as follows: Assemble 1 oz (30 g) flour, 3 pt (1.5 L) cold water, 1½ fl oz (45 mL) lemon juice, and 1 tbsp (15 mL) salt. Mix the flour with a little of the cold water to make a smooth paste, then stir this mixture into the rest of the cold water. Add the lemon juice and salt. Bring to a boil. Add the artichokes and simmer until just tender. Drain.

Maple Cinnamon Apples Sous Vide

YIELD: 8 OZ (240 G)

U.S.	METRIC	INGREDIENTS
8 oz	240 g	Apples, cut into small balls with parisienne scoop or into medium dice
2 tbsp	30 mL	Maple syrup
pinch	pinch	Salt
¼ tsp	1 mL	Cinnamon

PROCEDURE

1. Place all ingredients in a vacuum bag and seal on 90% vacuum.
2. Place the bag in a thermal circulator heated to 165°F (74°C). Cook for 30 minutes.
3. Use as garnish for soups, salads, desserts, or other dishes.

Per 1 ounce (28.35 g): Calories, 30; Protein, 0 g; Fat, 0 g (0% cal.); Cholesterol, 0 mg; Carbohydrates, 7 g; Fiber, <1 g; Sodium, 20 mg.

Asparagus Cream

YIELD: 1 LB (480 G)

U.S.	METRIC	INGREDIENTS	PROCEDURE
9 oz EP (see step 2)	250 g EP (see step 2)	Asparagus	
14 oz	400 g	Heavy cream	

1. Spray a 8 × 3 ³/₄-in. (20 × 9.5 cm) or comparable pan with nonstick spray and line with acetate or plastic wrap for molding the cream. Set aside.
2. Break off and discard the woody bottoms of the asparagus spears. Cut off the tips and reserve them for another use. Weigh the remaining portions of the spears to get the EP yield indicated at the left.
3. Cut these remaining portions of the asparagus into thin slices.
4. Combine the asparagus slices and heavy cream in a small saucepan. Bring to a simmer and cook until asparagus is tender, about 15 minutes.

U.S.	METRIC	INGREDIENTS	PROCEDURE
1 tbsp	15 mL	Lemon juice	
to taste	to taste	Salt	
to taste	to taste	White pepper	
0.18 oz	5 g	High-acyl gellan gum	

5. Purée the mixture in a blender until smooth.
6. Remove the mixture from the blender and scale out 17.6 oz (500 g) of the mixture. If necessary, add additional cream to bring the mixture to this weight.
7. Mix the lemon juice into the scaled purée and season with salt and pepper. Place this quantity back into the blender.
8. Turn the blender on low. When a vortex forms in the center, gradually sprinkle the gellan gum into this vortex to blend it evenly with the purée (a).
9. When all the gellan is added, turn the blender to high speed and blend for 2 minutes.
10. Pour the asparagus cream into the acetate-lined pan (b). Refrigerate until set.
11. For service, cut the cream into desired shapes. Warm in an oven at low heat.

Per 1 ounce (28.35 g): Calories, 90; Protein, <1 g; Fat, 9 g (90% cal.); Cholesterol, 35 mg; Carbohydrates, 2 g; Fiber, <1 g; Sodium, 15 mg.

Asparagus Cream

(a)

(b)

Cauliflower Press

YIELD: 1 LB 3 OZ (570 G)

U.S.	METRIC	INGREDIENTS
1 tbsp	15 mL	Shallots
1/2 tsp	2 mL	Garlic
1 tbsp	15 mL	Olive oil
1 tbsp	15 mL	Butter
8 oz	240 g	Cauliflower florets
to taste	to taste	Salt
to taste	to taste	Pepper
5 fl oz	150 mL	Chicken stock
5 fl oz	150 mL	Heavy cream
1 tsp	5 mL	Lemon zest, grated
		Sachet:
1/2	1/2	Bay leaf
1 tsp	5 mL	Anise seeds
2 sprigs	2 sprigs	Fresh thyme
2 sprigs	2 sprigs	Fresh tarragon
0.176 oz (5 g)	5 g	Agar-agar
2 1/2 tbsp	37 mL	Butter, unsalted
1/4 tsp	1 mL	Lemon juice
9 1/2 oz	285 g	Small cauliflower florets, blanched until al dente
1 tbsp	15 mL	Italian parsley, cut chiffonade
1 tsp	5 mL	Chives, cut fine
1 tsp	5 mL	Fresh tarragon, cut chiffonade

PROCEDURE

1. Sweat the shallots and garlic in the olive oil and butter over low heat.
2. Add the cauliflower, salt, and pepper. Cover and sweat for a few minutes.
3. Add the stock, cream, lemon zest, and sachet. Simmer until the cauliflower is tender. Strain, reserving the liquid.
4. Purée the drained cauliflower in a blender, adding just enough of the cooking liquid to make this possible.
5. Weigh 6.27 oz (175 g) of the purée and add to it 2.5 oz (71 g) of the reserved cooking liquid.
6. While beating with a wire whip, gradually add the agar-agar.
7. Place the cauliflower mixture in a stainless steel saucepan and add the butter and lemon juice. Add salt and pepper to taste.
8. Simmer 2 minutes, stirring constantly.
9. Add the cauliflower florets and herbs.
10. Pour the mixture into a flat mold, making sure there are no trapped air bubbles. Cover the surface with plastic wrap, press with a light weight, and chill.
11. Unmold and cut into desired shapes.

Per 1 ounce (28.35 g): Calories, 60; Protein, <1 g; Fat, 6 g (90% cal.); Cholesterol, 15 mg; Carbohydrates, 2 g;Fiber, <1 g; Sodium, 25 mg.

Cauliflower Press

SAUTÉING AND PAN-FRYING

According to the basic definitions given in Chapter 6, the main differences between sautéing and pan-frying are the amount of fat used and the cooking time. *Sautéing* means cooking quickly in a small amount of fat. The product is often tossed or flipped in the pan over high heat. *Pan-frying* means cooking in a larger amount of fat, usually for a longer time at lower heat, and the product is not tossed or flipped. In practice, the two methods are similar, and the distinction between them is hard to draw. Pan-fried vegetables are often given a coating, such as breading, that gives the cooked vegetable a crisp exterior that makes an appealing contrast with the tender vegetable inside.

   Both methods may be used for finish-cooking precooked or blanched vegetables as well as for completely cooking vegetables from the raw state. Sautéing in butter is especially popular for finishing precooked and chilled vegetables for service.

   Stir-frying is a quick-cooking technique used in Asian cookery. In effect, it is similar to sautéing, except the pan is left stationary and the items being cooked are stirred and flipped in hot fat with spatulas or other tools. For the basic stir-frying procedure, see page 502. This is the general procedure for stir-frying all foods, including meats. To use the procedure for vegetables only, omit steps 4, 5, and 6. The basic procedure for vegetables is illustrated by the recipe on page 324.

PROCEDURE for Sautéing Vegetables

This method is used for precooked or blanched vegetables and for tender, small-cut vegetables that cook quickly.

1. Collect all equipment and food products.

2. Prepare vegetables as required.

3. Place sauté pan on high heat.

4. When the pan is hot, add a small amount of clarified butter, oil, or other fat, enough to coat the bottom of the pan. (Clarified butter is used because the milk solids in whole butter burn quickly at the high heat necessary for sautéing.)

5. As soon as the fat is hot, add the vegetable. Do not overload the pan, or the temperature will be lowered too much and the vegetables will simmer instead of sauté.

6. After the heat has recovered, flip the pan a few times to turn and toss the vegetables (see Figure 11.4). Let the pan set again over the heat.

7. Continue to flip the vegetables as often as necessary for them to cook or heat evenly and become coated with the cooking fat. (Don't flip more than necessary, however.

Too much flipping may break fragile vegetables. Also, the heat must have time to recover between flips.)

8. As soon as the vegetables are cooked, or heated through if precooked, remove from the pan. Evaluate the quality of the cooked vegetable (see p. 285). Browning may or may not be desirable, depending on the vegetable and the particular preparation.

Flipping action of wrist

FIGURE 11.4 To flip foods in a sauté pan, give the handle a sharp twist upward with the wrist. Be sure to move the pan back far enough to catch the foods as they come down.

PROCEDURE for Pan-Frying Vegetables

Note: A griddle is often used for this procedure if only a small amount of fat is required.

1. Collect all equipment and food products.

2. Prepare vegetables as required. Preparation may include breading or dredging in flour.

3. Place a sauté pan or cast-iron skillet on moderately high heat. Add required amount of fat to the pan and let it heat.

4. Place prepared vegetables in the pan. Adjust the heat so the product cooks through with the desired amount of browning but without burning the outside.

5. Turn vegetables with a spatula and continue to cook until done.

6. Remove from pan. If necessary, drain on absorbent paper to eliminate excess fat.

7. Evaluate the quality of the cooked vegetable. In addition to the quality standards on p. 273, pan-fried vegetables that have been breaded or given another coating have a crisp exterior that has been attractively browned but not charred.

Zucchini Sauté Provençale 🌹

PORTIONS: 12   PORTION SIZE: 3¹/₂ OZ (100 G)

U.S.	METRIC	INGREDIENTS
3 lb	1.35 kg	Zucchini
3 fl oz	90 mL	Olive oil
3 oz	90 g	Shallots or onions, minced
2–3	2–3	Garlic cloves, chopped
to taste	to taste	Chopped parsley
to taste	to taste	Salt
to taste	to taste	White pepper

PROCEDURE

1. Wash and trim the zucchini. Cut crosswise into thin slices.
2. Heat the oil in two or three sauté pans (or sauté in several batches—do not overload the pans). Add the shallot or onion and the garlic. Sauté until soft but not browned.
3. Add the zucchini and sauté until slightly browned but still somewhat crisp.
4. Add the parsley and toss to mix. Season to taste.

Per serving: Calories, 80; Protein, 1 g; Fat, 7 g (72% cal.); Cholesterol, 0 mg; Carbohydrates, 5 g; Fiber 1 g; Sodium, 5 mg.

VARIATIONS

Cut the zucchini into other shapes, but keep them small enough to cook quickly. Examples: bâtonnet, julienne, dice, and shredded on a coarse grater.

Shredded Zucchini with Shallots

Shred the zucchini on a coarse grater. Sauté with shallots as in basic recipe, but without browning. Omit the garlic and parsley.

Zucchini with Tomatoes

Sauté as in the basic recipe. When half cooked, add 1 lb 4 oz (600 g) drained, chopped, canned tomatoes or fresh tomatoes concassé (p. 295). Finish cooking. Season with oregano and basil.

Zucchini Sauté Provençale

Pan-Fried Eggplant with Tomato Sauce

PORTIONS: 12   PORTION SIZE: 3¹/₂ OZ (100 G) EGGPLANT, 2 FL OZ (60 ML) SAUCE)

U.S.	METRIC	INGREDIENTS
3 lb 4 oz	1.5 kg	Eggplant
		Breading:
3 oz	90 g	Flour
³/₄ tsp	3 mL	Salt
¹/₄ tsp	1 mL	White pepper
8 fl oz	250 mL	Egg wash
10 oz	300 g	Bread crumbs
as needed	as needed	Oil for frying
1¹/₂ pt	750 mL	Tomato sauce

PROCEDURE

1. Wash and trim eggplants. Pare if skins are tough. Cut crosswise into ¹/₄-in. (0.5-cm) slices.
2. Hold in strongly salted cold water up to 30 minutes. (This step may be omitted, but it helps prevent darkening and eliminates some bitter flavors.)
3. Set up breading station, seasoning the flour with the salt and pepper.
4. Drain the eggplants and dry them well. Pass through Standard Breading Procedure (see p. 150).
5. Heat ¹/₄ in. (0.5 cm) oil in a heavy iron skillet or sauté pan. Pan-fry the breaded eggplant on both sides until browned. Remove from pan with slotted spatula and drain on absorbent paper.
6. Serve 2–3 slices per portion, depending on size. Nap each portion with 2 fl oz (60 mL) tomato sauce. Ladle the sauce in a band across the eggplant; do not cover completely.

Per serving: Calories, 260; Protein, 7 g; Fat, 13 g (44% cal.); Cholesterol, 70 mg; Carbohydrates, 30 g; Fiber, 4 g; Sodium, 490 mg.

VARIATIONS

Instead of Standard Breading Procedure, simply dredge slices in seasoned flour and pan-fry.

Pan-Fried Eggplant Creole

Use Creole sauce instead of tomato sauce.

Eggplant Parmigiana

Pan-fry as in the basic recipe. Top each fried slice with a thin slice of mozzarella cheese. Arrange in layers in a baking pan, covering each layer with tomato sauce and sprinkling with parmesan cheese. Bake 30 minutes at 350°F (175°C).

Stir-Fry of Asparagus, Green Beans, and Snow Peas

PORTIONS: 10   PORTION SIZE: 3¹/₂ OZ (105 G)

U.S.	METRIC	INGREDIENTS	PROCEDURE
12 oz	360 g	Asparagus, trimmed, lower ends peeled (p. 277)	1. Blanch the asparagus and green beans separately in boiling salted water until they are about three-fourths cooked and still crisp.
12 oz	360 g	Green beans, trimmed	2. Cut the asparagus and beans into 2-in. (5-cm) lengths.
10 oz	300 g	Snow peas, trimmed	3. Heat the oil in a sauté pan or wok over high heat. Add the snow peas and bell pepper and stir-fry.
6 oz	180 g	Red bell pepper, 2-in. (5-cm) batonnet	4. As soon as the peas are hot and coated with oil, add the green beans. Continue to stir-fry another 20–30 seconds.
1¹/₂ fl oz	45 mL	Vegetable oil	5. Add the asparagus. Stir-fry until all the vegetables are crisp-tender.
1 fl oz	30 mL	Soy sauce	6. Add the soy sauce and toss to mix.
1 tbsp	15 mL	Sesame oil	7. Add the sesame oil and toss quickly.
to taste	to taste	Salt	8. Add salt to taste.
1 tbsp	15 mL	Sesame seeds, toasted (optional)	9. Remove the vegetables from the pan or wok and serve immediately.
			10. If desired, sprinkle each portion lightly with toasted sesame seeds.

Per serving: Calories, 90; Protein, 3 g; Fat, 6 g (55% cal.); Cholesterol, 0 mg; Carbohydrates, 8 g; Fiber, 3 g; Sodium, 220 mg.

Stir-Fry of Asparagus, Green Beans, and Snow Peas

Corn with Poblanos

PORTIONS: 12   PORTION SIZE: 3 OZ (90 G)

U.S.	METRIC	INGREDIENTS	PROCEDURE
12 oz	360 g	Poblano chiles	1. Roast and peel the chiles as illustrated on page 290. Remove and discard the stem, seeds, and inner membranes.
2 lb	900 g	Corn, fresh or frozen	2. Cut the chiles into bâtonnet.
1¹/₂ fl oz	45 mL	Olive oil	3. If the corn is frozen, drop into boiling water until thawed, then drain.
8 oz	240 g	Onion, chopped fine	4. Heat the olive oil in a sauté pan over moderately high heat.
1 tsp	5 mL	Dried oregano	5. Add the onion and sauté until it is soft and beginning to brown.
to taste	to taste	Salt	6. Add the poblanos, corn, and oregano. Sauté until hot and well blended.
to taste	to taste	Pepper	7. Add salt and pepper to taste.

Per serving: Calories, 180; Protein, 6 g; Fat, 7 g (30% cal.); Cholesterol, 0 mg; Carbohydrates, 30 g; Fiber, 9 g; Sodium, 25 mg.

VARIATIONS

Gratin of Corn and Poblanos

Prepare as in the basic recipe. Transfer to a gratin dish and cover with a generous layer of Monterey jack or similar mild cheese. Place under a salamander or broiler until the cheese is melted and lightly browned.

Corn and Poblanos in Cream

Prepare as in the basic recipe, but in step 6 add 12 fl oz (360 mL) heavy cream. Simmer until the cream is thickened and reduced by about one-third.

Corn with Poblanos

Stir-Fried Mixed Vegetables

PORTIONS: 16   PORTION SIZE: 4 OZ (125 G)

U.S.	METRIC	INGREDIENTS	PROCEDURE
1¹/₂ lb	750 g	Chinese cabbage	1. Cut off the root end of the cabbage and separate the leaves. Cut out the thick center ribs, then cut them crosswise into 2-in. (5-cm) chunks. Cut the thin, leafy parts into shreds.
16	16	Dried black mushrooms	2. Soak the mushrooms in boiling water to cover. When soft, drain and squeeze dry, reserving the liquid. Discard the stems and cut the caps into julienne.
6 oz	175 g	Bamboo shoots, drained	3. Cut the bamboo shoots into thin slices.
4 oz	125 g	Celery	4. Cut the celery on the diagonal into thin slices.
4 oz	125 g	Carrots	5. Cut the carrots into julienne.
15-oz can	425-g can	Baby corn	6. Drain the baby corn.
4	4	Scallions	7. Slice the scallions into shreds at a sharp angle.
2–3 fl oz	60–90 mL	Oil	8. Heat the oil in a wok or large sauté pan.
1	1	Garlic clove, crushed (optional)	9. Add the garlic and cook about 15 seconds to flavor the oil, then remove and discard the garlic.
5 slices	5 slices	Fresh ginger root	10. Add the ginger and salt and let cook about 15 seconds.
1 tsp	5 mL	Salt	11. Add the cabbage ribs and stir-fry 1–2 minutes.
12 fl oz	350 mL	Water or chicken stock	12. Add the remaining vegetables, except the shredded cabbage leaves. Continue to stir-fry another 1–2 minutes. Add the shredded cabbage.
1 fl oz	30 mL	Soy sauce (optional)	13. Add the stock and soy sauce and continue to stir and cook until the vegetables are cooked but still crisp.
¹/₄ cup	25 g (60 mL)	Cornstarch	14. Mix the cornstarch with the cold water, then stir it, a little at a time, into the vegetables to thicken the sauce. Do not add it all at once because you may not need it all. The sauce should not be too thick but, rather, the consistency of a light velouté sauce.
3 fl oz	100 mL	Water, cold	
1 tsp	5 mL	Sesame oil (optional)	15. Stir in the sesame oil and serve at once.

Per serving: Calories, 90; Protein, 2 g; Fat, 4 g, (36% cal.); Cholesterol, 0 mg; Carbohydrates, 14 g; Fiber, 3 g; Sodium, 165 mg.

VARIATIONS

This is a basic procedure for stir-fried vegetables. One or two vegetables, or any harmonious assortment, can be cooked using the same recipe, merely substituting different vegetable ingredients.

Mixed Vegetables with Thai Green Curry

PORTIONS: 10   PORTION SIZE: 4 OZ (120 G)

U.S.	METRIC	INGREDIENTS	PROCEDURE
8 oz	240 g	Carrots	
8 oz	240 g	Green beans	
10 oz	300 g	Bok choy	
4 oz	120 g	Shiitake mushrooms	
4 oz	120 g	Scallions	
¹/₂ oz	15 g	Thai chiles or other small, fresh chiles	
1¹/₂ fl oz	45 mL	Oil	
1 pt	500 mL	Thai Green Curry Sauce (p. 210)	

PROCEDURE

1. Peel and trim the carrots. Slice on a sharp diagonal into long, oval slices. Blanch in boiling salted water until crisp-tender. Refresh in ice water.

2. Trim the green beans and cut into 2-in. (5-cm) pieces. Blanch until crisp-tender. Refresh in ice water.

3. Trim the bok choy. Cut crosswise into ¹/₂-in. (1-cm) slices.

4. Trim and discard the stems from the shiitake mushrooms. Slice into narrow strips.

5. Trim the root ends and tops of the scallions, reserving the white and some of the green. Cut on a sharp diagonal into 1-in. (2.5-cm) lengths.

6. Trim the chiles, discarding the stems and seeds. Chop coarsely.

7. Heat the oil in a large sauté pan or wok over high heat. Add the scallions, chiles, and mushrooms. Stir-fry 30 seconds.

8. Add the bok choy. Stir-fry until the bok choy leaves are wilted.

9. Add the carrots and green beans and continue to stir-fry until all the vegetables are hot.

10. Add the green curry sauce and bring to a boil. Toss to mix the sauce with the vegetables.

11. Serve immediately.

Per serving: Calories, 170; Protein, 3 g; Fat, 16 g (77% cal.); Cholesterol, 0 mg; Carbohydrates, 8 g; Fiber, 2 g; Sodium, 340 mg.

Mixed Vegetables with Thai Green Curry

Brussels Sprouts with Walnuts 🌰

PORTIONS: 12   PORTION SIZE: 3¹/₂ OZ (100 G)

U.S.	METRIC	INGREDIENTS	PROCEDURE
3 lb	1.5 kg	Brussels sprouts	1. Trim the bases of the sprouts and remove any damaged leaves. 2. Blanch the sprouts in a large quantity of boiling salted water until half to three-fourths cooked. 3. Drain and refresh the sprouts in ice water. Drain again. (If the sprouts are to be finished immediately, refreshing can be omitted.) 4. Cut the sprouts in half lengthwise.
2 oz	60 g	Butter	5. Heat the butter in a sauté pan large enough to hold the sprouts in a thin layer.
6 oz	180 g	Walnut pieces	6. Add the sprouts and the walnuts to the pan. Sauté until the sprouts are tender and lightly browned.
to taste	to taste	Salt	7. Add salt to taste.

Per serving: Calories, 183; Protein, 6 g; Fat, 15 g (65% cal.); Cholesterol, 10 mg; Carbohydrates, 12 g; Fiber, 5 g; Sodium, 60 mg.

VARIATION

For a vegan version of this recipe, substitute walnut oil for the butter.

Brussels Sprouts with Walnuts

BRAISING

Braising, as you know, is a slow, moist-heat cooking method using a small amount of liquid. When meats are braised, they are seared or browned in fat before liquid is added. Braised vegetables are not always cooked in fat before liquid is added, although some kind of fat is used in the preparation.

Braised vegetable preparations tend to be more complex than boiled or steamed vegetables, and the cooking times are longer. Unfortunately, there are so many variations of braised vegetable that it is not possible to prescribe a single basic procedure. Instead, we discuss the procedures in general terms and use the recipes to illustrate them.

Characteristics of Vegetable Braising Procedures

1. Fat is added to a braising or baking pan or a saucepan and heated. Finely diced mirepoix or other flavoring ingredients may be cooked briefly in the fat. The fat contributes to flavor and eating quality.

2. The vegetable (blanched or raw) is placed in the pan. It may or may not be cooked in the fat before the liquid is added, depending on the recipe.

3. Liquid is added—stock, water, wine, or a combination of liquids. The liquid generally covers the vegetable only partway.

4. The pot or saucepan is covered and the vegetable is cooked slowly in the oven or on the rangetop.

5. The flavorful cooking liquid is served with the vegetable. It is sometimes drained off and reduced over high heat before serving in order to concentrate flavor.

KEY POINTS TO REVIEW

- What are the steps in the procedure for sautéing vegetables? How is sautéing similar to stir-frying, and how is it different?

- What are the steps in the procedure for pan-frying vegetables?

- What are some common examples of braised vegetables? Describe the cooking procedure for each of these examples.

👨‍🍳 Braised Red Cabbage

PORTIONS: 12   PORTION SIZE: 5 OZ (150 G)

U.S.	METRIC	INGREDIENTS	PROCEDURE
3 lb	1.5 kg	Red cabbage	1. Remove the outer leaves of the cabbage and cut it into quarters. Remove the core and shred the cabbage with a knife (**Figure 10.7**) or a power shredder attachment. Do not chop; cabbage should be in long, fine shreds.
6 oz	175 g	Bacon, diced	
8 oz	250 g	Onions, sliced	
½ oz	15 g	Sugar	
12 fl oz	375 mL	White stock (chicken, pork, veal) or water	2. Render the bacon in a large, heavy pot. Add the onions and sugar and cook until the onion is soft.
8 oz	250 g	Apples (unpeeled), cored and diced	3. Add the cabbage and stir over heat until it is coated with fat.
2	2	Cloves	4. Add the stock, apples, and spices, tied in a cheesecloth bag. Cover and simmer until cabbage is nearly tender, about 30 minutes.
3	3	Whole allspice	
1 small	1 small	Cinnamon stick	
2 fl oz or more	60 mL or more	Red wine vinegar	5. Add the vinegar and red wine and simmer another 10 minutes. Remove spice bag.
4 fl oz	125 mL	Red wine (or more vinegar)	6. Taste and correct seasoning. If not tart enough or color is not red enough, add more vinegar.
to taste	to taste	Salt	
to taste	to taste	Pepper	

Per serving: Calories, 130; Protein, 3 g; Fat, 8 g (54% cal.); Cholesterol, 10 mg; Carbohydrates, 11 g; Fiber, 3 g; Sodium, 110 mg.

VARIATIONS

Substitute lard, salt pork, or chicken fat for the bacon. Vegetable oil may be used, but it does not contribute to flavor.

Eliminate cinnamon, cloves, and allspice. Add 1 tbsp (15 mL) caraway seeds to onions when sautéing them.

Braised Green or White Cabbage

Prepare as in the basic recipe, but season with 1 bay leaf, 6–8 parsley stems, 6 peppercorns, and a pinch of thyme instead of the cinnamon, cloves, and allspice. Omit sugar, apples, wine, and vinegar. Butter may be used as the cooking fat, if desired.

Braised Red Cabbage

Braised Sauerkraut ♥

PORTIONS: 12   PORTION SIZE: 4½ OZ (125 G)

U.S.	METRIC	INGREDIENTS
3 lb 4 oz	1.5 kg	Sauerkraut, canned or fresh
1 oz	30 g	Lard or bacon fat
8 oz	250 g	Onions, sliced
8 fl oz	250 mL	Dry white wine (optional)
1½ pt (approximately)	750 mL (approximately)	Chicken stock
3	3	Juniper berries
1	1	Bay leaves
1	1	Cloves
½ tsp	2 mL	Caraway or cumin seed
1	1	Garlic cloves
to taste	to taste	Salt

PROCEDURE

1. Rinse the sauerkraut in cold water. Drain and press out water. Taste and rinse again if still too briny. (See Appendix 2, p. 1026, for can sizes and substitutions.)

2. Heat the lard in a heavy pot and sauté the onions until soft. Add the sauerkraut, wine (if used), and enough stock to cover the sauerkraut by about three-fourths. Tie the spices and garlic in cheesecloth and add to the pot.

3. Cover and simmer 1½ hours on the rangetop or in a slow oven (300°F/150°C).

4. Remove the spice bag. Taste the sauerkraut and adjust seasoning.

Per serving: Calories, 50; Protein, 2 g; Fat, 2.5 g (38% cal.); Cholesterol, 5 mg; Carbohydrates, 7 g; Fiber, 3 g; Sodium, 790 mg

VARIATION

Choucroute Garni

Double quantities per portion. Cook a variety of fresh and smoked pork products and sausages in the sauerkraut. Add each item at the proper time so it is in the sauerkraut for its correct cooking time. Suggestions: fresh or smoked pork chops, slab bacon, bratwurst, frankfurters, smoked pork shoulder. Serve as a main course. Accompany with boiled potatoes.

> **SAUERKRAUT**
>
> Sauerkraut is popular not only in German-speaking countries but also in France, especially in Alsace, where it is usually served in the form of Choucroute Garni (see variation). The basic recipe here makes a mild-tasting sauerkraut closer to French rather than German style.

Fresh Sauerkraut ♥

U.S.	METRIC	INGREDIENTS
as desired (see step 1)	as desired (see step 1)	Green cabbage
as needed (see step 5)	as needed (see step 5)	Kosher salt

PROCEDURE

1. Sauerkraut can be made in as large a quantity as desired, depending on the containers available. For practicality, it is best to start with at least 5 lb (2.3 kg).

2. Trim and discard the outer leaves and any damaged areas from the cabbage.

3. Quarter the heads. Remove and discard the cores.

4. Slice the cabbage into fine shreds, preferably on a slicing machine.

5. Weigh the cabbage. Divide the weight by 40 to get the weight of salt needed.

6. Measure the salt and toss it with the shredded cabbage to mix evenly.

7. Pack the salted cabbage tightly into one or more nonreactive containers. Cover the top of the cabbage with a layer of plastic film and place a weight on top. If the cabbage is well packed, the juices drawn out by the salt should just cover the cabbage. (Note: The salt begins drawing out juices within minutes after being mixed with the cabbage.)

8. Place the cabbage in a cool place, preferably about 60°F (15°F), to ferment. After a day or two, the cabbage and liquid will begin to bubble slowly. Keep covered and let stand until the bubbling stops and the cabbage smells like sauerkraut. The fermentation may take from several days to about 2 weeks, depending on the temperature.

9. Discard any discolored cabbage from the top layer.

10. Refrigerate to stop the fermentation. The sauerkraut is ready for use.

Per 1 oz (28.35 g): Calories, 5; Protein, 0 g; Fat, 0 g (0% cal.); Cholesterol, 0 mg; Carbohydrates, 2 g; Fiber, 1 g; Sodium, 280 mg.

Fresh Raw Sauerkraut

Peas à la Française

PORTIONS: 16   **PORTION SIZE: 3 OZ (90 G)**

U.S.	METRIC	INGREDIENTS	PROCEDURE
3 oz	90 g	Butter	
2 oz	60 g	Onion, chopped, or whole tiny pearl onions, peeled	
2¹/₂ lb	1.1 kg	Peas, frozen	
8 oz	225 g	Lettuce, shredded	
2 tbsp	30 mL	Chopped parsley	
1 tsp	5 mL	Salt	
2 tsp	10 mL	Sugar	
4 fl oz	125 mL	Chicken stock or water, hot	
1 tbsp	15 mL	Beurre manié	

PROCEDURE

1. Heat the butter in a saucepan. Add the onions and sauté lightly.
2. Add the peas, lettuce, parsley, salt, and sugar. Cook over moderate heat, stirring a few times, until the vegetables begin to steam.
3. Add the stock or water. Bring to a boil, cover, and simmer over low heat or in the oven until peas are tender.
4. Stir in a little beurre manié to thicken the cooking liquid, and simmer another 2–3 minutes. Adjust seasoning. (For larger quantities, drain off liquid and thicken separately.)

Per serving: Calories, 100; Protein, 4 g; Fat, 5 g (43% cal.); Cholesterol, 15 mg; Carbohydrates, 11 g; Fiber, 4 g; Sodium, 250 mg.

Ratatouille

The method for this preparation is unlike that for the other braised vegetables in this section because no liquid is added. It is classified as a braised item because the vegetables are first sautéed in fat, then simmered in their own juices.

PORTIONS: 20   **PORTION SIZE: 4 OZ (125 G)**

U.S.	METRIC	INGREDIENTS
1 lb	500 g	Zucchini
1 lb	500 g	Eggplant
1 lb	500 g	Onions
4	4	Green bell peppers
4	4	Garlic cloves
2 lb	1 kg	Tomatoes (canned may be used if necessary)
6 oz, or more as needed	200 mL, or more as needed	Olive oil
¹/₂ cup	125 mL	Chopped parsley
1	1	Bay leaf
¹/₄ tsp	1 mL	Dried thyme
to taste	to taste	Salt
to taste	to taste	Pepper

PROCEDURE

1. Prepare the vegetables: Cut the zucchini into ¹/₂-in. (1-cm) slices. Peel the eggplant and cut into large dice. Slice the onions. Remove the cores and seeds of the peppers and cut into 1-in. (2.5-cm) dice. Chop the garlic. Peel and seed the tomatoes and cut into large dice (leave canned tomatoes whole; they will break up during cooking).
2. Sauté the zucchini in a little of the olive oil until it is about half cooked. Remove from pan.
3. Sauté the eggplant in olive oil until half cooked. Remove from pan.
4. Sauté the onions and peppers until half cooked. Add the garlic and sauté another minute.
5. Combine all vegetables and seasonings in brazier or heavy saucepan. Cover and cook in a slow oven (325°F/160°C) about 30 minutes, or until vegetables are tender and flavors are well blended. If the vegetables are too juicy, cook uncovered on a rangetop for a few minutes to reduce. Be careful not to scorch the vegetables on the bottom.
6. Adjust seasonings. Serve hot or cold.

Per serving: Calories, 110; Protein, 1 g; Fat, 9 g (67% cal.); Cholesterol, 0 mg; Carbohydrates, 9 g; Fiber, 2 g; Sodium, 5 mg.

VARIATION

For a ratatouille with a different texture and appearance, cut all vegetables into medium dice.

Ratatouille

BAKING

You could, if you wished, cook carrots by placing them in a pot of boiling water, placing the pot in a hot oven, and cooking until tender. This is not baking, however. It's plain old simmering. You'd just be using the heat of the oven rather than the rangetop to simmer the water.

When we talk about baking vegetables, we usually mean one of two things:

1. Cooking starchy vegetables, such as potatoes, winter squash, and sweet potatoes, as well as root vegetables and other moist, dense-textured vegetables such as tomatoes, beets, eggplant, onions, and turnips, from the raw to the finished state. Starchy vegetables are baked because the dry heat produces a desirable texture. Baked potatoes, for example, do not have the same texture as boiled or steamed potatoes.

   Vegetables that are cut before baking may become browned on the cut surfaces or only on the edges. A degree of browning enhances appearance, and the caramelization of sugars creates a rich, complex flavor. The term *roasted* is often used for this type of baked vegetable preparation.

   In theory, any vegetable with enough moisture can be baked like potatoes, but the drying effects of the oven and the long cooking time make it undesirable for most small vegetables, such as peas and green beans.

2. Finishing certain vegetable combinations, sometimes known as *casseroles*. The vegetables in these items are usually parcooked by simmering or steaming before they are baked.

   Vegetable casseroles are baked for either of two reasons:

   - The slow, all-around heat allows the product to cook undisturbed. The agitation and stirring of rangetop cooking is not always desirable. Baked beans could be finished on top of the range, but they would be mushier and more broken. Custard-based timbales would be pourable, not firmly set.

   - The dry heat produces desirable effects, such as browning and caramelizing of sugars. For example, you could put a pan of candied sweet potatoes in a steamer, but the moist heat would not allow a glaze to form.

PROCEDURE for Baking Vegetables

1. Collect all equipment and food products.
2. Prepare vegetables as required.
3. Place in appropriate pan and set in preheated oven.
4. Bake to desired doneness
5. Evaluate the quality of the cooked vegetable (see page 273). Browning, if any, should not be excessive, as this could create burnt flavors.

Roasted Onion Purée

YIELD: 2 LB (1 KG)

U.S.	METRIC	INGREDIENTS
3 lb	1.5 kg	Onions, large, whole
1 pt	500 mL	Heavy cream
to taste	to taste	Salt
to taste	to taste	White pepper
to taste	to taste	Nutmeg

PROCEDURE

1. Leave the onions whole; do not peel. Place them on a sheet pan or baking pan and bake at 350°F (175°C) until soft, about 45–60 minutes.
2. Peel the onions. Purée the pulp using a food processor.
3. Mix in the cream. Bring to a simmer and cook until slightly thickened.
4. Season to taste with salt, white pepper, and nutmeg.
5. Serve as an accompaniment for roasted or grilled meats.

Per serving: Calories, 70; Protein, 1 g; Fat, 6 g (73% cal.); Cholesterol, 21 mg; Carbohydrates, 4 g; Fiber, 1 g; Sodium, 5 mg.

VARIATION

Caramelized Roasted Onion Purée

In place of the large onions, use small onions. Peel them, cut in halves or quarters, coat them with oil, and roast until browned. Purée as in basic recipe.

Roasted Onion Purée

Baked Acorn Squash ♥ 🌼

PORTIONS: 12   PORTION SIZE: ¹/₂ SQUASH

U.S.	METRIC	INGREDIENTS
16	6	Acorn squash, small
as needed	as needed	Butter, melted
2¹/₂ oz	75 g	Brown sugar
1¹/₄ tsp	6 mL	Salt
1 fl oz	30 mL	Sherry (optional)

Per serving: Calories, 130; Protein, 2 g; Fat, 2 g (13% cal.); Cholesterol, 5 mg; Carbohydrates, 28 g; Fiber, 3 g; Sodium, 270 mg.

PROCEDURE

1. Wash and cut squash in half lengthwise. Scrape out seeds. (If using large squash, cut into portion sizes.)
2. Brush cut surfaces and cavity with melted butter. Place close together, cut side down, on baking sheet. (This helps squash cook faster without drying by retaining steam.)
3. Bake at 350°F (175°C) until almost tender, about 30–40 minutes.
4. Turn the squash cut side up and brush again with butter. Sprinkle the cavities with sugar and salt. Add a few drops of sherry to each if desired.
5. Bake 10–15 minutes more, or until surface is glazed.

VARIATIONS

Hubbard, buttercup, and other winter squash varieties may be cut into portion sizes and baked as in basic recipe.

Gingered Squash

Mix 1¹/₂ tsp (7 mL) ground ginger with the sugar in the basic recipe.

Puréed Squash

Bake cut Hubbard squash until tender. Remove from shell and purée in food mill. Add butter, salt, and pepper to taste.

PROCEDURE VARIATION: COMBI OVEN

1. Wash and cut squash in half lengthwise. Scrape out seeds. (If using large squash, cut into portion sizes.)
2. Brush cut surfaces and cavity with melted butter. Sprinkle the cavities with sugar and salt. Add a few drops of sherry to each if desired. Place close together, cut side up, on baking sheet.
3. Bake at 350°F (175°C) in combi mode for 8 minutes. With a pastry brush, brush all of the cut surfaces of the acorn squash with the melted sugar/butter mixture that pools in the squash's cavity.
4. Keep the oven in combi mode, and increase to maximum browning level (0 percent humidity).
5. Bake 7–9 minutes more, or until surface is glazed.

Baked Acorn Squash

Roasted Winter Vegetables

PORTIONS: 16   PORTIONS SIZE: 4 OZ (125 G)

U.S.	METRIC	INGREDIENTS
12 oz	375 g	Carrots, peeled
12 oz	375 g	Celery root, peeled
8 oz	250 g	Turnips, peeled
8 oz	250 g	Parsnips, peeled
12 oz	375 g	Waxy potatoes, peeled
8 oz	250 g	Butternut squash, peeled and seeded
12	12	Shallots, peeled
12	12	Garlic cloves, peeled
4 fl oz	125 mL	Olive oil
1½ tsp	7 mL	Dried thyme
1½ tsp	7 mL	Coarse salt
1 tsp	5 mL	Coarsely ground black pepper

PROCEDURE

1. Cut the carrots, celery root, turnips, parsnips, potatoes, and squash into 1-in. (2.5-cm) dice.
2. Place these cut vegetables, plus the shallots and garlic cloves, in a baking pan.
3. Pour the olive oil over the vegetables and sprinkle with the thyme, salt, and pepper. Toss or mix until the vegetables are well coated with oil. Add more oil if necessary.
4. Bake at 375°F (190°C) about 45 minutes, or until the vegetables are tender and lightly browned. Turn or stir the vegetables several times during baking so they cook evenly. Do not allow them to become too browned, or they may be bitter.

Per serving: Calories, 120; Protein, 1 g; Fat, 7 g (51% cal.); Cholesterol, 0 mg; Carbohydrates, 14 g; Fiber, 3 g; Sodium, 230 mg.

VARIATIONS

Vegetable proportions may be varied as desired. Other vegetables, such as sweet potatoes, stalk celery, onions, and rutabagas, may be added.

Roasted Onions

Substitute 4 lb (2 kg) onions, sliced ¼ in. (5 mm) thick, for all the vegetables in the basic recipe. Bake as in the basic recipe, cooking until the onions are browned and caramelized. Onions lose a lot of moisture during baking, so total yield is only about 2¼ lb (1.1 kg).

Roasted Summer Vegetables

Omit the parsnips, turnips, celeriac, and butternut squash. Substitute an assortment of summer vegetables in desired proportions, such as eggplant, summer squash, fennel, bell peppers, cherry tomatoes, and baby turnips. Season with fresh chopped basil and parsley.

PROCEDURE VARIATION: COMBI OVEN

Bake at 365°F (185°C) in combi mode with 85 percent browning level for about 30 minutes, or until the vegetables are tender and lightly browned. Turn or stir the vegetables several times during baking so they cook evenly.

Roasted Winter Vegetables

Roasted Beets with Beet Greens

PORTIONS: 12   PORTION SIZE: 3 OZ (90 G)

U.S.	METRIC	INGREDIENTS	PROCEDURE
3 lb	1.5 kg	Beets, large (without tops)	1. Trim the beets as necessary, but leave the root and 1–2 in. (3-5 cm) of the stems attached. Wash well. 2. Place in a baking pan and roast at 375°F (190°C) until tender, about 1 hour for large beets (**Figure 11.5**). Test by piercing with a thin-bladed knife. 3. Cool and refrigerate until needed.
1 lb	480 g	Beet greens	4. Trim the beet greens, discarding the stems. Wash well in several changes of water, and drain. 5. Blanch the greens in boiling salted water. Drain and refresh in ice water. Drain again. Squeeze out excess water. 6. Chop the greens coarsely.
2 oz	60 g	Butter	7. Peel the roasted beets (**Figure 11.6**). Cut them into small dice.
to taste	to taste	Salt	8. Heat the butter in a sauté pan. Add the diced beets and chopped greens and sauté until hot.
to taste	to taste	Pepper	9. Season to taste with salt and pepper.

Per serving: Calories, 100; Protein, 3 g; Fat, 4 g (33% cal.); Cholesterol, 10 mg; Carbohydrates, 15 g; Fiber, 4 g; Sodium, 230 mg.

Roasted Beets with Beet Greens

FIGURE 11.5 Roast beets, uncovered, until tender.

FIGURE 11.6 With a paring knife, pull the peels from the beets.

Roasted Garlic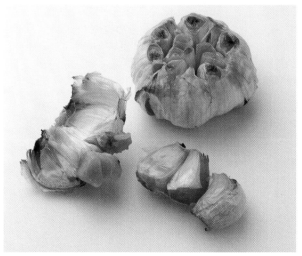

YIELD: APPROXIMATELY 6 OZ (175 G) GARLIC PULP

U.S.	METRIC	INGREDIENTS
6	6	Garlic heads, whole
1 fl oz	30 mL	Olive oil

PROCEDURE

1. Preheat an oven to 400°F (200°C).
2. Rub the heads of garlic with olive oil.
3. Place on a sheet pan in the oven. Roast about 30 minutes, or until soft.
4. Remove from the oven and cool slightly.
5. For roasted garlic pulp, cut the heads in half crosswise and squeeze out the pulp.
6. For roasted garlic cloves to use as garnish, separate the cloves. Serve peeled or unpeeled.

Per 1 ounce: Calories, 90; Protein, 2 g; Fat, 5 g (48% cal.); Cholesterol, 0 mg; Carbohydrates, 10 g; Fiber, 1 g; Sodium, 5 mg.

VARIATION

For whole roasted heads of garlic to use as garnish, cut off the tops (the pointed end) of the heads before rubbing with oil.

Roasted Garlic

Roasted Cauliflower with Almonds and Raisins

YIELD: 1 LB (480 G)

U.S.	METRIC	INGREDIENTS
1 lb 12 oz	840 g	Cauliflower florets
1 tbsp	15 mL	Olive oil
1/4 tsp	1 mL	Curry powder
to taste	to taste	Salt
to taste	to taste	Pepper
1 tbsp	15 mL	Shallots, chopped fine
1 1/2 tsp	7 mL	Olive oil
1/4 cup	60 mL	Golden raisins, soaked in warm water and drained
1 fl oz	30 mL	Chicken stock or water
1/4 cup	60 mL	Toasted sliced almonds
1 tbsp	15 mL	Italian parsley, cut chiffonade

PROCEDURE

1. Mix the cauliflower with the olive oil, curry powder, salt, and pepper.
2. Roast in an oven heated to 350°F (175°C) until the cauliflower is lightly browned at the edges.
3. Sweat the shallots in the second quantity of olive oil.
4. Add the raisins and cauliflower and cook for 1 minute.
5. Add the stock to moisten. Add the almonds and the parsley. Mix to combine.
6. Adjust the seasonings with additional salt and pepper if necessary.

Per 1 ounce (28.35 g): Calories, 40; Protein, 1 g; Fat, 2 g (45% cal.); Cholesterol, 0 mg; Carbohydrates, 4 g; Fiber, 1 g; Sodium, 10 mg.

Glazed Sweet Potatoes ♥

PORTIONS: 12   PORTION SIZE: 5 OZ (150 G)

U.S.	METRIC	INGREDIENTS	PROCEDURE
4 lb	1.8 kg	Sweet potatoes	1. Scrub the sweet potatoes and boil or steam until nearly tender. Do not overcook.
			2. Spread the potatoes on a sheet pan to cool.
			3. Peel the potatoes when they are cool enough to handle. Remove dark spots. Cut into neat, uniform pieces for easy portioning. Arrange in a buttered baking pan.
3 fl oz	90 mL	Water	4. Place the water, syrup, and sugar in a saucepan. Stir over heat until sugar is dissolved. Add the remaining ingredients and boil until the mixture is reduced to about 12 fl oz (350–400 mL) and forms a heavy syrup.
6 fl oz	175 mL	Light corn syrup or maple syrup	
3 oz	90 g	Brown sugar	5. Pour the syrup over the potatoes.
4 fl oz	125 mL	Orange juice	6. Bake at 350°F (175°C) until potatoes are thoroughly cooked and glazed, about 45–60 minutes. Baste with the syrup several times during baking.
1 fl oz	30 mL	Lemon juice	
1 oz	30 g	Butter	
1/2 tsp	2 mL	Cinnamon	
1/8 tsp	0.5 mL	Ground cloves	
1/4 tsp	1 mL	Salt	

Per serving: Calories, 190; Protein, 2 g; Fat, 2 g (9% cal.); Cholesterol, 5 mg; Carbohydrates, 44 g; Fiber, 2 g; Sodium, 100 mg.

Vanilla Roasted Beets

YIELD: APPROX. 12 OZ (360 G)

U.S.	METRIC	INGREDIENTS	PROCEDURE
4	4	Medium red or golden beets (about 4 oz each)	1. Peel the beets and cut into medium dice.
1/2	1/2	Vanilla bean	2. Split the vanilla bean and scrape out the seeds.
2 tsp	2 tsp	Vegetable oil	3. Combine the oil, vanilla seeds, salt, and pepper. Toss with the beets.
to taste	to taste	Salt	4. Place in a small baking pan, cover with foil, and bake at 300°F (150°C) until tender.
to taste	to taste	Black pepper	

Per 1 ounce (28.35 g): Calories, 25; Protein, <1 g; Fat, 1 g (36% cal.); Cholesterol, 0 mg; Carbohydrates, 4 g; Fiber, <1 g; Sodium, 30 mg.

Moussaka

PORTIONS: 16   PORTION SIZE: 9 OZ (250 G)

U.S.	METRIC	INGREDIENTS	PROCEDURE
1 lb	450 g	Onions, small dice	1. Sauté the onion and garlic in the olive oil until soft. Remove with a slotted spoon.
3	3	Garlic cloves, chopped	2. Add the meat to the pan and brown lightly.
2 fl oz	60 mL	Olive oil	3. Return the onion and garlic to the pot and add the tomato, wine, parsley, oregano, and cinnamon. Simmer, uncovered, until the liquid is reduced and the mixture is thick.
3¹/₂ lb	1.6 kg	Ground lamb or beef	
2 lb 4 oz	1 kg	Tomatoes, canned or fresh, peeled and chopped, with juice	
4 fl oz	100 mL	Red wine	4. Season to taste with salt and pepper.
2 tbsp	30 mL	Chopped parsley	
1¹/₂ tsp	7 mL	Dried oregano	
¹/₄ tsp	1 mL	Cinnamon	
to taste	to taste	Salt	
to taste	to taste	Pepper	
4 lb	1.8 kg	Eggplant	5. Peel the eggplant if the skin is tough. Cut into ¹/₂-in. (1-cm) slices.
as needed	as needed	Olive oil	6. Fry the eggplant slices in olive oil until tender. Set aside and season with salt.
to taste	to taste	Salt	
1 qt	1 L	Béchamel, cold	7. Season the béchamel (which should be quite thick when cold) with a little salt, white pepper, and nutmeg.
to taste	to taste	Salt	
to taste	to taste	White pepper	8. Beat the eggs and mix into the béchamel.
to taste	to taste	Nutmeg	
4	4	Eggs	
as needed	as needed	Olive oil	9. Oil the bottom of a hotel pan or other pan measuring 12 × 20 in. (30 × 50 cm) with olive oil. Sprinkle lightly with bread crumbs.
as needed	as needed	Dry bread crumbs	
2 oz	60 g	Romano or parmesan cheese, grated	10. Arrange the eggplant slices in the pan so they completely cover the bottom. Push them together as necessary.
			11. Put the meat mixture on the eggplant in a smooth layer.
			12. Pour the béchamel over the top and sprinkle with the grated cheese.
			13. Bake at 350°F (175°C) until hot and the top is golden, about 45–60 minutes.
			14. Cut into squares to serve.

Per serving: Calories, 470; Protein, 24 g; Fat, 33 g (62% cal.); Cholesterol, 140 mg; Carbohydrates, 20 g; Fiber, 4 g; Sodium, 580 mg.

BROILING AND GRILLING

Grilled quick-cooking vegetables such as peppers, zucchini, large mushroom caps, and eggplant are pleasant accompaniments to grilled and roasted meats and poultry. Cut the vegetables into broad slices, brush with oil, and grill until lightly cooked and lightly browned. Heavy browning may produce an unpleasant burned taste. Grilled vegetables are often dressed with vinaigrette.

Broiling is also used to finish cooked or partially cooked vegetables by browning or glazing them on top. Bread crumbs are sometimes used to give a pleasing brown color and to prevent drying. Casseroles or gratin dishes that do not brown sufficiently in the oven may be browned for a few seconds under the broiler or salamander.

PROCEDURE for Broiling or Grilling Vegetables

1. Collect equipment and food supplies.
2. Prepare the vegetables as necessary, including cutting them into required shapes and seasoning or marinating them.
3. Preheat the broiler or grill.
4. If necessary, brush the grill with a wire brush to clean it of any charred food particles. Lightly oil the cleaned grill rods.
5. Place the vegetables directly on the grill or broiler grate. Alternatively, place tender vegetables on broiler platters or sheet pans and set under the broiler. Cook the vegetables to the desired doneness and color, turning them as necessary (Figure 11.7).
6. Remove from broiler or grill. Evaluate the quality of the cooked vegetables (see p. 273). Depending on the exact procedure used, grilled vegetables should have a lightly browned exterior, grill marks when appropriate, and a pleasantly charred flavor.
7. Serve immediately.

FIGURE 11.7 Grilling vegetables.

Grilled Vegetable Kebabs

PORTIONS: 12   PORTION SIZE: 3 OZ (90 G)

U.S.	METRIC	INGREDIENTS	PROCEDURE
6 oz	180 g	Zucchini, trimmed	1. Cut the zucchini and yellow squash into 12 equal slices each.
6 oz	180 g	Yellow summer squash, trimmed	
6 oz	180 g	Red or orange bell peppers, cut into 1½-in. (2.5-cm) squares	2. Arrange the vegetables on 12 bamboo skewers (see Note). Give each skewer an equal arrangement of vegetable pieces.
12 oz	360 g	Red onion, large dice	3. Place the skewers in a single layer in a hotel pan.
12	12	Mushroom caps, medium	
12 fl oz	360 mL	Olive oil	4. Mix together the oil, garlic, herbs, salt, and pepper to make a marinade.
½ oz	15 g	Garlic, crushed	5. Pour the marinade over the vegetables, turning them to coat completely.
1½ tsp	7 mL	Dried rosemary	
½ tsp	2 mL	Dried thyme	6. Marinate 1 hour. Turn the skewers once or twice during marination to ensure the vegetables are coated.
2 tsp	10 mL	Salt	
½ tsp	2 mL	Black pepper	7. Remove the skewers from the marinade and let the excess oil drip off.

8. Broil the skewers until the vegetables are lightly charred on the surface but still somewhat crisp in the center.
9. Serve immediately.

Per serving: Calories, 50; Protein, 1 g; Fat, 3 g (53% cal.); Cholesterol, 0 mg; Carbohydrates, 5 g; Fiber, 1 g; Sodium, 40 mg.

Note: Soaking bamboo skewers in water ahead of time helps delay their charring.

VARIATIONS

Other assortments of vegetables may be used, and the skewers may be made larger if desired. Any vegetables you want to be completely cooked when served, such as cauliflower florets, should be blanched and cooled before being skewered, as they will not cook completely on the skewer.

Grilled Vegetable Medley

YIELD: ABOUT 3 LB (1.5 KG)   PORTIONS: 9   PORTION SIZE: 5 OZ (150 G)

U.S.	METRIC	INGREDIENTS
3–3¹/₂ lb	1.5–1.75 kg	Assorted vegetables:
		Small eggplants
		Zucchini
		Yellow summer squash
		Bell peppers
		Radicchio
		Large onions
as needed	as needed	Olive oil
to taste	to taste	Salt
as needed	as needed	Balsamic vinegar

PROCEDURE

1. Prepare the vegetables: Trim the stem ends of the eggplants and cut them lengthwise into thick slices. If they are very small, just cut them in half lengthwise. Trim the stem ends of the zucchini and yellow squash. Cut lengthwise into thick slices. Core and seed the peppers and cut into quarters lengthwise. Remove any bruised outer leaves of the radicchio and cut in halves or quarters through the base, leaving the core in to hold the leaves together. Cut the onion into thick slices, holding the rings of each slice together with a bamboo skewer.
2. Brush the vegetables with olive oil and sprinkle them with salt.
3. Grill the vegetables over medium heat, turning as necessary, until they are tender and lightly grill-marked. Cooking time will vary by vegetable. Regulate the heat or distance from the flame so the vegetables cook without browning too much.
4. Remove from the grill and brush with a little balsamic vinegar, and, if desired, a little more olive oil. Serve warm.

Per serving: Calories, 180; Protein, 1 g; Fat, 16 g (78% cal.); Cholesterol, 0 mg; Carbohydrates, 9 g; Fiber, 2 g; Sodium, 10 mg.

VARIATIONS

Other vegetables and vegetable assortments may be grilled in the same manner. Suggestions include large mushroom caps, Belgian endive, blanched potatoes, fennel, asparagus, leeks, and scallions. Grilled vegetables may be served with various sauces, such as aïoli, sauce Vierge, salsa cruda, and vinaigrette variations.

Grilled Vegetable Medley

Broiled Tomato Slices

PORTION: 10   PORTION SIZE: ABOUT 2 SLICES, 3–4 OZ (100 G)

U.S.	METRIC	INGREDIENTS
2¹/₂ lb	1.1 kg	Tomatoes
2 fl oz	60 g	Melted butter or olive oil
to taste	to taste	Salt
to taste	to taste	White pepper
		Topping (optional):
1 cup	100 g	Dry bread crumbs
4 fl oz	100 g	Melted butter or olive oil
1 oz	30 g	Onion, minced very fine

PROCEDURE

1. Wash the tomatoes, cut out the core ends, and slice crosswise into ¹/₂-in. (1-cm) slices.
2. Place the slices in a single layer on an oiled baking sheet.
3. Drizzle melted butter or oil over the tomatoes and sprinkle with salt and pepper.
4. Place in a broiler, 4 in. (10 cm) from the heat, and broil just until bubbling and hot but still firm enough to hold shape.
5. Serve 2 slices per portion, depending on size.
6. To use optional ingredients, cook tomatoes halfway. Combine topping ingredients and sprinkle over tomatoes. Brown under broiler.

Per serving: Calories, 60; Protein, 1 g; Fat, 5 g (65% cal.); Cholesterol, 10 mg; Carbohydrates, 5 g; Fiber, 1 g; Sodium, 55 mg.

VARIATIONS

Herbed Broiled Tomatoes

Top tomatoes with ¹/₄ cup (60 mL) chopped parsley and ¹/₂ tsp (2 mL) dried basil or oregano before broiling, or mix herbs with crumb topping.

Parmesan Broiled Tomatoes

Add ¹/₂ cup (125 mL) grated Parmesan cheese to crumb topping.

DEEP-FRYING

Potatoes (covered in the next chapter) and onion rings are the most popular fried vegetables, but many others may be fried, too.

Deep-fried vegetables may be divided into five categories:

1. Vegetables dipped in batter and fried.

2. Vegetables breaded and fried.

3. Vegetables fried without a coating.

Potatoes are the obvious example. Other starchy vegetables, such as sweet potatoes, may be fried without breading or batter if they are cut thin to reduce cooking time. The sugar in them burns easily if they are cooked too long.

Thin slices and shavings of vegetables, deep-fried until light and crisp, make an attractive and interesting garnish for many dishes. Root vegetables, such as beets, celery root, and parsnips, can be sliced thin and fried like potato chips (slice long roots like parsnips lengthwise). Other vegetables, such as leeks and celery, can be cut into thin shreds or julienne and fried. These may be dusted in flour before frying.

4. Small vegetables or cuts mixed with a batter and dropped with a scoop into hot fat. The term *fritter* is used for this preparation, as well as for that in category 1.

5. Croquettes: thick vegetable purées or mixtures of small pieces of vegetable and a heavy béchamel or other binder, formed into shapes, breaded, and fried.

Before continuing with this section, you should also review the discussion of breadings and batters in Chapter 7 (pp. 150-152). You should understand these procedures and techniques well in order to produce quality breaded and battered vegetables. In addition, you should observe the General Guidelines for Deep-Frying, which apply to all foods.

In addition to the batters included in the recipe for Onion Rings below, the batters in the recipes for Fish and Chips (p. 433) and Shrimp and Vegetable Tempura (p. 637) can be used for coating a variety of vegetables.

GENERAL GUIDELINES for Deep-Frying

1. Fry at proper temperatures.

Most foods are fried at 350°–375°F (175°–190°C). Excessive greasiness in fried foods is usually caused by frying at too low a temperature.

2. Don't overload the baskets.

Doing so greatly lowers the fat temperature.

3. Use good-quality fat.

The best fat for frying has a high smoke point (the temperature at which the fat begins to smoke and to break down rapidly).

4. Replace 15–20 percent of the fat with fresh fat after each daily use.

This extends frying life.

5. Discard spent fat.

Old fat loses frying ability, browns excessively, and imparts off-flavors.

6. Avoid frying strong- and mild-flavored foods in the same fat, if possible.

French fries should not taste like fried fish.

7. Fry as close to service as possible.

Do not leave foods in the basket above the fry kettle, and do not hold under heat lamps for more than a few minutes. The foods' moisture quickly makes the breading or coating soggy.

8. Protect fat from its enemies:

Heat. Turn the fryer off or to a lower holding temperature (200°–250°F/95°–120°C) when not in use.

Oxygen. Keep fat covered between services, and try to aerate the fat as little as possible when filtering.

Water. Remove excess moisture from foods before frying. Dry baskets and kettle thoroughly after cleaning. Keep liquids away from the fryer to prevent accidental spills.

Salt. Never salt foods over the fat.

Food particles. Shake loose crumbs off breaded items before placing over the fat. Skim and strain the fat frequently.

Detergent. Rinse baskets and kettle well after cleaning.

PROCEDURE for Deep-Frying Vegetables

1. Collect all equipment and food products.

2. Preheat fryer to proper temperature.
   Most vegetables are fried at 325°–350°F (160°–175°C).

3. Prepare food items as required. Apply breading or batter if necessary.

4. Place proper amount of food in fryer. Do not overload.

5. Fry to desired doneness.

6. Remove food from fryer and let fat drain from it.

7. Evaluate the quality of the finished item. In addition to the general standards of quality for cooked vegetables on page 273, also refer to the standards for fried foods below.

8. Serve at once, or, if necessary, hold uncovered in a warm place for the shortest possible time.

Standards of Quality for Deep-Fried Foods

High quality in a deep-fried product is characterized by the following properties:

Minimal fat absorption

Minimal moisture loss (that is, not overcooked)

Attractive golden color

Crisp surface or coating

No off-flavors imparted by the frying fat

VEGETABLES FOR DEEP-FRYING

Most vegetables large enough to coat with breading or batter may be fried. Tender, quick-cooking vegetables can be fried raw. Others may be precooked by simmering or steaming briefly to reduce the cooking time they need in the frying fat.

Raw vegetables for frying in breading or batter:

Eggplant	Onion rings	Tomatoes
Mushrooms	Peppers	Zucchini

Blanched or precooked vegetables for frying in breading or batter:

Artichoke hearts	Carrots	Fennel
Asparagus	Cauliflower	Okra
Beans, green and yellow	Celery	Parsnips
Broccoli	Celery root	Turnips
Brussels sprouts	Cucumbers	

KEY POINTS TO REVIEW

- What are the two basic kinds of baked vegetable preparation?

- What are the steps in the procedure for broiling or grilling vegetables?

- What are the basic kinds of deep-fried vegetable preparations?

- What are the steps in the procedure for deep-frying vegetables?

 Onion Rings

PORTIONS: 10   PORTION SIZE: 3 OZ (90 G), 8–10 PIECES

U.S.	METRIC	INGREDIENTS	PROCEDURE
1	1	Egg yolks, beaten	1. Combine the egg yolks and club soda in a bowl.
8 fl oz	240 mL	Club soda	2. Sift together the flour, baking powder, and salt.
5 oz	150 g	Flour	3. Add the dry ingredients to the liquid and mix to make a smooth batter.
1 tsp	5 mL	Baking powder	
¼ tsp	1 mL	Salt	
1	1	Egg whites	4. Whip the egg whites to soft peaks.
			5. Fold into the batter.
1 lb 8 oz	720 g	Onions, large	6. Peel the onions and cut crosswise into ¼-in. (0.5-cm) slices. Separate into rings (save unusable pieces for another purpose).
as needed	as needed	Flour	7. Place the onions in cold water, if they are not used immediately, to maintain crispness.
			8. Drain and dry the onions thoroughly.
			9. Dredge with flour and shake off excess. (This step isn't always necessary, but it helps the batter adhere.)
			10. Dip a few pieces at a time in the batter and fry in deep fat (350°F/175°C) until golden brown.
			11. Drain and serve immediately.

Per serving: Calories, 150; Protein, 3 g; Fat, 7 g (40% cal.); Cholesterol, 20 mg; Carbohydrates, 21 g; Fiber, 2 g; Sodium, 85 mg.

VARIATIONS

Beer Batter

Substitute light beer for the club soda.

Other Fried Vegetables

Any of the vegetables on the list at the beginning of this section may be fried in this batter.

Onion Rings

Vegetable Fritters

PORTIONS: 10   PORTION SIZE: 3 OZ (90 G), 2 PIECES

U.S.	METRIC	INGREDIENTS	PROCEDURE
		Batter:	1. Combine the eggs and milk.
3	3	Eggs, beaten	2. Mix together the flour, baking powder, salt, and sugar. Add to the milk and eggs and mix until smooth.
8 fl oz	240 mL	Milk	
8 oz	240 g	Flour	3. Let the batter stand for several hours in a refrigerator.
1 tbsp	15 mL	Baking powder	
¹/₂ tsp	2 mL	Salt	
¹/₂ oz	15 g	Sugar	
12 oz EP	350 g EP	Vegetables: Choice of corn, cooked diced carrots, baby lima beans, diced asparagus, diced celery or celery root, turnip, eggplant, cauliflower, zucchini, parsnips	4. Stir the cold, cooked vegetables into the batter.
			5. Drop with a No. 24 scoop into deep fat at 350°F (175°C). Hold the scoop just above the hot fat when dropping. Fry until golden brown.
			6. Drain well and serve.

Per serving: Calories, 140; Protein, 4 g; Fat, 6 g (37% cal.); Cholesterol, 45 mg; Carbohydrates, 19 g; Fiber, 1 g; Sodium, 230 mg.

VARIATIONS

For lighter fritters, beat egg whites separately and fold into batter.

Fruit Fritters

Increase sugar to 1 oz (30 g). Use fresh, frozen, or canned fruits such as blueberries, diced pineapple, or apple. Fruit must be well drained. Dust each portion with powdered sugar at service time. (Batter may be seasoned with cinnamon, vanilla, brandy, or other appropriate flavoring.)

Chiles Rellenos

PORTIONS: 16   PORTION SIZE: 1 PEPPER

U.S.	METRIC	INGREDIENTS	PROCEDURE
16	16	Chiles poblanos (see Note)	1. Char the chiles over a gas flame until the skin is blackened. Rub off the blackened skin under running water.
3 lb (approximately)	1.4 kg (approximately)	Picadillo (p. 510) (see Note)	2. Slit one side of each pepper and remove the seeds, but be careful to keep the peppers intact.
			3. Stuff the peppers with the picadillo.
12	12	Egg yolks	4. Beat the egg yolks and water slightly, then mix in the flour and salt.
1 fl oz	30 mL	Water	
1 oz	30 g	Flour, sifted	5. Whip the whites until they form soft peaks. Fold them into the yolk mixture.
¹/₂ tsp	2 mL	Salt	
12	12	Egg whites	6. Carefully dust the filled peppers with flour, then dip in the egg batter. Deep-fry at 350°F (175°C) until lightly browned. (*Hint:* Carefully lower each pepper into the fat with the slit side up. If the slit tends to open, spoon a little of the batter over the slit. This helps keep the opening sealed and the filling inside the pepper.)
as needed	as needed	Flour for dredging	
3–4 pt	1.5–2 L	Tomato Broth for Chiles Rellenos (p. 206)	7. For each portion, ladle 3–4 fl oz (90–125 mL) broth into a broad serving bowl or soup plate. Place 1 chile in the center of the bowl and serve at once.

Per serving: Calories, 430; Protein, 24 g; Fat, 30 g (62% cal.); Cholesterol, 210 mg; Carbohydrates, 17 g; Fiber, 2 g; Sodium, 460 mg.

Note: Anaheim peppers or frying peppers may be used if poblanos are not available, but the results will not be as flavorful. The exact amount of filling needed depends on the size of the peppers.

VARIATIONS

For cheese-filled chiles, use chunks of American Muenster or Monterey jack cheese instead of the picadillo.

For baked chiles rellenos, omit the egg batter and simply bake the stuffed chiles in a casserole until they are heated through. Serve with the tomato broth as in the basic recipe.

Chile Relleno

Suggested Vegetable Seasonings, Flavorings, and Combinations

Asparagus	Lemon juice, brown butter, mustard sauce, Parmesan cheese; hard-cooked egg, peas, artichokes, mushrooms
Beans, green	Dill, basil, tarragon, oregano, garlic, brown butter, soy sauce; almonds, sesame seed, onion, tomato, celery, mushrooms, bacon
Beans, lima	Oregano, sage, thyme, sour cream, cheddar cheese; corn, peas, onions, mushrooms, pimiento, bacon
Beets	Lemon, allspice, caraway, cloves, dill, ginger, horseradish, bay leaf, orange, sour cream, onion
Broccoli	Lemon, mustard sauce, almonds, buttered and toasted bread crumbs, hard-cooked egg
Brussels sprouts	Caraway, dill, parmesan cheese, cheddar cheese, chestnuts
Cabbage	Caraway, celery seed, dill, mustard, nutmeg, garlic; bacon, ham, carrots, onion
Carrots	Parsley, dill, fennel, tarragon, ginger, nutmeg, bay leaves, caraway, mint, orange; celery, peas, zucchini
Cauliflower	Dill, nutmeg, mustard, curry, cheese, tomato sauce; hard-cooked egg, peas, almonds
Celery	Parsley, tarragon, onion, green or red pepper, potatoes
Corn	Chili powder, mild cheddar or jack cheese, tomato, bacon, lima beans
Cucumber	Dill, garlic, mint, tarragon; peas
Eggplant	Garlic, marjoram, oregano, parsley, Parmesan cheese; tomato, chopped walnuts
Mushrooms	Nutmeg, parsley, lemon, paprika, dill, sherry, Parmesan cheese, cayenne, heavy cream; peas, spinach, artichokes, green beans
Okra	Garlic, coriander, sage; tomatoes, corn
Onions	Nutmeg, sage, thyme, cheese sauce, sour cream; peas
Peas	Mint, basil, dill, sage; mushrooms, pearl onions, turnips, potatoes, carrots, water chestnuts, Jerusalem artichokes
Spinach	Nutmeg, garlic, heavy cream; mushrooms, hard-cooked egg, cheese
Squash, summer (including zucchini)	Cumin, basil, oregano, mustard seed, rosemary, garlic, Parmesan cheese, parsley; tomato, carrots (with zucchini), onion, almonds, walnuts
Squash, winter	Cinnamon, nutmeg, allspice, cloves, ginger; apples, bacon, pecans
Sweet potatoes	Allspice, cinnamon, cloves, nutmeg, ginger, brandy, orange; almonds, apples, bananas
Tomatoes	Basil, bay leaf, garlic, celery seed, oregano, thyme, rosemary, chili powder; peppers, black olives
Turnips	Parsley, chives, nutmeg; mushrooms, potatoes, peas

ADDITIONAL RECIPES

These additional recipes may be found on your CulinarE-Companion recipe management program:

Artichokes Clamart; Braised Celery; Corn Pudding; Eggplant Sichuan Style; Elote con Queso; Gratin of Fennel; Green Beans with Roasted Peppers and Bacon; Lecsó; Sautéed Mushrooms; Southwestern Corn and Pinto Bean Gratin; Spaghetti Squash with Tomato Confit; Spinach Timbales; Sunchoke Galettes.

QUESTIONS FOR DISCUSSION

1. Which vegetables would you simmer uncovered?

Asparagus	Cauliflower
Green beans	Peas
Beets	Sweet potatoes
Brussels sprouts	Rutabagas
Carrots	Turnips

2. Why are greens such as spinach not well suited to cooking in a compartment steamer?

3. In the recipe for Peas, Carrots, and Pearl Onions (p. 309), why could you not save a step and cook the three vegetables together in one pot?

4. Why is it important to drain vegetables well before combining with a cream sauce?

5. Which of the two methods for making glazed root vegetables (see Glazed Root Vegetables, p. 315, and Glazed Carrots, p. 315) might be more appropriate for à la carte service, or cooking to order? Why?

6. We have learned that green vegetables should be cooked in a neutral liquid because acids destroy green pigments. But the recipe for artichokes says to cook them with lemon juice. What's going on here?

7. Describe briefly how you would make breaded, fried onion rings rather than onion rings with batter.

12

POTATOES

The eating habits of most nations place a great deal of importance on a category of foods we call starches. In fact, for a large portion of the world's peoples, starch is the mainstay of the diet and supplies most of the day's calories. In North America and Europe, the most important starches are potatoes, rice, pasta, and bread. It is true we do not depend on these high-carbohydrate foods as much as many of the world's people, who eat far less meat than we do. Nevertheless, even as diet fashions and fads come and go, starches appear at nearly all our meals.

Because we eat them often and have devised a great many ways of preparing them, starchy foods require extra study beyond that which we give other vegetables. In this chapter, we turn our attention primarily to the preparation of our most important vegetable, the potato. Grains and other starches are discussed in Chapter 13.

AFTER READING THIS CHAPTER, YOU SHOULD BE ABLE TO

1. Classify potatoes into two types, describe the general properties of each type, and identify the most suitable cooking method for each type.

2. Identify characteristics of high-quality potatoes, and describe how to store them.

3. Cook potatoes by boiling and steaming.

4. Prepare potato purée.

5. Cook potatoes by baking, sautéing, pan-frying, and deep-frying.

UNDERSTANDING POTATOES

In classical cuisine, the potato is one of the most important of all foods. To many of us today, potatoes are considered an ordinary and humble food. Escoffier, however, treated the potato with great respect. His *Guide Culinaire* lists more than 50 potato preparations, far more than for any other vegetable or starch.

Considering how important the potato is in the cuisines of Europe and North America, it is surprising to think this vegetable wasn't widely used until the last half of the eighteenth century. Although the potato had been brought to Europe from the New World a few hundred years earlier, it wasn't until then that an army pharmacist named Antoine-Auguste Parmentier began promoting its use. To this day, many classical recipes featuring the potato are called *Parmentier* (par mawn tyay).

Botanically, the potato is a tuber, which is an enlarged underground stem with buds (or eyes) that become new shoots. Traditional main courses in western cooking feature a protein item, one or more vegetables, and a starch. The potato is, of course, a vegetable, but because of its high starch content it usually serves the same function on the menu as grains and other starchy foods.

Potatoes are traditionally classified as either starchy, low-moisture varieties or as waxy, high-moisture varieties. The following section summarizes the main characteristics of these two categories as well as traditional uses for each. Today, many potato varieties are available that were largely unknown not long ago. As always, chefs love to experiment with new foods and to find new uses for them. For example, they don't feel limited to russets for baking. So, after reading about the two main categories of potatoes below, continue to the next section for an introduction to some of the many varieties available today.

TYPES

Potatoes are classified according to their starch content. The amount of starch determines the use for which they are usually considered most suitable. Keep in mind that these categories are only general. Within each group is a range of starch and moisture content. For example, different varieties of **waxy potatoes** have different moisture content, depending not only on the variety of potato but also on the growing and storage conditions.

1. Waxy potatoes.

High moisture content, high sugar content, low starch content.

Usually small and round in shape, but some varieties can be large, and some may be elongated. Flesh is white, yellow, or even blue or purple. Skin is white, red, yellow, or blue.

Hold shape well when cooked. Firm, moist texture. Use for boiling whole, for salads, soups, hash browns, and any preparation where the potato must hold its shape.

Do not use for deep-frying. High sugar content will cause dark streaks and poor texture.

Waxy potatoes, clockwise from top left: small and large red-skinned potatoes, white potatoes, large and small yellow potatoes.

Mashed or Whipped Potatoes

PORTIONS: 12   **PORTION SIZE: 5 OZ (150 G)**

U.S.	METRIC	INGREDIENTS
4¹/₂ lb	2 kg	Potatoes
3 oz	90 g	Butter
¹/₂ cup	125 mL	Light cream, hot
as needed	as needed	Milk, hot
to taste	to taste	Salt
to taste	to taste	White pepper

PROCEDURE

1. Peel and eye the potatoes and cut them into uniform sizes. Simmer in water to cover until tender.
2. Drain well and let the potatoes steam dry for a few minutes.
3. Pass the potatoes through a food mill or ricer into the bowl of a mixer. *Alternative method:* Place potatoes in mixer with paddle attachment. Mix until well broken up. Replace paddle with whip and beat until well puréed. Do not overwhip, or potatoes will become pasty.
4. Beat in butter, then cream.
5. Add enough hot milk to bring potatoes to proper consistency. They should be soft and moist, but firm enough to hold their shape, not runny.
6. Add salt and white pepper to taste.
7. If desired, whip briefly at high speed until potatoes are light and fluffy. Do not overwhip.

Per serving: Calories, 190; Protein, 3 g; Fat, 8 g (36% cal.); Cholesterol, 20 mg; Carbohydrates, 29 g; Fiber, 3 g; Sodium, 65 mg.

VARIATIONS

Garlic Mashed Potatoes

Method 1: Simmer 3–4 whole, peeled cloves of garlic with the potatoes. Purée the garlic and the potatoes together.

Method 2: Purée 1 head roasted garlic (p. 334) and mix into the potatoes before adding cream.

Ancho Mashed Potatoes

Add Ancho Sauce (p. 207) to taste to whipped potatoes before adding milk. Reduce the quantity of milk as necessary to achieve the proper consistency.

Duchesse Potatoes

PORTIONS: 12   **PORTION SIZE: 4 OZ (100 G)**

U.S.	METRIC	INGREDIENTS
3¹/₂ lb	1.5 kg	Potatoes, peeled and quartered
2 oz	50 g	Butter, melted
to taste	to taste	Salt
to taste	to taste	White pepper
to taste	to taste	Nutmeg
5	5	Egg yolks
as needed	as needed	Egg wash (optional)

PROCEDURE

1. Steam the potatoes or simmer them in water until tender. Drain in a colander and let dry in an oven several minutes.
2. Pass the potatoes through a food mill or ricer.
3. Add butter and mix to a smooth paste. Season to taste with salt, pepper, and just a little nutmeg (the potatoes should not taste strongly of nutmeg).
4. If the potatoes are very moist, stir over a low flame to stiffen. They must be much stiffer than mashed potatoes.
5. Add the egg yolks (off the fire) and beat until smooth.
6. Put the mixture in a pastry bag with a star tube and bag out into desired shapes on sheet pans or as platter borders (see **Figure 12.1** on p. 354). Cone-shaped spiral mounds are most popular for individual portion service.
7. If desired, brush lightly with egg wash for greater browning.
8. At service time, place potatoes in hot oven (400°–425°F/200°–230°C) until lightly browned. Platter borders may be browned under the salamander.

Per serving: Calories, 150; Protein, 3 g; Fat, 6 g (34% cal.); Cholesterol, 95 mg; Carbohydrates, 23 g; Fiber, 2 g; Sodium, 45 mg.

VARIATION

Duchesse potato mixture is also used as the base for Potato Croquettes (p. 362). It is considered one of the basic hot kitchen preparations.

FIGURE 12.1 Using the pastry bag: duchesse potatoes.

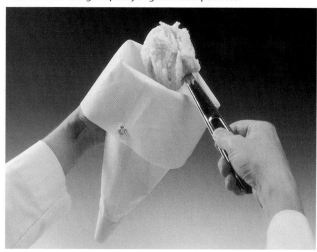

(a) Turn down the top of the pastry bag as shown. Slip your hand under this collar and hold the top open with your thumb and forefinger while you fill it with duchesse potato mixture.

(b) Turn the top of the bag up again and gather the loose top together as shown. Hold the bag shut with your thumb and forefinger. To force out the potatoes, squeeze the top of the bag in the palm of your hand. Use your free hand to guide the tip or hold the item being filled or decorated. You can make potato croquettes quickly by forcing out the potato mixture in long strips, using a large plain tube. Cut the strips into 2-in. (5-cm) lengths with a knife.

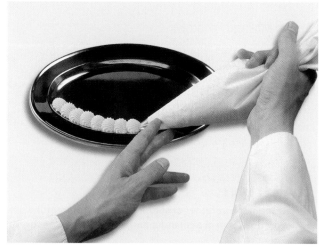

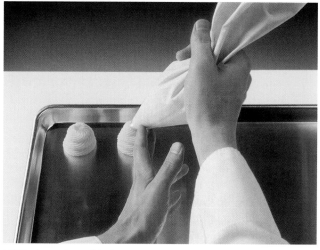

(c) Duchesse potatoes are often used to decorate platters, as in this illustration. This technique is also used in decorating cakes and desserts with icing, whipped cream, or meringue.

(d) Single portions of duchesse potatoes are usually piped out into a tall spiral shape. They are then browned in the oven.

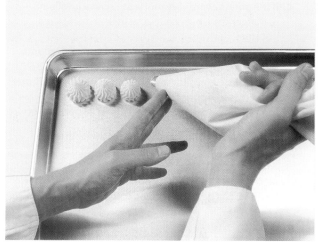

(e) Dauphine and Lorette potatoes may be bagged out into many shapes, such as these small stars. Some cookies are also shaped this way.

BAKING

Preparing baked potatoes is a simple procedure that is widely misunderstood and therefore needlessly complicated. Properly baked potatoes are white, fluffy, mealy, and steamy, and they have a dry skin that crackles slightly when pressed. Poorly baked potatoes, unfortunately common, are gray and soggy and have a damp, soft skin.

Russet potatoes are most often used for baked potatoes. However, many varieties are now available, some of which are excellent for baking and yield different flavors and textures. Don't be afraid to experiment with some of the varieties listed on pages 347–348.

PROCEDURE for Baking Potatoes

1. For standard baked potatoes, select russets or other regularly shaped starchy potatoes.

2. Scrub well and pierce the ends with a fork or skewer so steam can escape.

3. For crisp skins, rub lightly with oil. For more tender skins, leave dry.

4. Place on sheet pans or on sheet pan racks in a preheated 400°F (200°C) oven and bake until done, about 1 hour. To test doneness, squeeze gently. Done potatoes yield to gentle pressure.

   Note: Using sheet pan racks eliminates the hard spot that forms where the potato is in contact with the sheet pan.

5. Remove from oven.

6. To hold for service, keep warm and uncovered so the potatoes will not be made soggy by trapped steam. Hold no more than 1 hour, if possible, though they will keep longer with some loss of quality.

   Note that nothing was said about wrapping potatoes in foil. Foil-wrapped potatoes do not bake but rather steam in their own moisture. The texture of a steamed potato is entirely different from that of a baked potato. Save yourself the trouble and expense of wrapping in foil and serve a better product.

 Baked Potatoes ♥

PORTION SIZE: 1 POTATO

U.S.	METRIC	INGREDIENTS
as needed	as needed	Idaho or baking potatoes
as needed	as needed	Vegetable oil (optional)

PROCEDURE

1. Scrub the potatoes well and pierce the ends with a skewer or fork to allow steam to escape.

2. Lightly oil the potatoes if a crisp skin is desired. Leave them dry for a somewhat softer skin.

3. Place on sheet pan in 400°F (200°C) oven. Bake until done, about 1 hour. Test for doneness by squeezing a potato gently.

Per serving: Calories, 130; Protein, 3 g; Fat, 0 g (0% cal.); Cholesterol, 0 mg; Carbohydrates, 31 g; Fiber, 3 g; Sodium, 10 mg.

Stuffed Baked Potatoes

PORTIONS: 10   **PORTION SIZE: 1 POTATO**

U.S.	METRIC	INGREDIENTS
10	10	Baking potatoes, 7–8 oz (200–225 g) each
2 oz	60 g	Butter, melted
4 oz	100 mL	Light cream or milk, hot
to taste	to taste	Salt
to taste	to taste	White pepper

PROCEDURE

1. Bake the potatoes according to the basic method.
2. Remove from the oven. Cut a slice off the top of each potato and scoop out the pulp, leaving a shell about $1/4$-in. ($1/2$-cm) thick.
3. Pass the pulp through a food mill or ricer. Beat in the butter and enough cream or milk to make a smooth purée. Season to taste. (Note that this preparation is basically the same as whipped potatoes.)
4. Fill the potato shells with the purée, using a pastry bag or kitchen spoon. (A pastry bag is faster and neater.) Place them on a baking sheet.

U.S.	METRIC	INGREDIENTS
3 tbsp	45 mL	Dry bread crumbs
1 oz	30 g	Butter, melted
3 tbsp	45 mL	Parmesan cheese, grated

5. Mix the bread crumbs and melted butter until all the crumbs are moistened. Then mix in the parmesan cheese and top the potatoes with this mixture.
6. Place in hot oven (400°F/200°C) until potatoes are heated through and tops are browned, about 15 minutes.

Per serving: Calories, 270; Protein, 5 g; Fat, 10 g (32% cal.); Cholesterol, 30 mg; Carbohydrates, 42 g; Fiber, 4 g; Sodium, 140 mg.

VARIATIONS

For each variation listed, add the indicated ingredients to the potato purée mixture. Proportions are for $2^1/2$–3 lb (1.1–1.4 kg) purée.

1. 2 oz (60 g) grated parmesan cheese
2. 8 oz (225 g) minced onion, sautéed in butter
3. 4 oz (100 g) cooked ham, small dice
   4 oz (100 g) mushrooms, chopped and sautéed in butter
4. 8 oz (225 g) bacon, diced and cooked crisp
   1 green pepper, chopped and sautéed in butter or bacon fat

Macaire Potatoes

Scoop out the pulp completely and discard the skins. Mash the pulp with a kitchen fork or break it up with the paddle of a mixer. Omit the melted butter and cream or milk. Instead, mix in 7 oz (200 g) soft butter. Season. Form into small cakes and pan-fry in clarified butter until golden brown on both sides.

Oven Roast Potatoes ♥

PORTIONS: 12   **PORTION SIZE: 4 OZ (125 G)**

U.S.	METRIC	INGREDIENTS
5 lb AP	2.25 kg AP	Potatoes
as needed	as needed	Vegetable oil or olive oil
to taste	to taste	Salt
to taste	to taste	White pepper

PROCEDURE

1. Peel and eye potatoes. Cut into 25 uniform portions and trim pieces to shape. Save the trimmings for other use.
2. Dry the potatoes well and rub with oil. Place in oiled baking pan and season with salt and pepper.
3. Place in 400°F (200°C) oven and bake until browned and cooked through, about 1 hour. Halfway through baking time, turn potatoes and brush with additional oil.

Per serving: Calories, 160; Protein, 3 g; Fat, 2.5 g (14% cal.); Cholesterol, 0 mg; Carbohydrates, 32 g; Fiber, 3 g; Sodium, 10 mg.

PROCEDURE VARIATION: COMBI OVEN

After step 2 in the procedure:

3. Place in combi oven at 240°F (115°C) in steam mode and steam until cooked mostly through, about 10 minutes, depending on size.
4. Raise temperature to 375°F (190°C) and continue to cook in combi mode for 25 minutes. Halfway through baking time, turn the potatoes and brush with additional oil.

Olive Oil and Garlic Confit Red Potatoes

YIELD: 2 LB (960 G)

U.S.	METRIC	INGREDIENTS	PROCEDURE
8 fl oz	240 mL	Olive oil	1. Combine olive oil, garlic, and herbs in a small saucepan. Heat gently over low heat for 15 minutes to infuse the garlic and herb flavors into the oil.
4	4	Garlic cloves, crushed	
3–4 sprigs	3–4 sprigs	Fresh thyme	
1 sprig	1 sprig	Fresh rosemary	
2 lb	960 g	Red potatoes, small, sliced $1/4$ in. (6 mm) thick (see note)	2. Put the potatoes in a small baking pan or sauté pan. Pour the oil over the potatoes. The oil should come about halfway up the potatoes.
to taste	to taste	Salt	3. Cover the pan and bake in an oven at 275°F (135°C) until the potatoes are tender. Turn the potatoes frequently as they bake. Total baking time is about 30–40 minutes.
to taste	to taste	Pepper	4. Drain the oil from the potatoes and reserve for another use (such as browning or reheating potatoes for service).
			5. Season the potatoes to taste with salt and pepper.

Per 1 ounce (28.35 g): Calories, 35; Protein, <1 g; Fat, 1.5 g (39% cal.); Cholesterol, 0 mg; Carbohydrates, 5 g; Fiber, <1 g; Sodium, 5 mg.

Note: If desired, cut potatoes into medium dice instead of slices.

BAKED "EN CASSEROLE"

A number of preparations call for potatoes baked in a baking pan or casserole, with or without liquid added. The best-known is scalloped potatoes. A characteristic of most of these preparations is that they are baked uncovered at least part of the time so a brown crust forms on top. (Note that two versions of Gratin Dauphinoise are included here. The first is a modern version, while the second is a more traditional version.)

Scalloped Potatoes

PORTIONS: 12   PORTION SIZE: 5 OZ (150 G)

U.S.	METRIC	INGREDIENTS	PROCEDURE
$2^{1}/_{2}$ pt	1.25 L	Milk	1. Make a thin white sauce (béchamel) using the ingredients listed (see p. 181). Keep hot while preparing the potatoes.
$1^{1}/_{2}$ oz	45 g	Butter	
$1^{1}/_{2}$ oz	45 g	Flour	
1 tsp	5 mL	Salt	
to taste	to taste	White pepper	
3 lb 12 oz	1.75 kg	Potatoes	2. Peel and eye the potatoes. Cut into slices $1/8$ in. (3 mm) thick.
			3. Place the potatoes in a buttered baking pan, making several layers.
			4. Pour in the white sauce. Lift the potatoes slightly so the sauce can run between the layers.
			5. Cover with foil or greased paper and place in oven at 350°F (175°C) for 30 minutes.
			6. Uncover and continue to bake until top is lightly browned and potatoes are tender.

Per serving: Calories, 200; Protein, 6 g; Fat, 6 g (27% cal.); Cholesterol, 20 mg; Carbohydrates, 31 g; Fiber, 2 g; Sodium, 80 mg.

Note: Unthickened milk may be used instead of a thin white sauce, but the milk is more likely to curdle. The roux helps prevent curdling.

VARIATIONS

Scalloped Potatoes with Onions

Add 10 oz (300 g) sliced onions to baking pan with the potatoes.

Scalloped Potatoes with Ham

Add $1^{1}/_{4}$ lb (700 g) diced ham.

Gratin Dauphinoise I

PORTIONS: 12   PORTION SIZE: 4 OZ (125 G)

U.S.	METRIC	INGREDIENTS
2¹/₂ lb	1.25 kg	Firm-fleshed potatoes
1 tsp	5 mL	Salt
¹/₈ oz	3 g	Garlic, chopped
1 pt	500 mL	Cream
8 fl oz	250 mL	Milk
to taste	to taste	White pepper

PROCEDURE

1. Peel the potatoes and cut into thin slices (as for potato chips). Do not place the sliced potatoes in water.

2. Toss the potatoes in the salt and garlic and arrange in an even layer in a buttered half-size hotel pan or a roasting pan of equivalent size.

3. Mix the cream and milk and lightly season with pepper. Bring to a boil.

4. Pour over the potatoes, then place the pan in the oven at 400°F (200°C) for about 30 minutes, or until the potatoes are tender and the sauce is thick.

Per serving: Calories, 300; Protein, 4 g; Fat, 23 g (67% cal.); Cholesterol, 85 mg; Carbohydrates, 21 g; Fiber, 2 g; Sodium, 490 mg.

Gratin Dauphinoise

Gratin Dauphinoise II

PORTIONS: 15   PORTION SIZE: 5 OZ (150 G)

U.S.	METRIC	INGREDIENTS
3 lb	1.4 kg	Potatoes
to taste	to taste	Salt
to taste	to taste	White pepper
to taste	to taste	Nutmeg
¹/₂ lb	225 g	Gruyère cheese, grated
1 pt	500 mL	Milk
1 cup	250 mL	Heavy cream
3	3	Egg yolks

PROCEDURE

1. Peel and eye the potatoes. Cut into very thin slices.

2. Place some of the potatoes in a layer in a buttered baking pan. Season with salt, pepper, and a very small amount of nutmeg. Sprinkle with a little of the cheese. Repeat until all the potatoes and about three-fourths of the cheese are used up.

3. Combine the milk and half the cream and heat to a simmer. Beat the egg yolks with the remaining cream. Slowly stir in the hot milk mixture.

4. Pour the milk mixture over the potatoes. Top with remaining cheese.

5. Bake uncovered at 350°F (175°C) until done, 45–60 minutes.

Per serving: Calories, 220; Protein, 8 g; Fat, 13 g (53% cal.); Cholesterol, 85 mg; Carbohydrates, 18 g; Fiber, 1g; Sodium, 80 mg.

VARIATION

Savoyarde Potatoes

Prepare as above, but use chicken stock instead of milk.

Boulangère Potatoes

PORTIONS: 12   PORTION SIZE: 5 OZ (150 G)

U.S.	METRIC	INGREDIENTS
1 lb 4 oz	600 g	Onions, sliced
2¹/₂ oz	75 g	Butter or fat drippings from roast (see Note)
3 lb 12 oz	1.75 kg	Potatoes, peeled and cut into thick slices
1 pt	500 mL	Stock, chicken or lamb (if available), hot
to taste	to taste	Salt
to taste	to taste	Pepper

PROCEDURE

1. Sauté the onions in butter or fat until they are translucent and just beginning to brown.
2. Add the potatoes and toss until coated with fat.
3. Place in a baking pan or in a roasting pan under a partially cooked roast. Pour in the hot stock. Season.
4. Bake 1–1¹/₂ hours at 350°F (175°C) or at the roasting temperature of lamb, until potatoes are done. Add more stock during cooking if necessary to keep potatoes from drying out.

Per serving: Calories, 160; Protein, 3 g; Fat, 5 g (27% cal.); Cholesterol, 15 mg; Carbohydrates, 28 g; Fiber, 3 g; Sodium, 55 mg.

Note: Boulangère potatoes may be cooked separately, but they are usually cooked with a roast, especially leg of lamb (see p. 488).

If the potatoes are cooked with a roast, they must be added to the pan at the right time so they will be done at the same time as the meat.

Boulangère Potatoes

Potatoes au Gratin

PORTIONS: 12   PORTION SIZE: 6 OZ (175 G)

U.S.	METRIC	INGREDIENTS
3 lb 12 oz	1.75 kg	Potatoes
1 qt	1 L	Cheddar Cheese Sauce (p. 182), hot
¹/₃ cup	80 mL	Dry bread crumbs
1 tsp	5 mL	Paprika
1 oz	30 g	Butter, melted (optional)

PROCEDURE

1. Scrub the potatoes and simmer or steam them until tender but still firm.
2. Drain and spread on sheet pan to cool.
3. When the potatoes are cool enough to handle, peel and cut them into uniform ³/₈-in. (1-cm) dice.
4. Combine with the hot cheese sauce in a baking pan.
5. Mix the bread crumbs and paprika and sprinkle over the potatoes. Drizzle the butter evenly over the top.
6. Bake at 350°F (175°C) about 30 minutes, or until hot and browned.

Per serving: Calories, 190; Protein, 7 g; Fat, 6 g (29% cal.); Cholesterol, 20 mg; Carbohydrates, 26 g; Fiber, 2 g; Sodium, 150 mg.

Note: Cream sauce may be used instead of cheese sauce. Grated cheese (cheddar or parmesan) may be sprinkled over the potatoes before topping them with bread crumbs.

PROCEDURE VARIATION: COMBI OVEN

1. Scrub the potatoes, peel and cut them into uniform ³/₈-in. (1-cm) dice.
2. Place in a 4-inch-deep hotel pan and steam them in a combi oven in steam mode at 240°F (115°C) until tender but still firm, about 10 minutes.
3. Combine with the hot cheese sauce in the same baking pan.
4. *(continued)*
5. Mix the bread crumbs and paprika and sprinkle over the potatoes. Drizzle the butter evenly over the top.
6. Bake at 375°F (190°C) in combi-mode with maximum browning (0% humidity) about 15 minutes, or until hot and browned.

DEEP-FRYING

All the rules of deep-frying that you learned in Chapter 11 apply to potatoes. Review pages 339–340 to refresh your memory.

There are two kinds of deep-fried potato preparations:

1. Potatoes fried raw.

These are potatoes that are simply cut into shapes and deep-fried until golden and crisp. They include all the varieties of French fries as well as potato chips.

Russet or Idaho potatoes are most suitable for frying because of their high starch content and their regular shape, which permits less trimming loss.

2. Preparations made from cooked, puréed potatoes.

Most of these products are made from duchesse potato mixture. They include potato croquette variations, Dauphine potatoes, and Lorette potatoes.

Starchy potatoes are used for these recipes, as they are for duchesse potatoes, because they make a good dry, mealy purée.

FRENCH FRIES

Because French fries, or deep-fried potatoes, are one of the most popular items in North American food service, you must know how to prepare them well. Most French fries served are made from blanched, frozen product, but it is important to know how to make them from fresh potatoes.

The recipe on page 361 gives the complete procedure for preparing French fries. Note that they are fried in two stages. It is possible to cook them in one step, but this is impractical in a volume operation because of the long cooking time. The more common practice is to blanch them in frying fat. This is done at a lower temperature so they cook through without browning. They are then drained and refrigerated until service time. Portions can then be finished to order in a few minutes.

Frozen products have been prepared through step 5 in the recipe and then frozen. To use them, simply begin with step 6.

KEY POINTS TO REVIEW

- What are the steps in the procedure for baking potatoes?
- What are the two basic categories of sautéed or pan-fried potato preparations?
- What are the steps in the procedure for making French fries?
- What are potato croquettes? How are they made?

 French Fries

PORTIONS: AS NEEDED (2¹/₄ LB AP/1 KG AP POTATOES YIELDS ABOUT 1 LB/450 G COOKED POTATOES)

U.S.	METRIC	INGREDIENTS
as needed	as needed	Idaho potatoes

PROCEDURE

1. Peel and eye the potatoes.

2. Cut the potatoes into strips ³/₈ in. (1 cm) square and about 3 in. (7.5 cm) long. (See **Figure 7.11** for cutting procedure.) Hold the cut potatoes in cold water until needed, to prevent discoloration.

3. Line sheet pans with several layers of brown paper and have them ready by the deep fryer.

4. Drain and dry the potatoes well. Deep-fry in fat heated to 325°F (160°C) until they are just beginning to turn a pale golden color. At this point, they should be cooked through and soft.

5. Remove the potatoes from the fryer and turn them out onto the sheet pans in a single layer to drain. Refrigerate.

6. At service time, fry the potatoes in small quantities in fat heated to 350°–375°F (175°–190°C) until brown and crisp.

7. Drain well. Salt them lightly away from the fryer, or let customers salt their own. Serve immediately.

Per 3.2 oz (90.7 g): Calories, 290; Protein, 4 g; Fat, 15 g (46% cal.); Cholesterol, 0 mg; Carbohydrates, 36 g; Fiber, 3 g; Sodium, 200 mg.

VARIATIONS

Pont-Neuf Potatoes

Prepare as in basic recipe, but cut the potatoes in thicker strips, about ¹/₂ in. (1.25 cm) square or slightly larger. Blanching time will be slightly longer.

Allumette Potatoes (Shoestring or Matchstick Potatoes)

Cut the potatoes into thin strips, slightly less than ¹/₄ in. thick (about 0.5 cm). Because they are so thin, they are usually fried in one step (without blanching) until very crisp.

Straw Potatoes

Cut into very thin strips, about ¹/₈ in. (3 mm) thick. Fry in one step in hot fat (375°F/190°C).

Steakhouse Fries

Scrub but do not peel potatoes. Cut in half lengthwise, then cut each half lengthwise into 4–6 wedges, depending on size. Prepare as in basic recipe.

Potato Chips

Cut potatoes into very thin slices, less than ¹/₈ in. (3 mm) thick. Fry in one step in hot fat (375°F/190°C).

Waffle or Gaufrette Potatoes

Set the fluted blade of a mandoline to cut very thin slices. Cut potatoes into round slices, turning the potato about 90 degrees between slices so you cut waffle shapes (see Figure 12.2). Fry like potato chips.

FIGURE 12.2 Gaufrette Potatoes.

(a) Set the fluted blade of the mandoline so it cuts very thin slices.

(b) Slice the potatoes, turning the potato about 90 degrees between slices so the ridges on the two sides of each slice cross each other.

(c) You may need to adjust the thickness of the cut after the first slice or two. The slices should be thin enough to have holes.

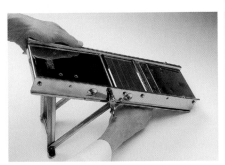

Potato Croquettes 🌹

These are made from duchesse potato mixture. The procedure for duchesse potatoes is repeated here for the sake of convenience.

PORTIONS: 10   PORTION SIZE: 3 PIECES, 1¹/₂ OZ (40 G) EACH

U.S.	METRIC	INGREDIENTS	PROCEDURE
3¹/₂ lb	1.5 kg	Potatoes, peeled and quartered	1. Steam the potatoes or simmer in salted water until tender. (Steaming is preferable because it results in a drier product.) Drain in a colander (if simmered) and let dry in an oven a few minutes.
2 oz	50 g	Butter	2. Pass the potatoes through a food mill or ricer.
to taste	to taste	Salt	3. Add the butter and mix to a smooth paste. Season to taste with salt, pepper, and nutmeg.
to taste	to taste	White pepper	
to taste	to taste	Nutmeg	4. Set over moderate heat and stir the mixture to dry it out well. If it is not dry enough, the croquettes will not hold their shape. *Alternative method:* Add enough cornstarch or dry potato starch to absorb excess moisture and stiffen the mixture.
5	5	Egg yolks	5. Remove from the fire, add the egg yolks, and beat in thoroughly.
			6. To shape croquettes, two methods are available: (a) Spread the mixture to cool in a pan, cover with plastic or buttered paper to keep a crust from forming, and refrigerate. Dust hands with flour and shape potatoes by hand into cylinders the shape of corks, about 2 in. (5 cm) long and about 1¹/₂ oz (40 g) each. (b) Dust sheet pans with flour. Place the warm potato mixture in a pastry bag fitted with a large, plain tip. Bag out the potatoes into long strips on the pans (see **Figure 12.1**). With a knife, cut the strips into 2-in. (5-cm) lengths.
		Standard Breading Procedure:	7. Set up a breading station and pass the potatoes through Standard Breading Procedure (p. 151).
as needed	as needed	Flour	8. At service time, fry croquettes in deep fat at 350°F (175°C) until golden brown. Drain well.
as needed	as needed	Egg wash	9. Serve immediately, 3 pieces per portion.
as needed	as needed	Bread crumbs	

Per serving: Calories, 443; Protein, 9 g; Fat, 20 g (42% cal.); Cholesterol, 155 mg; Carbohydrates, 54 g; Fiber, 3 g; Sodium, 315 mg.

VARIATIONS

Other shapes may be used as desired. Add 4 oz (110 g) of any one of the following to the potato mixture:

> Grated cheese
> Minced, sautéed onion
> Chopped, sautéed mushrooms
> Chopped ham
> Finely chopped nuts

Berny Potatoes

Shape into small balls. Bread with finely slivered almonds instead of bread crumbs. (In classical cuisine, minced truffles are added to the potato mixture.)

Dauphine Potatoes

Method 1: For each pound of duchesse or croquette potato mixture, add ¹/₃ lb (150 g) Pâte à Choux or cream puff paste (p. 1000), made without sugar and with half the amount of butter. To fry, bag out into desired shapes onto greased brown paper. Slide into hot fat. Remove paper when potatoes float loose.

Method 2: Hold pastry bag over deep fryer. Force out potato mixture and cut off short lengths with the back of a knife, letting them drop into the hot fat.

Potato Croquettes

Lorette Potatoes

Prepare like Dauphine potatoes and add 1 oz grated parmesan cheese per pound of mixture (60 g per kg). Shape as desired (the classic shape is a small crescent) and fry without breading.

SAUTÉING AND PAN-FRYING

The procedures for sautéing and pan-frying potatoes are basically the same as for other vegetables (p. 321).

There are many sautéed and pan-fried potato preparations. Some are made with raw potatoes, others with precooked or blanched potatoes. Many of these recipes are especially useful because they are excellent ways to utilize leftover boiled potatoes.

This group of recipes may be divided into two categories based on production technique.

1. **Potatoes mixed or tossed while cooking.**

   The procedure for sautéing vegetables, page 321, is used for these preparations. The potatoes are cut into pieces or into small shapes and cooked in a small amount of fat. They are turned or tossed in the pan so they brown on all sides. This category includes rissolé, parisienne, noisette, château, and American fried or home-fried potatoes.

2. **Potatoes cooked and served in compact cakes.**

   The procedure for pan-frying vegetables (see p. 321) is the basic method used for these preparations. The potatoes are not mixed while cooking but are made into cakes, which are browned on both sides. This category includes hash browns and variations as well as potato pancakes and macaire potatoes (see p. 356).

Rissolé or Cocotte Potatoes

PORTIONS: 12   PORTIONS SIZE: 4 OZ (125 G)

U.S.	METRIC	INGREDIENTS
7 lb	3.5 kg	Potatoes (see Note)
5 oz	150 g	Clarified butter
to taste	to taste	Salt
to taste	to taste	White pepper

PROCEDURE

1. Peel and eye the potatoes and trim or tourné them into small oval shapes about 1¹/₂ in. (4 cm) long. (See **Figure 7.21** for technique.)
2. Place the potatoes in a saucepan, cover with salted water, and bring to a boil. Reduce heat and simmer 7–8 minutes, or until about three-fourths cooked.
3. Drain and turn out onto a sheet pan to steam dry.
4. Heat the clarified butter in a large sauté pan. Add the potatoes and sauté over moderate heat until golden brown and fully cooked. (Potatoes may also be browned in deep fat if speed of service is critical.)
5. Season with salt and pepper.

Per serving: Calories, 290; Protein, 4 g; Fat, 11g (34% cal.); Cholesterol, 30 mg; Carbohydrates, 45 g; Fiber, 4 g; Sodium, 125 mg.

Note: The quantity of potatoes needed varies depending on how heavily they are trimmed. Save trimmings for purées or other uses.

VARIATIONS

Alternative Method: Potatoes may be sautéed without prior blanching. Sauté over low heat and keep covered during the first half or three-fourths of the cooking period so the potatoes cook fully without overbrowning. Potatoes cooked this way absorb more butter than those that are blanched first. Thus they are tastier, but they also are higher in calories.

Parisienne and Noisette Potatoes

Prepare as in basic recipe, but cut potatoes with a melon ball cutter. Parisienne potatoes are cut with a scoop slightly larger than 1 in. (about 3 cm). Noisette potatoes are smaller than 1 in. (about 2.5 cm). Blanch 3–5 minutes or cook from raw state.

Château Potatoes

Prepare as in basic recipe, but tourné the potatoes into larger ovals, about 2 in. (5 cm) long.

Potato Rods Rissolé

Cut potatoes into rod shapes, using a ¹/₂-in. (12-mm) brass tube (see photo, p. 885). Prepare as in basic recipe.

Hash Brown Potatoes

PORTIONS: 12   PORTION SIZE: 4¹/₂ OZ (125 G)

U.S.	METRIC	INGREDIENTS
3 lb 12 oz	1.7 kg	Boiled potatoes, cooled and peeled
5 oz	140 g	Oil, clarified butter, or a mixture of oil and clarified butter
to taste	to taste	Salt
to taste	to taste	Pepper

PROCEDURE

1. Chop the potatoes into small pieces.
2. Ladle a thin layer of oil or butter into a well-seasoned 6-in. (15-cm) sauté pan and set it over high heat. A griddle may also be used for this preparation.
3. When the fat is hot, add 1 portion of potatoes and flatten them into a round cake. Shake the pan back and forth to keep the potatoes from sticking.
4. When the potatoes are well browned on the bottom, flip them over or turn them with an offset spatula. Try to keep the potato cake unbroken. Season with salt and pepper.
5. When the second side is done, tilt the pan to drain off any excess fat for reuse, holding the potatoes in the pan with the spatula. Slide the potatoes out of the sauté pan onto a plate.
6. Repeat with remaining portions.

Per serving: Calories, 180; Protein, 2 g; Fat, 7 g (35% cal.); Cholesterol, 0 mg; Carbohydrates, 27 g; Fiber, 2 g; Sodium, 5 mg.

VARIATIONS

Any of the following can be added to the potatoes to vary the flavor: chopped parsley, chives, bacon, hard-cooked egg, grated cheese, and garlic.

Rösti Potatoes

Shred boiled potatoes on the large holes of a hand grater, or use a machine. Prepare as in basic recipe. Potatoes should be very crisp on the outside but soft inside. (These are often called hash brown potatoes but, strictly speaking, hash browns are made with chopped potatoes, as hash means "to chop.")

Lyonnaise Hash Browns

Combine 10 oz (300 g) onion, chopped and sautéed in butter, with the chopped or shredded potatoes before pan-frying.

Lyonnaise Potatoes

PORTIONS: 12   PORTION SIZE: 4¹/₂ OZ (125 G)

U.S.	METRIC	INGREDIENTS
3 lb 4 oz	1.5 kg	Boiled potatoes, cooled and peeled
12 oz	350 g	Onions
4 oz	110 g	Clarified butter, vegetable oil, or mixture of oil and butter
to taste	to taste	Salt
to taste	to taste	White pepper

PROCEDURE

1. Cut the potatoes into slices about ¹/₄ in. (0.5 cm) thick.
2. Peel the onions, cut in half lengthwise, and slice into julienne.
3. Heat half the fat in a sauté pan and sauté the onions until they are golden. Remove from the pan with a slotted spoon and set aside.
4. Put the rest of the fat into the pan. Set the pan on high heat and add the potatoes.
5. Sauté the potatoes, tossing them in the pan until well browned on all sides.
6. Add the onions and continue to sauté for another minute, or until onions and potatoes are well mixed and the flavors are blended.
7. Season to taste.

Per serving: Calories, 190; Protein, 2 g; Fat, 9 g (42% cal.); Cholesterol, 25 mg; Carbohydrates, 26 g; Fiber, 2 g; Sodium, 95 mg.

Note: This preparation may be made on a griddle instead of in a sauté pan.

VARIATIONS

Home Fries or American Fries

Prepare as in basic recipe, but omit onions.

Potatoes O'Brien

Cook 5 oz (150 g) diced bacon until crisp. Remove bacon from pan. Sauté 5 oz onion (150 g), cut in fine dice, and 5 oz (150 g) green bell pepper, cut in fine dice, in bacon fat. Sauté 3 lb 4 oz (1.5 kg) diced potatoes as in basic recipe and add vegetables.

Add the crisp bacon and 2 oz (60 g) diced pimiento to finish, and season to taste.

Lyonnaise Potatoes

Potato Pancakes

PORTIONS: 10   **PORTION SIZE: 2 PANCAKES, ABOUT 2 OZ (60 G) EACH**

U.S.	METRIC	INGREDIENTS	PROCEDURE
3 lb	21.35 kg	Potatoes	
8 oz	225 g	Onions	
1	1	Lemons	
3	3	Eggs	
2 tbsp	30 mL	Chopped parsley (optional)	
1 tsp	5 mL	Salt	
1/4 tsp	1 mL	White pepper	
1 oz or more	30 g or more	Flour (see Note)	
as needed	as needed	Oil for pan-frying	

PROCEDURE

1. Peel the potatoes and onions. Grate them together into a stainless-steel bowl. Juice the lemons, add the juice to the potatoes to prevent discoloration, and toss to mix.
2. Place the potatoes in a china cap and squeeze out the excess liquid. Hold the liquid and let the starch settle out. Drain the liquid from the starch.
3. Return the potatoes to a stainless-steel bowl and add the potato starch.
4. Beat in the eggs, parsley, salt, and pepper.
5. Stir in enough flour to bind the potato mixture. (If the batter is too thin, the pancakes will fall apart in the pan. Test-fry a little first, and add more flour if necessary.)
6. Pour about 1/4 in. (1/2 cm) oil into a heavy iron skillet. Heat the oil over moderately high heat. The oil should reach about 325°F/160°C.
7. Measuring with a solid kitchen spoon, place portions of the batter in the pan to make individual pancakes.
8. Pan-fry, turning once, until golden brown on both sides.
9. Remove from the pan with a slotted spoon or spatula and drain briefly on absorbent paper.
10. *Alternative method:* Lightly brown in oil and place in one layer on a sheet pan. Finish in the oven (375°F/190°C) until brown and crisp.

Per serving: Calories, 220; Protein, 5 g; Fat, 10 g (40% cal.); Cholesterol, 65 mg; Carbohydrates, 29 g; Fiber, 3 g; Sodium, 260 mg.

Note: Matzoh meal or dried potato starch may be used instead of flour for binding the batter.

ADDITIONAL RECIPES

These additional recipes may be found on your CulinarE-Companion recipe management program:

Anna Potatoes; Potato and Eggplant Stew with Cilantro; Potatoes with Bacon and Onions; Potatoes with Chickpeas and Chiles; Roasted New Potatoes with Herbs and Garlic; and Scalloped Potatoes with Cheese.

TERMS FOR REVIEW

waxy potato

russet

all-purpose

new potato

solanine

duchesse potatoes

QUESTIONS FOR DISCUSSION

1. True or false: French fries made from fresh potatoes are always better than French fries made from frozen, blanched potato strips. Explain.
2. If mature, starchy potatoes are best for puréeing (mashed, duchesse, etc.), then why doesn't everyone use russets or Idahos, which are the starchiest?
3. Why is it not a good idea to put parisienne potatoes on your menu unless you are also serving a puréed potato product?
4. Many of the potato recipes in this chapter do not indicate what type of potato to use. For those recipes, indicate whether you would select all-purpose, russet, or waxy potatoes.

13

LEGUMES, GRAINS, PASTA, AND OTHER STARCHES

This chapter continues the discussion of starchy foods. The previous chapter focuses on a fresh vegetable, the potato, one of the most important starches on European and North American menus. This chapter, by contrast, discusses preparations based on dried foods: legumes and grains.

For most of human history—and prehistory—grains have been the most important source of nutrients and calories to sustain life, and this remains true today in many parts of the world. For example, in parts of Asia, rice is eaten at nearly every meal. In Japan, the standard words for *breakfast*, *lunch*, and *dinner* can be translated as "morning rice," "noon rice," and "evening rice."

In many regions, dried beans, eaten together with rice, are a more common source of protein than meat or fish. Dried legumes are actually matured, dried versions of some of the fresh vegetables discussed in earlier chapters and are not immediately thought of as starches. However, they do have a high starch content and, because of their dried nature, are handled much like grains.

Pastas and noodles are important starchy foods made from grains, usually wheat, but also rice and other grains. In this chapter we look primarily at the noodle products inherited from Italian cuisine, but we also learn about noodles from other cuisines.

AFTER READING THIS CHAPTER, YOU SHOULD BE ABLE TO

1. Distinguish the major types of dried legumes.

2. Cook dried legumes.

3. Distinguish the major types of rice.

4. Distinguish the major types of other grains used in food service.

5. Prepare grains by simmering and by the pilaf and risotto methods.

6. Distinguish major kinds and shapes of commercial pasta, and determine their quality.

7. Prepare fresh and commercial pasta products, and list the steps involved in the alternate steam-table method of its preparation.

DRIED LEGUMES

A **legume** is a plant that bears seed pods that split along two opposite sides when ripe. Legumes include beans, peas, lentils, and a number of other plants. In culinary usage, we use the word to refer to the seeds from these pods, especially when they are mature and dried. (Do not confuse the English meaning of *legume* with that of the French word *legume*, which means "vegetable.")

Dried beans and peas have been used as food for thousands of years, and they continue to be important foods today. In fact, with today's increased interest in healthful eating and in vegetables of all sorts, as well as a greater knowledge of cuisines from around the world, many more interesting varieties of beans are widely available now than only a few years ago.

Legumes are high in protein and, thus, are important in vegetarian diets. They are rich in B vitamins and minerals. Some legumes, like the soybean, are also rich in fat.

TYPES AND VARIETIES

The three most important types of dried legumes in Western kitchens are kidney beans, peas, and lentils. Several unrelated legumes, including chickpeas, fava beans, soybeans, and lima beans, play smaller roles on the menu. The following descriptions include most of the commonly available beans.

Kidney Beans

Most of the many-colored beans in the illustrations are types of kidney bean. These are all varieties of one species of plant—the same plant that gives us the common green bean. Their flavors and textures vary slightly, but their cooking and handling characteristics are similar, although some may require longer cooking times than others.

A subgroup of this family is sometimes called **haricot beans** (*haricot* is the French word for "bean"). These are all varieties of green bean (*haricots verts*) that are allowed to ripen until the seeds are mature and dry. The members of this group are white beans of various sizes. The term *kidney bean* is then used for the remaining beans in this family, which have colors other than white.

The most common kinds of kidney bean and haricot bean are summarized in Table 13.1.

Peas

Dried *green* and *yellow* peas are the same peas we eat as a fresh vegetable, but they are left on the vine until mature and dry. They are usually split, with the hull removed, in order to speed cooking time, although whole peas are also available.

Split peas cook quickly without preliminary soaking.

Black-eyed peas and *pigeon peas* are popular in the southern United States as well as in parts of Africa and the Caribbean. They are not related to green or yellow peas nor to kidney beans but, like regular peas, are often sold fresh in the pod as well as dried. Black-eyed peas are small, white, kidney-shaped beans with a black spot where the bean attaches to the pod. Pigeon peas are small and round or oval, with beige skin flecked with brown.

Lentils

Lentils are small, lens-shaped legumes. They have a shorter cooking time than kidney beans, even when whole, and do not need soaking. If desired, however, they may be soaked, resulting in an even shorter cooking time.

The most prized lentils are the tiny *green lentils*, commonly known as *Le Puy* lentils because the best ones are grown in Le Puy, France. (**Note:** Only lentils actually grown in Le Puy should be given this name. If grown elsewhere, they are simply called *green lentils*.) They have a dark green or gray-green hull and, unlike other lentils, keep their shape fairly well when cooked.

The larger *brown lentils* are the most common lentils in Western kitchens. They range in color from medium brown to greenish-brown.

Red lentils have long been a mainstay in India and have become popular in the West only recently. They are tiny, salmon-pink lentils that have been split, with the dark hull removed. Red lentils turn yellow when cooked and break apart to form a purée.

Yellow lentils are small split lentils similar to red lentils, but less often seen.

TABLE 13.1 The Kidney Bean/Haricot Bean Family

Name	Description
Haricot beans	
Navy bean	small, oval white bean
Pea bean	similar to but smaller than navy bean
Rice bean	tiny white bean slightly larger than a grain of cooked rice
Great Northern bean	medium-small white bean
White kidney bean, cannellini	large, white kidney-shaped bean
Soissons (swah sohn)	medium white bean
Flageolet	small, pale green bean, harvested while immature and then dried
Tarbais	large, white bean, broad and flat; larger than cannellini
Kidney beans	
Red kidney bean	dark red-brown medium bean with tough skin
Pink kidney bean	pink-red bean similar to red kidney
Pinto bean	medium kidney bean with tan skin splotched with pink-brown
Black bean or turtle bean	small, black, oval kidney bean
Cranberry bean or borlotti bean	medium kidney bean, pale pink skin spotted with red
Brown or Swedish bean	small, oval, brown-skinned bean
Calypso bean	medium oval bean, white with black patches
Appaloosa bean	medium kidney-shaped bean with black and white skin

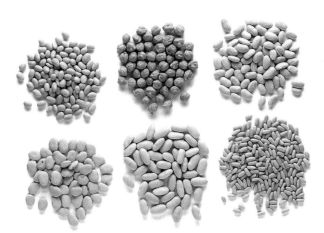

Top row: navy beans, garbanzo beans or chickpeas, Great Northern beans. Bottom row: baby lima beans, cannellini beans or white kidney beans, rice beans.

Top row: black turtle beans, dried fava beans. Bottom row: Swedish brown beans, calypso beans, flageolet beans.

Top row: red kidney beans, pink beans, appaloosa beans. Bottom row: cranberry beans or borlotti, Christmas lima beans, pinto beans.

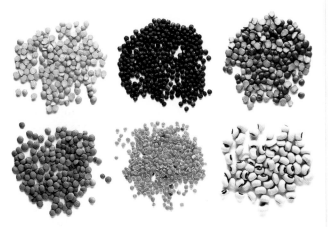

Top row: yellow split peas, green lentils, green split peas. Bottom row: brown lentils, red lentils, black-eyed peas.

Other Legumes

A number of other legumes are also important.

Lima beans, also called *butter beans*, are flat, broad beans ranging from creamy white to pale pastel green in color, in a range of sizes from large to small. They have a starchy texture and a distinctive flavor unlike that of any of the kidney beans. Like peas, they are often harvested when they are immature and moist and eaten as a fresh vegetable.

Chickpeas, also called *garbanzos* or, in Italian, *ceci* (chay chee), are round, hard, yellow-brown beans about twice the size of peas and best known as the main ingredient in hummus (p. 742). They stay firm and whole when cooked and have a nutty flavor. Dried chickpeas are also ground into flour.

Fava beans are broad, flat beans, the matured dried form of the vegetable discussed on pages 278–279. The brilliant green of the immature vegetable turns a creamy brown color when the bean is dried. The skin of the dried fava is tough and the texture is starchy.

Soybeans are one of the world's most important bean crops because of their high protein content as well as their rich oil content. Fresh soybeans are used as a vegetable (p. 292). Dried soybeans are very hard and require a long cooking time, but they are not often cooked and eaten directly. Rather, they are used to make a variety of other foods, such as soy sauce, soybean paste or *miso*, and tofu or bean curd. These products are discussed in Chapter 26.

Mung beans are small, round beans with a dark green skin and white interior. They are often sprouted to make the bean sprouts widely used in Chinese cooking. As dried legumes, they are available whole or hulled and split. The split form has a much shorter cooking time. Split mung beans break apart when cooked to make a purée. Mung beans belong to the same family as pigeon peas and black-eyed peas.

Adzuki beans are small, oval beans with a thick red skin. They are used primarily in Asian cooking in sweet as well as savory dishes.

Dal, or *dhal*, is the generic term in India for dried legume. Legumes used in Indian cuisine include *moong dal* (mung bean), *masoor dal* (red lentil), *urad dal* (a tiny, white split lentil), and *chana* (chickpea).

COOKING LEGUMES

Because dried beans, peas, and lentils are dry and hard, they must be rehydrated—that is, they must absorb water—in order to be made edible. Thus, the primary cooking method used to prepare dried legumes is simmering. Once the beans are cooked and tender, they can be finished in a variety of ways. The recipes in this section give a sample of ways to prepare legumes. Additional recipes are included in Chapter 9 (Soups), Chapter 21 (Salads and Salad Dressings), and Chapter 26 (Cooking for Vegetarian Diets).

PROCEDURE for Preparing Dried Legumes

1. Pick over to remove any foreign particles. Rinse well.

2. Soak overnight in 3 times the volume of water. (Split peas and some lentils do not require soaking. Check package directions.)

3. Drain the beans. Place them in a pot and add fresh cold water to cover by 1–2 inches (2.5–5 cm).

4. Bring to a boil. Reduce heat and simmer, covered, until tender. Do not boil, or the legumes may toughen. Some beans require up to 3 hours of simmering.

5. Check texture for doneness. A properly cooked bean is soft and creamy in texture, not hard or grainy. Unless a purée or soup is desired, kidney beans should remain intact, not cooked so long they are falling apart. Other peas and beans will form a purée or remain whole, depending on the type.

6. If you forget to soak beans overnight, an alternative method can be used. Put the beans in a cooking pot with 3 times their volume of cold water. Bring to a boil. When water just starts to boil, cover tightly and remove from the heat. Let stand 1 hour. Then proceed with step 3 above.

KEY POINTS TO REVIEW

• What are the three major types of dried legumes? Give examples of each.

• In addition to the three main types of legumes, what are six other important legumes used in the kitchen? Describe them.

• What are the steps in the basic procedure for preparing dried legumes?

🎩 White Beans, Bretonne Style

YIELD: ABOUT 6 LB (3 KG)   **PORTIONS: 20**   **PORTION SIZE: 5 OZ (150 G)**

U.S.	METRIC	INGREDIENTS	PROCEDURE
2 lb	1 kg	Dried white beans	1. Soak the beans overnight in cold water.
1	1	Carrot, small	2. Drain the beans and place in a pot with enough water to cover by 1 in. (2.5 cm). Add the carrot, celery, onion, and sachet. Simmer until the beans are tender but not soft or broken (1–3 hours, depending on the beans). Drain, but save the liquid. Discard the vegetables and sachet.
1 stem	1 stem	Celery	
1	1	Onion, small, peeled	
		Sachet d'épices:	
1	1	Bay leaf	
6–8	6–8	Parsley stems	
3–4	3–4	Peppercorns	
1	1	Clove	
pinch	pinch	Thyme	
4 oz	125 g	Butter	3. Heat the butter in a large saucepot or brazier. Sauté the onion and garlic until soft. Add the tomatoes and cook a few minutes to reduce liquid.
8 oz	250 g	Onion, diced	
2	2	Garlic cloves, chopped	4. Add the beans to this mixture and stir carefully. Simmer until heated through and flavors are blended. If too dry, add some of the bean cooking liquid. Add pan drippings, if you are using them. Season to taste.
1 lb	500 g	Canned tomatoes, with juice, coarsely chopped	
as needed	as needed	Pan juices from roast lamb (optional; see Note)	
to taste	to taste	Salt	
to taste	to taste	Pepper	

Per serving: Calories, 200; Protein, 11 g; Fat, 5 g (22% cal.); Cholesterol, 10 mg; Carbohydrates, 29 g; Fiber, 7 g; Sodium, 90 mg.

Note: This preparation is often served with roast leg of lamb or other lamb roast. If so, the pan drippings may be used to flavor the beans.

White Beans, Bretonne Style

Baked Beans, New England Style

PORTION: 10   PORTION SIZE: 4¹/₂ OZ (125 G)

U.S.	METRIC	INGREDIENTS	PROCEDURE
1 lb	450 g	Dried beans, navy or Great Northern	1. Soak the beans overnight in enough water to cover by 2 in. (5 cm).
		Bouquet garni:	2. Place the beans and liquid in a pot and add the bouquet garni. Bring to a boil and skim foam. Reduce heat to a simmer. Cover and simmer 45 minutes–1 hour, or until beans are just tender but not soft. Add more water if necessary during cooking.
1	1	Bay leaf	
6–8	6–8	Parsley stems	3. Drain the beans, reserving the cooking liquid. Discard the bouquet garni.
¹/₄ tsp	1 mL	Dried thyme	
a few	a few	Celery tops	
1¹/₂ cup	120 mL	Molasses	4. Mix the molasses, brown sugar, dry mustard, salt, and 1 pt (500mL) of the bean cooking liquid. If there is not enough bean liquid, add water to make up the difference.
1 oz	30 g	Brown sugar	
1¹/₂ tsp	7 mL	Dry mustard	5. Mix the beans, molasses mixture, and salt pork in a 2-qt (2-L) pot or deep baking pan.
1¹/₂ tsp	7 mL	Salt	
4 oz	110 g	Salt pork, medium dice	6. Bake, covered, at 300°F (150°C) for 2–2¹/₂ hours. Add more liquid if necessary during baking.

Per serving: Calories, 290; Protein, 10 g; Fat, 10 g (31% cal.); Cholesterol, 10 mg; Carbohydrates, 41 g; Fiber, 7 g; Sodium, 520 mg.

VARIATIONS

Michigan Baked Beans

Reduce molasses to 2 tbsp (30 mL) and add 1 cup (240 mL) tomato sauce or tomato purée.

Frijoles de la Olla (Mexican Pinto Beans)

PORTIONS: 16–20   PORTION SIZE: 4 OZ (125 G)

U.S.	METRIC	INGREDIENTS	PROCEDURE
1¹/₂ lb	750 g	Dried pinto beans or pink beans	1. Combine the beans, water, onion, garlic, and jalapeño in a pot. Bring to a boil, reduce heat, and simmer, covered, 1¹/₂ hours. Check the pot from time to time and add more water, if needed, to keep the beans covered.
3 qt	3 L	Water, cold (see Note)	
6 oz	175 g	Onion, sliced thin	
1–2	1–2	Garlic cloves, chopped	
1	1	Jalapeño or other green chile, chopped (optional)	2. Add the lard and salt. Continue to simmer until the beans are tender. Do not let the beans go dry. There should always be some broth. Add hot water if necessary.
2 oz	60 g	Lard or rendered pork fat	
2 tsp	10 mL	Salt	3. The beans will hold refrigerated for several days.

Per serving: Calories, 170; Protein, 8 g; Fat, 4 g (20% cal.); Cholesterol, 5 mg; Carbohydrates, 27 g; Fiber, 9 g; Sodium, 290 mg.

Note: The beans may be soaked overnight, if desired (although many Mexican cooking authorities feel the results are not as good). If they are soaked, reduce the water for cooking to 1¹/₂ pt (750 mL).

VARIATIONS

Frijoles Refritos

For the quantity of beans in the basic recipe, make in at least 3 batches. Mash the beans coarsely. Heat 2 oz (60 g) lard in a large sauté pan. Add 2 oz (60 g) chopped onion and fry until soft, but do not brown. Add one-third of the cooked, mashed beans (about 1¹/₂ lb/750 g, including broth) to the pan. Stir and mash the beans over heat until the beans start to dry out and pull away from the sides of the pan. Roll the mass out of the pan like an omelet. Sprinkle with grated cheese (mild cheddar or Monterey jack) and serve with tortilla chips.

Vegetarian Pinto Beans

Omit the lard or pork fat from the basic recipe.

Black Bean Cakes with Salsa 🌹

YIELD: 20 CAKES, 2 OZ (60 G) EACH   PORTIONS: 10   PORTION SIZE: 2 CAKES, 1½ FL OZ (45 ML) SALSA

U.S.	METRIC	INGREDIENTS	PROCEDURE
1 fl oz	30 mL	Olive oil	1. Heat the olive oil in a sauté pan over moderately low heat.
1 lb	500 g	Onion, cut brunoise	2. Add the onions and garlic and sweat until very soft. Do not brown.
2–4	2–4	Garlic cloves, chopped	
1–2	1–2	Jalapeños, seeded and cut brunoise	3. Add the jalapeño and ground cumin. Cook slowly another few minutes.
2 tsp	10 mL	Ground cumin	4. Add the beans and oregano. Cook until heated through.
2 lb	1 kg	Cooked black beans (see Note)	
1 tsp	5 mL	Dried oregano	
to taste	to taste	Salt	5. Transfer the mixture to a food processor and process to form a coarse purée. The mixture should be thick enough to hold its shape, but not too dry. If necessary, moisten with a little water.
to taste	to taste	Pepper	
3 fl oz	100 mL	Olive oil	6. Taste and adjust the seasoning with salt and pepper.
1 pt	450 mL	Salsa Cruda (p. 207)	7. Divide the mixture into 2-oz (60-g) portions. Form into small, flat cakes.
			8. Brown the cakes lightly on both sides in hot olive oil in a sauté pan. They will be very soft; handle carefully.
			9. Serve 2 cakes per portion with 1½ fl oz (45 mL) salsa.

Per serving: Calories, 260; Protein, 9 g; Fat, 12 g (41% cal.); Cholesterol, 0 mg; Carbohydrates, 30 g; Fiber, 9 g; Sodium, 180 mg.

Note: Canned black beans may be used, but they should be drained and rinsed.

Black Bean Cakes with Salsa

Green Lentils with Celery Root and Mustard ♥ 🌹

YIELD: 3 LB 12 OZ (1.8 KG)   PORTIONS: 12   PORTION SIZE: 5 OZ (150 G)

U.S.	METRIC	INGREDIENTS	PROCEDURE
1 lb 4 oz	600 g	Green (Le Puy) lentils	1. Rinse and drain the lentils.
1	1	Onion, small, whole	2. Place the lentils, onion, and sachet in a saucepan and add enough water to cover the lentils by 2 in. (5 cm).
		Sachet d'épices (p. 158):	3. Bring to a simmer and cook until tender but still firm. This may take from 10 minutes to more than 30, depending on the lentils. Add more water if necessary if the cooking time is long.
¹/₄ tsp	1 mL	Dried thyme	
6	6	Peppercorns	
1	1	Bay leaf	4. Drain the lentils, reserving the cooking liquid. Discard the onion and sachet.
8–10	8–10	Parsley stems	
1	1	Whole cloves	
1	1	Garlic clove	
4 oz	120 g	Celery root, small dice	5. Sweat the celery root, onion brunoise, and carrot in the olive oil 1 minute.
1¹/₂ oz	45 g	Onion, brunoise	6. Add a little of the lentil cooking liquid to moisten the vegetables and continue to cook until the celery root is tender. Add more liquid from time to time as needed.
1¹/₂ oz	45 g	Carrot, brunoise	
1 fl oz	30 mL	Olive oil	
1 fl oz	30 mL	Dijon-style mustard	7. When the vegetables are tender, stir in the mustard.
to taste	to taste	Salt	8. Add the lentils and stir. Cook just until the lentils are hot.
to taste	to taste	Pepper	9. Add salt and pepper to taste.

Per serving: Calories, 200; Protein, 13 g; Fat, 2.5 g (11% cal.); Cholesterol, 0 mg; Carbohydrates, 32 g; Fiber, 8 g; Sodium, 90 mg.

Green Lentils with Celery Root and Mustard

Hoppin' John ♥

PORTIONS: 12   PORTION SIZE: 6 OZ (180 G)

U.S.	METRIC	INGREDIENTS	PROCEDURE
1 lb	450 g	Dried black-eyed peas	1. Clean, soak, and cook the peas according to the procedure for cooking dried legumes (p. 370). Drain. Reserve the cooking liquid.
8 oz	225 g	Bacon, chopped	2. Render the bacon in a heavy saucepot, cooking until crisp.
8 oz	225 g	Onion, chopped fine	3. Remove the bacon with a slotted spoon and set aside.
¹/₂ tsp	2 mL	Red pepper flakes	4. Add the onion and the red pepper flakes to the pot. Cook until the onion is soft.
as needed	as needed	Water or chicken stock	5. Measure the reserved cooking liquid from step 1. Add enough water or chicken stock to measure 2 qt (2 L).
1 lb	450 g	Long-grain rice	6. Add the liquid, the cooked beans, and the rice to the pot. Bring to a boil, reduce heat, and cover. Cook at a low simmer until the rice is tender, 15 minutes or a little longer.
to taste	to taste	Salt	7. Remove from the heat and let stand, covered, 5 minutes.
			8. Fluff the rice with a fork while stirring in the reserved bacon, and salt to taste.

Per serving: Calories, 210; Protein, 9 g; Fat, 2 g (9% cal.); Cholesterol, 5 mg; Carbohydrates, 37 g; Fiber, 5 g; Sodium, 100 mg.

Purée of Flageolet Beans with Garlic

YIELD: 2 LB (1 KG)   PORTIONS: 8   PORTION SIZE: 4 OZ (125 G)

U.S.	METRIC	INGREDIENTS	PROCEDURE
12 oz	375 g	Dried flageolet beans	1. Rinse and drain the beans. Soak overnight in enough cold water to cover by several inches (5 cm or more).
1	1	Sachet d'épices (p. 158)	
1¹/₂ pt	750 mL	Water or light vegetable stock	2. Drain. Add the sachet and the water or stock. Simmer until the beans are tender, about 45 minutes. Discard the sachet.
3–4	3–4	Garlic cloves, peeled and chopped	3. Mash the chopped garlic with the salt.
1 tsp	5 mL	Salt	4. Drain the beans. Place the beans in a food processor with the garlic and olive oil. Blend to a purée.
6 fl oz	180 mL	Olive oil	5. With the motor running, pour in the lemon juice. The purée should have the consisency of soft mashed potatoes.
1¹/₂ fl oz	45 mL	Lemon juice	
to taste	to taste	Additional salt	6. Adjust seasoning to taste with salt and pepper if necessary.
to taste	to taste	Pepper	7. Serve warm as a vegetable accompaniment (reheat as necessary) or cold as a dip.

Per serving: Calories, 320; Protein, 9 g; Fat, 22 g (60% cal.); Cholesterol, 0 mg; Carbohydrates, 24 g; Fiber, 8 g; Sodium, 290 mg.

VARIATIONS

Other beans, such as white kidney beans and pinto beans, may be substituted.

Flageolet Beans with Wilted Arugula

Cook the beans as in the basic recipe, but do not purée. Use 1–2 cloves garlic and 2 fl oz (60 mL) olive oil. Chop the garlic and sauté in the olive oil. Add 8 oz (250 g) coarsely chopped arugula and sauté just until wilted. Add this mixture to the drained, hot beans. Season to taste.

Masoor Dal (Red Lentils with Spices)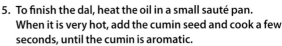

PORTIONS: 12   PORTION SIZE: 8 FL OZ (240 ML)

U.S.	METRIC	INGREDIENTS	PROCEDURE
1 lb 4 oz	575 g	Red lentils	1. Sort, rinse, and drain the lentils.
2¹/₂ qt	2.5 L	Water	2. Put the lentils and water in a saucepot. Bring to a boil. Skim off any froth that rises to the top.
4	4	Garlic cloves, chopped	
4 thin slices	4 thin slices	Fresh ginger root	3. Add the garlic, ginger, cilantro, turmeric, and cayenne. Simmer until the lentils are very tender and falling apart, about 1¹/₂ hours.
2 tbsp	30 mL	Chopped fresh cilantro	
2 tsp	10 mL	Turmeric	
¹/₂ tsp	2 mL	Cayenne	
1¹/₂ fl oz	45 mL	Lemon juice	4. Add the lemon juice and salt to taste.
to taste	to taste	Salt	
2 fl oz	6 mL	Vegetable oil	5. To finish the dal, heat the oil in a small sauté pan. When it is very hot, add the cumin seed and cook a few seconds, until the cumin is aromatic.
2 tsp	10 mL	Cumin seed	
			6. Pour the oil and seeds into the dal and stir in.

Per serving: Calories, 200; Protein, 12 g; Fat, 5 g (21% cal.); Cholesterol, 0 mg; Carbohydrates, 30 g; Fiber, 8 g; Sodium, 20 mg.

Masoor Dal

Cranberry Beans with Asparagus

PORTIONS: 12   PORTION SIZE: 5 OZ (150 G)

U.S.	METRIC	INGREDIENTS	PROCEDURE
1½ lb	700 g	Asparagus	1. Peel and trim the bottom of the asparagus stalks (see p. 277).
			2. Cook the asparagus in boiling salted water until just tender. Drain, cool in cold water, and drain again.
			3. Cut the asparagus into pieces about 1½ inches (4 cm) long.
2 fl oz	60 mL	Olive oil	4. Heat the olive oil in a sauté pan over moderate heat.
4 oz	125 g	Shallots, chopped fine	5. Add the shallots and garlic. Cook until tender and lightly browned.
¼ oz	7 g	Garlic, chopped fine	6. Add the beans to the pan. Toss over heat until the beans are hot.
3 lb	1.4 kg	Cranberry beans, cooked or canned, drained	7. Add the asparagus and herbs. Heat another minute or two, until the asparagus is hot.
2 tbsp	30 mL	Chopped fresh parsley	8. Add salt to taste.
1 tbsp	15 mL	Chopped fresh basil	
1 fl oz	30 mL	Red wine vinegar	9. Just before serving, stir in the vinegar and the cheese.
⅓ cup	80 mL	Grated parmesan cheese	

Per serving: Calories, 180; Protein, 10 g; Fat, 5 g (24% cal.); Cholesterol, 0 mg; Carbohydrates, 25 g; Fiber, 10 g; Sodium, 50 mg.

Cranberry Beans with Asparagus

GRAINS

Wheat and rice are without doubt the world's two most important grains. Unlike rice, wheat is rarely cooked and eaten as a whole grain. Rather, its primary use is to be ground into flour for making breads and other baked goods. The use of grain flours in baked goods is the subject matter of the last seven chapters of this book. In this section, we look at other uses of grains in food-service kitchens.

TYPES OF GRAINS

Grains are the edible seeds of various members of the grass family. Each seed consists of four parts:

1. The husk—an inedible fibrous outer layer that is removed during processing
2. The endosperm—the starchy mass that forms most of the kernel
3. The bran—a tough but edible layer covering the endosperm
4. The germ—the tiny embryo that forms the new plant when the seed sprouts

A product labeled *whole grain* consists of the **endosperm**, **bran**, and **germ**. The grain may be polished or milled to remove the bran and germ. White rice and other polished grains are only the endosperm.

Although rice appears on menus more often than other grain starches, several other grains can also be served as side dishes to add variety to your menu. The most popular of these grains are described following the section on rice.

Top row: basmati rice, glutinous rice, plain long-grain rice. Bottom row: Japanese short-grain rice, jasmine rice, Arborio rice.

Top row: true wild rice, Wehani rice, cultivated wild rice. Bottom row: parboiled or converted rice, brown long-grain rice.

Rice

Regular milled white rice has been milled to remove the outer bran coating. This process removes some vitamins and minerals, but it produces a white, lighter-textured product most people prefer. White rice appears in several forms:

Enriched rice has received a coating of vitamins to compensate for some of the nutrients lost in milling.

Short-grain and **medium-grain** rice have small, round kernels that become sticky when cooked. They are used for such preparations as rice pudding and rice molds. In addition, the regular boiled rice used in Japanese cuisine for everyday eating and for making sushi is short-grain rice.

Long-grain rice has long, slender grains that stay separate and fluffy when properly cooked. It is used for side dishes, entrées, casseroles, and so on.

Parboiled or **converted** rice is a specially processed long-grain rice. It has been partially cooked under steam pressure, redried, and then milled or polished. This process results in a higher vitamin and mineral content, compared with regular milled white rice.

Parboiled rice is the most widely used in food service. The grains stay firm, separate, and light, and the product holds well in the steam table without becoming mushy or sticky. However, the flavor and texture are not like those of regular long-grain rice, so it is not preferred by all customers.

Converted rice takes slightly more liquid and time to cook.

Instant rice has been precooked and dried so it can be prepared quickly. It does not hold well after cooking, and the grains quickly lose their shape and become mushy.

Brown rice has the bran layer left on, giving it a light brown color, a slightly coarse, crunchy texture, and nutty flavor. Brown rice is available as short, medium, or long grain. Brown rice takes about twice as long to cook as white rice.

Arborio rice is one of several Italian varieties of a type of short-grain rice essential for making the highest-quality risotto (see p. 386). It is the variety most often found in North America and the one specified in recipes. Two other varieties, less widely available, used for risotto are *carnaroli* and *vialone nano*.

Basmati rice is an extra-long-grain rice widely used in India and surrounding countries. It has a distinctive nutty flavor. Brown basmati rice is also available.

Jasmine rice is a long-grain white rice from Thailand and other parts of Southeast Asia. It is fragrant, a little like basmati rice but more delicate or floral.

Wehani rice is another aromatic rice, red in color, with a rich, earthy flavor.

Wild pecan rice is a cultivated, not wild, long-grain rice from Louisiana. It is aromatic, with a nutty flavor, from which it gets the name *pecan*.

Glutinous rice, also called **sticky rice** and *sweet rice*, is a sweet-tasting short-grain rice that becomes quite sticky and chewy when cooked. It is used for a number of special dishes, including desserts, in Chinese and Japanese cuisines. It is often cooked by soaking and steaming rather than boiling (see the recipe on p. 441). Contrary to what you may read elsewhere, it is *not* the rice used for sushi, which is made with regular Japanese short-grain rice.

Wild rice is not a type of rice but an unrelated grain, so it is discussed with specialty grains below.

Corn

Unlike other grains, which have a husk covering each seed, **corn** has a set of husks covering the entire seed head, or ear. Also unlike other grains, corn is eaten as a fresh vegetable, although different varieties are grown as grain and as a vegetable.

Corn as a grain is not often cooked whole. More often, it is ground into **cornmeal** and cooked into a porridge or used in baked goods. *Meal* can be defined as a coarsely ground grain, as distinguished from *flour*, which is finely ground grain. Common cornmeal is yellow or white, depending on the variety of corn it is made from.

Polenta is Italian-style cornmeal. Polenta has become popular in North America in recent years. Its preparation and uses are explained and illustrated on pages 390–391.

Hominy is corn that has been treated with lye. When it is cracked into a coarse meal, it becomes **grits**, popular in the southern United States and wherever the foods of the South are appreciated. Hominy in whole-grain form is known in Mexican cuisine as **pozole** (poh so hleh). It requires several hours of simmering.

Blue corn, usually available as blue cornmeal, is derived from early varieties of corn grown by Native Americans.

Wheat

The most common use of wheat is to be made into flour. The milling process for white flour separates the bran and germ. **Wheat germ** and **wheat bran** can be purchased separately. They are usually used as additions to baked goods and some other dishes to enrich their nutritional content and to add flavor interest.

Whole wheat grains that have been cut into smaller pieces are called **cracked wheat**. This product is often added to breads and also can be cooked like pilaf (see p. 384).

Wheatberries are the whole grain minus the hulls. They are generally cooked by boiling or simmering, but cooking time can be several hours. Soaking overnight reduces the cooking time to about 1 hour.

Bulgur is a type of cracked wheat that has been partially cooked or parched. It is usually available in coarse, medium, and fine granulations. Its cooking time is shorter than regular cracked wheat and, in fact, the fine granulations can be prepared simply by pouring boiling water over them and letting them stand for $^1/_2$ hour. This type of bulgur is often served cold, mixed with lemon juice, olive oil, chopped scallions, and fresh herbs.

Green wheat is wheat that is harvested while immature and then dried. It can be cooked like cracked wheat.

Couscous (koose koose) is not actually a grain, although it resembles one. It is made from semolina wheat, a variety of high-protein wheat, and is sort of a granular pasta. See the discussion of couscous in the pasta section of this chapter (p. 397).

Top row: Egyptian green wheat, hulled wheatberries. Bottom row: couscous, kasha, farro.

Other Grains

Wild rice is not actually rice but rather the seed of an unrelated grass native to the northern United States and Canada. The grains are long, slender, hard, and dark brown or nearly black in color. Because of its unique nutty flavor, scarcity, and high price, wild rice is considered a luxury food.

Wild rice is now widely cultivated, but the cultivated type is slightly different from that harvested in the wild. Grains of cultivated wild rice are generally larger and firmer, but the texture of the cooked rice is coarser and the flavor less complex. Cultivation has helped reduce the price of wild rice, however.

Farro (far oh) is a wheatlike grain that may be an ancestor of modern wheat. It has been used in the Mediterranean region for thousands of years and is still widely known in Tuscany and other parts of Italy. Farro has a flavor similar to that of wheat. It is higher in protein than wheat and can often be eaten by people who have wheat allergies. In North America, farro is known as **spelt**, although the Italian name is catching on. (Some sources argue that spelt and farro are different grains, but according to the International Plant Genetic Resources Institute, "the only registered varieties of farro belong to *T. spelta* or spelt.")

Kamut (kah moot), like farro, is an ancient relative of wheat. It is similar to spelt in composition and flavor.

Buckwheat is technically not a grain because it is the seed not of a grass but of a plant with branched stems and broad, arrow-shaped leaves. Whole buckwheat is often ground into flour. When the grains are crushed into coarse pieces, they are called **buckwheat groats** and can be cooked like rice. Toasted buckwheat is called **kasha**. The toasting gives it a nutty flavor. Kasha is popular in Eastern European and Jewish cooking. Kasha is also cooked like pilaf (p. 383).

Barley is usually purchased as **pearled barley**, which has been milled to remove the outer bran layers. It is commonly used in soups, but it can also be cooked by the pilaf method and served like rice, although it has a longer cooking time.

Oats are most familiar in North America as a breakfast food. **Steel-cut oats** are whole grains that have been cut into small pieces, somewhat resembling cracked wheat. They are usually cooked as a porridge. **Rolled oats** are whole grains that have been steamed until soft and then flattened between rollers. This processing reduces their cooking time considerably. If they are cooked and not just softened during the steaming process, they become **instant oats**. These need no additional cooking, only reconstituting with boiling water.

Millet is a small, round yellow grain that is an important food source in much of Africa and Asia. It has a high protein content and a mild flavor. Millet is often used as bird seed in North America. It can be cooked like rice.

Quinoa (keen wah) is a grain native to the South American Andes that has only recently become an occasional feature of North American menus. Quinoa is high in good-quality protein and lower in carbohydrates than other grains. It is a tiny, round grain with an ivory color and a mild, delicate flavor. When cooked, the germ of the grain unwinds, making it look as though each grain has a tail. Before cooking, quinoa must be washed and rinsed well to remove a bitter coating that occurs naturally on the grain.

Triticale is a high-protein hybrid of wheat and rye. It is often ground into flour, but it can also be cooked whole like rice. Triticale has a nutty, sweet flavor.

Amaranth is a tiny, yellow-brown seed with a somewhat spicy, nutty flavor when cooked. It contains high-quality protein and thus is useful in vegetarian diets.

Flaxseeds are technically not a grain because they are not seeds of a grass. Flax has recently gained popularity because it is thought to have some health benefits due to its beneficial fiber as well as a high quantity of omega-3 fatty acids (see p. 94). The seeds are used mostly in small quantities in breads and in commercial breakfast cereal preparations. Consuming them in large quantities can cause digestive problems.

Top row: quinoa, triticale, pearl barley. Bottom row: blue cornmeal, pozole, bulgur wheat.

STORING AND HANDLING GRAINS

Store raw grains at room temperature in a dark, dry place and in a tightly sealed container to keep out moisture and insects. Milled grains such as white rice that have had the germ removed will keep for many months. Whole grains are somewhat more perishable because the fat content of the germ can become rancid.

Depending on the source, whole grains may need to be picked over like dried beans (see p. 369) to remove foreign matter such as tiny stones or bits of soil. In addition, whole grains usually should be washed and drained before cooking.

Rice, our most commonly cooked grain, should be rinsed in cold water before boiling or steaming. This removes the excess starch that makes it sticky. The rice industry recommends *not* washing rice because it removes some of the vitamin coating of enriched rice. But that's probably a small price to pay for a more attractive product. This is a decision you will have to make in your own operation.

Do not buy low-grade rice, which tends to be dirty, or rice that has been coated with talc.

Rice cooked by the pilaf method (p. 383) does not need to be washed (unless it is dirty) because the fat coating each kernel helps keep the grains separate and reduces stickiness.

Converted rice and instant rice do not need to be washed.

KEY POINTS TO REVIEW

- What are the four parts of a grain kernel? Which of these parts are included in products labeled whole grain?

- What are the most important kinds of rice used in the kitchen? Describe them.

- What are the most important corn products used in the kitchen?

- In addition to rice and corn products, what other important whole-grain items are used as ingredients in the kitchen? Describe each.

- How should grains and grain products be stored?

COOKING GRAINS

Most grains are cooked by one of three cooking methods: the simmering method, the pilaf method, and the risotto method.

Simmering Methods

The most common method for cooking rice and other whole grains is to place the washed grain in a heavy pot with the right amount of water or other liquid to hydrate it, bring it to a simmer, cover, and cook slowly until all the water is absorbed. The exact amount of liquid needed varies considerably, depending on these factors:

1. The type of grain, its age, and its moisture content
2. Tightness or looseness of the cover (degree of moisture loss during cooking)
3. Desired moistness of the finished product

Because of all these factors, it is difficult to be precise when determining how much liquid to use. It is better to add too much liquid than too little. With too little moisture, the grain will not cook to tenderness. If a little too much is added, remove the pan from the heat, keeping it tightly covered, and let it stand for a few minutes. Liquid remaining in the bottom of the pan is likely to be absorbed. If there is much excess liquid, drain it off.

A second method, called the *pasta method*, can be used. It is so called because, like pasta, the item is cooked in a large quantity of water and drained. This method is good for producing separate, unsticky grains. However, some nutrients are lost in the cooking water, so chefs disagree about the value of the method.

PROCEDURE for Simmering Rice and Other Grains

1. For whole grains, wash the grain in cold water as necessary. Drain.

2. Combine the grain with the proper amount of water or other liquid in a heavy saucepot. Bring to a boil. Stir.

3. Cover and cook over very low heat for the proper cooking time, depending on the grain.

4. Test for doneness. Cook a few additional minutes if necessary.

5. Remove from the heat. Drain excess liquid if necessary. Let stand, covered, to allow moisture to be absorbed uniformly by the grain.

6. For rice and any grains that stick together, fluff with a fork and turn out into a hotel pan to let steam escape and stop the cooking.

VARIATION: PASTA METHOD

1. Drop the washed, drained grain into a large pot of boiling salted water.

2. When just tender, pour into a strainer and drain well.

3. Place in a hotel pan. Cover and steam dry in oven 5–10 minutes, or leave uncovered and place in a steamer to steam dry.

Boiled and Steamed Rice ♥ 🌹

YIELD: ABOUT 3 LB (1.4 KG) PORTIONS: 10 PORTION SIZE: 4¹/₂–5 OZ (140 G)
12 4 OZ (115 G)
16 3 OZ (90 G)

PROPORTIONS	U.S.	METRIC
Regular long-grain white rice		
Rice	1 lb	475 g
Water	1 qt	1 L
Salt	1 tsp	5 mL
Butter	1 oz	30 g
Parboiled long-grain rice		
Rice	1 lb	475 g
Water	4¹/₂ cups	1.1 L
Salt	1 tsp	5 mL
Butter	1 oz	30 g
Medium-grain white rice		
Rice	18 oz	525 g
Water	1 qt	1 L
Salt	1 tsp	5 mL
Butter	1 oz	30 g
Brown rice		
Rice	12 oz	350 g
Water	1 qt	1 L
Salt	1 tsp	5 mL
Butter	1 oz	30 g

Per 4¹/₂–5 oz (140 g) serving: Calories 200; Protein, 4 g; Fat, 2.5 g (12% cal.); Cholesterol, 5 mg; Carbohydrates, 39 g; Fiber 1 g; Sodium, 260 mg.

PROCEDURES

Rangetop

1. Wash rice in cold water until water runs clear (optional step; see p. 379 for note on washing rice).
2. Combine all ingredients in a heavy pot. Bring to boil. Stir. Cover and cook over very low heat.

   Cooking times:
   Long- and medium-grain:
    15–20 minutes
   Parboiled: 20–25 minutes
   Brown: 40–45 minutes

3. Test rice for doneness. Cook 2–4 minutes more if necessary.
4. Turn rice out into a hotel pan. Fluff with fork or slotted spoon to let steam escape.

Oven

1. Wash rice in cold water until water runs clear (optional step; see p. 379 for note on washing rice).
2. Bring salted water to boil. Combine all ingredients in a shallow steamer pan. Cover with foil or tight lid. Place in 375°F (175°C) oven.

   Cooking times:
   Long- and medium-grain:
    25 minutes
   Parboiled: 30–40 minutes
   Brown: 1 hour

3. Test rice for doneness. Bake 2–4 minutes more if necessary.
4. Fluff rice with fork or slotted spoon to let steam escape.

Steamer

1. Wash rice in cold water until water runs clear (optional step; see p. 379 for note on washing rice).
2. Bring salted water to boil. Combine all ingredients in a shallow steamer pan. Place uncovered pan in steamer for cooking time recommended by equipment manufacturer.

   Cooking times depend on type of steamer.

3. Test rice for doneness. Steam 2–4 minutes more if necessary.
4. Fluff rice with fork or slotted spoon to let steam escape.

Simmering Meals and Cereals

Procedures for cooking meals and cereals such as polenta and oatmeal differ somewhat from procedures for cooking whole grains. Cracked or flaked cereals such as rolled oats consist of large particles. They are usually stirred into boiling water. Because of the size of the particles, there is little danger of lumping.

Granular meals such as cornmeal can be stirred into boiling water if care is taken to add the grain slowly and to stir constantly and vigorously while doing so, in order to avoid lumps. See the recipe for polenta (p. 390). Alternatively, combine the meal with cold liquid, stir, and bring to a simmer while continuing to stir. Mixing the meal with cold liquid separates the granules to prevent lumping.

PROCEDURE for Simmering Whole, Cracked, or Flaked Cereals

1. Measure the correct amount of liquid into a pot and bring to a boil.

2. Measure the correct amount of meal or cereal.

3. Add the cereal slowly to the boiling liquid, stirring constantly.

4. Stir until some thickening takes place. Depending on the grain, continue to stir constantly, or stir only occasionally. Some cereals, such as oatmeal, become gummy with excessive stirring.

5. Reduce heat to a slow simmer and cook to desired doneness and consistency.

6. To prevent drying, keep the cooked grain covered until serving.

VARIATION: SIMMERING MEALS AND GRANULAR CEREALS

The procedure is the same as above, except the cereal is mixed with cold liquid. This separates the grains to prevent lumping. Place the mixture in a pot, bring to a simmer, and cook as in the basic procedure.

The Pilaf Method

The pilaf method is equivalent to braising. The grain is first sautéed in fat and then cooked in liquid—preferably in the oven for uniform heating—until the liquid is absorbed (see Figure 13.1). The fat helps keep the grains separate and adds flavor.

FIGURE 13.1 Making rice pilaf.

(a) Sweat the onion or shallot.

(b) Add the rice and sauté briefly.

(c) Pour in the hot stock or other liquid.

(d) Bring to a boil and cover. Cook at low heat on top of the stove or in the oven for the required time.

(e) The finished pilaf.

It is normal to measure rice by volume when making pilaf, as the proportions are based on volume measure. One pint of raw rice weighs about 14 ounces, or 1 pound measures about 2¼ cups (1 L weighs about 875 g; or 1 kg measures 1.15 L). Regarding exact measurements, see the note following the pilaf recipe.

PROCEDURE for Cooking Grain by the Pilaf Method

1. Heat the desired fat (such as butter or olive oil) in a heavy pan. Add chopped onion or other aromatic vegetable, if desired, and sauté until soft but not browned.

2. Add the grain. Stir to coat the grains with fat.

3. Cook the grain in the fat, stirring, to toast the grain lightly.

4. Add the proper amount of hot liquid.

5. Bring to a simmer, stirring occasionally.

6. Cover tightly. Cook on the stovetop or, preferably, in an oven, for the correct length of time, depending on the grain.

7. Remove from the heat and let stand, covered, to allow the moisture to be absorbed uniformly by the grain.

The Risotto Method

Risotto is a classic Italian preparation made by a special procedure that is like neither the boiling method nor the pilaf method. After sautéing the rice, add a small amount of hot stock or other liquid and stir until the liquid is absorbed. Repeat this procedure until the rice is cooked but still firm. Risotto should be served quickly, as it does not hold well. The finished product has a creamy consistency due to the starch that is cooked out of the rice. The grains are not fluffy and separate.

The word *risotto* comes from the Italian word *riso*, meaning "rice." Other grains can be cooked using the same method, although strictly speaking they should not be called *risotto*. In Italy, farro cooked by the risotto method is called *farrotto*, and orzo pasta cooked this way is called *orzotto*. In the English-speaking world, however, the word *risotto* is more likely to be understood and is often used on menus, but with the grain specified if any type other than rice is used. For example, farrotto might be called *farro risotto*.

PROCEDURE for Cooking Grain by the Risotto Method

1. Heat the desired fat (such as butter or olive oil) in a heavy pan. Add chopped onion or other aromatic vegetable, if desired, and sauté until soft but not browned.

2. Add the grain. Stir to coat the grains with fat.

3. Cook the grain in the fat, stirring, to toast the grain lightly.

4. Add a small amount of boiling liquid. Cook slowly, stirring, until the liquid is absorbed by the grain.

5. Add a second small quantity of liquid and repeat the procedure.

6. Continue adding a small quantity of liquid at a time, stirring constantly, and waiting until the liquid is absorbed before adding more.

7. Stop adding liquid when the grain is tender but still firm. It should be moist and creamy but not runny.

KEY POINTS TO REVIEW

- What are the steps in the basic procedure for simmering rice and other grains?

- How is the procedure for cooking cracked and flaked cereals different from cooking meals or granular cereals?

- What are the steps in the procedure for cooking grain by the pilaf method?

- What are the steps in the procedure for cooking grain by the risotto method?

 # Rice Pilaf

YIELD: ABOUT 3 LB (1.4 KG)   PORTIONS: 10   PORTION SIZE: 5 OZ (150 G)
    12   4 OZ (125 G)
    16   3 OZ (90 G)

U.S.	METRIC	INGREDIENTS
2 oz	60 g	Butter
3 oz	90 g	Onions, fine dice
1 pt (see Note)	500 mL (see Note)	Long-grain rice
1¹/₂–2 pt (see Note)	750 mL–1 L (see Note)	Chicken stock or water, boiling
to taste	to taste	Salt

PROCEDURE

1. Heat the butter in a heavy saucepan. Add the onion and sauté until it begins to soften. Do not brown.

2. Add the rice, without washing. Stir over heat until the rice is completely coated with butter.

3. Pour in the boiling liquid. Return the liquid to a boil with the rice. Taste and adjust seasonings; cover tightly.

4. Place in a 350°F (175°C) oven and bake 18–20 minutes, or until liquid is absorbed and rice is dry and fluffy. Taste the rice and, if it is not done, replace in oven 3–5 minutes.

5. Turn out into a hotel pan and fluff the rice with a fork. This releases steam and prevents further cooking. Keep hot for service.

6. If desired, additional raw butter may be stirred into finished rice.

Per 5-oz (150-g) serving: Calories, 190; Protein, 4 g; Fat, 5 g (23% cal.); Cholesterol, 15 mg; Carbohydrates, 33 g; Fiber, 1 g; Sodium, 50 mg.

Note: Rice for pilaf is measured by volume rather than by weight. Use 1¹/₂–2 times its volume in stock or water (1³/₄ times is the normal proportion for long-grain rice). For example, use 2 pints liquid per 1 pint of rice (1 L liquid per 0.5 L rice) if you desire a moister product or if you are using parboiled rice. Use 1¹/₂ pints (0.75 L) liquid if you desire a drier product and if your cover is tight enough to retain most of the steam.

VARIATIONS

Tomato Pilaf

Prepare as in the basic recipe, using 12–16 oz (375–500 mL) chicken stock and 1¹/₂ lb (700 g) chopped tomatoes with juice.

Spanish Rice

Prepare like Tomato Pilaf, but use bacon fat and sauté 6 oz (175 g) diced green bell pepper, 1 crushed garlic clove, and 1 tbsp (15 mL) paprika with the onion.

Turkish Pilaf

Sauté ¹/₄ tsp (1 mL) turmeric with the rice. To finished rice, add 4 oz (125 g) tomato concassé or drained, chopped canned tomatoes, 4 oz (125 g) cooked peas, and 4 oz (125 g) raisins (soaked and drained). Let stand 10–15 minutes before serving.

Cracked Wheat Pilaf

Prepare as in the basic recipe, using cracked wheat instead of rice.

Orzo Pilaf

Prepare as in the basic recipe, using orzo (rice-shaped pasta) instead of rice.

Barley Pilaf

Prepare as in the basic recipe, using pearled barley instead of rice. Use 2¹/₂ pints (1.25 L) stock and bake 45 minutes. Mushrooms are often added to barley pilaf.

Additions to Rice Pilaf

Pimiento	Spinach, chopped
Chopped nuts	Mushrooms
Celery, diced	Olives, chopped or sliced
Carrot, diced or grated	Ham, diced or cut julienne
Scallions	Raisins or currants
Peas	Water chestnuts
Green bell pepper, diced	Bacon

Basic Fried Rice 🌐🍽

PORTIONS: 16   PORTION SIZE: 6 OZ (175 G)

U.S.	METRIC	INGREDIENTS
4 lb	1.8 kg	Cooked rice, cold
4–6 fl oz	125–175 mL	Oil
1 lb	450 g	Cooked meat (cut into shreds) or seafood (flaked or sliced)
3 oz	90 g	Scallions, sliced thin
1 lb	450 g	Vegetables (see Variations), shredded or small dice
4–6 fl oz	125–175 mL	Soy sauce (optional)
to taste	to taste	Salt
4–8	4–8	Eggs, beaten

PROCEDURE

1. Break up the rice to remove all lumps.
2. Divide the rice and other ingredients into two or more batches, depending on the size of the pan or wok. Do no more than 1–2 lb (0.5–1 kg) rice at one time. If you fry too much at once, it will not fry properly.
3. Heat a small amount of oil in the wok. Add the meat and stir-fry 1–2 minutes.
4. Add the scallions and stir-fry 1 minute.
5. Add any raw vegetables and stir-fry until almost done.
6. Add the rice and stir-fry until it is hot and lightly coated with oil.
7. Add any cooked vegetables and mix in.
8. Add soy sauce, if used, and salt.
9. Add beaten egg and mix in. Stir-fry lightly to cook the egg, then serve.

Per serving: Calories, 280; Protein, 11 g; Fat, 10 g (33% cal); Cholesterol, 70 mg; Carbohydrates, 35 g; Fiber, 2 g; Sodium, 35 mg.

VARIATIONS

The quantities given in the basic recipe are only guidelines, but rice should be the predominant ingredient. You can omit the meat or fish items.

For plain rice, you can omit the vegetables, too.

Eggs can be omitted, or they can be added to fried rice in several other ways:

1. Remove the meat and vegetables from the pan when they are cooked. Add the egg to the pan and scramble. Add the rice, return the meat and vegetables to the pan, and continue with the recipe.
2. In step 9, push the rice to the sides of the pan. Add the egg to the well in the middle. When it starts to set, gradually mix in the rice.
3. Scramble the eggs separately and add to the rice at the end.
4. Mix the raw beaten egg with the cold cooked rice before cooking.

Suggested Ingredients

Meats: cooked pork, beef, chicken, duck, ham, bacon, Chinese sausage
Seafood: shrimp (diced or whole), crab, lobster
Vegetables: bamboo shoots, bean sprouts, celery, peas, mushrooms, onions, peppers, water chestnuts

Bulgur Pilaf with Lemon ♥

PORTIONS: 12   PORTION SIZE: 2¹/₂ OZ (75 G)

U.S.	METRIC	INGREDIENTS
1 oz	30 g	Butter or vegetable oil
4 oz	125 g	Onion, chopped fine
8 oz	250 g	Bulgur, coarse
4 tsp	20 mL	Grated lemon zest
1¹/₂ pt	750 mL	Chicken stock or vegetable stock, hot
to taste	to taste	Salt
to taste	to taste	Pepper
4 tbsp	60 mL	Chopped chives

PROCEDURE

1. Heat the butter in a saucepan. Add the chopped onion and sauté gently without browning until the onion is soft.
2. Add the bulgur. Stir to coat with butter. Stir over heat 1 minute to lightly toast the grain.
3. Add the grated lemon zest and stir to mix.
4. Stir in the hot stock. Add salt and pepper to taste. Bring to a simmer.
5. Cover the pot and cook over low heat or in an oven heated to 350°F (175°C) until the bulgur is tender, about 20 minutes.
6. Uncover and fluff the grain with a kitchen fork. Add the chives and toss to mix in.

Per serving: Calories, 90; Protein, 3 g; Fat, 2 g (20% cal.); Cholesterol, 25 mg; Carbohydrates, 15 g; Fiber, 4 g; Sodium, 25 g.

Risotto alla Parmigiana 🍽️

PORTIONS: 10   PORTION SIZE: 5 OZ (150 G)

U.S.	METRIC	INGREDIENTS
1 oz	30 g	Butter
1 fl oz	30 mL	Vegetable oil
1 oz	30 g	Onion, chopped fine
1 lb	450 g	Italian Arborio rice
1½ qt (approximately)	1.4 L (approximately)	Chicken stock, hot

PROCEDURE

1. Heat the butter and oil in a large, straight-sided sauté pan. Add the onion and sauté until soft. Do not brown.
2. Add the rice and sauté until well coated with the fat.
3. Using a 6-oz (150-mL) ladle, add one ladle of stock to the rice. Stir the rice over medium heat until the stock is absorbed and the rice is almost dry.
4. Add another ladle of stock and repeat procedure. Do not add more than one ladleful of stock at a time.
5. Stop adding stock when the rice is tender but still firm. It should be very moist and creamy, but not runny. The cooking should take about 30 minutes.

U.S.	METRIC	INGREDIENTS
1 oz	30 g	Butter
3 oz	90 g	Parmesan cheese, grated
to taste	to taste	Salt

6. Remove from the heat and stir in the raw butter and parmesan cheese. Salt to taste.

Per serving: Calories, 260; Protein, 7 g; Fat, 11 g (38% cal.); Cholesterol, 25 mg; Carbohydrates, 34 g; Fiber, 3g; Sodium, 210 mg.

VARIATIONS

Restaurant Method: Prepare basic pilaf (p. 384) using 1 lb Italian Arborio rice to 1 qt chicken stock (500 g rice to 1 L stock). To finish for service, place desired number of portions in a sauté pan and moisten with additional stock. Simmer until slightly moist and creamy, as in basic recipe. Finish with raw butter and parmesan cheese.

Risotto Milanese

Prepare as in basic recipe, but add ¼–½ tsp (1–2 mL) saffron soaked in 1 cup (200 mL) stock near the end of cooking.

Risotto with Mushrooms

Add 4–8 oz (100–200 g) mushrooms, chopped and sautéed in butter, near the end of cooking time.

Risi Bisi

Add 1 lb (450 g) cooked peas and ¼ cup (60 mL) chopped parsley to basic risotto. (This is not authentic Risi Bisi, which is considered a thick soup in Italy. However, it is similar.)

Risotto Milanese

FIGURE 13.2 Making risotto.

(a) Sauté the onion or shallot until soft.

(b) Add the rice. Sauté until coated with the fat.

(c) Add a ladleful of stock to the rice. Stir until the liquid is absorbed.

(d) Continue adding stock and stirring until the rice is cooked.

(e) For Risotto Milanese, add saffron steeped in hot stock near the end of the cooking period.

(f) Finish by stirring in parmesan cheese and butter.

Arroz à la Mexicana

PORTIONS: 16   PORTION SIZE: 4¹/₂ OZ (125 G)

U.S.	METRIC	INGREDIENTS
1¹/₂ lb	700 g	Long-grain rice
3 fl oz	90 mL	Oil
12 oz	350 g	Tomato purée
3 oz	90 g	Onion, chopped fine
2	2	Garlic cloves, mashed to a paste
3¹/₂ pt	1.75 L	Chicken stock
1 tbsp	15 mL	Salt

PROCEDURE

1. Rinse the rice well to remove excess starch. Soak in cold water at least 30 minutes. Drain well.
2. Heat the oil in a pot and add the rice. Stir over moderate heat until it begins to brown lightly.
3. Add the tomato purée, onion, and garlic. Cook until the mixture is dry. Be careful not to let it burn.
4. Add the chicken stock and salt. Stir. Simmer, uncovered, over medium heat until most of the liquid is absorbed.
5. Cover, turn the heat to very low, and cook 5–10 minutes, or until the rice is tender.
6. Remove from the heat and let it stand, without removing the cover, 15–30 minutes before serving.

Per serving: Calories, 230; Protein, 4 g; Fat, 6 g (24% cal.); Cholesterol, 5 mg; Carbohydrates, 39 g; Fiber, 1 g; Sodium, 440 mg.

VARIATION

Arroz Verde

Omit the tomato purée. Purée the onion and garlic in a blender along with the following: 6 fl oz (175 mL) water, 3 tbsp (45 mL) chopped fresh cilantro leaves, ³/₄ cup (45 g or 200 mL) chopped parsley, and 3 oz (90 g) green chiles (or part green chiles and part green bell peppers). Use this purée in place of the tomato purée. Reduce the quantity of stock to 3 pt (1.5 mL). You may use water instead of stock.

Arroz à la Mexicana

Barley with Wild Mushrooms and Ham

PORTIONS: 10   PORTION SIZE: 3¹/₂ OZ (110 G)

U.S.	METRIC	INGREDIENTS
1 oz	30 g	Dried porcini mushrooms
4 fl oz	125 mL	Water, hot
1 fl oz	30 mL	Oil
4 oz	125 g	Onions, small dice
4 oz	125 g	Celery, small dice
10 oz	300 g	Barley
1 pt 4 fl oz (or more; see step 5)	600 mL (or more; see step 5)	Brown stock, chicken stock, or vegetable broth
4 oz	125 g	Cooked ham, small dice
to taste	to taste	Salt

PROCEDURE

1. Soak the dried porcini in hot water until soft. Drain the mushrooms and squeeze them out, reserving all the soaking liquid. Strain or decant the liquid to remove any sand or grit. Chop the mushrooms.
2. Heat the oil in a heavy pot. Add the onion and celery. Sauté briefly.
3. Add the barley and sauté briefly, as for making rice pilaf.
4. Add the stock and mushroom liquid. Bring to a boil. Stir in the chopped mushrooms and the ham. Add salt to taste.
5. Cover tightly. Cook on top of the stove over low heat or in the oven at 325°F (160°C) until the barley is tender and the liquid is absorbed, 30–45 minutes. Check periodically during cooking to make sure that enough liquid remains, and add more stock or water if necessary. Some barley needs more liquid to fully hydrate.

Per serving: Calories, 170; Protein, 7 g; Fat, 6 g (30% cal.); Cholesterol, 10 mg: Carbohydrates, 24 g; Fiber, 6 g; Sodium, 125 mg.

VARIATIONS

Add other vegetables to the barley, such as diced carrots, turnips, fennel, or parsnips. Add them at the same time as the onion and celery, or cook them separately and add them at the end of cooking.

Barley with Wild Mushrooms and Ham

Paella

PORTIONS: 16 PORTION SIZE: SEE PROCEDURE

U.S.	METRIC	INGREDIENTS
2	2	Chickens, 2¹/₂–3 lb (1.1–1.4 kg) each
as needed	as needed	Olive oil
8 oz	225 g	Chorizo sausage (see Note)
2 lb	900 g	Lean pork, cut into large dice
16	16	Large shrimp, peeled and deveined
2 lb	900 g	Squid, cleaned (p. 605), cut into rings
2	2	Red bell peppers, large dice
2	2	Green bell peppers, large dice
16	16	Small clams
16	16	Mussels
8 fl oz	250 mL	Water
as needed	as needed	Chicken stock
1 tsp	5 mL	Saffron
12 oz	350 g	Onion, small dice
6	6	Garlic cloves, minced
2 lb	900 g	Tomatoes, chopped
2 tsp	10 mL	Dried rosemary
2 lb	900 g	Short-grain rice, such as Italian Arborio
2 tsp	10 mL	Salt
to taste	to taste	Pepper
4 oz	125 g	Cooked green peas
16	16	Lemon wedges

PROCEDURE

1. Cut each chicken into 8 pieces.
2. In a large sauté pan, brown the chicken in olive oil. Remove and set aside.
3. Using additional oil as needed, briefly sauté the sausage, pork, shrimp, squid, and peppers. Do each ingredient separately, then remove to separate containers.

4. Combine the clams and mussels with the water in a covered pot. Steam just until they open.
5. Remove the shellfish and set them aside. Strain the liquid, then add enough chicken stock to measure 2 qt (2 L).
6. Add the saffron to the stock mixture.

7. In the skillet used for browning the meats, sauté the onion and garlic until soft. Use additional olive oil if necessary.
8. Add the tomatoes and rosemary. Cook until most of the liquid has evaporated and the tomatoes form a rather dry paste.
9. Add the rice and stir. Add the chicken, sausage, pork, squid, and peppers.
10. Bring the stock mixture to a boil in a separate pot, then add to the rice and stir. Add salt and pepper to taste.
11. Bring to a simmer, cover, and put in a 350°F (175°C) oven for 20 minutes. (This dish is traditionally made uncovered on top of the stove, but making it in the oven is more practical for restaurants because it requires less attention.)

12. Remove the pan from the oven. Check the moisture level and add more stock, if necessary. It should be quite moist but not soupy.
13. Sprinkle the peas over the rice. Arrange the shrimp, clams, and mussels on top. Cover loosely and let stand 10 minutes to heat the shellfish.
14. For each portion, allow 8 oz (225 g) rice and vegetables, 1 shrimp, 1 clam, 1 mussel, 1 piece of chicken, and at least 1 piece each of pork, sausage, and squid. Garnish each portion with 1 lemon wedge.

Per serving: Calories, 630; Protein, 52 g; Fat, 22 g (32% cal.); Cholesterol, 260 mg; Carbohydrates, 52 g; Fiber, 5 g; Sodium, 630 mg.

Note: If chorizo is not available, use pepperoni or other hard, spicy sausage. You may cut the sausage into ¹/₂-oz (15-g) chunks before sautéing or cut them up just before serving.

Paella

Polenta

YIELD: ABOUT 5 LB (2.5 KG)

U.S.	METRIC	INGREDIENTS
5 pt	2.5 L	Water
1 tbsp	15 mL	Salt
1 lb	500 g	Polenta (Italian coarse-grained yellow cornmeal)

PROCEDURE

1. Bring the water and salt to a boil in a saucepot.
2. Very slowly sprinkle the cornmeal into the boiling water, stirring constantly. This must be done slowly and carefully to avoid lumps (see **Figure 13.3**).
3. Cook over low heat, stirring almost constantly. The polenta will become thicker as it cooks and eventually start to pull away from the sides of the pot. This will take 20–30 minutes.
4. Lightly moisten a large, flat surface, such as a wooden board or a platter.
5. Pour the polenta onto this board or platter. Serve immediately, hot, or let cool and use in any of a number of ways, including the variations below.

Per 1 ounce (28.35 g): Calories, 20; Protein, 0 g; Fat, 0 g (0% cal.); Cholesterol, 0 mg; Carbohydrates, 4 g; Fiber, 0 g; Sodium, 90 mg.

VARIATIONS

Freshly made hot polenta is good with many kinds of stews and other braised dishes that provide plenty of flavorful juices for the polenta to soak up. It is also served with grilled dishes.

Polenta con Sugo di Pomodoro

Serve hot polenta with tomato sauce or Meat Sauce (p. 399).

Polenta con Salsicce

Serve hot polenta with pork sausages cooked with tomatoes or tomato sauce.

Polenta al Burro e Formaggio

Stir 6 oz (175 g) fresh butter and 2–3 oz (60–90 g) grated parmesan cheese into hot polenta as soon as it is cooked.

Polenta Fritta or Grigliata

Let polenta cool and cut it into slices ½ in. (1 cm) thick. Pan-fry in oil until a thin crust forms. Alternatively, heat slices on a grill or broiler until hot and lightly grill-marked.

Polenta Grassa

This can be prepared in two ways.

1. Pour a layer of hot polenta into a buttered baking dish. Cover with sliced fontina cheese and dot with butter. Cover with another layer of polenta, then another layer of cheese and butter. Bake until very hot.
2. Prepare as in the first method, but instead of the hot, freshly made polenta, use cold polenta cut into thin slices.

Polenta Pasticciata

Prepare Meat Sauce (p. 399), using pork sausage in addition to the beef. Also, add sautéed sliced mushrooms to the sauce. Cut cold polenta into thin slices. Fill a baking pan with alternating layers of polenta slices, meat sauce, and parmesan cheese. Bake until hot.

Polenta Pasticciata

FIGURE 13.3 Making polenta.

(a) Slowly sprinkle the polenta into the simmering water, stirring constantly to avoid lumps.

(b) Simmer while stirring until the polenta reaches the desired texture. Serve at once, or proceed to the next step.

(c) Pour the polenta into a sheet pan and cool.

(d) Cut into desired shapes.

Grits with Cheddar Cheese

PORTIONS: 12   PORTION SIZE: 6 OZ

U.S.	METRIC	INGREDIENTS
2 cups	500 mL	Hominy grits
2 qt	2 L	Water
1 tsp	5 mL	Salt
8 oz	250 g	Cheddar cheese, grated

PROCEDURE

1. Stir the grits into the water in a saucepan.
2. If you are using stone-ground grits, let them settle, then skim off any hulls that float to the surface.
3. Bring the mixture to a boil while stirring.
4. Simmer the grits, stirring every 5–10 minutes. If you are using quick-cooking grits, they will be done in 7–10 minutes but can be cooked longer if a thicker product is desired. If you are using regular grits, cooking time is about 45 minutes, and you may have to add water as they thicken. In any case, continue to cook until creamy, adding water as necessary.
5. Stir in salt to taste.
6. Add the cheese and stir until it is melted.

Per serving: Calories, 170; Protein, 7 g; Fat, 7 g (35% cal.); Cholesterol, 20 mg; Carbohydrates, 22 g; Fiber, 1 g; Sodium, 310 mg.

VARIATIONS

For regular grits, omit the cheese. Serve topped with a pat of butter. If desired, substitute milk for 1/3 of the water.
Grits can be poured into a pan, chilled, cut into shapes, and browned lightly in butter.

Grits

Wheatberries with Pecans and Poblanos

PORTIONS: 12   PORTION SIZE: 4 OZ (125 G)

U.S.	METRIC	INGREDIENTS
1 lb 8 oz	750 g	Whole wheatberries, rinsed and soaked overnight in cold water (see Note)
2 qt	2 L	Water, cold
2 fl oz	60 mL	Olive oil
3 oz	90 g	Chopped pecans
3 oz	90 g	Roasted poblano chile, diced
to taste	to taste	Salt

PROCEDURE

1. Drain the soaking water from the wheatberries. Add the wheatberries to the cold water in a pot. Bring to a boil. Reduce heat to a simmer, cover, and simmer until the wheatberries are tender but still slightly crunchy, about 1 hour. Remove from heat and allow to stand, covered, 10 minutes. Drain.

2. Heat the olive oil in a sauté pan. Add the pecans and diced poblanos. Sauté about 1 minute.

3. Add the cooked wheatberries. Toss over heat until the mixture is hot. Season to taste.

Per serving: Calories, 280; Protein, 8 g; Fat, 10 g (31% cal.); Cholesterol, 0 mg; Carbohydrates, 42 g; Fiber, 7 g; Sodium, 0 mg.

Note: This recipe was developed using soft wheatberries with the bran left on. Other types of wheatberry may be used, but the yields and cooking times will vary. For example, the white wheatberries shown in the illustration on page 378 cook in less than 30 minutes and yield 3 times their dry weight (about 4½ lb/2.25 kg for this recipe). When using a new type of product, test cooking time and yield with a small quantity before adapting it to production.

VARIATIONS

Substitute any green chile or any sweet bell pepper for the poblanos.

Brown Rice, Barley, Farro, or Cracked Wheat with Pecans and Poblanos

Substitute cooked brown rice, cooked barley, cooked farro, or cracked wheat pilaf for the cooked wheatberries.

Wheatberries with Pecans and Poblanos

Farrotto with Pecorino Cheese

PORTIONS: 10   PORTION SIZE: 4 OZ (1150 G)

U.S.	METRIC	INGREDIENTS	PROCEDURE
1 lb	450 g	Farro	1. Soak the farro in cold water 45 minutes. Drain.
			2. Boil the farro in boiling salted water 20 minutes. Drain. At this point, the farro should be about half-cooked.
2 fl oz	60 mL	Olive oil	3. Heat the olive oil in a sauté pan over moderate heat.
1 oz	30 g	Onion, chopped fine	4. Sweat the onion in the oil until soft.
1¹/₂ qt (approximately)	1.5 L (approximately)	Vegetable stock, hot	5. Add the farro. Stir and cook over moderate heat 2 minutes.
			6. Ladle in 4 fl oz (125 mL) stock. Stir over moderate heat until the stock is absorbed and the farro is almost dry.
			7. Add another ladleful of stock and repeat the procedure.
			8. Stop adding stock when the farro is tender.
1 oz	30 g	Butter	9. Stir in the butter and cheese. Serve immediately.
4 oz	120 g	Pecorino cheese, grated	

Per serving: Calories, 260; Protein, 8 g; Fat, 12 g (39% cal.); Cholesterol, 15 mg; Carbohydrates, 34 g; Fiber, 0 g; Sodium, 102.7 mg.

Farrotto with Pecorino Cheese

Kasha Pilaf with Egg

PORTIONS: 10   PORTION SIZE: 4 OZ (125 G)

U.S.	METRIC	INGREDIENTS	PROCEDURE
2 cups	250 mL	Buckwheat groats (kasha)	1. Put the buckwheat in a sauté pan over moderate heat.
2	2	Eggs, lightly beaten	2. Add the eggs and stir vigorously so all the grains are coated with egg.
2 oz	60 g	Chicken fat or butter	3. Continue to cook, stirring, until the grains are dry.
1 qt	1 L	Water or chicken stock	4. Add the fat and stir.
2 tsp	10 mL	Salt	5. Add the water or stock, salt, and pepper. Bring to a boil, cover, and set over low heat to simmer 10 minutes, or until the liquid is absorbed.
¹/₂ tsp	2 mL	Pepper	

Per serving: Calories, 180; Protein, 5 g; Fat, 7 g (35% cal.); Cholesterol, 50 mg; Carbohydrates, 24 g; Fiber, 2 g; Sodium, 480 mg.

Kasha Pilaf with Egg

PASTA, NOODLES, AND DUMPLINGS

Macaroni products, or pastas, are popular alternatives to other starch foods. The word **pasta** is Italian for "paste," so called because pasta is made from a mixture of wheat flour and water and, sometimes, eggs.

Not so many years ago, many of us knew only spaghetti with tomato sauce and elbow macaroni with cheese, among all pasta products. Today, thanks to the influence of Italian cooks, we have a choice of a great variety of pasta dishes.

In addition, noodle products play an important role in other cuisines, notably those of Asia. These include not only wheat noodles but also noodles made of rice and other starches. Although most of our attention here is devoted to Italian-style pastas, we take a look at these other products as well.

ITALIAN-STYLE PASTAS

Italian pastas have spread beyond their original borders to become one of the most popular foods in North America and Europe. Please note that when we use the term *Italian-style pasta* we are referring only to the noodle products themselves, not to the dishes prepared from them. Appearing on menus are many dishes that are made with Italian-style pastas but that are not recognizably Italian in any way, as they use ingredients from other cuisines, including those of Asia and Latin America. Adopted by chefs from many cultures, Italian pastas have become citizens of the world.

Kinds, Characteristics, and Quality Factors

Commercial dried pasta is made from dough that has been shaped and dried. To refer to this category of food, we sometimes use the term **macaroni**, meaning any dried pasta made from flour and water. These include spaghetti, lasagna, elbow macaroni, and many other shapes.

The best dried macaroni pastas are made from **semolina**, a high-protein flour from the inner part of durum wheat kernels. Lower-quality products are made from farina, a softer flour.

Specialty pastas include ingredients in addition to semolina and water. Whole wheat pasta may be made with all whole wheat flour or a mixture of semolina and whole wheat. Other grain flours, including buckwheat and farro (spelt), are used in other specialty items.

Spinach pasta is perhaps the most popular of the pastas that contain vegetable purées. Others include red peppers, hot chile, seaweed, beet, tomato, and pumpkin.

Pasta containing *squid ink* is black in color and goes well with seafood sauces.

When purchasing macaroni products (unflavored), look for a good yellow color, not gray-white. The product should be very hard, brittle, and springy, and it should snap with a clean, sharp-edged break. When cooked, it should be firm and hold its shape well. Poor-quality pastas are soft and pasty when cooked.

In addition to pastas made of flour and water, dried **egg pastas** are also available. They contain at least 5.5 percent egg solids in addition to the flour and water. They are usually sold as flat noodles of various widths.

Fresh egg pasta is made from flour and eggs and, sometimes, a small quantity of water and/or oil. Use a regular all-purpose or bread flour. Hard semolina flour, used for factory-made spaghetti and macaroni, is not appropriate for fresh egg pasta. Softer flour makes a more tender pasta. Soft egg noodle products are also available fresh and frozen from manufacturers. They take less time to cook than dried macaroni products.

Other flours, such as whole wheat flour, can be used to make fresh noodles. When you are making them yourself, you can experiment with ingredients. Keep in mind that flours other than wheat flour don't form much gluten (see p. 897), so they should be mixed with some wheat flour. Buckwheat flour added to white flour makes especially tasty noodles that are featured in a northern Italian classic called *Pizzoccheri* (peet so kerry; p. 405).

Vegetable purées and other flavoring ingredients are often added to fresh egg pasta. Spinach, tomato, beet, mushrooms, fresh herbs, dried chile, grated lemon zest, saffron, and squid ink are among the many possibilities.

Shapes and Their Uses

Pasta is made in hundreds of shapes and sizes. Each shape is appropriate for different preparations because of the way different kinds of sauce cling to them or the way their textures complement the texture of the topping. The illustration shows some of the most popular kinds. Table 13.2 describes the most common shapes and gives suggestions for use.

Remember that fresh egg pasta and factory-made spaghetti and macaroni are different products. It makes no sense to say that one type is better than the other. Italian cooks use fresh and dried pasta in different ways, with different recipes for each type. Factory pasta has a chewy, robust texture, good with robust sauces, while fresh egg pasta is tender and more delicate. Fresh egg pasta absorbs sauces more deeply than factory macaroni products. In general, factory-made pasta is ideal for olive oil–based sauces, and fresh, homemade pasta is better with butter or cream-based sauces.

TABLE 13.2 Commercial Pasta Shapes and Uses

Name	Description	Suggested Uses
Spaghetti	Long, round	With great variety of sauces, especially tomato sauces
Spaghettini	Thin, long, round	Like spaghetti, especially with olive oil and seafood sauces
Vermicelli	Very thin	With light, delicate sauces and, broken, in soups
Linguine	Looks like slightly flattened spaghetti	Like spaghetti; popular with clam sauces
Perciatelli, bucatini	Looks like thick, hollow spaghetti	Like spaghetti, but can handle heavy, chunky sauces
Fusilli	Long, shaped like a corkscrew	Thick, creamy sauces
Macaroni	Long, hollow tubes	Especially good with hearty meat sauces
Elbow macaroni	Short, bent macaroni	Cold, in salads; baked, in casseroles
Penne or mostaccioli	Hollow tubes, cut diagonally; may be smooth or ridged	Baked, with meat sauce or with tomato sauce and cheese; freshly cooked, with tomato sauce
Ziti	Short, hollow tubes, cut straight	
Rigatoni	Larger tubes, with ridges	
Manicotti (sometimes called cannelloni, which are actually rolled from fresh egg noodle dough)	Large hollow tubes, sometimes with ridges	Stuff with cheese or meat filling
Orecchiette	Little ears	Chunky vegetable sauces
Rotelle, ruote	Wheels	Chunky tomato, meat, or vegetable sauces; in soups
Radiatore	Radiators; curled, ruffled shapes	Cold, in salads; hot, with chunky sauces
Fettuccine	Flat egg noodles	Rich cream sauces or meat sauces
Tagliatelle	Wide, flat egg noodles	
Lasagna	Broad, flat noodles, often with rippled edges	Baked with meat, cheese, or vegetable fillings
Conchiglie	Shells	With seafood or meat sauces; small sizes can be used in salads
Bow ties or farfalle	Look like bow ties	With sauces containing chunks of meat, sausage, or vegetable
Pastina (little pasta)		In soups; cold, in salads; buttered, as a side dish
Ditalini	Very short, hollow tubes	
Orzo	Rice-shaped	
Stelline	Tiny stars	
Acini di pepe	Peppercorns	
Pepe bucato	Peppercorns with holes	
Rotelline	Little wheels	
Semi di melone	Melon seeds	

Top row: elbow macaroni, pepe bucato, radiatore, ziti, conchiglie. Middle row: fettuccine; spaghettini; fusilli; (three small piles, from top to bottom) orzo, stelline, and ditalini; lasagna; spaghetti; (two small piles, from top to bottom) gemelli and rigatoni. Bottom row: bow ties (farfalle), penne, manicotti.

OTHER NOODLE PRODUCTS

Asian countries have a wide variety of noodle products. We can divide these into two categories: wheat noodles and noodles made of other starches.

Noodles Made of Wheat

Most of the Asian wheat noodles we know in the West are from China and Japan.

Chinese noodles are made from either flour and water or flour, water, and egg. If they contain egg, they are usually labeled as *egg noodles*. Flat noodles come in a variety of widths, from very thin to nearly 1 inch (2.5 cm) wide. Round noodles come in a variety of thicknesses, from thin vermicelli to thicker, spaghetti-like noodles.

Cantonese noodles are a special type of Chinese wheat noodle made by stretching a single large piece of dough in one length until it is as thin as spaghetti. Some restaurants feature a skilled noodle maker performing this amazing feat in the dining room. Cantonese noodles are available fresh or dried.

Japanese wheat noodles come in several varieties. **Udon** are thick, white noodles made from wheat flour. They are available fresh in vacuum packs or dried. **Somen** are thin, white wheat noodles usually packaged in small bundles. Tamago somen are made with egg in addition to wheat flour and water (*tamago* means "egg").

Soba are thin noodles made with buckwheat in addition to wheat flour. A special variety of buckwheat noodle is **chasoba**, made with powdered green tea in addition to the buckwheat.

In spite of the *soba* in the name, **chukasoba** contain no buckwheat. They are wheat noodles made with flour and water, with an alkali such as sodium carbonate added to the water. The noodles are yellow in color, leading some people to think they are egg noodles, but they contain no egg. These noodles are used in the popular ramen dishes familiar in the West from the single-portion packages of noodles and soup stock, found in nearly every supermarket.

Noodles Made of Other Starches

With the increased popularity of Southeast Asian and Chinese cuisines in the West, **rice noodles** have become familiar. Rice noodles are available as very fine, almost hairlike noodles called **rice vermicelli**, and as flat noodles of various widths. Rice noodles are sometimes known as **rice sticks**.

Rice vermicelli are often cooked by deep-frying the dry noodles, without using any water. The noodles puff up and become crisp and tender. Rice vermicelli can also be broken apart and stir-fried, as long as enough liquid is added to the stir-fry to rehydrate them.

Rice noodles of all types are not usually boiled in water because they become too soft and sticky. Rather, they are covered with hot water and soaked until tender. This takes from a few minutes to about an hour, depending on the thickness of the noodle and the temperature of the water. The noodles are drained and added to stir-fried dishes and soups during the last minutes of cooking.

Bean thread noodles, also called **cellophane noodles**, are made with mung bean starch. They are very thin noodles that resemble rice vermicelli. Like rice noodles, they are either deep-fried or soaked in hot water until tender and then added to soups and braised dishes.

Couscous

Couscous is a kind of granular pasta made from semolina flour (see sidebar). It is cooked by soaking and then steaming, using a fairly time-consuming process. **Instant couscous** is prepared by simply adding the dry product to hot or boiling water and letting it stand 5 minutes. The accompanying procedures outline the methods for making both classic and instant couscous.

COUSCOUS

Couscous is a classic dish of North Africa, especially Morocco and Algeria. The word refers to both the grainlike product and the dish made from it. After soaking, couscous is steamed in the perforated top section of a double kettle called a *couscousière* (koos koos yair), while a spicy stew cooks in the lower section.

Rather than being made from a smooth dough like noodles are, couscous is made by sprinkling water into a bowl of semolina flour and stirring with the hand, forming tiny granules of dough. The granules are then sifted out and dried, and the process is repeated.

A related product called *Israeli couscous* is made in the lands at the eastern end of the Mediterranean. Israeli couscous is made from regular wheat rather than semolina, and it is formed into larger, spherical grains. It is typically served in place of rice.

PROCEDURE for Steaming Traditional Couscous

1. Place the couscous in a bowl and add enough cold water to cover it by several inches (cm). Stir the couscous and then drain off the excess water through a fine sieve. Smooth the couscous in the bowl and let stand 15 minutes, allowing the couscous to absorb the moisture that coats it.

2. With wet hands, stir and rub the grains to break up all lumps.

3. Line the top of a couscousière with a double layer of cheesecloth. Place the couscous inside. If you don't have a couscousière, select a colander that will fit over a large saucepan or similar pot. Line with cheesecloth and put the couscous in it.

4. Set the couscousière top or colander over a simmering stew (or simmering water). Steam, uncovered, 20 minutes.

5. Empty the couscous into a hotel pan and spread it into a flat layer. Sprinkle with a little salt and just enough water to moisten it slightly. Coat your hands with oil and stir and rub the couscous to break up all lumps. Let stand 10 minutes. At this point, if the couscous feels dry, sprinkle with a little more water and stir.

6. Give the couscous a second steaming by repeating step 4.

7. Turn out the finished couscous into a hotel pan or other container for service. Stir lightly to break up any lumps.

PROCEDURE for Preparing Instant Couscous

1. Measure equal parts by volume dry instant couscous and water.

2. Place the water in a saucepan and bring to a boil. Add ½ tsp (2 mL) salt and 2 tsp (10 mL) butter per pint (0.5 L) water.

3. When the water boils, add the couscous and stir. Remove from the heat, cover, and let stand 5 minutes.

4. Before serving, stir with a fork to break up lumps.

COOKING PASTA

Doneness

Pasta should be cooked **al dente**, or "to the tooth." This means cooking should be stopped when the pasta still feels firm to the bite, not soft and mushy. Much of the pleasure of eating pasta is its texture (that's why there are so many shapes), and this is lost if it is overcooked.

Testing Doneness and Serving

Many suggestions have been made for testing doneness, but none is more reliable than breaking off a very small piece and tasting it. As soon as the pasta is al dente, the cooking must be stopped at once. Half a minute extra is enough to overcook it.

Cooking times differ for every shape and size of pasta. Timing also depends on the kind of flour used and the moisture content. Times indicated on packages are often too long.

Fresh egg pasta, if it has not been allowed to dry, takes only 1 to 1^1/$_2$ minutes to cook after the water returns to a boil.

Italian practice is to toss the pasta with the sauce the minute it is drained. The sauce immediately coats all surfaces of the pasta, and cheese, if there is any, melts in the heat of the boiling hot noodles. If you are attempting to serve an authentic Italian pasta dish, follow this practice rather than simply topping the pasta with the sauce.

Pasta is best if cooked and served immediately. Whenever possible, you should try to cook pasta to order. Fresh pasta, in particular, cooks so quickly there is little reason to cook it in advance. In volume operations, however, commercial pasta may have to be cooked ahead of time. The following procedures can be used for quantity cookery.

Yields

One pound (450 g) uncooked dried pasta yields about 3 pounds (1.4 kg) cooked pasta. This is enough for 4–6 main-course portions or 8–10 side-dish or first-course portions.

One pound (450 g) uncooked fresh pasta yields 2–2^1/$_2$ pounds (900–1,100 g) cooked pasta.

PROCEDURE for Cooking Pasta in Large Quantities

1. Use at least 4 quarts boiling salted water per pound of pasta (4 L per 500 g). Use about 1^1/$_2$ tablespoons (25 g) salt per 4 quarts (4 L) water.

2. Have the water boiling rapidly and drop in the pasta. As it softens, stir gently to keep it from sticking together and to the bottom.

3. Continue to boil, stirring a few times.

4. As soon as the pasta is al dente, drain it immediately in a colander and rinse with cold running water until completely cooled. Otherwise, it would continue to cook and become too soft. (If you are cooking just a few portions to serve immediately, just drain well and do not rinse. Sauce and serve without a moment's delay.)

If the pasta is to be used cold in a salad, it is ready to be incorporated into the recipe as soon as it has cooled.

5. If the pasta is to be held, toss gently with a small amount of oil to keep it from sticking.

6. Measure portions into mounds on trays. Cover with plastic film and refrigerate until service time. (Do not store pasta in cold water. The pasta will absorb water and become soft, as though it had been overcooked.)

7. To serve, place the desired number of portions in a china cap and immerse in simmering water to reheat. Drain, plate, and add sauce.

ALTERNATIVE METHOD: Steam Table Service

Pasta gradually becomes soft and mushy when kept hot for service, but it will hold reasonably well for 30 minutes. It will not be as good as if freshly cooked, however. This method should not be used unless cooking pasta to order is not possible in a particular food-service operation.

1. Follow steps 1 to 3 above.

2. Drain the pasta while still slightly undercooked. Rinse briefly in cool water, enough to stop the cooking and

rinse off starch but not enough to cool the pasta. Pasta should still be quite warm.

3. Transfer the pasta to a steam table pan and toss with oil to prevent sticking.

4. Hold for up to 30 minutes.

Italian Tomato Sauce for Pasta

YIELD: 1¹/₂ QT (1.5 L)   PORTIONS: 16   PORTION SIZE: 3 FL OZ (ML)

U.S.	METRIC	INGREDIENTS
8 fl oz	240 mL	Olive oil (see Note)
4 oz	110 g	Onion, chopped fine
4 oz	110 g	Carrot, chopped fine
4 oz	110 g	Celery, chopped fine
3 lb	1360 g	Canned whole tomatoes
1	1	Garlic cloves, minced
¹/₂ oz	15 g	Salt
1¹/₂ tsp	7 mL	Sugar

PROCEDURE

1. Heat the olive oil in a large saucepot. Add the onions, carrots, and celery and sauté lightly for a few minutes. Do not let the vegetables brown.
2. Add remaining ingredients. (See Appendix 2 for can sizes and substitutions.) Simmer, uncovered, about 45 minutes, until reduced and thickened.
3. Pass through a food mill. Taste and adjust seasonings.
4. For service, this sauce should be tossed with the freshly cooked spaghetti or other pasta in a bowl before being plated, rather than simply ladled over the pasta.

Per serving: Calories, 190; Protein, 1 g; Fat, 18 g (82% cal.); Cholesterol, 0 mg; Carbohydrates, 8 g; Fiber, 2 g; Sodium, 660 mg.

Note: The quantity of olive oil may seem high, but it is only 1 tbsp for a 3-oz portion (15 mL per 90 g). These are typical proportions for basic tomato sauce in Italy, where sauces are not used in such large quantities as in North America. The olive oil is intended to be a major ingredient, not just a sautéing medium for the mirepoix, so use a good, flavorful oil. If a lower-fat sauce is desired, oil may be cut in half. Except for meat sauce, most Italian sauces are cooked less than American-style tomato sauce and have fewer ingredients. As a result, they have a more pronounced fresh tomato taste.

VARIATIONS

Omit onion, carrot, and celery. Reduce oil to 4 fl oz (120 mL). Add fresh chopped parsley and basil to taste.

Meat Sauce

Brown 1 lb (500 g) ground beef, ground pork, or a mixture of beef and pork, in oil or rendered pork fat. Add 4 fl oz (120 mL) red wine, 1 qt (1 L) tomato sauce, 1 pt (500 mL) beef or pork stock, and parsley, basil, and oregano to taste. Simmer 1 hour, uncovered.

Tomato Cream Sauce

Use 4 oz (110 g) butter instead of the olive oil in the basic recipe. At service time, add 1 cup heavy cream per quart of tomato sauce (250 mL per L). Bring to simmer and serve.

Tomato Sauce with Sausage

Slice 1¹/₂ lb (700 g) fresh Italian sausage and brown in oil. Drain and add to basic tomato sauce. Simmer 20 minutes.

Tomato Sauce with Sausage and Eggplant

Prepare like Tomato Sauce with Sausage, but use 12 oz (350 g) each sausage and peeled, diced eggplant.

Tomato Sauce with Ham and Rosemary

Cook 8 oz (220 g) ham, cut into fine dice, and 1 tbsp (15 mL) dried rosemary leaves in a little olive oil for a few minutes. Add to basic tomato sauce (after it has been passed through the food mill) and simmer 5 minutes.

Pesto (Fresh Basil Sauce)

YIELD: ABOUT 3 CUPS (750 ML)   PORTIONS: 12   PORTION SIZE: 2 OZ (60 ML)

U.S.	METRIC	INGREDIENTS
2 qt	2 L	Fresh basil leaves
1¹/₂ cups	375 mL	Olive oil
2 oz	60 g	Walnuts or pine nuts (pignoli)
6	6	Garlic cloves
1¹/₂ tsp	7 mL	Salt
5 oz	150 g	Parmesan cheese, grated
1¹/₂ oz	50 g	Romano cheese, grated

PROCEDURE

1. Wash the basil leaves and drain well.
2. Put the basil, oil, nuts, garlic, and salt in a blender or food processor. Blend to a paste, but not so long that the mixture is smooth. It should have a slightly coarse texture.
3. Transfer the mixture to a bowl and stir in the cheese.
4. To serve, cook pasta to order according to the basic procedure. Just before the pasta is done, stir a little of the hot cooking water into the pesto to thin it, if desired. Toss the drained pasta with the pesto and serve immediately. Pass additional grated cheese.

Per 1 fl oz (29.57 mL): Calories, 350; Protein, 8 g; Fat, 35 g (88% cal.); Cholesterol, 15 mg; Carbohydrates, 3 g; Fiber, 1 g; Sodium, 550 mg.

Fresh Egg Pasta

YIELD: 1¹/₂ LB (700 G)

U.S.	METRIC	INGREDIENTS
1 lb	450 g	Bread flour
5	5	Eggs
¹/₂ fl oz	15 mL	Olive oil
pinch	pinch	Salt

Per 1 ounce (28.35 g): Calories, 90; Protein, 4 g; Fat, 2 g (20% cal.); Cholesterol, 45 mg; Carbohydrates, 14 g; Fiber, 0 g; Sodium, 20 mg.

PROCEDURE

1. Mound the flour on a work surface. Make a well in the center and add the eggs, oil, and salt.
2. Working from the center outward, gradually mix the flour into the eggs to make a dough.
3. When it is firm enough to knead, begin kneading the dough, incorporating more flour. If the dough is still sticky when all the flour is incorporated, add more flour, a little at a time. Knead well for at least 15 minutes.
4. Cover the dough and let it rest at least 30 minutes.
5. Cut the dough into 3–5 pieces. Set the rollers of a pasta machine at the widest opening. Pass the pieces of dough through the machine, folding them in thirds after each pass and dusting them lightly with flour to keep them from getting sticky. Continue passing each piece through the machine until it is smooth. See **Figure 13.4**.
6. Working with one piece of dough at a time, decrease the width between the rollers one notch and pass the dough through them again. After each pass, turn the rollers one notch narrower, dust the dough with flour, and pass it through again. Continue until the dough is as thin as desired. The pasta is now ready to cut into desired shapes and to cook. See below for cutting instructions.

VARIATIONS

Cutting Instructions

Fettuccine or Tagliatelle: Roll dough thin and cut with wide cutting rollers.

Taglierini: Roll dough thin and cut with narrow cutting rollers.

Pappardelle: Cut by hand, using a fluted cutting wheel, into long noodles about ³/₄ in. (18 mm) wide.

Tonnarelli: Roll dough to the same thickness as the width of the narrow cutting roller. Cut with the narrow cutting rollers. The result is like square spaghetti.

Bow ties: Cut into rectangles about 1¹/₂ × 3 in. (4 × 8 cm). Pinch in the middle to make a bow.

Lasagne: Cut by hand into broad strips about 8–12 in. (20–30 cm) long.

Spinach Pasta

Clean 1 lb (450 g) AP spinach, discarding stems. Simmer 5 minutes in salted water. Drain, rinse in cold water, and squeeze dry. Chop as fine as possible. Incorporate in basic pasta recipe, adding it to the flour at the same time as the eggs. Reduce the quantity of eggs to 4.

Other Colored Pastas

Other colored vegetables, in small quantities, cooked until tender and puréed or chopped fine, can be substituted for spinach to color pasta. For example, experiment with beets, red bell peppers, and carrots.

Herb Pasta

Roll out regular egg pasta as in step 6. When the sheets of pasta are almost as thin as desired, Sprinkle one sheet of dough with coarsely chopped fresh herbs. Top with a second sheet to enclose the herbs and continue to roll to desired thinness.

Whole Wheat Pasta

Substitute whole wheat flour for half of the white flour.

Buckwheat Pasta

In place of the 1 lb (450 g) white flour, use 10 oz (280 g) buckwheat flour and 6 oz (180 g) white flour. Omit the olive oil.

FIGURE 13.4 Working with fresh egg pasta.

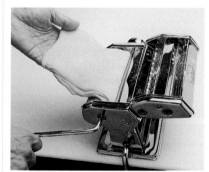

(a) Set the rollers of the machine at their widest setting. Pass the piece of dough through the rollers, fold in thirds, and repeat until the dough is smooth.

(b) Decrease the opening between the rollers one notch at a time and pass the dough through them to roll to desired thickness.

(c) Pass the rolled-out dough through the appropriate cutters to make pasta of desired size and shape.

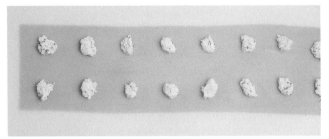

(d) To make ravioli, lay a thin sheet of pasta on the workbench. Deposit portions of fitting on the dough using a spoon, small scoop, or pastry bag.

(e) Cover with another sheet of pasta.

(f) Press down between the mounds of filling to seal the layers of pasta together. Try to remove air bubbles from between the layers.

(g) Cut out the ravioli with cutters, or cut them apart with a pastry wheel.

Ravioli with Cheese Filling

PORTIONS: 10   PORTION SIZE: 5 OZ (150 G) UNCOOKED OR APPROXIMATELY 7 OZ (200 G) COOKED

U.S.	METRIC	INGREDIENTS	PROCEDURE
1 lb 3 oz	560 g	Ricotta cheese	1. Mix together the ricotta, parmesan, egg yolks, parsley, and seasonings.
3^1/$_2$ oz	100 g	Parmesan cheese, grated	
2	2	Egg yolks	
5 tbsp	20 g	Chopped parsley	
^{1}/$_4$ tsp	1 mL	Nutmeg	
to taste	to taste	Salt	
to taste	to taste	White pepper	
1^1/$_2$ lb	800 g	Fresh pasta	2. Roll the pasta into thin sheets.

3. Over half the pasta, make small mounds of cheese filling, about 1 tsp (5 mL) each, arranging them in a checkerboard pattern 1^1/$_2$–2 in. (4–5 cm) apart.

4. Lay the remaining pasta over the top and press down between the mounds of cheese to seal well (see **Figure 13.4**). While doing this, try to avoid sealing large air bubbles inside the ravioli. Note: If the pasta is fresh and moist, the layers will seal together if pressed firmly. If it is dry, moisten the bottom layer lightly between the mounds of cheese, using a brush dipped in water. Do not get the pasta too wet.

5. Cut the ravioli apart with a pastry wheel. Check each piece to be sure it is completely sealed.

6. The cheese filling does not keep well, so the ravioli should be cooked the same day they are made. They can be held briefly on sheet pans covered with dry, clean towels; turn them over from time to time so they do not stick. Alternatively, cook them immediately in boiling salted water, keeping them slightly underdone. Drain and rinse under cold water, drain, and toss with oil or melted butter. Spread in a single layer on a sheet pan and refrigerate. They can then be reheated to order by sautéing them briefly in butter or oil or by dipping them in boiling water.

7. Serve with your choice of sauce, such as tomato sauce, meat sauce, Bolognese sauce, tomato cream sauce, or just melted butter and parmesan cheese.

Per serving: Calories, 410; Protein, 21 g; Fat, 17 g (38% cal.); Cholesterol, 205 mg; Carbohydrates, 42 Fiber, 1 g; Sodium, 270 mg.

Fettuccine Alfredo 🌹

PORTIONS: 10   **PORTION SIZE: 6–7 OZ (175–200 G)**

U.S.	METRIC	INGREDIENTS	PROCEDURE
1 cup	250 mL	Heavy cream	1. Combine 1 cup (250 mL) cream and the butter in a sauté pan. Bring to a simmer, reduce by one-fourth, and remove from heat.
2 oz	60 g	Butter	
1½ lb	700 g	Fresh fettuccine	2. Drop the noodles into boiling salted water, return to a full boil, and drain. The noodles must be slightly undercooked because they will cook further in the cream mixture.
1 cup	250 mL	Heavy cream	3. Put the drained noodles in the pan with the hot cream and butter mixture. Over low heat, toss the noodles with two forks until they are well coated.
6 oz	175 g	Freshly grated parmesan cheese	4. Add the remaining 1 cup (250 mL) cream and the cheese, and toss to mix well. (If the noodles seem dry at this point, add a little more cream.)
to taste	to taste	Salt	5. Add salt and pepper to taste.
to taste	to taste	Pepper	6. Plate and serve immediately. Offer additional grated cheese at the table.

Per serving: Calories, 500; Protein, 17 g; Fat, 32 g (56% cal.); Cholesterol, 195 mg; Carbohydrates, 35 g; Fiber, 1 g; Sodium, 430 mg.

VARIATIONS

Fettuccine with Vegetables I (Fettuccine Primavera)

Fresh, lightly cooked vegetables can be added to fettuccine to make a great variety of dishes. In the basic recipe, use about half the quantity of cream. Select 4–6 fresh vegetables, cut them into appropriately small sizes and shapes, cook them al dente, and add them to the pasta when it is being tossed in the cream. The following are examples of appropriate vegetables:

Mushrooms	Tiny green beans
Peas	Asparagus
Broccoli	Artichoke hearts
Red or green bell pepper	Zucchini

Small quantities of finely diced ham, prosciutto, or bacon can also be added as a flavor accent.

Fettuccine with Vegetables II

Prepare like Fettuccine with Vegetables I, but omit all butter and cream. Instead, toss the freshly cooked fettuccine and cooked vegetables with olive oil. Add parmesan cheese as desired.

Fettuccine Bolognese

Serve the freshly cooked fettuccine with Bolognese Sauce (p. 403) instead of the cream sauce.

Fettuccine with Seafood

Use half the quantity of cream and cheese in the basic recipe. Prepare like Fettuccine with Vegetables I, adding only 1–3 types of vegetables. At the same time, add the desired quantity of cooked seafood, such as shrimp, scallops, crab, or lobster. For a fuller flavor, reduce a small amount of fish stock and white wine with the cream in the first step.

Fettuccine with Gorgonzola

Prepare as in the basic recipe, except use light cream instead of heavy cream in the first step. Omit the second quantity of heavy cream, and instead add 6 oz (175 g) gorgonzola cheese (Italian blue cheese). Reduce the quantity of parmesan cheese to 2 oz (60 g).

Fettuccine with Vegetables II

Bolognese Sauce (Ragù Bolognese)

YIELD: 1 QT (1 L)

U.S.	METRIC	INGREDIENTS	PROCEDURE
1 fl oz	30 mL	Vegetable oil	1. Heat the oil and butter in a heavy saucepot over moderate heat. Add the onion, celery, and carrot. Sweat the vegetables until they just begin to soften.
2^1/$_2$ oz	75 g	Butter	
3 oz	90 g	Onion, chopped fine	
3 oz	90 g	Celery, chopped fine	
3 oz	90 g	Carrot, chopped fine	2. Add the ground beef, along with a little salt and pepper. Stir to break up lumps, and cook until the meat has all lost its red color, but do not brown.
1^1/$_2$ lb	720 g	Ground beef, preferably chuck	
to taste	to taste	Salt	
to taste	to taste	Pepper	
12 fl oz	360 mL	Milk	3. Add the milk and nutmeg. Simmer slowly until the milk has almost completely reduced.
^{1}/$_4$ tsp	1 mL	Nutmeg	4. Add the wine. Continue to simmer until the wine has almost completely reduced.
1 pt	500 mL	Dry white wine	
1 lb 8 oz	720 g	Canned Italian-style tomatoes, chopped, with their juice	5. Stir in the tomatoes. Simmer over low heat, barely bubbling, about 3 hours, or until the sauce is quite thick. Stir from time to time as it cooks.
to taste	to taste	Salt	6. Taste and adjust the seasonings with salt and pepper.
to taste	to taste	Pepper	

Per fl oz: Calories, 90; Protein, 5 g; Fat, 5 g (53% cal.); Cholesterol, 20 mg; Carbohydrates, 2 g; Fiber, 0 g; Sodium, 70 mg.

Linguine with White Clam Sauce 🍽️

PORTIONS: 10   PORTION SIZE: APPROXIMATELY 12 OZ (350 G)

U.S.	METRIC	INGREDIENTS	PROCEDURE
4 dozen	4 dozen	Cherrystone clams	1. Open the clams. Strain and reserve 1 pt (500 mL) of their juice. Chop the clams coarsely.
1 cup	250 mL	Olive oil	2. Heat the olive oil in a large sauté pan. Add the garlic and brown it very lightly. Do not let it get too brown, or it will be bitter.
4–6	4–6	Garlic cloves, sliced thin	3. Add the red pepper and then, very carefully, add the wine. (If the pan is very hot, you may want to cool it a little first to prevent dangerous spattering when the liquid is added.) Reduce the wine by half.
^{1}/$_2$ tsp	2 mL	Red pepper flakes	
^{1}/$_2$ cup	125 mL	Dry white wine (optional)	
2 tsp	10 mL	Dried oregano	4. Add the reserved clam juice and reduce by half.
			5. Add the oregano.
2 lb	900 g	Linguine	6. Drop the linguine into boiling, salted water and boil al dente. Drain and plate.
4 tbsp	60 mL	Chopped parsley	7. While the linguine is boiling, add the chopped clams and the parsley to the olive oil mixture. Heat gently, just until the clams are hot. Do not overcook them, or they will be tough.
to taste	to taste	Black pepper	8. Add pepper to taste. (Because clams are salty, the sauce will probably not need any salt, but taste to make sure.)
			9. Spoon the sauce over the hot linguine and serve at once.
			10. Many people prefer this dish without parmesan cheese, but provide it on the side for those who want it.

Per serving: Calories, 600; Protein, 21 g; Fat, 24 g (36% cal.); Cholesterol, 25 mg; Carbohydrates, 74 g; Fiber, 7 g; Sodium, 40 mg.

Spaghetti Carbonara

PORTIONS: 10   PORTION SIZE: 11 OZ (330 G)

U.S.	METRIC	INGREDIENTS	PROCEDURE
2 fl oz	60 mL	Olive oil	1. Heat the oil in a sauté pan over moderate heat. Add the pancetta and sauté until most of the fat has rendered and it begins to crisp.
12 oz	360 g	Pancetta, cut into short bâtonnet	
2	2	Garlic cloves, chopped fine	2. Add the garlic and cook another few seconds.
3 fl oz	90 mL	Dry white wine	3. Add the wine. Reduce by three-fourths.
2 lb	1 kg	Spaghetti	4. While the pancetta is cooking, boil the spaghetti. Drain.
4	4	Whole eggs, preferably pasteurized, lightly beaten	5. Reduce the heat under the sauté pan to low and add the spaghetti. Toss to coat with the fat from the pancetta.
5 oz	150 g	Parmesan cheese, grated	6. Add the eggs, cheese, pepper, and parsley. Mix well over low heat until the spaghetti is well coated and the eggs have begun to coagulate. Heat only until the pasta has a creamy appearance. Do not cook so long that the eggs set hard.
1/4 tsp or to taste	1 mL or to taste	Black pepper	
1 oz	30 g	Chopped parsley	7. Taste for seasonings. Because of the salt in the pancetta, you may not need salt, but add a little if needed.
as needed	as needed	Salt	8. Serve immediately.

Per serving: Calories, 620; Protein, 25 g; Fat, 25 g (37% cal.); Cholesterol, 125 mg; Carbohydrates, 71 g; Fiber, 4 g; Sodium, 940 mg.

SPAGHETTI CARBONARA

According to legend, spaghetti carbonara originated outside Rome as a hearty meal for coal miners (Italian for "coal" is *carbone*) or charcoal makers. Whatever its origins, the dish has become popular both inside and outside Italy. Although most versions served in North American restaurants contain cream, in Italy, the authentic spaghetti carbonara is made without cream, like the version here. In addition, the authentic version is made with *guanciale*(gwan chah leh), or cured pork jowl, but pancetta is a good substitute if guanciale is not available.

Spaghetti Carbonara

Macaroni and Cheese

PORTIONS: 15   **PORTION SIZE: 6 OZ (175 G)**

U.S.	METRIC	INGREDIENTS
1 lb	450 g	Elbow macaroni
1 qt	1 L	Medium Béchamel, hot (p. 181)
1 tsp	5 mL	Dry mustard
dash	dash	Tabasco
1 lb	450 g	Cheddar cheese, grated
		Garnish:
as needed	as needed	Bread crumbs
as needed	as needed	Paprika

PROCEDURE

1. Cook macaroni according to basic method for boiling pasta. Drain and rinse in cold water.
2. Flavor the béchamel with the dry mustard and Tabasco.
3. Mix the macaroni with the cheese. Combine with the béchamel.
4. Pour into a buttered half-hotel pan. Sprinkle with bread crumbs and paprika.
5. Bake at 350°F (175°C) until hot and bubbling, about 30 minutes.

Per serving: Calories, 330; Protein, 14 g; Fat, 17 g (46% cal.); Cholesterol, 50 mg; Carbohydrates, 31 g; Fiber, 1 g; Sodium, 290 mg.

Note: Cheese sauce may be used instead of béchamel. If you do so, reduce grated cheese to 4 oz (100 g) or omit.

Pizzoccheri

PORTIONS: 12   **PORTION SIZE: 8 OZ (240 G)**

U.S.	METRIC	INGREDIENTS
1¹/₂ lb	700 g	Buckwheat pasta dough (p. 400)
4 oz	120 g	Butter
2 oz	60 g	Garlic, crushed
1 lb	450 g	Small, waxy potatoes, cut into ¹/₄-in. (6-mm) slices
2 lb	900 g	Swiss chard, stalks only, cut into 2-in. (5-cm) pieces
8 oz	240 g	Taleggio cheese, sliced into small pieces
4 oz	120 g	Parmesan cheese, grated

PROCEDURE

1. Prepare the pasta: Roll the buckwheat pasta dough into sheets slightly thicker than for fettuccine. Cut the sheets into strips 1 in. (2.5 cm) wide, then cut the strips diagonally into pieces about 3 in. (8 cm) long.
2. Prepare the garlic butter: Heat the butter in a small saucepan and add the garlic. Cook until the garlic is golden brown, then strain the butter and discard the garlic.
3. Drop the potatoes and chard into a large pot of boiling salted water. Cook at a slow boil just until the potatoes are tender.
4. When the potatoes are cooked, drop the buckwheat noodles into the water with the potatoes. Boil until the pasta is just cooked but al dente. Drain immediately in a colander.
5. Transfer the mixture to a large, buttered gratin dish or to several small gratin dishes.
6. Pour the garlic butter over the mixture and toss gently to coat the noodles, potatoes, and chard.
7. Add the Taleggio and parmesan cheeses and mix in gently.
8. Bake at 400°F (200°C) until the mixture is hot and bubbling and the top is lightly browned, about 10 minutes.

Per serving: Calories, 380; Protein, 18 g; Fat, 18 g (42% cal.); Cholesterol, 135 mg; Carbohydrates, 38 g; Fiber, 4 g; Sodium, 610 mg.

Pizzoccheri

Whole Wheat Maltagliate with Porcini and Peas

PORTIONS: 8   **PORTION SIZE: 8 OZ (240 G)**

U.S.	METRIC	INGREDIENTS	PROCEDURE
1¹/₂ lb	700 g	Whole wheat pasta dough (p. 400)	1. Prepare the pasta: Roll out the dough into sheets as for making fettuccine. By hand, cut the pasta into triangles about 3 in. (8 cm) across. (Maltagliate means "badly cut," referring to the irregular shapes of the pasta.)
2 oz	60 g	Dried porcini mushrooms	2. Put the dried mushrooms in a bowl and cover them with hot water. Let stand 30 minutes.
			3. Lift the mushrooms out of the water, squeezing them dry and letting the water run back into the bowl.
			4. Cut the mushrooms into ¹/₂-in. (1-cm) pieces.
			5. Strain the soaking liquid through a paper filter to remove sand or soil.
6 fl oz	180 mL	Olive oil	6. Heat the olive oil in a sauté pan. Add the garlic and cook until soft.
1 tbsp	15 mL	Finely chopped garlic	7. Add the mushrooms. Stir to coat with oil.
4 tbsp	60 mL	Chopped fresh parsley	8. Add the soaking liquid, parsley, and peas. Simmer until almost all the liquid has evaporated.
1 lb	450 g	Peas, fresh or frozen	
2 oz	60 g	Butter	9. In a separate pot, boil the pasta in salted water until al dente.
to taste	to taste	Black pepper	10. Drain and immediately add to the peas.
4 oz	120 g	Parmesan cheese, grated	11. Add the butter and pepper. Toss to coat the noodles.
			12. Add the grated cheese and toss to mix.
			13. Serve immediately.

Per serving: Calories, 600; Protein, 23 g; Fat, 36 g (53% cal.); Cholesterol, 160 mg; Carbohydrates, 50 g; Fiber, 7 g; Sodium, 360 mg.

Whole Wheat Maltagliate with Porcini and Peas

Pad Thai

PORTIONS: 2   PORTION SIZE: 8 OZ

U.S.	METRIC	INGREDIENTS	PROCEDURE
4 oz	240 g	Thai flat rice noodles	1. Soak the rice noodles in warm water for about 1 hour, or until they are soft but not mushy.
2 tsp	10 mL	Tamarind paste	2. Mix the tamarind paste, water, fish sauce, brown sugar, and chili sauce. Set aside.
2 tbsp	30 mL	Water	
1 tbsp	15 mL	Fish sauce (nam pla)	
1 tbsp	15 mL	Brown sugar	
1 tsp	5 mL	Thai roasted red chili paste (nam prik pow)	
1 tsp	5 mL	Oil	3. Heat the oil in a wok or large sauté pan over moderately high heat.
1	1	Egg, beaten	4. Add the egg. Stir and cook until the egg is scrambled, then remove it with a perforated spoon.
1 tbsp	15 mL	Oil	5. Add the second quantity of oil to the pan. Add the garlic and scallions and stir-fry just until they start to brown.
1 tsp	5 mL	Garlic, chopped fine	6. Add the tofu and stir-fry 30 seconds.
1	1	Scallions, sliced	7. Drain the noodles. Add the noodles and the tamarind mixture to the pan. Begin to stir-fry, mixing the noodles with the other ingredients. Continue to stir-fry until the noodles are just tender.
3 oz	90 g	Firm tofu, bâtonnet	
3 oz	90 g	Bean sprouts	
1 tbsp	15 mL	Cilantro, chopped	8. Add the bean sprouts and cilantro and mix with the noodles just long enough for the sprouts to heat through.
		Garnish:	9. Transfer the noodles to one or more large serving bowls or portion into individual bowls.
1 oz	120 g	Peanuts, coarsely chopped	
1	1	Scallions, sliced	10. Sprinkle with the peanuts, scallions, and cilantro. Arrange the lime wedges on the side.
1½ tsp	7 mL	Cilantro, chopped	
2	2	Lime wedges	

Per serving: Calories, 1,120; Protein, 38 g; Fat, 54 g (42% cal.); Cholesterol, 425 mg; Carbohydrates, 131 g; Fiber, 7 g; Sodium, 900 mg.

VARIATIONS

Vegan Pad Thai

Omit the egg and, if desired, increase the tofu. Substitute soy sauce for the fish sauce.

PAD THAI

In the West, pad thai is perhaps the most famous Thai dish and is on nearly every Thai restaurant menu. In Thailand, however, this dish of stir-fried noodles is ordinary luncheon food, served by street vendors and simple diners. There are thousands of versions of this dish, many of which incorporate ingredients not included in the version presented here, including both dried and fresh shrimp, salted turnip, banana flower, and Chinese chives.

Rice Sticks, Singapore Style

YIELD: 2 QT (2 L)   PORTIONS: 10   PORTION SIZE: 8 OZ (225 G)

U.S.	METRIC	INGREDIENTS	PROCEDURE
1 oz	30 g	Dried Chinese black mushrooms	1. Soak the dried mushrooms in warm water until soft. 2. Remove the mushrooms from the water and squeeze them dry. Cut off and discard the stems. 3. Cut the mushroom caps into julienne.
1 tbsp	15 mL	Soy sauce	4. Mix the soy sauce, water or stock, and curry powder.
4 fl oz	120 mL	Water or chicken stock	
5 tbsp	75 mL	Madras curry powder	
1 lb	450 g	Thin rice noodles (rice sticks)	5. Soak the noodles in warm, not hot, water until softened, about 20 minutes. Drain.
1 tbsp	15 mL	Vegetable oil	6. Heat the oil in a wok or large sauté pan. Add the egg and swirl it to cover the bottom of the pan in a thin layer. 7. As soon as the egg is set, remove it to a cutting board. Cut into thin shreds.
3	3	Eggs, beaten	
2 fl oz	60 mL	Vegetable oil	8. Heat the rest of the oil in the same wok or pan until very hot. 9. Add the scallions, garlic, ginger, and salt. Stir-fry 1 minute. 10. Add the shrimp and stir-fry until the shrimp is about half cooked. 11. Add the bean sprouts, peppers, mushrooms, and meat. Continue to stir-fry until the sprouts and peppers are cooked but still somewhat crisp. 12. Add the noodles and continue to stir-fry until the items are well mixed and hot. 13. Add the curry mixture. Quickly stir and toss the mixture to distribute it evenly. Continue to stir-fry until the liquid is absorbed. 14. Return the shredded omelet to the pan and toss to mix in. 15. Serve immediately.
4	4	Scallions, cut diagonally into thin shreds	
1 tbsp	15 mL	Finely chopped garlic	
1 tbsp	15 mL	Finely chopped fresh ginger root	
1 tsp	5 mL	Salt	
8 oz	225 g	Small shrimp, peeled and deveined	
8 oz	225 g	Mung bean sprouts	
4 oz	110 g	Red bell peppers, cut julienne	
4 oz	110 g	Cooked pork or chicken, cut julienne	

Per serving: Calories, 320; Protein, 13 g; Fat, 10 g (28% cal.); Cholesterol, 115 mg; Carbohydrates, 45 g; Fiber, 3 g; Sodium, 500 mg.

SINGAPORE NOODLES

The city-nation of Singapore sits at the tip of the Malay Peninsula, which juts into the Indian Ocean. Situated as it is between India and China, Singapore has always been an important stop on trade routes between East and West. Thus, it is not surprising that its cooking shows the influences of many regions, such as curry from India and rice noodles from South China.

That's a good story, but there isn't much evidence that the dish called Rice Stick Singapore Style, or Singapore Noodles, was invented in Singapore. Rather, it was probably invented in Chinese restaurants in Europe in the 1970s and quickly became popular around the world. Today, it is one of the most popular dishes in Chinese restaurants from Hong Kong to London.

Use a good brand of Madras curry powder to create the true flavor of this dish.

Rice Sticks, Singapore Style

DUMPLINGS

Dumplings are starch products made from soft doughs or batters and cooked by simmering or steaming. They are served as side dishes and in soups and stews. Many national cuisines have their own kinds of dumpling. (For Chinese filled dumplings or wontons, see p. 441.)

KEY POINTS TO REVIEW

- What are the major kinds and shapes of commercial pasta?

- What are the quality factors to look for in commercial pasta?

- How should pasta be cooked for à la carte service?

- What procedure should be used if pasta is to be cooked ahead in quantity?

- How are rice noodles prepared for cooking?

Spaetzle

PORTIONS: 15   PORTION SIZE: 4 OZ (125 G)

U.S.	METRIC	INGREDIENTS
6	6	Eggs
1¹/₂ cups	375 mL	Milk or water
1 tsp	5 mL	Salt
¹/₈ tsp	0.5 mL	Nutmeg
¹/₈ tsp	0.5 mL	White pepper
1 lb or more	450 g or more	Flour
as needed	as needed	Butter, for service

PROCEDURE

1. Beat the eggs in a bowl and add the milk or water, salt, nutmeg, and pepper.
2. Add the flour and beat until smooth. You should have a thick batter. If it is too thin, beat in a little more flour.
3. Let the batter stand 1 hour before cooking to relax the gluten.
4. Set a colander or perforated hotel pan (or a spaetzle machine, if available) over a large pot of boiling salted water (see **Figure 13.5**). The colander should be high enough so the steam doesn't cook the batter in the colander.
5. Place the batter in the colander and force it through the holes with a spoon or plastic scraper.
6. After the spaetzle float to the top of the water, let them simmer 1–2 minutes, then remove them with a skimmer. Cool quickly in cold water and drain well.
7. Cover and refrigerate until service.
8. Sauté portions to order in butter until hot. Serve immediately.

Per serving: Calories, 260; Protein, 7 g; Fat, 15 g (52% cal.); Cholesterol, 120 mg; Carbohydrates, 24 g; Fiber, 1 g; Sodium, 310 mg.

FIGURE 13.5 Making spaetzle.

(a) Force the batter through the holes of the perforated pan into simmering water.

(b) Remove the spatzle from the simmering water with a skimmer and drop into ice water.

Potato Dumplings

PORTIONS: 10   PORTION SIZE: 5 OZ (150 G)

U.S.	METRIC	INGREDIENTS	PROCEDURE
2½ lb	1.1 kg	Boiled potatoes, peeled, cold	1. Grate the potatoes into a mixing bowl. 2. Add the flour and salt and mix lightly until just combined.
12 oz	350 g	Flour	3. Add the eggs and mix well to form a stiff dough. Work in more flour if necessary.
2 tsp	10 mL	Salt	
2	2	Eggs	4. Divide the dough into 20 equal portions. Roll each piece into a ball. Refrigerate 1 hour. Dumplings may be made ahead up to this point.
4 oz	125 g	Butter	5. Heat the butter in a sauté pan and add the bread crumbs. Sauté for a few minutes, until the crumbs are toasted and brown. Set aside.
4 oz	125 g	Dry bread crumbs	6. Place the dumplings in a pot of boiling salted water. Stir so they rise to the top and don't stick to the bottom of the pan. Simmer 10 minutes. 7. Remove with a slotted spoon and place in a single layer in a hotel pan (or onto serving plates). 8. Top with the toasted buttered bread crumbs. Serve 2 pieces per order. (Dumplings may also be served with melted butter or pan gravy.)

Per serving: Calories, 360; Protein, 8 g; Fat, 11 g (28% cal); Cholesterol, 65 mg; Carbohydrates, 57 g; Fiber, 3 g; Sodium, 680 mg.

VARIATIONS

One or more of the following may be added to the dough: ¼ cup (60 mL) chopped parsley; 4 oz (125 g) diced bacon, cooked crisp; 2 oz (60 g) onion, chopped fine and sautéed in butter or bacon fat.

For the cork-shaped dumplings shown on p. 519, cut cylinders out of the cooked dumplings using a round cutter. Brown the ends in butter.

Potato Gnocchi with Tomato Sauce

PORTIONS: 16   PORTION SIZE: 4½ OZ (140G)

U.S.	METRIC	INGREDIENTS	PROCEDURE
4 lb	2 kg	All-purpose potatoes (see Note)	1. Wash the potatoes, but do not peel. Boil until tender. 2. Peel the potatoes while they are still hot, and force them though a food mill.
1 lb	500 g	Flour	3. Add about thrvee-fourths of the flour to the potatoes and knead to make a soft, sticky mixture. Continue to work in more flour to form a soft, smooth dough. It should still be somewhat sticky. You may not need all the flour. 4. Divide the dough into smaller pieces. Roll each piece into a sausage shape about ½ in. (1.25 cm) thick. Cut into pieces about ¾ in. (2 cm) long. 5. To shape the gnocchi, pick up one piece of the dough and press it with your fingertip against the times of a fork. Then flip the piece with the finger and allow it to drop on the worktable. This will give the piece grooves on one side and an indentation on the other side.
3 pt	1.5 L	Italian Tomato Sauce for Pasta (p. 399)	6. Drop the gnocchi into a large quantity of boiling salted water. When they float to the surface, let them boil 10–15 seconds, then remove with a skimmer or slotted spoon.
1½ cup	350 mL	Grated parmesan cheese	7. Plate the gnocchi. Top each portion with 2 fl oz (60 mL) tomato sauce and 1 tbsp (15 mL) grated parmesan cheese.

Per serving: Calories, 380; Protein, 10 g; Fat, 17 g (40% cal.); Cholesterol, 5 mg; Carbohydrates, 47 g; Fiber, 4 g; Sodium, 680 mg.

Note: All-purpose potatoes give the best results in this recipe. If you are using either very starchy potatoes, such as russets, or very waxy potatoes, add 2 beaten eggs to the mixture in step 3 to help the gnocchi hold together when cooking.

VARIATIONS

Gnocchi may be served with other pasta sauces, such as pesto, or simply with melted butter and grated cheese.

Potato Gnocchi with Tomato Sauce

ADDITIONAL RECIPES

These additional recipes may be found on your CulinarE-Companion recipe management program:

Baked Lasagne (Lasagne al Forno); Chickpeas in Spicy Tomato Sauce; Fettuccine with Chiles and Grilled Chicken; Pasta e Fagioli; Rigatoni or Penne with Sausage and Clams; Spaghetti Puttanesca; Vegetable Ravioli in Lemongrass Broth.

TERMS FOR REVIEW

legume	**germ**	**bulgur**	**rice noodles**
haricot bean	**parboiled or converted rice**	**pasta**	**couscous**
lentil	**Arborio rice**	**commercial dried pasta**	**al dente**
dal	**polenta**	**macaroni**	**dumplings**
endosperm	**hominy**	**semolina**	
bran		**egg pastas**	

QUESTIONS FOR DISCUSSION

1. Describe how to prepare dried beans, lentils, and peas for cooking.

2. What is the main difference between cooking dried kidney beans and dried lentils?

3. Describe the three basic methods for cooking grains.

4. Should rice be washed before cooking? always, sometimes, or never? Discuss.

5. Can wild rice and long-grain rice be cooked together to decrease the portion cost of wild rice? Explain.

6. What factors determine how much water is needed to cook rice?

7. Describe two ways in which rice noodles are cooked or prepared.

8. Describe the procedure for making cheese ravioli, starting with a freshly made piece of pasta dough.

14

Cooking Methods for Meat, Poultry, and Fish

All cooking involves one or more methods of applying heat to foods in order to change them in ways we can control. We learned the definitions of the basic cooking methods in Chapter 6, and we looked at these methods in more detail in the discussion of cooking vegetables in Chapter 11. But the changes we look for in meats, poultry, and fish are different from those we look for in vegetables, so it is helpful to take another look at these techniques.

We begin with the dry-heat methods, then the dry-heat methods with fat, and finally the moist-heat methods. Each section in this chapter includes a general discussion, guidelines on how to evaluate the quality of products cooked by that method, and a detailed procedure. Each procedure is illustrated by a recipe using that technique.

The procedures outlined in this chapter are general ones, since they are used for so many different products. Be aware that they might be modified slightly in specific recipes. In Chapters 16, 18, and 20, we discuss some of the most important of these variations. In this chapter, we study the cooking methods in their most basic form. Review the procedures in this chapter as necessary as you prepare recipes later in the book.

Information in Chapters 15, 17, and 19 includes guidelines on how to select appropriate cooking methods for various meat cuts and poultry and seafood items.

AFTER READING THIS CHAPTER, YOU SHOULD BE ABLE TO

1. Explain the procedure for roasting and baking meats, poultry, and seafood, and evaluate foods cooked by this method.

2. Explain the procedure for barbecuing meats, poultry, and seafood, and evaluate foods cooked by this method.

3. Explain the procedure for grilling and broiling meats, poultry, and seafood, and evaluate foods cooked by this method.

4. Explain the procedure for sautéing meats, poultry, and seafood, and evaluate foods cooked by this method.

5. Explain the procedure for pan-frying meats, poultry, and seafood, and evaluate foods cooked by this method.

6. Explain the procedure for deep-frying meats, poultry, and seafood, and evaluate foods cooked by this method.

7. Explain the procedure for simmering and submersion-poaching meats, poultry, and seafood, and evaluate foods cooked by this method.

8. Explain the procedure for shallow-poaching meats, poultry, and seafood, and evaluate foods cooked by this method.

9. Explain the procedure for steaming meats, poultry, and seafood, and evaluate foods cooked by this method.

10. Explain the procedure for braising meats, poultry, and seafood, and evaluate foods cooked by this method.

11. Explain basic sous vide procedures, as they are applied to meats, poultry, and seafood.

ROASTING AND BAKING

To *roast* and to *bake* both mean to cook foods by surrounding them with hot, dry air, usually in an oven. Cooking on a spit in front of an open fire may also be considered roasting.

The term *roasting* usually applies to meats and poultry. The term *baking* usually applies to breads, pastries, vegetables, and fish. It is a more general term than roasting, although, in practice, there is little or no difference in actual technique, and the terms are often interchangeable (except for breads and pastries).

1. Cooking *uncovered* is essential to roasting. Covering holds in steam, changing the process from dry-heat to moist-heat cooking, such as braising or steaming.

2. Meat and poultry are usually roasted on a rack (or, in the case of rib roasts, on their own natural rack of bones). The rack prevents the meat from simmering in its own juices and fat. It also allows hot air to circulate around the product.

3. When roasting in a conventional oven, the cook should allow for uneven temperatures by occasionally changing the position of the product. The back of the oven is often hotter because heat is lost at the door.

4. A roast may be browned by another cooking method, such as pan-frying or broiling, before being placed in the oven. This technique is most useful for small poultry and small cuts of meat, which may not brown sufficiently in the oven due to their short roasting times.

5. *Rotisserie-roasting*, also called *spit-roasting*, is unlike oven-roasting in that radiant heat from electric or gas elements or an open fire, rather than hot air in a closed space, is the primary heat source. Nevertheless, spit roasting is considered the original roasting method. Before modern times, most roasts were cooked on a spit in front of an open fire.

   Modern rotisserie equipment is discussed on page 43. Poultry or large cuts of meat are fixed on spits, in front of the heating elements. Electric motors turn the spits so that the items roast evenly on all sides. Drip pans catch fat and juices that drop from the meat. The items may be basted with drippings or seasoned liquids as they roast to flavor them and give them a good color.

In principle, roasting meats, poultry, and seafood is a simple procedure. The prepared item is placed in an oven at a selected temperature, and it is removed when done. What could be easier?

However, there are many variables, and chefs often disagree about proper roasting procedures, especially when it comes to the fine points. The following are some of the major variations in the basic roasting procedure.

SEASONING

Salt added to the surface of an item just before roasting penetrates only a fraction of an inch during cooking. The same is true of the flavors of herbs, spices, and aromatics. In the case of most poultry (except large turkeys and capons) and smaller cuts of meat, such as beef tenderloin and rack of lamb, the seasoned, browned skin or crust that forms during roasting is an important part of the flavor of the finished dish. Although opinions vary, many chefs advocate seasoning such roasts immediately before roasting so the salt doesn't have time to draw moisture to the surface, which inhibits browning. Alternatively, season well in advance so the salt has time to penetrate into the roast. Brining (p. 149) is another option for seasoning some roasts.

In the case of large roasts, such as beef ribs and steamship rounds, there is so little crust in proportion to meat that seasoning before roasting has little effect. Also, if the surface of the roast is mostly fat covering or bone, the seasoned fat and bones may not even be served, so the seasoning has little effect.

With roasts of any size, two alternatives to seasoning just before roasting are often used:

- Marinate the product or apply seasonings in advance, to give the time for flavors to penetrate. See pages 148–149 for a discussion of marinades and dry seasoning rubs. See also the discussion of brining on page 149. Brining is most suitable for pork and poultry. It is not often used for red meats.

- Serve the product with a flavorful sauce, gravy, or jus. The sauce serves as a seasoning and flavoring.

Another way to add flavor to roasts is to smoke-roast them. Commercial smoker ovens roast meats in the same way as conventional ovens, except they also have a smoke-generating unit that passes smoke through the oven chamber, flavoring foods as they cook. The flavor of wood smoke in cooked meats is so popular that some restaurants have even installed wood-burning hearth ovens to bake and roast meats, pizzas, and other items. The use of smoke is discussed in more detail in the next section on barbecuing.

Stovetop smoke-roasting is an alternative to smoker ovens. The procedure is explained on page 539. Examples of smoke-roasted fish and poultry can be found on pages 545 and 620.

TEMPERATURE

Low-Temperature Roasting

As we discuss on page 472, it was once thought that starting the roast at a high temperature "seals the pores" by searing the surface, thus keeping in more juices.

We now know this is not the case. Repeated tests have shown that *continuous roasting at a low temperature* gives a superior product with four distinct advantages:

1. Less shrinkage
2. More flavor, juiciness, and tenderness
3. More even doneness from outside to inside
4. Greater ease in carving

Low roasting temperatures generally range from 250° to 325°F (120° to 165°C), depending on two factors:

1. *The size of the item*. The larger the cut, the lower the temperature. This ensures the outer portion is not overcooked before the inside is done.
2. *The operation's production schedule*. Lower temperatures require longer roasting times, which may or may not be convenient for a particular operation.

Searing

If a well-browned, crusted surface is desired for flavor or for appearance, as when the roast is to be carved in the dining room, a roast may be started at high temperature (400°–450°F/200°–230°C) until it is browned. The temperature should then be lowered to the desired roasting temperature and the meat roasted until done, as for low-temperature roasting.

Alternatively, small roasts that may not brown adequately by the time they are done may be seared in a pan on the stove before being roasted. This is especially useful for small, whole roasted poultry.

High-temperature Roasting

Very small items that are to be roasted rare may be cooked at a high temperature, from 375° to 450°F (190° to 230°C). The effect is similar to that of broiling: a well-browned, crusted exterior and a rare interior. The meat is in the oven for so short a time that shrinkage is minor. Examples of items that may be roasted at a high temperature are rack of lamb, beef tenderloin, squab, game birds, and breast of duck.

Whole duck roasted well done is often roasted at a high temperature, because its fat content gives some protection from drying. Great care must be taken to avoid overcooking, however.

Convection Ovens and Combi Ovens

If a convection oven is used for roasting, the temperature should be reduced about 50°F (25°–30°C). Many chefs prefer not to use convection ovens for large roasts because the drying effect of the forced air seems to cause greater shrinkage. On the other hand, convection ovens are effective in browning and are good for high-temperature roasting.

Combi ovens are capable of adding steam to the baking compartment, so it can be argued that cooking meat in combi mode is not true roasting because moist heat is used, at least for part of the cooking time. Nevertheless, combi ovens can give results similar to roasting, but in less time and with less weight loss and more moisture retention. See the variations following recipes on pages 418 and 490 for examples of adapting roasting recipes to the combi oven.

FAT SIDE UP OR FAT SIDE DOWN

Roasting meats fat side up provides continuous basting as the fat melts and runs down the sides. This method is preferred by perhaps the majority of chefs, although there is not complete agreement.

In this book, we use the fat-side-up method. In the classroom, you should be guided by the advice of your instructor.

BASTING

Basting is unnecessary if the meat has a natural fat covering and is roasted fat side up. For lean meats, barding has the same effect. **Barding** is covering the surface of the meat with a thin layer of fat, such as sliced pork fatback or bacon.

If a roast is basted by spooning pan drippings over it, use only the fat. Fat protects the surface of the roast from drying, while moisture washes away protective fat and allows drying. Juices used in basting will not soak into the meat.

Basting with drippings or juices may be used to increase the appetite appeal of the roast because it enhances browning. Gelatin and other solids dissolved in the juices are deposited on the surface of the meat, helping form a flavorful brown crust. This does not increase juiciness, however. Some cookbooks claim basting forms a waterproof coating that seals in juices, but this is not the case.

Basting sometimes produces more tender roasts for an unexpected reason: Frequent basting interrupts and slows the cooking. Every time the oven door is opened, the temperature in the oven drops considerably, so the roasting time is longer and more connective tissue breaks down. Thus, it is not the basting but the lower temperature that increases tenderness.

USE OF MIREPOIX

Mirepoix is often added during the last part of the roasting time to flavor the roast and to add extra flavor to the pan juices.

Many chefs feel, however, that mirepoix adds little flavor, if any, to the roast and that it is actually harmful because the moisture of the vegetables creates steam around the roast. Mirepoix can be more easily added when the gravy is being made. If no gravy or juice is to be served, mirepoix may not be needed at all.

The use of mirepoix is more important for poultry and white meats—veal and pork—which, because they are usually cooked well done, lose more juices and need a good gravy or jus to give them moistness and flavor.

GRAVY AND JUS

The general procedures for making pan gravy are given in Chapter 8 (p. 198). Read or review this section if necessary. The recipe for roast chicken in this chapter includes the procedure for making pan gravy. The procedure for making jus, given in Chapter 16 in the recipe for roast prime rib of beef au jus, is the same, except no roux or other thickening agents are used. In other words, use the methods for making pan gravy (p. 198), but eliminate steps 5 and 6 from Method 1 and step 3 from Method 2.

STANDARDS OF QUALITY FOR ROASTED ITEMS

1. The food should be cooked to the desired degree of doneness, as indicated by a thermometer inserted into the interior. Even foods cooked well done should have a moist interior, not overcooked and dry.

2. The surface of the food should be attractively browned. Red meats should be well browned. White meats should be a rich golden to golden-brown. Poultry should have a well-browned, crisp skin.

3. Gravy made from the pan drippings should have an attractive color, a full, rich flavor, and the correct texture. Unthickened jus should have body and not be watery, and thickened gravy should have a good nappé texture and not be thick and pasty.

BASIC PROCEDURE for Roasting

The following procedure applies primarily to meats and whole poultry. When fish and cut-up poultry items are baked, some variations in the procedure are applied. These variations are detailed in Chapters 18 and 20.

1. Collect all equipment and food supplies. Select roasting pans that have low sides (so moisture vapor does not collect around the roast) and that are just large enough to hold the roast. If pans are too large, drippings will spread out too thin and burn.

2. Prepare or trim the item for roasting. Heavy fat coverings on meat should be trimmed to about ½ inch (1 cm) thick.

3. If desired, season the item several hours ahead or the day before.

4. Preheat the oven to the desired temperature. Refer to the discussion of roasting temperatures above.

5. Optional step. Sear the meat under a broiler or in hot fat on the stove. This technique is useful for small items that may not brown enough during the roasting time.

6. Place the item on a rack in the roasting pan. In the case of meat cuts, the item should be fat side up. The rack holds the roast out of the drippings. Bones may be used if no rack is available. Bone-in rib roasts need no rack because the bones act as a natural rack.

7. Insert a meat thermometer (clean and sanitary) so the bulb is in the center of the meat, not touching bone or fat. (Omit this step if you are using an instant-read thermometer.)

8. Do not cover or add water to the pan. Roasting is a dry-heat cooking method.

9. Place the item in a preheated oven.

10. Roast to desired doneness, allowing for carryover cooking.

11. If desired, add mirepoix to the pan during the last half of the cooking period.

12. Remove the roast from the oven and let stand in a warm place 15 to 30 minutes. This allows the juices to be reabsorbed through the meat so less juice is lost when the meat is sliced. Also, resting the meat makes slicing easier.

13. If the roast must be held, place it in an oven or warmer set no higher than the desired internal temperature of the roast.

14. While the roast is resting, prepare jus or pan gravy from the drippings (following the procedure on p. 198). Mirepoix may be added to the drippings now if it was not added in step 8.

15. Slice the roast as close as possible to serving time. In almost all cases, slice the meat against the grain for tenderness.

16. Evaluate the finished product (see above).

Roast Chicken with Natural Gravy

PORTIONS: 8   PORTION SIZE: ¼ CHICKEN PLUS 2 FL OZ (60 ML) GRAVY

U.S.	METRIC	INGREDIENTS
2	2	Chickens, 3–3½ lb (1.4–1.6 kg) each
to taste	to taste	Salt
to taste	to taste	Pepper
as needed	as needed	Oil or butter
		Mirepoix:
2 oz	60 g	Onion, large dice
1 oz	30 g	Carrots, large dice
1 oz	30 g	Celery, large dice

PROCEDURE

1. Remove giblets from chickens. Check inside cavities to make sure they are well cleaned. Reserve giblets for other use.
2. Season the insides of the chickens with salt and pepper.
3. Truss the chickens (see **Figure 17.3**).
4. Rub the outside of the chickens with oil or butter (butter promotes faster browning). Season the skin with salt and pepper, as it will be served with the meat.
5. Place the mirepoix in a roasting pan. Place a rack over the mirepoix, and place the chickens breast down on the rack. (*Note*: If desired, the mirepoix may be added to the pan later in the roasting process.)
6. Place the chickens in an oven preheated to 450°F (230°C). After 15 minutes (not longer), turn the heat down to 325°F (165°C).
7. When the chickens have been in the oven 45–60 minutes, turn them breast side up. Baste with the fat in the roasting pan and finish roasting. Total cooking time is about 1½ hours.
8. Remove the chickens from the roasting pan and hold them in a warm place for service.

U.S.	METRIC	INGREDIENTS
1 qt	1 L	Strong chicken stock
⅔ oz (8 tsp)	20 g	Cornstarch or arrowroot
4 tsp	20 mL	Water or stock, cold
to taste	to taste	Salt
to taste	to taste	Pepper

9. Set the roasting pan on the range over high heat and cook until all the moisture has cooked off ("clarify the fat") and the mirepoix is well browned but not burned. Pour off the fat.
10. Add the strong stock to deglaze the pan. Boil until the gravy is reduced by about one-third. Degrease carefully. See **Figure 14.1**.
11. Stir the starch with the cold water or stock. Stir it into the gravy. Bring to a boil and simmer until thickened. The stock will reduce further.
12. Strain into a bain-marie, using a china cap lined with cheesecloth. Season carefully with salt and pepper.
13. Quarter the chickens, or carve them as shown in **Figure 14.2**. Serve a quarter-chicken with 2 oz (60 mL) gravy.

Per serving: Calories, 350; Protein, 45 g; Fat, 17 g (45% cal.); Cholesterol, 105 mg; Carbohydrates, 2 g; Fiber, 0 g; Sodium, 140 mg.

VARIATIONS

Roast Herbed Chicken

Place 3–4 parsley stems and a pinch each of tarragon and marjoram in the cavity of each bird.

After turning the chicken breast-up in the roasting pan, rub the skin with chopped parsley, tarragon, and marjoram.

Roast Chicken with Gravy

Save 1⅓ oz (40 g) of the fat skimmed from the juices and make a blond roux with the fat and 1⅓ oz (40 g) bread flour. Beat the roux into the juices and simmer until thickened.

Roast Chicken with Cream Gravy

Prepare as in basic recipe, but use only 1 pt (500 mL) chicken stock. Boil until reduced by one-third and strain (step 10). Add 10 fl oz (300 mL) hot milk and thicken with 2½ oz (75 g) blond roux. Finish with 1½ fl oz (45 mL) heavy cream.

PROCEDURE VARIATION: COMBI OVEN

Substitute the following steps for steps 5, 6, and 7 in the main recipe:

5. Place the mirepoix in a roasting pan. Place a rack over the mirepoix, and place the chickens breast up on the rack.
6. Place the chickens in a combi-oven preheated to 325°F (163°C) in combi mode.
7. After 20 minutes (not longer), turn the heat up to 385°F (195°C) and increase to maximum browning level (reduce humidity to 0 percent). Continue roasting until chicken is finished, basting with the fat in the roasting pan. Total cooking time is about 50 minutes.

Roast Chicken with Natural Gravy, Whipped Potatoes, Glazed Root Vegetables, Roasted Garlic, Peas

FIGURE 14.1 Preparing pan gravy.

(a) Remove the rack of cooked poultry from the roasting pan.

(b) If the mirepoix is not already browned, set it over a burner to brown it. Degrease the pan.

(c) Deglaze the pan with stock.

(d) Bring to a simmer and cook until reduced by about one-third. Stir in a starch slurry or a roux to thicken. Simmer until further reduced and the desired texture is reached.

(e) Strain the gravy.

FIGURE 14.2 Carving roast chicken.

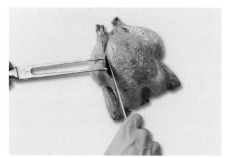

(a) Place the chicken on a clean, sanitary cutting board. Cut through the skin between the leg and the breast sections.

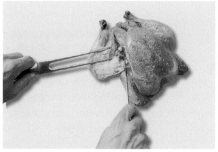

(b) Pull the leg away from the body of the chicken with the fork. Using the knife, cut between the thigh bone and the hip to separate the leg completely.

(c) Hold the chicken steady by bracing the backbone with the fork. Cut through the breast between the two halves, just to one side of the keel bone.

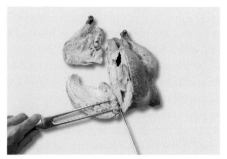

(d) Pull the breast section away from the bone. With the knife, cut through the joint where the wing bone is attached to the body. Separate the breast and wing section completely from the carcass. Repeat steps (a) through (d) on the other side of the chicken.

(e) Cut the wings from the breast portions and cut the drumsticks apart from the thighs.

REGIONAL BARBECUE STYLES

Variations of barbecue are popular worldwide, and barbecue styles—in terms of the foods cooked, seasonings, sauces, and even the woods for the fire—vary considerably. Within North America, a great range of styles exist. In the United States, main styles can be divided into four regions, but even within these regions styles vary. The heartland of American barbecue is in the southern states, from the Atlantic coast to Texas. Canada, similarly, enjoys many barbecue styles across the country, and even within the same province.

For many people across the continent, the barbecue styles of **Kansas City** are what they think of when they think of barbecue. Both pork and beef are widely used, cooked with a dry rub. The sauces are thick, tomato-based, sweet, and spicy. Most of the barbecue sauces sold in supermarkets are of this style. Sauces are most often served at the table as an accompaniment, not applied during cooking. The barbecue sauces of St. Louis, on the opposite side of the state of Missouri, are usually milder and not as thick. Pork ribs are especially popular in St. Louis.

Further south, **Memphis** barbecue features mostly pork shoulder and pork ribs. Ribs are cooked "dry"—that is, with a spice rub and no sauce—and "wet"—basted with a sauce during and after cooking. The sauce is a tomato-based sauce but thinner than Kansas City sauce. Pork shoulder is shredded and served as "pulled pork" in sandwiches and other dishes.

Texas barbecues vary considerably across this large region, but beef is the most popular meat. In East Texas, pork shares the menu with beef. Both are cooked with hickory wood, and both are served chopped, with sweet, spicy, tomato-based sauce. In Central Texas, meats are cooked with a dry rub, and they are sliced rather than chopped for service. Sauce may be served on the side, but the flavor of the meat itself is considered most important. West Texas barbecue cooks use mesquite wood, which gives the meat a stronger, smokier flavor, and the dry rubs and sauces are enlivened with chiles and other spices.

Carolina barbecue is the best known of the barbecues of the Southeast. Even in this smaller area, however, styles vary. According to some experts, there are as many as five distinct styles in North and South Carolina. Pork is the most popular meat, and in the eastern part of the region, whole hogs are barbecued. The meat, whether the whole hog or just the shoulder, is usually served pulled or chopped. Sauces are usually thin, vinegar-based mixtures, although in the western part of the region, tomato may also be added. In parts of South Carolina, mustard is added to the vinegar-based sauce.

Barbecue styles popular in **Canada** developed in part from the cooking of the native peoples and in part from European and American practices. As in the United States, styles vary greatly across the country. Of particular note is the whole salmon cooked on cedar planks in British Columbia.

BARBECUING

As indicated in Chapter 6, to **barbecue** means to cook with dry heat created by the burning of hardwood or by the hot coals of this wood. According to this definition, barbecuing is a roasting technique, and we could treat it as a variation of the roasting procedure we have just discussed. However, barbecuing has become such an important cooking method in North America that we devote a separate section to it.

Original barbecue was done in a hole or pit dug in the ground. After a wood fire was built in the hole, the foods were placed inside and the hole covered to hold in the smoke and heat. We still use the term *pit* for some kinds of barbecue equipment.

Most barbecue equipment falls into three categories:

1. *Indirect cooking with a wood fire.* The wood is burned in a separate chamber, and the smoke and heat flow through the compartment holding the food. Foods should be placed on racks or suspended in the oven so that smoke can contact all surfaces. In this type of barbecue, cooking temperatures are usually maintained at around 200°–250°F (93°–120°C). This is an ideal temperature range for cooking tougher cuts of meat, such as beef brisket and pork shoulder, to tenderness over a period of hours.

2. *Indirect cooking with electric or gas heat.* In this equipment, the heat source is generally in the same compartment as the food, but thermostatic controls enable the cook to keep the temperature at the same low range, 200°–250°F (93°–120°C). The equipment also contains a means of heating hardwood chips to provide smoke. This type of unit is the most versatile for commercial food service. Although it doesn't fit the strict definition of barbecue, because the heat is not provide by a wood fire, the presence of wood smoke provides almost identical results, and the heat is more controllable.

3. *Direct cooking with a wood fire.* The food is placed directly above the wood coals. This is similar to a wood- or charcoal-burning grill, except that the unit is covered to hold in the heat and smoke, making it more like an oven. Cooking temperatures are higher than in an indirect-cooking unit, usually around 300°–350°F (150°–175°C). The higher heat gives results closer to standard roasting than to traditional slow-cook barbecue.

In addition to these three categories of equipment, the combi oven also needs mention. As discussed on page 41, some combi ovens with a smoke option are available. The combination of wood smoke with the controlled heat and humidity of combi ovens can produce tender, slow-cooked items in shorter times than conventional barbecue equipment.

Even if you have a combi oven without a smoke option, consider this equipment as a useful alternative to barbecue. Items such as pork shoulder and ribs, coated with a rub and cooked to tenderness in a combi, can have nearly all the eating qualities of barbecue except for the smoke flavor. Perhaps include natural smoke flavoring as an ingredient in the rub or in the accompanying barbecue sauce.

The long, slow cooking of barbecue is ideally suited for tougher cuts of meat, such as pork shoulder and beef brisket, which can become meltingly tender during the extended roasting period. Tender items, such as poultry and seafood, can also be barbecued, although of course cooking times are shorter, and the items do not have as much time to absorb the flavor of the wood smoke. Seafood is not as often cooked by barbecuing. When it is, the direct heat method (the third option described) is most often used.

RUBS AND MOPS

There is more to a traditional barbecue than long, low-temperature roasting with wood smoke. Blends of seasonings and flavorings applied before, during, and after cooking account for much of the popularity of barbecue.

Before cooking, foods are often coated with a rub and allowed to stand for several hours or overnight to absorb the flavors. A rub is a blend of herbs and spices, plus salt and sometimes sugar. The blend may be a *dry rub*, made of dried herbs and spices, or a *wet rub*, made of dried spices plus ingredients,

like fresh garlic and chiles, that supply enough moisture to make a paste. An example of a dry rub can be found on page 422. For a wet rub, see the jerk recipe on page 554.

During cooking, the food is sometimes basted with a highly seasoned sauce called a mop. Mops usually contain vinegar or other acidic ingredients in addition to spices. Sugar is usually avoided in mops because it burns too easily.

After cooking, the food may be glazed with a barbecue sauce, or it may be served dry, with barbecue sauce on the side.

Seasonings and basting sauces are an important part of barbecue, as discussed in the Rubs and Mops sidebar. Regional traditions call for specific meats as well as different seasoning combinations and even different woods for producing smoke. More information on regional barbecue styles can be found in the sidebar. Various regional barbecue sauces can be found in Chapter 8.

STANDARDS OF QUALITY FOR BARBECUED ITEMS

1. Foods should be cooked to the correct degree of doneness. For most barbecued items, this means well done and extremely tender. Barbecued meats and poultry easily pull away from bone.

2. The interior of barbecued foods should have a moist texture and not be overcooked and dry.

3. The exterior of barbecued items should have an attractive, browned or reddish-brown color created by long cooking with smoke and by the spice rub (if used).

4. Meats and poultry cooked for extended periods with smoke have a detectable **smoke ring,** a red color that extends from to surface to about $1/4$ to $1/2$ inch (6–12 mm) into the interior, depending on how long the item was smoked (see photo).

5. Barbecued foods should have a noticeably smoky flavor.

FIGURE 14.3 Smoke ring on pork shoulder.

BASIC PROCEDURE for Barbecuing

1. Collect all equipment and food supplies.

2. Prepare or trim the item. Heavy fat coverings on meat should be trimmed to about ½ inch (1 cm) thick.

3. If using a rub, prepare the spice mixture. Coat the meat items evenly on all sides with the rub (**Figure 14.4**). Cover and refrigerate for 24 hours.

4. Before cooking, remove the items to be barbecued from the refrigerator and let stand at room temperature for about 1 hour.

5. Select the desired hardwood (choices include hickory, mesquite, pecan, and fruit tree woods such as apple and cherry) and prepare it according the instructions for your equipment.

6. Preheat the smoking equipment to the desired temperature (200°–250°F or 93°–120°C is a typical temperature range for low-temperature barbecuing).

7. Arrange the meats or other items on racks in the smoker. Allow enough space between items to permit the smoke to circulate well and contact all surface areas.

8. If a mop is used (see sidebar), baste the items with the liquid at intervals during the cooking period. Note that a mop is not the same as a barbecue sauce. Unlike barbecue sauces, mops contain no sugar, which would burn too easily.

9. Barbecue until the items are tender. For items containing bones, the bones should easily pull free from the meat. Total time can range from a few hours for tender items to more than 10 or 12 hours for whole pork shoulder or beef brisket cooked at the low end of the temperature range.

10. Barbecue sauces may be used to glaze items such as spareribs and brisket. Apply the sauce to the items and return the meat to the smoker to dry out the sauce or, alternatively, place the ribs in a broiler or on a grill just long enough to lightly brown the sauce. Do not allow the sauce to burn.

11. Evaluate the finished product (see p. 421).

FIGURE 14.4 Applying a dry rub to a rack of spareribs.

 # Barbecued Spareribs

YIELD: 7.5 LB (3.6 KG)   PORTIONS: 10   PORTION SIZE: 12 OZ (360 G)

U.S.	METRIC	INGREDIENTS
9 lb	4.1 kg	Fresh pork spareribs
3 oz, or as needed	90 g, or as needed	Spice Rub I (below)
2¹/₂ pt (approx.)	1.2 L (approx.)	Barbecue Sauce (p. 213), Kansas City Barbecue Sauce (p. 214), or Chili Barbecue Sauce (p. 215)

PROCEDURE

1. Rub the ribs with the spice rub so they are completely coated with a thin layer.
2. Refrigerate overnight.
3. Preheat barbecue equipment to 200°F (93°C). Place the ribs on racks in the smoker with plenty of space between them.
4. Roast until the meat is tender and can be pulled away from the bones, approximately 4¹/₂ hours.
5. Remove the ribs from the smoker. Brush both sides of each rack with some of the barbecue sauce. Place on a grill or under a broiler just until the sauce is lightly browned. Turn the ribs over to brown the other side.
6. If desired, repeat step 5 with a second coating of barbecue sauce.
7. Serve in multirib portions, or cut into separate ribs for easier eating. Serve additional barbecue sauce in a small ramekin on the side.

Per serving: Calories, 730; Protein, 47 g; Fat, 54 g (68% cal.); Cholesterol, 190 mg; Carbohydrates, 11 g; Fiber, 2 g; Sodium, 1010 mg.

VARIATION

If barbecue equipment is not available, the ribs may be baked in a conventional oven at 300°F (150°C). Cooking time is approximately 2¹/₂ hours. In a combi oven, using 60 percent humidity, cooking time is about 1¹/₂ hours.

Barbecued Spareribs

Spice Rub I

YIELD: APPROX. 6 OZ (180 G)

U.S.	METRIC	INGREDIENTS
1¹/₂ oz	45 g	Paprika
1 oz	30 g	New Mexico chili powder
2 tbsp	30 mL	Dried oregano
2 tbsp	20 mL	Dried thyme
4 tsp	20 mL	Ground coriander
4 tsp	20 mL	Ground cumin
2 oz	60 g	Salt
2 tbsp	30 mL	Black pepper

PROCEDURE

1. Combine all ingredients and mix well.
2. Store in a tightly sealed container in a dark place.

Per 1 oz: Calories, 50; Protein, 2 g; Fat, 2 g (27% cal.); Cholesterol, 0 mg; Carbohydrates, 10 g; Fiber, 6 g; Sodium, 3720 mg.

KEY POINTS TO REVIEW

- What are the steps in the basic roasting procedure?
- What are the factors to consider when determining at what temperature to use when roasting?
- What standards can be used to evaluate the quality of roast items?
- What are the steps in the basic barbecuing procedure?
- What standards can be used to evaluate the quality of barbecued items?

BROILING AND GRILLING

To **broil** means to cook with radiant heat from above. To **grill** means to cook on an open grid over a heat source. Broiling and grilling are dry-heat cooking methods, which use very high heat to cook meat quickly. Properly broiled meats have a well-browned, flavorful crust on the outside, and the inside is cooked to the desired doneness and still juicy.

Broiling and grilling are especially well suited to tender meats cooked to the rare or medium-done stage. The intense heat browns the surface of the item quickly, and the item is removed from the heat before it is overcooked and dry.

Items cooked well done or medium-well—items such as pork, veal, poultry, and seafood—must be cooked using lower heat so that they have time to cook through before the surface becomes too charred. Be very careful not to let such foods overcook, because they can quickly dry out in the intense heat.

Many seafood items are too delicate to be placed directly on the grill bars. Put these items in hinged hand racks to protect them during cooking and keep them from breaking apart. To broil tender fish fillets, such as cod and sole, dip them in oil or melted butter, place them on heated sizzler platters, and place under the broiler. Cook without turning over.

TEMPERATURE CONTROL

The object of broiling is not just to cook the meat to the desired doneness but also to form a brown, flavorful, crusty surface.

The goal of the broiler cook is to create the right amount of browning—not too much or too little—by the time the inside is cooked to the desired doneness. The key to accomplishing this is to broil the item at the right temperature.

In general, the shorter the cooking time, the higher the temperature, or else the meat won't have time to brown. The longer the cooking time, the lower the temperature, or the meat will brown too much before the inside is done.

Cooking time depends on two factors:

1. The desired doneness
2. The thickness of the cut

For example, a well-done steak should be cooked at a lower heat than a rare one. A thin steak cooked rare must be broiled at a higher temperature than a thick one cooked rare.

To control the cooking temperatures on a grill, set different zones for different temperatures and grill meats in the appropriate area. Broilers can also be adjusted for different heat zones. In addition, the rack of a broiler is adjustable. Move it up (closer to the heat) for higher temperatures and down for lower temperatures.

SEASONING

As with roasting, chefs disagree on when to season. Some feel that meats should not be seasoned before broiling. This is because salt draws moisture to the surface and retards browning. Others feel that seasoning before broiling improves the taste of the meat because the seasonings become part of the brown crust rather than something sprinkled on afterward.

Generally, if you have a professional broiler that has been properly preheated, it is not difficult to brown meat that has been salted. Low-powered broilers such as those found in home kitchens, however, do not get as hot. In such cases, it is better to salt after broiling, not before.

One way around this problem is to serve the meat with a seasoned butter (p. 192). Another option is to marinate the meat in seasoned oil 30 minutes or more before broiling. Be sure to drain marinated meats well before placing them on the broiler.

STANDARDS OF QUALITY FOR BROILED AND GRILLED ITEMS

1. Foods should have a well-browned but not burned or excessively charred surface, showing noticeable grill marks.
2. The surface of the foods should look moist, not dried out.
3. Foods should be cooked to the correct degree of doneness. Even well-done items should not be overcooked and dried out or tough.
4. Food should have a crisp surface and moist interior.
5. Foods cooked on a grill should have a somewhat smoky, "chargrilled" flavor. This flavor is created by fats dripping onto the heat source and creating smoke. Foods from a wood-fired grill should have an even more noticeable smoky taste. Broiled foods, on the other hand, are cooked under, not over, the heat source, so lack this chargrilled flavor, but they should still have the flavor of well-browned foods.
6. If the food item was marinated or seasoned before grilling, the flavor of the seasonings should be in balance and not overpower the flavor of the meat, poultry, or seafood item.

BASIC PROCEDURE for Broiling and Grilling

In a broiler, the heat source is above the food. In a grill, the heat is below the food. Except for this difference, the basic procedure is the same for both.

Make sure you understand how to test broiled meats for doneness (p. 473-475) before starting.

1. Collect and prepare all equipment and food supplies. Trim excess fat from meats to avoid flare-ups that can char the meat too much and coat it with smoky residue. If necessary, score the fatty edges of thin cuts of meat to prevent curling. Meat, poultry, and some fish items may be marinated in advance to flavor them.
2. Preheat the broiler or grill.
3. Thoroughly clean the grill bars with a stiff brush. The equipment may be cleaned before preheating, but preheating burns away much of the encrusted food, and the remaining ash can then be brushed off.
4. Season the grill bars by brushing them lightly with oil. This can be done by rubbing them with a rolled cloth dipped in oil, using tongs to hold the cloth. Keep the grill brush and oiled cloth handy to repeat the cleaning and oiling process as necessary.
5. Brush the food with oil and season it, or, if it has been marinated, remove it from the marinade and let the excess drip off. (The oil helps prevent sticking and keeps the product moist. This step may be unnecessary for meats high in fat. Using too much oil can cause grease fires.)
6. Place the item on the broiler or grill. Cook until the item is about one-fourth done. To grill-mark the item, pick it up with tongs and rotate it about 90 degrees to give the surface a cross-hatch pattern (see Figure 14.5).
7. When about half done, turn the item over with tongs. Cook the second side until the item is cooked to the desired doneness.
8. Remove from the broiler or grill. Evaluate the finished product (see above).
9. Serve immediately.

FIGURE 14.5 Grill-marking steaks.

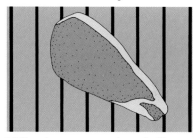

(a) Place the meat on a preheated grill at an angle as shown.

(b) When the meat is about one-fourth done, turn the meat 60–90 degrees, as shown. Do not turn it over.

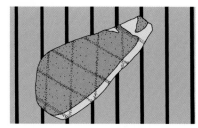

(c) When the steak is about half done, turn it over. The grill marks will appear as shown.

Broiled Strip Loin Steak Maître d'Hôtel

YIELD: 4 PORTION

U.S.	METRIC	INGREDIENTS	PROCEDURE
4	4	Strip loin steaks, boneless, 10–12 oz (300–350 g) each	1. Trim steaks as necessary, leaving a thin layer of fat on the edge.
to taste	to taste	Salt	2. Season the steaks to taste with salt and pepper. Brush lightly with oil.
to taste	to taste	Black pepper	3. Place the steaks on a preheated grill or broiler.
as needed	as needed	Vegetable oil	4. When the steak is about one-fourth done, turn it 60–90 degrees to grill-mark it (see **Figure 14.5**).
			5. When the steak is half done, turn it over and complete the cooking to the desired doneness. If turning with a fork, pierce the fat, not the meat, or juices will be lost.
2 oz	60 g	Maître d'Hôtel Butter (p. 192)	6. Remove the steaks from the broiler and immediately place on a hot plate. Top each steak with a ¹/₂-oz (15-g) slice of seasoned butter.

Per 8 ounces: Calories, 550; Protein, 60 g; Fat, 33 g (55% cal.); Cholesterol, 185 mg; Carbohydrates, 0 g; Fiber, 0 g; Sodium, 250 mg.

VARIATIONS

Other steaks may be cooked by the same method, including rib steak, rib-eye steak, tenderloin, club steak, porterhouse, T-bone, and sirloin. Chopped beef patties may also be prepared using this recipe.

Other seasoned butters and sauces make good accompaniments to broiled steaks, including:

Garlic Butter	Chasseur Sauce	Mushroom Sauce (brown)
Béarnaise Sauce	Anchovy Butter	Madeira Sauce
Bercy Sauce (brown)	Foyot Sauce	Bordelaise Sauce
Périgueux Sauce	Lyonnaise Sauce	Choron Sauce
Marchand de Vin Sauce		

Broiled Lamb Chops

Prepare as for broiled steaks, using rib, loin, or shoulder chops.

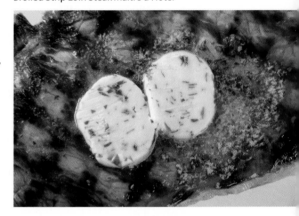

Broiled Strip Loin Steak Maître d'Hôtel

SAUTÉING

To **sauté** is to cook quickly in a small amount of fat. If you review the general definitions of sautéing and pan-frying in Chapter 6, you will see the differences between these methods are largely a matter of degree. Sautéing uses high heat and a small amount of fat and is usually used for small pieces of food. Pan-frying uses moderate heat, a moderate amount of fat, and is usually employed for larger items, such as chops. But at what point does moderate heat become high heat and a small amount of fat become a moderate amount of fat? It is impossible to draw an exact dividing line between sautéing and pan-frying.

Nevertheless, there are important differences between these two basic cooking methods, and so we devote separate discussions to them. For the purposes of our discussion, consider the following important differences between sautéing and pan-frying:

- Pan-fried foods are generally larger and/or thicker than sautéed food. This means they need lower heat and a longer cooking time than sautéed foods in order to achieve doneness with the proper amount of browning.

- Sautéed foods are often finished with a sauce made by deglazing the sauté pan. Pan-fried foods are not.

- Pan-fried foods are often breaded or given another coating before cooking. Sautéed foods may be dredged in flour, but they are not breaded.

Each time you cook a piece of meat, poultry, or seafood, you must judge how much heat and how much fat to use to do the job best. This depends on the kind of food and the size of the pieces.

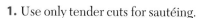

GUIDELINES for Sautéing Meats, Poultry, and Seafood

1. Use only tender cuts for sautéing.

2. Smaller or thinner pieces of meat require higher heat. The object is to brown or sear the meat in the time it takes to cook it to the desired doneness. Very small or thin pieces cook in just a few moments.

3. If large or thick items are browned over high heat, it may be necessary to finish them at lower heat or in an oven (uncovered) to avoid burning them.

4. Do not overload the pan, and do not turn or toss the food more than necessary. This will cause the temperature to drop too much, and the items will simmer in their juices rather than sauté.

5. Use clarified butter or oil or a mixture of the two for sautéing. Whole butter burns easily.

6. Dredging foods in flour promotes even browning and helps prevent sticking. Flour the items immediately before cooking, not in advance, or the flour will get pasty. Also, shake off excess flour before adding meat to the pan.

DEGLAZING THE PAN

A sauce made by deglazing the pan often accompanies sautéed meats. To **deglaze** means to swirl a liquid in a sauté or other pan to dissolve cooked particles of food remaining on the bottom. (Review discussions of deglazing in Chapter 8, p. 177 and p. 180.) The deglazing liquid can be used to flavor a sauce in one of two ways:

1. *Combine the reduced deglazing liquid with a prepared sauce.* The deglazing liquid adds flavor and color to the sauce.

2. *Use the deglazing liquid to make a freshly prepared sauce.* Add stock or other liquids and other flavoring and thickening ingredients and finish the sauce as indicated in the recipe.

The procedure for making a sauce by deglazing is outlined in the general sautéing procedure on p. 428.

STANDARDS OF QUALITY FOR SAUTÉED ITEMS

1. The item should be cooked to the desired degree of doneness. Items cooked well done should still be juicy, not overcooked and dry.

2. The surface of the food should be evenly browned, with no burned spots and no pale, unbrowned spots.

3. The item should retain its correct, natural shape and not be curled due to uneven cooking. It should appear plump and juicy, not shrunken.

4. The flavor of the seasonings should be balanced so that they enhance the flavor of the main item and not overpower it. The amount of salt added should be just enough, not too much.

5. The caramelized flavor of seared or browned foods should be well developed, but there should be no scorched or burned flavors.

6. All added ingredients should be cooked properly so that there are no raw flavors.

7. The sauce should be well seasoned and flavorful.

8. The sauce should be properly thickened (by added thickeners, by reduction, or both), neither too thin nor too thick.

9. There should be enough sauce so that each bite of the main item can be moistened with a little of the sauce.

10. The sauce should have an attractive and appropriate color.

11. Main item, garnish, and sauce should be served hot on a clean, hot plate.

12. All elements, including sauce, should be arranged on the plate in a neat, attractive fashion. The main item should be most prominent and not covered by other ingredients.

PRESENTATION SIDE

The side of a piece of meat, poultry, or fish that is browned first generally browns more attractively than the side that is browned second. This is because the juices that are forced upward to the surface while the first side is cooking tend to mar the appearance when the item is turned over and the second side is browned. Therefore, the most attractive side of a food item, known as the *presentation side*, should be browned first. For poultry pieces, this is usually the skin side. For fish fillets, the presentation side is usually the side that was against the backbone, not the side that had the skin on it.

After sautéing, the item is plated presentation side up.

BASIC PROCEDURE for Sautéing

1. Collect all equipment and food supplies. Select a pan large enough so that the food items can be placed in the pan in a single layer and not overlap.

2. Prepare food items as required. Meats, poultry and fish to be sautéed should be seasoned before cooking. If desired, they may also be dredged in flour. This should be done only at the last minute before cooking

3. Place the sauté pan over high heat until the pan is hot.

4. Add just enough fat to the hot pan to cover the bottom with a thin coating.

5. Add the meat to the pan, presentation side down (see sidebar). Do not overcrowd the pan.

6. For flat pieces of food, allow the items to sauté undisturbed until the first side is browned. Then turn over with tongs or spatula to brown the second side (**Figure 14.6**). For small pieces of food (such as thin strips of meat or poultry or small items like shrimp), allow to cook undisturbed until the heat recovers and the bottoms of the foods are browned. Then toss or flip the pan to turn over the items. Toss only as often as necessary to brown all sides evenly.

7. Remove the items from the pan and keep warm. If necessary, transfer the food to the oven to finish cooking.

8. Drain excess fat, if any, from the sauté pan.

9. Optional step: Add any sauce ingredients to be sautéed, such as shallots or mushrooms, as indicated in the recipe. Sauté them as necessary.

10. Add liquid for deglazing, such as wine or stock. Simmer while swirling and scraping the pan to release food particles on the bottom so they can dissolve in the liquid. Reduce the liquid.

11. Add a prepared sauce or other sauce ingredients, and finish the sauce as indicated in the recipe.

12. Serve the meat with the sauce, or return the meat to the sauce in the pan to reheat briefly and coat it with the sauce. Do not let the meat cook in the sauce.

13. Plate the hot item with hot sauce on a hot plate. If any sauce drips onto the rim of the plate, wipe it clean with a clean, hot cloth dampened with hot water and rung out.

14. Evaluate the finished product (see p. 427).

FIGURE 14.6 Sautéing chicken.

(a) After browning the first side of the chicken, turn each piece over to brown the second side.

(b) To prepare the sauce, sauté the ingredients, such as shallots.

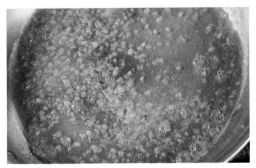

(c) Deglaze the pan by adding liquid as called for in the recipe.

(d) Finish the sauce as indicated in the recipe.

STIR-FRYING

The Chinese technique of stir-frying is very much like sautéing, except that in sautéing, the food items are often tossed by flipping the pan, while in stir-frying, the pan is left stationary and the foods are tossed with spatulas or other tools. Although true Chinese stir-frying is done in a round-bottomed pan, called a **wok**, set over a special burner, you can use the same technique with a standard sauté pan.

A procedure for stir-frying can be found on page 502.

 # Sautéed Chicken with Tomatoes and Mushrooms

PORTIONS: 4   PORTION SIZE: 1/4 CHICKEN

U.S.	METRIC	INGREDIENT	PROCEDURE
1	1	Chickens, about 2³/₄ lb (1.3 kg) each	1. Cut the chicken into 8 pieces, as shown in **Figure 17.5**.
to taste	to taste	Salt	2. Season the chicken lightly with salt and pepper.
to taste	to taste	Pepper	3. Pour about ¹/₈ in. (3 mm) oil into a sauté pans over moderate heat.
as needed	as needed	Vegetable oil	4. Sauté the chicken pieces until they are cooked through. Regulate the heat so they brown well but do not scorch.
			5. Remove the chicken pieces from the sauté pans and keep them hot. (*Alternative method*: Brown the chicken pieces well in the sauté pans, transfer to a sheet pan or hotel pans, without crowding them, and finish cooking in a 375°F [190°C] oven.)
²/₃ oz (4 tsp)	20 g	Butter	6. Drain the oil from the sauté pans and discard. Add the butter to the pans over moderately high heat.
1¹/₂ oz	45 g	Shallots, chopped fine	7. Add the shallots, garlic, and mushrooms. Sauté until tender and lightly browned.
1	1	Garlic clove, chopped fine	8. Add the wine. Reduce by half.
¹/₂ lb	240 g	Mushrooms, sliced or quartered (see Note)	9. Stir in the tomatoes and demi-glace. Return the mixture to a boil.
3¹/₂ fl oz	100 mL	White wine	10. Stir in the chopped parsley. Season to taste with salt.
¹/₂ lb	240 g	Tomato concassé	11. Spoon the tomato mixture onto plates and top with the chicken. (If the chicken has cooled, it can be reheated in the tomato mixture, but do not let it cook further.)
4 fl oz	120 mL	Demi-glace	
2 tsp	10 mL	Chop ped parsley	
to taste	to taste	Salt	

Per serving: Calories, 470; Protein, 72 g; Fat, 28 g (53% cal.); Cholesterol, 140 mg; Carbohydrates, 9 g; Fiber, 2 g; Sodium, 370 mg.

Note: Use white button mushrooms or a mixture of mushrooms, such as shiitake, portobello, and cremini.

Sautéed Chicken with Tomatoes and Mushrooms

KEY POINTS TO REVIEW

- What are the steps in the basic broiling and grilling procedures?
- What standards can be used to evaluate the quality of broiled and grilled items?
- What are the steps in the basic sautéing procedure?
- What standards can be used to evaluate the quality of sautéed items?

PAN-FRYING

Pan-frying is the second of three dry heat cooking methods using fat that we are discussing in this chapter. We just explained the technique of sautéing, which uses just enough fat to coat the bottom of the pan. Later we discuss deep-frying, in which food is completely submerged in hot fat. In terms of the amount of fat used, pan-frying lies between these two methods. To **pan-fry** is to cook foods in an uncovered pan in a moderate amount of fat, generally enough to come half to two-thirds up the sides of the food.

Pan-fried items are generally larger and thicker than sautéed items. Therefore, they must be cooked over lower heat for a longer time, so that the surface doesn't become too browned before the items are cooked to proper doneness.

Foods to be pan-fried are typically breaded, dredged in flour, or, less frequently, dipped in batter. Review the explanation of the Standard Breading Procedure in Chapter 7 (pp. 150–152).

BASIC PROCEDURE for Pan-Frying

1. Collect all equipment and food supplies. Select a heavy pan, such as a cast-iron skillet or heavy sautoir that is large enough to hold the food items in one layer without touching.
2. If foods are to be breaded, set up a breading station as explained on p. 151.
3. Prepare meats, poultry, or seafood items as required. Trim off fat and silverskin from meats. Preparation may include breading, dredging with flour, or coating in batter. (Breading may be done in advance, but do not dredge or batter until just before cooking.)
4. Heat a moderate amount of fat in a sauté pan or skillet until hot. The fat should be about 350°F (175°C), or hot enough to start to brown a piece of bread in about 45 seconds.

FIGURE 14.7 Pan-frying breaded veal cutlets.

5. Add the food items to the pan. Be careful when placing the food in the pan to avoid splashing hot fat. Do not overcrowd the pan (**Figure 14.7**).
6. Brown the items on one side. Turn with tongs or a slotted spatula and brown the other side. Larger pieces may need to be finished at reduced heat after browning. If required, they may be removed from the fat, placed on sheet pans or sizzler platters, and finish cooking, uncovered, in the oven.
7. Drain the items on clean paper or cloth towels to absorb fat. Evaluate the finished product (see p. 431).
8. Serve immediately.

STANDARDS OF QUALITY FOR PAN-FRIED ITEMS

1. Breaded and battered foods should have a crust that is uniformly golden or golden-brown and crisp. Foods dredged in flour should also have an attractively browned surface. Degree of browning depends on cooking time. Thin items cooked for a short time will not be as browned as thicker items.

2. Foods should be cooked to the correct degree of doneness. The interior should be juicy, not overcooked and dry.

3. Foods should be well drained of frying oil and should not be greasy or heavy.

4. Foods should be served hot.

5. Any sauce or garnish served with the item should enhance it both in terms of flavor and color. Sauce should not cover or destroy the crisp surface of the fried item.

 # Breaded Veal Cutlets

PORTIONS: 8   PORTION SIZE: 4 OZ (125 G)

U.S.	METRIC	INGREDIENTS
2 lb	1 kg	Veal cutlets (scaloppine): 24 pieces, 4 oz (125 g) each (See **Figure 15.15** for preparation of veal.)
to taste	to taste	Salt
to taste	to taste	Pepper
		Standard Breading Procedure (see Note):
2 oz	60 g	Flour
2	2	Eggs
4 fl oz	125 mL	Milk
12 oz	375 g	Bread crumbs, dry or fresh
as needed	as needed	Oil or clarified butter, or a mixture of oil and clarified butter
4 oz	125 g	Butter

Per serving: Calories, 550; Protein, 31 g; Fat, 38 g (63% cal.); Cholesterol, 165 mg; Carbohydrates, 19 g; Fiber, 1 g; Sodium, 380 mg.

Note: Quantities given for breading materials are only guidelines. You may need more or less, depending on the shapes of the meat pieces, the care used in breading, and other factors. In any case, you will need enough so even the last piece to be breaded can be coated easily and completely.

PROCEDURE

1. Lightly flatten each piece of veal with a meat mallet. Do not pound too hard, or you may tear the meat.

2. Season the meat with salt and pepper and pass through Standard Breading Procedure (see p. 151).

3. Heat about $1/4$ in. (5 mm) oil or butter in a large sauté pan or cast-iron skillet. Place the cutlets in the pan and pan-fry until golden brown. Turn and brown the other side. Remove from the pan and place on hot plates.

4. Heat the butter in a small saucepan or sauté pan until lightly browned. Pour $1/2$ oz (15 g) brown butter over each portion.

Veal Cutlet, Viennese-Style

VARIATIONS

Veal Cutlet Sauté Gruyère

Top each cooked cutlet with 1 or 2 thin slices of tomato and 1 slice of Gruyère cheese. Pass under a broiler to melt cheese. Serve with tomato sauce placed under the cutlet or in a ribbon (cordon) around the cutlet.

Schnitzel à la Holstein

Top each portion with a fried egg and 4 anchovy fillets placed around the edge of the egg.

Veal Cutlet Viennese-Style (Wiener Schnitzel)

Top each cutlet with 1 peeled lemon slice and 1 anchovy fillet rolled around a caper. Garnish the plate with chopped hard-cooked egg white, sieved egg yolk, and chopped parsley.

Veal Parmigiana

Top each cutlet with 2 fl oz (60 mL) tomato sauce, 1 slice of mozzarella cheese, and 2 tbsp (30 mL) parmesan cheese. Pass under a broiler to melt cheese.

Veal Cordon Bleu

Use 2 thin 2-oz (60-g) cutlets per portion. Sandwich 1 thin slice ham and 1 thin slice of Swiss cheese between 2 cutlets. Pound edges lightly to seal. Bread and fry as in basic recipe.

Breaded Pork Cutlets

Cutlets from pork leg or loin may be breaded and pan-fried like veal. They must be cooked well done.

DEEP-FRYING

To **deep-fry** means to cook a food submerged in hot fat. Commercial deep-fryers are equipped with thermostats to control the temperature of the fat, and with baskets to lower food into the fat and to remove and drain them when fried. Deep-frying can also be done in a pot on the stove-top. In this case, it is necessary to use an appropriate thermometer to ensure correct cooking temperature.

The guidelines and procedures in this section apply primarily to poultry and seafood items. Although it is possible to deep-fry meats, it is done less frequently than for poultry and fish. (Breaded veal and pork cutlets may be an exception, but they are more often pan-fried than deep-fried.) Deep-fried vegetable and potato items are discussed in Chapters 11 and 12.

Poultry and fish items are almost always breaded or dipped in batter before frying. This forms a protective coating between food and fat and helps give the product crispness, color, and flavor. Obviously, the quality of the breading or batter affects the quality of the finished product (see Chapter 7, pp. 150–152).

Breaded items may be placed in frying baskets and lowered into the hot fat. Battered items, by contrast, must be dropped by hand into the hot fat, because the raw batter would stick to the baskets. For example, to fry shrimp in batter, hold the shrimp by the tail, dip in batter, then lower into the hot fat, holding the tail for a few moments until a crust forms on the batter. If the item is dropped too soon, it will sink and stick to the bottom of the fryer.

GUIDELINES for Deep-Frying Seafood, Poultry, and Meat

1. Review the general deep-frying guidelines on page 339.
2. Select tender foods for deep-frying. Tougher items are not suitable for deep-frying, because of the short cooking times.
3. Fry at proper temperatures. Most poultry and seafood items are fried at 325°–375°F (165°–190°C). Excessive greasiness in fried foods is usually caused by frying at too low a temperature.
4. Season foods before applying a coating of breading or batter.
5. Breaded items may be placed in frying baskets and lowered into the hot fat (the *basket method*). Battered items, on the other hand, are fried without baskets, using the *swimming method.* The items must be dropped by hand into the hot fat, because the raw batter would stick to the baskets. Lower the battered item into the fat, holding it for a moment until a crust starts to form on the batter. If the item is dropped too soon, it will sink and stick to the bottom of the fryer.
6. Have clean, paper-lined pans at the fry station to receive foods that have just been fried, to drain off excess fat.
7. Fry poultry and seafood items in separate fryers if possible, so that fish flavors aren't transferred to other fried foods.

BASIC PROCEDURE for Deep-Frying

1. Collect all equipment and food supplies
2. Heat the fat to the proper temperature. If you are not using a thermostatically controlled deep-fryer, check the temperature of the fat with an appropriate thermometer.
3. Season the food items and coat with the desired coating (breading or batter).
4. Add the food to the hot fat, using either the basket method or the swimming method, whichever is appropriate.
5. If necessary, turn the food items as they cook so that they brown evenly on all sides.
6. Remove the food from the fat and place on clean paper to drain.
7. Evaluate the finished product (see below).
8. Serve immediately, while the food is hot.

STANDARDS OF QUALITY FOR DEEP-FRIED FOODS

High quality in a deep-fried product is characterized by the following properties:

Minimal fat absorption

Minimal moisture loss (that is, not overcooked)

Attractive golden color

Crisp surface or coating

No off-flavors imparted by the frying fat

 # Fish and Chips

PORTIONS: 12   PORTION SIZE: 8 OZ (240 G) FISH, 6 OZ (180 G) FRENCH FRIES

U.S.	METRIC	INGREDIENTS	PROCEDURE
		Batter (see Note):	1. Mix together the flour, salt, and pepper in a bowl.
7¹/₂ oz	225 g	Flour	2. Gradually beat in the beer to make a smooth batter. This step must be done just before cooking, or the batter will lose its carbonation. The carbonation from the beer provides leavening.
¹/₂ tsp	2 mL	Salt	
¹/₄ tsp	1 mL	White pepper	
12 fl oz	360 mL	Beer	
12	12	Cod fillets, 8 oz (240 g) each	3. Dredge the fillets in flour. Shake off excess.
as needed	as needed	Flour for dredging	4. One at a time, dip the fillets in the batter and then drop into a deep fryer heated to 350°F (175°C). Fry until golden.
4¹/₂ lb	2.2 kg	French Fries (p. 361)	5. Drain on absorbent paper.
24 fl oz	7200 mL	Rémoulade Sauce (p. 216) or Tartar Sauce (p. 216)	6. Serve with French fries on plates or in paper-lined baskets. Serve each portion with 2 fl oz (60 mL) rémoulade or tartar sauce in a cup.

Per serving: Calories, 1040; Protein, 46 g; Fat 72 g (62% cal.); Cholesterol, 130 mg; Carbohydrates, 50 g; Fiber, 3 g; Sodium, 970 mg.

Note: The quantity of batter needed varies greatly, depending on the fish. Thin fillets need more batter per unit of weight of fish because there is more surface area per unit of weight.

VARIATIONS

Substitute club soda for the beer in the batter.
Instead of the cod, use another lean, white fish, such as haddock, flounder, halibut, sole, or snapper.

Fisherman's Platter

Instead of using only one type of fish, use a variety of fish and seafood. For example: two kinds of fish fillets plus squid rings and shrimp.

Fish and Chips

 ## KEY POINTS TO REVIEW

- What are the steps in the basic pan-frying procedure?

- What standards can be used to evaluate the quality of pan-fried items?

- What are the steps in the basic deep-frying procedure?

- What standards can be used to evaluate the quality of deep-fried items?

SIMMERING AND SUBMERSION POACHING

To **simmer** and to **poach** both involve cooking foods submerged in water or in a seasoned or flavored liquid. The two methods are almost identical except for temperature. The two methods can be summarized as follows:

Simmering

- Cooking liquid is bubbling gently at a temperature of about 185° to 200°F (85° to 93°C).
- Used primarily to tenderize tougher meats and poultry by long, slow cooking. Also used to cook some shellfish items, such as lobster.

Poaching

- Cooking liquid is not bubbling but is barely moving, at a temperature of about 160° to 185°F (71 to 85°C).
- Used primarily for cooking seafood and some tender cuts of poultry and meat, to retain moisture and develop a light, subtle flavor.

A third cooking method in which foods are cooked submerged in a liquid is **boiling** at a temperature of 212°F (100°C). Boiling is a common cooking method for vegetables and starches but not for protein foods such as meat, poultry, and fish. The high temperature toughens the proteins, and the rapid bubbling breaks up delicate foods. Some traditional recipe names, such as "Boiled Beef," are misleading, because meat should not be boiled.

Much of the reason for the popularity of roasted, grilled, and fried meats and poultry is due to the kind of flavor produced by browning with dry heat. Simmered meats lack this flavor component. However, simmering is used effectively for less tender cuts for which browning is not desired or not appropriate. Popular examples of simmered meats are cured products such as ham and corned beef, fresh or cured tongue, fresh beef brisket, white stews such as veal blanquette, and flavorful mature poultry items such as stewing hen or fowl.

LIQUIDS AND FLAVORING INGREDIENTS FOR SIMMERED AND POACHED FOODS

Simmering Less Tender Meats and Poultry

- *For fresh meats and poultry*, use enough liquid to cover the meat completely, but don't use too much, as flavors will be diluted. Water is the main cooking liquid, but other liquids, such as wine, can be added to flavor the meat. Add herbs, spices, and a generous amount of mirepoix to give a good flavor to the meat.
- *For cured meats*, especially those that are heavily salted or smoked, use a generous amount of water to help draw excess salt or smoky flavor from the meat. In some cases, such as country hams, the water may even have to be changed during cooking to remove salt from the meat. Heavily seasoned cured meats, such as corned beef, are often simmered in pure, unseasoned water, but milder cured meats may be simmered with mirepoix and herbs. Do not add salt, however, because cured meats already contain a great deal of salt.

Poaching Tender Meats and Poultry

Use a flavorful stock appropriate to the item, such as chicken stock for fowl and beef stock for beef tenderloin. Add appropriate herbs, spices, and mirepoix or other aromatics.

Poaching Fish and Seafood

Fish is usually poached in court bouillon, fish stock or fumet, or wine. **Court bouillon** (koor bwee yohn) may be defined as water containing seasonings, herbs, and usually an acid, used for cooking fish. The name means "short broth" in French, so called because it is made quickly, unlike stocks.

Court bouillon is discussed in more detail in Chapter 20 (see p. 638).

STANDARDS OF QUALITY FOR SIMMERED AND POACHED ITEMS

1. Foods should be cooked to the correct degree of doneness:
   - Poached tender poultry items and whole fish should be fully cooked but still moist and tender, not overcooked or dry.
   - Poached meats should be cooked to the correct degree of doneness.
   - Tougher meat and poultry items should be cooked until fork-tender but still moist.

2. Flavoring ingredients in the poaching liquid should enhance the flavor of the main item, not cover it up or conflict with it.

3. After cooking, the item should be well drained. Poaching liquid should not be transferred to the plate.

4. Poached items should be served hot, on hot plates.

5. Sauces should enhance the flavor of the item, not cover it up. Most poached items have a fairly delicate flavor, so in most cases they should not be served with strongly flavored sauces.

BASIC PROCEDURES for Simmering and Poaching

1. Collect all equipment and food supplies.
2. Prepare meat, poultry, or seafood for cooking. This may include cutting, trimming, tying, or blanching.
3. Prepare the cooking liquid:
   - Review the discussion of cooking liquids above.
   - For fresh meats and poultry, start with boiling, seasoned liquid.
   - For cured and smoked meats, start with cold, unsalted liquid to help draw out some of the salt from the meats.
   - For seafood: Start shellfish, small fish, and portion cuts in hot liquid. Start large, whole fish in cold liquid to cook more evenly and to avoid sudden contractions that would split the skin and spoil the appearance.
   - In all cases, use enough liquid to cover the food completely.

FIGURE 14.8 Simmering beef brisket.

4. Place the meat in the cooking liquid and return (or bring) to a boil (**Figure 14.8**).
5. Reduce heat to the proper simmering or poaching temperature. Do not allow the liquid to boil after the food has been added. Skim the surface, if necessary.
6. Cook until the food is done, skimming as necessary.
   - Simmered meat and poultry are cooked until the meat is tender, skimming as necessary. To test for doneness, insert a kitchen fork into the meat. The meat is tender if the fork slides out easily. This is called fork-tender.
   - Poached tender poultry items should be fully cooked. Tender meats should be cooked to the desired degree of doneness. Both can be tested with an instant-read thermometer. Cooked poultry items should feel firm to the touch and the flesh should be opaque throughout.
   - Fish should feel firm to the touch and the flesh should be opaque. Crustacean shells should have turned bright red or pink. Clams, oysters, and mussels should be open.
7. If the food is to be served hot, remove from the liquid and serve immediately. If it is to be served cold or to be reheated later, cool it in its cooking liquid to retain moisture. Cool rapidly in a cold-water bath, as for stocks.
8. Evaluate the finished product (see above).

Simmered Fresh Beef Brisket

PORTIONS: 8   PORTION SIZE: 6 OZ (180 G)

U.S.	METRIC	INGREDIENTS
5 lb	2.4 kg	Fresh beef brisket, well trimmed
		Mirepoix:
4 oz	125 g	Onion, coarsely chopped
2 oz	60 g	Carrot, coarsely chopped
2 oz	60 g	Celery, coarsely chopped
1	1	Garlic cloves
1	1	Bay leaf
¹/₄ tsp	1 mL	Peppercorns
1	1	Whole cloves
6	6	Parsley stems
to taste	to taste	Salt

PROCEDURE

1. Place beef in a stockpot with enough boiling water to cover. Return the water to a boil, reduce heat to a simmer, and skim the scum carefully.
2. Add the mirepoix and seasonings.
3. Simmer until the meat is tender when tested with a fork.
4. Transfer the meat to a steam-table pan and add enough of the broth to barely cover (to keep the meat moist), or cool the meat with some of the broth in a cold-water bath and refrigerate.
5. To serve, cut the meat into thin slices across the grain. Slice at an angle to make the slices broader. Serve each portion with Horseradish Sauce (p. 184), prepared horseradish, or mustard and with boiled vegetables, such as carrots, potatoes, or turnips.
6. Strain the broth and save for soups or sauces. If desired, use some of the broth to make horseradish sauce to accompany the meat.

Per serving: Calories, 280; Protein, 35 g; Fat, 15 g (49% cal.); Cholesterol, 110 mg; Carbohydrates, 0 g; Fiber, 0 g; Sodium, 90 mg.

VARIATIONS

Beef tongue (fresh, cured, or smoked), beef shank, various cuts of beef chuck, beef short ribs, fresh or smoked ham, pork shoulder, and lamb shoulder or leg may be cooked using the same method.

Simmered Pork Shoulder with Cabbage

Cook fresh or smoked pork shoulder or pork butt as in basic recipe. Cut 1 lb 10 oz (780 g) cabbage (for 8 portions) into wedges and simmer in some of the pork broth. Serve each portion of meat with a cabbage wedge. For 8 portions, 6 oz (180 g) each, use about 7¹/₂ lb (3.6 kg) bone-in, skin-on shoulder.

Simmered Pork Shoulder, Braised Red Cabbage, Kasha Pilaf with Parsley, Roasted Onions

Simmered Beef Brisket with Vegetables, Horseradish Sauce, and Deep-Fried Julienne of Fresh Horseradish

SIMMERED STEWS

The term **stewing** means cooking small pieces of meat by simmering or braising (a composite method that includes both browning and simmering). Stews cooked by braising are covered in the next section. See also the discussion of stewing and braising on pages 442–444.

One difference between stews and many other simmered meats is that stews are served in a sauce or gravy made of the cooking liquid.

SHALLOW POACHING

The **shallow-poaching** method is used primarily for fish, but it is also sometimes used for chicken breasts. The procedure is based on one of the great dishes of classical cuisine: sole or other white fish poached in fumet and white wine. If well prepared, it can be one of the most exquisite dishes on the menu.

This method of preparation is best for lean, delicate white fish, such as sole, halibut, turbot, haddock, cod, pike, and perch. It is also used for salmon and trout. The fish is always served with a sauce made from the **cuisson**—that is, the poaching liquid.

The procedure and recipe given here are for fillets of sole au vin blanc (in white wine). This is the basic preparation, and most other classical poached fish recipes are variations on it. Many of the variations involve only different garnishes.

For a recipe that uses the shallow-poaching method to prepare chicken breasts, see p. 568.

Because of the delicacy of flavors, this preparation requires good-quality fish or poultry item and well-made stock. If wine is used, it should have a good flavor. A cheap, bad-tasting wine will spoil the dish.

The basic classical procedure is detailed in this section. Using this procedure as a pattern, the technique of shallow poaching can also be used for other fish preparations, substituting other liquids for the fumet, wine, velouté, and cream.

STANDARDS OF QUALITY FOR SHALLOW-POACHED ITEMS

1. Items should be cooked just to doneness and be moist and tender.
2. The sauce should have a light, nappé texture and not be heavy or pasty. Its color should be attractive and appropriate to the ingredients. Sauces based on velouté are typically ivory in color.
3. The sauce should have a refreshing flavor from the acids in the poaching liquid (usually wine) and complement the flavor of the food item.

BASIC PROCEDURE for Shallow Poaching

1. Collect all equipment and food supplies. Select a pan just large enough to hold the fish portions in a single layer. This will enable you to use a minimum amount of poaching liquid. Also, use a pan with low, sloping sides. This makes it easier to remove fragile cooked fish from the pan.
2. Butter the bottom of the pan and sprinkle with chopped shallots or other aromatics.
3. Arrange the portions of fish or poultry in the pan in a single layer. Season them lightly.
4. Add enough cooking liquid (usually stock and white wine) to cover the items no more than about halfway. Use no more liquid than necessary so the flavor will be more concentrated and less reduction will be required later. (Also, keep in mind that the food items will release additional liquid as they cook.)
5. Cover the items with a piece of buttered parchment or other paper and cover the pan with a lid. The paper holds in the steam to cook the top of the fish. It is sometimes omitted if the pan has a tight lid, but it does help the fish cook more evenly. (Some chefs like to cut a vent hole in the center of the parchment, but this is not necessary, because the steam can easily vent from around the edges.)
6. Bring the liquid just to a simmer and finish poaching in the oven at moderate heat. Thin fish fillets will cook in just a few minutes. Poaching may be done on top of the range, but the oven provides more even, gentle heat from both top and bottom.
7. Carefully remove the cooked items from the pan to a clean, hot dish and ladle a little of the cooking liquid over them to help keep them moist. Cover and keep warm.
8. Reduce the cuisson over high heat to about one-fourth its volume. (If more liquid drains from the cooked items during this time, drain and add it to the cuisson while it is reducing.)
9. Finish the sauce as indicated in the recipe. For example, add fish velouté and heavy cream and bring to a boil. Adjust seasoning with salt, white pepper, and lemon juice.
10. Strain the sauce.
11. Evaluate the finished product (see above).
12. Arrange the fish or poultry on plates for service, coat with the sauce, and serve immediately.

VARIATIONS in Sauce Production

Using a prepared velouté makes the sauce production very quick, and the procedure can easily be used for cooking to order.

An alternative method may be used if no velouté is available or if a large quantity of fish is being poached for banquet service:

1. Use a larger quantity of fumet and wine for cooking the fish and reduce it by only about half, depending on the amount of sauce needed.

2. Thicken the liquid with roux or beurre manié and simmer until no raw starch taste remains.

3. Finish the preparation as in the basic method.

Another popular method uses no starch thickener. Instead, the reduced poaching liquid is lightly bound with heavy cream or raw butter (monter au beurre).

• To bind the sauce with cream, add about 2 oz (60 mL) heavy cream per portion to the reduced cooking liquid and continue to reduce until the sauce is lightly thickened.

• To bind with butter, whip raw butter into the reduced cooking liquid as for monter au beurre (p. 177). Use about ¹⁄₂ oz (15 g) butter or more per portion. If you use a larger quantity of butter, you are essentially making beurre blanc (p. 191).

A further variation is known as **à la nage**, which means "swimming." To serve poached seafood à la nage, reduce the cooking liquid only slightly, season, and strain it carefully. If desired, enrich the liquid with a very small quantity of butter. Serve the seafood with the liquid in a soup plate or other plate deep enough to hold the juices.

The procedure of shallow poaching in wine and stock or fumet can be adapted to other preparations. Use other liquids as directed in the recipe. To finish the sauce, reduce the cuisson and, instead of finishing the sauce with velouté and cream, finish it as directed in the individual recipe.

Sole Vin Blanc (Poached Fillets of Sole in White Wine Sauce)

PORTIONS: 8   PORTION SIZE: 4 OZ (125 G) FISH, 2 FL OZ (60 ML) SAUCE

U.S.	METRIC	INGREDIENTS	PROCEDURE
16	16	Sole fillets, 2 oz (60 g) each	1. Fold the fillets in half or roll them up, starting with the large end (see **Figure 14.9**). Be sure the skin side of the fillet is on the inside of the fold or the roll (called paupiette).
1 oz	30 g	Butter	2. Butter the inside of a shallow pan or pans just large enough to hold the fish in a single layer. Sprinkle with the shallots. Lay the fillets on top of the shallots in a single layer.
1 oz	30 g	Shallots, fine dice	
3¹⁄₂ fl oz	100 mL	White wine	3. Pour the wine into the pan and add enough stock to cover the fish halfway.
7 fl oz or as needed	200 mL or as needed	Fish stock	4. Butter a piece of parchment or waxed paper cut the same size as the pan. Cover the fish closely with it, buttered side down (**Figure 14.10**). Cover the pan.
			5. Set the pan on the range and bring just barely to a simmer. Do not boil.
			6. Place the pan in an oven heated to 350°F (175°C) and cook the fish until it is just barely done, about 5 minutes.
			7. Remove the fish to a holding dish, moisten with a little of the cooking liquid, cover, and keep warm.
1 pt	500 mL	Fish Velouté (p. 182)	8. Reduce the poaching liquid over high heat to about one-fourth its volume. If more liquid collects under the fish as it stands, add this to the rest.
1	1	Egg yolks	9. Add the velouté and bring to a simmer. Reduce to about 10 fl oz (300 mL).
2¹⁄₂ fl oz	75 mL	Heavy cream	
4 tsp	20 g	Butter	10. Beat the egg yolk, then mix in the cream to make a liaison.
to taste	to taste	Salt	11. Whip in a little of the hot velouté to temper the liaison. Stir the tempered liaison into the sauce.
to taste	to taste	White pepper	12. Swirl in the raw butter (monter au beurre). Season to taste with salt, pepper, and a few drops of lemon juice if necessary.
as needed	as needed	Lemon juice	13. To serve, place 2 fillets on a dinner plate and coat with 2 fl oz (60 mL) sauce.

Per serving: Calories, 260; Protein, 25 g; Fat, 14 g (51% cal.); Cholesterol, 125 mg; Carbohydrates, 4 g; Fiber 0 g; Sodium, 190 mg.

Paupiettes of Sole Dugléré, Buttered Spinach, Steamed Potatoes

FIGURE 14.9 Rolling and folding sole fillets.

(a) To make paupiettes, or rolled fillets of sole, lay the fillets on the work surface skin side up. Starting at the large end, roll tightly. As the fish cooks, the connective tissues on the skin side shrink and keep the roll tight. If you roll the fillet skin side out, it unrolls as it cooks.

VARIATIONS

Any lean white fish may be poached using this recipe or any of the variations below. For example:

Halibut	Cod	Perch	Haddock
Turbot	Pike	Scallops	

For a sauce that is lower in fat, omit the liaison and the final monter au beurre. Finish the sauce with the desired amount of heavy cream, 6–12 fl oz (175–350 mL).

À la Carte Service, Fast Method: Poach fish as in the basic recipe. Plate the fish as soon as it is cooked. Coat it with a pre-prepared white wine sauce. The poaching liquid may be reused throughout the service period and then used to make velouté for the next day's sauce. Other sauces based on fish velouté or béchamel may be used instead of white sauce:

Bercy	Normandy	Nantua
Herb	Mushroom	Mornay

(b) If the fillets are to be poached flat or folded, first make a series of very shallow cuts in the skin side as shown. This helps keep the tissues from shrinking and deforming the fillet. Fold so the flesh side is out (skin side on the inside).

Glazed Poached Fish

Prepare as in the basic recipe. Immediately before serving, fold 1 pt (500 mL) hollandaise and 8 fl oz (250 mL) heavy cream, whipped, into the sauce. Coat the fish and brown it quickly under the salamander.

Poached Fish Bonne Femme

Add 1¹/₂ lb (700 g) sliced mushrooms to the shallots in the poaching pan. Poach as in the basic recipe. Omit egg yolks. Do not strain the sauce.

Poached Fish Dugléré

Add 1¹/₂ lb (400 g) tomato concassé (p. 295) and 4 tbsp (60 mL) chopped parsley to the pan when poaching the fish. Poach as in the basic recipe. Omit liaison, but stir 5 oz (150 g) raw butter into the sauce before serving. Do not strain.

Poached Fish Mornay

Strain the reduced poaching liquid (after step 8) and add a little of it to a thick Mornay Sauce for Glazing (p. 182), just enough to thin it to the desired consistency. Coat the fish with the sauce and brown under the salamander or broiler.

Note: Scallops are often prepared à la Mornay and served in scallop shells with a duchesse potato border. The dish is called Coquille St Jacques Mornay. (St Jacques is the French term for "scallop," and coquille means "shell.")

Poached Fish Florentine

Prepare like Poached Fish Mornay, but place the cooked fish on beds of cooked, buttered spinach before coating with sauce.

Sole Paupiettes with Seafood Mousseline

Before rolling, spread the fillets with a thin layer of seafood mousseline (p. 865).

Seafood à la Nage

Double the quantity of fish stock. Omit the velouté, liaison, and final butter. After poaching, reduce the cooking liquid only slightly. Season carefully and strain. Serve the seafood with the broth.

STEAMING

To **steam** is to cook foods by putting them in direct contact with steam. Steaming is a popular method for cooking vegetables, as discussed on pages 306–307. In Western cuisines, it is less often used for meats, poultry, and fish, although it is more common in some Asian cuisines (as in the Chinese recipe in this section). This cooking method is best suited to foods that are naturally tender. It is not usually appropriate for foods that must be tenderized by long, slow cooking.

Small items, such as small whole fish, shellfish, small chicken breasts, and small meat items such as Pearl Balls are the best choices for steaming. These cook in a short time in a steamer. Monitor the cooking time carefully so that they don't overcook.

Quantity steaming is most easily done in compartment steamers. Individual portions can be steamed on a rack over simmering liquid in a covered pot. This method gives you the opportunity to use liquids other than plain water to create steam. Adding spices, aromatics, wine, and similar ingredients can give a subtle flavoring to the steamed item. The effect is slight, however, so this technique is best used when steaming mild-flavored foods.

STANDARDS OF QUALITY FOR STEAMED ITEMS

1. Items should be cooked just to doneness and be moist and tender. Flesh appears opaque rather than translucent. Clam and oyster shells are open and the edges of the flesh are curled. Shrimp and lobster shells are pink or red.

2. Items appear plump, not shrunken, and the surface is free of coagulated albumen. Steaming too long can shrink and toughen protein items.

3. Items should have a good natural color. Short cooking times and lack of contact with cooking liquids help preserve color.

4. Flavors should be delicate. If any aromatics were used in steaming, their flavor should be subtle and not overpowering.

BASIC PROCEDURE for Steaming

1. Collect all equipment and food supplies.
2. Prepare the seafood, poultry, or meat items for cooking. Trim as needed.
3. If using a compartment steamer, preheat the unit. If using a rack over a boiling liquid, select a pan large enough to hold the items without crowding, so steam can circulate. Add the liquid and bring to a boil.
4. Place the food in the unit. Cover or close the steamer.
5. Steam to the correct doneness. Observe written cooking times, but check the items for doneness early, as overcooking can happen very quickly.
6. Remove the items from the steamer rack. Evaluate the finished product.
7. Serve hot, on hot plates, with appropriate sauce, garnish, and accompaniments.

 Pearl Balls

YIELD: ABOUT 20 PIECES

U.S.	METRIC	INGREDIENTS	PROCEDURE
1 cup	250 mL	Glutinous rice (also called sweet rice)	1. Wash the rice in several changes of cold water. Drain. Add enough fresh water to cover by 1 in. (2.5 cm). Let soak at least 30 minutes.
1¹/₂ tbsp	22 mL	Cornstarch	2. Mix together the cornstarch and water.
1 tbsp	15 mL	Water, cold	3. Combine all ingredients except the glutinous rice and mix together.
1 lb	450 g	Ground pork	4. Form the meat mixture into small meatballs, about 1 oz (30 g) each.
2	2	Scallions, minced	5. Drain the rice. Roll the balls in the rice so they are well coated.
4	4	Water chestnuts, minced	6. Line a rack or perforated steamer pan with cheesecloth. Arrange the meatballs in the pan, allowing about ¹/₂ in. (1–2 cm) between them.
1 tsp	5 mL	Minced fresh ginger root	
1	1	Eggs, beaten	7. Steam 30–45 minutes, or until the rice is translucent and the pork is done.
1 tbsp	15 mL	Soy sauce	
1 tbsp	15 mL	Sherry or Shaoxing wine	
1 tsp	5 mL	Sugar	
³/₄ tsp	3 mL	Salt	

Per 1 piece: Calories, 70; Protein, 5 g; Fat, 3.5 g (49% cal.); Cholesterol, 25 mg; Carbohydrates, 3 g; Fiber, 0 g; Sodium, 145 mg.

VARIATIONS

Fried Pork Balls

Omit the rice coating and cook the meatballs by deep-frying them.

Wontons

The pork mixture can be used for wonton filling. Put a small spoonful of meat in the center of a wonton skin. Moisten the edges of the skin with beaten egg, and then fold the skin in half to make a triangle (or, if you are using round wonton skins, a semicircle) enclosing the filling. Moisten one of the two corners (on the folded edge) with egg, then twist the wonton to bring the two corners together. Press the corners together to seal. Makes 60 or more wontons, depending on size. Wontons can be cooked by simmering, steaming, or deep-frying. They are often served in chicken broth as wonton soup.

Pearl Balls

BRAISING

Braising is a combination of dry-heat and moist-heat cooking methods. To **braise** meats and poultry, first brown or sear them in fat or in a hot oven, then simmer in a flavorful liquid until tender.

The popularity of properly braised items is due to the flavor imparted by the browning and by the sauce made from the braising liquid. Clearly, the quality of a braised meat depends greatly on the quality of the stock the meat is cooked in. Other liquids used in braising include wine, marinades, tomato products, and, occasionally, water.

Note that this cooking method applies primarily to meat and poultry. So-called **braised fish** is actually cooked by a variation of the baking or roasting method. This procedure is discussed on page 616.

BRAISES AND STEWS

- A **braise** consists of large pieces of food, at least portion-size, cooked by the braising method—first dry heat, then moist heat.
- A **stew** consists of small pieces of food, bite-sized or slightly larger, cooked either by the braising method—first dry heat, then moist heat—or by the simmering method—moist heat only.

TYPES OF BRAISED MEAT AND POULTRY DISHES

1. *Large cuts.* A large cut of meat braised whole, sliced, and served with a sauce or gravy is sometimes called a **pot roast**.

2. *Individual portion cuts.* Meats and poultry may be cut into portion sizes before braising instead of afterward. When portion cuts of beef round are braised in a brown sauce, the process is sometimes called *swissing*, and the product is called *Swiss steak*.

    Other braised portion-cut meats include short ribs, lamb shanks, and pork chops. Cut-up chicken is often cooked by braising.

3. *Stews.* Stews are made of meats and poultry cut into small pieces or cubes. Most stews are made by braising, but some are cooked by simmering only, without first browning or searing the meat.

    Stews are usually made with enough liquid or gravy to cover the meat completely while cooking. However, so-called dry stews are braised in their own juices or in a very little added liquid.

    Brown stews are made by browning the meat thoroughly before simmering. A **fricassée** (free kah say) is a white stew made by cooking white meat in fat over low heat without letting it brown, then adding liquid (see p. 575). Compare this to a **blanquette** (blawn ket), which is a white stew made by simmering the meat in stock without first cooking it in fat. The cooking method for blanquettes, therefore, is simmering rather than braising.

    Note: This use of the term *fricassée* is its traditional or classical usage. Today the word is used for many kinds of stews.

    Many other dishes can be classified as braised stews, even if we don't normally think of them that way. Chili, for example, is a braised dish made of finely cut or ground beef or pork. Even meat sauce for spaghetti (p. 399) is actually a braised meat or a stew.

    Many chefs prefer to use the term *braising* only for large cuts of meat, and they use the term *stewing* for small cuts (see sidebar). However, the basic cooking method—using first dry heat, then moist heat—is the same for both large and small cuts.

GUIDELINES for Braising Meats and Poultry

The basic principle of braising is a combination of searing or browning and then simmering. This process accomplishes two things: it cooks the meat, and it produces a sauce. (You will use some of your sauce-making techniques when you braise meats.)

Before giving basic procedures that apply to most popular braised meats, we discuss factors that affect the quality of the finished product.

1. Seasoning.

The meat may be seasoned before browning, or it may receive its seasonings from the cooking liquid while braising. But remember that salt on the surface of meat retards browning. Also, herbs may burn in the high heat necessary for browning.

Marinating the meat for several hours or even several days before browning is an effective way to season meat because the seasonings have time to penetrate it. The marinade is often included as part of the braising liquid.

2. Browning.

Dry the meat thoroughly before browning. Small pieces for stew may be dredged in flour for better browning. In general, red meats are well browned; white meats and poultry are browned less heavily, usually until they are golden.

3. **Amount of braising liquid.**

The amount of liquid to be added depends on the type of preparation and on the amount of sauce required for serving. Do not use more liquid than necessary, or the flavors will be less rich and less concentrated.

Pot roasts usually require about 2 oz (60 mL) sauce per portion, and this determines the amount of liquid needed. The size of the braising pot used should allow the liquid to cover the meat by one-third to two-thirds. The top portion of the meat not covered by liquid is cooked with the aid of the steam held in the covered pot.

Stews usually require enough liquid to cover the meat.

Some items are braised with no added liquid. They are browned, then covered, and the item cooks in its own moisture, which is trapped by the pan lid. Pork chops are frequently cooked in this way. If roasted, sautéed, or pan-fried items are covered during cooking, they become, in effect, braised items.

4. **Vegetable garnish.**

Vegetables to be served with the meat may be cooked along with the meat or cooked separately and added before service.

If the first method is used, the vegetables should be added just long enough before the end of cooking for them to be cooked through but not overcooked.

5. **Stovetop or oven.**

Braising may be done on the range or in the oven. Oven-braising has three major advantages:

- Uniform cooking. The heat strikes the braising pot on all sides, not just the bottom.
- Less attention required. Foods braise at a low, steady temperature without having to be checked constantly.
- Range space is free for other purposes.

6. **Adjusting the sauce.**

Braising liquids may be thickened by a roux either before cooking (Method 2) or after cooking (Method 1). In some preparations, the liquid is left unthickened or is naturally thick, such as tomato sauce.

In any case, the sauce may require further adjustment of its consistency by

- Reducing
- Thickening with roux, beurre manié, or other thickening agent
- The addition of a prepared sauce, such as demi-glace or velouté

STANDARDS OF QUALITY FOR BRAISED ITEMS

1. Braised items should be fork-tender. The interior should be juicy.

2. Although tender, braised foods should maintain their shape and not be falling apart.

3. Braised foods should have a well-developed, concentrated flavor due to long cooking in flavorful liquids. Thin flavor may be caused by insufficient cooking or cooking at the wrong temperature, so liquids didn't have time to reduce and concentrate.

4. Braised foods should have an attractive color developed from proper browning or searing at the beginning of the cooking process and from long cooking in concentrated liquids.

BASIC PROCEDURES **for Braising**

METHOD 1: BRAISING IN UNTHICKENED LIQUID

1. Collect all equipment and food supplies.

2. Cut or trim the meat as required. Dry it thoroughly. For stews, the meat may be dredged with flour.

3. Brown the meat thoroughly on all sides in a heavy pan with a small amount of fat, or in an oven.

4. Remove the meat from the pan and brown mirepoix in the fat left in the pan.

5. Return the meat to the pan and add the required amount of liquid.

6. Add a sachet or other seasonings and flavorings.

7. Bring the liquid to a simmer, cover the pot tightly, and simmer in the oven or on top of the range until the meat is tender.

   Oven braising provides more uniform heat. Temperatures of 250° to 300°F (120° to 150°C) are sufficient to maintain a simmer. Do not let boil.

8. Remove the meat from the pan and keep it warm.

9. Prepare a sauce or gravy from the braising liquid. This usually includes the following:

   • Skim fat.

   • Prepare a brown roux with this fat or with another fat if desired.

   • Thicken the braising liquid with the roux. Simmer until the roux is cooked thoroughly.

   • Strain and adjust seasonings.

10. Combine the meat (sliced or whole) with the sauce.

11. Evaluate the finished product (see p. 443).

METHOD 2: BRAISING IN THICKENED LIQUID

1. Collect all equipment and food supplies.

2. Prepare the meat for cooking, as required.

3. Brown the meat thoroughly in a heavy pan with fat or in a hot oven.

4. Remove the meat from the pan (if required) and brown mirepoix in remaining fat.

5. Add flour to make a roux. Brown the roux.

6. Add stock to make a thickened sauce. Add seasonings and flavorings.

7. Return the meat to the pan. Cover and simmer in the oven or on the range until the meat is tender.

8. Adjust the sauce as necessary (strain, season, reduce, dilute, etc.).

9. Evaluate the finished product (see p. 443).

FIGURE 14.11 Preparing braised meats (Method 2).

(a) Brown the meat well in a heavy pan.

(b) Remove the meat from the pan. Add the mirepoix to the pan and brown well.

(c) Add flour to the mirepoix. Stir to combine the flour with the fat and brown the roux.

(d) Add stock to the pan and whip to combine with the roux. Add seasonings and flavorings and return the meat to the pan. For portion cuts, there should be just enough liquid to cover the meat.

 Braised Short Ribs

PORTIONS: 12   PORTION SIZE: 10 OZ (300 G) MEAT WITH BONE, 2 FL OZ (60 ML) SAUCE

U.S.	METRIC	INGREDIENTS	PROCEDURE
12	12	Short rib sections, 10 oz (300 g) each	1. Dry the meat so it will brown more easily.
4 oz	125 mL	Oil	2. Heat the oil in a heavy brazier until very hot. Brown the meat well on both sides. Remove it and set aside.
5 oz	150 g	Onion, medium dice	3. Add the onion, celery, and carrot to the fat in the pan and sauté until browned.
2¹/₂ oz	75 g	Celery, medium dice	4. Stir in the bread flour to make a roux. Cook until the roux is browned.
2¹/₂ oz	75 g	Carrot, medium dice	5. Stir in the stock and tomato purée and simmer until the sauce thickens. Add the bay leaves and season to taste with salt and pepper.
2¹/₂ oz	75 g	Bread flour	6. Return the short ribs to the pan. Cover and braise in the oven at 300°F (150°C) until tender, about 2 hours.
2¹/₂ pt	1.25 L	Brown stock	7. Transfer the short ribs to a hotel pan for service.
2¹/₂ oz	75 mL	Tomato purée	8. Strain the sauce (optional). Degrease. Adjust the seasoning and consistency and pour over the short ribs.
1	1	Bay leaves	
to taste	to taste	Salt	
to taste	to taste	Pepper	

Per serving: Calories, 299; Protein, 29 g; Fat, 17 g (52% cal.);
Cholesterol, 55 mg; Carbohydrates, 6 g; Fiber, 0 g; Sodium, 70 mg.

VARIATIONS

Short ribs and the variations that follow may be braised in a prepared brown sauce or espagnole instead of a specially made sauce. Omit steps 4, 5, and 6.

If desired, use 8 oz (250 g) chopped canned tomatoes instead of the purée. Alternatively, use 1 oz (30 g) tomato paste; add the paste to the browned mirepoix and continue to brown until the paste turns a rusty brown color.

Swiss Steak

Instead of the short ribs, use beef round steaks, about 5–6 oz (150–180 g) per portion. Omit celery and carrot if desired.

Swiss Steaks in Tomato Sauce

Reduce flour to 1¹/₄ oz (38 g). For braising liquid, use 1 pt 4 fl oz (625 mL) brown stock, 1 lb 4 oz (625 g) chopped canned tomatoes with their juice, and 10 oz (300 g) tomato purée. Season with bay leaf, oregano, and basil. After removing cooked steaks, reduce sauce to desired consistency. Do not strain. Garnish each portion with chopped parsley.

Swiss Steaks with Sour Cream

Prepare as in the basic Swiss Steak recipe, above. When steaks are cooked, finish the sauce with 8 oz (250 g) sour cream, 1¹/₄ fl oz (37 mL) Worcestershire sauce, and 1 tbsp (15 g) prepared mustard.

Swiss Steaks in Red Wine Sauce

Prepare as in the basic Swiss Steak recipe, above, but add 8 fl oz (250 mL) dry red wine to the braising liquid.

Braised Oxtails

Allow 1 lb (500 g) oxtails per portion. Cut into sections at joints.

Braised Lamb Shanks

Allow 1 lamb shank per portion. Add chopped garlic to mirepoix if desired.

Braised Short Ribs with Steamed Broccoli and Mashed Potatoes

SOUS VIDE COOKING

Sous vide cooking, introduced on page 7, is such a new science that chefs are only beginning to explore its possibilities. Techniques will surely evolve and change in years to come.

Some of the main applications of sous vide cooking as it is practiced today are detailed in the sidebar. Please note temperature ranges are approximate. Immersion circulators (p. 48) are extremely accurate, and chefs may specify temperatures to within a fraction of a degree to get the precise results they want. Although not as precise as immersion circulators, combi ovens (p. 41) are also used for some sous vide cooking.

Tender Meats and Poultry

Tender meats and poultry are usually cooked in a water bath heated to the exact doneness temperature desired. The lamb loin described above is a typical example. Cooking temperatures usually range from 140° to 149°F (60°–65°C), although higher or lower temperatures are also used, and cooking times may range from 20 to 60 minutes, or sometimes longer.

Remember that if the cooking temperature is within the food danger zone (p. 17), cooking time must be counted as part of the four-hour rule (p. 22). For this reason, tender meats and poultry are almost always finished and served immediately after cooking and not chilled and stored for later use.

Tough Meats

Tough meats can be cooked to tenderness, while retaining more moisture than if they were braised or simmered. Cooking temperatures range from 149° to 158°F (65°–70°C). These temperatures are much lower than usual braising temperatures, but they are high enough to break down connective tissue. At the higher end of this range, meats become falling-apart tender, while at the lower end they retain more of their shape and have a firmer texture while still being tender and juicy.

Because it takes time to break down connective tissue, often 12 to 48 hours, tough meats are usually cooked to doneness, then immediately chilled and refrigerated for later use.

Because of the long cooking times, avoid cooking temperatures below 149°F (65°C). Higher temperatures provide more safety from bacterial growth.

Fish and Seafood

Fish are naturally tender and have even more delicate connective tissue than tender meats. Cooking procedures are similar to those for tender meats, but often with even lower temperatures and/or shorter cooking times. Typical cooking temperatures may range from 122° to 140°C (50°–60°C), or sometimes higher, with cooking times as short as 10 to 15 minutes. Use only the freshest, cleanest fish, and finish and serve it immediately after cooking.

Shellfish such as lobster and shrimp can be toughened by the higher heat of traditional cooking techniques, so sous vide cooking can be a benefit. Cooking temperature for these items is typically around 140°F (60°C).

TEMPERATURE PROBES FOR SOUS VIDE

To measure the interior temperature of an item cooked sous vide, it is necessary to insert a probe into the food without puncturing the bag in a way that would let air in. To do this, use a needle probe especially designed for sous vide, and stick a piece of *closed-cell foam tape* to the bag in a spot where you want to insert the probe. You can then safely insert the probe through the tape and bag and into the meat.

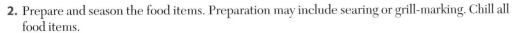

BASIC PROCEDURES **for Sous Vide Cooking**

1. Collect all equipment and food supplies.

2. Prepare and season the food items. Preparation may include searing or grill-marking. Chill all food items.

3. Place the food items in appropriate plastic bags designed for vacuum sealing. Add any flavoring ingredients indicated in the recipe. If more than one item is in a bag, space the items so that they are not touching.

4. Using a chamber-type vacuum machine, evacuate the air from the bag, using the correct pressure setting for that item, and seal the bag.

5. Chill the bag briefly.

6. Completely submerge the bag in a controlled water bath kept at the correct cooking temperature for that item (using an immersion circulator, **Figure 14.12**), and cook for the desired time or until the interior of the food reaches the correct temperature (see Temperature Probes for Sous Vide sidebar).

7. If the food is to be served as soon as it is cooked (typically for foods with short cooking times), remove the items from the bag, finish as directed in the recipe, and serve immediately. If the item is to be stored for later use, proceed to step 8.

8. For items to be stored, remove the bag from the water bath and immediately place it in an ice bath to chill it quickly (the ice bath should be at least 50 percent ice to be most effective). The items must be chilled so that their interior temperature reaches 37.4°F (3°C) in 2 hours or less.

9. Remove from the ice bath, label correctly (product name, time and date of processing, and date to be discarded), and either freeze or store in a refrigerator at 37.4°F (3°C) or lower.

10. To serve, reheat in an immersion circulator until the interior temperature reaches the desired level. Remove from the bag and finish as desired. Serve at once.

FIGURE 14.12 Immersion circulator.

Beef Short Ribs Sous Vide with Bordelaise Sauce

PORTIONS: 1   PORTION SIZE: 5 OZ (150 G)

U.S.	METRIC	INGREDIENTS
5 oz	150 g	Beef short ribs, boneless, in one piece
to taste	to taste	Salt
to taste	to taste	Pepper

PROCEDURE

1. Review the guidelines for safe sous vide cooking on page 119.
2. Season the meat with salt and pepper. Place the meat in a plastic bag appropriate for sous vide cooking and seal it under vacuum.
3. Place the sealed meat in a hot bain-marie heated with an immersion circulator to 158°F (70°C) and cook 20 hours. At the end of cooking time, the interior of the meat should be 158°F (70°C).
4. If the meat is not to be finished and served immediately, chill it as quickly as possible in an ice bain-marie or blast cooler. The interior of the meat must reach 37°F (3°C) or colder in less than 90 minutes.

U.S.	METRIC	INGREDIENTS
as needed	as needed	Oil
1 fl oz	30 mL	Bordelaise Sauce (p. 191; omit the marrow garnish)

5. For finishing and serving, rewarm the meat, still sealed in its plastic bag, in a warm bain-marie at 140°F (60°C).
6. Open the bag and remove the meat.
7. Heat a thin film of oil in a sauté pan. Brown the meat lightly on top and bottom.
8. Serve the meat with the sauce around it.

Per serving: Calories, 290; Protein, 21 g; Fat, 22 g (3% cal.); Cholesterol, 70 mg; Carbohydrates, 2 g; Fiber, 0 g; Sodium, 60 mg.

VARIATION

PROCEDURE VARIATION: COMBI OVEN

If an immersion circulator is not available, cook in a combi oven in steam mode at 158°F (70°C) for 18–20 hours.

Beef Short Ribs Sous Vide with Bordelaise Sauce

KEY POINTS TO REVIEW

- What are the steps in the basic steaming procedure?
- What standards can be used to evaluate the quality of steamed items?
- What are the steps in the basic braising procedure?
- What is the difference between braises and stews?
- What standards can be used to evaluate the quality of braised items?
- What are the steps in the basic sous vide cooking procedures?

TERMS FOR REVIEW

roast	deglaze	shallow-poaching	fricassée
bake	pan-fry	cuisson	blanquette
barbecue	deep-fry	à la nage	sous vide
broil	simmer	steam	
grill	poach	braise	
sauté	court bouillon	pot roast	

QUESTIONS FOR DISCUSSION

1. List four advantages of roasting at a low temperature.

2. What is the purpose of basting?

3. Which steaks require the highest broiler temperature, thick ones or thin ones? steaks to be cooked rare or steaks to be cooked well done?

4. Why is it important not to overload the pan when sautéing foods?

5. What is meant by the term *presentation side?*

6. What is the difference between simmering and submersion poaching?

15

UNDERSTANDING MEATS AND GAME

Meat is muscle tissue. It is the flesh of domestic animals (cattle, hogs, and lambs) and of wild game animals (such as deer). As a cook, chef, or food-service operator, you will spend more of your time and money on meats than on any other food.

It is important, then, to understand meats thoroughly in order to cook them well and profitably. Why are some meats tender and some tough? How can you tell one cut from another when there are so many? How do you determine the best way to cook each cut?

In order to answer questions like these, it is helpful to start at the most basic level of composition and structure. We then proceed to discuss grading and inspection, basic cuts, and appropriate cooking and storage methods. In addition, we discuss the characteristics of variety meats and of popular game meats. Only then can we best approach the individual cooking methods and recipes presented in the following chapters.

AFTER READING THIS CHAPTER, YOU SHOULD BE ABLE TO

1. Describe the composition and structure of meat, and explain how they relate to meat selection and cooking methods.

2. Explain the use of the federal meat inspection and grading system in selecting and purchasing meats.

3. Explain the effect of aging on meat, and identify the two primary aging methods.

4. Identify the primal cuts of beef, lamb, veal, and pork, and list the major fabricated cuts obtained from each of them.

5. Select appropriate cooking methods for the most important meat cuts, based on the meat's tenderness and other characteristics.

6. Prepare variety meats.

7. Identify the characteristics of game meats, and select the appropriate cooking methods for them.

8. Determine doneness in cooked meat.

9. Store fresh meat and frozen meat to gain the maximum shelf life.

COMPOSITION, STRUCTURE, AND BASIC QUALITY FACTORS

COMPOSITION

Muscle tissue consists of three major components: water, protein, and fat.

WATER

Water is about 75 percent of muscle tissue. With such a high percentage of water, you can see why *shrinkage* can be a big problem in cooking meat. Too much moisture loss means dry meat, loss of weight, and loss of profit.

PROTEIN

Protein is an important nutrient and the most abundant solid material in meat. About 20 percent of muscle tissue is protein.

As we learned in Chapter 6, protein *coagulates* when it is heated. This means it becomes firmer and loses moisture. **Coagulation** is related to doneness. In dry-heat cooking, when protein has coagulated to the desired degree, the meat is said to be "done." Doneness is discussed later in this chapter.

After protein has coagulated, applying higher heat toughens it.

FAT

Fat accounts for up to 5 percent of muscle tissue. Of course, more fat may surround the muscles. A beef carcass can be as much as 30 percent fat.

Because of health and dietary concerns, many meat animals are being bred and raised with a lower fat content than in past years. Nevertheless, a certain amount of fat is desirable for three reasons:

1. **Juiciness.**

   Marbling is fat deposited within the muscle tissue. The juiciness we enjoy in well-marbled beef is due more to fat than to moisture.

   Surface fat protects the meat—especially roasts—from drying out during cooking as well as in storage. Adding surface fats where they are lacking is called *barding*.

2. **Tenderness.**

   Marbling separates muscle fibers, making them easier to chew.

3. **Flavor.**

   Fat is perhaps the main source of flavor in meat. A well-marbled Prime (top grade) steak tastes "beefier" than the same cut of a lower grade.

CARBOHYDRATE

Meat contains a very small amount of carbohydrate. From the standpoint of nutrition, its quantity is so small that it is insignificant. It is important, however, because it plays a necessary part in the complex reaction, called the *Maillard reaction* (see p. 108), that takes place when meats are browned by roasting, broiling, or sautéing. Without these carbohydrates, the desirable flavor and appearance of browned meats would not be achieved.

STRUCTURE

MUSCLE FIBERS

Lean meat is composed of long, thin muscle fibers bound together in bundles. These determine the *texture* or *grain* of a piece of meat. Fine-grained meat is composed of small fibers bound in small bundles. Coarse-textured meat has large fibers.

Feel the cut surface of a tenderloin steak, and compare its smooth texture to the rough cut surface of brisket or bottom round.

CONNECTIVE TISSUE

Muscle fibers are bound together in a network of proteins called **connective tissue**. Each muscle fiber also is covered in a sheath of connective tissue.

It is important for the cook to understand connective tissue for one basic reason: *Connective tissue is tough*. To cook meats successfully, you should know:

- Which meats are high in connective tissue and which are low.
- What are the best ways to make tough meats tender.

1. **Meats are highest in connective tissue if they come from more muscular regions or from older animals.**

   - Muscles that are more exercised—muscles in the legs, for example—have more connective tissue than muscles in the back.
   - Meat from a young animal, such as veal, is more tender than meat from a young steer, which, in turn, is more tender than meat from an old bull or cow. (Young animals have connective tissue, too, but it becomes harder to break down as the animal ages.)

2. **Meats high in connective tissue can be made more tender by using proper cooking techniques.**

   There are two kinds of connective tissue: **collagen**, which is white in color, and **elastin**, which is yellow.

   - **Collagen.**
     Long, slow cooking in the presence of moisture breaks down or dissolves collagen by turning it into gelatin and water. Of course, muscle tissue is about 75 percent water, so moisture is always present when meats are cooked. Except for very large roasts, however, long cooking by a dry-heat method has the danger of evaporating too much moisture and drying out the meat. Therefore, *moist-heat cooking methods at low temperatures are most effective for turning a meat high in connective tissue into a tender, juicy finished product*.

     Other factors also help tenderize collagen:

     > *Acid* helps dissolve collagen. Marinating meat in an acid mixture, or adding an acid such as tomato or wine to the cooking liquid, helps tenderize it.

     > *Enzymes* are naturally present in meats. They break down some connective tissue and other proteins as meat ages (see "Aging," pp. 455–456). These enzymes are inactive at freezing temperatures, slow-acting under refrigeration, active at room temperature, and destroyed by heat above 140°F (60°C).

     > *Tenderizers* are enzymes such as papain (extracted from papaya) that are added to meats by the cook or injected into the animal before slaughter. Exercise care when using enzyme tenderizers. Too long an exposure at room temperature can make the meat undesirably mushy.

   - **Elastin.**
     Older animals have a higher proportion of elastin than younger animals.

     Elastin is not broken down in cooking. Tenderizing can be accomplished only by *removing the elastin* (cutting away any tendons) and by mechanically breaking up the fibers, as in:

     > Pounding and cubing (cubed steaks)

     > Grinding (hamburger)

     > Slicing the cooked meat very thin against the grain (as in London broil)

INSPECTION AND GRADING

Cooks and food-service operators in Canada are assisted in their evaluation of meats by a federal inspection and grading system.

INSPECTION

1. **Inspection** is a *guarantee of wholesomeness*, not of quality or tenderness. It means the animal was not diseased and the meat is clean and fit for human consumption.
2. That the meat passed inspection is indicated by a round stamp with a crown and the number of the processing plant (Figure 15.1).
3. Inspection is required by law. All meat must be inspected.

The following discussion explains the above three points in more detail.

The Canadian Food Inspection Agency (CFIA) is responsible for carrying out meat inspection. All meat that is shipped from province to province or is exported to other countries is required to be federally inspected. This includes about 90 percent of all meats processed in Canada. The remainder is subject to provincial inspection regulations, which are similar to the federal regulations. The CFIA also inspects all meat imported into Canada. Meats that are provincially inspected but not federally inspected cannot be shipped across provincial borders.

Beginning 24 hours before slaughter, all livestock process at federally registered plants undergoes a two-step inspection process. First, all live animals are inspected in order to separate any animal that may be diseased or otherwise unsuitable for slaughter. Certified veterinarians examine these separated animals for wholesomeness.

After slaughter, the internal organs and lymph nodes of each animal are inspected, and any animal found to be diseased is removed from further processing. Condemned carcasses are disposed of as required by the CFIA.

QUALITY GRADING

1. **Grading** is a quality designation.
2. Quality grading is indicated by a maple leaf stamp (Figure 15.2(a)).
3. **Yield grading** indicates the proportion of lean meat in a carcass. It is indicated by a triangular stamp. Figure 15.2(b) shows a yield grade stamp.
4. Grading is not required by law.

Quality grading is based on the texture, firmness, and color of the lean meat, the age or maturity of the animal, and the marbling (the fat within the lean).

All these factors must be considered together. For example, old, tough meat can still have marbling, but it would rate a low grade because of the other factors.

BEEF GRADES

The Canadian Beef Grading Agency is responsible for grading Canadian beef. This private, nonprofit organization administers the grades according to federal standards. Grading is voluntary, and producers must pay a fee for the service.

There are 13 Canadian beef grades. The best grades, beginning with the highest, are Prime, AAA, AA, and A. All Prime and A grades must be from young animals. The meat must be firm, fine grained, and well developed. The muscle tissue must be bright red. The fat must be white and firm. Carcasses that meet these initial requirements are then divided into grades based on the amount of marbling, as follows:

Canada Prime—slightly abundant marbling or higher
Canada AAA—small marbling or higher
Canada AA—at least slight marbling, but less than small
Canada A—at least trace marbling, but less than slight

About 90 percent of all graded Canadian beef carcasses fall within these four grades. Prime and A grades are indicated by a red grade stamp.

Beef carcasses graded as prime or one of the A grades are also yield graded, as follows:

Canada 1—59 percent or more lean, usable meat
Canada 2—54–58 percent lean, usable meat
Canada 3—53 percent or less lean, usable meat

FIGURE 15.1
Canada inspection stamp for meat.
Courtesy of the Canadian Food Inspection Agency.

FIGURE 15.2 Canada quality grade stamp (a) and yield grade stamp (b) for meat.
Courtesy of the Canadian Food Inspection Agency.

(a)

(b)

Canada B grades are, starting with the highest, B1, B2, B3, and B4. These grades are also given to young animals that do not meet the requirements for the A grades. They are indicated by a blue stamp.

The lowest grades are D1, D2, D3, D4, and E (there are no C grades). These grades are all from mature animals. The meat is used primarily for ground beef and processed products. They are indicated by a brown stamp.

VEAL GRADES

Veal carcasses are divided into 10 grades: four A grades (A1 through A4), four B grades (B1 through B4), and two C grades (C1 and C2). Grades are based on muscling, meat color, and fat cover. The A grades are the highest, with the best muscling and some creamy white fat. Color ranges from light pink (A1) to red (A4).

PORK GRADES

Pork is not graded the way other meats are. Within the industry, carcasses are yield graded based on the ratio of fat to lean. The carcasses are not stamped, however, and the meat is trimmed at the processing plant before shipment. Thus, the grading system is of concern only to producers in that it is used to determine the price the packer pays the producer.

LAMB GRADES

Most lamb produced and sold in Canada is ungraded. Nevertheless, a grading system has been established, so you may find lamb graded Canada AAA, the highest grade. Other grades are Canada C1, Canada C2, Canada D1, and Canada D4. Grades are determined based on the maturity of the carcass, the condition of the break joint in the leg (the joint in young carcasses breaks easily and shows evidence of a blood supply), the color of the muscle (young carcasses have pink to light red meat, not dark red), the color of the fat (white rather than yellow), and the overall muscling and fat cover.

AGING

GREEN MEAT

Soon after slaughter, an animal's muscles stiffen due to chemical changes in the flesh. This stiffness, called *rigor mortis*, gradually disappears. Softening takes 3–4 days for beef, less time for smaller carcasses like veal, lamb, and pork. This softening is caused by enzymes in the flesh.

Green meat is meat that has not had enough time to soften. It is tough and relatively flavorless. Because it takes several days for meats to reach the kitchen from the slaughterhouse, green meat is seldom a problem with commercially available meats, except when meat is frozen while still green. The problem is sometimes encountered with game killed for home consumption, if the hunter cuts and freezes the meat when it is too fresh.

AGED MEAT

Enzyme action continues in muscle tissue even after meat is no longer green. This tenderizes the flesh even more and develops more flavor. Holding meats in coolers under controlled conditions to provide time for this natural tenderizing is called **aging**.

Beef and lamb can be aged because high-quality carcasses have enough fat cover to protect them from bacteria and from drying. Veal has no fat cover, so it is not aged. Pork does not require aging.

Aging does not mean just storing meat in the refrigerator. *There is a difference between aged meat and old meat*. Conditions must be carefully controlled so the meat becomes naturally tender without spoiling. There are two primary methods used for aging:

1. **Wet aging.**
   Today, most wholesale meat carcasses are broken down into smaller cuts and enclosed in plastic vacuum packs. These packs are usually known by the trade name **Cryovac®**. The air- and moistureproof packaging protects the meat from bacteria and mold, and it prevents weight loss due to drying. (However, Cryovac-aged meats often lose more weight in cooking than do dry-aged meats.) Vacuum-pack meats must be refrigerated.

2. Dry aging.

Dry aging is the process of storing meats, usually large cuts, under carefully controlled conditions. The meat is not packaged or wrapped, and it is exposed to air on all sides. Temperature, humidity, and air circulation are precisely controlled to prevent spoilage. Ultraviolet lights are sometimes used in aging coolers to kill bacteria.

Dry-aged meat can lose up to 20 percent of its weight through moisture loss, depending on the size of the cut and how long it is aged. Consequently, dry aging is a more expensive process than wet aging. Dry-aged meats are usually available from specialty purveyors only, and at a higher price than wet-aged meats. Many customers are willing to pay a premium for fine dry-aged steaks because they are considered the best for flavor and texture.

Aging increases tenderness and flavor. An off taste is not characteristic of aged meat. *If a meat smells or tastes spoiled, it probably is.* Sometimes meats in vacuum packs have a musty aroma when first opened, but this disappears quickly.

Aging costs money. Storage costs, weight loss due to drying, and heavier trimming due to dried and discolored surfaces all add to the price of aged meat (although wet aging costs less than dry aging). As a meat purchaser, you must decide how much quality is worth how much cost for your particular establishment.

KEY POINTS TO REVIEW

- What are the three main components of meat?

- What is connective tissue? How does it affect our choice of cooking techniques for meat?

- What is the difference between inspection and grading?

- What are the best two grades of beef? of lamb? of pork? of veal?

- What are the two types of aging? What effect does aging have on meat?

UNDERSTANDING THE BASIC CUTS

The following discussion of meat cuts focuses on the four primary meat categories in the wholesale and retail markets: beef, lamb, veal, and pork. Keep in mind, however, that game animals, discussed later in the chapter, have the same bone and muscle structure and are generally divided into the same or similar cuts as nongame animals.

Meat cuts are based on two factors:

1. The muscle and bone structure of the meat.
2. Uses of and appropriate cooking methods for various parts of the animal.

Food-service suppliers may follow a set of specifications called *Institution Meat Purchase Specifications* (IMPS). The *Canadian Meat Council* (CMC) has published a manual that includes the IMPS numbers and names of cuts. The CMC manual also includes cuts specific to Canada. All cuts are described in detail and listed by number. This simplifies purchasing, as you can order by number exactly the cut you want.

IMPS/ NAMPCLASSIFICATIONS

The IMPS/CMC system assigns a series of numbers to each major category of meat, as detailed in Table 15.1. Beef, for example, is the 100 series. This means all large beef cuts, from whole carcass to primals and prepared roasts, are assigned a three-digit number from 100 to 199. Portion-size and smaller cuts of beef, such as steaks and stew meat, are assigned a four-digit number, also beginning with 1. For example, a whole beef rib, roast-ready, has the number 109; a beef rib steak, bone in, is 1103.

Note that variety meats and processed meat products, such as cured and smoked meats and sausages, are also categorized (see Table 15.1).

AVAILABLE FORMS: CARCASSES, PARTIAL CARCASSES, PRIMALS, AND FABRICATED CUTS

Beef, lamb, veal, and pork may be purchased in some or all of these forms. Mutton and goat are also given classification numbers, as indicated in Table 15.1, but they have minimal importance in North American food service and are not covered here.

CARCASSES

The carcass is the whole animal, minus the entrails, head, feet, and hide (except pork, from which only the entrails and head are removed). Whole carcasses are rarely purchased by food-service operators because of the skill and labor required in cutting and because of the problem of total utilization.

SIDES, QUARTERS, FORESADDLES, HINDSADDLES

These represent the first step in breaking down a carcass.

Again, these larger cuts are no longer frequently used in food service. Fewer establishments cut their own meats.

1. Beef is split first through the backbone into sides. Sides are divided between the 12th and 13th ribs into forequarter and hindquarter.

2. Veal and lamb are not split into sides but are divided in half into foresaddle and hindsaddle, or the front half and the hind half. The cut is made between the 11th and 12th ribs.

3. Pork carcasses are not divided in this way. They are cut directly into primal cuts.

PRIMAL OR WHOLESALE CUTS

These are the primary divisions of quarters, foresaddles, hindsaddles, and carcasses. These cuts, called **primal cuts**, are still used, to some extent, in food service, because they

1. Are small enough to be manageable in many food-service kitchens.

2. Are still large enough to allow a variety of cuts for different uses or needs.

3. Are easier to utilize completely than quarters or halves.

Each primal may be **fabricated**, or cut up and trimmed, in several ways. Primal cuts are always the starting point for smaller cuts. For this reason, it will benefit you to be able to identify each one. Study the charts and photos in Figures 15.3 through 15.6. Learn the names of the primals, their location on the carcass, and the most important cuts that come from each. Then, whenever you work with a piece of meat, try to identify it exactly and match it with its primal cut.

FABRICATED CUTS

Primal cuts are fabricated into smaller cuts for roasts, steaks, chops, cutlets, stewing meat, ground meat, and so forth, according to individual customer requirements and, if applicable, IMPS/CMC specifications.

The amount of trim and exact specifications can have many variations. For example, a beef primal rib can be trimmed and prepared for roasting at least nine ways.

Portion-controlled cuts are ready-to-cook meats cut according to customer's specifications. Steaks and chops are ordered either by weight per steak or by thickness. Portion-controlled cuts require the least work for the cook of all meat cuts. They are also the most expensive per pound of all categories of cuts.

BONE STRUCTURE

Knowing the bone structure of meat animals is essential for:

1. **Identifying meat cuts.**
   The distinctive shapes of the bones are often the best clue to the identification of a cut. Note how the shapes of the bones in the photographs in Figures 15.3 through 15.6 help your recognition.

2. **Boning and cutting meats.**
   Bones are often surrounded by flesh. You need to know where they are even if you can't see them.

3. **Carving cooked meats.**
   Same reason as number 2.

TABLE 15.1
IMPS/CMC Meat Categories

Series Number	Series Name
100	Fresh Beef
200	Fresh Lamb and Mutton
300	Fresh Veal and Calf
400	Fresh Pork
500	Cured, Cured and Smoked, and Cooked Pork Products
600	Cured, Dried, and Smoked Beef Products
700	Variety Meats and Edible Byproducts
800	Sausage Products
11	Fresh Goat

FIGURE 15.3 Beef.

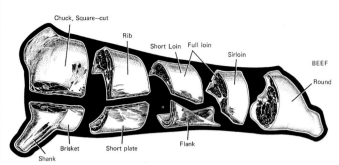

(a) Primal (wholesale) beef meat cuts
Courtesy National Livestock and Meat Board

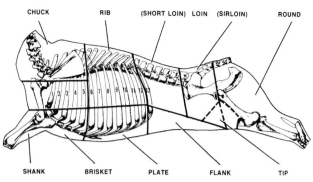

(b) Primal (wholesale) beef cuts and their bone structure
Courtesy National Livestock and Meat Board

Beef chuck, boneless, separated into
blade, clod, and arm (IMPS/CMC 115)

Beef rib, roast ready (IMPS/CMC 109)
(Reprinted with permission of John Wiley & Sons, Inc.)

Beef rib steak (IMPS/CMC 1103)

Beef loin (IMPS/CMC 172)
(Reprinted with permission of John Wiley & Sons, Inc.)

Beef short loin (IMPS/CMC 174)
(Reprinted with permission of John Wiley &
Sons, Inc.)

Beef porterhouse steak
(IMPS/CMC 1173)

Beef T-bone steak (IMPS/CMC 1174)

Beef tenderloin, trimmed (IMPS/CMC 189A)

Beef strip loin (IMPS/CMC 175)
(Reprinted with permission of John Wiley & Sons, Inc.)

Beef outside (bottom) round
(IMPS/CMC 170)

Beef round steak
(IMPS/CMC 1170)

Beef inside (top) round
(IMPS/CMC 168)

Beef knuckle, untrimmed
(IMPS/CMC 167)
(Reprinted with permission
of John Wiley & Sons, Inc.)

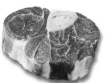

Beef shank, cross-cuts
(IMPS/CMC 117)

Beef flank steak
(IMPS/CMC 193)

FIGURE 15.4 Veal.

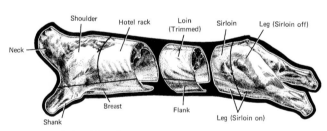

(a) Primal (wholesale) veal meat cuts
Courtesy National Livestock and Meat Board

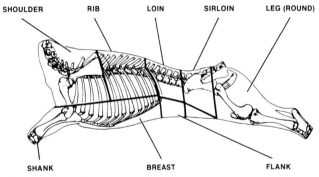

(b) Primal (wholesale) veal cuts and their bone structure
Courtesy National Livestock and Meat Board

Veal rib roast (IMPS/CMC 306)

Veal breast (IMPS/CMC 313)

FIGURE 15.5 Lamb.

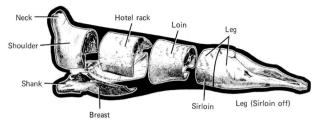

(a) Primal (wholesale) lamb meat cuts
Courtesy National Livestock and Meat Board

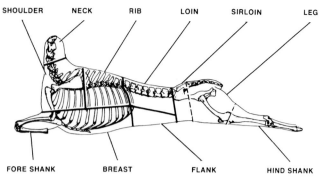

(b) Primal (wholesale) lamb cuts and their bone structure
Courtesy National Livestock and Meat Board

Lamb loin roast
(IMPS/CMC 232A)

Lamb arm chop
(IMPS/CMC 1207)

Lamb shoulder blade chop
(IMPS/CMC 1207)

Lamb rib chop
(IMPS/CMC 1204)

Lamb loin chop
(IMPS/CMC 1232A)

Lamb, square-cut shoulder, whole
(IMPS/CMC 207)

Lamb, whole leg
(IMPS/CMC 233)

Lamb, boneless shoulder, rolled and tied
(IMPS/CMC 208)

Lamb foreshank
(IMPS/CMC 210)

FIGURE 15.6 Pork.

(a) Primal (wholesale) pork meat cuts
Courtesy National Livestock and Meat Board

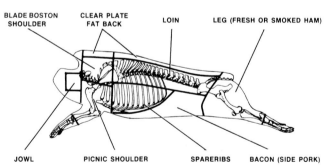

(b) Primal (wholesale) pork cuts and their bone structure
Courtesy National Livestock and Meat Board

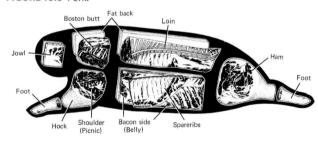

Pork tenderloin
(IMPS/CMC 415)

Pork shoulder butt
(IMPS/CMC 406)

Full pork loin (includes ribs)
(IMPS/CMC 410)

Pork loin chops (IMPS/CMC 1410)

Pork rib half and loin half roasts
(IMPS/CMC 410)

Study the chart of the beef skeleton in Figure 15.7 and learn the names of the major bones. Then compare the charts in Figures 15.3 through 15.6. You will see the bone structures for all the animals are identical (except for pork, which has more than 13 ribs). Even the names are the same.

The photographs in Figures 15.3 through 15.6 depict typical primal and fabricated cuts of beef, lamb, veal, and pork (courtesy National Livestock and Meat Board and National Pork Producers Council).

MEAT-CUTTING TERMINOLOGY

Although the public refers to retail meat cutters as butchers, the meat industry uses this term another way. To **butcher** means to kill and dress a meat animal. To **fabricate** means to cut raw meat into smaller pieces.

A third term, **carve**, also means to cut meat, but this always refers to cooked meat.

FIGURE 15.7 Beef bone structure.

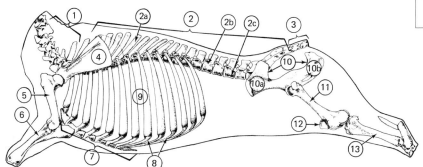

1. Neck bone	**3. Tailbone**	**8. Rib cartilage**	**11. Leg or round bone**
2. Backbone	**4. Blade bone**	**9. Ribs**	**12. Kneecap**
2a. Feather bone	**5. Arm bone**	**10. Pelvis**	**13. Hindshank bone**
2b. Finger bone	**6. Foreshank bone**	10a. Hip bone	
2c. Chine bone	**7. Breastbone**	10b. Rump or aitch bone	

BEEF, LAMB, VEAL, AND PORK CUTS

Beef Primal Cuts and Fabricated Cuts

Primal	Major Bones	Common Fabricated Cuts	Primary Cooking Methods
Front quarter			
Chuck (square cut)	Ribs 1–5	Shoulder clod	Moist heat
	Blade bone	Triangle	
	Backbone (including chine and feather bones)	Boneless inside chuck	
		Chuck tender	
	Neck bones	Chuck short ribs	
	Arm bone	Cubed (delicatized) steaks	
		Stew meat	
		Ground chuck	
Brisket	Rib bones	Boneless brisket and corned beef brisket	Moist heat
	Rib cartilage	Ground beef	
	Breastbone	Stew meat	Moist heat
Shank	Shankbone	Ground beef	

Note: Square-cut chuck, brisket, and shank, in one piece, are called *cross-cut chuck*.

Primal	Major Bones	Common Fabricated Cuts	Primary Cooking Methods
Rib	Ribs 6–12	Rib roasts (prime rib)	Dry heat
	Backbone (chine and feather bones)	Rib steaks	
		Short ribs	Moist heat
Short plate	Rib bones	Short ribs	Moist heat
	Tip of breastbone	Stew meat	
	Rib cartilage	Ground beef	

(continued)

Beef Primal Cuts and Fabricated Cuts *(continued)*

Primal	Major Bones	Common Fabricated Cuts	Primary Cooking Methods
Hindquarter			
(Full loin)		Full tenderloin (to have tenderloin in one piece, it must be stripped out of loin before loin is split into short loin and sirloin)	Dry heat
Short loin	Rib 13	Club (wing) steaks	Dry heat
	Backbone (chine, feather bones, finger bones; see Note 1)	T-bone steaks	
		Porterhouse steaks	
		Strip loin	
		Strip loin steaks	
		Short tenderloin	
Sirloin	Backbone	Top sirloin steak, boneless	Dry heat
	Hip bone (part of pelvis)	Bottom sirloin steak, boneless	
		Tenderloin butt	
Flank	Tip of rib 13	Flank steak	Moist heat (exception: flank steak cooked as London broil)
		Ground beef	
Hip	Round (leg) bone	Sirloin tip (Knuckle)	Moist heat and dry heat
	Aitch bone (part of pelvis)	Inside (top) round	
	Shankbone	Outside (bottom) round	
	Tailbone	Eye of round (part of outside round)	
		Rump	
		Hind shank	

Note 1: Finger bones are the short horizontal bones attached to those chine bones that have no ribs attached. They are stems of the Ts in T-bones.

Lamb Primal Cuts and Fabricated Cuts

Primal	Major Bones	Common Fabricated Cuts	Primary Cooking Methods
Foresaddle (front half)			
Shoulder	Ribs 1–6	Shoulder roasts	Moist heat and dry heat
	Arm	Shoulder chops	
	Blade	Stew meat	
	Backbone (chine and feather bones)	Ground lamb	
	Neck bones		
Breast and shank	Rib bones	Riblets	Moist heat
	Rib cartilage	Breast	
	Breastbone	Stew meat	
	Shankbone	Ground lamb	
Hotel rack	Ribs 7–11 or 7–13	Rib roasts (rack)	Dry heat
	Backbone	Crown roast	
		Ribs, chops	
Hindsaddle (hind half)			
Loin (with or without flank)	Backbone (chine, feather bones, finger bones)	Loin roast	Dry heat
		Loin chops	
Leg	Backbone	Leg roast	Dry heat
	Tailbone	Leg chops	
	Pelvis	Sirloin chops	
	Round bone	Shank	Moist heat
	Hindshank		

Note: Rack and loin attached are called *whole loin double* or *back*: used mostly for chops. The whole loin may be divided after the 13th rib to yield a rib or rack section containing ribs 7–13.

Veal Primal Cuts and Fabricated Cuts

Primal	Major Bones	Common Fabricated Cuts	Primary Cooking Methods
Foresaddle (front half)			
Shoulder (square cut)	Ribs 1–6 Blade bone Backbone (chine and feather bones) Neck bones Arm bone	Shoulder roasts Shoulder chops Shoulder clod steaks Cubed steaks Stew meat Ground veal	Moist heat and dry heat (square cut)
Breast	Rib bones Rib cartilage Breastbone	Boneless breast Cubed steaks Ground veal	Moist heat
Shank	Shankbone	Shank cross-cut (osso buco)	Moist heat
Hotel rack	Ribs 7–11 or 7–13 Backbone (chine and feather bones)	Rib roast Rib chops	Dry heat and moist heat

Note: Rack plus connecting portions of breast is called a *bracelet* or *rib and flank, double.*

Primal	Major Bones	Common Fabricated Cuts	Primary Cooking Methods
Hindsaddle			
Loin (with or without flank)	Ribs 12 and 13 or no ribs Backbone (chine, feather bones, finger bones)	Saddle (loin roast) Loin chops	Dry heat and moist heat
Leg	Backbone Tailbone Pelvis (hip bone, aitch bone) Round bone Hindshank	Leg roast Scaloppine or cutlets Shank cross-cut (osso buco)	Dry heat Moist heat

Note: Hotel rack and loin attached are called *back* or *whole loin*; used mostly for chops. The whole loin may be divided after the 13th rib to yield a rib or rack section containing ribs 7–13.

Pork Primal Cuts and Fabricated Cuts

Primal	Major Bones	Common Fabricated Cuts	Primary Cooking Methods
Shoulder picnic	Shoulder (arm) bone Shankbone	Fresh and smoked picnic Hocks Ground pork Sausage meat	Moist heat
Shoulder blade	Blade bone (rib bones, back and neck bones are removed)	Butt steaks Shoulder roasts Daisy (smoked) Ground pork Sausage meat	Dry heat and moist heat
Loin	Rib bones (see Note) Backbone (chine, feather bones, finger bones) Hip bone	Loin roast Loin chops (rib end and tenderloin end) Boneless loin Country-style ribs Canadian-style bacon (smoked)	Dry heat and moist heat
Ham	Aitch bone Leg bone Hindshank bone	Fresh ham Smoked ham Ham steaks	Dry heat and moist heat
Belly	None	Bacon	Dry heat and moist heat
Spare ribs	Rib bones Breastbone	Side ribs (belly ribs or spareribs)	Moist heat
Fatback and clear plate	None	Fresh and salt fatback Salt pork Lard	(Used as cooking fats)
Jowl	None	Jowl bacon	Moist heat
Feet	Foot bones		Moist heat

Note: Pork has more than 13 ribs (unlike beef, lamb, and veal) due to special breeding to develop long loins.

SELECTING MEATS FOR YOUR OPERATION

DECIDING WHICH FORMS TO PURCHASE

Whether you buy whole carcasses, fabricated cuts, or anything in between depends on four factors:

1. How much meat-cutting skill you or your staff has.

2. How much work and storage space you have.

3. Whether or not you can use all cuts and lean trim on your menu.

4. Which form gives you the best cost per portion after figuring in labor costs.

Meat purveyors can usually cut meat more economically than food-service operators can because they deal in large volume. Carcasses or primal cuts cost less per pound than fabricated cuts, but they have more waste (fat and bone) and require more labor (which costs money). However, some operators still do some of their own cutting, depending on how they answer the four questions above. They feel cutting their own meat gives them greater control over quality.

Some compromises are available. If you want the quality of freshly cut steaks, for example, you might buy boneless strip loins and cut your own steaks to order. You need not buy primal loins.

SPECIFICATIONS

When buying meat, you must indicate the following specifications:

1. **Item name.**
   Include IMPS/CMC number, if applicable.
   Example: 173 Beef Short Loin, Regular

2. **Grade.**
   Example: Canada AAA

3. **Weight range for roasts and large cuts.**
   Portion weight or thickness (not both) for steaks and chops.

4. **State of refrigeration.**
   Chilled or frozen.

5. **Fat limitations, or average thickness of surface fat.**
   Example: 2 cm average, 2.5 cm maximum.
   (This does not apply to veal.)

FIGURE 15.8 The radura is the international symbol for irradiation.

Meat purchasers may also have to choose whether or not to purchase irradiated meat. **Irradiation** is the process of exposing foods to radiation in order to kill bacteria, parasites, and other potentially harmful organisms. Irradiation does not harm the meat, make it radioactive, or change its structure, flavor, or nutritional value. In Canada, the irradiation of ground beef and poultry is awaiting approval from Health Canada. Currently, only potatoes, onions, wheat flour, whole wheat flour, and whole or ground spices and dehydrated seasonings are approved for irradiation and sale. The Canadian Food Inspection Agency (CFIA) requires that prepackaged wholly irradiated foods include labeling along with a statement that the product has been irradiated and the international symbol for irradiation, the radura (see Figure 15.8).

Some operators refuse to purchase irradiated foods because they or their customers have concerns about their health effects. The procedure has generated much controversy for other reasons as well. For example, some see the availability of the process as an excuse to avoid normal sanitation procedures. Nevertheless, there is so far no evidence that these foods are harmful for human beings to eat.

FABRICATING MEAT

Even though few operations today purchase large cuts, such as primals, and break them down in-house, you still need to know a number of trimming and fabricating techniques to finish or modify the fabricated cuts you purchase. The illustrations in this section demonstrate important procedures. These procedures are used for recipes in Chapter 16.

One term you will encounter often when trimming meat is **silverskin**, a thin layer or membrane of connective tissue that often covers the surface of a muscle. For braised meats, it is not always necessary to remove silverskin, unless it is very heavy, because slow cooking breaks down the collagen of the tissue. However, for roasts, sautés, and grills of tender meats, it should be removed for two reasons: (1) It is tough and would be unpleasantly chewy in the cooked product; (2) It usually shrinks when cooked, making the meat deform or curl.

To remove silverskin:

1. Hold the blade of the knife parallel to the silverskin and perpendicular to the grain of the meat.

2. Insert the tip of the blade just under the silverskin.

3. Hold the knife so the edge of the blade angles slightly upward. Carefully slip the blade under the silverskin in the direction of the grain of the meat while holding the meat steady with your other hand. (Angling the blade upward keeps it from digging into the meat.)

4. Repeat until all silverskin is removed.

This technique is illustrated in Figures 15.9, 15.15, and 15.16.

FIGURE 15.9 Preparing beef tenderloin.

(a) A whole, untrimmed beef tenderloin.

(b) Pull off the heavy fat from the outside of the tenderloin, freeing it with a knife as necessary.

(c) Separate the strip of gristly meat, or chain, from the side of the tenderloin. Use this piece for ground meat.

(d) Carefully remove the silverskin.

(e) The fully trimmed tenderloin before cutting.

(f) Cut into steaks of the desired size.

(g) This tenderloin has been cut into a variety of steaks as a demonstration. From left to right: four fillet steaks, two large pieces for Châteaubriand, two tournedos, four fillets mignons. In front: trimmings from both ends.

FIGURE 15.10 Shaping medallions.

FIGURE 15.11 Preparing a leg of lamb for roasting.

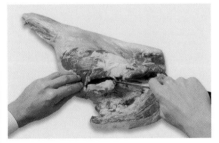

(a) Begin by removing the hip and tail bones.

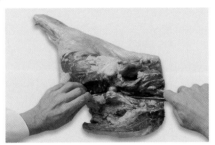

(b) With a sharp-pointed boning knife, cut along the hip bone to separate bone from meat. Always cut against the bone.

(c) Continue until the hip bone and tailbone are completely removed. Note the round ball joint at the end of the leg bone in the center of the meat.

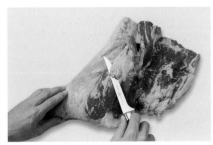

(d) Trim off excess external fat, leaving a thin covering.

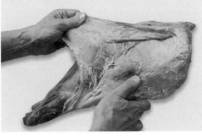

(e) Pull off the skin or fell on the outside of the leg.

(f) Full leg of lamb, ready for roasting. The end of the shankbone and part of the shank meat have been removed.

(g) The leg may be tied into a more compact shape.

(h) The sirloin portion may be cut off and used for another purpose, such as shish kebabs.

FIGURE 15.12 Preparing a rack of lamb for roasting.

(a) Begin by cutting down on both sides of the feather bones all the way to the chine bone.

(b) If a meat saw is available, turn the rack over and cut through the rib bones where they attach to the chine bone.

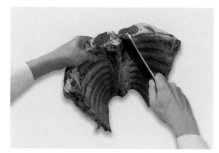

(c) If a meat saw is not available, use a cleaver. Stand the roast on end and cut through the rib bones where they join the chine bone. This separates one rack.

(d) Repeat the procedure on the other side of the chine.

(e) The two halves are separated from the chine.

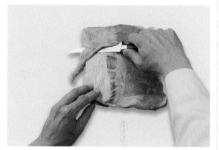

(f) Trim excess fat from the top of the meat, leaving a thin protective covering. During this step, you should also remove the shoulder blade cartilage, which is embedded in the layers of fat.

(g) To trim the fat and meat from the ends of the bones (called *frenching* the bones), first cut through the fat in a straight line down to the bone, keeping the cut about 1 in. (2.5 cm) from the tip of the eye muscle.

(h) Score the membrane covering the rib bones. Pull and cut the layer of fat from the bones.

(i) The roast is trimmed and ready to cook.

FIGURE 15.13 Butterflying and stuffing a pork loin.

(a) To butterfly, hold the knife blade parallel to the table and cut through the center of the loin as shown. Do not cut all the way through, but leave the meat attached at one edge.

(b) Open the cut meat and spread the filling on the bottom half.

(c) Fold the top half over the filling to reform the loin.

FIGURE 15.14 Tying a roast.

(a) Tie a length of butcher's twine tightly around one end of the roast. After tying the knot securely, twist the length of twine into a loop as shown.

(b) Pass the loop over the end of the roast.

(c) Position the loop an inch or two (2.5–5 cm) from the first loop and pull the loose end to tighten.

(d) Continue making loops and tightening them until the whole length of meat is tied.

(e) Turn the meat upside down. Pass the loose end of the twine under the last loop and wind it once around as shown.

(f) Repeat step (e) with each of the loops. When you reach the end of the roast, tie it off securely to the short length of twine from the knot made in the first step.

FIGURE 15.15 Trimming, cutting, and pounding veal for scaloppine.

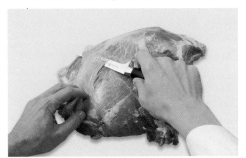

(a) Remove all tendons and connective tissue (silverskin) from the veal. Slip the point of a thin boning knife under the skin. Angle the edge of the blade upward against the skin and cut it away carefully without cutting through the meat.

(b) Holding the blade of the knife at an angle if necessary to get a broader slice, cut across the grain of the meat as shown to make thin slices.

(c) Broader slices can be cut from narrower pieces of meat by butterflying. Cut the slice almost through the meat but . . .

(d) . . . leave it attached.

(e) Then cut a second slice the same way, but cut all the way through.

(f) A butterflied scaloppine is twice as large as a single slice. Unfortunately, it has a seam in the center that often detracts from the appearance of the finished dish, unless the veal is breaded or covered with a topping.

(g) If desired, pound the cutlet to an even thickness with a cutlet mallet. This helps disguise the seam in a butterflied cutlet.

FIGURE 15.16 Trimming a pork tenderloin.

FIGURE 15.17 For stuffed pork chops, cut a pocket in the chops as shown.

FIGURE 15.18 Larding meat using a larding needle.

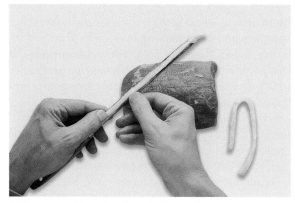

(a) Cut a strip of fatback to fit inside the needle.

(b) Insert the needle through the meat. Pull out the needle, holding the strip of fat so it stays inside the meat.

COOKING AND HANDLING MEATS

TENDERNESS AND APPROPRIATE COOKING METHODS

The heat of cooking affects tenderness in two ways:

1. It tenderizes connective tissue if moisture is present and cooking is slow.
2. It toughens protein. Even meats low in connective tissue can be tough and dry if cooked at excessively high heats for too long.

THE PRINCIPLES OF LOW-HEAT COOKING

1. High heat toughens and shrinks protein and results in excessive moisture loss. Therefore, low-heat cooking should be the general practice for most meat cooking methods.
2. Broiling seems to be a contradiction to this rule. The reason carefully broiled meat stays tender is that it is done quickly. It takes time for the heat to be conducted to the interior of the meat, so the inside never gets very hot. Meat broiled to the point of being well done, however, is likely to be dry.

MILK-FED, GRAIN-FED, OR GRASS-FED

The properties of meats are determined, in part, by the diet of the animals. Most of the beef on the market in North America is grain-fed, even though grass, not grain, is the natural diet of cattle. Feeding cattle grain enables producers to raise and fatten them for market more quickly than letting them browse on grass does. Grain-fed beef is tender and has more marbling than grass-fed beef, and it is preferred by most North American consumers. Grass-fed or pastured beef is usually perceived as less tender and less juicy, although it is lower in saturated fat and may have more health benefits. Its flavor is often described as "beefier" than that of grain-fed beef. Grass-fed beef is common in some other countries, such as the beef-eating and beef-producing nation of Argentina. In North America, producers of pastured beef are campaigning for more consumer recognition.

The effect of diet can be seen in other meat animals. Traditionally, the highest-quality veal is considered to be so-called milk-fed veal, more properly called *formula-fed*. The meat is light pink in color and mild and delicate in flavor. Calves fed solid food or allowed into a pasture have darker, more reddish meat with a somewhat beefier flavor. Ethical objections exist about the raising of formula-fed veal because the animals are penned and not allowed much movement. As for the flavor of milk-fed versus pastured veal, this is a matter of personal preference.

The youngest lamb is called *spring lamb*. It is slaughtered before it begins a diet of solid food, and its meat is light in color and delicate in flavor. Older lamb is darker in color and has a more pronounced flavor. After the age of one year, this meat is no longer called lamb but *mutton*, and it has a still darker color and stronger flavor. Little mutton is sold in North America. (In some markets the name mutton may also be used for goat meat, although this is not traditional English usage.)

3. Roasts cooked at low temperatures have better yields than those roasted at high heat—that is, they shrink less and lose less moisture.

4. Because both liquid and steam are better conductors of heat than air, moist heat penetrates meat quickly. Therefore, to avoid overcooking, meat should be simmered, never boiled.

BREAKING DOWN CONNECTIVE TISSUE

Remember that connective tissue is highest in muscles that are frequently exercised and in mature animals.

Look again at the primary cooking methods (column 4) in the table of meat cuts (p. 461-462). You should detect a pattern of tender cuts, cooked primarily by dry heat; slightly less tender cuts, cooked sometimes by dry and sometimes by moist heat; and least tender cuts, cooked almost always by moist heat.

The concept of moist-heat cooking needs further explanation as it applies to breaking down connective tissue in meat. The usual explanation of the effect of moist heat on connective tissue is that heat breaks down collagen in the presence of moisture. But meat is about 75 percent water, so **moisture is always present**. Collagen breaks down because of long, slow cooking, no matter what cooking method is used.

The catch is that, for small cuts of meat, dry-heat cooking methods are usually short, quick methods. Cooking must be short, in part because too long an exposure to dry heat results in excessive moisture loss from the product. The terms **moist-heat cooking method** and **dry-heat cooking method** refer to the way in which heat is transferred from the heat source to the food, whether by dry means, like hot air or radiation, or moist means, like steam or simmering liquid. Because the product is surrounded by moisture when it is simmered, steamed, or braised, moist-heat cooking methods promote moisture retention, not moisture loss, so cooking time can be as long as desired.

A tough steak on the grill or in the oven doesn't have enough time to become tender before it is dried out. On the other hand, large cuts of less tender meat can be roasted successfully because they are too large to dry out during a long roasting time. A 40-pound (18-kg) roast steamship round of beef can be tender because it takes hours to cook even to the rare stage. A grilled steak cut from the same round, however, is likely to be tough.

To summarize: Long, slow cooking tenderizes collagen. Moist-heat methods are most suitable for long, slow cooking. Dry-heat methods usually are short, quick cooking methods, suitable only for tender cuts, except when larger items are roasted for a relatively long time. The following list summarizes the cooking characteristics of the major cuts.

1. **Rib and loin cuts.**

   Always the most tender cuts, used mostly for roasts, steaks, and chops.

   Beef and lamb. Because these meats are often eaten rare or medium done, the rib and loin are used almost exclusively for roasting, broiling, and grilling.

   Veal and pork. Pork is generally eaten well done, and veal is most often eaten well done, although many people prefer it slightly pink in the center. Therefore, these meats are occasionally braised, not to develop tenderness but to help preserve juices. Veal chops, which are very low in fat, may be broiled if great care is taken not to overcook them and dry them out. A safer approach is to use a method with fat, such as sautéing or pan-frying, or to use moist heat.

2. **Hip, leg or round.**

   Beef. The cuts of the hip are less tender and are used mostly for braising.

   Top grades, such as Canada Prime, Canada AAA, U.S. Prime, and U.S. Choice, can also be roasted. The roasts are so large that, roasted at low temperatures for a long time, the beef's own moisture helps dissolve collagen. Inside round (top round) is favored for roasts because of its size and relative tenderness.

   Beef hip is very lean. It is best roasted rare. Lack of fat makes well-done round taste dry.

Veal, lamb, and pork. These meats are from young animals and therefore tender enough to roast.

Legs make excellent roasts because large muscles with few seams and uniform grain allow easy slicing and attractive portions.

Figure 15.19 shows the muscle structure of the round in cross section. A center-cut steak from a whole round of beef, lamb, veal, or pork has this same basic structure.

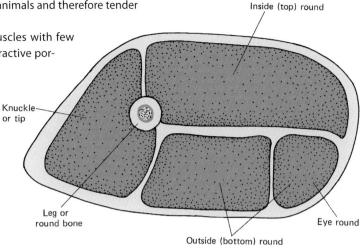

FIGURE 15.19 Location of the muscles in a whole center-cut round steak of beef, veal, lamb, or pork.

3. Chuck or shoulder.

Beef. Beef chuck is a tougher cut that is usually braised. Although chuck is not the ideal choice for braising if uniform slices are desired, it makes braised dishes of excellent eating quality. Its connective tissue is easily broken down by moist cooking, yielding moist, tender meat with abundant gelatin content.

Veal, lamb, and pork. These are most often braised but are young enough to be roasted or cut into chops for broiling. Shoulder roasts are not the most desirable because they consist of many small muscles running in several directions. Therefore, they do not produce attractive, solid slices.

4. Shanks, breast, brisket, and flank.

These are the least tender cuts, even on young animals, and are almost always cooked by moist heat.

Shanks are desirable for braising and simmering because their high collagen content is converted into gelatin that gives body to braising liquids and good eating quality to the meat.

Beef flank steaks can be broiled (as London broil) if they are cooked rare and cut across the grain into thin slices. This cuts the connective tissue into chewable pieces (see mechanical tenderization, p. 453).

5. Ground meat, cubed steaks, and stew meat.

These can come from any primal cut. They are usually made from trimmings, although whole chucks are sometimes ground into chopped meat. Ground meat and cubed steaks can be cooked by dry or moist heat because they are mechanically tenderized. Stew meat is, of course, cooked by moist heat.

OTHER FACTORS INFLUENCING CHOICE OF COOKING METHODS

1. Fat content.

Meats high in fat, such as Prime beef or lamb, are generally cooked without added fat, such as by roasting or broiling.

Meats low in fat, such as veal, are often cooked with added fat to prevent dryness. Sautéing, pan-frying, or braising is generally preferable to broiling for veal chops that are cooked well done.

Fat can be added to lean meats in two ways:

- **Barding**. Tying slices of fat, such as pork fatback, over meats with no natural fat cover to protect them while roasting.

- **Larding**. Inserting strips of fat with a larding needle into meats low in marbling.

These two techniques were developed in Europe when meats were much leaner and not as tender. They are not often used with today's tender, grain-fed meats. These techniques are useful, however, when cooking lean game, such as venison.

2. Developing tenderness is not the only goal of cooking.

Other goals are:

- Developing flavor.
- Preventing excessive shrinkage and nutrient loss.
- Developing appearance.

You must often compromise to get a balanced result. For example, preliminary browning of a roast at high heat increases shrinkage but may be desirable for some roasts to develop flavor and appearance.

SEARING AND "SEALING"

Searing

Searing meats at high heat creates desirable flavor and color by browning the surfaces. It was long believed that searing the surface of meat "seals the pores," keeping in juices.

This does not actually happen. Meat does not have pores but rather an open network of fibers. Think of the surface of a steak as resembling the cut end of a thick rope. There are no pores to seal. It is true that heavy browning creates a kind of crust on the surface of the meat, but this crust is no more waterproof than an unbrowned surface.

You can easily demonstrate this. Place a steak or chop on a hot griddle or grill and sear it well. Turn it over and continue cooking. As it cooks, you will see meat juices driven up through the seared top surface. You will continue to hear a sizzling sound, which is the sound of moisture escaping from the meat and quickly vaporizing. Remove the finished steak from the grill and let it set on a plate for a few minutes, and you will see a small pool of juices collect. Everyone who has cooked a steak has seen this demonstration that searing doesn't seal.

Roasts cooked from the start at a low temperature retain more juices than roasts that are seared at high heat first.

Steaks, chops, and cutlets cooked quickly at high heat retain more moisture at first because the intense heat instantly evaporates the juices from the surface of the meat and forces internal juices further into the meat. This permits browning, because moisture creates steam and inhibits browning. However, overcooked steaks are dry whether or not they were seared.

Blanching and "Sealing"

Dropping meat into boiling water doesn't seal the pores, either. What actually happens is this: Many proteins dissolve in cold water. When heated, these proteins coagulate and become froth or scum on the surface of the water. When meat is placed into boiling water, some of the protein coagulates inside that meat, and not as much is carried out of the meat with the lost moisture. Prolonged cooking shrinks meat as much if started in boiling water as if started in cold water.

COOKING FROZEN MEATS

Some sources recommend cooking some meats from the frozen state, without thawing, in order to eliminate drip loss that occurs during defrosting. However, it is usually better to thaw before cooking because of the following reasons:

1. Frozen meats lose no moisture from defrosting but lose more during cooking. The total loss is about the same as for thawed meats. Besides, the perception of juiciness depends as much or more on fat content than on moisture content.

2. Cooking frozen meats complicates the cooking process and requires adjustments in procedure. It is possible for roasts to be cooked on the outside but still frozen in the center. Frozen steaks, too, are more difficult to cook evenly than thawed steaks. Thawed meats, on the other hand, are handled like fresh meats.

3. Cooking frozen meats requires extra energy, and energy is expensive. A hard-frozen roast may take three times as long to cook as a thawed roast.

DONENESS

DEFINITIONS

The meaning of the term **doneness** depends on whether the cooking method uses dry heat or moist heat.

1. Dry heat.

Meat is "done" when the proteins have reached the desired degree of coagulation (see p. 108), as indicated by internal temperature.

2. Moist heat.

Meat is "done" when connective tissues have broken down enough for the meat to be palatable. With a few exceptions, meat cooked by moist heat is always well done.

DRY-HEAT COOKING

The object of dry-heat cooking is to achieve the desired degree of doneness (protein coagulation) while preserving natural tenderness and juiciness.

Degree of Doneness

As meat cooks, its pigments change color. These color changes indicate degrees of doneness.

Red meat (beef and lamb) changes from red to pink to gray or gray-brown.

- Rare: browned surface; thin layer of cooked (gray) meat; red interior
- Medium: thicker layer of gray; pink interior
- Well done: gray throughout

(Of course, there are stages in between.)

White meat (veal and pork) changes from pink or gray-pink to white or off-white. It is generally cooked well done, although many cuts of veal may be considered done when still slightly pink in the center.

Trichinosis is a disease caused by a parasite that lives in the muscle tissue of hogs and some wild animals. In Canada and the United States, cases of trichinosis from pork are virtually nonexistent, but in countries in which this disease is a problem, pork must be cooked long enough to eliminate this danger. This parasite is killed at 137°F (58°C), but, to be safe, pork should be cooked to at least 145° (63°C). At this stage, pork is only medium to medium-well done. Some people are happy to eat pork that is still pink in the center, but most people prefer it to be cooked slightly more than this. On the other hand, it is not necessary to cook pork to 185°F (85°C), as older guidelines said. At this temperature, pork is overcooked and dry. For diners who avoid any trace of pink in pork, perhaps the best doneness range is 160°–170°F (71°–77°C). The Canadian Food Inspection Agency stipulates a temperature of 160°F (71°C) for safety from salmonella and other pathogens.

Testing Doneness

Determining doneness is one of the most difficult and critical aspects of meat cooking. Anyone can put a steak on the grill or a roast in the oven. But it takes experience and skill to take it off the fire at the right time.

Color change cannot be used to test doneness because it would be necessary to cut the meat. Piercing the meat and examining the color of the juices is not a reliable method.

Internal Temperature

Testing the interior of meat with a meat thermometer is the most accurate method of testing doneness. Thermometers are of two types: *standard*, which are inserted before roasting and left in the roast; and *instant-read*, which are inserted at any time, read as soon as the needle stops moving, and pulled out. Whatever thermometer you use, make sure it is *clean and sanitary* before inserting it in the meat.

The tip of the thermometer should be inserted into the center of the thickest part of the flesh, not touching fat or bone. Table 15.2 gives internal temperatures of meats at various degrees of doneness.

TABLE 15.2 Interior Temperatures of Cooked Meats

Meat	Medium-Rare	Medium	Well Done
Beef	145°F (63°C)	160°F (71°C)	170°F (77°C)
Lamb	145°F (63°C)	160°F (71°C)	170°F (77°C)
Veal	—	160°F (71°C)	170°F (77°C)
Pork	—	—	160°F (71°C)
All ground meat and meat mixtures			160°F (71°C)

Please note that the table indicates government safety standards for meat doneness. The temperatures in the table may not be in agreement with popular preferences. For example, the temperature indicated for medium, 160°F (71°C), is closer to what many people think of as well done. Although preferences vary in different regions of Canada, for many people, rare meat is cooked to around 130°F (54°C), medium to 140°–145°F (60°–63°C), and well done to 160°F (71°C). These temperatures are below what is considered safe, but those who like their steaks rare are not likely to be swayed by this argument and will continue to request meat done to their liking. Each food-service operator must decide whether to please their customers or to follow food safety guidelines in the cooking of meat. In any case, whether or not 145°F (63°C) is the lowest safe temperature for cooking most meats, it is not really accurate to call it *rare*.

Carryover Cooking

Internal temperature continues to rise even after the meat is removed from the oven. This is because the outside of roasting meat is hotter than the inside. This heat continues to be conducted into the meat until the heat is equalized throughout the roast.

Carryover cooking can raise internal temperatures from 5°F (3°C) for small cuts to as much as 25°F (14°C) for very large roasts, such as a steamship round. The usual range is 10°–15°F (6°–8°C) for average roasts. Exact temperature change depends on the size of the cut and on the oven temperature.

Remove roasts from the oven when internal temperature is 10°–15°F (6°–8°C) below the desired reading. Let the roast stand 15–30 minutes before slicing. For example, a beef rib roast cooked rare should be removed from the oven when the thermometer reads 115°–120°F (46°–49°C). Carryover cooking will bring the temperature to 130°F (54°C) after the roast has stood for 30 minutes.

Touch

The small size of steaks and chops makes using a thermometer impractical. The cook must depend on his or her sense of touch.

Meat gets firmer as it cooks. Pressing it lightly with the finger indicates its doneness. Press the center of the lean part, not the fat.

Rare. Feels soft, gives to pressure, though not as soft and jellylike as raw meat.

Medium. Feels moderately firm and resilient, springs back readily when pressed.

Well done. Feels firm, does not give to pressure.

Time-Weight Ratio

Many charts give roasting times per pound of meat. However, these can be approximate only and should be used in estimating and planning cooking times, not in determining doneness.

Many factors other than weight and oven temperature determine cooking time:

1. Temperature of the meat before roasting.
2. Amount of fat cover (fat acts as an insulator).
3. Bones (bones conduct heat faster than flesh, so boneless roasts cook more slowly than bone-in roasts of the same weight).
4. Size, type, and contents of the oven.
5. Number of times the oven door is opened.
6. Shape of the cut (a flat or a long, thin cut cooks more quickly per pound than a round, compact cut).

You can see why roasting requires experience and judgment. To be really accurate and useful, a complete roasting chart that took all variables into consideration, including all meat cuts, sizes, oven temperatures, and so on, would be the size of a small book.

Point 6 above is a key point. It is the *thickness* of a cut, not its *weight*, that determines cooking time—the time needed for the heat to penetrate to the center. Half a pork loin roasts in about the same time as a whole pork loin, even though it weighs half as much. The thickness is the same.

Perhaps the most useful roasting time charts are those you make yourself. When you regularly roast the same cuts in the same way with the same equipment and find they always take the same length of time, you may use those times as indicators of doneness. Many food-service operators have developed charts based on their own practices, and the correct times are indicated on their individual recipe cards.

MOIST-HEAT COOKING

Meat cooked by moist heat is cooked well done and actually beyond well done. Doneness is indicated by tenderness, not by temperature.

Piercing with a meat fork is the usual test for doneness. When the prongs of the fork go in and slide out easily, the meat is done.

Low temperatures—no higher than simmering—are essential to avoid toughening protein in moist-cooked meats. Oven temperatures of 250°–300°F (120°–150°C) are usually sufficient to maintain a simmer.

JUICINESS

Three main factors determine the juiciness—or, more accurately, the perception of juiciness—in cooked meat. Despite the myths about basting with stock and about searing meat to "seal in the juices," the following are the only factors that have any significant effect on juiciness.

1. **Internal fat.**
   Fat makes meat taste juicy. This is why well-marbled meats taste juicier than lean meats. We understand the health effects of too much fat in the diet, but there is no getting around the fact that high fat content makes meat taste juicier. When lean meats are cooked, other measures (such as using sauces and, especially, avoiding overcooking) are used to increase palatability.

2. **Gelatin.**
   This factor is most important in braised meats. Gelatin, converted from connective tissue, helps bind water molecules and hold them in the meat. Also, the texture of the gelatin improves the texture of the meat in the mouth. This is why braised beef shank tastes so much juicier than braised outside round.

3. **Protein coagulation.**
   As you know, as protein coagulates or is cooked, it breaks down and begins to lose water. The more it is cooked, the more it contracts and forces out moisture. No matter how much you try to sear to "seal in the juices," this moisture will be lost. The only way to minimize the loss is to avoid overcooking.

KEY POINTS TO REVIEW

- What are the primal cuts of beef? of lamb? of veal? of pork? What are the main fabricated cuts from each of these primal cuts?

- How do you determine the most appropriate cooking methods for the various fabricated cuts of meat?

- How can you tell when meat is done?

COOKING VARIETY MEATS

Variety meats, also known as offal, include the organs, glands, and other meats that don't form a part of the dressed carcass of the animal.

For cooking purposes, we can divide the most popular variety meats into two groups:

Glandular Meats	Muscle Meats
Liver	Heart
Kidneys	Tongue
Sweetbreads	Tripe
Brains	Oxtails

Glandular meats do not consist of muscle tissue like regular meats but instead are internal organs or glands. This fact is important for two reasons.

First, because they do not consist of bundles of muscle fibers, the texture of glandular meats is unlike that of regular meats. Because they are not muscle tissue, they are naturally tender and do not need long, slow cooking like muscular variety meats do. If organ meats are dry and tough, it is usually because they have been overcooked.

Second, glandular meats are much more perishable than muscle meats. While some muscle meats, especially beef, benefit from aging, organ meats must be very fresh to be of the best quality. Liver, sweetbreads, and brains must be used within a day or two after purchase. If brains or sweetbreads must be kept longer, they should be blanched as described below so they will keep another day or two.

Heart, tongue, oxtails, and tripe are made of muscle tissue, just like other meats from the carcass. They are all tough, however, and must be cooked for a long time by simmering or braising in order to be made tender.

LIVER

Calf's liver is the most prized because it is tender and delicate in flavor. It is easily recognized by its pale, pinkish color. Most calf's liver is served pan-fried, sautéed, or broiled.

Beef liver is darker in color (see accompanying photo), stronger in flavor, and tougher than calf's liver. It is also pan-fried or broiled, and it is frequently braised.

Pork liver is also available, but it is used mostly in pâtés and sausages.

Top: milk-fed calf's-liver slice;
Bottom: beef-liver slice

Preparation

- Remove outer skin.
- Slice on the bias about ¹/₄ inch (0.5 cm) thick. Slicing is easier if the liver is partially frozen.
- Remove tough membranes.

Cooking

Cook to order. Do not cook ahead.

- To broil: Brush with (or dip in) oil or melted butter. Broil according to basic procedure for meats.
- To pan-fry, griddle, or sauté: Dredge in seasoned flour. Cook in desired fat over moderately high heat.
- *Do not overcook*, unless customer requests well done. To be moist, liver must be slightly pink inside. Liver cooked well done is very dry.
- Serve with bacon, French-fried or smothered onions, or seasoned butter.

KIDNEYS

Veal and lamb kidneys are the most popular, especially in the more upscale restaurants. They are usually prepared by sautéing and broiling. Beef kidneys are tougher and more strongly flavored. They are often cooked by braising and served in specialty items, like steak and kidney pie. Pork livers are not often used.

Veal kidneys weigh 8 to 12 oz (225 to 350 g) each. Lamb kidneys are very small, $1\frac{1}{2}$ to 3 oz (40 to 85 g) each. If you purchase whole lamb or veal carcasses, you will find a pair of kidneys inside the cavity, attached to the small of the back in the region of the tenderloin and surrounded by a heavy layer of fat or suet.

Preparation

If the kidney is encased in fat, pull the fat away with your hands and use a knife to cut it away from the core area where the ducts emerge from inside the kidney.

Lamb kidneys are usually broiled and served two or three per portion, or as part of a mixed grill. Butterfly them by splitting them almost in half, starting at the curved or convex side. Spread them open and skewer them to hold them open during cooking.

Veal kidneys can be broiled like lamb kidneys, but they are most often cut up, sautéed, and served in a sauce. To prepare them for sautéing, first split them in half. Remove the white ducts from the center. Then cut into large dice or thick slices.

Cooking

There are two main pitfalls to cooking kidneys. First, they become tough and rubbery if overcooked. Properly cooked, they are pink in the middle and still tender and juicy. Cooking time is very short.

Second, they have a high moisture content, which can interfere with proper sautéing. Make sure the pan is very hot before adding the kidneys, and do not overcrowd the pan. Failure to do this results in kidneys that are boiled in their juices rather than sautéed.

To avoid overcooking when sautéing over high heat, do not try to brown the kidneys too heavily. Brown them only lightly and remove them from the pan when they are still somewhat rare. Set them aside while you deglaze the pan and prepare the sauce. During this time, some juices will be released from the kidneys. Drain this juice and add it to the sauce if desired, or discard it if you feel the flavor is too strong. Finally, add the kidneys to the sauce and warm them gently. Do not let them simmer long. Serve at once.

Left: lamb kidney. Right: veal kidney

SWEETBREADS

Sweetbreads are the thymus glands of calves and young beef animals. (The gland gradually disappears as the animal matures.) They are considered a delicacy and are often expensive. Sweetbreads are mild in flavor and delicate in texture. They are usually braised or breaded and sautéed in butter.

Before cooking, sweetbreads should be prepared according to the following procedure (see Figure 15.20):

1. Soak in several changes of cold water for several hours or overnight. This removes blood, which would darken the meat when cooked.

2. Blanch in simmering salted water for 10 minutes. Some chefs like to add a little lemon juice or vinegar to the water to preserve whiteness and make the meat firmer.

3. Refresh under cold water and peel off membranes and connective tissue.

4. Press between two trays, with a light weight on top, and refrigerate for several hours. If desired, wrap in cheesecloth before pressing, as shown in Figure 15.20.

5. Prepare for cooking:
   - For braising, leave whole or cut into large dice.
   - For breading and sautéing, split in half horizontally. Pass through Standard Breading Procedure (p. 151) or dredge in flour.

BRAINS

Brains are not a popular item, but they are delicate in both flavor and texture. Calf's brains are the most frequently used.

FIGURE 15.20 Preparing sweetbreads.

(a) Raw sweetbreads.

(b) After the sweetbread has been blanched, peel off the membrane.

(c) Wrap the sweetbreads in clean cheesecloth.

(d) Tie the ends securely.

(e) Place in a hotel pan or other flat pan and top with another pan.

(f) Place weights in the top pan and refrigerate for several hours.

Brains are very perishable and should be cooked as soon as possible. They are also fragile and must be handled carefully.

Brains must be pre-prepared according to the following procedure. They may then be served hot with black butter (p. 190) or cooled and then dipped in batter, deep-fried, and served with tomato sauce.

1. Soak in fresh water, as for sweetbreads.
2. Peel off outer membrane (this may be done before or after poaching).
3. Poach 20 minutes in court bouillon made of 1 oz (25 mL) lemon juice or vinegar per pint (500 mL) of salted water, plus a bouquet garni.
4. Drain and serve immediately, or cool in fresh, cold water.

HEART

Heart, usually from veal or beef, is very tough and lean. It can be braised or simmered, or it may be ground and added to chopped meat for casserole dishes and meatloaf.

Before cooking, trim coarse fibers and veins inside and at top.

TONGUE

Cooked beef tongue is popular as a cold, sliced meat for sandwiches. It may be fresh, cured, or smoked. Veal and lamb tongues are also available.

Tongue is almost always cooked by simmering. After simmering, remove the skin and trim the gristle at the base of the tongue before slicing.

OXTAILS

Oxtails contain flavorful meat and a rich gelatin content, making them highly desirable for soups and stews.

To disjoint oxtails, cut into sections at the joints with a French knife or butcher knife. Do not use a cleaver, or you may splinter the bones.

TRIPE

Tripe is the muscular stomach lining of meat animals. Although lamb and pork tripe are sometimes available, beef tripe is by far the most widely used. Because cattle have four stomachs, there are four kinds of beef tripe. Honeycomb tripe, from the second stomach, is the kind most widely available. Other kinds, however, can be substituted in recipes that call for honeycomb tripe. In France, another type of beef tripe, known as **gras-double**, is popular; it is smooth rather than honeycombed.

Tripe

Most tripe that comes from the market has been partially cooked, but it still requires several hours of simmering to be made tender. Undercooked tripe is chewy and somewhat rubbery, but tripe that has simmered long enough is tender, with a pleasant gelatinous texture.

To prepare, first remove any lumps of fat by pulling or cutting them off. Next, blanch the tripe, if desired. Although it is already partially cooked when purchased, blanching freshens it. Place it in a pot with cold, salted water. Bring to a boil, simmer 5–10 minutes, drain, and rinse under cold water.

OTHER VARIETY MEATS

Intestines

The most common use for intestines is to make sausage casings. These are discussed in Chapter 27.

Chitterlings are pork intestines that are treated like tripe. They are blanched or simmered, and then braised or fried. Chitterlings are generally available in 10-lb (4.5-kg) pails. Because they shrink a great deal when simmered, this quantity yields only 3 lb (1.3 kg) or less of finished product.

Caul

Pig's **caul** is a fatty membrane covering the animal's stomach. It looks somewhat like a delicate piece of lace. Its main uses are to line terrines and to wrap forcemeats and other foods so they hold their shape during cooking and do not dry out. Sausage patties wrapped in caul are called *crépinettes* (see p. 843). The advantage of using caul instead of fatback to line terrines is that the caul is so thin it melts away almost completely during cooking.

Feet

Feet are exceptionally rich in gelatin. For this reason, they are added to soups, stews, and stocks to add richness and body. Indeed, some stews made with feet, such as Tripes à la Mode de Caen, may be so rich in gelatin that not only do they solidify when cold but they can even be unmolded and sliced like cold cuts.

Pig's feet are readily available in most markets. Calf's feet and ox's feet are also available, but often only on the wholesale market. The feet from older animals have less gelatin. If a recipe calls for a calf's foot but none is available, in most cases you can substitute two pig's feet.

KEY POINTS TO REVIEW

- What is the difference between glandular variety meats and muscular variety meats? List the most important types of each.

- Muscular variety meats are nearly always cooked by what cooking methods? Why?

- What are the most appropriate cooking methods for liver?

- How are sweetbreads prepared for cooking?

GAME AND SPECIALTY MEATS

The term **game** is used to refer to poultry and meat animals normally found in the wild. However, most of the "wild" game that has become so popular on restaurant menus is actually from farm-raised animals. Venison farms, in particular, have become numerous and productive, supplying a growing demand.

Farm-raised game birds are discussed along with other poultry in Chapter 17. This section is concerned with furred game.

Although a great variety of game, large and small, can be found on hunters' tables, the supply of game for the restaurant and retail markets is more limited. Venison, the most popular game item, is the main subject of this section. Other products, such as boar and hare, are occasionally available as well. In addition, domestic rabbit is considered here, although its meat has little in common with true game.

Note that the term *venison* is sometimes used in a broader sense to mean meat from deer, elk, moose, caribou, antelope, and pronghorn. However, when any of these meats is offered for sale, the name of the animal must appear on the packaging.

The French terms for game meats are often used on menus and in cooking manuals and references. To clarify these terms, a list of those most commonly used follows:

Chevreuil: often translated as "venison" the term refers specifically to the roe deer, the most prized European variety.

Cerf: red deer; often farm raised.

Daim: fallow deer; often farm raised.

Marcassin: young boar, especially under six months of age.

Sanglier: boar.

Lapin: rabbit.

Lapereau: young rabbit.

Lièvre: hare.

Levraut: young hare.

Venaison: usually translated as "venison"; the term in fact refers to the meat of any game animal.

VENISON

Several varieties of deer are raised on farms for use as meat, including the red deer and the smaller fallow deer. Deer meat is typically called **venison**. An important advantage of farm-raised venison, besides its year-round availability, is that the cook can be assured it is from young, tender animals. In the wild, young animals less than two years old are likely to have tender meat, but the meat rapidly becomes tough as the animal matures and ages. The tradition of marinating game for several days in strong wine marinades originates, in large part, from efforts to tenderize hunted game enough to make it palatable.

Marination, Flavor, and Tenderness

The first thing to be said about farm-raised venison is that it is milder in flavor than venison hunted in the wild. It has little, if any, of the strong, gamy flavor usually associated with wild game. In fact, a farm-raised venison steak tastes rather like an especially flavorful lean cut of beef. Those who enjoy strong, gamy flavors may even find farmed venison a little bland. Although it does have some tenderizing effect, marination is not necessary for commercially raised venison because the meat is already tender. Nevertheless, marinating is widely used as a flavoring technique. Much of the flavor traditionally associated with venison, in fact, is due less to its gaminess than to the red wine marinades that were invariably used.

To retain more of the natural flavor of the meat, cook it without marination, or let it marinate for only a short period (30 minutes to 3 or 4 hours) with the desired seasonings and flavoring ingredients. Modern quick marinades are often simple and may contain only a few ingredients.

Fat Content

Venison, like other game, is very low in fat. This makes it especially popular with health-conscious diners. The meat is likely to become dry unless the cook takes great care.

The loin and leg, being tender, are best cooked by dry-heat methods and served rare or medium done. If cooked longer, they will dry out. Roast these cuts whole, either bone-in or deboned, or cut them into steaks, cutlets, and medallions, and sauté, pan-fry, or broil them, taking care not to overcook.

Whole leg of venison, completely boned, seamed, and vacuum packed, is available. Weights range from 5 to 10 pounds (2 to 4.5 kg). Whole bone-in saddle weighs 5 to 20 pounds (2.3 to 9 kg), while the loin muscle weighs about half that after boning and trimming.

Tougher cuts, chiefly the shoulder, neck, and breast, are braised, stewed, or made into ground meat or sausage. These cuts are also lean, but because they are higher in connective tissue and gelatin, they take more readily to stewing and braising.

To generalize, farm-raised venison can be treated like very lean beef. Take care not to cook it to the point of dryness.

BOAR

Boar is a type of wild pig. Its meat is somewhat similar to pork, except it is leaner and its flavor fuller and richer. Boar is now raised commercially on a few farms and is available in limited quantities.

BONE STRUCTURE OF GAME

The bone and muscle structure of furred game such as venison and elk is the same as that of familiar meats such as beef and lamb. The carcasses are also broken down and fabricated in the same ways. After you have become familiar with the charts and diagrams on pages 458–461, you can apply the same cuts to venison and other large game.

Unlike those larger game animals, however, rabbit is cut differently, so separate illustrations are provided in this section beginning on page 482.

Boar is somewhat more difficult to cook than venison and other game because, like pork, it must be cooked until well done. At the same time, it is leaner and less tender than domestic pork, so it tends to be somewhat dry and chewy. Special care must be taken to cook it adequately without overcooking. Because boar is usually tougher than farm-raised venison, its legs or hams are better suited for braising or slow roasting, while the loins can be used for roasts or cut into medallions and sautéed.

Traditionally, boar is handled much like venison, and typical recipes call for red wine marinades. Although marinating a white meat in red wine may seem strange at first, this treatment actually works very well with boar. The red wine accentuates the more pronounced flavor of boar (as compared to pork) and makes it taste more like game.

OTHER LARGE GAME

Other meats are sometimes found in food service kitchens. *Elk*, *caribou*, *moose*, and *antelope* are all similar to venison and are handled in much the same way. The first three of these, especially moose, are larger than deer, so it may be necessary to allow for longer cooking times when using venison recipes for them.

Buffalo, or *American bison*, is raised on ranches in western Canada and United States and handled like beef. Flavor and cooking characteristics are similar to those of beef, but the meat is somewhat richer in flavor and has less fat and cholesterol than beef.

RABBIT

Domestic rabbit is a versatile meat that can be cooked in most of the same ways as chicken. In fact, in some countries it is classified as poultry. Some typical recipes for rabbit are included in Chapter 16, but nearly any chicken recipe can be used for domestic rabbit as well. In addition, many recipes for veal or pork are adaptable to rabbit.

Rabbit's light, delicate meat is often compared to chicken, but there are differences. It is somewhat more flavorful than chicken, with a mild but distinctive taste that is not exactly like that of other poultry or meat. Also, it is very lean (more like chicken or turkey breast than legs) and can become dry if overcooked.

Rabbit takes well to marination; it can also be cooked without prior marination. Either way, it can be cooked by long, slow simmering, braising, or stewing, or it can be quickly cooked by sautéing, grilling, or roasting.

The structure of rabbit, of course, is like that of other land mammals rather than like that of poultry. Cutting methods divide the meaty hind legs, the bonier forelegs, and the choice saddle or back section (*râble* in French). The whole carcass, cut up, is used for stews and sautés, while the saddle alone is often roasted. It may be boned or bone-in. (See Figure 15.21.)

Small rabbits, 3 pounds (1.5 kg) or less, are the best for cooking. Mature rabbits, weighing 4 to 5 pounds (about 2 kg), tend to be tougher and drier.

HARE

Hare is a wild cousin of the rabbit. (Please note that rabbits and hares are different animals. The American jackrabbit, for example, is actually a hare, not a rabbit.) Unlike domestic rabbit, with its light-colored, delicate meat, hare has flesh that is dark reddish-brown and gamy.

Hares 7–8 months old and weighing about 6 pounds (2.7 kg) make the best eating. Larger ones, over 8 pounds (3.6 kg), are likely to be tough and stringy.

Because its structure is the same, hare is cut the same way as rabbit.

Roast Saddle of Hare

Like other game, hare is very lean and therefore becomes dry if overcooked. If roasted, it should be removed from the oven while rare or at least still pink. Rare roast hare has an attractive, deep red color. A typical classic preparation of saddle of hare is as follows. Note that this is also the classic treatment for roast venison.

1. Marinate the saddle of hare in a red wine marinade (such as the venison marinade on p. 497).

INSPECTION OF GAME

The Canadian Food Inspection Agency (CFIA) oversees inspection of game offered for sale. Just as for domesticated meats, federal and provincial inspection of farm-raised game for wholesomeness is mandatory. Game may be imported only from countries that have approved inspection systems. Wild game that can be hunted legally under federal or provincial authority may not be sold.

FIGURE 15.21 Cutting rabbit for stews and sautés.

(a) Cut off the hind legs, separating them at the hip joint.

(b) Cut off the forelegs by cutting under the shoulder blade.

(c) Cut off the hip bone.

(d) Cut through the backbone to separate the bony rib section from the meaty loin or saddle.

(e) Carefully separate the remaining rib bones from the loin and remove them. The saddle can then be cut crosswise through the backbone into pieces if desired.

(f) This is the cut-up rabbit, with the forelegs and rib section on the left, the saddle in the center, and the hind legs and hip bone on the right.

2. Brown it on top of the stove and roast it rare or medium done, about 15 minutes at 425°F (220°C).

3. Remove the loin muscles from the bone and cut lengthwise into thin slices. Remove the tenderloins from the underside of the saddle and leave whole or slice as desired.

4. Serve with a poîvrade sauce (p. 187).

STORAGE OF MEATS

The quality of a finished meat product depends not only on proper selection and cooking of the meat but also on its proper storage. Fresh meat is highly perishable. The high cost of meat makes it essential to avoid spoilage.

FRESH MEATS

1. Check purchases on arrival to ensure the purchased meat is of good quality.

2. Do not wrap tightly. Bacteria and mold thrive in moist, stagnant places. Air circulation inhibits their growth. Store meat loosely arranged on pans or racks to allow air circulation between pieces, but cover cut surfaces to prevent excessive drying.

3. Do not open vacuum-packed meats until ready to use.

4. Store at 32°–36°F (0°–2°C). Meat does not freeze until about 28°F (–2°C).

5. Keep meats separated in the cooler (or, even better, in separate coolers) and on the work-table to avoid cross-contamination.

6. Use as soon as possible. Fresh meats keep well only two to four days. Ground meats keep even less well because so much surface area is exposed to bacteria. Cured and smoked products may keep up to one week. For these reasons, frequent deliveries are better than long storage.

7. Do not try to rescue meats that are going bad by freezing them. Freezing will not improve the quality of spoiling meat.

8. Keep coolers clean.

Frozen Meats

1. Wrap frozen meats well to prevent freezer burn.

2. Store at 0°F (−18°C) or colder.

3. Rotate stock—first in, first out. Frozen meats do not keep indefinitely. Recommended shelf life at 0°F (−18°C) for beef, veal, and lamb: 6 months; for pork: 4 months (pork fat turns rancid easily in the freezer).

4. Defrost carefully. Tempering in the refrigerator is best. Defrosting at room temperature encourages bacterial growth.

5. Do not refreeze thawed meats. Refreezing increases loss of quality.

6. Keep freezers clean.

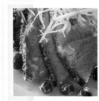

KEY POINTS TO REVIEW

- What are the most important kinds of game used in food service?

- How does the fat content of game meats affect its cooking qualities?

- What are the proper ways to store fresh meats? frozen meats?

TERMS FOR REVIEW

coagulation	aging	irradiation	gras-double
marbling	Cryovac®	silverskin	caul
connective tissue	dry aging	barding	game
collagen	primal cuts	larding	venison
elastin	fabricated (cuts)	doneness	boar
inspection	portion-controlled cuts	carryover cooking	hare
grading	butcher	variety meats	
yield grading	fabricate	sweetbreads	
green meat	carve	tripe	

QUESTIONS FOR DISCUSSION

1. Many people assume that the leaner a meat is, the better it is. Do you agree? Explain.

2. What is connective tissue? Why is it important for the cook to understand connective tissue?

3. Flank steak (beef) is high in connective tissue, yet it is often broiled and served in thin slices as London broil. How is this possible?

4. Why are portion-controlled meats so widely used in food service, even though their per-pound cost is higher?

5. Can you explain why veal loin, a tender cut, is sometimes braised, while veal shoulder, a less tender cut, is sometimes roasted?

6. Which of the following cuts are you more likely to braise? Which might you roast?

Beef chuck	Ground pork	Beef rib
Corned beef brisket	Veal rib	Lamb leg
Lamb shanks	Beef strip loin	Pork shoulder

7. You wish to cook a roast rib of beef to a final internal temperature of 145°F (63°C). Why, then, would you remove the roast from the oven when the temperature on the meat thermometer reads 130°F (54°C)?

8. Why are weight-time roasting charts inadequate for determining the doneness of roast meats?

9. Describe the fat content of game meats such as venison, boar, and elk. Explain how the fat content affects how these meats are handled and cooked.

10. How does farm-raised venison differ from wild venison?

16

COOKING MEATS AND GAME

This chapter presents a collection of recipes to give you practice with basic cooking methods as they apply to beef, lamb, veal, pork, and game such as venison. Be sure you have a thorough understanding of the cooking methods described in Chapter 14, and review that material as often as necessary.

It is also important that you have read and understood the basic material in Chapter 15, especially the sections on matching particular cuts to appropriate cooking methods and on testing for doneness. If necessary, please review those sections.

The procedures described in Chapter 14 are general. Be aware they may be modified slightly in specific recipes. Some of the most important of these variations are discussed at the beginning of each section in this chapter. In addition, your instructors may wish to show you variations or methods that differ from those presented here.

Each time you prepare one of these recipes, you should be thinking not just about the product you are making but also about the techniques you are using and how they can be applied to other products. It is helpful to compare the recipes in each section, how they are alike and how they are different. This way you will be learning to cook, not just to follow recipes.

AFTER READING THIS CHAPTER, YOU SHOULD BE ABLE TO

1. Cook meats by roasting, baking, and barbecuing.

2. Cook meats by broiling, grilling, and pan-broiling.

3. Cook meats by sautéing and pan-frying.

4. Cook meats by simmering, submersion poaching, steaming, and sous vide.

5. Cook meats by braising.

ROASTING, BAKING, AND BARBECUING

The roasting and barbecuing procedures are explained in detail in Chapter 14 (pp. 414–424). Note that the core recipe illustrating the roasting procedure is Roast Chicken. If you have completed that unit, you had the opportunity to practice this cooking method with an inexpensive product. In this chapter, the first recipe illustrating the roasting method involves greater food cost, so it is especially important to review the procedure thoroughly. Note also that, unlike the roast chicken in Chapter 14, the beef roast recipe here includes the procedure for making an unthickened *jus.* If a thickened gravy is desired, a separate recipe follows.

The remaining recipes in this section present a cross section of roasted meats, illustrating variations on the roasting method. In addition, barbecue recipes are also placed in this section, as barbecuing is low-temperature roasting with smoke.

Among the other recipes here are meats that are marinated before roasting, as well as one recipe for a brined roast. Both marinades and brines are discussed in Chapter 7, and a detailed procedure for brining is found on page 149. Review this material as necessary.

 # Roast Rib of Beef au Jus

YIELD: 5 LB (2.3 KG) BONELESS, TRIMMED MEAT PORTIONS: 12 PORTION SIZE: 6¹/₂ OZ (190 G), 1¹/₂ OZ (50 ML) JUS
                                   10 8 OZ (225 G), 1¹/₂ OZ (50 ML) JUS

U.S.	METRIC	INGREDIENTS	PROCEDURE

10 lb	4.5 kg	Beef rib, roast ready, bone in (one half of an average-size rib roast)	1. Place the meat fat side up in a roasting pan.
			2. Place meat in a preheated 300°F (150°C) oven. Roast until rare or medium done, as desired, allowing for carryover cooking. Internal temperature (tested with an instant-read thermometer):
			    Rare: 120°F (49°C)
			    Medium 130°F (54°C)
			    (Outer slices will be cooked more than center.)
			    Roasting time will be at least 3–4 hours.
			3. Remove the meat from the pan and let stand in a warm place 30 minutes before carving.

		Mirepoix:	4. Drain off all but 1–2 oz (50 g) of the fat from the roasting pan. Be careful to retain any juices in the pan. Add the mirepoix to the pan (see **Figure 16.1**).
4 oz	125 g	Onion	5. Set the pan over high heat and cook until mirepoix is brown and moisture has evaporated, leaving only fat, mirepoix, and browned drippings.
2 oz	60 g	Carrot	6. Pour off any excess fat.
2 oz	60 g	Celery	
1 qt	1 L	Brown stock	7. Pour about 8 fl oz (250 mL) stock into the roasting pan to deglaze it. Stir over heat until brown drippings are dissolved.
to taste	to taste	Salt	8. Pour the deglazing liquid and mirepoix into a saucepot with the remaining stock. Simmer until mirepoix is soft and liquid is reduced by about one-third.
to taste	to taste	Pepper	9. Strain through a china cap lined with cheesecloth into a bain-marie. Skim fat carefully. Season to taste with salt and pepper. 10. For service, stand the roast on its widest end. Cut down beside the bones to free the meat, and slice the meat across the grain. 11. Serve each portion with 1¹/₂ oz (50 mL) jus.

Per serving: Calories, 810; Protein, 52 g; Fat, 65 g (74% cal.); Cholesterol, 180 mg; Carbohydrates, 0 g; Fiber, 0 g; Sodium, 150 mg.

VARIATIONS

Roast Rib-Eye Roll, Top Round, Sirloin, or Strip Loin

These cuts may be roasted by the same procedure; roast them on a rack.

Roast Beef with Gravy

Roast the desired cut of beef according to the basic recipe. Prepare gravy according to the following recipe.

FIGURE 16.1 Preparing pan jus.

(a) After removing the cooked meat from the roasting pan, degrease the pan. Add mirepoix to the pan and brown on the stovetop or in the oven.

(b) Deglaze with brown stock.

(c) Pour the mirepoix and deglazing liquid into a saucepan. Simmer for the desired time.

(d) Strain through a fine chinois or a china cap lined with cheesecloth.

Roast Beef Gravy ♥

YIELD: ABOUT 1½ PT (750ML)   PORTIONS: 12   PORTION SIZE: 2 FL OZ (60 ML)

U.S.	METRIC	INGREDIENTS	PROCEDURE
as needed	as needed	Pan drippings from roast beef (previous recipe)	1. After removing the roast, add the mirepoix to the drippings in the roasting pan.
		Mirepoix:	2. Set the pan over high heat and cook until mirepoix is brown and moisture has evaporated, leaving only fat, mirepoix, and browned drippings (see **Figure 16.1**).
4 oz	125 g	Onion	
2 oz	60 g	Carrot	
2 oz	60 g	Celery	3. Pour off and save the fat.
1 qt	1 L	Brown stock	4. Deglaze the pan with some of the stock. Pour the deglazing liquid and mirepoix into a saucepot with the remaining stock. Add the tomato purée. Bring to a boil and reduce heat to a simmer.
2 oz	60 g	Tomato purée	
2 oz	60 g	Flour	5. Make a brown roux with the flour and 2 oz (60 g) of the reserved fat. Cool the roux slightly and beat it into the simmering stock to thicken it.
to taste	to taste	Salt	
to taste	to taste	Pepper	6. Simmer 15–20 minutes, or until all raw flour taste is cooked out and the liquid is reduced slightly.
to taste	to taste	Worcestershire sauce	7. Strain through a china cap into a bain-marie.
			8. Season to taste with salt, pepper, and Worcestershire sauce.

Per serving: Calories. 35; Protein, 1 g; Fat, 1.5 g (40% cal.); Cholesterol, 5 mg; Carbohydrates, 4 g; Fiber, 0 g; Sodium, 10 mg.

VARIATION

Jus Lié

Omit tomato purée and roux. Thicken liquid with ¾ oz (25 g) cornstarch or arrowroot blended with 2 fl oz (50 mL) cold water or stock.

Roast Stuffed Shoulder of Lamb

PORTIONS: 10   PORTION SIZE: 5 OZ (150 G) MEAT AND STUFFING, 2 FL OZ (60 ML) GRAVY

U.S.	METRIC	INGREDIENTS	PROCEDURE
		Stuffing:	1. Sauté the onion and garlic in oil until soft. Remove from heat and cool.
4 oz	125 g	Onion, fine dice	
1 tsp	5 mL	Garlic, chopped fine	2. Combine onion and garlic with the remaining stuffing ingredients and mix lightly.
2 fl oz	60 g	Olive oil, vegetable oil, or butter	
3 oz	100 g	Soft, fresh bread crumbs (about 2 cups)	
1/3 cup	80 mL	Chopped parsley	
1/2 tsp	2 mL	Dried rosemary	
1/4 tsp	1 mL	Black pepper	
1/2 tsp	2 mL	Salt	
1	1	Egg, beaten	
1	1	Boneless lamb shoulder, about 4 lb (1.8 kg)	3. Lay the lamb shoulder out flat, fat side down (see **Figure 16.2**).
as needed	as needed	Oil	4. Spread the lamb with the stuffing and roll it up. Tie the roll tightly.
as needed	as needed	Salt	5. Rub the meat with oil, salt, pepper, and rosemary.
as needed	as needed	Pepper	6. Place the meat on a rack in a roasting pan. Insert a meat thermometer into the thickest part of the meat (not into the stuffing).
as needed	as needed	Dried rosemary	7. Place in a 325°F (165°C) oven. Roast the meat about 1 1/2 hours.
		Mirepoix:	8. Place the mirepoix in the bottom of the roasting pan. Baste the meat with fat and continue to roast until the thermometer reads 160°F (71°C). Total cooking time is about 2 1/2 hours.
4 oz	125 g	Onion, chopped	
2 oz	60 g	Carrot, chopped	9. Remove the roast from the pan and let stand in a warm place.
2 oz	60 g	Celery, chopped	
2 oz	60 g	Flour	10. Set the roasting pan over high heat to clarify the fat and finish browning the mirepoix. Drain off about three-fourths of the fat.
1 qt	1 L	Brown beef stock or lamb stock	11. Add the flour to the pan to make a roux, cooking it until it is brown.
4 oz	125 g	Tomatoes, canned	12. Stir in the stock and tomatoes and bring to a boil. Simmer, stirring, until the gravy is thickened and reduced to about 1 1/2 pt (750 mL).
to taste	to taste	Salt	13. Strain and skim excess fat.
to taste	to taste	Pepper	14. Season to taste with salt and pepper.
			15. Slice the roast crosswise, so each slice contains stuffing in the center. When slicing, be careful to keep the slices from falling apart. Serve each portion with 2 oz (60 mL) gravy.

Per serving: Calories, 390; Protein, 31 g; Fat, 25 g (57% cal.); Cholesterol, 120 mg; Carbohydrates, 11 g; Fiber, 1 g; Sodium, 275 mg.

VARIATIONS

Roast Boneless Shoulder of Lamb

Roast a tied, boneless lamb shoulder as in basic recipe, without stuffing.

Roast Leg of Lamb

Prepare leg of lamb for roasting as shown in Figure 15.11. Rub with oil, salt, pepper, rosemary, and garlic. Roast as in basic recipe (without stuffing) to rare, medium, or well-done stage. Leg of lamb may be served with natural juices (au jus) instead of thickened gravy, if desired. An 8 lb (3.6 kg) AP leg of lamb yields about 3 1/2 lb (1.6 kg) cooked meat. Yield is less if cooked well done. See Figure 16.3 for carving technique.

Roast Leg of Lamb Boulangère

About 1 1/2 hours before lamb is done, transfer the meat to a rack over a pan of Boulangère Potatoes (p. 358) and finish cooking.

Roast Stuffed Shoulder of Lamb

FIGURE 16.2 Stuffing a lamb shoulder.

(a) Bone out the shoulder. To remove the blade bone, slide the knife against the bone to separate it from the meat.

(b) Spread the boned shoulder with the stuffing.

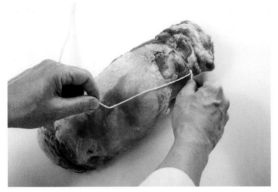

(c) Roll the shoulder and tie it securely. (See Figure 15.14 for tying procedure.)

FIGURE 16.3 Carving a leg of lamb. Hams and other leg roasts may be carved using the same basic technique shown here.

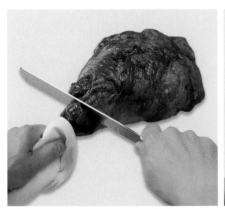

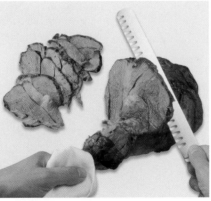

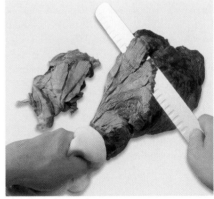

(a) Place the roast on a clean, sanitary cutting board. Begin by making a vertical cut through to the bone about 1 in. (2.5 cm) from the end of the shank meat. The small collar of shank meat forms a guard to protect the hand in case the knife slips.

(b) Using long, smooth strokes, cut thin slices on a slight bias as shown.

(c) When slices become too large, angle the knife. When the top of the roast is completely sliced, turn over and repeat the procedure on the bottom of the roast.

Roast Rack of Lamb

PORTIONS: 8   PORTION SIZE: 2 CHOPS, 1 FL OZ (30 ML) JUS

U.S.	METRIC	INGREDIENTS	PROCEDURE
2	2	Racks of lamb, 8 ribs each	1. Prepare lamb for roasting as shown in **Figure 15.12**.
to taste	to taste	Salt	2. Place any trimmed-off bones in the bottom of a roasting pan. Place the meat fat side up on top of the bones. Season with salt, pepper, and thyme.
to taste	to taste	Pepper	
to taste	to taste	Dried thyme	3. Place in a hot oven (450°F/230°C) and roast to desired doneness. Rack of lamb is usually roasted rare or medium. Test doneness with a meat thermometer or by the touch method, as for steaks. Total time is about 30 minutes.
			4. Remove the lamb from the roasting pan and hold in a warm place. Leave the bones in the pan.
			5. Set the roasting pan over moderate heat to caramelize the juices and clarify the fat. Pour off the fat.
2	2	Garlic cloves, chopped	6. Add the garlic to the pan and cook 1 minute.
1 pt	500 mL	White or brown veal stock	7. Deglaze the pan with the stock and reduce by half. Strain, degrease, and season to taste.
			8. Cut the meat between the ribs into chops. Serve 2 chops per portion with 1 oz (30 mL) jus.

Per serving: Calories, 280; Protein, 19 g; Fat, 22 g (72% cal.); Cholesterol, 75 mg; Carbohydrates, 0 g; Fiber 0 g; Sodium, 70 mg.

VARIATIONS

Rack of Lamb aux Primeurs (with Spring Vegetables)

Place the racks on one or two heated serving platters. Garnish the platters with an assortment of spring vegetables, cooked separately: tournéed carrots, tournéed turnips, buttered peas, green beans, rissolé potatoes. Pour the jus into a warm gooseneck or sauceboat. Carve and serve the meat, vegetables, and jus in the dining room.

Rack of Lamb Persillé

Prepare as in basic recipe. Combine the ingredients for Persillade, listed below. Before carving and serving, spread the top (fat side) of each rack with 1 tbsp (15 g) soft butter. Pack the persillade onto the top of the racks and brown under the salamander.

Persillade

4 cloves garlic, minced
2 oz (60 g) fresh bread crumbs (about 1 cup)
1/3 cup (80 mL) chopped parsley

PROCEDURE VARIATION: COMBI OVEN

In place of step 3 in the main recipe, substitute the following step:

3. Place in a hot combi oven (375°F) in combi-mode with maximum browning level and roast to desired doneness. Rack of lamb is usually roasted rare or medium. Test doneness with a meat thermometer or by the touch method, as for steaks. Total time is about 25 minutes.

Roast Rack of Lamb; White Beans Bretonne; Steamed Brussels Sprout Leaves

Roast Brined Pork Loin with Date and Gorgonzola Stuffing

PORTIONS: 8   **PORTION SIZE: APPROX. 5 OZ (150 G)**

U.S.	METRIC	INGREDIENTS
6 oz	180 g	Pitted dates, chopped
1 tsp	5 mL	Butter
1/8 tsp	0.5 mL	Dried rosemary
3 fl oz	90 mL	Water
2 oz	60 g	Gorgonzola cheese
3 lb	1.5 kg	Pork loin, center cut, boneless
2 qt, or as needed	2 L, or as needed	Brine (p. 149)
to taste	to taste	Salt
to taste	to taste	Pepper
to taste	to taste	Ground coriander

PROCEDURE

1. In a saucepan, briefly cook the dates in butter until slightly softened.
2. Add the rosemary and water. Cook until the mixture forms a thick paste. Cool.
3. When the date mixture is completely cool, add the gorgonzola and mix in well.
4. If the loin has a fat covering, trim the excess fat from it, leaving only a thin covering. If the loin was purchased without a fat covering, trim off all silverskin.
5. Marinate in enough brine to cover, refrigerated, 24 hours.
6. Stuff the meat. This can be done in either of two ways:
   a. Butterfly the meat as shown in **Figure 15.13**. Spread the date mixture inside and refold the meat to enclose the stuffing. Tie the meat well (**Figure 15.14**).
   b. Poke a hole lengthwise through the center of the loin with a clean, sanitized spoon handle or similar rod-shaped item. Force the date mixture into the hole so it fills the hole completely.
7. If the loin has no fat covering and was trimmed of silverskin in step 4, oil the surface well. Season the surface of the meat lightly with salt, pepper, and ground coriander. Hold in the refrigerator until ready to cook.
8. Preheat the oven to 425°F (220°C). Place the roast in the oven. After 10 minutes, reduce the heat to 325°F (160°C). (The initial roasting at high heat is optional; its purpose is to help brown the roast.) Continue to roast until the internal temperature is 150°F (65°C).
9. Remove from the heat and let stand in a warm place at least 10 minutes before slicing.

Per serving: Calories, 420; Protein, 38 g; Fat, 21 g (18% cal.); Cholesterol, 120 mg; Carbohydrates, 19 g; Fiber, 2 g; Sodium, 540 mg.

Roast Brined Pork Loin with Date and Gorgonzola Stuffing, Spaetzle, Glazed Carrots

Barbecued Pork Shoulder

YIELD: APPROX. 4 LB (2 KG) TRIMMED, BONELESS MEAT

U.S.	METRIC	INGREDIENTS
8 lb	4 kg	Pork shoulder or butt
2 oz, or as needed	60 g, or as needed	Spice Rub I (p. 423) or Spice Rub II (p. 493)

PROCEDURE

1. Trim excess external fat from the meat.
2. Rub the meat with the spice rub so it is completely coated with a thin layer.
3. Refrigerate the meat overnight.
4. Cook the meat on racks in a smoke roaster at 225°F (107°C) until very tender, about 8 hours. Turn the meat over once per 1–2 hours so that it cooks evenly. When done, the blade bone should be loose enough to be pulled out (see photo), and the meat should be tender enough to pull apart with a fork.
5. The pork may be cut from the bones and sliced or cut into chunks and served with a barbecue sauce on the side, or it may be shredded (see photo), mixed with a little barbecue sauce, and used as a sandwich filling. For barbecue sauces, see pp. 213–215.

Per serving: Calories, 60; Protein, 7 g; Fat, 3 g (49% cal.); Cholesterol, 25 mg; Carbohydrates, 0 g; Fiber, 0 g; Sodium, 140 mg.

VARIATION

Barbecued Shoulder of Boar

Substitute shoulder of boar for the pork shoulder.

Finished shoulder

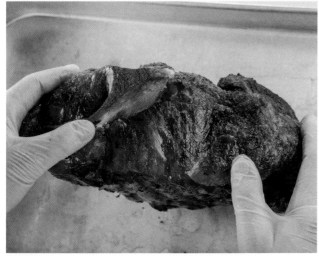

Loose blade bone

Pulling pork

Barbecued Beef Brisket

YIELD: APPROX. 5 LB (2.25 KG)

U.S.	METRIC	INGREDIENTS
8 lb	3.6 kg	Fresh beef brisket, well trimmed, but with a thin layer of fat on top
2 oz, or as needed	60 g, or as needed	Spice Rub I (p. 423)

PROCEDURE

1. Rub the spice mixture evenly over the brisket.
2. Refrigerate the brisket overnight.
3. Place the brisket, fat side up, in a barbecue cooker heated to 225°F (107°C). Cook the meat until tender, approximately 8 hours.
4. Cut across the grain into thin slices, holding the knife at an angle to make broader slices. Serve with desired barbecue sauce.

Per 1 ounce (28.35 g): Calories, 60; Protein, 9 g; Fat, 1.5 g (23% cal.); Cholesterol, 30 mg; Carbohydrates, 0 g; Fiber, 0 g; Sodium, 105 mg.

Spice Rub II

YIELD: APPROX. 7 OZ (200 G)

U.S.	METRIC	INGREDIENTS
1 1/2 oz	45 g	Paprika
1 1/2 oz	45 g	Chili powder
2 tsp	10 mL	Dry mustard
2 tbsp	30 mL	Onion powder
1 tsp	5 mL	Celery seed
2 tsp	10 mL	Dried thyme
1 oz	30 g	Sugar
2 oz	60 g	Salt
2 tbsp	30 mL	Black pepper

PROCEDURE

1. Combine all ingredients and mix well.
2. Store in a tightly sealed container in a dark place.

Per 1 oz (28.35 g): Calories, 70; Protein, 2 g; Fat, 2 g (26% cal.); Cholesterol, 0 mg; Carbohydrates, 14 g; Fiber, 5 g; Sodium, 3250 mg.

Roast Pork Tenderloin with Kalbi Marinade

PORTIONS: 10   **PORTION SIZE: 5 OZ (150 G)**

U.S.	METRIC	INGREDIENTS	PROCEDURE
8 fl oz	250 mL	Soy sauce, preferably Korean or Japanese	1. In a nonreactive container large enough to hold the pork tenderloins, combine the soy sauce, brown sugar, garlic, scallions, sesame oil, and pepper to make a marinade.
2 oz	60 g	Brown sugar	
1 tbsp	15 mL	Garlic, crushed	
4	4	Scallions, sliced	
2 fl oz	60 mL	Sesame oil	
$^1/_4$ tsp	1 mL	Black pepper	
4 lb	2 kg	Pork tenderloins	2. Trim all fat and silverskin from the tenderloins (see **Figure 15.16**).
1 fl oz	30 mL	Vegetable oil	3. Add the tenderloins to the marinade, turning them so they are coated on all sides. Cover and refrigerate overnight.
			4. Remove the tenderloins from the marinade. Blot them dry on clean towels, making sure no particles of scallion or garlic cling to them, as these can burn during cooking.
			5. Heat the oil in a sauté pan over moderately high heat. Brown the meat lightly on all sides. Keep in mind the sugar in the marinade can burn easily, so watch the meat carefully as it browns.
			6. Place the pan in a preheated 425°F (220°C) oven for about 15 minutes, or until a meat thermometer indicates an interior temperature of 150°F (66°C) or desired doneness.
			7. Remove the tenderloins from the oven. Let stand 10 minutes in a warm place, then slice across the grain into medallions.

Per serving: Calories, 240; Protein, 37 g; Fat, 8 g (32% cal.); Cholesterol, 105 mg; Carbohydrates, 1 g; Fiber, 0 g; Sodium, 250 mg.

KALBI

The word *kalbi* is Korean and refers to ribs, usually beef ribs. In Korean cuisine, the meat is sliced, marinated in a sweetened soy sauce mixture, and grilled. In Western kitchens, the marinade for the ribs has become popular and is adapted to a variety of dishes not typically Korean, such as the pork tenderloin recipe here.

Roast Kalbi Pork with Hoisin Sauce

Glazed Ham with Cider Sauce

PORTIONS: 12   PORTION SIZE: 5 OZ (150 G) HAM, 2 FL OZ (60 ML) SAUCE

U.S.	METRIC	INGREDIENTS	PROCEDURE
7 lb	3.2 kg	Smoked ham (bone-in, skin-on)	1. Place the ham(s) in a stockpot with enough water to cover. Bring to a boil and reduce heat to a simmer. Simmer 1 hour. Drain. 2. Cut off skin and excess fat. Leave fat covering about $1/2$ in. (1 cm) thick. Score the fat with a knife.
$1^1/2$–2 tbsp	22–30 mL	Prepared mustard	3. Place the ham fat side up in a roasting or baking pan. Spread with a thin layer of prepared mustard. Mix the sugar and cloves and sprinkle over the ham.
3 oz	90 g	Brown sugar	
$1/8$ tsp	0.5 mL	Ground cloves	4. Bake at 350°F (175°C) about 1 hour. (Caution: Sugar burns easily, so check ham after 30–45 minutes.)
$1^1/2$ pt	750 mL	Apple cider	5. Place cider, raisins, sugar, nutmeg, and lemon rind in a saucepan and simmer 5 minutes.
4 oz	125 g	Raisins, seedless (optional)	6. Mix cornstarch with a little cold water or cold cider and stir into the sauce. Simmer until thickened.
$1^1/2$ oz	45 g	Brown sugar	7. Add salt to taste.
$1/4$ tsp	1 mL	Nutmeg	8. Slice ham (as for leg of lamb, **Figure 16.3**). Serve 5-oz (150-g) portion with 2 oz (60 mL) sauce on the side.
$1/2$ tsp	2 mL	Grated lemon rind	
3 tbsp	22 g	Cornstarch	
to taste	to taste	Salt	

Per serving: Calories, 300; Protein, 32 g; Fat, 7 g (21% cal.); Cholesterol, 70 mg; Carbohydrates, 25 g; Fiber, 1 g; Sodium, 1680 mg.

Note: The amount of cooking required depends on the type of ham. Aged country hams must be soaked 24 hours in cold water, scrubbed, and simmered about 20 minutes per pound (500 g). Hams labeled tenderized or ready to cook may be baked without simmering (starting with step 2) or just blanched before baking (place in cold water, bring to a boil, and drain).

VARIATIONS

If using fully cooked, boneless, skinless ham, skip steps 1 and 2. You will need less total weight of ham. Allow about $4^1/2$ lb (2.2 kg) for one recipe.

Ham with Brown Cider Sauce

When ham is baked, drain fat from pan and deglaze with 12 fl oz (375 mL) cider. Add $1^1/2$ pt (750 mL) demi-glace or espagnole and simmer until reduced and thickened. Flavor to taste with mustard and a little sugar.

Fruit-Glazed Ham

Omit mustard-sugar glaze. During last half of baking, spoon fruit preserves (apricot, pineapple, or peach) over ham to glaze.

Glazed Ham with Cider Sauce

Home-Style Meatloaf

PORTIONS: 12   PORTION SIZE: 4 OZ (125 G)

U.S.	METRIC	INGREDIENTS
8 oz	250 g	Onions, fine dice
4 oz	125 g	Celery, fine dice
1 oz	30 mL	Oil
6 oz	180 g	Soft bread crumbs
6 fl oz	180 mL	Tomato juice, stock, or milk
1¼ lb	625 g	Ground beef
1¼ lb	625 g	Ground pork
1¼ lb	625 g	Ground veal
4 oz	125 g	Beaten egg
1½ tsp	7 mL	Salt
¼ tsp	1 mL	Black pepper
1½ pt	750 mL	Tomato sauce, Spanish sauce, or Creole sauce

PROCEDURE

1. Sauté the onions and celery in oil until tender. Remove from pan and cool thoroughly.
2. In a large bowl, soak the bread crumbs in the juice, stock, or milk.
3. Add the sautéed vegetables and the meat, eggs, salt, and pepper. Mix gently until evenly combined. Do not overmix.
4. Form the mixture into 1 or 2 loaves in a baking pan, or fill loaf pans with the mixture.
5. Bake at 350°F (175°C) 1–1½ hours, or until done. Test with a meat thermometer for an internal temperature of 165°F (74°C).
6. For service, cut the loaves into 4-oz (125 g) slices. Serve with 2 oz (60 mL) sauce per portion.

Per serving: Calories, 360; Protein, 27 g; Fat, 21 g (53% cal.); Cholesterol, 135 mg; Carbohydrates, 16 g; Fiber, 2 g; Sodium, 680 mg.

VARIATIONS

Home-Style All-Beef Meatloaf

In place of the mixture of beef, pork, and veal, use 3¾ lb (1.875 kg) ground beef.

Italian-Style Meatloaf

Add the following ingredients to the basic mix:

> 2 tsp (10 mL) chopped garlic, sautéed with the onion
> ½ oz (15 g) parmesan cheese
> ⅓ cup (75 mL) chopped parsley
> ¾ tsp (3 mL) basil
> ½ tsp (2 mL) oregano

Salisbury Steak

Divide meat mixture for all-beef meatloaf into 6-oz (175 g) portions. Form into thick, oval patties and place on a sheet pan. Bake at 350°F (175°C) about 30 minutes.

Baked Meatballs

Divide basic meat mixture or Italian-Style Meatloaf mixture into 2½-oz (75 g) portions using a No. 16 scoop. Form into balls and place on sheet pans. Bake at 350°F (175°C). May be served with tomato sauce over pasta.

Loin or Rack of Venison Grand Veneur

PORTIONS: 8   **PORTION SIZE:** ¼ LOIN OR RACK (2 CHOPS)

U.S.	METRIC	INGREDIENTS	PROCEDURE
2	2	Loins or racks of venison (2½–3 lb/ 1.1–1.4 kg each)	1. Trim the venison, removing all silverskin. Because venison is very lean, there will be very little fat to remove.
2 qt	2 L	Red Wine Marinade for Game (below)	2. Marinate the venison for 2 days, using enough marinade to cover the meat completely. (The quantity indicated is approximate.)
12 oz (approx.)	350 g (approx.)	Pork fatback for barding	3. Cut the fatback into thin sheets on a slicing machine. Bard the venison by covering the meat with the sheets of fat and tying them in place.
			4. Roast the meat at 450°F (230°C) until rare, 30–45 minutes.
1 pt	500 mL	Poîvrade Sauce (p. 187)	5. When the meat is done, set it aside in a warm place for 15 minutes. Degrease the roasting pan and deglaze it with a little of the marinade. Reduce slightly and strain it into the poîvrade sauce.
1 oz	30 g	Red-currant jelly	6. Heat the sauce with the jelly until the jelly is melted and dissolved.
3 fl oz	90 mL	Heavy cream	7. Temper the cream with a little of the sauce, then add to the rest of the sauce.
1 lb	480 g	Chestnut purée (fresh or canned), thinned to a soft texture with a little demi-glace and cream	8. Cut the meat into chops, or else cut the meat from the bones in one piece and slice it into medallions.
			9. Serve each portion with 2 oz (60 mL) sauce and garnish with 2 oz (60 g) chestnut purée.

Per serving: Calories, 990; Protein, 67 g; Fat, 65 g (60% cal.); Cholesterol, 345 mg; Carbohydrates, 8 g; Fiber, 1 g; Sodium, 200 mg.

VARIATION

Leg of Venison Grand Veneur

Leg of venison can be prepared and served in the same way. A whole leg of venison weighing 4–5 lb (about 2 kg) yields 8–10 portions. This larger cut should be marinated slightly longer, 2–3 days.

Red Wine Marinade for Game

YIELD: 2 QT (2 L)

U.S.	METRIC	INGREDIENTS	PROCEDURE
4 oz	125 g	Carrot, chopped fine	1. Combine all ingredients in a nonreactive container (e.g., stainless steel, glass, plastic; do not use aluminum).
4 oz	125 g	Onion, chopped fine	
4	4	Garlic cloves, crushed	2. Marinate meat as desired or as indicated in recipe. Marinating times may vary from a few hours to several days. After marinating, use the liquid as a cooking medium and as the base for a sauce.
25	25	Parsley stems	
2 tsp	10 mL	Dried thyme	
4	4	Bay leaves	
2 tsp	10 mL	Dried ground sage	
1 tsp	5 mL	Peppercorns, crushed	
4	4	Cloves	
8 fl oz	250 mL	Red wine vinegar	
2 qt	2 L	Red wine	

Per serving: Calories, 25; Protein, 0 g; Fat, 0 g (0% cal.); Cholesterol, 0 mg; Carbohydrates, 23 g; Fiber, 5 g; Sodium, 200 mg.

KEY POINTS TO REVIEW

- What are the steps in the procedure for roasting a beef rib roast?

- What are the steps in preparing pan jus for roast meats?

- How is the procedure for preparing beef jus different from preparing roast beef gravy?

- What are the steps in the carving technique for bone-in roast leg of lamb?

BROILING, GRILLING, AND PAN-BROILING

The basic procedures for broiling and grilling are explained on pages 424–425. Review this material carefully before beginning this section. In addition to the core recipe for grilled steak on page 426, this chapter includes a range of grilled and broiled meats to give you additional practice with variations on the basic method.

SAUCES AND ACCOMPANIMENTS FOR GRILLED AND BROILED MEATS

Many kinds of sauces and accompaniments are appropriate for grilled meats, including compound butters; butter sauces such as béarnaise; brown sauce variations such as Bercy, mushroom, and bordelaise; tomato sauce variations; and salsas and relishes. For other examples, see the recipes in this section. Note that, unlike pan sauces made by deglazing sauté pans (see p. 427), all these sauces are prepared in advance because broiling or grilling does not give you the opportunity to deglaze a pan.

Part of the appeal of broiled and grilled meats is their brown, crisp surface. For this reason, it is best not to cover the item with the sauce. Also, less sauce is usually served with grilled items than with sautéed items. Serve the sauce on the side or around the meat or, at most, in a thin ribbon across only part of the meat.

Similarly, vegetables and accompaniments for broiled and grilled meats should, in most cases, not be heavily sauced. Grilled vegetables are often good choices as accompaniments.

PAN-BROILING

In the absence of broiling or grilling equipment, similar results can be achieved by cooking meats using the pan-broiling method. Like grilling, this method involved dry heat from below, except that the meat is cooked in a heavy pan rather than on an open grid over a heat source.

Pan-broiling is an especially useful technique for very thin steaks, called **minute steaks.** Broiling such thin steaks (minute steaks) to the rare stage is difficult because the heat is not high enough to form a good brown crust without overcooking the inside. Pan-broiling in a heavy iron skillet is an answer to this problem.

PROCEDURE for Pan-Broiling Meats

1. Preheat an iron skillet over a high flame until it is very hot. Do not add fat. (The pan should, of course, be well seasoned.)

2. If desired, brush the meat lightly with oil. Place the meat in the pan, presentation side down. Adjust the heat as necessary for proper cooking. Thin cuts should be cooked over very high heat in order to create enough browning in the short time they will cook. Thicker items should be cooked over slightly lower heat so that they will not brown too much during the longer cooking time.

3. During cooking, pour off fat that accumulates in the pan, if necessary.

4. When the bottom of the meat is brown and the meat is cooked halfway, turn it over.

5. Cook the second side until the meat is cooked to the desired doneness.

6. Remove from the pan and serve immediately.

London Broil

PORTIONS: 12   PORTION SIZE: 5 OZ (150 G)

U.S.	METRIC	INGREDIENTS
5 lb	2.4 kg	Flank steak
		Marinade:
8 fl oz	240 mL	Vegetable oil
1 fl oz	30 mL	Lemon juice
1 tsp	5 mL	Salt
1 tsp	5 mL	Black pepper
1/2 tsp	2 mL	Dried thyme
1 1/2 pt	720 mL	Mushroom Sauce (brown) (p. 187)

PROCEDURE

1. Trim all fat and connective tissue from the beef.
2. Combine the marinade ingredients in a hotel pan. Place the steaks in the pan and turn them so they are coated with oil. Cover and refrigerate at least 2 hours.
3. Remove the meat from the marinade and place in a preheated broiler or grill. Broil at high heat 3–5 minutes on each side, until well browned outside but rare inside (see Note).
4. Remove from broiler and let rest 2 minutes before slicing.
5. Slice the meat very thin on a sharp angle across the grain (see **Figure 16.4**).
6. Weigh 5-oz (150-g) portions. Serve each portion with 2 oz (60 mL) sauce.

Per serving: Calories, 520; Protein, 41 g; Fat, 37 g (65% cal.); Cholesterol, 110 mg; Carbohydrates, 3 g; Fiber, 0 g; Sodium, 370 mg.

Note: Flank steak should be broiled rare. If cooked well done, it will be tough and dry.

VARIATIONS

Thick-cut steaks from the round or chuck are sometimes used for London broil.

Teriyaki-Style London Broil

Marinate the steaks in a mixture of the following ingredients: 10 fl oz (300 mL) Japanese soy sauce, 3 oz (100 mL) vegetable oil, 2 oz (60 mL) sherry, 3 oz (90 g) chopped onion, 1 tbsp (15 g) sugar, 1 tsp (5 mL) ginger, 1 crushed garlic clove. Marinate at least 4 hours or, preferably, overnight. Broil as in basic recipe.

FIGURE 16.4 Slicing London broil flank steak.

(a) Holding the knife at a sharp angle, slice the meat in very thin slices across the grain. Use a table fork or kitchen fork to hold the meat steady. Some chefs slice the meat toward the fork.

(b) Others prefer to slice away from the fork. The result is the same.

Beef Fajitas

PORTIONS: 12   **PORTION SIZE: 4 OZ (150 G), PLUS GARNISH**

U.S.	METRIC	INGREDIENTS	PROCEDURE
3 fl oz	90 mL	Vegetable oil	1. To make a marinade, mix together the oil, lime juice, cumin, chili powder, garlic, salt, and pepper.
6 fl oz	180 mL	Lime juice	
4 tsp	20 mL	Ground cumin	2. Place the meat in a nonreactive container with the marinade, turning the meat so it is coated on all sides. Refrigerate 2–4 hours.
4 tsp	20 mL	Chili powder	
4	4	Garlic cloves, chopped fine	
4 tsp	20 mL	Salt	3. Remove the meat from the marinade. Grill the meat on both sides on a grill or under a broiler until it is lightly charred and rare to medium done, as desired.
2 tsp	10 mL	Pepper	
4 lb	1.9 kg	Beef skirt steak or flank steak, trimmed of fat	
1 fl oz	30 mL	Vegetable oil	4. While the beef is grilling, heat the oil in a large sauté pan over high heat.
1 lb 8 oz	700 g	Bell peppers, assorted colors, cut into strips	5. Add the peppers and sauté briefly, keeping them slightly crisp.
1 lb 8 oz	700 g	Onion, in thick slices	6. Remove the peppers from the pan, and then sauté the onions in the same way.
			7. When the beef is done, remove it from the grill and let it rest 5 minutes.
			8. Slice the meat across the grain into thin slices.
24, or as desired	24, or as desired	Flour tortillas, steamed to soften	9. Serve the meat strips and vegetables on a hot plate or sizzle platter. Serve the tortillas and the garnishes on the side in separate containers. Diners make their own soft tacos by rolling meat, vegetables, and choice of condiments in tortillas.
as desired	as desired	Garnishes:	
		Guacamole or sliced avocado	
		Sour cream	
		Shredded lettuce	
		Diced tomato	
		Salsa Cruda (p. 207)	

Per serving: Calories, 740; Protein, 44 g; Fat, 22 g (27% cal.); Cholesterol, 60 mg; Carbohydrates, 88 g; Fiber, 6 g; Sodium, 830 mg.

Broiled Smoked Pork Chop with Flageolet Beans and Wilted Arugula

PORTIONS: 15   **PORTION SIZE: 1 8-OZ (250-G) PORK CHOP, ABOUT 5 OZ (150 G) VEGETABLE**

U.S.	METRIC	INGREDIENTS	PROCEDURE
15	15	Smoked pork loin chops, bone in, about 8 oz (250 g) each	1. Brush the chops very lightly with oil. Place on a preheated grill or broiler until grill-marked on both sides and heated through. Smoked pork chops are fully cooked, so it is necessary only to heat them through.
as needed	as needed	Vegetable oil	
4 lb 12 oz	2.2 kg	Flageolet Beans with Wilted Arugula (see Note)	2. Serve each chop with 5 oz (150 g) of the beans with arugula.

Per serving: Calories, 620; Protein, 57 g; Fat, 26 g (39% cal.); Cholesterol, 110 mg; Carbohydrates, 26 g; Fiber, 9 g; Sodium, 3110 mg.

Note: The quantity of beans and arugula required for this recipe is equal to the yield of the recipe on page 375.

Broiled Smoked Pork Chop with Flageolet
Beans and Wilted Arugula

Grilled Marinated Pork Tenderloin with Sweet Potato Purée and Warm Chipotle Salsa

PORTIONS: 10   PORTION SIZE: 5 OZ (150 G) MEAT

U.S.	METRIC	INGREDIENTS
4 lb	2 kg	Pork tenderloin
		Wet rub:
2 oz	60 g	Onion, chopped fine
1	1	Garlic clove, chopped fine
2 tbsp	30 mL	Powdered red New Mexico chile (see Note)
1 tsp	5 mL	Salt
1/2 tsp	2 mL	Dried oregano
1/4 tsp	1 mL	Ground cumin
1/8 tsp	0.5 mL	Cinnamon
2 fl oz	60 mL	Lime juice
1 fl oz	30 mL	Olive oil
1	1	Garlic clove, unpeeled
1 lb	500 g	Plum tomatoes or other small tomatoes
2	2	Whole chipotle chiles in adobo (canned)
1/2 tsp	2 mL	Salt
2–3 tsp	10–15 mL	Sauce from the canned chiles
3 lb	1.5 kg	Sweet potatoes
to taste	to taste	Salt

PROCEDURE

1. Trim fat and membranes from the tenderloins (see **Figure 15.16**).

2. Mix together the onion, garlic, powdered chile, salt, oregano, cumin, cinnamon, lime juice, and oil.

3. Coat the meat with this mixture. Wrap and refrigerate several hours or overnight.

4. Roast the garlic and tomatoes in a preheated 450°F (230°C) oven for 10 minutes.

5. Remove the skins from the tomatoes and garlic. Place the tomatoes and garlic in a blender.

6. Carefully cut open the chiles. Scrape out and discard the seeds. Chop the chiles.

7. Add the chiles, salt, and the sauce from the chiles to the blender. Blend to make a coarse purée. Add more salt if needed.

8. Bake the sweet potatoes at 400°F (200°C) until soft. Cut in half and scoop out the flesh. Pass through a food mill to purée. Season lightly.

9. Scrape the onions and garlic off the meat (they will burn if left on).

10. Grill the meat until just medium-well to well done, as desired (internal temperature at least 145°F (63°C). Be careful not to overcook, or the meat will be dry.

11. To serve, place 3 oz (90 g) sweet potato purée on the plate. Slice the meat across the grain into medallions. Arrange 5 oz (150 g) meat on top of the sweet potato. Drizzle with 1 1/2 oz (45 mL) tomato chipotle salsa.

Per serving: Calories, 410; Protein, 41 g; Fat, 10 g (22% cal.); Cholesterol, 110 mg; Carbohydrates, 38 g; Fiber, 5 g; Sodium, 630 mg.

Note: For a slightly different flavor, or if powdered New Mexico chile is not available, use a regular chili powder blend.

Grilled Marinated Pork Tenderloin with Sweet Potato Purée and Warm Chipotle Salsa

SAUTÉING AND PAN-FRYING

This section includes recipes for both sautéed and pan-fried items. Review pages 427–428 for an explanation of the sautéing procedure, and pages 430–431 for the pan-frying procedure, as it is applied to meats, poultry, and seafood.

When you have become comfortable with the basic sautéing and pan-frying techniques, you will learn to adapt them and modify them as needed for different preparations. As you look at recipes for sautéing meats, poultry, and fish in this and later chapters, you will find recipes that seem to be slightly different from the basic procedures on p. 428. For example, a few recipes appear to be sautés except that the pan is not deglazed to make a sauce. In such cases, you may be using steps 1–7 of the basic sautéing procedure and finishing the dish without a sauce or with a sauce prepared in advance. Be prepared for adaptations and variations of the basic sautéing and pan-frying methods

STIR-FRYING

As is mentioned in Chapter 14, the technique of stir-frying, used for many Asian dishes, is similar in many ways to sautéing. A detailed procedure for stir-frying is given below. Several recipes in this chapter as well as in Chapters 18 and 20 illustrate how this procedure is put into practice.

BASIC PROCEDURE for Stir-Frying

1. Heat a wok or sauté pan over high heat until very hot.
2. Add a small quantity of oil and let it heat.
3. Add seasonings for flavoring the oil—one or more of the following: salt, garlic, ginger root, scallions.
4. If meat, poultry, or seafood items are part of the dish, add them at this point. As when sautéing, do not overload the pan. Leave the food pieces untouched for a few moments so they begin to brown properly. Then stir and toss them with a spatula so they sear and cook evenly.
5. If any liquid seasoning for the meat, such as soy sauce, is used, add it now, but only in small quantities, so the meat continues to fry and does not start to simmer or stew.
6. Remove the meat from the pan or leave it in, depending on the recipe. If a small quantity of quick-cooking vegetables is used, the meat can sometimes be left in the pan and the vegetables cooked with it. Otherwise, remove the meat when it is almost done and keep it on the side while cooking the vegetables.
7. Repeat steps 2 and 3 if necessary.
8. Add the vegetables to the pan and stir-fry. If more than one vegetable is used, add the longer-cooking ones first and the quicker-cooking ones last.
9. Some dishes are dry-fried, meaning prepared without liquid or sauce. In this case, simply return the meat item, if any, to the pan to reheat with the vegetables, then serve. Otherwise, proceed to the next step.
10. Add liquid ingredients, such as stock or water, and continue to cook and stir until the vegetables are almost cooked.
11. Add the meat item, which was removed in step 6, to the pan to reheat.
12. Optional but widely used step: Add a mixture of cornstarch and water to the pan and cook until lightly thickened.
13. Serve at once.

KEY POINTS TO REVIEW

- What kinds of sauces are appropriate for grilled meat items, and how should these sauces be served?

- What are the steps in the basic procedure for pan-broiling meats?

- What are the steps in the basic procedure for stir-frying meats?

Veal Scaloppine alla Marsala

PORTIONS: 10   PORTION SIZE: 4 OZ (125 G)

U.S.	METRIC	INGREDIENTS	PROCEDURE
2¹/₂ lb	1.25 kg	Small veal scaloppine: 20 pieces, 2 oz (60 g) each (see **Figure 15.15** for preparation of veal)	1. Lightly flatten each piece of veal with a meat mallet. Do not pound hard, or you may tear the meat.
to taste	to taste	Salt	2. Dry the meat, season it with salt and pepper, and dredge it in flour. Shake off excess. (Do not do this step until immediately before cooking.)
to taste	to taste	White pepper	
for dredging	for dredging	Flour	3. Heat the oil in a large sauté pan until very hot. Add the veal and sauté over high heat just until lightly browned on both sides. (If necessary, sauté the meat in several batches.)
2 fl oz	60 mL	Oil	
			4. Remove the meat from the pan and drain the excess oil.
4 fl oz	125 mL	Marsala wine	5. Add the Marsala to the pan and deglaze.
8 fl oz	250 mL	Strong white stock, veal or chicken (see Note)	6. Add the stock and reduce over high heat by about half.
2 oz	60 g	Butter, cut in pieces	7. Add the pieces of butter and swirl the pan until they are melted and blended with the sauce.
2 tbsp	30 mL	Chopped parsley	8. Add the veal to the pan and bring just to the simmer. Turn the meat to coat it with the sauce.
			9. Serve immediately, 2 pieces per portion, sprinkled with chopped parsley.

Per serving: Calories, 360; Protein, 27 g; Fat, 26 g (65% cal.); Cholesterol, 115 mg; Carbohydrates, 2 g; Fiber, 0 g; Sodium, 120 mg.

Note: Brown sauce may be used instead of white stock. However, the white stock makes a more delicate product without masking the flavor of the veal.

VARIATIONS

Veal Scaloppine with Sherry

Substitute sherry for the Marsala.

Veal Scaloppine à la Crème

Prepare as in basic recipe, but omit the wine. Deglaze the pan with the stock. Add 1 cup (250 mL) heavy cream and reduce until thickened. Omit the butter. Season the sauce with a few drops of lemon juice. Taste carefully for salt.

Veal Scaloppine with Lemon

Substitute 3 fl oz (90 mL) lemon juice for the 4 fl oz (125 mL) wine. After plating, top each piece of veal with 1 lemon slice and sprinkle with chopped parsley.

Veal Scaloppine with Mushrooms and Cream

Prepare as for Veal Scaloppine à la Crème, but sauté ¹/₂ lb (250 g) sliced mushrooms in butter in the sauté pan before deglazing.

Tournedos Vert-Pré

PORTIONS: 1   **PORTION SIZE: 5–6 OZ (150–175 G)**

U.S.	METRIC	INGREDIENTS
1 oz	30 g	Clarified butter
2	2	Tournedos (see Note), 2¹/₂–3 oz (75–90 g) each
2 slices	2 slices	Maître d'Hôtel Butter (p. 192)
as needed	as needed	Allumette Potatoes (p. 361)
as needed	as needed	Watercress

PROCEDURE

1. Heat the butter in a small sauté pan over moderately high heat.
2. Place the tournedos in the pan and cook until well browned on the bottom and about half cooked.
3. Turn the meat over and continue to cook until rare or medium done, according to customer's request.
4. Place the tournedos on a hot dinner plate and top each with a slice of maître d'hôtel butter. Garnish the plate with a portion of allumette potatoes and a generous bunch of watercress. Serve immediately, while the butter is still melting.

Per serving: Calories, 640; Protein, 27 g; Fat, 59 g (83% cal.); Cholesterol, 185 mg; Carbohydrates, 0 g; Fiber, 0 g; Sodium, 410 mg.

Note: Tournedos (TOOR-nuh-doe; singular form: one tournedos) are small tenderloin steaks cut about 1¹/₂ in. (4 cm) thick. The same recipe may be used for fillet steaks, which are larger but thinner cuts from the tenderloin. See Figure 15.9 for cutting tenderloin.

VARIATIONS

Tournedos Béarnaise

Pan-fry tournedos as in basic recipe and serve with béarnaise sauce (p. 195).

Tournedos Bordelaise

Pan-fry as in basic recipe. Top each steak with 1 slice poached beef marrow and coat lightly with bordelaise sauce (p. 187).

Tournedos Chasseur

Pan-fry as in basic recipe. Plate the steaks and deglaze the sauté pan (drained of cooking fat) with ¹/₂ fl oz (15 mL) white wine. Add 2 fl oz (60 mL) chasseur sauce (p. 187), bring to a simmer, and pour around the tournedos.

Tournedos Rossini

Pan-fry as in basic recipe. Set the tournedos on croûtons (rounds of bread cut the same size as the steaks and fried in butter until golden).

Top each steak with 1 slice pâté de foie gras (goose liver pâté) and 1 slice truffle (if available). Coat lightly with Madeira sauce (p. 187).

Tournedos Rossini; Berny Potatoes; Braised Lettuce

Calf's Liver Lyonnaise

PORTIONS: 10   **PORTION SIZE: 1 SLICE LIVER, 1 1/2 OZ (50 G) ONION GARNISH**

U.S.	METRIC	INGREDIENTS
2 lb	1 kg	Onions
3 oz	90 g	Butter
1 cup	250 mL	Demi-glace or strong brown stock
to taste	to taste	Salt
to taste	to taste	Pepper
10 slices	10 slices	Calf's liver, 1/2 in. (6 mm) thick, about 4 oz (125 g) each
as needed	as needed	Salt
as needed	as needed	Pepper
as needed	as needed	Flour
as needed	as needed	Clarified butter or oil

PROCEDURE

1. Peel and slice the onions.
2. Heat the butter in a sauté pan and add the onions. Sauté them over medium heat until tender and golden brown.
3. Add the demi-glace or stock and cook a few minutes, until the onions are nicely glazed. Season to taste.
4. Place in a bain-marie and keep warm for service.

5. Season the liver and dredge in flour. Shake off excess flour.
6. Pan-fry the liver in butter or oil over moderate heat until browned on both sides and slightly firm to the touch. Do not overcook or use high heat.
7. Serve each portion with 1 1/2 oz (45 g) onion mixture.

Per serving: Calories, 310; Protein, 24 g; Fat, 19 g (54% cal); Cholesterol, 445 mg; Carbohydrates, 13 g; Fiber, 1 g; Sodium, 250 mg.

Calf's Liver Lyonnaise

Thai Green Curry of Pork with Vegetables 🍽️

PORTIONS: 12   **PORTION SIZE: 7 OZ (200 G)**

U.S.	METRIC	INGREDIENTS
2 fl oz	60 mL	Vegetable oil
2 lb	1 kg	Boneless pork loin, cut into thin slices
1 lb	500 g	Bok choy, cut into 1-in. (2.5-cm) pieces
10 oz	300 g	Yellow summer squash, sliced
8 oz	250 g	Green bell pepper, medium dice
6 oz	180 g	Shiitake mushrooms, sliced
1 tbsp	15 mL	Garlic, chopped
12	12	Scallions, sliced
18 fl oz	550 mL	Thai Green Curry Sauce (p. 210), hot

PROCEDURE

1. Heat half the oil in a large sauté pan or wok.
2. Over high heat, stir-fry the pork just until it loses its pink color. Cook it in several batches if necessary. Remove from pan.
3. Add the rest of the oil to the pan.
4. With the pan still over high heat, add the bok choy, squash, peppers, mushrooms, garlic, and scallions. Stir-fry for a few minutes, keeping the vegetables crisp.
5. Return the pork to the pan and add the sauce. Simmer until the pork is cooked through.
6. Serve immediately with steamed rice.

Per serving: Calories, 250; Protein, 18 g; Fat, 19 g (64% cal.); Cholesterol, 40 mg; Carbohydrates, 6 g; Fiber, 2 g; Sodium, 310 mg.

Note: This method of making a curry is designed for advance preparation and quick, last-minute cooking. A more traditional method is to make the sauce as an integral sauce. Stir-fry the meat and vegetables, add curry paste (see sauce recipe, p. 184), and cook until aromatic. Add the coconut milk, other liquids, flavorings, and any other ingredients indicated in the sauce recipe, and finish cooking.

Thai Green Curry of Pork with Vegetables

Stir-Fried Beef with Bell Peppers

PORTIONS: 8   PORTION SIZE: 4 OZ (125 G)

U.S.	METRIC	INGREDIENTS
1 lb 4 oz	600 g	Flank steak
2 fl oz	60 mL	Soy sauce
1 tbsp	15 mL	Sherry or Shaoxing wine
2¹/₂ tsp	12 mL	Cornstarch
3	3	Bell peppers, 2 or 3 colors
2 slices	2 slices	Fresh ginger root
1	1	Garlic cloves, sliced
1 oz	30 g	Scallion, sliced
1¹/₂–2 fl oz	45–60 mL	Oil
¹/₄ tsp	1 mL	Salt
1 fl oz	30 mL	Chicken stock

PROCEDURE

1. Cut the flank steak lengthwise (with the grain) into strips 2 in. (5 cm) wide. Then cut the strips crosswise into very thin slices. (This is easier if the meat is partially frozen.)
2. Toss the meat with the soy sauce, sherry, and cornstarch. Let marinate 30 minutes or longer.
3. Core and seed the peppers. Cut them into strips ¹/₄ in. (6 mm) wide.
4. Have the ginger, garlic, and scallions ready in separate containers.
5. Stir-fry the beef in 2 or more batches, depending on the size of the pan or wok. Use a little of the oil for each batch, as needed.
6. As each batch of the beef is cooked, remove it from the pan and set it aside.
7. Heat additional oil in the pan and add the salt, ginger, garlic, and scallions. Stir-fry for a few seconds to develop flavor.
8. Add the peppers and stir-fry until lightly cooked but still crisp.
9. Add the stock and toss the vegetables a few times.
10. Return the meat to the pan. Toss the meat with the vegetables until it is hot and evenly combined with the peppers. Serve at once.

Per serving: Calories, 180; Protein, 16 g; Fat, 11 g (54% cal.); Cholesterol, 35 mg; Carbohydrates, 5 g; Fiber, 1 g; Sodium, 525 mg.

VARIATIONS

Other vegetables may be used instead of the peppers, such as celery, broccoli, snow peas, green beans, asparagus, mushrooms, bok choy. Or use 2 or 3 fresh vegetables, plus water chestnuts and/or bamboo shoots.

Chicken or pork may be used instead of beef. If chicken is used, cut it into medium dice or bâtonnet. Also, reduce the quantity of soy sauce to avoid discoloring the light meat of the chicken.

CHINESE STIR-FRIES

Chinese restaurants in North America often finish stir-fried dishes by adding a quantity of commercially prepared sauce to the mixture. In traditional Chinese cooking, however, the sauce is more often made as part of the stir-frying procedure. Far less liquid is used, so the finished dish has much less sauce than the Chinese-style dishes that most North Americans are familiar with. The recipe included here is an example of this drier style of stir-frying.

Stir-Fried Beef with Bell Peppers

SIMMERING, SUBMERSION POACHING, STEAMING, AND SOUS VIDE

Meats are not often simmered. Part of the reason simmered meats are not as popular as meats cooked in other ways may be that they lack the kind of flavor produced by browning with dry heat.

However, as explained in Chapter 14, simmering can be used for less tender cuts for which browning is not desired or not appropriate. Popular examples of simmered meats are cured products such as cured meats and simmered stews.

The core recipe illustrating the basic procedure in Chapter 14 is a fresh meat product. In this chapter, you will find a basic recipe for a cured meat item, corned beef. So, while the recipe follows the basic procedure, the cooking liquid differs, as explained on page 434. Please note that the recipe name "New England Boiled Dinner," is traditional. The meat is not actually boiled but simmered.

The term **stewing** means cooking small pieces of meat by simmering or braising (a composite method that includes both browning and simmering). Stews cooked by braising are covered in the next section. See also the discussion of stewing and braising on pages 442–444.

One difference between stews and many other simmered meats is that stews are served in a sauce or gravy made of the cooking liquid. The procedure for making simmered stews is illustrated by several recipes, including a classic dish called a *blanquette* on page 508. The French term blanquette comes from the word *blanc,* meaning "white," since this is a white, not a browned, stew.

Submersion poaching, more common with fish and poultry, is rarely used for meats. One exception is the recipe for poached beef tenderloin on page 512, which uses flavorful stock and garnish to complement the subtle flavor of the meat.

The steaming procedure for meats is illustrated by the recipe for Pearl Balls in Chapter 14. In this chapter, steaming comes into play in the recipe for Lamb Tagine, in which the meat is cooked partly by simmering liquid and partly by steam trapped in the cooking utensil.

Finally, the sous vide recipe in Chapter 14 is supplemented by an additional recipe in this section.

New England Boiled Dinner

PORTIONS: 8   **PORTION SIZE: 3 OZ (90 G) MEAT, PLUS ASSORTED VEGETABLES**

U.S.	METRIC	INGREDIENTS	PROCEDURE
2 lb 12 oz	1.4 kg	Corned beef brisket	1. Trim excess fat from corned beef if necessary. 2. Place the beef in a stockpot or steam kettle. Cover with cold water. 3. Bring to a boil and reduce heat to a simmer. Simmer until the meat feels tender when pierced with a fork. Cooking time will be 2–3 hours. 4. To hold and serve hot, place the cooked meat in a steam-table pan and add some of the cooking liquid to keep it moist.
1 heads	1 heads	Green cabbage	5. Prepare the vegetables: Cut each cabbage into 8 wedges; pare the turnips and carrots and cut them into serving-size pieces; peel the onions; scrub the beets and potatoes.
1 lb	500 g	Turnips	
1 lb	500 g	Carrots	6. Cook the cabbage, turnips, carrots, onions, and potatoes separately in a little of the beef cooking liquid.
24	24	Pearl onions	
16	16	Baby beets	7. Steam the beets, then peel them.
16	16	Small red-skinned potatoes	
as needed	as needed	Horseradish Sauce (p. 184) or prepared horseradish	8. To serve, cut the meat across the grain into slices, holding the knife at an angle to get broader slices. Serve with horseradish sauce or prepared horseradish.

Per serving: Calories, 700; Protein, 35 g; Fat, 26 g (46% cal.); Cholesterol, 75 mg; Carbohydrates, 81 g; Fiber, 17 g; Sodium, 2240 mg.

VARIATION

To serve the corned beef cold, cool the beef in some of its cooking liquid, to keep it moist, in a cold-water bath. When cool, remove from the liquid and refrigerate, covered. Cold corned beef may be reheated in its cooking liquid.

New England Boiled Dinner

 # Blanquette of Veal

PORTIONS: 12   **PORTION SIZE: 5 OZ (150 G)**

U.S.	METRIC	INGREDIENTS	PROCEDURE
5 lb	2.4 kg	Boneless, trimmed veal breast, shoulder, or shank	1. Cut the veal into 1-in. (2.5- cm) dice. 2. Blanch the meat: Place in a saucepot and cover with cold water. Bring to a boil, drain, and rinse the meat under cold water (see Note).
1	1	Medium onion stuck with 2 cloves	3. Return the meat to the pot and add the onion stuck with cloves, bouquet garni, and salt.
1	1	Bouquet garni	4. Add enough stock to just cover the meat.
2 tsp	10 mL	Salt	5. Bring to a boil, skim, cover, and lower heat to a slow simmer.
2$^{1}/_{2}$ pt (approximately)	1.2 L (approximately)	White veal stock	6. Simmer until meat is tender, about 1$^{1}/_{2}$ hours. Skim when necessary. 7. Strain the stock into another pan. Reserve the meat and discard the onion and bouquet garni.
		Roux:	8. Reduce the stock to about 20 fl oz (600 mL).
2 oz	60 g	Butter, clarified	9. Meanwhile, prepare a white roux with the butter and flour. Beat into the stock to make a velouté sauce and simmer until thickened and no raw flour taste remains.
2 oz	60 g	Flour	
		Liaison:	
3	3	Egg yolks	10. Remove the sauce from the heat. Beat the egg yolks and cream together, temper with a little of the hot sauce, and stir into the sauce. (Strain the sauce if desired.)
8 fl oz	250 mL	Heavy cream	
to taste	to taste	Lemon juice	11. Combine the sauce and meat. Heat gently; do not boil.
pinch	pinch	Nutmeg	12. Season to taste with a few drops of lemon juice, a pinch of nutmeg and white pepper, and more salt if needed.
pinch	pinch	White pepper	

Per serving: Calories, 350; Protein, 35 g; Fat, 21 g (55% cal.); Cholesterol, 230 mg; Carbohydrates, 4 g; Fiber, 0 g; Sodium, 550 mg.

Note: Blanching eliminates impurities that discolor the sauce. This step can be omitted, but the product will have a less attractive appearance.

VARIATIONS

Blanquette of Lamb

Prepare as in basic recipe, using lamb shoulder or shank. If desired, use white lamb stock.

Blanquette of Pork

Prepare as in basic recipe, using pork shoulder or butt. If desired, use white pork stock.

Blanquette of Veal with Herb Pasta (p. 400) and Vegetables

Irish Lamb Stew

U.S.	METRIC	INGREDIENTS
3¹/₂ lb	1.75 kg	Lean, boneless lamb shoulder or shanks
3 pt (approximately)	1.5 L (approximately)	Water or white lamb stock
1	1	Small onion stuck with 2 cloves
		Sachet:
1	1	Bay leaf
1 clove	1 clove	Garlic
4	4	Whole peppercorns
6	6	Parsley stems
¹/₄ tsp	1 mL	Dried thyme
to taste	to taste	Salt
1 lb	500 g	Onions, sliced thin
8 oz	250 g	Leeks (white part), sliced
2 lb	1 kg	Potatoes, peeled and sliced thin
as needed	as needed	Chopped parsley

PROCEDURE

1. Cut meat into 1-in. (2¹/₂-cm) cubes.
2. Bring the water to a boil in a large, heavy saucepot. Add the lamb. There should be just enough liquid to cover the meat; add more liquid if necessary.
3. Return to a boil, reduce heat to a simmer, and skim the scum carefully.
4. Add the onion stuck with cloves, the sachet ingredients tied in a piece of cheesecloth, and salt to taste. Simmer 1 hour.
5. Add the onions, leeks, and potatoes. Continue to simmer until the meat is tender and the vegetables are cooked. The potatoes should break down somewhat and thicken the stew.
6. Remove and discard the sachet and the onion stuck with cloves. Correct the seasoning.
7. Garnish each portion with chopped parsley.

Per serving: Calories, 200; Protein, 18 g; Fat, 7 g (33% cal.); Cholesterol, 60 mg; Carbohydrates, 14 g; Fiber, 2 g; Sodium, 55 mg.

VARIATION

Carrots and white turnips may be cooked with the stew or cooked separately and added as a garnish.

Tripes à la Mode de Caen

U.S.	METRIC	INGREDIENTS
4 lb 8 oz	2.2 kg	Beef tripe
2	2	Calf's feet (see Note)
8 oz	250 g	Onion, medium dice
6 oz	185 g	Carrot, sliced
6 oz	185 g	Leek, sliced
		Sachet:
12	12	Peppercorns, lightly crushed
2	2	Bay leaf
12	12	Parsley stems
¹/₂ tsp	2 mL	Dried thyme
4	4	Whole cloves
2 pt	1 L	Dry white wine
1 pt	500 mL	White stock
to taste	to taste	Salt
3 oz	90 mL	Calvados (apple brandy)

PROCEDURE

1. Trim all fat from the tripe. Put the tripe in a pot of cold water and bring it to a boil. Simmer 5 minutes. Drain and rinse in cold water. Cut the tripe into 1¹/₂-in. (4-cm) squares.
2. Cut the feet into pieces with a meat saw as necessary so they fit into the braising pan.
3. Combine all the ingredients, except the Calvados, in a braising pan or other heavy pot. Salt lightly. Bring to a boil, cover tightly, and put in an oven at 325°F (160°C). Cook 5 hours or longer, until the tripe is very tender.
4. Remove the feet and bone them out. Dice the skin and meat and return it to the pot. Discard the bone, fat, and connective tissue.
5. Stir in the Calvados. Adjust the seasoning. Simmer a few minutes to blend in the flavor of the Calvados.
6. Serve with boiled potatoes.

Per serving: Calories, 640; Protein, 159 g; Fat, 28 g (40% cal.); Cholesterol, 345 mg; Carbohydrates, 8 g; Fiber, 1 g; Sodium, 200 mg.

Note: If calf's feet are not available, substitute twice the number of pig's feet. Do not omit, or the tripe stew will not have enough gelatin to give it the proper texture.

This dish is from the Normandy region of France, famed for, among other things, its apples. The traditional recipe calls for hard cider, but white wine is an acceptable substitute.

Tripes à la Mode de Caen

Shredded Pork (Carnitas)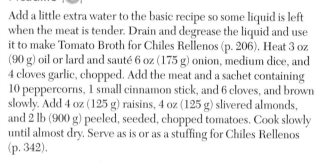

YIELD: ABOUT 3¹/₂ LB (1.6 KG)

U.S.	METRIC	INGREDIENTS
6 lb	2.8 kg	Pork butt or shoulder, boned
1	1	Onion, medium, cut in half
1	1	Garlic clove, chopped
1 tbsp	15 mL	Salt
¹/₄ tsp	1 mL	Pepper
1 tsp	5 mL	Dried oregano
1 tsp	5 mL	Cumin seeds

PROCEDURE

1. Remove most of the large chunks of fat from the pork, leaving a little of it on. Cut the meat into strips measuring 1–2 in. (2.5–5 cm).

2. Put the pork in a large pot with the rest of the ingredients. Add water to barely cover the meat.

3. Bring to a boil, reduce heat, and simmer slowly, uncovered, until all the liquid has evaporated. By this time, the meat should be tender. If it is not, add more water and continue to cook until it is.

4. Remove the onion and discard it.

5. Lower the heat and let the meat cook in the rendered fat, stirring from time to time, until browned and very tender. Shred the meat slightly.

6. Serve as a snack, an appetizer, or a filling for tortillas, either as it is or moistened with any of the sauces in this section or with guacamole.

Per ounces: Calories, 390; Protein, 31 g; Fat, 28 g (67% cal.); Cholesterol, 120 mg; Carbohydrates, 0 g; Fiber, 0 g; Sodium 590 mg.

VARIATIONS

Picadillo

Add a little extra water to the basic recipe so some liquid is left when the meat is tender. Drain and degrease the liquid and use it to make Tomato Broth for Chiles Rellenos (p. 206). Heat 3 oz (90 g) oil or lard and sauté 6 oz (175 g) onion, medium dice, and 4 cloves garlic, chopped. Add the meat and a sachet containing 10 peppercorns, 1 small cinnamon stick, and 6 cloves, and brown slowly. Add 4 oz (125 g) raisins, 4 oz (125 g) slivered almonds, and 2 lb (900 g) peeled, seeded, chopped tomatoes. Cook slowly until almost dry. Serve as is or as a stuffing for Chiles Rellenos (p. 342).

Shortcut Picadillo

Instead of preparing Shredded Pork, use 5 lb (2.3 kg) raw ground pork. Sauté it with the onion and garlic in the picadillo recipe, then proceed as directed with the rest of the recipe.

PROCEDURE VARIATION: COMBI OVEN

In place of the procedure in the main recipe, use the following procedure:

1. Remove most of the large chunks of fat from the pork, leaving a little of it on.

2. Grate the onion and garlic together. Combine the grated onion and garlic with the seasonings, and rub over the outside of the pork butt/shoulder.

3. Place the roast in a 225°F (107°C) combi oven in combi-mode and cook until done, by checking internal temperature, about 2¹/₂ hours.

4. Shred the meat slightly. Serve as a snack, an appetizer, or a filling for tortillas, either as it is or moistened with any of the sauces in this section or with guacamole.

Lamb Tagine with Chickpeas

PORTIONS: 12   **PORTION SIZE: 10 OZ (285 G) MEAT, VEGETABLES, AND SAUCE**

U.S.	METRIC	INGREDIENTS
3 fl oz	90 mL	Vegetable oil
8 oz	240 g	Onion, chopped fine
1/2 oz	15 g	Garlic, crushed
1 tsp	5 mL	Ground ginger
1 tsp	5 mL	Turmeric
1 oz	30 g	Fresh cilantro, chopped
2 tsp	10 mL	Salt
1 tsp	5 mL	Pepper
5 1/2 lb	2.5 kg	Lamb shoulder, boneless, trimmed of fat, cut into 1 1/2-in. (4-cm) pieces
1 pt	500 mL	Water
1 1/2 lb	675 g	Chickpeas, cooked or canned, drained
1 1/2 lb	675 g	Carrots, cut into 1-in. (2.5-cm) pieces
6 oz	180 g	Mediterranean-type olives, such as Kalamata, pitted
3 fl oz	90 mL	Lemon juice
to taste	to taste	Salt

PROCEDURE

1. In a heavy braising pan or casserole or in the base of a tagine (see sidebar), mix together the oil, onion, garlic, ginger, turmeric, cilantro, salt, and pepper.
2. Add the lamb to the pan and toss with the spice mixture to coat all the pieces.
3. Add the water. Bring to a boil, and then cover tightly. Simmer slowly on top of the stove or in an oven heated to 325°F (165°C) for 1 1/2 hours.
4. Add the chickpeas and carrots to the pan. If necessary, add a little more water if the tagine is becoming dry.
5. Return to the heat and cook until the carrots and meat are very tender.
6. Add the olives and lemon juice and stir. Simmer another 5 minutes.
7. Remove the meat and vegetables with a slotted spoon.
8. Degrease the cooking liquid.
9. Reduce the cooking liquid over moderate heat until it has the consistency of thick gravy.
10. Season the liquid with salt if necessary. Pour over the meat and vegetables.

Per serving: Calories, 620; Protein, 43 g; Fat, 40 g (57% cal.); Cholesterol, 145 mg; Carbohydrates, 24 g; Fiber, 7 g; Sodium, 640 mg.

TAGINES

A *tagine* (tah zheen) is a type of stew originating in North Africa. It is traditionally cooked in an earthenware pot also called a *tagine*. In Morocco, the traditional tagine consists of a round, shallow base and a cone-shaped lid. The stew is usually cooked on the stovetop over low heat. The dish is usually made with poultry or less expensive cuts of meat, such as lamb neck or shoulder, which are made tender by long, slow cooking.

Traditional spices used to flavor tagines include cinnamon, saffron, ginger, turmeric, cumin, paprika, and black pepper, Vegetables, fruits, nuts, and legumes may also be added to the stew.

The lamb tagine in this book is shown in a classic tagine pot to illustrate the appearance of this vessel. In a commercial kitchen, you are more likely to cook such stews in a brazier or other heavy pan.

Lamb Tagine with Chickpeas

Poached Beef Tenderloin with Beef Short Rib Ravioli in Morel Consommé

PORTIONS: 4   PORTION SIZE: 4 OZ (125 G) MEAT, 3 OZ (90 G) RAVIOLI, 3 MUSHROOMS, 3 FL OZ (90 ML) CONSOMMÉ

U.S.	METRIC	INGREDIENTS	PROCEDURE
8 oz	250 g	Trimmed meat from Braised Short Ribs (p. 445)	1. Prepare the ravioli filling: Shred the meat and place it in a bowl. Add the chopped parsley and cheese. Mix well.
2 tbsp	30 mL	Chopped parsley	2. If the meat mixture is dry, moisten with a little of the short rib braising liquid or brown stock.
1 tbsp	15 mL	Grated parmesan cheese	3. Roll out the pasta and, using the meat mixture as a filling, make ravioli following the procedure illustrated on page 401.
8 oz	500 g	Fresh Egg Pasta (p. 400)	4. If desired, cook the ravioli in advance. Simmer in salted water until just tender. Drain. Rinse briefly with cold water to stop the cooking. Toss with a little vegetable oil to keep the ravioli from sticking. Refrigerate, covered, until needed.
12	12	Dried morel mushrooms	5. Place the mushrooms in a bowl and add hot water to cover. Let soak until soft.
as needed	as needed	Hot water	6. Drain the mushrooms, squeezing them lightly. Strain and reserve the soaking liquid.
12 fl oz	375 mL	Consommé (p. 232)	7. Combine the soaking liquid with an equal volume of the consommé. Bring to a simmer. Add the mushrooms and cook until they are tender.
			8. Drain the mushrooms. Strain the cooking liquid again and add it to the consommé.
1 lb	500 g	Beef tenderloin, completely trimmed of fat and silverskin, in 1 or 2 pieces	9. Season the beef with salt and pepper.
			10. Bring the stock to a boil in a pot large enough to hold both the stock and the beef.
to taste	to taste	Salt	11. Lower the beef into the stock. Adjust the heat and cook at a slow simmer until the meat reaches the desired doneness, as determined by a meat thermometer. The temperature at the center of the meat should be 120°F (49°C) for rare, 130°F (54°C) for medium. Cooking time will be 20–30 minutes for rare, slightly longer for medium.
to taste	to taste	Pepper	
1 qt	1 L	Brown stock	12. Remove the meat from the liquid and let rest in a warm place about 15 minutes.
			13. Reserve the stock for another use, such as for making the next batch of consommé.
			14. While the meat is cooking, bring the consommé to a simmer and hold.
			15. If the ravioli were cooked in advance, reheat them by dropping them for a moment into boiling water. Drain. If they were not cooked ahead, cook them now in simmering salted water until just tender.
			16. Slice the meat. Arrange it in heated broad soup plates with the ravioli and the morels. Ladle 3 fl oz (90 mL) consommé into each bowl.

Per serving: Calories, 680; Protein, 46 g; Fat, 40 g (55% cal.); Cholesterol, 210 mg; Carbohydrates, 29 g; Fiber, 1 g; Sodium, 180 mg.

Beef Tenderloin Sous Vide

PORTIONS: 2   PORTION SIZE: APPROX. 6 OZ (180 G)

U.S.	METRIC	INGREDIENTS	PROCEDURE
2 sprigs	2 sprigs	Fresh thyme	1. Wrap the thyme, bay leaf, and parsley stems in a small piece of plastic wrap. Cut the ends off the parcel so that it is open at the ends.
1	1	Bay leaf	
4	4	Parsley stems	2. Place the tenderloin and the parcel of herbs in a vacuum bag. Vacuum at 100% and seal.
1 piece	1 piece	Beef tenderloin, center cut, trimmed, about 13 oz (400 g) each	3. Cook the beef in a thermal circulator heated to 127°F (53°C) for 45 minutes.
			4. Remove the beef from the bag. Discard the herb parcel. Dry the beef on paper towels.
to taste	to taste	Salt	5. Season the beef with salt and pepper.
to taste	to taste	Pepper	6. Heat the oil in a small sauté pan. Add the beef and brown evenly on all sides.
1 tbsp	15 mL	Vegetable oil	
1 oz	30 g	Butter	7. Reduce heat to low. Add the butter, garlic, shallot, and thyme to the pan. Continue to cook the beef, turning frequently to cook evenly on all sides and basting with the butter.
2 cloves	2 cloves	Garlic, crushed	
1 oz	30 g	Shallot, sliced	8. The beef was cooked to the rare stage in the immersion circulator, so to serve it rare, stop cooking after a few minutes. (If medium done beef is desired, the best method is to cook to a higher internal temperature in the immersion circulator. Alternatively, you could cook it slightly longer in the sauté pan at this stage.)
3 sprigs	3 sprigs	Fresh thyme	
			9. Remove from the pan and let rest 10 minutes before slicing for service.

Per serving: Calories, 370; Protein, 35 g; Fat, 23 g (56% cal.); Cholesterol, 125 mg; Carbohydrates, 4 g; Fiber, <1 g; Sodium, 180 mg.

BRAISING

Braising is sometimes referred to as a combination cooking procedure, because combines dry heat—for browning the item—with moist heat—for cooking the item to tenderness in a liquid and at the same time making a sauce. A detailed guide to the braising procedure is given in Chapter 14. Because these procedures have many steps, it is a good idea to review them carefully before continuing with this section.

As Chapter 14 explains, there are three main types of meat items cooked by the braising procedure: large cuts, individual portion cuts, and stews. The core recipe used to illustrate braising is for individual portion cuts, in this case short ribs. In this section, you will find recipes for large cuts and stews as well. The first two recipes are for a classic beef pot roast and a classic beef stew. The pot roast is cooked in an unthickened liquid using the first braising method (p. 444), and the stew is cooked in a thickened liquid using the second braising method.

One special type of stew, the classical *fricassée*, is described in Chapter 14, but a more complete explanation is left for Chapter 18, where a recipe for Chicken Fricassée is included. This same procedure and recipe, however, can be used for making fricassees of white meats such as veal and pork, as noted later in this section (p. 517). Because the procedure can be given in a short space, it is included here as well, for convenience.

PROCEDURE for Making Classical Fricassées

1. Follow Braising Method 2, p. 444, *except*:
   - Do not brown the meat. Cook it gently in the fat without browning.
   - Add flour to the meat in the pan and make a blond roux.

2. Finish the sauce with a liaison of egg yolks and cream.

KEY POINTS TO REVIEW

- How is the procedure for simmering cured meats, such as corned beef, different from simmering fresh meats?

- What are the steps in the procedure for making a classic meat blanquette?

- What are the steps in the procedure for making a classic meat fricassee?

Beef Pot Roast

PORTIONS: 12   PORTION SIZE: 4 OZ (125 G) MEAT, 2 FL OZ (60 ML) SAUCE

U.S.	METRIC	INGREDIENTS	PROCEDURE
4 lb 12 oz	2.3 kg	Beef shoulder clod or bottom round, well trimmed (see Note)	1. Dry the meat so it will brown more easily. Heat the oil in a brazier over high heat and brown the meat well on all sides. Remove from pan. (*Alternative method:* Brown meat in a very hot oven.)
2 fl oz	60 mL	Oil	
		Mirepoix:	2. Add the mirepoix to the brazier and brown it.
4 oz	125 g	Onion, medium dice	3. Add the tomato product, the stock, and the sachet. Bring to a boil, cover, and place in an oven preheated to 300°F (150°C), or just hot enough to maintain a simmer.
2 oz	60 g	Celery, medium dice	
2 oz	60 g	Carrot, medium dice	
3 oz	90 g	Tomato purée	4. Braise the meat until tender, 2–3 hours.
		or	5. Remove meat from pan and keep warm for service in a covered pan. Discard sachet. (See alternative method of service given below.)
6 oz	180 g	Tomatoes, canned	
2½ pt	1.25 L	Brown stock	
		Sachet:	
1	1	Bay leaf	
pinch	pinch	Dried thyme	
6	6	Peppercorns	
1	1	Garlic clove	
2 oz	60 g	Flour	6. Skim the fat from the braising liquid and reserve 2 oz (60 g) of it.
			7. Make a brown roux with the flour and the reserved fat. Cool the roux slightly.
			8. Bring the braising liquid to a simmer and beat in the roux. Simmer the sauce at least 15–20 minutes, or until thickened and reduced slightly.
			9. Strain the sauce and adjust the seasonings.
			10. Slice the meat across the grain. The slices should not be too thick. Serve each 4-oz (125-g) portion with 2 fl oz (60 mL) sauce.

Per serving: Calories, 320; Protein, 38 g; Fat, 15 g (45% cal.); Cholesterol, 90 mg; Carbohydrates, 4 g; Fiber, 0 g; Sodium, 70 mg.

Note: Other cuts of beef from the round, or from the chuck or brisket, may be used instead of bottom round. Braised round makes the best slices, but it tends to be dry. Chuck (shoulder) and brisket are moister when braised because they have a higher fat content.

VARIATIONS

Alternative Method of Service: Cool beef as soon as it is cooked. For service, slice cold meat on an electric slicer and arrange in hotel pans. Add sauce, cover pans, and reheat in oven or steamer. Individual portions may also be reheated to order in the sauce.

Braised Beef Jardinière

Garnish the finished product with 8 oz (250 g) each carrots, celery, and turnips, all cut bâtonnet and boiled separately, and 8 oz (250 g) pearl onions, boiled and sautéed until brown.

Braised Lamb Shoulder

Prepare boned, rolled shoulder of lamb according to the basic recipe. Use either regular brown stock or brown lamb stock.

Beef Stew

PORTIONS: 12   PORTION SIZE: 8 OZ (250 G)

U.S.	METRIC	INGREDIENTS	PROCEDURE
3 lb	1.5 kg	Beef chuck, boneless and well trimmed of fat	1. Cut the meat into 1-in. (2.5-cm) cubes.
2 fl oz	60 mL	Oil	2. Heat the oil in a brazier until very hot. Add the meat and brown well, stirring occasionally to brown all sides. If necessary, brown the meat in several small batches to avoid overcrowding the pan.
8 oz	250 g	Onion, fine dice	
1 tsp	5 mL	Chopped garlic	3. Add the onion and garlic to the pan and continue to cook until onion is lightly browned.
2 oz	60 g	Flour	4. Add the flour to the meat and stir to make a roux. Continue to cook over high heat until the roux is slightly browned.
4 oz	125 g	Tomato purée	
1 qt	1 L	Brown stock	5. Stir in the tomato purée and stock and bring to a boil. Stir with a kitchen spoon as the sauce thickens.
		Sachet:	
1	1	Bay leaf	6. Add the sachet. Cover the pot and place in an oven at 325°F (165°C). Braise until the meat is tender, 1½–2 hours.
pinch	pinch	Thyme	
small sprig	small sprig	Celery leaves	
8 oz	250 g	Celery, EP	7. Cut the celery and carrots into large dice.
12 oz	375 g	Carrots, EP	8. Cook the celery, carrots, and onions separately in boiling salted water until just tender.
8 oz	250 g	Small pearl onions, EP	
4 oz	125 g	Tomatoes, canned, drained, and coarsely chopped	9. When meat is tender, remove the sachet and adjust seasoning. Degrease the sauce.
4 oz	125 g	Peas, frozen, thawed	10. Add celery, carrots, onions, and tomatoes to the stew.
to taste	to taste	Salt	11. Immediately before service, add the peas. Alternatively, garnish the top of each portion with peas. Season with salt and pepper.
to taste	to taste	Pepper	

Per serving: Calories, 240; Protein, 27 g; Fat, 9 g (34% cal.); Cholesterol, 60 mg; Carbohydrates, 13 g; Fiber, 2 g; Sodium, 150 mg.

Note: For more elegant service, remove the cooked meat from the sauce before adding the vegetables. Strain the sauce and pour it back over the meat.

VARIATIONS

Vegetables for garniture may be varied as desired.

Beef Stew with Red Wine

Prepare as in basic recipe, but use 1¼ pt (625 mL) dry red wine and 12 fl oz pt (375 mL) brown stock instead of 1 qt (1 L) brown stock.

Boeuf Bourguignon

Prepare Beef Stew with Red Wine, using rendered salt pork or bacon fat instead of oil. (Cut the pork into bâtonnet shapes, sauté until crisp, and save the cooked pork for garnish.) Increase garlic to 1 tbsp (15 mL). Omit vegetable garnish (celery, carrots, pearl onions, tomatoes, and peas) indicated in basic recipe, and substitute lardons (cooked salt pork or bacon pieces), small mushroom caps browned in butter, and boiled pearl onions browned in butter. Serve with egg noodles.

Navarin of Lamb (Brown Lamb Stew)

Prepare as in basic recipe, using lamb shoulder instead of beef chuck. Increase garlic to 1 tbsp (30 mL).

Brown Veal Stew

Prepare as in basic recipe, using veal shoulder or shank.

Brown Veal Stew with White Wine

Prepare Brown Veal Stew, replacing 8 fl oz (250 mL) stock with white wine.

Beef Pot Pie

Fill individual casserole dishes with stew and vegetable garnish. Top with pie pastry (p. 984). Bake in a hot oven (400°–450°F/200°–225°C) until the crust is brown.

Chile con Carne

PORTIONS: 12   PORTION SIZE: 8 OZ (250 G)

U.S.	METRIC	INGREDIENTS
1 lb 4 oz	600 g	Onion, small dice
10 oz	300 g	Green bell pepper, small dice
1/2 oz	15 g	Garlic, chopped
2 fl oz	60 g	Oil
2 1/2 lb	1.2 kg	Ground beef
3 lb 4 oz	1.4 kg	Canned tomatoes
5 oz	150 g	Tomato paste
1 1/4 pt	625 mL	Brown stock
1 1/2 oz	45 g	Chili powder
to taste	to taste	Salt
to taste	to taste	Pepper

PROCEDURE

1. Sauté the onion, pepper, and garlic in oil in a heavy saucepot until tender but not browned. Remove from the pot.
2. Add the meat to the pot and brown over high heat, breaking it up with a spoon as it browns. Drain off the fat.
3. Return the vegetables to the pot and add the remaining ingredients.
4. Simmer uncovered until the chili is reduced to the desired thickness, 45–60 minutes. Stir occasionally during the cooking period.

Per serving: Calories, 310; Protein, 20 g; Fat, 19 g (54% cal.); Cholesterol, 55 mg; Carbohydrates, 16 g; Fiber, 4 g; Sodium, 380 mg.

VARIATION

Chile with Beans

Add 2 lb (1 kg) (drained weight) cooked or canned and drained kidney beans or pinto beans about 15 minutes before the end of cooking.

Game Chile

In place of the ground beef, use ground bison, venison, elk, caribou, or boar.

Texas Red

YIELD: 6 LB 12 OZ (3 KG)   PORTIONS: 12   PORTION SIZE: 9 OZ (250 G)

U.S.	METRIC	INGREDIENTS
6 lb	2.75 kg	Lean beef, preferably chuck or shank, well trimmed
2 fl oz	60 mL	Vegetable oil
1 oz	30 g	Garlic, chopped fine
5 oz	140 g	Chili powder
1 1/2 tbsp	22 mL	Ground cumin
1 1/2 tbsp	22 mL	Dried oregano
1–2 tsp	5–10 mL	Cayenne
2 qt	2 L	Brown stock or beef broth
2 oz	60 g	Cornmeal or masa harina
4 fl oz	120 mL	Cold water
to taste	to taste	Salt

PROCEDURE

1. Cut the beef into 1-in. (2.5-cm) cubes.
2. In a brazier, brown the beef in the vegetable oil over high heat. Brown a little at a time so as not to overcrowd the pan.
3. Add the garlic, chili powder, cumin, oregano, and cayenne to the beef over moderate heat. Stir and cook a few minutes, until the spices are aromatic.
4. Add the stock. Bring to a boil, cover, and place in an oven at 325°F (165°C) for 1 1/2–2 hours, or until the beef is tender.
5. Degrease the cooking liquid carefully.
6. Mix the cornmeal or masa harina with the cold water to make a smooth paste.
7. Stir the cornmeal mixture into the beef mixture.
8. Simmer 30 minutes to thicken the cooking liquid.
9. Season to taste with salt.

Per serving: Calories, 380; Protein, 46 g; Fat, 16 g (38% cal.); Cholesterol, 90 mg; Carbohydrates, 12 g; Fiber, 5 g; Sodium, 200 mg.

VARIATION

Texas Short Ribs

In place of the lean beef, use 7 lb 8 oz (3.4 kg) beef short ribs. After the ribs are tender, remove them from the braising liquid and carefully degrease the liquid. Add just enough cornmeal or masa harina to lightly thicken the liquid, about half the amount in the basic recipe.

Texas Red

Braised Beef with Ancho Chiles

YIELD: 3 LB 12 OZ (1.7 KG)   PORTIONS: 10   PORTION SIZE: 6 OZ (170 G) MEAT, 2 FL OZ (60 ML) JUS

U.S.	METRIC	INGREDIENTS
6 lb	2.75 kg	Beef chuck, well trimmed, in large pieces
as needed	as needed	Vegetable oil
1 lb	900 g	Onions, sliced
4 oz	120 g	Slab bacon, in 1 piece
4	4	Whole dried ancho chiles, stems and seeds removed (see Note)
1 oz	30 g	Garlic, coarsely chopped
1/4 tsp	1 mL	Ground cumin
1/4 tsp	1 mL	Dried thyme
3/4 tsp	3 mL	Dried oregano
2 oz	60 g	Tomato paste
3 pt	1.5 L	White stock
1 tbsp	15 mL	Salt
1 tsp	5 mL	Pepper

PROCEDURE

1. In a braising pan, brown the beef well in vegetable oil. Brown just a little of the meat at a time if necessary to prevent overcrowding the pan. Remove the meat from the pan.
2. Using additional oil as needed, brown the onions over moderate heat until they are well browned but not scorched.
3. Return the beef to the pot and add the remaining ingredients.
4. Bring to a boil, cover, and place in an oven at 325°F (165°C). Braise about 2 hours, or until the meat is very tender.
5. Remove the meat from the braising liquid and keep warm. Discard the bacon.
6. Degrease the cooking liquid.
7. Remove the chiles from the liquid. Place the chiles, skin side down, on a cutting board, and carefully scrape the soft pulp from the inside surface. Discard the thin, transparent skins. Stir the pulp back into the cooking liquid.
8. Reduce the liquid over moderate heat to about 1 1/2 pt (700 mL).
9. Taste and add salt and pepper if needed.

Per serving: Calories, 410; Protein, 56 g; Fat, 16 g (36% cal.); Cholesterol, 110 mg; Carbohydrates, 9 g; Fiber, 3 g; Sodium, 920 mg.

Note: If whole chiles are not available, or if it is necessary to save the labor of scraping the pulp of the chiles after cooking (step 7), substitute 2 1/2 tbsp (40 mL) ground ancho chiles for the whole chiles.

Braised Beef with Ancho Chiles on Grits with Cheddar, Topped with Fried Parsley

Veal Fricassée
Pork Fricassée

See Chicken Fricassée variations, page 576.

Saltimbocca alla Romana

PORTIONS: 8   PORTION SIZE: 2 PIECES

U.S.	METRIC	INGREDIENTS
16	16	Veal scaloppine, 1 1/2–2 oz (45–60 g) each
to taste	to taste	Salt
to taste	to taste	White pepper
16	16	Thin slices of prosciutto, about the same diameter as the scaloppine
16	16	Fresh sage leaves
2 oz	60 g	Butter
6 fl oz	175 mL	White wine

PROCEDURE

1. Pound the scaloppine with a cutlet pounder. Season with salt and white pepper. Put 1 slice of prosciutto and 1 sage leaf on top of each and fasten with a toothpick.
2. Sauté briefly in butter on both sides.
3. Add the wine and continue to cook until the meat is done and the wine is partly reduced, no more than 5 minutes.
4. Remove the meat from the pan and serve, ham side up, with a spoonful of the pan juices over each.

Per serving: Calories, 320; Protein, 28 g; Fat, 21 g (60% cal.); Cholesterol, 115 mg; Carbohydrates, 0 g; Fiber, 0 g; Sodium, 800 mg.

Osso Buco

PORTIONS: 12 PORTION SIZE: 1 PIECE, APPROXIMATELY 8 OZ (240 G) COOKED WEIGHT, PLUS 1¹/₂ FL OZ (45 ML) SAUCE

U.S.	METRIC	INGREDIENTS	PROCEDURE
12 pieces	12 pieces	Veal shank, cut crosswise into thick slices, about 11 oz (320–350 g) each	1. Tie each piece of veal shank tightly around the circumference to hold the meat to the bone.
to taste	to taste	Salt	2. Lightly season the meat with salt and pepper.
to taste	to taste	Pepper	3. Heat the oil in a brazier just large enough to hold the pieces of veal shank in a single layer.
2 fl oz	60 mL	Vegetable oil	4. Dredge each piece of meat in flour to coat completely, then shake off the excess.
as needed	as needed	Flour	5. Brown the meat on all sides in the oil.
1¹/₂ oz	45 g	Clarified butter	6. Remove the meat from the pan and set it aside.
		Mirepoix:	7. Drain off excess oil from the pan, then add the clarified butter.
8 oz	250 g	Onion, small dice	8. Add the mirepoix and garlic. Brown lightly.
4 oz	125 g	Carrot, small dice	
4 oz	125 g	Celery, small dice	
¹/₄ oz	7 g	Garlic, chopped	
8 fl oz	250 ml	Dry white wine	9. Return the meat to the pan and add the white wine. Simmer about 10 minutes.
1 lb	500 g	Canned plum tomatoes, coarsely chopped, with their juice	10. Add the tomatoes, lemon zest, stock, and sachet. Bring to a simmer, cover, and place in an oven at 325°F (160°C). Braise until the meat is very tender, 1¹/₂–2 hours.
2–3 strips	2–3 strips	Lemon zest, in wide strips	11. Remove the meat from the pan with a slotted spoon and set aside in a warm place.
1 pt	500 mL	Brown stock	12. Strain and degrease the sauce. Reduce the sauce, if necessary, to about 1¹/₂ fl oz (45 mL) per portion. The flour on the meat should have been enough to thicken the sauce to a very light consistency.
		Sachet:	
8	8	Parsley stems	
¹/₄ tsp	1 mL	Dried thyme	
1	1	Bay leaf	
		Gremolada (optional):	13. The traditional but optional seasoning finish for osso buco is called gremolada. Mix together the lemon zest, parsley, garlic, and anchovy. Just before serving, sprinkle the gremolada over the osso buco and turn the meat so it gets an even coating of the mixture.
1¹/₂ tsp	7 mL	Grated lemon zest	
1¹/₂ tbsp	22 mL	Chopped parsley	
¹/₂ tsp	2 ml	Garlic, chopped fine	14. Serve 1 piece of veal shank per portion, along with 1¹/₂ fl oz (45 mL) sauce. Osso buco is traditionally served with Risotto Milanese (p. 386) as well.
2	2	Anchovy filets, chopped fine (optional)	

Per serving: Calories, 360; Protein, 42 g; Fat, 14 g (11% cal.); Cholesterol, 170 mg; Carbohydrates, 10 g; Fiber, 2 g; Sodium, 260 mg.

Osso Buco

Sauerbraten

PORTIONS: 12   PORTION SIZE: 4 OZ (125 G), 2 FL OZ (60 ML) SAUCE

U.S.	METRIC	INGREDIENTS	PROCEDURE
5 lb	2.5 kg	Beef bottom round, trimmed (see Note)	1. Place the trimmed beef in a nonmetallic crock or barrel.
		Marinade:	2. Add all the marinade ingredients to the crock. If the meat is not completely covered by the liquid, add equal parts vinegar and water until it is. Cover.
1 pt	500 mL	Red wine vinegar	
1 pt	500 mL	Water	3. Refrigerate 3–4 days. Turn the meat in the marinade every day.
8 oz	250 g	Onion, sliced	
4 oz	125 g	Carrot, sliced	
1	1	Garlic cloves, chopped	
1 oz	30 g	Brown sugar	
1	1	Bay leaves	
2	2	Whole cloves	
1/2 tsp	2 mL	Peppercorns, crushed	
1 tsp	5 mL	Salt	
as needed	as needed	Vegetable oil, if needed for browning meat	4. Remove the meat from the marinade. Dry it thoroughly with paper towels.
			5. Brown the meat on all sides. This may be done on the range in an iron skillet, on a very hot griddle, under the broiler, or in a brazier in a hot oven.
			6. Place the meat in a braising pan. Strain the marinade. Add the vegetables to the meat and enough of the liquid to cover the meat by half. Cover and braise in a 300°F (150°C) oven until the meat is tender, 2–3 hours.
			7. Remove the meat from the braising liquid and transfer to a hotel pan. Set aside.
4 fl oz	125 mL	Red wine	8. Strain 1 qt (1 L) of the braising liquid into a saucepan and skim off fat. Bring to a boil. Reduce to about 1½ pt (750 mL).
2 oz	60 g	Gingersnap crumbs	9. Add the wine and boil another 2–3 minutes.
			10. Reduce heat to a simmer and stir in the gingersnap crumbs. Simmer another 3–4 minutes. Remove from heat and let stand 5 minutes to allow the crumbs to be completely absorbed.
			11. Slice the meat across the grain. Serve 4 oz (125 g) meat per portion, overlapping the slices on the plate. Ladle 2 oz (60 mL) sauce over the meat.

Per serving: Calories, 260; Protein, 37 g; Fat, 8 g (32% cal.); Cholesterol, 80 mg; Carbohydrates, 4 g; Fiber, 0 g; Sodium, 110 mg.

Note: If you are preparing this item in large quantities and are using whole bottom round (called gooseneck), separate the eye of round from the bottom round and cut the bottom round in half lengthwise, so two pieces are about the size of the eye of round.

Brisket or chuck may also be used for sauerbraten. They do not make attractive slices, but the eating quality is very good.

VARIATION

Sauerbraten with Sour Cream Gravy

Marinate and braise meat as in basic recipe. Prepare gravy through step 8. Thicken the sauce with a roux made of 2 oz (60 g) butter or beef fat, 2 oz (60 g) flour, and 1 oz (30 g) sugar. Cook the roux until well browned and use it to thicken the sauce. Omit wine and add 4 oz (125 mL) sour cream.

Sauerbraten with Red Cabbage and Potato Dumplings

Veal Curry with Mangos and Cashews

PORTIONS: 12   PORTION SIZE: 8 OZ (250 G)

U.S.	METRIC	INGREDIENTS	PROCEDURE
4¹/₂ lb	2.25 kg	Boneless, lean veal (shoulder, shank, or breast)	1. Cut the veal into 1-in. (2.5-cm) cubes.
4 fl oz	125 mL	Oil	2. Heat the oil in a brazier over medium heat. Add the meat and cook it in the fat, stirring occasionally, until seared on all sides but only lightly browned.
1 lb 4 oz	600 g	Onions, medium dice	3. Add the onions and garlic to the pan. Sauté until softened, but do not brown.
1 tbsp	15 mL	Garlic, chopped	4. Add the spices and salt and stir. Cook 1 minute.
2¹/₂ tbsp	37 mL	Curry powder (see Note)	
1¹/₂ tsp	7 mL	Ground coriander	
1¹/₂ tsp	6 mL	Paprika	
¹/₂ tsp	2 mL	Ground cumin	
¹/₂ tsp	2 mL	Pepper	
¹/₄ tsp	1 mL	Cinnamon	
1	1	Bay leaves	
1 tsp	5 mL	Salt	
2 oz	60 g	Flour	5. Stir in the flour to make a roux and cook another 2 minutes.
1 qt	1 L	White stock	6. Add the stock and tomatoes. Bring to a boil while stirring.
5 oz	150 g	Tomato concassé	7. Cover and simmer slowly in the oven (300°F/150°C) or on top of the range until the meat is tender, 1–1¹/₂ hours.
4 fl oz	125 mL	Heavy cream, hot	8. Degrease, discard the bay leaf, and add the cream. Adjust the seasonings.
2	2	Mangos	9. Peel the mangos with a paring knife or vegetable peeler. Cut the mango flesh from the stone in thick slices. Cut into medium dice.
as needed	as needed	Boiled or steamed rice	
2 oz	60 g	Cashews, coarsely chopped	10. To serve, place a bed of rice on a plate. Spoon the curry onto the center of the rice. Top with diced mango. Sprinkle with chopped cashews and chopped parsley.
1 tbsp	15 mL	Chopped parsley	

Per serving: Calories, 430; Protein, 31 g; Fat, 26 g (64% cal.); Cholesterol, 95 mg; Carbohydrates, 18 g; Fiber, 2 g; Sodium, 330 mg.

Note: If desired, increase curry powder to taste and omit other spice (except bay leaf).

VARIATIONS

In place of the mango and cashews, serve meat curries with an assortment of other condiments, such as raisins, chutney, peanuts, chopped scallions or onions, diced pineapple, diced banana, diced apple, shredded coconut, and poppadums.

Lamb Curry

Substitute lean boneless lamb shoulder, breast, or leg for the veal.

Veal Curry with Mangos and Cashews

Carbonnade à la Flammande

PORTIONS: 16   PORTION SIZE: 6–7 OZ (175–200 G)

U.S.	METRIC	INGREDIENTS	PROCEDURE
3 lb	1.4 kg	Onions	1. Peel the onions. Cut them into small dice.
as needed	as needed	Beef fat or vegetable oil	2. Cook the onions in a little fat over moderate heat until golden. Remove from the heat and set aside.
6 oz	175 g	Flour	3. Season the flour with salt and pepper. Dredge the meat in the flour. Shake off the excess flour.
2 tsp	10 mL	Salt	
1 tsp	5 mL	Pepper	4. Brown the meat well in a sauté pan. Do a little at a time to avoid overcrowding the pan. As each batch is browned, add it to the pot with the onions.
5 lb	2.3 kg	Beef chuck, 1-in. (2.5-cm) dice	
2½ pt	1.25 L	Dark beer	5. Deglaze the sauté pan with the beer and add it to the pot. Add the stock, sachet, and sugar.
2½ pt	1.25 L	Brown stock	
		Sachet:	6. Bring to a boil, cover, and transfer to the oven. Cook at 325°F (160°C) until very tender, 2–3 hours.
2	2	Bay leaves	
1 tsp	5 mL	Dried thyme	7. Degrease. Adjust the consistency of the sauce. If it is too thin, reduce over moderately high heat. If it is too thick, dilute with brown stock.
8	8	Parsley stems	
8	8	Peppercorns	8. Taste and adjust the seasonings. Serve with plain boiled potatoes.
1 tbsp	15 mL	Sugar	

Per serving: Calories, 450; Protein, 30 g; Fat, 29 g (52% cal.); Cholesterol, 100 mg; Carbohydrates, 19 g; Fiber, 1 g; Sodium, 290 mg.

ADDITIONAL RECIPES

These additional recipes may be found on your CulinarE-Companion recipe management program:

Baked Pork Chops with Prune Stuffing; Beef Stroganoff; Boeuf à la Mode (Beef Braised in Red Wine); Braised Sweetbreads; Broiled Lamb Kidneys with Bacon; Costolette di Vitello Ripiene alla Valdostana (Veal Cutlets Val d'Aosta-Style); Dillkött; Grilled Chopped Lamb "Steaks" with Rosemary and Pine Nuts; Grilled Loin of Elk; Grilled Venison with Lime Butter; Hungarian Goulash (Veal, Beef, or Pork); Lamb Vindaloo; Lombatine di Maiale alla Napoletana (Braised Pork Chops Naples-Style); Medallions of Venison Poîvrade with Cassis; Pork Chops Charcutière; Provençal Beef Stew; Rabbit with Mustard; Roast Loin of Pork with Sage and Apples; Roast Loin of Rabbit with Risotto; Sautéed Veal Sweetbreads with Shiitake Mushrooms and Port Wine Sauce; Shish Kebab; Steak en Chevreuil; Swedish Meatballs; Thyme-Scented Medallions of Lamb with Balsamic Glaze.

QUESTIONS FOR DISCUSSION

1. Why does the recipe for Roast Brined Pork Loin with Date and Gorgonzola Stuffing (p. 491) include no pan gravy or jus?

2. The recipes for Roast Pork Tenderloin with Kalbi Marinade (p. 494) and Loin or Rack of Venison Grand Veneur (p. 497) instruct you to dry the meat with paper towels before roasting. Why?

3. In the recipe for Home-Style Meatloaf (p. 496), why are the sautéed vegetables cooled after cooking in step 1?

4. When plating a grilled steak with a sauce, how should the sauce be added to the plate?

5. You have the following ingredients and would like to make a stir-fried dish with them: pork loin, bok choy, red pepper flakes, soy sauce, fresh ginger, garlic, cornstarch, water, oil, and salt. Explain what steps you would use to make this dish.

6. Why is the menu term "New England Boiled Dinner" inaccurate?

7. What is the difference between Veal Blanquette and Veal Fricassée?

UNDERSTANDING POULTRY AND GAME BIRDS

The versatility, popularity, and relatively low cost of poultry items make them ideal for all kinds of food-service operations, from high-end restaurants to cafeterias and fast-food restaurants. Also, chicken and turkey are popular among diet-conscious people because they are lower in fat and cholesterol than other meats. Game birds, such as pheasant, are also increasing in popularity and availability because they are now raised domestically by many producers. Farm-raised game birds are similar, in many ways, to chicken, so learning techniques for cooking and handling chicken teaches you a great deal about handling these other birds as well.

Learning about poultry is, in some ways, easier than learning about meats like beef and lamb. Because chickens, turkeys, and other poultry are much smaller, they are not cut up in such detail.

However, poultry has its own cooking problems, so it is important to observe both the similarities and the differences between meat and poultry.

AFTER READING THIS CHAPTER, YOU SHOULD BE ABLE TO

1. Explain the differences between light meat and dark meat, and describe how these differences affect cooking.

2. Describe four techniques that help keep chicken or turkey breast moist while roasting.

3. Define the following terms used to classify poultry: kind, class, and style.

4. Identify popular types of farm-raised game birds and the cooking methods appropriate to their preparation.

5. Store poultry items.

6. Determine doneness in cooked poultry, both large roasted birds and smaller birds.

7. Truss poultry for cooking.

8. Cut up chicken into parts.

COMPOSITION AND STRUCTURE

The flesh of poultry and game birds is muscle tissue, as is the flesh of beef, lamb, veal, pork, and game. Its composition and structure are essentially the same as those of meat.

Review the section on meat composition and structure (Chapter 15, pp. 450–483). Remember that muscle tissue is composed of

- Water (about 75 percent)
- Protein (about 20 percent)
- Fat (up to 5 percent)
- Other elements, including carbohydrate, in small quantities

Remember that muscles consist of muscle fibers held together in bundles by connective tissue.

MATURITY AND TENDERNESS

We learned in Chapter 15 that the tenderness of a piece of meat—or poultry—is related to connective tissue and that connective tissue increases with

- Use or exercise of the muscle.
- Maturity or age of the animal or bird.

1. Use or exercise is of less concern in poultry. Most poultry is so young that it is relatively tender throughout. However, there are differences, discussed in the next section, between **light meat** and **dark meat**.
2. **Maturity** is a major consideration when selecting poultry. Young, tender birds are cooked by dry-heat methods, such as broiling, frying, and roasting, as well as by moist-heat methods. Older, tougher birds need slow, moist heat to be made palatable.

   Maturity is the major factor in categorizing each kind of poultry (see p. 527).

   Skin color is determined by diet and is not related to the flavor or tenderness of the poultry.

FREE-RANGE CHICKENS

Most chickens on the market are produced by large operations that house their poultry indoors in carefully controlled environments and feed them scientifically monitored diets. This process enables the industry to raise healthy chickens quickly and in large numbers to meet the great demand. Many people feel these chickens lack flavor because they are not allowed to move around outdoors. Some farmers, in response, offer **free-range** chickens, which are allowed to move around freely and eat outdoors in a more natural environment.

It is important to note that there is no legal definition of *free-range,* and that free-range chickens are considerably more expensive than ordinary chickens. Many people, however, feel free-range chickens are more flavorful and worth the extra cost. Because quality varies from producer to producer, it is necessary to do careful taste-testing to determine whether you want to purchase free-range poultry for your operation.

A term related to free-range is **organic**, which has recently been defined by the USDA as food produced without using most conventional pesticides, fertilizers made with synthetic ingredients or sewage sludge, bioengineering, or ionizing radiation. See page 275 for additional information on organic foods.

LIGHT MEAT AND DARK MEAT

Poultry is not divided into as many small cuts as are meats. Chicken and turkey, however, are usually thought of as consisting of two kinds of parts, depending on the color of the meat. These color differences reflect other differences:

FREE-RANGE VERSUS CAGE-FREE

Some poultry products, especially eggs, are marketed as "cage-free." How does this differ from free-range?

Cage-free simply means the birds aren't kept in cages, but they *are* kept indoors. To be called free-range, they must be allowed outdoors.

"Light meat"—breast and wings

> Less fat
>
> Less connective tissue
>
> Cooks faster

"Dark meat"—legs (drumsticks and thighs)

> More fat
>
> More connective tissue
>
> Takes longer to cook

Duck, goose, and squab have all dark meat, but the same differences in connective tissue hold true.

The dark color of dark meat is due to a protein called *myoglobin*. This protein stores oxygen for muscles to use during periods of great activity. The breast muscles of birds are used for flying, and because chickens and turkeys rarely, if ever, fly, these muscles don't need a great deal of myoglobin. In flying birds, such as ducks, the breast muscles have more myoglobin and thus are darker. Active muscles, in addition to being darker, also have more connective tissue.

The cook must observe these differences when preparing poultry.

1. **Cooking whole birds.**

   Everyone has tasted chicken or turkey breast so dry it was difficult to swallow. In fact, light meat is overcooked more often than not because it cooks faster than the legs and is done first. In addition, the breast has less fat than the legs, so it tastes much drier when cooked (or overcooked).

   A major problem in roasting poultry is cooking the legs to doneness without overcooking the breast. Chefs have devised many techniques to help solve this problem. Here are some of them.

   - Roasting breast-down for part of the roasting period. Gravity draws moisture and fat to the breast rather than away from it.

   - Basting with fat only, not with water or stock. Fat protects against drying, but moisture washes away protective fat.

   - Barding, or covering the breast with a thin layer of pork fat. This is usually done with lean game birds.

   - Separating breast from leg sections and roasting each for a different time. This is often done with large turkeys.

2. **Cooking poultry parts.**

   Many recipes have been devised especially for certain poultry parts, such as wings, drumsticks, and boneless chicken breasts. These recipes take into account the different cooking characteristics of each part. For example, flattened boneless chicken breasts can be quickly sautéed and remain juicy and tender. Turkey wings, when braised, release enough gelatin to make a rich sauce.

   Many of these items have especially high customer appeal, especially boneless chicken breast, and are served in the most elegant restaurants.

   Several of the chicken and turkey recipes in Chapter 18 are for specific parts. Those that use cut-up whole chickens can easily be adapted for specific parts. For example, you may want to buy whole chickens, braise the leg sections, and reserve the breasts for other preparations.

INSPECTION AND GRADING

Like meat, poultry is subject to federal **inspection** and **grading**. (*Note*: Unlike those for meats, poultry inspection and grading stamps are not stamped on the birds but instead are printed on tags and packing cases.)

FIGURE 17.1 Inspection stamps for poultry. *Courtesy of the Canadian Food Inspection Agency.*

FIGURE 17.2 Grade stamp for poultry. *Courtesy of the Canadian Food Inspection Agency.*

INSPECTION

1. A guarantee of wholesomeness (fit for human consumption).
2. Indicated by a round stamps (Figure 17.1).
3. Required by Canadian law.

GRADING

1. Based on quality.
2. Indicated by a leafstamp and letter grade (Figure 17.2).
3. Not required by law but widely practiced in Canada and the United States.

Canadian poultry grades are Canada Grade A, Canada Utility, and Canada Grade C (A being the best). The grade stamps are colored red for Grade A, blue for Utility, and brown for Grade C. (In limited circumstances, the stamps may be colored black. For example, when non-prepackaged items are shipped in a carton, it is permissible to print the carton in black rather than in color.) Grades are based on

Shape of carcass (lack of defects)	Pinfeathers (present or absent)
Amount of flesh	Skin tears, cuts, broken bones
Amount of fat	Blemishes and bruises

Most poultry used in food service is Grade A. Some utility-grade poultry is also used, although most utility- and nearly all C-grade poultry is used by canners and processors.

CLASSIFICATION AND MARKET FORMS

The following terms are used to classify poultry:

Kind—the species, such as chicken, turkey, or duck.

Class—the subdivision of kind, depending on age and sex.

Style—the amount of cleaning and processing.

Live: almost never purchased in food service.

Dressed: killed, bled, and plucked. Also rarely seen in food service.

Ready to cook: dressed and eviscerated, with head and feet removed.

- Whole
- Cut up, or parts

State of refrigeration—chilled or frozen.

Table 17.1 describes the kinds and classes of domestic poultry. **Chicken** is the most common kind of poultry in the kitchen. As indicated in the table, age or maturity determines the differences among the classes of chicken. **Rock Cornish game hens** (usually called Cornish hens), **broilers**, and **fryers** are young, tender chickens suitable for sautéing, broiling, or frying, while **roasters** and **capons** are larger chickens that are usually roasted. Older **hens** and (rarely marketed) **roosters** must be simmered or braised to make them tender.

In addition to the common classes of chicken listed in the table, you may also find **poussin** (poo-san) on the market in certain localities. Similar to Rock Cornish game hen but often smaller, a poussin is a young chicken weighing 1 pound (450 g) or less. It commands a fairly high price because of the special techniques required for raising it.

Turkeys are larger birds that are usually roasted, although the cooking of turkey parts is increasingly common. For example, legs may be stewed or braised for special dishes (see, for example, Mole Poblano, p. 573), while breasts are cut into cutlets or scaloppine and sautéed like veal cutlets.

Ducks and **geese** also are usually roasted, although duck parts are sometimes cooked separately. Boneless breast of duck is sautéed or broiled and served rare, sliced into small medallions, and the legs may be braised. Ducks and geese have a thick layer of fat under the skin.

TABLE 17.1 Domestic Poultry Classes and Characteristics

Kind /Class	Description	Age	Weight Range
Chicken			
Rock Cornish game hen	Special breed of young chicken, very tender and delicate.	5 weeks or less	$^3/_4$–2 lb (0.34–0.9 kg)
Broiler or fryer	Young chicken of either sex. Tender flesh and flexible cartilage. Smooth skin.	6–12 weeks	Broiler: $1^1/_2$–$2^1/_2$ lb (0.7–1.1 kg) Fryer: $2^1/_2$–$3^1/_2$ lb (1.1–1.6 kg)
Roaster	Young chicken of either sex. Tender flesh and smooth skin, but less flexible cartilage.	3-5 months	$3^1/_2$–5 lb (1.6–2.3 kg)
Capon	Castrated male chicken. Flesh very tender and well flavored. Large breast. Expensive.	Under 8 months	5–8 lb (2.3–3.6 kg)
Hen or fowl	Mature female. Tough flesh and coarse skin. Hardened breastbone cartilage.	Over 10 months	$3^1/_2$–6 lb (1.6–2.7 kg)
Cock or rooster	Mature male. Coarse skin. Tough, dark meat.	Over 10 months	4–6 lb (1.8–2.7 kg)
Turkey			
Fryer-roaster	Young bird of either sex. Tender flesh, smooth skin, and flexible cartilage.	Under 16 weeks	4–9 lb (1.8–4 kg)
Young turkey (hen or tom)	Young turkeys with tender flesh but firmer cartilage.	5–7 months	8–22 lb (3.6–10 kg)
Yearling turkey	Fully matured turkey that is still reasonably tender.	Under 15 months	10–30 lb (4.5–14 kg)
Mature turkey or old turkey (hen or tom)	Old turkey with tough flesh and coarse skin.	Over 15 months	10–30 lb (4.5–14 kg)
Duck			
Broiler or fryer duckling	Young tender duck with soft bill and windpipe.	Under 8 weeks	2–4 lb (0.9–1.8 kg)
Roaster duckling	Young tender duck with bill and windpipe just starting to harden.	Under 16 weeks	4–6 lb (1.8–2.7 kg)
Mature duck	Old duck with tough flesh and hard bill and windpipe.	Over 6 months	4–6 lb (1.8–2.7 kg)
Goose			
Young goose	Young bird with tender flesh.	Under 6 months	6–10 lb (2.7–4.5 kg)
Mature goose	Tough old bird.	Over 6 months	10–16 lb (4.5–7.3 kg)
Guinea			
Young guinea	Domestic relative of the pheasant. Tender.	3–6 months	$^3/_4$–$1^1/_2$ lb (0.34–0.7 kg)
Mature guinea	Tough.	Up to 12 months	1–2 lb (0.45–0.9 kg)
Pigeon			
Squab	Very young pigeon with light, tender meat.	3–4 weeks	Under 1 lb (0.45 kg)
Pigeon	Older pigeon with tough, dark meat.	Over 4 weeks	1–2 lb (0.45–0.9 kg)

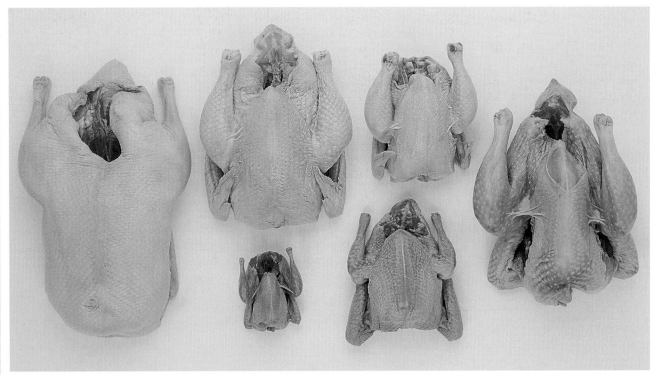

Clockwise from left: duckling, free-range chicken, poussin, guinea fowl, squab, quail

Compared with chicken and turkey, they have a low yield. For example, a 4-pound duck yields about 1 pound raw lean meat, and a 4-pound chicken yields about 2 pounds raw lean meat.

Most ducks marketed in North America are a breed called **White Pekin**; this includes the well-known Long Island duck. A specialty item available in some markets is **magret** (mah-gray). This is the boneless breast of a breed of duck called **moulard**. It is thicker and meatier than the breast of a regular Pekin.

Guineas are a domestically raised descendant of the pheasant. They taste like a flavorful chicken and are usually cooked and handled like young chickens.

Squabs are young, domestically raised pigeons, usually weighing less than 1 pound (450 g). Their rich, dark meat, which is usually cooked slightly rare to avoid dryness, has a slightly gamy flavor that combines well with flavorful brown sauces.

GAME BIRDS AND SPECIALTY PRODUCTS

In recent years, the availability of such birds as quail and squab has increased dramatically, and they are seen regularly on restaurant menus. The poultry items discussed in this section are classified as game birds, but they are all, in fact, raised domestically. While farm-raised pheasants and partridge lack the full gamy flavor of their wild cousins, they do have a richer, more gamelike taste compared to chicken. With bland, factory-raised chickens dominating the market, cooks and eaters are turning more and more to exotic poultry and are willing to pay the higher price.

Traditionally, true wild game is hung and allowed to age, usually before plucking and dressing. The purpose is essentially the same as for aging beef, namely to allow the natural enzymes in the meat to tenderize it and to develop flavor. Often, game is hung until it becomes high, to the point where spoiled meat is mistaken for aged meat. With today's farm-raised game birds, this procedure is not appropriate. Anyway, most customers prefer a fresh taste to a strong, gamy one.

Quail are small, weighing 4–5 ounces (110–140 g) each. A normal main-course portion is two birds. They have meaty breasts for their size, but not much meat on the legs. Quail are richly flavored without being gamy. The French name is *caille*.

Partridges are about the size of Rock Cornish game hens, weighing about 1 pound (450 g) each. It is important to look for young, tender birds because mature partridge is likely to be tough. They have excellent flavor, but they are not as delicate as squab or pheasant. The French names are *perdreau* (young partridge) and *perdrix* (mature partridge).

Pheasant is a popular game bird, and farm-raised pheasant is widely available. Most pheasant sold weighs from 2 to 2^1/$_2$ pounds (900 to 1200 g), but young pheasant weighing 1 pound (450 g) or less is also available. This bird has delicate, light-colored meat with subtle flavor similar to that of chicken. Most recipes for chicken are also suitable for pheasant, but the simplest preparations are usually the best, because the flavor stands well on its own and is easily covered by too many spices. Pheasant can be dry if overcooked. The French name is *faisan*.

Many varieties of **wild duck** are eaten, but mallard is the most common. Farm-raised mallards weigh from 1^1/$_2$ to 3 pounds (700 to 1400 g). Unlike domestic duck, wild duck is very lean. It has dark, flavorful flesh.

Handling game birds is easy if you remember that their structure is basically the same as the structure of chickens. All the cutting and trussing techniques you learn for chicken can be applied to these other birds.

Because farm-raised game birds are usually young and tender, they can be roasted, sautéed, grilled, and barbecued. The most important thing to remember about them is they are usually very lean. Therefore, they are best served slightly rare. If cooked to well done, they become dry. This is especially true of wild duck, which is almost inedible if overcooked. Wild duck is usually left rarer than the other birds discussed here. Its meat is then red and juicy.

Pheasant is also very dry if well done. Its light-colored meat is best if still slightly pink at the bone. Quail doesn't become as dry, but it too has the best flavor if still slightly pink at the bone.

Another category of farm-raised birds is the category technically known as **ratites**. **Ostrich** and **emu** are the most familiar members of this category. Although these meats have lost some popularity in recent years, they may still be available in some markets. The meat of both birds is lean and red and resembles venison or very lean beef in appearance, although it is slightly lighter in color than venison.

Because ostrich and emu are so lean, they are best cooked to the medium rare or, at most, medium stage to avoid dryness. Recommended safe internal temperature is 155°–160°F (68°–71°C). Grilling, sautéing, and pan-frying are the best cooking methods for small, tender cuts, while larger tender cuts can be roasted. Moist-heat methods, especially braising, are sometimes recommended for less tender cuts, but this often results in excessively dry meat because it is so lean. Take care to avoid overcooking if you braise ostrich or emu. Another option is to grind the less tender cuts. Mixed with seasonings and added moisture, ground emu and ostrich can make excellent burgers, meatballs, and meatloaf.

Ostrich and emu are best cooked like other lean red meats and game. Recipes for venison, in particular, are often excellent when applied to these meats, as are recipes for grilled or sautéed beef. Ostrich producers often recommend cooking their product like veal. This may be slightly misleading because veal is often cooked medium well or well done. Nevertheless, grilled, sautéed, and roast veal recipes can often be used for ostrich and emu as well, as long as the meat is not overcooked. Two recipes developed specifically for ostrich or emu are included in Chapter 18 as a sample. For other cooking ideas, look for appropriate recipes in Chapter 16 based on the guidelines just described.

KEY POINTS TO REVIEW

- How do the differences between light meat and dark meat affect how you cook different poultry parts?

- What are four ways to keep breast meat from becoming too dry when cooking whole poultry?

- What is the significance of inspection and grading of poultry products?

- What do the terms *kind, class,* and *style* mean with respect to poultry? What are the major kinds and classes of domestic poultry used in food service?

- What are the most important kinds of farm-raised game birds used in food service? Describe each kind.

HANDLING AND STORAGE

FRESH POULTRY

1. Fresh poultry is extremely perishable. It should arrive well chilled, ideally at 32°F (0°C) and kept well chilled until used.

2. Ideally, use poultry within 24 hours of receiving. Never hold it for more than 4 days.

3. Poultry often carries salmonella bacteria. Wash all equipment and cutting surfaces after handling poultry to avoid contamination of other foods.

Frozen Poultry

1. Store frozen poultry at 0°F (–18°C) or lower until ready to thaw.

2. Thaw in original wrapper in refrigerator, allowing 1 to 2 days for chickens, 2 to 4 days for larger birds. If pressed for time, thaw in cold, running water in original wrapper.

3. Do not refreeze thawed poultry.

DONENESS

Domestic poultry is almost always cooked well done (except squab and sautéed or grilled duck breast). Many cooks, however, cannot tell the difference between well done and over-cooked. Chicken and turkey are low in fat, so they quickly become dry and unpalatable when overcooked. Even duck and goose, which are very fatty, taste dry and stringy if cooked too long.

Skilled chefs with years of experience can often tell the doneness of a roast chicken or turkey just by looking at it. Nevertheless, the only way to be certain of proper doneness is to use a thermometer.

FOR LARGE ROASTED BIRDS

Internal temperature, as tested with a thermometer, is the most accurate guide to doneness. The thermometer should be inserted into the thickest muscle of the inner part of the thigh, away from the bone. The thigh is tested rather than the breast because the thigh is the last part of the bird to become fully cooked.

The recommended safe internal temperature (see p. 26) for roast whole poultry is 185°F (85°C). This temperature provides a margin of safety to allow for the difficulty of measuring poultry temperatures accurately. (Most customers do not like to see red or pink at the bone joints of chicken and turkey.) Depending on the size of the bird, removing it from the oven when the thermometer reads between 170° and 175°F (77° and 79°C) should result in a final temperature of 185°F (85°C) after carryover cooking.

FOR SMALLER BIRDS, COOKED BY ANY METHOD

Doneness of smaller birds is determined in the following ways.

1. Looseness of joints. The leg moves freely in its socket.

2. Clear juices. Juices inside the cavity of a roasted bird are clear yellow rather than cloudy and red or pink.

3. Flesh separating from bone. Muscles begin to pull away from bones, especially breast-bone and leg bones. Excessively shrunken flesh means it's overcooked and dry.

4. Firmness to touch. Test with finger pressure as you would a steak (see p. 474). This method is especially useful for sautéed boneless chicken breasts.

5. As for large roasted birds, testing with a properly calibrated thermometer is the most accurate way to determine doneness. Minimum temperature for whole roasted birds, again, is 185°F (85°C). For poultry pieces, final temperature should be 165°F (74°C).

Not recommended: Do not test by piercing deeply with a fork and twisting the flesh. Too many valuable juices will be lost.

TRUSSING METHODS

Trussing means tying the legs and wings against the body to make a compact, solid unit. It has two main purposes:

1. Even cooking. Extended legs and wings cook too quickly.
2. More attractive appearance, especially when presented or served whole or carved in the dining room.

One of many trussing methods is illustrated in Figure 17.3. Your instructor may wish to show you other methods.

FIGURE 17.3 Trussing chicken.

(a) Place the chicken breast up, with the neck end toward you. Tuck the first joint of the wings behind the back.

(b) Press the legs forward and down against the body.

(c) Pass the center of a length of twine under the hip bone just ahead of the tail.

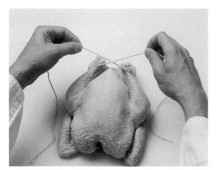

(d) Bring the twine up and across the ends of the legs.

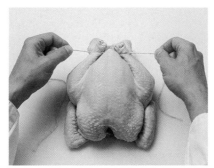

(e) Pass the twine under the ends of the legs as shown and pull tight.

(f) Bring the ends of the twine toward the neck end of the bird. Pull firmly on the twine while pressing on the breast portion with the thumbs as shown.

(g) Tie the twine tightly.

(h) The stub of the neck holds the twine in place, preventing it from slipping behind the back.

(i) The trussed chicken.

CUTTING UP CHICKEN

There are many ways to cut up chickens. Every chef has his or her preferred methods. Some of these methods are illustrated, step by step, in Figures 17.4, 17.5, and 17.6. These show how to split a chicken for broiling and how to cut whole chickens into quarters and eighths, for both bone-in parts and semi-boneless pieces. Cutting chicken apart at the joints is also known as *disjointing*.

As for meats, it is important to know the bone structure of chicken in order to cut it up. The best way to learn this is to practice cutting chickens.

FIGURE 17.4 Splitting chicken for broiling.

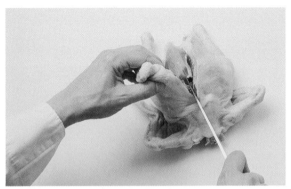

(a) Hold the chicken up by the tail. Cut through the bones to one side of the backbone, all the way to the neck.

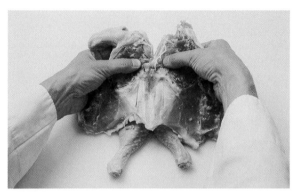

(b) Split the chicken open.

(c) Cut off the backbone as shown.

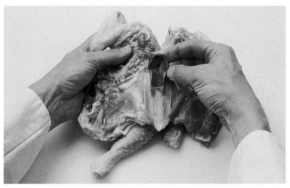

(d) Pull out the breastbone or keel bone. This helps the chicken lie flat and cook evenly.

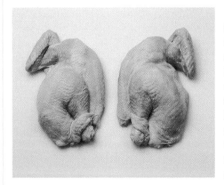

(e) For a portion size of one-half chicken, cut the chicken in half down the center of the breast. Make a split in the skin below the leg and slip the end of the leg through it, as shown, to hold the chicken in shape.

(f) Portion-size Cornish game hens are left whole.

FIGURE 17.5 Cutting chicken into quarters and eighths, bone in.

(a) Place the chicken on the cutting board breast up. Split the chicken down the center of the breast with a heavy knife, as shown.

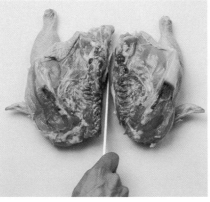

(b) Spread the chicken open and cut through the bones on one side of the backbone.

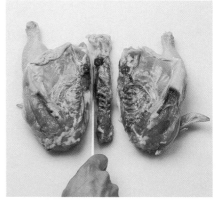

(c) Cut off the backbone completely. Save for stocks.

(d) Cut through the skin between the leg and the breast.

(e) Pull the leg back and cut off the entire leg section. Repeat with the other half. The chicken is now in quarters.

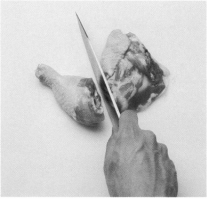

(f) To cut into eighths, cut the drumstick and thigh apart at the joint.

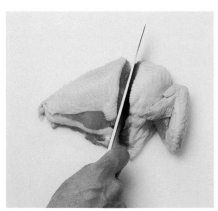

(g) Cut the breast and wing quarter into two equal pieces. (Another method is simply to cut off the wing.)

(h) The chicken cut into eighths. Note that the first joint of each wing has been cut off.

FIGURE 17.6 Cutting up chickens, semi-boneless.

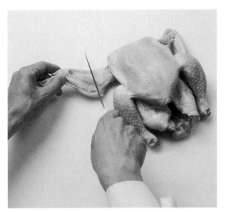

(a) Cut off the wings at the second joint. Save for stocks.

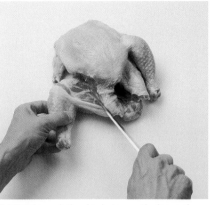

(b) Cut through the skin between the leg and the body.

(c) Turn the chicken on its side and pull the leg back. Carefully start to cut the flesh from the bone, being sure to get the "oyster," the little nugget of tender meat in the hollow of the hip bone. Cut through the ligaments at the hip joint.

(d) Holding the chicken steady with the knife, pull off the leg. Repeat with the other leg.

(e) Turn the breast portion upright. Cut down along one side of the ridge of the breastbone to separate the breast meat from the bone.

(f) Continue the cut along the wishbone to the wing joint.

(g) Holding the chicken by the wing, cut through the wing joint.

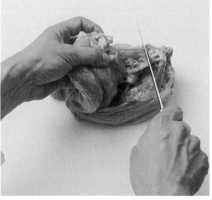

(h) Holding the carcass steady with the knife, pull back on the wing and breast meat.

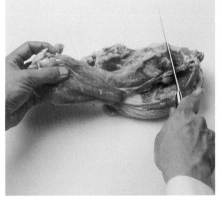

(i) Pull the breast meat completely off the bone. Be sure to hold on to the small "tenderloin" muscle inside the breast so it doesn't separate from the rest of the meat. Repeat with the other side.

(j) If desired, remove the thigh bone. Cut down along both sides of the bone to separate it from the meat.

(k) Lift out the bone and cut it off at the joint.

(l) For a neater appearance, chop off the end of the wing bone with the heel of the knife.

(m) A semi-boneless breast with the wing bone left in is known as a **suprême** or an airline breast. The wing bone may be left as is or frenched (meat scraped off), as in the illustration.

(n) The cut-up chicken. From left: breast portions without and with wing bone; leg portions without and with thigh bone; wing sections and carcass for stockpot. The drumstick and thigh (bone-in) may be cut apart at the joint, as in Figure 17.5.

KEY POINTS TO REVIEW

- How is fresh poultry properly stored? How is frozen poultry stored and handled?
- How is doneness of poultry determined?
- Why is whole poultry often trussed before cooking?
- What methods are used for cutting up poultry? Describe as thoroughly as possible how to cut chicken into bone-in and semi-boneless parts.

TERMS FOR REVIEW

light meat	free-range	grading	trussing
dark meat	organic	magret	suprême
maturity	inspection	ratite	

QUESTIONS FOR DISCUSSION

1. Why are hens or fowl not roasted in commercial kitchens?

2. Why is the breast section so often dry when whole chickens and turkeys are roasted? Can you suggest ways to remedy this problem?

3. Give a brief description of each of the following classes of poultry. Be sure to mention relative tenderness and approximate size.

Capon	Roaster	Yearling turkey
Roaster duckling	Young tom turkey	Hen or fowl
Broiler/fryer	Rock Cornish hen	

4. How should fresh and frozen poultry be stored?

5. Describe five methods for determining doneness in poultry items.

6. What is the purpose of trussing poultry?

7. Why are most game birds better if not cooked until well done?

18

COOKING POULTRY AND GAME BIRDS

The muscle tissue of domestic poultry and game birds, as we have said, has basically the same structure as the muscle tissue of meat animals. In particular, the breast meat of chicken and turkey is so similar to veal that they are interchangeable in many recipes.

Imaginative cooks realize that even when two meats are quite different—chicken and beef, for example—interesting new dishes can be made with substitutions that might seem unusual. For example, it is possible to make a delicious chili from chicken or turkey meat, a preparation that has the added advantage of having a lower food cost than beef chili.

Be sure to review the basic cooking procedures in Chapter 14 before proceeding with any of the recipes in this chapter. Also, please review the discussion of light and dark meat (pp. 524–525) in the previous chapter, as well as the methods for testing doneness (p. 530).

AFTER READING THIS CHAPTER, YOU SHOULD BE ABLE TO

1. Cook poultry by roasting and baking.
2. Cook poultry by broiling and grilling.
3. Cook poultry by sautéing, pan-frying, and deep-frying.
4. Cook poultry by simmering, submersion poaching, and steaming.
5. Cook poultry by braising.
6. Identify the safety, quality, and practicality concerns associated with preparing dressings and stuffings.
7. List basic ingredients for dressings and stuffings.
8. Prepare dressings and stuffings.

ROASTING AND BAKING

The general procedures for roasting and baking are explained in detail in Chapter 14. Review pages 415 and 469–470 for a discussion of low-temperature roasting. However, there are some differences in the ways poultry items are handled. The guidelines below should be observed.

Remember that poultry items are almost always cooked well done (except for squab and game birds).

Please note that the core recipe for roast chicken is found in Chapter 14 following the basic roasting procedure.

SEASONING AND BASTING

1. Seasonings and, if desired, a little mirepoix or a bouquet garni should be placed inside the cavity. When roasting whole birds, you need to season the skin only if it is to be served and eaten, as the seasonings will not penetrate the skin.

2. Oil the skin before roasting to help in browning and to protect against drying. The skin may be basted with fat during roasting, but this is unnecessary if the bird is roasted breast down (and turned breast up just at the end of the roasting period to brown).

   Basting is beneficial for large turkeys, which must be subjected to dry heat for several hours. If you baste large poultry during roasting, do it every 20–30 minutes. More frequent basting results in the loss of a great deal of heat from the oven because the door is opened so often.

3. Basting is unnecessary for duck and goose, which have a great deal of fat under the skin. These birds are usually roasted breast up for that reason.

TEMPERATURE

Selection of roasting temperature depends on the product being roasted.

1. *Low-temperature roasting* is best for large items such as turkeys and capons. It results in a tender, juicy product. Review pages 415–416 and 469–470 for a discussion of low-temperature roasting.

   Large turkeys may be roasted at 250°–325°F (120°–165°C). For most operations, however, cooking times at the lower end of this range are too long. In addition, if a turkey is stuffed (see pp. 581–582), it is not advisable to roast it at a very low temperature because the interior temperature will take too long to rise, providing a good breeding ground for bacteria.

   Some recipes call for starting large turkeys at a high temperature for 15–30 minutes in order to brown them. This is usually not necessary because they will likely brown anyway during the long cooking time.

   Smaller items, such as roasting chickens, are usually roasted at 325°–375°F (165°–190°C). Small items roasted at low temperatures may not brown well by the time they are done. In such cases, the heat can be turned up for a few minutes when they are almost done in order to brown them.

2. The *searing* method may be used for chickens under 5 pounds (2 kg) and for baked chicken parts. Start roasting in the oven at 450°F (230°C) for 15 minutes, then reduce the oven temperature to 250°–325°F (120°–160°C). These small items cook so quickly that continuous roasting at a low temperature produces very little browning.

   Ducks and geese also may be started at a high temperature in order to melt off some of the heavy fat layer under the skin and to make the skin brown and crisp.

3. *High-temperature roasting* is used for small items such as squab and game birds, which are often served rare. Cornish hens and other small poultry under 3 pounds (1.35 kg) may also be flash-roasted. Great care is necessary when roasting at high temperatures, however, because the poultry will quickly become overcooked if left in the oven only a few minutes too long. For example, a 2½-pound (1.2-kg) chicken or guinea hen roasted at 450°F (230°C) may be perfectly cooked and nicely browned after 45 minutes but overcooked and dry if left another 10 minutes.

   Ducks may also be roasted at a continuously high temperature (400°–425°F/ 200°–220°C) because their fat content protects them from drying. Great care should be used to prevent overcooking, however, because this can happen very quickly at these temperatures. (Goose should not be roasted at a continuously high temperature because of its larger size and the length of time it takes to cook.)

BAKED POULTRY

Roasting and baking are the same process. Cutting up the chicken doesn't change the cooking method. Baked chicken or turkey parts are treated like roasted poultry.

Chicken parts are sometimes coated with seasoned crumbs or flour and rolled in fat before baking. Such products are sometimes misleadingly called oven-fried because of their resemblance to breaded fried chicken.

POÊLÉS

A **poêlé** (pwah lay) is a classical preparation for white meats and poultry in which the item is cooked with a matignon (see sidebar on p. 548) in a covered container and basted with butter before and during cooking. Because the container is covered, the procedure is not a dry-heat method and, therefore, is not a genuine roasting procedure. Nevertheless, poêléing is usually translated as "butter-roasting" and is traditionally discussed along with other roasting procedures. (Escoffier also refers to a poêlé as a roast.) In this book, we follow this tradition and include the method here rather than with moist-heat procedures.

The recipe on page 548 illustrates the classical procedure for a poêlé, following Escoffier's method. Several alternatives to this procedure are indicated in the recipe notes and in the sidebar on page 548.

SMOKE-ROASTING

Rangetop smoke-roasting, also called **pan-smoking**, is a procedure done in a closed container, using wood chips to make smoke. You can use this procedure not only for small, tender, quick-cooking poultry and meat items, but also for fish fillets and some vegetables. A recipe for pan-smoked chicken breast is on page 545.

THE POÊLE

The term *poêlé* is derived from the word for a type of cooking utensil, the *poêle* (pwahl). At one time, this term was used for the kind of covered casserole used to cook meats and poultry by the method described in the text. In French kitchens today, however, the poêle is a type of sauté pan or frying pan. Consequently, in modern French, the verb poêler usually means "to sauté or pan-fry."

In this text, as in most English-speaking kitchens, we use the word *poêlé* to refer to the classical procedure described in the text. To avoid confusion, however, it is helpful to be aware of how the word is used in modern French. In fact, many familiar cooking terms have slightly different meanings in French. For example, *casserole*, a term used in the preceding paragraph, refers in French to what we would call a saucepan.

BASIC PROCEDURE Rangetop Smoke-Roasting

1. Collect all equipment and food supplies.
2. Place a layer of fine hardwood chips or shavings on the bottom of a hotel pan (see Figure 18.1). Disposable pans may be used for light smoking.
3. Place a rack in the pan over the chips and lay the seasoned food items on the rack. Cover tightly with a second hotel pan or with aluminum foil.

4. Place on the cooktop (making sure the ventilating hood is on!) over moderate heat. Smoke will begin rising from the wood chips.
5. After about 5 minutes, remove the food items from the smoke-roaster and, if necessary, complete the cooking in the oven. Leaving the food in the smoke too long results in a strong, bitter taste.

FIGURE 18.1 Rangetop smoke-roasting.

(a) Place hardwood chips or sawdust in a disposable hotel pan. Place over moderately high heat and heat until the wood begins to smoke.

(b) Place the items to be cooked on a rack and set the rack over the chips so the food is not touching the chips. Cover tightly with another pan and cook for the desired time.

 Roast Turkey with Giblet Gravy

YIELD: ABOUT 7 LB (3.5 KG) SLICEABLE MEAT   PORTIONS: 22   PORTION SIZE: 5 OZ (150 G)
                                                     28   4 OZ (125 G)
                                                     36   3 OZ (100 G)

U.S.	METRIC	INGREDIENTS	PROCEDURE
1	1	Turkey, dressed, 20 lb (10 kg)	1. Remove the giblets from the cavity of the turkey. Check inside the turkey to make sure it is well cleaned. Lock the wings in place by twisting the wing tips behind the back of the turkey.
to taste	to taste	Salt	
to taste	to taste	Pepper	2. Season the inside of the turkey with salt and pepper. Rub the skin thoroughly with oil.
as needed	as needed	Oil	

3. Place the turkey on one side in a roasting pan, on a rack if possible (see Note).

4. Place in an oven preheated to 325°F (165°C). (Lower temperatures are preferable if production schedule permits. See p. 538 for explanation.)

5. Roast 1 1/2 hours. Turn the turkey on the other side. Roast another 1 1/2 hours. Baste turkey with drippings (fat only) every 30 minutes.

6. While turkey is roasting, place turkey heart, gizzard, and neck in a saucepan. (Reserve the liver for another use, or add it to bread dressing.) Cover the giblets with water and simmer until very tender, 2–3 hours. Reserve the broth and the giblets for gravy.

		Mirepoix:	
8 oz	250 g	Onions, chopped medium	
4 oz	125 g	Carrots, chopped medium	
4 oz	125 g	Celery, chopped medium	

7. Turn the turkey breast-up. Place the mirepoix in the pan.

8. Return the turkey to the oven and continue to roast. Baste occasionally by spooning the fat in the pan over the turkey.

9. Allowing for carryover cooking, the turkey is done when a thermometer inserted into the thickest part of the inside of the high reads 165°–175°F (74°–79°C). Total roasting time is about 5 hours. (See p. 530 for a discussion of how to determine doneness.)

10. Remove the turkey from the roasting pan and let stand in a warm place at least 15 minutes before carving.

3 qt	3 L	Chicken stock, hot	
6 oz	175 g	Bread flour	
to taste	to taste	Salt	
to taste	to taste	Pepper	

11. Drain off and save the fat from the roasting pan.

12. Set the roasting pan with the mirepoix and drippings on the range to reduce the moisture and brown the mirepoix. Brown lightly if a light gravy is desired. Brown heavily if a dark gravy is desired.

13. Deglaze the pan with about 1 qt (1 L) of the chicken stock. Pour into a saucepan with the rest of the stock and the giblet broth (from step 6). Bring to a simmer. Degrease well.

14. Make a blond roux with the flour and 6 oz (175 g) of the fat from the roasting pan. Beat the roux into the gravy to thicken it.

15. Simmer at least 15 minutes, or until the gravy is smooth and no raw flour taste remains. Strain and season.

16. Chop or dice the giblets very fine and add to the gravy.

17. Slice the turkey and serve desired portion with 2 fl oz (60 mL) gravy. See **Figure 18.2** for slicing techniques. For quantity service, see procedures demonstrated in **Figure 18.3**.

Per 5 ounces (141.75 g): Calories, 460; Protein, 68 g; Fat, 16 g (33% cal.); Cholesterol, 135 mg; Carbohydrates, 6 g; Fiber, 0 g; Sodium, 170 mg.

Note: Because of the difficulty of handling large turkeys, many chefs do not use a rack. Also, large turkeys are easier to turn if they are placed first on one side, then on the other, rather than breast-down.

    For more even cooking, separate leg and thigh sections from breast section. Roast dark and light meat on separate pans, if many turkeys are being prepared. Remove each part from oven when done.

VARIATIONS

Roast Capons and Large Chickens

Prepare like roast turkey. Reduce roasting time, depending on size of bird. A 6-lb (2.7-kg) bird requires about 3 hours at 325°F (165°C).

Roast Turkey, Chicken, or Capon with Cream Gravy

Prepare as in basic recipe, but use half stock and half milk for the gravy instead of all stock. When gravy is finished, add 8 oz (250 mL) heavy cream.

FIGURE 18.2 Carving roast turkey.

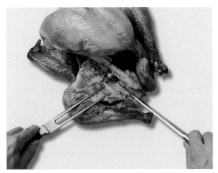

(a) Place the turkey on a clean, sanitary cutting board. Cut through the skin between the leg and the body. Pull the leg outward with a fork. The leg should pull off easily, but use the knife as necessary to separate the thigh from the hip.

(b) Cut the drumstick and thigh apart at the joint. Repeat with the other leg.

(c) Cut the meat from the drumstick and thigh in thin slices.

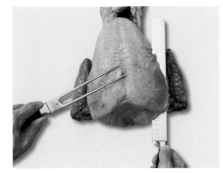

(d) Make a horizontal cut just above the wing, cutting all the way through to the bones of the body cavity. This cut helps the breast slices separate evenly.

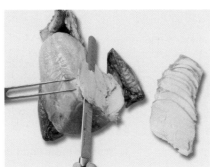

(e) With long, smooth strokes, cut the breast into thin slices.

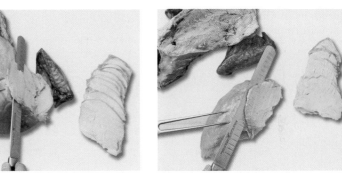

(f) When the slices become too large, change the angle of the knife slightly as shown. Continue until both sides of the breast are completely sliced.

(g) An alternative method is to cut off an entire half-breast in one piece. This piece can then be sliced as shown. Cut across the grain, holding the knife at an angle to get broader slices.

FIGURE 18.3 Quantity service of roast turkey and dressing.

(a) Using a scoop, place measured portions of baked dressing in hotel pans, as shown, or on sheet pans.

(b) Place measured portions of dark meat slices on top of the mounds of dressing.

(c) Place light meat slices on last, using the best-looking slices for the tops. Cover the pans. Refrigerate if they are to be held for later service. To serve, ladle a small amount of stock over the portions to keep them moist, and reheat, covered, in a moderate oven. Ladle 2 fl oz (60 mL) gravy over the portions when served.

Baked Chicken

PORTIONS: 12   **PORTION SIZE:** ¼ CHICKEN

U.S.	METRIC	INGREDIENTS	PROCEDURE
4 oz	125 g	Flour	1. Combine the flour and seasonings in a pan.
2½ tsp	12 mL	Salt	2. Dry the chicken pieces with paper towels if they are wet. Dredge in the seasoned flour.
¼ tsp	1 mL	White pepper	
1 tsp	5 mL	Paprika	3. Dip the chicken in the fat so that all sides are coated. Let excess drip off.
¼ tsp	1 mL	Dried thyme	
7½ lb	3.5 kg	Fryer chicken parts or quarters (see Note)	4. Arrange the chicken pieces on a sheet pan or in baking pans, skin side up. If using both dark and light meat parts, place them on separate pans.
8 oz	250 g	Melted butter, oil, or a mixture of butter and oil	5. Bake the chicken at 350°F (175°C) until done, about 1 hour.

Per serving: Calories, 450; Protein, 38 g; Fat, 30 g (61% cal.); Cholesterol, 150 mg; Carbohydrates, 6 g; Fiber, 0 g; Sodium, 590 mg.

Note: If you are starting with whole chickens, cut them into quarters or eighths, as shown in Figure 17.5. Any chicken parts may be used for this recipe. For example, you might bake just the legs and use the breasts and wings for other preparations.

VARIATIONS

Baked Herbed Chicken

Add 1½ tsp (7 mL) dried tarragon, ½ tsp (2 mL) dried marjoram, 1 tbsp (15 mL) fresh chives, and 1 tbsp (15 mL) dried parsley to the flour mixture. Omit paprika.

Baked Rosemary Chicken

Prepare as in basic recipe. After placing chicken parts in baking pans, sprinkle with dried rosemary, about 2 tsp (20 mL) for 12 portions. Fifteen minutes before chicken is done, sprinkle with 3–4 tbsp (45–60 mL) lemon juice.

Baked Chicken Parmesan

Instead of flour for dredging, use 2 oz (50 g) grated parmesan cheese mixed with ¾ cup (75 g) fine dry bread crumbs. Season as in basic recipe.

Quail Baked With Prosciutto And Herbs

PORTIONS: 8   **PORTION SIZE: 2 QUAIL**

U.S.	METRIC	INGREDIENTS	PROCEDURE
2 oz	60 g	Butter	1. Melt the butter. Add the herbs and seasonings.
4 tsp	20 mL	Sage, crumbled	2. Brush the quail heavily with the herb butter. Wrap a slice of prosciutto around the breast of each quail.
1 tsp	5 mL	Dried rosemary, crumbled	
2 tbsp	30 mL	Chopped parsley	3. Place the quail in a casserole just large enough to hold them. Cover and bake at 400°F (200°C) until done, about 30 minutes.
to taste	to taste	Salt	
to taste	to taste	Pepper	
16	16	Quail, whole	
16	16	Prosciutto slices, paperthin	
2 fl oz	60 mL	Balsamic vinegar or red wine vinegar	4. When they are done, remove the quail from the casserole and keep them warm. Without degreasing it, deglaze the pan with the vinegar and reduce au sec. Add the chicken stock and bring to a boil.
4 fl oz	125 mL	Chicken stock	5. Place 1 tbsp (15 mL) sauce on each plate and place 2 quail on top. Garnish with your choice of vegetable accompaniment.

Per serving: Calories, 500; Protein, 47 g; Fat, 34 g (61% cal.); Cholesterol, 185 mg; Carbohydrates, 1 g; Fiber, 0 g; Sodium, 900 mg.

VARIATIONS

Chicken, squab, and guinea hens can be cooked using this recipe. Cut them into pieces, brush each piece with the herb butter, and wrap with a slice of prosciutto. If desired, the skin can be removed before the pieces are brushed with the herb butter.

Roast Squab with Mushrooms

PORTIONS: 4   PORTION SIZE: 1 SQUAB

U.S.	METRIC	INGREDIENTS	PROCEDURE
		Stuffing:	1. Chop the stuffing ingredients fine and sauté them together lightly in the butter. Cool and reserve.
2 oz	60 g	Bacon	
4	4	Squab livers	
2 oz	60 g	Carrots	
4 oz	120 g	Onion	
2	2	Garlic cloves	
1 tsp	5 mL	Dried rosemary	
1 oz	30 g	Butter	
4 oz	120 g	Mushrooms	2. Cut the mushrooms and shallots into small dice.
1 oz	30 g	Shallots	3. Sauté the mushrooms and shallots in butter until nearly dry. Reserve.
1 oz	30 g	Butter	
4	4	Squabs	4. Season the squabs inside and out with salt and pepper.
to taste	to taste	Salt	5. Stuff them with the stuffing mixture and truss them for roasting. Caution: If you are not stuffing the birds immediately before roasting, the stuffing must be chilled before being placed in the birds to avoid spoilage or bacterial growth.
to taste	to taste	Pepper	
as needed	as needed	Clarified butter for browning the squabs	
			6. In a sauté pan, brown the squabs on all sides in clarified butter.
			7. In a roasting pan or in the same sauté pan, put the squabs on their sides and brush with some fresh butter. Put into a very hot oven (500°F/260°C) and roast them 5 minutes on one side. Turn them to the other side and roast another 5 minutes. Finally, turn them breast-up for another 5–6 minutes. Baste with butter during the roasting period.
			8. Remove them from the oven. They should be slightly rare. Cut off the legs, and if they are too rare, return them to the oven for another minute. Cut off each side of the breast in one piece, keeping the skin intact. Cover and keep warm.
1 pt	500 mL	White stock	9. Chop the carcass and put the carcass and stuffing into a pan with the stock. Boil 5 minutes, pressing down on the solids to help extract flavors.
1 oz	30 g	Butter	
to taste	to taste	Salt	10. Strain through a fine sieve, pressing down on the solids to squeeze out as much liquid as possible. Degrease the liquid and reduce by one-half. Monter au beurre with the last quantity of butter and adjust the seasonings.
to taste	to taste	Pepper	
4 small bunches	4 small bunches	Microgreens or watercress	11. Reheat the mushroom mixture from step 3.
			12. For each portion, spoon a little of the mushrooms onto the center of the plate.
			13. Arrange the breast pieces and legs of the squab on top of the mushrooms.
			14. Spoon the sauce around the squab.
			15. Garnish with a small bunch of microgreens or watercress.

Per serving: Calories, 1090; Protein, 49 g; Fat, 94 g (77% cal.); Cholesterol, 375 mg; Carbohydrates, 13 g; Fiber, 2 g; Sodium, 520 mg.

Roast Duckling à l'Orange

PORTIONS: 8   PORTION SIZE: ¼ DUCKLING

U.S.	METRIC	INGREDIENTS	PROCEDURE
2	2	Ducklings, about 5 lb (2.3 kg) each	1. Make sure the ducklings are well cleaned. Use the neck, gizzard, and heart in preparing the sauce (step 5); save the liver for another use.
to taste	to taste	Salt	2. Season the cavities with salt and pepper.
to taste	to taste	Pepper	3. Place the ducks breast-up on a rack in a roasting pan. (Caution: Do not use baking sheets or shallow pans. The pans must be deep enough to hold the large quantity of fat from the ducks.)
			4. Place the pan in an oven heated to 450°F (230°C). After 15 minutes, reduce heat to 375°F (190°C) and continue roasting until the ducks are done, about 1½ hours in all. (See discussion of roasting temperatures on pp. 538–539.)
1½ pt	720 mL	Demi-glace	5. While the duck is roasting, prepare the sauce. Add the duck neck and giblets (except liver) to the demi-glace and simmer slowly about 1 hour, or until reduced to 1 pt (500 mL). Strain. Keep the sauce hot.
3	3	Oranges	6. Peel the zest from the oranges in long strips. Cut them into julienne (or use a zester). Blanch the zests in boiling water 5 minutes. Drain and cool in ice water. Drain again and reserve the zest.
1½ fl oz	45 mL	Lemon juice	
3 oz	90 g	Sugar	7. Squeeze enough juice from the oranges to measure 4 fl oz (120 mL). Add to the lemon juice and set aside.
2 fl oz	60 mL	Wine vinegar, red or white	8. Make a gastrique (p. 185): Heat the sugar in a small saucepan over low heat until it melts. As soon as it turns a golden caramel color, remove from the heat, cool briefly, and add the vinegar and fruit juices. Simmer until the caramel is dissolved and the liquid is reduced by half.
2 fl oz	60 mL	Orange liqueur, such as Curaçao (optional)	9. Add the demi-glace and bring to a simmer.
			10. If the sauce is too thin, adjust the consistency by adding a little more demi-glace and reducing. The sauce should be just thick enough to coat lightly. It should not be heavy.
			11. If desired, add the liqueur and simmer a few minutes.
4 fl oz	120 mL	Chicken stock or white wine	12. When the ducklings are done, remove them from the pan. (Check to be sure they are done by pouring the juices in the cavity into a pan. There should be no trace of pink in the juices.)
			13. Drain off all the fat from the roasting pan.
			14. Deglaze the pan with the stock or wine, stirring to dissolve all the brown drippings.
			15. Reduce the liquid by half. Degrease and strain into the sauce.
			16. Cut the ducklings into 8 pieces each.
as desired	as desired	Garnish: peeled orange sections (optional)	17. To serve, place 1 breast piece and 1 leg piece on each plate, garnish the top with some of the blanched orange peels, and ladle on 2 fl oz (60 mL) sauce. Garnish the plate with orange sections, if desired.

Per serving: Calories, 900; Protein, 46 g; Fat, 71 g (71% cal.); Cholesterol, 215 mg; Carbohydrates, 20 g; Fiber, 1 g; Sodium, 200 mg.

VARIATIONS

Ducklings may be roasted following steps 1–4 of the basic recipe and served without a sauce, with a pre-prepared Bigarade Sauce (p. 187), or simply with applesauce (p. 213). Special seasonings create different products, such as the following recipe:

Roast Duckling, Bohemian Style

Season the cavity of each duckling with salt, pepper, and about ½ tsp (2 mL) caraway seeds. Place on rack in roasting pan and season the breast with salt, pepper, and another ½ tsp (2 mL) caraway. Roast as in the basic recipe. Serve without sauce, but with Sauerkraut (p. 328) or Braised Red Cabbage (p. 327) and Potato Dumplings (p. 410) or Spaetzle (p. 409).

Roast Goose

Roast goose at a continuous low temperature of 350°F (175°C). Because goose is much larger than duckling, the roasting time is long enough for the skin to crisp and brown at a low roasting temperature. Plan 20–25 minutes per pound for roasting time, but determine doneness by minimum internal cooking temperature of 165°F (74°C).

Pan-Smoked Spiced Chicken Breasts with Fruit Salsa

PORTIONS: 12   PORTION SIZE: 1 CHICKEN BREAST, 3 OZ (90 G) VEGETABLE GARNISH, 2 OZ (60 G) SALSA

U.S.	METRIC	INGREDIENTS	PROCEDURE
2 tbsp	30 mL	Paprika	1. Combine the paprika, cumin, thyme, coriander, salt, and pepper.
1¹/₂ tsp	7 mL	Ground cumin	
³/₄ tsp	4 mL	Dried thyme	2. Coat the chicken breasts with the spice mixture.
1¹/₂ tsp	7 mL	Ground coriander	3. Brush lightly with oil. Allow to marinate, refrigerated, 3–4 hours.
1¹/₂ tsp	7 mL	Salt	
³/₄ tsp	4 mL	Pepper	4. Set up a smoke-roasting system as shown in **Figure 18.1**. Heat the pan of wood chips or sawdust on top of the stove until smoke appears. Lay the chicken breasts on the rack, cover, and turn the heat to medium-low. Smoke-roast 10 minutes.
12	12	Chicken breasts, boneless and skinless, about 5 oz (150 g) each	
as needed	as needed	Vegetable oil	5. Transfer the pan to an oven preheated to 400°F (200°C) and roast another 10 minutes.
2 lb 4 oz	1.1 kg	Wheatberries with Pecans and Poblanos (p. 389)	6. For each portion, place a 3-oz (90-g) portion of wheatberries on the center of a plate. Slice a chicken breast on the diagonal and arrange the slices, overlapping, on top of the wheatberries. Spoon 2 oz (60 g) salsa next to the chicken and wheatberries. Garnish with cilantro.
24 oz	725 g	Fruit Salsa (p. 215)	
as needed	as needed	Cilantro sprigs for garnish (optional)	

Per serving: Calories, 200; Protein, 29 g; Fat, 6 g (27% cal.); Cholesterol, 80 mg; Carbohydrates, 7 g; Fiber, 1 g; Sodium, 360 mg.

Pan-Smoked Spiced Chicken Breasts with Fruit Salsa

Stuffed Chicken Legs with Pecan Butter

PORTIONS: 6   **PORTION SIZE: 1 CHICKEN LEG**

U.S.	METRIC	INGREDIENTS	PROCEDURE
1 oz	30 g	Shallots, chopped fine	1. Sweat the shallots and garlic in the butter. Do not let them brown. Cool thoroughly.
1 small clove	1 small clove	Garlic, chopped fine	
1/4 oz	7 g	Butter	
2 oz	60 g	Pecans, chopped fine	2. Combine the cooled shallot mixture, pecans, soft butter, bread crumbs, salt, and pepper in a bowl. Mix together.
2 oz	60 g	Butter, soft	
1 tbsp	15 ml	Parsley, chopped	
1/2 oz	15 g	Dry bread crumbs, preferably panko (see Note)	
to taste	to taste	Salt	
to taste	to taste	Pepper	
6	6	Large chicken legs (drumstick and thigh in one piece)	3. Partially bone the chicken legs as shown in **Figure 18.4**, removing the thigh bone but leaving in the drumstick bone.
to taste	to taste	Salt	4. Stuff the legs with the pecan mixture as shown in the illustration.
to taste	to taste	Pepper	5. Arrange the legs, skin side up, in a pan (such as a baking pan or sauté pan) just large enough to hold them. It is not necessary to tie them if baked this way; they should retain their shape and hold the stuffing.
			6. Bake at 400°F (200°C) until their internal temperature reaches 165°F (74°C).
			7. If the chicken needs more color, brown under a salamander or broiler.
			8. Remove the chicken from the pan and keep warm while you prepare the pan juices.
8 fl oz	240 mL	Chicken stock	9. Degrease the pan in which the chicken was cooked. Deglaze with the stock. Reduce by half.
8 fl oz	240 mL	Demi-glace	10. Add the demi-glace. Reduce by one-third. Strain and season to taste.
to taste	to taste	Salt	11. Plate the chicken legs. If desired, cut them in half and arrange to display the stuffing.
to taste	to taste	Pepper	12. Moisten each portion with about 1 fl oz (30 mL) of the pan juices.

Per serving: Calories, 570; Protein, 47 g; Fat, 40 g (64% cal.); Cholesterol, 185 mg; Carbohydrates, 4 g; Fiber, 1 g; Sodium, 230 mg.

Note: Coarse, panko-style bread crumbs should be used to create a light texture for the stuffing. Each chicken leg will hold about 1 oz (30 g) stuffing made with panko. If finer crumbs are used, the amount of stuffing may need to be increased, as the stuffing will be denser.

VARIATIONS

Chicken breasts can be used instead of legs. Flatten the breasts with a meat mallet, stuff, and roll up starting from a long side.

Many mixtures can be used for stuffing in place of the pecan butter. The following are a few suggestions:

Mushroom duxelles

Mixture of dried fruits, such as prunes and apricots, soaked and simmered until soft, mixed with browned almond slices and sautéed minced shallot.

Lightly steamed mixed vegetables julienne or brunoise, seasoned with fresh herbs and mixed with grated parmesan cheese.

Stuffed Chicken Legs with Pecan Butter

FIGURE 18.4 Boning and stuffing a chicken leg.

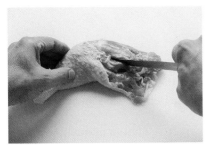

(a) Make a shallow cut along the thigh bone.

(b) Continue to cut along the bone to free it from the meat.

(c) The thigh bone is completely freed from the meat.

(d) Grasp the end of the thigh bone with one hand and wrap the fingers of the other hand around the bone near the joint. Pulling on the bone as shown separates some of the drumstick meat from the bone, making more room for stuffing.

(e) Cut the thigh bone from the drumstick bone at the joint. The bones are shown here as separated.

(f) Force some of the stuffing into the space in the drumstick created in step (d).

(g) Wrap the thigh meat around the stuffing.

(h) Legs stuffed in this manner, if carefully handled, can be baked without tying and will retain their shape. Note that the end of the drumstick bone has also been cut off for a neater, more compact appearance.

Chicken Poêlé

PORTIONS: 4   PORTION SIZE: ¼ CHICKEN, 2 FL OZ (60 ML) SAUCE

U.S.	METRIC	INGREDIENTS	PROCEDURE
1	1	Whole chicken, about 4 lb (2 kg)	1. Season the cavity of the chicken with salt and pepper. Stuff the herbs into the cavity.
to taste	to taste	Salt	2. Truss the chicken. Season the skin with additional salt and pepper
to taste	to taste	Pepper	
3 large sprigs	3 large sprigs	Parsley	
1 sprig	1 sprig	Fresh thyme	
2 sprigs	2 sprigs	Fresh tarragon	
		Raw matignon:	3. Place the matignon in the bottom of a brazier or casserole just large enough to hold the chicken.
4 oz	120 g	Onion, brunoise	
2 oz	60 g	Carrot, brunoise	4. Place the chicken, breast up, on the matignon and baste generously with butter.
2 oz	60 g	Celery, brunoise	
1 oz	30 g	Lean ham, cut paysanne	5. Cover the brazier or casserole and place in an oven preheated to 350°F (175°C). Let cook 1 hour. Two or three times during this period, uncover and brush the chicken with the butter and fat at the bottom of the pan.
2 oz, or more as needed	60 g, or more as needed	Butter, melted	
			6. After 1 hour of cooking, uncover the pan so the chicken will brown slightly. Continue to cook until the chicken is done, about 30 minutes more.
1 fl oz	30 mL	Madeira	7. Remove the chicken to a platter or another pan, cover, and keep warm while preparing the sauce.
1 pt	500 mL	Brown stock	8. Add the Madeira to the matignon and juices in the pan and simmer until the liquid is reduced slightly.
to taste	to taste	Salt	
to taste	to taste	Pepper	9. Add the brown stock. Bring to a boil and reduce by half.

10. Strain the liquid through a china cap lined with several layers of cheesecloth (see Note).

11. Degrease the liquid (see Note). Season to taste with salt and pepper.

12. Carve the chicken. Plate as desired. Serve the juices in a sauceboat on the side.

Per serving: Calories, 750; Protein, 71 g; Fat, 46 g (56% cal.); Cholesterol, 250 mg; Carbohydrates, 8 g; Fiber, 1 g; Sodium, 540 mg.

Note: The liquid should not be degreased earlier than this step. The fat absorbs some of the flavors of the matignon and chicken during cooking, and some of these flavors are transferred back to the stock when it is reduced, resulting in a more flavorful sauce.

Escoffier directs that the sauce be strained before being degreased, so this recipe follows the classical procedure. Many chefs prefer to leave the vegetables in the sauce. If they were cut carefully, this is an attractive and flavorful variation. If desired, the sauce may be thickened lightly with a slurry of arrowroot or cornstarch.

MATIGNON

A classical matignon is a mixture of aromatic ingredients used to add flavor to a meat or other product. It is similar to a mirepoix, but with several important differences. First, a pork product, usually ham, is added to the vegetables. Second, in most cases the mixture is sweat in butter and deglazed with Madeira or another wine.

Although a matignon is usually cooked in butter before use, Escoffier specifically directs that a raw matignon be used for the poêlé procedure, so we follow those instructions in the basic recipe included here. The matignon cooks fully during the poêléing process and contributes its flavor to the poultry and juices. Nevertheless, when preparing this dish, some chefs prefer to sweat the matignon in butter before adding it to the pan.

Chicken Poêlé

BROILING AND GRILLING

Tender, young poultry items may be cooked on the grill or broiler using the basic procedure explained on pages 424–425.

Use lower temperatures than for meats. The outside can be burned easily before the inside is cooked through. Poultry skin, in particular, browns and then burns very easily. For quantity production, broiled chicken is sometimes finished in the oven on sheet pans, preferably on racks.

Start poultry pieces skin side down. This helps keep flavorful juices from dripping out. Brush generously with melted butter or other fat before and during broiling.

Because the skin of broiled chicken is often eaten, it may be seasoned before cooking.

Large or thick poultry items are not well suited to broiling and grilling because it takes so long for the heat to penetrate to the center and cook them through. For example, turkey thighs could be broiled if you kept the heat low enough, but, in practice, this is rarely done. On the other hand, if you boned them out and flattened them lightly with a cutlet mallet, you would make them more suitable for broiling and grilling.

Grilled meat and poultry items are fairly simple, straightforward dishes, without the variety of ingredients and components you find in, for example, stews. However, there are several ways you can give variety to grilled poultry, including the following:

1. **Marinate the poultry or rub it with seasonings before cooking.**
   Keep in mind that marinade ingredients such as sugar and tomato burn easily, so use these with care. Also, herbs on the surface of the poultry burn easily. Charred rosemary can give a pleasant aroma to the food, but herbs such as parsley, if used in large quantity and allowed to burn, may taste like burned leaves.

2. **Baste with seasoned butter, marinade, or other flavorings during broiling.**
   Again, be careful with ingredients that burn easily. Use them only toward the end of cooking.

3. **Serve with an appropriate sauce or seasoned butter.**
   Flavored butters can be placed on top of grilled poultry, but sauces should be underneath or on the side so they don't detract from the crisp, browned skin.

4. **Select vegetable garnishes for variety and interest.**
   Well-chosen and carefully plated garnishes should be thought of as part of the whole presentation, not just something served on the side.

KEY POINTS TO REVIEW

- What techniques are used for seasoning and basting roast poultry?
- What are the factors to consider when determining the temperature at which to roast poultry?
- What are the steps in the procedure for range-top smoke roasting?
- What poultry items are best suited to broiling and grilling?
- What are four ways to give flavor variety to broiled and grilled poultry?

 # Broiled Chicken

PORTIONS: 10   PORTION SIZE: ½ CHICKEN

U.S.	METRIC	INGREDIENTS
5	5	Broiler chickens, about 2 lb (900 g) each
4 fl oz	125 mL	Melted butter or oil
to taste	to taste	Salt
to taste	to taste	Pepper

PROCEDURE

1. Split the chickens in half and prepare for broiling, as shown in **Figure 17.4** (if chickens are large, cut into quarters).
2. Brush the chickens on both sides with melted butter. Season with salt and pepper.
3. Place the chickens skin side down on broiler rack (or on grill). Broil at moderately low heat until the chicken is half cooked and well browned on one side.
4. Turn the chickens over. (Use tongs or fork, but do not pierce the flesh with the fork.) Continue to broil until the chicken is done (no pink near thigh bone) and well browned on second side. (See first variation below for oven method.)
5. Remove from broiler. To serve, place half a chicken on dinner plate, skin side up.

Per serving: Calories, 600; Protein, 59 g; Fat, 39 g (60% cal.); Cholesterol, 215 mg; Carbohydrates, 0 g; Fiber, 0 g; Sodium, 270 mg.

VARIATIONS

Broiled Chicken, Quantity Method

For large quantities, or for bigger chickens (quartered instead of halved), use slightly higher heat (set the broiler rack higher), and broil until browned on both sides, as in basic recipe. Remove from broiler and place in baking pans or on sheet pans. Finish cooking in the oven at 325°–350°F (165°–175°C).

Broiled Rock Cornish Game Hen

Remove backbones from hens and prepare for broiling (see Figure 17.4). Broil as in the basic recipe for any of the variations. Serve 1 hen per portion.

Broiled Tarragon Chicken

Before brushing with melted butter, rub each chicken half with ¼ tsp (1 mL) dried tarragon and 1–2 tsp (5–10 mL) chopped fresh parsley. Broil as in basic recipe.

Broiled Deviled Chicken (Poulet à la Diable)

When the chicken is almost completely cooked, brush the tops (skin side) lightly with about 1 tbsp (15 mL) mustard per portion and sprinkle with 1 oz (30 g) bread crumbs per portion. Return to the broiler to brown the crumbs. Serve with 2 fl oz (60 mL) Sauce Diable (p. 187) per portion.

Broiled Deviled Chicken

Chicken Paillard with Grilled Vegetables

PORTIONS: 12   PORTION SIZE: 1 CHICKEN BREAST, ABOUT 6 OZ (175 G)

U.S.	METRIC	INGREDIENTS	PROCEDURE
12	12	Boneless, skinless chicken breasts, about 6 oz (175 g) each	1. Place each chicken breast between sheets of plastic film. With a meat mallet, carefully pound to a uniform thickness of about 1/4 in. (6 mm).
3	3	Garlic cloves, chopped	2. Combine the garlic, rosemary, salt, and pepper. Lightly rub the flattened chicken breasts on both sides with the mixture. Sprinkle both sides with the lemon juice, then with the olive oil. Let marinate 2–4 hours in the refrigerator.
1 tbsp	15 mL	Fresh rosemary, chopped	
1 1/2 tsp	7 mL	Salt	
1/2 tsp	2 mL	Pepper	3. Preheat a grill or broiler to very hot. Place the chicken breasts on the grill, skin side (that is, the side that had the skin on) down, and grill until about one-fourth done. Rotate on the grill, as in **Figure 14.5**, to mark. Continue to cook until about half done. Turn over and continue to grill until just cooked through.
3 fl oz	90 mL	Lemon juice	
3 fl oz	90 mL	Olive oil	
as desired	as desired	Grilled Vegetable Medley (p. 338)	4. Plate and serve at once with an assortment of grilled vegetables. Garnish each portion with a sprig of rosemary.
12	12	Fresh rosemary sprigs	

Per serving: Calories, 250; Protein, 34 g; Fat, 11 g (41% cal.); Cholesterol, 95 mg; Carbohydrates, 1 g; Fiber, 0 g; Sodium, 370 mg.

Chicken Paillard with Grilled Vegetables

Grilled Spiced Squab with Couscous and Fig Compote

PORTIONS: 12   PORTION SIZE: 1 SQUAB, PLUS GARNISH

U.S.	METRIC	INGREDIENTS	PROCEDURE
12	12	Whole squab	1. Split the squab for broiling, as shown in **Figure 17.4**, leaving the birds in one piece.
1 tbsp	15 mL	Salt	
1 tbsp	15 mL	Quatre Épices (p. 843)	2. Combine the salt, spices, thyme, and pepper. Rub the squab with the spice mixture. Refrigerate several hours.
3/4 tsp	3 mL	Dried thyme	
3/4 tsp	3 mL	Black pepper	3. Brush the squab lightly with olive oil. Place on the grill or broiler, starting skin side down. Grill or broil, turning as necessary, until the breast portion is medium rare.
as needed	as needed	Olive oil	
3 pt	750 mL	Water	4. Combine the water and salt in a saucepan and bring to a boil. Stir in the couscous. Cover and let stand 5 minutes. Remove the cover and fluff with a kitchen fork.
1 1/2 tsp	7 mL	Salt	
12 oz	375 g	Instant couscous (uncooked)	5. Wash and drain the salad greens. Toss with the vinaigrette. Place a small mound of greens at the top of each dinner plate. Top with halved cherry tomatoes.
1 lb	500 g	Tender, young salad greens, such as mesclun	
6 fl oz	175 mL	Basic Vinaigrette (p. 660) made with red wine vinegar and olive oil	6. At the bottom center of each plate, place a small mound of couscous. In the center of the plate, above the couscous, place 1 oz (30 g) fig compote.
36	36	Cherry tomatoes, cut in half	7. Split the squab in half. Place 1 half-squab on each side of the couscous.
12 oz	360 g	Fig Compote (p. 217)	

Per serving: Calories, 1300; Protein, 69 g; Fat, 94 g (68% cal.); Cholesterol, 325 mg; Carbohydrates, 42 g; Fiber, 5 g; Sodium, 1180 mg.

VARIATION   Grilled Spiced Cornish Hen

Prepare and grill Cornish hens using the same procedure as in the basic recipe, but make sure the chicken is more thoroughly cooked.

Grilled Chicken with Garlic and Ginger

PORTIONS: 8   PORTION SIZE: APPROXIMATELY 12 OZ (375 G), BONE IN

U.S.	METRIC	INGREDIENTS	PROCEDURE
6–8 lb	2.8–3.6 kg	Chicken parts or halves, or whole Cornish hens	1. Cut the chickens into parts or halves for broiling. If using Cornish hens, split down the back and flatten, or cut into halves.
6 fl oz	180 mL	Lemon or lime juice	2. Combine the lemon or lime juice, oil, ginger, garlic, salt, and pepper to make a marinade.
2 fl oz	60 mL	Vegetable oil	3. Marinate the chicken 3–4 hours.
1 oz	30 g	Fresh ginger root, grated	
2 tsp	10 mL	Garlic, chopped fine	
to taste	to taste	Salt	
to taste	to taste	Pepper	
as needed	as needed	Melted butter	4. Remove the chicken from the marinade. Broil or grill the chicken until done. Baste with the marinade several times during cooking.
to taste	to taste	Ginger Oil (p. 216; optional)	5. Shortly before the chicken is done, brush the pieces once or twice with melted butter.
			6. If desired, drizzle a few drops of ginger oil around the chicken after plating.

Per serving: Calories, 430; Protein, 42 g; Fat, 27 g (59% cal.); Cholesterol, 140 mg; Carbohydrates, 1 g; Fiber, 0 g; Sodium, 150 mg.

VARIATIONS

Substitute any of the following for the chicken parts in the above recipe or in any of the following variations:

Chicken or turkey brochettes (cubes of boneless meat on skewers)
Turkey paillards (broad slices of turkey breast pounded thin)
Boneless chicken breast
Chicken or turkey thighs, boned out and lightly pounded
Turkey cutlets (thick slices of turkey breast)

Southwestern Grilled Chicken

Substitute the following ingredients for the marinade in the basic recipe:

4 oz	120 g	Tomato, puréed
4 tsp	20 mL	Very finely chopped jalapeño
4 tbsp	60 mL	Chopped cilantro
3 fl oz	90 mL	Red wine vinegar

Grilled Chicken Oriental Style

Substitute the following ingredients for the marinade in the basic recipe. Brush the chicken with the marinade frequently during grilling.

8 fl oz	250 mL	Soy sauce
2 fl oz	60 mL	Rice wine or sherry
2 fl oz	60 mL	Chicken stock or water
1 fl oz	30 mL	Lemon juice
2 tsp	10 mL	Sugar
4 tsp	20 mL	Ginger root, grated

Spicy Barbecue-Style Grilled Chicken

Substitute the following ingredients for the marinade in the basic recipe. Toss the chicken pieces by hand so they are well coated.

2 fl oz	60 mL	Worcestershire sauce
1 fl oz	30 mL	Red wine vinegar
4 tsp	20 mL	Paprika
2 tsp	10 mL	Chili powder
1 tsp	5 mL	Dry mustard
1 tsp	5 mL	Crushed garlic
1/2 tsp	2 mL	Black pepper
1 tsp	5 mL	Salt

PROCEDURE VARIATION: COMBI OVEN

In place of steps 4, 5, and 6 in the main recipe, substitute the following steps:

4. Remove the chicken from the marinade. Place the chicken skin-side down on a preheated charbroiler or grill to mark the surface. Baste with the marinade while grill-marking the chicken.

5. Finish cooking the chicken in a combi oven preheated to 325°F (162°C) in combi mode, for about 10 minutes. (If you are cooking only breast pieces, instead use a temperature of 180°F (82°C). Cooking time at this lower temperature will be about 25 minutes.)

6. Shortly before the chicken is done, brush the pieces once or twice with melted butter.

7. If desired, drizzle a few drops of ginger oil around the chicken after plating.

Grilled Chicken Marinated in Yogurt and Spices

Substitute the following ingredients for the marinade in the basic recipe. Marinate the chicken overnight.

8 oz	250 g	Unflavored yogurt
2 oz	60 g	Onion, grated
1/2 tsp	2 mL	Crushed garlic
1/2 tsp	2 mL	Grated ginger root
1 tsp	5 mL	Ground cumin
1/2 tsp	2 mL	Ground cardamom
pinch	pinch	Mace
1/2 tsp	2 mL	Cayenne
1/2 tsp	2 mL	Black pepper
1 tsp	5 mL	Salt

TANDOORI CHICKEN

The *tandoor* is a clay oven widely used in northern India. With wood or coal as a fuel, it cooks foods quickly at a high temperature. Tandoori chicken is especially popular in Indian restaurants in North America and Europe. In the original version, whole chickens are marinated in yogurt and spices, usually strongly colored with a red food coloring, then impaled on skewers and lowered into the tandoor. Western kitchens approximate this dish by cooking the chicken on a grill or in a very hot oven.

The marinade for Grilled Chicken Marinated in Yogurt and Spices (above) is similar to typical marinades for tandoori chicken, but without the red coloring.

Grilled Quail Marinated in Soy Barbecue Sauce

PORTIONS: 6   PORTION SIZE: 2 QUAIL

U.S.	METRIC	INGREDIENTS	PROCEDURE
12	12	Quail	1. Remove the backbones and split open the quail, as shown in **Figure 17.4**. Keep each quail in one piece.
1 pt	500 mL	Soy Barbecue Sauce (p. 215)	2. Marinate the quail in the barbecue sauce 2 hours. Drain.
			3. Place the quail on a preheated broiler or grill. Cook, turning as necessary, until they are browned and the breast meat is medium done. Do not overcook, or the breast meat will be dry.

Per serving: Calories, 790; Protein, 58 g; Fat, 53 g (81% cal.); Cholesterol, 215 mg; Carbohydrates, 14 g; Fiber, 0 g; Sodium, 1780 mg.

Grilled Ostrich or Emu with Adobo Spices

PORTIONS: 8   PORTION SIZE: 4 OZ (125 G)

U.S.	METRIC	INGREDIENTS	PROCEDURE
		Spice rub (see Note):	1. Combine the ingredients for the spice rub.
1 tbsp	15 mL	Powdered ancho or pasilla chiles	2. Rub the spices over both sides of the steaks to lightly but evenly coat them.
1/2 tsp	2 mL	Salt	3. Refrigerate 1 hour or longer.
1/4 tsp	1 mL	Dried oregano	4. Grill or broil until medium done. Minimum internal cooking temperature is 155°F (68°C) (see p. 26).
1/4 tsp	1 mL	Ground cumin	5. Cut into thin slices across the grain.
1/4 tsp	1 mL	Black pepper	
8	8	Ostrich or emu steaks or filets, 4 oz (125 g) each	

Per serving: Calories, 130; Protein, 25 g; Fat, 3 g (21% cal.); Cholesterol, 85 mg; Carbohydrates, 0 g; Fiber, 0 g; Sodium, 230 mg.

Note: The quantities given for the spice rub are enough to mildly spice the meat. For a spicier product, double the quantities.

Jerk Spice Mixture

YIELD: 1 CUP (240 ML)

U.S.	METRIC	INGREDIENTS
4 tbsp	60 mL	Ground allspice
4 tbsp	60 mL	Dried thyme
3 tbsp	45 mL	Black pepper
1 tbsp	15 mL	Nutmeg
1 tbsp	15 mL	Cinnamon
2 oz	60 g	Salt

PROCEDURE

1. Combine all ingredients and mix well.
2. Store in an airtight container until needed.

Per serving: Calories, 45; Protein, 1 g; Fat, 1 g (16% cal.); Cholesterol, 0 mg; Carbohydrates, 11 g; Fiber, 5 g; Sodium, 4930 mg.

Note: Traditional Jamaican jerk relies on fresh Scotch bonnet chiles rather than dried chiles for its heat. Therefore, this formula doesn't include cayenne or other chile spices. If you wish a spice blend that can be used to give a jerk flavor to foods without a complete marination, add 1¹/₂ tsp (7 mL) cayenne, or to taste, to the spice mix. In addition, 2 tbsp (30 mL) each garlic powder and onion powder may be added to the mix if fresh garlic and onion will not be used in cooking.

JAMAICAN JERK

Jerk is a traditional Jamaican cooking style in which meats are marinated in a mixture that includes allspice and Scotch bonnet peppers, as well as a variety of other spices, before barbecuing or grilling over a smoking fire. Pork, goat, and chicken are the most popular foods cooked by this method, but beef, fish, and sausage are also flavored and grilled this way.

Jerk Chicken

PORTIONS: 8   PORTION SIZE: 10 OZ (300 G)

U.S.	METRIC	INGREDIENTS
1 cup	240 mL	Jerk Spice Mix (above)
4–8	4–8	Scotch bonnet chiles, stemmed, seeded, and chopped (see Note)
8	8	Garlic cloves, chopped
8	8	Scallions, sliced
6 oz	180 g	Onion, chopped
1 oz	30 g	Brown sugar
4 fl oz	120 mL	Lime juice
3 fl oz	90 mL	Vegetable oil
2 fl oz	60 mL	Soy sauce
5 lb	2.4 kg	Chicken parts

PROCEDURE

1. Combine the spice mix, chiles, garlic, scallions, sugar, lime juice, oil, and soy sauce in a blender. Blend to a paste. (Do half of the ingredients at a time if you have a small blender, and combine the two batches.)

2. Place the chicken in a nonreactive container. Add the spice mixture and stir so all pieces of chicken are coated.
3. Marinate, refrigerated, at least 1 hour but up to 48 hours.
4. Before cooking, remove from the marinade and let stand at room temperature 20–30 minutes.
5. Grill over low heat until done. Alternatively, cook in a smoker oven at 300°F (150°C).

Per serving: Calories, 360; Protein, 39 g; Fat, 21 g (54% cal.); Cholesterol, 125 mg; Carbohydrates, 2 g; Fiber, 0 g; Sodium, 170 mg.

Note: Wear gloves when handling the chiles. They are extremely hot and can burn the skin. Reduce or increase the quantity of chiles according to taste and tolerance for spicy-hot foods.

Jerk Chicken with Grilled Pineapple

Grilled Spiced Turkey Burger

PORTIONS: 10   **PORTION SIZE: 5 OZ (150 G)**

U.S.	METRIC	INGREDIENTS	PROCEDURE
6 oz	180 g	Onion, chopped fine	1. Sweat the onions in the oil until soft. Cool completely.
1 tbsp	15 mL	Vegetable oil	2. Combine the onions, turkey, salt, spices, and water in a bowl. Toss gently until uniformly mixed.
3 lb	1.5 kg	Ground turkey	
1 tbsp	15 mL	Salt	3. Divide the mixture into 5-oz (150-g) portions (or as desired). Form each portion into a thick patty.
1 tbsp	15 mL	Curry powder	
1½ tsp	7 mL	Grated lemon zest	4. Grill or broil until just well done, but avoid overcooking, which makes the burger dry.
½ tsp	2 mL	Pepper	
¼ tsp	1 mL	Cinnamon	
½ tsp	2 mL	Ground coriander	
¼ tsp	1 mL	Ground cumin	
¼ tsp	1 mL	Ground cardamom	
4 fl oz	125 mL	Water	
as desired	as desired	Tomato Raisin Chutney (p. 208) or other chutney as desired	5. Plate the burgers. Top each with a spoonful of chutney (or place the chutney on the side) and a small mound of greens. Alternatively, serve the burger and garnish as a sandwich on whole-grain bread.
as desired	as desired	Microgreens or cilantro leaves	

Per serving: Calories, 250; Protein, 27 g; Fat, 14 g (52% cal.); Cholesterol, 100 mg; Cholesterol, 2 g; Fiber, 1 g; Sodium, 800 mg.

Grilled Spiced Turkey Burger; Tomato Raisin Chutney; Curry Oil

SAUTÉING, PAN-FRYING, AND DEEP-FRYING

Because chicken and turkey are lean, tender meats, cooking in fat is an appropriate and popular way to prepare them. The procedures for sautéing and pan-frying are explained in detail in Chapter 14 (pp. 426–428 and 430–431). Deep-frying procedures are explained on pp. 432–433. For the stir-fried chicken recipe on p. 564, it will be helpful to review the basic stir-frying procedure on p. 502. Also, please note the guidelines below that apply particularly to poultry and game bird items.

Tender game birds and specialty poultry items may also be cooked by sautéing or pan-frying. For most game birds, only the breasts are usually cooked by these methods. The legs are small and have more connective tissue, so they require longer cooking. They are often braised or roasted until tender and served as garnish for the breast, either bone-in or as boneless meat.

For lean items, such as squab, partridge, and quail, the breasts are best if not cooked well done but rather kept somewhat pink inside, or even rare, to preserve moisture. Breast of pheasant and guinea may also be served with a little pink in the interior, although because this meat is so similar to the white meat of chicken, many customers may prefer it well done.

Dark red poultry, such as ostrich and emu steaks and breast of duck, are also lean and most often served medium to medium rare. Remember, however, that the minimum safe temperature for ostrich and emu is 155°F (68°C). Refer to page 26. Duck breasts present a special case for pan-frying because of the heavy layer of fat between the skin and the meat. Pan-fried duck breasts are started skin side down and cooked until much of the fat is rendered and the skin is crisp. This takes several minutes, or most of the cooking time. To finish, they are turned over and cooked skin side up for just a few moments, until they reach the desired doneness.

SAUTÉING

1. Boneless chicken breasts, thin slices of turkey breast, and other quick-cooking items are ideal for sautéing.

2. Larger items, such as bone-in chicken cut into eighths, must be cooked to lower heat if they are to cook to doneness by sautéing because they need longer cooking times. Such items are often browned by sautéing and then finished in the oven. Breasts of game birds, on the other hand, may be cooked rare or medium and can thus be easily cooked from start to finish on the stovetop.

3. In a classical chicken sauté, after pan is deglazed and the sauce is made, the chicken is sometimes reheated briefly in the sauce. If the chicken is allowed to cook for a longer time in the sauce, the dish becomes a braised item rather than a sauté. Recipes for this kind of preparation are included under Braising, page 575.

PAN-FRYING

1. Pan-fried chicken is usually breaded or floured before cooking for even browning and crispness.

2. About ¼ inch (½ cm) or more of fat is needed in the pan to pan-fry chicken.

3. The side that will face up on the plate should be browned first for best appearance. This is called the **presentation side**. For chicken pieces, it is usually the skin side.

4. After browning on all sides over moderately high heat, lower the heat so the chicken cooks to doneness without overbrowning. Pan-fried chicken takes 30–45 minutes to cook.

DEEP-FRYING

1. The procedure for deep-frying is like that for pan-frying, except the item doesn't have to be turned because it is submerged in the hot fat. Review page 432 for deep-frying instructions.

2. Pieces from small chickens (under $2^{1}/_{2}$ lb/l kg) are best for deep-frying. Larger pieces require such a long cooking time that the surface may brown too much.

   If necessary, fried items may be finished in the oven.

3. Fry chicken at 325°–350°F (160°–175°C) for even cooking.

 Pan-Fried Chicken

PORTIONS: 12   **PORTION SIZE: $^{1}/_{2}$ CHICKEN**
      24                         $^{1}/_{4}$ **CHICKEN**

U.S.	METRIC	INGREDIENTS
6	6	Chickens, $2^{1}/_{2}$ lb (1.1 kg) each
8 oz	225 g	Flour
$2^{1}/_{2}$ tsp	12 mL	Salt
$^{1}/_{2}$ tsp	2 mL	White pepper
as needed	as needed	Oil

Per $^{1}/_{2}$ chicken: Calories, 820; Protein, 75 g; Fat, 51 g (57% cal.); Cholesterol, 235 mg; Carbohydrates, 11 g; Fiber, 0 g; Sodium, 580 mg.

PROCEDURE

1. Cut chickens into 8 pieces, as shown in **Figure 17.5**.
2. Place the flour in a small hotel pan and season with salt and pepper.
3. Pour about $^{1}/_{4}$ in. (6 mm) oil into enough heavy iron skillets to hold all the chicken pieces in a single layer. Heat over moderately high heat.
4. Dredge the chicken pieces in the seasoned flour and shake off excess.
5. Place the pieces skin side down in the hot oil. Let the pieces fall away from you to avoid splashing hot oil on yourself.
6. Fry the chicken until golden brown on the bottom. Turn the pieces with tongs and brown the other side.
7. Lower the heat slightly to avoid overbrowning. Continue to cook the chickens, turning once or twice more, until cooked through. Breast meat cooks faster than leg meat—remove it when it is done. Total cooking time will be 20–40 minutes, depending on the size of the chickens and the temperature of the fat.
8. Remove the chicken from the pan and drain well. Place on hot dinner plates or hold for service in counter pans. Do not cover pans or hold too long, or chicken will lose its crispness.

VARIATIONS

For slightly crustier, browner chicken, dip in milk before dredging in flour.

Alternative Method, Quantity Service: Brown chickens in hot oil as in basic recipe. Place on sheet pans or in baking pans skin side up, and finish cooking in a 350°F (175°C) oven.

Country-Style Fried Chicken

Fry chickens as in basic recipe. For 12 portions: Pour all but 2 oz (60 g) fat from the pans. Add 2 oz (60 g) flour and make a blond roux. Stir in $2^{1}/_{2}$ pt (1.2 L) milk and bring to a boil. Stir constantly as the gravy thickens. Simmer a few minutes to eliminate all raw starch taste, and season with salt and white pepper. Adjust the consistency with stock, water, or additional milk if necessary. Strain. Serve the chicken with gravy and mashed potatoes.

Garlic Fried Chicken

PORTIONS: 4   **PORTION SIZE: ¹/₂ CHICKEN**
       8   ¹/₄ **CHICKEN**

U.S.	METRIC	INGREDIENTS	PROCEDURE
2	2	Chickens, 2¹/₂ lb (1.1 kg) each	1. Cut the chickens into 8 pieces as shown in **Figure 17.5**.
6–8 cloves	6–8 cloves	Garlic, chopped fine	2. Combine the garlic, milk, salt, white pepper, and lemon juice in a nonreactive container large enough to hold the chicken. (The milk will curdle slightly because of the lemon juice.)
12 fl oz	360 mL	Milk	
2 tsp	10 mL	Coarse salt	3. Add the chicken pieces and turn them in the mixture so that they are all coated.
¹/₄ tsp	1 mL	White pepper	
1 fl oz	30 mL	Lemon juice	4. Cover tightly and let marinate in the refrigerator 48 hours.
4 oz, or as needed	120 g, or as needed	Flour	5. Place the flour in a small hotel pan.
			6. Remove the chicken from the marinade and, a few pieces at a time, dredge well with the flour to coat them completely.
as needed	as needed	Oil	7. Pour about ¹/₄ in. (6 mm) oil in enough heavy iron skillets to hold all the chicken pieces in a single layer. Heat over moderately high heat.
			8. Place the pieces skin side down in the hot oil. Let the pieces fall away from you to avoid splashing hot oil on yourself.
			9. Fry the chicken until golden brown on the bottom. Turn the pieces with tongs and brown the other side.
			10. Lower the heat slightly to avoid overbrowning. Continue to cook the chickens, turning once or twice more, until cooked through. Breast meat cooks faster than leg meat—remove it when it is done. Total cooking time will be 20–40 minutes, depending on the size of the chickens and the temperature of the fat.
			11. Remove the chicken from the pan and drain well. Place on hot dinner plates or hold for service in counter pans. Do not cover pans or hold too long, or chicken will lose its crispness.

Per serving: Calories, 430; Protein, 40 g; Fat, 21 g (44% cal.); Cholesterol, 150 mg; Carbohydrates, 16 g; Fiber, <1 g; Sodium, 560 mg.

VARIATION

Buttermilk Fried Chicken: Substitute buttermilk for the regular milk in the basic recipe, and omit the lemon juice.

Pan-frying chicken

Sautéed Boneless Breast of Chicken with Mushroom Sauce

PORTIONS: 10   **PORTION SIZE: 1 CHICKEN BREAST, ABOUT 4 OZ (125 G), 2 FL OZ (60 ML) SAUCE**

U.S.	METRIC	INGREDIENTS	PROCEDURE
2 oz	60 g	Clarified butter	1. Add enough clarified butter to a sauté pan to just cover the bottom with a thin film. Place on the range over moderate heat.
10	10	Boneless, skinless chicken breasts from 5 chickens, 3¹/₂ lb (1.6 kg) each	2. While the pan is heating, season the chicken breasts and dredge in flour. Shake off excess flour.
to taste	to taste	Salt	3. Place the breasts in the hot pan, presentation side (that is, the side that had the skin) down.
to taste	to taste	White pepper	4. Sauté over moderate heat until lightly browned and about half cooked. The heat must be regulated so the chicken doesn't brown too fast.
2 oz	60 g	Flour for dredging	
to taste			5. Turn the chicken over and complete the cooking.
to taste			6. Remove the chicken from the pan and place on hot dinner plates for service. Keep warm.
10 oz	300 g	White mushrooms, sliced	7. Add the mushrooms to the pan and sauté briefly. After a few seconds, before the mushrooms start to darken, add the lemon juice. Toss the mushrooms in the pan as they sauté.
1 fl oz	30 mL	Lemon juice	8. Add the suprême sauce to the pan and simmer a few minutes, or until reduced to the proper consistency. (The juices from the mushrooms will dilute the sauce.)
2¹/₂ cups	600 mL	Suprême Sauce (p. 183), hot	9. Ladle 2 fl oz (60 mL) sauce over each portion and serve immediately.

Per serving: Calories, 360; Protein, 36 g; Fat, 20 g (51% cal.); Cholesterol, 145 mg; Carbohydrates, 8 g; Fiber, 1 g; Sodium, 200 mg.

VARIATIONS

Alternative (Quick) Method: Sauté the chicken as in basic recipe. Plate and ladle 2 fl oz (60 mL) prepared Mushroom Sauce (p. 184, made with suprême sauce as a base) over each portion.

Other sauces based on chicken stock may be used in place of mushroom sauce to serve with sautéed chicken breasts, including suprême, aurora, Hungarian, and ivory (p. 184).

Sautéed Chicken Breast with Ivory Sauce, Squash Purée, and Mixed Vegetables

Deep-Fried Chicken

PORTIONS: 12   PORTION SIZE: ¹/₂ **CHICKEN**

U.S.	METRIC	INGREDIENTS
6	6	Chickens, about 2 lb (900 g) each
		Standard Breading Procedure (see Note):
4 oz	125 g	Flour
1 tsp	5 mL	Salt
¹/₂ tsp	2 mL	White pepper
1	1	Eggs
1¹/₄ cups	300 mL	Milk
12 oz	375 g	Dry bread crumbs

PROCEDURE

1. Cut chickens into 8 pieces, as shown in **Figure 17.5**.
2. Set up breading station: seasoned flour, egg wash, and crumbs (see pp. 150–151).
3. Pass the chicken through the Standard Breading Procedure.
4. Heat the fat in a deep fryer to 325°–350°F (165°–175°C).
5. Fry the chicken until golden brown and cooked through. Fry light meat and dark meat pieces in separate baskets, as the light meat cooks faster.
6. Remove from the fat, drain well, and serve immediately.

Per serving: Calories, 880; Protein, 63 g; Fat, 58 g (61% cal.); Cholesterol, 205 mg; Carbohydrates, 22 g; Fiber, 1 g; Sodium, 520 mg.

Note: Quantities given for breading materials are only guidelines. You may need more or less depending on the shapes of the chicken pieces, the care used in breading, and other factors. In any case, you need enough so even the last piece to be breaded can be coated easily and completely.

VARIATIONS

Alternative Method: For larger chickens or for quantity service, brown the chicken in the deep fryer. Drain, place on sheet pans, and finish in the oven at 350°F (175°C).

Fried Chicken Maryland

Fry the chicken as in basic recipe. Serve each portion with the following sauce and garnish:

2 fl oz (60 mL) cream sauce, suprême sauce, or horseradish sauce made with béchamel, placed on the plate under the chicken

2 strips crisp bacon, placed in a cross on top of the chicken

2 Corn Fritters (p. 342)

2 banana quarters, breaded and fried

Fried Chicken Breast Strips

Cut boneless, skinless chicken breasts into strips about ³/₄ in. (2 cm) wide. Bread and fry as in basic recipe. Serve with desired dipping sauce, such as barbecue sauce or blue cheese dressing.

Quail with Balsamic Glaze

PORTIONS: 12 (SEE NOTE)   **PORTION SIZE: 1 QUAIL PLUS GARNISH**

U.S.	METRIC	INGREDIENTS	PROCEDURE
1 pt	500 mL	Balsamic vinegar	1. Combine the balsamic vinegar and rosemary in a stainless-steel pan. Reduce to about 2 oz (60 mL), or until syrupy. Strain.
1/2 tsp	2 mL	Rosemary, dried	
1 fl oz	30 mL	Olive oil	2. In another saucepan or sauté pan, heat the olive oil and add the garlic. Cook briefly.
4	4	Garlic cloves, chopped	3. Add the chard leaves. Stir while cooking over high heat to wilt the greens and evaporate excessive moisture. Cook just until tender.
1 1/2 lb	750 g	Swiss chard, greens only (reserve the thick stalks for another use)	4. Season to taste with cayenne, salt, and pepper.
to taste	to taste	Cayenne	
to taste	to taste	Salt	
to taste	to taste	Pepper	
12	12	Quail	5. Cut each quail into 4 semi-boneless parts, as shown in **Figure 17.6**. Leave the third wing joint in the breast sections. Season lightly with salt and pepper.
to taste	to taste	Salt	6. Heat the oil in sauté pan or pans. Add the legs and pan-fry until well browned and slightly crisp on the outside.
to taste	to taste	Pepper	7. When the legs are almost cooked, add the breast sections, skin side down, and pan-fry until medium rare, turning as necessary to brown both sides.
3 fl oz	90 mL	Oil	8. Remove the quail from the pan and degrease. Return to the pan and add the balsamic vinegar (not the reduced vinegar from step 1). Cook over moderate heat, turning the pieces until they are lightly glazed.
4 fl oz	125 mL	Balsamic vinegar	9. For each portion, spoon 1 fl oz (30 mL) pepper coulis onto a plate.
12 fl oz	375 mL	Red Bell Pepper Coulis (p. 204)	10. Make a bed of Swiss chard on the center of the plate. Arrange 2 leg portions and 2 breast portions on top of the chard.
			11. Place dots of the balsamic reduction around the quail.

Per serving: Calories, 600; Protein, 37 g; Fat, 36 g (54% cal.); Cholesterol, 110 mg; Carbohydrates, 33 g; Fiber, 5 g; Sodium, 200 mg.

Note: One quail per portion is suitable for the first course. For the main course, use 2 quail per portion.

Quail with Balsamic Glaze, served with a corn and black-eyed pea mixture

Quail with Creole Spices

PORTIONS: 12 AS FIRST COURSE   **PORTION SIZE:** 1 QUAIL, 2 OZ (60 G) VEGETABLES, 1¹⁄₂ OZ (45 G) RICE
6 AS MAIN COURSE   2 QUAIL, 4 OZ (125 G) VEGETABLES, 3 OZ (90 G) RICE

U.S.	METRIC	INGREDIENTS	PROCEDURE
12	12	Quail	1. Remove the backbones from the quail, spread the birds open, and flatten them as illustrated in **Figure 17.4(f)**.
1¹⁄₂ tsp	7 mL	Paprika	2. Mix together the paprika, cayenne, oregano, thyme, pepper, and salt.
¹⁄₄ tsp	1 mL	Cayenne	
¹⁄₄ tsp	1 mL	Dried oregano	3. Lay out the quail skin side up and sprinkle them generously with the spice mixture.
¹⁄₄ tsp	1 mL	Dried thyme	
¹⁄₄ tsp	1 mL	Pepper	4. Heat the oil in one or more sauté pans over moderate heat. Pan-fry the quail on both sides until well browned and cooked through.
¹⁄₂ tsp	2 mL	Salt	
2 fl oz	20 mL	Vegetable oil	5. Remove the quail from the pan and keep warm.
			6. Pour off all but about 1 fl oz (30 mL) fat from the pan.
6 oz	180 g	Onion, chopped fine	7. Add the onion, garlic, celery, and green pepper to the pan. Sauté over medium heat until the onion just starts to brown.
1	1	Garlic clove, chopped fine	
4 oz	125 g	Celery, small dice	8. Add the tomatoes, tomato purée, and thyme. Simmer 10 minutes, or until the juices are somewhat reduced and thickened.
4 oz	125 g	Green bell pepper, small dice	
1 lb	500 g	Tomatoes, canned, chopped, with their juice	9. Stir in the parsley.
1 oz	30 g	Tomato purée	10. Add salt and pepper to taste.
1 tsp	5 mL	Dried thyme	11. To serve, place a scoop of hot rice on a dinner plate. Spoon vegetables around the rice. Lean the quail against the mound of rice.
2 tbsp	30 mL	Chopped parsley	
to taste	to taste	Salt	
to taste	to taste	Pepper	
18 oz	550 g	Steamed white rice	

Per serving: Calories, 300; Protein, 23 g; Fat, 16 g (48% cal.); Cholesterol, 85 mg; Carbohydrates, 16 g; Fiber, 2 g; Sodium, 221 mg.

Quail with Creole Spices

Tangerine-Marinated Breast of Duck

PORTIONS: 12   PORTION SIZE: 6 OZ (180 G)

U.S.	METRIC	INGREDIENTS
12	12	Boneless duck breast halves, skin on, about 6 oz (180 g) each
1¹/₂ oz	45 mL	Shallots, chopped fine
1 fl oz	30 mL	Soy sauce
4 fl oz	120 mL	Tangerine juice
¹/₂ tsp	2 mL	Grated tangerine zest
¹/₄ tsp	1 mL	Cinnamon
¹/₄ tsp	1 mL	Crushed star anise
1 lb 8 oz	720 g	Wheatberries with Pecans (p. 389, prepared without poblanos)
as needed	as needed	Tangerine suprêmes (use procedure shown in **Figure 7.22**)
12 fl oz	360 mL	Oriental Vinaigrette (p. 662; optional)

PROCEDURE

1. With a sharp knife, score the duck skin in a diamond pattern. Cut through the skin and into the layer of fat but not into the meat. (This is easiest to do if the duck is chilled.)
2. In a nonreactive container, combine the shallots, soy sauce, tangerine juice and zest, cinnamon, and star anise and mix together.
3. Add the duck breasts, turning them to coat on both sides. Refrigerate 2–4 hours.
4. Remove the breasts from the marinade. Pat them dry with clean towels. Discard the marinade and any pieces of star anise clinging to the meat.
5. Pan-fry the breasts skin side down in a sauté pan over moderately high heat about 6 minutes, or until the skin is well browned and crisp and much of the fat has rendered. Turn over and pan-fry another 4 minutes. The breasts should still be rare.
6. Remove from the pans and let rest 5 minutes in a warm place.
7. Place 2 oz (60 g) wheatberries on each plate.
8. Cut each duck breast diagonally into thin slices. Fan the slices out, leaning them against the mound of wheatberries.
9. Top the wheatberries with a few tangerine suprêmes.
10. If desired, drizzle a little vinaigrette onto the plate around the duck breast.

Per serving: Calories, 340; Protein, 32 g; Fat, 16 g (43% cal.); Cholesterol, 165 mg; Carbohydrates 16 g; Fiber, 3 g; Sodium, 510 mg.

Tangerine-Marinated Breast of Duck

Chicken Stir-Fry with Walnuts

PORTIONS: 12   PORTION SIZE: 6 OZ (175 G)

U.S.	METRIC	INGREDIENTS	PROCEDURE
2¹/₂ lb	1.2 kg	Boneless, skinless chicken meat	1. Remove all fat from the chicken. Cut into ¹/₂-in. (1-cm) dice.
1 lb	450 g	Green or red bell peppers	2. Cut the peppers in half and remove the core and seeds. Cut into strips ¹/₄ in. (0.5 cm) wide.
1 lb	450 g	Scallions	3. Cut off the roots and withered parts of the green tops of the scallions. Split the scallions in half lengthwise and cut into 1-in. (2.5-cm) pieces. Combine with the peppers.
3 tbsp	45 mL	Cornstarch	4. Stir the cornstarch with the soy sauce until smooth.
4 fl oz	125 mL	Soy sauce	5. Have the remaining ingredients ready in separate containers. Everything must be ready before starting to cook because cooking takes only a few minutes.
4 oz	125 g	Walnut pieces or unsalted peanuts	
¹/₄ tsp	1 mL	Ground ginger	
¹/₈ tsp	0.5 mL	Cayenne	
1¹/₂ cups	350 mL	Chicken broth or water	
2 fl oz	60 mL	Oil	6. Heat half the oil in a large sauté pan or skillet until very hot, almost smoking.
			7. Add the peppers and scallions and sauté rapidly about 2 minutes, or until the vegetables are only slightly cooked. Remove from the pan.
			8. Add the remaining oil to the pan and again get it very hot.
			9. Add the chicken and sauté rapidly until no longer pink. If the chicken sticks to the pan, use a spatula to stir.
			10. Add the nuts, ginger, and cayenne and sauté another minute.
			11. Quickly stir the cornstarch mixture (the starch settles out) and add it and the stock or water to the pan. Stir to deglaze the pan and bring to a simmer.
			12. Add the sautéed vegetables and simmer just until heated through. Correct the seasonings.
			13. Serve immediately with boiled or steamed rice.

Per serving: Calories, 270; Protein, 22 g; Fat, 16 g (53% cal.); Cholesterol, 60 mg; Carbohydrates, 10 g; Fiber, 2 g; Sodium, 610 mg.

Chicken Stir-Fry with Walnuts

Vietnamese Stir-Fried Chicken with Chile

PORTIONS: 10   **PORTION SIZE: 6 OZ (180 G) CHICKEN AND VEGETABLES, 6 OZ (180 G) RICE NOODLES**

U.S.	METRIC	INGREDIENTS	PROCEDURE
3 lb	1.4 kg	Chicken breast, boneless and skinless	1. Cut the chicken breasts into 1-in. (2.5-cm) pieces.
6 fl oz	180 mL	Nuoc nam (Vietnamese fish sauce) or nam pla (Thai fish sauce)	2. In a bowl, toss the chicken with the fish sauce, garlic, and pepper. Marinate 15–20 minutes.
3	3	Garlic cloves, chopped fine	
¹/₂ tsp	2 mL	Black pepper	
1 lb	450 g	Rice sticks (p. 397)	3. Soak the rice sticks in warm water 20 minutes. Drain.
4 fl oz	120 mL	Nuoc Cham (p. 209)	4. Drop the rice sticks into boiling water. As soon as the water returns to a boil, drain.
			5. Toss the rice sticks with the nuoc cham. Keep warm while cooking the chicken.
3 fl oz	90 mL	Vegetable oil	6. Heat the oil in a sauté pan or wok over high heat.
6	6	Thai green chiles or serrano chiles, seeded and sliced thin	7. Drain the chicken, reserving the marinade. Add the chicken to the pan. Stir-fry until the chicken is about half cooked.
12	12	Scallions, sliced	8. Add the chiles, scallions, and reserved marinade. Continue to stir-fry until the chicken is nearly done.
8 oz	240 g	Snow peas, trimmed and blanched	9. Add the snow peas and stir-fry 1 minute, or until the snow peas are just cooked and still crisp.
6 oz	180 g	Tomato concassé	10. Add the tomato. Stir-fry just until the tomato is hot.
2 tbsp	30 mL	Sesame seeds, toasted	11. Place the rice noodles in the center of dinner plates.
1 pt	500 mL	Nuoc Cham	12. Top with the chicken mixture.
			13. Sprinkle with sesame seeds.
			14. Serve with additional nuoc cham in small ramekins on the side.

Per serving: Calories, 410; Protein, 38 g; Fat, 16 g (35% cal.); Cholesterol, 110 mg; Carbohydrates, 28 g; Fiber, 2 g; Sodium, 3320 mg.

Vietnamese Stir-Fried Chicken with Chile

SIMMERING, SUBMERSION POACHING, STEAMING, AND SOUS VIDE

Simmering and poaching are both methods of cooking in a liquid. These cooking procedures are explained in detail in Chapter 14 (pp. 434–435). The major difference between them, as you should recall, is the temperature. In simmering, the liquid is a little below the boiling point and bubbling very gently. In poaching, the temperature is even lower, and the liquid is not really bubbling.

For the recipe for Chicken Blanquette on page 569, it may be helpful to review the discussion of simmered stews in Chapters 14 and 16 and the recipe for classic Veal Blanquette on p. 508. Note that the first variation following the chicken blanquette recipe is made starting with raw chicken, using the same procedure as for Veal Blanquette, while the main recipe is a shortcut technique made with cooked chicken.

This section includes an example of the steaming method as applied to poultry, in this case whole duck as prepared in a classic Chinese recipe. Review the basic steaming procedure explained on page 440.

Finally, review the information on sous vide cooking on pages 117–119 and 446–447 before preparing the sous vide recipe in this section.

SIMMERING

1. The simmering method is used to cook fowl and other tough items that require long cooking in moist heat to be made tender. Cooking time is about $2^{1}/_{2}$ hours.
2. The cooking liquid is usually water seasoned with salt and, most often, mirepoix and herbs as well.
3. Simmered fowl yields a rich, flavorful broth. The meat can be used for soups, creamed dishes, casseroles, salads, and similar preparations.
4. Start the fowl in cold water if a flavorful soup is your main objective. Start with hot water to retain more flavor in the meat.

POACHING

1. The poaching method is used to gently cook tender poultry in order to retain moisture and to develop a light, subtle flavor. Cooking time is usually short because the product is naturally tender.
2. The cooking liquid is usually stock, sometimes with the addition of wine and other flavorings and seasonings. Cold liquid is added to the poultry product in the pan to cover partway, and the pan must be covered to retain steam. Covering also helps prevent drying and discoloration.
3. After cooking, the liquid may be used to make a sauce, such as suprême sauce, to serve with the cooked product.
4. It is important to drain the poultry well after cooking, because any remaining liquid may spoil the appearance of the sauce on the plate.
5. Poaching may be done on the rangetop or in the oven. Oven poaching provides more even heat.

Simmered Fowl

This preparation and the variation that follows are usually not served as is. Instead, they are the basis for other recipes that call for simmered or cooked chicken or turkey and for chicken or turkey stock or velouté. The cooked meat can also be used for chicken or turkey salads.

YIELD: 4¹/₂–5 LB (2–2.3 KG) COOKED MEAT

U.S.	METRIC	INGREDIENTS	PROCEDURE
3	3	Fowls, about 5 lb (2.3 kg) each	1. Truss the fowls. This step is optional but recommended because it keeps the chicken from falling apart, especially if you are cooking more than one in the pot.
		Mirepoix:	
8 oz	250 g	Onion, coarsely chopped	2. Place the fowls in a stockpot.
4 oz	125 g	Celery, coarsely chopped	3. Add boiling water to cover and return to a boil. Skim the scum carefully.
4 oz	125 g	Carrots, coarsely chopped	4. Add the mirepoix and sachet.
		Sachet:	5. Simmer until the fowls feel tender when pressed on the thigh, about 2¹/₂ hours.
1	1	Bay leaf	6. If the meat and broth are to be used immediately for another preparation, remove the fowl from the liquid, place in another pan, and keep covered until needed. Strain the broth.
6	6	Parsley stems	
¹/₄ tsp	1 mL	Peppercorns	
2	2	Whole cloves	7. If the meat and broth are not needed right away, leave the fowls in the broth and cool quickly in a cold-water bath, as for cooling stocks (p. 161). When completely cool, remove the fowls and refrigerate, covered. Strain the broth and refrigerate.
1 tsp	5 mL	Salt	
			8. To use, disjoint the fowls, remove all bones and skin, and dice or cut as required.

Per 1 ounce: Calories, 70; Protein, 10 g; Fat, 2.5 g (36% cal.); Cholesterol, 30 mg; Carbohydrates, 0 g; Fiber, 0 g; Sodium, 65 mg.

VARIATION

Simmered Chicken or Turkey

Simmer young chickens or turkey as in basic recipe. They may be whole or disjointed. A 3-lb (1.4-kg) chicken will take about 45–60 minutes to cook completely. Turkey will take about 1¹/₂–3 hours, depending on size. Do not overcook young poultry, and do not let the liquid boil.

Poached Chicken Breast Princesse

PORTIONS: 8   **PORTION SIZE:** ½ **CHICKEN BREAST, 2 FL OZ (60 ML) SAUCE, PLUS GARNISH**

U.S.	METRIC	INGREDIENTS	PROCEDURE
8	8	Boneless, skinless half-breasts from 12 chickens, 3 lb (1.2 kg) each	1. Select a baking pan just large enough to hold the chicken breasts in a single layer. Butter the inside of the pan and sprinkle in the shallots.
as needed	as needed	Butter	2. Season the chicken breasts with salt and pepper. Place them in the pan, presentation side (that is, the side that had the skin on) up.
1 oz	30 g	Shallots, brunoise	
as needed	as needed	Salt	3. Sprinkle with the lemon juice and add enough chicken stock to barely cover the chicken.
as needed	as needed	White pepper	
4 tsp	20 mL	Lemon juice	4. Cover the chicken with a buttered piece of parchment or waxed paper (see **Figure 14.10**).
1 pt (approximately)	500 mL (approximately)	Chicken stock, cold	5. Bring to a simmer on top of the stove. Finish poaching in a 325°F (165°C) oven or over low heat on the stove. Cooking time will be 5–10 minutes.
			6. Remove the chicken breasts from the liquid. Place them in a hotel pan, cover, and keep them warm.
		Beurre manié:	7. Reduce the poaching liquid over high heat to about 14 fl oz (400 mL).
1 oz	30 g	Butter, softened	8. Knead the butter and flour together to make a beurre manié (p. 174).
1 oz	30 g	Flour	
7 fl oz	200 mL	Heavy cream, hot	9. With a wire whip, beat the beurre manié into the simmering stock to thicken it. Simmer a minute to cook out any starchy taste.
to taste	to taste	Salt	10. Add the hot cream to the sauce. Season to taste.
24	24	Asparagus tips, cooked, hot	11. Place each chicken breast, well drained, on a plate, and coat with 2 fl oz (60 mL) sauce. Garnish with 3 asparagus tips. Serve immediately.

Per serving: Calories, 330; Protein, 37 g; Fat, 17 g (47% cal.); Cholesterol, 140 mg; Carbohydrates, 6 g; Fiber, 1 g; Sodium, 140 mg.

VARIATIONS

Alternative Method: Poach the chicken as in the basic recipe. Plate immediately and coat with prepared suprême sauce. Save poaching liquid for next day's sauce.

Other sauces may be used to coat poached chicken breasts, including:

Allemande	Aurora	Hungarian
Ivory	Mushroom	

Poached Chicken Breast Florentine

Poach the chicken as in the basic recipe. Place each portion on a bed of buttered spinach, well drained. Coat with Mornay sauce. Optional: Sprinkle with parmesan cheese and brown under the broiler.

Poached Chicken
Breast Princesse

Chicken Blanquette I

PORTIONS: 12   **PORTION SIZE: 5 OZ (150 G)**

U.S.	METRIC	INGREDIENTS	PROCEDURE
2^1/$_2$ lb	1.2 kg	Cooked chicken meat, 1-in. (2^1/$_2$-cm) dice	1. Combine the chicken with the sauce and bring to a simmer.
1 pt 4 fl oz	600mL	Chicken velouté	2. Remove from the heat. Beat the egg yolks and cream together. Temper with a little of the hot velouté sauce, and stir into the sauce.
		Liaison:	
3	3	Egg yolks	3. Return the pot to the heat and bring to just below the simmer. Do not boil.
8 fl oz	240 mL	Heavy cream	
to taste	to taste	Lemon juice	4. Season to taste with a few drops of lemon juice and a pinch each of nutmeg, white pepper, and salt.
pinch	pinch	Nutmeg	
pinch	pinch	White pepper	
pinch	pinch	Salt	

Per serving: Calories, 240; Protein, 21 g; Fat, 16 g (60% cal); Cholesterol, 135 mg; Carbohydrates, 3 g; Fiber, 0 g; Sodium, 160 mg.

VARIATIONS

Chicken Blanquette II

Follow the recipe for Veal Blanquette, page 508. Use 6 lb (3 kg) disjointed chicken (raw) in place of the 5 lb (2.4 kg) veal. Use chicken stock instead of veal stock.

Chicken Blanquette à l'Ancienne (Ancient Style)

Garnish each portion with 2 cooked pearl onions and 1 cooked mushroom cap, fluted if possible (see p. 287).

Chicken Blanquette Brunoise

Add to the sauce 2 oz (60 g) each carrot, celery, and leeks or onions, all cut brunoise and sautéed lightly in butter.

Chicken Blanquette Argenteuil

Garnish each portion with 3 cooked asparagus tips.

Chicken or Turkey Pot Pie

PORTIONS: 6   **PORTION SIZE: 3 OZ (90 G) MEAT, 2 OZ (60 G) VEGETABLES, 4 FL OZ (125 ML) SAUCE**

U.S.	METRIC	INGREDIENTS	PROCEDURE
1 lb 2 oz	550 g	Cooked chicken or turkey meat (light and dark meat)	1. Cut the chicken or turkey into ^{3}/$_4$-in. (1-cm) dice.
3 oz	90 g	Potatoes, medium dice	2. Cook the vegetables separately in boiling salted water. Drain and cool.
3 oz	90 g	Carrots, medium dice	3. Season the velouté to taste with salt and pepper. Add the tarragon.
3 oz	90 g	Tiny white onions, peeled	
3 oz	90 g	Peas	4. Prepare the pastry and cut out circles to cover the tops of casserole serving dishes. You will need about 2 oz (60 g) pastry per portion.
3 cups	750 mL	Chicken velouté	
to taste	to taste	Salt	5. Divide the light and dark meat evenly among individual serving casseroles (3 oz/90 g per portion).
to taste	to taste	Pepper	
^{1}/$_2$ tsp	2 mL	Dried tarragon	6. Divide the vegetables evenly among the casseroles (about ^{1}/$_2$ oz/15 g per portion of each vegetable).
6	6	Flaky pie pastry covers (p. 984)	7. Ladle about 4 fl oz (125 mL) velouté into each casserole.
			8. Top the dishes with the pastry. Cut holes in the centers to allow steam to escape.
			9. Place the dishes on a sheet pan. Bake at 400°F (200°C) until the crust is well browned.

Per serving: Calories, 520; Protein, 25 g; Fat, 32 g (55% cal.); Cholesterol, 80 mg; Carbohydrates, 34 g; Fiber, 5 g; Sodium, 430 mg.

VARIATIONS

Vegetable ingredients may be varied as desired. Other vegetables that may be used include celery, mushroom caps, and lima beans.

Chicken or Turkey Stew

Prepare the meat, vegetables, and velouté as in the basic recipe. Omit pastry. Combine the ingredients in a saucepot and bring to a simmer. Hold for service.

Oyako Donburi

PORTIONS: 8   PORTION SIZE: SEE PROCEDURE

U.S.	METRIC	INGREDIENTS
1 qt	1 L	Dashi
2¹/₂ fl oz	75 mL	Soy sauce
2¹/₂ oz	75 g	Sugar
1 fl oz	30 mL	Sake (optional)
3 pt	1.5 L	Raw Japanese short-grain rice
12 oz	360 g	Boneless, skinless chicken meat
8	8	Scallions
8	8	Eggs

PROCEDURE

1. Combine the dashi, soy sauce, sugar, and sake in a saucepan. Bring to a simmer to dissolve the sugar. Remove from heat and reserve.

2. Cook the rice by following steps 1–4 in the recipe for sushi rice, page 652. Keep it hot.

3. Cut the chicken into strips 1 in. (2¹/₂ cm) wide, then slice diagonally ¹/₄ in. (6 mm) thick.

4. Trim the roots and the coarser greens from the scallions, leaving the tender green parts intact. Cut diagonally into ¹/₂-in. (1-cm) pieces.

5. For each portion, put 1¹/₂ oz (45 g) chicken, 1 sliced scallion, and 4¹/₂ fl oz (125 mL) dashi mixture in a small sauté pan. Simmer until the chicken is nearly done.

6. Break 1 egg into a bowl. Mix lightly but do not beat.

7. Pour the egg in a stream around the chicken in the sauté pan. Continue to simmer until the egg is half set.

8. Put 1¹/₂ cups (375 mL) hot rice in a large, deep soup bowl.

9. When the egg is nearly set, give the egg and chicken mixture a light stir and pour the contents of the sauté pan over the rice.

Per serving: Calories, 570; Protein, 23 g; Fat, 8 g (13% cal.); Cholesterol, 240 mg; Carbohydrates, 98 g; Fiber, 4 g; Sodium, 600 mg.

VARIATION

Tendon

Omit the chicken, scallions, and egg in the recipe above. Reduce the dashi to 1 pt 500mL) and double the amount of sake. Top each bowl of rice with 1 piece of shrimp tempura (p. 637) and 2 pieces of vegetable tempura. Pour about 2¹/₂ fl oz (75 mL) hot dashi mixture over the rice and serve. (The name of this dish comes from the first syllables of tempura and donburi.)

OYAKO

The name of this dish, oyako, means "parent and child," referring to the chicken and eggs. A donburi is a type of serving bowl, and the word also refers to foods served in this type of bowl, generally rice with toppings and sauce.

Oyako Donburi

Red-Cooked Chicken

PORTIONS: 16   **PORTION SIZE:** ¹/₈ CHICKEN

U.S.	METRIC	INGREDIENTS
2	2	Chickens, about 4 lb (1.8 kg) each
2 cloves	2 cloves	Star anise
3 slices	3 slices	Fresh ginger root
1 tbsp	15 mL	Sichuan peppercorns
¹/₂ pt	250 mL	Soy sauce
2 pt	1 L	Water or chicken stock
1 oz	30 g	Sugar
2	2	Scallions
2 fl oz	60 mL	Sherry or Shaoxing wine

Per serving: Calories, 260; Protein, 29 g; Fat, 14 g (49% cal.); Cholesterol, 90 mg; Carbohydrates, 3 g; Fiber, 0 g; Sodium, 990 mg.

VARIATIONS

The star anise and Sichuan peppercorns may be omitted for a simpler version of this dish.

Other meats (using cuts appropriate for simmering) may be cooked this way, including pork, beef, tripe, and duck.

PROCEDURE

1. Cut each chicken into eighths.
2. Tie the star anise, ginger root, and peppercorns in a cheesecloth bag.
3. Combine the soy sauce and water or stock in a pot and add the spice bag, sugar, scallions, and sherry. Bring to a boil.
4. Add the chicken. Simmer until tender.
5. Serve the chicken hot or cold. If it is to be served cold, cool it and store it in the cooking liquid. If desired, combine part of the cooking liquid with additional soy sauce to taste, reduce to a glaze, and coat the chicken.

Red-Cooked Chicken

COOKING STYLES IN CHINESE DISHES

A popular misconception in the Western world about Chinese cooking is that nearly all dishes are stir-fried. An important advantage of stir-frying if cooking fuel is scarce is that, although preparation times are long, due to all the required cutting and slicing, cooking times are short. Once the mise en place is done, stir-fried dishes can be sent to the table in a matter of minutes.

Although stir-frying is an important technique in China, many other cooking techniques are also used, especially simmering and steaming. Two of the recipes in this section, Tea-Smoked Duck and Red-Cooked Chicken, are typical examples of steaming and simmering.

Although both these recipes begin with whole poultry, the birds are cut into small pieces before they are served. In a typical Chinese meal, meats, fish, and vegetables are not so much main dishes as they are accompaniments to rice or, sometimes, noodles. Portion sizes of protein items are small, and the dishes are served family-style in the center of the table. Each diner takes a small quantity of the desired dishes to eat between bites of rice.

Because China is so large and has such an array of climates, there is no single cooking style. The styles of Beijing in the north, Guangdong (Canton) in the southeast, and Sichuan (Szechwan) in the interior, are perhaps as different as the styles of Germany, France, and Italy.

Tea-Smoked Duck 🌐

YIELD: 1 DUCK

U.S.	METRIC	INGREDIENTS	PROCEDURE
3 tbsp	45 mL	Coarse salt	1. Toast the salt and peppercorns in a dry skillet over moderate heat, until peppercorns are fragrant.
1 tbsp	15 mL	Sichuan peppercorns	2. Cool the mixture, then crush with a rolling pin.
1	1	Duck, about 5 lb (2.3 kg)	3. Clean the duck well, removing excess fat. Flatten the duck slightly by pressing down on the breastbone to break it.
			4. Rub the duck inside and out with the salt and peppercorn mixture.
			5. Put the duck in a hotel pan, weight it, and refrigerate 1–2 days.
6	6	Scallions, trimmed	6. Rinse the duck.
4 slices	4 slices	Ginger root	7. Put the scallions and ginger slices in the cavity.
			8. Steam the duck 1–1¹/₂ hours, or until tender.
3 oz	90 g	Raw rice	9. Line a large wok or other heavy pan with aluminum foil.
¹/₂ cup	125 mL	Brown or black tea leaves	10. Mix together the rice, tea leaves, and sugar. Put the mixture in the bottom of the wok.
2 oz	60 g	Sugar	11. Put the duck on a rack over the tea mixture and cover the pan tightly.
			12. Set the pan over high heat 5 minutes, then over moderate heat 20 minutes. Turn the heat off and let stand another 20 minutes without uncovering.
			13. Cool the duck. Chop it into pieces measuring 1–2 in. (3–5 cm), bones and all. Alternatively, bone it out and cut the meat into strips 1 in. (2.5 cm) wide. This dish is normally served at room temperature.

Per 1/6 recipe: Calories, 520; Protein, 30 g; Fat, 43 g (75% cal.); Cholesterol, 130 mg; Carbohydrates, 2 g; Fiber, 1 g; Sodium, 2969 mg.

VARIATIONS

For spicier duck, add 1 tsp (5 mL) five-spice powder to the dry marinade after toasting.

Crispy Duck

This variation may be made with smoked duck or with steamed but unsmoked duck (step 8). When the duck is cool, cut it into quarters. You may bone it if desired, but try to keep it in its original shape. Deep-fry until the skin is crisp. Drain, cut up, and serve at once. (Optional step: Rub cornstarch into the skin before deep-frying.)

Mole Poblano de Pollo *or* de Guajolote

PORTIONS: 16 **PORTION SIZE: 3 FL OZ (90 ML) SAUCE, CHICKEN OR TURKEY QUANTITY VARIABLE**

U.S.	METRIC	INGREDIENTS	PROCEDURE
15	15	Mulato chiles (see Note)	1. Remove and discard the seeds and stem ends of the chiles. Grind the chiles to a powder.
1¹/₂ oz	45 g	Sesame seeds	2. Grind the sesame seeds in a spice grinder or with a mortar and pestle. Set them aside and grind the almonds in the same way.
4 oz	125 g	Almonds	
3	3	Tortillas	3. Fry the tortillas in the fat about 30 seconds. Drain and reserve the fat for step 6. Break the tortillas into pieces.
6 oz	175 g	Lard or rendered chicken, turkey, or pork fat	4. Put the ground sesame, ground almonds, tortillas, cloves, cinnamon, pepper, and coriander into the container of a blender.
¹/₄ tsp	1 mL	Ground cloves	5. Peel the tomatoes if they are fresh. Add the tomatoes and the garlic to the blender. Blend to a smooth purée. If the mixture is too thick to blend, add a little chicken or turkey broth or water.
¹/₂ tsp	2 mL	Cinnamon	
¹/₂ tsp	2 mL	Black pepper	
¹/₄ tsp	1 mL	Ground coriander	6. Heat the reserved fat from step 3 in a saucepot over moderate heat. Add the powdered chiles and cook about 30 seconds. Be careful not to let the chile powder burn.
8 oz	225 g	Tomatoes, canned or fresh	
4	4	Garlic cloves, chopped	7. Add the purée from the blender. Cook 5 minutes, stirring constantly. The mixture will be very thick.
1 oz	30 g	Bitter (unsweetened) chocolate, grated or broken into pieces	8. Add the chocolate. Stir constantly until the chocolate is completely blended in. The sauce may be prepared to this point 1–2 days ahead of time and held in the refrigerator.
10–14 lb	4.5–6.5 kg	Chicken (pollo) or turkey (guajolote), disjointed	9. Put the poultry, onion, carrot, garlic, peppercorns, and salt in a large pot. Add water to cover.
6 oz	175 g	Onion, chopped	10. Simmer until the poultry is tender.
2 oz	60 g	Carrot, chopped	11. Remove the poultry from the broth and set aside to keep warm.
1	1	Garlic clove	12. Strain the broth. Measure 3 pt (1.5 L) broth and stir it into the chile sauce base. Simmer slowly 30–45 minutes, or until the flavors are blended and the mixture has the consistency of a light sauce. (Reserve the remaining broth for another use.)
8	8	Peppercorns	
4 tsp	20 mL	Salt	
as needed	as needed	Water	
as needed	as needed	Lard	13. Heat the lard in a sauté pan and brown the cooked poultry pieces lightly. (This step is optional.)
to taste	to taste	Salt	14. Add the poultry to the sauce and simmer a few minutes until quite hot.
			15. Adjust the seasoning with salt, if necessary, and serve.

Per serving: Calories, 510; Protein, 38 g; Fat, 36 g (63% cal.); Cholesterol, 120 mg; Carbohydrates, 9 g; Fiber, 4 g; Sodium, 420 mg.

Note: Instead of mulato chile peppers, you may use ancho or pasilla chiles or a mixture of different kinds. If none of these is available, you may substitute about 1 cup (125 g) chili powder.

MOLE POBLANO

The most internationally well known examples of Mexican cooking—enchiladas, burritos, tacos, and tamales—represent only a small part of the varied cooking styles of Mexico. Beyond these few dishes, mole poblano is one of the few dishes widely known outside its home.

The Spanish word *salsa* means, literally, "sauce," and includes a wide variety of cooked and uncooked sauces and condiments. In Mexican cuisine, a mole (mo-lay) is a more specific kind of

sauce, one that is cooked, contains chiles in addition to spices, and is usually thickened with corn, ground seeds, or nuts. It is a more complex preparation than most salsas. The best-known mole, mole poblano, meaning mole from the town of Puebla, contains bitter chocolate in addition to its other seasonings.

Mole does not mean chocolate sauce, and simply putting chocolate in a sauce does not make it a mole. In addition, many moles contain no chocolate.

Stuffed Chicken Roulade Sous Vide

YIELD: APPROX. 1¹/₂ LB (720 G)

U.S.	METRIC	INGREDIENTS	PROCEDURE
		Chicken sausage:	1. Review sausage forcemeats on pages 882–884.
8 oz	240 g	Boneless, skinless chicken thigh meat, diced	2. Keeping the mixture cold at all times, grind the sausage ingredients together finishing with the small die of the grinder.
4 oz	120 g	Fatback or fatty bacon, diced	3. Transfer the mixture to the cold bowl of a mixer and mix together at medium speed until smooth and slightly sticky.
³/₄ tsp	3 mL	Salt	
¹/₄ tsp	1 mL	Pepper	4. Chill the mixture until you are ready to fill the chicken breast.
1 tsp	5 mL	Garlic, chopped fine	
¹/₂ tsp	2 mL	Fennel seed	
¹/₈ tsp	0.5 mL	Red chile flakes	
2 tbsp	30 mL	Chopped parsley	
1 fl oz	30 mL	Cold water	
1	1	Double chicken breast, boneless, skin on	5. Place the chicken breast skin side down between two sheets of plastic wrap and pound lightly until the meat is of even thickness all over.
to taste	to taste	Salt	6. Remove the top piece of plastic wrap. Form the sausage meat into a log shape the length of the chicken breast. Place it down the middle of the breast and roll up. (See the illustrations of chicken galantine on p. 864 for the procedure.)
to taste	to taste	Pepper	
			7. Wrap tightly in plastic wrap.
			8. Place in a vacuum bag. Vacuum on high and seal.
			9. Cook for 50–60 minutes in an immersion circulator heated to 160°F (71°C).
			10. Remove from the bag and dry with paper towels.
1 tbsp	15 mL	Vegetable oil	11. Heat the oil over medium heat and brown the roulade well on all sides.
			12. Let rest 10 minutes before slicing.

Per 1 oz (28.35 g): Calories, 70; Protein, 4 g; Fat, 6 g (77% cal.); Cholesterol, 20 mg; Carbohydrates, 0 g; Fiber, 0 g; Sodium, 90 mg.

Chicken Roulade with an herbed velouté and root vegetables paysanne on a root vegetable purée

BRAISING

A moist-heat cooking method, braising may be used to tenderize tough poultry products. Also, as for veal and pork, it can be used to provide moistness and flavor to tender poultry items. Coq au vin, the well-known braised chicken in red wine, was originally made with a tough old rooster (coq), but today the same recipe is applied to tender young chicken.

The braising method is explained in detail in Chapter 14. Review pages 441–444 before beginning this section. Poultry products are braised using the same procedures as for meats, except that mirepoix is frequently omitted. Other flavoring ingredients may be used instead, depending on the recipe. Method 1 of the basic braising procedure is used more often than Method 2.

A variation on the second braising method is used for making white stews called *fricassees.* The recipe for Chicken Fricassée in this section illustrates the following basic procedure. Note that this procedure is the same as the second braising method (p. 444) except that the meat and roux are not browned.

BASIC PROCEDURE for Making Classic Fricassées

1. Collect all equipment and food supplies.

2. Prepare the meat for cooking, as required.

3. Add the fat to the pan and heat over moderated heat.

4. Add the meat to the fat and sear it gently on all sides, but do not brown it.

5. Remove the meat from the pan (if required). If mirepoix or other aromatic vegetables are used, add them to the fat in the pan and sweat them without browning.

6. Add flour to the fat in the pan. Make a blond roux.

7. Add stock to make a thickened sauce. Add seasonings and flavorings.

8. Return the meat to the pan. Cover and simmer in the oven or on the range until the meat is tender.

9. Adjust the sauce as necessary (strain, season, reduce, dilute, etc.).

BRAISED "SAUTÉS"

If you review the procedure for sautéing meats (p. 428), you will see that if the product is not completely cooked when browned in step 6 and then finished by simmering it in the sauce in step 11, the result is a braised item. This procedure is sometimes used for classical "sautés," which then become braises rather than true sautés. The recipe for Chicken Chasseur on page 577 is an example. (An alternative method is to finish cooking the chicken or meat in a covered pan in the oven while you are making the sauce. This is also braising because the cover holds in moisture.)

KEY POINTS TO REVIEW

- What poultry items are best for sautéing?

- What is the difference between sautéing and pan-frying, when applied to poultry?

- What poultry items are best for deep-frying? At what temperature should they be fried?

- What are the differences between simmering and poaching? Describe each procedure as it is applied to poultry items.

- What is a classical sauté, and why does it sometimes become a braised item rather than a sautéed one? How would cook the item if you wanted it to be a true sauté rather than a braise?

- What are the steps in the basic procedure for making classic fricassees?

 # Chicken Fricassée

PORTIONS: 12   PORTION SIZE: ¼ CHICKEN, 3 FL OZ (90 ML) SAUCE

U.S.	METRIC	INGREDIENTS	PROCEDURE
3	3	Chickens, 2½–3 lb (1.1–1.4 kg) each	1. Cut the chickens into 8 pieces each. Season with salt and white pepper.
to taste	to taste	Salt	2. Melt the butter in a brazier over moderate heat.
to taste	to taste	White pepper	3. Add chicken and onion. Sauté very lightly so the chicken is seared on all sides, but do not brown.
3 oz	90 g	Butter	
6 oz	175 g	Onion, cut brunoise	
3 oz	90 g	Flour	4. Add the flour and stir so it combines with the fat to make a roux. Cook another 2 minutes, without browning.
3 pt (approximately)	1.5 L (approximately)	Chicken stock	5. Gradually stir in enough stock to cover the chicken. Bring to a simmer, stirring, until the sauce thickens.
		Sachet:	6. Add the sachet.
1	1	Bay leaf	7. Cover and place in a slow oven (300°F/150°C) or over very low heat on the range. Cook until tender, 30–45 minutes.
1	1	Small piece of celery	
4	4	Parsley stems	
¼ tsp	1 mL	Thyme	
		Liaison:	8. Remove the chicken from the sauce and keep it warm in a covered pan.
3	3	Egg yolks, beaten	9. Degrease the sauce. Reduce it over high heat to proper thickness. You should have about 1 qt (1 L) sauce. Strain through cheesecloth.
8 fl oz	240 mL	Heavy cream	
1 tbsp	15 mL	Lemon juice	10. Combine the egg yolks and cream. Temper with a little hot sauce, and add the liaison to the sauce. Bring to just below the simmer. Do not boil.
to taste	to taste	Salt	
to taste	to taste	White pepper	11. Season to taste with lemon juice, salt, white pepper, and nutmeg. Pour the sauce over the chicken.
to taste	to taste	Nutmeg	

Per serving: Calories, 440; Protein, 39 g; Fat, 27 g (56% cal.); Cholesterol, 175 mg; Carbohydrates, 8 g; Fiber, 0 g; Sodium, 180 mg.

VARIATIONS

Chicken Fricassée with Tarragon

Add 1½ tsp (7 mL) dried tarragon to the sachet.

Chicken Fricassée à l'Indienne

Add 2 tbsp (30 mL) curry powder when making the roux.

Fricassée of Turkey Wings

Prepare as in the basic recipe, using 1 or 2 turkey wings per portion, depending on size. Cut large turkey wings into 2 pieces.

Veal Fricassée

Prepare as in the basic recipe, using 5 lb (2.3 kg) boneless veal shoulder, cut into large dice. Use white veal stock.

Pork Fricassée

Use 5 lb (2.3 kg) boneless, diced pork and pork, veal, or chicken stock.

Fricassée à l'Ancienne
Fricassée Brunoise
Fricassée Argenteuil

Use the same garnishes as for the corresponding Chicken Blanquette variations (p. 569).

Chicken Fricassée Brunoise with Cauliflower Press

Chicken Chasseur

PORTIONS: 10 PORTION SIZE: 1/2 **CHICKEN, 3 FL OZ (90 ML) SAUCE**

U.S.	METRIC	INGREDIENTS	PROCEDURE
5	5	Chickens, 2–2¹/₄ lb (0.9–1 kg) each (see Note)	1. Cut the chickens into 8 pieces. Season with salt and pepper.
to taste	to taste	Salt	2. Heat the oil in a brazier or large sauté pan. Brown the chicken well on all sides.
to taste	to taste	Pepper	3. Remove the chicken pieces from the pan. Cover and keep them hot.
2 fl oz	60 mL	Oil	
2 oz	60 g	Shallots or onions, cut brunoise	4. Add the shallots and mushrooms to the pan and sauté lightly without browning.
8 oz	250 g	Mushrooms, sliced	5. Add the white wine and reduce by three-fourths over high heat.
8 fl oz	250 mL	White wine	
1¹/₂ pt	750 mL	Demi-glace	6. Add the demi-glace and tomatoes and bring to a boil. Reduce slightly. Season with salt and pepper.
8 oz	250 g	Tomato concassé, fresh	
		or	7. Place the chicken in the sauce. Cover and simmer slowly on the stove or in the oven at 325°F (165°C), 20–30 minutes, or until done.
4 oz	125 g	Drained, chopped canned tomatoes	
to taste	to taste	Salt	8. When the chicken is done, remove it from the pan and reduce the sauce slightly over high heat. Add the chopped parsley and check the seasonings.
to taste	to taste	Pepper	
2 tbsp	30 mL	Chopped parsley	9. Serve ¹/₂ chicken (2 pieces dark meat and 2 pieces light meat) per portion. Cover with 3 fl oz (90 mL) sauce.

Per serving: Calories, 780; Protein, 76 g; Fat, 47 g (55% cal.); Cholesterol, 250 mg; Carbohydrates, 72 g; Fiber, 1 g; Sodium, 290 mg.

Note: Large chickens may be used, if desired. For 3¹/₂-lb (1.6-kg) chickens, use ¹/₄ chicken per portion (1 piece dark meat and 1 piece light meat).

VARIATIONS

Alternative Method: Brown chickens as in the basic recipe. Drain excess fat. Add 1 qt (1 L) prepared chasseur sauce and finish cooking the chickens as in the basic method.

Chicken Bercy

Method 1: Prepare as in the basic recipe, but omit the mushrooms and tomato.

Method 2: Brown chickens as in the basic recipe. Add 1 qt (1 L) prepared Bercy sauce and simmer the chickens until done.

Chicken Portugaise

Method 1: Prepare as in the basic recipe, but omit the mushrooms and wine. Use 4 oz (125 g) onions, cut brunoise, and add 1 tsp (5 mL) chopped garlic. Substitute tomato sauce for the demi-glace.

Method 2: Brown chicken as in the basic recipe. Add 1 qt (1 L) portugaise sauce and simmer the chicken until done.

Chicken Hongroise

Prepare as in the basic recipe, but sauté chicken only lightly. Do not brown. Omit mushrooms and wine. Use Hungarian (hongroise) sauce instead of demi-glace. When the chicken is cooked, add 4–6 fl oz (125–175 mL) heavy cream (tempered or heated) to the sauce. Omit parsley garnish. Serve with rice pilaf.

Chicken Chasseur

Coq au Vin

PORTIONS: 12   PORTION SIZE: ¼ CHICKEN, 2½ FL OZ (75 ML) SAUCE

U.S.	METRIC	INGREDIENTS	PROCEDURE
12 oz	350 g	Salt pork or slab bacon	1. Cut the salt pork or bacon into bâtonnet shapes, 1 × ¼ × ¼ in. (2 × ½ × ½ cm).
1 fl oz	30 mL	Oil	2. Place the bacon pieces in a saucepan. Cover with cold water. Bring to a boil and drain.
			3. Add the oil to a large sauté pan and place over moderate heat. When hot, add the blanched bacon. Sauté until lightly browned. Remove with a slotted spoon and set aside.
3	3	Chickens, 3½ lb (1.6 kg) each, cut into 8 pieces	4. Increase the heat to high. Add the chickens to the fat remaining in the pan and brown well on all sides. Remove the chicken from the pan.
24	24	Tiny white onions, peeled and parboiled	5. Add the onions and mushrooms to the pan and sauté until browned. Remove with a slotted spoon and set aside with the bacon pieces. Pour off the fat from the pan.
1½ lb	700 g	Small mushroom caps	
1 qt	1 L	Dry red wine	6. Add the wine and stock to the pan and bring to a boil.
1 pt	500 mL	Chicken stock	7. Add the sachet.
		Sachet:	8. Return the chicken to the pan. Bring the liquid back to a boil. Cover and cook in a 300°F (150°C) oven or over very low heat on top of the stove until chicken is done, 30–40 minutes.
½ tsp	2 mL	Dried thyme	
1	1	Bay leaf	
4	4	Large garlic cloves, crushed	
		Beurre manié:	9. Remove the chicken from the cooking liquid and place in a heated pan or on a serving platter. Garnish with the mushrooms, onions, and bacon pieces.
2 oz	60 g	Butter, softened	10. Degrease the cooking liquid carefully.
2 oz	60 g	Flour	11. Place over high heat and boil until the liquid is reduced to about 1 qt (1 L).
to taste	to taste	Salt	12. Mix the butter and flour to make beurre manié. Beat in the beurre manié a little at time, just enough to thicken the sauce lightly.
as needed	as needed	Chopped parsley	13. Add salt to taste.
			14. Strain the sauce over the chicken and garnish.
			15. At service time, sprinkle each portion with a little chopped parsley.

Per serving: Calories, 690; Protein, 50 g; Fat, 41 g (54% cal.); Cholesterol, 165 mg; Carbohydrates, 10 g; Fiber, 2 g; Sodium, 480 mg.

Coq au Vin

Chicken Couscous

PORTIONS: 8   PORTION SIZE: 9 OZ (270 G) STEW, 4¹/₂ OZ (135 G) COUSCOUS

U.S.	METRIC	INGREDIENTS
1 lb	480 g	Couscous, regular or instant

PROCEDURE

1. Review the procedures for preparing couscous on page 397. If you are using instant couscous, do not prepare until just before serving. If you are using regular couscous, do step 1 of the couscous procedure. (*In the rest of this recipe, couscous preparation is distinguished from the chicken stew preparation by being printed in italics.*)

U.S.	METRIC	INGREDIENTS
1 oz	30 g	Oil or clarified butter
3 lb 8 oz	1.7 kg	Chicken, cut into eighths (**Figure 17.5**)
12 oz	360 g	Onions, sliced
¹/₂ tsp	2 mL	Ground ginger
¹/₂ tsp	2 mL	Turmeric
1	1	Cinnamon stick
2 tsp	10 mL	Salt
¹/₂ tsp	2 mL	Pepper
1 pt	0.5 L	Water

2. In the bottom of a couscousière (see p. 397) or large, heavy saucepan, heat the oil over moderate heat.
3. Add the chicken, onions, ginger, turmeric, cinnamon stick, salt, and pepper. Cook over moderate heat, stirring occasionally, 15 minutes, taking care not to let any ingredients scorch.
4. Add the water. Cover and simmer 45 minutes.
5. *If you are using standard couscous, prepare it for steaming by performing steps 2 and 3 of the procedure on page 397.*

U.S.	METRIC	INGREDIENTS
3 oz	90 g	Raisins
1 lb	480 g	Carrots, peeled and sliced
1 lb	480 g	Zucchini, in 1-in. (2.5-cm) slices
1 lb 8 oz	720 g	Tomatoes, peeled and quartered

6. Soak the raisins in hot water until soft. Drain and set aside.
7. Add the carrots to the chicken stew.
8. *If you are using standard couscous, perform steps 4 and 5 of the procedure on page 397 while the stew is simmering. Steam the couscous in the top section of the couscousière or in a colander set over the stew.*
9. Add the zucchini, tomatoes, and raisins to the stew. Simmer an additional 20 minutes.
10. *If you are using standard couscous, steam it for the last time (step 6 on p. 397). If you are using instant couscous, prepare it following the procedure on page 397.*

U.S.	METRIC	INGREDIENTS
1 oz	30 g	Butter
to taste	to taste	Salt

11. Turn out the couscous onto a serving platter or into a hotel pan. Toss with the butter.
12. Taste the stew and add more salt if necessary. Remove the cinnamon stick.
13. For platter presentation, make a well in the center of the couscous and spoon the stew into the center. For single portions, make a ring of couscous on a plate and spoon a portion of stew into the center.

Per serving: Calories, 620; Protein, 38 g; Fat, 21 g (30% cal.); Cholesterol, 95 mg; Carbohydrates, 70 g; Fiber, 8 g; Sodium, 730 mg.

Chicken Couscous

Duck Confit

PORTIONS: 8 (SEE NOTE)

U.S.	METRIC	INGREDIENTS
8 lb	3.6 kg	Duck parts, preferably legs
1 oz	30 g	Salt
1 tsp	5 mL	White pepper
1/2 tsp	2 mL	Nutmeg
1/2 tsp	2 mL	Powdered bay leaf
pinch	pinch	Ground cloves
as needed	as needed	Extra duck fat

PROCEDURE

1. Trim off excess fat from the duck and reserve.

2. If you are using duck legs (the ideal part for this preparation), leave the thigh attached to the drumstick.

3. Rub the duck pieces with the salt and spices. Refrigerate overnight.

4. Render the trimmed fat plus as much extra fat as needed. You will need enough rendered fat to completely cover the duck pieces.

5. Put the duck and the rendered fat in a brazier, large saucepan, or casserole. Simmer gently in the fat over low heat or in a 300°F (150°C) oven until very tender, 1 1/2–2 hours.

6. Remove the cooked duck from the fat and pack into a clean crock or other container. Pour the melted fat over the meat so it is completely covered, but be careful not to pour in any of the juices. Refrigerate. (The degreased juices may be used for another purpose, such as cooking beans.)

7. Remove duck pieces and use as needed. For best storage, remaining pieces must be kept covered by the fat. For serving, the confit is usually browned in a little of the fat until it is heated through and the skin is crisp. Serve with such accompaniments as braised cabbage, cooked white beans, or sautéed potatoes with garlic, or on a bed of salad greens.

Per serving: Calories, 620; Protein, 35 g; Fat, 52 g (77% cal.); Cholesterol, 155 mg; Carbohydrates, 0 g; Fiber, 0 g; Sodium, 1480 mg.

Note: The exact weight of the finished confit can vary considerably.

CONFIT OF DUCK AND GOOSE

Confit means "preserved." Confit of duck and goose originated as a byproduct of the production of foie gras, as a way to make use of and preserve the meat of birds raised and fed for their enlarged, fattened livers. After the parts are cooked by the procedure in the recipe on this page, they are packed in crocks and enough fat is poured over them to seal them from the air, thus preserving them for a time.

Today, of course, refrigeration makes this method of preserving unnecessary. But confit is more popular than ever because of the tenderness and flavor the cooking method yields. In the case of ducks, the legs are usually made into confit, while the boneless breasts are reserved for pan-frying and serving rare. A special breed of duck called *moulard*, with a large, meaty breast, is used for foie gras production. It is this duck whose legs are traditionally made into confit. However, the legs of any domestic duck can be used in this recipe.

Duck Confit with Walnut Mesclun Salad

DRESSINGS AND STUFFINGS

Stuffing chickens and turkeys is usually not practical in production kitchens. Baking the stuffing separately gives better results, for these reasons:

1. **Safety.**

   Stuffing inside a bird is an ideal breeding ground for bacteria that cause food poisoning.

2. **Quality.**

   Additional roasting time is needed to heat the stuffing through. The result is often overcooked poultry.

3. **Efficiency.**

   Filling poultry with stuffing and removing it after roasting is impractical, time consuming, and messy.

   Stuffing that is baked separately is usually called *dressing*.

   Is poultry ever stuffed? Yes. Small birds served whole as one or two portions can be stuffed, and often are. Stuffed Cornish hens or small game birds such as quail are popular items.

BASIC INGREDIENTS OF DRESSINGS

1. Starch base, such as bread or rice.

2. Aromatic vegetables, generally onions and celery.

3. Fat, such as butter or chicken fat, for sautéing the vegetables and for providing richness.

   Dressings for chicken and turkey, which are lean, may require more fat than dressings for duck and goose, which are fatty.

4. Liquid, usually stock, to provide moisture.

5. Seasonings, herbs, and spices.

6. Eggs, sometimes added as a binder but not always necessary.

7. Other ingredients for flavor, character, and bulk, such as:

Sausage	Chestnuts
Oysters	Fruits
Giblets	Nuts

GUIDELINES for Making Dressings

1. All ingredients that require cooking must be completely cooked before combining with other dressing ingredients. They will cook very little more during baking or roasting.

2. Cool all ingredients before combining to avoid growth of dangerous bacteria.

3. Never let baked or unbaked dressing stay in the Food Danger Zone (see p. 17) longer than 1 hour.

   - Refrigerate unbaked dressing if it is not to be baked immediately.
   - Hold baked dressing above 135°F (57°C) for service, or chill as rapidly as possible.
   - Reheat baked dressing rapidly in oven or steamer to an internal temperature of 180°F (82°C).

4. Bake dressing in shallow pans (2 in./5 cm deep) for rapid cooking, to get it above the Danger Zone quickly.

5. Do not overmix bread dressings, or they will become pasty. Toss ingredients together lightly.

6. For light texture, do not pack dressings into baking pans; instead, spoon loosely.

7. If you stuff poultry instead of baking the dressing separately, fill the birds loosely. Do not pack. Stuffings expand during cooking.

Basic Bread Dressing

YIELD: ABOUT 4 LB (2 KG)

U.S.	METRIC	INGREDIENTS
1 lb	500 g	Onion, small dice
¹/₂ lb	250 g	Celery, small dice
¹/₂ lb	250 g	Fat such as butter, chicken fat, or bacon fat
2 lb	1 kg	White bread, 2 days old
1 oz	30 g	Chopped fresh parsley
1 tsp	5 mL	Sage
¹/₂ tsp	2 mL	Dried thyme
¹/₂ tsp	2 mL	Dried marjoram
¹/₂ tsp	2 mL	White pepper
2 tsp	10 mL	Salt
1–2 pt	0.5–1 L	Chicken stock, cold

PROCEDURE

1. Sauté the onion and celery lightly in the fat until tender but not browned. Cool thoroughly.
2. Cut the bread into small cubes. If desired, crusts may be trimmed first.
3. Combine the bread and cooked vegetables in a large stainless-steel bowl. Add the herbs and seasonings and toss gently until all ingredients are well mixed.
4. Add the stock a little at a time, and mix the dressing lightly after each addition. Add just enough to make the dressing slightly moist, neither dry nor soggy. Adjust the seasonings.
5. Place in a greased baking pan and bake at 375°F (190°C), until hot at the center, about 1 hour.

Per 1 ounce: Calories, 70; Protein, 1 g; Fat, 3.5 g (47% cal.); Cholesterol, 10 mg; Carbohydrates, 8 g; Fiber, 0 g; Sodium, 175 mg.

VARIATIONS

Sausage Dressing

Cook 1 lb (500 g) crumbled pork sausage meat, drain, and cool. Use some of the drained fat to cook the vegetables for the dressing. Add the cooked sausage to the dressing before adding the stock.

Chestnut Dressing

Reduce the bread to 1¹/₂ lb (750 g). Add 1 lb (500 g) cooked, coarsely chopped chestnuts to the dressing before adding the stock.

Mushroom Dressing

Cook 2 lb (1 kg) sliced mushrooms with the onion and celery. Proceed as in the basic recipe.

Giblet Dressing

Add ¹/₂ lb (250 g) cooked, chopped chicken or turkey gizzards and hearts to the dressing before adding the stock.

Cornbread Dressing

Substitute cornbread for all or part of the white bread in Basic Bread Dressing or Sausage Dressing.

ADDITIONAL RECIPES

These additional recipes may be found on your CulinarE-Companion recipe management program:

Arroz con Pollo (Spanish Rice with Chicken); Braised Duckling with Sauerkraut; Chicken alla Cacciatora; Chicken Braised with Vinegar; Chicken Breasts Parmesan; Chicken Teriyaki; Paprika Chicken; Pheasant en Cocotte; Salmis of Partridge; Turkey Scallopine with Shiitake Mushrooms and Roasted Shallots.

QUESTIONS FOR DISCUSSION

1. Describe the three roasting methods discussed in this chapter: low-temperature roasting, searing, and high-temperature roasting. When is each used?

2. True or false: Chicken should be broiled at a lower temperature than steaks. Explain your answer.

3. What is meant by the term ***presentation side***?

4. Why is it difficult to cook large chicken pieces by deep-frying? How can this problem be solved?

5. What are the differences between simmering and poaching as applied to poultry?

6. Give three reasons for baking dressing in a separate pan rather than stuffing it into roast poultry.

UNDERSTANDING FISH AND SHELLFISH

At one time, fresh fish was enjoyed only in limited areas—along the seacoast and, to a lesser extent, around lakes and rivers. Today, thanks to modern refrigeration and freezing technology, fish products are enjoyed much more widely.

For the cook, the difficulties of understanding fish and shellfish are, in some ways, the reverse of those for meat. With meat, we are presented with only a few animals but a bewildering array of cuts from each. With fish, we see only a few cuts but hundreds of species, each with its own characteristics and cooking requirements.

For this reason, it is especially important that students learn the basic principles of structure, handling, and cooking so they can utilize the many varieties of seafood in a systematic way.

Fish products are divided into two categories: **fin fish**, or fish with fins and internal skeletons, and **shellfish**, or fish with external shells but no internal bone structure. Because these two categories have many differences, it is helpful to look at them separately, as we do in this chapter.

AFTER READING THIS CHAPTER, YOU SHOULD BE ABLE TO

1. Explain how the cooking qualities of fish are affected by its lack of connective tissue.

2. Determine doneness in cooked fish.

3. Demonstrate the appropriate cooking methods for fat and lean fish.

4. List seven basic market forms of fish.

5. Dress and fillet round fish and flatfish.

6. List and describe common varieties of saltwater and freshwater fin fish used in North American food service.

7. Identify the characteristics of fresh fish, and contrast them with characteristics of not-so-fresh fish.

8. Store fish and fish products.

9. Understand the popular varieties of shellfish, and discuss their characteristics.

10. Outline the special safe handling and cooking procedures for shellfish.

11. Open clams and oysters, split lobsters, and peel and devein shrimp.

FIN FISH

COMPOSITION AND STRUCTURE

The edible flesh of fish, like that of meat and poultry, consists of water, proteins, fats, and small amounts of minerals, vitamins, and other substances. The differences, however, are perhaps more important than the similarities.

Fish has very little connective tissue. This is one of the most important differences between fish and meat. It means that

1. *Fish cooks very quickly*, even at low heat (just enough heat to coagulate the proteins).
2. *Fish is naturally tender*. Toughness is the result not of connective tissue but of the toughening of the protein by high heat.
3. *Moist-heat cooking methods* are used not to create tenderness but to preserve moistness and provide variety.
4. *Cooked fish must be handled very carefully* or it will fall apart.

SPECIAL PROBLEMS IN COOKING FISH

DONENESS AND FLAKING

When fish is cooked, the flesh breaks apart into its natural separations. This is called **flaking**. Most books, somewhat misleadingly, say that fish is done when it flakes easily. Unfortunately, some cooks interpret this as "nearly falling apart." Because fish continues to cook in its retained heat even when removed from the fire, it is often dreadfully overcooked by the time it reaches the customer. *Fish is very delicate and is easily overcooked*.

Observe these tests for doneness:

1. The fish just *separates into flakes*—that is, it is beginning to flake but does not yet fall apart easily.
2. If bone is present, the flesh separates from the bone, and the bone is no longer pink.
3. The flesh has turned from translucent to opaque (usually white, depending on the kind of fish).

Remember, the major flaw in fish preparation is *overcooking*.

COOKING FAT FISH AND LEAN FISH

The fat content of fish ranges from 0.5 percent to 20 percent.

Lean fish are those that are low in fat. Examples: flounder, sole, cod, red snapper, bass, perch, halibut, pike.

Fat fish are those that are high in fat. Examples: salmon, tuna, trout, butterfish, mackerel.

Cooking Lean Fish

Because lean fish has almost no fat, it can easily become dry, especially if overcooked. It is often served with sauces to enhance moistness and give richness.

Moist-heat methods. Lean fish is especially well suited to poaching. This method preserves moistness.

Dry-heat methods. Lean fish, if it is broiled or baked, should be basted with butter or oil. Take special care not to overcook it, or the fish will be dry.

Dry-heat methods with fat. Lean fish may be fried or sautéed. The fish gains palatability from the added fat.

Cooking Fat Fish

The fat in these fish enables them to tolerate more heat without becoming dry.

Moist-heat methods. Fat fish, like lean fish, can be cooked by moist heat. Poached salmon and trout are very popular.

Dry-heat methods. Fat fish are well suited to broiling and baking. The dry heat helps eliminate excessive oiliness.

Dry-heat methods with fat. Large fat fish, like salmon, and stronger-flavored fish, like bluefish and mackerel, may be cooked in fat, but care should be taken to avoid excessive greasiness. Smaller ones, like trout, are often pan-fried. Drain the fish well before serving.

CUTTING FISH

MARKET FORMS

Fish are available in several forms, as illustrated in Figure 19.1. Or they may be cut by the cook into these forms, depending on how they are to be cooked.

BUYING PROCESSED FISH VERSUS CUTTING THEM YOURSELF

Most food-service establishments purchase fish in the forms in which they intend to cook them. They find it less expensive to pay the purveyor to do the cutting than to hire and train the personnel and to allocate the storage and workspace to do it in-house.

Some restaurants still buy whole fish. Here are a few reasons why they do:

1. Their clienteles demand it. Some high-priced luxury restaurants stake their reputations on using only the freshest, most unprocessed ingredients. They are able to charge enough to cover their high labor costs.
2. They are located in the heart of a fresh fish market, where fresh whole fish, delivered daily, are economical. They can best take advantage of seasonal bargains.
3. They are high-volume specialty restaurants and find it more economical to clean the fish themselves and watch the market for the best prices every day.
4. They make fish stocks and use the bones.
5. They serve the whole fish. Examples: sautéed or poached trout presented whole; whole cold poached fish as a buffet display.

Your purchasing decisions will depend on what you plan to do with the fish and what forms are most economical for those purposes.

FIGURE 19.1 Market forms of fish.

(a) Whole or round: completely intact, as caught.

(b) Drawn: viscera removed.

(c) Dressed: viscera, scales, head, tail, and fins removed.

(d) Steaks: cross-section slices, each containing a section of backbone.

(e) Fillets: boneless sides of fish, with skin on or off.

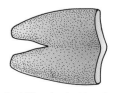

(f) Butterflied fillets: both sides of a fish still joined, but with bones removed.

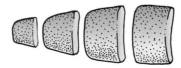

(g) Sticks or tranches: cross-section slices of fillets.

FIGURE 19.2 Dressing a fish.

(a) Scale the fish. Lay the fish flat on the work surface. Rub a scaling tool or the back of a knife against the scales from tail to head. Repeat until all the scales are removed. Rinse. (Exceptions: Trout, with very tiny scales, and scaleless fish, like catfish, are not scaled.)

DRESSING AND FILLETING

Although most of you will work with ready-to-cook fish products, you should know how to clean and fillet whole fish.

1. **Dressing.**
   Figure 19.2 illustrates how to dress a whole fish.

2. **Filleting.**
   There are two basic shapes of fish: **flatfish** (like flounder and sole) and **round fish** (like cod and trout). They are filleted differently. Flatfish have four fillets; round fish have two. Figures 19.3 and 19.4 show the two methods for filleting these fish.

   After a large fish, such as salmon, is filleted, it may be necessary for some recipes to cut the fillets into thinner slices, or escalopes. Figure 19.5 shows this procedure.

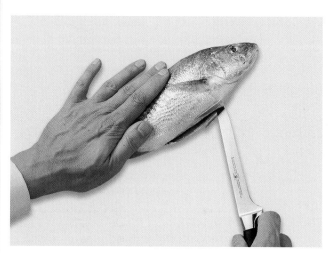

(b) Eviscerate. Slit the belly and pull out the viscera. Rinse the cavity.

(c) Cut off the tail and fins. Scissors are easiest to use.

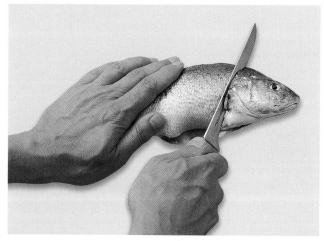

(d) Remove the head. Cut through the flesh just behind gills. Cut or break the backbone at the cut and pull off the head.

(e) The fish is dressed.

FIGURE 19.3 Filleting flatfish.

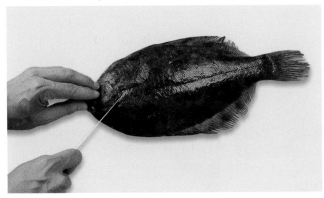

(a) Use a thin-bladed, flexible knife. Cut off the head, just behind the gills. (This step is optional.)

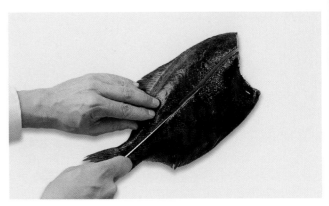

(b) Make a cut from head to tail just to one side of the center line, down to the backbone.

(c) Turn the knife so it is almost parallel to the table. Making long, smooth cuts, cut horizontally against the backbone toward the outer edge of the fish. Gently separate the fillet from the bone.

(d) Remove the fillet completely. Repeat to remove the three remaining fillets

(e) To skin, place the fillet skin side down on the work surface with the tail pointing toward you. Holding the skin at the tail end, slide the knife between the skin and flesh, scraping against the skin to avoid cutting into the fillet. Note: Dover or English sole is skinned before filleting. Cut through the skin at the tail. Holding the tail with one hand, peel off the skin toward the head. Caution: Do not do this with flounder. You will tear the flesh.

FIGURE 19.4 Filleting round fish.

(a) Cut into the top of the fish along one side of the backbone from head to tail. Cut against the bone with smooth strokes of the knife to separate the flesh from the bone.

(b) Cut under the flesh toward the tail; detach it.

(c) Cut along the curved rib bones and finish detaching the fillet at the head end. Turn the fish over and repeat to remove the second fillet. Lightly run your finger along the flesh side of the fillets to see if any bones remain in them. Using a needlenose pliers, pull out any you find. Skin the fillets as for flatfish.

FIGURE 19.5 Cutting escalopes of salmon.

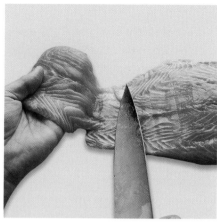

(a) Holding the knife at a sharp angle, cut a thin slice of the fillet, slicing toward the tail or thin end.

(b) The cut slice. Weigh the slice to check for accuracy of cutting, then continue making slices.

KEY POINTS TO REVIEW

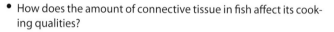

- How does the amount of connective tissue in fish affect its cooking qualities?

- How is fish tested for doneness?

- What are some examples of fat fish and lean fish? What are appropriate cooking methods for each type?

- What are the basic market forms of fish?

- How is filleting flatfish different from filleting round fish?

SEAFOOD AND SUSTAINABILITY

In Chapter 1 we discussed the concept of sustainable agriculture (see p. 6). Similar principles apply to both farmed and wild-caught seafood. In order to be considered sustainable, a seafood species should meet these criteria:

- The population is abundant enough so that fishing won't endanger its numbers.

- The species is harvested in such a way that the accidental harvest of other species, especially endangered species, is limited. (Seafood that is accidentally caught in the process of harvesting a targeted species is called bycatch.)

- Fishing of the species is well managed. Fishing plans should be carefully managed based on current research.

- The species is harvested in a way that limits damage to habitat and to other species.

Rankings of seafood species change from year to year. Some species become threatened, while others may experience population growth. Chefs and food service operators should always be aware of the current rankings of the fish they serve and avoid species that are not recommended.

VARIETIES AND CHARACTERISTICS

Hundreds of varieties of fish are eaten around the world. However, relatively few species account for the majority of the fish used in food service in Canada and the United States. Following are the most common varieties that are used fresh. Some are illustrated in photographs.

There is a significant difference in flavor between freshwater fish and saltwater fish, because saltwater fish, as might be expected, have more salt in their flesh. The following survey of fish varieties divides them into these two basic categories. However, some fish may spend part of their lives in the ocean and part in fresh water. Salmon, for example, lives in the ocean but migrates up rivers to lay eggs, or *spawn*. Fish that live in salt water but spawn in fresh water are called **anadromous**. Shad and arctic char are two other examples of anadromous fish. On the other hand, fish that live in fresh water but swim downstream to spawn in the ocean are called **catadromous**. Eel is a catadromous fish.

In the following listing, fish are categorized by where they spend most of their adult lives. Thus, anadromous fish such as salmon are listed with saltwater fish, while the catadromous fish such as eel are listed with freshwater fish.

SALTWATER FISH—FLATFISH

These are all popular in commercial kitchens. Flatfish have lean, white flesh and a mild, delicate flavor. They are all very flat, oval in shape, with both eyes on one side of the head.

You should note that, in North American waters, there are no true sole. All fish marketed as sole, such as lemon sole, gray sole, and Pacific Dover sole, are actually flounder. True Dover sole, from Atlantic waters off Europe, is the only actual sole in this list. Also, be aware that both true Dover sole and the flounder called Pacific Dover sole are both sometimes called English sole.

Flounder

Type: Lean.

Varieties: Winter flounder, lemon sole, gray sole, Pacific Dover sole, sand dab.

Characteristics: White flesh; fine flakes; mild, sweet flavor.

Weight: 1/2 to 5 lb (0.2 to 2.3 kg).

Flounder

Sole (Dover sole or English sole)

Type: Lean.

Characteristics: Narrower, more elongated than flounder. Flesh similar to flounder, but firmer in texture. One of the most prized of all fish. Expensive.

Weight: 1 to 2 lb (0.5 to 1 kg).

Dover sole

Halibut

Type: Lean.

Characteristics: Looks like a giant flounder, with thicker flesh, delicate flavor. Cut into steaks and fillets.

Weight: 4 to 100 lb or more (2 to 45 kg).

Turbot

Type: Lean.

Characteristics: Large, broad flatfish. White, firm, delicate flesh.

Weight: 1 to 25 lb (0.5 to 11 kg).

Halibut
Courtesy of Alaska Seafood Marketing Institute

SALTWATER FISH—ROUND FISH

Turbot

Arctic Char

Type: Fat. Anadromous.

Characteristics: Similar to both salmon and trout, to which it is related. Flesh is similar to salmon but with less fat, although with more fat than trout.

Weight: Usually around 4 lb (1.8 kg), but can be as large as 25 lb (11 kg).

Black Sea Bass

Type: Lean.

Characteristics: Small, black-skinned fish with firm, delicate, sweet white flesh.

Weight: Up to 3 lb (1.5 kg).

Bluefish

Type: Fat.

Characteristics: Flavorful, oily flesh that is bluish when raw, grayish when cooked. Abundant and inexpensive.

Weight: 1 to 10 lb (0.5 to 4.5 kg).

Black sea bass

Bluefish

Chilean Sea Bass

Type: Fat.

Varieties: This fish is not a bass; the name is a marketing term for the Patagonian toothfish. Due to its popularity, it has been seriously overfished and its populations have declined dramatically. Fishing is highly regulated, and imports to North America must be labeled as legally caught, although there is much illegal fish on the market. Many chefs have agreed to stop serving this fish in order to slow or halt the illegal fishing that threatens its survival.

Characteristics: White, oily flesh with large flakes, firm texture, and mild flavor.

Weight: Usually around 20 lb (9 kg), although it can be much larger.

Cod

Cod

Type: Lean.

Varieties: Small, young cod is called scrod.

Characteristics: Lean, white, delicately flavored flesh with large flakes. One of the most widely used food fish in North America. Most fish sticks and similar items are made from cod.

Weights: Scrod: 1 to 2$^1/_2$ lb (0.5 to 1 kg). Cod: 2$^1/_2$ to 25 lb and up (1 to 11 kg).

Escolar

Type: Fat.

Characteristics: A firm, white, oily fish containing a fat that is not metabolized by the human body. Because of this fat, some people become ill when eating this fish. The fat seems to act like a laxative. In the United States, the Food and Drug Administration banned escolar in 1990, but its sale is once again legal. Food-service operators who serve escolar should advise customers of its potential dangers.

Weight: 65 to 110 lb (30 to 35 kg).

Grouper

Grouper

Type: Lean.

Characteristics: Many varieties with varying shape and skin color. Firm white fish, similar in texture and flavor to red snapper. Tough skin.

Weight: Up to 700 lb (300 kg), but most groupers on the market weigh 5 to 15 lb (2.3 to 7 kg).

Haddock

Type: Lean.

Varieties: Finnan haddie is smoked haddock, not a separate kind of fish.

Characteristics: Similar to cod, but generally smaller.

Weight: 1 to 5 lb (0.5 to 2.3 kg).

Herring

Herring

Type: Fat.

Characteristics: Small, full-flavored, oily fish. Most herring sold to food-service and retail establishments is pickled or smoked rather than fresh, but fresh herring is sometimes available. Smoked herring are sometimes called kippers. Fresh herring are excellent butterflied or left whole and grilled. Because of their high fat content, they do not keep well, so they should be checked carefully for freshness when purchased.

Weight: Up to 8 oz (240 g).

Jack

Blue runner, a type of jack

Jack

Type: Fat.

Varieties: Members of the jack family include kingfish, blue runner, yellowtail, amberjack, and golden thread. The best-known jack is the pompano, listed separately.

Characteristics: Smooth, shiny skin; firm, oily flesh; strong flavor. Some varieties are much stronger in flavor than others.

Weight: Varies greatly depending on variety.

John Dory

Also known as St. Peter's fish, St Pierre.

Type: Lean.

Characteristics: Identified by the characteristic black spot ("St. Peter's thumbprint") on each side of the body behind the head. Firm, sweet, white flesh with fine flakes; broad, thin fillets.

Weight: About 2 lb (900 g) average.

Mackerel

Type: Fat.

Varieties: Spanish and Boston mackerel are the most common small varieties. King mackerel is larger, usually cut into steaks.

Characteristics: Fat, firm flesh with rich flavor and slightly dark color.

Weight: $^1/_2$ to 5 lb (0.2 to 2.3 kg).

Boston mackerel

Spanish mackerel

Mahi-Mahi

Also known as dorado and dolphinfish (not related to the mammal called dolphin).

Type: Lean.

Characteristics: Firm, fine-textured, pinkish flesh with rich, sweet taste. Becomes very dry when overcooked, so best cooked with moist heat or with fat or served with a sauce.

Weight: 5 to 40 lb (2.3 to 18 kg).

Mahi-mahi

Monkfish

Also known as lotte, anglerfish, bellyfish.

Type: Lean.

Characteristics: Large, ugly fish, but only tail is used. White, very firm flesh with fine texture, somewhat like lobster. Rich flavor. Holds up well in soups and stews, but dries out easily if cooked dry without fat. As with skate and shark, the skeleton of monkfish is cartilage rather than bone.

Weight: 5 to 50 lb (2.3 to 23 kg).

Ocean Perch

Type: Lean.

Varieties: Many. Red-skinned varieties (redfish) are especially popular and more expensive.

Characteristics: Mild, somewhat bony.

Weight: Depends on variety, but typically about 1 lb (0.5 kg).

Monkfish

Orange Roughy

Type: Lean.

Characteristics: A South Pacific fish available in some markets. Texture and flavor similar to red snapper.

Weight: $^3/_4$ to 2 lb (0.3 to 1 kg).

Pompano

Porgy

Red mullet

Red snapper

Salmon

Sardine

Pompano

Type: Fat.

Characteristics: Small fish with rich, sweet-flavored flesh. Expensive. A variety of jack (see entry above).

Weight: $3/4$ to 2 lb (0.3 to 1 kg).

Porgy

Type: Lean.

Characteristics: Small, oval fish. Sweet and mild, but quite bony.

Weight: Up to 2 lb (1 kg).

Red Mullet

Also known as rouget barbet (roo-zhay bar-bay), rouget.

Type: Lean.

Varieties: A member of the goatfish family. In addition to the Mediterranean variety discussed here, similar varieties are found in North American waters.

Characteristics: Mediterranean fish resembling a small red snapper. Rich flavor, with a slight taste of shrimp or lobster. Always cooked with skin on, because much of the desired flavor comes from the skin.

Weight: Usually 12 oz (350 g) or less.

Red Snapper

Type: Lean.

Characteristics: Firm, delicate, sweet white flesh with large flakes. Large, coarse bones. Skin is red. Highly prized, and popular in restaurants.

Weight: 1 to 15 lb (0.5 to 7 kg).

Salmon

Type: Fat. Anadromous.

Varieties: Atlantic, king, chinook, sockeye, coho, chum, humpback.

Characteristics: Rich pink to red flesh, with somewhat meaty texture and flavor. One of the most prized of all fish. Much is canned or smoked.

Weight: 4 to 25 lb (2 to 11 kg).

Sardine

See Herring. True sardines are closely related to herring, but are scarce. Most fish sold as sardines are actually small herring. Usually sold canned.

Shad

Type: Fat. Anadromous.

Characteristics: Oily, rich flavor and many bones in several rows in each fillet. Its roe (egg sacs) is especially prized. Fresh shad is highly seasonal (February to June).

Weight: 1¹/₂ to 5 lb (0.7 to 2.3 kg).

Shark

Type: Fat.

Varieties: Mako shark is the variety usually used.

Characteristics: Firm texture, similar to swordfish, but softer and a little moister and with finer grain; less expensive. Has cartilage skeleton, no bones. Usually cut into steaks.

Weight: 25 to 40 lb (11 to 18 kg).

Skate

Also known as ray.

Type: Lean.

Characteristics: May be sold whole, but often only the triangular wings are marketed. Skeleton is cartilage, not bone; tough skin, gray on top, white on bottom. Flesh consists of sweet, white, gelatinous strips of meat extending the width of the wings; two fillets per wing, one above and one below the cartilage.

Weight: Some varieties weigh up to a ton, but most sold for food weigh 1 to 4 lb (0.5 to 2 kg) per wing.

Skate wings, top view (left) and bottom view (right)

Striped Bass

Type: Lean.

Varieties: Wild striped bass is highly prized but available in only small numbers in recent years. Farmed striped bass, developed in response to demand, is a cross between white bass and wild striped bass.

Characteristics: Firm, white, delicately flavored fish with large flakes.

Weight: 1 to 10 lb (0.5 to 4.5 kg).

Striped bass

Sturgeon

Type: Fat. Anadromous

Varieties: White sturgeon is found along the Pacific coast of North America and is also farm-raised. Atlantic sturgeon is found along the East Coast of North America. Sturgeon is also the source of the most prized caviar (see p. 745).

Characteristics: Firm, meaty, oily flesh. Mild flavor.

Weight: White sturgeon: 10 to 15 lb (4.5 to 6.8 kg). Atlantic sturgeon: 60 to 80 lb (27 to 36 kg).

Swordfish

Type: Fat.

Characteristics: Very large, fat fish with dense, meaty, not flaky, texture. High yield. Sold mostly as steaks. Expensive.

Weight: Up to 1,000 lb (450 kg).

Tilefish

Type: Lean.

Characteristics: Firm, sweet, mild white flesh, pinkish when raw. Used mostly for steaks, but smaller ones can be filleted or poached whole.

Weight: 4 to 8 lb (1.8 to 3.6 kg).

Tilefish

Triggerfish

Triggerfish

Type: Lean.

Characteristics: Tough-skinned fish with firm, meaty, white to gray-white flesh. Low yield because of large head.

Weight: About 2 lb (900 g) average.

Tuna

Type: Fat.

Varieties: Yellowfin and bluefin have red flesh, gray when cooked. Albacore (more often canned than used fresh) has pink flesh, off-white when cooked. Some other varieties are also available. Note that bluefin tuna populations have decreased drastically. Many fisheries experts advocate limiting or stopping fishing for this species, to allow populations to recover.

Characteristics: Meaty texture and appearance. Belly cuts much fattier than back ("loin"). Red-fleshed varieties often served raw as sashimi or sushi. When cooked, usually cut into steaks and grilled. Should not be cooked well done, or will be very dry.

Weight: Depends on variety. May be several hundred pounds (100 kg and up).

Wahoo or Ono

Type: Fat.

Characteristics: Firm flesh similar in texture to mackerel and tuna, to which it is related. Pale pink flesh turns white when cooked. The biology of this fish is not well known, and there are concerns about overfishing. It is a popular sport fish.

Weight: 11 to 80 lb (5 to 35 kg), but can be much larger.

Weakfish

Weakfish

Also called sea trout, but unrelated to trout.

Type: Lean.

Characteristics: Mild, light-gray flesh with soft texture.

Weight: Up to 10 lb (4.5 kg).

Whiting

Whiting

Type: Lean.

Characteristics: Fragile, white flesh with mild flavor. Fillets must be handled carefully or they will break up.

Weight: $1/4$ to 3 lb (0.1 to 1.4 kg).

FRESHWATER FISH

Catfish

Type: Fat to somewhat lean, depending on origin.

Varieties: Bullhead is similar, but it is not the same species.

Characteristics: Firm flesh with abundant flavor. Layer of fat directly under skin. Catfish has no scales and is usually skinned before frying or pan-frying. Farmed catfish is milder and usually leaner than wild catfish.

Weight: 1 to 8 lb (0.5 to 3.6 kg).

Catfish

Eel

Type: Fat. Catadromous.

Characteristics: A long, slender, snakelike fish with a slippery skin. Flesh is firm, mild, and oily. Must be alive until shortly before cooking or flesh will be mushy. Skin before cooking and fillet or cut into 1¹/₂-in. (4-cm) lengths. Popular in Japanese restaurants (where it is called unagi), served filleted and grilled, basted with a soy-based glaze.

Weight: Typically 2 to 4 lb (900 g to 1.8 kg), but can be much smaller or larger.

Eel

Perch

Type: Lean.

Varieties: Yellow perch is the most common. Walleyed pike, or walleye, is actually a perch, not a pike. Zander (French: sandre) is a popular European perch sometimes mislabeled in North America as European walleye, although it is different from, and usually less expensive than, true walleye.

Characteristics: Mild-flavored, flaky white flesh with firm texture and fine grain.

Weight: ¹/₂ to 5 lb (0.2 to 2.3 kg).

Perch

Pike

Type: Lean.

Varieties: Northern pike is most common in North America. Walleyed pike or walleye is not a pike but a perch.

Characteristics: Firm white flesh similar to perch but not as thick, and with many small bones.

Weight: 2 to 12 lb (1 to 5.4 kg).

Swai

Type: Fat.

Characteristics: A Southeast Asian fish related to catfish. Milder and sweeter in taste and more delicate in texture than North American river catfish. It is also called *iridescent shark,* but it is not a shark.

Weight: Large adults can weigh as much as 95 lb (43 kg), but swai is usually sold as frozen fillets weighing 3 to 11 oz (85 to 310 g) each.

Tilapia

Type: Lean.

Characteristics: Nearly always from aquafarms, because wild tilapia often taste muddy. Firm, mild white flesh.

Weight: Up to 3 lb (1.4 kg); usually about 1¹/₂ lb (700 g).

Tilapia

Trout

Type: Fat.

Varieties: Lake trout, river trout, brook trout, rainbow trout, steelhead trout, many local varieties.

Characteristics: Soft, fine-textured flesh with rich, delicate flavor. Color of flesh may be white, pink, or reddish.

Weight: Lake trout: 4 to 10 lb (2 to 4.5 kg). Others: ¹/₂ to 3 lb (0.2 to 1.4 kg).

Trout

Whitefish

Type: Fat.

Characteristics: Flaky, white flesh with somewhat sweet flavor.

Weight: 1¹/₂ to 6 lb (0.7 to 2.7 kg).

Whitefish

Zander (see Perch)

HANDLING AND STORAGE

Fish and shellfish are among the most perishable foods you will handle. It is especially important to store them carefully and use them quickly. The fishy taste that turns many people away from fish is actually a sign of decomposition. Fresh fish tastes and smells sweet and fresh. Guidelines for checking fish quality are summarized in Table 19.1.

TABLE 19.1 Checklist for Fish Freshness

Characteristics	Fresh Fish	Not-So-Fresh Fish
Overall appearance	Good appearance, clear slime, moist and pliable fins	Dry or cloudy appearance, dried fins, bruises or cuts in flesh
Odor (see Note)	Fresh and mild, no off-odors	Strong fishy odor
Eyes	Clear, shiny, bulging	Cloudy, sunken
Gills	Red or pink	Gray or brown
Texture of flesh	Firm, elastic	Soft, dents easily
Scales	Shiny, tight on skin	Loose, not shiny
Belly cavity	No belly burn; natural-colored flesh, pink or red along backbone (see Note)	Belly burn: browned, decomposed flesh along ribs and backbone

Note: Because most fish is not purchased whole or dressed but as fillets, steaks, or other portions, odor must be your primary check for freshness.
Belly burn results when the intestines are left in the cavity too long, resulting in spoilage.

STORING FRESH FISH

Objectives

1. To maintain temperature of 30° to 34°F (–1° to 1°C).

2. To keep the fish moist.

3. To prevent fish odors and flavors from transferring to other foods.

4. To protect the delicate flesh from being bruised or crushed.

Methods

1. On crushed ice—the preferred method. Use drip pans to allow for drainage of melted ice. Change ice daily. Cover container or store in separate box away from other foods. Whole fish should be **drawn**—that is, viscera removed (see Figure 19.1)—as soon as possible because the entrails deteriorate quickly. Whole or drawn fish are not wrapped. Cut fish (fillets, steaks, portions) should be wrapped or left in original moistureproof wrap.

2. In refrigerated box at 30° to 34°F (–1° to 1°C) if crushed ice storage is not available or practical. Wrap all fish or leave in original moistureproof wrap.

Storage Time

Fresh fish may be stored 1 or 2 days. If it must be kept longer, you may (1) wrap and freeze it immediately, or (2) cook and then refrigerate it for later use in recipes calling for cooked fish.

Check stored fish for freshness just before you use it. Even if it was fresh when received, it may not be fresh after a few days in storage.

Inspection

The Canadian Food Inspection Agency (CFIA) conducts a program of fish and fish product inspection to promote the safety of fish and shellfish products produced in Canada and imported into Canada. The agency sets the guidelines for product quality and establishes standards for fish and seafood processing plants, for fishing boats, and for equipment used to handle and store fish. These standards cover all aspects of the harvesting and processing of fish products. In order to be approved by the CFIA, a processing facility must use procedures that follow the principles of HACCP (see p. 29). For domestic fish products, the "Canada Inspected" logo is indicated by a round stamp with a maple leaf (Figure 19.6). Standards cover

FIGURE 19.6 Canadian inspection logo. Courtesy of the Canadian Food Inspection Agency.

all aspects of the harvesting and processing of fish products. The program is required for fish that is imported, exported, or shipped between provinces. In addition, the provinces—in particular, Nova Scotia, Prince Edward Island, British Columbia, and Yukon—have important regulatory and licensing functions governing fishing, fish farming, and the sale of seafood within their borders.

FROZEN, CANNED, AND OTHER PROCESSED FISH

Frozen Fish

Frozen seafood products account for more of the fish served today than do fresh. If it were not for the wide availability of frozen seafood products, commercial kitchens would serve much less fish than they do.

Checking Quality

1. Frozen products should be frozen, not thawed, when received.

2. Look for fresh, sweet odor or none at all. Strong, fishy odor means poor handling.

3. Items should be well wrapped, with no freezer burn.

4. Some frozen fish is glazed with a thin layer of ice to prevent drying. Check for shiny surface to make sure glaze has not melted off or evaporated.

Storage

1. Store at 0°F (–18°C) or colder.

2. Keep well wrapped to prevent freezer burn.

3. Maximum storage time:

    Fat fish: 2 months.

    Lean fish: 6 months.

4. Rotate stock—first in, first out.

Thawing and Handling

1. Frozen raw fish.

    - Thaw in refrigerator, never at room temperature. Allow 18 to 36 hours, depending on size. Alternative method, if pressed for time: Keep in original moistureproof wrapper and thaw under cold running water.

    - Small pieces (fillets, steaks, portions) up to 8 oz (250 g) can be cooked from frozen state to make handling easier and to prevent excessive drip loss. Large fish should be thawed for more even cooking from surface to interior.

    - Fillets or other portions to be breaded or prepared in some other way before cooking may be partially thawed (for example, for a few seconds in a microwave) and then prepped and cooked. They will be easier to handle than if fully thawed.

    - Handle thawed fish as you would fresh fish.

    - Do not refreeze.

2. Breaded and battered fish, fully prepared entrées, and other frozen, prepared fish items.

    - Read and follow package directions.

    - Most of these items are cooked from the frozen state, usually in the deep fryer, oven, microwave, or steamer.

Canned Fish

1. Check cans for signs of damage. Discard swollen cans (or return to the supplier).

2. Store, like other canned goods, in a cool, dry place.

3. Opened canned fish should be placed in covered containers, labeled with the contents and date, and refrigerated. It will keep for 2 or 3 days.

KEY POINTS TO REVIEW

- What are the four main kinds of saltwater flatfish used in food service? List examples of saltwater round fish and freshwater fish and briefly describe each.

- What are the key signs of freshness in fish?

- How should fresh fish be stored? How should frozen fish be stored and thawed?

SHELLFISH

Shellfish are distinguished from fin fish by their hard outer shells and their lack of backbone or internal skeleton.

There are two classifications of shellfish:

1. **Mollusks** are soft sea animals that fall into three main categories:
   - **Bivalves**, which have a pair of hinged shells (such as clams and oysters).
   - **Univalves**, which have a single shell (such as abalone and conch).
   - **Cephalopods** (such as octopus, squid, and cuttlefish).

   From the scientist's point of view, there are other mollusks as well, but they do not concern us here.

2. **Crustaceans** are animals with segmented shells and jointed legs.

MOLLUSKS

The most important mollusks in commercial kitchens are oysters, clams, mussels, scallops, squid, and octopus.

OYSTERS

Characteristics

St. James River oysters

1. Oysters have rough, irregular shells. The bottom shell is slightly bowl-shaped. The top shell is flat.
2. The flesh of the oyster is extremely soft and delicate and contains a high percentage of water.
3. Oysters are available all year, even in months without an *R* in their names, but they are at their best in the fall, winter, and spring.
4. There are four main varieties in Canada and the United States, depending on their origin. Note in particular that the dozens of varieties of Eastern oysters are all the same species; their flavors differ, however, depending on the environment in which they grew.

   Eastern: Known by many local names, depending on their place of origin, such as Bluepoint (Long Island), Box Oyster (Long Island), Chesapeake Bay, Chincoteague (Virginia), Cotuit (Nantucket), Kent Island (Maryland), Malpeque (Prince Edward Island, Canada), Patuxent (Maryland), Apalachicola (Florida), Breton Sound (Louisiana), Wellfleet (Massachusetts).

   Olympia: Very small, from the Pacific coast.

   Belon: European oyster now grown in North America. Shells are flatter than those of Eastern oysters. More properly called ***European flat oysters***, reserving the term ***Belon*** for those flat oysters grown near the mouth of the Loire River in France. Prized for their intense, briny flavor.

   Japanese or Pacific: Usually large oysters from the Pacific coast. The much smaller kumamoto oyster is of the same species.

Pine Island oysters

Market Forms

1. Live, in the shell.
2. Shucked—fresh or frozen. Shucked oysters are graded by size as follows.

Grade	Number per Gallon (3.8 liters)
Extra Large or Counts	160 or fewer
Large or Extra Selects	161–210
Medium or Selects	211–300
Small or Standards	301–500
Very Small	Over 500

3. Canned—rarely used in food service.

Checking Freshness

1. Oysters in the shell must be alive to be good to eat. Tightly closed shells, or shells that close when jostled, indicate live oysters. Discard dead ones.

2. Live or shucked oysters should have a very mild, sweet smell. Strong odors indicate spoilage.

Opening Oysters

1. Scrub shells thoroughly before opening.

2. Oysters to be served raw must be opened in a way that leaves the bottom shell intact and the tender oyster undamaged. The technique illustrated in Figure 19.7 is one common way of opening oysters. Your instructor may wish to show you another method.

3. Oysters to be cooked may be opened by spreading them on a sheet pan and placing them in a hot oven just until the shells open. Remove from shells and cook immediately. Discard any that do not open.

Storage

1. Keep live oysters in a cold, wet place in the cartons or sacks in which they arrived. They should keep at least 1 week.

2. Store fresh shucked oysters in their original container in the refrigerator at 30° to 34°F (–1° to 1°C). They will keep up to 1 week.

3. Keep frozen oysters in the freezer at 0°F (–18°C) or colder, until ready for use. Thaw in refrigerator 24 hours or more, depending on size of container.

Cooking Oysters

1. Cook just enough to heat through to keep oysters juicy and plump. Overcooking makes them shrunken and dry.

2. Cooking methods: Poaching, deep-frying, baking on the half-shell with toppings, in soups and stews.

CLAMS

Characteristics

There are two major kinds of clams from the east coast of North America: hard shell and soft shell. The West Coast also has some local varieties.

1. *Hard-shell clams* or *quahogs*. These go by different names, depending on size.

Littlenecks are the smallest. They are the most tender for eating raw or for steaming.

Cherrystones are medium-sized, and perhaps the most common. They can be eaten raw and are good for steaming, though tougher than littlenecks.

Chowders, the largest, are also called *quahogs* in the Northeast. Rather tough, they are chopped for cooking in chowders or cut into strips for frying.

FIGURE 19.7 Opening oysters.

(a) Examine the shell to see that it is tightly closed, indicating a live oyster. Rinse the shell under cold running water. Hold oyster in left hand, as shown. (Left-handers will hold oyster in right hand.) Hold the oyster knife near the tip as shown. Insert the knife between the shells near the hinge.

(b) Twist the knife to break the hinge.

(c) Slide the knife under the top shell and cut through the adductor muscle (which closes the shells) near the top shell. Try not to cut the flesh of the oyster, or it will lose plumpness. Remove the top shell.

(d) Carefully cut the lower end of the muscle from the bottom shell to loosen oyster. Remove any particles of shell from the oyster before serving.

Clams, clockwise from top left: steamers, littlenecks, cherrystones, chowder clams.

Cockles

2. *Soft-shell clams*. These are sometimes called *longnecks* because of the long tube that protrudes from between the shells. They have very thin shells that do not close completely. They are also called *steamers* because the usual way to serve them is to steam them and serve them with their own broth and with melted butter for dipping.

3. **Cockles**. These are not actually clams. They are from a different family, even though they look like tiny clams that are usually no more than 1 in. (2.5 cm) across. They can be cooked like clams and are almost always served in the shell.

Market Forms (Hard-Shell Clams)

1. Live, in the shell.

2. Shucked, fresh or frozen.

3. Canned, whole or chopped.

Checking Freshness

Same as for oysters. Clams in the shell must be alive. Live and shucked clams should smell fresh.

Opening Clams

1. Scrub shells thoroughly before opening.

2. Hard-shell clams are sometimes sandy inside, and soft-shell clams nearly always are. They can be flushed as follows:

   • Make a salt brine, using ¹/₃ cup salt per gallon of water (20 mL salt per liter).

   • Soak the clams in the brine 20 minutes.

   • Drain and repeat until the clams are free of sand.

   • Some chefs put cornmeal in the water and refrigerate the clams in it for a day. The clams eat the cornmeal and expel the sand.

   • Rinse in fresh water before using.

3. Opening hard-shell clams is different from opening oysters. This technique is illustrated in Figure 19.8.

4. Like oysters, clams to be cooked may be opened by spreading on sheet pans and placing in a hot oven just until the shells open. Discard any that do not open because they were likely dead before cooking.

Storage

Same as for oysters.

Cooking Clams

1. Clams become tough and rubbery if overcooked. Cook just enough to heat through.

2. When steaming clams in the shell, steam just until shells open.

3. Cooking methods: Steaming, poaching, deep-frying, baking on the half-shell with toppings, simmering in soups and chowders.

MUSSELS

Characteristics

The most common mussels resemble small black or dark-blue clams. Their shells are not as heavy as clamshells. Their flesh is yellow to orange in color and firm but tender when cooked. Mussels are harvested worldwide and are also extensively farmed.

Green mussels, from New Zealand and Southeast Asia, are larger and have a lighter-colored gray or tan shell with a green edge. Green mussels often command premium prices.

FIGURE 19.8 Opening clams.

(a) Examine the shell to see that it is tightly closed, indicating a live clam. Rinse the shell under cold running water. Avoid jostling the clam too much, or it will "clam up" tighter. Hold clam in left hand as shown (or in right hand if you are left-handed). Place the sharp edge of the clam knife against the crack between the shells.

(b) Squeeze with the fingers of the left hand, forcing the knife between the shells.

(c) Change the angle of the blade as shown in the illustration and slide the knife against the top shell to cut the adductor muscles (clams have two; oysters have only one). Be careful not to cut or pierce the soft clam.

(d) Open the clam and finish detaching the meat from the upper shell.

(e) Cut the muscles against the lower shell to loosen the clam completely. Discard the top shell. Remove any particles of shell from the clam before serving.

Market Forms

Most mussels are sold alive in the shell. Many are also sold shucked and packed in brine.

Checking Freshness

1. Like oysters and clams, mussels must be alive to be good to eat. Check for tightly closed shells or shells that just close when jostled.

2. Discard any mussels that are very light in weight or seem to be hollow. Also, discard any that are much too heavy—they are probably full of sand.

Mussels, left to right: green, blue (wild), blue (farm-raised)

Cleaning

1. Clean shells thoroughly:
   - Scrub well under cold running water.
   - Scrape off barnacles, if any, with a clam knife.
   - Remove the beard, a fibrous appendage protruding from between the shells (see Figure 19.9). Do not do this until just before cooking, because it may kill the shellfish.

New Zealand green lips

2. Mussels may be sandy inside if not commercially grown. They may be soaked in brine and flour or cornmeal like clams (see previous section) to rid them of sand.

FIGURE 19.9 When cleaning mussels, pull off the fibrous beard that extends from between the shells.

Storage

Keep refrigerated at 32°–35°F (0°–2°C) and protect from light. Store in original sack and keep sack damp.

Cooking

Unlike oysters and clams, mussels are almost never served raw. They are usually steamed and served in their cooking broth, in soups, or chilled and served with mayonnaise-type sauces. Cook only until shells open and mussels are heated through. Do not overcook. Discard any that are not open after cooking.

SCALLOPS

Characteristics

1. Scallops are almost always sold shucked. The only part we usually eat is the adductor muscle, which closes the shell. If live scallops in the shell are available, leave the orange, crescent-shaped coral attached to the adductor muscle when shucking.

2. There are two main kinds of scallop:

    Bay scallops: Small, with delicate flavor and texture; expensive; 32 to 40 per lb (70 to 88 per kg) on average.

    Sea scallops: Larger, not as delicate as bay scallops, but still tender unless overcooked; 10 to 15 per lb (22 to 33 per kg) on average.

3. Scallops are creamy white in color and have a sweet flavor.

4. They are available all year.

Scallops with roe or coral

Market Forms

1. Fresh, shucked. Sold by volume or weight.

2. Frozen.

    • IQF (individually quick frozen).

    • In 5-lb (2.3-kg) blocks.

Checking Freshness

A sweet, clean smell is a sign of freshness. Strong fishy odor or a brownish color is a sign of age or spoilage.

Handling

1. Shucked scallops can be cooked without further preparation. They are improved, however, if you pull off the small, tough tendon or sinew on the side of each, as shown in Figure 19.10.

2. Large sea scallops are sometimes cut into smaller pieces before cooking.

Storage

Keep scallops covered and refrigerated at 30° to 34°F (–1° to 1°C). Do not let them rest directly on ice, or they will lose flavor and become watery.

Cooking Scallops

Scallops are cooked in almost every way that fish are cooked. The most popular methods are sautéing, deep-frying, broiling, and poaching.

FIGURE 19.10 Removing the tendon from the side of a scallop.

CEPHALOPODS

Cephalopods are classified as mollusks, even though they have no external shell. The term *cephalopod* means "head-foot," referring to the fact that these animals have tentacles, or "legs," attached to the head and surrounding the mouth. The most important cephalopods in the North American kitchen are squid and octopus. A third type, cuttlefish, is similar to squid but is usually seen only in limited markets.

SQUID

Squid are usually referred to on menus by their Italian name, **calamari**. They are soft-bodied animals somewhat resembling octopus, but they have 10 tentacles, 2 of them longer than the others.

Squid must be skinned and eviscerated. The head, beak, and the internal plastic-like quill are discarded. The hollow body and tentacles are eaten. Figure 19.11 illustrates how to clean squid.

Somewhat chewy, squid are cut up and either fried quickly or simmered about 45 minutes in a seasoned liquid or sauce.

Note: The similar cuttlefish has a shorter, thicker body than does squid. Instead of a thin, transparent interior quill, it has a hard, chalky cuttlebone.

OCTOPUS

Octopus (the name means "eight feet") range in size from less than an ounce to many pounds. All sizes are firm-textured, even chewy, but the larger sizes are usually considered too tough to eat. Because of its texture, octopus requires either mechanical tenderization (such as pounding with a mallet) or long, slow cooking in a court bouillon.

Like squid, octopus is cleaned by cutting off the tentacles, discarding the head and beak, and eviscerating the body cavity. Pull the skin off the body pouch and tentacles; it may be necessary to parboil the octopus a few minutes to loosen the skin. The skin is reddish gray, turning purple-red when cooked.

FIGURE 19.11 Cleaning squid.

(a) Pull off the head. The interior organs will come out with it.

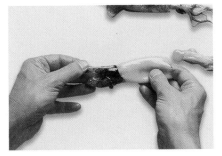
(b) Pull off the skin.

(c) Pull out the plastic-like quill from the body sac. Rinse out the sac to clean it well.

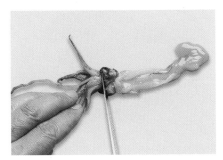
(d) Cut off the tentacles just above the eyes. Discard the head and organs.

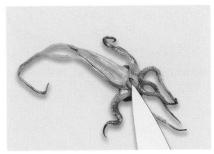
(e) Be sure to remove the hard beak, which is found at the center of the tentacle cluster, as shown by the tip of the knife in this picture.

(f) The body sac may be left whole for stuffing or cut into rings for frying, sautéing, or stewing.

CRUSTACEANS

The most important crustaceans in commercial kitchens are lobsters, rock lobsters or langoustes, shrimp, and crabs.

LOBSTERS

Characteristics

FIGURE 19.12 Splitting a lobster for broiling.

1. The northern lobster is perhaps the most prized of all shellfish. It has a large, flexible tail, four pairs of legs, and two large claws. Its shell is dark green or bluish green but turns red when cooked.

2. Meat from the tail, claws, and legs is eaten. It is white and sweet, with a distinctive taste. Claw meat is considered especially good. The coral (roe or eggs), which is dark green when raw and red when cooked, and the light green tomalley (liver) in the thorax or body portion are also eaten.

3. Lobsters are classified by weight.

Chicken	1 lb (450 g)
Quarters	1¼ lb (575 g)
Halves or Selects	1½ to 2¼ lb (675 to 1025 g)
Jumbos	over 2½ lb (1130 g)

(a) Place the lobster on its back on a cutting board. With a firm thrust of a French knife, pierce the head to kill the lobster quickly.

(b) Bring the knife down firmly through the center of the lobster to split it in half.

4. Lobsters weighing over 20 lb (9 kg) have been caught, but they are rare.

5. Yield: A 1-lb lobster yields about ¼ lb cooked meat, or a 500-g lobster yields about 125 g cooked meat.

6. Sometimes customers request female lobsters in order to get the coral, so you should be able to tell females from males. Look at the pairs of tiny legs (called swimmerets) under the tail. If the pair closest to the front is soft and flexible, the lobster is female. If it is hard, the lobster is male.

Market Forms

1. Live.
   Most live lobsters have both claws intact. A lobster with only one claw is called a *cull.* A *pistol* is missing both claws.

(c) With the hands, crack the back of the shell by spreading the lobster open.

(d) Pull out and discard stomach, a sac just behind the eyes.

(e) If desired, remove the tomalley for use in the crumb stuffing.

(f) With a sharp blow of the back of the knife, crack the claws.

(g) The lobster is ready for broiling. If it is broiled as shown, the end of the tail should be weighted to keep it from curling.

(h) You may also split the tail all the way through and curl up the two sides as shown. In this position, weighting the tail is not necessary. Note the claws have been broken off and placed beside the lobster.

FIGURE 19.13 Cutting a lobster for sautés and stews.

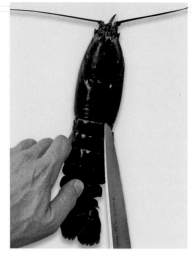

(a) Place the lobster on the cutting board. Pierce the head with a firm thrust of the knife point to kill the lobster quickly.

(b) Cut off the legs and claws.

(c) Remove the tail section from the thorax, either by breaking it off or by inserting the knife behind the thorax as shown and cutting through the flesh.

(d) Cut the thorax in half lengthwise.

(e) Remove and discard the stomach, a sac just behind the eyes.

(f) Remove the tomalley and coral for use in the sauce to accompany the lobster.

2. Cooked meat, fresh or frozen.

Hot pack refers to canned, retorted (heated) lobster that is shelf-stable.

Cold pack refers to canned lobster that has not been retorted and so must be kept frozen.

A *popsicle pack* is a whole cooked lobster packed with brine in a plastic sleeve and frozen.

The term *CK*, or *claws and knuckles*, indicates cooked meat from claws and legs only, no tail meat.

(g) Cut the tail into pieces where the segments join. This is a small lobster. Large tails should be cut into more pieces (at least four or five) so each piece is not too large.

Checking Freshness

1. Live lobster must be alive when cooked. This is indicated by movement of the legs and claws and by a tightly curled tail.
2. If the lobster is dead when cooked, the meat will fall apart. If it is alive, the meat will be firm and the tail will spring back when straightened.
3. Sleepers (dying lobsters, alive but sluggish and unable to hold up their claws) should not be purchased. Already purchased lobsters in your storage that become sleepers should be cooked immediately so the meat will still be usable.
4. Cooked lobster meat (fresh or thawed) should smell fresh and sweet.

Handling

1. Live lobsters are either cooked live or cut up before cooking (as for broiling or sautéing). Live lobsters are plunged headfirst into boiling water to kill them instantly. They are then simmered 5 to 6 minutes per lb (500 g). If served hot, they are drained well and split in half, and the claws are cracked.

(h) The cut-up lobster, ready to cook.

2. Splitting and cutting up live lobsters is necessary for certain preparations. Figure 19.12 shows splitting a lobster for broiling. Figure 19.13 shows cutting up a lobster for sautés and stews. Note that the methods are somewhat different.

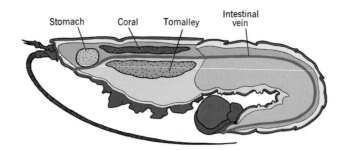

FIGURE 19.14 Cross-section of a female lobster showing the location of the stomach, tomalley, coral, and vein.

In addition to the many terms in the text, you might also see the following terms in connection with lobster:

Berried lobster: A female lobster with eggs under its tail. By law, these lobsters must be returned to the sea and not harvested.

Crusher: The larger of the two claws.

Hard-shell: A lobster whose shell has fully hardened after molting (see Shedder).

Pincher: The smaller of the two claws.

Shedder: A lobster that is in the molting stage. Lobsters must shed their shells and then grow new ones in order to increase size.

Soft-shell: A lobster after it has molted, before the new shell has hardened.

3. The thorax section contains four parts you should recognize. Figure 19.14 shows their location.

- The stomach, or sac, located just behind the eyes, is often sandy and should be discarded.
- The **tomalley**, or liver, is pale green. It is eaten plain or removed, mashed, and added to sauces that accompany the lobster.
- The **coral**, or roe, is red when cooked and dark green when raw. It is present only in females. Like the tomalley, it is considered a delicacy.
- The intestinal vein may be removed from the split lobster if it contains dark material, but is often left in if it is clear.

Storage

1. Live lobsters can be kept in two ways:
   - Packed in moist seaweed or in moist, heavy paper, kept in a cool place.
   - In salt water. Special lobster tanks are used in restaurants for display and so customers can select their own lobster. Air must be bubbled through the water to keep the lobsters alive.
2. Cooked lobster meat must be covered and refrigerated at 30° to 34°F (−1° to 1°C). It is very perishable and should be used in 1 or 2 days.

Cooking Lobsters

1. Lobster meat becomes tough if cooked at too high a temperature or for too long. Boiling too long will also make the meat dry. Whole lobster is usually cooked by dropping into boiling water and then simmering 5 to 6 minutes per lb (500 g). For jumbo lobsters, reduce the cooking time slightly.
2. Whole or cut-up lobster can be cooked by simmering in water or court bouillon, simmering in sauce or soup, sautéing, baking, or broiling.

ROCK LOBSTERS

Characteristics

1. Rock lobsters are also known as *spiny lobsters* or *langoustes*. They are warm-water relatives of northern lobsters but have no claws. Only the tails are marketed, sold as lobster tails.
2. The flesh of the rock lobster tail is similar to that of the northern or Maine lobster, but it is drier and coarser, with less flavor.
3. Rock lobster tails weigh 2–12 oz (60–340 g).
4. **Langoustines**, or **langostinos**, are smaller relatives of the rock lobster. These small shellfish are often marketed as **rock shrimp**. When out of the shell, they look like shrimp, but their flavor is milder and sweeter than shrimp. **Scampi** (plural form of *scampo*) refers not to shrimp but to a variety of langoustine from Italian waters. The name *scampi* is often used, incorrectly, for large shrimp broiled with butter and garlic.

Market Forms

Nearly all rock lobster tails are sold IQF (individually quick frozen).

Handling and Cooking

1. Rock lobsters are handled and cooked much like northern lobsters. The most common cooking methods are steaming, simmering, and broiling.

2. Tails to be broiled will be moister if poached 5 minutes before splitting and broiling.

3. Tails steamed or simmered whole should have the shells split before serving as a convenience to the customer.

SHRIMP

Characteristics

1. Shrimp are small crustaceans that look somewhat like tiny, clawless lobsters. Only the tail is marketed and eaten, as a rule.

2. Shrimp come in many varieties, depending on where they are caught, but the particular variety is usually of little importance to the cook.

3. Shrimp are classified by count per pound—the higher the count, the smaller the shrimp. (For example, 16/20 means 16 to 20 per pound.) Classification systems differ by market—that is, in different markets a given size shrimp may have different names.

4. Large shrimp are more expensive per pound but require less work to peel and devein.

5. Yield: 1 lb raw shrimp (tails) in the shell yield about $1/2$ lb peeled, cooked shrimp (500 g raw yields 250 g peeled, cooked).

6. The term **prawn** is sometimes used for large shrimp, sometimes for langoustines (see above). Use of the term varies from region to region.

Market Forms

1. *Green shrimp* are raw shrimp in the shell.

   Fresh: Not widely available, except near source of supply.

   Frozen: In 5-lb (2.3-kg) blocks.

2. P/D (peeled, deveined): Usually IQF (individually quick frozen).

3. PDC (peeled, deveined, and cooked): Usually IQF.
   Note: IQF shrimp are usually glazed (see p. 599).

Checking Freshness

1. Frozen shrimp should be solidly frozen when received.

2. Glazed shrimp should be shiny, with no freezer burn.

3. All shrimp should smell fresh and sweet. A strong fishy or iodine smell indicates age or spoilage.

Storing

1. Like other frozen fish, shrimp should be kept frozen at 0°F (–18°C) or lower until ready for use.

2. Thaw in refrigerator, allowing sufficient slack time.

3. Fresh or thawed shrimp in the shell are stored on crushed ice, like whole fish.

4. Peeled shrimp lose soluble nutrients and flavor when stored unwrapped on ice. They should be wrapped before placing on ice or covered and simply refrigerated.

FIGURE 19.15 Peeling and deveining shrimp.

(a) Pull off the legs with your forefinger.

(b) Peel back the shell as shown and remove.

(c) For deep-fried and broiled shrimp, leave on the tail section of the shell for appearance. This also gives you something to hold when dipping the shrimp in batter.

Handling

1. Shrimp served hot must normally be peeled and deveined before cooking. Figure 19.15 shows how.
2. Shrimp to be served cold may be peeled after cooking to preserve flavor.
3. Large shrimp are sometimes butterflied, as shown in Figure 19.15(f). This is done for appearance (it makes shrimp seem larger, with more surface area for breading) and to speed cooking by reducing thickness.

Cooking Shrimp

Like most shellfish, shrimp become tough and rubbery when cooked at too high a heat. Shrimp can be cooked by simmering, deep-frying, sautéing, broiling, and baking.

CRABS

Six kinds of crab are important in commercial kitchens.

1. **Alaskan king crab.**
   Largest of the crabs, weighing 6 to 20 lb (2.7 to 9 kg). The meat can be removed in large chunks, making it especially attractive to serve in restaurants. It is expensive.
2. **Alaskan snow crab.**
   Smaller than the king crab. Often used as a less expensive substitute.
3. **Dungeness crab.**
   Another West Coast crab, weighing 1 1/2 to 4 lb (0.7 to 1.8 kg). The meat is very sweet.
4. **Blue crab.**
   Small crab from the East Coast, weighing about 5 oz (150 g). Most frozen crabmeat is from blue crabs.
5. **Soft-shell crab.**
   Actually a molting blue crab, harvested before the new shell has hardened. It is sautéed or fried and eaten shell and all; only the gills and head are removed.
   Soft-shell crabs must be cleaned before being cooked. The procedure is shown in Figure 19.16.
6. **Stone crab.**
   Popular in the Southeastern United States. Only the claws are eaten.

Market Forms

1. Live. Crabs taste best when fresh, but very few (except soft-shell crabs) are purchased live because of the labor required to pick the meat. An average blue crab yields less than 1 oz (30 g) meat.

(d) For most other preparations, remove the tail section of the shell.

(e) With a paring knife, make a shallow cut down the back of the shrimp and pull out the intestinal vein, which is just below the surface.

(f) To butterfly shrimp, make the cut in step (e) deeper so the shrimp can be spread open as shown.

FIGURE 19.16 Cleaning soft-shell crabs.

(a) Soft-shell crabs, seen from the bottom and top.

(b) Cut off the head just behind the eyes.

(c) Pull out the stomach sac.

(d) Pull back one side of the soft top shell to reveal the feathery gills.

(e) Pull off the gills. Repeat on the other side.

(f) Cut off the apron on the underside. The wide apron indicates this is a female crab. Males have a much narrower apron.

2. Cooked, frozen, in the shell.

    King crab legs, whole and split.

    Snow and stone crab claws.

    Soft-shell crabs, whole.

3. Cooked, frozen meat. All varieties.

Freshness and Storage

1. Live crabs should be kept alive until cooked. They are packed in damp seaweed and kept cool.

2. Frozen crabmeat should be treated like any other frozen fish. It is very perishable when thawed.

Handling and Cooking

1. Hard-shell crabs are picked of their meat after cooking.

- Simmer 10 to 15 minutes in salted water ($1/2$ cup salt/gal or 30 mL/L). Cool rapidly in ice water.
- Break off the pointed shell on the underside (called the apron) and remove the top shell.
- Remove and discard the spongy gills and the stomach, which is just behind the eyes.
- Pick out the body meat.
- Crack the legs and claws with a mallet or the back of a heavy knife and pick out the meat.

Blue crab

Dungeness crab

Alaskan king crab legs

Alaskan snow crab legs

2. Soft-shell crabs.
   * With a knife or shears, cut off the head section, just behind the eyes (Figure 19.16).
   * Lift the pointed, outside corners of the top shell and pull out the spongy gills.
   * Cut off the apron, the small flap on the underside.
   * Dredge the crab in flour for sautéing, or bread or batter it for deep-frying.

3. Frozen crabmeat.
   * Crabmeat is usually watery. You may need to squeeze out excess moisture before cooking. Whenever possible, save the liquid for use in sauces and soups.
   * Frozen crabmeat is already cooked. It needs only to be heated through to be prepared for serving.

CRAYFISH

Crayfish or *crawfish* (the preferred term in the southern United States) are freshwater relatives of the lobster. Not long ago, they were used almost exclusively in Southern regional cuisine and in French restaurants. With the spreading popularity of Southern cuisines, including Creole and Cajun, they have become more widely available.

Crayfish are marketed live or frozen (peeled tail meat or whole).

MISCELLANEOUS SEAFOOD

Several other seafood items play a role in food-service kitchens. They are classified with fish, even though some of them spend part or all of their lives on land, like frogs and snails.

Snails or escargots are popular hors d'oeuvres when baked in their shells with a highly seasoned butter. Fresh snails require long, slow cooking before being baked with escargot butter, but canned snails are fully cooked and ready to prepare. Canned snails can be improved, however, by first simmering them in white wine and seasonings.

Frogs' legs are often compared to chicken in taste and texture (but then, so are a lot of things). Only the hind legs are used, and they are sold in pairs. They may be sautéed, deep-fried, or poached and served with a sauce.

Surimi is a processed seafood product made by grinding lean, inexpensive white fish, combining it with flavorings, and forming it into various shapes. The most popular shapes are shredded crab and crab leg segments, but other items, such as lobster claws, are also available. Coloring is added for a realistic appearance. Originally from Japan, surimi has found growing acceptance in North America because of the reasonably good quality of the product and the high cost of crab and lobster.

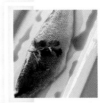

KEY POINTS TO REVIEW

- What are the four most important kinds of shelled mollusk in food service? Describe the characteristics and handling of each kind.

- What are the two most important kinds of cephalopod in food service? Describe the characteristics and handling of each kind.

- What are five kinds of crustacean used in food service? Describe the characteristics and handling of each kind.

TERMS FOR REVIEW

fin fish	anadromous	crustacean	scampi
shellfish	catadromous	cockle	prawn
flaking	drawn	calamari	crayfish
lean fish	mollusks	tomalley	surimi
fat fish	bivalve	coral	
flatfish	univalve	langoustine/langostino	
round fish	cephalopod	rock shrimp	

QUESTIONS FOR DISCUSSION

1. Fish has very little connective tissue. How does this affect the ways in which you handle it and cook it?

2. Based on what you learned about fat and lean fish and about individual species of fish, suggest at least one cooking method for each of the following.

Mackerel	Salmon
Cod	Perch
Flounder	Swordfish
Trout	Whitefish
Red snapper	Halibut

3. List and describe the major market forms of fresh fish.

4. What are the differences between filleting flatfish and round fish? Describe or demonstrate.

5. You have just received delivery of fresh whole red snapper and fresh cod fillets. What should you check before accepting the shipment? After accepting the fish, what do you do with them?

6. You are making a casserole of shrimp with a prepared Newburg sauce and frozen, glazed cooked shrimp. When you add the frozen shrimp to the sauce, it thins out so much it is no longer usable. Why did this happen, and how could you have prevented it?

7. What is the most important indication of freshness in fresh oysters, clams, lobsters, and crabs? Describe how you would preserve this freshness.

8. How does opening oysters differ from opening clams, and how is it similar?

9. What happens to most kinds of shellfish when they are overcooked?

20

COOKING FISH AND SHELLFISH

Because most fish and shellfish have little connective tissue, the procedures for cooking them are somewhat different from those for cooking meats and poultry. When we cook meats, we are often concerned with tenderizing the product. Fish is naturally tender, however, and cooking—especially overcooking—is more likely to toughen the product and dry it out.

In this chapter, we apply the basic cooking procedures explained in Chapter 14 to the preparation of fish dishes. Our goals are to cook the product just to doneness, to preserve moisture and texture, and to preserve and enhance natural flavors.

Of course, your study of these procedures depends on your understanding of the basic information on fish in the previous chapter. In particular, you may want to review the sections on testing for doneness and on fat and lean fish and their appropriate cooking methods.

AFTER READING THIS CHAPTER, YOU SHOULD BE ABLE TO

1. Cook fish and shellfish by baking.

2. Cook fish and shellfish by broiling and grilling.

3. Cook fish and shellfish by sautéing and pan-frying.

4. Cook fish and shellfish by deep-frying.

5. Cook fish and shellfish by submersion poaching, shallow poaching, and simmering.

6. Cook fish and shellfish by steaming, sous vide, and mixed cooking techniques.

7. Prepare dishes made of raw seafood.

BAKING

Whole fish and fish portions may be cooked by baking in an oven. Although large whole fish may be baked, this is usually not practical in volume food service. The method is more often used with steaks and fillets and with small fish. Baking is also a popular method for preparing shellfish such as stuffed clams or oysters.

Baking is often combined with other cooking methods. For example, partially broiled fish can be finished by baking. Baked fish casseroles are usually made with cooked fish.

Whole fish or portions of fish baked in the oven may also be referred to as roasted, as is currently the fashion. The term *roasting* may be applied both to basic baking and to moist baking, discussed below.

The basic roasting and baking procedures are explained in detail in Chapter 14, and you should review pages 414–417 as necessary before continuing with this section. However, baking fish presents some unique problems unlike those of roasting meat and poultry. The guidelines and procedures below outline the techniques used for baking fish.

In addition, you should review the procedure for pan smoking (range-top smoke roasting) on page 539 before working with the recipe for Pan-Smoked Salmon on page 620.

GUIDELINES for Baking Fish

1. Fat fish are better for baking because they are less likely to dry out.

2. Lean fish may be baked successfully if great care is taken not to overcook them. Basting with butter or oil helps prevent drying.

3. In most cases, baking temperatures are from 350° to 400°F (175° to 200°C). Large fish are best baked at the lower end of this range so they bake more evenly.

   It is also possible to bake thin fillets or slices of fillets (cut like scaloppine) at very high temperatures. Great care must be used in this case because the fish may cook in a minute or less, and a few seconds too long might ruin it. The effect of the high heat is almost like broiling, and the normal guidelines for broiling thin fish fillets should be followed, as explained in the next section.

4. It is not possible to give specific baking times because fish vary so much in shape and composition. Different ingredients and types of baking pans also affect the baking time. The following guideline is helpful, however: Measure the thickness of the fish at the thickest point. At 400°F (200°C), baking time is about 10 minutes per inch (2.5 cm) of thickness.

5. Serving baked fish with a sauce or seasoned butter enhances its moistness and improves palatability. Serving with lemon also enhances the fish.

6. If fish is baked with a moist topping or sauce, strictly speaking, it is no longer being cooked by a dry-heat method. However, because the basic procedure is the same, the fish is treated as if it were.

PROCEDURE for Baking Fish

1. Collect all equipment and food supplies.

2. Prepare and season fish (whole, steaks, fillets) as directed in the recipe.

3. Place the fish on oiled or buttered baking sheets. Brush tops with oil or butter.

   Alternative method: Dip fish in oil or melted butter to coat both sides. Place on baking sheets.

4. Apply toppings, if desired. Examples: seasoned bread crumbs, lemon slices, mushrooms or other vegetable garnish, and sauces.

5. Bake at 350° to 400°F (175° to 200°C) until done. If the fish is lean and has no moist topping, baste with oil or butter during baking.

MOIST BAKING OR BRAISING

As we already noted above, if fish is baked with moist ingredients or with liquids, strictly speaking, the method is no longer a dry-heat method, although such preparations are included in this section. In French cooking, baking fish—usually whole fish or large pieces—with vegetables and liquid is called *braising*. To avoid confusion with the braising method as applied to meats, however, we avoid that term and refer to such dishes as *baked*.

This procedure is the same as the basic procedure for baking fish described above, with the following special features:

1. The baking pan should be just large enough to hold the fish, so you will not need too much liquid.

2. The bottom of the pan is buttered or oiled, then covered with a layer of sliced or chopped vegetables, such as carrots, onions, shallots, and mushrooms. The vegetables may be raw or first sautéed gently in butter or oil. The fish is then placed on top of the vegetables.

3. Frequently, a small amount of liquid, such as equal parts wine and fish stock, is added. Just enough liquid is used to cover the fish about halfway or less. During baking, the fish is basted with this liquid.

4. The fish may be baked uncovered or covered only lightly. It should not be covered tightly, however, because the liquid must be able to reduce somewhat so it will become more concentrated and more flavorful. Remember also that liquid will be released from the fish, diluting the cooking liquid.

5. For service, the fish is removed from the dish. The liquid is strained, degreased, reduced, and finished in various ways, such as by adding butter, cream, or velouté sauce.

You can see that this is a sort of combination technique. Because it often uses wine and other liquids, the method is, in some ways, similar to shallow poaching, as explained on page 437.

 # Baked Cod Fillets Portugaise

PORTIONS: 12   PORTION SIZE: 5 OZ (150 G) FISH, 2 FL OZ (60 ML) SAUCE

U.S.	METRIC	INGREDIENTS	PROCEDURE
12	12	Cod fillets, 5-oz (150-g) portions	1. Place cod fillets on a well-oiled baking sheet or baking pan, flesh side up (that is, skin side down).
1¹⁄₂ fl oz	45 mL	Lemon juice	2. Brush the fish lightly with lemon juice. Then brush generously with the butter or oil and season lightly with salt and pepper.
4 fl oz	120 mL	Melted butter or oil	
to taste	to taste	Salt	3. Place the pan in a preheated 350°F (175°C) oven until done, about 10–15 minutes.
to taste	to taste	White pepper	4. Halfway through the cooking time, check the fish and, if the tops appear to be drying out, brush with more butter or oil.
1¹⁄₂ pt	720 mL	Portugaise Sauce (p. 189)	5. Serve each portion with 2 fl oz (60 mL) sauce. Nap the sauce across the center of the portion. Do not cover the entire fillet. For an alternative plating, see photo.

Per serving: Calories, 170; Protein, 16 g; Fat, 10 g (53% cal.); Cholesterol, 60 mg; Carbohydrates, 4 g; Fiber, 1 g; Sodium, 250 mg.

VARIATIONS

Many other fish may be baked according to the basic recipe, such as

Haddock (fillets or steaks)   Pike (fillets)
Halibut (steaks or fillets)   Salmon (fillets or steaks)
Snapper (fillets)   Perch (fillets)
Bluefish (fillets)   Swordfish (steaks)
Bass (fillets)   Flounder (fillets)
Mackerel (fillets)   Whitefish (fillets)

Other appropriate sauces may be used, such as

Melted butter   Mustard (for strong-flavored fish only,
Beurre noisette   such as mackerel or bluefish)
Maître d'hôtel butter   Curry (not for salmon or other fat fish)
Tomato and tomato-based
sauces such as Creole
(not for salmon or for very
delicate fish like flounder)

Baked Cod Fillet Portugaise with a
bouquet of vegetables

Baked Fish with Tomatoes and Mushrooms

PORTIONS: 10   PORTION SIZE: 1 FISH

U.S.	METRIC	INGREDIENTS	PROCEDURE
10	10	Small whole fish, about 12 oz (375 g) each (see Note)	1. Scale and clean the fish, but leave the heads on. Season the fish inside and out with salt and pepper, and put a small pinch of thyme and a sprig of parsley in the cavity of each.
to taste	to taste	Salt	
to taste	to taste	Pepper	
pinch	pinch	Dried thyme	
10	10	Parsley sprigs	
as needed	as needed	Olive oil	2. Select as many baking pans as necessary to just hold the fish in a single layer. Oil the pans with a little olive oil.
8 oz	250 g	Onion, small dice	3. Sauté the onions and shallots in a little olive oil about 1 minute. Add the mushrooms and sauté lightly.
1 oz	30 g	Shallots, minced	
8 oz	250 g	Mushrooms, chopped	4. Put the sautéed vegetables and the tomatoes in the bottoms of the baking pans.
1 lb	500 g	Tomato concassé	5. Put the fish in the pans. Oil the tops lightly. Pour in the wine.
8 fl oz	250 mL	Dry white wine	

6. Bake at 400°F (200°C) until the fish is done. The time will vary but will average 15–20 minutes. Baste often with the liquid in the pan.

7. Remove the fish and keep them warm until they are plated.

8. Remove the vegetables from the pans with a slotted spoon and check for seasonings. Serve a spoonful of the vegetables with the fish, placing it under or alongside each fish.

9. Strain, degrease, and reduce the cooking liquid slightly. Just before serving, moisten each portion with 1–2 tbsp (15–30) mL of the liquid.

Per serving: Calories, 350; Protein, 55 g; Fat, 9 g (24% cal.); Cholesterol, 120 mg; Carbohydrates, 6 g; Fiber, 1 g; Sodium, 210 mg.

Note: Many types of fish can be used, including sea bass, red snapper, porgy, perch, and trout. As an alternative, use thick steaks or thick pieces of fillet from larger fish such as cod or tilefish.

VARIATIONS

Baked Fish à la Menagère

Use butter instead of olive oil. Substitute sliced leeks for part or all of the onion. Omit the tomatoes. Add 4 oz (125 g) sliced carrot and cook it with the leek. Slice the mushrooms instead of chopping them, and add them to the pan raw. Add 1 cup (250 mL) fish stock along with the wine. After straining and reducing the cooking liquid in step 9, thicken it very lightly with a little beurre manié. Enrich the sauce with a little raw butter or cream.

Baked Fish with Tomatoes and Mushrooms

Baked Pike Fillets English Style

PORTIONS: 12   PORTION SIZE: 1 FILLET, 5 – 6 OZ (150–180 G)

U.S.	METRIC	INGREDIENTS	PROCEDURE
12	12	Pike fillets, 5–6 oz (150–180 g) each	1. Brush the fillets lightly with lemon juice and season with salt and pepper.
1 fl oz	30 mL	Lemon juice	2. Dip both sides of each fillet in the butter and then in the bread crumbs. Press the crumbs on lightly so they adhere.
to taste	to taste	Salt	
to taste	to taste	White pepper	3. Place the fillets flesh side up (skin side down) on a sheet pan (if cooking in quantity) or on sizzler platters (if cooking to order).
5 oz	150 g	Melted butter	
12 oz, or as needed	360 g, or as needed	Bread crumbs, dry	4. Place in an oven heated to 400°F (200°C) and bake until fish is done and crumbs are lightly browned.
12 fl oz	360 mL	Herb Butter Sauce (p. 192, optional)	5. Place the fish on hot plates. Spoon a thin band of butter sauce onto the plate partway around the fish.
12	12	Lemon wedges	6. Place a lemon wedge on each plate and serve immediately.

Per serving: Calories, 350; Protein, 33 g; Fat, 14 g (36% cal.); Cholesterol, 155 mg; Carbohydrates, 22 g; Fiber, 1 g; Sodium, 370 mg.

Baked Clams Oreganata

PORTIONS: 10   PORTION SIZE: 3 CLAMS (APPETIZER PORTION)

U.S.	METRIC	INGREDIENTS	PROCEDURE
30	30	Cherrystone clams	1. Open the clams (see **Figure 19.9** for technique). Catch the juice in a bowl.
			2. Remove the clams from the shell. Place them in a strainer over the bowl of juice. Let them drain 15 minutes in the refrigerator. Save the 30 best half-shells.
			3. Chop the clams into small pieces.
2 fl oz	60 mL	Olive oil	4. Heat the oil in a sauté pan. Add the onion and garlic. Sauté about 1 minute, but do not brown.
1 oz	30 g	Onions, shallots, or scallions, chopped fine	
1 tsp	5 mL	Finely chopped garlic	5. Add half the clam juice and reduce by three-fourths over high heat.
1 fl oz	30 mL	Lemon juice	6. Remove from the heat and add the lemon juice, crumbs, parsley, oregano, and white pepper. Mix gently to avoid making the crumbs pasty.
10 oz	300 g	Fresh bread crumbs	
1 tbsp	15 mL	Chopped parsley	
3/4 tsp	3 mL	Dried oregano	7. Taste and adjust seasonings if necessary. (Clams are usually very salty.)
1/8 tsp	0.5 mL	White pepper	8. Cool the mixture. Mix in the chopped clams.
1/3 cup	25 g	Parmesan cheese	9. Fill the 30 clamshells with the mixture. Sprinkle with parmesan cheese and (very lightly!) with paprika.
as needed	as needed	Paprika	10. Place on a sheet pan and refrigerate until needed.
10	10	Lemon wedges	11. For each order, bake 3 clams in a hot oven (450°F/230°C) until they are hot and the top is brown.
			12. Garnish with lemon wedge.

Per serving: Calories, 180; Protein, 10 g; Fat, 8 g (41% cal.); Cholesterol, 20 mg; Carbohydrates, 16 g; Fiber, 1 g; Sodium, 230 mg.

Note: Clams and oysters are often baked on a bed of rock salt to hold them steady. The rock salt also holds heat well.

Baked Clams Oreganata

Pan-Smoked Salmon Fillet with Pepper Salad

PORTIONS: 12   PORTION SIZE: 4 OZ (125 G) FISH, 4 OZ (125 G) GARNISH

U.S.	METRIC	INGREDIENTS	PROCEDURE
3 lb	1.5 kg	Salmon fillets	1. Cut the salmon fillets into 4-oz (125-g) portions.
1 fl oz	30 mL	Vegetable oil	2. Brush the salmon lightly with oil.
3/4 tsp	3 mL	Ground coriander	3. Combine the spices and salt. Sprinkle over the fish to coat them in a light, even layer.
3/4 tsp	3 mL	Ground cumin	
1/4 tsp	1 mL	Ground cloves	4. Set up a smoke-roasting system as shown in **Figure 4.1**. Heat the pan of wood chips or sawdust on top of the stove until smoke appears. Lay the salmon fillets on the rack, cover, and turn the heat to medium-low. Smoke-roast for 2 minutes.
1/2 tsp	2 mL	Ground fennel	
1/2 tsp	2 mL	Black pepper	
1 tsp	5 mL	Salt	
			5. Transfer the pan to an oven preheated to 400°F (200°C) and continue roasting another 8–10 minutes, or until the fish is just cooked through.
3 lb	1.5 kg	Roasted Pepper Salad (p. 687)	6. Arrange the salmon fillets and the pepper salad on plates. Serve immediately.

Per serving: Calories, 370; Protein, 29 g; Fat, 25 g (60% cal.); Cholesterol, 70 mg; Carbohydrates, 9 g; Fiber, 2 g; Sodium, 470 mg.

Pan-Smoked Salmon Fillet with Pepper Salad

Baked Oysters with Balsamic Vinegar, Arugula, and Pine Nuts

PORTIONS: 12   PORTION SIZE: 3 OYSTERS (APPETIZER PORTION)

U.S.	METRIC	INGREDIENTS	PROCEDURE
36	36	Oysters	1. Open the oysters as shown in **Figure 19.8**. Arrange the oysters on a sheet pan.
1/2 oz	15 mL	Balsamic vinegar	2. Top each oyster with a few drops of balsamic vinegar, then with a little chopped arugula, a few pine nuts, and, finally, about 1/4 tsp (1.2 mL) olive oil. If the oysters are not salty, you may wish to add a little salt.
1 oz	30 g	Arugula, chopped	
1 oz	30 g	Pine nuts, toasted	
1 1/2 fl oz	45 mL	Olive oil	
			3. Bake at 450°F (230°C) for a few minutes, just until the oysters are hot. Serve immediately.

Per serving: Calories, 70; Protein, 3 g; Fat, 6 g (77% cal.); Cholesterol, 15 mg; Carbohydrates, 2 g; Fiber, 0 g; Sodium, 75 mg.

Roasted Monkfish with Spinach and White Beans

PORTIONS: 12   PORTION SIZE: 6–8 OZ (180–240 G) FISH, 3 OZ (90 G) BEANS, 2 OZ (60 G) SPINACH

U.S.	METRIC	INGREDIENTS	PROCEDURE
6–12	6–12	Garlic cloves	1. Cut the garlic cloves lengthwise into slivers.
12	12	Monkfish tails, boned, skinned, and trimmed, 6–8 oz (180–240 g) each	2. Lard the monkfish with the garlic by making slits in the fish with the point of a paring knife and inserting the slivers in them.
to taste	to taste	Salt	3. Season the fish with salt and pepper. Dust the fish lightly with flour and shake off excess.
to taste	to taste	White pepper	
as needed	as needed	Flour	4. Heat the butter and oil in a sauté pan. Brown the monkfish on all sides.
3 oz	90 g	Butter	
3 fl oz	90 mL	Olive oil	5. Add the wine and transfer the pan to an oven heated to 400°F (200°C). Roast 4–5 minutes, or until the fish is cooked. (*Note:* The pan must not be too large, or too much liquid will evaporate.)
12 fl oz	360 mL	White wine	
			6. Remove the fish from the pan and keep warm.
2 lb 4 oz	1.1 kg	Cooked white kidney beans, cannellini beans, or navy beans	7. Add the beans to the pan and set over moderate heat. Cook until the beans are hot and the liquid, if any, is reduced so that the beans are quite moist but without excess liquid. Adjust the seasonings. Add a little butter or olive oil if desired.
to taste	to taste	Butter or olive oil	
1 lb 8 oz	720 g	Cooked spinach, buttered	8. Spoon 3 oz (90 g) beans onto the center of each dinner plate and spread them out slightly.
			9. Place 2 oz (60 g) spinach in a small mound in the center of the beans.
			10. Cut each monkfish tail into 3 pieces and arrange on or around the spinach.

Per serving: Calories, 450; Protein, 36 g; Fat, 18 g (36% cal.); Cholesterol, 65 mg; Carbohydrates, 30 g; Fiber, 11 g; Sodium, 135 mg.

Roasted Monkfish with Spinach and White Beans

BROILING AND GRILLING

Broiled and grilled seafood items, like grilled meats, are increasingly popular. Customers perceive them as simpler and more healthful and, because they are prepared quickly, as fitting the faster pace of modern life.

Grilled dishes are, in their purest form, simple and straightforward, but they lend themselves to many variations in presentation. By varying sauces, vegetable accompaniments, and garnishes, you can offer a great assortment of grilled fish on the menu. And, because the sauces and accompaniments are generally prepared ahead of time, these dishes are still quick to prepare, even if the presentation is elaborate.

A slightly crisped, browned, or grill-marked surface is important to the appeal of grilled or broiled fish. Do not cover the item with sauce, and do not serve too much sauce. A small piece of seasoned butter or a drizzle of a flavorful condiment or sauce can be used to decorate the top of the item. In most other cases, it is better to place sauces on the side. Appropriate healthful sauces include vegetable coulis and salsas. For richer dishes, beurre blanc is a good accompaniment, especially for lean fish like halibut.

Another popular approach is to serve the fish without a sauce, except perhaps for a small dab of a condiment such as a relish or chutney, and to complement the fish with an attractive variety of vegetables.

Carefully review the explanation and the basic grilling and broiling procedure explained on pages 424–425. In addition, observe the following guidelines:

GUIDELINES for Broiling or Grilling Fish

1. Because of the intense heat of the broiler, great care is needed to avoid overcooking the fish.

2. Select appropriate fish for broiling or grilling. Small slices and thin fillets are usually too delicate to be placed directly on the grid of a broiler or grill. Arrange them on oiled sheet pans, broiler platters, or other baking dishes or pans, and cook under a broiler. Alternatively, place fish in hinged hand racks to protect them and to help keep them from breaking apart. Whole fish or thick slices of firm fish such as salmon, shark, and swordfish can be placed directly on the grill if care is taken to avoid breaking them up. Be sure the grid is clean and oiled to prevent sticking.

3. Fat fish is best for broiling because it doesn't get as dry as lean fish. However, all fish, fat or lean, should be coated with a fat before broiling to reduce drying.

4. Lean fish may be dredged in flour before dipping in oil or melted butter. The flour helps form a flavorful browned crust.

5. Instead of being dredged with flour, fish may be coated with fat and then with bread crumbs or cornmeal. Use

caution, however, as bread crumbs burn very easily. Use this technique only with items that cook quickly.

6. If the fish has an attractive skin (such as red snapper and black sea bass), the skin may be left on and used as the presentation side. This enhances appearance and also helps hold the delicate flesh together. Make sure the fish is thoroughly scaled. To prevent splitting during cooking, score the skin with a sharp knife, making a series of parallel cuts across the fish or cutting in a diamond pattern for larger fish. For small fillets, scoring may not be necessary.

7. Broil fish to order and serve immediately.

8. Broiled fish may be garnished lightly with paprika if more color is desired, but don't overdo it. A common fault in broiling or baking fish is coating them with a heavy layer of paprika, which ruins their delicate flavor.

9. Thick cuts should be turned once during broiling in order to cook evenly. Thin pieces may be arranged on an oiled pan and broiled on one side only. Lobster is also broiled without turning.

KEY POINTS TO REVIEW

- What are the steps in the basic procedure for baking fish? Describe guidelines for success in baking fish.

- What is braising of fish, and how does it compare to baking?

- What are the steps in the basic procedure for broiling or grilling fish? Describe guidelines for success in broiling and grilling fish.

Broiled Fish Steaks Maître d'Hôtel

PORTIONS: AS NEEDED   **PORTION SIZE: 5–6 OZ (150–175 G)**

U.S.	METRIC	INGREDIENTS	PROCEDURE
as needed	as needed	Fish steaks, 5–6 oz (150–175 g) each (see Note)	1. Season the steaks with salt and pepper.
to taste	to taste	Salt	2. Place the oil or melted butter in a small pan. Dip both sides of the steaks in it to coat completely.
to taste	to taste	White pepper	3. Place on the rack of a preheated broiler. Broil under medium heat until half cooked. Turn over with a spatula. At this point, it may be necessary to brush the tops of the steaks with more oil or butter if they are becoming dry.
as needed	as needed	Oil or melted butter	4. Complete the cooking on the second side.
as needed	as needed	Maître d'Hôtel Butter (p. 192)	5. Plate the fish. Place a slice of seasoned butter on top of each steak. Garnish the plate with a lemon wedge. Serve immediately.
as needed	as needed	Lemon wedges	

Per serving: Calories, 320; Protein, 27 g; Fat, 22 g (63% cal.); Cholesterol, 105 mg; Carbohydrates, 2 g; Fiber, 1 g; Sodium, 2460 mg.

Note: Salmon, tuna, and swordfish steaks are ideal for broiling, but they are also expensive. Other fish steaks that may be broiled include cod, haddock, halibut, king mackerel, and large bluefish.

Fillets may also be broiled using this recipe if they are thick or firm enough to avoid breaking up on the grill.

VARIATIONS

For lean white fish (halibut, cod, etc.), dredge in flour and shake off excess before dipping in the melted butter or oil. Broil as in basic recipe.

Other compound butters may be used in place of maître d'hôtel butter.

For fat fish, omit the butter and serve the fish with a small quantity of flavorful vinaigrette.

Grilled Mahi-Mahi with Fruit Salsa

PORTIONS: 12   **PORTION SIZE: 6 OZ (180 G)**

U.S.	METRIC	INGREDIENTS	PROCEDURE
3 fl oz	90 mL	Lemon juice	1. In a nonreactive container, mix together the lemon juice, salt, white pepper, and cayenne.
1 tsp	5 mL	Salt	2. Place the fish fillets in the mixture, turning them to coat all sides. Let stand 15 minutes.
1/4 tsp	1 mL	White pepper	3. Remove the fish from the marinade. Dip the fillets in the oil and remove, letting the excess oil drip off.
1/8 tsp	0.5 mL	Cayenne	4. Grill the fillets on a preheated, seasoned grill or broiler. Grill until just done. Do not overcook, or the fish will be dry.
12	12	Mahi-mahi fillets, 6 oz (180 g) each	
4 fl oz, or as needed	125 mL, or as needed	Vegetable oil	
1 lb 2 oz	600 g	Fruit Salsa (p. 215)	5. Plate the fillets. Next to each fillet, place 1 1/2 oz (50 g) fruit salsa.

Per serving: Calories, 210; Protein, 32 g; Fat, 6 g (27% cal.); Cholesterol, 125 mg; Carbohydrates, 5 g; Fiber, 1 g; Sodium, 170 mg.

Grilled Tuna with Balsamic Vinaigrette and Sweet-Sour Baby Onions

PORTIONS: 4   **PORTION SIZE: 5–6 OZ (150–180 G) TUNA**

U.S.	METRIC	INGREDIENTS	PROCEDURE
1 fl oz	30 mL	Balsamic vinegar	1. Make a vinaigrette (see p. 660 for procedure) with the vinegar, water, herbs, salt, pepper, and olive oil.
1/2 oz	15 mL	Water	2. Mix in the diced red pepper.
1/4 tsp	1 mL	Parsley, chopped	
1/4 tsp	1 mL	Chives, chopped	
to taste	to taste	Salt	
to taste	to taste	White pepper	
2 1/2 fl oz	75 mL	Olive oil	
1 oz	30 g	Roasted red pepper (p. 290), small dice	
4	4	Tuna steaks, 5–6 oz (150–180 g) each	3. Season the tuna steaks with a little salt and brush with a little of the vinaigrette.
to taste	to taste	Salt	4. Grill the steaks until rare or medium rare.
8 oz	240 g	Cipolline in Agrodolce (p. 314)	5. For each portion, arrange the onions in the center of the plate.
			6. Cut the steak in half to display the interior. Arrange the two halves on top of the onions.
			7. Spoon a little vinaigrette around the tuna.

Per serving: Calories, 450; Protein, 36 g; Fat, 26 g (54% cal.); Cholesterol, 65 mg; Carbohydrates, 14 g; Fiber, 1 g; Sodium, 220 mg.

Grilled Tuna with Balsamic Vinaigrette and Sweet-Sour Baby Onions

Broiled Salmon in Escabeche

PORTIONS: 12   **PORTION SIZE: 5–6 OZ (150–180 G)**

U.S.	METRIC	INGREDIENTS	PROCEDURE
4 fl oz	120 mL	Lime juice	1. In a nonreactive container, mix together the lime juice, salt, pepper, ground cumin, and ground allspice.
1 tsp	5 mL	Salt	2. Place the salmon steaks in the mixture, turning them to coat both sides.
1/2 tsp	2 mL	Pepper	3. Refrigerate 30–60 minutes.
1/2 tsp	2 mL	Ground cumin	
1/2 tsp	2 mL	Ground allspice	
12	12	Salmon steaks, 5–6 oz (150–180 g) each	
2 fl oz	60 mL	Vegetable oil	4. Heat the oil in a sauté pan over moderate heat.
12 oz	360 g	Onion, sliced 1/8 in. (3 mm) thick	5. Add the onion. Sauté until lightly browned.
6	6	Garlic cloves, peeled and halved	6. Add the garlic and jalapeños. Sauté 2–3 minutes.
4	4	Pickled jalapeños, sliced	7. Add the sachet, vinegar, and water. Simmer over low heat until reduced by half.
		Sachet:	8. Remove the salmon from the marinade and pat dry with clean towels.
6	6	Whole cloves	
1/2 tsp	2 mL	Whole allspice	
1 tsp	5 mL	Peppercorns	
1/2 tsp	2 mL	Cumin seed	
3	3	Bay leaves, broken	
1 tsp	5 mL	Dried oregano	
1-in. piece	2–3 cm piece	Cinnamon stick	
6 fl oz	180 mL	Cider vinegar	
1 pt	500 mL	Water	
as needed	as needed	Vegetable oil	9. Dip the salmon in oil and remove, letting the excess oil drip off.
as needed	as needed	Cilantro leaves	10. Broil the steaks on a preheated, seasoned broiler or on a grill. Broil until just done. Do not overcook, or the fish will be dry.
			11. Serve each steak with about 1 1/2 fl oz (45 mL) onion, spice, and vinegar mixture. Top with a few cilantro leaves.

Per serving: Calories, 260; Protein, 31 g; Fat, 12 g (43% cal.); Cholesterol, 80 mg; Carbohydrates, 5 g; Fiber, 0 g; Sodium, 150 mg.

VARIATION

For a more traditional escabeche, add the fish to the hot vinegar mixture and cool. After 1 hour, refrigerate the fish in the liquid. To serve, rewarm to room temperature or slightly warmer.

ESCABECHE

The word *escabeche* (ess ka bey chay) is Spanish for "brine," and *en escabeche* means "pickled." Many foods en escabeche are popular in the cuisines of Mexico, especially seafood and vegetables. The method originated as a way to preserve foods by pickling them in vinegar and spices. Today, with refrigeration available, they are enjoyed for their flavors more than their keeping qualities.

In a true escabeche, the food is left in the vinegar mixture for a time at room temperature or refrigerated long enough for it to develop a pickled taste. It is later served at room temperature or rewarmed. In addition, fish prepared this way is often pan-fried in oil before being pickled. The accompanying recipe is modified from the traditional methods for a lighter flavor. The fish is grilled rather than pan-fried, and the spice and vinegar mixture is served with it as a sauce rather than being used to pickle it.

Broiled Salmon in Escabeche

Broiled Lobster

PORTIONS: 1   PORTION SIZE: 1 LOBSTER

U.S.	METRIC	INGREDIENTS	PROCEDURE
1	1	Live lobster, 1–1¹/₂ lb (450–700 g)	1. Split the lobster as shown in **Figure 19.13**. Remove and discard the stomach (just behind the eyes) and the vein that runs through the tail. The liver and coral may be left in or removed and added to the stuffing (step 3), as desired.
1 tsp	5 mL	Finely chopped shallot	2. Sauté the shallot in the butter just until it starts to become tender.
1 tbsp	15 g	Butter	3. Optional step: Chop the lobster coral and liver (tomalley) and add to the pan. Sauté just until firm, 10–20 seconds.
1 oz	30 g	Dry bread crumbs	4. Add the bread crumbs and brown them lightly in the butter. Remove from the heat.
3 tbsp	45 mL	Chopped parsley	5. Add the parsley. Season the crumbs with salt and pepper.
to taste	to taste	Salt	
to taste	to taste	Pepper	
as needed	as needed	Melted butter	6. Place the lobster shell side down on a small sheet pan or in a shallow baking pan. Fill the body cavity with the crumb mixture. Do not put the crumbs over the tail meat.
			7. Brush the tail well with melted butter.
			8. Place a few of the legs on top of the stuffing. If the lobster was split by the first method, shown in **Figure 19.13(g)**, weight the end of the tail to keep it from curling.
			9. Place the lobster under the broiler at least 6 in. (15 cm) from the heat. Broil until the crumbs are well browned.
			10. At this point, the lobster will probably not be completely cooked, unless it is very small and the broiler heat very low. Place the pan with the lobster in a hot oven to finish cooking.
2 fl oz	60 mL	Melted butter	11. Remove the lobster from the heat and serve immediately with a small cup of melted butter and with lemon garnish.
as needed	as needed	Lemon wedges	

Per serving: Calories, 760; Protein, 12 g; Fat, 71 g (82% cal.); Cholesterol, 210 mg; Carbohydrates, 22 g; Fiber, 1 g; Sodium, 1090 mg.

VARIATION   Broiled Rock Lobster Tail

Rock lobster is usually dry if broiled like lobster. A better method is to poach it in salted water (see p. 638) until just cooked. Then split the tails, brush with butter, and run under the broiler 1–2 minutes.

Broiled Shrimp, Scampi Style

PORTIONS: 10   PORTION SIZE: 4¹/₂ OZ (125 G)

U.S.	METRIC	INGREDIENTS	PROCEDURE
50	50	Shrimp, size 16/20	1. Peel, devein, and butterfly the shrimp as shown in **Figure 19.16**. Leave tails on.
			2. Place shrimp in individual service casserole dishes or in a shallow baking pan, tails up and cut side down. (Shrimp will curl more when cooked, so tails stand up as shrimp are broiled.)
			3. Keep refrigerated until needed.
6 oz	175 g	Butter	4. Heat butter and oil in a saucepan until the butter is melted.
¹/₂ cup	125 mL	Oil, preferably olive oil (see Note)	5. Add the garlic, lemon juice, parsley, salt, and pepper.
1 tbsp	15 mL	Garlic, very finely chopped	6. Pour the butter sauce over the shrimp.
1 fl oz	30 mL	Lemon juice	7. Place under the broiler at medium heat. Broil until the tops are lightly browned. (Don't worry if the tips of the tails burn a little; this is normal.)
2 tbsp	30 mL	Chopped parsley	8. Transfer the shrimp to the oven above the broiler for a few minutes to finish cooking.
to taste	to taste	Salt	
to taste	to taste	Pepper	

Per serving: Calories, 250; Protein, 7 g; Fat, 25 g (88% cal.); Cholesterol, 95 mg; Carbohydrates, 1 g; Fiber, 0 g; Sodium, 210 mg.

Note: All butter may be used, instead of a mixture of butter and oil. Or, if you are using a good-quality olive oil, use more oil and less butter or all olive oil.
Serve this dish with rice or with plenty of bread to soak up the flavorful butter.

(continues on next page)

VARIATIONS

The shrimp can be marinated 1–2 hours in the oil, chopped garlic, lemon juice, and seasonings. Add the butter at cooking time. Or omit the butter when cooking, and serve small cups of garlic butter on the side.

Shrimp Brochettes

Marinate the shrimp as indicated above. Put the shrimp on skewers and broil, basting several times with the marinade and melted butter.

Broiled Scallops

Place the scallops in individual service casseroles (5–6 oz/ 150–175 g per portion). Top each portion with 1 tbsp (15 mL) dry bread crumbs. Pour the butter sauce over the scallops and broil as in basic recipe.

Broiled Fish Fillets or Steaks with Garlic Butter

Use fillets or steaks of any lean, white fish. Place fish on sheet pans and prepare according to procedure for Broiled Scallops, using the bread crumbs.

Oysters Casino

PORTIONS: 12   PORTION: 3 OYSTERS (APPETIZER PORTION)

U.S.	METRIC	INGREDIENTS	PROCEDURE
36	36	Oysters	1. Open oysters as shown in **Figure 19.8**. Discard top shell.
			2. Place oysters on a sheet pan or in a shallow baking pan (see Note).
¹/₂ lb	225 g	Butter	3. Place the butter in the bowl of a mixer and beat with the paddle attachment until soft and smooth.
2 oz	60 g	Green bell pepper, chopped fine	4. Add the green pepper, pimiento, shallots, parsley, and lemon juice. Mix until evenly combined. Season to taste with salt and pepper. (Casino butter can be rolled in parchment, refrigerated or frozen, and sliced to order.)
1 oz	30 g	Pimiento, chopped fine	
1 oz	30 g	Shallots, chopped fine	
¹/₄ cup	60 mL	Chopped parsley	
1 fl oz	30 mL	Lemon juice	
to taste	to taste	Salt	
to taste	to taste	White pepper	
9 strips	9 strips	Bacon	5. Cook the bacon in the oven or on the griddle until about half cooked. Drain.
			6. Cut each strip into 4 pieces.
			7. Place about 2 tsp (10 mL) butter mixture on top of each oyster.
			8. Top each oyster with a piece of bacon.
			9. Run the oysters under the broiler until the bacon is brown and the oysters are hot. Do not overcook.

Per serving: Calories, 610; Protein, 26 g; Fat, 51 g (75% cal.); Cholesterol, 230 mg; Carbohydrates, 12 g; Fiber, 0 g; Sodium, 860 mg.

Note: Broiled or baked oysters and clams are often placed on beds of rock salt to hold them steady.

VARIATIONS

Clams Casino

Prepare as in basic recipe, using cherrystone or littleneck clams.

Oysters Casino

SAUTÉING AND PAN-FRYING

As in meat and poultry cookery, the exact distinction between sautéing and pan-frying fish is impossible to draw. For many purposes, the two terms are used interchangeably. For example, many fish preparations that do not involve deglazing the pan are commonly referred to as sautéed. In many ways, they seem more like pan-fried items, except that only a small amount of fat is used. You should review the basic procedures in Chapter 14 (pp. 426–431), but be aware that there are many variations on these procedures.

A classic method for sautéing fish is called **à la meunière** (mun yair). In this preparation, the product is dredged in flour and sautéed in clarified butter or oil. It is then plated and sprinkled with lemon juice and chopped parsley, and freshly prepared hot brown butter (beurre noisette) is poured over it. When the hot butter hits the lemon juice, it creates a froth. The fish should then be served at once. The basic procedure for sautéing à la meunière is given below.

Other pan-fried or sautéed fish preparations may call for Standard Breading Procedure (p. 151) or for dredging the fish with a product other than flour, such as cornmeal. Also, a variety of garnishes may be used.

The procedures and variations just described apply to most popular sautéed and pan-fried fish recipes. In general, because most types of fin fish are so delicate, especially if filleted, they do not lend themselves to a great many sautéing variations. Rather, variety is created with accompaniments, sauces, and garnishes. On the other hand, firm shellfish, like shrimp and scallops, are easy to sauté, and there is a greater variety of recipes for them.

GUIDELINES for Sautéing and Pan-Frying Fish and Shellfish

1. Lean fish are especially well suited to sautéing because the cooking method supplies fat the fish lack.

   Fat fish may also be sautéed, as long as you take care not to get them too greasy.

2. Sautéed fish is usually given a coating of flour, breading, or other starchy product before sautéing. This forms a crust that browns attractively, enhances the flavor, and helps hold the fish together and prevent sticking.

3. Fish may be soaked in milk briefly before dredging in flour. This helps the flour form a good crust.

4. Clarified butter and oil are the preferred fats for sautéing and pan-frying. Whole butter is likely to burn, unless the fish items are very small.

5. Use a minimum of fat. About 1/8 inch (3 mm), or enough to cover the bottom of the pan, is enough.

6. Observe the guidelines for the basic sautéing procedure (p. 427). In particular, be sure the pan is hot before adding the fish to it. After the item has begun to cook, adjust the heat as necessary. Small items, such as shrimp and scallops, are sautéed over high heat. Larger items, such as whole fish or thick steaks, require lower heat to cook evenly.

7. Very large fish may be browned in fat and then finished in the oven, uncovered.

8. Brown the most attractive side—the presentation side—first. For fillets, this is usually the flesh side or the side against the bone, not the skin side.

9. Handle fish carefully during and after cooking to avoid breaking the fish or the crisp crust.

10. Sauté or fry to order and serve immediately.

PROCEDURE for Cooking Fish à la Meunière

1. Collect all equipment and food supplies.

2. Heat a small amount of clarified butter in a sauté pan.

3. Season the fish and dredge in flour. Shake off excess.

4. Place the fish in the pan, presentation side down.

5. Sauté the fish, turning once with a spatula, until both sides are brown and the fish is just cooked through.

6. Remove the fish from the pan with a spatula and place on a serving plate, presentation side up.

7. Sprinkle the fish with lemon juice and chopped parsley.

8. Heat some raw butter in the sauté pan until it turns light brown. Pour it over the fish immediately.

9. Serve at once.

PROCEDURE VARIATION

Instead of sprinkling the lemon juice and parsley on the fish, deglaze the pan with lemon juice after the butter turns brown in step 8. Be careful, as the hot butter may spatter when the lemon juice hits it. Then add the parsley to the lemon butter. Pour this mixture over the fish.

 Fillets of Sole Meunière

PORTIONS: 10   PORTION SIZE: 4 OZ (125 G)

U.S.	METRIC	INGREDIENTS	PROCEDURE
20	20	Sole fillets, 2 oz (60 g) each	1. Have all ingredients ready, but do not season and flour the fish until immediately before cooking.
to taste	to taste	Salt	2. Unless you are cooking to order, use as many sauté pans as necessary to hold all the fillets, or cook them in several batches. Place the sauté pans over medium heat so they will be ready as soon as the fish is floured.
to taste	to taste	White pepper	
3 oz	90 g	Flour	
6 oz	175 g	Clarified butter or oil, or a mixture of butter and oil	3. Season the fillets with salt and pepper. Place the clarified butter in the hot pans to heat. Dredge the fish in flour and shake off excess. Place the fish in the pans flesh side (presentation side) down.
			4. Sauté until lightly browned. Turn over with a spatula and brown the other side. Be careful not to break the fillets when turning.
			5. Remove the fillets from the pan with a spatula, being careful not to break them. Plate the fish on hot dinner plates.
1 fl oz	30 mL	Lemon juice	6. Sprinkle the fish with lemon juice and chopped parsley.
¼ cup	60 mL	Chopped parsley	7. Heat the butter in small saucepan or sauté pan until it turns light brown (beurre noisette).
5 oz	150 g	Butter	8. Pour the hot butter over the fish.
20	20	Slices of peeled lemon	9. Quickly place a lemon slice on top of each fillet and serve immediately.

Per serving: Calories, 370; Protein, 20 g; Fat, 29 g (71% cal.); Cholesterol, 130 mg; Carbohydrates, 7 g; Fiber, 0 g; Sodium, 370 mg.

VARIATIONS

Other white fish fillets, as well as shellfish such as scallops and shrimp, may be cooked by the same procedure.

Placing the fish in milk before dredging in flour helps form an attractive, well-browned crust. However, the fish must be drained well before flouring or the flour coating might become heavy and pasty.

Fillets of Fish Doré

Sauté the fish as in the basic recipe, but serve without the lemon juice, chopped parsley, and beurre noisette. Garnish the plate with lemon and parsley sprig. (Doré means "golden.")

Trout Meunière

Prepare whole, drawn trout as in the basic recipe. Dip the fish in milk before dredging in flour to form a better crust.

Fish Sauté Amandine

Prepare as in the basic recipe. Brown sliced almonds in the butter used for garnishing. Omit garnish of lemon slices, and garnish plate with lemon wedges.

Fish Sauté Grenobloise

Prepare as in the basic recipe. Garnish the fish with capers and diced, peeled lemon sections in addition to the chopped parsley before pouring on the brown butter.

Sautéed Soft-Shell Crabs

Prepare as in the basic recipe. Serve 2 per portion. Chopped parsley and lemon slices may be omitted.

Scallops Meunière

Spicy Shrimp or Scallop Sauté

PORTIONS: 10   PORTION SIZE: 4 OZ (125 G)

U.S.	METRIC	INGREDIENTS
1 tsp	5 mL	Paprika
1/4 tsp	1 mL	Cayenne
1/4 tsp	1 mL	Black pepper
1/4 tsp	2 mL	White pepper
1/4 tsp	1 mL	Dried thyme
1/4 tsp	1 mL	Dried basil
1/4 tsp	1 mL	Dried oregano
1/2 tsp	2 mL	Salt
2 1/2 lb	1.25 kg	Peeled, deveined shrimp or scallops
6 oz	175 g	Onion, sliced
1	1	Garlic clove, chopped
as needed	as needed	Clarified butter

PROCEDURE

1. Mix together the spices, herbs, and salt.
2. If the shrimp or scallops are wet, dry them with paper towels. If you are using sea scallops and they are large, cut them into halves or quarters.
3. Toss the shrimp or scallops with the dry seasonings.
4. Sauté the onion and garlic in a little clarified butter until they are tender and only lightly browned. Remove them from the pan and set them aside.
5. Add a little more butter to the pan and sauté the seafood just until it is cooked.
6. Return the onion and garlic to the pan and toss to combine. Serve immediately, accompanied by white rice.

Per serving: Calories, 160; Protein, 18 g; Fat, 9 g (50% cal.); Cholesterol, 185 mg; Carbohydrates, 2 g; Fiber, 0 g; Sodium, 390 mg.

VARIATION

For spicy fish fillets, season fish well with the dry seasoning mix in the basic recipe. Then dredge them in flour and sauté as for meunière.

Escalope of Salmon with Red Wine Sauce

PORTIONS: 12   PORTION SIZE: 4 OZ (120 G) SALMON, 1 1/2 FL OZ (45 ML) SAUCE

U.S.	METRIC	INGREDIENTS
4 lb (approx.)	2 kg (approx.)	Salmon fillet (see Note)
as needed	as needed	Clarified butter
18 fl oz	540 mL	Beurre Rouge for Fish (p. 192)

PROCEDURE

1. Following the procedure in **Figure 19.5**, cut twelve 4-oz (120 g) escalopes from the fillet.
2. Heat a thin film of oil over high heat in as many sauté pans as needed to hold the salmon.
3. Sauté the salmon escalopes about 1–2 minutes per side. Remove and drain briefly on paper towels to absorb excess butter.
4. Plate the salmon. Spoon 1 1/2 fl oz (45 mL) of sauce around each escalope.

Per serving: Calories, 590; Protein, 36 g; Fat, 46 g (71% cal.); Cholesterol, 200 mg; Carbohydrates, 1 g; Fiber, 0 g; Sodium, 390 mg.

Note: The weight given for the salmon includes skin, which is discarded after the escalopes are cut. If you are using skinless fillets, the weight needed is less.

VARIATIONS

Escalope of Salmon with Herb Cream

In place of the beurre rouge, serve the fish with Herbed Cream Sauce (p. 205).

Escalope of Salmon with Red Wine Sauce

Crab Cakes with Roasted Pepper Rémoulade

PORTIONS: 8   PORTION SIZE: 4 OZ (125 G)

U.S.	METRIC	INGREDIENTS
1¹/₂ lb	720 g	Crabmeat
2 oz	60 g	Fresh bread crumbs
3 oz	90 g	Mayonnaise
2	2	Eggs, beaten
1 tsp	5 mL	Prepared mustard
1 tsp	5 mL	Worcestershire sauce
1 tsp	5 mL	Salt
¹/₄ tsp	1 mL	White pepper
1¹/₂ tbsp	22 mL	Chopped parsley
4	4	Scallions, chopped fine
as needed	as needed	Clarified butter, for cooking
1¹/₂ oz	45 g	Roasted red bell peppers, peeled, cored, and seeded (see **Figure 10.14**)
8 fl oz	240 mL	Rémoulade Sauce (p. 216)
8	8	Lemon wedges

PROCEDURE

1. Pick over the crabmeat to remove any bits of shell.
2. Mix together the bread crumbs, mayonnaise, eggs, mustard, Worcestershire sauce, salt, pepper, parsley, and scallions. Fold in the crabmeat.
3. Form by hand into round cakes. For each portion, allow 1 large cake, about 4 oz (125 g), or 2 small cakes, about 2 oz (60 g) each (see Note).
4. Pan-fry the cakes in butter until browned on both sides and cooked through.
5. Cut the roasted peppers into brunoise. Mix into the rémoulade.
6. Serve each crab cake with 1 fl oz (30 mL) sauce and 1 lemon wedge.

Per serving: Calories, 440; Protein, 21 g; Fat, 36 g (72% cal.); Cholesterol, 145 mg; Carbohydrates, 7 g; Fiber, 0 g; Sodium, 1070 mg.

Note: This mixture has very little bread filler, so it may be some what difficult to handle. If desired, add more bread crumbs to make a firmer mixture that will pack more easily into cakes. The texture of the crabmeat may also affect the texture of the cakes and the quantity of bread crumbs needed.

Cornmeal-Crusted Soft-Shell Crabs

PORTIONS: 10   PORTION SIZE: 1 CRAB

U.S.	METRIC	INGREDIENTS
10	10	Soft-shell crabs
8 oz	250 mL	Milk
5 oz	150 g	Cornmeal
4 fl oz	125 mL	Oil
6 oz	175 g	Butter
to taste	to taste	Salt
5 fl oz	150 mL	Rémoulade Sauce (p. 216)

PROCEDURE

1. Clean the crabs as shown in **Figure 19.17**.
2. Dip the crabs in milk. Drain, then dredge in cornmeal. Discard leftover milk.
3. Heat the oil and butter in enough sauté pans to hold the crabs in a single layer (or pan-fry in separate batches). Place the crabs in the pan upside down and cook over moderate heat until lightly browned. Salt lightly to taste while the crabs are cooking. Turn over and brown the other side. Remove from the pan and drain a few seconds on paper towels to remove excess fat.
4. Serve each crab with 1 tbsp (15 mL) rémoulade sauce.

Per serving: Calories, 420; Protein, 16 g; Fat, 27 g (57% cal.); Cholesterol, 120 mg; Carbohydrates, 30 g; Fiber, 2 g; Sodium, 500 mg.

Cornmeal-Crusted Soft-Shell Crab on a bed of sautéed Swiss chard, with corn, roasted cherry tomatoes, Corn Purée, and sautéed polenta squares

Peppered Haddock with Garlic Mashed Potatoes and Parsley Sauce

PORTIONS: 12   PORTION SIZE: 5 OZ (150 G) PLUS GARNISH

U.S.	METRIC	INGREDIENTS	PROCEDURE
10 fl oz	300 mL	Olive oil	1. **Prepare the sauce:** Combine the oil, lemon juice,
1 fl oz	30 mL	Lemon juice	chopped parsley, and salt in a blender. Process until
1/2 cup	125 mL	Chopped parsley	the parsley is puréed.
1/2 tsp	2 mL	Salt	
3 lb 12 oz	1.8 kg	Haddock fillets, cut into 5-oz (150-g) portions	2. Coat the fish fillets evenly with a light sprinkling of crushed peppercorns. Season with salt.
2 tbsp	30 mL	Crushed black peppercorns	3. Heat the olive oil in as many sauté pans as necessary to hold the fish in a single layer.
1 tsp	5 mL	Salt	4. Place the fish in the pans, presentation side down, and
2 fl oz	60 mL	Olive oil	sauté over moderate heat until lightly browned and
2 lb 4 oz	1.1 kg	Garlic Mashed Potatoes (p. 353)	about half cooked. Turn over and finish the cooking.

5. Place a 3-oz (90-g) portion of potatoes in the center of each plate. Top with the fish fillet. Drizzle about 1 fl oz (30 mL) sauce in a circle around the fish.

Per serving: Calories, 480; Protein, 28 g; Fat, 33 g (62% cal.); Cholesterol, 95 mg; Carbohydrates, 17 g; Fiber, 2 g; Sodium, 420 mg.

VARIATIONS

Other firm-fleshed white fish, such as cod, sea bass, striped bass, red snapper, or grouper, may be substituted.

Peppered Haddock with Purée of Flageolet Beans

Substitute Purée of Flageolet Beans with Garlic (p. 375) for the potatoes.

Peppered Haddock with Garlic Mashed Potatoes, Parsley Sauce, and Fried Parsley

Shrimp and Cucumber in Thai Red Curry

PORTIONS: 12   **PORTION SIZE: 6 OZ (180 G)**

U.S.	METRIC	INGREDIENTS
2 fl oz	60 mL	Vegetable oil
2¹/₂ lb	1.2 kg	Shrimp, medium to large, peeled and deveined
12 oz	375 g	Cucumber, peeled, seeded, medium dice
18 fl oz	550 mL	Thai Red Curry Sauce (p. 210)

PROCEDURE

1. Heat the oil in a sauté pan or wok over high heat.
2. Add the shrimp and stir-fry until well seared.
3. Add the cucumber and stir-fry 1 minute.
4. Add the curry sauce. Cook a few minutes longer, or until the shrimp are completely cooked.
5. Serve with rice.

Per serving: Calories, 200; Protein, 17 g; Fat, 16 g (65% cal.); Cholesterol, 140 mg; Carbohydrates, 2 g; Fiber, 2 g; Sodium, 450 mg.

Note: This method of making a curry is designed for advance preparation and quick, last-minute cooking. A more traditional method is to make the sauce as an integral sauce. Stir-fry the shrimp, add curry paste (see sauce recipe, p. 184), and cook until aromatic. Add the coconut milk, other liquids, flavorings, and any other ingredients indicated in the sauce recipe, and finish cooking.

Shrimp and Cucumber in Thai Red Curry

Skate with Caper Butter

PORTIONS: 12   **PORTION SIZE: 4 OZ (125 G)**

U.S.	METRIC	INGREDIENTS
12	12	Skate fillets, 4 oz (125 g) each
to taste	to taste	Salt
to taste	to taste	White pepper
3 fl oz	90 mL	Vegetable oil
3 fl oz	90 mL	Clarified butter
as needed	as needed	Flour
1 oz	30 g	Shallots, chopped fine
6 fl oz	180 mL	White wine
1 fl oz	30 mL	Lemon juice
12 tbsp	180 mL	Capers, drained
12 oz	360 g	Butter, cut into small pieces
to taste	to taste	Salt

PROCEDURE

1. Season the fillets with salt and white pepper.
2. Heat the oil and clarified butter in a sauté pan over moderate heat.
3. Dredge the fillets in flour and shake off excess.
4. Sauté the fish, browning it lightly on both sides, until done.
5. Place the fillets on hot dinner plates.
6. Drain the butter and oil from the sauté pan, leaving a thin film on the bottom.
7. Add the shallots and sauté just until they are soft.
8. Add the wine. Reduce until about 1 fl oz (30 mL) remains.
9. Add the lemon juice and capers.
10. Whip in the raw butter as for making beurre blanc (p. 191).
11. Season the caper butter to taste with salt.
12. Pour the butter over the fish fillets.

Per serving: Calories, 450; Protein, 2 g; Fat, 37 g (90% cal.); Cholesterol, 115 mg; Carbohydrates, 5 g; Fiber, 0 g; Sodium, 510 mg.

VARIATION

For a different kind of caper butter, plate the sautéed skate and top each portion with 1 tbsp (15 mL) capers. Pour hot beurre noisette over the fish, using 1–2 tbsp (15–30 mL) butter per portion.

DEEP-FRYING

Deep-frying is perhaps the most popular method of preparing fish in North America. While fried fish may not be the most subtle or refined preparation, it can be of very high quality if the fish is fresh and not overcooked, the frying fat is of good quality, and the item is served without delay after cooking.

Guidelines and procedures for deep-frying meat, poultry, and fish are explained in Chapter 14, which also includes a core recipe for the British popular classic of Fish and Chips, batter-coated fried fish fillets with French fries. Review pages 432–433 as needed before continuing with this section. Note also that the batter recipes on page 341 can also be used for seafood.

Lean fish—either small whole fish or small portions such as fillets or sticks—and shellfish such as shrimp, clams, oysters, and scallops are best for deep-frying.

Fish to be fried is breaded or battered to protect it from the frying fat and to protect the frying fat from it. Also, the breading or batter provides a crisp, flavorful, and attractive coating.

Frozen breaded fish portions are widely used. They should be fried without thawing.

Fried fish is usually served with lemon and/or a cold sauce such as tartar, rémoulade, or cocktail sauce on the side.

Breading procedures are discussed in detail in Chapter 7, and general guidelines for deep-frying are outlined in Chapter 11. There is no need to repeat them here, but you should review those sections if necessary.

KEY POINTS TO REVIEW

- What are the steps in the basic procedure for sautéing fish à la meunière?

- What are the guidelines for success in sautéing and pan-frying fish?

- What kinds of fish and seafood items are most appropriate for deep-frying?

- Why is fish usually breaded or battered before being deep-fried?

Fried Breaded Fish Fillets

PORTIONS: 12   PORTION SIZE: 4 OZ (125 G)

U.S.	METRIC	INGREDIENTS	PROCEDURE
		Standard Breading Procedure (see Note):	1. Set up a breading station (see p. 151): Place the flour in one pan, the eggs beaten with milk in a shallow bowl, and the bread crumbs in another pan.
2 oz	60 g	Flour	
2	2	Whole eggs, beaten	
4 fl oz	120 mL	Milk	
10 oz	300 g	Dry bread crumbs	
12	12	4-oz (125-g) fillets of lean, white fish, such as haddock, perch, pike, bass, sole, or flounder	2. Season the fish lightly with salt and white pepper.
			3. Bread the fish fillets by passing them through the flour, egg wash, and crumbs. Press the crumbs on firmly. (See p. 151 for detailed breading instructions.)
to taste	to taste	Salt	4. Fry the fillets until golden brown in deep fat heated to 350°F (175°C).
to taste	to taste	White pepper	
12	12	Parsley sprigs	5. Drain and serve immediately. Garnish each portion with a parsley sprig and lemon wedge. Serve with 1 fl oz (25 mL) tartar sauce.
12	12	Lemon wedges	
12 fl oz	300 mL	Tartar sauce	

Per serving: Calories, 490; Protein, 25 g; Fat 36 g (66% cal.); Cholesterol, 115 mg; Carbohydrates, 16 g; Fiber, 1 g; Sodium, 430 mg.

Note: Quantities given for breading materials are only guidelines. You may need more or less, depending on the shapes of the fish pieces, the care used in breading, and other factors. In any case, you will need enough so even the last piece to be breaded can be coated easily and completely.

VARIATIONS

Breaded fish fillets may also be pan-fried in butter or oil (see previous section).

Fried Breaded Scallops

Prepare as in the basic recipe. Use wire baskets in the breading station to simplify the procedure, as explained on page 150.

Fried Breaded Shrimp

Peel, devein, and butterfly shrimp, as shown in Figure 19.16. Leave tails on. Bread and fry as in the basic recipe.

Fried Oysters or Clams

Prepare like scallops.

Cod Cakes

PORTIONS: 12   PORTION SIZE: 2 CAKES, 2¹/₂ OZ (75 G) EACH

U.S.	METRIC	INGREDIENTS	PROCEDURE
2 lb	900 g	Cooked cod	1. Flake the fish until it is well shredded.
2 lb	900 g	Potato purée (p. 352)	2. Combine with the potato, egg, and egg yolk. Mix well.
2¹/₂ oz (1¹/₂ eggs)	70 g (1¹/₂ eggs)	Whole eggs, beaten	
²/₃ oz (1 yolk)	20 g (1 yolk)	Egg yolks, beaten	3. Season to taste with salt, pepper, and a little ground ginger.
to taste	to taste	Salt	
to taste	to taste	White pepper	4. Scale the mixture into 2¹/₂-oz (75-g) portions. Shape into round, slightly flattened cakes.
pinch	pinch	Ground ginger	
		Standard Breading Procedure:	5. Pass the cakes through the Standard Breading Procedure (p. 151).
as needed	as needed	Flour	
as needed	as needed	Egg wash	6. Deep-fry at 350°F (175°C) until golden brown.
as needed	as needed	Bread crumbs	
as desired	as desired	Tomato sauce or tartar sauce	7. Serve 2 cakes per portion. Accompany with tomato sauce or tartar sauce.

Per serving: Calories, 280; Protein, 23 g; Fat, 6 g (19% cal.); Cholesterol, 110 mg; Carbohydrates, 33 g; Fiber, 2 g; Sodium, 360 mg.

VARIATION

Salmon or Tuna Cakes

Prepare as in the basic recipe, using well-drained canned salmon or tuna. If you have duchesse potato mixture on hand, simply combine equal parts fish and duchesse mixture. No additional eggs are needed, as the potatoes already contain eggs.

Salmon Cakes with Mixed Greens

Deep-Fried Calamari with Spicy Tomato Sauce and Aïoli

PORTIONS: 12   PORTIONS SIZE: 6 OZ (175 G) SQUID, 2 FL OZ (60 ML) EACH SAUCE

U.S.	METRIC	INGREDIENTS	PROCEDURE
4¹⁄₂ lb	2.25 kg	Small cleaned squid (see **Figure 19.12**)	1. Slice the body sacs of the squid crosswise into rings. Leave the tentacle sections whole, or cut them in half if they are large.
4 fl oz	125 mL	Lemon juice	2. Combine the squid with the lemon juice and marinate, refrigerated, 2–3 hours.
1 lb	500 g	Flour	3. Mix the flour and salt.
3 tbsp	45 mL	Salt	4. Drain the squid and dry on clean towels. Immediately before cooking, toss the squid with the flour, then shake in a large strainer to remove excess flour.
			5. Deep-fry at 350°F (175°C) just until lightly golden. Remove from the fryer and drain.
1¹⁄₂ pt	750 mL	Italian Tomato Sauce for Pasta (p. 399)	6. Heat the tomato sauce and season to taste with the hot pepper sauce.
to taste	to taste	Hot red pepper sauce	7. Heap the fried squid in the center of the serving plates, accompanied by 2 fl oz (60 mL) of each of the sauces in small cups. If desired, top with a little deep-fried parsley.
1¹⁄₂ pt	750 mL	Aïoli I (p. 216) or Aïoli II (p. 665)	
as desired	as desired	Optional garnish: Deep-fried parsley leaves	

Per serving: Calories, 700; Protein, 30 g; Fat, 50 g (64% cal.); Cholesterol, 395 mg; Carbohydrates, 33 g; Fiber, 2 g; Sodium, 1710 mg.

VARIATION

For other versions of deep-fried calamari, bread the squid (see Standard Breading Procedure, p. 151) or dip in batter, using one of the batter recipes for Onion Rings (p. 341); the recipe for Beer Batter is recommended. Follow steps 6–8 in the onion ring recipe for battering and frying.

Deep-Fried Calamari with Spicy Tomato Sauce and Aïoli

Shrimp and Vegetable Tempura

PORTIONS: 8   PORTION SIZE: 3 SHRIMP, PLUS VEGETABLES

U.S.	METRIC	INGREDIENTS	PROCEDURE
24	24	Large shrimp	1. Peel the shrimp, leaving the tails attached. Devein and butterfly them.
2	2	Green bell peppers	2. Core and seed the peppers. Cut each one lengthwise in 8 wedges or strips.
10 oz	300 g	Sweet potatoes	3. Peel the sweet potatoes. Cut into slices about $1/6$ in. (4 mm) thick.
16	16	Small mushrooms (or halves or quarters of large mushrooms)	4. Clean the mushrooms and trim the bottoms of the stems.
12 oz	375 g	Cake flour or other low-gluten flour	5. Sift the flour into a mixing bowl.
12 fl oz	375 mL	Water, ice cold	6. Mix together the water and egg yolks.
3	3	Egg yolks	7. Mix the liquid into the flour until just combined. Do not worry about a few lumps. The batter should be somewhat thinner than pancake batter.
12 fl oz	375 mL	Dashi (p. 167)	8. Make a dipping sauce by combining the dashi, soy sauce, and mirin.
2 1/2 fl oz	75 mL	Soy sauce	
1 1/2 fl oz	50 mL	Mirin (sweet rice wine)	
as needed	as needed	Flour for dredging	9. Divide the shrimp and vegetables equally into 16 portions. Fry the vegetables first, then the shrimp, by dredging with flour, shaking off the excess, then dipping in the batter and dropping into clean frying fat at 350°F (175°C). Fry just until lightly golden.
4 oz	125 g	Daikon (large white Japanese radish), grated	
1 1/2 tbsp	22 mL	Grated fresh ginger root	10. Drain and serve at once. Tempura is traditionally served on a bamboo tray covered with a clean piece of absorbent paper. Accompany with about 2 fl oz (60 mL) dipping sauce in a shallow bowl. Put a small mound each of grated daikon and grated ginger on each serving tray. The diner mixes these to taste into the dipping sauce.

Per serving: Calories, 380; Protein, 11 g; Fat 12 g (28% cal.); Cholesterol, 115 mg; Carbohydrates, 56 g; Fiber, 3 g; Sodium, 560 mg.

TEMPURA

Tempura (TEM-poo-rah) is not a native Japanese dish but rather originated with the deep-fried dishes Portuguese traders and missionaries introduced to Japan in the sixteenth century. Japanese cooks transformed the dish by perfecting a particularly light, lacy batter and by serving the tempura with a dipping broth based on dashi (p. 167) and soy sauce and flavored with grated ginger and daikon radish.

POACHING AND SIMMERING

The procedures in this section all involve cooking seafood in a liquid at temperatures lower than boiling. Basic techniques and guidelines for *simmering* and *submersion poaching* are explained in detail on pages 434–435 of Chapter 14. *Shallow poaching* is discussed on pages 437–438.

SUBMERSION POACHING AND SIMMERING IN COURT BOUILLON

Recall that the difference between simmering and poaching is primarily one of temperature. Simmered foods are cooked at about 185°–200°F (85°–94°C), while poached foods are cooked at 160°–185°F (71°–85°C). The lower temperature of poaching better preserves the texture of delicate fish items. Simmering is used primarily for shellfish such as lobster, while fin fish are rarely simmered.

As explained in Chapter 14, the liquid typically used for poaching fish is called **court bouillon**. As defined there, court bouillon consists of water containing seasoning, herbs, and usually an acid. Refer to the recipes on pages 639–641.

In quantity food service, this method is perhaps used most often for cooking large whole fish to be decorated and served cold on a buffet. Slightly higher simmering temperatures are used for cooking crustaceans, such as lobster, crab, and shrimp.

SHALLOW POACHING

Shallow poaching is a technique used primarily for fish. Procedures and guidelines are explained at length in Chapter 14, followed by a classic recipe for fish poached in white wine and fumet. Review this section as necessary.

Glazing

Shallow-poached fish is sometimes glazed before serving. This is done as follows:

1. According to the particular recipe, combine the finished sauce with egg yolk, hollandaise sauce, and/or lightly whipped cream. Alternatively, combine the reduction of the cooking liquid with Mornay sauce instead of fish velouté.

2. Coat the fish with the sauce and run the plate or platter under the salamander or broiler for a few seconds, or until the sauce is golden brown.

    Note: It's a good idea to test a little of the sauce under the salamander before coating the fish to make sure it will brown.

TRUITE AU BLEU

The famous preparation called **truite au bleu** (blue trout) in French and *forelle blau* in German is made by poaching trout that are alive until cooking time. The fish must be alive and must not be washed in order for the fish to turn blue. Live fish have a protective slippery coating on the skin, and the blue color results from the vinegar in the court bouillon reacting with this coating.

GUIDELINES for Poaching Fish in Court Bouillon

1. Both fat and lean fish may be cooked by this method.

2. Seasoned liquid for cooking fish may be as simple as salted water. More often, however, it contains flavoring ingredients such as spices, herbs, and mirepoix, and acid ingredients such as lemon juice, vinegar, and white wine.

3. Cook flavoring ingredients in court bouillon to extract the flavors before cooking the fish.

4. To cook seafood in court bouillon, follow the basic procedures for simmering and poaching detailed on pages 434–435. Cooking temperature is 160°–180°F (70°–80°C), well below boiling. A temperature of 160°F (70°C) is sufficient to cook fish, and it reduces the likelihood of overcooking. Higher temperatures are harmful to the delicate texture and flavor of fish.

    Lobsters, crabs, and shrimp may be cooked at a simmer because their textures are less fragile. The terms *boiled lobster* and *boiled fish* are often used but are inaccurate. Lobster and fish should never be boiled.

5. Start shellfish, small fish, and portion cuts in hot liquid to preserve flavors. Start large fish in cold liquid to cook more evenly and to avoid sudden contractions that would split the skin and spoil the appearance.

6. Special fish poachers with racks are best for poaching. They allow the fish to be removed from the liquid without damage. If these utensils are not available, wrap the fish in cheesecloth so it can be lifted out easily, or tie the fish loosely to a board.

7. Serve poached fish with an appropriate sauce, such as hollandaise for hot fish and a mayonnaise-based sauce for cold fish. Mild vinaigrettes go well with both hot and cold poached fish.

Ordinary Court Bouillon for Fish

YIELD: 1 GAL (4 L)

U.S.	METRIC	INGREDIENTS
1 gal	4 L	Water
8 fl oz	250 mL	White vinegar, wine vinegar, or lemon juice
8 oz	250 g	Onions, sliced
4 oz	125 g	Celery, sliced
4 oz	125 g	Carrots, sliced
2 oz	60 g	Salt
1/4 tsp	2 mL	Peppercorns, crushed
1	1	Bay leaf
1/2 tsp	1 mL	Dried thyme
10–12	10–12	Parsley stems

PROCEDURE

1. Combine all ingredients (**Figure 20.1**) in a stockpot or saucepot and bring to boil.
2. Reduce heat and simmer 30 minutes.
3. Strain and cool.

Per 1 fl oz (29.57 mL): Calories, 0; Protein, 0 g; Fat, 0 g (0% cal.); Cholesterol, 0 mg; Carbohydrates, 0 g; Fiber, 0 g; Sodium, 170 mg.

FIGURE 20.1   Court bouillon ingredients.

Court Bouillon with White Wine

YIELD: 1 GAL (4 L)

U.S.	METRIC	INGREDIENTS
2 qt	2 L	Water
2 qt	2 L	White wine
4 oz	125 g	Onions, sliced
4 oz	125 g	Celery, coarsely chopped
4 oz	125 g	Leeks, coarsely chopped
1	1	Bay leaf
1/2 tsp	2 mL	Peppercorns, crushed
1/2 tsp	2 mL	Dried thyme
10–12	10–12	Parsley stems
2	2	Whole cloves
1 oz	30 g	Salt

PROCEDURE

1. Combine all ingredients in a stock pot and bring to a boil.
2. Reduce heat and simmer about 20 minutes.
3. Strain and cool.

Per 1 ounce: Calories, 40; Protein, 0 g; Fat, 0 g (0% cal.); Cholesterol, 0 mg; Carbohydrates, 2 g; Fiber, 0 g; Sodium, 274 mg.

Poached Whole Fish

PORTIONS: 10   PORTION SIZE: 4 OZ (125 G)

U.S.	METRIC	INGREDIENTS	PROCEDURE
5 lb	2.5 kg	Drawn fish	
		or	
4 lb	2 kg	Dressed fish (1 large or 2 or more smaller fish; see Note)	
3 qt or as needed	3 L or as needed	Court bouillon, cold	

PROCEDURE

1. Place the clean dressed or drawn fish on the lightly oiled rack of a fish poacher. If a rack is unavailable, wrap the fish in cheesecloth or tie it loosely on a board so it can be lifted out when cooked.

2. Place the fish in the poaching pan and pour in enough court bouillon to cover the fish completely.

3. Set the pan over moderately low heat and slowly bring barely to a simmer.

4. Reduce heat to very low and cook the fish at below the simmering point until done. The fish will feel firm, not mushy, at its thickest part, and the backbone, as seen inside the cavity, will no longer be pink. Total cooking time will vary from 5 to 20 minutes, depending on the size of the fish and the exact cooking temperature. In general, plan on about 8–10 minutes for every inch of thickness at the thickest point.

Suggested sauces:
Hollandaise
Mousseline
Beurre noisette
Herb vinaigrette

5. For serving hot: Remove fish from liquid, drain well, and serve immediately with choice of sauce.

For serving cold: Add ice to the court bouillon to stop the cooking. Cool the fish rapidly in the liquid and refrigerate. Drain when chilled.

Per serving: Calories, 220; Protein, 47 g; Fat, 2 g (9% cal.); Cholesterol, 140 mg; Carbohydrates, 0 g; Fiber, 0 g; Sodium, 320 mg.

Note: For attractive presentations, such as for buffet work, fish is often poached with the head on. Suggested fish for poaching whole:

Haddock	Red snapper	Striped bass
Cod	Salmon	Trout

VARIATION

Poached Fish Steaks

Prepare as in basic recipe, except start with boiling court bouillon. Drain the cooked fish, remove skin and center bone, and serve immediately with selected sauce. Suggested fish steaks: cod, haddock, halibut, turbot, salmon.

"Boiled" Shellfish (Lobster, Crab, Shrimp)

Prepare as for fish steaks using salted water, court bouillon, or acidulated water (4 oz lemon juice and 1/2 oz salt per quart of water/125 mL lemon juice and 15 g salt per liter). Water may simmer when cooking shellfish.

Poaching whole fish

Scallops and Shrimp à la Nage ♥

PORTIONS: 12   PORTION SIZE: 8 OZ (240 G)

U.S.	METRIC	INGREDIENTS	PROCEDURE
9 oz	270 g	Carrots, trimmed and peeled	1. Cut the carrots, leeks, and celery into julienne.
9 oz	270 g	Leeks, trimmed and cleaned	2. Sweat the vegetables in butter for a few minutes.
6 oz	180 g	Celery, trimmed	3. Add the court bouillon and simmer until the vegetables are cooked but still slightly crisp. Remove them from the court bouillon with a slotted spoon.
1½ oz	45 g	Butter	
3 pt	1.5 L	Court Bouillon with White Wine (p. 639), strained	
1 lb 8 oz	720 g	Shrimp, peeled and deveined	4. Return the court bouillon to a simmer. Add the shrimp and scallops. Poach them gently until they are just cooked, about 5 minutes.
1 lb 8 oz	720 g	Scallops	5. Remove the seafood from the court bouillon with a slotted spoon and keep warm.
to taste	to taste	Salt	6. Bring the court bouillon to a boil and reduce by one-third.
			7. Season and strain through cheesecloth or a fine chinois.
2 tbsp	30 mL	Chives, chopped	8. Divide the shrimp, scallops, and vegetables among broad soup plates or other appropriate dishes for service.
			9. Pour the hot court bouillon (reheated if necessary) over them. Sprinkle with chives and serve immediately.

Per serving: Calories, 220; Protein, 20 g; Fat, 4 g (16% cal.); Cholesterol, 115 mg; Carbohydrates, 11 g; Fiber, 1 g; Sodium, 740 mg.

VARIATIONS

A small amount of crème fraîche or herb butter may be stirred into the broth after it is reduced in step 6.

Salmon à la Nage

Omit the shrimp and scallops. For each portion, use one boneless, skinless salmon fillet, 4–5 oz (125–150 g).

Scallops and Shrimp à la Nage

Zuppa di Vongole 🍽️

PORTIONS: 8

U.S.	METRIC	INGREDIENTS
7¹/₂ lb	3.5 kg	Small clams, such as littlenecks
8 fl oz	250 mL	Water

PROCEDURE

1. Scrub the clams under cold water to remove sand and grit from the shells.
2. Put the clams and water in a heavy covered pot and heat gently, just until the clams open. Set the clams aside. Strain and reserve the liquid.
3. Depending on how you wish to serve them, you can leave the clams in the shell or shell all but 4–6 of them per portion to use as garnish.

U.S.	METRIC	INGREDIENTS
3 fl oz	90 mL	Olive oil
2¹/₂ oz	75 g	Onion, small dice
1–3	1–3	Garlic cloves, chopped
3 tbsp	45 mL	Chopped parsley
6 fl oz	175 mL	White wine
12 oz	350 g	Canned plum tomatoes, with juice, coarsely chopped

4. Heat the olive oil in a large pot. Sauté the onion until soft but not brown.
5. Add the garlic and cook another minute.
6. Add the parsley and the wine and boil for 1 minute.
7. Add the tomatoes and the reserved clam juice. Simmer 5 minutes.
8. Taste for seasoning and adjust if necessary.
9. Add the clams and reheat them gently. Do not overcook, or the clams will be tough.
10. Serve with plenty of crusty bread for dipping in the broth.

Per serving: Calories, 680; Protein, 93 g; Fat, 18 g (25% cal.); Cholesterol, 240 mg; Carbohydrates, 22 g; Fiber, 1 g; Sodium, 470 mg.

VARIATIONS

Zuppa di Cozze

Substitute mussels for the clams.

Zuppa di Frutti di Mare

Use a mixture of clams, mussels, squid (cut up), and shrimp (shelled). Keep all the items separate. Cook the clams and mussels as in the basic recipe. Add the squid at the same time as the tomatoes and broth. Simmer slowly, covered, until tender. Add the shrimp and cook just 1 minute before adding the clams and mussels.

Zuppa di Pesce

Use a mixture of shellfish and fin fish, as desired. Add each type of fish just long enough before the end of cooking so it cooks through without overcooking.

Zuppa di Vongole

STEAMING, SOUS VIDE, AND MIXED COOKING TECHNIQUES

The recipes in this section are placed here because they are difficult to classify as one of the basic cooking methods, although they are all based on moist-heat methods. As preparation for working with this section, you should review the explanation of steaming methods in Chapter 14 (pp. 440–441). That discussion is concerned primarily with steaming in a compartment steamer and steaming on a rack above a simmering liquid. In this section, you will find that a number of other techniques also use trapped steam to cook seafood.

Most of the recipes in this section share two characteristics:

1. The item cooks in its own juices and, usually, a small amount of added liquid.

2. The item is served with its flavorful cooking liquid.

In some cases, enough liquid is added to barely cover, and the item simmers. In other cases, little liquid is added and the item cooks in the steam trapped by the pot lid.

The French term **étuver** (ay too vay) is used for this kind of procedure, in which the item cooks slowly in very little liquid. The word is usually translated as "stew," but this may be misleading. More precisely, it means "to cook or steam in its own juices" or "to sweat."

In addition, this section includes two traditional recipes for dishes made with cooked seafood.

Finally, this section also contains recipes for fish cooked sous vide. It is essential that you read the discussion of sous vide cooking, and especially the safety precautions, on pages 117–119 before you try this recipe.

VARIATIONS

Note the following three variations represented by the recipes in this section:

1. The product is cooked for a few minutes in fat over low heat, along with mirepoix or onion, to begin extracting juices. A little liquid is then added, the pot is covered, and the item is cooked. Example: Fisherman's Stew (p. 648).

2. The product is sautéed over high heat. Then other ingredients and liquids are added and the item is cooked, covered, over low heat. Example: Lobster à l'Americaine (p. 644).

3. The product is simply placed in a pot with liquids and flavoring ingredients. The pot is covered, and the item is steamed or simmered. Example: Moules Marinière (p. 645).

COOKING EN PAPILLOTE

An unusual version of the third variation above is called cooking **en papillote** (on poppy-yote), or in paper. The fish item, plus flavoring ingredients and sauce, is tightly enclosed in a piece of parchment so steam cannot escape. When the paper package is heated, the item steams in its own moisture. All the juices, flavors, and aromas are held inside the paper, which is not opened until it is placed before the customer.

Sometimes a starch-thickened sauce is used in cooking fish en papillote. In this case, the fish is usually precooked (poached) so it will not exude juices that would dilute and spoil the sauce. The problem with this method is that the fish is often overcooked by the time it reaches the customer.

STOVETOP STEAMING

The simplest method of steaming is to place the item on a rack above simmering liquid, cover the pan, and steam until done. Because it does not come in contact with seasoned liquid or fat during cooking, steamed seafood is usually delicate in flavor, even more so than submersion-poached seafood. For this reason, this method is usually used when the cook wants to emphasize the pure, natural taste of the best-quality seafood.

Three methods are available to season and flavor steamed seafood:

1. Apply salt and other seasonings directly to the item.

2. Season the steaming liquid with aromatic herbs and other ingredients. The steam rising from the liquid carries a little of the aroma to the fish. This method gives only the subtlest and most delicate flavors to fish and is most effectively used with mild-tasting seafood.

3. Serve the seafood with an appropriate sauce. Accompaniments as simple as melted butter and lemon wedges are classic accompaniments to steamed seafood.

KEY POINTS TO REVIEW

- What are the steps in the basic procedure for poaching fish in court bouillon (submersion poaching)? Describe guidelines for success in poaching fish in court bouillon.

- What tools and procedures are used for steaming fish? How is fish cooked en papillote?

COMPARTMENT STEAMING

Two precautions should be observed if you cook fish and shellfish in a compartment steamer:

1. Watch the cooking time carefully. Fish cooks quickly, especially in the high heat of a steamer, and is easily overcooked.

2. Avoid pressure-steaming fish and shellfish, if possible. The high temperatures toughen fish protein very quickly. Lobster tails and similar items can become rubbery.

Lobster à l'Americaine

PORTIONS: 2   **PORTION SIZE:** ½ **LOBSTER**

U.S.	METRIC	INGREDIENTS	PROCEDURE
1	1	Live lobster, about 1½ (700 g)	1. Cut up the lobster as shown in **Figure 19.14.**
1 oz	30 g	Butter, softened	2. Remove the tomalley (liver) and coral (if any). Mash them in a small bowl with the soft butter.
2 fl oz	60 mL	Oil	3. Heat the oil in a sauté pan and add the lobster pieces. Sauté over high heat until the shells turn red.
1 tbsp	15 mL	Finely chopped shallot	
½ tsp	2 mL	Finely chopped garlic	4. Drain off the oil by tilting the pan and holding the lobster in with the pan lid.
2 fl oz	60 mL	Brandy	5. Add the shallot and garlic to the pan. Sauté for a few seconds.
6 fl oz	200 mL	White wine	6. Remove from the heat (to avoid burning yourself if the brandy flares up) and add the brandy. Return to the heat and add the wine, fish stock, tomato, parsley, tarragon, and cayenne.
4 fl oz	125 mL	Fish stock	
4 oz	125 g	Tomato concassé (p. 295)	
		or	7. Cover the pan and simmer until the lobster is cooked, about 10–15 minutes.
2 oz	60 g	Tomato purée	8. Remove the lobster from the cooking liquid and place it on a serving platter or in broad soup plates for service. The meat may be left in the shell or removed from the shell, as desired.
1 tbsp	15 mL	Chopped parsley	
¼ tsp	1 mL	Dried tarragon	9. Reduce the cooking liquid over high heat to about 6 oz (175 mL).
pinch	pinch	Cayenne	10. Remove from the heat and stir in the mixture of butter, tomalley, and coral from step 2. Heat the sauce gently for a minute, but do not boil or it will curdle. Adjust the seasoning.
			11. Strain the sauce and pour it over the lobster. Serve immediately.

Per serving: Calories, 340; Protein, 7 g; Fat, 26 g (67% cal.); Cholesterol, 50 mg; Carbohydrates, 5 g; Fiber, 1 g; Sodium, 230 mg.

VARIATIONS

Lobster Newburg

Prepare through step 4. Omit remaining ingredients. Instead, add 1 tbsp (15 mL) brandy, 3 tbsp (45 mL) sherry or Marsala or Madeira wine, and 3 fl oz (100 mL) fish stock. Cover and simmer as in basic recipe. Remove lobster meat from shells and discard shells. Reduce cooking liquid by half and add 1 cup (250 mL) heavy cream or light cream sauce. Reduce the sauce slightly and finish by adding the mixture of butter, tomalley, and coral (step 10 of basic recipe). If desired, flavor with more sherry. Pour sauce over the lobster meat.

Shrimp à l'Americaine or Shrimp Newburg

Shrimp may be cooked using the main recipe or the variation. They should be shelled before combining with the finished sauce.

Moules Marinière (Steamed Mussels)

PORTIONS: 10 **PORTION SIZE: APPROX. 12 OZ (360 G)**

U.S.	METRIC	INGREDIENTS
7 lb	3.2 kg	Mussels, in shells
3 oz	90 g	Shallots or onions, chopped fine
6	6	Parsley stems
1/4 tsp	1 mL	Pepper
1 cup	250 mL	White wine
1/4 cup	60 mL	Chopped parsley
3 oz	90 g	Butter
to taste	to taste	Salt
to taste	to taste	Lemon juice

PROCEDURE

1. Scrub the mussels well with a stiff brush and remove the beards. Clean them well by soaking them according to the procedure given in Chapter 19.
2. Place the mussels in a stockpot or large saucepot. Add the shallots or onions, parsley stems, pepper, and wine.
3. Cover the pot and set it over moderately high heat. Cook until the mussels open, about 5 minutes.
4. Drain the mussels and strain the liquid through cheesecloth into a broad saucepan. Bring to a boil.
5. Add the parsley and butter. Swirl the liquid in the pan until the butter is melted. Season to taste with salt and a few drops of lemon juice.
6. For service, remove the top shells of the mussels (or leave them on, if desired). Place the mussels in broad soup plates and pour the sauce over them.

Per serving: Calories, 220; Protein, 20 g; Fat, 11 g (44% cal.); Cholesterol, 65 mg; Carbohydrates, 8 g; Fiber, 0 g; Sodium, 540 mg.

VARIATION

Steamed Mussels (without wine)

Substitute water for the wine and add 2 fl oz (60 mL) lemon juice. Increase the onion or shallot to 6 oz (175 g) and add 3 oz (90 g) sliced celery.

Mussels in Cream

Prepare the basic recipe. Reduce the cooking liquid by half and add 1 cup (250 mL) heavy cream or a liaison of 2 egg yolks and 1 cup (250 mL) heavy cream.

Moules Marinière

Mackerel en Papillote

PORTIONS: 1   PORTION SIZE: 4 OZ (125 G)

U.S.	METRIC	INGREDIENTS
1	1	Mackerel fillet, 4 oz (125 g) (see Note)
2 tsp	10 mL	Melted butter
to taste	to taste	Salt
to taste	to taste	White pepper
2 tsp	10 mL	Chopped parsley
pinch	pinch	Dried marjoram
1 tsp	5 mL	Very finely chopped shallots
2	2	Thin lemon slices

PROCEDURE

1. Cut out a piece of parchment in a heart shape, as shown in **Figure 20.2**. (Foil may be used instead of parchment.) The piece must be big enough to hold the fish and still have room for crimping the edges. Oil the parchment and place on the workbench oiled side down. (If using foil, place it oiled side up.)

2. Place the fillet on one side of the heart. Brush with melted butter and sprinkle with salt, pepper, parsley, marjoram, and chopped shallot. Lay the lemon slices on top.

3. Fold and crimp the parchment, as shown in the illustration, to enclose the fish tightly.

4. Place the folded package in a sauté pan or, if several orders are being done at once, on a sheet pan. Set on the range to start the cooking.

5. As soon as the paper begins to puff, place the pan in a hot oven (450°F/230°C). Bake until the parchment is puffed and browned, 5–8 minutes. (If the paper doesn't brown, you may run it under the broiler for a second.)

6. Serve immediately. The parchment should be cut open in front of the customer.

Per serving: Calories, 290; Protein, 20 g; Fat, 23 g (71% cal.); Cholesterol, 85 mg; Carbohydrates, 1 g; Fiber, 0 g; Sodium, 150 mg.

Note: Pompano or bluefish may be used.

FIGURE 20.2 Preparing foods en papillote.

(a) Cut out a heart-shaped piece of parchment by folding a parchment sheet in half and cutting half a heart from the folded side. Oil or butter the parchment and place on the work surface, oiled side down.

(b) Place the fish fillet or other item plus any sauce, topping, or seasoning, on one side of the heart.

(c) Fold over the other half of the heart. Starting at the top of the fold, make a small crimp in the edges as shown.

(d) Continue crimping around the edge. Each crimp holds the previous one in place.

(e) When you reach the bottom of the heart, fold the point under to hold it in place. The papillote is now ready for cooking.

Steamed Sea Bass with Garlic and Ginger

PORTIONS: 10   **PORTION SIZE: 5–6 OZ (150–180 G)**

U.S.	METRIC	INGREDIENTS	PROCEDURE
10	10	Sea bass fillets, skin on, 5–6 oz (150–180 g) each	1. With the point of a sharp knife, lightly score the skin of the sea bass, making 3 or 4 diagonal cuts across the fillets.
10	10	Scallions, trimmed, halved lengthwise	2. Arrange the scallion halves on the rack of a steamer.
10 thin slices	10 thin slices	Fresh ginger root, peeled	3. Place the fillets, skin side up, on top of the scallions.
4–5	4–5	Garlic cloves	4. Cut the ginger slices into thin shreds.
			5. Cut the garlic cloves into paper-thin slices.
			6. Scatter the ginger and garlic on top of the fillets.
			7. Pour water into the bottom of a steamer and bring to a boil.
			8. Place the steamer rack holding the fish in the steamer, cover, and steam just until the fish is done, about 4–8 minutes, depending on the thickness of the fish.
5	5	Scallions, sliced thin	9. Remove the fish and the scallions from the steamer. Discard the scallion halves.
5 fl oz	150 mL	Soy sauce	10. Place the fillets on hot plates.
5 fl oz	150 mL	Peanut oil	11. Sprinkle the tops of the fillets with the scallion slices.
			12. Pour 1 tbsp (15 mL) soy sauce over each portion.
			13. Heat the peanut oil until it is very hot, almost smoking.
			14. Pour 1 tbsp (15 mL) oil over each portion.
			15. Serve immediately.

Per serving: Calories, 280; Protein, 29 g; Fat, 17 g (54% cal.); Cholesterol, 60 mg; Carbohydrates, 4 g; Fiber, 1 g; Sodium, 1130 mg.

VARIATION

Instead of finishing with soy sauce and hot oil, serve with Oriental Vinaigrette (p. 662).

Fisherman's Stew

PORTIONS: 10   **PORTION SIZE: SEE STEP 9**

U.S.	METRIC	INGREDIENTS
2 lb	900 g	Fish steaks or fillets (see Note for suggested fish)
10	10	Clams, in shell
20	20	Mussels, in shell (or 10 more clams)
5	5	Lobster tails, small (or 10 large shrimp)
4 fl oz	125 mL	Olive oil
8 oz	250 g	Onions, sliced
8 oz	250 g	Leeks, cut julienne
2 tsp	10 mL	Chopped garlic
1/4 tsp	1 mL	Fennel seed
12 oz	350 g	Tomato concassé or drained, chopped canned tomato
2 qt	2 L	Fish stock
4 fl oz	100 mL	White wine (optional but recommended)
2	2	Bay leaves
2 tbsp	30 mL	Chopped parsley
1/4 tsp	1 mL	Dried thyme
2 tsp	10 mL	Salt
1/4 tsp	1 mL	Pepper
20–30	20–30	French bread slices, dry or toasted

PROCEDURE

1. Cut the fish into 3-oz (90-g) serving pieces.
2. Scrub the clams and mussels well.
3. Cut the lobster tails in half lengthwise with a heavy chef's knife. Remove the intestinal vein.
4. Heat the oil in a heavy saucepot or large straight-sided sauté pan.
5. Add the onions, leeks, garlic, and fennel seed. Sweat them in the oil for a few minutes.
6. Add the pieces of fish and the lobster tails (or shrimp). Cover and cook over low heat for a few minutes to begin extracting juice from the fish.
7. Remove the cover and add the clams and mussels.
8. Add the tomato, fish stock, wine, bay leaves, parsley, thyme, salt, and pepper. Cover and bring to a boil. Reduce heat and simmer 5–10 minutes, or until the clams and mussels are open.
9. To serve, place 2 or 3 thin slices of French bread in the bottoms of soup plates. For each portion, place 1 piece of fish, 1 clam, 2 mussels, and 1/2 lobster tail in each plate. Ladle 8 fl oz (250 mL) broth and vegetables over the fish.

Per serving: Calories, 470; Protein, 32 g; Fat, 26 g (51% cal.); Cholesterol, 90 mg; Carbohydrates, 25 g; Fiber, 2 g; Sodium, 910 mg.

Note: Any firm fish may be used, such as halibut, cod, haddock, sea bass or striped bass, red snapper, or mackerel. Avoid delicate fish like flounder or sole, which break up easily during cooking. If desired, reduce or eliminate shellfish from the recipe and increase the quantity of fish.

Fisherman's Stew

Sea Bass Sous Vide with Asian Sweet-and-Sour Sauce

PORTIONS: 4   PORTION SIZE: 1 FISH FILLET, 6 OZ (180 G)

U.S.	METRIC	INGREDIENTS
1 oz	30 g	Fresh ginger root
as needed	as needed	Water
4	4	Sea bass fillets, skin on, 6 oz each
to taste	to taste	Salt

PROCEDURE

1. Review the guidelines for safe sous vide cooking on pages 117–119.
2. Peel and grate the fresh ginger. Gather the grated ginger in a small square of cheesecloth and squeeze out the juice into a small ramekin. Discard the solids. Dilute the juice with approximately 3 times its volume of water.
3. Brush the skin side of the fillets lightly with the ginger juice.
4. Sprinkle both sides of the fillets lightly with salt.
5. Vacuum-pack each fillet in a plastic bag.
6. Cook at 140°F (60°C) for 12 minutes in a thermal bath (see p. 117).

U.S.	METRIC	INGREDIENTS
2 fl oz	60 mL	Oil
6 fl oz	180 mL	Asian Sweet-and-Sour Sauce (p. 212)
1 fl oz	30 mL	Basil Oil (p. 216) or Basil Sauce (p. 218), if possible made with Thai basil
as needed	as needed	Thai basil or cilantro leaves

7. Remove the fillets from their bags. Heat the oil in a sauté pan over high heat. Carefully place the fillets in the pan, skin side down, and cook just until the skin is lightly browned, about 1 minute.
8. For each portion, spoon a pool of 1½ fl oz (45 mL) sauce in the center of each plate. Spread the sauce a little with the back of a spoon. Place a fillet on the sauce, skin side up. Drizzle a little basil oil or basil sauce around the outside of the sweet-sour sauce.
9. Garnish the top of the fish with a few leaves of Thai basil or cilantro. If desired, deep-fry the herb garnish.

Per serving: Calories, 480; Protein, 33 g; Fat, 28 g (53% cal.); Cholesterol, 75 mg; Carbohydrates, 22 g; Fiber, 2 g; Sodium, 125 mg.

PROCEDURE VARIATION: COMBI OVEN

If an immersion circulator thermal bath is not available, cook the vacuum-packed fish in a combi oven at 140°F (60°C) in steam mode for 12 minutes.

Sea Bass Sous Vide with Asian Sweet-and-Sour Sauce

Butter-Poached Salmon Sous Vide

PORTIONS: 2   **PORTION SIZE: 6 OZ (180 G)**

U.S.	METRIC	INGREDIENTS	PROCEDURE
6 oz	180 g	Butter	1. Combine the butter, thyme, fennel seed, bay leaf, and peppercorns in a small saucepan. Heat to melt the butter.
¹/₂ tsp	2 mL	Fresh thyme	2. Clarify the butter (see p. 190 for procedure) and strain out the seasonings.
¹/₄ tsp	1 mL	Fennel seeds	
¹/₂	¹/₂	Bay leaf	
8	8	Peppercorns	
2	2	Salmon fillets, 6 oz each	3. Season the salmon fillets with salt and white pepper.
to taste	to taste	Salt	4. Place the salmon fillets in vacuum bags, add the clarified butter, and vacuum seal at 95 percent vacuum.
to taste	to taste	White pepper	5. Cook the salmon in an immersion circulator heated to 128°F (53°C) for 25 minutes, until the internal temperature of the salmon is 128°F (53°C).
			6. If using immediately, carefully remove the salmon from the bag, and place the salmon on a towel to remove excess butter. Serve.
			7. If storing for later service, plunge the sealed bags of salmon in ice water until the salmon has cooled to at least 41°F (5°C). Refrigerate.

Per serving: Calories, 520; Protein, 39 g; Fat, 41 g (71% cal.); Cholesterol, 195 mg; Carbohydrates, 0 g; Fiber, 0 g; Sodium, 85 mg.

Butter Poached Salmon Sous Vide on Orange and Tarragon Fennel Sous Vide, with Red Pepper Coulis, Asparagus Cream, and a salad of shaved fennel, asparagus, and red onion with lemon and olive oil

SEAFOOD SERVED RAW

Raw oysters and clams have long been enjoyed in North American restaurants as well as in homes. Smoked salmon, too, has always been a popular raw seafood item, although many people don't think of it as raw fish because it has been processed.

More recently, with the popularity of Japanese sushi and sashimi, many more people enjoy eating items made with raw fish, and chefs are experimenting with new ways to serve raw seafood items. Because of the danger of contamination from polluted fishing waters, however, many health officials advise against serving raw seafood. Nevertheless, these items remain popular. If you choose to serve raw seafood, observe the following guidelines:

1. Use only the freshest fish.

2. Buy the fish from a reliable purveyor.

3. Use only saltwater seafood from clean waters. Do not use freshwater fish, which is more likely than saltwater fish to contain parasites. To destroy possible parasites in saltwater fish, freeze it, and then thaw it before use.

4. Observe the strictest sanitation procedures.

5. Keep the fish cold. Handle it as little as possible.

Four kinds of raw fish recipes are included in this section. **Fish tartare** is a mixture of chopped raw fish mixed with various condiments and seasonings. **Fish carpaccio** is a dish consisting of very thin slices of firm, meaty fish such as tuna, served with various garnishes and usually with a piquant sauce such as a vinaigrette. *Carpaccio* (karpachio), is the Italian word for a dish of thin slices of raw beef, served the same way. **Seviche**, or *ceviche* (seh vee chay, or, in Spanish, seh bee chay), is a preparation, native to Latin America, of raw seafood marinated in an acid mixture. The acid coagulates the protein so the texture of the raw fish resembles that of cooked fish. It is common to say the acid "cooks" the fish. However, because the food is not subjected to heat, it is important to still think of the item as raw, for sanitation reasons. **Sushi** is seasoned Japanese short-grain rice garnished with one or more foods, usually, but not always, raw fish.

Sushi has become especially popular in Western countries, but there is much confusion about it. Sushi is often equated with raw fish, but in fact the term refers to cooked rice lightly flavored with seasoned vinegar. The rice is served with various garnishes, which include not only raw fish (the most popular garnish) but also cooked seafood and many kinds of vegetables and pickles.

KEY POINTS TO REVIEW

- What are five safety guidelines for serving raw fish items?

- What is seviche? How is it made?

- What is the basic meaning of the term *sushi*? How is sushi rice made?

- What is seafood tartare, and how is it made?

- What is tuna carpaccio, and how is it made?

Tuna Carpaccio

PORTIONS: 8   **PORTION SIZE: 2¹/₂ OZ (75 G)**

U.S.	METRIC	INGREDIENTS	PROCEDURE
1 lb 4 oz	600 g	Yellowfin tuna or other red-fleshed tuna	1. On a slicing machine, or with a very sharp knife, cut the tuna into slices weighing 2¹/₂ oz (75 g) each. 2. Place each slice between two sheets of plastic film. Carefully pound each slice until it is paper thin, being careful to keep it of even thickness. 3. If the carpaccio is not to be served immediately, keep it between the layers of plastic film and refrigerate.
8–12 oz	240–360 g	Micro greens or baby arugula	4. Lay each portion of tuna flat in the center of a cold plate. 5. Toss the greens with just enough of the vinaigrette to coat them lightly.
8–12 fl oz	240–360 mL	Mustard Vinaigrette (p. 660)	6. Arrange the greens in a mound in the center of the sheet of tuna. 7. Drizzle the remaining vinaigrette over the tuna.

Per serving: Calories, 270; Protein, 18 g; Fat, 22 g (72% cal.); Cholesterol, 35 mg; Carbohydrates, 1 g; Fiber, 1 g; Sodium, 280 mg.

Sushi Rice

YIELD: ABOUT 2¹/₂ LB (1.1 KG)

U.S.	METRIC	INGREDIENTS	PROCEDURE
3 cups	750 mL	Japanese short-grain rice (see Note)	1. Wash the rice in several changes of cold water. Drain well.
3¹/₂ cups	875 mL	Water, cold	2. Put the drained rice in a heavy saucepan and add the measured water. Cover tightly and let stand at least 30 minutes.
			3. With the cover in place, set the pan over high heat and bring to a boil. When the water boils, reduce the heat to medium and let cook until all the water is absorbed. Do not remove the cover to check, but listen to the sounds. The bubbling will stop, and there will be a faint hissing sound.
			4. Reduce the heat to very low and cook another 5 minutes. Then remove from heat and let stand at least 15 minutes before removing the cover. You now have the basic white rice that is eaten with Japanese meals.
3–4 fl oz	100–125 mL	Sushi vinegar (see Note)	5. In Japan, mixing in the vinegar is done in special wooden tub used only for this purpose (to avoid off flavors). The advantage of wood is that it absorbs excess moisture. If you use a nonabsorbent mixing bowl, transfer the rice to a clean bowl whenever the mixing bowl becomes coated with moisture. Using a wooden paddle or plastic spatula, break up the hot rice to get rid of all lumps. At the same time, fan the rice to cool it.
			6. When the rice is slightly warm to the touch, begin adding the sushi vinegar. Add a little at a time while mixing gently. The rice is ready when it has a glossy appearance and a very mild taste of the vinegar. The vinegared rice is best if used within 2–3 hours, and it must not be refrigerated.

Per 1 ounce: Calories, 40; Protein, 1 g; Fat, 0g (0% cal.); Cholesterol, 0 mg; Carbohydrates, 9 g; Fiber, 0 g; Sodium, 105 mg.

Note: Do not confuse Japanese short-grain rice with glutinous rice, which is an entirely different product.
Sushi vinegar is commercially available, but it can also be made in the kitchen. Combine 1 pint (500 mL) Japanese rice vinegar, 8 oz (250 g) sugar, and 4 oz (125 g) salt. Heat and stir until the sugar and salt are dissolved, then cool.

VARIATIONS

Nigirizushi (Finger Sushi)

1. Prepare wasabi (green horseradish) by mixing wasabi powder with a little water to form a thick paste. Let stand, covered, a few minutes to allow flavor to develop.

2. Prepare sushi toppings by cutting very fresh fish fillets (use saltwater fish or smoked salmon) into slices about 1¹/₂ × 2¹/₂ in. (4 × 6 cm). Tuna, the most popular fish for sushi, is tender and is usually cut about ¹/₄ in. (6 mm) thick. Other fish are cut thinner.

3. Wet your hands with cold water to keep the rice from sticking to them, then pick up about 2 tbsp (30 mL) sushi rice. Shape it into firm oval about 1¹/₂ in. (4 cm) long (see **Figure 20.3**). Pick up a slice of fish in one hand. Dip a finger of the other hand in the wasabi and spread a very small amount on the underside of the fish slice. Drape the fish over the rice, with the wasabi underneath next to the rice, and press it gently but firmly in place. Serve with soy sauce for dipping.

Chirashizushi (Scatter Sushi)

Fill a serving bowl half full of sushi rice. Carefully and attractively arrange an assortment of raw fish, cooked shrimp, crabmeat, and neatly cut vegetables such as snow peas, cucumbers, carrots, mushrooms, and pickled ginger on top of the rice.

Kappa-maki (Cucumber Roll)

To make rolled sushi, you will need a special bamboo mat called a sudare (see **Figure 20.3**). You could also use a sheet of parchment, but the roll will be harder to make.

1. Peel a cucumber. Cut it in half and scrape out the seeds. Cut lengthwise into julienne.

2. Cut a sheet of nori (a type of seaweed for rolled sushi) in half crosswise. Toast it by passing it briefly above a burner flame, being careful not to burn it.

3. Put the bamboo mat on the table in front of you with the bamboo strips horizontal. Put the half-sheet of nori on the mat, smooth side down.

4. Cover the two-thirds of the nori closest to you with a layer of sushi rice about ¹/₄ in. (6 mm) thick.

5. Spread a light streak of wasabi from right to left across the middle of the rice.

6. Lay strips of cucumber evenly on top of the strip of wasabi.

7. Lift the edge of the mat closest to you and roll up firmly. This is best done by lifting the mat with the thumbs while holding the cucumber in place with the fingers. Press the roll in the mat gently but firmly to make it tight.

8. Wipe the blade of a very sharp knife, then cut the roll in half crosswise. Do not saw the roll but cut it cleanly with a single stroke. Wiping the blade on a damp cloth after every cut, cut each half-roll into 3 or 4 pieces.

Tekka-maki (Tuna Roll)

Make kappa-maki, but instead of cucumber, use raw tuna cut into bâtonnet strips.

FIGURE 20.3 Making sushi.

(a) For nigirizushi, wet your hands with cold water to keep the rice from sticking to them. Form a bit of rice into a small oval in one hand.

(b) In the other hand, pick up a slice of fish and smear a dab of wasabi on the bottom of it.

(c) Top the rice with the fish, wasabi side against the rice, and press in place in the palm of one hand with two fingers of the other.

(d) Turn it over and press the same way. Then press the sides with the thumb and forefinger to finish shaping it.

(e) For rolled sushi, put a half-piece of nori on the sudare (bamboo mat). Wetting your hands with cold water to keep the rice from sticking to them, spread a layer of sushi rice over the bottom two-thirds of the nori.

(f) Lay strips of the filling across the middle of the rice.

(g) Holding the filling in place with the fingers, lift the corner of the mat with the thumbs and roll up.

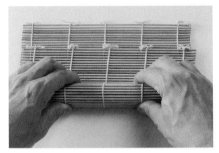

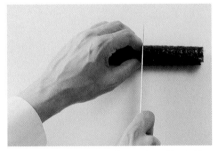

(h) Press the mat firmly and evenly to make a tight roll.

(i) Using a dampened knife, cut the roll in half with a single forward stroke.

(j) Wipe the blade between cuts, and cut each half-roll into 3 or 4 pieces.

SUSHI: A JAPANESE CLASSIC

Fish and rice dominate the cuisine of the island nation of Japan, and these two staples are combined in sushi, a classic Japanese preparation that is especially popular in the West. Sushi originated many years ago as a way of preserving fish by salting it and packing it with vinegared rice. When the fish was eaten, the rice was discarded.

In Japan, it takes many years of apprenticeship to become a sushi master, or *itamae* (ee-tah-mah-eh). A master is skilled in preparing rice properly, making rice fingers of just the right size and compactness for nigirizushi, and especially in selecting, judging, and cutting the many varieties of fresh seafood used as toppings.

When one is eating nigirizushi, it is considered improper to dip the rice ball into soy sauce. The large amount of soy sauce absorbed by the rice masks the fresh flavor of the toppings and makes the rice ball break up. Instead, turn the sushi over and dip the seafood topping lightly in the sauce.

Incidentally, in Japanese, an initial S sound, as in the word *sushi,* may change to a Z sound when it is in the middle of a compound word, as in *nigirizushi.*

Tuna Tartare

YIELD: APPROX. 1 LB 12 OZ (850 G)   **PORTIONS 8**   **PORTION SIZE: 3¹/₂ OZ (105 G)**

U.S.	METRIC	INGREDIENTS
1 lb 8 oz	720 g	Sashimi-quality tuna, well trimmed
1 oz	30 g	Shallots, minced
2 tbsp	30 mL	Parsley, chopped
2 tbsp	30 mL	Fresh tarragon, chopped
2 tbsp	30 mL	Lime juice
1 fl oz	30 mL	Dijon-style mustard
2 fl oz	60 mL	Olive oil
to taste	to taste	Salt
to taste	to taste	White pepper

PROCEDURE

1. Mince the tuna with a knife. The most effective way to do this is to cut it into thin slices, cut the slices into thin strips, and cut the strips into very small dice. (The fish may also be chopped with a food processor, although the result is not as good as hand cutting. Pulse the machine, and be careful not to process the fish to a paste.) Do not pass the fish through a grinder, as this produces an undesirable texture.

2. Mixed the chopped tuna with the remaining ingredients shortly before serving, seasoning to taste with salt and pepper. Do not mix too long in advance, as the acidity of the lime juice coagulates the protein as for seviche (p. 651).

3. Tuna tartare may be served very simply mounded on a small plate or in more elaborate ways (see variations).

Per serving: Calories, 200; Protein, 21 g; Fat, 12 g (54% cal.); Cholesterol, 35 mg; Carbohydrates, 2 g; Fiber, 0 g; Sodium, 130 mg.

VARIATIONS

Omit the shallots, parsley, tarragon, lime juice, and mustard. Add some chopped chives and a dash of hot pepper sauce.

Place a round biscuit cutter in the center of a plate. Fill it with the tartare and smooth the top. Remove the cutter (see **Figure 20.4**). Garnish and decorate as desired. For example, place a fresh herb leaf or dab of caviar on top, and place spoonfuls of caviar around the disk of tartare. With a squeeze bottle or spoon, place a ring of vinaigrette on the plate around the tartare.

Salmon Tartare

Substitute salmon for the tuna. Be sure all bones, skin, and sinews are removed from the salmon. In addition, cut away the dark, fatty tissue that runs down the centerline of the fillet on the skin side.

Tartare of Salmon and Sea Bass

In place of the tuna, use 12 oz (340 g) sea bass and 12 oz (340 g) salmon.

Tuna Tartare

FIGURE 20.4 Shaping food with a ring mold.

(a) Position a ring mold on the plate. Fill with the food and level the top with a palette knife or spatula.

(b) Carefully lift off the mold.

Scallop Seviche

PORTIONS: 12   PORTION SIZE: 4 OZ (125 G)

U.S.	METRIC	INGREDIENTS	PROCEDURE
2 lb	1 kg	Sea scallops (see Note)	1. Cut the scallops into quarters vertically, then slice crosswise $^1/_4$ in. (6 mm) thick.
8 fl oz	250 mL	Lime juice	2. In a nonreactive container, mix together the scallops, lime juice, and onion.
4 oz	125 g	Red onion, chopped fine	3. Refrigerate about 12 hours, or until the scallops have the texture of cooked scallops.
1	1	Jalapeño, seeded and chopped fine	4. Drain the scallops and onions, discarding the lime juice.
6 tbsp	90 mL	Chopped cilantro	5. Gently toss the scallops with the jalapeño, cilantro, avocado, and tomato.
2	2	Avocados, medium dice	6. Add salt to taste.
8 oz	250 g	Tomato, peeled and seeded, small dice	7. Mix in the olive oil.
to taste	to taste	Salt	
4 fl oz	125 mL	Olive oil	
as needed	as needed	Lettuce leaves	8. To serve, line small bowls or cups with lettuce leaves and spoon in the seviche.

Per serving: Calories, 220; Protein, 14 g; Fat, 15 g (61% cal.); Cholesterol, 25 mg; Carbohydrates, 8 g; Fiber, 2 g; Sodium, 130 mg.

Note: As an alternative to sea scallops, use bay scallops and leave them whole.

Scallop Seviche

ADDITIONAL RECIPES

These additional recipes may be found on your CulinarE-Companion recipe management program:

Baked Stuffed Mackerel; Broiled Mako Shark Steaks with Browned Garlic Vinaigrette; Pan-Fried Catfish with Shrimp Etouffée; Pesce con Salsa Verde; Sautéed Scallops with Tomato, Garlic, and Parsley; Seafood Casserole au Gratin; Seafood Newburg.

TERMS FOR REVIEW

à la meunière	**étuver**	**fish tartare**	**seviche**
court bouillon	**en papillote**	**fish carpaccio**	**sushi**
truite au bleu			

QUESTIONS FOR DISCUSSION

1. What major precaution must be taken when baking or broiling lean fish?
2. Describe the procedure for cooking fish à la meunière.
3. Which side of a fish fillet is the presentation side?
4. What techniques can you use for lifting whole poached fish out of court bouillon without breaking it or damaging its appearance?
5. What temperatures are best for poaching fish?
6. What techniques can be used to make a sauce that is suitable for glazing poached fish?
7. Discuss the precautions a cook should take to ensure that raw fish is safe to eat.

21

SALAD DRESSINGS AND SALADS

In the days before modern refrigeration, the pantry was the storeroom where food products were kept before being brought into the kitchen. Because this room was cooler than the kitchen, it was especially suited as a work area for the production of cold food, especially aspics, chaud-froids, and other elaborate buffet preparations. In kitchens around the world, this department is often referred to by its French name, *garde manger* (gard mawn zhay).

Today, the pantry is the department responsible for cold foods and related items. This does not mean that no cooking is done in the pantry. On the contrary, garde manger chefs must be masters of a wide range of cooking techniques. In addition, they must have artistic judgment as well as the patience and dexterity to perform a great many hand operations quickly and efficiently.

This chapter deals with two groups of items prepared in the pantry: salads and salad dressings.

AFTER READING THIS CHAPTER, YOU SHOULD BE ABLE TO

1. Identify the major salad dressing ingredients.

2. Prepare the following: oil and vinegar dressings, mayonnaise and mayonnaise-based dressings, cooked dressings, and specialty dressings.

3. Identify and describe five salad types, and select appropriate recipes for use as appetizer, accompaniment, main course, separate course, or dessert salad.

4. Identify a dozen popular salad greens, list six categories of other salad ingredients, and recognize several examples from each category.

5. Judge the quality of fruits, and complete the pre-preparation procedures for fruit.

6. Identify the four basic parts of a salad.

7. Prepare and arrange salads that achieve maximum eye appeal.

8. Set up an efficient system for producing salads in quantity.

9. Prepare the following types of salads: green, vegetable, bound, fruit, combination, and gelatin.

10. Set up a successful salad bar and buffet service.

SALAD DRESSINGS

Salad dressings are liquids or semiliquids used to flavor salads. They are sometimes considered cold sauces, and they serve the same functions as sauces—that is, they flavor, moisten, and enrich.

Most of the basic salad dressings used today can be divided into three categories:

1. Oil and vinegar dressings (most unthickened dressings).

2. Mayonnaise-based dressings (most thickened dressings).

3. Cooked dressings (similar in appearance to mayonnaise dressings, but more tart and with little or no oil content).

A number of dressings have as their main ingredient such products as sour cream, yogurt, and fruit juices. Many of these are designed specifically for fruit salads or for low-calorie diets.

Most salad dressings are *emulsions,* either temporary or permanent. As background to the study of salad dressings, review the discussion of emulsions in Chapter 6 (p. 109).

INGREDIENTS

Because the flavors of most salad dressings are not modified by cooking, their quality depends directly on the quality of the ingredients.

Most salad dressings are made primarily of an oil and an acid, with other ingredients added to modify the flavor or texture.

OILS

Kinds

Corn oil is widely used in dressings. It has a light golden color and is nearly tasteless, except for a mild cornmeal-type flavor.

Cottonseed oil, *soybean oil*, *canola oil*, and *safflower oil* are bland, nearly tasteless oils. *Vegetable oil* or *salad oil* is a blend of oils and is popular because of its neutral flavor and relatively low cost.

Peanut oil has a mild but distinctive flavor and may be used in appropriate dressings. It is somewhat more expensive.

Olive oil has a distinctive, fruity flavor and aroma and a greenish color. The best olive oils are called *virgin* or *extra-virgin*, which means they are made from the first pressing of the olives. Because of its flavor, olive oil is not an all-purpose oil but may be used in specialty salads such as Caesar salad.

Walnut oil has a distinctive flavor and a high price. It is occasionally used in fine restaurants featuring specialty salads. Other nut and seed oils, such as *hazelnut oil* and *grapeseed oil*, are sometimes used.

Quality Factors

All-purpose oils for dressings should have a mild, sweet flavor. Strongly flavored oils can make excellent salad dressings but are not appropriate with every food.

Winterized oil should be used with dressings that are to be refrigerated. These oils have been treated so they remain a clear liquid when chilled.

Rancidity is a serious problem with oils because even a hint of a rancid flavor can ruin an entire batch of dressing. A thin film of oil, such as might be left on containers through careless washing, becomes rancid very quickly. Clean all dressing containers thoroughly, and never pour a fresh batch into a jar containing older dressing.

VINEGAR

Kinds

Cider vinegar is made from apples. It is brown in color and has a slightly sweet apple taste.

White or *distilled vinegar* is distilled from grain alcohol and purified so that it has a neutral flavor.

Wine vinegar may be white or red, and it has, naturally, a winy flavor.

Flavored vinegars have had another product added to them, such as tarragon, garlic, or raspberries.

Sherry vinegar is made from sherry wine and, consequently, has the distinctive flavor of that wine.

Balsamic vinegar is a special wine vinegar aged in wooden barrels (see sidebar). It is dark brown in color and has a noticeably sweet taste.

Other specialty vinegars include malt vinegar, rice vinegar, and vinegars flavored with fruits.

Quality Factors

Vinegars should have a good, clean, sharp flavor for their type.

Strength of acidity determines the tartness of the vinegar—and of the dressing made from it. Most salad vinegars are about 5 percent acidity, but some range as high as 7 or 8 percent. Read the label for this information. Vinegar that is too strong should be diluted with a little water before it is measured for a recipe.

White vinegar is used when a completely neutral flavor is desired for a dressing. Other vinegars are used for their characteristic flavors. Wine vinegars are usually preferred for the best-quality oil-and-vinegar dressings.

LEMON JUICE

Fresh lemon juice may be used in place of or in addition to vinegar in some preparations, when its flavor is desired.

VERJUS

Verjus, also called *verjuice*, is the unfermented juice of unripe green grapes. It has a high acidity level, although not as high as vinegar. Like lemon juice, it can be used in place of vinegar in salad dressings.

EGG YOLK

Egg yolk is an essential ingredient in mayonnaise and other emulsified dressings. For safety, pasteurized eggs should be used (see p. 766 and Appendix 5, p. 1030), and the finished product should be refrigerated to guard against spoilage.

SEASONINGS AND FLAVORINGS

Nearly any herb or spice can be used in salad dressings. Fresh herbs are preferable to dried herbs as flavorings, especially when the dressings are used for simple, light mixed green salads. Remember that dried herbs and spices need extra time to release their flavors if they are not heated in the product. This is why most dressings are best made at least two or three hours before serving. Review Chapter 6 to refresh your memory on the use of herbs and spices.

Other ingredients added for flavoring include mustard, ketchup, Worcestershire sauce, and cheeses.

A note on blue cheese and **Roquefort cheese**: Many restaurants sell "Roquefort dressing" that is actually blue cheese dressing. Roquefort is a brand name for a special kind of blue cheese made in Roquefort, France. It is made of sheep's milk, has a distinctive taste, and is expensive. Do not use the term *Roquefort* for blue cheese dressings unless you are actually using this brand of cheese.

BALSAMICO TRADIZIONALE (TRADITIONAL BALSAMIC VINEGAR)

True balsamic vinegar is made by small artisan producers, as distinguished from the industrial product found in most kitchens and supermarkets. It is made not from wine or wine vinegar but from grape juice, usually from white Trebbiano grapes, although four other grapes are permitted by Italian law. Balsamico tradizionale is aged in a series of small wooden barrels for at least 10 but as long as 50 years. The result is an intensely flavorful, thick, almost syrupy, dark brown liquid.

Because of the small production and the long aging, true balsamic vinegar is very expensive, the oldest bottlings being among the most expensive foods anywhere.

The familiar inexpensive balsamico vinegars most of us are familiar with are made in large quantities from wine vinegar and caramelized sugar, sometimes with the addition of a quantity of aged balsamico. The quality of inexpensive balsamic vinegars ranges from terrible to good. The better ones can be excellent salad ingredients. True balsamico tradizionale is too costly to be mixed with salad dressings. It is generally used by itself as a condiment, measured out in mere drops.

OIL-AND-VINEGAR DRESSINGS

Basic **vinaigrette**, the first recipe in this section, is a simple mixture of oil, vinegar, and seasonings. It can be used as is, but it is usually the base for other dressings, such as the variations that follow.

The ratio of oil to vinegar in a basic vinaigrette is 3 parts oil to 1 part vinegar. This is not a divine law, however, and the proportions may be changed to taste. Different acids (primarily vinegars and citrus juices) and different oils (such as some very flavorful olive oils) have varying effects on the balance of flavors. Taste the vinaigrette and adjust the ratio as necessary. In general, less oil makes the dressing more tart, while more oil makes it taste milder and oilier.

A very strong vinegar, more than 5 percent acid, may have to be diluted with water before being measured and added to the recipe.

For guidelines in the preparation of vinaigrettes, review the discussion of temporary emulsions on page 110. The emulsion in the basic vinaigrette recipe holds only a short time because the formula contains no stabilizers, with the minor exception of a little pepper. To make a good emulsion, mix some mustard with the vinegar, as in the first variation of the basic recipe.

 # Basic Vinaigrette

YIELD: 1 QT (1 L)

U.S.	METRIC	INGREDIENTS	PROCEDURE
1 cup	250 mL	Wine vinegar	1. Mix the vinegar, salt, and white pepper until the salt is dissolved.
1 tbsp	15 mL	Salt	
1 tsp	5 mL	White pepper	2. Using a wire whip, a mixing machine, or a blender, begin adding the oil a few drops at a time. Gradually increase the oil to a thin stream. (See discussion of temporary emulsions, p. 110.)
3 cups	750 mL	Salad oil, olive oil, or part salad oil and part olive oil	3. Mix again before using. (The best way to re-emulsify a separated vinaigrette is to put it in a blender and spin at high speed until it is recombined.)

Per 1 fl oz (29.57 mL): Calories, 180; Protein, 0 g; Fat, 21 g (100% cal.); Cholesterol, 0 mg; Carbohydrates, 0 g; Fiber, 0 g; Sodium, 220 mg.

VARIATIONS

Mustard Vinaigrette

Add 1–2 oz (30–60 g) prepared mustard (French or Dijon type) to the basic recipe. Mix with the vinegar in step 1.

Herbed Vinaigrette

Add to the basic recipe or to the Mustard Vinaigrette variation ½ cup or 1 oz (30 g) chopped parsley, 1 tsp (5 mL) chopped fresh basil, 1 tsp (5 mL) chopped fresh marjoram or oregano, and 2 tsp (10 mL) chopped chives. If fresh herbs are not available, use half their volume of dried herbs.

Lemon Vinaigrette

In place of the wine vinegar in the basic recipe or the Mustard Vinaigrette variation, use 2½ fl oz (75 mL) wine vinegar and 5½ fl oz (175 mL) fresh lemon juice.

Balsamic Vinaigrette

Use balsamic vinegar in place of half to three-quarters of the wine vinegar in the basic recipe.

Italian Dressing

Use all or part olive oil. Add to the basic recipe 1½ tsp (7 mL) minced garlic, 1 tbsp (15 mL) dried oregano, and ¼ cup (60 mL) chopped parsley.

Piquante Dressing

Add to the basic recipe 2 tsp (10 mL) dry mustard, 2 tbsp (30 mL) finely chopped onion, and 2 tsp (10 mL) paprika.

Chiffonade Dressing

Add to the basic recipe the following ingredients, all chopped fine: 2 hard-cooked eggs, 4 oz (125 g) cooked or canned red beets (drained), 2 tbsp (30 mL) chopped parsley, and 1 oz (30 g) onion or scallions.

Avocado Dressing

Add 1 lb (500 g) puréed avocado to the basic recipe or to Herbed Vinaigrette. Beat until smooth. Increase salt to taste.

Blue Cheese or Roquefort Vinaigrette

Mix 4 oz (125 g) crumbled blue cheese or Roquefort cheese and 4 fl oz (125 mL) heavy cream in a mixer with a paddle attachment or by hand in a stainless-steel bowl. Gradually beat in 1½ pt (750 mL) Basic Vinaigrette.

American French or Tomato French Dressing

YIELD: 2 QT (2 L)

U.S.	METRIC	INGREDIENTS
4 oz	125 g	Onion
12 oz	375 mL	Cider vinegar
2¹/₂ cups	625 mL	Ketchup
4 oz	125 g	Sugar
1 tsp	5 mL	Mashed garlic
1 tbsp	15 mL	Worcestershire sauce
1 tsp	5 mL	Paprika
¹/₄ tsp	1 mL	Hot pepper sauce (such as Tabasco)
¹/₂ tsp	2 mL	White pepper
1 qt	1 L	Salad oil

PROCEDURE

1. Grate the onion on a hand grater or grind in food chopper.
2. Combine all ingredients except the oil in a stainless-steel bowl.
3. Mix with a wire whip until well combined and sugar is dissolved.
4. Using a wire whip or a mixing machine, gradually beat in the oil.
5. Beat or stir again before serving.

Per 1 fl oz (29.57 mL): Calories, 140; Protein, 0 g; Fat, 14 g (86% cal.); Cholesterol, 0 mg; Carbohydrates, 5 g; Fiber, 0 g; Sodium, 115 mg.

Sauce Gribiche

YIELD: 1¹/₂ PT (750 ML)

U.S.	METRIC	INGREDIENTS
6	6	Hard-cooked egg yolks
1¹/₂ tbsp	45 mL	Prepared mustard, French or Dijon-style
to taste	to taste	Salt
to taste	to taste	Pepper
1 pt	475 mL	Olive oil
as needed	as needed	Wine vinegar or water
2 tbsp	30 mL	Chopped parsley
2 tbsp	30 mL	Chopped fresh chervil
2 tbsp	30 mL	Chopped fresh tarragon
2 tbsp	30 mL	Capers
2 tbsp	30 mL	Cornichons or sour gherkin pickles, chopped
6	6	Hard-cooked egg whites, cut julienne

PROCEDURE

1. Mash the yolks or force them through a sieve into a bowl.
2. Add the mustard, salt, and pepper, and mix well.
3. As when making mayonnaise (see p. 664), very gradually beat in the olive oil a few drops at a time at first. Because the cooked yolks do not enable a stable emulsion like raw yolks do, the mixture curdles easily. When it does, beat in a little hot water. Continue adding oil alternately with hot water. The mixture should have the consistency of a thin mayonnaise.
4. Add just enough vinegar to give the dressing a slightly tart taste.
5. Stir in the herbs, capers, cornichons, and egg whites.
6. Taste and add more salt if necessary.

Per 1 fl oz (29.57 mL): Calories, 180; Protein, 2 g; Fat, 19 g (96% cal.); Cholesterol, 50 mg; Carbohydrates, 0 g; Fiber, 0 g; Sodium, 70 mg.

Oriental Vinaigrette

YIELD: 1¹/₂ PT (750 ML)

U.S.	METRIC	INGREDIENTS
³/₄ cup	200 mL	Rice vinegar or white vinegar
¹/₄ cup	60 mL	Soy sauce
1 tbsp	15 mL	Grated fresh ginger root
2 tsp	10 mL	Pepper
¹/₄ tsp	1 mL	Crushed garlic
¹/₂ tsp	2 mL	Hot pepper sauce (such as Tabasco)
1³/₄ cups	425 mL	Salad oil
¹/₄ cup	60 mL	Sesame oil
as needed	as needed	Salt

PROCEDURE

1. Combine all ingredients except the oils and salt in a bowl and mix well.
2. Taste the dressing and add salt if necessary (the soy sauce may contain enough salt).
3. Using a wire whip or a mixing machine, gradually beat in the salad oil and then the sesame oil.
4. Mix or stir again before using.

Per 1 fl oz (29.57 mL): Calories, 160; Protein, 0 g; Fat, 18 g (100% cal.); Cholesterol, 0 mg; Carbohydrates, 0 g; Fiber, 0 g; Sodium, 150 mg.

Reduced-Fat Vinaigrette

YIELD: 1 QT (1 L)

U.S.	METRIC	INGREDIENTS
1 cup	250 mL	Wine vinegar
1 pt	500 mL	Jus Lié (p. 186) made with white stock, vegetable stock, or vegetable juice
1 tbsp	15 mL	Salt
1 tsp	5 mL	White pepper
1 cup	250 mL	Salad oil, olive oil, or part salad oil and part olive oil

PROCEDURE

1. Mix the vinegar, jus lié, salt, and white pepper until the salt is dissolved.
2. Using a wire whip or a mixing machine, mix in the oil.
3. Mix or stir again before using.

Per 1 fl oz (29.57 mL): Calories, 60; Protein, 0 g; Fat, 7 g (100% cal.); Cholesterol, 0 mg; Carbohydrates, 0 g; Fiber, 0 g; Sodium, 220 mg.

Fat-Free Vinaigrette

YIELD: 10 FL OZ (300 ML)

U.S.	METRIC	INGREDIENTS
8 oz	250 g	Onions, whole, with peel
2 tbsp	30 mL	Prepared mustard, French or Dijon style
2 fl oz	60 mL	Wine vinegar
4 fl oz	125 mL	Vegetable stock or juice
2 tbsp	30 mL	Chopped parsley
to taste	to taste	Salt
to taste	to taste	Black pepper

PROCEDURE

1. Roast the onions at 350°F (175°C) until they are completely soft, 45 to 60 minutes.
2. Cool until they are cool enough to handle. Peel.
3. Purée the onions in a blender or food processor.
4. Add the mustard, vinegar, and stock. Blend to mix well.
5. Stir in the chopped parsley.
6. Season to taste with salt and pepper.

Per 1 fl oz (29.57 mL): Calories, 15; Protein, 0 g; Fat, 0 g (0% cal.); Cholesterol, 0 mg; Carbohydrates, 3 g; Fiber, 0 g; Sodium, 80 mg.

VARIATIONS

Substitute chicken stock for the vegetable stock.

Fat-Free Roasted Garlic Vinaigrette

Substitute roasted garlic for the roasted onions in the basic recipe.
See the variations following Basic Vinaigrette for other flavoring ideas.

EMULSIFIED DRESSINGS

Mayonnaise is the most important emulsified dressing. It is sometimes used by itself as a salad dressing, but more often it serves as the base for a wide variety of other dressings. Mayonnaise-based dressings are generally thick and creamy. In fact, many of them are made with the addition of sour cream.

Emulsified French dressing is similar to basic French dressing, except egg yolk is added to keep the oil and vinegar from separating. Its preparation is similar to that of mayonnaise. Emulsified French dressing is given a red-orange color and a subtle flavoring through the addition of Spanish paprika.

PREPARATION OF MAYONNAISE

Good-quality prepared mayonnaise is readily available on the market, and few establishments make their own. But it is such a basic preparation and, like the mother sauces you studied in Chapter 8, the foundation of many others. Therefore, it is important to know how to make it.

Homemade mayonnaise is not as stable as the commercial product, which is prepared with special equipment that creates a finer emulsion and which may have added stabilizers to increase its shelf life. Also, the commercial product is usually less expensive. Nevertheless, making mayonnaise in your operation takes only minutes with a power mixer, and by carefully selecting your ingredients you can make a superior-tasting product.

To make mayonnaise, you must observe several conditions in order to get an emulsion. Study the guidelines on p. 664 before proceeding with the recipe.

Standards of Quality for Vinaigrette and Mayonnaise-Based Salad Dressings

Vinaigrettes

1. Oil and vinegar should be in emulsion at the time the dressing is used. If the temporary emulsion has broken, the dressing should be reemulsified. A properly emulsified dressing of the right texture clings lightly to the greens when tossed without looking excessively oily.

2. Oil and vinegar should be in proper balance, so that the dressing doesn't taste too oily or too acidic.

3. Flavors of additional ingredients should be pleasing and in good balance with each other and with the flavors of the oil and vinegar.

Mayonnaise-Based Dressings

1. Dressing should be pourable, not too thick. Excessively thick dressings are difficult to mix and can damage delicate greens.

2. The dressing should have an appropriate level of acidity to balance the richness and creaminess of the mayonnaise and of dairy ingredients (if any).

3. The flavors of additional ingredients should be pleasing and in good balance.

GUIDELINES for Making Mayonnaise

1. **Use fairly bland ingredients if the mayonnaise is to be used as a base for other dressings.**
   The mayonnaise will be more versatile as a base if it has no strong flavors. Olive oil and other ingredients with distinctive flavors may be used for special preparations.

2. **Use the freshest eggs possible for the best emulsification. For safety, use pasteurized eggs.**

3. **Have all ingredients at room temperature.**
   Cold oil is not easily broken into small droplets, so it is harder to make an emulsion.

4. **Beat the egg yolks well in a bowl.**
   Thorough beating of the yolks is important for a good emulsion.

5. **Beat in the seasonings.**
   It is helpful to add a little of the vinegar at this time as well. The emulsion will form more easily because the acidity of the vinegar helps prevent curdling of the egg yolk proteins. Also, the vinegar helps disperse the spices and dissolve the salt.

6. **Begin to add the oil very slowly, beating constantly.**
   It is critical to add the oil slowly at first, or the emulsion will break. When the emulsion has begun to form, the oil may be added more quickly. But never add more oil at once than the amount of mayonnaise that has already formed in the bowl, or the emulsion may break.

7. **Gradually beat in the remaining oil alternately with the vinegar.**
   The more oil you add, the thicker the mayonnaise gets. Vinegar thins it. Add a little vinegar whenever the mayonnaise gets too thick to beat.

   Beating with a power mixer using the wire whip attachment makes a more stable emulsion than beating by hand.

8. **Add no more than 8 ounces (240 mL) oil per large egg yolk, or no more than 1 quart (950 mL) per 4 yolks.**
   The emulsion may break if more oil is added than the egg yolks can handle.

9. **Taste and correct the seasonings.**
   Finished mayonnaise should have a smooth, rich, but neutral flavor, with a pleasant tartness. Its texture should be smooth and glossy, and it should be thick enough to hold its shape.

10. **If the mayonnaise breaks, it can be rescued.**
    Beat an egg yolk or two or some good prepared mayonnaise in a bowl, and very slowly begin to beat in the broken mayonnaise, as in step 6. Continue until all the mayonnaise has been added and re-formed.

    Alternative repair method: Place the broken mayonnaise in a blender and spin until the emulsion is re-formed.

FIGURE 21.1 Making mayonnaise by hand.

(a) Whip the egg yolks until light. Whip in this first quantity of vinegar and seasonings.

(b) Slowly pour in the oil in a thin stream, whipping constantly.

(c) The finished mayonnaise should be thick enough to hold its shape.

 Mayonnaise

YIELD: 1 QT (1 L)

U.S.	METRIC	INGREDIENTS	PROCEDURE
4	4	Egg yolks, pasteurized	1. Review the guidelines for making mayonnaise on page 664.
1 tbsp	15 mL	Vinegar	2. Place the egg yolks in the bowl of a mixer and beat with the whip attachment until well beaten.
1 tbsp	15 mL	Water	
1 tsp	5 mL	Salt	3. Add the first quantity of vinegar and the water and beat well.
1 tsp	5 mL	Dry mustard	4. Mix the dry ingredients and add to the bowl. Beat until well mixed.
pinch	pinch	Cayenne	
25 fl oz	850 mL	Salad oil	5. Turn the mixer to high speed. Very slowly, almost drop by drop, begin adding the oil. When the emulsion forms, you can add the oil slightly faster.
2 tbsp	30 mL	Vinegar	
1¹/₂–2 tbsp	25–30 mL	Lemon juice	6. When the mayonnaise becomes thick, thin with a little of the second quantity of vinegar.

7. Gradually beat in the remaining oil alternately with the vinegar. (If the emulsion breaks, see step 10 in the basic procedure to repair it.)

8. Adjust the tartness and the consistency by beating in a little lemon juice.

Per 1 fl oz **(29.57 mL):** Calories, 220; Protein, 0 g; Fat, 25 g (100% cal.); Cholesterol, 25 mg; Carbohydrates, 0 g; Fiber, 0 g; Sodium, 75 mg.

Mayonnaise-Based Dressings

For each of the following dressings, add the listed ingredients to 1 qt (1 L) mayonnaise, as indicated.

Thousand Island Dressing

8 fl oz (250 mL) chili sauce, 1 oz (30 g) minced onion, 2 oz (60 g) finely chopped green bell pepper, 2 oz (60 g) chopped drained pimiento, and (optional ingredient) 1¹/₂ chopped hard-cooked eggs.

Louis Dressing

Prepare Thousand Island Dressing without the chopped eggs. Add 8 fl oz (250 mL) heavy cream.

Russian Dressing

8 fl oz (250 mL) chili sauce or ketchup, ¹/₄ cup (60 mL) drained horseradish, 1 oz (30 g) minced onion, and (optional ingredient) ¹/₂ cup (125 mL) lumpfish or whitefish caviar.

Chantilly Dressing

8 fl oz (250 mL) heavy cream, whipped. Fold the whipped cream into the mayonnaise carefully to retain volume. Do this as close as possible to service time.

Blue Cheese Dressing

12 fl oz (60 mL) white vinegar, 1 tsp (5 mL) Worcestershire sauce, a few drops of hot red pepper sauce, and 8 oz (250 g) crumbled blue cheese. Thin to desired consistency with 4–8 fl oz (125–250 mL) heavy cream or half-and-half. Variation: Substitute sour cream for up to half of the mayonnaise.

Ranch Dressing

1¹/₂ pt (750mL) sour cream, 2¹/₂ cups (625mL) buttermilk, 4 fl oz (125 mL) wine vinegar, 3 fl oz (90 mL) lemon juice, 3 fl oz (90 mL) Worcestershire sauce, 3 tbsp (45 mL) chopped parsley, 2 tbsp (30 mL) chopped chives, 3 crushed garlic cloves, 2 chopped scallions, 1 fl oz (30 mL) prepared mustard, 1¹/₂ tsp (7 mL) celery seed.

Aïoli II

Mash 1–2 oz (30-60 g) garlic with the salt in the basic recipe. Add this to the egg yolks. Use olive oil or half olive oil and half salad oil. For another version of Aïoli, see page 216.

Emulsified French Dressing

YIELD: 1 QT (1 L)

U.S.	METRIC	INGREDIENTS	PROCEDURE
1	1	Eggs, preferably pasteurized	1. Place the eggs in the bowl of a mixer and beat with the whip attachment until well beaten.
1½ tsp	7 mL	Salt	2. Mix the dry ingredients and add to the bowl. Beat until well mixed.
1½ tsp	7 mL	Paprika	3. Turn the mixer to high speed. Very slowly begin adding the oil, as when making mayonnaise.
1½ tsp	7 mL	Dry mustard	
¼ tsp	1 mL	White pepper	4. When the dressing becomes thick, thin with a little of the vinegar.
1½ pt	700 mL	Salad oil	5. Gradually beat in the remaining oil alternately with the vinegar.
4 fl oz	125 mL	Cider vinegar	6. Beat in the lemon juice.
2 fl oz	60 mL	Lemon juice	
as needed	as needed	Vinegar, lemon juice, or water	7. The dressing should be pourable, not thick like mayonnaise. If it is too thick, taste for seasonings first. If the dressing is not tart enough, thin with a little vinegar or lemon juice. If it is tart enough, thin with water.

Per 1 fl oz (29.57 mL): Calories, 190; Protein, 0 g; Fat, 21 g (100% cal.); Cholesterol, 5 mg; Carbohydrates, 0 g; Fiber, 0 g; Sodium, 110 mg.

Caesar Dressing

YIELD: 1 QT (1 L)

U.S.	METRIC	INGREDIENTS	PROCEDURE
25	25	Anchovy fillets (see Note)	1. Mash the anchovies and garlic together to make a paste.
2 tsp	10 mL	Crushed garlic	
4	4	Eggs, pasteurized	2. Place the eggs in the bowl of a mixer and whip with the whip attachment until well beaten.
3 fl oz	90 mL	Lemon juice	3. Add the anchovy and garlic paste and the first quantity of lemon juice. Whip until well mixed.
2½ cups	600 mL	Olive oil	4. With the mixer on high speed, slowly begin adding the oil, as when making mayonnaise.
3 fl oz	90 mL	Lemon juice	5. When the dressing becomes thick, add a little of the remaining lemon juice.
2 oz	60 g	Parmesan cheese, grated	6. Gradually beat in the rest of the oil alternating with the rest of the lemon juice.
to taste	to taste	Salt	7. Mix in the parmesan cheese and salt.

Per 1 fl oz (29.57 mL): Calories, 170; Protein, 2 g; Fat, 18 g (93% cal.); Cholesterol, 30 mg; Carbohydrates, 1 g; Fiber, 0 g; Sodium, 160 mg.

Note: Anchovies are a main ingredient in traditional Caesar salads but may be omitted according to taste.

KEY POINTS TO REVIEW

- What kinds of oils and vinegars are used in salad dressings? What are their quality factors?

- In addition to oil and vinegar, what are the other major ingredients in most salad dressings?

- How do you make a basic vinaigrette?

- How do you make mayonnaise?

OTHER DRESSINGS

Cooked salad dressing is similar in appearance to mayonnaise, but it has a more tart flavor, while mayonnaise is richer and milder. Cooked dressing is made with little or no oil and with a starch thickener. It may be made in the kitchen or purchased already prepared. Formerly, it was little used in commercial kitchens because of its strong flavor and tartness, but now it is preferred to mayonnaise in some regions.

You will find in many cookbooks a great variety of dressings based on neither mayonnaise nor oil and vinegar. They include dressings based on sour cream and on fruit juice and yogurt (for fruit salads), and low-calorie dressings that appeal to the dieter. The important thing to remember is that these dressings should have well-balanced flavors with a pleasant tartness, and they should harmonize with and complement the salad with which they are served.

Cooked Salad Dressing ♥ 🌸

YIELD: 1 QT (1 L)

U.S.	METRIC	INGREDIENTS
2 oz	60 g	Sugar
2 oz	60 g	Flour
1 tbsp	15 mL	Salt
1 tbsp	15 mL	Dry mustard
1/8 tsp	0.5 mL	Cayenne
2	2	Eggs
2	2	Egg yolks
1 1/2 pt	750 mL	Milk
2 oz	60 g	Butter
6 fl oz	180 mL	Cider vinegar

PROCEDURE

1. Mix the sugar, flour, salt, mustard, and cayenne in a stainless-steel bowl.
2. Add the eggs and yolks and beat until smooth.
3. Place the milk in a saucepan and bring to a simmer. Be careful not to scorch it.
4. Gradually beat about half the milk into the egg mixture. Then return the mixture to the saucepan.
5. Cook over low heat, stirring constantly, until very thick and no raw flour taste remains.
6. Remove from heat and stir in the butter.
7. When the butter is melted and mixed in, stir in the vinegar.
8. Immediately transfer the dressing to a stainless-steel container. Cover and cool.

Per 1 fl oz (29.57 mL): Calories, 50; Protein, 2 g; Fat, 3 g (49% cal.); Cholesterol, 35 mg; Carbohydrates, 5 g; Fiber, 0 g; Sodium, 250 mg.

Low-Fat Buttermilk Yogurt Dressing ♥ 🌸

YIELD: 1 PT 12 FL OZ (850 ML)

U.S.	METRIC	INGREDIENTS
1 pt	500 mL	Fat-free or low-fat yogurt, unflavored
8 fl oz	250 mL	Buttermilk
1 tbsp	15 mL	Prepared mustard, French or Dijon style
1 fl oz	30 mL	Wine vinegar
1 fl oz	30 mL	Lemon juice
1 fl oz	30 mL	Worcestershire sauce
1 1/2 tbsp	22 mL	Chopped parsley
1 tsp	5 mL	Finely chopped garlic
1/2 oz	15 g	Shallots, chopped fine
1/2 tsp	2 mL	Celery seed
4 tsp	20 mL	Sugar
to taste	to taste	Salt
to taste	to taste	Pepper

PROCEDURE

1. Mix all ingredients until uniformly blended.

Per 1 fl oz (29.57 mL): Calories, 15; Protein, 1 g; Fat, 0 g (0% cal.); Cholesterol, 0 mg; Carbohydrates, 3 g; Fiber, 0 g; Sodium, 45 mg.

SALADS

Because the number and variety of salad combinations is nearly endless, it is helpful to divide salads into categories in order to understand how they are produced. For the pantry chef, the most useful way to classify salads is by ingredients: green salads, vegetable salads, fruit salads, and so on. This is because production techniques are slightly different for each kind. We use this classification when we discuss specific recipes later in this chapter.

Before the pantry chef can produce the salads, first he or she must decide exactly what salads should be made. Therefore, you should know what kinds of salad are best for which purposes. For this reason, salads are also classified according to their function in the meal. Keep in mind that there are no exact dividing lines between the types of salad discussed here. For example, a salad suitable as the first course of a dinner may also be an excellent main course on a luncheon menu.

TYPES OF SALADS

Today, the variety of salads on offer seems to be greater than ever in memory. Restaurants that once listed no more than two or three salads on their menu now devote an entire page to the category. New kinds of salad fill bin after bin in the prepared-food sections of supermarkets and delicatessens.

At the same time, more traditional salads have not lost their importance. In schools, hospitals, nursing homes, neighborhood diners, and mom-and-pop restaurants, cooks who never heard of mesclun still must know how to clean a head of iceberg lettuce and how to prepare flavored gelatins.

The following classification of salad types describes the roles salads fill in modern menus. These categories apply to both traditional and modern recipes. Examples of both are included later in this chapter.

APPETIZER SALADS

Many establishments serve salads as a first course, often as a substitute for a more elaborate first course. Not only does this ease the pressure on the kitchen during service but it also gives the customers a satisfying food to eat while their dinners are being prepared.

In addition, more elaborate composed salads are popular as appetizers (and also as main courses at lunch) in many elegant restaurants. These often consist of a poultry, meat, or fish item, plus a variety of vegetables and garnishes, attractively arranged on a bed of greens.

Appetizer salads should stimulate the appetite. This means they must have fresh, crisp ingredients; a tangy, flavorful dressing; and an attractive, appetizing appearance.

Preportioned salads should not be so large as to be filling, but they should be substantial enough to serve as a complete course in themselves. (Self-service salad bars, of course, avoid this problem.) Tossed green salads are especially popular for this reason, as they are bulky without being filling.

The combination of ingredients should be interesting, not dull or trite. Flavorful foods like cheese, ham, salami, shrimp, and crabmeat, even in small quantities, add appeal. So do crisp raw or lightly cooked vegetables. A bowl of poorly drained iceberg lettuce with a bland dressing is hardly an exciting way to start a meal.

Attractive arrangement and garnish are important because visual appeal stimulates the appetite. A satisfying, interesting starter puts the customer in a good frame of mind for the rest of the meal.

ACCOMPANIMENT SALADS

Salads can also be served with the main course. They serve the same function as other side dishes (vegetables and starches).

Accompaniment salads must balance and harmonize with the rest of the meal, like any other side dish. For example, don't serve potato salad at the same meal at which you are serving French fries or another starch. Sweet fruit salads are rarely appropriate as accompaniments, except with such items as ham or pork.

Side-dish salads should be light and flavorful, not too rich. Vegetable salads are often good choices. Heavier salads, such as macaroni or high-protein salads containing meat, seafood, cheese, and so on, are less appropriate, unless the main course is light. Combination salads with a variety of elements are appropriate accompaniments to sandwiches.

MAIN-COURSE SALADS

Cold salad plates have become popular on luncheon menus, especially among nutrition- and diet-conscious diners. The appeal of these salads is in variety and freshness of ingredients.

Main-course salads should be large enough to serve as a full meal and should contain a substantial portion of protein. Meat, poultry, and seafood salads, as well as egg salad and cheese, are popular choices.

Main-course salads should offer enough variety on the plate to form a balanced meal, both nutritionally and in flavors and textures. In addition to the protein, a salad platter should offer a variety of vegetables, greens, and/or fruits. Examples are chef's salad (mixed greens, raw vegetables, and strips of meat and cheese), shrimp or crabmeat salad with tomato wedges and slices of avocado on a bed of greens, and cottage cheese with an assortment of fresh fruits.

The portion size and variety of ingredients give the chef an excellent opportunity to use imagination and creativity to produce attractive, appetizing salad plates. Attractive arrangements and good color balance are important.

SEPARATE-COURSE SALADS

Many fine restaurants serve a refreshing, light salad after the main course. The purpose is to cleanse the palate after a rich dinner and to refresh the appetite and provide a pleasant break before dessert.

Salads served after the main course were the rule rather than the exception many years ago, and the practice deserves to be more widespread. A diner who may be satiated after a heavy meal is often refreshed and ready for dessert after a light, piquant salad.

Separate-course salads must be very light and in no way filling. Rich, heavy dressings, such as those made with sour cream and mayonnaise, should be avoided. Perhaps the ideal choice is a few delicate greens, such as Bibb lettuce or Belgian endive, lightly dressed with vinaigrette. Fruit salads are also popular choices.

DESSERT SALADS

Dessert salads are usually sweet and may contain items such as fruits, sweetened gelatin, nuts, and cream. They are often too sweet to be served as appetizers or accompaniments and are best served as dessert or as part of a buffet or party menu.

INGREDIENTS

Freshness and variety of ingredients are essential for high-quality salads. Lettuce, of course, is the first choice for most people, but many other foods can make up a salad.

The following tables list, by category, most of the ingredients used in popular salads. You will be able to think of others. Add them to the lists as they occur to you or as they are suggested by your instructor. The lists will be useful when you are creating your own salad ideas.

Following these lists are detailed descriptions of two groups of food that have not been covered in previous chapters and belong especially in the pantry: salad greens and fresh fruits.

SALAD GREENS

Iceberg lettuce	Dandelion greens
Romaine lettuce	Watercress
Boston lettuce	Arugula
Bibb or limestone lettuce	Radicchio
Loose-leaf lettuce	Mesclun
Escarole	Tatsoi

Chicory or curly endive
Frisée
Belgian endive
Chinese cabbage or celery cabbage
Spinach

Mâche
Microgreens
Sprouts
Edible flowers

VEGETABLES, RAW

Avocado
Bean sprouts
Broccoli
Cabbage, white, green, and red
Carrots
Cauliflower
Celery
Celeriac (celery root)

Cucumbers
Sunchokes (Jerusalem artichokes)
Kohlrabi
Mushrooms
Onions and scallions
Peppers, red, green, and yellow
Radishes
Tomatoes

VEGETABLES, COOKED, PICKLED, AND CANNED

Artichoke hearts
Asparagus
Beans (all kinds)
Beets
Carrots
Cauliflower
Corn
Cucumber pickles (dill, sweet, etc.)

Hearts of palm
Leeks
Olives
Peas
Peppers, roasted and pickled
Pimientos
Potatoes
Water chestnuts

STARCHES

Dried beans (cooked or canned)
Potatoes

Macaroni products
Grains

Bread (croutons)

FRUITS, FRESH, COOKED, CANNED, OR FROZEN

Apples
Apricots
Bananas
Berries
Cherries
Coconut
Dates
Figs
Grapefruit

Grapes
Kiwi fruit
Kumquats
Mandarin oranges and tangerines
Mangoes
Melons
Nectarines
Oranges
Papayas

Peaches
Pears
Persimmons
Pineapple
Plums
Prunes
Pomegranates
Prickly pear
Raisins

PROTEIN FOODS

Meats (beef, ham)
Poultry (chicken, turkey)
Fish and shellfish (tuna, crab,
   shrimp, lobster, salmon,
   sardines, anchovies, herring,
   any fresh cooked fish)
Salami, prosciutto, luncheon meats, etc.

Bacon
Eggs, hard-cooked
Cheese, cottage
Cheese, aged or cured types

MISCELLANEOUS

Gelatin (plain or flavored)

Nuts

LETTUCE AND OTHER SALAD GREENS

Iceberg Lettuce

The most popular salad ingredient. Firm, compact head with crisp, mild-tasting pale green leaves. Valuable for its texture because it stays crisp longer than other lettuces. Can be used alone but is best mixed with more flavorful greens, such as romaine, because it lacks flavor itself. Keeps well.

Romaine or Cos Lettuce

Elongated, loosely packed head with dark green, coarse leaves. Crisp texture, with full, sweet flavor. Keeps well and is easy to handle. Essential for Caesar salad. For elegant service, the center rib is often removed.

Boston Lettuce

Small, round head with soft, fragile leaves.Deep green outside shading to nearly white inside. The leaves have a rich, mild flavor and delicate, buttery texture. Bruises easily and does not keep well. Cup-shaped leaves excellent for salad bases.

Bibb or Limestone Lettuce

Similar to Boston lettuce, but smaller and more delicate. A whole head may be only a few inches (less than 10 cm) across. Color ranges from dark green outside to creamy yellow at the core. Its tenderness, delicate flavor, and high price make it a luxury in some markets. The small, whole leaves are often served by themselves, with a light vinaigrette dressing, as an after-dinner salad.

Loose-Leaf Lettuce

Forms bunches rather than heads. Soft, fragile leaves with curly edges. May be all green or with shades of red. Wilts easily and does not keep well, but is inexpensive and gives flavor, variety, and interest to mixed green salads.

Escarole or Broad-Leaf Endive

Broad, thick leaves in bunches rather than heads. Texture is coarse and slightly tough, and flavor is somewhat bitter. Mix with sweeter greens to vary flavor and texture, but do not use alone because of the bitterness. Escarole is frequently braised with olive oil and garlic and served as a vegetable in Italian cuisine.

Chicory or Curly Endive

Narrow, curly, twisted leaves with firm texture and bitter flavor. Outside leaves are dark green; core is yellow or white. Attractive when mixed with other greens or used as a base or garnish, but may be too bitter to be used alone.

Frisée

Frisée is the same plant as curly endive or chicory, but it is grown in a way that makes it more tender and less bitter. Except for the outer layer, the leaves are pale yellow, slender, and feathery, with a distinct but mild taste.

Belgian Endive or Witloof Chicory

Narrow, lightly packed, pointed heads resembling spearheads, 4–6 inches (10–15 cm) long. Pale yellow-green to white in color. Leaves are crisp, with a waxy texture and pleasantly bitter flavor. Usually expensive. Often served alone, split in half or into wedges, or separated into leaves, accompanied by a mustard vinaigrette dressing.

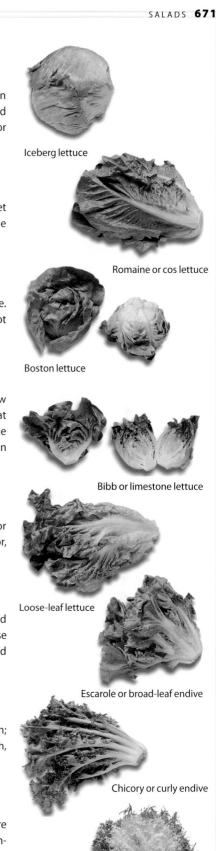

Iceberg lettuce

Romaine or cos lettuce

Boston lettuce

Bibb or limestone lettuce

Loose-leaf lettuce

Escarole or broad-leaf endive

Chicory or curly endive

Frisée

Belgian endive or witloof chicory

Chinese cabbage

Chinese Cabbage

Elongated, light green heads with broad, white center ribs. Available in two forms: narrow, elongated head, often called **celery cabbage**, and thicker, blunt head, called **napa cabbage**. Tender but crisp, with a mild cabbage flavor. Adds excellent flavor to mixed green salads. Also used extensively in Chinese cooking.

Spinach

Spinach

Small, tender spinach leaves are excellent salad greens, either alone or mixed with other greens. A popular salad is spinach leaves garnished with sliced raw mushrooms and crisp, crumbled bacon. Spinach must be washed thoroughly, and the coarse stems must be removed.

Watercress

Most commonly used as a garnish, watercress is also excellent in salads. Small, dark green, oval leaves with a pungent, peppery flavor. Remove thick stems before adding to salads.

Watercress

Arugula

Also known as **rugula** or **rocket**, these pungent, distinctively flavored greens are related to mustard and watercress. They are tender and perishable, and they often are sandy, so they must be washed carefully. Arugula was once found almost exclusively in Italian restaurants, but it has since become more widely available and popular.

Arugula or rocket

Mesclun

Mesclun is a mixture of tender baby lettuces. It is available as a mixture, but some chefs prefer to buy individual baby lettuces and make their own mixture.

Baby Lettuces

The small, tender leaves that make up a mesclun mix are also available separately. These include baby Bibb (both red and green), baby romaine, baby red oak leaf, and lola rossa (a red lettuce with ruffled leaves).

Mesclun

Sprouts

Sprouts are young plants that have just emerged from their seeds, before the true leaves develop. Sprouts from mung beans are commonly used in Chinese cooking. Alfalfa, daikon radish, and mustard sprouts are often used in delicate salads. Alfalfa sprouts have a mild flavor, while radish and mustard sprouts have a peppery flavor.

Lola rossa

Microgreens

These are the first true leaves that develop after a seed sprouts. Tiny herb leaves and tiny leaves from lettuce and other salad greens, younger and smaller than baby lettuces, are used mostly as garnish for other dishes, both hot and cold.

Tatsoi

Tatsoi is a small, round, dark green leaf. Its flavor has a pleasant bite similar to that of arugula, watercress, and other members of the mustard family. It is sometimes included in mesclun mixtures, although it is not actually a lettuce.

Red oak leaf

Alfalfa sprouts

Microgreens

Tatsoi

Mizuna

Mizuna, also known as *Japanese mustard greens*, is a dark green leaf with jagged edges resembling dandelion leaves. It has a mild, mustardy taste.

Mâche

Also called *corn salad*, *lamb's lettuce*, *lamb's tongue*, and *field salad*, mâche is a small, very tender green with spoon-shaped leaves. It has a delicate, nutty flavor.

Radicchio

Radicchio (ra dik ee oh), a red-leafed Italian variety of chicory, has creamy white ribs or veins and generally comes in small, round heads. It has a crunchy texture and a slightly bitter flavor. Radicchio is expensive, but only a leaf or two are needed to add color and flavor to a salad.

Treviso

Treviso is a red-leafed plant like radicchio, but with elongated leaves somewhat like Belgian endive. Like radicchio and endive, it belongs to the chicory family and has a slightly bitter flavor.

Dandelion Greens

The familiar lawn ornament is also cultivated for use in the kitchen. Only young, tender leaves may be used. Older leaves are coarse and bitter, though cultivated varieties are milder than wild dandelion. Best in spring.

Precleaned, Precut Salad Greens

Precut greens are sold in large, sealed plastic bags. They save labor costs in large operations but are more perishable than unprocessed greens. Keep refrigerated, and do not open until ready to use. Unopened bags will keep for two or three days. Taste before serving to make sure the greens do not have too much antioxidant on them, making them bitter.

FRESH FRUITS: EVALUATING AND PREPARING

The following is a summary of the most commonly available fresh fruits. Emphasis is on the qualities to look for when purchasing them and on how to trim and prepare the fruit for use. In addition, identification information is included for certain exotic items. Nearly everyone knows what apples, bananas, and strawberries are, but not everyone can identify a persimmon or a passion fruit. Trimming yields are also given. How to use these percentages is explained on pages 82–83.

Apples

Mature apples have a fruity aroma, brown seeds, and a slightly softer texture than unripe fruit. Overripe or old apples are soft and sometimes shriveled. Avoid apples with bruises, blemishes, decay, or mealy texture. Summer varieties (sold until fall) do not keep well. Fall and winter varieties keep well and are available for a longer period. Apples with a good acid content are usually better for cooking than bland eating varieties like Red Delicious. Granny Smith and Golden Delicious are widely used for cooking. To prepare, wash; pare if desired. Quarter and remove core, or leave whole and core with a special coring tool. Use a stainless-steel knife for cutting. After paring, dip in solution of lemon juice (or other tart fruit juice) or ascorbic acid to prevent browning.

Percentage yield: 75%

Apricots

Only tree-ripened apricots have sufficient flavor, and they keep for a week or less under refrigeration. They should be golden yellow, firm, and plump, not mushy. Avoid fruit that is soft, blemished, or decayed.

Wash, split in half, and remove pit. Peeling is not necessary for most purposes.

Percentage yield: 94%

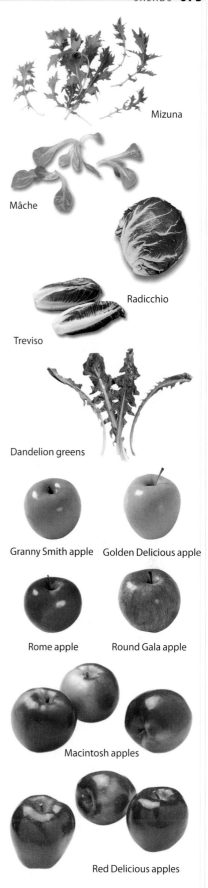

Mizuna

Mâche

Radicchio

Treviso

Dandelion greens

Granny Smith apple   Golden Delicious apple

Rome apple   Round Gala apple

Macintosh apples

Red Delicious apples

Apricots

Bananas

Bananas

Look for plump, smooth bananas without bruises or spoilage. All bananas are picked green, so you don't need to avoid unripe fruit. Avoid overripe fruit, however.

Ripen at room temperature for three to five days; fully ripe fruit is all yellow with small brown flecks and no green. Do not refrigerate, or fruit will discolor. Peel and dip in fruit juice to prevent browning.

Percentage yield: 70%

Berries

This category includes blackberries, blueberries, cranberries, black currants (cassis), red currants, white currants, lingonberries, raspberries, and strawberries. Berries should be full, plump, and clean, with bright, fully ripe color. Watch for moldy or spoiled fruits. Wet spots on carton indicate damaged fruit.

Refrigerate in original container until ready to use in order to reduce handling. Except for cranberries, berries do not keep well. Sort out spoiled berries and foreign materials. Wash with gentle spray and drain well. Remove stems from strawberries. Red currants for garnishing are often left on the stem. Handle berries carefully to avoid bruising.

Percentage yield: 92–95%

Blackberries

Cherries

Look for plump, firm, sweet, juicy cherries. Bing or black cherries should be uniformly dark to almost black.

Refrigerate in original container until ready to use. Just before use, remove stems and sort out damaged fruit. Rinse and drain well. Pit with a special pitting tool.

Percentage yield: 82% (pitted)

Blueberries

Coconuts

Shake to hear liquid inside; fruits with no liquid are dried out. Avoid cracked fruits and fruits with wet eyes.

Pierce eye with ice pick or nail and drain liquid. Crack with hammer and remove meat from shell (easier if placed in 350°F/175°C oven 10–15 minutes first). Peel brown skin with paring knife or vegetable peeler.

Percentage yield: 50%

Cranberries

Figs

Calimyrna figs, also called *Smyrna figs*, are light green; Black Mission figs and Black Spanish figs (also called *Brown Turkey*) are purple. All figs are sweet when ripe, and soft and delicate in texture. They should be plump and soft, without spoilage or sour odor.

Keep refrigerated (although firm, unripe figs can be left at room temperature, spread in one layer, for a few days to ripen slightly). Rinse and drain, handling carefully. Trim off hard stem ends.

Percentage yield: 95% (80–85% if peeled)

White currants

Raspberries

Strawberries

Cherries

Coconut

Figs

Calimyrna figs

Black mission figs

Grapefruit

Select fruit that is heavy for its size and has a firm, smooth skin. Avoid puffy, soft fruits and those with pointed ends, which have low yield and a lot of rind. Cut and taste for sweetness.

For sections and slices, peel with a chef's knife, removing all white pith (see p. 146). Free sections from membrane with a small knife.

Percentage yield: 45–50% (flesh without membrane); 40–45% (juiced)

Grapefruit

Red grapefruit

Grapes

Look for firm, ripe, good-colored fruits in full bunches. Grapes should be firmly attached to stems and should not fall off when shaken. Watch for rotting or shriveling at stem ends.

Refrigerate in original container. Wash and drain. Except for seedless varieties, cut in half and remove seeds with the point of a paring knife.

Percentage yield: 90%

Grapes

Guava

There are many varieties of these small, tropical fruits. They may be round, oval, or pear-shaped, with aromatic flesh that may be green, pink, yellow, red, or white, sometimes full of seeds and sometimes nearly seedless. The flavor is complex and ranges from sweet to sour. Select tender fruits with a full aroma.

Cut in half and scoop out the flesh. For many uses, the flesh is puréed in a food processor or blender, seeds and all. Alternatively, cut into dice or other shapes as desired.

Percentage yield: 80%

Guava

Kiwi Fruit

Kiwis are firm when unripe; they become slightly softer when ripe but do not change color significantly. Common kiwis have green flesh; golden kiwis are also available. Allow them to ripen at room temperature. Avoid fruits with bruises or soft spots.

Pare thin outer skin. Cut crosswise into slices.

Percentage yield: 80%

Kiwi fruit

Kumquats

These look like tiny, elongated oranges, about the size of a medium olive. The skin and even the seeds can be eaten. In fact, the skin is sweet, while the flesh and juice are tart. Avoid soft or shriveled fruit. Kumquats keep well and are usually in good condition in the market.

Wash, drain well, and cut as desired.

Percentage yield: 95–100%

Kumquats

Lemons and Limes

Look for firm, smooth skins. Colors may vary: Limes may be yellow, and lemons may have green on the skin.

Cut in wedges, slices, or other shapes for garnish, or cut in half crosswise for juicing. Wash first if using the zest.

Percentage yield: 40–45% (juiced)

Lemons

Limes

Litchis (or Lychees)

This Chinese fruit is about the size of a walnut or Ping-Pong ball. Its rough, leathery outer skin, which ranges from reddish to brown, is easily peeled away to reveal aromatic, juicy white flesh that surrounds an inedible pit. Look for heavy, plump fruit with good color.

Peel, cut in half, and remove the seed.

Percentage yield: 50%

Litchis

Mangoes

This tropical fruit comes in two main types: oval, with a skin that ranges from green to orange to red, and kidney-shaped, with skin that is more uniformly yellow when ripe. Mangoes have a thin but tough skin and yellow to yellow-orange flesh that is juicy and aromatic. Fruit should be plump and firm, with clear color and no blemishes. Avoid rock-hard fruit, which may not ripen properly.

Mangoes

Cantaloupe

Let ripen at room temperature until slightly soft. Peel and cut flesh away from center stone, or cut in half before peeling, working a thin-bladed knife around both sides of the flat stone.

Percentage yield: 75%

Melons

Look for the following characteristics when selecting melons:

Cantaloupes: Smooth scar on stem end, with no trace of stem (called full slip, meaning the melon was picked ripe). Yellow rind, with little or no green. Heavy, with good aroma.

Honeydew: Good aroma, slightly soft, heavy, creamy white to yellowish rind, not too green. Large sizes have best quality.

Crenshaw, Casaba, Persian, Canary, Santa Claus: Heavy, with a rich aroma and slightly soft blossom end.

Watermelon: Yellow underside, not white. Firm and symmetrical. Large sizes have best yield. Velvety surface, not too shiny. When cut, look for hard, dark brown seeds and no white heart (hard white streak running through center).

To prepare hollow melons, wash, cut in half, and remove seeds and fibers. Cut into wedges and cut flesh from rind, or cut balls with ball cutter. For watermelon, wash, cut in half or into pieces, and cut balls with ball cutter, or cut flesh from rind and remove seeds.

Percentage yield: Watermelons: 45%; others: 50–55%

Honeydew

Nectarines

See Peaches and Nectarines.

Oranges, Mandarins, and Tangerines

To buy high-quality oranges, use the same guidelines as for grapefruit. Mandarins may feel puffy, but they should be heavy for their size. Unusual varieties include blood oranges, with dark red flesh and juice and intense flavor, and Seville oranges, with tart rather than sweet flesh. Seville oranges are prized for making marmalade.

Peel mandarins by hand and separate the sections. For juicing, cut oranges in half crosswise. For sections, see Grapefruit.

Percentage yield: 60–65% (sections with no membranes); 50% (juiced)

Canary melon

Papayas

Papayas are pear-shaped tropical fruits with a mild, sweet flavor and slightly floral aroma. The flesh is yellow or pinkish, depending on the variety, and the center cavity holds a mass of round, black seeds. Papayas may weigh from less than 1 pound to several pounds (less than 500 g to more than 1 kg) each. Their skin is green when unripe, becoming yellow as they ripen. For best quality, select fruits that are firm and symmetrical, without bruises or rotten spots. Avoid dark green papayas, which may not ripen properly.

Let ripen at room temperature until slightly soft and nearly all yellow, with only a little green. Wash. Cut in half lengthwise and scrape out the seeds. Peel, if desired, or serve like cantaloupe.

Percentage yield: 65%

Crenshaw melon

Piel de sapo melon

Watermelon

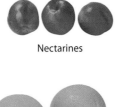

Nectarines

Tangerines

Papaya

Oranges

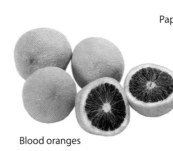

Blood oranges

Passion Fruit

These are tropical fruits about the size of eggs, with a brownish purple skin that wrinkles when ripe. (There is also a yellow-skinned variety.) They are mostly hollow when ripe, with juice, seeds, and a little flesh inside. The tart juice has an intense, exotic flavor and aroma greatly prized by pastry chefs. Select fruits that are large and heavy for their size. If they are smooth, let ripen at room temperature until the skin is wrinkled.

To use, cut in half, taking care not to lose any juice. Scrape out the seeds, juice, and pulp. Seeds can be eaten, so do not discard. If you need only the juice, it is much more economical to buy the frozen juice, as fresh fruits are expensive.

Percentage yield: 40–45%

Passion fruits

Peaches and Nectarines

Peaches should be plump and firm, without bruises or blemishes. Avoid dark green fruits, which are immature and will not ripen well. Avoid fruits refrigerated before ripening, as they may be mealy. Select freestone varieties of peaches. Clingstone varieties require too much labor (they are used primarily for canning).

Let ripen at room temperature, then refrigerate. Peel peaches by blanching in boiling water 10–20 seconds, until skin slips off easily, and cool in ice water. (Nectarines do not need to be peeled.) Cut in half, remove pits, and drop into fruit juice, sugar syrup, or ascorbic acid solution to prevent darkening.

Percentage yield: 75%

Peaches

Pears

Pears should be clean, firm, and bright, with no blemishes or bruises.

Pears for eating raw should be fully ripe and aromatic. For cooking, they are better if slightly underripe, as fully ripe pears are very soft when cooked. Wash, pare, cut in halves or quarters, and remove core. To prevent browning, dip in fruit juice.

Percentage yield: 75% (peeled and cored)

Bartlett pears

Persimmons

Persimmons are orange-red fruits available in two varieties. The most common is Hachiya, which is shaped somewhat like a large acorn (about 8 oz/250 g each). It is extremely tannic when unripe, making it nearly inedible until it ripens to a soft, jellylike mass. Ripe persimmons are sweet, juicy, and mild but rich in flavor. The other variety, Fuyu, is smaller and more squat in shape. It lacks the tannin content of Hachiya persimmons and can be eaten even when not fully ripe. Select plump persimmons with good red color and stem cap attached.

Ripen at room temperature until very soft, then refrigerate. Remove stem cap, cut as desired, and remove seeds, if there are any.

Percentage yield: 80%

Bosc pears

Butter French pear
Courtesy of the California
Pear Advisory Board

Pineapples

Pineapples should be plump and fresh-looking, with an orange-yellow color and abundant fragrance. Avoid soft spots, bruises, and dark, watery spots.

Store at room temperature for a day or two to allow some tartness to disappear, then refrigerate. Pineapples may be cut in many ways. For slices, chunks, and dice, cut off top and bottom and pare the rough skin from the sides, using a stainless-steel knife. Remove all eyes. Cut into quarters lengthwise and cut out the hard center core. Slice or cut as desired.

Percentage yield: 50%

Comice pear
Courtesy of the California
Pear Advisory Board

Seckel pear
Courtesy of the California
Pear Advisory Board

Pineapple

Hachiya persimmons

Taylor gold pear
Courtesy of the California
Pear Advisory Board

Starcrimson pear
Courtesy of the California
Pear Advisory Board

Forelle pear
Courtesy of the California
Pear Advisory Board

Prune plums

Santa Rosa plums

Red plums

Black freestone plums

Pomegranates

Prickly pears

Quinces

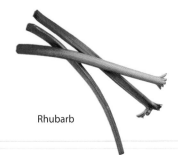

Rhubarb

Star fruit

Plums

Plums should be plump and firm but not hard, with good color and no blemishes.

Wash, cut in half, and remove pits, or serve whole.

Percentage yield: 95% (pitted only)

Pomegranates

The pomegranate is a subtropical fruit about the size of a large apple. It has a dry red skin or shell enclosing a mass of seeds. Each seed is surrounded by a small sphere of juicy, bright red pulp. Pomegranates are used mostly for their red, tart-sweet juice. The seeds, with their surrounding pulp, can also be used as an attractive garnish for desserts and even meat dishes. Look for heavy fruits without bruises. When squeezed, they should yield to gentle pressure; if they are too hard, they may be dried out.

To prepare, lightly score the skin without cutting into the seeds and carefully break the fruit into sections. Separate the seeds from the membranes. Juicing is difficult. Some methods crush the seeds and make the juice bitter. This method makes a better juice: Roll the whole pomegranate on the countertop under the palm of the hand to break the juice sacs. Then pierce a hole in the side and squeeze out the juice.

Percentage yield: 55%

Prickly Pears or Cactus Pears

This is a barrel-shaped fruit about the size of a large egg. Its skin color ranges from magenta to greenish red, and it has a bright pinkish red, spongy interior with black seeds. The pulp is sweet and aromatic, but with a mild flavor. Good-quality fruits are tender but not mushy, with a good skin color, not faded. Avoid fruits with rotten spots.

If the fruit is firm, allow to ripen at room temperature, then refrigerate. As it is the fruit of a cactus, thorns grow on the skin. These are removed before shipping, but small, hard-to-see thorns may remain. To avoid getting stung, hold the fruit with a fork while you slice off the top and bottom. Still holding it with a fork, pare the sides with a knife and discard the peels without touching them. Cut or slice the pulp as desired, or force it through a sieve to purée it and remove the seeds.

Percentage yield: 70%

Quinces

Quinces grow in temperate climates and were once very popular in Europe and North America. Many old, neglected quince trees remain in New England and elsewhere. The fruit resembles a large, yellow, lumpy pear with a smooth or slightly downy skin. The raw fruit is never eaten, as it is dry and hard. When cooked (usually stewed or poached in a sugar syrup), it becomes aromatic, flavorful, and sweet, and the color of the flesh turns slightly pink. The fruit keeps well. Select fruit with good color and free of bruises or blemishes.

Cut, pare, and core like apples or pears, then cook.

Percentage yield: 75%

Rhubarb

Rhubarb is a stem, not a fruit, but it is used like a fruit. Buy firm, crisp, tender rhubarb with thick stalks, not thin and shriveled.

Cut off all traces of leaf, which is poisonous. Trim root end, if necessary. Peel with a vegetable peeler if desired, or omit this step if the skin is tender. Cut into desired lengths.

Percentage yield: 85–90% (if purchased without leaves)

Star Fruit or Carambola

The star fruit is a shiny, yellow, oblong fruit with five ridges running the length of the fruit, so it forms stars when sliced crosswise. It is fragrant, ranging from tart to sweet, with a crisp texture. Look for full, firm fruits. Avoid fruits with ribs that have browned and shrunk.

Wash and slice crosswise.

Percentage yield: 99%

KEY POINTS TO REVIEW

- What are the five main types of salad, as categorized by their place in menus? What kinds of salad are appropriate for each category?

- What are the main categories of salad ingredients? Give examples of each.

- What are the most important varieties of salad greens? Describe them.

- What fruits are used in the kitchen? How is each kind judged for quality? What pre-preparation techniques are used for each?

ARRANGEMENT AND PRESENTATION

THE STRUCTURE OF A SALAD

A plated salad may have as many as four parts: base, body, dressing, and garnish. All salads have body, and most have dressing, but base and garnish are parts of only some salads, as you will see in the following discussion.

Of course this discussion refers only to individual plated salads. When we use the term *salad* to refer to a bulk mixture, as in "two pounds of potato salad," references to the **four parts of a salad** do not apply.

Base or Underliner

A scoop of potato salad looks bare when served by itself on a salad plate as a side dish. Placing it on a bed of lettuce leaves makes it more appealing and also emphasizes its identity as a salad. Although most tossed green salads and many composed salads are presented without an underliner, bound salads and some other vegetable salads may be more attractive and appetizing when served on a bed of leafy greens.

Cup-shaped leaves of iceberg or Boston lettuce make attractive bases. They give height to salads and help confine loose pieces of food.

A layer of loose, flat leaves (such as romaine, loose-leaf, or chicory) or of shredded lettuce may be used as a base. This kind of base involves less labor and food cost, as it is not necessary to separate whole cup-shaped leaves from a head.

Body

This is the main part of the salad and, as such, receives most of our attention in this chapter.

Garnish

A garnish is an edible decorative item added to a salad for eye appeal, though it often adds to the flavor as well. It should not be elaborate or dominate the salad. Remember this basic rule of garnishing: Keep it simple.

Garnish should harmonize with the rest of the salad ingredients and, of course, be edible. It may be mixed with the other salad ingredients (for example, shreds of red cabbage mixed into a tossed green salad), or it may be added at the end.

Often, the main ingredients of a salad form an attractive pattern in themselves, and no garnish is necessary. In the case of certain combination salads and other salads with many ingredients or components, there may be no clear distinction between a garnish and an attractive ingredient that is part of the body. In general, if a salad is attractive and balanced without an added garnish, don't add one.

Nearly any of the vegetables, fruits, and protein foods listed on pages 669–670, cut into simple, appropriate shapes, may be used as garnish.

Dressing

Dressing is a seasoned liquid or semiliquid added to the body of the salad for flavor, tartness, spiciness, and moistness.

The dressing should harmonize with the salad ingredients. In general, use tart dressings for green salads and vegetable salads and use slightly sweetened dressings for fruit salads. Soft, delicate greens like Boston or Bibb lettuce require a light dressing. A thick, heavy dressing will turn them to mush.

Dressings may be added at service time (as for green salads), served separately for the customer to add, or mixed with the ingredients ahead of time (as in potato salad, tuna salad, egg salad, and so on). A salad mixed with a heavy dressing, like mayonnaise, to hold it together is called a ***bound salad***.

Remember: Dressing is a *seasoning* for the main ingredients. It should accent their flavor, not overpower or drown them. Review the rules of seasoning in Chapter 6.

ARRANGING THE SALAD

Perhaps even more than with most other foods, the appearance and arrangement of a salad are essential to its quality. The colorful variety of salad ingredients gives the chef an opportunity to create miniature works of art on the salad plate.

Unfortunately, it is nearly as difficult to give rules for arranging salads as it is for painting pictures because the principles of composition, balance, and symmetry are the same for both arts. It is a skill you must develop an eye for, by experience and by studying good examples.

GUIDELINES for Arranging Salads

1. **Keep the salad off the rim of the plate.**
   Think of the rim as a picture frame and arrange the salad within this frame. Select the right plate for the portion size, not too large or too small.

2. **Strive for a good balance of colors.**
   Plain iceberg lettuce looks pale and sickly all by itself, but it can be enlivened by mixing in darker greens and perhaps a few shreds of carrot, red cabbage, or other colored vegetable. On the other hand, don't go overboard. Sometimes just a few shades of green create a beautiful effect. Too many colors might look messy.

3. **Height helps make a salad attractive.**
   Ingredients mounded on the plate are more interesting than if they are spread flat. Lettuce cups as bases add height. Often just a little height is enough. Arrange ingredients like fruit wedges or tomato slices so they overlap or lean against each other rather than lie flat on the plate.

4. **Cut ingredients neatly.**
   Ragged or sloppy cutting makes the whole salad look sloppy and haphazard.

5. **Make every ingredient identifiable.**
   Cut every ingredient into large enough pieces that the customer can recognize each immediately. Don't pulverize everything in the buffalo chopper or VCM. Bite-size pieces are the general rule, unless the ingredient can be cut easily with a fork, such as tomato slices. Seasoning ingredients, like onion, may be chopped fine.

6. **Keep it simple.**
   A simple, natural arrangement is pleasing. An elaborate design, a gimmicky or contrived arrangement, or a cluttered plate is not pleasing. Besides, elaborate designs take too long to make.

PROCEDURE **for Quantity Salad Production**

When salads are made in quantity, an assembly-line production system is most efficient. Figure 21.2 illustrates this technique.

Remember the rules of safe food handling. Most salad ingredients are eaten without further cooking, so it is essential to avoid cross-contamination. Use sanitary tools and work surfaces. Wash hands properly before beginning work. Local health laws may require the use of gloves whenever you are handling ready-to-eat foods.

1. Prepare all ingredients. Wash and cut greens. Prepare cooked vegetables. Cut all fruits, vegetables, and garnish. Mix bound and marinated salads (egg salad, potato salad, three-bean salad, etc.). Have all ingredients chilled.

2. Arrange salad plates on worktables. Line them up on trays for easy transfer to refrigerator.

3. Place bases or underliners on all plates.

4. Arrange body of salad on all plates.

5. Garnish all salads.

6. Refrigerate until service. Do not hold more than a few hours, or the salads will wilt. Holding boxes should have high humidity.

7. Do not add dressing to green salads until service, or they will wilt.

FIGURE 21.2 Efficient production of salads in quantity.

(a) Prepare all ingredients ahead. Arrange cold salad plates on trays for easy refrigeration.

(b) Place lettuce bases on all plates.

(c) Place body of salad (in this case, potato salad) on all plates.

(d) Garnish all salads. Refrigerate until service.

RECIPES AND TECHNIQUES

Thorough mise en place is extremely important in salad-making. Little cooking is involved, but a great deal of time-consuming handwork is. Salads can be made quickly and efficiently only if the station is set up properly.

GREEN SALADS

Principles

Salad greens must be fresh, clean, crisp, cold, and well drained, or the salad will be of poor quality. Good greens depend on proper preparation.

Moisture and air are necessary to keep greens crisp.

1. Leaves wilt because they lose moisture. Crispness can be restored by washing and refrigeration. The moisture that clings to the leaves after thorough draining is usually enough. Too much water drowns them and dissolves out flavor and nutrients.

2. Air circulation is essential for the greens to breathe. Do not seal washed greens too tightly or pack them too firmly. Refrigerate in colanders covered with clean, damp towels, or in specially designed perforated plastic bins. These protect from drying while allowing air circulation.

Browning or rusting occurs when cut greens are held too long. This can be partially avoided by rinsing them in a mild antioxidant and by using stainless-steel knives. Better yet, plan purchasing and production so you don't need to hold them too long.

BASIC Procedure for Making Green Salads

1. **Wash greens thoroughly.**

   Remove core from iceberg lettuce by striking the core gently against a flat surface, such as a cutting board or the side of a vegetable sink, and twisting it out. Do not smash it, or you'll bruise the entire head. Cut through the core of other greens or separate the leaves so all traces of grit can be removed. Wash in several changes of cold water, until completely clean. For iceberg lettuce, run cold water into the core end (after removing core), then turn over to drain.

2. **Drain greens well.**

   Lift greens from the water and drain in a colander. Tools and machines are available that quickly spin-dry greens. Poor draining results in a watered-down dressing and a soupy, soggy salad.

3. **Crisp the greens.**

   Refrigerate greens in a colander covered with damp towels or in a perforated storage bin to allow air circulation and complete drainage.

4. **Cut or tear into bite-size pieces.**

   Many people insist on tearing leaves instead of cutting, but this is a slow method if you have a large quantity to do. Also, you are more likely to crush or bruise the leaves.

   Use sharp stainless-steel knives for cutting. Bite-size pieces are important as a convenience to the customer. It is difficult to eat or cut large leaves with a salad fork.

5. **Mix the greens.**

   Toss gently until uniformly mixed. Nonjuicy raw vegetable garnish such as green pepper strips or carrot shreds may be mixed in at this time. Just make sure the vegetables are not cut into compact little chunks that will settle to the bottom of the bowl. Broad, thin slices or shreds stay better mixed.

   For tossed salads to be served immediately, add the dressing to the greens in the bowl. Toss to coat the greens with the dressing. Plate (step 6) and serve immediately. For pre-plated salads, proceed with steps 6 through 9.

6. **Plate the salads (including underliners, if used).**

   Cold plates, please. Don't use plates right out of the dishwasher.

   Avoid plating salads more than an hour or two before service, or they are likely to wilt or dry.

7. **Garnish.**

   Exceptions: (a) Garnish that is tossed with the greens in step 5. (b) Garnish that will not hold well (croutons will get soggy, avocado will discolor, etc.). Add these at service time.

8. **Refrigerate.**

9. **Add dressing immediately before service, or serve it on the side.**

Standards of Quality for Green Salads

1. Greens should be fresh and bright, not wilted or browned.

2. Except for some specialty salads, greens should be in bite-size pieces.

3. Greens should have been properly dried before dressing and plating. Water clinging to improperly dried greens dilutes the dressing and makes the salad excessively wet.

4. Tossed salads should have just enough dressing to lightly coat all the greens. Dressing should not pool on the bottom of the plate.

5. Dressing should be appropriate to the greens. Light, delicate greens require a lighter dressing, usually a vinaigrette. Sturdier greens, such as romaine, can handle heavier dressings.

KEY POINTS TO REVIEW

- What are the four parts of a salad? Do all salads have all four parts?
- What are the basic guidelines for arranging salads?
- What are the steps in the basic procedure for making green salads?

Mixed Green Salad

PORTIONS: 12   PORTION SIZE: 2¹/₂–3 OZ (70–90 G)

U.S.	METRIC	INGREDIENTS
2 lb 4 oz	1 kg	Assorted salad greens
8-12 fl oz	250-375 mL	Basic Vinaigrette or variation (p. 660)

PROCEDURE
1. Review guidelines and procedure for preparing green salads (p. 682).
2. Wash and drain the greens thoroughly. Chill in refrigerator.
3. Cut or tear the greens into bite-size pieces.
4. Place the salad greens in a large mixing bowl and toss gently until uniformly mixed.
5. Immediately before service, add the dressing and toss to coat all the leaves with the dressing.
6. Place on cold salad plates and serve immediately.

Per serving: Calories, 130; Protein, 1 g; Fat, 13 g (91% cal.); Cholesterol, 0 mg; Carbohydrates, 2 g; Fiber, 2 g; Sodium, 160 mg.

VARIATIONS

Any combination of salad greens may be used. When using head lettuces, make allowances for variations in trimming yields. Plan on an EP weight of 2¹/₂–3 oz (70–90 g) per portion.

Vegetable ingredients, if they are not juicy, may be tossed with the greens. See page 669 for a listing. Shredded carrot and red cabbage are useful because a small amount gives an attractive color accent.

Garnishes may be added after the salads are plated, such as

Tomato wedges	Pepper rings
Cherry tomatoes	Red onion rings
Cucumber slices	Croutons
Radishes	Hard-cooked egg wedges or slices

Service variation: Instead of tossing the salads with the dressing, plate the greens and hold for service. Ladle dressing over salads just before service, or serve dressing in a separate container.

Spinach Salad

PORTIONS: 12   PORTION SIZE: 3 OZ (90 G)

U.S.	METRIC	INGREDIENTS
1¹/₂ lb	700 g	Spinach leaves, trimmed (no stems)
6 oz	175 g	Bacon
8 oz	225 g	Fresh white mushrooms
3	3	Hard-cooked eggs

PROCEDURE
1. Wash the spinach leaves in several changes of cold water until there is no trace of sand on them. Drain well. Chill in the refrigerator.
2. Cook the bacon until crisp on a griddle or in the oven on a sheet pan. Drain and let cool.
3. Crumble the bacon.
4. Wash the mushrooms and dry them well. Trim the bottoms of stems. Cut the mushrooms into thin slices.
5. Chop the eggs coarsely.
6. Place the spinach in a large bowl. Tear large leaves into smaller pieces. Smaller leaves may be left whole.
7. Add the mushrooms. Toss to mix thoroughly.
8. Portion the salad onto cold salad plates.
9. Sprinkle the salad with the chopped eggs.
10. Hold for service in refrigerator.
11. At serving time, sprinkle with the crumbled bacon.
12. Serve with a vinaigrette variation or with emulsified French dressing.

Per serving: Calories, 60; Protein, 5 g; Fat, 3.5 g (50% cal.); Cholesterol, 55 mg; Carbohydrates, 3 g; Fiber, 2 g; Sodium, 130 mg.

Note: Bacon may be added to salads when they are assembled (step 9). However, it will be less appetizing because the fat congeals in the refrigerator. For best quality, cook the bacon as close to serving time as possible.

Caesar Salad (Method 1: Tableside Preparation)

PORTIONS: 8   **PORTION SIZE: 4 OZ (125 G) PLUS DRESSING**

U.S.	METRIC	INGREDIENTS	PROCEDURE
2 lb	1 kg	Romaine leaves	1. Wash and drain the greens thoroughly. Chill in the refrigerator.
4 oz	125 g	White bread	2. Trim the crusts from the bread. Cut the bread into small cubes measuring about $^3/_8$ in. (1 cm).
1–1½ fl oz	30–45 mL	Olive oil	3. Heat a thin layer of olive oil in a sauté pan over moderately high heat. Add the bread cubes and sauté in the oil until golden and crisp. Add more oil as needed.
			4. Remove the croutons from the pan and hold for service. Do not refrigerate.
1–2	1–2	Garlic cloves	5. Have all ingredients prepared ahead of time and arranged on a cart in the dining room.
4–8	4–8	Anchovy fillets	6. Ask the customers how much garlic they would like. Depending on their answer, either rub the bowl with a cut clove of garlic and remove it, or leave it in the bowl and crush it with the anchovies.
8 fl oz	250 mL	Olive oil	
2	2	Eggs, pasteurized	
2½ fl oz	75 mL	Lemon juice	7. Ask the customers how many anchovies, if any, they would like.
1 oz	30 g	Parmesan cheese, grated	8. Mash the garlic and anchovies to a paste in the salad bowl.
to taste	to taste	Salt	9. Beat in about half the olive oil.
			10. Add the greens and toss to coat with the oil mixture.
			11. Break the egg over the bowl and drop it in. Toss the lettuce well.
			12. Add the lemon juice, the rest of the oil, the parmesan cheese, and a little salt. Toss again until well mixed.
			13. Add the croutons and toss a final time.
			14. Plate and serve.

Per serving: Calories, 370; Protein, 6 g; Fat, 33 g (80% cal.); Cholesterol, 55 mg; Carbohydrates, 12 g; Fiber, 2 g; Sodium, 270 mg.

VARIATIONS

Caesar Salad (Method 2: Pantry Preparation)

Prepare the croutons and salad greens as in the basic recipe. Toss the greens with Caesar Dressing (p. 666). Plate and garnish with croutons.

Grilled Chicken Caesar

Top Caesar salads with sliced, grilled chicken breast.

Caesar Salad

Garden Salad ♥ 🌹

PORTIONS: 12   **PORTION SIZE: 3 OZ (90 G) PLUS GARNISH**

U.S.	METRIC	INGREDIENTS
1 lb 12 oz	800 g	Mixed salad greens (see Note)
4 oz	125 g	Cucumbers
2 oz	60 g	Celery
2 oz	60 g	Radishes
2 oz	60 g	Scallions
2 oz	60 g	Carrots
12 oz	350 g	Tomatoes

PROCEDURE

1. Wash and drain the greens thoroughly. Chill in the refrigerator.
2. Score the cucumbers lengthwise with a fork (see p. 284), or peel them if they are waxed. Cut into thin slices.
3. Cut the celery into thin slices on the bias.
4. Trim the radishes and cut into thin slices.
5. Trim the roots and wilted tops of the scallions. Cut in half crosswise. Then slice lengthwise into thin shreds.
6. Trim and peel the carrots. Shred on a medium grater.
7. Remove the core end of the tomatoes. Cut into wedges, 8–10 per tomato, depending on size.
8. Cut or tear the lettuce and other greens into bite-size pieces.
9. Place all ingredients except tomatoes in a large mixing bowl. Toss until evenly mixed.
10. Plate the salads on cold plates or bowls.
11. Garnish with tomato wedges.
12. Hold for service in refrigerator.
13. Serve with an appropriate dressing.

Per serving: Calories, 25; Protein, 2 g; Fat, 0.5 g (14% cal.); Cholesterol, 0 mg; Carbohydrates; 5 g; Fiber, 2 g; Sodium, 25 mg.

Note: Include some firm-textured, crisp lettuce in the mixed greens, such as romaine or iceberg.

VEGETABLE, GRAIN, LEGUME, AND PASTA SALADS

Principles

Vegetable salads are salads whose main ingredients are vegetables other than lettuce or other leafy greens. Some vegetables are used raw, such as celery, cucumbers, radishes, tomatoes, and green peppers. Some are cooked and chilled before including in the salad, such as artichokes, green beans, beets, and asparagus. See page 669 for lists of vegetables that can be used.

Starchy items such as grains, pastas, and dried legumes can also form the body of a salad. These ingredients usually have a bland, flat taste, so they are enhanced by a well-seasoned, tart dressing. Raw or cooked vegetables are usually added to the starch item to enhance the color, flavor, and nutritional balance of the salad. Depending on the proportion of vegetables and starch item, it is not always possible to classify the salad as a vegetable or starch salad. However, the following guidelines apply to the preparation of all these salads.

In addition, protein items such as poultry, meat, seafood, and cheese may be added to vegetable and starch salads.

Some bound salads, discussed on page 694, could be considered as a subcategory of vegetable and starch salads. However, many bound salads have a protein item as a main ingredient, so we discuss them as a separate category. There is no exact dividing line between these types. Coleslaw with mayonnaise dressing, for example, may be considered to be in either category.

GUIDELINES **for Making Vegetable, Legume, Grain, and Pasta Salads**

1. Neat, accurate cutting of ingredients is important because the shapes of the vegetables add to eye appeal. The design or arrangement of a vegetable salad is often based on different shapes, such as long, slender asparagus and green beans, wedges of tomato, slices of cucumber, strips or rings of green pepper, and radish flowers.

2. Cut vegetables as close as possible to serving time, or they may dry or shrivel at the edges.

3. Cooked vegetables should have a firm, crisp texture and good color. Mushy, overcooked vegetables are unattractive in a salad. See Chapter 10 for vegetable cooking principles.

4. After cooking, vegetables must be thoroughly drained and chilled before being included in the salad.

5. Starches, pastas, and legumes should be cooked until completely tender, but not overcooked. Starches absorb liquid from the dressing, so they may become mushy if they were overcooked. Undercooked grains and dried beans may be unpleasantly firm when cooled.

6. Vegetables are sometimes marinated, or soaked in a seasoned liquid, before being made into salads, as for Mixed Bean Salad (p. 691). The marinade is usually some form of oil and vinegar dressing that also serves as the dressing for the salad. Do not plate marinated salads too far ahead of time, or the lettuce base will wilt. Use crisp, sturdy greens (such as iceberg, romaine, or chicory) as bases, as they do not wilt as quickly.

7. Grains and pastas may also be marinated for a short time, but avoid marinating for more than a few hours, as they are likely to absorb too much liquid and become very soft. This is especially true of cooked pasta. Legumes should not be marinated long either, but for the opposite reason. The acid in the marinade may toughen the proteins in the beans or lentils.

 # Coleslaw

PORTIONS: 12   **PORTION SIZE: 3 OZ (100 G)**

U.S	METRIC	INGREDIENTS	PROCEDURE
12 fl oz	375 mL	Mayonnaise	1. Combine the mayonnaise, vinegar, sugar, salt, and pepper in a stainless-steel bowl. Mix until smooth.
1 fl oz	30 mL	Vinegar	
1/2 oz	15 g	Sugar (optional)	2. Add the cabbage and mix well.
1 tsp	5 mL	Salt	3. Taste and, if necessary, add more salt and/or vinegar.
1/4 tsp	1 mL	White pepper	
2 lb EP	1 kg EP	Cabbage, shredded	
12	12	Lettuce cups	4. Arrange the lettuce leaves as underliners on cold salad plates.
			5. Using a No. 12 scoop, place a mound of coleslaw in the center of each plate.
			6. Hold for service in refrigerator.

Per serving: Calories, 230; Protein, 2 g; Fat, 24 g (89% cal.); Cholesterol, 25 mg; Carbohydrates, 5 g; Fiber, 2 g; Sodium, 270 mg.

VARIATIONS

1. Use Cooked Salad Dressing (p. 667) instead of mayonnaise. Reduce or omit vinegar.
2. Substitute sour cream for half of the mayonnaise.
3. Substitute heavy cream for 4 fl oz (125 mL) mayonnaise.
4. Substitute lemon juice for the vinegar.
5. Use 8 fl oz (250 mL) basic vinaigrette and omit mayonnaise and vinegar. Flavor with 1 tsp (5 mL) celery seed and 1/2 tsp (2 mL) dry mustard.
6. Add 1 tsp (5 mL) celery seed to the basic mayonnaise dressing.

Mixed Cabbage Slaw

Use half red cabbage and half green cabbage.

Carrot Coleslaw

Add 8 oz (250 g) shredded carrots to the basic recipe. Reduce cabbage to 1 lb 12 oz (850 g).

Garden Slaw

Add the following ingredients to the basic recipe: 4 oz (125 g) carrots, shredded; 2 oz (60 g) celery, chopped or cut julienne; 2 oz (60 g) green bell pepper, chopped or cut julienne; 1 oz (30 g) scallions, chopped. Reduce cabbage to 1 lb 12 oz (850 g).

Coleslaw with Fruit

Add the following ingredients to the basic recipe: 2 oz (60 g) raisins, soaked in hot water and drained; 4 oz (125 g) unpeeled apple, cut in small dice; 4 oz (125 g) pineapple, cut in small dice. Use sour cream dressing (dressing variation 2 above) and use lemon juice instead of vinegar.

Coleslaw

Roasted Pepper Salad

PORTIONS: 12 **PORTION SIZE: 3¹/₂ OZ**

U.S.	METRIC	INGREDIENTS
1 lb 12 oz	840 g	Red bell peppers
1 lb 12 oz	840 g	Yellow bell peppers
10 fl oz	300 mL	Vinaigrette (p. 660), made with olive oil and red wine vinegar
¹/₂ oz	15 g	Fresh basil, chopped

PROCEDURE

1. Roast and peel the peppers, following the procedure illustrated on page 290.
2. Cut the peppers into strips ¹/₂-in. (1 cm) wide.
3. Mix the peppers, vinaigrette, and basil. Check for seasonings and add salt and pepper if desired, although none may be needed if the dressing is well seasoned.

Per serving: Calories, 190; Protein, 1 g; Fat, 17 g (83% cal.); Cholesterol, 0 mg; Carbohydrates, 9 g; Fiber, 2 g; Sodium, 180 mg.

Mixed Vegetable Salad with Pasta

PORTIONS: 12 **PORTION SIZE: 4 OZ (125 G)**

U.S.	METRIC	INGREDIENTS
12 oz	350 g	Cooked ditalini pasta, cold
8 oz	225 g	Cooked chickpeas or other dried beans, cold
6 oz	175 g	Zucchini, medium dice, raw or blanched
6 oz	175 g	Green beans, cooked, cut into ¹/₂-in. (1-cm) lengths
4 oz	125 g	Red onions, small dice
3 oz	90 g	Small pitted black olives
3 oz	90 g	Celery, medium dice
2 oz	60 g	Green bell pepper, medium dice
2 oz	60 g	Red bell pepper, medium dice
2 tbsp	30 mL	Capers, drained
2 oz	60 g	Parmesan cheese, grated
12 fl oz	350 mL	Italian Dressing (p. 660)
12	12	Lettuce leaves for underliners
12	12	Tomato wedges or cherry tomatoes

PROCEDURE

1. Combine the pasta, beans, vegetables, and cheese in a large bowl. Toss to mix.
2. No more than 1–2 hours before service, add the dressing and toss.
3. Arrange the lettuce leaves on cold salad plates.
4. Just before service, place a 4-oz (125-g) portion of the salad on each lettuce leaf.
5. Garnish each salad with a tomato wedge.

Per serving: Calories, 290; Protein, 6 g; Fat, 22 g (67% cal.); Cholesterol, 5 mg; Carbohydrates, 18 g; Fiber, 4 g; Sodium, 420 mg.

VARIATIONS

Add 8 oz (225 g) diced or sliced salami, pepperoni, or mozzarella cheese to the salad mixture.

Carrot Salad

PORTIONS: 12   **PORTION SIZE: 3 OZ (100 G)**

U.S.	METRIC	INGREDIENTS	PROCEDURE
2¹/₂ lb	1.2 kg	Carrots	1. Peel the carrots. Shred them on a coarse grater.
6 fl oz	180 mL	Mayonnaise	2. Combine the mayonnaise and vinaigrette. Mix until smooth.
4 fl oz	125 mL	Vinaigrette	3. Add the carrots and mix. Season to taste with salt.
to taste	to taste	Salt	
12	12	Lettuce cups	4. Arrange the lettuce cups as underliners on cold salad plates.
6	6	Pitted black olives	5. Using a No. 12 scoop, place a mound of carrot salad in each lettuce cup.
			6. Cut the olives in half lengthwise. Garnish the top of each salad with an olive half.

Per serving: Calories, 200; Protein, 1 g; Fat, 18 g (79% cal.); Cholesterol, 10 mg; Carbohydrates, 10 g; Fiber, 3 g; Sodium, 20 mg.

VARIATIONS

Carrot Raisin Salad

Simmer 4 oz (125 g) raisins in water 2 minutes. Cool, then drain. Mix raisins with the carrots.

Carrot Pineapple Salad

Mix 6 oz (180 g) drained pineapple tidbits with the carrots.

Carrot Celery Salad

Reduce the carrots to 1 lb 12 oz (850g). Mix 12 oz (375 g) celery (cut julienne) or celery root (shredded) with the carrots.

Celery Salad

Use celery or celery root instead of carrots in basic recipe. Cut stalk celery into thin slices instead of shredding it. Add 1 tbsp (15 mL) French or Dijon-type mustard to the dressing.

Carrot Salad

Greek Salad

PORTIONS: 12

U.S.	METRIC	INGREDIENTS	PROCEDURE
		Vinaigrette:	1. Prepare vinaigrette, using the ingredients listed and following the procedure on p. 660.
1¹/₂ fl oz	45 mL	Lemon juice	
¹/₂ tsp	2 mL	Salt	
¹/₈ tsp	0.5 mL	Pepper	
1 tbsp	15 mL	Dried oregano	
4¹/₂ fl oz	135 mL	Olive oil	
12 oz	360 g	Cucumber, cut in half lengthwise and sliced	2. Combine the cucumber, tomatoes, cheese, onion, and olives in a bowl.
1 lb 8 oz	720 g	Tomatoes, large dice	3. Add the vinaigrette and toss to mix.
9 oz	270 g	Feta cheese, crumbled	4. Divide the lettuce among individual serving bowls (1¹/₂ oz or 45 g per portion).
6 oz	180 g	Red onion, in thin slices	5. Spoon the vegetable and cheese mixture onto the lettuce.
4 oz	120 g	Pitted Kalamata olives	
1 lb 2 oz	540 g	Romaine lettuce, cut into bite-size pieces	

Per serving: Calories, 210; Protein, 5 g; Fat, 18 g (74% cal.); Cholesterol, 20 mg; Carbohydrates, 9 g; Fiber, 2 g; Sodium, 500 mg.

VARIATIONS

Instead of dicing the tomatoes, cut them into wedges. Garnish each plated salad with 2 or 3 wedges.

Heirloom Tomato Salad

PORTIONS: 12   **PORTION SIZE: ABOUT 4 OZ (125 G)**

U.S.	METRIC	INGREDIENTS
3 lb	1.5 kg	Assorted ripe heirloom tomatoes, 3 or 4 varieties (see Procedure)
12 small bunches	12 small bunches	Mixed baby greens
6 fl oz	180 mL	Extra-virgin olive oil
3 fl oz	90 mL	Balsamic vinegar
to taste	to taste	Coarse salt
to taste	to taste	Black pepper

PROCEDURE

1. Select the tomatoes: Use 3 or 4 types of ripe, flavorful tomatoes of varied colors (yellow, green, orange, different shades of red), depending on availability. Include at least one type of small tomato, such as cherry tomato.
2. Prepare the tomatoes: Peel large tomatoes by blanching them 10 seconds in boiling water, cutting out the core end, and pulling off the skins. Small tomatoes with tender skins can be left unpeeled. Cut large tomatoes into slices. Cut small tomatoes into quarters or halves or leave whole, depending on size.
3. Arrange the tomatoes attractively on large salad plates.
4. Place a small bunch of greens on each plate.
5. Drizzle the tomatoes with a little olive oil and then a little vinegar.
6. Sprinkle lightly with coarse salt and black pepper.

Per serving: Calories, 150; Protein, 1 g; Fat, 14 g (82% cal.); Cholesterol, 0 mg; Carbohydrates, 6 g; Fiber, 1 g; Sodium, 100 mg.

VARIATIONS

Instead of the oil and vinegar, drizzle the salad with Balsamic Vinaigrette (p. 660) or Mustard Vinaigrette (p. 660).

Insalata Caprese

Use only one variety of large, red tomato. Peel and slice. Alternate overlapping slices of tomato and fresh mozzarella on a plate or platter. Sprinkle with fresh basil, cut chiffonade, coarse salt, and pepper. Drizzle with olive oil.

Modern Caprese

For a modern interpretation of Caprese Salad, Place a rectangle of Balsamic Sheet (p. 222) on a rectangular plate, as shown in the illustration. Brush tomato wedges and halved cherry tomatoes lightly with olive oil and arrange on the sheet. Place a few wedges of fresh mozzarella between the tomatoes and add a few lightly oiled fresh greens. Place a small spoonful of Olive Oil Powder (p. 222) at each end and garnish the salad with Basil Caviar (p. 221).

Modern Caprese

Pacific Rim Salad with Beef

PORTIONS: 12   PORTION SIZE: 5 OZ (150 G)

U.S.	METRIC	INGREDIENTS	PROCEDURE
1 lb	450 g	Bean sprouts	1. Mix the bean sprouts, snow peas, carrots, water chestnuts, scallions, and almonds. Hold in the refrigerator.
8 oz	225 g	Snow peas, trimmed and blanched	
4 oz	120 g	Carrots, cut julienne	
6 oz	175 g	Water chestnuts, sliced	
2 oz	60 g	Scallions, chopped	
2 oz	60 g	Slivered almonds, toasted	
12 oz	350 g	Cold roast beef	2. Slice the beef about $1/4$ in. (6 mm) thick, then cut into thin strips.
14 fl oz	400 mL	Oriental Vinaigrette (p. 662)	3. No more than about 2 hours before serving, marinate the beef in the vinaigrette 30–60 minutes.
12 oz	350 g	Chinese cabbage, shredded	4. Add the mixed vegetables to the beef and vinaigrette mixture. Toss to mix well.
24	24	Orange wedges	5. Place a bed of shredded Chinese cabbage on salad plates.
			6. Portion the salad mixture onto the plates.
			7. Garnish each salad with 2 orange wedges.

Per serving: Calories, 330; Protein, 11 g; Fat, 27 g (72% cal.); Cholesterol, 25 mg; Carbohydrates, 13 g; Fiber, 4 g; Sodium, 200 mg.

VARIATIONS

Substitute chicken, turkey, ham, or duck meat for the beef.

Pacific Rim Salad with Beef

Mixed Bean Salad with Olives and Tomatoes

PORTIONS: 12   PORTION SIZE: 4 OZ (125 G)

U.S.	METRIC	INGREDIENTS
8 oz	250 g	Chickpeas, cooked or canned, drained
8 oz	250 g	Red kidney beans, cooked or canned, drained
8 oz	250 g	Navy or white kidney beans, cooked or canned, drained
10 oz	300 g	Tomatoes, peeled, seeded, small dice
2 oz	60 g	Green olives, pitted, sliced
2 oz	60 g	Black olives, pitted, sliced
2 oz	60 g	Red onion, chopped fine
8 fl oz	250 mL	Mustard Vinaigrette (p. 660)
2 tbsp	30 mL	Chopped parsley
12	12	Lettuce leaves

PROCEDURE

1. Combine the chickpeas and beans in a bowl. Add the tomatoes, olives, and onion. Toss gently.
2. Add the vinaigrette. Mix.
3. Cover and refrigerate 2–4 hours.
4. Before serving, mix in the chopped parsley.
5. Arrange the lettuce leaves on cold salad plates.
6. Mound the salad mixture on the lettuce leaves.

Per serving: Calories, 220; Protein, 5 g; Fat, 16 g (62% cal.); Cholesterol, 0 mg; Carbohydrates, 17 g; Fiber, 6 g; Sodium, 320 mg.

VARIATION

White Bean Salad

Instead of the mixture of beans, use only white kidney beans.

Mixed Bean Salad with Olives and Tomatoes

Panzanella

PORTIONS: 10   PORTION SIZE: 4 OZ (125 G)

U.S.	METRIC	INGREDIENTS
12 oz	375 g	Firm, white Italian bread
1¹/₂ lb	750 g	Ripe tomatoes, peeled
1¹/₂ oz	45 g	Red onion, chopped fine
¹/₂ oz	15 g	Fresh basil, torn into small pieces
4 fl oz	125 mL	Olive oil
2 fl oz	60 mL	Red wine vinegar
¹/₈ tsp	0.5 mL	Red pepper flakes
to taste	to taste	Salt
to taste	to taste	Pepper

PROCEDURE

1. Use a hearty, firm bread for this recipe. If necessary, let it sit out to dry for a few hours, or dry it slightly in an oven.
2. Tear the bread into bite-size pieces. Place in a large bowl.
3. Chop the tomatoes coarsely into ¹/₂-in. (1-cm) pieces. Add to the bowl.
4. Add the remaining ingredients. Toss to mix.
5. Let stand about 1 hour, or until the bread has absorbed the juices and softened.

Per serving: Calories, 200; Protein, 4 g; Fat, 12 g (53% cal.); Cholesterol, 0 mg; Carbohydrates, 20 g; Fiber, 2 g; Sodium, 200 mg.

VARIATION

Add either or both of the following ingredients to the mixture: 8 oz (250 g) peeled, seeded cucumber, chopped; 4 oz (125 g) hearts of celery, sliced.

Quinoa Salad with Bell Peppers

PORTIONS: 12   PORTION SIZE: 4¹/₂ OZ (135 G)

U.S.	METRIC	INGREDIENTS	PROCEDURE
9 oz	275 g	Quinoa	1. Rinse the quinoa thoroughly in cold water. Drain.
1¹/₄ pt	625 mL	Water	2. Combine with the water and salt. Bring to a boil. Lower heat, cover, and simmer slowly until the grain is cooked, about 15 minutes.
¹/₄ tsp	1 mL	Salt	3. Spread the cooked grain in a shallow pan to cool.
4 oz	125 g	Red bell pepper, small dice	4. Combine the cooled quinoa, bell peppers, scallions, cucumber, and apricots in a bowl.
4 oz	125 g	Green bell pepper, small dice	5. Add the dressing and toss to mix.
2 oz	60 g	Scallion, chopped fine	6. Add salt to taste.
6 oz	180 g	Cucumber, peeled and seeded, small dice	
3 oz	90 g	Dried apricots, chopped fine	
8 fl oz	250 mL	Italian Dressing (p. 660)	
to taste	to taste	Salt	
12	12	Lettuce or radicchio leaves	7. Arrange the lettuce leaves on cold salad plates.
			8. Mound the salad mixture on the lettuce leaves.

Per serving: Calories, 220; Protein, 4 g; Fat, 14 g (56% cal.); Cholesterol, 0 mg; Carbohydrates, 21 g; Fiber, 2 g; Sodium, 200 mg.

VARIATIONS

This salad can also be made with brown rice, farro, or couscous.

Quinoa Salad with Bell Peppers

Lentil Salad

PORTIONS: 10   PORTION SIZE: 4 OZ (125 G)

U.S.	METRIC	INGREDIENTS	PROCEDURE
12 oz	375 g	Green lentils	1. Pick over, rinse, and drain the lentils.
1 qt	1 L	Water	2. Place them in a saucepan with the water. Bring to a boil, reduce heat, and simmer until the lentils are just tender and not falling apart, about 25 minutes.
			3. Drain the lentils and place them in a bowl.
4 oz	125 g	Celery, cut brunoise	4. While the lentils are still warm, add the celery, carrot, onion, parsley, olive oil, and lemon juice. Toss to mix.
4 oz	125 g	Carrot, cut brunoise	5. Season to taste with salt and pepper.
1¹/₂ oz	45 g	Red onion, cut brunoise	
1 oz	30 g	Parsley, chopped	
4 fl oz	145 mL	Olive oil	
2 fl oz	60 mL	Lemon juice	
to taste	to taste	Salt	
to taste	to taste	Pepper	
12	12	Lettuce leaves	6. Arrange the lettuce leaves on salad plates.
			7. Serve the lentil salads warm or cold. Mound the salad on the lettuce leaves.

Per serving: Calories, 200; Protein, 9 g; Fat, 11 g (44% cal.); Cholesterol, 0 mg; Carbohydrates, 23 g; Fiber, 6 g; Sodium, 35 mg.

Broccoli, Prosciutto, and Sunflower Salad

PORTIONS: 12   PORTION SIZE: 4¹/₂ OZ (140 G)

U.S.	METRIC	INGREDIENTS
4 oz	125 g	Raisins
2 lb	1 kg	Broccoli florets
4 oz	125 g	Prosciutto, sliced thin, chopped
3 oz	90 g	Sunflower seeds, toasted (shelled)
3 oz	90 g	Shallot or red onion, chopped fine
6 fl oz	180 mL	Emulsified French Dressing (p. 666)
4 fl oz	125 mL	Heavy cream

PROCEDURE

1. Place the raisins in a bowl and add warm water to cover. Let stand 1 hour.
2. Drain.
3. Combine the raisins, broccoli, prosciutto, sunflower seeds, and shallot or onion in a bowl. Toss to mix evenly.
4. Combine the dressing and cream. Add to the broccoli and toss to mix.

Per serving: Calories, 240; Protein, 7 g; Fat, 19 g (67% cal.); Cholesterol, 15 mg; Carbohydrates, 14 g; Fiber, 3 g; Sodium, 280 mg.

Broccoli, Prosciutto, and
Sunflower Salad

Tabbouleh

PORTIONS: 12   PORTION SIZE: 4 OZ (125 G)

U.S.	METRIC	INGREDIENTS
12 oz	375 g	Bulgur wheat, fine or medium texture
1 lb	500 g	Cucumber
1¹/₂ tsp	7 mL	Coarse salt
1¹/₂ oz	45 g	Parsley, chopped
1¹/₂ oz	45 g	Scallions, sliced thin
6 oz	180 g	Tomato, peeled, seeded, and chopped
3 fl oz	90 mL	Lemon juice
3 fl oz	90 mL	Olive oil
to taste	to taste	Salt
to taste	to taste	Pepper
12	12	Lettuce leaves

PROCEDURE

1. Place the bulgur wheat in a bowl. Pour over it about twice its volume of boiling water.
2. Cover and let stand until completely cool. The bulgur should have absorbed most or all of the water and should be tender enough to eat. If any liquid remains, drain it and squeeze out the grain gently. Fluff with a fork.
3. Peel the cucumbers and quarter them lengthwise. Scoop out and discard the seeds. Slice the cucumbers ¹/₄ in. (6 mm) thick.
4. Toss with the coarse salt in a bowl and let stand 30 minutes.
5. Rinse, drain, and pat dry.
6. Mix the bulgur, cucumbers, parsley, scallions, tomato, lemon juice, and olive oil.
7. Add salt and pepper to taste.
8. Arrange the lettuce leaves on cold salad plates.
9. Mound the tabbouleh on the lettuce leaves.

Per serving: Calories, 170; Protein, 4 g; Fat, 7 g (36% cal.); Cholesterol, 0 mg; Carbohydrates, 24 g; Fiber, 6 g; Sodium, 240 mg.

BOUND SALADS

Principles

Bound salads are mixtures of foods held together, or bound, with a dressing, usually a thick dressing such as mayonnaise. The term bound is used most often for traditional mixtures of cooked protein, starch, and vegetable items with mayonnaise, such as chicken salad, tuna salad, egg salad, and potato salad.

As noted in the discussion of vegetable and starch salads on page 685, there is no exact dividing line between that category and bound salads, so you should keep in mind the guidelines for preparing both kinds of salads when preparing bound salads.

Some of these salads, mainly those made with protein items, are also used as sandwich fillings. Sandwich filling ingredients must usually be chopped fine or cut into small dice to be practical for this use. Plated salads, by contrast, may contain larger cuts if desired.

Popular choices for cooked salads are the following:

Chicken	Lobster
Turkey	Eggs
Ham	Potatoes
Tuna	Pastas
Salmon	Rice
Crab	Mixed vegetables
Shrimp	

GUIDELINES for Making Bound Salads

1. Cooked ingredients must be thoroughly cooled before being mixed with mayonnaise, and the completed salad mixture must be kept chilled at all times. Mayonnaise-type salads are ideal breeding grounds for bacteria that cause food poisoning.

2. Bound salads are good ways to use leftovers such as chicken, meat, or fish, but the ingredients must have been handled according to the rules of good sanitation and food handling. The product will not be cooked again to destroy any bacteria that might grow in the salad and cause illness.

3. Potatoes for salads should be cooked whole, then peeled and cut, in order to preserve nutrients.

4. Except in the case of sandwich fillings, don't cut ingredients too small, or the final product will be like mush or paste, with no textural interest.

5. Crisp vegetables are usually added for texture. Celery is the most popular, but other choices are green peppers, carrots, chopped pickles, onions, water chestnuts, and apples. Be sure the flavors go together, however.

6. Bland main ingredients, such as potatoes and some seafoods, may be marinated in a seasoned liquid such as vinaigrette before being mixed with the mayonnaise and other ingredients. Any marinade not absorbed should be drained first to avoid thinning the mayonnaise.

7. Fold in thick dressings gently to avoid crushing or breaking the main ingredients.

8. Bound salads are usually portioned with a scoop. This has two advantages: (a) It provides portion control. (b) It gives height and shape to the salad.

9. For plated salads, serve on a base of greens, and choose attractive, colorful garnishes when appropriate. A scoop of potato or chicken salad looks pale and uninteresting when plated without a base or garnish.

Chicken or Turkey Salad

PORTIONS: 12   **PORTION SIZE: 3¹/₂ OZ (100 G)**

U.S.	METRIC	INGREDIENTS	PROCEDURE
1 lb 14 oz	875 g	Cooked chicken or turkey, ¹/₂-in. (1-cm) dice	1. Combine all ingredients in a mixing bowl. Toss gently until thoroughly mixed.
6 oz	175 g	Celery, ¹/₄-in. (0.5-cm) dice	
8 fl oz	240 mL	Mayonnaise	
1 fl oz	30 mL	Lemon juice	
to taste	to taste	Salt	
to taste	to taste	White pepper	
12	12	Lettuce cups	2. Arrange lettuce as underliners on cold salad plates.
as needed	as needed	Parsley or watercress sprigs	3. Using a No. 10 scoop, place a mound of chicken salad on each plate. Garnish with parsley or watercress.
			4. Hold for service in refrigerator.

Per serving: Calories, 200; Protein, 15 g; Fat, 22 g (74% cal.); Cholesterol, 60 mg; Carbohydrates, 2 g; Fiber, 1 g; Sodium, 170 mg.

VARIATIONS

If desired, increase the quantity of celery and decrease the quantity of chicken by the same amount.

Add any of the following ingredients to the basic recipe:

   3 oz (90 g) broken walnuts or pecans
   3 hard-cooked eggs, chopped
   4 oz (115 g) seedless grapes, cut in half, and 1¹/₂ oz (45 g) chopped or sliced almonds
   4 oz (115 g) drained, diced pineapple
   4 oz (115 g) diced avocado
   8 oz (225 g) peeled, seeded, diced cucumber, substituted for 8 oz (225 g) of the celery
   4 oz (115 g) sliced water chestnuts

Egg Salad

Substitute 16–18 diced hard-cooked eggs for the chicken in the basic recipe.

Tuna or Salmon Salad

Substitute 1 lb 12 oz (820 g) drained, flaked canned tuna or salmon for the chicken in the basic recipe. Add 1 oz (30 g) chopped onion. Optional ingredient: 2 oz (50 g) chopped pickles or drained capers.

Chicken Salad

Macedoine of Vegetables Mayonnaise

PORTIONS: 12   PORTION SIZE: 4 OZ (125 G)

U.S.	METRIC	INGREDIENTS	PROCEDURE
1 lb	500 g	Cooked carrots, 1/4-in. (0.5-cm) dice	1. Chill all ingredients before combining.
1 lb	500 g	Cooked white turnips, 1/4-in. (0.5-cm) dice	2. Place vegetables and mayonnaise in a bowl and mix until evenly combined. Use just enough mayonnaise to bind. Season to taste with salt and white pepper.
8 oz	250 g	Cooked green beans, sliced in 1/4-in. (0.5-cm) pieces	
8 oz	250 g	Cooked green peas	
8 fl oz or as needed	250 mL or as needed	Mayonnaise	
to taste	to taste	Salt	
to taste	to taste	White pepper	
12	12	Lettuce cups	3. Place lettuce bases as underliners on cold salad plates.
12	12	Tomato wedges	4. Using a No. 10 scoop, place a mound of salad on each plate. Garnish with 1 tomato wedge.

Per serving: Calories, 180; Protein, 3 g; Fat, 14 g (68% cal.); Cholesterol, 10 mg; Carbohydrates, 12 g; Fiber, 4 g; Sodium, 150 mg.

Macedoine of Vegetables Mayonnaise

Ham Salad

PORTIONS: 25   PORTION SIZE: 3 1/2 OZ (105 G)

U.S.	METRIC	INGREDIENTS	PROCEDURE
1 1/2 lb	725 g	Cooked smoked ham, small dice	1. Combine ham, celery, pickles, onion, mayonnaise, and vinegar in a mixing bowl. Toss gently until evenly mixed. Adjust seasonings.
8 oz	240 g	Celery, small dice	
4 oz	120 g	Chopped pickles (sweet or dill) or drained pickle relish	2. Refrigerate until ready for use.
1 oz	30 g	Onion, chopped fine	
8 fl oz	250 mL	Mayonnaise	
1 fl oz	30 mL	Vinegar	
12	12	Lettuce leaves for underliners	3. Serve a 3 1/2-oz (105-g) portion on a bed of lettuce.
24	24	Tomato wedges (optional)	4. If desired, garnish with tomato wedges.

Per serving: Calories, 150; Protein, 1 g; Fat, 15 g (87% cal.); Cholesterol, 5 mg; Carbohydrates, 4 g; Fiber, 1 g; Sodium, 170 mg.

VARIATIONS

Ham Salad Spread

Grind the ham, or chop very fine. Chop the celery very fine. Use for sandwiches.

Deviled Ham

Grind the ham. Add 1 1/2 tbsp (22 mL) prepared mustard and 1/2 tsp (2 mL) hot red pepper sauce to the basic recipe. Increase the onion to 2 oz (60 g). Use as a canapé spread.

Corned Beef Salad

Substitute corned beef for the ham in the basic recipe.

Macaroni and Ham Salad

Reduce the ham to 6 oz (180 g) and add 1 1/2 lb (725 g) cooked, drained, chilled elbow macaroni. Omit the pickles and add 3 oz (90 g) green pepper, cut in small dice.

Potato Salad

PORTIONS: 12   **PORTION SIZE: 4 OZ (125 G)**

U.S.	METRIC	INGREDIENTS	PROCEDURE
2¹/₂ lb AP	1.25 kg AP	Waxy potatoes (see Note)	1. Scrub the potatoes. Steam or boil until tender, but do not overcook.
			2. Drain the potatoes. Leave in the colander or spread out on a sheet pan until cool enough to handle.
6 fl oz	180 mL	Basic Vinaigrette (p. 660)	3. Peel the warm potatoes. Cut into ¹/₂-in. (1-cm) dice.
³/₄ tsp	3 mL	Salt	4. Combine the dressing, salt, and pepper. Add the potatoes and mix carefully to avoid breaking or crushing them.
¹/₈ tsp	0.5 mL	White pepper	5. Marinate until cold. For the purpose of food safety, chill the potatoes in the refrigerator before proceeding with the next step.
6 oz	180 g	Celery, small dice	6. If any vinaigrette has not been absorbed by the potatoes, drain it off.
2 oz	60 g	Onion, chopped fine (Optional ingredients— see Variations below)	7. Add the celery, onion, and, if desired, any of the optional ingredients listed below. Mix gently.
8 fl oz	250 mL	Mayonnaise	8. Add the mayonnaise. Mix carefully until evenly blended.
			9. Keep refrigerated until ready to use.
12	12	Lettuce cups	10. Arrange the lettuce as underliners on cold salad plates.
24	24	Pimiento strips	11. Using a No. 10 scoop, place a 4-oz (125-g) mound of potato salad on each plate.
			12. Garnish each salad with 2 strips pimiento placed crosswise on top.
			13. Hold for service in refrigerator.

Per serving: Calories, 290; Protein, 2 g; Fat, 24 g (74% cal.); Cholesterol, 10 mg; Carbohydrates, 17 g; Fiber, 2 g; Sodium, 360 mg.

Note: See pages 346–347 for explanation of potato types. Do not use starchy, mealy potatoes for salad because they will not hold their shape.

VARIATIONS

Optional ingredients, to be added in step 7:

2–3 hard-cooked eggs, diced
1 oz (30 g) green bell peppers, small dice
1 oz (30 g) pimientos, small dice
2 oz (60 g) chopped pickles or capers or sliced olives
2 tbsp (30 mL) chopped parsley

Vinaigrette marination (steps 4–5) may be omitted if necessary. In this case, chill the potatoes before mixing with the dressing. Add 1 fl oz (30 mL) vinegar to the mayonnaise and check carefully for seasonings. Refrigerate 2 hours or more before serving.

French Potato Salad

PORTION: 12   **PORTION SIZE: 4 OZ (125 G)**

U.S.	METRIC	INGREDIENTS	PROCEDURE
3¹/₂ lb	1.75 kg	Waxy potatoes	1. Scrub the potatoes. Steam or boil until tender, but do not overcook.
			2. Drain the potatoes. Leave in the colander or spread out on a sheet pan until cool enough to handle.
4 fl oz	125 mL	Salad oil	3. Peel the potatoes while still hot. Cut into slices ¹/₄-in. (0.5-cm) thick or into ¹/₂-in. (1-cm) dice.
3 fl oz	90 mL	Wine vinegar (white or red)	4. Mix the potatoes with the remaining ingredients. Allow to stand at least 15 minutes while the potatoes absorb the dressing.
2 oz	60 g	Onions or shallots, chopped fine	
¹/₈ tsp	0.5 mL	Garlic, chopped fine	
2 tbsp	30 mL	Chopped parsley	5. Serve warm or cold. This salad is a popular accompaniment to hot cooked sausages.
1¹/₂ tsp	7 mL	Dried tarragon	
to taste	to taste	Salt	
to taste	to taste	Pepper	

Per serving: Calories, 180; Protein, 2 g; Fat, 9 g (46% cal.); Cholesterol, 0 mg; Carbohydrates, 22 g; Fiber, 2 g; Sodium, 5 mg.

VARIATION

Hot German Potato Salad

Omit oil and tarragon from the basic recipe. Cook 4 oz (125 g) diced bacon until crisp. Add the bacon, the bacon fat, and 4 fl oz (125 mL) hot chicken stock to the dressing ingredients. (More stock may be needed if the potatoes absorb a great deal.) Place the mixed salad in a hotel pan, cover, and heat in a 300°F (150°C) oven about 30 minutes. Serve hot.

French Potato Salad

Dilled Shrimp Salad

PORTION: 12   **PORTION SIZE: 3¹/₂ OZ (100 G)**

U.S.	METRIC	INGREDIENTS	PROCEDURE
1¹/₂ lb	700 g	Cooked, peeled, deveined shrimp	1. Cut the shrimp into ¹/₄-in. (0.5-cm) pieces. (If the shrimp are very small, leave them whole.)
12 oz	350 g	Celery, small dice	2. Combine the celery and shrimp in a bowl.
8 fl oz	250 mL	Mayonnaise	3. Mix the mayonnaise, lemon juice, dill, and salt.
1 tbsp	15 mL	Lemon juice	4. Add the dressing to the shrimp mixture. Mix in thoroughly.
1 tsp	5 mL	Dried dill weed (or 2 tbsp/30 mL chopped fresh dill)	
¹/₄ tsp	1 mL	Salt	
12	12	Lettuce cups	5. Arrange the lettuce leaves as underliners on cold salad plates.
24	24	Tomato wedges	6. Using a No. 10 scoop, place a mound of shrimp salad on each plate.
			7. Garnish with tomato wedges, using 2 per salad.

Per serving: Calories, 200; Protein, 12 g; Fat, 15 g (67% cal.); Cholesterol, 115 mg; Carbohydrates, 5 g; Fiber, 1 g; Sodium, 300 mg.

VARIATIONS

Crab or Lobster Salad

Prepare as in the basic recipe, using crab or lobster meat instead of shrimp.

Crab, Shrimp, or Lobster Louis

Use Louis Dressing (p. 665) instead of the mixture of mayonnaise, lemon juice, and dill. Serve on shredded lettuce. If food cost permits, omit celery and increase shellfish to 2¹/₄ lb/1 kg.

Rice and Shrimp Salad

Reduce shrimp in the basic recipe to 1 lb (450 g), and add 2 lb (900 g) cooked rice.

Curried Rice Salad with Shrimp

Prepare Rice and Shrimp Salad, but omit the dill. Instead, flavor the dressing with ¹/₂ tsp (2 mL) curry powder heated lightly in 1 tsp (5 mL) oil and cooled. Optional: Substitute diced green bell pepper for half the celery.

FRUIT SALADS

Principles

As their name indicates, **fruit salads** have fruits as their main ingredients. They are popular as appetizer salads, as dessert salads, and as part of combination luncheon plates, often with a scoop of cottage cheese or other mild-tasting protein food.

GUIDELINES for Making Fruit Salads

1. Fruit salads are often arranged rather than mixed or tossed because most fruits are delicate and easily broken. An exception is the Waldorf salad, made of firm apples mixed with nuts, celery, and a mayonnaise-based dressing.

2. Broken or less attractive pieces of fruit should be placed at the bottom of the salad, with the more attractive pieces arranged on top.

3. Some fruits discolor when cut and should be dipped into an acid such as tart fruit juice. See pages 673–679 for pre-preparation guidelines for individual fruits.

4. Fruits do not hold as well as vegetables after being cut. If both vegetable and fruit salads are being prepared for a particular meal service, the vegetable salads should usually be pre-pared first.

5. Drain canned fruits well before including them in the salad, or the salad will be watery and sloppy. The liquid from the canned fruit may be reserved for use in fruit salad dressing or other preparations.

6. Dressings for fruit salads are often slightly sweet, but a little tartness is usually desirable as well. Fruit juices are often used in dressings for fruit salad.

KEY POINTS TO REVIEW

- What are the guidelines for making vegetable, legume, grain, and pasta salads?

- What is a bound salad? What are the guidelines for making bound salads?

- What are the guidelines for making fruit salads?

Waldorf Salad

PORTIONS: 12   **PORTION SIZE: 3 OZ (90 G)**

U.S.	METRIC	INGREDIENTS	PROCEDURE
6 fl oz	180 mL	Chantilly Dressing (p. 665)	1. Prepare the dressing. Place it in a large stainless-steel bowl and have it ready in the refrigerator. (See Note.)
2 lb AP	900 g AP	Crisp, red eating apples	2. Core the apples and dice them to ¹/₂ in. (1 cm) without peeling them.
8 oz	225 g	Celery, small dice	3. As soon as the apples are cut, add them to the dressing and mix in to prevent darkening.
2 oz	55 g	Walnuts, coarsely chopped	4. Add the celery and walnuts. Fold in until evenly mixed.
12	12	Lettuce cups	5. Arrange the lettuce bases as underliners on cold salad plates.
1 oz	30 g	Chopped walnuts (optional)	6. Using a No. 12 scoop, place a mound of salad on each plate.
			7. If desired, garnish each salad with about 1 tsp (5 mL) chopped nuts.
			8. Hold for service in refrigerator.

Per serving: Calories, 150; Protein, 1 g; Fat, 12 g (69% cal.); Cholesterol, 10 mg; Carbohydrates, 11 g; Fiber, 2 g; Sodium, 40 mg.

Note: Plain mayonnaise may be used instead of Chantilly dressing.

VARIATIONS

Any of the following ingredients may be added to the basic Waldorf mixture. If any of these changes is made, the item should no longer be called simply Waldorf Salad. Change the menu name to indicate the product contains other ingredients. For example: Pineapple Waldorf Salad or Apple Date Salad.

8 oz (225 g) diced pineapple
4 oz (100 g) chopped dates, substituted for the walnuts
4 oz (100 g) raisins, plumped in hot water and drained
1 lb (450 g) shredded cabbage or Chinese cabbage, substituted for the celery

Arugula, Citrus, and Fennel Salad

PORTIONS: 12   PORTION SIZE: 4 OZ (125 G)

U.S.	METRIC	INGREDIENTS
1 fl oz	30 mL	Lime juice
2 fl oz	60 mL	Orange juice
1/2 oz	15 g	Shallots, chopped fine
1 tsp	5 mL	Grated fresh ginger root
2 tsp	10 mL	Grated lime zest
3 fl oz	90 mL	Olive oil
to taste	to taste	Salt
2 lb	1 kg	Grapefruit
1 lb	500 g	Fennel, trimmed
6 oz	180 g	Arugula

PROCEDURE

1. Mix the lime juice, orange juice, shallots, ginger, and zest.
2. Whip in the olive oil to make a vinaigrette.
3. Add salt to taste.

4. Peel and section the grapefruit according to the procedure illustrated on page 146. You should have about 1 lb (500 g) grapefruit sections.
5. Cut the fennel bulb in half vertically. Lay the halves on the cutting board, cut side down, and cut vertically into thin slices.
6. Trim the stems from the arugula. Tear into pieces.
7. Just before serving, toss together the grapefruit, fennel, and arugula.
8. Mound on cold salad plates.
9. Drizzle 1 tbsp (15 mL) vinaigrette over each portion.

Per serving: Calories, 90; Protein, 1 g; Fat, 7 g (66% cal.); Cholesterol, 0 mg; Carbohydrates, 7 g; Fiber, 2 g; Sodium, 25 mg.

Arugula, Citrus, and Fennel Salad

Thai Papaya and Mango Salad

PORTIONS: 12   PORTION SIZE: 4 OZ (125 G)

U.S.	METRIC	INGREDIENTS
6 fl oz	180 mL	Lime juice
1 fl oz	30 mL	Nam pla (Thai fish sauce) or nuoc nam (Vietnamese fish sauce)
2	2	Serrano chiles, seeded, chopped fine
2	2	Scallions, chopped fine
1 oz	30 g	Brown sugar or raw sugar
1 1/2 lb	750 g	Mango, medium dice
1 1/2 lb	750 g	Papaya, medium dice
12 oz	360 g	Napa cabbage, shredded
4 tbsp	60 mL	Chopped peanuts

PROCEDURE

1. Mix the lime juice, fish sauce, chiles, scallions, and sugar. Stir until the sugar is dissolved.

2. Mix the mango and papaya.
3. Arrange 1 oz (30 g) cabbage on each salad plate.
4. Top with a mound of the fruit mixture.
5. Drizzle each portion with 4 tsp (20 mL) dressing.
6. Sprinkle each portion with 1 tsp (5 mL) chopped peanuts.

Per serving: Calories, 100; Protein, 2 g; Fat, 1.5 g (13% cal.); Cholesterol, 0 mg; Carbohydrates, 21 g; Fiber, 3 g; Sodium, 240 mg.

Thai Papaya and Mango Salad

Pear Salad

YIELD: APPROX. 1 LB (480 G)

U.S.	METRIC	INGREDIENTS	PROCEDURE
		Pear mixture:	1. Toss together all the ingredients for the pear mixture.
1	1	Pear, peeled and cut paysanne	
1/2 cup	120 mL	Celery, sliced very thin	
4 tbsp	60 mL	Golden raisins	
4 tbsp	60 mL	Red onion, sliced very thin and then cut into 1/2-in. pieces, rinsed in cold water	
2 tsp	10 mL	Parsley, chopped very fine	
2 tsp	10 mL	Capers, chopped fine	
		Honey vinaigrette:	2. Mix together the vinegar, honey, shallot, and garlic in a small bowl.
2 tbsp	30 mL	Sherry vinegar	3. Very slowly add the olive oil while beating with a wire whip (see p. 660 for vinaigrette procedures).
4 tsp	20 mL	Honey	
1 tsp	5 mL	Shallot, chopped very fine	4. Season the vinaigrette to taste with salt, pepper, and fresh lemon juice.
1/2 tsp	2 mL	Garlic, chopped very fine	
4 tbsp	60 mL	Olive oil	5. Add the vinaigrette to the pear mixture and toss carefully to coat the salad ingredients.
to taste	to taste	Salt	
to taste	to taste	Black pepper	
to taste	to taste	Lemon juice	

Per 1 oz (28.35 g): Calories, 50; Protein, 0 g; Fat, 3.5 g (63% cal.); Cholesterol, 0 mg; Carbohydrates, 6 g; Fiber, <1 g; Sodium, 15 mg.

Apple Fennel Salad

YIELD: APPROX. 15 OZ (450 G)

U.S.	METRIC	INGREDIENTS	PROCEDURE
7 oz	210 g	Granny Smith apple, peeled and cut julienne	1. Toss together all ingredients except the parsley and chives.
3 oz	90 g	Fennel bulb, shaved as thin as possible (as on a slicing machine)	2. Just before serving, mix in the parsley and chives.
3 oz	90 g	Red onion, sliced very thin, rinsed and drained	
2 tbsp	30 mL	Cider vinegar	
1 tbsp	15 mL	Honey	
2 tbsp	30 mL	Olive oil	
to taste	to taste	Salt	
to taste	to taste	Black pepper	
1 tbsp	15 mL	Italian parsley, cut chiffonade	
1 tsp	5 mL	Chives, cut fine	

Per 1 oz (28.35 g): Calories, 30; Protein, 0 g; Fat, 1.5 g (45% cal.); Cholesterol, 0 mg; Carbohydrates, 4 g; Fiber, <1 g; Sodium, 0 mg.

Pickled Apple

YIELD: 12 OZ (360 G) [1 AP APPLE = 8 OZ; EP = 75%]

U.S.	METRIC	INGREDIENTS
¹/₂ cup	120 mL	Water
3¹/₂ oz	105 g	Sugar
¹/₂ cup	120 mL	Cider vinegar
		Sachet:
3–4	3–4	Peppercorns
¹/₄ tsp	1 mL	Mustard seed
¹/₄ tsp	1 mL	Coriander seed
¹/₄ tsp	1 mL	Salt
2	2	Granny Smith apples, peeled, cored, and cut into small dice
1 tbsp	15 mL	Italian parsley, cut chiffonade (p. 146)

PROCEDURE

1. Combine the water, sugar, vinegar, sachet, and salt in a saucepan. Bring to a boil.
2. Pour the liquid and the sachet over the apples in a nonreactive container.
3. Let stand, refrigerated, 3–4 hours or overnight.
4. Immediately before serving, drain the apples and toss with the parsley.

Per 1 oz (28.35 g): Calories, 50; Protein, 0 g; Fat, 0 g (0% cal.); Cholesterol, 0 mg; Carbohydrates, 12 g; Fiber, 0 g; Sodium, 50 mg.

COMPOSED SALADS

Principles

Composed salads are salads made by arranging two or more elements attractively on a plate. They are called composed because the components are arranged on the plate rather than being mixed together. One or more of the elements may be mixed or tossed salads, but the individual mixed salads are arranged on the plate with other components for the final presentation.

Because they are more elaborate and can be substantial in size, composed salads are usually served as main courses or first courses rather than as accompaniments or side dishes.

There are so many kinds of composed salad that guidelines for preparing this category of salad are very general.

GUIDELINES for Preparing Composed Salads

1. Observe the guidelines for preparing each of the salad components. For example, if one of the components is a mixed green salad, observe the guidelines for preparing green salads.

2. Prepare and season each component separately, and evaluate it for flavor and quality. If one or more of the components is a salad, dressing may be added to each salad component separately, or in some cases dressing may be added to the entire salad just before serving.

3. Arrangements may be plated ahead of time only if the components will hold well. Add delicate items just before serving.

4. If any of the components is to be served hot or warm, prepare and add that item just before serving.

5. Flavors, textures, and colors of all components should harmonize or provide pleasing contrast. See the discussion of flavor building in Chapter 6.

6. Observe the general concepts of plating and presentation discussed in Chapter 29.

Chef's Salad

PORTIONS: 12

U.S.	METRIC	INGREDIENTS
3 lb	1.4 kg	Mixed salad greens, washed, trimmed, and crisped
12 oz	350 g	Turkey breast, cut into thin strips
12 oz	350 g	Pullman ham, cut into thin strips
12 oz	350 g	Swiss cheese, cut into thin strips
24	24	Tomato wedges or cherry tomatoes
24	24	Hard-cooked egg quarters
12	24	Radishes
4 oz	115 g	Carrots, cut bâtonnet
12	12	Green bell pepper rings

PROCEDURE

1. Place the greens in cold salad bowls, approximately 4 oz (125 g) per portion.
2. Arrange the turkey, ham, and cheese strips neatly on top of the greens. Keep the items separate—do not mix them all together.
3. Arrange the remaining items attractively on the salad.
4. Hold for service. If salads must be held for over 1 hour, they should be covered so the meats and cheese don't dry out.
5. Serve with any appropriate salad dressing on the side in a separate container.

Per serving: Calories, 400; Protein, 37 g; Fat, 25 g (54% cal.); Cholesterol, 485 mg; Carbohydrates, 10 g; Fiber, 3 g; Sodium, 570 mg.

VARIATIONS

Other vegetable garnish may be used in addition to or in place of the items in the basic recipe. See lists on pages 669–670.

Roasted Beet Salad with Gorgonzola

PORTIONS: 12   PORTION SIZE: 5 OZ (150 G)

U.S.	METRIC	INGREDIENTS
1 lb 8 oz	750 g	Red beets, medium to large
1 lb 8 oz	750 g	Yellow beets, small

PROCEDURE

1. Wrap the red beets in foil. Wrap the yellow beets in a separate foil package.
2. Bake at 400°F (200 °C) until the beets are tender, about 1 hour.
3. Cool the beets slightly. Trim the root and stem ends, and pull off the peels.
4. Cut the red beets crosswise into thin slices.
5. Cut the yellow beets vertically into quarters.

U.S.	METRIC	INGREDIENTS
12 oz	375 g	Mesclun
12 fl oz	375 mL	Mustard Vinaigrette (p. 660)
6 oz	180 g	Gorgonzola cheese, crumbled

6. Arrange the sliced red beets in circles on cold salad plates.
7. Toss the mesclun with half the vinaigrette.
8. Drizzle the remaining vinaigrette over the sliced beets.
9. Place a small mound of greens in the center of each plate.
10. Arrange the quartered yellow beets around the greens.
11. Sprinkle the tops of the salads with the crumbled gorgonzola.

Per serving: Calories, 290; Protein, 6 g; Fat, 26 g (76% cal.); Cholesterol, 15 mg; Carbohydrates, 12 g; Fiber, 3 g; Sodium, 590 mg.

Roasted Beet Salad
with Gorgonzola

Salade Niçoise

PORTIONS: 12

U.S.	METRIC	INGREDIENTS
1¹/₂ lb	1.4 kg	Waxy potatoes, scrubbed
1¹/₂ lb	1.4 kg	Green beans, washed and trimmed
1 lb	900 g	Mixed salad greens, washed, trimmed, and crisped
30 oz	850 g	Canned tuna, solid pack or chunk
12	12	Anchovy fillets
24	24	Olives, black or green
24	24	Hard-cooked egg quarters
48	48	Tomato wedges
2 fl oz	30 mL	Chopped parsley
		Vinaigrette:
1 pt	500 mL	Olive oil
4 fl oz	125 mL	Wine vinegar
¹/₂ tsp	2 mL	Garlic, chopped fine
1¹/₂ tsp	7 mL	Salt
¹/₄ tsp	1 mL	Pepper

PROCEDURE

1. Cook the potatoes in boiling salted water until just tender. Drain and let cool. Peel. Cut into thin slices. Hold in refrigerator, covered.
2. Cook the beans in boiling salted water. Drain and cool under cold running water. Cut into 2-in. (5-cm) pieces. Hold in refrigerator.
3. Line cold salad bowls or plates with the lettuce leaves (see Note).
4. Combine the potatoes and green beans. Divide the mixture among the salad bowls, about 3 oz (90 g) per portion.
5. Drain the tuna and break it into chunks. Place a 1¹/₂-oz (50-g) portion in the center of each salad.
6. Arrange the anchovy fillets, olives, egg quarters, and tomato wedges attractively on the salads.
7. Sprinkle the salads with chopped parsley.
8. Hold for service in refrigerator.
9. Combine the dressing ingredients and mix well. Just before service, mix again and dress each salad with 1¹/₂ fl oz (50 mL) dressing.

Per serving: Calories, 710; Protein, 37 g; Fat, 53 g (67% cal.); Cholesterol, 440 mg; Carbohydrates, 22 g; Fiber, 5 g; Sodium, 890 mg.

Note: Salade Niçoise (nee-swahz) may be plated on large platters or in bowls to serve 2–6 portions each.

Salade Niçoise

Rohkostsalatteller (German Vegetable Salad)

PORTIONS: 16   PORTION SIZE: SEE PROCEDURE

U.S.	METRIC	INGREDIENTS	PROCEDURE
6 fl oz	175 mL	White wine vinegar	1. Make a dressing by mixing the vinegar, sour cream, salt, sugar, and chives. Set aside.
1 pt	500 mL	Sour cream	
2 tsp	10 mL	Salt	
1/2 tsp	2 mL	Sugar	
2 tbsp	30 mL	Chopped chives	
1 lb	450 g	Carrots	2. Peel the carrots. Shred them on a coarse grater.
2 tbsp	30 mL	Horseradish, well drained	3. Mix the carrots with the horseradish, then with 6 fl oz (175 mL) sour cream dressing, or just enough to bind. Season to taste with salt.
to taste	to taste	Salt	
1 lb 6 oz	625 g	Cucumbers	4. Peel the cucumbers. Cut them into thin slices. Toss with the coarse salt and let stand 1–2 hours.
1 tbsp	15 mL	Coarse salt	5. Press the juices out of the cucumbers. Rinse off excess salt and drain.
2 fl oz	60 mL	White wine vinegar	6. Mix the vinegar, water, sugar, dill, and white pepper.
3 fl oz	90 mL	Water	7. Mix this dressing with the cucumbers. If necessary, add salt to taste.
1 tbsp	15 mL	Sugar	
2 tsp	10 mL	Fresh dill weed, chopped	
pinch	pinch	White pepper	
1 lb 4 oz	575 g	Celery root	8. Peel the celery root. Grate it on a coarse grater. Immediately mix with the lemon juice.
1 1/2 fl oz	50 mL	Lemon juice	9. Mix in the cream. Season with salt and white pepper.
5 fl oz	150 mL	Heavy cream	
to taste	to taste	Salt	
to taste	to taste	White pepper	
2 lb	900 g	Bibb or Boston lettuce greens	10. If necessary, thin the remaining sour cream dressing with a little water until it is the consistency of heavy cream.
32	32	Tomato wedges	11. Toss the greens with the dressing. Plate in the center of large salad plates.
			12. Around the outside edge of each plate, arrange 2 tomato wedges and about 1 oz (30 g) each of the carrot, cucumber, and celery salads.

Per serving: Calories, 140; Protein, 3 g; Fat, 10 g (58% cal.); Cholesterol, 25 mg; Carbohydrates, 13 g; Fiber, 3 g; Sodium, 730 mg.

Rohkostsalatteller

Cobb Salad

PORTIONS: 12 LUNCH ENTRÉES

U.S.	METRIC	INGREDIENTS
1 lb 8 oz	720 g	Tomatoes, peeled and seeded
1 lb 8 oz	720 g	Chicken breast, cooked
6	6	Hard-cooked eggs
1 lb 2 oz	540 g	Roquefort or other blue cheese
24 strips	24 strips	Bacon, crisp
4	4	Avocados
12 oz	360 g	Romaine
12 oz	360 g	Iceberg lettuce
6 oz	180 g	Watercress
6 oz	180 g	Frisée
1¹/₂ pt	720 mL	Mustard Vinaigrette (p. 660)

PROCEDURE

1. Cut the tomatoes into small dice.
2. Cut the chicken into small dice.
3. Chop the hard-cooked eggs.
4. Crumble the blue cheese.
5. Crumble the bacon.
6. Cut the avocados into small dice. Do not do this until just before serving, so it does not discolor before it is served.
7. Cut the salad greens into bite-size pieces and place in a bowl.
8. Add half the vinaigrette to the greens and toss.
9. Place the greens in individual salad bowls.
10. Arrange the tomatoes, chicken, eggs, cheese, bacon, and avocado on top of the greens.
11. Drizzle the remaining vinaigrette over the salads. Serve immediately.

Per serving: Calories, 880; Protein, 39 g; Fat, 78 g (77% cal.); Cholesterol, 210 mg; Carbohydrates, 12 g; Fiber, 6 g; Sodium, 1760 mg.

COBB SALAD

The first Cobb salad was made by Robert Cobb, owner of the Brown Derby restaurant in Hollywood, California. According to the story, late one night in 1938, Mr. Cobb was hungry for a snack and found a variety of ingredients in the refrigerator, including avocado, hard-cooked egg, tomato, and Roquefort cheese. He chopped them up, put them in a salad, and a legendary dish was born. After he put it on the restaurant's menu, it quickly became popular, and other restaurants picked up the idea.

Stuffed Tomato Salad with Tuna

PORTIONS: 12   **PORTION SIZE: 1 TOMATO**

U.S.	METRIC	INGREDIENTS
12	12	Tomatoes, small (about 4 oz/120 g each)
to taste	to taste	Salt
1 lb 8 oz	720 g	Tuna Salad (p. 695)
12	12	Lettuce leaves for underliners
12	12	Small parsley sprigs

PROCEDURE

1. Wash the tomatoes and remove the core at the stem end.
2. Set the tomatoes on the cutting board, stem end down. Cut into eighths to within ¹/₂ inch (1 cm) of the bottom—that is, leave the sections attached at the bottoms.
3. Carefully spread the sections apart slightly. Sprinkle the insides with salt and turn upside down on a sheet pan to drain 15–20 minutes.
4. Fill the tomatoes with tuna salad, using 2 oz (60 g) per portion.
5. Arrange the lettuce leaves on salad plates and place a filled tomato on top of each. Garnish the tops with parsley.
6. Serve at once, or hold for service in the refrigerator.

Per serving: Calories, 160; Protein, 10 g; Fat, 11 g (61% cal.); Cholesterol, 15 mg; Carbohydrates, 6 g; Fiber, 2 g; Sodium, 210 mg.

VARIATIONS

Tomatoes may be stuffed with any of the following:
Chicken Salad and variations (p. 695)   Ham Salad (p. 696)
Egg Salad (p. 695)   Cottage cheese
Dilled Shrimp Salad (p. 699) and variations

Alternative Method: Tomatoes may be hollowed out for stuffing by cutting off the tops and scooping out the insides, being careful not to pierce the sides. Large tomatoes may be cut in half and hollowed out.

Salad of Seared Sea Scallops with Oriental Vinaigrette

PORTIONS: 10   **PORTION SIZE: 3 OZ (90 G) SCALLOPS, 1¹/₂ OZ (45 G) GREENS, 1¹/₂ FL OZ (45 ML) VINAIGRETTE**

U.S.	METRIC	INGREDIENTS
1 lb	500 g	Mesclun or other delicate mixed greens
1 pt	500 mL	Oriental Vinaigrette (p. 662)
2 lb	1 kg	Sea scallops
as needed	as needed	Butter
30	30	Orange segments (free of membranes)

PROCEDURE

1. Wash and drain the salad greens.
2. Prepare the vinaigrette.
3. Trim the scallops by removing the small, tough side muscle. If any scallops are very large, cut them in half crosswise. Dry them well.
4. Heat a little butter in a nonstick sauté pan. Sear a few scallops at a time until they are browned on top and bottom.
5. Toss the mesclun with half the vinaigrette.
6. Mound the mesclun in the center of the plates.
7. Arrange the orange segments so they are leaning against the mound of salad greens. Use 3 per portion.
8. Arrange the scallops around the salads.
9. Drizzle the remaining vinaigrette around the scallops.

Per serving: Calories, 360; Protein, 9 g; Fat, 33 g (80% cal.); Cholesterol, 20 mg; Carbohydrates, 9 g; Fiber, 2 g; Sodium, 480 mg.

Salad of Seared Sea Scallops with Oriental Vinaigrette

Goat Cheese and Walnut Salad

PORTIONS: 12

U.S.	METRIC	INGREDIENTS	PROCEDURE
6 oz	180 g	Belgian endive or radicchio	1. Trim, wash, and drain the salad greens.
9 oz	270 g	Arugula	2. Tear into bite-size pieces. Toss together.
6 oz	180 g	Bibb lettuce	
9 oz	270 g	Romaine lettuce	
1 cup	240 mL	Bread crumbs, dry	3. Mix the crumbs, herbs, and pepper.
1 tbsp	15 mL	Dried thyme	4. Slice the cheese into 1-oz (30-g) pieces. Roll the
1 tbsp	15 mL	Dried basil	pieces in the seasoned crumbs to coat them.
1¹/₂ tsp	7 mL	Black pepper	
1 lb 8 oz	720 g	Fresh goat's milk cheese, preferably in log shape	
4¹/₂ oz	135 g	Walnut pieces	5. At service time, arrange the cheese pieces on a sheet pan. Bake at 425°F (220°C) for 10 minutes.
8 fl oz	240 mL	Vinaigrette (p. 660) made with red wine vinegar and olive oil	6. At the same time, toast the walnuts in a dry sauté pan or in the oven with the cheese.
			7. Toss the greens with the vinaigrette and arrange on cold plates. Top each plate of greens with 2 pieces of cheese and sprinkle with walnuts.

Per serving: Calories, 460; Protein, 17 g; Fat, 40 g (75% cal.); Cholesterol, 50 mg; Carbohydrates, 13 g; Fiber, 3 g; Sodium, 530 mg.

VARIATIONS

Mixed Green Salad with Blue Cheese and Walnuts

Omit the goat cheese and herbed crumbs from the basic recipe. Instead, sprinkle the salads with crumbled gorgonzola, stilton, Roquefort, or other blue cheese.

Goat Cheese and Walnut Salad

GELATIN SALADS

Principles

Gelatin salads have a distinguished history. Their ancestors are aspics, the highly ornamented appetizers and elaborate buffet pieces made with meat and fish stocks rich in natural gelatin extracted from bones and connective tissue. Aspics are part of the glory of classical cuisine and still an important part of modern buffet work.

It's no longer necessary to extract gelatin from bones in your kitchen. Purified, granular gelatin and gelatin sheets have long been available for use in the pantry. Many excellent gelatin-based salads can be made with little labor using these products. However, most gelatin products today are made with sweetened prepared mixes whose high sugar content and heavy reliance on artificial color and flavor make their appropriateness as salads somewhat questionable. (Often, in a cafeteria line, you will see in the salad section little squares of gelatin with a lettuce leaf underneath and a dab of mayonnaise on top, and in the dessert section the identical product, without the lettuce leaf and with a dab of whipped cream in place of the mayo.)

Nevertheless, as a professional cook, you need to know how to prepare these products because many customers expect them. You should also know how to prepare salads using unflavored gelatin, relying on fruit juices and other ingredients for flavor. Unflavored gelatin is especially valuable for preparing molded vegetable salads because shredded cabbage and other vegetables make a poor combination with highly sweetened dessert gelatin.

GUIDELINES for Making Gelatin Salads

1. It is important to use the right amount of gelatin for the volume of liquid in the recipe. Too much gelatin makes a stiff, rubbery product. Too little makes a soft product that will not hold its shape.

   Basic proportions for unflavored gelatin are $2\frac{1}{2}$ ounces dry gelatin per gallon (19 g per L) liquid, but you will almost always need more than this because of acids and other ingredients in the recipe. Basic proportions for sweetened, flavored gelatin are 24 ounces per gallon (180 g per L) liquid.

   Acids, such as fruit juices and vinegar, weaken the gelatin set, so a higher proportion of gelatin to liquid is needed, sometimes as much as 4 ounces or more per gallon (30 g per L). The setting power is also weakened by whipping the product into a foam and by adding a large quantity of chopped foods. It is impossible to give a formula for how much gelatin to use, as it varies with each recipe. Test each recipe before using it.

2. Gelatin dissolves at about 100°F (38°C), but higher temperatures will dissolve it faster.

   To dissolve unflavored gelatin, stir it into cold liquid to avoid lumping and let it stand 5 minutes to absorb water. Then heat it until dissolved, or add hot liquid and stir until dissolved.

   To dissolve sweetened, flavored gelatin, stir it into boiling water. It will not lump because the gelatin granules are held apart by sugar granules, much the way starch granules in flour are held separate by the fat in a roux.

3. To speed setting, dissolve the gelatin in up to half the liquid and add the remainder cold to lower the temperature. For even faster setting, add crushed ice in place of an equal weight of cold water. Stir until the ice is melted.

4. Do not add raw pineapple or papaya to gelatin salads. These fruits contain enzymes that dissolve the gelatin. If cooked or canned, however, these fruits may be included.

5. Add solid ingredients when the gelatin is partially set—that is, when thick and syrupy. This will help keep them evenly mixed rather than floating or settling.

6. Canned fruits and other juicy items must be well drained before being added, or they will dilute the gelatin and weaken it.

7. For service, pour into pans and cut into equal portions when set, or pour into individual molds.

8. To unmold gelatin:

   • Run a thin knife blade around the top edges of the mold to loosen.

   • Dip the mold into hot water 1 or 2 seconds.

   • Quickly wipe the bottom of the mold and turn it over onto the salad plate (or invert the salad plate over the mold and flip the plate and mold over together). Do not hold in the hot water for more than a few seconds, or the gelatin will begin to melt.

   • If the gelatin doesn't unmold after a gentle shake, repeat the procedure. You may also wrap a hot towel (dipped in hot water and wrung out) around the mold until it releases, but this is more time-consuming.

9. Refrigerate gelatin salads until service to keep them firm.

Jellied Fruit Salad

PORTIONS: 12   **PORTION SIZE: 4 OZ (125 G)**

U.S.	METRIC	INGREDIENTS	PROCEDURE
10 oz	300 g	Canned pineapple cubes, with juice	1. Drain the pineapple and reserve the juice. You should have about 6 oz (180 g) drained fruit.
4 oz	120 g	Grapefruit sections	2. Cut the grapefruit and orange sections into $^1/_2$-in. (1-cm) dice. (See p. 146 for cutting citrus sections.)
6 oz	180 g	Orange sections	
4 oz	120 g	Grapes	3. Cut the grapes in half. Remove seeds, if any.
			4. Place the fruit in a colander or strainer over a bowl and hold in the refrigerator.
1 oz	30 g	Unflavored gelatin	5. Stir the gelatin into the cold water and let stand at least 5 minutes.
4 fl oz	120 mL	Water, cold	
as needed	as needed	Fruit juice: grapefruit, orange, or pineapple	6. Add enough fruit juice (or part juice and part water) to the liquid from the pineapple to measure 1$^3/_4$ pt (825mL).
3 oz	90 g	Sugar	7. Bring the fruit juice to a boil in a stainless-steel pan. Remove from heat.
$^1/_8$ tsp	0.5 g	Salt	8. Add the sugar, salt, and softened gelatin. Stir until gelatin and sugar are dissolved.
3 tbsp	45 mL	Lemon juice	9. Cool the mixture. Add the lemon juice.
			10. Chill until thick and syrupy but not set.
			11. Fold the drained fruits into the gelatin mixture.
			12. Pour into individual molds. Chill until firm.
12	12	Lettuce leaves for underliners	13. Line cold salad plates with lettuce leaves.
6 fl oz	180 mL	Chantilly Dressing (p. 665)	14. Unmold the salads.
			15. Place a gelatin salad on each plate. Hold for service in the refrigerator.
			16. At service time, top each salad with 1 tbsp (15 mL) dressing.

Per serving: Calories, 170; Protein, 3 g; Fat, 10 g (49% cal.); Cholesterol, 15 mg; Carbohydrates, 20 g; Fiber, 1 g; Sodium, 55 mg.

Basic Flavored Gelatin with Fruit ♥

PORTIONS: 25   **PORTION SIZE: 4 OZ (125 G)**

U.S.	METRIC	INGREDIENTS	PROCEDURE
12 oz	375 g	Flavored gelatin mix	1. Place the gelatin in a bowl.
1 qt	1 L	Water, boiling	2. Pour in the boiling water. Stir until dissolved.
1 qt	1 L	Water or fruit juice, cold	3. Stir in the cold water or juice.
			4. Chill until thick and syrupy but not set.
2 lb	1 kg	Fruit, well drained	5. Fold the fruit into the gelatin mixture.
			6. Pour into molds or into a half-hotel pan.
			7. Chill until firm.
			8. Unmold. If using a hotel pan, cut 5 × 5 into portions.

Per serving: Calories, 70; Protein, 1 g; Fat, 0 g (0% cal.); Cholesterol, 0 mg; Carbohydrates, 18 g; Fiber, 0 g; Sodium, 40 mg.

VARIATIONS

The possible combinations of fruits and flavored gelatin are nearly limitless. The following suggestions are only a few possibilities. Note: When using canned fruits, use the syrup from the fruits as part of the liquid in step 3.

  Black cherry-flavored gelatin; Bing cherries
  Raspberry-flavored gelatin; peach slices or halves
  Strawberry, raspberry, or cherry-flavored gelatin; canned fruit cocktail
  Orange-flavored gelatin; equal parts sliced peaches and pears
  Cherry-flavored gelatin; equal parts crushed pineapple and Bing cherries
  Lime-flavored gelatin; grapefruit sections or pear halves or slices

SALAD BARS AND BUFFET SERVICE

Salad bars are frequent fixtures in restaurants and are popular with both customer and restaurateur. Diners enjoy customizing their own salads with selections from a large bowl of greens, smaller containers of assorted condiments, and a variety of dressings. The restaurateur likes salad bars because they take some pressure off the dining room staff during service. Many restaurants have designed unique salad bars that have become almost a trademark. Others may not have salad bars as part of their regular meal service but rely on them for efficiency at certain times, such as weekend brunch.

For successful salad bar service, it is important to keep several points in mind:

1. Keep the salad bar attractive and well stocked from the beginning until the end of service. Refill containers before they begin to look depleted, wipe the edges of dressing containers, and clean up debris scattered by customers.

2. Keep the components simple but attractive. Elaborately arranged salad bowls lose their effect as soon as two or three customers have dug into them.

3. Select a variety of condiments to appeal to a variety of tastes. Try both familiar and unusual items to make your salad bar stand out. There is no reason to restrict the choices to the same old stuff everyone else is serving.

   There are two basic kinds of salad bar condiments:

   - *Simple ingredients*. Nearly any item in the salad ingredient list on pages 669–670 might be selected. Your choice will depend on balance of flavors and colors, customer preference, and cost.

   - *Prepared salads*. Marinated vegetable salads, such as three-bean salad, and cooked salads, like macaroni salad, are especially suitable. The choice is large.

4. Arrange the salad bar in the following order (see Figure 21.3):

   - Plates.
   - Mixed greens.
   - Condiments (put the expensive ones at the end).
   - Dressings.
   - Crackers, breads, etc., if desired.

5. Make sure your setup conforms to your state health department regulations.

6. Some portion control can be achieved by selecting the right size plates, condiment servers, and dressing ladles.

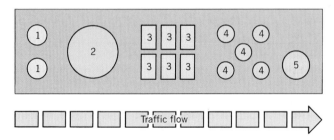

FIGURE 21.3 Suggested arrangement of a salad bar.
Key: (1) plates; (2) large bowl of salad greens; (3) condiments;
(4) dressings; (5) crackers, breads, etc.

KEY POINTS TO REVIEW

- What are the guidelines for making arranged or composed salads?

- What are the guidelines for making gelatin salads?

- What guidelines should be kept in mind when planning and setting up a salad bar?

ADDITIONAL RECIPES

These additional recipes may be found on your CulinarE-Companion recipe management program:

Chicken Breast Salad with Walnuts and Blue Cheese; Cucumbers and Onions in Sour Cream; Fruit Salad Dressing; Honey Lemon Dressing; Mushrooms à la grecque; Sour Cream Fruit Salad Dressing; Taco Salad; Vegetable Chopped Salad with Kidney Beans and Provolone; Wheatberry Salad with Mint.

TERMS FOR REVIEW

winterized oil	appetizer salad	four parts of a salad
strength of acidity	accompaniment salad	vegetable salad
verjus	main-course salad	bound salad
Roquefort cheese	separate-course salad	fruit salad
vinaigrette	dessert salad	composed salad
temporary emulsion	full slip	gelatin salad

QUESTIONS FOR DISCUSSION

1. List three or four salads that may be served as appetizers, as accompaniments, as main dishes, as separate-course salads, and as desserts. Give reasons for your choices.

2. What is the effect of salad dressing on the crispness of salad greens, and what are some ways to solve this problem?

3. You are asked to prepare 250 Waldorf salads for a banquet. Explain the procedure you will use. List each step, from raw ingredients to plated salads. (You may refer to the recipe on p. 700.)

4. How can you ensure salad greens will be crisp?

5. You are making mixed green salads and have the following ingredients to choose from. Which would you toss together, and which would you add after plating or at service time? Why?

Iceberg lettuce	Chicory
Shredded red cabbage	Avocado slices
Carrot strips or shreds	Tomato wedges
Watercress	Romaine lettuce
Sliced celery	

6. You are preparing tossed green salads, potato salads, and avocado and grapefruit salads for luncheon service. How will you plan your preparation—that is, what will you do first, second, and so on?

7. You are trying a new recipe for a molded vegetable salad using unflavored gelatin. After evaluating the flavor, you decide it isn't tart enough and more vinegar should be added. Should you make any other adjustments?

8. When you are making mayonnaise, you should take a number of precautions to make sure a good emulsion is formed. Name as many as you can. If you forget one of these and your mayonnaise breaks, what can you do?

SANDWICHES

The sandwich is a favorite and convenient lunchtime food. It is quickly made and served, and it is adaptable to so many variations that it satisfies nearly every taste and nutrition requirement.

Sandwiches have long been the domain of the pantry department, along with salads and other cold preparations. However, when you consider that the most popular sandwich today is the hamburger, you realize that sandwich preparation is as much the responsibility of the short-order cook as it is of the pantry cook.

Preparing hot and cold sandwiches to order is one of the fundamental skills required in modern food service. In this chapter, we start by looking at the fundamentals of sandwich making, the basic ingredients, and basic sandwich types. We then look at the setup of the sandwich station and methods for efficient production.

AFTER READING THIS CHAPTER, YOU SHOULD BE ABLE TO

1. Select, store, and serve fresh, good-quality breads for sandwiches.

2. Use sandwich spreads correctly.

3. Identify the most popular types of sandwich filling.

4. Set up an efficient sandwich station.

5. Prepare the major types of sandwiches to order.

6. Prepare sandwiches in quantity.

BREADS

One of the functions of the bread in a sandwich is to provide an edible casing for the food inside. Ideally, though, the bread should do more than this. Good-quality breads provide variety, texture, flavor, and eye appeal to sandwiches, as well as bulk and nutrients.

TYPES

Pullman or sandwich loaves of white bread are most frequently used for simple sandwiches. These are long, rectangular loaves that provide square slices of specified thickness, from $3/8$ to $5/8$ inch (10 to 16 mm) thick.

Commercial sandwich bread should be of fine rather than coarse texture and firm enough to accommodate spreads well. Supermarket white bread is unsuitable because it is too soft for spreading and for holding most fillings, and it becomes pasty in the mouth.

Because of its neutral flavor, white bread is suitable for the largest variety of fillings.

Other kinds of breads add variety and interest, provided they harmonize with the filling. The following are some possibilities:

Rolls, including hard and soft rolls, hamburger and hot dog rolls, long rolls for submarine sandwiches

French or Italian bread and rolls, including sourdough and ciabatta, split horizontally

Whole wheat

Cracked wheat

Rye and pumpernickel

Pita bread

Raisin bread

Cinnamon bread

Fruit and nut breads

Focaccia

STORAGE

Fresh bread is essential for top-quality sandwiches. Stale or dry bread is undesirable. The following measures can be taken to ensure freshness.

1. Daily delivery, or delivery as frequent as possible, depending on your location. Bread stales rapidly, and day-old bread has lost much of its freshness.

2. Keep bread tightly wrapped in moistureproof wrapping until it is used. This prevents drying and guards against absorption of odors.

Breads made of various flours: top row: multigrain, white pan loaf, marble rye, whole wheat, and olive loaf. Bottom row: raisin swirl, pumpernickel, and Jewish rye.

Breads in various shapes: Top row: Pullman loaf, baguette, home-style, bâtard, sub roll, and hearth. Middle row: Cuban, kaiser, hot dog, New England hot dog, hamburger roll, ciabatta, and steak roll. Bottom row: naan, flour tortillas, whole wheat wrap, and pita.

3. French bread and other hard-crusted breads should not be wrapped, or the crusts will soften. These breads stale rapidly and should be used the day they are baked.

4. Store at room temperature, away from ovens or hot equipment. Do not refrigerate, because refrigerated bread becomes stale faster.

5. If bread must be kept more than one day, it may be frozen. Thaw frozen bread without unwrapping.

6. Day-old bread may be used for toasting without loss of quality.

SPREADS

PURPOSES OF SPREADS

1. To protect the bread from soaking up moisture from the filling.

2. To add flavor.

3. To add moisture or mouthfeel.

BUTTER

Butter should be soft enough to spread easily without tearing the bread. It may be softened by whipping in a mixer or by simply letting it stand at room temperature for half an hour.

Whipping gives the butter greater volume, and this cuts food cost. However, whipped butter does not keep as well because the incorporated air speeds the development of rancidity.

Some operators whip a small amount of water or milk into the butter. This increases both spreadability and volume. However, it adds nothing to the quality of the sandwich and increases the likelihood of soaking the bread.

Margarine is sometimes used instead of butter, if food costs require it or if customers request it.

Flavored butters, such as those listed on page 192, may be used with appropriate fillings.

MAYONNAISE

Mayonnaise is often preferred to butter as a spread because it contributes more flavor. However, it does not protect the bread from moisture as well as butter does.

Because of the danger of food-borne disease, sandwiches made with mayonnaise should be served immediately or refrigerated at once and kept refrigerated until served.

OTHER SPREADS

Although butter and mayonnaise are the most widely used spreads, almost any food of a spreadable consistency can be used to add flavor interest to sandwiches, including Pesto (p. 399), Tapenade (p. 754), Liver Pâté (p. 757), Hummus (p. 742), Babaganouj (p. 742), Guacamole (p. 743), and Romesco (p. 742). Be aware, however, that most of these do not protect the bread from soaking up moisture from the filling. Moist spreads are best if applied just before serving.

FILLINGS

The filling is the heart of the sandwich. As we have already said, nearly any kind of food may be served between two slices of bread. The following are possible fillings that may be used separately or in combination.

MEATS AND POULTRY

Most meats for sandwiches are precooked, though some are cooked to order. Sliced meats dry out and lose flavor, so avoid slicing farther ahead than necessary, and keep sliced meats covered or wrapped.

Leftovers may be used, but only if they are of good quality and have been properly handled and stored to avoid contamination.

Thin slices are more tender, and sandwiches made with them are easier to eat. Also, many thin slices make a thicker sandwich than one or two thick slices of the same total weight.

1. Beef	2. Pork products	3. Poultry	4. Sausage products
Sliced roast beef, hot or cold	Roast pork	Turkey breast	Salami
Hamburger patties	Barbecued pork	Chicken breast	Frankfurters
Small steaks	Ham, all kinds		Bologna
Corned beef	Bacon		Liverwurst
Pastrami	Canadian bacon		Luncheon meats
Tongue, fresh or smoked			Grilled sausages

CHEESE

Like meats, cheese dries out rapidly when unwrapped and sliced. When slicing is done ahead, the slices should remain covered until service time. See Chapter 25 for a summary of cheese varieties.

The most popular sandwich cheeses are:

Cheddar types	Process cheese	Cream cheese, mascarpone
Swiss types	Provolone	Cheese spreads

FISH AND SHELLFISH

Most seafood fillings for sandwiches are highly perishable and should be kept well chilled at all times.

Some popular seafood fillings are:

Tuna	Shrimp	Fried fish portions
Sardines	Anchovies	Grilled or pan-fried fish fillets
Smoked salmon and lox		

BOUND SALADS

Refer to page 694 for preparation of bound salads. The most popular salads for sandwich fillings are tuna salad, egg salad, chicken or turkey salad, and ham salad.

VEGETABLE ITEMS

Lettuce, tomato, and onion are indispensable in sandwich production. In addition, nearly any vegetable used in salads may also be included in sandwiches. See page 670 for a listing. Grilled vegetables are popular not only in vegetarian sandwiches but also as part of the filling in meat sandwiches.

MISCELLANEOUS

Peanut butter	Fruits, fresh or dried
Jelly, jam, and preserves	Nuts (such as sliced almonds)
Egg	

TYPES OF SANDWICHES

COLD SANDWICHES

1. **Simple cold sandwiches** are those made with two slices of bread or two halves of a roll, a spread, and a filling. They are called *simple* because they are made with just two slices of bread, not because they are necessarily simple in construction. Simple cold sandwiches range from a single slice of cheese or meat between two slices of buttered bread to complex constructions like the submarine sandwich (also called a *hero sandwich* or *grinder*), a long Italian roll filled with salami, ham, capocollo, mortadella or bologna, provolone cheese, peppers, onions, olives, tomatoes, and more.

   Most popular sandwiches fall into this category.

2. **Multidecker sandwiches** are made with more than two slices of bread (or rolls split into more than two pieces) and with several ingredients in the filling.

   The **club sandwich** is a popular multidecker sandwich made with three slices of toast and filled with sliced chicken or turkey breast, mayonnaise, lettuce, tomato, and bacon. It is cut into four triangles, as shown in Figure 22.1.

3. **Open-faced sandwiches** are made with a single slice of bread, like large canapés, which is what they are. Also as in canapés, the filling or topping should be attractively arranged and garnished. Canapé ingredients and method are discussed in Chapter 23.

4. **Tea sandwiches** are small, fancy sandwiches generally made from light, delicate ingredients and bread trimmed of crust. They are often cut into fancy shapes. Fillings and spreads can be the same as those for canapés.

5. **Wraps** are sandwiches in which the fillings are wrapped, like a Mexican burrito, in a large flour tortilla or similar flatbread. They may be served whole or cut in half if large.

HOT SANDWICHES

1. **Simple hot sandwiches** consist of hot fillings, usually meats but sometimes fish, grilled vegetables, or other hot items, between two slices of bread or two halves of a roll. They may also contain items that are not hot, such as a slice of tomato or raw onion on a hamburger.

   Hamburgers and hot dogs and all their variations are the most popular hot sandwiches.

2. **Open-faced hot sandwiches** are made by placing buttered or unbuttered bread on a serving plate, covering it with hot meat or other filling, and topping with a sauce, gravy, cheese, or other topping. Some versions are browned under the broiler before serving. This type of sandwich is eaten with a knife and fork.

3. **Grilled sandwiches**, also called *toasted sandwiches*, are simple sandwiches buttered on the outside and browned on the griddle, in a hot oven, or in a panini grill (see sidebar, p. 720). Sandwiches containing cheese are popular for grilling.

4. **Deep-fried sandwiches** are made by dipping sandwiches in beaten egg and, sometimes, in bread crumbs, and then deep-frying. This type of sandwich is often cooked on a griddle or in a hot oven instead, as deep-frying makes it greasy.

5. If wraps are classified as cold sandwiches, then hot burritos, quesadillas, and filled enchiladas could be considered hot sandwiches. **Pizzas** could be considered open-faced hot sandwiches, and we can define them as thin sheets of lean bread dough baked with a topping. In fact, one could classify any food made with a dough product filled or topped with another food product as a sandwich. Dough products, whether breads, puff pastry, tortillas, or egg roll skins, make useful and versatile bases or cases for foods, whether or not the result resembles something we might think of as a sandwich. Several of these items (enchiladas, quesadillas, and egg rolls) are considered in other chapters of this book. Pizzas are included in this chapter for convenience. The recipe on page 732 gives the basic procedure for baking pizzas. For other flavors, vary the toppings as desired, using the same technique for baking.

FIGURE 22.1 Cutting a club or multidecker sandwich.

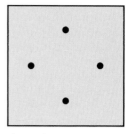

(a) Place four picks in the sandwich in the locations shown by the dots in the illustrations.

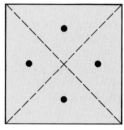

(b) Cut the sandwiches into quarters from corner to corner. Plate the sandwich with the points up.

Panini grill

PANINI

The basic meaning of the Italian word **panino** (plural: *panini*) is "roll," as in bread roll or dinner roll—literally, "little bread." By extension, panini are sandwiches made of small rolls plus fillings. The filling for a typical panino is a slice or two of cured meat, such as salami, mortadella (Italian bologna), or prosciutto. Cheese may also be used, as well as a variety of other items, such as sliced tomato and canned tuna. In summary, Italian panini are most often small, simple, cold sandwiches.

Outside of Italy, panino has come to mean something quite different. It is usually a hot grilled sandwich, often with a fairly substantial filling. What distinguishes panini from other grilled sandwiches is that they are made on a special griddle called a panini grill or panini press, which grills the sandwich on both sides at once while compressing or squeezing the bread and filling together. The griddle surfaces may be smooth or grooved. Grooved griddles toast grill marks onto the bread for the typical panino look.

Grilled panini often contain cheese as one of the filling ingredients. The melted cheese as well as the crisp toasted crusts are part of the attraction of this type of sandwich.

If a sandwich press is not available, you can create a similar kind of sandwich by toasting it on a standard flat griddle or grooved griddle, putting a grill brick on the sandwich to compress it, and turning it over when heated halfway through to toast both sides.

For pressed sandwiches, avoid making the fillings too thick. Unlike for grilled sandwiches, it is not always necessary to butter or oil the outside surfaces of the bread. The moisture and fat in the filling may be enough to moisten the bread as it grills. Adding more sometimes makes the sandwich too greasy.

KEY POINTS TO REVIEW

- What kinds of spread are used for sandwiches? What is their purpose?

- What categories of ingredients are used for sandwich fillings? List as many examples of each category as possible.

- What are the major kinds of hot and cold sandwiches?

MAKING SANDWICHES

The preparation of sandwiches requires a great deal of handwork. Many individual motions may be required, especially if the sandwiches are multideckers or have several ingredients. Whether you are making sandwiches in quantity or to order, your goal must be to reduce your motions to make the production as efficient and quick as possible.

SETTING UP THE STATION FOR PREPARED-TO-ORDER SANDWICHES

A station setup depends on the menu and on the available equipment and space, so there is no single correct way to set up.

Any setup involves two elements: ingredients and equipment.

Ingredients

This phase of the setup has two parts:

1. **Prepare ingredients.**
   Mix fillings, prepare spreads, slice sandwich meats and cheeses, separate lettuce leaves, slice tomatoes, prepare garnishes, and so on. In other words, have everything ready ahead of time, so nothing is left to do but assemble the ingredients.

2. **Arrange or store ingredients for maximum efficiency.**
   To reduce your movements to a minimum, the ideal setup has everything you need within easy reach of both hands. Depending on the kitchen layout, this may not be possible, especially if the sandwich menu is large. But try to get as close to it as possible.
   Arrange ingredients so you can use both hands. For example, while the left hand reaches for the bread, the right hand reaches for the butter spreader. Then, while the right hand puts the spreader back, the left reaches for the sliced ham. The right hand, on its way back from the butter, picks up a slice of cheese and so on. On a busy sandwich station, every second counts.

Two other considerations are important while we're talking about ingredients:

1. **Sanitation.**
   Because cold sandwiches are subjected to a lot of handling and are not cooked, it is especially important that ingredients be properly refrigerated and protected at all times. A refrigerated table—sort of a cold version of a steam table—is usually used. Refrigerated drawers or under-the-counter reach-ins are used for less frequently needed items.
   Wash hands carefully and thoroughly before beginning work. In addition, be aware of local sanitation requirements. In many places, laws require that workers wear gloves when handling ready-to-eat foods (see p. 22).

2. **Portion control.**
   Sliced items are portioned by the count and by weight. If portioning is by the count, you must take care, during pre-prep, to slice to the proper thickness. If done by weight, each portion can be placed on squares of waxed paper and stacked in a container.

Equipment

The equipment needed for a sandwich station depends, of course, on the menu and the size of the operation.

1. *Storage equipment* for ingredients includes refrigeration equipment for cold ingredients and a steam table for hot ingredients, such as roasted meats.

2. *Hand tools* are basic requirements for sandwich making and are often the only tools necessary. These include spreaders, spatulas, and knives, including a serrated knife and a sharp chef's knife for cutting the finished sandwich. A cutting board, of course, is also required. A power slicer may be necessary for any slicing not done ahead.

3. *Portion control equipment* includes scoops for fillings and a portion scale for other ingredients.

4. *Cooking equipment* is necessary for most hot sandwiches. Griddles, grills, broilers, and deep fryers are all used for cooking sandwich ingredients to order. Microwave ovens are sometimes used to heat ingredients or finished sandwiches.

SETTING UP AND PREPARING SANDWICHES IN QUANTITY

Once the ingredients are prepared and the hand tools assembled, all that's needed for a complete sandwich station is a large table.

Assembly-line production is the most efficient method because it simplifies movements. This is the same method applied to producing salads in quantity in Chapter 21.

PROCEDURE for Making Simple Cold Sandwiches in Quantity

1. Prepare and assemble all ingredients.

2. Assemble necessary equipment, including wrapping materials.

3. Arrange bread slices in rows on the tabletop.

4. Spread each slice with butter or whatever spread is required.

5. Place fillings evenly and neatly on alternate slices, leaving the other slices plain. Fillings should not hang over the edges of the bread. If the filling is spreadable, spread it evenly to the edges. See Figure 22.2 for spreading technique.

6. Top the filled slices with the plain buttered slices.

7. Stack two or three sandwiches and cut with a sharp knife.

8. To hold, do one of the following:
   - Wrap separately in plastic, waxed paper, or sandwich bags.
   - Place in storage pans, cover tightly with plastic wrap, and cover with clean, damp towels. The towels must not touch the sandwiches; their purpose is to provide a moisture barrier to help prevent drying.

9. Refrigerate immediately and hold until served.

FIGURE 22.2 Spread sandwiches efficiently with three quick strokes as shown.

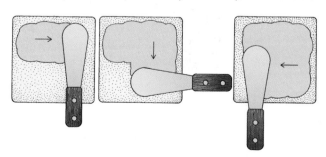

SERVICE

With a few exceptions, such as hamburgers and hot dogs, sandwiches are cut before serving. Cutting serves two purposes: It makes the sandwich easier to handle and eat, and it makes possible a more attractive presentation.

The first purpose is usually accomplished by simply cutting the sandwich in half or, if it is very large or thick, into thirds or quarters.

The second purpose can be served by displaying the cut edges rather than the crust edges to the outside. If the sandwich was neatly made of good ingredients and tastefully garnished, it will be appetizing and attractive. Little purpose is served by cutting and arranging the sandwich in complicated ways.

Hamburgers and other uncut sandwiches may be presented open-face to display the attractive ingredients. For example, a hamburger version often called a *California burger* is presented with the meat on the bottom half of the bun and, alongside, a lettuce leaf and a slice of tomato on the top half of the bun.

Hot sandwiches prepared to order offer the same challenges for creative and attractive presentation as other hot foods, as discussed in Chapter 29. The most important factors in the presentation are using good, fresh ingredients and preparing them with care, using proper techniques for cooking meats, poultry, fish, vegetables, and any other components of the sandwich.

COLD SANDWICH COMBINATIONS

The following suggestions are only a few of the many sandwiches that can be made from the ingredients listed in the first part of this chapter.

1. Roast beef on rye bread spread with a mixture of softened cream cheese and horseradish.
2. Beef tongue, lettuce, and tomato on onion roll; mayonnaise.
3. Bologna, provolone cheese, tomatoes, and chopped pimiento on hard roll; mayonnaise.
4. Liverwurst, onion slices, and sour pickles on pumpernickel; mayonnaise or butter.
5. Corned beef or ham, Swiss cheese, mustard, and dill pickle slices on rye; mayonnaise or butter.
6. Corned beef, coleslaw (well drained), and Swiss cheese on rye; mayonnaise or butter.
7. Ham, salami, tomato slice, Russian dressing, and lettuce on rye toast.
8. Chicken or turkey, ham, Swiss cheese, and lettuce on white or whole wheat toast; mayonnaise.
9. Chicken and cucumber slices on whole wheat; mayonnaise.
10. Turkey, bacon, Swiss cheese, and lettuce on white or whole wheat toast; mayonnaise.
11. Deviled ham, pineapple slice, and lettuce on white toast.
12. Tuna salad, lettuce, tomato, and shredded cheddar cheese on white toast.
13. Sardines and onion slices on dark rye spread with cream cheese; served open-face.

KEY POINTS TO REVIEW

- How is a typical sandwich station set up and organized? What equipment is used?

- What are the steps in the procedure for making simple cold sandwiches in quantity?

California Burger

YIELD: 1 SANDWICH

U.S.	METRIC	INGREDIENTS
1	1	Hamburger patty, 4 oz (125 g)
1	1	Hamburger roll
as needed	as needed	Butter
2 tsp	10 mL	Mayonnaise
1	1	Lettuce leaf
1	1	Thin slice of onion (optional)
1	1	Tomato slice

PROCEDURE

1. Cook the hamburger patty on a griddle or grill to desired doneness.
2. While the meat is cooking, prepare the roll. Butter the bottom half very lightly. Spread the top half with mayonnaise.
3. Place the halves of the roll side by side on a serving plate.
4. On the top half, place the lettuce leaf, the onion slice (if used), and the tomato slice.
5. When the hamburger patty is cooked, place it on the bottom half of the roll. Serve immediately, open faced.

Per serving: Calories, 480; Protein, 24 g; Fat, 32 g (61% cal.); Cholesterol, 85 mg; Carbohydrates, 23 g; Fiber, 2 g; Sodium, 370 mg.

VARIATIONS

California Cheeseburger

Prepare as in the basic recipe, except place a slice of cheddar or American cheese on the hamburger patty 1 minute before it is done. Cook until the cheese melts.

Cheeseburger (Plain)

Omit mayonnaise, lettuce, onion, and tomato, but add the slice of cheese as in California Cheeseburger.

Cheeseburger with Bacon

Prepare like a cheeseburger, but place 2 half-strips of cooked bacon on the cheese.

California Cheeseburger Deluxe

Prepare like a California Cheeseburger, but place 2 half-strips of bacon on the cheese.

California Burger

Submarine Sandwich

YIELD: 1 SANDWICH

U.S.	METRIC	INGREDIENTS
1	1	Submarine roll
2 tbsp	30 mL	Mayonnaise
1 oz	30 g	Salami, cut in thin slices
1 oz	30 g	Ham, cut in thin slices
1 oz	30 g	Bologna, cut in thin slices
1 oz	30 g	Provolone cheese, cut in thin slices
2	2	Tomato slices
2	2	Onion slices, very thin
3	3	Green bell pepper rings

PROCEDURE

1. Split the roll horizontally, but leave it hinged on one side.
2. Spread the roll with mayonnaise.
3. Arrange the meats and cheese in the sandwich in layers. If the slices of meat are too wide to fit, fold them in half.
4. Arrange the tomato, onion, and pepper slices on top of the meats and cheese.
5. Close the sandwich. Leave it whole or cut it in half for service.
6. Serve the sandwich with mustard and olives or pickles on the side.

Per serving: Calories, 770; Protein, 28 g; Fat, 51 g (60% cal.); Cholesterol, 85 mg; Carbohydrates, 50 g; Fiber, 5 g; Sodium, 1650 mg.

Club Sandwich

YIELD: 1 SANDWICH

U.S.	METRIC	INGREDIENTS
3 slices	3 slices	White bread, toasted
as needed	as needed	Mayonnaise
2 leaves	2 leaves	Lettuce
2 slices	2 slices	Tomato, about 1/4 in. (0.5 cm) thick
3 strips	3 strips	Bacon, cooked crisp
2 oz	60 g	Sliced turkey or chicken breast

PROCEDURE

1. Place the toast slices on a clean work surface. Spread the tops with mayonnaise.
2. On the first slice, place 1 lettuce leaf, then 2 slices of tomato, then 3 strips of bacon.
3. Place the second slice of toast on top, spread side down.
4. Spread the top with mayonnaise.
5. On top of this, place the turkey or chicken, then the other lettuce leaf.
6. Top with the third slice of toast, spread side down.
7. Place frilled picks on all 4 sides of the sandwich, as shown in **Figure 22.1**.
8. Cut the sandwich from corner to corner into 4 triangles. Each triangle will have a pick through the center to hold it together.
9. Place on a plate with the points up. The center of the plate may be filled with potato chips, French fries, or other garnish or accompaniment.

Per serving: Calories, 580; Protein, 32 g; Fat, 25 g (40% cal.); Cholesterol, 75 mg; Carbohydrates, 53 g; Fiber, 3 g; Sodium, 910 mg.

VARIATION

Bacon, Lettuce, and Tomato Sandwich (BLT)

Using only 2 slices of toast, prepare the basic recipe through step 3. Omit remaining ingredients. Cut sandwich in half diagonally for service.

Club Sandwich

Reuben Sandwich

YIELD: 1 SANDWICH

U.S.	METRIC	INGREDIENTS
2 slices	2 slices	Dark rye bread
4 tsp	20 mL	Russian or Thousand Island dressing
2 oz	60 g	Corned beef, sliced very thin
1 oz	30 g	Sauerkraut, well drained
1 oz	30 g	Swiss cheese (1 or 2 slices)
as needed	as needed	Butter

PROCEDURE

1. Place the slices of bread on a clean work surface.
2. Spread each slice with about 2 tsp (10 mL) dressing.
3. On one of the slices, place the corned beef, then the sauerkraut, then the cheese.
4. Place the second slice of bread on top, spread side down.
5. Butter the top of the sandwich and place buttered side down on a preheated griddle. Immediately butter the other side of the sandwich, which is now on top. (This method is less messy than buttering both sides before placing it on the griddle.)
6. Griddle the sandwich, turning once, until browned on both sides and hot through.
7. Cut the sandwich into halves and serve immediately.

Per serving: Calories, 590; Protein, 25 g; Fat, 39 g (59% cal.); Cholesterol, 105 mg; Carbohydrates, 35 g; Fiber, 4 g; Sodium, 1580 mg.

Grilled Cheese Sandwich 🌹

YIELD: 1 SANDWICH

U.S.	METRIC	INGREDIENTS
1 slice (1 oz)	1 slice (30 g)	Cheddar or American cheese cut to the size of the bread
2 slices	2 slices	White bread
as needed	as needed	Butter

PROCEDURE

1. Place the slice of cheese between the slices of bread.
2. Butter the outsides of the sandwich and place on a griddle preheated to 350°–375°F (175°–190°C).
3. Cook until golden brown on one side. Turn over and cook until the second side is golden brown and the cheese starts to melt.
4. Remove the sandwich from the griddle. Cut in half diagonally and serve immediately.

Per serving: Calories, 360; Protein, 13 g; Fat, 20 g (49% cal.); Cholesterol, 55 mg; Carbohydrates, 34 g; Fiber, 1 g; Sodium, 580 mg.

VARIATIONS

Grilled Ham and Swiss Sandwich

Make the sandwich with a ¹/₂-oz (15-g) slice of Swiss cheese and a 1-oz (30-g) slice of ham. Griddle as in basic recipe.

Croque Monsieur

Make like the Grilled Ham and Swiss Sandwich, above, except use 1 oz (30 g) Gruyère cheese and 1¹/₂ oz ham, sliced thin. As a variation, a croque monsieur may also be coated in Mornay Sauce (p. 182) and heated in an oven until lightly browned.

Grilled Cheese and Bacon Sandwich

Make the sandwich with 1 oz (30 g) cheddar or American cheese and 2 strips of crisp cooked bacon. Griddle as in the basic recipe.

Monte Cristo Sandwich

YIELD: 1 SANDWICH

U.S.	METRIC	INGREDIENTS
2 slices	2 slices	White bread
as needed	as needed	Butter
1 oz	30 g	Sliced turkey or chicken breast
1 oz	30 g	Sliced ham
1 oz	30 g	Sliced Swiss cheese
1	1	Egg, beaten
2 tbsp	30 mL	Milk

PROCEDURE

1. Place the bread on a clean work surface. Spread the tops with butter.
2. Place the turkey, ham, and cheese slices on the buttered side of one of the pieces of bread. Top with the remaining slice of bread, buttered side down.
3. Secure the sandwich with 2 picks placed in opposite corners.
4. Beat the egg and milk together.
5. Dip the sandwich in the batter until it is completely coated and the liquid has partially soaked into the bread.
6. Fry the sandwich in deep fat at 375°F (190°C) until golden brown.
7. Cut in half and serve immediately.

Per serving: Calories, 570; Protein, 35 g; Fat, 30 g (48% cal.); Cholesterol, 285 mg; Carbohydrates, 37 g; Fiber, 1 g; Sodium, 650 mg.

VARIATION

Alternative Method: Omit the picks and cook on a griddle until browned on both sides.

Monte Cristo Sandwich

Turkey BLT Wrap

YIELD: 1 SANDWICH

U.S.	METRIC	INGREDIENTS
1	1	Flour tortilla, 10 in. (25 cm) in diameter
1/2 oz	15 g	Mayonnaise
1/2–1 oz	15–30 g	Lettuce leaves
3	3	Tomato slices, thin
1 oz	30 g	Cooked bacon, crumbled
2 oz	60 g	Turkey breast, in thin slices

PROCEDURE

1. Spread the tortilla with the mayonnaise, leaving a 1/2-in. (15-mm) border around the outside unspread. (For a richer, moister sandwich, double the quantity of mayonnaise.)
2. Arrange the lettuce leaves in the center of the tortilla, leaving a border of about 2 in. (5 cm) uncovered.
3. Arrange the tomato, bacon, and turkey on top of the lettuce.
4. To roll the wrap, first fold the uncovered rim of the tortilla on opposite edges of the circle toward the center. This closes the ends of the roll to hold in the filling. Then roll the tortilla tightly. The mayonnaise on the edge of the tortilla helps seal it closed.
5. Serve whole or cut in half on the diagonal.

Per serving: Calories, 810; Protein, 32 g; Fat, 35 g (52% cal.); Cholesterol, 75 mg; Carbohydrates, 42 g; Fiber, 3 g; Sodium, 920 g.

FIGURE 22.3 Making a wrap sandwich.

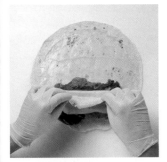   

(a) Spread the tortilla with the desired spread. Arrange the fillings toward one edge.

(b) Fold over the sides of the tortilla so that the ends of the wrap will be closed.

(c) Roll up.

(d) The finished, cut Turkey BLT Wrap.

Vegetarian Wrap with White Beans

YIELD: 1 SANDWICH

U.S.	METRIC	INGREDIENTS
1	1	Flour tortilla, 10 in. (25 cm) in diameter
as needed	as needed	Olive oil
1/2–1 oz	15–30 g	Lettuce leaves
3 oz	90 g	White Bean Salad (p. 691)
1 1/2 oz	45 g	Cooked rice, white or brown, cold
1/2 oz	15 g	Green bell pepper, diced

PROCEDURE

1. Lay the tortilla flat on the workbench. Brush it very lightly with olive oil.
2. Arrange the lettuce leaves in the center of the tortilla, leaving a border of about 2 in. (5 cm) uncovered.
3. Place the bean salad in a sieve to drain excess vinaigrette, reserving the vinaigrette. Mix together the bean salad and the rice. If the mixture is dry, add enough of the dressing back to the mixture so that it is just sufficiently moistened.
4. Mix in the diced pepper.
5. Place the bean mixture on the tortilla in an oblong mound. Roll and wrap the tortilla tightly around the filling as in the recipe for Turkey BLT Wrap.

Per serving: Calories, 510; Protein, 13 g; Fat, 21 g (38% cal.); Cholesterol, 0 mg; Carbohydrates, 65 g; Fiber, 6 g; Sodium, 360 mg.

Falafel and Roasted Vegetables in Pita

YIELD: 1 SANDWICH

U.S.	METRIC	INGREDIENTS
1	1	Pita
1 fl oz	30 mL	Tahini Yogurt Dressing (p. 728)
2 oz	60 g	Falafel (p. 824), 1-oz (30-g) balls, hot
2 oz	60 g	Roasted Summer Vegetables (p. 332), hot

PROCEDURE

1. The sandwich may be made by splitting open the pita to make a pocket or by wrapping the whole pita around the filling. If you are splitting open the pita, spread the inside with 1 tbsp (15 mL) dressing.
2. Fill with the falafel and vegetables.
3. Drizzle the remaining dressing over the filling.
4. Serve immediately.

Per serving: Calories, 350; Protein, 12 g; Fat, 11 g (28% cal.); Cholesterol, 20 mg; Carbohydrates, 53 g; Fiber, 5 g; Sodium, 660 mg.

VARIATION

Traditional Falafel in Pita

Omit the roasted vegetables. Use two 2-oz (60-g) falafel patties or four 1-oz (30-g) balls. Fill the pita with the falafel, ½ oz (15 g) shredded lettuce, ½–1 oz (15–30 g) diced tomato, and the tahini yogurt dressing.

Traditional Falafel in Pita

Southwestern Grilled Salmon Sandwich

YIELD: 1 SANDWICH

U.S.	METRIC	INGREDIENTS
1	1	Salmon fillet, about 4 oz (120 g)
1 tbsp	15 mL	Chili Marinade (p. 728)
1	1	Ciabatta roll or 5-in. (13-cm) length of French baguette
2	2	Tomato slices
1½ oz	45 g	Avocado, sliced
1–2	1–2	Leaf lettuce leaves

PROCEDURE

1. Rub the salmon fillet on both sides with the marinade. Let stand 10 minutes.
2. Grill or broil the salmon until just cooked through.
3. Split the roll or baguette in half horizontally.
4. Place the salmon on the bottom half.
5. Top with the tomato slices and the avocado slices.
6. Place a lettuce leaf and the top half of the bread on top.
7. To serve, leave the sandwich whole or cut it in half.

Per serving: Calories, 650; Protein, 39 g; Fat, 21 g (29% cal.); Cholesterol, 80 mg; Carbohydrates, 75 g; Fiber, 8 g; Sodium, 1210 mg.

Southwestern Grilled Salmon Sandwich

Tahini Yogurt Dressing 🌐 ❤ 🌹

YIELD: 10 FL OZ (300 ML)

U.S.	METRIC	INGREDIENTS
8 oz	240 g	Plain yogurt
1 oz	30 g	Tahini (sesame paste)
1 fl oz	30 mL	Lemon juice

PROCEDURE

1. Mix all ingredients together.
2. Refrigerate until needed.

Per 1 ounce (29.57 mL): Calories, 30; Protein, 1 g; Fat, 2 g (60% cal.); Cholesterol, 5 mg; Carbohydrates, 2 g; Fiber, 0 g; Sodium, 10 mg.

Chili Marinade ❤ 🌹

YIELD: ABOUT 6 OZ (180 G)

U.S.	METRIC	INGREDIENTS
2 oz	60 g	Chili powder
1 tbsp	15 mL	Dried oregano
1/2 tsp	2 mL	Ground cloves
1 oz	30 g	Garlic, crushed
2 tsp	10 mL	Salt
1 oz	30 g	Brown sugar
4 fl oz	120 mL	Red wine vinegar

PROCEDURE

1. Combine all ingredients except the vinegar and grind together to make a uniform mixture.
2. Add the vinegar and mix well.
3. Refrigerate until needed.

Per 1 ounce (28.35 g): Calories, 60; Protein, 2 g; Fat, 1.5 g (19% cal.); Cholesterol, 0 mg; Carbohydrates, 12 g; Fiber, 4 g; Sodium, 870 mg.

Barbecued Pork or Beef Sandwich

PORTIONS: 20   PORTION SIZE: 1 SANDWICH WITH 4 OZ (115 G) FILLING

U.S.	METRIC	INGREDIENTS
3 lb	1.3 kg	Cooked pork or beef
2 1/4 pt	1 L	Barbecue Sauce (p. 213)
20	20	Hamburger rolls
as needed	as needed	Butter

PROCEDURE

1. Using a slicing machine or chef's knife, cut the meat into very thin slices.
2. Combine the meat and sauce in a saucepan. Simmer uncovered over low heat for 10–15 minutes, until the meat has absorbed some of the flavor of the sauce and the liquid has reduced and thickened slightly.
3. Keep the meat hot for service.
4. For each order, butter a hamburger roll. Place a 4-oz (115-g) portion of the meat mixture on the bottom half of the roll. Close the sandwich and serve immediately.

Per serving: Calories, 400; Protein, 20 g; Fat, 24 g (54% cal.); Cholesterol, 65 mg; Carbohydrates, 26 g; Fiber, 2 g; Sodium, 300 mg.

Smoked Ham and Taleggio Panino

YIELD: 1 SANDWICH

U.S.	METRIC	INGREDIENTS
2 tbsp	30 mL	Mayonnaise
1 tsp	5 mL	Lemon juice
1/4 tsp	1 mL	Grated lemon zest
2 slices	2 slices	Firm country bread
2 thin slices	2 thin slices	Smoked ham
2 thin slices	2 thin slices	Tomato
2 slices	2 slices	Taleggio cheese
3–4 leaves	3–4 leaves	Arugula

PROCEDURE

1. Mix together the mayonnaise, lemon juice, and zest.
2. Lay the bread on the work surface.
3. Spread the two slices of bread with the mayonnaise.
4. Top one of the slices with the ham.
5. Place the tomato slices on top of the ham.
6. Top the ham with the slices of cheese.
7. Top the cheese with the arugula leaves.
8. Place the second slice of bread on the sandwich, mayonnaise side down. Press the sandwich together firmly.
9. Cook in a panino press or on a griddle until the cheese is melted. If using a griddle, place a grill brick on the sandwich and turn it over when half done.
10. Cut in half to serve.

Per serving: Calories, 770; Protein, 30 g; Fat, 42 g (50% cal.); Cholesterol, 65 mg; Carbohydrates, 66 g; Fiber, 3 g; Sodium, 2030 mg.

Smoked Ham and Taleggio Panino

Eggplant, Roasted Pepper, and Fontina Panino

YIELD: 1 SANDWICH

U.S.	METRIC	INGREDIENTS
2 slices	2 slices	Eggplant, about 3/8 in. (1 cm) thick, peeled
as needed	as needed	Olive oil
1/2	1/2	Red bell pepper, roasted and peeled (p. 290)
1	1	Ciabatta roll or other large, flat dinner roll or bun
as needed	as needed	Olive oil
2 slices	2 slices	Fontina cheese

PROCEDURE

1. Brush both sides of the eggplant slices with oil.
2. Cook on a grill or griddle or in a sauté pan until just tender.
3. Remove the seeds, stem, and inner membranes from the pepper. Cut it in into two pieces and lay the pieces out flat.
4. Cut the roll in half horizontally.
5. Brush both halves with olive oil.
6. Lay the eggplant slices on the bottom half.
7. Top with the pepper and then the fontina cheese.
8. Place the top half of the roll in place and press together.
9. Cook in a panino press or on a griddle until the cheese is melted. If using a griddle, place a grill brick on the sandwich and turn it over when half done.
10. Cut in half to serve.

Per serving: Calories, 640; Protein, 27 g; Fat, 26 g (36% cal.); Cholesterol, 65 mg; Carbohydrates, 75 g; Fiber, 8 g; Sodium, 1210 mg.

Grilled Vegetable Sandwich with Goat Cheese and Sun-Dried Tomatoes

YIELD: 8 SANDWICHES

U.S.	METRIC	INGREDIENTS
8 oz	250 g	Zucchini, trimmed
8 oz	250 g	Eggplant, trimmed
8 oz	250 g	Bell peppers, any color
8 oz	250 g	Onion, large
as needed	as needed	Olive oil
as needed	as needed	Salt
as needed	as needed	Balsamic vinegar
(see step 8)	(see step 8)	Focaccia
4–6 oz	125–180 g	Sun-dried tomatoes, packed in oil, drained, cut julienne
8 oz	250 g	Fresh, soft goat cheese, sliced

PROCEDURE

1. Cut the zucchini lengthwise into slices about $1/4$ in. (6 mm) thick.
2. If the eggplant is large, with a thick skin, peel it. Cut large eggplant crosswise into slices $1/4$ in. (6 mm) thick. Cut small eggplants lengthwise to make larger slices.
3. Core and seed the peppers. Cut into quarters lengthwise.
4. Cut the onion crosswise into slices about $1/3$ in. (8 mm) thick. Hold the rings of each slice together with a bamboo skewer.
5. Brush the vegetables with oil and sprinkle them with salt.
6. Grill the vegetables over medium heat, turning as necessary, until they are tender and lightly grill-marked. Cooking times will vary for different vegetables. Regulate the heat or the distance from the flame so the vegetables cook without browning too much.
7. Remove from the grill and brush with a little balsamic vinegar.
8. Cut the focaccia into $3^{1}/_{2} \times 5$ in. (9 × 13 cm) rectangles. If the focaccia is thick, split the rectangles in half horizontally. If it is thin, use 2 rectangles per portion.
9. Arrange the grilled vegetables on half of the rectangles.
10. Arrange the tomatoes on top of the vegetables.
11. Top with the goat cheese and the rest of the bread rectangles.
12. For service, serve whole or cut in half at an angle.

Per serving: Calories, 180; Protein, 8 g; Fat, 10 g (47% cal.); Cholesterol, 15 mg; Carbohydrates, 17 g; Fiber, 3 g; Sodium, 640 mg.

Grilled Vegetable Sandwich with Goat Cheese and Sun-Dried Tomatoes

Grilled Portobello and Boursin Sandwich

YIELD: 1 SANDWICH

U.S.	METRIC	INGREDIENTS
1 tbsp	15 mL	Olive oil
1¹/₂ tsp	7 mL	Balsamic vinegar
¹/₄ tsp	1 mL	Prepared mustard, French or Dijon style
pinch	pinch	Salt
1	1	Portobello mushroom cap, large
1 oz	30 g	Boursin or other soft, creamy cheese
1	1	Whole-grain roll or bun, split in half
6–8	6–8	Arugula leaves, torn

PROCEDURE

1. With a wire whip, mix together the oil, vinegar, mustard, and salt to make a vinaigrette.
2. Brush the mushroom cap with about two-thirds of the vinaigrette. Reserve the rest of the vinaigrette for step 6.
3. Place the mushroom cap on a grill or broiler grid, stem side down. Grill until about half cooked.
4. Turn the mushroom over so the bottom of the cap faces up. Top with the cheese.
5. Grill until the mushroom is tender and the cheese is partly melted.
6. Toss the arugula with the rest of the vinaigrette.
7. Place the mushroom cap, cheese side up, on the bottom half of the roll.
8. Top with the arugula and the top of the bun.

Per serving: Calories, 390; Protein, 7 g; Fat, 28 g (64% cal.); Cholesterol, 35 mg; Carbohydrates, 28 g; Fiber, 4 g; Sodium, 560 mg.

VARIATION   Grilled Portobello and Tomato Sandwich

Omit the cheese. Spread the bread with mayonnaise and top with the grilled mushroom, 2 tomato slices, and the arugula.

Grilled Chicken Aïoli Sandwich with Tomato and Avocado

YIELD: 1 SANDWICH

U.S.	METRIC	INGREDIENTS
1	1	Boneless skinless chicken breast, about 5 oz (150 g)
¹/₄ tsp	1 mL	Finely chopped garlic
pinch	pinch	Salt
pinch	pinch	Pepper
1¹/₂ tsp	7 mL	Lemon juice
1¹/₂ tsp	7 mL	Olive oil
2 slices	2 slices	Hearty country-style whole-grain bread, sliced ¹/₂ in. (1 cm) thick
1 tbsp	15 mL	Aïoli (p. 216 or p. 665)
2	2	Tomato slices
1 oz	30 g	Avocado, sliced
1–2	1–2	Leaf lettuce leaves

PROCEDURE

1. Place the chicken breast between two sheets of plastic film. With a meat mallet, gently pound to a uniform thickness of about ¹/₂ in. (6 mm).
2. Combine the garlic, salt, pepper, lemon juice, and olive oil.
3. Lightly coat the chicken with this mixture. Let marinate about 2 hours in the refrigerator.
4. Preheat a grill or broiler to very hot.
5. Grill the chicken until cooked through.
6. Spread the bread slices with aïoli.
7. Top one of the slices with the chicken, then the tomatoes, the avocado, and the lettuce. Top with the remaining slice of bread.
8. For service, cut in half.

Per serving: Calories, 690; Protein, 38 g; Fat, 31 g (41% cal.); Cholesterol, 100 mg; Carbohydrates, 64 g; Fiber, 5 g; Sodium, 810 mg.

Grilled Chicken Aïoli Sandwich with Tomato and Avocado

Pizza Margherita 🍽️ 🥬

YIELD: ONE 12-INCH (30-CM) PIZZA

U.S.	METRIC	INGREDIENTS
12 oz	360 g	Fresh, ripe plum tomatoes or canned Italian-style plum tomatoes
1 tbsp	15 mL	Olive oil
to taste	to taste	Salt
12 oz	360 g	French Bread dough (p. 919), fermented
6	6	Fresh basil leaves, torn in half
4 oz	120 g	Fresh mozzarella cheese, preferably Italian buffalo-milk mozzarella, sliced
1 tbsp	15 mL	Parmesan cheese (optional)
1½ tbsp	22 mL	Olive oil

PROCEDURE

1. If using fresh tomatoes, peel, seed, and chop them. If using canned, chop and drain them.
2. Combine the tomatoes and oil in a saucepan. Cook uncovered over moderate heat until the tomatoes are no longer watery.
3. Add salt to taste.
4. Cool.
5. Flatten the dough and roll it out into a circle. Drape the dough over the backs of the hands and carefully stretch the circle to a diameter of 12 in. (30 cm). Leave the dough thicker around the rim than in the center.
6. Flour a peel so the dough does not stick. Lay the circle of dough on it.
7. Spread the tomatoes over the dough, leaving the rim uncovered.
8. Distribute the basil and the mozzarella cheese over the pizza.
9. If using the parmesan, sprinkle it over the pizza.
10. Drizzle with olive oil.
11. Slide directly onto the hearth of a deck oven preheated to 500°F (260°C). Bake until the dough is browned and the cheese is melted.
12. Using the peel, remove from the oven and serve immediately.

Per pizza: Calories, 1450; Protein, 49 g; Fat, 63 g (39% cal.); Cholesterol, 90 mg; Carbohydrates, 170 g; Fiber, 9 g; Sodium, 1880 mg.

VARIATION

Pizza Marinara

Omit the mozzarella cheese and the basil. Increase the tomatoes to 2 lb (900 g). Increase the first quantity of olive oil to 1½ tbsp. In step 2, cook the tomatoes with 3 cloves garlic, sliced thin, and ½ tsp (2 mL) fresh chopped oregano or ¼ tsp dried oregano.

Pizza Margherita before baking

Pizza Margherita after baking

ASSOCIAZIONE VERACE PIZZA NAPOLETANA

Naples, Italy, takes pride in being the birthplace of pizza. Today the Associazione Verace Pizza Napoletana sets down the rules by which its members must abide if they wish to claim they serve authentic Neapolitan pizza. The rules state that the dough must contain only flour, water, salt, and natural yeast and that it must be made by hand or in an approved mixer. The pizza must be shaped by hand and baked on the hearth of a wood-burning oven. Toppings are limited to a list of approved ingredients.

Two pizzas—Margherita, topped with tomatoes, basil, and buffalo mozzarella, and Marinara, topped with tomatoes, garlic, oregano, and olive oil—are considered the original and true Neapolitan pizzas.

Pizza has grown beyond its origins to become an international favorite. In North America, most of the pizzas served make no claim to being authentically Italian and instead sport toppings such as barbecued chicken, beef with peppers and taco seasonings, and smoked salmon with artichokes, in addition to old favorites like sausage and pepperoni.

TERMS FOR REVIEW

pullman loaf	**club sandwich**	**wrap**	**pizza**
simple sandwich (hot or cold)	**open-faced sandwich (hot or cold)**	**grilled sandwich**	**panino**
multidecker sandwich	**tea sandwich**	**deep-fried sandwich**	

QUESTIONS FOR DISCUSSION

1. If you cannot get daily bread delivery, what are some measures you can take to ensure that the bread in the sandwiches you serve is always fresh?

2. What precautions must you take when using mayonnaise as a sandwich spread?

3. Briefly describe the setup of a short-order sandwich station.

4. How does a setup for preparing sandwiches in quantity differ from a short-order sandwich setup?

5. Why are most sandwiches cut before serving?

23

HORS D'OEUVRES

In addition to salads and salad dressings, the pantry or garde manger department is generally responsible for the small food items known as *appetizers* or *hors d'oeuvres*. The function of these foods is to enliven the appetite before dinner, often to the accompaniment of drinks, so they are generally small in size and spicy or piquant in flavor.

There is some confusion as to the distinction, if any, between the terms *appetizer* and *hors d'oeuvre*. In general, the first course of a multicourse meal is called an **appetizer**, and a finger food item served at receptions and with cocktails is called an **hors d'oeuvre**. In some regions, however, the terms are used interchangeably.

This chapter deals not with first courses but primarily with the kinds of food usually encountered away from the dinner table—foods that are the special domain of the pantry chef. They include finger foods such as canapés and relishes, indispensable accompaniments to receptions and parties.

AFTER READING THIS CHAPTER, YOU SHOULD BE ABLE TO

1. Name and describe the two principal methods of serving hors d'oeuvres at a reception.

2. Prepare canapés.

3. Prepare hors d'oeuvre cocktails and relishes.

4. Prepare dips.

5. Prepare a variety of other hors d'oeuvres, including antipasti, bruschette, and tapas.

6. Name and describe the three types of sturgeon caviar, and name and describe three other kinds of caviar.

SERVING HORS D'OEUVRES

This chapter focuses on foods not served as part of a normal lunch or dinner menu. These foods are usually small items that may be served at a formal reception preceding a meal, as part of a separate event not connected with a meal, or simply as accompaniments to beverages at an informal gathering.

The two most common ways of serving hors d'oeuvres are butler-style and buffet-style.

In **butler-style service**, the hors d'oeuvre selections are offered to guests by service staff carrying small trays as they pass among the assembled group. Several points should be kept in mind when planning this style of service:

- Each item should be small enough to be eaten in one or two bites.

- Each item should be easily handled by the guest. Canapés or other foods that can be picked up without soiling the fingers are ideal. For foods with a moist or oily surface, offer picks that guests can use to handle the food without touching it. Cocktail napkins should always be offered. Foods requiring a plate are better served on a cocktail buffet.

- Strive for simple, attractive arrangements on the tray. The foods should look appetizing even when only a few items are left. Trays that no longer look appealing should be brought back to the kitchen or pantry area to be refreshed and refilled.

- Ideally, each tray should hold only one food selection. For simple items, two or three selections can be presented on a single tray, but avoid making the tray too complex or overloaded.

- Avoid presenting hot and cold items on the same tray. Cold items are easiest to serve. Hot items should be brought back to the kitchen or pantry area as soon as they are no longer warm.

- For items to be offered with dips, a small bowl of the dip can be presented on the same tray as the individual hors d'oeuvres.

In **buffet-style service**, hors d'oeuvres are arranged attractively on one or more tables, and guests help themselves. Small plates can be offered on buffet tables, so it is not necessary to confine the food selections to finger foods. Cold hors d'oeuvres are usually presented arranged on trays, which can be easily replaced when they are depleted. Canapés and similar items are best arranged in neat rows, circles, or other arrangements. Raw vegetables may be piled in neat stacks, with bowls of dips placed among them. Hot items should be offered in chafing dishes, which keep them hot.

Buffet presentation is discussed in more detail in Chapter 29.

For both butler and buffet service, hors d'oeuvres must be prepared and garnished with care so that they are attractive and appealing. Remember that their function is to enliven the appetite, and to do this they must look good. Select garnishes in attractive colors that go together, and cut all ingredients, including garnish, neatly and uniformly.

Finally, hors d'oeuvres are also served informally as accompaniments to beverages. This category is often known as *bar food* and may be served one item at a time, as ordered by the customer, or presented on a buffet. The classic bar food is the popular Spanish-style tapas, discussed on page 745.

CANAPÉS

Canapés may be defined as bite-size, open-faced sandwiches.

Canapés are perhaps the most traditional and also the most modern of hors d'oeuvres. This is no doubt because they are so varied and so versatile. Because they consist of tiny portions of food presented on bases of bread, toast, or pastry, they are perfect finger food, easily handled and easily eaten. Nearly any food that can be served in a small portion can be served as a canapé topping. The variety of possible combinations is nearly unlimited.

Most canapés consist of three parts: base, spread, and garnish.

BASE

Canapé bases may be made from several items. The following are suggestions:

Bread cutouts

Toast cutouts

Crackers

Melba toasts

Tiny unsweetened pastry shells (short dough, phyllo dough, or other pastry)

Profiteroles (miniature unsweetened cream puff shells (p. 1001)

Toasted pita wedges

Tortilla chips or cups

Tiny biscuits, split in half if necessary

Polenta cutouts

Miniature pancakes

Many of these items, such as crackers and Melba toasts, can be purchased ready-made, but bread and toast cutouts are the most widely used and offer the lowest food cost, though they require more labor.

Untoasted bread for canapés should be firm enough to allow the finished product to be handled easily. It may be cut thick and flattened slightly with a rolling pin to make firmer. Toast is, of course, firmer, and it gives a pleasing texture and crispness to canapés.

"OUTSIDE THE WORK"

The French expression *hors d'oeuvre* literally translates as "outside the work," meaning "apart from the main meal or main part of the meal." In French, the term is not spelled or pronounced with an s at the end to make it plural, so you will often see the plural form spelled the same as the singular. The term has been thoroughly adopted into English, however, and in English-language dictionaries, the plural is spelled and pronounced with a final *s*.

PROCEDURE for Preparing Canapés from Toast

METHOD 1

1. You may use ready-sliced bread (after trimming the crusts), but it is usually most efficient to use long, unsliced pullman loaves. Cut the crusts from all sides (save for bread crumbs). Cut the bread horizontally into slices ¼ inch (6 mm) thick, as shown in Figure 23.1.

2. Toast the slices in the oven or in a large toaster.

3. Let the toasts cool.

4. Cover with a thin, even layer of the chosen spread and cut into desired shapes with a knife (see Figure 23.2). Make the cuts neat and uniform.

   Alternatively, cut the toasts into desired shapes with small cutters and reserve the trim for bread crumbs. Spread each cutout with desired topping. (This method is more time-consuming, but it may be used with round or odd-shaped cutters if you want to save food cost by not losing spread on unused trimmings.)

5. Garnish the cutouts as desired.

METHOD 2

1. Cut bread slices as in step 1 above.

2. Cut into desired shapes. Brush both sides of each cutout with melted butter and arrange on sheet pans. Place in a hot oven (450°F/230°C) until golden brown and very crisp, 6 to 8 minutes.

3. Let the bases cool.

4. Assemble the canapés.

   This method is more costly but gives a crisper base that holds up better with a moist spread.

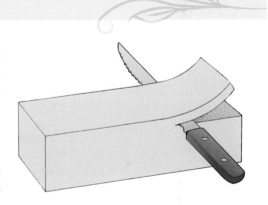

FIGURE 23.1 For canapés, trim the crusts from a pullman loaf. With a serrated knife, cut the loaf horizontally into thin slices.

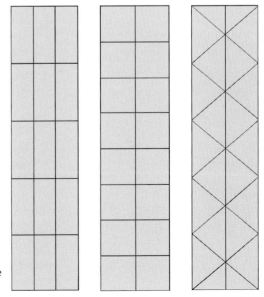

FIGURE 23.2 Bread slices for canapés can be cut into several basic shapes with no waste.

SPREAD

Canapé spreads may be as simple as butter or softened cream cheese, but it is better to use a more highly flavored spread because sharp or spicy flavors are better for stimulating the appetite.

The spread should be thick enough to cling well to the base and so the garnish sticks to it without falling off.

Spreads may be divided into three basic categories, as follows.

Flavored Butters

Basic procedures for making flavored or compound butters are explained in Chapter 8 (see recipes on p. 192). Most flavored butters are made simply by blending the flavoring ingredients with the softened butter until completely mixed. Solid ingredients should be puréed or chopped very fine so the butter can be spread smoothly.

Proportions of flavoring ingredients to butter can be varied widely, according to taste. For example, to make anchovy butter, you could double the quantity of anchovies indicated in the recipe on page 192 to get a stronger flavor, or you could decrease it to get a milder flavor. Because of this variability, and because the basic procedure is so simple, you should be able to make many flavored butters without individual recipes. Use the recipes in Chapter 8 and the following list as guides. Popular and versatile flavors for butter spreads include

Lemon	Caviar	Shrimp
Parsley	Mustard	Olive
Tarragon	Horseradish	Shallot or scallion
Chive	Pimiento	Curry
Anchovy	Blue cheese	Caper

Flavored Cream Cheese

Flavored cream cheese spreads are made like flavored butters, except cream cheese is substituted for the butter. Alternatively, use a mixture of cream cheese and butter, well blended. Flavor variations are the same as those listed for butter.

In addition, cream cheese is often blended with sharper, more flavorful cheeses that have been mashed or grated. Adding cream cheese to firmer cheese helps make the latter more spreadable. A liquid such as milk, cream, or port wine may be added to make the mixture softer. Such cheese spreads are often flavored with spices and herbs such as paprika, caraway seeds, dry mustard, parsley, or tarragon.

Meat or Fish Salad Spreads

You can use many cold meat or fish mixtures, such as cooked salads, to make canapé spreads. Popular examples include tuna salad, salmon salad, shrimp salad, chicken salad, deviled ham, and liver pâté.

To convert a salad recipe (see pp. 695, 696, and 698) to a spread recipe, you may need to make one or more of the following modifications:

1. Chop the solid ingredients very fine, or grind or purée them, so the mixture is spreadable and not chunky.

2. Do not add the liquid ingredients and mayonnaise all at once. Add them a little at a time, just until the mixture reaches a thick, spreadable consistency.

3. Check the seasonings carefully. You may want to increase the seasonings to make the spread more stimulating to the appetite.

GARNISH

The garnish of a canapé is any food item or combination of items placed on top of the spread. It may be a major part of the canapé, such as a slice of ham or cheese, or it may be a small tidbit selected for color, design, texture, or flavor accent, such as a pimiento cutout, a slice of radish, a caper, or a dab of caviar. Even the spread can be used as a garnish. For example, you may

GUIDELINES for Assembling Canapés

1. **Good mise en place is essential.**
   Preparing thousands of canapés for large functions can be tedious work, so it is essential that all bases, spreads, and garnishes be prepared ahead of time in order that final assembly may go quickly and smoothly.

2. **Assemble as close as possible to serving time.**
   Bases quickly become soggy, and spreads and garnishes dry out easily. As trays are completed, they may be covered lightly with plastic and held for a short time under refrigeration. Be sure to observe all rules for safe food handling and storage, as you learned in Chapter 2.

3. **Select harmonious flavor combinations in spreads and garnish.**
   For example, caviar and chutney or anchovy and ham are not appealing combinations, but these combinations are:

   Mustard butter and ham

   Lemon butter and caviar

   Pimiento cream cheese and sardine

   Horseradish butter and smoked salmon or smoked tongue

   Tuna salad and capers

   Anchovy butter, hard-cooked egg slice, and olive

4. **Be sure that at least one of the ingredients is spicy or pronounced in flavor.**
   A bland canapé has little value as an appetizer.

5. **Use high-quality ingredients.**
   Canapés can be a good way to utilize leftovers, but only if the leftovers have been carefully handled and stored to retain freshness.

6. **Keep it simple.**
   Simple, neat arrangements are more attractive than elaborate, overworked designs. Besides, you don't have time to get too fancy. Be sure the canapés hold together and do not fall apart in the customers' hands.

7. **Arrange the canapés carefully and attractively on trays.**
   Much of the attraction of canapés is eye appeal, and the customer never sees just one at a time, but a full tray. Each tray should carry an assortment of flavors and textures, so there is something for every taste.

   Figure 23.3 shows a tray of simple, attractive canapés.

FIGURE 23.3 Assorted canapés, from left: gravlax with mustard cream cheese; chicken liver pâté with mustard butter and black olive; caviar, red onion, and sour cream in red potato; herbed Boursin cheese with almonds; beef tenderloin with horseradish cream cheese and capers.

make a canapé with a mustard butter spread and a slice of ham, then decorate the ham with a border or design of mustard butter piped on with a paper cone.

Here are some of the many food items that may be used alone or in combination to decorate canapés:

Vegetables, Pickles, and Relishes

Radish slices	Pickled onions	Cherry tomato slices or halves
Olives	Chutney	Watercress leaves
Pickles	Asparagus tips	Marinated mushrooms
Capers	Cucumber slices	Parsley
Pimiento		

Fish

Smoked oysters and clams	Shrimp	Crab meat
Smoked salmon	Rolled anchovy fillets	Lobster chunks or slices
Smoked trout	Caviar	Sardines
Herring	Salmon or tuna flakes	

Meats

Ham	Smoked tongue
Salami	Roast beef
Chicken or turkey breast	

Other

Cheese

Hard-cooked egg slices

KEY POINTS TO REVIEW

- What are the two styles of hors d'oeuvre service? Describe each.

- What are canapés? What are the three components of canapés?

- What are the guidelines for assembling canapés?

COCKTAILS

The term **cocktail** is used not only for alcoholic beverages and vegetable and fruit juices but also for a group of appetizers made of seafood or fruit, usually with a tart or tangy sauce. Such cocktails are always served well chilled, often on a bed of crushed ice.

Oysters and clams on the half-shell are popular seafood cocktails, as are shrimp, crabmeat, lobster, and firm, flaked white fish with an appropriate sauce. Recipes for a standard tomato-based cocktail sauce and for a lighter sauce called *mignonette* are included in Chapter 8.

Fresh oysters and clams on the half-shell should be opened just before they are served (see pp. 600–604) and arranged on flat plates, preferably on a bed of ice. Provide cocktail sauce in a small cup in the center or at the side of the plate. Lemon wedges should also be provided.

Cocktails of shrimp and other cooked seafood are generally served in a stemmed glass or in a small, cup-shaped bowl, which may be nestled in a bed of ice. The cocktail sauce may be put in the glass first and the seafood then arranged on top, partially immersed. Or the cocktail sauce may be added to the seafood as a topping. A third alternative is to serve the sauce separately in a small cup, as for raw oysters. Garnish the dish attractively with lettuce or other salad greens and with lemon wedges.

Fruit cups served as cocktails should be pleasantly tart and not too sweet. Many fruit salads (see Chapter 21) may be served as cocktails. Adding fresh lemon or lime juice to fruit mixtures or serving with a garnish of lemon or lime wedges provides the necessary tartness. A simple wedge of melon with lime is a refreshing cocktail.

A few drops of a flavored liqueur can also be used to perk up the flavor of a fruit cocktail.

RELISHES

The term **relish** covers two categories of foods: raw vegetables and pickled items.

RAW VEGETABLES

Raw vegetables are also known as **crudités** (croo dee tays; cru in French means "raw").

Any vegetable that can be eaten raw may be cut into sticks or other attractive, bite-size shapes and served as a relish. Most popular are celery, carrots, and radishes. Other good choices are green and red peppers, zucchini, cucumbers, scallions, cauliflower and broccoli florets, peeled broccoli stems, peeled kohlrabi, cherry tomatoes, and Belgian endive leaves. Crudités are often served with an appropriate dip (see the next section).

Raw vegetables must be served crisp and well chilled, just as in salads. Use the freshest, most attractive vegetables possible. If they are a little wilted, they can be recrisped by holding them for a short time in ice water. Serving vegetables embedded in crushed ice will maintain their crispness.

An imaginative pantry chef can make an attractive, colorful bouquet of raw vegetables.

PICKLED ITEMS

A wide variety of items such as dilled cucumber pickles, gherkins, olives, watermelon pickles, pickled peppers, spiced beets, and other preserved vegetables and fruits are served as relishes. These items are rarely made in-house but are purchased already prepared. Like raw vegetables, they should be served chilled.

DIPS

Savory **dips** are popular accompaniments to potato chips, crackers, and raw vegetables.

Proper consistency is important for any dip you prepare. It must not be so thick it cannot be scooped up without breaking the chip or cracker, but it must be thick enough to stick to the items used as dippers. Proper consistency means thickness *at serving temperature*. Most dips become thicker when held in the refrigerator.

Many mixtures used as spreads (see section on canapés) can also be used as dips. Thin or soften them by adding a little mayonnaise, cream, or other appropriate liquid.

The recipes here are examples of typical dips. Many other sauces and salad dressings can be used as dips. Salsas (p. 206) and aïoli (pp. 216 and 665) are two popular examples. Review the cold sauces in Chapter 8.

Blue Cheese Dip

YIELD: 1 QT (1 L)

U.S.	METRIC	INGREDIENTS
12 oz	375 g	Cream cheese
5 fl oz	150 mL	Milk
6 oz	175 g	Mayonnaise
1 fl oz	30 mL	Lemon juice
1 oz	30 g	Onion, minced
1/2 tsp	2 mL	Hot red pepper sauce
1/2 tsp	2 mL	Worcestershire sauce
10 oz	300 g	Blue cheese, crumbled

PROCEDURE

1. In a mixer with the paddle attachment, beat the cream cheese at low speed until soft and smooth.
2. With the machine running, slowly beat in the milk.
3. Add the rest of the ingredients and blend in well.
4. Taste and adjust seasonings. Chill.

Per 1 fl oz (29.57 mL): Calories, 110; Protein, 3 g; Fat, 11 g (86% cal.); Cholesterol, 20 mg; Carbohydrates, 1 g; Fiber, 0 g; Sodium, 190 mg.

VARIATION

Cheddar Cheese Dip

Substitute grated sharp cheddar cheese for the blue cheese.
If desired, add chopped chives.

Garlic Cheese Dip

Add mashed garlic to taste to Cheddar Cheese Dip.

Bacon Cheese Dip

Add crumbled crisp bacon to Cheddar Cheese Dip.

Cheese and Chile Dip

Flavor Cheddar Cheese Dip with canned green chiles, chopped.

Romesco

YIELD: 1 PT (500 ML)

U.S.	METRIC	INGREDIENTS	PROCEDURE
1 oz	30 g	White bread	1. Sauté the bread in the olive oil until golden.
¹/₂ fl oz	15 mL	Olive oil	2. Toast the almonds in an oven at 350°F (175°C) about 15 minutes or in a skillet over moderately low heat until light golden, not dark brown. Remove the almonds from the pan as soon as they are golden so they do not brown further.
3 oz	90 g	Blanched almonds	
2 tsp	10 mL	Garlic, chopped	3. Combine the bread, almonds, and garlic in a food processor. Grind until fine.
10 oz	300 g	Tomatoes, peeled	4. Add the tomatoes, paprika, and cayenne. Process to a paste.
2 tsp	10 mL	Spanish paprika	
¹/₈ tsp	0.5 mL	Cayenne	5. With the machine running, gradually add the vinegar and then the oil in a slow stream.
1¹/₂ fl oz	45 mL	Red wine vinegar	6. Adjust the seasonings with salt and pepper. Add more vinegar if required. The sauce should not be too acidic but should have a definite sharpness.
3–4 fl oz	90–120 mL	Olive oil	
to taste	to taste	Salt	
to taste	to taste	Pepper	

Per 1 fl oz (29.57 mL): Calories, 100; Protein, 2 g; Fat, 9 g (80% cal.); Cholesterol, 0 mg; Carbohydrates, 3 g; Fiber, 1 g; Sodium, 15 mg.

Hummus (Chickpea Dip)

YIELD: 1 QT (1 L)

U.S.	METRIC	INGREDIENTS	PROCEDURE
1 lb	500 g	Cooked or canned chickpeas, drained	1. Purée the chickpeas with the tahini, garlic, lemon juice, and olive oil.
8 oz	250 g	Tahini (sesame paste)	2. If necessary, thin the purée with a little water or additional lemon juice, depending on the taste.
¹/₄ oz	8 g	Garlic, crushed	
4 fl oz	125 mL	Lemon juice	3. Season with salt to taste and with cayenne.
1 fl oz	30 mL	Olive oil	4. Chill at least 1 hour to allow the flavors time to blend.
to taste	to taste	Salt	
pinch	pinch	Cayenne	
1–2 fl oz	30–50 mL	Olive oil	5. Spoon the hummus into serving bowls. Drizzle additional olive oil over each bowl before serving.

Per 1 fl oz (29.57 mL): Calories, 80; Protein, 3 g; Fat, 6 g (63% cal.); Cholesterol, 0 mg; Carbohydrates, 5 g; Fiber, 2 g; Sodium, 0 mg.

VARIATION

Babaganouj

Substitute eggplant purée for the chickpeas. Make the eggplant purée as follows: Toast whole eggplants under a broiler, over a gas burner, or directly on a flattop range until the skin is charred and the eggplant is soft. Peel off the charred skin under running water and cut off the tops. Remove large clumps of seeds, if desired. Let stand in a china cap or sieve to let excess moisture drain, then purée the pulp. Reduce the lemon juice to 2–3 fl oz (60–90 mL) and the tahini to 4 oz (125 g). Double the olive oil.

Hummus

Guacamole

YIELD: APPROXIMATELY 1 QT (1 L)

U.S.	METRIC	INGREDIENTS
4	4	Ripe avocados, medium size
2 oz	60 g	Onion, grated
1	1	Small, hot green chile such as jalapeño, minced
1 fl oz	30 mL	Lime or lemon juice
1 fl oz	30 mL	Olive oil
to taste	to taste	Salt
12 oz	375 g	Fresh tomato (optional)

PROCEDURE

1. Pit and peel the avocados. Mash the pulp coarsely. The purée should be slightly lumpy rather than smooth.
2. Mix in the onion, minced chile, lime or lemon juice, olive oil, and salt to taste.
3. If desired, peel, seed, and dice the tomato and mix it into the avocado.
4. Cover tightly with plastic wrap placed on the surface of the guacamole. This is to protect it from air, which will darken it. For the same reason, guacamole should not be made too long before serving time. Chill the guacamole until ready to serve.

Per 1 fl oz (29.57 mL): Calories, 45; Protein, 1 g; Fat, 4 g (75% cal.); Cholesterol, 0 mg; Carbohydrates, 2 g; Fiber, 2 g; Sodium, 0 mg.

Note: If fresh chiles are not available, use canned chiles or a few dashes of hot red pepper sauce.

VARIATION

Sour Cream Avocado Dip

Use 1 oz (30 g) onion. Omit the olive oil and tomato. Mash the avocado to a smooth purée. Add $\frac{1}{2}$–$\frac{3}{4}$ pt (250–375 mL) sour cream.

Guacamole

Parmesan Dip

YIELD: 1 PT (480 ML)

U.S.	METRIC	INGREDIENTS
1 pt	480 mL	Mayonnaise
3$\frac{1}{2}$ oz	105 g	Parmesan cheese, grated
4$\frac{1}{2}$ tsp	22 mL	White vinegar
4 tsp	20 mL	Sugar
4 tsp	20 mL	Dijon mustard
1 tsp	5 mL	Garlic powder
1$\frac{1}{2}$ tsp	7 mL	Paprika
3 tbsp	45 mL	Water

PROCEDURE

1. Combine the mayonnaise, cheese, vinegar, sugar, mustard, garlic powder, and paprika in a bowl. Mix thoroughly.
2. Stir in a little water to thin to desired consistency.

Per 1 fl oz (29.57 mL): Calories, 240; Protein, 3 g; Fat, 24 g (92% cal.); Cholesterol, 15 mg; Carbohydrates, 2 g; Fiber, 0 g; Sodium, 280 mg.

MISCELLANEOUS HORS D'OEUVRES

A great variety of other foods, both hot and cold, can be served as hors d'oeuvres. If they are to be served away from the dinner table, it is best if they can be eaten with the fingers or speared with a pick. At a reception at which many hors d'oeuvres are served, it is all right if a few of them must be eaten with forks from small plates, but finger food is much easier for the guests, who are likely to be standing and holding a wineglass or cocktail glass while eating.

Of course, there are thousands of hors d'oeuvre recipes, including many adapted from the cuisines of other lands. Those included here are a sampling of some popular types.

Many of the recipes given elsewhere in this book can be adapted as hors d'oeuvre recipes. In most cases, unit size or portion size should be decreased. For example, meatballs should be made small enough to be eaten in one or two bites. Among the items most readily adapted are the following (check the Index for page numbers):

Barbecued Spareribs	Seviche
Shish Kebab	Vegetable Fritters
Deep-Fried Chicken (using wing sections)	Many salads
	Quiche
Baked Clams Oreganata	Cheese Wafers and Straws
Oysters and Clams Casino	Sushi
Broiled Shrimp, Scampi Style	Gravlax
Broiled Scallops	Smoked Salmon
Fried Breaded Scallops, Shrimp, Oysters, or Clams	Smoked Trout
Steamed Mussels (served on the half-shell, with a sauce)	Smoked Duck
	Pâtés, Terrines, and Galantines
Cod Cakes or other fish cakes	Ham Mousse
Fish Tartares	Foie Gras Mousse and Terrine

In addition to the categories of hors d'oeuvres introduced so far, the following additional terms merit brief explanation.

ANTIPASTO

Italian cuisine is particularly rich in hors d'oeuvres, or antipasti, as they are called (singular form: **antipasto**). Many books give a recipe for a mixed salad called *antipasto*. This is misleading, however, because the Italian term does not refer to a specific recipe but to any typically Italian hors d'oeuvre, hot or cold.

Many Italian-style restaurants offer a cold antipasto plate or platter comprising an assortment of flavorful tidbits. Typical components include the following:

Cured meats, such as salami, prosciutto, bologna, and boiled ham.

Seafood items, especially canned or preserved items such as sardines, anchovies, and tuna.

Cheeses, such as provolone and mozzarella.

Hard-cooked eggs and stuffed eggs.

Relishes, such as raw carrots, celery, fennel, radishes, cauliflower, and tomatoes, and cooked or pickled items, such as olives, artichoke hearts, small hot peppers, and onions.

Cooked dried beans and other firm vegetables in a piquant vinaigrette.

BRUSCHETTA

In recent years, one of the most popular hors d'oeuvres is the Roman garlic bread called **bruschetta** (broo sket ta; plural: *bruschette*, pronounced broo sket teh). Bruschetta, in its most basic form, is a slice of Italian bread that is toasted, rubbed with crushed garlic, and drizzled with olive oil. When made with good bread and extra-virgin olive oil, this simple hors d'oeuvre is fine on its own. But it is no doubt the variety of toppings that can be served on this garlic-bread base that accounts for its popularity.

Perhaps the best-known topping is a seasoned mixture of diced fresh tomatoes and olive oil, but this is far from the only possibility. Meats, cheeses, cooked dried beans, and raw and cooked vegetables are all appropriate. In fact, the bruschetta is a type of antipasto, and many of the items listed in the preceding section may also be served as bruschetta topping. Several variations are included among the recipes in this chapter.

Bruschette are similar to canapés in that they are toast with topping. However, they are larger and heartier than typical canapés and can be served as the appetizer course of a dinner.

TAPAS

Another type of hors d'oeuvre that has become increasingly popular is the Spanish **tapa**. The variety of tapas is so great that it is difficult to give a definition, except to say that a tapa is a small food item intended to be eaten with wine or other drinks, usually in taverns and bars. The Spanish term literally means "lid," and the original tapa is thought to be a small bit of food, perhaps a slice of cured ham, placed on top of a glass of sherry.

As varied as they are, tapas have a few points in common. They are served in small portions and are intended to be eaten immediately and quickly. In most cases, they are served on small plates. Many can be eaten by hand, but many others, especially those served in a sauce, are eaten with a fork. Often they are the same kinds of foods that can be served as first courses of a dinner, but presented in smaller portions.

CAVIAR

Caviar is the salted roe, or eggs, of the sturgeon. In the United States and Canada, any product labeled simply *caviar* must come from sturgeon. Roe from any other fish must be labeled as such (for example, *whitefish* caviar).

The most important caviar-producing countries have traditionally been Russia and Iran, both of which border the Caspian Sea, where the sturgeon are harvested. Sturgeon live in other waters as well, including North American waters. Production of North American caviar has been growing in recent years, in part because of difficulties obtaining caviar from the Caspian.

To categorize it further, caviar is given the name of the species of sturgeon it is taken from. The *beluga* is the largest and scarcest sturgeon, and it yields the largest and most expensive eggs. Next in size are *osetra* and *sevruga*.

Although the larger eggs are generally more expensive, size or price alone does not necessarily indicate quality. All three types of caviar vary considerably in quality. The only sure way to determine the quality of a particular tin or jar of caviar is to taste it.

Good-quality caviar is made up of shiny, whole eggs, with few, if any, broken. It should not have a strong, fishy smell, and it should not look watery or oily.

Caviar made with a relatively low proportion of salt is labeled *malassol*, which means "little salt." Malassol caviar is considered to be of better quality than the more highly salted varieties. However, for those who must restrict their salt intake, there is no such thing as low-sodium caviar. Even malassol is salty.

Caviar is either fresh or pasteurized. Fresh caviar in an unopened tin will keep for a few weeks, as long as it is kept cold. Once opened, it begins to deteriorate quickly and should be eaten the same day, if possible. Pasteurized caviar is of lower quality because it has been heat-treated. Unopened, it will keep much longer than fresh caviar, but once opened, it too should be eaten as quickly as possible, or within a few days.

SPANISH TAPAS TRADITIONS

Normal lunchtime in Spain is two or three o'clock, while dinner usually isn't eaten before ten o'clock at night. What better way to spend some of those long hours after work and before dinner than to meet friends at a local tavern and enjoy a glass or two of sherry? People go to tapas bars not so much to eat but to talk and to enjoy each other's company. But of course lunch was hours earlier, and dinner will be hours later, so having a few bites to eat with your wine is almost a necessity.

Out of these traditional dining habits grew the sociable custom of tapas, which has spread around the world.

Sturgeon caviar

Salmon caviar

Tobiko

Other Caviars

Roe from other fish, including trout, salmon, whitefish, and lumpfish, is also salted to make caviar. Once again, quality varies considerably, ranging from good to nearly inedible. Those of better quality are useful alternatives to expensive sturgeon caviar. Three popular types are **salmon caviar**, large red eggs, often paired with smoked salmon or other salmon products in appetizer dishes; **bleak roe**, tiny, mild-tasting, orange eggs from a fish native to the Baltic; and **tobiko**, tiny orange eggs from Japanese flying fish, introduced to the West in Japanese sushi restaurants.

Serving Caviar

The best caviar should be served ice-cold and as simply as possible. The traditional caviar accompaniments of chopped onion, lemon, chopped egg, and sour cream may be all right for inexpensive roes, but they overwhelm the delicate flavor of top-quality beluga, osetra, and sevruga. Spoons for eating caviar should be made of bone, porcelain, mother of pearl, or even plastic, but not metal, which reacts with the caviar to give it an unpleasant taste.

For more elaborate presentations, caviars a notch or two below the best quality are more economical choices. Foods served with caviar should be fairly mild or delicate. Such items as mild smoked salmon, oysters, and potatoes are often used.

AMUSE BOUCHE

An **amuse bouche** (ah mews boosh) is a tiny appetizer or hors d'oeuvre offered, compliments of the chef and usually in more expensive restaurants, to guests seated at their tables, either before or after they have ordered from the menu. It is an opportunity to showcase an aspect of the chef's cooking style and talent and to welcome the guests.

Nearly anything that can be served in a tiny portion can be served as an amuse bouche, including salads, soups (in espresso cups), canapés, and little portions of meat, fish, or vegetable with a few drops of sauce and a garnish. Usually chefs don't use a separate category of recipe for these items but borrow something from one of their regular menus, preferably not a dish on the menu that evening, and give it a different presentation, garnish, or sauce.

KEY POINTS TO REVIEW

- What are hors d'oeuvre cocktails? What are relishes?

- What is the proper consistency for dips?

- What do the following terms mean: antipasto; bruschetta; tapas? Give examples of each.

- What does the term *caviar,* when used by itself, mean? What are the three types of sturgeon caviar? Give the names and descriptions of three other types of caviar.

Spinach Boreks 🍽️ 🌹

YIELD: 50 PIECES

U.S.	METRIC	INGREDIENTS
2 lb	900 g	Spinach
4 oz	100 g	Butter
4 oz	100 g	Onions, chopped fine
1 oz	30 g	Scallions, chopped fine
1 oz	30 g	Fresh dill weed, chopped
1 lb	450 g	Feta cheese, crumbled
to taste	to taste	Salt
to taste	to taste	Pepper
25 sheets	25 sheets	Phyllo dough (about 1 lb/450 g)
8 oz approximately	225 g approximately	Melted butter

PROCEDURE

1. Trim, wash, and steam or boil the spinach just until it is thoroughly wilted.
2. Drain, cool under cold running water, and squeeze dry. Chop fine.
3. Heat the butter in a sauté pan. Sauté the onions and scallions over low heat until soft.
4. Remove from the heat and add the spinach and dill. Mix to coat the spinach lightly with butter.
5. Mix in the cheese.
6. Season to taste with salt and pepper.
7. Thaw the phyllo if it is frozen. Unwrap and unfold the stack of sheets and cut them in half lengthwise. Keep them covered to prevent drying.
8. Taking one sheet at a time, brush each lightly with melted butter. Fold the sheet in half lengthwise and butter it again.
9. Put a small mound (about $1/2$–$2/3$ oz/15–20 g) of the spinach mixture toward the bottom of the strip and a little to one side, as shown in **Figure 23.4**.
10. Fold into triangular packets as indicated in the illustration.
11. Arrange the triangles on baking sheets with the loose ends of the phyllo on the bottom. Brush the tops with melted butter.
12. Bake at 375°F (190°C) until golden brown and crisp, 20–25 minutes.
13. Serve warm.

Per 1 piece: Calories, 110; Protein, 2 g; Fat, 8 g (69% cal.); Cholesterol, 25 mg; Carbohydrates, 6 g; Fiber, 1 g; Sodium, 210 mg.

VARIATION

Many other fillings can be baked wrapped in phyllo, as long as they are not too juicy. The following are some suggestions:

Diced ham, cheddar cheese, and prepared mustard

Gruyère cheese, blue cheese, walnuts

Feta and cream cheese (2 parts feta to 1 part cream cheese) mixed together, plus 1 egg yolk per 12 oz (350 g) cheese

Sautéed mushrooms and onions, crumbled bacon, parsley, parmesan cheese

Diced cooked chicken, mozzarella cheese, sun-dried tomatoes, basil

Crabmeat, sautéed shellfish, cream cheese, hot pepper sauce

Ratatouille (p. 329)

FIGURE 23.4 Folding phyllo dough triangles.

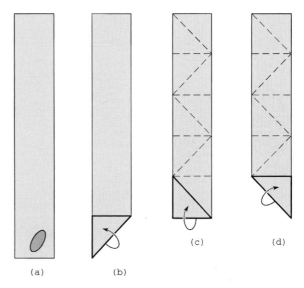

(a)   (b)   (c)   (d)

(**a**) Place the filling at the bottom of the strip of buttered phyllo dough and a little to one side.
(**b**) Fold the bottom corner over at a 45-degree angle to form a triangle.
(**c, d**) Continue folding the triangle as shown in the diagram.

Chicken Satay

YIELD: ABOUT 24 PIECES, 1 OZ (30 G) EACH

U.S.	METRIC	INGREDIENTS
		Marinade:
1 stalk	1 stalk	Lemongrass
1 tbsp	15 mL	Peanut butter
1¹/₂ oz	45 g	Shallots, chopped fine
1¹/₂ tsp	7 mL	Brown sugar
1 tsp	5 mL	Salt
¹/₂ tsp	2 mL	Ground cumin
¹/₄ tsp	1 mL	Ground coriander
¹/₄ tsp	1 mL	Cinnamon
¹/₄ tsp	1 mL	Turmeric
1 tbsp	15 mL	Oil
1¹/₂ lb	720 g	Chicken breasts, boneless, skinless, trimmed of all fat
1 pt, or as needed	500 mL, or as needed	Indonesian Peanut Sauce (p. 211)

PROCEDURE

1. Trim and discard the lower end of the lemongrass. Cut the tender bottom portion of the stalk into paper-thin slices. Discard the top.
2. Mix the lemongrass, peanut butter, shallots, sugar, salt, spices, and oil into a coarse paste. If necessary, add a few drops of water.
3. Cut the chicken breast lengthwise into thin strips. You should have about 24 pieces.
4. Mix the chicken with the marinade so all pieces are coated. Refrigerate 3 hours.
5. While the chicken is marinating, soak bamboo skewers in water at least 3 hours to keep them from burning during grilling.
6. Thread each chicken strip onto a skewer, leaving the top half of the skewer empty so it can be used as a handle.
7. Grill the chicken skewers about 2 minutes on each side, or until done.
8. Serve with the sauce for dipping.

Per 1 piece: Calories, 110; Protein, 9 g; Fat, 7 g (55% cal.); Cholesterol, 17.5 mg; Carbohydrates, 4 g; Fiber, 1 g; Sodium, 240 mg.

CHICKEN SATAY

Satay, or saté, is a snack food, often sold by street vendors, popular throughout Southeast Asia. Although it originated in Indonesia, where there are many dozens of versions, it can also be found in Malaysia, Singapore, Thailand, the Philippines, and neighboring countries. Satay consists of cubes or strips of meat or poultry that are marinated, threaded on skewers, and grilled. It is usually served with some kind of spicy sauce.

The marinade for the satay included here is inspired by the famous satays of Singapore.

Chicken Satay

Rumaki

YIELD: 60 PIECES

U.S.	METRIC	INGREDIENTS
15	15	Chicken livers
8 fl oz	250 mL	Soy sauce
1 clove	1 clove	Garlic, cut in half
3 slices	3 slices	Fresh ginger root
15 (approx.)	15 (approx.)	Water chestnuts
30 strips	30 strips	Bacon

PROCEDURE

1. Trim the chicken livers and cut them into quarters.
2. Marinate the livers overnight with the soy sauce, garlic, and ginger.
3. Cut each water chestnut into 3–5 pieces to make the same number as liver pieces.
4. Cut each slice of bacon in half.
5. Wrap each liver piece together with a water chestnut piece in a piece of bacon and secure with a pick.
6. Bake at 400°F (200°C) until the bacon is cooked, about 15 minutes. Brush with the marinade 2 or 3 times during baking.

Per 1 piece: Calories, 62; Protein, 2 g; Fat, 5 g (79% cal.); Cholesterol, 25 mg; Carbohydrates, 1 g; Fiber, 0 g; Sodium, 390 mg.

Smoked Trout Rollups

YIELD: 24 PIECES

U.S.	METRIC	INGREDIENTS	PROCEDURE
10 oz	300 g	Smoked trout fillets (p. 834), skinless	1. Carefully examine the trout fillets and remove all bones.
			2. Shred the trout, breaking up all lumps.
2 oz	30 g	Cream cheese	3. In a bowl, mix the cream cheese and first quantity of mayonnaise until softened and smooth.
3 tbsp	15 mL	Mayonnaise	
1 oz	30 g	Shallot, chopped fine	4. Add the trout and shallot. Mix well.
2 fl oz	75 mL	Mayonnaise	5. Mix the mayonnaise and mustard.
1 tsp	7 mL	Dijon-style mustard	6. Lay 1 crêpe on a work surface, good side down.
6	6	Crêpes (p. 750), 6 in. (15 cm) in diameter	7. Drop 2 tsp (10 mL) of the mayonnaise mixture on the crêpe and spread it over the crêpe to cover it completely in a thin layer.
5	5	Cornichons (see Note)	8. Portion $1\frac{1}{2}$ oz (45 g) of the trout mixture and spread it in a wide band across the center of the crêpe.
			9. Roll up the crêpe, pressing gently (to avoid tearing the crêpe) but firmly to make a tight roll.
			10. Repeat with the remaining crêpes.
			11. Slice the cornichons on the diagonal into thin slices.
			12. Trim the ends off the crêpe rolls, then cut the rolls crosswise into fourths.
			13. Push the point of a cocktail pick through a cornichon slice and stick the pick into the center of a rollup. Repeat with the remaining rollups.

Per 1 piece: Calories, 80; Protein, 3 g; Fat, 5 g (65% cal.); Cholesterol, 30 mg; Carbohydrates, 3 g; Fiber, 0 g; Sodium, 150 mg.

Note: Cornichons are tiny, sour cucumber pickles packed in vinegar, usually flavored with tarragon.

VARIATION

Smoked Trout Purses

Instead of making rollups, cut a circle from the center of a crêpe using a 4-in. (10-cm) cutter. Place about 2 tsp (10–12 g) of the trout mixture in the center of the circle. (Omit the mustard mixture.) Gather the edges of the circle over the top of the trout and tie with a length of chive. (You will need 30–36 crêpes for the quantity of trout in the basic recipe.)

Beef Teriyaki Skewers

YIELD: ABOUT 50 PIECES

U.S.	METRIC	INGREDIENTS	PROCEDURE
		Marinade:	1. Combine the sake, mirin, soy sauce, and sugar in a saucepan. Bring to a boil to dissolve the sugar.
$1\frac{1}{2}$ fl oz	45 mL	Sake	
3 fl oz	90 mL	Mirin (sweet Japanese rice wine)	2. Cool thoroughly, then chill.
$3\frac{1}{2}$ fl oz	105 mL	Soy sauce	
1 tbsp	15 mL	Sugar	
1 lb 8 oz	720 g	Beef tenderloin, or other tender beef cut, trimmed of all fat and silverskin	3. Cut the beef into $\frac{3}{4}$-in. (2-cm) cubes.
			4. Mix the beef with the marinade so all of it is coated. Refrigerate 3–4 hours.
			5. Impale each cube of meat on the end of a bamboo skewer.
			6. Grill or broil the meat to the desired doneness. Serve immediately.

Per 1 piece: Calories, 20; Protein, 2 g; Fat, 1 g (52% cal.); Cholesterol, 5 mg; Carbohydrates, 0 g; Fiber, 0 g; Sodium, 20 mg.

VARIATION

For more authentic Japanese teriyaki, do not marinate the beef. Instead, when the meat is about half cooked, begin basting frequently with the marinade to give it a light glaze.

Crêpes

YIELD: ABOUT 36 CRÊPES

U.S.	METRIC	INGREDIENTS	PROCEDURE
8 oz	250 g	Flour	1. Sift the flour and salt into a bowl.
1 tsp	5 g	Salt	2. Add the eggs to the flour and mix with a wire whip.
6	6	Eggs, lightly beaten	3. Gradually mix in the milk and the butter. The batter should be about the consistency of heavy cream. Strain to remove any lumps.
1 pt	500 mL	Milk	
1½ oz	50 g	Clarified butter or beurre noisette (p. 189)	4. Let the batter rest 2 hours before frying.
as needed	as needed	Oil for seasoning pan	5. Rub a 6- or 7-in. (15–18-cm) crêpe pan or skillet with oil (**Figure 23.5**). Heat the pan over moderately high heat until it is very hot.
as needed	as needed	Clarified butter for frying	6. Brush the pan lightly with clarified butter and pour off any excess.
			7. Remove the pan from the heat and pour in 3–4 tablespoons (45–60 mL) of the batter. Very quickly tilt the pan to cover the bottom with a thin layer. Immediately dump out any excess batter, as the crêpe must be very thin.
			8. Return to the heat for 1–1½ minutes, until the bottom is lightly browned. Flip over and brown the second side. The second side will brown in only a few spots and will not be as attractive as the first. The first side should always be the visible side when the crêpe is served.
			9. Slide the crêpe onto a plate. Repeat with the remaining batter. Butter the pan lightly when necessary.
			10. Cover the finished crêpes and refrigerate until needed.

Per 1 crêpe: Calories, 70; Protein, 2 g; Fat, 4 g (53% cal.); Cholesterol, 40 mg; Carbohydrates, 6 g; Fiber, 0 g; Sodium, 90 mg.

VARIATION

Dessert Crêpes

Sift 2 oz (60 g) sugar with the flour and salt.

FIGURE 23.5 Preparing crêpes.

(a) Brush the hot pan lightly with clarified butter.

(b) Ladle in 3–4 tablespoons (45–60 mL) of the batter. Very quickly tilt the pan to cover the bottom with a thin layer. Immediately dump out any excess batter.

(c) Cook until the crêpe is lightly browned on the bottom.

(d) Flip the crêpe over.

(e) Brown the other side.

Cucumber Cups with Dilled Shrimp Salad

YIELD: 24 PIECES

U.S.	METRIC	INGREDIENTS
2, or as needed (see step 2)	2, or as needed (see step 2)	Long, slender English, cucumbers about 6 oz (180 g) each
as needed	as needed	Salt
12 oz	360 g	Dilled Shrimp Salad (p. 699)
24 tiny sprigs	24 tiny sprigs	Fresh dill

PROCEDURE

1. Channel the cucumbers as shown in **Figure 10.9(b)**. Alternatively, peel the cucumbers with a vegetable peeler.
2. Trim off the ends of the cucumbers, then cut the cucumbers into slices $3/4$ in. (1 cm) thick. (You need enough cucumbers to make 24 slices. Use an additional cucumber if necessary.)
3. With a parisienne scoop, hollow out each slice to make a shallow cup.
4. Sprinkle the hollowed side of each slice lightly with salt, then turn upside down on a sheet pan lined with paper towels to drain. Refrigerate 1 hour.
5. Turn the cucumber slices hollow side up. Fill each with $1/2$ oz (15 g) shrimp salad.
6. Decorate the top of each with a tiny sprig of dill.
7. Refrigerate until service time.

Per 1 piece: Calories, 30; Protein, 2 g; Fat, 2.5 g (75% cal.); Cholesterol, 15 mg; Carbohydrates, 0 g; Fiber, 0 g; Sodium, 40 mg.

VARIATION

Cucumber Boats with Dilled Shrimp Salad

Peel the cucumbers, cut in half lengthwise, and scoop out the seeds. Cut crosswise at a slight angle into pieces about $3/4$ in. (2 cm) wide. Fill with shrimp salad as in basic recipe.

Cucumber Boats and Cups with Dilled Shrimp Salad

Belgian Endive with Herbed Chèvre 🌹

YIELD: 24 PIECES

U.S.	METRIC	INGREDIENTS
2–3 heads	2–3 heads	Belgian endive
10 oz	300 g	Fresh goat cheese
1 oz	30 g	Parsley, chopped
1 tbsp	15 mL	Fresh chervil, chopped
1 tbsp	15 mL	Chives, chopped
as needed	as needed	Water or heavy cream

PROCEDURE

1. Cut off the bases of the Belgian endive and separate the leaves. Select the 24 best and largest leaves. (Use the small inner leaves for salads or another purpose.)
2. Blend the cheese and herbs to a soft paste. If the paste is too stiff to force through a pastry bag, soften it by mixing in a little water or heavy cream.
3. Put the herbed cheese in a pastry bag fitted with a small star tip.
4. Pipe about 2 tsp (10 mL) of the cheese into the hollow side of each Belgian endive leaf at the base end.

Per 1 piece: Calories, 35; Protein, 2 g; Fat, 2.5 g (74% cal.); Cholesterol, 5 mg; Carbohydrates, 0 g; Fiber, 0 g; Sodium, 45 mg.

Belgian Endive with Herbed Chèvre

Chicken and Sweet Potato Fritters

YIELD: APPROXIMATELY 24 PIECES

U.S.	METRIC	INGREDIENTS	PROCEDURE
2	2	Egg yolks	1. Make the batter: Beat the egg yolks and water. Add the flour all at once and stir briefly to make a loose, rather lumpy batter. Do not overmix.
13 fl oz	390 mL	Water, ice cold	
6 oz	180 g	Flour, sifted	
1 lb	480 g	Chicken breast, boneless and skinless	2. Cut the chicken breast into thin strips about 1 in. (2.5 cm) long.
2 fl oz	60 mL	Soy sauce	3. Mix the chicken with the soy sauce and ginger. Marinate 20 minutes.
1 tsp	5 mL	Fresh ginger, peeled and grated	4. Peel and grate the sweet potatoes on a coarse grater. Hold in cold water until ready to cook.
8 oz	240 g	Sweet potatoes	5. Drain the chicken meat and dry on clean towels.
as needed for dusting	as needed for dusting	Flour	6. Drain the sweet potatoes and dry on clean towels. Mix with the chicken.

7. Dust the chicken mixture lightly with flour and toss to coat. Add about three-fourths of the batter and mix gently. If the mixture is dry, add the rest of the batter.

8. Heat a deep fryer to 350°F (175°C). Drop small spoonfuls of the chicken mixture into the fat and fry just until light golden brown. Remove from the fat and drain well.

9. Serve immediately. If a dipping sauce is desired, serve with soy sauce flavored with a little wasabi (Japanese horseradish). (Note: The illustration shows wasabi and grated radish for mixing into the soy sauce for dipping.)

Per 1 piece: Calories, 90; Protein, 6 g; Fat, 3.5 g (33% cal.); Cholesterol, 30 mg; Carbohydrates, 10 g; Fiber, 1 g; Sodium, 35 mg.

Chicken and Sweet Potato Fritters

Spiced Shrimp with Smoked Paprika

YIELD: 26–30 PIECES

U.S.	METRIC	INGREDIENTS
1 lb	480 g	Peeled, deveined shrimp, size 26–30
3 tbsp	15 mL	Lemon juice
1 1/2 tsp	7 mL	Olive oil
1/2 tsp	2 mL	Peppercorns, crushed
1 clove	1 clove	Garlic, crushed
1 oz	30 g	Onion, sliced
1	1	Bay leaf
1/4 tsp	1 mL	Cayenne
1 tbsp	15 mL	Salt
as needed	as needed	Water
1 fl oz	30 mL	Olive oil
1 tbsp	15 mL	Lemon juice
1 tbsp	15 mL	Smoked paprika

PROCEDURE

1. Combine the shrimp, lemon juice, oil, peppercorns, garlic, onion, bay leaf, cayenne, and salt in a stainless-steel saucepan. Add just enough water to cover the shrimp.

2. Bring to a boil, then reduce to a slow simmer. Cook 2–4 minutes, or until just cooked through.

3. Remove the pan from the heat and set in an ice bath to cool quickly. Refrigerate.

4. Mix the olive oil, lemon juice, and paprika in a bowl large enough to hold the shrimp.

5. Drain the shrimp and dry on clean towels.

6. Add the shrimp to the bowl with the olive oil mixture. Toss until all the shrimp are well coated with the mixture.

7. Arrange the shrimp on a platter and serve with cocktail picks. Alternatively, place on round toast cutouts and serve as canapés.

Per 1 piece: Calories, 30; Protein, 4 g; Fat, 1.5 g (46% cal.); Cholesterol, 30 mg; Carbohydrates, 0 g; Fiber, 0 g; Sodium, 55 mg.

Spiced Shrimp with Smoked Paprika

Mushrooms Stuffed with Tapenade

YIELD: 50 PIECES

U.S.	METRIC	INGREDIENTS
8 oz	250 g	Pitted Mediterranean or Greek black olives
1 oz	30 g	Capers, drained
1 oz	30 g	Anchovies, drained
1 oz	30 g	Tuna, drained
1 tsp	5 mL	Dijon-style mustard
2¹/₂ fl oz	75 mL	Olive oil
1 tsp	5 mL	Lemon juice
2 tbsp	30 mL	Chopped parsley
pinch	pinch	Dried thyme
to taste	to taste	Salt
to taste	to taste	Pepper
50	50	Small to medium fresh white mushrooms
as needed	as needed	Pimiento

PROCEDURE

1. In a blender or food processor, purée the olives, capers, anchovies, tuna, mustard, oil, lemon juice, and herbs. Season to taste.
2. Chill the mixture several hours to let the flavors blend.
3. Remove the stems from the mushrooms and clean the caps.
4. Fill each cap with a small spoonful of the tapenade.
5. Garnish the top of each with a dot of pimiento.

Per 1 piece: Calories, 30; Protein, 1 g; Fat, 25 g (74% cal.); Cholesterol, 0 mg; Carbohydrates, 1 g; Fiber, 0 g; Sodium, 110 mg.

Deviled Eggs

YIELD: 50 PIECES

U.S.	METRIC	INGREDIENTS
25	25	Hard-cooked eggs
¹/₂ pt	250 mL	Mayonnaise
1 fl oz	30 mL	Lemon juice or vinegar
1 tsp	5 mL	Dry mustard
1 tsp	5 mL	Worcestershire sauce
to taste	to taste	Salt
to taste	to taste	White pepper
to taste	to taste	Cayenne
as needed	as needed	Assorted garnish, such as:
		Chopped parsley
		Tiny dill sprigs
		Capers
		Diced pimiento
		Sliced stuffed olives
		Paprika
		Red or black caviar

PROCEDURE

1. Halve the eggs lengthwise and remove the yolks.
2. Mash the yolks or force them through a sieve.
3. Add the remaining ingredients (except the garnish) and mix to a smooth paste.
4. Using a pastry bag with a star tip, fill the egg white halves.
5. Using a variety of garnishes, decorate the top of each egg.

Per 1 piece: Calories, 70; Protein, 3 g; Fat, 7 g (84% cal.); Cholesterol, 110 mg; Carbohydrates, 0 g; Fiber, 0 g; Sodium, 45 mg.

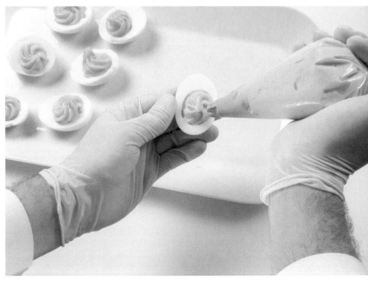

Filling deviled eggs with a pastry bag

VARIATIONS

Vary the flavor of stuffed eggs by adding any of the following ingredients to the egg yolk mixture in the basic recipe.

Anchovy: 2–3 oz (60–90 g) anchovy paste.

Curry: 2 tbsp (30 mL) curry powder, heated very gently with a little oil and cooled.

Blue cheese: 6 oz (175 g) mashed blue cheese.

Parmesan: 3 oz (90 g) grated parmesan cheese.

Tarragon: Use tarragon vinegar in the filling and add 2 tsp (10 mL) dried tarragon.

Tuna: 6 oz (175 g) well-mashed, drained tuna.

Miniature Gougère Puffs

YIELD: ABOUT 80 PIECES

U.S.	METRIC	INGREDIENTS
1 lb 4 oz	600 g	Éclair Paste (p. 1000)
4 oz	120 g	Gruyère cheese, grated
as needed	as needed	Egg wash

PROCEDURE

1. Mix the éclair paste and the cheese.
2. Using a pastry bag with a small, plain tip, pipe small mounds of about 1 tbsp (15 mL) each onto sheet pans lined with parchment. See page 1001 for guidelines on making tiny cream puff shells or profiteroles, which is essentially what these are.
3. Bake at 400°F (200°C) until puffed and brown, 20–30 minutes.
4. Make a little slit in the side of each to allow steam to escape. Put the puffs in a warm oven until they are dry.
5. Serve warm or at room temperature.

Per 1 piece: Calories, 25; Protein, 1 g; Fat, 1.5 g (63% cal.); Cholesterol, 15 mg; Carbohydrates, 1 g; Fiber, 0 g; Sodium, 25 mg.

Egg Rolls or Spring Rolls

YIELD: 16 PIECES

U.S.	METRIC	INGREDIENTS
1 oz	30 mL	Oil
6 oz	175 g	Cooked meat or poultry, cut fine julienne
6 oz	175 g	Chinese cabbage, shredded
3	3	Scallions, shredded
6 oz	175 g	Bean sprouts
5	5	Dried black mushrooms (stems discarded), soaked in boiling water, cut julienne
2 oz	60 g	Bamboo shoots, cut julienne
1¹/₂ oz	45 g	Raw shrimp, chopped
2 tsp	10 mL	Soy sauce
1 tsp	5 mL	Sherry or Shaoxing wine
3 fl oz	100 mL	Chicken stock
1¹/₂ tsp	7 mL	Cornstarch
1 tbsp	15 mL	Water, cold
16	16	Egg roll skins or spring roll skins
as needed	as needed	Egg, beaten

PROCEDURE

1. Heat the oil in a wok or sauté pan. Stir-fry the meat, cabbage, scallions, bean sprouts, black mushrooms, bamboo shoots, and shrimp.
2. Add the soy sauce, sherry, and stock. Continue to stir and cook for another 1–2 minutes.
3. Mix the cornstarch with the cold water. Stir into the vegetable mixture and cook until reduced and thickened.
4. Remove from the heat and adjust the seasonings. Cool thoroughly.
5. Lay an egg roll skin on the worktable with one of the corners toward you. Spoon about 1¹/₂ oz (45 g) filling onto the lower half of the skin in a sort of sausage shape. (See **Figure 23.6.**)
6. Fold the lower corner of the skin (the corner pointing at you) over the filling so it is covered. Then start to roll it up like a cylinder, giving it just a half turn.
7. Brush the left and right corners with a little beaten egg. Fold one corner over the filling, then the other, pressing down to seal. At this point, it should look like an open envelope.
8. Brush the top corner with beaten egg. Roll into a firm, compact cylinder. Seal the top corner well.
9. Repeat with the remaining skins and filling.
10. Deep-fry the egg rolls until the skins are crisp and brown. (Note: Egg roll skins are heavier than spring roll skins. They must be thoroughly fried or they will be doughy.) Drain and serve at once, with a little hot mustard or bottled duck sauce for dipping.

Per 1 roll: Calories, 180; Protein, 8 g; Fat, 8 g (38% cal.); Cholesterol, 25 mg; Carbohydrates, 21 g; Fiber, 1 g; Sodium, 240 mg.

FIGURE 23.6 Rolling egg rolls.

(a) Lay an egg roll skin on the bench with one corner toward you. Place the filling as shown.

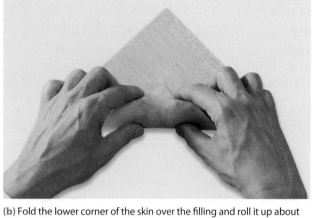

(b) Fold the lower corner of the skin over the filling and roll it up about one-third of the way.

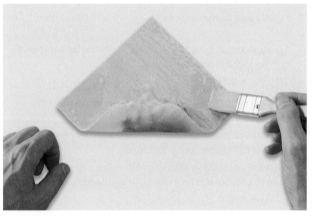

(c) Brush the left and right corners of the skin with beaten egg.

(d) Fold the corners over the filling so the skin resembles an open envelope.

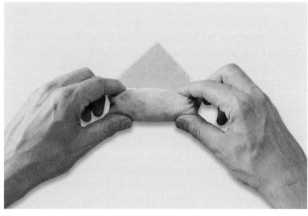

(e) Brush the top edge with egg and roll up tightly.

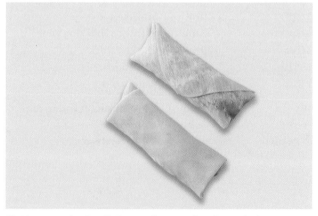

(f) When completely rolled, egg rolls are ready to be cooked.

Black Bean Quesadillas

YIELD: 48 PIECES

U.S.	METRIC	INGREDIENTS
12 oz	360 g	Black turtle beans, cooked or canned, drained
1–2 fl oz	30–60 mL	Water or bean cooking broth
to taste	to taste	Salt
to taste	to taste	Cayenne or ground chipotle pepper
12	12	Flour or corn tortillas
6 tbsp	90 mL	Salsa Cruda (p. 207)
1 tbsp	15 mL	Jalapeño or Serrano chile, seeded and chopped fine
2 tbsp	30 mL	Chopped cilantro
4 oz	120 g	Monterey jack or mild cheddar cheese, grated

PROCEDURE

1. Mash the beans coarsely.
2. Add just enough water or bean broth to moisten the beans slightly. They should not be dry.
3. Add salt and cayenne or chipotle to taste.
4. Top 1 tortilla with about 2 oz (60 g) bean mixture and spread it to cover the tortilla. Repeat this process until 6 tortillas are topped with the bean mixture.
5. For each of the 6 tortillas, sprinkle 1 tbsp (15 mL) salsa over the beans, then $1/2$ tsp (2.5 mL) chopped chile and 1 tsp (5 mL) cilantro.
6. Lightly oil a hot griddle or sauté pan and place the tortilla on it.
7. Sprinkle about $2/3$ oz (20 g) cheese over the filling in each tortilla and top with a second tortilla.
8. When the cheese is soft, flip the tortilla sandwich and heat the other side.
9. Remove from the heat and cut into 8 wedges.

Per 1 piece: Calories, 110; Protein, 4 g; Fat, 4 g (30% cal.); Cholesterol, 0 mg; Carbohydrates, 17 g; Fiber, 2.4 g; Sodium, 210 mg.

Chicken Liver Pâté

YIELD: 2$1/2$ LB (1.2 KG)

U.S.	METRIC	INGREDIENTS
2 lb	1 kg	Chicken livers
to taste	to taste	Salt
as needed	as needed	Milk
4 oz	125 g	Onion, chopped
3 oz	90 g	Butter
$1/2$ tsp	2 mL	Dried oregano
$1/4$ tsp	1 mL	White pepper
pinch	pinch	Nutmeg
pinch	pinch	Ground ginger
pinch	pinch	Ground cloves
1 tsp	5 mL	Salt
12 oz	375 g	Cream cheese
1–2 fl oz	30–60 mL	Brandy, Madeira, or port (optional)
to taste	to taste	Salt

PROCEDURE

1. Trim fat and sinews from the livers.
2. Sprinkle the livers lightly with salt. Add milk to cover and let stand overnight, refrigerated. This step is optional, but it results in a slightly milder flavor and lighter color.
3. Sauté the onion lightly in the butter until tender but not brown.
4. Add the livers (drained and rinsed), herbs, and spices. Brown the livers lightly and cook them until they are still slightly pink in the center. Remove from the heat and cool.
5. Grind the livers and onions in a grinder or buffalo chopper.
6. Add cream cheese and continue to process to obtain a uniformly mixed paste.
7. Add brandy or wine to taste. Add more salt if necessary.
8. Pack the mixture into containers and chill overnight.

Per 1 ounce (28.35 g): Calories, 60; Protein, 3 g; Fat, 5 g (74% cal.); Cholesterol, 75 mg; Carbohydrates, 1 g; Fiber, 0 g; Sodium, 105 mg.

Bruschette (Basic)

YIELD: 24 PIECES

U.S.	METRIC	INGREDIENTS
24 slices	24 slices	Italian or French bread, 3–4 in. (8–10 cm) wide and ¹/₂ in. (6 mm) thick
4–6	4–6	Garlic cloves, cut in half
8–12 fl oz	240–360 mL	Extra-virgin olive oil
to taste	to taste	Salt
to taste	to taste	Pepper

PROCEDURE

1. Toast the bread on a grill or under a broiler until golden brown.
2. While the bread is still hot, rub one side of each slice with a cut clove of garlic.
3. Brush or drizzle olive oil generously over each slice. The toast should be partly soaked with the oil.
4. Add salt and pepper to taste.
5. Serve warm.

Per 1 piece: Calories, 140; Protein, 2 g; Fat, 10 g (65% cal.); Cholesterol, 0 mg; Carbohydrates, 10 g; Fiber, 1 g; Sodium, 115 mg.

VARIATIONS

Bruschette with Tomato and Basil

Toast the bread and rub with garlic as in the basic recipe. Cut 2 lb (1 kg) ripe plum tomatoes in half and remove the seeds. Cut into medium dice. Tear 16 fresh basil leaves into small pieces and mix with the tomatoes. Top the toasts with the tomatoes. Drizzle with olive oil and add salt and pepper to taste.

Bruschette with Parmesan Cheese

Toast the bread and rub with garlic as in the basic recipe. Using a vegetable peeler, cut cheese, preferably a good-quality Parmigiano-Reggiano, into thin shavings. While the toast is still hot, top each piece with about ¹/₄ oz (7 g) cheese and drizzle with olive oil. Sprinkle with black pepper.

Bruschette with White Beans and Prosciutto

Toast the bread and rub with garlic as in the basic recipe. Mix 1 lb 4 oz (600 g) cooked or canned and drained white kidney beans, 2 oz (60 g) red onion cut in small dice, 6 oz (180 g) prosciutto, sliced thin and chopped, and 4 fl oz (120 mL) olive oil. Add salt and pepper to taste. Top toasts with this mixture.

Bruschette, clockwise from left: Bruschetta with White Beans and Prosciutto, Bruschetta with Parmesan Cheese, Bruschetta with Tomato and Basil

Vegetable Samosas

YIELD: 48 PIECES

U.S.	METRIC	INGREDIENTS	PROCEDURE
6 oz	180 g	Bread flour	1. In the bowl of a mixer fitted with the paddle attachment, combine the two flours and the oil. Mix until well blended.
4 oz	120 g	Whole wheat flour	
3 tbsp	45 mL	Vegetable oil	2. Add the water. Mix at low speed until a dough forms.
6 fl oz	180 mL	Water	3. Remove the dough from the mixer and knead it until smooth.
			4. Rub the dough lightly with oil to keep a skin from forming, wrap in plastic, and let rest 30 minutes.
			5. Divide the dough into 24 equal pieces, about $^2/_3$ oz (20 g) each. Roll each piece into a ball.
1 lb 2 oz	500 g	Potato Filling for Samosas (p. 760), cooled	6. Roll out each ball of dough into a circle 4 in. (10 cm) in diameter. Cut the circles in half to make two semicircles.
			7. Moisten one half of the cut edge of a semicircle with water, then shape the dough into a cone. Move the moistened edge to overlap the other half of the cut edge and press together to seal.
			8. Fill the cone with about $^1/_3$ oz (10 g) potato filling.
			9. Moisten the inside of the open end of the filled cone with water, then press the edges of the dough together in a flat seam. Seal the seam by pressing with a fork.
			10. Repeat with the remaining dough and filling.
			11. Deep-fry at 350°F (175°C) until browned.
			12. Serve warm. Offer them plain or with Cucumber Raita (p. 209) as a dip.

Per 1 piece: Calories, 50; Protein, 1 g; Fat, 2.5 g (41% cal.); Cholesterol, 0 mg; Carbohydrates, 7 g; Fiber, 0 g; Sodium, 20 mg.

Note: These instructions are for small, hors d'oeuvre-size samosas. For larger samosas, divide the dough into 12 pieces, about $1^1/_3$ oz (40 g) each, and make the circles $5^1/_2$ in. (14 cm) across. Increase the amount of filling.

VARIATION

Meat Samosas

Instead of the potato filling, fill the samosas with Meat Filling for Samosas.

Vegetable Samosas

Potato Filling for Samosas

YIELD: 1 LB 2 OZ (500 G)

U.S.	METRIC	INGREDIENTS
1 lb	480 g	Boiled potatoes, cooled and peeled
1 fl oz	30 mL	Oil
1/2 tsp	2 mL	Cumin seeds
3 oz	90 g	Onion, small dice
1/4 tsp	1 mL	Turmeric
1/2 tsp	2 mL	Salt
1/8 tsp	0.5 mL	Pepper
1/4 tsp	1 mL	Ground coriander
1/8 tsp	0.5 mL	Ground cardamom
pinch	pinch	Cinnamon
pinch	pinch	Ground cloves
1/8 tsp	0.5 mL	Cayenne
1 tbsp	30 mL	Lemon juice

PROCEDURE

1. Mash the potatoes coarsely. They should be slightly lumpy, but the lumps should be small so the filling fits easily into samosas.
2. Heat the oil in a sauté pan over moderate heat. When the oil is hot, add the cumin seeds. Cook just until the seeds darken slightly and become aromatic.
3. Add the onion and turmeric. Sauté until the onion is soft.
4. Add the potatoes and all the remaining ingredients. Cook, stirring often, until the potatoes are hot and all the seasonings are well blended.
5. Taste and add more salt if necessary.
6. Cool before using to fill samosas. (The potatoes can also be served as a side dish, in which case they should be served hot.)

Per serving: Calories, 40; Protein, 1 g; Fat, 1.5 g (33% cal.); Cholesterol, 0 mg; Carbohydrates, 6 g; Fiber, 1 g; Sodium, 55 mg.

Spicy Meat Empanadas

YIELD: 24 PIECES

U.S.	METRIC	INGREDIENTS
8 oz	240 g	Bread flour
4 oz	120 g	Cake flour
1 1/2 oz	45 g	Lard or vegetable shortening
3/4 tsp	3 mL	Salt
6 fl oz	180 mL	Warm water
18 oz	480 g	Picadillo (p. 510)

PROCEDURE

1. Sift the two flours together into a mixing bowl.
2. Add the lard or shortening and blend it thoroughly with the flour.
3. Dissolve the salt in the water and add it to the flour mixture. Mix to make a moderately stiff dough. Knead the dough for a few minutes, until smooth.
4. Let the dough rest 30 minutes.
5. Divide the dough into 24 equal parts by rolling it into a long cylinder and cutting it into equal segments.
6. Roll each piece into a ball. Cover and let rest 15–30 minutes.
7. For each empanada, flatten a ball of dough and roll out into a circle 4 in. (10 cm) in diameter.
8. Brush the rim of the circle lightly with water.
9. Place about 3/4 oz (20 g) picadillo on one side of the circle. Fold the other side over it to make a turnover. Press the dough around the filling to force out air. Seal the edge well by crimping with the tines of a fork.
10. Deep-fry at 375°F (190°C) until golden brown.
11. Serve immediately. If desired, accompany with Guacamole (p. 743) or Salsa Cruda (p. 207).

Per 1 piece: Calories, 130; Protein, 6 g; Fat, 6 g (43% cal.); Cholesterol, 15 mg; Carbohydrates, 12 g; Fiber, 1 g; Sodium, 140 mg.

VARIATION

Empanadas with Roasted Poblanos and Cheese

Roast and peel poblano peppers as shown on page 290. Dice the poblanos. In place of the picadillo, fill the empanadas with poblanos and grated Monterey jack or mild cheddar cheese.

Cauliflower Pakoras

YIELD: 30–40 PIECES

U.S.	METRIC	INGREDIENTS	PROCEDURE
8 oz	240 g	Chickpea flour	1. Sift together into a bowl the chickpea flour, pepper, cayenne, cumin, turmeric, baking powder, and salt.
1/4 tsp	1 mL	Black pepper	
1/4 tsp	1 mL	Cayenne	
1/2 tsp	2 mL	Ground cumin	2. Gradually stir in the water until the mixture forms a thick batter about the consistency of ketchup. You may not need all the water.
1/4 tsp	1 mL	Turmeric	
1/2 tsp	2 mL	Baking powder	
1/2 tsp	2 mL	Salt	
12 fl oz	360 mL	Water	
1 1/2 lb	720 g	Cauliflower, trimmed	3. Divide the cauliflower into florets no bigger than 1 in. (2.5 cm) across.
			4. Dip the florets into the batter and deep-fry at 360°F (180°F) until lightly browned and tender.

Per 1 piece: Calories, 50; Protein, 2 g; Fat, 3 g (49% cal.); Cholesterol, 0 mg; Carbohydrates, 5 g; Fiber, 1 g; Sodium, 45 mg.

VARIATIONS

Other vegetables, such as peppers, onion rings, potatoes (thin slices), broccoli, eggplant, and okra, may be substituted for the cauliflower.

Cauliflower Pakoras

ADDITIONAL RECIPES

These additional recipes may be found on your CulinarE-Companion recipe management program:

Brandade de Morue; Dates Stuffed with Gorgonzola; Meat Filling for Samosas; Profiteroles with Ham Salad or Deviled Ham; Prosciutto and Melon Balls.

TERMS FOR REVIEW

appetizer	**canapé**	**dip**	**caviar**
hors d'oeuvre	**cocktail**	**antipasto**	**amuse bouche**
butler-style service	**relish**	**bruschetta**	
buffet-style service	**crudité**	**tapa**	

QUESTIONS FOR DISCUSSION

1. What is butler-style service? Describe at least five guidelines to keep in mind when planning butler-style service of hors d'oeuvres.

2. How can you avoid soggy canapé bases?

3. In order to use a ham salad recipe to make a ham spread, you may need to change the recipe slightly. What are three modifications you may need to make?

4. What is the difference between a fruit salad and a fruit cocktail?

5. What is the proper consistency or thickness for a dip?

6. List six or more items that might appear on a cold antipasto platter.

7. Describe how to prepare the simplest type of bruschetta.

8. Describe the best way to serve the best-quality caviar.

24

BREAKFAST PREPARATION

When we speak of breakfast cookery, we are not just talking about a particular meal. We are referring to a particular small group of foods that appears on perhaps every breakfast menu. These items not only appear on breakfast menus but also are popular for brunches, snacks, and late suppers. Many establishments offer a breakfast menu all day long.

Eggs, of course, are the most popular breakfast food, and they are the primary subject of this chapter. In addition, the chapter examines the preparation of other breakfast staples: pancakes, waffles, French toast, and breakfast meats.

AFTER READING THIS CHAPTER, YOU SHOULD BE ABLE TO

1. Describe the composition of eggs and the major differences among grades.

2. Store eggs properly.

3. Prepare the following egg items: hard-, medium-, and soft-cooked eggs; poached eggs; fried eggs; baked eggs; scrambled eggs; omelets; entrée soufflés; and savory custards.

4. List the key differences between waffle batter and pancake batter, and prepare each.

5. Prepare French toast, and identify the common variations possible by changing the basic ingredients.

6. Prepare each of the two general types of breakfast cereal.

7. Identify the three most common breakfast meats and prepare them.

EGGS

Contrary to popular opinion, there is no law that says one must have eggs or cereal or pancakes or pastries for breakfast and must not have shrimp curry or chili or spaghetti and meatballs. Although most of us would think these last suggestions rather strange for the morning meal, there is probably no food that someone, somewhere, does not enjoy for breakfast. No doubt many Japanese, who have soybean soup, sour pickles, and rice for their first meal of the day, think Western breakfast habits are strange.

However, the egg remains a favorite breakfast food, even as we become more adventurous and explore ethnic cuisines. For such apparently simple items, eggs are used in many ways in the kitchen and require special study. We examine not only the usual breakfast preparation but other egg dishes as well, such as soufflés and custards.

UNDERSTANDING EGGS

COMPOSITION

A whole egg consists primarily of a yolk, a white, and a shell. In addition, it contains a membrane that lines the shell and forms an air cell at the large end, and two white strands called chalazae that hold the yolk centered. Figure 24.1 is a cross-sectional diagram that shows the location of these features.

1. The **yolk** is high in both fat and protein, and it contains iron and several vitamins. Its color ranges from light to dark yellow, depending on the diet of the chicken.

2. The **white** is primarily albumin protein, which is clear and soluble when raw but white and firm when coagulated. The white also contains sulfur.
   The white has two parts: a thick portion that surrounds the yolk and a thinner, more liquid portion outside of this.

3. The **shell** is not the perfect package, in spite of what you may have heard. Not only is it fragile but it is also porous, allowing odors and flavors to be absorbed by the egg and allowing the egg to lose moisture even if unbroken.

BROWN EGGS

Many consumers like to buy brown-shelled eggs, even pay more for them, because they think they are more nutritious, more flavorful, or somehow more natural. In fact, shell color is determined by the breed of the hen and has no relation to the flavor, purity, or nutritional value of the egg.

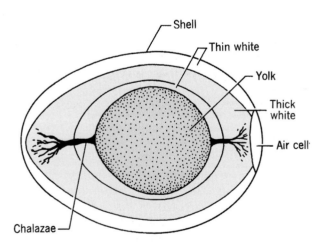

FIGURE 24.1 The parts of an egg. The diagram shows, in simplified form, the location of the parts of an unbroken egg, as described in the text.

GRADES AND QUALITY

Grades

In Canada, eggs are graded for quality at grading stations registered with the Canadian Food Inspection Agency (CFIA). The four grades are A, B, C, and Canada Nest Run.

The best grade (A) has a firm yolk and white that stand up high when broken onto a flat surface and do not spread over a large area. In the shell, the yolk is well centered, and the air sac is small.

As eggs age, they lose density. The thin part of the white becomes larger, and the egg spreads over a larger area when broken. Also, the air sac becomes larger as the egg loses moisture through the shell. Figure 24.3 shows the differences among grades A, B, and C.

In Canada, only grade A eggs are sold for retail use. Canada B eggs are sold for commercial baking, with their grade clearly marked. Canada C and Canada Nest Run are used for commercial processing. Grading is indicated by a maple leaf stamp (Figure 24.2).

FIGURE 24.2 Canada grade stamp for eggs.

Courtesy of the Canadian Food Inspection Agency.

FIGURE 24.3 Egg grades.
Courtesy of the USDA.

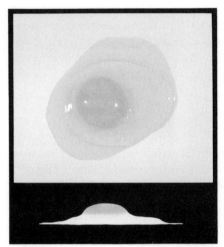

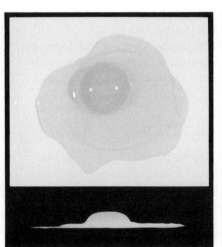

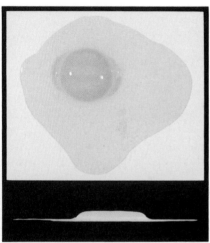

(a) Grade A.

(b) Grade B.

(c) Grade C eggs, as seen from the top and side. Note how the white and yolk lose thickness and spread more in the lower grades.

Maintaining Quality

Proper storage is essential for maintaining quality. Eggs keep for weeks if held at 36°F (2°C) but lose quality quickly if held at room temperature. In fact, they can lose a full grade in one day at warm kitchen temperatures. There's no point in paying for Grade A eggs if they are Grade B or C by the time you use them.

Store eggs away from foods that might pass on undesirable flavors or odors.

Grades and Use

One glance at Figure 24.3 will show you why Grade A is the best to use for fried or poached eggs. Lower grades spread too much to produce a high-quality product.

For hard-cooked eggs, use Grade A eggs that have been held a few days in the refrigerator. Very fresh eggs are difficult to peel when cooked in the shell.

Grade B eggs are suitable for use in baking. If you are certain they have developed no strong flavors, they may be used for scrambled eggs, where the firmness of the whole egg is less important.

SIZE

Eggs are also graded by size. Table 24.1 gives the minimum weight per egg (including shell) according to size category.

Most food-service operations use large eggs, and recipes in most books are based on this size.

TABLE 24.1 Egg Size Classifications

Size	Minimum Weight per Egg	
	Canada	U.S.
Jumbo	70 g or more	2$\frac{1}{2}$ oz or more
Extra-large	64 g	2$\frac{1}{2}$ oz
Large	56 g	2 oz
Medium	49 g	1$\frac{1}{2}$ oz
Small	42 g	1$\frac{1}{4}$ oz
Peewee	less than 42 g	less than 1$\frac{1}{4}$ oz

MARKET FORMS

1. **Fresh eggs or shell eggs.**
   These are most often used for breakfast cookery and are the main subject of this section.

2. **Frozen eggs.**
   - Whole eggs
   - Whites
   - Yolks
   - Whole eggs with extra yolks

   Frozen eggs are usually made from high-quality fresh eggs and are excellent for use in scrambled eggs, omelets, French toast, and in baking. They are pasteurized and are usually purchased in 30-pound (13.6-kg) cans. These take at least 2 days to thaw at refrigerator temperatures.

3. **Dried eggs.**
   - Whole eggs
   - Yolks
   - Whites

   Dried eggs are used primarily for baking. They are not suggested for use in breakfast cookery.
   Unlike most dehydrated products, dried eggs are not shelf-stable and must be kept refrigerated or frozen, tightly sealed.

SANITATION

In recent years, cases of salmonella food poisoning have been caused by raw or undercooked eggs. As a result, cooks have been made more aware of egg-related sanitation concerns. Pasteurized egg products are used in more operations. For a more detailed discussion of eggs and food safety, see Appendix 5, page 1030.

EGG SUBSTITUTES

Egg yolks, in addition to being high in fat, are also high in cholesterol. Efforts to reduce cholesterol in the diet have led to the development of commercial egg substitutes. These are of two types:

1. Egg substitutes that can be used to make such dishes as scrambled eggs, omelets, and custards are made of pasteurized egg whites with the addition of a blend of ingredients to substitute for the yolks, such as vegetable oil, milk solids, vegetable gums, salt, emulsifiers, and vitamin additives. They are sold in bulk liquid form, usually frozen, and can be substituted, ounce for ounce, for whole liquid eggs in most egg preparations.

2. Eggless egg substitutes contain no egg product. They are made of flours or other starches, plus vegetable gums and stabilizers, and, sometimes, soy protein. They are intended for use in baked goods only and are not suitable for use in breakfast egg preparations or custards. If they contain no milk products (read ingredient lists on individual products), they may be used in vegan diets.

GENERAL COOKING PRINCIPLES

The most important rule of egg cookery is simple: *Avoid high temperatures and long cooking times*. In other words, do not overcook. This should be a familiar rule by now.
Overcooking produces tough eggs, causes discoloration, and affects flavor.

Coagulation

Eggs are largely protein, so the principle of coagulation (p. 108) is important to consider.

Eggs coagulate at the following temperatures:

Whole eggs, beaten	about 156°F (69°C)
Whites	140°–149°F (60°–65°C)
Yolks	144°–158°F (62°–70°C)
Custard (whole eggs plus liquid)	175°–185°F (79°–85°C)

Note that whites coagulate or cook before yolks do. This is why it is possible to cook eggs with firm whites but soft yolks.

Note also that when eggs are mixed with a liquid, they become firm at a higher temperature. However, 185°F (85°C) is still much lower than the temperature of a sauté pan or skillet over high heat. As the temperature of coagulation is reached, the eggs change from semiliquid to solid, and they become opaque. If their temperature continues to rise, they become even firmer. ***An overcooked egg is tough and rubbery***. Low temperatures produce the best-cooked eggs.

If egg-liquid mixtures such as custards and scrambled eggs are overcooked, the egg solids separate from the liquids, or **curdle**. This is often seen as tough, watery scrambled eggs.

Sulfur

The familiar green ring you often see in hard-cooked eggs is caused by cooking at high temperatures or cooking too long. The same green color appears in scrambled eggs that are overcooked or held too long in the steam table.

This ring results when the sulfur in the egg whites reacts with the iron in the yolk to form iron sulfide, a compound that has a green color and a strong odor and flavor. The best way to avoid green eggs is to use ***low temperatures and short cooking and holding times***.

Foams

Whipped egg whites are used to give lightness and rising power to soufflés, puffy omelets, cakes, some pancakes and waffles, and other products. The following guidelines will help you handle beaten egg whites properly (see Figure 24.4).

1. **Fat inhibits foaming.**
   When separating eggs, be careful not to get any yolk in the whites. Yolks contain fats. Use very clean equipment when beating whites.

2. **Mild acids help foaming.**
   A small amount of lemon juice or cream of tartar gives more volume and stability to beaten egg whites. Use about 2 teaspoons cream of tartar per pound of egg whites (20 mL per kg).

3. **Egg whites foam better at room temperature.**
   Remove them from the cooler 1 hour before beating.

4. **Do not overbeat.**
   Beaten egg whites should look moist and shiny. Overbeaten eggs look dry and curdled and have lost much of their ability to raise soufflés and cakes.

5. **Sugar makes foams more stable.**
   When making sweet puffed omelets and dessert soufflés, add some of the sugar to the partially beaten whites and continue to beat to proper stiffness. (This will take longer than when no sugar is added.) The soufflé will be more stable before and after baking.

FIGURE 24.4 Whipping egg whites.

(a) The whites are just beginning to whip into a foam.

(b) The whites have reached the soft-peak stage.

(c) The whites have reached the firm-peak stage. Note the smooth texture. Whipping beyond this stage will cause the foam to break.

KEY POINTS TO REVIEW

- What are the three components of a whole shell egg? Describe each of these components.

- How are eggs graded for size and quality? What are the grades?

- How should egg products be stored?

- What are five guidelines to keep in mind when whipping egg white foams?

COOKING EGGS

SIMMERING IN THE SHELL

The term *hard-boiled egg* is not a good one to use because eggs should be simmered instead of boiled.

Eggs may be simmered in water to the soft-, medium-, or hard-cooked stage according to the following methods.

Standards of Quality for Eggs Simmered in the Shell

1. Cooked to the proper degree of hardness or softness. The yolk should be still liquid for soft-cooked eggs, soft but no longer liquid for medium-cooked eggs, and fully set with a uniform color throughout for hard-cooked eggs. The whites should be completely set, even for soft-cooked eggs, but not rubbery.

2. No green ring around the yolk.

PROCEDURES for Simmering Eggs in the Shell

METHOD 1

1. Collect equipment and food items.

2. Bring eggs to room temperature by (a) removing them from the cooler 1 hour before cooking, or (b) placing them in warm water for 5 minutes and draining. Cold eggs are more likely to crack when placed in boiling water.

3. Place eggs in boiling water and return the water to a simmer.

4. Simmer, do not boil, for the required time:

Soft-cooked	3–4 minutes
Medium-cooked	5–7 minutes
Hard-cooked	12–13 minutes

   Exact cooking time depends on temperature of eggs, size of eggs, and amount of water used.

5. Drain immediately and cool under cold running water to stop the cooking. Cool just a few seconds if eggs are to be served hot. Cool further if they are to be held for later use.

6. To peel, crack the shell and pull it away, starting at the large end (where the air sac is located). For easier peeling, peel while still warm, and hold under running water to help loosen shell. Very fresh eggs are hard to peel. Eggs for cooking in the shell should be several days old.

METHOD 2

1. Collect equipment and food items.

2. Place eggs in saucepan and cover with cold water.

3. Bring water to a boil.

4. Reduce heat and simmer for the required time:

Soft-cooked	1 minute
Medium-cooked	3–5 minutes
Hard-cooked	8–9 minutes

METHOD 3; FOR HARD-COOKED EGGS ONLY

Proceed as in Method 2, but remove pan from heat and cover as soon as it comes to a boil. Let stand off heat 20 minutes.

POACHING

The principles of cooking eggs in the shell are applicable to poached eggs. The only difference between the two items is the shell.

This difference, of course, complicates the cooking process, as emphasized in the following procedure. The object is to keep the eggs egg-shaped—that is, in a round, compact mass rather than spread all over the pan.

Standards of Quality for Poached Eggs

1. Bright, shiny appearance.

2. Compact, round shape, not spread or flattened.

3. Firm but tender whites; warm, liquid yolks.

PROCEDURE for Poaching Eggs

1. Collect equipment and food items.

2. Use the freshest Grade A eggs whenever possible for best results. These maintain their shape best because the yolks and whites are firm.

3. If eggs are not very fresh, add 1 teaspoon salt and 1 tablespoon distilled vinegar per quart of water (5 mL salt and 15 mL vinegar per L). The vinegar helps coagulate the egg white faster so it keeps a better shape.

Vinegar is not necessary if very fresh eggs are used. Omit in this case because whites will be tougher and not as shiny if cooked with vinegar.

4. Bring water to a simmer.

If water is boiling, eggs will toughen and may be broken up by the agitation.

If water is not hot enough, eggs will not cook quickly enough and will spread.

5. Break eggs, one at a time, into a dish or a small plate and slide into the simmering water. Eggs will hold their shape better if they slide in against the edge of the pan.

6. Simmer 3–5 minutes, until whites are coagulated but yolks are still soft.

7. Remove eggs from pan with slotted spoon or skimmer.

8. To serve immediately, drain very well. For better appearance, trim ragged edges.

9. To hold for later service, plunge immediately into ice water to stop the cooking. At service time, reheat briefly in hot water.

 # Eggs Benedict

YIELD: 1 PORTION (SEE NOTE)

U.S.	METRIC	INGREDIENTS
¹/₂	¹/₂	English muffin
as needed	as needed	Butter
1	1	Egg, fresh Grade A
1 slice	1 slice	Back bacon or ham, cooked (about 2 oz/60 g per slice)
1¹/₂ fl oz	50 mL	Hollandaise Sauce (p. 194)

PROCEDURE

1. Toast the muffin half. Spread it with butter and place on a serving plate.

2. Poach the egg according to the basic procedure given in this section.

3. While the egg is poaching, heat the back bacon or ham for 1 minute on a hot griddle or in a sauté pan. Place the meat on the toasted muffin.

4. Drain the poached egg well and place it on the back bacon.

5. Ladle hollandaise over the top. Serve immediately.

Per serving: Calories, 660; Protein, 19 g; Fat, 58 g (79% cal.); Cholesterol, 480 mg; Carbohydrates, 15 g; Fiber, 1 g; Sodium, 1260 mg.

Note: To prepare Eggs Benedict in quantity, the eggs may be poached ahead of time, cooled in ice water, and refrigerated. At service time, reheat the eggs in simmering water 30–60 seconds. Drain, plate, and serve.

VARIATIONS

Eggs Florentine

Instead of the muffin and bacon, place the egg on a bed of hot, buttered cooked spinach (about 2 oz/60 g). Cover with Mornay sauce instead of hollandaise. Optional: Sprinkle with parmesan cheese and brown under the salamander or broiler.

Eggs Bombay

Instead of the muffin and bacon, place the egg on a bed of hot rice pilaf (about 2 oz/60 g). Cover with curry sauce instead of hollandaise.

FRYING

Fried eggs are an especially popular breakfast preparation. They should always be cooked to order and served immediately. For best quality, observe each step in the following procedure.

The choice of cooking fat is a matter of taste and budget. Butter has the best flavor, but margarine or oil may be used. Use bacon fat only if that flavor is desired by the customer.

PROCEDURE for Frying Eggs to Order

1. Collect all equipment and food items.

   Eggs may be fried in small, individual sauté pans (omelet pans) or on the griddle. Griddled eggs are not as attractive because they tend to spread more. See page 773 for the procedure for conditioning sauté pans to avoid sticking.

2. Select very fresh Grade A eggs for best results.

3. Add about ⅛ inch (2 mm) fat to the sauté pan and set it over moderate heat, or preheat the griddle to 325°F (165°C) and ladle on a small quantity of fat. Too much fat will make the eggs greasy. Not enough will cause them to stick, unless a pan with a nonstick coating is used.

4. Break the eggs into a dish. This lessens the chance of breaking the yolks.

5. When the fat is hot enough so a drop of water sizzles when dropped into it, slide the eggs into the pan (or onto the griddle).

   If the fat is not hot enough, the eggs will spread too much and may stick. If it is too hot, the eggs will become tough or even crisp.

6. Reduce heat to low (if using sauté pan) and cook the eggs to order as indicated below. See Figures 24.5 and 24.6 for flipping and turning techniques.

FIGURE 24.5 Flipping eggs in a pan.

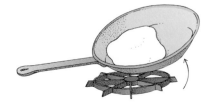

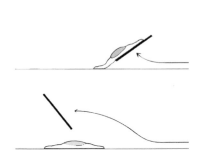

(a) Lift the handle of the pan and slide the eggs to the far edge with a quick jerk.

(b) With a quick flip of the wrist, as shown by the arrow, turn the eggs over. Do not flip the eggs too hard, or the yolks may break when they land.

FIGURE 24.6 When frying eggs on a griddle, turn them with one smooth motion of the spatula, as shown. The left corner of the egg never actually leaves the surface of the griddle.

- **Sunny side up**. Cook slowly without flipping until white is completely set but yolk is still soft and yellow. Heat must be low, or bottom will toughen or burn before top is completely set.

- **Basted**. Do not flip. Add a few drops of water to pan and cover so steam cooks the top. A thin film of coagulated white will cover the yolk, which should remain liquid. Note: This preparation is sometimes called country style. The term basted is used because the same effect may be achieved by spooning hot fat over the egg as it fries. This method may make the eggs excessively greasy, however.

- **Over easy**. Fry and flip over. Cook just until the white is just set but the yolk is still liquid.

- **Over medium**. Fry and flip over. Cook until the yolk is partially set.

- **Over hard**. Fry and flip over. Cook until the yolk is completely set.

Standards of Quality for Fried Eggs

1. White should be shiny, uniformly set, and tender, not browned, blistered, or crisp at edges.

2. Yolk should be unbroken (unless requested by customer) and set properly according to desired doneness. Sunny-side-up yolks should be yellow and well rounded. In other styles, the yolk is covered with a thin layer of coagulated white.

3. Relatively compact, standing high. Not spread out and thin.

KEY POINTS TO REVIEW

- What are the procedures for simmering eggs in the shell?

- What are the steps in the procedure for poaching eggs?

- What are the steps in the procedure for frying eggs to order?

Huevos Rancheros

YIELD: 1 PORTION

U.S.	METRIC	INGREDIENTS	PROCEDURE
as needed	as needed	Vegetable oil	1. Heat a thin layer of oil in a sauté pan.
1	1	Corn tortilla	2. Fry the tortilla briefly in the oil, turning it once, until softened.
2	2	Eggs	3. Remove from the pan and drain on paper towels.
4 fl oz	125 mL	Salsa Roja (p. 206)	4. Fry the eggs sunny side up or basted, following the procedure on page 770.
1 oz	30 g	Monterey jack or fresh Mexican white cheese, grated or crumbled	5. Place the tortilla on a warm dinner plate.
			6. Top with the eggs.
			7. Ladle the sauce over the whites of the eggs, leaving the yolks uncovered.
			8. Top with the grated cheese.
			9. If desired, add a portion of Frijoles Refritos (p. 372) to the plate.

Per serving: Calories, 400; Protein, 22 g; Fat, 27 g (60% cal.); Cholesterol, 450 mg; Carbohydrates, 19 g; Fiber, 3 g; Sodium, 340 mg.

BAKED EGGS

Baked eggs are baked in individual serving dishes. They are also called *shirred eggs* and *eggs en cocotte*.

They may also be baked with or garnished with a variety of meats and sauces, as indicated in the variations that follow.

Baked Egg Variations

Any of the following may be placed in the buttered egg dish before adding the egg:

Ham or back bacon, thin slice, lightly browned on griddle or in sauté pan

Bacon, cooked crisp, 3 or 4 half-strips

Corned beef hash, beef hash, or ham hash

Cheese, such as cheddar, Swiss, or Gruyère, grated

Diced chicken in cream sauce

Tomato concassé, sautéed in butter

PROCEDURE for Making Baked Eggs

1. Collect equipment and food items.

2. Butter individual-portion ramekins or baking dishes.

3. Break eggs into dish.

4. Place in oven at 350°F (175°C) and cook to desired doneness. Unless otherwise requested, they should be soft-cooked, with the whites set and the yolks barely set, still partly liquid.

5. Serve in the same dish or ramekin.

Baked eggs with sausages

Any of the following may be placed on top of the eggs, either before or after baking. Place solid garnish to one side. Spoon sauces around the outside. Do not cover the yolk.

Heavy cream, hot	Sautéed chicken livers and brown sauce
Brown sauces such as bordelaise, Madeira, or demi-glace	Small grilled sausages
Tomato sauce	Mushrooms sautéed in butter or cream sauce
Soubise sauce	Asparagus tips

Standards of Quality for Baked Eggs

1. Whites should be set and milky-looking, with no part of the white still liquid.
2. Unless the customer requests them otherwise, yolks should be barely set, still partly liquid.
3. Additional ingredients should be in appropriate proportion to the egg, enough to add a pleasing flavor component but not so much as to overwhelm the egg.

SCRAMBLED EGGS

Like other egg preparations, scrambled eggs are best if cooked to order. However, they may be made in larger quantities. They should be undercooked if they are to be held for volume service, as they will cook more in the steam table.

If scrambled eggs must be held over 30 minutes, they will be more stable if the eggs are mixed with a medium white sauce (béchamel) before cooking. Use about 8 ounces sauce per quart of eggs (250 mL per L).

Do not overcook scrambled eggs or hold them too long. Overcooked eggs are tough and watery, and they eventually turn green in the steam table.

Scrambled eggs should be soft and moist, unless the customer requests "scrambled hard."

PROCEDURE for Scrambling Eggs

1. Collect equipment and food items.
2. Break eggs into a stainless-steel bowl and beat until well blended. Season with salt and white pepper. Do not use aluminum, which may discolor the eggs.
3. If desired, add a small amount of milk or cream, about 1–1½ tablespoons (15–20 mL) for 2 eggs, or 8–12 ounces per quart of eggs (250–375 mL per L).
   Too much liquid may make cooked eggs watery, and it dilutes the flavor. Heavy cream adds richness but also adds cost.
4. Heat butter in a small sauté pan (for cooking to order) or in a large skillet, as for fried eggs.
   Note: Steam kettles or tilting skillets may be used for scrambling large quantities of eggs.
5. When fat is just hot enough to make a drop of water sizzle, pour in eggs.
6. Cook over low heat, stirring gently from time to time as the eggs coagulate. Lift portions of coagulated egg so uncooked egg can run underneath.
   Too much stirring breaks up eggs into very small particles.
   Do not let the eggs brown. Keep heat low.
7. When eggs are set but still soft and moist, remove from heat. Turn out onto plate or into steam table pan.

Additions to Scrambled Eggs

Flavor variations may be created by adding any of the following ingredients to scrambled eggs before serving:

Chopped parsley and/or other herbs	Sautéed diced onion and green bell pepper
Grated cheese (cheddar, Swiss, parmesan)	Diced smoked salmon
Diced ham	Sliced cooked breakfast sausage
Crumbled bacon	

Scrambled Eggs in Breakfast Sandwiches

The breakfast sandwich is a popular item for people on the go. Fillings generally contain scrambled eggs, often with one or more of the additions listed in the previous section, plus breakfast sausage, and served hot. Wrap sandwiches (p. 719) are especially popular because they are easiest to eat while engaged in another activity. Refer to the principles of sandwich making discussed in Chapter 22.

An example of a typical breakfast wrap is a flour tortilla wrapped around a filling of a few pieces of breakfast sausage and a portion of scrambled eggs mixed with grated cheese and sautéed, diced onion and green pepper. For a traditional bread sandwich, enclose this same filling between two pieces of bread, omitting the sausage and including a slice of ham.

Standards of Quality for Scrambled Eggs

1. Soft, moist, creamy texture, not hard, rubbery curds (unless the customer requests "scrambled hard").

2. Uniform yellow color. No browned bits (from excessively high heat) or greenish tinge (from overcooking and long holding at high temperature).

OMELETS

Making omelets is like riding a bicycle. When you are learning, it seems difficult, and you can't imagine how anyone can do it. But once you have mastered the technique, it seems easy, and you don't understand how anyone could have trouble doing it.

We are talking about the plain or French omelet. There are several kinds, as described here, but the French omelet remains the standard. Making it is a technique worth mastering.

French Omelet

Omelets may be described as sophisticated scrambled eggs. The first part of the technique is similar to that for making scrambled eggs. But the similarities end there, and the omelet emerges from the pan not as a shapeless pile of curds but an attractive oval with a light, delicate texture.

Two elements are necessary for making omelets:

1. Moderately high heat. This seems like a contradiction to our basic principle of low-temperature egg cookery. But the omelet cooks so fast that its internal temperature never has time to get too high.

2. A conditioned omelet pan. First, the pan must have sloping sides and be the right size so the omelet can be shaped properly. Second, it must be well seasoned or conditioned to avoid sticking.

PROCEDURE for Conditioning an Omelet Pan

The following method is only one of many. Your instructor may show you another. The object is to seal the surface of the metal with a layer of baked-on oil.

1. Rub the clean pan with a thin film of vegetable oil.

2. Set the pan over moderately high heat until it is very hot.

3. Remove from heat and let cool.

4. Do not scour the pan or wash with a detergent after use. Rub with salt, which will scour the pan without harming the primed surface. Rinse only after pan has cooled, or wipe with a clean towel.

5. Reseason as often as necessary, or after each day's use.

PROCEDURE for Making a French Omelet

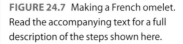

See Figure 24.7 for illustration of technique.

1. Collect all equipment and ingredients.

2. Beat 2 or 3 eggs in a small bowl just until well mixed. Do not whip until frothy. Season with salt and pepper.

   If desired, 1 tablespoon (15 mL) water may be added to make the omelet lighter.

   For extended service, beat a large quantity of eggs. Measure each portion with a ladle.

3. Place an omelet pan over moderately high heat.

4. When the pan is hot, add about 1 tablespoon (15 mL) clarified butter and swirl it around to coat the inside of the pan. Give it a second to get hot.

   Raw butter may be used, but great care is necessary to keep it from burning. Milk solids in the butter will brown, detracting from the appearance of the finished omelet.

5. Add the eggs to the pan. They should begin to coagulate around the edges and on the bottom in a few seconds.

6. With one hand (the left, if you are right-handed), vigorously shake the pan back and forth. At the same time, stir the eggs with a circular motion with the bottom side of a fork, but do not let the fork scrape the pan.

   This is the difficult part. The most common errors are not shaking and stirring vigorously enough and using heat that is too low. The purpose of this action is to keep the eggs in motion so they coagulate uniformly.

7. Stop shaking and stirring when the eggs are almost set but still very moist. If you continue stirring, you will have scrambled eggs instead of an omelet.

8. Tilt the handle up and shake the pan so the omelet slides to the opposite side of the pan and begins to climb up the opposite slope.

9. For a filled omelet, spoon the filling across the center of the egg, perpendicular to the handle.

10. With the fork, fold the sides of the omelet over the center. The omelet should now be resting in the corner of the pan and have an approximately oval shape.

11. Grasp the handle of the pan with your palm underneath and tilt the omelet out onto a plate so it inverts and keeps an oval shape.

    The whole procedure should take less than 1 minute.

    The finished omelet should be moist on the inside, tender on the outside, and yellow or only slightly browned.

FIGURE 24.7 Making a French omelet. Read the accompanying text for a full description of the steps shown here.

(a) As soon as the eggs are added to the hot pan, shake the pan back and forth with one hand and stir the eggs in a circular motion with a fork.

(b) When the eggs are almost set, tilt the pan and shake the eggs down to the opposite side of the pan. Rapping the handle sharply helps move the eggs.

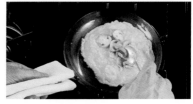

(c) Spoon the filling, if used, across the center.

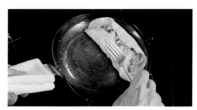

(d) Fold over the side of the omelet to make an oval shape.

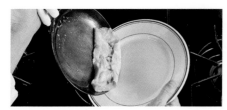

(e) Grasp the handle of the pan with your palm underneath and tilt the omelet onto a plate.

(f) The finished omelet should have a neat, oval shape. The color should be a bright, uniform yellow, with no browning (although some chefs prefer a light browning).

Suggested Omelet Fillings

Cheese

Sautéed or creamed mushrooms

Creamed or curried chicken

Creamed or buttered spinach

Sautéed onions, with or without bacon

Sautéed onions and diced potatoes

Seafood Newburg or seafood in a cream sauce

Red caviar

Thick Spanish Sauce (p. 189)

Ratatouille (p. 329)

American-Style or Folded Omelet

This style of omelet is often called a *French omelet*, but it is not a French omelet. It was probably devised by cooks who hesitated to tackle the French method.

It is made somewhat like a French omelet, except low heat is used and the eggs are not stirred or agitated. Instead, the edges of the cooked portion are lifted with a fork or spatula, allowing the uncooked portion to flow underneath. The finished omelet may be folded in half or like a French omelet.

The advantage of this method is that it is easier to learn.

The disadvantages are that the omelet is not as light or delicate in texture and the method is much slower.

Fluffy Omelet or Soufflé Omelet

These omelets are made by beating the egg whites separately and folding them into the beaten yolks, which may have some milk added. The mixture is poured into a hot, buttered omelet pan, and the omelet is finished in the oven. It is folded in half for service.

Fluffy omelets are not often made in food service because of the time they take to make.

Frittatas and Other Flat Omelets

A **frittata** is a flat omelet that originated in Italy. The same basic techniques are used for many popular American preparations. Flat omelets are made by mixing beaten eggs with a variety of ingredients, such as vegetables, meats, or cheese, and cooking the mixture over low heat without stirring. To finish, it is either flipped over or run under the broiler or into the oven until the top is set.

A popular American frittata (actually derived from the Chinese egg foo yung) is the Western omelet, containing diced sautéed onion, green bell pepper, and ham.

Standards of Quality for Omelets

French omelets

1. Neat oval shape.

2. Creamy interior, not dry.

3. Properly seasoned with salt.

3. If a filling is used, the filling should be hot.

Fluffy omelets

1. Light, puffy texture; completely set but with a moist, tender, not rubbery, interior.

2. Properly seasoned with salt.

3. Puffy, not sunken, appearance on the plate.

Flat omelets

1. Fully cooked, hot, and completely set but with a moist, tender, not rubbery, interior.

2. Properly seasoned with salt.

3. Interior garnish is flavorful and properly seasoned.

Summer Squash, Spinach, and Leek Frittata

YIELD: 4 PORTIONS

U.S.	METRIC	INGREDIENTS
4 oz	125 g	Leeks, white part and a little of the green, trimmed and cleaned
10 oz	300 g	Yellow summer squash or zucchini
1 oz	30 g	Butter
8 oz	250 g	Spinach leaves (no stems)
6	6	Eggs
to taste	to taste	Salt
to taste	to taste	Pepper
¹/₂ oz	15 g	Butter

PROCEDURE

1. Split the leeks in half lengthwise, and then slice them crosswise into thin slices.
2. Trim and slice the squash.
3. Heat the butter in a sauté pan over moderate heat.
4. Add the leeks and sauté until wilted.
5. Add the squash and sauté until just tender.
6. Remove from the pan and cool.
7. Blanch the spinach in boiling water until wilted.
8. Drain and cool the spinach in cold water. Drain again and squeeze dry.
9. Chop the spinach coarsely and mix it with the squash.
10. Beat the eggs and add them to the vegetables.
11. Add salt and pepper to taste.
12. Heat the butter over moderate heat in a well-seasoned or, preferably, a nonstick 10-in. (25-cm) sauté pan (see Note).
13. Add the egg mixture. Immediately lower the heat as low as possible. Cover loosely.
14. Cook slowly until the eggs are mostly set but creamy in the middle.
15. Place the pan under the broiler until the eggs are set.
16. Slide the frittata onto a plate.
17. Cut into 4 wedges. Serve immediately.

Per serving: Calories, 230; Protein, 12 g; Fat, 17 g (65% cal.); Cholesterol, 340 mg; Carbohydrates, 9 g; Fiber, 3 g; Sodium, 220 mg.

Note: To make individual portions to order, cook one-fourth of the egg mixture in a 6-in. (15-cm) sauté pan.

Summer Squash, Spinach, and Leek Frittata

SOUFFLÉS

Soufflés are not normally featured on breakfast menus. However, they are important basic egg preparations with which you should be familiar.

Amateur cooks often consider soufflés difficult to make. Actually, they are relatively easy preparations. Many restaurants have no difficulty turning out large numbers of soufflés to order. The only hard part is making sure the waiter picks up the order when it is ready.

A standard entrée soufflé consists of three elements:

1. Base—usually a heavy béchamel sauce.
2. Flavor ingredient—cheese, vegetables, seafood, etc.
3. Egg whites, beaten.

GENERAL PROCEDURE for Preparing Entrée Soufflés

1. Prepare a heavy béchamel sauce.
2. Combine the sauce with egg yolks.
3. Prepare the flavor ingredients—grate cheese, cook and chop vegetables, and so on.
4. Combine the base and the flavor ingredients.
5. Beat egg whites and fold in.
6. Bake in a soufflé dish that has been buttered and dusted with parmesan cheese.
7. Serve immediately.

À LA CARTE SERVICE

Prepare through step 4 and hold in refrigerator.

If several flavors are offered, prepare a single large batch of base and keep the flavor ingredients separate.

For each order, beat egg whites and combine with measured amount of base.

Standards of Quality for Soufflés

1. Well risen above the rim of the baking dish.
2. The top should be level, not tilted; that is, the soufflé should be evenly risen on all sides.
3. Moist interior.
4. Light, puffy texture.
5. Properly seasoned.

KEY POINTS TO REVIEW

- What are the steps in the procedure for making baked eggs?
- What are the steps in the procedure for making scrambled eggs?
- What are the steps in the procedure for making a French omelet?
- What are the steps in the procedure for making entrée soufflés?

Cheese Soufflé

PORTIONS: 12 **PORTION SIZE: 4 OZ (125 G)**

U.S.	METRIC	INGREDIENTS	PROCEDURE
as needed	as needed	Butter	1. Select three 1¹⁄₂-qt (1¹⁄₂-L) soufflé dishes (4 portions each) or two 2-qt (2-L) dishes (6 portions each). Butter the insides of the dishes well. Sprinkle with cheese or crumbs so the bottom and sides are completely coated.
as needed	as needed	Parmesan cheese or dry bread crumbs	
		Roux:	2. Make a white roux with the butter and flour. Cook the roux a few minutes.
2¹⁄₂ oz	75 g	Butter	
2¹⁄₂ oz	75 g	Flour	3. Beat in the hot milk. Bring to a boil while stirring. Cool and stir until very thick and smooth.
1¹⁄₂ pt	750 mL	Milk, hot	
1¹⁄₂ tsp	7 mL	Salt	4. Remove from the heat. Stir in the salt, pepper, cayenne, and nutmeg.
1 tsp	5 mL	White pepper	
pinch	pinch	Cayenne	
pinch	pinch	Nutmeg	
12	12	Egg yolks	5. Add the egg yolks to the hot sauce and quickly mix in with a wire whip.
10 oz	300 g	Gruyère cheese, coarsely grated (see Note)	6. Stir in the cheese.
12–15	12–15	Egg whites	7. Beat the egg whites with the salt until they form stiff peaks. (The larger number of egg whites will make a lighter soufflé.)
¹⁄₄ tsp	1 mL	Salt	
			8. Fold the egg whites into the cheese mixture.
			9. Pour the mixture into the prepared soufflé dishes.
			10. Place the dishes in a preheated 375°F (190°C) oven. Bake 40 minutes without opening the oven door. After this time, check for doneness by *very gently* shaking the dishes. If the centers are firm and do not jiggle, the soufflés are done. If necessary, bake another 5–10 minutes.
			11. Remove from oven and serve *immediately*.

Per serving: Calories, 290; Protein, 17 g; Fat, 21 g (65% cal.); Cholesterol, 265 mg; Carbohydrates, 8 g; Fiber, 0 g; Sodium, 600 mg.

Note: Other cheeses may be used: sharp cheddar, Swiss, a mixture of Swiss and Gruyère, or a mixture of Swiss or Gruyère and parmesan.

VARIATIONS

À la Carte Service: Prepare the basic recipe through step 6. Chill mixture quickly and hold in refrigerator. For each order, scale 3¹⁄₂ oz (100 g) of the mixture. Beat 1 egg white and fold in. Bake in an individual soufflé dish 20–30 minutes.

Spinach Soufflé

Reduce cheese to 5 oz (150 g). Add 5 oz (150 g) well-drained, chopped cooked spinach.

Spinach and Ham Soufflé

Add 2 oz (60 g) ground or finely chopped ham to Spinach Soufflé.

Mushroom Soufflé

Reduce cheese to 5 oz (150 g). Add 4 oz (125 g) cooked chopped mushrooms.

Other Vegetable Soufflés

Follow the procedure for Spinach Soufflé, using chopped cooked vegetables such as broccoli, asparagus, or carrots.

Salmon Soufflé

Make the sauce base with milk plus liquid from canned salmon. Add 1¹⁄₂ oz (45 g) tomato paste to the base. Reduce cheese to 4 oz (125 g) and add 8 oz (250 g) flaked cooked or canned salmon.

CUSTARDS

A **custard** is a liquid that is thickened or set by the coagulation of egg protein.

There are two basic kinds of custard:

1. *Stirred custard* is stirred as it cooks and remains pourable when done.
2. *Baked custard* is not stirred and it sets firm.

One basic rule governs the preparation of both custards: Do not heat custards higher than an internal temperature of 185°F (85°C).

This temperature, as you know, is the point at which egg-liquid mixtures coagulate. If they are heated more than this, they are likely to curdle. An overbaked custard becomes watery because the moisture separates from the toughened protein.

Most custards are sweet. These preparations are covered in the baking and dessert section of this book.

The **quiche** (keesh), which is a custard baked in a pastry shell, is probably the most popular form of savory custard. The following recipe illustrates the technique for preparing savory custards.

Standards of Quality for Quiches

1. The filling must be completely set with soft, moist texture, not overcooked and rubbery.
2. Top should be smooth and level, with no large cracks.
3. Interior garnish should be flavorful, properly seasoned, and properly cooked.
4. Crust should be completely cooked and golden, with a tender, flaky texture.

Quiche au Fromage (Cheese Tart)

YIELD: FOUR 8-INCH (20-CM) TARTS PORTIONS: 24   PORTION SIZE: ¹/₆ OF TART
16   ¹/₄ OF TART

U.S.	METRIC	INGREDIENTS	PROCEDURE
2 lb	900 g	Mealy Pie Dough (p. 984)	1. Scale the dough into 4 pieces, 8 oz (225 g) each.
			2. Roll the dough into 4 circles, ¹/₈ in. (3 mm) thick.
			3. Fit the dough into four 8-in. (20-cm) pie pans or tart pans.
			4. Hold the pie shells in the refrigerator until needed (see Note).
1 lb	450 g	Swiss or Gruyère cheese, grated	5. Sprinkle 4 oz (110 g) cheese into the bottom of each tart shell.
12	12	Eggs, beaten	6. Beat together the eggs, cream, milk, and seasonings. Pour into the tart shells.
1 pt	500 mL	Heavy cream	7. Place the tarts in a 375°F (190°C) oven on the bottom shelf or, if using a deck oven, directly on the deck.
2 pt	950 mL	Milk	
2 tsp	10 mL	Salt	8. Bake until the filling is set, about 20–30 minutes.
¹/₄ tsp	1 mL	White pepper	9. Serve hot or cold. Cut into wedges of desired size.
¹/₈ tsp	0.5 mL	Nutmeg	

Per 1/6 tart: Calories, 370; Protein, 12 g; Fat, 30 g (69% cal.); Cholesterol, 155 mg; Carbohydrates, 18 g; Fiber, 3 g; Sodium, 450 mg.

Note: Pastry shells may be partially baked before filling if uncooked bottoms tend to be a problem. This is sometimes the case if you are using shiny aluminum pie pans or if the bottom heat of the oven isn't strong enough. See page 987 for procedure.

VARIATIONS

Quiche Lorraine

Dice 1 lb (450 g) bacon strips and cook until crisp. Drain and add to pie shell in step 5. Omit cheese or leave it in, as desired. (Quiche Lorraine was originally made without cheese.)

Onion Quiche

Sauté 2 lb (900 g) sliced onions very slowly in 2 oz (60 g) butter until golden and tender. Cool and add to empty pie shells. Reduce cheese to 8 oz (225 g).

Spinach Quiche

Sauté 3 oz (90 g) chopped onion in 3 oz (90 g) butter until soft. Add 1¹/₂ lb (700 g) cooked, drained chopped spinach. Sauté until

most of the liquid evaporates. Cool and add to empty pie shell. Omit cheese.

Mushroom Quiche

Sauté 2 lb (900 g) sliced mushrooms and 3 oz (90 g) chopped onion in 3 oz (90 g) butter. Add 1 tbsp (15 mL) lemon juice to keep the mushrooms white. Cook until juices evaporate. Cool and add to the empty pie shell. Omit cheese.

Seafood Quiche

Substitute 8 oz (225 g) cooked diced shrimp and 8 oz (225 g) cooked diced crabmeat for the cheese. Add 3 fl oz (90 mL) sherry and 2 oz (60 g) tomato paste to the egg mixture.

BREAKFAST BREADS, CEREALS, AND MEATS

Bread items probably play a more important role at breakfast than even eggs. Hardly an order of eggs is sold without an order of toast on the side. And for the diner who prefers a continental breakfast, coffee and a bread item such as a roll or pastry constitute the entire breakfast.

Except for toast, few breakfast breads are prepared to order. Most operations purchase such items ready-made. These products include muffins, doughnuts, Danish pastries, sweet rolls, and regional favorites such as bagels and cornbread.

In this section, we consider three items that are made to order: pancakes, waffles, and French toast. You may not think of pancakes and waffles as breads, but they are actually a form of quick bread, a category of foods we consider in more detail in the baking section of this book.

PANCAKES AND WAFFLES

Waffles and pancakes, also called *griddle cakes* and *hot cakes*, are made from pourable batters. Pancakes are made on a griddle, while waffles are made on a special tool called a *waffle iron*.

Both items should be cooked to order and served hot. Waffles lose their crispness very quickly, and pancakes toughen as they are held. However, batters may be prepared ahead and are often mixed the night before.

Serve with butter and with maple syrup or syrup blends (pure maple syrup is expensive). Other condiments that may accompany these items are fruit syrups, jams and preserves, applesauce, and fruits such as strawberries or blueberries.

INGREDIENTS AND PROCEDURES

Compare the basic pancake and waffle recipes and note how much alike the batters are, with important exceptions:

1. Waffle batter contains more fat.
2. Waffle batter contains less liquid, so it is slightly thicker.
3. Waffles are given extra lightness when the egg whites are beaten separately and folded into the batter. (Some recipes omit this step.)

A standard-size pancake requires $1/4$ cup (60 mL) batter. The amount of batter needed for waffles depends on the size of the waffle iron.

PRE-PREPARATION FOR VOLUME SERVICE

Pancake and waffle batters leavened by *baking powder only* may be mixed the night before and stored in the cooler. Some rising power may be lost, so baking powder may have to be increased.

Batters leavened by baking soda should not be made too far ahead because the soda will lose its power. Mix dry ingredients and liquid ingredients ahead and combine just before service.

Batters using beaten egg whites and baking powder may be partially made ahead, but *incorporate the egg whites just before service*.

Buttermilk Pancakes

YIELD: 3¹/₂ PT (1.75 L) BATTER, ENOUGH FOR 25–30 LARGE OR 50 MEDIUM PANCAKES

U.S.	METRIC	INGREDIENTS	PROCEDURE
1 lb	500 g	Pastry flour or all-purpose flour	1. Sift together the flour, sugar, salt, baking powder, and baking soda.
2 oz	60 g	Sugar	
1 tsp	5 mL	Salt	
1 tbsp	15 mL	Baking powder	
1¹/₂ tsp	7 mL	Baking soda	
4	4	Eggs, beaten	2. Mix the beaten eggs, buttermilk, and butter or oil.
1 qt	1 L	Buttermilk	3. Add the liquid ingredients to the dry ingredients. Mix just until the dry ingredients are thoroughly moistened. Do not overmix. (Note: Buttermilk makes a thick batter. Thin with a little skim milk or water if the batter seems too thick.)
4 fl oz	125 mL	Melted butter or oil	
			4. Depending on the size pancake desired, measure 1–2 fl oz (30–60 mL) portions of the batter onto a greased, preheated griddle (375°F/190°C), allowing space for spreading.
			5. Griddle the pancakes until the tops are full of bubbles and begin to look dry and the bottoms are golden brown. Turn and brown the other side.
			6. Remove from the griddle and serve.

Per 2-ounce (56.7-g) pancake: Calories, 120; Protein, 4 g; Fat, 5 g (36% cal.); Cholesterol, 40 mg; Carbohydrates, 16 g; Fiber, 0 g; Sodium, 250 mg.

Waffles

YIELD: 3¹/₂ PT (1.75 L) BATTER

U.S.	METRIC	INGREDIENTS	PROCEDURE
1 lb 4 oz	625 g	Pastry flour or all-purpose flour	1. Sift together the flour, salt, and baking powder.
1 tsp	5 mL	Salt	
2 tbsp	30 mL	Baking powder	
6	6	Egg yolks, beaten	2. Mix the egg yolks, milk, and butter or oil.
1¹/₂ pt	750 mL	Milk	3. Add the liquid ingredients to the dry ingredients. Mix just until the dry ingredients are thoroughly moistened. Do not overmix.
8 fl oz	250 mL	Melted butter or oil	
6	6	Egg whites	4. Whip the egg whites until they form soft peaks. Add the sugar and whip until stiff peaks form.
2 oz	60 g	Sugar	5. Fold the egg whites into the batter.
			6. Pour enough batter onto a lightly greased, preheated waffle iron to almost cover the surface with a thin layer. Close the iron.
			7. Cook waffles until signal light indicates they are done or until steam is no longer emitted.
			8. Remove from the iron and serve.

Per 1 fl oz (29.57 mL) batter: Calories, 80; Protein, 2 g; Fat, 4 g (45% cal.); Cholesterol, 30 mg; Carbohydrates, 9 g; Fiber, 0 g; Sodium, 100 mg.

Buckwheat Pancakes

YIELD: 2 QT (2 L) BATTER, ENOUGH FOR ABOUT 30 LARGE OR 50 MEDIUM PANCAKES

U.S.	METRIC	INGREDIENTS	PROCEDURE
6 oz	180 g	Buckwheat flour	1. Sift together the flours, baking powder, and salt.
8 oz	240 g	Pastry flour or all-purpose flour	
4 tsp	20 mL	Baking powder	
1 tsp	5 mL	Salt	
4	4	Egg yolks, beaten	2. Mix the egg yolks, milk, melted butter, and honey.
2 pt	1 L	Milk	3. Add the liquid ingredients to the dry ingredients. Mix just until the dry ingredients are thoroughly moistened. Do not overmix.
4 oz	120 g	Melted butter	
5 fl oz	150 mL	Honey	
4	4	Egg whites	4. Whip the egg whites until they form soft peaks.
			5. Fold the egg whites into the batter.
			6. Depending on the size pancake desired, measure 1–2-oz (30–60-g) portions of the batter onto a greased, preheated griddle (375°F/190°C), allowing space for spreading.
			7. Griddle the pancakes until the tops are full of bubbles and begin to look dry and the bottoms are golden brown. Turn and brown the other side.
			8. Remove from the griddle and serve.

Per 2-ounce (56.7-g) pancake: Calories, 120; Protein, 3 g; Fat, 4.5 g (35% cal.); Cholesterol, 35 mg; Carbohydrates, 16 g; Fiber, 1 g; Sodium, 140 mg.

FRENCH TOAST

French toast in different versions is popular in many regions, and it has the advantage of being an excellent way to utilize day-old bread.

Basic French toast consists of slices of bread dipped in a batter of eggs, milk, a little sugar, and flavorings. French toast is cooked on a griddle like pancakes.

Variations may be created by changing the basic ingredients:

Bread. White pullman bread is standard. Specialty versions can be made with French bread, rich egg bread, or whole-grain breads.

Batter. Milk is the usual liquid, mixed with egg in various proportions. Deluxe versions may include cream or sour cream.

Flavorings. Vanilla, cinnamon, and nutmeg are popular choices. Other possibilities are grated lemon and orange rind, ground anise, rum, and brandy.

The most common fault in making French toast is not soaking the bread long enough to allow the batter to penetrate. If the bread is just dipped in the batter, the final product is just dry bread with a little egg on the outside.

French toast is dusted with powdered sugar and served, like pancakes, with accompanying butter, syrups, preserves, or fruits.

Cinnamon Raisin French Toast

PORTIONS: 6 PORTION SIZE: 2 SLICES

U.S.	METRIC	INGREDIENTS
6	6	Eggs
12 fl oz	375 mL	Milk
4 fl oz	125 mL	Heavy cream
4 oz	125 g	Sugar
1 tbsp	15 mL	Vanilla extract
1 tsp	5 mL	Cinnamon
$^1/_8$ tsp	0.5 mL	Nutmeg
$^1/_8$ tsp	0.5 mL	Salt
12 slices	12 slices	Cinnamon raisin bread, $^3/_4$ in. (2 cm) thick
1$^1/_2$ oz	45 g	Butter
as needed	as needed	Confectioners' sugar

PROCEDURE

1. Beat together the eggs, milk, cream, sugar, vanilla, cinnamon, nutmeg, and salt until the sugar is dissolved.

2. Soak the bread in the egg mixture until the bread is soaked through, but do not leave so long that it falls apart.

3. For each portion, heat 1$^1/_2$ tsp (7 mL) butter in a sauté pan large enough to hold 2 slices of bread.

4. Put 2 slices of bread in the pan and brown the bottoms.

5. Turn the bread over and cook 30 seconds.

6. Transfer the sauté pan to an oven preheated to 375°F (190°C). Bake 10 minutes. The bread should be cooked through and slightly puffed up.

7. Plate, dust with confectioners' sugar, and serve immediately.

Per serving: Calories, 470; Protein, 15 g; Fat, 23 g (43% cal.); Cholesterol, 260 mg; Carbohydrates, 51 g; Fiber, 2 g; Sodium, 400 mg.

VARIATION

The amount of batter in the recipe is enough to soak about 24 thin slices of sliced bread. Regular slices of bread are thin enough to be cooked from start to finish on a griddle or in a sauté pan.

Cinnamon Raisin French Toast

Standards of Quality for Pancakes and French Toast

1. Moist and tender, not dry or tough.

2. Golden brown exterior.

3. Pancakes and waffles should not taste strongly of baking powder (excessive baking powder gives a soapy taste).

4. Pancakes and waffles should contain no lumps of dry flour.

5. Interior of French toast should be moist and eggy, not merely dry bread.

CEREALS

HOT COOKED CEREALS

Cooked cereals are of two types:

1. Whole, cracked, or flaked cereals, such as oatmeal (rolled oats), Scotch oatmeal (cracked oats), and cracked wheat. The particles are large and can be added to boiling water without lumping.

2. Granular cereals, such as farina and cornmeal. The particles are small and tend to lump when added to boiling water.

For more information on cooking grains, see Chapter 13.

PROCEDURE for Cooking Whole, Cracked, or Flaked Cereals

1. Collect equipment and ingredients.

2. Measure the correct amount of water and salt into a pot and bring to a boil. Read package directions for quantities.

   Using milk or part milk makes a richer cereal, but a more expensive one. Be careful not to scorch the milk if you use it.

3. Measure the correct amount of cereal.

4. Add the cereal slowly, stirring constantly.

5. Stir until some thickening takes place, then stop stirring. Too much stirring makes cereal gummy.

6. Reduce heat to a slow simmer, cover, and cook until desired doneness and consistency are reached. Cooking times vary greatly.

7. Keep covered until service to prevent drying.

PROCEDURE for Cooking Granular Cereals

The procedure is the same as above, except the cereal is mixed with a little cold water before being added to boiling water. This separates the grains and prevents lumping. The cold water must be calculated as part of the total amount of liquid. Alternatively, mix the cereal with all the cold liquid, then bring to a boil and simmer until done.

COLD CEREALS

Cold, dry cereals are purchased ready prepared and need no preparation by the kitchen. Like hot cereals, they are served with accompanying milk or cream, sugar, and, sometimes, fruit such as berries or sliced bananas.

BREAKFAST MEATS

Meats and meat cooking methods are covered in previous chapters, but we mention them again because three meats in particular—bacon, sausage, and ham—appear on most breakfast menus.

Although most breakfast meats are traditionally pork products, items like turkey bacon and other non-pork products are also available. In addition, meat substitutes made from soy can be used on vegetarian menus.

BACON

Bacon is a cured, smoked pork product. It is available in whole slabs but is almost always purchased sliced. Thickness of slices is specified by number of slices per pound, usually 18–22 (40–48 per kg).

Low-temperature cooking applies to bacon as well as to other meats. Bacon is about 70 percent fat and shrinks a great deal. However, cooking at low temperatures minimizes shrinkage. The oven is most often used for cooking bacon in quantity, though a griddle or sauté pan may also be used.

To cook in the oven, lay out the bacon strips on sheet pans in a single layer, or, even better, on racks over sheet pans. (Bacon may be purchased already laid out on parchment.) Bake at 300°–350°F (150°–175°C) until about three-fourths done. Remove from the oven, being careful not to spill the hot fat. Finish individual portions to order on the griddle or in the oven, cooking them until crisp.

HAM

Ham for breakfast service is almost always precooked. Slices in 3- to 4-ounce (90- to 115-g) portions need only be heated and browned slightly on a griddle or under the broiler.

Back bacon, also called Canadian bacon, is boneless pork loin that is cured and smoked like ham. It is handled like ham in the kitchen.

SAUSAGE

Breakfast sausage is simply fresh pork that has been ground and seasoned. It is available in three forms: patties, links, and bulk.

Because it is fresh pork, sausage must be cooked well done. This does not mean, however, that it should be cooked until it is just hard, dry, shrunken little nuggets, as it often is.

Most kitchens cook sausages by the same methods as bacon. For volume service, sausage is partially cooked in the oven and then finished to order. Link sausages hold better than patties because the links are protected from drying by their casings.

KEY POINTS TO REVIEW

- How does pancake batter differ from waffle batter?

- How can a basic French toast recipe be modified to create variety?

- What are the steps in the procedure for cooking whole, cracked, or flaked cereals? What are the steps in the procedure for cooking granular cereals?

- How are bacon, ham, and sausages prepared for breakfast service?

TERMS FOR REVIEW

yolk	**sunny side up**	**over hard**	**custard**
white	**basted**	**baked egg**	**quiche**
shell	**over easy**	**frittata**	**breakfast sausage**
curdle	**over medium**	**soufflé**	

QUESTIONS FOR DISCUSSION

1. Which grade of egg would you choose to prepare poached eggs? hard-cooked eggs? fried eggs? scrambled eggs? Why?

2. Is it possible to prepare hard-cooked eggs in a pressure steamer? Give reasons for your answer.

3. When separating eggs, many chefs advise breaking them one by one over a small bowl, then transferring each white to the larger bowl as it is separated. Can you give a reason for this advice?

4. Give two reasons for being careful not to add too much vinegar to the poaching water for eggs.

5. In the recipe for waffles, what is the purpose of beating the sugar into the egg whites rather than combining it with the other dry ingredients?

6. What precautions might you take if you were making French toast from thick slices of French bread?

25

DAIRY AND BEVERAGES

Milk is one our most versatile foods, not only as a beverage but even more importantly as a cooking and baking ingredient and as a raw material for the production of a wide array of other foods, including cream, butter, and hundreds of kinds of cheese.

This chapter identifies and describes the many dairy products used regularly in the kitchen and gives guidelines for handling, storing, and cooking with them. The last section of the chapter describes the many types of cheese produced in North America and Europe and gives examples of each type.

The chapter concludes with a discussion of coffee and tea beverages, including guidelines for their production.

AFTER READING THIS CHAPTER, YOU SHOULD BE ABLE TO

1. Describe the major milk, cream, and butter products.
2. Explain why milk curdles and why it scorches, and identify the steps to take to prevent curdling and scorching.
3. Whip cream.
4. Describe the most important kinds of cheese used in the kitchen.
5. Store and serve cheese properly.
6. Cook with cheese.
7. Prepare coffee and tea.

DAIRY PRODUCTS

MILK AND CREAM

CATEGORIES AND DEFINITIONS

When we talk about milk and cream used in food service, we are nearly always talking about milk from cows. Milk from other animals, including goats, sheep, and water buffalo, is used to make some cheeses, but most of the liquid milk we see, except for a small amount of goat milk, is milk from dairy cows.

Milk is used as a beverage and also in cooking. Similarly, other milk products, including cream, butter, and cheese, are eaten as purchased and also used in cooking.

Pasteurization

Liquid milk, directly as it comes from the cow and before anything is done to it, is called *raw milk*. Because raw milk may contain disease-causing bacteria or other organisms, it is almost always **pasteurized** before being sold or before being processed into other products. Pasteurized milk has been heated to 161°F (72°C) and held at this temperature for 15 seconds to kill disease-causing organisms, and then quickly chilled. By law, all Grade A liquid milk and cream must be pasteurized. (Grades B and C are used in food processing and industrial uses and are rarely seen in food service or in the retail market.)

Even after pasteurizing, milk and cream are highly perishable products. Some cream products are **ultra-pasteurized** to extend their shelf life. By heating the product to a much higher temperature (275°F/135°C) for 4 seconds, this process kills not only disease-causing bacteria but nearly all organisms that cause spoilage. Ultra-pasteurized products must still be refrigerated both before and after opening.

Ultra-high-temperature (UHT) pasteurization involves even higher temperatures. The resulting product is packed into sterile cartons. If the cartons are unopened, the milk will keep at room temperature for up to 10 months. Once opened, the milk must be refrigerated like regular pasteurized milk. UHT milk has a somewhat cooked taste and is better suited to cooking than for drinking as a beverage.

Fresh Milk Products

Whole milk is fresh milk as it comes from the cow, with nothing removed and nothing (except vitamin D) added. It contains about 3.5 percent fat (known as **milk fat** or **butterfat**), 8.5 percent nonfat milk solids, and 88 percent water.

Skim or **nonfat milk** has had most or all of the fat removed. Its fat content is 0.5 percent or less.

Two percent milk has a fat content between 1.9 percent and 2.1 percent, while partially skimmed milk has a fat content of between 0.9 percent and 1.1 percent. Both these products must contain at least 8.5 percent milk solids (other than milk fat), the same as for whole milk. Both these products are usually fortified with vitamins A and D and extra nonfat milk solids.

Fortified nonfat or *low-fat milk* has had substances added to increase its nutritional value, usually vitamins A and D and extra nonfat milk solids.

Flavored milks, such as *chocolate milk*, have had flavoring ingredients added. A label such as *chocolate milk drink* or *chocolate-flavored drink* indicates the product does not meet the standards for regular milk. Read ingredient labels.

Except, of course, for nonfat milk, natural liquid milk contains fat, which, because it is lighter than water, will gradually separate and float to the top in the form of cream. **Homogenized milk** has been processed so the cream doesn't separate. This is done by forcing the milk through very tiny holes, which breaks the fat into particles so small they stay distributed in the milk. Nearly all liquid milk on the market has been homogenized.

Fresh Cream Products

Whipping cream has a fat content of 32 to 40 percent. Whipping cream labeled *ultra-pasteurized* keeps longer than regular pasteurized cream. Pure ultra-pasteurized cream does

not whip as well as regular pasteurized cream, so additives such as vegetable gums are added to it to make it more whippable.

Double cream is richer than regular whipping cream. It contains 40 percent or more milk fat.

Table cream, also called coffee cream, contains between 16 and 32 percent fat, usually about 18 percent.

Half-and-half has a fat content of 10 to 12 percent.

Fermented Milk and Cream Products

Sour cream has been cultured or fermented by added lactic acid bacteria, which makes it thick and slightly tangy in flavor. It has about 18 percent fat. Low-fat and fat-free sour cream are also available. These products are thickened with stabilizers.

Crème fraîche (kremm fresh) is a slightly aged, cultured heavy cream. It is widely used for sauce making in Europe because of its pleasant, slightly tangy flavor and its ability to blend easily into sauces. Unlike regular heavy cream, it usually doesn't require tempering and can be added directly to hot sauces. It is available commercially but is expensive. A close approximation can be made by warming 1 quart (1 L) heavy cream to about 100°F (38°C), adding 1½ ounces (50 mL) buttermilk, and letting the mixture stand in a warm place until slightly thickened, 6 to 24 hours.

Buttermilk is fresh, liquid milk, usually skim milk or low-fat milk, which has been cultured or soured by bacteria. It is usually called *cultured buttermilk* to distinguish it from the original buttermilk, which was the liquid left after butter making. Buttermilk is used in recipes calling for sour milk.

Yogurt is milk (whole or low-fat) cultured by special bacteria. It has a custard-like consistency. Most yogurt has additional milk solids added, and some of it is flavored and sweetened.

Milk Products with Water Removed

Evaporated milk is milk, either whole or skim, that has at least 50 percent of the water removed and is then sterilized and canned. Evaporated milk has a somewhat cooked flavor. Evaporated milk made from skim or low-fat milk is labeled as such.

Condensed milk is whole milk that has had about 60 percent of the water removed and is heavily sweetened with sugar. It is available canned and in bulk.

Dried whole milk is whole milk that has been dried to a powder. **Nonfat dry milk** is skim milk that has been dried in the same way. Both are available in regular form and in instant form, which dissolves in water more easily.

ARTIFICIAL DAIRY PRODUCTS

A wide variety of imitation cream and dessert topping products are made from various fats and chemicals, which are listed on the label. They are used in some institutions because they keep longer and are generally less expensive than dairy products. Some people feel they are acceptable, but many find their flavors objectionable. These products may leave a waxy mouthfeel.

PROBLEMS IN COOKING MILK AND CREAM PRODUCTS

Curdling

Curdling is a process by which milk proteins solidify and separate from the whey. Curdling is usually caused by acids, tannins, salt, and heat. The mild acids in many vegetables and the tannins in potatoes are often enough to curdle milk.

Starches partially stabilize milk and cream. This is why it is possible to make soups and sauces with both milk or cream and acid ingredients. Avoid combining milk or cream with strong acids unless a starch is present.

Reducing temperatures and cooking times also helps. Curdling is more likely at high heat or with prolonged cooking.

Do not add salt, or salt only lightly, unless the milk has been stabilized by starch.

When adding milk or cream to a hot liquid, heat it first in a separate pot, or temper it by stirring a little of the hot liquid into it first.

Reconstituted dry milk is more likely to curdle than fresh milk.

FIGURE 25.1 Whipping cream.

(a) The cream has begun to thicken.

(b) The cream has reached the soft-peak stage. Stop at this stage if the cream is to be folded into a batter or other mixture.

(c) The cream has reached the firm-peak stage. Whipping beyond this stage causes the cream to break or separate.

EUROPEAN BUTTERS

As explained in the text, typical European butters have a higher fat content and lower moisture content than American butters.

A second difference is in the type of cream used. Most North American butters are made from fresh cream and are called *sweet cream butter*. By contrast, many European butters are made from slightly cultured cream, somewhat like crème fraîche (p. 789). This gives the butter a more complex, richer flavor.

Scorching

Scorching occurs when milk that is being heated coagulates on the bottom of the pan due to high heat. This deposit is likely to burn if cooking continues.

To avoid scorching, heat milk in a double boiler, steamer, or steam-jacketed kettle rather than over direct heat.

Skin Formation

The formation of scum or skin on top of heated milk or milk sauces is caused by coagulation of proteins in contact with air. Prevent it by covering the utensil or by coating the surface with a layer of melted fat.

WHIPPING CREAM

Cream with a fat content of 32 percent or more can be whipped into a foam (see Figure 25.1). One quart or liter of cream produces up to 2 quarts or liters of whipped cream.

For the best results, observe the following guidelines:

1. Have cream and all equipment well chilled.

2. Do not sweeten until the cream is whipped. Sugar decreases stability and makes the cream harder to whip. Use powdered sugar instead of granulated sugar for best stability.

3. Do not overwhip. Stop beating when the cream forms stiff peaks. If it is whipped longer, it first becomes granular and then turns into butter and whey.

4. Cream to be folded into other ingredients should be underbeaten because the action of folding it in whips it more and may overwhip it.

BUTTER

BUTTER CHARACTERISTICS AND GRADES

Fresh **butter** in Canada and the United States consists of about 80 percent milk fat. The remainder is milk solids and water.

In Europe, butter often has a higher fat content, usually around 82 percent. Some manufacturers in North America have begun processing and selling European-style butters with this higher fat content. They are intended to replicate the qualities of European butters in cooking and baking.

In Canada, butter is graded for flavor, body, color, and salt content. Grading is mandatory. Grades are Canada 1, Canada 2, and Canada 3. Most food service operations use Canada 1.

Most butter on the market is lightly salted. A maximum of 2 percent salt is permitted. **Unsalted butter** is more perishable but has a fresher, sweeter taste.

Because of its flavor, butter is the preferred cooking fat for most purposes. It has no equal in sauce making and is used as a sauce itself, as discussed in Chapter 8.

Clarified butter (see p. 189 for production procedure) is used as a cooking fat more often than whole butter because the milk solids in whole butter burn easily.

The smoke point of butterfat is only 300°–350°F (150°–175°C), so another product, such as vegetable oil, should be used when high cooking temperatures are required.

STORING

Have you ever been served butter that tasted like onions? Butter absorbs odors and flavors easily, so it should be kept well wrapped and away from foods that might transfer odors to it in the refrigerator.

Best storage temperature is 35°F (2°C).

MARGARINE

Margarine is a manufactured product meant to resemble butter in taste, texture, and appearance. It is made from vegetable and animal fats, plus flavoring ingredients, emulsifiers, coloring agents, preservatives, and added vitamins. Liquid vegetable oils for margarine are hydrogenated or partially hydrogenated to make them into solids. Like butter, it is about 80 percent fat. Diet and low-fat margarines have less fat and a much higher water content, so they can't be used in cooking like butter or regular margarine.

Flavors of different brands should be evaluated carefully because they vary considerably. Margarine should have a reasonably clean, fresh flavor, although you should not expect even the best to taste like high-grade butter.

Margarines that include an emulsifier called *lecithin* foam and brown like butter when heated. Those without lecithin do not. In all other respects, margarine is handled and stored like butter.

KEY POINTS TO REVIEW

- What are three types of pasteurization? Describe them.

- What are the major milk, cream, and butter products?

- What is curdling? How can it be prevented? How can scorching of milk products be prevented during cooking?

- What guidelines should be observed when whipping cream?

CHEESE

COMPOSITION

Cheese is a food produced by separating milk solids from whey by curdling or coagulation. This curdling is brought about by introducing selected bacteria or an enzyme called *rennet* into the milk (see sidebar). The resulting curds are drained, processed, and cured or aged in a variety of ways. The liquid that drains from the curds is called whey.

Processing techniques are so numerous that from a single basic ingredient (milk from cows, sheep, or goats) it is possible to produce hundreds of kinds of cheese, from cottage cheese to parmesan, from cheddar to Swiss, from blue to Limburger. Variables that produce these differences include the type of milk used, the method of curdling and the temperatures during curdling, the method of cutting and draining the curd, the way the curds are heated, pressed, or handled, and all the conditions of ripening or curing.

Ripening is the process that converts freshly made curds into distinctive, flavorful cheeses. Ripening is brought about by certain bacteria or molds that are introduced during manufacture. Much of a cheese's final character is determined by its ripening agent and the way it acts on the cheese.

Cheeses can be classified by the way in which they are ripened.

Bacteria ripened from inside, such as cheddar, Swiss, Gouda, and parmesan. This category includes most hard-ripened and hard grating cheeses. These ripen uniformly throughout the interior.

Washed-rind cheeses, such as Limburger and Liederkranz. These are so called because the surface of each cheese is periodically washed with a salt brine solution during the first stages of aging and ripening. Washed-rind cheeses usually have an orange or reddish rind that is thinner than the mold rind of mold-ripened cheeses. They usually become softer as they ripen. Many washed-rind cheeses have a pungent aroma when ripe.

Blue-veined cheeses, such as Roquefort and Stilton. These contain mold cultures that spread through the interior of the cheese.

RENNET AND VEGETARIAN DIETS

The most commonly used curdling agent in cheese production is rennet, as the text explains. Most traditional cheeses are made with rennet extracted from the stomach lining of calves. As a result, vegetarians usually avoid such cheeses. However, rennet derived from a plant base is also available. Cheeses made with a vegetable-based rennet can be included in the diets of lacto-vegetarians. In addition, some cheeses are made without rennet but with other curdling agents, such as citric acid or other mild acid.

Top row: Morbier, cheddar, and fontina.
Bottom row: Tilsit, baby Gouda, and Emmentaler.

Washed-rind cheeses, clockwise from left: Époisses, Livarot, Limburger, and Reblochon.

Blue-veined cheeses, clockwise from left: Stilton, Cabrales, Roquefort, and Bleu d'Auvergne.

Top row: Brie, Explorateur, Taleggio.
Bottom row: Camembert, Edel de Cleron.

Unripened cheeses. Back row: queso blanco, ricotta, fresh mozzarella. Front row: string cheese, cream cheese, fresh chèvre, and Boursin.

Mold-rind cheeses, such as Brie, Camembert, and St. André. These have a mold crust or rind that is white and velvety when the cheese is young but that may darken or become mottled with orange or brown as the cheese ages.

Unripened cheeses, such as cottage, cream, and baker's cheese.

The three major components of cheese are water, fat, and protein. The water content of cheese ranges from about 80 percent for a fresh, soft cheese like cottage cheese to about 30 percent for a very hard, aged cheese like parmesan.

The fat content of cheese, when it is listed on a label, generally refers to the percentage of solids. In other words, if a cheddar cheese has a 50 percent fat content, this means the cheese would be 50 percent fat if all the moisture were removed. In fact, the cheese may have a moisture content of about 40 percent, and its actual fat content may be about 30 percent of the total.

Double-crème (at least 60 percent fat) and **triple-crème** (at least 75 percent fat, dry weight) are very rich cheeses. Most of these styles of cheese originated in France, but they have become popular and are now made in many countries. Most of them fall into the unripened, soft-ripened, or blue-veined categories, discussed on the next page.

VARIETIES

Hundreds, possibly even thousands, of cheeses are produced in North America and Europe. The following is a representative sampling of most of the types of cheese commonly available. See also the accompanying photographs.

Unripened Cheeses

These are soft, white, freshly made cheeses.

Cottage cheese is a moist, loose-curd cheese that may or may not have cream added. ***Baker's cheese*** or ***pot cheese*** is similar but drier. Baker's cheese is used in cheesecakes and pastry.

Ricotta cheese is sometimes called ***Italian cottage cheese***, but it is smoother, moister, and sweeter than cottage cheese. Traditional ricotta is made from whey, but most ricotta in the North American market is made from fresh milk.

Cream cheese is a smooth, mild cheese with a high fat content. It is extensively used in making sandwiches, canapés, and hors d'oeuvres and in baking.

Neufchâtel is similar to cream cheese, but it has less fat. An Italian cream cheese called *mascarpone* is very soft and rich, and it looks almost like whipped cream. It has a slightly tangy taste that goes well with fruits as a dessert.

Mozzarella is a soft, mild cheese made from whole milk or part skim milk. It has a stringy texture that comes from being pulled and stretched during production. It is widely used in pizzas and Italian-style dishes. The freshly made mozzarella one finds in Italian neighborhoods is moister and more tender than the packaged varieties.

Mozzarella di bufala, made from the milk of water buffaloes, is imported from Italy and is available in some areas at a somewhat high price. It is much softer and more delicate in texture than regular mozzarella, and it has a slightly acidic flavor that is refreshing.

Feta is a crumbly, curdy cheese that originated in Greece and other Balkan countries. Instead of being aged or cured, it is pickled in brine. This, plus the fact that it is generally made from goat's or sheep's milk, gives it a distinctive and salty flavor.

Buffalo mozzarella

Feta

Semisoft Cheeses

Bel Paese and *fontina* from Italy, *Port Salut* from France, and American *muenster* and *brick* cheeses are the best known of a large group of cheeses that range from bland and buttery when young to more earthy and full-flavored when older. They are often used as dessert cheeses and as hors d'oeuvres.

France produces many washed-rind soft cheeses with orange rinds and with flavors that range from mild to pungent. Among the better-known ones are *Pont l'Évêque* and *Livarot* from Normandy and *Munster* from Alsace. (These cheeses might also be categorized as soft-ripened cheeses, below, although they may not become as soft and runny as the washed-rind cheeses discussed in the next section.)

Port Salut

Soft-Ripened Cheeses

These cheeses ripen from the outside toward the center. When very young, they are firm and cakey and have little flavor. As they mature, they gradually become softer and, when fully ripe, may be actually runny. The ripening starts just inside the rind and spreads to the center.

Soft-ripened cheeses include two categories: mold-ripened and washed-rind or bacteria-ripened.

Brie and *Camembert* from France are ripened by mold. They are made in flat, round shapes and are covered with a crust that varies in color from white to straw. When ripe, these cheeses are creamy and flavorful, but they develop a sharp odor of ammonia when overripe.

Many rich double- and triple-crème cheeses fall into this category, including *Explorateur*, *Brillat-Savarin*, *St. André*, *Boursault*, and *Boursin* (which may be flavored with pepper or with garlic and herbs). *Chaource* is similar in texture and appearance to a double-crème cheese but is actually closer in composition to Brie, having a fat content of 45 to 50 percent.

Liederkranz, made in the United States, and its Belgian cousin, *Limburger*, are ripened by bacteria rather than mold, but they also become softer as they age. They are widely misunderstood because of their aroma. Actually, when not overripe, these cheeses are not nearly as strong as most people expect, and they have a pleasant, smooth texture.

Brie

Cheddar

Époisses from Burgundy, France, is one of the great soft-ripened washed-rind cheeses. Small cheeses are packed individually in round wooden boxes. They are so soft when ripe that they are left in their boxes and served with a spoon.

Hard Cheeses

These are cured cheeses with a firm texture and varying degrees of mildness or sharpness, depending on their age.

Cheddar is an English invention, but American versions are so popular in the United States that it is often thought of as a distinctly American cheese. It ranges in flavor from mild to sharp and in color from light yellow to orange. Cheddar is eaten as is and is also widely used in cooking. *Colby* and *Monterey jack* are similar to very mild cheddars. Monterey jack is usually sold when quite young. In this case, it is more like American muenster and belongs in the semisoft category.

Swiss-type cheeses are also popular. They are produced in many countries, but the original Swiss cheese from Switzerland, *Emmentaler*, is perhaps the most flavorful. These are very firm, slightly rubbery cheeses with a nutty taste. Their large holes are caused by gases formed during ripening. *Gruyère* is another Swiss-type cheese from either Switzerland or France. It has smaller holes and a sharper, earthier flavor. Gruyère is important in cooking, and both it and Emmentaler are widely used for sauces, soufflés, fondue, and gratinéed items. Other cheeses related to Swiss are *Comté* from France, *Appenzeller* and *Raclette* from Switzerland, and *Jarlsberg* from Norway.

Edam and *Gouda* are the familiar round Dutch cheeses with the yellow and red wax rinds. Hard in texture, with a mellow, nut-like flavor, they are often seen on buffet platters and among dessert cheeses.

Provolone is an Italian cheese that resembles mozzarella when very young, but it becomes sharper as it ages. It is also available smoked.

Clockwise from top: Emmentaler, gorgonzola, provolone, Gruyère, Gouda, Locatelli Romano, Pont l'Évêque, baby Gouda; Lower center: Livarot.

Blue-Veined Cheeses

These cheeses owe their flavor and appearance to the blue or green mold that mottles their interiors. The most famous of the blue cheeses is *Roquefort*, made in France from sheep's milk and cured in limestone caves near the town of Roquefort. *Stilton*, from England, is a mellower, firmer blue cheese that the English call "Roquefort with a college education." Italy's *gorgonzola* is a soft, creamy cheese with an unmistakable pungency. Spanish *Cabrales* is a mellow but intensely flavorful blue cheese. Blue cheeses made in Denmark and in the United States are also widely used.

Less widely known but worth seeking out are a number of special blues, including *Bleu de Bresse*, *Fourme d'Ambert*, and *Pipo Crem'* from France, *Saga* from Denmark, and *Bavarian Blue* and *Blue Castello* from Germany. The last four of these are double- or triple-crèmes.

Jarlsberg

Edam

Blue Stilton

Gorgonzola

Goat Cheeses

Cheeses made from goat's milk are produced in dozens of varieties in France, where this type of cheese, called **chèvre** (shev r', or shev), is very popular. It has also become well known in the United States, which now has many producers. With a few exceptions, most goat cheeses are small, ranging in size from tiny buttons to logs, cakes, cones, and pyramids weighing up to about 6 ounces (170 g).

Chèvre

Fresh, unaged chèvres are the most popular and the mildest in flavor. Their paste is very white, with a soft but interestingly dry texture. They have a distinctive peppery, slightly acidic taste. The most widely available fresh French chèvre is probably the cylindrical *Montrachet*, either plain or with a coating of edible ash. Other fresh goat cheeses, both domestic and imported, may be available in different localities, and many have no name other than *chèvre*.

As goat cheese ages it becomes firmer, and the peppery, acidic flavor becomes stronger. Cheeses 2 or 3 months old can be quite powerful, while the youngest might taste almost like the unaged ones. Some names of chèvres are *Boucheron*, *Banon* (wrapped in chestnut leaves), *Pyramide*, *Crottin de Chavignol*, *Chabis*, and *Rocamadour*. *Saint-Marcellin* is made of part goat's milk and part cow's milk. Shaped into small disks, it softens rather than hardens as it ripens.

Goat cheeses, top row: Bucheron, Humboldt Fog, and Valençay; Bottom row: Banon and button chèvres.

Hard Grating Cheeses

The hard grating cheeses, typified by Italian *parmesan*, are called *grana* cheeses, referring to their grainy textures. The best of all granas is called *Parmigiano-Reggiano*. It is the true parmesan, aged at least 2 years, and it is very expensive. It is imitated widely around the world, and the imitations vary from bad to very good. Another Italian grana is *Romano*. Italian Romanos are made with sheep's milk, but American versions are usually made with cow's milk. Romano is stronger and saltier than parmesan.

These cheeses are often sold already grated. This is a convenience for commercial kitchens, of course, but, unfortunately, pre-grated cheese has much less flavor than freshly grated cheese. A merchant was once arrested for selling what he claimed was grated parmesan cheese but was actually grated umbrella handles. A large share of pre-grated cheese sold today resembles grated umbrella handles in flavor.

Hard cheeses, clockwise from top left: Parmigiano-Reggiano, Caerphilly, dry jack, Pecorino Romano, and aged Gouda.

Process Cheeses

Up to now, we have been talking about so-called natural cheeses, made by curdling milk and ripening the curds. **Process cheese**, by contrast, is manufactured by grinding one or more natural cheeses, heating and blending them with emulsifiers and other ingredients, and pouring the mixture into molds to solidify. Process cheese is a uniform product that does not age or ripen like natural cheese. Thus, it keeps very well. It is usually mild in flavor and gummy in texture.

Because of its melting quality and low price, it is often used in cooking. However, it is not as good a value as its price implies. Because it is relatively flavorless, you have to use much more of it to get the same flavor as from a smaller quantity of sharp cheddar.

In addition to its price and keeping qualities, the chief advantages of process cheese are that it melts easily and that its blandness appeals to many people who don't like more flavorful cheese.

In North America, most process cheeses are made from cheddar, while European process cheeses more often contain Swiss-type cheeses. Among them is a process cheese called *Gruyère*, which bears little resemblance to true *Gruyère*.

Process cheese food and *process cheese spread* contain a lower percentage of cheese and more moisture than a product labeled simply *process cheese*. *Cold pack* or *club cheese*, on the other hand, is not heated and pasteurized like process cheese but is simply ground and mixed with flavorings and seasonings to a spreadable consistency. Some brands are fairly flavorful.

North American Artisan Cheeses

Many of the fine cheeses from France, Italy, and other European countries, including many of those already listed, are handmade in small batches. By contrast, until recently, nearly all the cheese produced in North America was made by large industrial concerns.

In recent years, the production of small-batch artisan cheeses in North America has sky-rocketed. This development has been stimulated, no doubt, by the adoption of the European practice of offering a cheese course after dinner in fine dining establishments. Increased interest in specialty cheeses has created a new market for these products. North American chefs have made their cheese menus into showcases for fine local cheeses.

The American Cheese Society offers the following definitions of limited-production cheeses:

Specialty cheese is cheese of limited production, with particular attention to natural flavor and texture profiles. Specialty cheeses may be made from all types of milk and may include flavorings.

Artisan or **artisanal cheese** is cheese produced primarily by hand, in small batches, with particular attention to the tradition of the cheese-maker's art and using as little mechanization as possible.

Farmstead cheese is cheese made with milk from a farmer's own herd or flock on the farm where the animals are raised. No milk from an outside source may be used.

European cheeses are the product of hundreds of years of tradition and include nearly every type of cheese imaginable, so it is natural that North American producers looked to European cheeses as models. At the same time, the best producers have also tried to develop distinctive products that set new standards of quality. Today, cheese producers all across Canada and the United States produce cheeses that compare in quality with the best from Europe.

It is impossible in a short space to list all the fine artisan and farmstead cheeses in North America. The following are just small random samples of the hundreds of varieties from many regions.

Fresh unripened cheeses

Mozzarella Company

Mascarpone, Texas

Shepherd's Way Ricotta
(sheep's milk), Minnesota

Silani Sweet Cheese Ricotta,
Ontario

Soft washed-rind cheeses

Madawaska, Québec

Clandestin, Québec

Le Douanier, Québec

Colo Rouge, Colorado

Mold-rind soft-ripened

Blythedale Farm Brie and
Camembert, Vermont

Groupe Fromage Côté Triple
Crème, Quebec

Bittersweet Plantation Fleur-de-Teche
Triple Cream, Louisiana

Marquis de Temiscouata, Québec

Semi-firm and hard cheeses

Roth Kase Gruyère, Wisconsin

Gort's Gouda, British Columbia

Sylvan Star Gouda, Alberta

Mona (sheep and cow's
milk blend), Wisconsin

Contomme, Québec

Carr Valley Canaria, Wisconsin

McCadam Cheddar, New York

Blue cheeses

Bleu Benedictin, Québec

Dragon's Breath, Nova Scotia

Firefly Farms Mountain
Top Blue, Maryland

Point Reyes Blue, California

Hubbardston Blue, Massachusetts

Goat cheeses, fresh and aged

Sentinelle, Québec

Emissaire de Notre Dame,
Québec

Fromagerie Bergeron Patte
Blanche, Québec

Capriole Pipers Pyramid, Indiana

STORAGE AND SERVICE

Storing

Keeping qualities of cheese vary considerably. In general, *the firmer and more aged the cheese, the longer it will keep*. Cottage cheese should be used within a week, while a whole, uncut parmesan may keep a year or more.

Soft-ripened cheeses like Brie, Camembert, and Liederkranz deteriorate rapidly once they reach maturity. They are difficult cheeses to purchase because in their whole lifespan there may be only one week when they are neither underripe nor overripe.

Other ripened cheeses are not as fussy, as long as you store them under refrigeration and well wrapped to prevent drying. Cut cheeses dry especially quickly, so they must be wrapped in plastic at all times.

Serving

Serve cheese at room temperature. This is the single most important rule of cheese service. Only at room temperature will the full flavors develop. (This does not apply to unripened cheese like cottage cheese.)

Cut cheese just before service to prevent drying. Better yet, set out whole cheeses and large pieces when possible so portions can be cut to order by the customer or service personnel.

COOKING WITH CHEESE

Three varieties of cheese account for the majority of cheese used in cooking. *Cheddar* is the most frequently used in North American dishes, especially in sauces, as a casserole ingredient, and as a melted or gratinéed topping. *Swiss-type* cheeses are used more often in European-style dishes. Emmentaler and Gruyère are essential ingredients in fondue, Mornay sauce, gratinéed dishes, soufflés, and quiches. *Parmesan-type* cheeses are used in grated form for toppings and for seasoning and flavoring purposes.

GUIDELINES for Cooking with Cheese

1. Use low temperatures. Cheese contains a high proportion of protein, which toughens and becomes stringy when heated too much. Sauces containing cheese should not be boiled.

2. Use short cooking times, for the same reasons. Cheese should be added to a sauce at the end of cooking. Stirring it into the hot sauce off the heat is usually enough to melt it.

3. Grate cheese for faster and more uniform melting.

4. Aged cheeses melt and blend into foods more easily than young cheeses.

5. Aged cheeses add more flavor to foods than young, mild cheeses, so you need less of it.

KEY POINTS TO REVIEW

- What are five categories of cheese based on type of ripening? Give examples of each category.

- How should cheese be stored and served?

- What guidelines should be observed when cooking with cheese?

Welsh Rabbit

PORTIONS: 6   PORTION SIZE: 4 OZ (125 G)

U.S.	METRIC	INGREDIENTS	PROCEDURE
2$^{1}/_{4}$ tsp	11 mL	Worcestershire sauce	1. Mix the Worcestershire sauce and spices in a heavy saucepan.
$^{1}/_{2}$ tsp	2 mL	Dry mustard	
a few grains	a few grains	Cayenne	2. Add the beer or ale. Heat almost to a simmer.
5 fl oz	150 mL	Beer or ale	3. Set the pan over very low heat. Add the grated cheese, a little at a time. Stir constantly. Continue to stir over low heat until the mixture is smooth and thick.
1 lb 4 oz	600 g	Sharp cheddar cheese, grated	4. Remove from heat. The mixture may be kept warm in a steam table or bain-marie, but it is better if served immediately.
6 slices	6 slices	White bread	5. Toast the bread.
			6. For each portion, place a slice of hot toast on a plate. Ladle 4 fl oz (125 mL) cheese mixture over the toast. Serve.

Per serving: Calories, 470; Protein, 26 g; Fat, 31 g (60% cal.); Cholesterol, 95 mg; Carbohydrates, 19 g; Fiber, 1 g; Sodium, 750 mg.

Note: This dish is sometimes called Welsh Rarebit, although Rabbit is the original name.

Welsh Rabbit

Sirniki (Russian Fried Cheese Cakes)

PORTIONS: 6   PORTION SIZE: 3$^{1}/_{2}$ OZ (100 G)

U.S.	METRIC	INGREDIENTS	PROCEDURE
1 lb 4 oz	600 g	Pot cheese	1. Place the pot cheese in a strainer lined with cheesecloth. Fold the overhanging cloth over the top of the cheese so it is covered. Set the strainer over a bowl and refrigerate 24 hours to drain the cheese.
1$^{1}/_{2}$ oz	45 g	Bread flour	2. Force the cheese through a sieve or food mill into the bowl of a mixer.
2	2	Egg yolks	3. Add the flour, egg yolks, salt, and sugar.
$^{1}/_{8}$ tsp	0.5 mL	Salt	4. With the paddle attachment, mix until smooth.
$^{1}/_{2}$ oz	15 g	Sugar	5. Divide the dough into 3 parts. Roll each part into a cylinder about 3 in. (7.5 cm) thick. Wrap in plastic film. Refrigerate 2 hours or more.
as needed	as needed	Butter for frying	6. At service time, cut the cheese rolls into cakes about $^{3}/_{4}$ in. (2 cm) thick.
			7. Heat about $^{1}/_{8}$ in. (3 mm) butter in a heavy sauté pan. Pan-fry the cakes over low heat until golden brown on both sides. Turn very carefully with a spatula.
as needed	as needed	Confectioners' sugar	8. Plate the cakes and sprinkle lightly with confectioners' sugar. Place about 1$^{1}/_{2}$ tbsp (22 mL) sour cream on the plate next to the cake. Serve immediately.
5 fl oz	1140 mL	Sour cream	

Per serving: Calories, 350; Protein, 18 g; Fat, 23 g (63% cal.); Cholesterol, 130 mg; Carbohydrates, 12 g; Fiber, 0 g; Sodium, 480 mg.

Note: This dish is often served as a main course for brunch or lunch. In this case, portion sizes may be increased. Sirniki may also be served with strawberry or other preserves.

Swiss Fondue

PORTIONS: 4   PORTION SIZE: 8 OZ (250 G)

U.S.	METRIC	INGREDIENTS
1	1	Garlic clove
1 pt	500 mL	Dry white wine
1 lb	500 g	Swiss Emmentaler cheese, or half Emmentaler and half Gruyère, grated
1 tsp	5 mL	Cornstarch
3 tbsp	45 mL	Kirsch (see Note)
to taste	to taste	Salt
to taste	to taste	White pepper
to taste	to taste	Nutmeg
2	2	Small loaves French bread, cut into bite-size pieces

PROCEDURE

1. Cut the garlic clove in half. Rub the inside of a 1½-qt (1.5-L) fondue pot or casserole with the garlic.

2. Add the wine to the pot and set over moderate heat. Heat the wine until it is hot but not simmering. Do not boil.

3. Add the cheese to the wine, about one-fourth at a time. Stir well between each addition.

4. Dissolve the cornstarch in the kirsch. Stir into the cheese mixture. Stir over very low heat until smooth and slightly thickened.

5. Season to taste with salt, white pepper, and just a trace of nutmeg.

6. Set the casserole over a chafing dish heating element for service. Serve the bread cubes in baskets. To eat fondue, the diner spears a cube of bread on a special fondue fork and swirls it in the cheese mixture, which is kept hot over the heating element.

Per serving: Calories, 870; Protein, 43 g; Fat, 35 g (36% cal.); Cholesterol, 105 mg; Carbohydrates, 70 g; Fiber, 3 g; Sodium, 1000 mg.

Note: Kirsch is a white (that is, clear) alcoholic beverage distilled from cherries. While it is traditional in Swiss fondue, it may be omitted if unavailable. In this case, dissolve the cornstarch in cold water, or mix it with the grated cheese.

Gorgonzola "Truffle"

YIELD: APPROX. 6 OZ   PORTIONS: NUMBER OF PIECES VARIABLE, DEPENDING ON SIZE

U.S.	METRIC	INGREDIENTS
4 oz	120 g	Gorgonzola cheese
2 oz	60 g	Cream cheese
¼ tsp	1 mL	Fresh thyme, chopped fine
1 tsp	5 mL	Shallot, chopped fine
¼ tsp	1 mL	Garlic, chopped fine
to taste	to taste	Kosher salt
to taste	to taste	Black pepper
		Standard Breading Procedure:
as needed	as needed	Flour
as needed	as needed	Egg wash
as needed	as needed	Bread crumbs, preferably fine panko

PROCEDURE

1. Combine the cheeses, thyme, shallot, and garlic in a small bowl and mix well with a spatula. Add salt and pepper to taste and mix again until evenly blended.

2. Using a small round scoop, portion the cheese mixture into balls of the desired size and drop onto sheet pans lined with parchment. Freeze slightly to make the mixture firmer.

3. Roll the cheese portions into round balls.

4. Set up a breading station as described on p. 151. Bread the cheese balls by passing them through the flour, egg wash, and crumbs.

5. Double-bread the balls by dipping them a second time into the egg wash and then into the crumbs.

6. Place the balls again onto a sheet pan and freeze.

7. Deep-fry at 360°F (182°C) until light golden brown.

Per 1 oz (28.35 g): Calories, 320; Protein, 10 g; Fat, 18 g (51% cal.); Cholesterol, 90 mg; Carbohydrates, 28 g; Fiber, 2 g; Sodium, 340 mg.

Cheese Wafers

YIELD: ABOUT 75 WAFERS

U.S.	METRIC	INGREDIENTS
5 oz	500 g	Sharp cheddar cheese, grated
4 oz	250 g	Butter, softened
6 oz	375 g	Bread flour
1/4 tsp	2 mL	Salt
1/8 tsp	1 mL	White pepper

PROCEDURE

1. Combine all ingredients in the bowl of a mixer. Mix at low speed with the paddle attachment until the mixture forms a uniform dough.
2. Remove from the mixer and knead lightly on a floured board until the dough holds together well.
3. Divide the dough into 2 or 3 pieces. Roll each piece into a cylinder 1 in. (2.5 cm) in diameter. Wrap in waxed paper or plastic film and chill.
4. Slice the dough into thin rounds a little less than 1/4 in. (0.5 cm) thick. Place on greased baking sheets.
5. Bake at 425°F (220°C) about 10 minutes, or until crisp and lightly browned.
6. Serve hot or cold as an hors d'oeuvre or as a soup accompaniment.

Per 1 wafer: Calories, 30; Protein, 1 g; Fat, 2.5 g (65% cal.); Cholesterol, 5 mg; Carbohydrates, 2 g; Fiber, 0 g; Sodium, 40 mg.

VARIATIONS

Cheese Straws

Roll out the dough like pie dough, slightly less than 1/4 in. (0.5 cm) thick. Cut into strips, 1/4 × 3 in. (0.5 × 7.5 cm). Bake as in basic recipe.

Cheese Wafers

COFFEE AND TEA

COFFEE

Many people judge a restaurant by its coffee. Regardless of the quality of the food, one of the things they are most likely to remember about an establishment is whether the coffee is good or bad.

Whether or not that seems fair to you, it is at least a clear signal that you ought to learn to make coffee properly. Coffee making is a simple procedure. All you do is pass hot water through ground coffee. The care with which you perform this operation, paying attention to all the details, makes the difference between a rich, aromatic, satisfying beverage and a bitter, unpleasant liquid.

VARIETIES, ROASTS, AND BLENDS

Coffee roast varieties, top to bottom: city roast, medium dark roast, French roast

Coffee beans are harvested as berries from a tropical shrub. Two species of coffee account for nearly all the beans grown worldwide: *Arabica* coffees are delicate plants, difficult to grow. Nevertheless, they supply 65 to 75 percent of the world's coffee. Most of the finest coffees are from arabica beans. *Robusta* coffee plants are hardier and easier to grow. Most ordinary supermarket coffee is made from robusta beans, but robusta can also yield beans of very high quality.

Each coffee berry contains two seeds. The harvested berries are fermented and hulled, yielding green or gray-green coffee beans. The green beans are roasted to develop their flavor. The degree of roasting—light, medium, dark—affects the flavor. Most Americans drink medium roast, sometimes called *city roast*, while darker roasts, sometimes called *Viennese roast* (medium dark) and *French roast* (very dark), are popular in Europe and increasingly popular in North America. *Espresso roast* is also a dark roast and is brewed using a special process discussed below.

BASIC PRINCIPLES of Coffee Making

Coffee is made by extracting flavors from ground coffee beans by dissolving them in hot water. The essence of making good coffee is to extract enough of these solids to make a flavorful beverage, but not to brew so long as to extract those solids that make the coffee bitter.

With this principle in mind, study the following guidelines for making good coffee. The list is long, and every item is important.

1. Use fresh coffee.

Once it is ground, coffee loses flavor and aroma rapidly. To maintain freshness, store coffee, tightly sealed, in a cool, dry place. Even with the best storage, however, you should not use coffee more than a week old. Vacuum-packed coffee keeps longer, but it too deteriorates as soon as it is opened. If you can't grind your own coffee (some restaurants do), at least you can arrange for frequent delivery.

2. Use the right grind and the right brewing time.

A coarse grind requires more time for extraction than a fine grind. You must use the grind that is suited to your equipment.

Grind	Extraction Time
Fine or vacuum	2–4 minutes
Drip or urn	4–6 minutes
Regular (percolator)	6–8 minutes

3. Use the right proportions.

Always measure. Recommended proportions are 1 pound of coffee and 1³/₄ to 2¹/₂ gallons of water (500 g coffee and 7.5 to 10.5 L water), depending on the strength desired.

To make weaker coffee, add more hot water after removing the used grounds. Using more water while actually making the coffee extends the brewing time, resulting in overextraction and bitterness. In fact, many experts feel that passing no more than 2 gallons of water through 1 pound of ground coffee (8 L to 500 g) and then diluting to taste is the surest way to avoid bitterness.

Coffee strength is a matter of customer preference and varies from region to region. For example, people in New York generally prefer stronger coffee than people in Chicago. In some areas, the preferred ratio is 1 pound (500 g) coffee to 3 gallons (12 L) water.

4. Use fresh water.

Fresh, cold water brought to a boil contains dissolved air. Water that has been kept hot for a long time does not, so it tastes flat, and it makes flat-tasting coffee.

Tap water is usually best to use. Special filtration systems are available for tap water that has off flavors or is heavily chlorinated. Do not use chemically softened water.

5. Use water at the right brewing temperature: 195° to 200°F (90° to 93°C).

Water that is too hot extracts bitter solids. Water that is too cold does not extract enough flavor and yields coffee that is too cool for serving.

6. Use a good brewing procedure.

Most operations use either urns, for large volume, or automatic drip makers, which make one pot at a time, as shown in the photograph. These machines can make excellent coffee because water passes through the grounds only once.

Percolator-type coffee makers should not be used. They boil the coffee as it is being brewed and pass it through the grounds repeatedly.

7. Use clean equipment.

Urns and coffee makers must be cleaned every day. Coffee leaves oily deposits that quickly turn rancid or bitter and can ruin the next batch of coffee.

8. Use good filters.

Good filters are the only way to ensure sparkling, clear coffee. Most operations use paper filters, which are discarded after use. If cloth filters are used, they must be perfectly clean and free from odors.

9. Use proper holding procedures.

Proper holding temperature is 185°–190°F (85°–88°C). Higher temperatures decompose the coffee quickly. Lower temperatures mean the customer gets cold coffee.

Coffee made in carafe-type coffee makers is usually kept warm over electric burners.

Do not hold brewed coffee over heat longer than 30 minutes. After this time, loss of quality is considerable. If it must be held longer, transfer it as soon as it is brewed to preheated Thermos containers. Plan production so coffee is always fresh. Discard old coffee.

Decanter-type automatic drip coffee maker
Courtesy of Cecilware.

PROCEDURE for Making Coffee in an Urn

1. Be familiar with your equipment. Models differ in details.

2. Check to make sure the urn holds sufficient fresh water at the proper temperatures for brewing.

3. Fit the filter securely in place.

4. Spread a measured amount of coffee evenly in the filter. An even bed is necessary for uniform extraction.

5. Pass the correct amount of water through the ground coffee. If the urn is manual, pour the water slowly in a circular motion. If it is automatic, all you need to do is make sure the nozzle is in place.

6. Keep the top covered during brewing to retain heat.

7. Remove the filter with the used grounds as soon as brewing is complete. Leaving the grounds in the urn results in overextraction and bitterness.

8. Mix the coffee. Because the coffee at the bottom is stronger, you must draw out some of it—about 1 gallon (4 L) per pound (500 g) of coffee—and pour it back into the top of the urn.

9. Hold at 185°–190°F (85°–88°C) for up to 1 hour.

10. Clean the urn thoroughly after use.
    Using special urn brushes, clean the inside of the urn as well as inside spigots and glass gauges. Rinse and fill with several gallons of fresh water if the urn is to stand for a time. Empty and rinse with hot water before next use.

    Twice a week, clean thoroughly with urn cleaning compound, following manufacturer's instructions.

Twin urn
Courtesy of Cecilware.

Coffee is grown in many tropical countries, and each producing area is known for certain quality and flavor characteristics. Excellent coffees are grown in Colombia, Brazil, Venezuela, Mexico, Jamaica, Hawaii, Indonesia, and nations in Africa and the Middle East.

Most ground coffees are blends of several varieties. Blending enables the processor to combine desirable quantities from a number of beans to produce a well-balanced beverage.

Coffee may be purchased in whole bean or ground form. Whole beans stay fresh longer, but unopened vacuum packs of ground coffee keep well for a week or two and, for many establishments, are the easiest and most economical way to buy coffee. Ground coffee, once opened, should be kept in airtight containers and used within a few days. Better yet, buy ground coffee in premeasured packs suitable for your brewing equipment. Whole beans keep several weeks once opened and months in the freezer. The best practice, though, is to have frequent small deliveries so you always have the freshest coffee on hand.

Brewing procedures for fresh coffee are discussed in the procedures below.

In addition to the standard cup of regular hot coffee, the following coffee drinks are sometimes served.

1. **Instant coffee** is a powdered, soluble extract from coffee beans. To simplify somewhat, instant coffee is made by brewing regular coffee and drying it. In the process, the coffee loses a portion of its flavor and aroma. Most coffee lovers agree that it does not taste as good as freshly brewed coffee. Instant coffee is rarely used in food service.

2. **Decaffeinated coffee**. Caffeine is a chemical stimulant that occurs naturally in coffee, tea, and chocolate. Decaffeinated coffee is coffee from which the caffeine has been removed. Most decaffeination processes use chemical solvents, although the Swiss water process uses water only. Decaffeinated coffee is often specially requested by some customers. In the past, most restaurants offered decaffeinated coffee only in the instant form. Now, however, nearly all restaurants serve freshly brewed decaffeinated coffee.

3. **Espresso** or **expresso** is a strong, dark coffee made from beans roasted until they are almost black and ground to a powder. Espresso is served in small cups as an after-dinner beverage. In addition, several popular coffee drinks are made with brewed espresso, as discussed in the next section.

4. **Iced coffee** is made from double-strength brewed coffee to compensate for dilution by melting ice.

An espresso maker
Courtesy of Cecilware.

Espresso Drinks

Espresso and espresso drinks have grown in popularity in recent years, and specialty coffee shops have sprung up seemingly on almost every street corner.

Espresso is a concentrated coffee beverage brewed in special machines (see photo) from dark-roast, finely ground beans. For each drink, the ground coffee is packed into a small metal filter, which is then firmly attached to the machine. Water is forced through the grounds under high steam pressure, making a small cup of strong beverage topped with a layer of rich foam called the *crema*. A shot of espresso is about 2 fluid ounces (55 mL), and it is usually served in a tiny cup or as a double shot in a standard coffee cup.

In addition, the following drinks are made with brewed espresso:

Cappuccino (cap oochee no): equal parts espresso and frothy steamed milk.

Latte (lah tay): short for *caffè latte*, or "coffee milk." One part espresso to two (or more) parts steamed milk, without the layer of foam found on cappuccino. The French *café au lait* (cah fay oh lay; meaning "coffee with milk") is basically the same as a latte, but it may be made with strong regular dark roast coffee rather than espresso.

Macchiato (mahkee ah toe): espresso topped with a little frothed milk.

Americano: espresso diluted with hot water.

Breve (bray vay): espresso with steamed half-and-half.

Mocha (moh kah): espresso mixed with hot chocolate or cocoa, topped with whipped cream.

Specialty coffee shops also sell a great variety of flavored espresso coffee drinks, hot and cold, using flavored syrups such as caramel, vanilla, and mint.

TEA

Tea is one of the world's most popular beverages, and it is widely drunk even in coffee-drinking countries. In many regions, tea is a much more popular beverage in the home than in the restaurant. Part of the difference may be due to mishandling in the restaurant and indifference on the part of restaurateurs.

Food-service professionals would do well to pay more attention to tea. First of all, it is much less expensive than coffee to serve. One pound (500 g) of tea yields 200 servings, as compared with 40 servings from 1 pound (500 g) of coffee. Moreover, tea is one of the simplest of beverages to serve and does not require the equipment or the labor of coffee service.

VARIETIES

All the world's varieties of tea are produced from one species of evergreen shrub. Most of the differences among varieties are the results of growing conditions and modifications in processing techniques.

As in the case of coffee, different regions produce teas of different quality and flavor characteristics. Most of the tea consumed in North America is imported from India and Sri Lanka (Ceylon).

Variations in processing produce three categories of tea. *Black tea* is fermented by allowing the freshly harvested leaves to oxidize in a damp place. *Green tea* is dried without fermenting. *Oolong tea* is partially fermented to a greenish-brown color. Specialty teas and flavored teas are also available.

Tea varieties, from top to bottom: black tea, green tea, oolong tea.

Black teas are graded by leaf size according to a rather complicated system. This is important to remember because most people think of orange pekoe as a variety of tea, whereas it is actually a specific leaf size of any black variety.

After grading, teas are blended to ensure consistency and uniformity. A blend may contain as many as 30 individual teas.

Many excellent blends are available from numerous purveyors. A smart food-service operator would do well to shop around rather than serve the same mediocre blend the competition serves.

Herbal Teas and Specialty Tea Drinks

Herbal teas are beverages that are brewed like tea but that are made with herbs, spices, dried fruits, and other plant ingredients in place of tea leaves, or sometimes in addition to tea leaves. In other words, most herbal tea contains no actual tea. Hundreds of flavors and blends are available, far too many to list, and most purveyors have their own proprietary blends. Mint and chamomile are among the most popular herb teas.

Chai (rhymes with pie) is a spiced milk and tea blend that originated in India and has become popular in the West. *Chai* is the word for "tea" in several languages. The mixed beverage called *chai* is made of black tea, milk, spices such as cardamom, cinnamon, ginger, and black peppercorns, and sugar or another sweetener. In the West, chai is generally made from commercial mixes.

PACKAGING AND MARKET FORMS

Tea is packaged in bulk as loose tea and in tea bags of various sizes. Standard cup-size bags are packaged 200 to the pound (500 g), while the pot-size bag (that is, individual service pot) is packaged at 150 to 175 per pound (500 g). This is important for you to know if you are purchasing tea because the larger bags will not be as economical if the service in your establishment is by the cup.

Larger tea bags that contain 1 or 2 oz (30 to 60 g) of tea are available for brewing larger quantities, especially for iced tea.

Instant tea is a soluble extraction made by brewing a very strong tea, using lesser grades, and drying the liquid to obtain a powder. This product is used primarily for iced tea because the processing results in the loss of much of the flavor and aroma essential to a good hot tea.

PREPARING TEA

In most restaurants, it seems, when one orders tea, one receives a cold cup, a tea bag in a little package, and a pot of warm water that has been standing in an urn for hours. This is absolutely the worst possible way to serve tea, with the possible exception of brewing a large quantity and keeping it warm all day. No wonder most people don't order tea.

Here is the right way:

PROCEDURE for Making Hot Tea

1. Use proper proportions of tea and water. One teaspoon (5 mL) loose tea or one single-service tea bag makes a 6-oz (175-mL) cup.

2. Rinse the teapot with hot water to warm it. Use china, glass, or stainless steel. Other metals may give an off flavor.

3. Bring fresh, cold water to a boil. Water that has been kept warm for a time makes flat-tasting tea.

4. Place the loose tea or tea bag in the pot and pour the water directly over it.

5. Let the tea steep 3–5 minutes. Then remove the tea bag or strain off the tea from the loose leaves.

   Establishments specializing in tea service present the customer with the pot of tea and a pot of hot water so they can dilute the tea to taste.

6. Serve immediately. Tea does not hold well.

PROCEDURE for Making Iced Tea

The following method makes 1 gallon (4 L). The tea is brewed stronger to allow for melting ice.

1. Place 2 ounces (60 g) tea in a pot.

2. Bring 1 quart (1 L) water to a boil and pour over the tea.

3. Steep 5 minutes. Remove tea bags, or strain out loose leaves.

4. Add 3 quarts (3 L) cold tap water.

5. Hold at room temperature up to 4 hours. Refrigeration may make the tea cloudy.

6. Serve over ice.

KEY POINTS TO REVIEW

- What nine guidelines should be observed when making coffee?

- What is the procedure for making coffee in an urn?

- What is espresso? What popular drinks are made from espresso?

- What is the procedure for making hot tea? Iced tea?

TERMS FOR REVIEW

pasteurized

ultra-pasteurized

ultra-high-temperature (UHT) pasteurization

whole milk

milk fat

butterfat

skim (nonfat) milk

two percent milk

flavored milk

homogenized milk

whipping cream

double cream

table cream

half-and-half

sour cream

crème fraîche

buttermilk

yogurt

evaporated milk

condensed milk

dried whole milk

nonfat dry milk

curdling

scorching

butter

unsalted butter

clarified butter

margarine

ripening

double-crème cheese

triple-crème cheese

chèvre

process cheese

specialty cheese

artisan(al) cheese

farmstead cheese

instant coffee

decaffeinated coffee

espresso/expresso

iced coffee

cappuccino

latte

café au lait

macchiato

Americano

breve

mocha

herbal tea

chai

QUESTIONS FOR DISCUSSION

1. What is curdling, and how can you prevent it when cooking with milk?

2. What is washed-rind cheese? Describe the typical appearance of washed-rind cheese, and compare it to mold-rind cheese.

3. Why does cheese combine more smoothly with a sauce at low heat than at high heat?

4. Why is using the proper grind important in making coffee?

5. Describe the procedure for making coffee in an urn.

6. Describe the proper method for making tea.

26

COOKING
FOR VEGETARIAN DIETS

Vegetarian diners are an important and growing segment of the dining public. In the United States, for example, it is estimated that about 15 million people consider themselves vegetarians. Young people, in particular, embrace many forms of vegetarianism. In college dining rooms, the proportion of clients choosing vegetarian options may be as high as 40 percent. Clearly, food service cannot afford to ignore this segment.

It is important that cooks and chefs who want to please their customers know something about the needs of vegetarian diners. Beyond the financial benefits of serving foods that appeal to the widest range of customers, chefs find other benefits as well. Vegetarians are often more knowledgeable and enthusiastic about their dining choices because they have thought more about them. Accomplished cooks often say that meeting the challenge of cooking for knowledgeable diners is one of the most satisfying aspects of their jobs and that vegetarian menus give them new opportunities for creativity.

In addition to people who have a strong commitment to vegetarianism, many others eat meat on other occasions but choose vegetarian items simply because the choices are so appealing in a particular dining facility. Chefs who create satisfying, innovative choices for vegetarians often find that creating a good vegetarian menu selection is one of their best professional decisions.

AFTER READING THIS CHAPTER, YOU SHOULD BE ABLE TO

1. Describe the main types of vegetarian diets.

2. Describe complementary proteins and describe how to include them in the diet.

3. List three nutrients other than proteins that nonvegetarians get mostly from animal products, and describe how vegetarians can include these nutrients in their diet.

4. Name and describe five food types derived from soybeans.

5. Explain why refined sugar may not be permitted in a vegan diet.

6. List seven guidelines for building a vegetarian menu.

UNDERSTANDING VEGETARIAN DIETS

Preparing food for anyone who follows a restricted diet requires understanding the nature and limitations of that diet. Vegetarian diets present challenges because there are several types of vegetarianism.

TYPES OF VEGETARIAN DIET

A **vegetarian** diet is one consisting entirely or mostly of foods derived from plants. Most committed vegetarians fall into one of the following categories:

The **vegan** diet is the most restrictive form of vegetarianism. Vegans eat plant products only. All animal products, including dairy products and eggs, are off limits. Even foods that might sound safe are off limits to the strictest vegans. Examples of such foods include honey, because it comes from bees, and cane sugar, which may be refined with the use of animal products (more on this subject on p. 813). When preparing a vegetarian menu, the chef should keep in mind that a menu appropriate to a vegan diet has the broadest appeal because it can be eaten by all categories of vegetarians.

Lacto-vegetarians eat dairy products in addition to plant products but will not eat other animal products.

Ovo-vegetarians eat eggs in addition to plant products.

Lacto-ovo-vegetarians eat dairy and egg products as well as plant products.

Pesco-vegetarians eat fish and plant products but not meat or poultry. They may or may not eat dairy and egg products.

Vegetarianism may be based on strong ethical or moral beliefs or on health concerns. Naturally, the chef who cares for his or her customers is eager to respect these beliefs and concerns. Vegetarians may have chosen their diet based on deeply held ethical or religious beliefs and may be dedicated to following their diet rigorously.

In addition, many people choose vegetarianism for health reasons. Vegetarian diets are usually low in fat and cholesterol and in addition are free of the hormones and drugs often used in the raising of meat animals. Environmental concerns also lead some people to vegetarianism. Producing plant foods requires fewer natural resources than raising meat animals. Economic factors are yet another consideration in that vegetables and grains are, on average, much less expensive than meat, poultry, and seafood. Finally, some people are occasional vegetarians simply because they enjoy the food.

NUTRITIONAL CONSIDERATIONS

Because vegetarians eliminate major categories of food from their diet, nutrients otherwise obtained from animal products are lost and must be obtained from other foods. Refer to the MyPlate image and the various food pyramids on pages 97–99. Note that although dairy products, meats, fish, and eggs do not form the largest portions of the diet, they form an important part. When those foods are eliminated, the structure of the diet must be rebuilt, as in Figure 26.1, in order to ensure adequate nutrition.

Protein

The subject of complete proteins and complementary proteins was introduced in Chapter 5 (see p. 94). Because this subject is so important for vegetarian diets, it is discussed in greater detail here.

The major nutritional concern of a vegetarian diet is getting enough protein. Dairy products, eggs, and fish supply adequate amounts of good-quality protein, but vegans must plan their diet carefully in order to get adequate protein. Some plant products, such as grains, nuts, and dried beans, contain proteins. In the traditional food guide pyramid, dried beans and nuts are included in the meat group. However, with the important exception of soybeans and soy products such as tofu, most of these protein foods, **_when eaten alone_**, are not adequate for human nutrition.

AMINO ACIDS

These 9 compounds are called essential amino acids:

Histidine	Phenylalanine
Isoleucine	Threonine
Leucine	Tryptophan
Lysine	Valine
Methionine	

These 11 amino acids can be made by the body and so are called *nonessential amino acids*:

Aniline	Glutamine
Arginine	Glycine
Asparagine	Proline
Aspartic acid	Serine
Cysteine	Tyrosine
Glutamic acid	

It is not necessary to include nonessential amino acids in the diet.

When the body manufactures proteins, it puts together a chain of amino acids using those it has available, like a factory assembling an appliance out of parts. If it finds that one of the parts—in this case, amino acids—is missing, it takes apart the partial protein it has already assembled and sends the parts back to the supply room—the bloodstream.

This means that if one amino acid is in short supply, it limits the usefulness of those that are plentiful. An amino acid that is in short supply, thus limiting the usefulness of the others, is called a **limiting amino acid**.

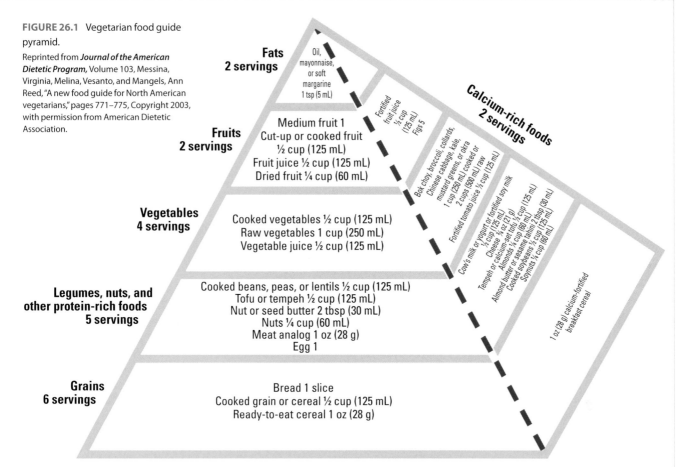

FIGURE 26.1 Vegetarian food guide pyramid.

Reprinted from *Journal of the American Dietetic Program,* Volume 103, Messina, Virginia, Melina, Vesanto, and Mangels, Ann Reed, "A new food guide for North American vegetarians," pages 771–775, Copyright 2003, with permission from American Dietetic Association.

Fats
2 servings

Oil, mayonnaise, or soft margarine 1 tsp (5 mL)

Fortified fruit juice ½ cup (125 mL) Figs 5

Calcium-rich foods
2 servings

Fruits
2 servings

Medium fruit 1
Cut-up or cooked fruit ½ cup (125 mL)
Fruit juice ½ cup (125 mL)
Dried fruit ¼ cup (60 mL)

Bok choy, broccoli, collards, Chinese cabbage, kale, mustard greens, or okra 1 cup (250 mL) cooked or 2 cups (500 mL) raw

Fortified tomato juice ½ cup (125 mL)

Cow's milk or yogurt or fortified soy milk ½ cup (125 mL)
Cheese ¾ oz (21 g)
Tempeh or calcium-set tofu ½ cup (125 mL)
Almonds ¼ cup (60 mL)
Almond butter or sesame tahini 2 tbsp (30 mL)
Cooked soybeans ½ cup (125 mL)
Soynuts ¼ cup (60 mL)

1 oz (28 g) calcium-fortified breakfast cereal

Vegetables
4 servings

Cooked vegetables ½ cup (125 mL)
Raw vegetables 1 cup (250 mL)
Vegetable juice ½ cup (125 mL)

Legumes, nuts, and other protein-rich foods
5 servings

Cooked beans, peas, or lentils ½ cup (125 mL)
Tofu or tempeh ½ cup (125 mL)
Nut or seed butter 2 tbsp (30 mL)
Nuts ¼ cup (60 mL)
Meat analog 1 oz (28 g)
Egg 1

Grains
6 servings

Bread 1 slice
Cooked grain or cereal ½ cup (125 mL)
Ready-to-eat cereal 1 oz (28 g)

Proteins are long chains of smaller compounds called **amino acids**. There are, in all, 20 amino acids that, when joined in various combinations, make up over 100,000 proteins in the human body. Eleven of these amino acids can be made in the body, so it is not necessary to include them in the diet. All remaining nine amino acids must be included in the diet in order for the body to make all the proteins it needs. These nine are called **essential amino acids**.

Any food protein that contains all nine essential amino acids is called a **complete protein**. Proteins found in meat, poultry, seafood, milk and milk products, and eggs are complete proteins.

Some plant foods, especially dried legumes, grains, nuts, and seeds, contain **incomplete proteins**. This means that one or more of the essential amino acids is either missing or is not present in high enough concentration. Soybeans, quinoa, and amaranth are unusual among grains and legumes in that they contain complete proteins.

The key to getting enough protein in a plants-only diet is to eat, in the course of each day, a balance of these foods, so that amino acids missing from one of these foods is supplied by another one of them. Such proteins are called **complementary proteins**. For example, kidney beans are high in the amino acids isoleucine and lysine, but low in some of the others. Millet is low in lysine but high in the amino acids that kidney beans are missing. So if both kidney beans and millet are eaten during the day, all the essential amino acids are included in the diet.

Including Complementary Proteins in the Diet

Contrary to what you might think after reading the preceding discussion, you do not have to be a biochemist to prepare vegetarian menus. A basic understanding of what foods go together to supply complete proteins is the best place to start and will carry you a long way.

The following pairings of food categories are the most useful complementary proteins for planning vegetarian diets:

Dried legumes plus grains

Dried legumes plus seeds and nuts

Grains plus milk products

The first two of these pairings are important in vegan diets. The third pairing can be included in the diet of lacto-vegetarians.

Examining the cultures and cuisines of other lands, we see these complementary protein groupings have long been a part of the staple diet of peoples with limited supplies of meat. Think, for example, of the beans and corn tortillas (dried legumes plus grains) of Mexico and the rice and dal (also grains plus dried legumes) of India. People who have long relied on these foods have found tasty and varied ways to prepare them. Studying traditional vegetarian cuisines is a useful way to learn how to include these items in your own menus.

Other Nutrients

In addition to protein, other nutrients normally found in animal products must be supplied in other ways in vegetarian diets:

Vitamin B$_{12}$. This vitamin is found only in animal foods, including milk and eggs. Vegans must obtain it from grain foods, such as breakfast cereals, that have been fortified with this vitamin, or else take vitamin supplements. Vegetarians who eat sufficient dairy products and eggs usually can get enough vitamin B$_{12}$.

Vitamin D. This vitamin is found in vitamin D-fortified milk, and it is created in the skin on exposure to sunlight. A vegan or other vegetarian who doesn't get enough exposure to sunlight can get this nutrient from vitamin-fortified cereals or some soy beverages.

Calcium. Dairy products are rich in calcium, but vegans and other vegetarians who don't consume dairy products must get calcium from other sources, including green leafy vegetables and dried legumes. Calcium supplements or calcium-fortified beverages may be necessary in the vegan diet.

KEY POINTS TO REVIEW

- What are the main types of vegetarian diet? What kinds of food can be eaten in each of these diets?

- What are complete proteins? What are complementary proteins? How can complementary proteins be included in the diet?

- In addition to protein, what other nutrients are of special concern to people who follow a vegetarian diet? How can these nutrients be included in the diet in healthy quantities?

MENUS FOR VEGETARIAN DIETS

To a meat-eater, the idea of a vegetarian diet might seem monotonous or boring, but the reality can be the opposite. A typical meat-based diet often features the same few meats—beef, pork, chicken, and occasionally fish—over and over, accompanied by the same few simple vegetable side dishes and potatoes. By contrast, a diet in which vegetables are the focus and not a routine side dish can feature dozens of fresh vegetables, a great variety of grains and legumes, and, for at least some vegetarians, dairy products and possibly eggs to round out the selection.

The appeal of vegetarian menus can be seen in many fine restaurants that feature tasting menus. At such restaurants, the dinner menu each evening may offer a multicourse vegetarian menu in addition to the meat and seafood options. It is common for a pair of diners, even nonvegetarians, to order one vegetarian menu and one nonvegetarian menu so they can sample all the chef's creations.

INGREDIENTS

Because animal protein is omitted from vegetarian diets, some important ingredients get more attention than they do in meat diets. Several categories of plant food take on the role of main courses.

Dried Legumes

Dried beans, peas, and lentils are some of the most important sources of protein for vegans and other vegetarians. These are listed and discussed in detail in Chapter 13. Refer to that information as necessary in your menu planning.

One legume not mentioned earlier is the peanut, which is not a nut but a legume with an unusual growing habit. As the plants grow, the branches bearing the pods bend to the ground so the pods develop underground. Although peanuts are used in the kitchen like other nuts, nutritionally they are similar to other legumes. Keep this in mind when planning complementary proteins.

Dried soybeans are high in protein, but they are rarely cooked and served like other dried beans. They are very hard and take hours of cooking if whole, and when cooked they do not appeal to most tastes. Furthermore, their protein is not as usable as the protein in products derived from soybeans, such as soy milk and tofu. These are so important that they are discussed separately below.

Grains

Like legumes, grains are discussed in detail in Chapter 13. They are sources of complementary protein when eaten with legumes or dairy products. Of the grains described, rice is no doubt the most important worldwide. Many vegans and other vegetarians use brown rice instead of white rice to take advantage of its vitamin and fiber content. Wheat and corn are also important protein foods. In the form of bread, noodles, and tortillas, these grains are especially versatile.

Don't overlook the other grains. Products such as farro, barley, millet, and triticale add variety to the menu. Quinoa and amaranth are especially valuable for vegan diets because they contain complete protein.

Nuts and Seeds

Nuts and seeds are rich in protein, so almonds, cashews, hazelnuts, pecans, walnuts, pistachios, pumpkin seeds, sesame seeds, and sunflower seeds are useful in vegetarian diets. They are high in fat, however, and can't be relied on as heavily as other vegetable proteins.

Nut butters, such as cashew butter and almond butter, are simply nuts or seeds that have been roasted and ground to a paste. They are a flavorful and attractive way to include these foods in a diet. Sesame butter, or tahini, is especially prominent in Mediterranean diets and used in many recipes. See, for example, the recipe for hummus on page 742, which uses legumes (chickpeas) and seeds (tahini) to supply complementary protein.

Soy Products

Borrowed from Asian cuisines, soy derivatives such as tofu have long been familiar in the West and are important sources of protein for many people.

Soy milk is made by soaking dried soybeans, draining them, grinding them, combining them with water, boiling, and then straining the resulting milky liquid.

Plain soy milk can be used in cooking and as a beverage. Many brands of flavored and sometimes sweetened soy beverages are also available, and these can also be used in coffee beverages and in desserts.

Tofu, or **bean curd**, is made by curdling soy milk, just as cheese is made by curdling dairy milk. The result is a soft, white cake with a mild, faintly beany flavor that is adaptable to many sauces and accompaniments.

Tofu is available in several varieties, described here. Most types are available packed in water in sealed plastic packages. It keeps well as long as it is unopened, but it should be used within a day or two of opening.

Japanese silken tofu is the softest type. Unlike other tofu varieties, the curd of silken tofu is not pressed to remove moisture. Silken tofu has the most delicate texture and also the mildest flavor. In Japan it is regularly eaten cold with a light dipping sauce, or it is diced and added as a garnish to miso soup.

Japanese cotton tofu is firmer than the silken variety and can stand up to more vigorous handling. It is often deep-fried or added to stir-fried dishes.

Chinese firm tofu is the firmest of these three tofu types and is the coarsest in texture. It can be fried, stir-fried, grilled, baked in casseroles, simmered in stews, and put on skewers.

Silken tofu

Firm tofu

When tofu is to be marinated, fried, or stewed, removing excess moisture first helps the tofu absorb the added flavors more readily. Place the cakes of tofu on several layers of paper towel or on a perforated sheet pan, cover with several more layers of paper towel, place a weighted tray or sheet pan on top, and let stand about 15 minutes.

To make tofu even firmer and less likely to break apart when cooked, press it as above and then deep-fry until it has a light golden color. Or drop it into simmering water and simmer 5 minutes. Precooking coagulates the proteins, making them firmer.

Tempeh is a fermented soybean product that originated in Indonesia. In addition to the original version made with only soybeans, it is also made from soy plus grains, including barley, millet, and rice. Tempeh is sold in cakes and has a dense, meaty texture. It can be sliced or diced and braised, sautéed, stir-fried, baked, and added to casseroles.

Miso, also called **bean paste**, is a paste made of fermented soybeans, sometimes with the addition of wheat, rice, or barley, depending on the variety. There are several types of miso ranging from light, sweet, and mild to dark, salty, and robust. The lightest, called *white miso* (*shiromiso* in Japanese), is yellow, with a sweet flavor and soft, moist texture. It is often used in soups and in dressings or sauces for vegetables (p. 312). Other light miso types may be darker yellow and saltier. *Red miso* (*akamiso*) is brown or reddish brown, salty, and fuller in flavor. It may be smooth or chunky. Red miso is used in soups and in cooked dishes rather than in uncooked dressings. The thickest, darkest, strongest-tasting miso is *hatcho miso*. It is dark brown and thick enough to cut, almost like fudge. It is used in soups and cooked dishes.

Textured vegetable protein (TVP) is made from defatted soy flour, processed and dried to give it a spongelike texture. It is available unflavored or flavored to resemble various meats. To prepare TVP, mix with water, let stand until softened, and then add to recipes as you would meat. Because it doesn't taste exactly like real meat, it is best used in highly flavored or seasoned dishes such as chili. TVP is high in protein and fiber and free of cholesterol. It is useful on some menus for its nutritional value, but some vegetarians avoid it because they are not interested in highly processed foods that pretend to be meat.

Dairy and Eggs

For lacto-ovo vegetarians, milk products and eggs are valuable sources of protein, vitamin D, calcium, and other nutrients. Cheeses and eggs, however, are high in fat and cholesterol, so it is a good idea not to rely on them for one's sole source of protein.

MODIFYING RECIPES

When modifying recipes for vegetarian menus, you must first determine your customers' needs. Do you want to build a purely vegan menu, or will a less restrictive vegetarian selection work? The presence of any egg or dairy product indicates that the recipe cannot be used on a vegan menu, although it is suitable for other vegetarians. The first step in modifying a recipe is to read the list of ingredients carefully to make sure they are all permitted. If any of the ingredients is a packaged or manufactured food, read the list of ingredients on the package as well. Remember:

- No animal product of any kind may be used as an ingredient in a recipe intended for vegans.
- Dairy products (such as milk, cream, butter, and cheese) and eggs may be included as ingredients in recipes for lacto-ovo vegetarians, and thus they may be labeled as vegetarian. However, they will not be eaten by vegans.
- Service staff must be adequately trained to accurately answer patrons' questions about the menu.

Sometimes a soup, salad, or side dish recipe consists primarily of plant products but contains one or more animal products, such as a stock. The following methods can be used to modify existing recipes to adapt them for vegetarian menus:

1. Remove meat, poultry, and fish stocks from recipes and substitute an equal quantity of water or vegetable stock.
2. Omit sauces based on meat, poultry, or fish stocks, and, for vegans, omit sauces using dairy products or eggs. Substitute another suitable sauce. Vinaigrettes, salsas, vegetable coulis, and flavored oils are often good sauces for vegetarian dishes.
3. For vegans, substitute oil for butter. Replace dairy products with soy milk or other dairy substitutes.

Sugar in Vegetarian Recipes

Refined white sugar and most brown sugar are processed either from sugar cane or sugar beets. Refineries that process sugar from cane often use charcoal made from animal bones as a filter. Sugar made from beets, however, is not made with bone char. Vegans and many other vegetarians often avoid all refined sugar because of the difficulty of determining the sugar's origin. For home use, they may use unbleached sugar or beet sugar, but away from home it is easier just to avoid sugar altogether.

Some of the recipes in this book labeled as vegetarian have sugar as an ingredient. *If you use any of these recipes on a vegetarian menu, make sure to use sugar from a purely vegetarian source, such as beet sugar or unbleached sugar.*

Other recipes in this book may appear to be vegetarian but are not labeled as such. For example, the recipe for Cocktail Sauce (p. 217) contains only plant products, and sugar is not listed as an ingredient. However, two of the ingredients, ketchup and chili sauce, contain sugar. Because there is no way to tell the source of the sugar, the recipe can't easily be adapted to a vegetarian menu. In this book, no recipe is labeled as vegetarian if one of the ingredients is a manufactured product that may contain sugar or other impermissible ingredients.

DEVELOPING APPEALING MENUS

One of the problems in developing vegetarian menu choices is constructing a plate that is satisfying as a main course. Appetizers or first courses are less of a problem. Vegetarians and nonvegetarians alike enjoy soups, salads, and vegetable dishes as first courses. Similarly, a succession of 4–6 small courses in a tasting menu can be pleasing and satisfying. A main course, however, must usually feel like a complete meal in its own right, and a selection of steamed vegetables on a plate usually will not satisfy.

Center of the Plate

In the world of food service, the expression ***center of the plate*** is often used to describe the meat, poultry, fish, or other principal item that is the focus of the main course. For vegetarian menus, we need to rethink the concept of center of the plate somewhat.

Begin by thinking about the categories of foods we rely on for complementary proteins: legumes, grains, seeds and nuts, and dairy products, and, if appropriate, soy products like tofu. Dishes made with these ingredients tend to be more substantial than simple vegetables and thus are often suitable for main courses, especially when combined with vegetables and perhaps a sauce or condiment.

Complex dishes—dishes that have had something done to them beyond simple boiling or steaming—are appealing vegetarian main courses. For example, a baked squash stuffed with a well-seasoned grain and vegetable mixture holds its own in the center of the plate. Regional cuisines give us other examples, such as a bean-filled enchilada served with rice and a spicy tomato sauce.

Always keep in mind the vegetable-cooking methods and guidelines explained in other chapters of this book. On a vegetarian menu, vegetables are no longer just side dishes but are the center of attention, so you must use your best skills to prepare them and make them attractive.

Some preparations work as main courses whether or not they are made with meat. These types of dishes can be prepared many ways using only plant products, with or without the use of dairy products or eggs. Examples follow:

Stews

Curries

Chili

Stir-fries served with rice or noodles

Risotto

Pasta

Pizza

Pot pies, or foods served inside dough or pastry

Casseroles

Gratins

Finally, remember that it is not always necessary to get all the elements of a well-balanced meal onto one plate. For example, a classic casserole of macaroni and cheese is not a balanced meal in itself, but when accompanied by a fresh green salad or vegetable salad, either as a first course or as a side dish, makes an attractive and reasonably balanced lunch. The objective is to balance the complete meal and not necessarily each individual plate.

Similarly, when you are pairing complementary protein items, such as beans and rice or beans and corn, putting both items on the same plate is not your only option. For example, you could place a scoop of rice in the center of a bowl of bean chili, but you might instead serve cornbread on the side. A meal featuring rice as the main dish, such as a vegetarian risotto, might start with a first course of bean salad or bean soup. Of course, with an à la carte menu you have no control over the guests' selections, but vegetarian diners are generally knowledgeable about their own diets and are happy to take responsibility for making wise choices.

Small Plates

Yet another option lies in the modern popularity of "small plates" menus. Inspired by such concepts as Spain's tapas (see p. 745), Greece's mezze, and China's dim sum, small plates give diners a chance to sample a wide variety of dishes and are increasingly seen in many restaurants. Menus feature small portions of a wide variety of dishes, often from many different cuisines, that can be enjoyed by single diners or shared by groups. For restaurants that want to offer a variety of foods with vegetarian appeal, such a menu concept gives unlimited opportunity without the necessity of designing main-course vegetarian dishes.

VEGETARIAN RECIPES IN THIS BOOK

The ingredient information and prep and cooking procedures in other chapters in this book—especially in the chapters on vegetables; potatoes; legumes, grains, pastas, and other starches; and salads—give you tools you can use to build creative and satisfying vegetarian menus. Many of the recipes in those chapters can be used by vegetarians without modification. In fact, this book contains more than 450 vegetarian recipes, comprising more than 275 main recipes and 175 additional recipe variations. The main recipes are labeled with a small icon, shown in the margin. Look for the icons in the Recipe Table of Contents and also next to the recipe titles in the body of the text.

In addition, examples of other vegetarian items suitable for center-of-the-plate presentations conclude this chapter. A few of these use recipes from other chapters and combine them with other elements to make a more complete meal.

GUIDELINES for Building a Vegetarian Menu

1. Offer dishes containing no ingredients of animal origin, to appeal to vegans, and offer dishes containing dairy products and eggs for vegetarians who eat those foods.

2. Select a variety of legumes, grains, and seeds, and use them in combinations that provide complementary proteins.

3. Complementary proteins don't need to be on the same plate. For example, you could offer a grain salad as a first course and a vegetable stew with beans as a main course.

4. Use a wide variety of vegetables.

5. Offer dishes made with tofu and other soybean derivatives.

6. To limit fat and cholesterol, consider using low-fat milk products, and use eggs sparingly.

7. Offer foods containing nutrients, such as vitamin B_{12}, that are of special concern to vegetarians (see p. 810).

8. Read the ingredient labels of all packaged food products to make sure they contain no ingredients of animal origin.

KEY POINTS TO REVIEW

- What foods are important sources of protein for vegetarians?

- What foods commonly used in vegetarian diets are derived from soybeans?

- What problems does sugar pose for vegan diets?

- What are some ways nonvegetarian recipes can be modified to fit vegetarian diets?

- What guidelines should be followed when building a vegetarian menu?

Yellow Split Pea and Vegetable Curry with Spiced Rice and Cucumber Raita

PORTIONS: 12 **PORTION SIZE:** 3¹/₂ OZ (100 G) RICE, 6 OZ (180 G) VEGETABLE CURRY, 2 OZ (60 ML) RAITA

U.S.	METRIC	INGREDIENTS	PROCEDURE
		Spiced rice:	1. Wash the rice in several changes of cold water. Drain. Add cold water to cover by 1–2 in. (3–5 cm) and soak 1 hour or longer.
1 lb	500 g	Long-grain brown rice, preferably basmati	
2 fl oz	60 mL	Vegetable oil	2. Heat the oil in a heavy saucepan.
4 oz	125 g	Onion, chopped	3. Add the onion, garlic, and ginger. Sauté until the onion is lightly browned.
1 tsp	5 mL	Finely chopped garlic	
1 tsp	5 mL	Finely chopped fresh ginger root	4. Add the spices.
¹/₄ tsp	1 mL	Ground cardamom	5. Drain the rice and add it to the pan. Stir to coat the grains with oil.
¹/₄ tsp	1 mL	Cinnamon	6. Add the water or stock and the salt. Bring to a boil.
¹/₈ tsp	0.5 mL	Ground cumin	7. Cover and simmer over low heat about 40 minutes, or until the rice is tender. Check after about 30 minutes to make sure the liquid hasn't all been absorbed before the rice is done. If it has, add a little more hot water.
¹/₈ tsp	0.5 mL	Ground cloves	
¹/₈ tsp	0.5 mL	Nutmeg	
¹/₈ tsp	0.5 mL	Cayenne	
1 qt	1 L	Water or vegetable stock, hot	
2 tsp	10 mL	Salt	
8 oz	250 g	Yellow split peas	8. Pick over the peas to remove stones and other foreign matter. Rinse and drain the peas.
1 qt	1 L	Water	9. Simmer the peas in the water until they are very tender and falling apart. Do not drain.
2 fl oz	60 mL	Vegetable oil	10. Heat the oil in a heavy saucepot.
8 oz	250 g	Onion, medium dice	11. Add the onion and garlic. Sauté until lightly browned.
2 tsp	10 mL	Finely chopped garlic	12. Stir in the curry powder and salt.
2 tbsp	30 mL	Curry powder	13. Add the carrots, parsnips, and eggplant. Sauté 2–3 minutes over medium heat.
2 tsp	10 mL	Salt	
8 oz	250 g	Carrots, thickly sliced	14. Add the cauliflower. Sauté another minute.
6 oz	180 g	Parsnips, large dice	15. Add the cooked peas. Simmer 15–20 minutes, or until the vegetables are nearly tender.
6 oz	180 g	Eggplant, peeled, large dice	16. Add more water if the mixture becomes too thick. The peas should have the consistency of a sauce.
8 oz	250 g	Cauliflower, broken into small florets	17. Add the broccoli and continue to simmer until all the vegetables are tender. Taste and add more salt if necessary.
8 oz	250 g	Broccoli, broken into small florets	
8 fl oz	250 mL	Plain yogurt	18. Stir in the yogurt and lemon or lime juice.
2 fl oz	60 mL	Lemon juice or lime juice	
1¹/₂ pt	750 mL	Cucumber Raita (p. 209)	19. For each portion, spoon rice to cover about two-thirds of a plate, leaving space at one side. Make a well in the center. Spoon a portion of the vegetable curry into the center of the rice.
as needed	as needed	Cilantro leaves	
as needed	as needed	Paprika	20. To the side of the rice, spoon a portion of the raita.
			21. Top the curry with a few cilantro leaves.
			22. Sprinkle a dash of paprika onto the center of the raita.

Per serving (excluding Raita): Calories, 370; Protein, 12 g; Fat, 13 g (30% cal.); Cholesterol, 5 mg; Carbohydrates, 55 g; Fiber, 6 g; Sodium, 1030 mg.

Per serving (Raita only): Calories, 30; Protein, 2 g; Fat, 2 g (53% cal.); Cholesterol, 0 mg; Carbohydrates, 2 g; Fiber, 0 g; Sodium, 210 mg.

VARIATION

Vegan Vegetable and Split Pea Curry

Omit the yogurt. Increase the lemon juice or lime juice to taste.

Noodle Bowl with Stir-Fried Vegetables, Tofu, and Peanuts

PORTIONS: 12   PORTION SIZE: 4 OZ (125 G) NOODLES, 6 OZ (180 G) VEGETABLES

U.S.	METRIC	INGREDIENTS	PROCEDURE
2 fl oz	60 mL	Vegetable oil	1. Heat the oil in a large sauté pan or wok over high heat.
4	4	Scallions, chopped	
2	2	Garlic cloves, chopped	2. Add the scallions, garlic, and ginger. Stir-fry 1 minute.
1 tsp	5 mL	Finely chopped ginger root	
6 oz	180 g	Carrots, cut julienne	3. Add the carrots, peppers, and mushrooms. Continue to stir-fry another minute.
6 oz	180 g	Red bell peppers, cut bâtonnet	
8 oz	250 g	Shiitake mushroom caps, cut bâtonnet	4. Add the bok choy and bean sprouts. Stir-fry until the vegetables are wilted but still crisp.
1 lb	500 g	Bok choy, cut into 1-in. (2.5-cm) pieces	5. Add the snow peas and continue to cook about 30 seconds.
4 oz	125 g	Mung bean sprouts	
8 oz	250 g	Snow peas, trimmed	6. Add the tofu and peanuts. Toss the mixture to blend, and cook until the tofu is hot.
1 lb 4 oz	625 g	Firm tofu, pressed (pp. 811–812), cut in ¹/₂-in. (1-cm) dice	
6 oz	180 g	Roasted peanuts, shelled, skinless	
3 floz	90 mL	Soy sauce	7. Pour in the soy sauce, hoisin sauce, stock or water, and oil. Toss to mix.
2 floz	60 mL	Hoisin sauce	
4 floz	125 mL	Vegetable stock or water	
1 tbsp	15 mL	Sesame oil or chili oil	
3 lb	1.5 kg	Cooked Chinese wheat noodles or egg noodles, hot	8. Put the noodles into individual serving bowls and top with the vegetable mixture with its liquid.

Per serving: Calories, 400; Protein, 19 g; Fat, 18 g (39% cal.); Cholesterol, 0 mg; Carbohydrates, 45 g; Fiber, 6 g; Sodium, 830 mg.

Noodle Bowl with Stir-Fried Vegetables, Tofu, and Peanuts

Zucchini and Eggplant Lasagne 🌸

PORTIONS: 12   PORTION SIZE: 8 OZ (250 G)

U.S.	METRIC	INGREDIENTS	PROCEDURE
1½ lb	750 g	Eggplant	1. Trim and peel the eggplant.
1½ lb	750 g	Zucchini	2. Trim the stem ends of the zucchini.
as needed	as needed	Salt	3. Cut the eggplant and the zucchini lengthwise into slices about ⅓ in. (8 mm) thick.
as needed	as needed	Olive oil	4. Salt the slices lightly on both sides and let stand 15 minutes.
			5. Pat dry. Brush both sides of each slice lightly with oil.
			6. Arrange the eggplant and zucchini on separate sheet pans.
			7. Bake the eggplant in an oven at 425°F (210°C) until the bottoms are brown. Turn over and bake another 15 minutes.
			8. At the same time, bake the zucchini until softened but still somewhat crisp, 5–10 minutes.
			9. Remove the vegetables from the oven and let cool. These slices will serve as noodles for the lasagna.
2 oz	375 g	Ricotta cheese	10. Mix the ricotta, parmesan, egg, parsley, salt, and pepper.
1 oz	30 g	Parmesan cheese, grated	
1	1	Egg, beaten	
2 tbsp	30 mL	Chopped parsley	
to taste	to taste	Salt	
to taste	to taste	Pepper	
1½ qt	1.5 L	Tomato Sauce for Pasta (p. 399), made without sugar	11. Ladle a little tomato sauce into a half-hotel pan, 10 × 12 in. (25 × 30 cm). Spread it across the bottom.
1 lb	500 g	Mozzarella cheese, sliced	12. Lay the eggplant slices in a single layer over the sauce.
2 tbsp	30 mL	Fresh basil, cut chiffonade	13. Add the ricotta mixture, carefully spreading it in an even layer over the eggplant.
2 oz	60 g	Parmesan cheese, grated	14. Add the zucchini in a single layer.
			15. Cover with half the remaining sauce.
			16. Arrange the mozzarella slices over the top. Sprinkle the mozzarella with a little of the basil.
			17. Cover with the remaining sauce.
			18. Sprinkle with the second quantity of parmesan cheese.
			19. Bake at 375°F (190°C) until hot and bubbling.

Per serving: Calories, 410; Protein, 15 g; Fat, 34 g (73% cal.); Cholesterol, 65 mg; Carbohydrates, 14 g; Fiber, 4 g; Sodium, 1030 mg.

Pinto Bean Enchiladas

PORTIONS: 12   PORTION SIZE: 2 ENCHILADAS

U.S.	METRIC	INGREDIENTS
as needed	as needed	Vegetable oil
24	24	Corn tortillas
2 qt	2 L	Ancho Sauce (made with water) or the variation with tomato (p. 207) or Salsa Roja (p. 206)
2 lb 4 oz	1.1 kg	Vegetarian Pinto Beans (p. 372), warmed
2 lb	1 kg	Monterey jack cheese, grated

PROCEDURE

1. Heat a thin layer of oil in a sauté pan over moderate heat.
2. One at a time, dip the tortillas in the hot oil and fry them a few seconds, first on one side, then the other, just until they are soft. The purpose is to soften them, so do not fry until crisp. Drain and lay them on a work surface or in a pan.
3. Pour 1–2 cups (250–500 mL) sauce in a shallow pan.
4. One at a time, dip the tortillas in the sauce to coat both sides and let the excess drip off. Lay the sauced tortilla on the work surface. Place 1¹/₂ oz (45 g) beans and ¹/₂ oz (15 g) grated cheese in the center of the tortilla and roll up.
5. Add more sauce to the pan as it gets used up. You will need about 1 qt (1 L) of the sauce for dipping.
6. Arrange the rolled tortillas against each other in a baking pan, putting the loose ends on the bottom. Alternatively, place them in individual gratin dishes, allowing 2 per portion.
7. Pour the remaining sauce over the tortillas, making sure to cover the ends so they don't dry out.
8. Top with the remaining cheese.
9. Bake at 375°F (190°C) until heated through, about 20 minutes.

U.S.	METRIC	INGREDIENTS
3 lb	1.5 kg	Arroz Verde or Arroz à la Mexicana (p. 388), made with water instead of stock
as desired	as desired	Guacamole (p. 743)
as desired	as desired	Salsa Cruda (p. 207)
as desired	as desired	Sour cream

10. Serve 2 enchiladas per portion with 4 oz (125 g) rice.
11. Serve with guacamole, salsa, and sour cream as desired.

Per serving: Calories, 830; Protein, 35 g; Fat, 41 g (43% cal.); Cholesterol, 70 mg; Carbohydrates, 89 g; Fiber, 17 g; Sodium, 1040 mg.

ENCHILADAS

The Spanish word *enchilar* means "to put chile on something." The word *enchilada* is short for *tortilla enchilada*, meaning a tortilla covered with chile sauce. In the non-Spanish-speaking world, an enchilada is a corn tortilla rolled around a filling. Strictly speaking, however, the tortilla must first be coated with chile sauce before being rolled. A plain, dry tortilla with a filling is more accurately called a soft taco.

There are two basic methods for making enchiladas. In both methods, part of the procedure is to soften the tortilla enough so it can be rolled without breaking.

Fry the tortilla just enough to soften it, then dip in sauce, fill, and roll up.

Or coat the tortilla in chile sauce, then fry briefly, fill, and roll up.

Once the tortilla is filled, it can be served immediately. For North American tastes, however, it is usually baked with additional sauce and sometimes cheese toppings.

Pinto Bean Enchiladas

Three-Bean Chili with Tostaditas 🌹

YIELD: 6 LB   PORTIONS: 12   PORTION SIZE: 8 OZ (250 G)

U.S.	METRIC	INGREDIENTS	PROCEDURE
8 oz	250 g	Black turtle beans	1. Sort, wash, and drain the black turtle beans. Soak them overnight in cold water. 2. Drain the beans. Place them in a heavy saucepot and add enough water to cover them by 1 in. (2.5 cm). 3. Bring to a boil, reduce heat to a simmer, and simmer until the beans are just tender. 4. Keep an eye on the beans and add a little more water whenever necessary to keep the beans covered with water.
8 oz	250 g	White kidney beans	5. Repeat steps 1–4 for the white beans. It is necessary to cook the beans separately because they may have different cooking times.
1	1	Bay leaf	
8 oz	250 g	Red kidney beans or pinto beans	6. Repeat steps 1–4 for the red kidney or pinto beans.
1	1	Bay leaf	
2 fl oz	60 mL	Vegetable oil	7. Heat the vegetable oil in a sauté pan over moderate heat.
2 tbsp	30 mL	Cumin seed	8. Add the cumin and cook in the oil until aromatic.
6 oz	180 g	Onion, medium dice	9. Add the onion, garlic, and spices. Sauté until the onion is lightly browned.
1 oz	30 g	Garlic, chopped fine	
¼ cup	60 mL	Chili powder	10. Add the tomatoes and salt. Simmer 15 minutes.
1½ tsp	7 mL	Chipotle chili powder	11. Combine all three pots of beans in one large pot and add the spice and tomato mixture. Simmer until the beans are very tender. As before, add water if the stew becomes too dry.
1 tbsp	15 mL	Paprika	
2 tbsp	30 mL	Dried oregano	
2 lb	1 kg	Tomatoes, canned, chopped, with their juice	12. Taste for seasonings and add more salt if necessary.
1 tsp	5 mL	Salt	
24	24	Tortillas	13. To make the tostaditas, cut the tortillas into quarters, and then cut each quarter into 2 wedges.
6 fl oz	180 mL	Salsa Cruda (p. 207)	14. Deep-fry the tortillas until they are crisp. Drain well. 15. To serve, ladle the chili into bowls. 16. Set the bowls on large dinner plates as underliners. Arrange the tostaditas around the bowls on the plates. 17. Place 1 tbsp (15 mL) salsa on top of each portion of chili.

Per serving: Calories, 400; Protein, 17 g; Fat, 10 g (22% cal.); Cholesterol, 0 mg; Carbohydrates, 65 g; Fiber, 16 g; Sodium, 580 mg.

Three-Bean Chili
with Tostaditas

Red Rice, Spinach with Tofu Dressing, and Eggplant Dengaku

PORTIONS: 10   PORTION SIZE: APPROXIMATELY 5 OZ (150 G) RICE, 3 OZ (90 G) SPINACH, 3 OZ (90 G) EGGPLANT

U.S.	METRIC	INGREDIENTS	PROCEDURE
³/₄ cup	180 mL	Azuki beans	1. Wash and drain the beans.
4¹/₂ cups	1.1 L	Water	2. Place the beans and the water in a saucepan and bring to a boil. Simmer 10 minutes, or until the beans are just soft enough to be crushed between the fingers but not completely cooked.
4 cups	1 L	Glutinous rice (see p. 377)	3. Drain, reserving the cooking water. Cool both the beans and the water to room temperature. Refrigerate the beans.
			4. Wash the rice in several changes of cold water. Drain.
			5. Add the red bean cooking water to the rice and soak overnight.
			6. The next day, drain the rice and mix it with the beans.
			7. Cover the rack of a steamer with several layers of cheesecloth.
			8. Place the rice and bean mixture on top of the cheesecloth in a layer no more than 1¹/₂ in. (4 cm) thick.
			9. Steam until the rice and beans are tender, about 40 minutes.
10 oz	280 g	Firm Japanese-style tofu	10. Simmer the tofu in water 2 minutes. Drain.
2 tbsp	30 mL	Sesame seeds	11. Wrap the tofu in cheesecloth or in a clean towel and weight with a dinner plate 30–60 minutes to remove excess moisture.
2 tsp	10 mL	Soy sauce	12. Lightly toast the sesame seeds in a dry skillet, and then grind them coarsely in a mortar or spice mill.
1 tsp	5 mL	Mirin (Japanese sweet rice wine)	13. Rub the tofu through a fine sieve.
1 tbsp	15 mL	Sugar	14. Mix the puréed tofu with the sesame seeds, soy sauce, mirin, and sugar.
1 lb 12 oz	850 g	Spinach, fresh, cooked and lightly squeezed (see p. 310)	15. Chop the pressed spinach into large pieces, and then mix with the tofu dressing.
to taste	to taste	Salt	16. Add salt to taste.
6 oz	180 g	White miso	17. In a stainless-steel bowl, combine the miso and egg yolk and stir until smooth.
1	1	Egg yolk	18. Stir in the sake, mirin, sugar, and dashi.
1 tbsp	15 mL	Sake (Japanese rice wine)	19. Set the bowl in a pan of simmering water and stir until thick.
1 tbsp	15 mL	Mirin (sweet Japanese rice wine)	20. Cool.
1 tbsp	15 mL	Sugar	21. Trim the ends off the eggplant. Cut crosswise into rounds 1 in. (2.5 cm) thick.
2 fl oz	60 mL	Vegetarian dashi (p. 167)	22. Brush the cut sides of the eggplant slices with oil.
2 lb	900 g	Eggplant	23. Grill, broil, or pan-fry the eggplant until tender.
			24. Spread a layer of the miso mixture on top of each of the rounds.
			25. Pass under the broiler or salamander just until the miso topping is lightly colored.
as needed	as needed	Gomashio (see Note)	26. Serve each of the three items in separate bowls, one of each per diner. Sprinkle the top of the red rice with a little gomashio as garnish.

Per serving: Calories, 430; Protein, 16 g; Fat, 4.5 g (9% cal.); Cholesterol, 20 mg; Carbohydrates, 81 g; Fiber, 10 g; Sodium, 770 mg.

Note: Gomashio, or sesame salt, is a mixture of toasted black sesame seeds and coarse salt. If it is not available, toast sesame seeds and mix with a little kosher salt.

Red Rice, Spinach with Tofu Dressing, and Eggplant Dengaku

Vietnamese Vegetable Pancakes 🍽 🌹

PORTIONS: 6   **PORTION SIZE: 1 PANCAKE**

U.S.	METRIC	INGREDIENTS	PROCEDURE
		Dipping sauce:	1. Combine the dipping sauce ingredients and stir until the sugar is dissolved.
4 fl oz	125 mL	Soy sauce	2. Set aside.
1 tsp	5 mL	Finely chopped garlic	
2 tsp	10 mL	Sugar	
1/2 tsp	2 mL	Black pepper	
1 tbsp	15 mL	Chopped fresh red chile	
2 tbsp	30 mL	Chopped peanuts	
1 tbsp	15 mL	Lime juice	
1/2 cup	120 mL	Rice flour	3. Whip together the rice flour, eggs, water, and salt to make a thin batter. Let stand 10 minutes.
3	3	Eggs, beaten	4. Strain to remove any lumps.
9 fl oz	270 mL	Water	
1/8 tsp	0.5 mL	Salt	
4 oz	125 g	Shiitake mushroom caps, cut bâtonnet	5. Sauté the mushrooms in hot oil until tender. Cool.
1 tbsp	15 mL	Vegetable oil	
2 fl oz	60 mL	Vegetable oil	6. In a well-seasoned 10-in. (25-cm) sauté pan or, preferably, a nonstick pan, heat 2 tsp (10 mL) oil until very hot.
6 oz	180 g	Snow peas, trimmed	7. Add about 2 1/2 fl oz (75 mL) of the batter and swirl it around to cover the bottom of the pan.
6 oz	180 g	Bean sprouts	8. Add one-sixth of the mushrooms and snow peas. Cover and cook 1 minute.
2 oz	60 g	Scallions, sliced thin	9. Remove the lid and add one-sixth of the bean sprouts and scallions.
			10. Cook uncovered until the pancake is crisp.
			11. Slide the pancake onto a dinner plate.
			12. Repeat with the remaining batter and vegetables.
			13. Serve with dipping sauce.

Per serving: Calories, 400; Protein, 17 g; Fat, 10 g (22% cal.); Cholesterol, 0 mg; Carbohydrates, 65 g; Fiber, 16 g; Sodium, 580 mg.

VARIATIONS

For a nonvegetarian version, serve the pancake with Nuoc Cham (p. 209) instead of the soy dipping sauce.

Vietnamese Vegetable Pancake

Winter Vegetable Gratin with Feta Cheese and Bulgur Pilaf

PORTIONS: 10   PORTION SIZE: 6¹/₂ OZ (185 G) VEGETABLES, 3 OZ (90 G) PILAF

U.S.	METRIC	INGREDIENTS	PROCEDURE
10 oz	300 g	Turnips, peeled	1. Cut the turnips, parsnips, and carrots into 1¹/₂-in. (4-cm) pieces.
10 oz	300 g	Parsnips, peeled	
10 oz	300 g	Carrots, peeled	2. Break or cut the cauliflower into small florets.
10 oz	300 g	Cauliflower, trimmed	3. Cut the Brussels sprouts in half lengthwise.
6 oz	180 g	Brussels sprouts, trimmed	4. In a large bowl, toss the vegetables, including the onions, with the oil, salt, pepper, and herbs to coat them.
4 oz	125 g	Pearl onions, peeled	
3 oz	90 g	Olive oil or melted butter	5. Transfer to a half-hotel pan, 10 × 12 in. (25 × 30 cm), or a gratin dish of the same capacity. (Alternatively, bake in single-serving gratin dishes.)
1 tsp	5 mL	Salt	
¹/₂ tsp	2 mL	Pepper	
1 tsp	5 mL	Dried thyme	6. Bake at 400°F (200°C) until the vegetables are lightly browned and tender. Stir them occasionally during cooking so they brown evenly.
1 tsp	5 mL	Dried oregano	
12 oz	375 g	Feta cheese, crumbled	7. Remove the pan from the oven and sprinkle the feta cheese over the vegetables.
4 oz	125 g	Olive oil or melted butter	8. Mix the oil or butter with the crumbs until they are evenly mixed.
8 oz	250 g	Fresh bread crumbs	
30 oz (1 recipe)	900 g (1 recipe)	Bulgur Pilaf with Lemon (p. 385), made with vegetable stock	9. Sprinkle the crumbs over the vegetables in an even layer.
			10. Return to the oven and bake until the top is browned. If necessary, place under a broiler or salamander to finish browning.
			11. Serve with bulgur pilaf.

Per serving: Calories, 490; Protein, 12 g; Fat, 31 g (55% cal.); Cholesterol, 35 mg; Carbohydrates, 44 g; Fiber, 9 g; Sodium, 830 mg.

Winter Vegetable Gratin with Feta Cheese and Bulgur Pilaf

Falafel

YIELD: ABOUT 2 LB 4 OZ (1.1 KG)

U.S.	METRIC	INGREDIENTS
2 lb	1 kg	Chickpeas, cooked or canned, drained
1 oz	30 g	Garlic, chopped fine
4 oz	125 g	Scallion, chopped fine
3 tbsp	45 mL	Chopped parsley
2 tsp	10 mL	Ground cumin
2 tsp	10 mL	Ground coriander
1/4 tsp	1 mL	Cayenne
2 tsp	10 mL	Salt
1 tsp	5 mL	Black pepper
1 oz	30 g	Tahini (sesame paste)
2	2	Eggs, beaten
2 oz	60 g	Soft bread crumbs

PROCEDURE

1. Process the chickpeas in a food processor until well chopped but not completely puréed.
2. Add the remaining ingredients and mix well.
3. To make large falafel, use a No. 20 scoop to portion the mix. Roll each portion into a ball and then flatten into a thick patty. Handle the portions quickly and lightly. If they are handled too much, a loose crust may form during frying that will flake off after draining.
4. To make small, appetizer-size falafel, use a No. 40 or 50 scoop to portion the mix, and roll into balls.
5. Deep-fry at 375°F (190°C) until browned and crisp.

Per serving: Calories, 70; Protein, 3 g; Fat, 3 g (36% cal.); Cholesterol, 10 mg; Carbohydrates, 9 g; Fiber, 2 g; Sodium, 150 mg.

Marinated, Fried Tempeh with Stir-Fried Vegetables

PORTIONS: 10   PORTION SIZE: 3 OZ (90 G) TEMPEH

U.S.	METRIC	INGREDIENTS
1 lb 14 oz	900 g	Tempeh
1 qt	1 L	Water
8 oz	240 g	Onion, sliced thin
6 slices	6 slices	Fresh ginger
1/2 oz	15 g	Brown sugar

PROCEDURE

1. Cut the tempeh into slices about 1/2 inch (12 mm) thick or slightly less.
2. Combine the water, onion, ginger, and brown sugar in a saucepan and bring to a boil.
3. Add the tempeh and simmer, covered, for 20 minutes. Remove the cover and continue to simmer until the liquid is nearly completely evaporated.
4. Remove the tempeh from the pan and discard the onion, ginger, and liquid.
5. Deep-fry or pan-fry the tempeh at 350°F (175°C) until golden.

U.S.	METRIC	INGREDIENTS
15 fl oz	450 mL	Indonesian Peanut Sauce (p. 211)
10 portions	10 portions	Stir-Fry of Asparagus, Green Beans, and Snow Peas (p. 323)
as desired	as desired	Steamed or boiled rice (p. 381)

6. Place 1 1/2 oz Peanut Sauce in the center of a plate and arrange the tempeh on top of it.
7. Place the stir-fried vegetables on one side of the tempeh and a mound of rice on the other side.

Per portion: Calories, 440; Protein, 24 g; Fat, 32 g (65% cal.); Cholesterol, 0 mg; Carbohydrates, 22 g; Fiber, 4 g; Sodium, 460 mg.

Black Bean Vegetable Burger

PORTIONS: 6   **PORTION SIZE: 5 OZ (150 G)**

U.S.	METRIC	INGREDIENTS	PROCEDURE
1 lb 8 oz	720 g	Cooked or canned black beans, well drained	1. Purée the beans in a food processor, leaving the purée somewhat coarse.
2¹/₂ oz	75 g	Tomato concassé	2. Add the tomato, onion, garlic, chiles, lime juice, cornmeal, flour, cumin, and salt. Purée until well blended.
³/₄ oz	22 g	Onion, chopped fine	
¹/₂ tsp	2 mL	Garlic, chopped fine	
¹/₂ oz	15 g	Jalapeños or other fresh green chiles, chopped fine	3. Refrigerate the mixture for 2 hours or more.
1 tbsp	15 mL	Lime juice	4. Divide the mixture into 5-oz (150-g) portions. Form into balls and flatten into patties about ¹/₂ inch thick.
1 oz	30 g	Cornmeal	
2 oz	60 g	Flour	5. Pan-fry or griddle over moderate heat about 4 minutes on each side.
1 tsp	5 mL	Ground cumin	
¹/₂ tsp	2 mL	Salt	
6	6	Hamburger buns	6. Arrange each burger on a hamburger bun with a slice of tomato and lettuce leaf.
6	6	Tomato slices	
6	6	Lettuce leaves	

Per portion: Calories, 330; Protein, 16 g; Fat, 2.5 g (7% cal.); Cholesterol, 0 mg; Carbohydrates, 62 g; Fiber, 20 g; Sodium, 410 mg.

Zucchini and Hummus Pita "Pizza"

PORTIONS: 1

U.S.	METRIC	INGREDIENTS	PROCEDURE
1	1	Pita, 6-7 in. (15-18 cm) in diameter	1. Brush the rim of the pita with olive oil.
as needed	as needed	Olive oil	2. Spread the hummus on the pita, but do not spread it all the way to the edge.
2 oz	60 g	Hummus (p. 742)	
2 oz	60 g	Zucchini, small dice	3. Toss the zucchini with the second quantity of olive oil. Arrange on top of the hummus.
¹/₂ tsp	2 mL	Olive oil	
1 oz	30 g	Kalamata or other pitted black olives, sliced	4. Sprinkle the olives and the feta cheese on top of the zucchini.
1 oz	30 g	Feta cheese, crumbled	5. Bake at 400°F (200°C) 12–15 minutes, until hot.

Per portion: Calories, 550; Protein, 17 g; Fat, 32 g (52% cal.); Cholesterol, 25 mg; Carbohydrates, 51 g; Fiber, 5 g; Sodium, 1110 mg.

Pulao with Chickpeas and Vegetables

YIELD: ABOUT 2 LB 4 OZ (1 KG)

U.S.	METRIC	INGREDIENTS	PROCEDURE
1 fl oz	30 ml	Oil or clarified butter	1. In a heavy saucepan, heat the oil or butter. Add the rice and sauté until it begins to brown lightly.
1 cup	240 mL	Rice	
18 fl oz	540 mL	Water	2. Add the water, cardamom seeds, raisins, cinnamon stick, salt, turmeric, and nuts. Stir.
2	2	Cardamom pods, peeled (see note)	
¾ oz	22 g	Raisins	3. Bring to a simmer, cover, and simmer slowly for 10 minutes.
1-in piece	2.5-cm piece	Cinnamon stick	
1 tsp	5 mL	Salt	
1 tsp	5 mL	Turmeric	
1 ½ oz	45 g	Pistachios or cashews	
4 oz	120 g	Chickpeas, cooked or canned, drained	4. Add the chickpeas, pepper, and peas. Cover and cook another 10 minutes, until the rice is tender.
4 oz	120 g	Red bell pepper	
3 oz	90 g	Peas, fresh or frozen	

Per 1 oz (28.35 g): Calories, 45; Protein, 1 g; Fat, 1.5 g (30% cal.); Cholesterol, 0 mg; Carbohydrates, 7 g; Fiber, <1 g; Sodium, 65 mg.

Note: Use only the tiny black seeds from the cardamom pods; discard the papery husks.

VARIATIONS

Other vegetables can be substituted for the bell pepper and peas.

Curry Pakora

PORTIONS: 8   PORTION SIZE: 4–5 PAKORAS, 4 FL OZ (120 ML) SAUCE, 4 OZ (120 G) RICE

U.S.	METRIC	INGREDIENTS	PROCEDURE
8 fl oz	240 mL	Plain yogurt (see note)	1. Combine the yogurt, water, and chickpea flour. Beat together until smooth.
1 ½ pt	720 mL	Water	
6 tbsp	90 mL	Chickpea flour	
2 tsp	10 mL	Oil or clarified butter	2. Heat the oil or butter in a saucepan over moderate heat. Add the mustard seed, ground cloves, cumin, and coriander. Cook until the spices are aromatic.
½ tsp	2 mL	Black or yellow mustard seeds	
⅛ tsp	0.5 mL	Ground cloves	
½ tsp	2 mL	Ground cumin	3. Add the liquid from step 1. Stir in the salt, turmeric, and cayenne.
½ tsp	2 mL	Ground coriander	
¾ tsp	3 mL	Salt	4. Bring to a boil. Reduce heat and simmer, stirring often, until lightly thickened. It should be about the consistency of heavy cream.
½ tsp	2 mL	Turmeric	
¼ tsp	1 mL	Cayenne	
30–40 pieces (1 recipe)	30–40 pieces (1 recipe)	Cauliflower Pakoras (p. 761)	5. Just before serving, add the pakoras to the sauce. Heat until simmering.
2 lb	960 g	Steamed or boiled rice (p. 381)	6. Serve the pakoras and sauce with rice.

Per portion: Calories, 400; Protein, 13 g; Fat, 16 g (36% cal.); Cholesterol, 10 mg; Carbohydrates, 54 g; Fiber, 5 g; Sodium, 610 mg.

Note: For vegan version, substitute almond milk or soy milk for the yogurt.

Tofu Scramble with Vegetables

YIELD: ABOUT 1 LB 12 OZ (840 G)

U.S.	METRIC	INGREDIENTS	PROCEDURE
1 lb	480 g	Firm tofu	1. Press and drain the tofu as explained on p. 811–812.
			2. Crumble the tofu into small pieces.
2 tbsp	30 mL	Olive oil	3. Heat the olive oil in a sauté pan over moderate heat. Add the garlic and sauté briefly.
1 clove	1 clove	Garlic, chopped fine	4. Add the mushrooms and sauté until lightly browned.
3 oz	90 g	Mushrooms, sliced	5. Add the tomato and sauté just until the tomato is hot.
8 oz	240 g	Tomato concassé	6. Add the crumbled tofu, soy sauce, salt and pepper. Stir and cook 5 minutes.
2 tsp	10 mL	Soy sauce	7. Add the spinach, stir, and cook until the spinach is wilted.
to taste	to taste	Salt	8. Serve immediately. Tofu scrambles are often served at breakfast as a vegan substitute for scrambled eggs.
to taste	to taste	Pepper	
8 oz	240 g	Spinach, stems removed	

Per 1 oz (28.35 g): Calories, 35; Protein, 3 g; Fat, 2.5 g (64% cal.); Cholesterol, 0 mg; Carbohydrates, 1 g; Fiber, <1 g; Sodium, 30 mg.

TERMS FOR REVIEW

vegetarian	pesco-vegetarian	incomplete protein	tempeh
vegan	limiting amino acid	complementary protein	miso
lacto-vegetarian	amino acid	soy milk	bean paste
ovo-vegetarian	essential amino acid	tofu	textured vegetable protein (TVP)
lacto-ovo-vegetarian	complete protein	bean curd	

QUESTIONS FOR DISCUSSION

1. Explain how complementary proteins enable vegans to get enough protein in their diet without eating animal products.

2. What vitamin is most difficult for vegans to get in their diet? Why is this so? What are alternative sources for this vitamin?

3. Considering them as sources of protein, what makes soybeans, amaranth, and quinoa unique among foods from plants?

4. Name and describe the three main types of tofu.

5. What is TVP? Describe how to prepare it for use in recipes.

6. Explain why some refined sugars are not suitable for use in vegetarian recipes. Which sugars are unsuitable, and which sugars can be used?

27

SAUSAGES AND CURED FOODS

Sausages have been popular since ancient times. They were first made to utilize and preserve trimmings and less desirable cuts from a meat carcass. Most people have heard the expression about using "every part of the hog except the squeal." The preparation of sausages is an important part of this process.

The French term **charcutier** (shar koo tyay; the feminine form is *charcutière* [shar koo tyair]) means "one who prepares and sells pork products." The art of the charcutier is called **charcuterie** (shar koo tree). We use this term more generally to refer to the production of sausages, pâtés, smoked ham, and other cured and smoked products.

The main part of this chapter is devoted to the production of fresh sausages, which are easily prepared in any kitchen with relatively simple equipment. We also introduce the subject of cured and smoked sausages with an overview of curing and smoking. Preparing smoked hams and cured, air-dried meat products is an advanced subject beyond the scope of this book. However, reading the first section of this chapter will enable you to understand how these items are produced. A small sampling of recipes gives you some experience with curing and smoking before you proceed to the main section on sausages.

AFTER READING THIS CHAPTER, YOU SHOULD BE ABLE TO

1. Prepare simple dry-cured and brine-cured foods.

2. Prepare simple smoked foods.

3. Prepare fresh, cured, and smoked sausages.

CURING AND SMOKING

Curing and smoking have been important methods for preserving foods, especially protein foods, since prehistoric times. Before the days of modern preservation techniques such as canning, freezing, and refrigeration, curing and smoking, as well as drying, enabled people to store food in times of abundance for use in times of scarcity. Today we use these techniques for their contribution to flavor perhaps even more than for their preservative qualities. We smoke foods because we like the taste, not because we must in order to preserve them.

Nevertheless, it is important to understand how curing works to preserve foods so we can avoid spoilage and food-borne disease that can result from improper curing or improper handling of cured foods.

INGREDIENTS FOR CURING FOODS

Ingredients used in cures fill two main functions: preserving the food, and changing its flavor and texture. The following discussion of cure ingredients and processes is somewhat simplified, but it introduces the most important principles.

Clockwise from top: table salt, curing salt, kosher salt

Salt

Salt has been one of the most important food preservatives throughout most of human history. When a food comes in direct contact with salt, a two-way process takes place in an effort to make the salt concentrations inside and outside the food more nearly equal. Moisture from inside the cell walls of the food is drawn out and dilutes the salt used for the cure. At the same time, dissolved salt is absorbed into the cells to increase the salt concentration there. The result is foods that have less moisture and are saltier. Both the moisture reduction and the increased salt content make the food less hospitable to bacteria that cause spoilage and disease.

Of course, the reduced moisture and the increased salt content also change the flavor and texture of the food.

Nitrites and Nitrates

Some foods, especially meats, are susceptible to contamination by the bacteria that cause botulism. Nitrites and nitrates are added to the cures for these foods to make them safe from botulism infection. **Sodium nitrite** ($NaNO_2$) is the most important of these chemicals. Even in the very small quantities in which it is used, sodium nitrite has strong preservative powers. In addition, it keeps meats red or pink, which is why products like cured ham, corned beef, and certain sausages have their characteristic color.

Nitrites gradually break down inside the cured foods, but by the time they lose their effectiveness, the curing and, in some cases, smoking procedures are finished, the food is cooked or refrigerated, and the food remains safe. However, when the food is raw and air-dried, as in the case of prosciutto and various salamis, a stronger chemical, **sodium nitrate** ($NaNO_3$) is used. Nitrates break down more slowly than nitrites and therefore are effective for a longer time. (Potassium nitrate, or saltpeter, is sometimes used in cures, but it is not as safe and is strongly regulated. It should be avoided.)

Nitrates and nitrites are the subject of controversy regarding their safety. Substances called **nitrosamines** form when foods containing nitrates or nitrites are subjected to very high heat, such as when bacon is fried. Nitrosamines are known to cause cancer. Using only nitrites to cure foods usually avoids this problem because the nitrites break down quickly and are not usually a factor when the food is cooked. Nitrates, on the other hand, because they remain in meats longer, should not be used for curing bacon for this reason.

Some people feel that all use of nitrites and nitrates should be avoided because of the nitrosamine factor. However, if only nitrites and not nitrates are used in foods that will be exposed to extreme heat, the risk is low—much lower than the risk of botulism. So far, we know of no adequate substitute for these chemicals in curing pork and other meats.

Two special mixtures are employed that make it easy to add nitrites and nitrates in very low but sufficient concentration:

1. **Prague Powder #1**, or curing salt, is a blend of 6 percent sodium nitrite and 94 percent sodium chloride, or regular table salt. It is colored pink so it will not be confused with regular salt, and thus it is often called **tinted curing mix (TCM)**. It is also sold under various trade names, such as InstaCure I.

2. **Prague Powder #2** is similar to Prague Powder #1 except that it contains nitrates in addition to nitrites. It is used in curing products that have a long curing and drying period, as explained above. Prague Powder #2 is not used in any of the recipes in this chapter.

Fish and seafood are usually cured without the use of nitrites. Fish is almost always cured under refrigeration, and the salt cure is sufficient to protect it, even when it is cold smoked (p. 832).

The quantity of nitrite to be added to meats depends on several factors, including the type of meat, the type of cure, and the length of curing time. For cooked sausage, the USDA recommends 156 parts nitrite per million parts meat. This is equivalent to 1 ounce (30 g) nitrite per 400 pounds (192 kg) meat. To translate this to quantities similar to those used in this chapter, 10 pounds (4.5 kg) cooked sausage requires about 0.025 ounces (0.7 g) nitrite. This is the amount of nitrite contained in 0.4 ounces or 2 teaspoons (11 g or 10 mL) Prague Powder #1.

Keep in mind, however, that not all the nitrite specified in a recipe may be absorbed into the meat. In the case of the basic dry cures and brine cures (described below), part of the curing medium is discarded after the cure is complete. Therefore, enough nitrite must be used so the portion that is absorbed is adequate for the cure. The only exception is the case of sausages, in which all of the curing medium is mixed directly with the meat. The recipes in this book contain appropriate quantities of nitrite to cure the product.

Please note that some ham, sausage, and bacon products sold commercially are labeled "uncured." This is legal labeling (see the discussion of cured sausages on page 835) but is somewhat misleading. These products have been cured with salt and other ingredients, but nitrates and nitrites are not used as part of the curing medium.

Sugars

Ordinary white sugar (sucrose) and other forms of sugar, including corn syrup, honey, and maple syrup, are used in some cures. Sugars modify the flavor of the food and counteract some of the harsh flavor of the highly concentrated salt in the cure. Also, because salt extracts so much moisture, sugar adds to the perception of moistness in the cured product. Using less sweet forms of sugar, such as corn syrup and dextrose, provides the advantages of sugar without adding too much sweetness.

Herbs, Spices, and Other Flavorings

Nearly any spice or flavoring that can be used in cooking can be used in curing. Traditional recipes, of course, use traditional seasonings. Most of our most familiar cured sausages, for example, are of European origin and use such seasonings as garlic, pepper, coriander, caraway, nutmeg, and dry mustard. Many chefs today are experimenting with unusual ingredients to give a modern accent to cured meats, poultry, fish, and sausages.

CURING METHODS

The two basic types of cure are **dry cures** and **brine cures**. With both of these methods, the food item remains in contact with the curing mixture, which contains salt plus any of the other ingredients discussed above, until the curing medium has penetrated the food uniformly. The difference between the methods is that in brines, the salt is dissolved in water, whereas in dry cures, it is not.

Dry Cures

In a dry cure, the cure ingredients (salt, plus seasonings and, in many cases, sugar and Prague Powder) are mixed together and packed or rubbed over the food product to coat it completely. The length of time required to dry-cure meats depends on their thickness. Lean meat 1 inch (2.5 cm) thick requires 3–8 hours, while a whole ham needs about 45 days. During a long cure, the food is repeatedly turned and rubbed with the cure mixture in order to maintain uniform contact. It is important that the entire surface area be coated.

The curing procedure for sausages could be considered a dry cure. However, in this case, the cure is mixed directly with the meat. Because no time is required for the cure to penetrate to the center of the meat, the curing process takes place quickly, although the products may be air-dried or smoked for a longer period before being consumed.

Brine Cures

A brine is a solution of salt and other curing ingredients in water. The simplest way to use a brine is to immerse the meat in the brine and let it soak until the cure is complete. Compared with dry cures, brines are especially useful for poultry items, which are difficult to coat evenly with a dry cure because of their shape. Items that float, such as poultry, must be held down with a weight so they are completely submerged in the brine.

Simple brine soaking is used for small meat items, but because the brine takes time to penetrate to the center of large items such as hams, another method is used to speed the process. Brine is pumped or injected into the meat to make sure it penetrates evenly. After injection, the meat may be soaked in brine as well. Commercial operations use a variety of high-speed equipment for injecting brine. In addition, small pumps are available for brining by hand.

The length of time required for brining depends on the size and thickness of the item. Of course, meats injected with brine need less time in the brine soak.

Fresh brine should be made for each batch of cured meats. Do not reuse brines because they are diluted and contaminated with juices from the first batch of meats.

SMOKING

Smoking has been used as a way of drying and preserving foods since prehistoric times. Smoking does have some preservative effects but, for modern cooking, it is more important for the flavors it gives to meats, poultry, and seafood. Even smoked cheeses and vegetables are enjoyed for their special flavors.

The first rule of smoking foods is ***do not smoke meats, poultry, and fish that have not been cured***. The reason is a matter of food safety. During smoking, foods spend time in the Food Danger Zone (p. 17)—that is, at a temperature favorable for the growth of bacteria. Without the preservative effects of curing, smoking could be unsafe. (This rule does not apply to **smoke roasting** and **barbecuing** [p. 539], which are more properly considered cooking methods rather than smoking methods because they take place at higher temperatures.)

After meats, poultry, and fish are cured, they should be allowed to dry slightly before being smoked. So air can circulate all around the foods, place them on racks or hang them from hooks under refrigeration until the surface is dry to the touch. This preliminary drying allows the smoke to penetrate the foods more effectively.

The two basic types of smoking are **cold smoking** and **hot smoking**. In cold smoking, the temperature inside the smokehouse is kept at or below 85°F (30°C). At these temperatures, the foods take on the flavor of the smoke but are not cooked.

In hot smoking, the temperature inside the smokehouse may be as high as 165°F (74°C) for sausage and meats, and as high as 200°F (93°F) for fish and poultry. These temperatures are high enough to cook the foods being smoked. Higher temperatures are usually avoided because they result in excessive shrinkage. Foods may be hot smoked until they reach an internal temperature of 150°–165°F (65°–74°C) to ensure they are fully cooked. Alternatively, they may be hot smoked for a shorter period and then poached until they reach this internal temperature. This second method is used when a less intense smoke flavor is desired. No matter which method is used, hot-smoked foods are always sold fully cooked.

A typical smoker consists of the following elements:

- An enclosed chamber for holding the foods to be smoked
- A source of smoke
- A means to circulate the smoke around the food and then to exhaust it
- A way of controlling the temperature inside the chamber

The smoke source consists of a receptacle for wood chips or sawdust plus a heating element, usually electric. If the smoker is to be used for cold smoking, the smoke generator should be outside the main chamber that holds the food. If the smoke generator is inside the food chamber, as in some less expensive smokers, the temperature will rise too high for cold smoking. This type of smoker is used only for hot smoking. Foods to be smoked are arranged on racks or hung from hangers with enough space between them so the smoke circulates freely around all surfaces. Commercially made smokers are safest to use, as improperly built smokers may present a fire hazard. If a smoker is used indoors, it is essential to provide for ventilation of the exhausted smoke to the outdoors.

Hickory is perhaps the most popular wood for smoking, but other hardwoods that may be used include oak, mesquite, and fruitwoods such as apple and cherry. Soft woods, like pine, are not used because they release bitter, tarry components when burned. It is important to use woods from a reliable source. Pressure-treated woods should never be used, as they contain toxic chemicals such as arsenic.

To summarize, the smoking process consists of the following steps:

1. Curing (dry cure or brine cure)
2. Air-drying
3. Smoking (hot smoking or cold smoking)

KEY POINTS TO REVIEW

- What are the main ingredients used in curing? What is the purpose of each of them?

- What are the two main types of cures?

- Should foods be cured before being smoked? Why or why not?

- What are the two basic kinds of smoking processes? Describe them.

Gravlax

YIELD: 1 LB 14 OZ (850 G) WITHOUT SKIN

U.S.	METRIC	INGREDIENTS
2¹/₂ lb	1.2 kg	Salmon fillet, skin on
4 oz	125 g	Coarse salt
4 oz	125 g	Sugar
¹/₄ tsp	1 mL	White pepper
2 oz	60 g	Fresh dill sprigs

PROCEDURE

1. Pass your fingertips over the surface of the salmon fillet to locate any bones. Pull them out with needlenose pliers.
2. Mix the salt, sugar, and pepper.
3. Select a stainless-steel, glass, ceramic, or other nonreactive pan to hold the salmon for curing. Sprinkle a little of the salt mixture on the bottom of the pan and lay the salmon on it skin side down. Cover the flesh side of the fillet completely with a layer of the salt mixture. Then top with the dill, again covering the fillet completely.
4. If you are doubling this recipe and curing 2 fillets, salt the second fillet in the same manner and invert it on top of the first so the dill is sandwiched between the fillets and the skin side of each fillet is toward the outside.
5. Cover the pan well and refrigerate 1 day. Turn the fillet or fillets over and refrigerate another day (for a total of 2 days). Note: Some instructions say to place a weight on the fish during the cure. This is optional. Weighting the fish produces a slightly drier, firmer finished product.
6. After 2 days, drain off any liquid that accumulated in the pan. Carefully scrape all the dill and curing mixture from the fish.
7. To serve, cut on a sharp diagonal—that is, with the knife almost parallel to the table—into broad, paper-thin slices (**Figure 27.1**).

Per 1 ounce (28.35 g): Calories, 70; Protein, 8 g; Fat, 4 g (50% cal.); Cholesterol, 20 mg; Carbohydrates, 1 g; Fiber, 0 g; Sodium, 240 mg.

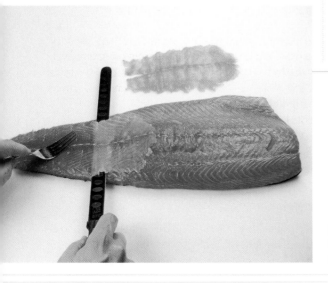

FIGURE 27.1 Slice gravlax and smoked salmon on the diagonal into paper-thin slices. Cut toward the tail end.

Smoked Salmon

YIELD: 1 LB 14 OZ (850 G) WITHOUT SKIN

U.S.	METRIC	INGREDIENTS
2¹/₂ lb	1.2 kg	Salmon fillet, skin on
6 oz	180 g	Coarse salt
3 oz	90 g	Sugar
2 tsp	10 mL	Coarse black pepper
1 tsp	5 mL	Dry mustard
¹/₂ tsp	2 mL	Ground allspice
¹/₄ tsp	1 mL	Cayenne
2 oz	60 g	Onion, chopped fine

PROCEDURE

1. Pass your fingertips over the surface of the salmon fillet to locate any bones. Pull them out with needlenose pliers.

2. Mix the salt, sugar, pepper, dry mustard, allspice, and cayenne.

3. Select a stainless-steel, glass, ceramic, or other nonreactive pan to hold the salmon for curing. Sprinkle a little of the salt mixture on the bottom of the pan and lay the salmon on it skin side down. Cover the flesh side of the fillet completely with a layer of the salt mixture. Then top with the chopped onion, distributing it evenly over the fillet.

4. If you are doubling this recipe and curing 2 fillets, salt the second fillet in the same manner and invert it on top of the first so the onion is sandwiched between the fillets and the skin side of each fillet is toward the outside.

5. Cover the pan well and refrigerate 12–24 hours.

6. Remove the fillet from the pan and rinse off all the salt mixture and onion. Place on a rack set on a sheet pan, skin side down, and allow to dry, uncovered, in the refrigerator until a thin, dry skin (called a pellicle) has formed on the surface of the flesh.

7. Cold smoke at 86°F (30°C).

8. To serve, cut on a sharp diagonal—that is, with the knife almost parallel to the table—into broad, paper-thin slices.

Per 1 ounce (28.35 g): Calories, 70; Protein, 8 g; Fat, 4 g (53% cal.); Cholesterol, 25 mg; Carbohydrates, 0 g; Fiber, 0 g; Sodium, 240 mg.

Smoked Trout

YIELD: 10 FILLETS, 6–7 OZ (180–200 G) EACH

U.S.	METRIC	INGREDIENTS
2 qt	2 L	Water
8 oz	250 g	Salt
2 oz	60 g	Light brown sugar
4	4	Bay leaves
2 tsp	10 mL	Black peppercorns
1 tsp	5 mL	Coriander seed
¹/₂ tsp	2 mL	Whole allspice
¹/₂ tsp	2 mL	Dry mustard
10	10	Trout fillets, about 8 oz (250 g) each

PROCEDURE

1. Prepare the brine: Combine the water, salt, sugar, bay leaves, peppercorns, coriander, allspice, and mustard in a pot. Bring to a simmer, stirring until the salt and sugar are dissolved.

2. Cool, then chill the brine.

3. Place the trout fillets in a stainless-steel, plastic, or other nonreactive pan in a single layer. Add enough cold brine to completely cover the fillets. Place a light weight on top of the fish to keep them submerged.

4. Refrigerate 6–8 hours.

5. Remove from the brine and rinse in cold water. Blot dry.

6. Arrange on racks and allow to dry several hours in the refrigerator.

7. Hot smoke at 185°F (85°C) until the internal temperature of the fish reaches 145°F (63°C), about 1–1¹/₂ hours. Cool, then refrigerate.

Per fillet: Calories, 320; Protein, 47 g; Fat, 12 g (36% cal.); Cholesterol, 135 mg; Carbohydrates, 2 g; Fiber, 0 g; Sodium, 3420 mg.

Smoked Duck

YIELD: 2 SMOKED DUCKS

U.S.	METRIC	INGREDIENTS
1½ gal	6 L	Water
12 oz	360 g	Salt
6 oz	180 g	Sugar
3 oz	90 g	Prague Powder #1
3	3	Bay leaves
2 tsp	10 mL	Onion powder
2	2	Small ducks, about 4 lb (1.8 kg) each

PROCEDURE

1. Prepare the brine: Heat the water in a pot until it is warm. Add the salt, sugar, and Prague Powder and stir until they are dissolved. Add the bay leaves and onion powder.
2. Cool, then chill the brine.

3. Place the ducks in a stainless-steel, plastic, or other nonreactive container. Pour enough brine over them to cover them completely. Weight them to keep them submerged.
4. Allow to cure in the refrigerator 2 days. (Note: Large ducks take 3–4 days to cure.)
5. Remove the ducks from the brine and rinse well.
6. Place on a rack and let dry in the refrigerator at least 8 hours.
7. Hot smoke at 185° (85°C) until the internal temperature of the duck reaches 165°F (74°F).

Per 1 duck, without skin: Calories, 510; Protein, 79 g; Fat, 17 g (32% cal.); Cholesterol, 305 mg; Carbohydrates, 4 g; Fiber, 0 g; Sodium, 4330 mg.

VARIATIONS

Smoked Turkey or Chicken

Brine turkey or chicken as in the basic recipe. Turkey requires 4–6 days to cure in the brine, depending on the size. Chickens require 2–4 days, depending on their size.

SAUSAGES

A **sausage** is a mixture of ground meat, usually pork, and seasonings stuffed into a casing. The term *sausage* may also be used for the meat mixture itself, without the casing. Reduced to its simplest form, sausage meat may be nothing more than ground pork seasoned with salt.

Although there are hundreds or even thousands of kinds of sausages, the majority are based on the same few basic principles. These principles are simple enough that we can make a wide variety of sausages in the kitchen and not have to rely entirely on commercial products. Furthermore, it is not necessary to restrict ourselves to traditional sausage recipes. Many chefs are experimenting with ingredients and seasonings to add variety to the menu.

CATEGORIES OF SAUSAGES

Sausages can be classified into three basic groups:

Fresh sausages

Cured sausages

Smoked sausages

A **fresh sausage**, by USDA definition, is one that contains no nitrates or nitrites. It is basically a mixture of ground meat, seasonings, and flavorings. Although they are often raw, fresh sausages may contain cooked ingredients, or they may even be fully cooked before being sold. Any fresh raw sausage containing pork, of course, must be fully cooked before being served and eaten.

A **cured sausage** is one that contains nitrites or nitrates of sodium. These chemicals help prevent spoilage and food-borne disease, as explained in the first part of this chapter.

Incidentally, they also keep the meat red or pink, even when cooked. Cured sausages may be sold raw or cooked, soft and moist like fresh sausages, semidried and firm, or dried and hard like salami. Pork salamis, which are Italian in origin, and similar cured, dried sausages are raw, but the curing, aging, and drying process renders them safe to eat. As explained on page 830, air-dried sausages meant to be eaten raw are made with nitrates in addition to nitrites for a longer lasting cure.

Smoked sausages may be hot smoked and, therefore, cooked, or cold smoked. Smoking may be light or heavy, depending on the sausage. Sausages, like other smoked meats, are cured before being smoked. The cure mixture is mixed directly with the sausage meat.

BASIC SAUSAGE INGREDIENTS

The basic ingredients of sausage meat are the following:

Lean pork

Pork fat, preferably hard fatback, ground with the meat

Salt

Spices, herbs, and other seasonings and flavorings

In the case of cured sausage, curing mixes containing nitrites or nitrates are added to the above list.

The Meat

Pork is the most commonly used meat in sausage making, but other meats or mixtures of meat may be used. Beef, veal, lamb, chicken, turkey, duck, liver, rabbit, and venison all find their way into sausages. Mixtures of pork plus one or more of these meats are often used. More exotic sausages may include such ingredients as sweetbreads and brains.

Certified pork—that is, pork certified to be free of trichinosis—is used for sausages meant to be air-dried and eaten uncooked.

The Fat

Pork fat or other fat, such as the beef fat used in all-beef sausages, is an important ingredient. Because our impression of juiciness in any cooked meat is largely due to the meat's fat content, some fat is included in sausage mixtures. Without it, the texture of the cooked sausage would be very dry.

In most traditional sausages, fat makes up 25 to 50 percent of the total weight, with 33 percent fat being the norm. In other words, proportions of fat to lean range from 3 parts lean plus 1 part fat (the leanest sausages) to 1 part lean plus 1 part fat (rich, fatty sausages). Varying the proportions changes the character of the sausage.

Hard fatback is preferred over other fats for pork sausage. Softer fats are more likely to melt out of the sausage during cooking. A quick and easy way to make pork sausages without worrying about the ratio of fat to lean is to use whole pork butt. The ratio of fat to lean in this cut is very good for sausages.

In today's diet-conscious atmosphere, it may make sense to try to create low-fat sausages. But be advised that extra care is required when making sausages with less than 25 percent fat. Lean sausages should never be overcooked, as overcooking makes them dry. Keep the meat mixture cold to avoid damage to the texture of the meat and fat. With care, it is possible to make tasty sausages with a fat content as low as 10 or 15 percent. One should not expect, however, that the eating qualities of lean sausages will be the same as those of fattier sausages.

Cereal ingredients and fillers (rice, barley, bread crumbs, and so on) can be used to help reduce fat content. Because these starches absorb and retain moisture, they enhance the total moisture content of low-fat sausages.

Seasonings

Herbs, spices, and other flavorings account for the primary differences among sausages. Many, if not most, of the world's fresh sausages are made of nothing more than ground pork and seasonings. It is the seasonings that give them their characteristic flavor. A glance at the recipes in this section will confirm this.

Some of the major spices and herbs used in sausage making are:

Allspice	Cumin	Paprika
Caraway seeds	Fennel seed	Parsley
Cayenne or hot red pepper	Ginger	Pepper, black and white
	Mace	
Cinnamon	Marjoram	Sage
Cloves	Mustard	Tarragon
Coriander	Nutmeg	Thyme

Other important ingredients include:

Garlic	Chives	Vinegar
Onion	Wine, white and red	Eggs
Shallots		

Standardized spice mixtures are often used to season sausages, pâtés, and similar items. One of the most common is **quatre épices** (French for "four spices"), a mixture that exists in many versions. It can usually be obtained commercially, or it can be homemade, using recipes such as those on page 843.

Cure Mixes

Sausages to be air-dried are cured with both nitrate and nitrite, as explained on page 830. For other cured sausages, only nitrite, usually in the form of Prague Powder #1, is used. Other ingredients, such as dextrose and additional salt, may be combined with the Prague Powder to make the cure mix. The cure mix may be incorporated into the sausage in one of two ways:

1. It may be mixed with the meat after it is diced. The diced meat is then chilled thoroughly before being ground.

2. It may be combined with the remaining spices and seasonings and mixed into the ground meat.

Refer to individual recipes for more detailed instructions on making and incorporating cure mixes.

Other Ingredients

A number of sausages are characterized by unusual or exotic ingredients. Some of these are traditional and time-honored, such as the black truffles included in some regional French sausages and the chestnuts or raisins in other specialty sausages.

Other unusual sausages are modern innovations by creative chefs. It is no longer uncommon, when reading today's restaurant menus, to come across sausages with such ingredients as sun-dried tomatoes or fresh vegetables like sweet bell peppers and spinach. In theory, there is no limit to what can be stuffed into sausage casings. The main requirement is that the ingredients complement or enhance one another, just as the meat, seasonings, and vegetable garnish on a dinner plate should complement one another.

CASINGS

Natural casings are made from the intestines of meat animals. *Sheep casings* are the smallest, ranging from 3/4 inch (18 mm) to more than 1 inch (25 mm) in diameter. They are used for breakfast links, frankfurters, and similar sausages. *Hog casings* are medium-sized, about 1–11/2 inches (3–4 cm) in diameter, depending on what part of the intestines they are taken from. They are used for many popular fresh sausages, such as Italian sausages and fresh bratwurst. *Beef casings* range in size from the so-called *beef round*, about 13/4 inches (45 mm) in diameter, through *beef middles*, 21/2 inches (63 mm), to the large *beef bung*, more than 4 inches (100 mm) in diameter.

Natural casings are often sold packed in salt. Because of the preservative effect of the salt, the casings keep indefinitely as long as they are refrigerated. Natural casings are easy to use if they are handled correctly. Before being stuffed, they must be untangled, rinsed, and flushed, and examined for holes according to the following procedure.

PROCEDURE for Preparing Natural Casings

1. Carefully remove the casings, one at a time, from the salt pack, and unravel them. Because a single casing may be 12 feet (4 m) long, it is easiest to do this on a large workbench. Separate the individual lengths and keep them separate in their own little stacks on the bench. When unraveling them, do not pull hard, because this may cause knotting.

   Unravel slightly more casing than you think you will need. It is easier to return unused casings to the salt pack than it is to separate and flush additional casings when you run out before you have stuffed your whole batch of meat.

2. Partially fill a large bowl with clean water and set it in a sink under the faucet. Take hold of the end of one casing and drop the rest of it into the bowl of water. Open the end of the casing and run cold water into it, enough to fill about 12 inches (30 cm) of it. Holding the casing at both ends of this "water sausage," allow the water to flush through the casing from one end to the other.

   This accomplishes two purposes. It rinses out the inside of the casing, and it identifies any holes that might be present. Pinpoint-size holes are no problem, but if a large hole is found, simply cut the casing in two at that point. Short pieces may be discarded for the sake of efficiency.

3. If you are making a small quantity of sausages and stuffing them immediately, each casing can be put on the stuffing horn as it is rinsed. If this is not the case, the casing must be stored for later use. Select a container with a cover and fill it about three-fourths full of cold water. Drop in the casing and let one end hang over the edge.

   Repeat with remaining casings. Fill the container to the top with cold water, cover, and refrigerate until needed. By letting the ends of the casings hang over the edge, you can remove one at a time from the container without tangling them. If the end of the casing dries out, simply cut it off.

Because the supply of natural casings is not nearly large enough to accommodate all the sausages produced, other types of casings have come into wide use. **Collagen casings** are molded from animal materials and are completely edible, like natural casings. Unlike natural casings, they are uniform in size, making portion control easier. Various types are manufactured for different uses. Some are used only for fresh sausages, as they are not strong enough to hold the weight of the sausages if they are hung for smoking. Other types are stronger and intended to be used for smoked sausages. Most collagen casings must be refrigerated to keep them from becoming dry and brittle. To use, dip them in water for a few seconds to soften them if they are dry, then put them on the sausage stuffing nozzle.

Synthetic fibrous casings are made from a plastic material and are not edible. They are widely used for salamis and luncheon meats, and the casing is peeled off before or after slicing. These casings are nonperishable and need no refrigeration. They must be soaked in water before using to make them flexible. Soaking time varies, and the manufacturer should specify the recommended time.

EQUIPMENT

Most fresh sausages can be made with no special equipment other than a meat grinder and a device for getting the meat into the casings.

The working parts of a meat grinder are a rotating blade and a selection of dies in various sizes. The size of the die determines the fineness of the grind. A screw forces the meat from the feed tube to the blade, and the blade chops the meat as it is forced through the holes of the die.

Before grinding meat, it is important to be sure of the following two points:

1. The equipment must be clean and sanitary. Make sure there is no trace of food from previous jobs hiding in the many nooks and crannies of the grinder, blades, and dies.

2. The equipment must be cold in order to avoid warming the meat. Refrigerate the grinder parts or set them in ice water before use.

The sausage stuffer is the second piece of equipment necessary for making sausages. The simplest stuffer resembles a funnel. The casings are pushed over the narrow end and the meat is pushed through the wide end by hand or with a wooden plunger. These are adequate for making a few pounds of sausage but are not suitable for larger batches.

Larger stuffers have detachable nozzles or horns for different casing sizes. The nozzle is attached to a cylindrical reservoir that holds the meat, which is pushed through by a piston. On smaller machines, the piston is simply pushed through by hand. These machines are suitable for small-scale production such as might be done in a small to medium-size restaurant. For large-scale commercial production, larger machines are used. The piston in a large machine is operated by means of a crank and a sequence of gears.

THE GRIND

The grind of the meat determines the texture of the sausage. We can divide sausage meat into two categories based on grind. Most common sausages are made simply of meat and fat ground to varying degrees of fineness or coarseness, mixed with seasonings, and stuffed into casings. We refer to these as **basic grind** sausages. Other sausages, including familiar ones like frankfurters, mortadella, and bologna, are made with meat and fat ground to a smooth purée and blended with a little liquid, such as water or milk. We refer to these as **emulsified** or **emulsion grind** sausages because of the added liquid content and the emulsion of the fat with the meat and liquid.

Basic Grind

The fineness or coarseness of the grind is an important characteristic of any sausage. For example, one identifying feature of Toulouse Sausages (p. 843) is their coarse texture. The meat is chopped by hand rather than ground. On the other hand, typical breakfast sausages have a fairly fine grind.

To grind meat, cut it into pieces small enough to fit easily into the feed tube of the grinder, then chill it well. For the best texture, start with the largest die, no matter what final grind is needed. Then grind with the next smaller die, until the desired texture is achieved. This process is called **progressive grinding**. Control the coarseness of the grind by selecting the proper die for the final grind, as indicated in the recipe. As indicated in the following procedure, the meat, fat, and seasonings are mixed until uniformly blended after the meat is ground.

The meat mixture must be kept cold during grinding. When the mixture gets too warm, the fat becomes soft and begins to lose its structure. As a result, it may melt out too readily when the sausage is cooked, resulting in excessive shrinkage, poor texture, and dryness. If the kitchen is warm, return the meat to the refrigerator to chill it thoroughly after cutting it up and before grinding it. If it must be ground more than once, return it to the refrigerator between grindings.

After the forcemeat is ground, it should be mixed to develop the texture. Unmixed sausage forcemeats are likely to be dry and crumbly after cooking. Mixing binds the forcemeats proteins, fats, and liquid together, resulting in a smoother, moister texture. Be sure to keep the mixture cold during mixing. The best way to do this is to use a mixer fitted with an ice bain-marie (a jacket surrounding the mixing bowl that can be filled with ice water).

Additional textural variation in basic grind sausages can be created by mixing chunks or dice of meat or other ingredients into a more finely ground forcemeat, as is often done in pâtés.

When following any sausage recipe, adhere closely to the grinding and processing directions in order to achieve the proper texture and character.

PROCEDURE for Making Fresh and Cured Basic Grind Sausages

1. Weigh the meats and fat. Cut them into chunks small enough to fit into the grinder.

2. If the sausages are to be cured, toss the cure mix with the diced meat at this point. Alternatively, combine the cure mix with the salt and spices in step 4.

3. Chill the meat well before proceeding.

4. Measure the salt and spices.

5. Grind the meat and fat, following the directions in the recipe to achieve the proper fineness or coarseness of grind.

6. Chill the meat well. Ideally, the meat should be chilled to 32°–35°F (0°–2°C) before mixing and stuffing. Best practice is to grind the meat directly into a bowl set in an ice bath.

7. Combine the meat, salt, and spices and mix thoroughly. For small quantities, this can be done by hand. The best procedure, however, is to use a mixer fitted with an ice bain-marie to keep the meat cold.

   If the recipe calls for cold water or other liquid, mix the salt and spices with the water first, then mix this combination with the meat. This facilitates a better distribution of the spices than mixing them dry with the meat.

   Many recipes say to refrigerate the meat overnight after it is mixed to allow it to absorb the seasonings and the flavors to blend. However, it is easier to stuff the sausages immediately after mixing. If the salted meat mixture is allowed to stand, it becomes firmer and thus harder to force through the stuffer. Stuffing immediately allows the casings to be filled more uniformly and with fewer air bubbles. Besides, the meat can absorb the sea-

sonings just as well in the casings as in the meat tub. If you do refrigerate the meat, mix it again until smooth before filling casings.

8. Test for seasonings. Do not taste the raw meat. Rather, make a small ball or patty and cook it in a small sauté pan or poach it in water. Then cool the meat slightly and taste. If more salt or other seasoning is needed, add it to the sausage mixture.

9. Place the meat in the stuffer, one handful at a time. Pack each handful firmly into the stuffer to eliminate air bubbles.

10. Slide the casings onto the nozzle. To help the casings slide on easily, moisten both them and the stuffer nozzle with water.

11. Stuff the sausages, following the directions for your equipment (see Figure 27.2). The sausages will pull the casings off the nozzle as the meat flows through it, but it is best to hold your hand at the end of the nozzle to help control the rate at which the casing is pulled from the nozzle. From time to time, as the casing is filled, it is necessary to push the bunched-up, unfilled casing toward the end of the nozzle so it will slide off more easily. Do not stuff the casings too tightly. If the meat is packed too tightly, it will be difficult to twist the sausage into links.

12. After all the meat is stuffed into casings, remove any air bubbles by pricking the casing and pressing the surface to expel the air.

13. Twist the sausage into links of uniform size.

FIGURE 27.2 Stuffing sausages.

(a) A small sausage stuffer fitted with a medium nozzle. This stuffer holds 5 lb (2.3 kg) meat at a time.

(b) Flush out the casings with fresh, cold water, while looking for holes.

(c) Slide the casings onto the proper size nozzle.

(d) When stuffing the casings, use one hand to guide the casing as it slides off the nozzle.

(e) Pinch and twist the sausage into links of the desired size.

Emulsified Grind

In the finest-textured sausages, the meat is actually puréed to a smooth paste. Puréed raw meat is capable of absorbing a good deal of moisture, and water or milk is usually added to improve the texture. Because of the addition of liquid to the meat purée, the resulting grind is called an *emulsified grind*. For sausages that include milk, a recipe may indicate either liquid milk or water plus dried milk solids. The dried milk is added at the same time as the spice mixture.

Because grinding meat to a smooth paste generates heat, water is added in the form of ice to keep the temperature low. If liquid milk is called for, freeze it into chunks.

Because a meat grinder is not capable of grinding sausage to this texture, the grinding procedure is done in two stages. First, grind the meat and fat separately using the fine die on the meat grinder. Keep the two separate and chill them after this grinding. Second, grind the meat to a smooth paste with a food chopper (buffalo chopper) or food processor. Place the lean meat in the bowl of the chopper and run it a few seconds. Then add the ice and continue to grind until smooth. Monitor the temperature carefully. Do not allow it to rise above 50°F (10°C). Add the fat and continue to grind to blend it in. Continue to monitor the temperature. Do not allow it to rise above 58°F (14°C).

Emulsified sausages are usually cooked in water after stuffing, then chilled in cold water. If they are to be smoked, they are smoked first and then cooked.

PROCEDURE for Making Fresh and Cured Emulsified Sausages

1. Keeping the meat and fat separate, grind them with the fine die of a meat grinder, following steps 1–6 of the basic sausage-making procedure. Chill well.

2. If indicated in the recipe, mix the seasonings with the lean meat. (Alternatively, add the seasonings at the same time as the ice in step 3.)

3. Place the lean meat in the bowl of a food chopper or food processor. Run the chopper a few turns, then add the ice. Continue to chop to a fine paste. Do not allow the temperature to rise above 50°F (10°C).

4. Add the fat and continue to grind to blend it in. Continue to monitor the temperature. Do not allow it to rise above 58°F (14°C).

5. Follow steps 8–12 in the procedure on page 840 to test the seasonings and stuff the casings.

SMOKING SAUSAGES

The procedure for smoking sausages is the same as for smoking other meats (see Figure 27.3). Refer to page 832 for basic information on smoking and follow the instructions in individual recipes. Note in particular the following points:

1. Only cured sausages should be smoked.
2. The sausages should be dried briefly before being smoked, like other smoked foods (see p. 832). Hang them so that air can circulate around them and allow to dry about 1½ hours or as indicated in the recipe.
3. Hot smoke or cold smoke as indicated in the recipe. Air-dried sausages to be eaten raw are cold smoked. Sausages that are sold cooked are generally hot smoked. After smoking, they are usually poached to an internal temperature of 160°–165°F (71°–74°C), then cooled quickly in ice water and blotted dry.

COOKING SAUSAGES

Although sausages are sometimes ingredients or components of more elaborate dishes, they are also popular as standalone menu items like other meats. Preparation of sausages before cooking is minimal. In most cases, sausages may be considered ready to cook—or, in the case of fully cooked sausages, ready to heat. Fresh raw sausages containing pork must be fully cooked before serving, like other fresh pork products. If a fresh sausage contains no pork it may be served somewhat less done. Removing it from the heat when still pink inside helps retain juices.

Fully cooked sausages need only be heated through before serving. Cooking times, however, are nearly the same as for raw sausages. In other words, the time it takes for the heat to penetrate to the center is about the same.

The following cooking methods are most often used for sausages.

FIGURE 27.3 Finished sausages before removing from the smoker.

Simmering

Place sausages in a pan with enough salted water to cover. Bring to a simmer and simmer until completely cooked. Do not let the water boil. This is likely to make the sausages burst or shrink excessively.

Time depends on the thickness of the sausage. Sausages in hog casings take about 20 minutes to cook; smaller ones may take as little as 10 minutes, larger ones 30 minutes or more.

Raw sausages may be simmered until cooked through, then finished by one of the following methods.

Sautéing and Pan-Frying

Sausages are sautéed or pan-fried using the same techniques as for other meats. For many kinds of sausage, the browning that results makes them more appetizing than simmered ones.

If the sausages are not raw, it is necessary only to cook them until they are lightly browned and heated through. If they are raw, lower heat is necessary so they have time to become completely cooked by the time they are browned.

Sautéing and pan-frying are used not only for sausages in casings but also for patties and *crêpinettes*, which are sausage patties wrapped in caul fat (see p. 843). Do not press on the patties with the spatula while cooking. This forces out juices and makes them dry.

Braising

Because fresh raw sausages may require long cooking times, braising is often the preferred cooking method. The sausages are browned by sautéing them over moderate heat for up to 5 minutes, then finished using a moist-heat cooking method. Cooking with moisture may take any of several forms, including the four methods listed below. These methods are suitable not only for fresh sausages but also for fully cooked and smoked sausages that are served hot:

- Covering the pan to hold in steam.
- Adding a small amount of liquid (water, stock, wine, and so on), covering the pan, and cooking until done.
- Glazing—that is, adding a small amount of stock and continuing to cook. The pan may be covered or left uncovered at the beginning, but the last part of cooking takes place with the cover off. The sausages are basted with the stock as it reduces, giving them a glaze. Any stock remaining in the pan after the sausages are cooked will be reduced and thickened, and it may be served with the sausages as a sauce. This cooking method is especially suitable for patties and crêpinettes.
- Cooking the sausages in a casserole or stew after they are browned. Sauerkraut, bean dishes, and gumbos are examples.

Broiling and Grilling

Brush the sausages with oil to prevent sticking, and broil or grill as for other meats. Moderate heat is usually best. High heat may brown the sausages too much before they are fully cooked, and it is more likely to cause splitting.

ABOUT THE SAUSAGE RECIPES

Before proceeding to the sausage recipes that follow, please take note of the following points:

1. Refer to the general procedures for making sausages on pages 840–841. Some of these points are not repeated in each recipe.
2. Ratios of fat to lean are not specified in most of the recipes. Refer to the explanation of fat-lean ratios on page 836 and adjust the amount of fat in the recipes as desired.

KEY POINTS TO REVIEW

- What are the three basic categories of sausages?
- What are the main ingredients of sausages?
- What are the steps in the procedure for preparing natural sausage casings?
- What are the steps in the procedure for making basic grind sausages?
- What are the steps in the procedure for making emulsified sausages?

Quatre Épices I

YIELD: 1¼ OZ (37 G)

U.S.	METRIC	INGREDIENTS
4 tbsp	25 g	Ground white or black pepper
2 tsp	4 g (10 mL)	Nutmeg
2 tsp	4 g (10 mL)	Ground cloves
2 tsp	4 g (10 mL)	Cinnamon

PROCEDURE

1. Combine the ingredients and mix well.
2. Store in a tightly sealed container.

Per 1 ounce (28.35 g): Calories, 90; Protein, 3 g; Fat, 2 g (13% cal.); Cholesterol, 0 mg; Carbohydrates, 19 g; Fiber, 8 g; Sodium, 10 mg.

Quatre Épices II

YIELD: 1¼ OZ (34 G)

U.S.	METRIC	INGREDIENTS
10 tsp	20 g	Ground white pepper
3 tsp	6 g (15 mL)	Nutmeg
3 tsp	6 g (15 mL)	Ground ginger
1 tsp	2 g (5 mL)	Ground cloves

PROCEDURE

1. Combine the ingredients and mix well.
2. Store in a tightly sealed container.

Per 1 ounce (28.35 g): Calories, 100; Protein, 3 g; Fat, 3 g (21% cal.); Cholesterol, 0 mg; Carbohydrates, 18 g; Fiber, 7 g; Sodium, 5 mg.

Pork Sausage

YIELD: 6 LB (3 KG)

U.S.	METRIC	INGREDIENTS
6 lb	3 kg	Pork and pork fat
2 tbsp	30 g	Salt
2 tsp	4 g (10 mL)	Quatre épices
6 fl oz	200 mL	Water, cold

PROCEDURE

1. Cut the meat and fat into cubes small enough to fit into the grinder. Chill thoroughly.
2. Grind the meat once with the large die and again with the small die. If necessary, chill the meat between grindings.
3. Mix the salt and spices with the water. Add to the ground meat and mix thoroughly.
4. Stuff into hog casings or sheep casings.

Per 1 ounce (28.35 g): Calories, 70; Protein, 5 g; Fat, 5 g (69% cal.); Cholesterol, 20 mg; Carbohydrates, 0 g; Fiber, 0 g; Sodium, 160 mg.

VARIATIONS

Toulouse Sausage

Use the ingredients specified in the recipe, but grind the meat coarsely. This can be done in three ways, with slightly different results for each method:

1. Chop the meat coarsely with a knife.
2. Grind only once, using the large die of the grinder.
3. Grind briefly in a food processor, just until achieving a coarse texture.

Stuff into hog casings.

Toulouse sausages should have a fat content of at least 33 percent—that is, 1 part fat to 2 parts lean.

Crêpinettes

Weigh out 3½-oz (100-g) portions of sausage meat. Shape them into oval patties. Cut squares of caul fat (p. 479) and wrap the sausage portions in the squares. Cook by pan-frying, glazing (see p. 842), or grilling.

Other ingredients are often mixed with the sausage meat. Classic additions include blanched, peeled pistachios (2–3 oz per lb sausage meat/125–175 g per kg); chopped, cooked chestnuts (4 oz per lb/250 g per kg); or fresh, diced truffle (whatever quantity the budget allows).

Other sausage mixtures, such as those in the other recipes in this section, may be used to make crêpinettes.

French Garlic Sausage

YIELD: 7 LB (3.5 KG)

U.S.	METRIC	INGREDIENTS
6 lb	3 kg	Pork and pork fat
1 lb	500 g	Pork rind (skin)
2 tbsp	30 g	Salt
1½ tsp	3 g (7 mL)	Black pepper
2 tsp	8 g	Crushed garlic
½ tsp	2 mL	Ground sage
½ tsp	2 mL	Dried marjoram
½ tsp	2 mL	Dried thyme
4 fl oz	125 mL	Dry white wine

PROCEDURE

1. Cut the meat and fat into cubes small enough to fit into the grinder. Chill thoroughly.
2. Simmer the pork rind in salted water at least 2 hours, or until very tender. Chill thoroughly.
3. Grind the meat once with the large die and again with the medium die. If necessary, chill the meat between grindings.
4. Repeat this grinding procedure with the pork rind. Mix the ground pork rind with the ground meat.
5. Mix the salt and spices with the wine. Add to the ground meat and mix thoroughly.
6. Stuff into hog casings.

Per 1 ounce (28.35 g): Calories, 70; Protein, 5 g; Fat, 5 g (69% cal.); Cholesterol, 20 mg; Carbohydrates, 0 g; Fiber, 0 g; Sodium, 150 mg.

Fresh Bratwurst

YIELD: 6 LB (3 KG)

U.S.	METRIC	INGREDIENTS
6 lb	3 kg	Pork and pork fat
2 tbsp	30 g	Salt
3 tsp	15 mL	White pepper
¼ tsp	1 mL	Mace
1 tsp	5 mL	Ground coriander
½ tsp	2 mL	Ground ginger
6 fl oz	200 mL	Water, cold

PROCEDURE

1. Cut the meat and fat into cubes small enough to fit into the grinder. Chill thoroughly.
2. Grind the meat once with the large die and again with the small die. If necessary, chill the meat between grindings.
3. Mix the salt and spices with the water. Add to the ground meat and mix thoroughly.
4. Stuff into hog casings.

Per 1 ounce (28.35 g): Calories, 70; Protein, 5 g; Fat, 5 g (69% cal.); Cholesterol, 20 mg; Carbohydrates, 0 g; Fiber, 0 g; Sodium, 160 mg.

VARIATION

Following the procedure on page 841, make an emulsified grind instead of a basic grind. Omit the water indicated in the recipe and add 12 oz (400 g) ice as indicated in the procedure.

Duck Sausage

YIELD: 6 LB (3 KG)

U.S.	METRIC	INGREDIENTS
6 lb	3 kg	Boneless duck meat and fat
2 tbsp	30 g	Salt
2 tsp	10 mL	Quatre épices
6 oz	185 g	Pine nuts, toasted, or pistachios, blanched and peeled
6 oz	185 g	Marinated sun-dried tomatoes, drained and chopped
½ tsp	2 mL	Dried thyme

PROCEDURE

1. Cut the meat and fat into cubes small enough to fit into the grinder. Chill thoroughly.
2. Grind the meat once with the medium die.
3. Mix the salt, spices, nuts, and sun-dried tomatoes. Add to the ground meat and mix thoroughly.
4. Stuff into small hog casings.

Per 1 ounce (28.35 g): Calories, 160; Protein, 3 g; Fat, 17 g (91% cal.); Cholesterol, 25 mg; Carbohydrates, 1 g; Fiber, 0 g; Sodium, 160 mg.

Hot Italian Sausage

YIELD: 6 LB (3 KG)

U.S.	METRIC	INGREDIENTS
6 lb	3 kg	Pork and pork fat
2 tbsp	30 g	Salt
2 tsp	10 mL	Black pepper
2 tsp	10 mL	Fennel seeds
4 tsp	20 mL	Paprika
2 tsp	10 mL	Crushed red pepper
1 tsp	5 mL	Ground coriander
2 tsp	10 mL	Sugar
6 fl oz	200 mL	Water, cold

PROCEDURE

1. Cut the meat into cubes small enough to fit into the grinder. Chill the meat thoroughly.
2. Grind once with the medium die.
3. Mix the salt and spices with the cold water. Add to the ground meat and mix thoroughly.
4. Stuff into hog casings.

Per 1 ounce (28.35 g): Calories, 70; Protein, 5 g; Fat, 5 g (69% cal.); Cholesterol, 20 mg; Carbohydrates, 0 g; Fiber, 0 g; Sodium, 160 mg.

VARIATIONS

Mild Italian Sausage

Omit the paprika, crushed red pepper, and coriander.

Spicy Garlic Sausage

Omit the fennel and coriander. Add 2 tsp (10 mL) dried oregano and 1–2 tsp (5–10 mL) chopped garlic.

Boudin Blanc (White Sausage)

YIELD: 7 LB (3.6 KG)

U.S.	METRIC	INGREDIENTS
1 lb 8 oz	750 g	Chicken breast, boneless and skinless
2 lb	1 kg	Lean white pork
1 lb 8 oz	750 g	Pork fat
12 oz	375 g	Onion, chopped
1 pt	500 mL	Milk
8 oz	250 g	Fresh bread crumbs
4	4	Eggs
3 tbsp	45 mL	Salt
2 tsp	10 mL	White pepper
1/2 tsp	2 mL	Mace
1 oz	30 g	Parsley, chopped (optional)

PROCEDURE

1. Cut the chicken, pork, and fat into cubes small enough to fit into the grinder. Combine the chicken with the lean pork, but keep the pork fat separate.
2. Grind the meat once with the large die, then once with the small die. Chill.
3. Grind the fat once with the large die, then once with the small die. Chill.
4. Simmer the onions in the milk until they are tender. Pour this mixture over the bread crumbs in a bowl. Cool, then set in the freezer until partly frozen.
5. Place the ground meat in the bowl of a food chopper or food processor. Chop a few seconds, then add the partially frozen milk mixture. Grind until the mixture is uniformly blended.
6. Beat together the eggs, salt, pepper, mace, and parsley (if used). Add to the meat in the grinder and continue to grind to a smooth paste. Check the temperature to make sure it does not rise above 50°F (10°C). If necessary, chill before continuing.
7. Add the fat to the grinder and continue to grind until the mixture is smooth and uniform. Do not let the temperature rise above 58°F (14°C).
8. Stuff into hog casings.
9. Poach the sausages by simmering them slowly in water until cooked through, about 20 minutes. Drain and plunge into ice water to cool. Drain again and refrigerate.

Per 1 ounce (28.35 g): Calories, 90; Protein, 4 g; Fat, 7 g (72% cal.); Cholesterol, 25 mg; Carbohydrates, 2 g; Fiber, 0 g; Sodium, 210 mg.

VARIATIONS

Other white meats, such as veal or rabbit, can be used instead of chicken.

Mortadella

YIELD: 5 LB (2.4 KG)

U.S.	METRIC	INGREDIENTS	PROCEDURE
2 lb 8 oz	1.2 kg	Pork, lean	1. Cut the pork into cubes small enough to fit into the grinder.
1 tsp	5 mL	Prague Powder #1	2. Toss the meat with the Prague Powder, salt, and wine, and refrigerate several hours or overnight.
5 tsp	25 mL	Salt	
1½ fl oz	45 mL	Dry white wine	3. Cut the pork fat into cubes. Refrigerate several hours or overnight.
1 lb 6 oz	660 g	Firm pork fat, such as fatback	
		Garnish:	4. Prepare the garnish. Cut the fatback into small dice. Blanch in boiling water 2 minutes. Drain. Refrigerate.
6 oz	180 g	Pork fatback	
2 oz	60 g	Pistachios, peeled	5. Pick over the pistachios to be sure they are all peeled and in good condition.
14 oz	400 g	Ice	6. Grind the lean pork using the small die of the grinder. Place it in a food chopper.
1½ tsp	7 mL	White pepper	
1½ tsp	7 mL	Paprika	7. Grind the fat from step 3 (not the fat for the garnish) through the small die and set it aside. Refrigerate it if the kitchen is warm, even though you will be using it in a few minutes.
1 tsp	5 mL	Mace	
1 tsp	5 mL	Quatre épices	
1 tsp	5 mL	Ground coriander	8. Start the food chopper and run a few seconds. Add the ice, the spices, and the milk powder and continue to chop the meat to a fine paste. Check the temperature of the meat. Do not let it rise above 50°F (10°C).
¼ tsp	1 mL	Ground cloves	
2 oz	60 g	Nonfat dry milk powder	
			9. Add the ground fat and continue to run the chopper. Stop the machine from time to time and stir briefly by hand if necessary for even mixing. Process to a fine, uniform paste. Do not let the temperature rise above 58°F (14°C).
			10. Remove the meat mixture from the chopper and place in a stainless-steel bowl set over ice. Add the garnish (diced pork and pistachios) and mix into the meat purée.
			11. Stuff into large beef casings or artificial casings.
			12. Allow to dry several hours in the refrigerator. This step is not essential, but it gives the sausage time to take on the desired pink color.
			13. Poach the sausage in water over very low heat, using a thermometer to ensure the temperature of the water stays at 165°F (74°C). Cook until the internal temperature of the sausage reaches 152°F (67°C).
			14. Remove the sausage from the cooking water and cool it in 2 steps. First, place it in room-temperature water 5–10 minutes. Next, place it a tub of cold running water until completely cooled.
			15. Refrigerate.

Per 1 ounce (28.35 g): Calories, 110; Protein, 4 g; Fat, 10 g (82% cal.); Cholesterol, 15 mg; Carbohydrates, 1 g; Fiber, 0 g; Sodium, 190 mg.

VARIATION

Bologna

Omit the garnish of diced fat and pistachios. Omit the wine. After stuffing the casings and drying the sausage, hot smoke at 165°F (74°C) for 1 hour. Remove from the smoker and immediately poach as in the basic recipe.

Smoked Garlic Sausage

YIELD: 5 LB 12 OZ (2750 G)

U.S.	METRIC	INGREDIENTS
6 lb	3 kg	Pork and pork fat
1¼ tsp	6 mL	Prague Powder #1
5 tsp	25 mL	Salt
1 tsp	5 mL	Sugar
1 tsp	5 mL	White pepper
1 tsp	5 mL	Quatre épices
2 tsp	10 mL	Crushed garlic
6 fl oz	180 mL	Dry white wine or water

PROCEDURE

1. Cut the pork and fat into cubes.
2. Mix together the Prague Powder, salt, and sugar.
3. Add this mixture to the diced meat and fat and mix well. Refrigerate several hours.
4. Grind the meat mixture using the medium die.
5. Add the remaining ingredients to the meat and mix thoroughly.
6. Stuff into large hog casings.
7. Hang the sausages in the refrigerator, if possible, or arrange on towel-covered sheet pans so they are not touching each other. Refrigerate 24 hours to dry the sausages before smoking.
8. Hot smoke at 160°F (71°C) for 1½ hours, then poach in water at 165°F (74°C) until the internal temperature is 152°F (67°C).
9. Cool the sausages in cold water, then drain, dry, and refrigerate.

Per 1 ounce (28.35 g): Calories, 70; Protein, 5 g; Fat, 5 g (68% cal.); Cholesterol, 20 mg; Carbohydrates, 0 g; Fiber, 0 g; Sodium,170 mg.

VARIATIONS

For a darker color and heavier smoke flavor, leave the sausages in the hot smoke until their internal temperature is 152°F (67°C), and do not poach them. Spray them with cold water after removing them from the smoker to keep them from shriveling.

Cured Garlic Sausage

Dry the sausages 24–48 hours (step 7). Omit the hot smoking and simply poach, cool, and refrigerate them as described in steps 8 and 9.

ADDITIONAL RECIPES

These additional recipes may be found on your CulinarE-Companion recipe management program:

Cajun-Style Sausage; Lamb Sausage; Veal or Beef Sausage.

TERMS FOR REVIEW

charcutier	tinted curing mix (TCM)	sausage	natural casing
charcuterie	Prague Powder #2	fresh sausage	collagen casing
sodium nitrite	dry cures	cured sausage	basic grind
sodium nitrate	brine cures	smoked sausage	emulsified grind
nitrosamine	cold smoking	certified pork	emulsion grind
Prague Powder #1	hot smoking	quatre épices	progressive grinding

QUESTIONS FOR DISCUSSION

1. Explain how salt helps preserve meats.
2. Nitrites and nitrites are both used to cure meats. Explain how their use differs. Against which food-borne disease are they important in protecting cured foods?
3. Explain why smoked meats are cured before smoking.
4. Describe the basic differences between fresh, cured, and smoked sausages.
5. What are the four basic kinds of ingredients in sausage meat? List ways this basic formula can be changed to create different types of sausages.
6. Why should sausage meat be kept cold when it is being ground and processed?
7. Describe the procedure for preparing natural sausage casings for stuffing.

28

PÂTÉS, TERRINES, AND OTHER COLD FOODS

The techniques and procedures presented in this chapter belong to the culinary department known as **garde manger** (gard mawn zhay), a term whose basic meaning is "larder" or "food storage place." As explained in Chapter 21 (p. 657), this area, because of its cooler temperature, was traditionally used for the preparation of cold foods. Thus, the work carried out in the storage area, or garde manger, became known by the same name.

The art of garde manger includes the techniques of cold food decoration, cold platter design and presentation, and the design and planning of buffets. Garde manger is an intricate and complex discipline that is the subject of whole books and of extended courses of study.

This book is primarily concerned with à la carte cooking. Buffet service is beyond its scope. Nevertheless, à la carte restaurants have inherited from classical garde manger a number of special food preparations that can be served in single portions as well as on butler platters. Pâtés, terrines, galantines, and mousses are not only ideal for buffets, they are also popular in many restaurants. This chapter serves as an introduction to these preparations.

AFTER READING THIS CHAPTER, YOU SHOULD BE ABLE TO

1. Prepare and use aspic jellies.
2. Prepare livers for use in forcemeats.
3. Prepare basic meat and poultry forcemeats.
4. Prepare pâtés and terrines using basic forcemeats.
5. Prepare galantines.
6. Prepare mousseline forcemeats and make terrines based on them.
7. Prepare specialty terrines and other molded dishes based on aspics and mousses.
8. Handle raw foie gras and prepare foie gras terrines.
9. Prepare baked liver terrines.
10. Prepare rillettes.

THE HANDLING AND SERVICE OF COLD DISHES

Because the dishes featured in this chapter are served cold, factors relating to their handling and service require special consideration. These have to do with sanitation and presentation.

SANITATION AND STORAGE

Hot foods that have been handled in a sanitary manner and that are served at or above a temperature that kills microorganisms can usually be considered safe and sanitary to eat. Cold foods, on the other hand, present special problems because they are stored and handled after cooking. During this time, they may be exposed to disease-causing organisms. Because these foods are not subjected to further cooking, the organisms will not be destroyed.

For this reason, it is particularly important to follow all the rules of safe food handling. Make sure tools, containers, and work surfaces are clean and sanitary. Keep ingredients refrigerated when they are not being worked on, and keep the finished product refrigerated until service time. Always observe the four-hour rule, as explained on page 22.

The length of time terrines and other cold foods can be stored in the refrigerator depends on the ingredients, the type of item, and the method of preparation. Uncut meat terrines sealed with a layer of fat (see p. 861) may keep as long as several weeks (although the quality may start to decline after a week or so), while seafood and vegetable terrines may keep no more than two or three days, or even less. Pâtés en croûte (see p. 858) do not keep as well as terrines because the pastry loses its freshness. Aspics should be kept covered or wrapped to prevent them from drying out.

PRESENTATION

Attractive plating or presentation of foods is, of course, always important, but it could be argued that it is even more important for cold foods than for hot foods. Foods presented hot and steaming, directly from the sauté pan or carving board, have an immediate appeal to the nose as well as the eye, but cold foods must rely more completely on visual impact to make their first impression.

Furthermore, because the urgency in getting the plate from the kitchen to the dining room before the food cools down is absent, the cook has more time to arrange cold foods on the plate. This does not mean, however, that the most elaborate or intricate presentation is the best. "Keep it simple" is a good rule of thumb. Food is not made more appetizing by excessive handling.

Arrangements should be kept neat, but this does not mean they must always be symmetrical or regular. As in the case of salads, a deliberate casualness in the arrangement can be appetizing when it suggests the dish has been freshly assembled with minimum handling and rushed to the table.

In the case of pâtés and terrines, careful handling is essential to the presentation. Slice these items carefully and plate each slice with the best side up. To make neat slices, use a sharp, thin-bladed slicing knife. Before each slice, wipe off any residue from the previous slice and dip the blade in hot water. Slice the pâté with a gentle sawing motion, using the full length of the blade. Don't force the knife straight down or make little jagged cuts; this will make the cut surface uneven rather than smooth.

If the cut end of the pâté has discolored somewhat from exposure to air, plate the first slice with this side down. In contrast to pâtés for buffet presentation, slices of pâté for à la carte service are often garnished with greens or other colorful items, which enliven the sometimes drab appearance of a plain meat pâté. In addition, greens and other vegetable garnish provide a pleasant flavor contrast to the somewhat rich, processed flavor of the pâté or terrine.

Tart or piquant garnishes and accompaniments, as well as tart sauces such as vinaigrette and mayonnaise variations, help counter the richness of pâtés, which are often rather fatty. This is why sour pickles and mustard are classic accompaniments for these foods.

Consideration must be given to the serving temperature of cold foods such as aspics, pâtés, and terrines. A common error is to serve these items too cold. At refrigerator temperature, their flavors are masked. Furthermore, their textures are too firm; the fat in pâtés and the gelatin in aspics are firmly congealed. A little warmer temperature is necessary to enable them to melt pleasantly in the mouth.

To bring them to suitable serving temperature, remove individual portions from refrigeration and let stand at room temperature about 5 or 10 minutes, but no longer. Remember the sanitation guidelines discussed earlier. This period is long enough to let them warm slightly but not long enough to give microorganisms time to start multiplying. Keep in mind, too, that this short period of tempering applies only to those portions to be served right away. Whole pâtés from which portions were cut, for example, should be returned immediately to refrigeration. Again, remember the four-hour rule.

ASPIC

Aspic jelly, or *gelée* (zhuh lay), is clarified stock that contains enough gelatin to solidify when cold. The gelatin may be naturally extracted from bones or added from a package. Good stock naturally contains a certain amount of gelatin but, in most cases, it must be supplemented with additional unflavored packaged gelatin.

Aspic jelly may be nearly colorless (white aspic) or various shades of amber. Most often, however, it must be crystal clear. This is achieved by clarifying it like a consommé. White or light-colored aspic is used when the natural colors of the foods and decorations must show through. Amber or golden aspic enhances the brown color of foods such as roasted meats and poultry.

Aspic is used as a coating for foods and as a binding ingredient. When it is used as a coating, it has three main purposes:

1. To protect foods from the air, which would dry them out and discolor them.

2. To improve appearance and give shine.

3. To add flavor. This last purpose is, of course, best accomplished if the stock is of high quality.

As a binding ingredient, aspic is used in mousses, terrines, and aspic molds, as discussed later in this chapter. *Note:* When aspic is used as a binding agent, it need not be perfectly clarified.

In addition, when congealed and chopped or cut into various shapes, aspic jelly is used as a garnish for platters or servings of pâtés, terrines, and other cold items.

PREPARING ASPIC JELLY

The best aspic is a well-made, naturally gelatinous stock. It has a superior texture and flavor, but it is time-consuming to make because a separate batch of stock must be made in addition to the normal stock production. Consequently, most aspics are made by reinforcing regular stock with gelatin. Aspic powders and mixes are available, but the flavor of aspic made from them does not compare with that made from stock. They can be useful in an emergency, however, or for pieces used purely for display or decoration.

Aspic Strength

As the previous discussion suggests, aspic has many uses. For some uses it must be stronger— that is, have a higher gelatin content—than for others. For our purposes in this chapter, we must be able to make two strengths of aspic:

Aspic for coating is used to glaze cold foods. It contains a moderate proportion of gelatin.

Aspic for slicing is used to line forms such as timbales, ring molds, and silicone forms. It also is used to bind the ingredients for aspic-based terrines. Aspic for this purpose must hold its shape at cool room temperature and must cut cleanly with a hot, sharp knife. It contains a high proportion of gelatin. However, it must not be tough or rubbery.

Classic Aspic Jelly

Prepare classic aspic jelly as you would white or brown veal stock, but with the addition of products that release a good deal of gelatin, such as split calves' feet or pigs' feet, pork skin, and veal knuckle bones. If enough of these items are used, the stock will contain enough natural gelatin to be used as an aspic jelly.

Follow basic stock-making procedures, except do not brown the added feet and pork skin when making brown stock. When the stock is made, use the following procedure to convert it to aspic jelly.

PROCEDURE for Preparing Classic Aspic Jelly for Coating

1. Test the stock for gelatin content. Ladle a small amount of cooled stock onto a small plate or saucer and refrigerate.

   • If the stock becomes firm, no additional gelatin is needed.

   • If it sets but is not firm enough, add about ½ oz or 2 tbsp (15 g) powdered gelatin per quart (L) of stock, or 2 oz (60 g) per gallon (4 L).

   • If it does not set at all (which is unlikely if the stock is properly made) but merely becomes thicker, add about 1 oz or 4 tbsp (30 g) powdered gelatin per quart (L), or 4 oz (125 g) per gallon (4 L). In this case, you will actually be making regular aspic jelly (see next section) rather than classic aspic jelly.

   Add the gelatin by first stirring it gradually into a little cold water, avoiding making lumps, and letting it soften. Then add the softened gelatin to the stock.

2. Clarify the stock as for consommé. For white aspic (made with white stock), omit the carrots in the mirepoix.

3. After the stock is clarified, remove all traces of fat. The aspic jelly is now ready to use.

PROCEDURE for Preparing Classic Aspic Jelly for Slicing

1. Prepare as for Classic Aspic Jelly for Coating, following the previous procedure, but add an additional 1 oz gelatin per quart (30 g per L) of stock. In other words:

   • If the stock sets firm but is not firm enough to hold its shape when sliced, add about 1 oz (30 g) powdered gelatin per quart (L), or 4 oz (125 g) per gallon (4 L).

   • If it does not set at all but merely becomes thicker, add about 2 oz or (60 g) powdered gelatin per quart (L), or 8 oz (250 g) per gallon (4 L).

2. Continue with steps 2 and 3 in the procedure above.

Regular Aspic Jelly

This is normal meat, poultry, or fish stock reinforced with gelatin and clarified. Regular stock rarely has enough natural gelatin to be used as a jelly, so extra gelatin must be added. To test the stock for gelatin content and to convert it to aspic jelly, follow the same procedure as for classic aspic jelly, described above. However, note that if the stock contains very little gelatin and stays watery when chilled, more than 4 oz (125 g) gelatin may be needed per gallon (4 L) if you are making aspic for coating. If you are making aspic for slicing, you will need about 8 oz (250 g) gelatin per gallon (4 L).

Aspic Powder

Aspic powder is unflavored gelatin mixed with a powdered stock base. To prepare it, follow the instructions on the container. Additional unflavored gelatin may be needed for some purposes.

USING ASPIC JELLY

Aspic is used to enhance both the appearance and the flavor of cold foods. For best results, the aspic, as well as the foods to be coated, should be prepared and handled in specific ways. The following sections offer general procedures for handling these products. Specific applications, such as recipes for aspic-based terrine molds, are included later in the chapter.

Aspic jelly must be cooled to just above congealing temperature before it is used to coat foods. If it is too warm, it will not have enough body to coat and will just run off.

Standards of Quality for Aspic Jelly and Aspic-Coated Items

1. The aspic should be perfectly clear, with no trace of cloudiness.

2. White aspic should be nearly colorless. Aspic made with brown stock may range from amber to golden brown.

3. The aspic should be set firmly enough to hold its shape but should melt in the mouth. It should not be rubbery.

4. The coating layer of aspic should be perfectly smooth, with no bubbles or unevenness. It should coat the food completely with a thin, uniform layer.

PROCEDURE for Cooling Aspic Jelly

1. If the jelly is congealed, it must first be melted. Set the pan or container of jelly in a hot-water bath. Stir it gently from time to time until it is completely melted.

2. Place the warm aspic jelly in a stainless-steel bowl.

3. At all times, be careful not to make any bubbles. Bubbles in the jelly may get transferred to the surface of your food item and mar its appearance.

4. Select a ladle that fits the curve of the bowl. Set the bowl in crushed ice, pushing it in so it sits in a well of ice. With the edge of the ladle against the inside of the bowl, rotate the bowl so the ladle continually scrapes the inside of the bowl. This method prevents the formation of lumps that occur when jelly touching the cold bowl solidifies too quickly.

5. Continue to rotate the bowl until the jelly is thick and syrupy but not yet set. The jelly is now ready for use. Remove from the ice bath and work quickly, because it will set very fast.

6. Remelt and recool the jelly as necessary.

PROCEDURE for Coating Foods with Aspic Jelly

1. Chill the food to be coated. For best results, the surface of the item should be as smooth and as free of fat as possible.

2. Place the item on a wire rack over a tray or sheet pan. Excess aspic that falls onto the tray can be remelted and reused.

3. Cool the aspic jelly according to the procedure above.

4. Use the aspic as soon as it is ready. Various methods can be used to coat foods with aspic, depending on the size and shape of the item.
   - For smooth, regularly shaped items, use a large ladle and nap them with a single smooth stroke, as illustrated in Figure 28.1. Working too slowly may produce an uneven, bumpy coat.
   - Large items and items with steep sides or irregular shapes are harder to coat. Using a ladle, coat the sides first and then the top for best results.
   - For small items, it may be more convenient to use a kitchen spoon than a ladle.
   - A pastry brush can be used to coat small items. A brush is often used for small portions, such as canapés, that need only a light glaze rather than a perfectly smooth coating of aspic.

5. Chill the items until the jelly is thoroughly set.

6. Repeat with additional coats, if necessary, until the aspic is of the desired thickness.

7. To decorate, dip pieces of decoration in liquid aspic and place on the product in the desired pattern. Some items appropriate to use for decorating aspic are

Leek leaves	Black olives	Tomato peels
Fresh herbs, especially flat-leaf parsley and tarragon	Truffles, real or artificial	Carrots

   As appropriate, cut the items for decoration into very thin slices and then into desired shapes. For most vegetable decorations, such as carrots and leek leaves, blanch to make them more limber and to intensify the color.

8. If decorations are used, cover the decorated item with a final layer of aspic jelly to protect the design.

9. Evaluate the quality of the finished item (see p. 852).

FIGURE 28.1 A smooth, regularly shaped item can be covered with an even layer of aspic using one stroke of the ladle.

PROCEDURE for Lining a Mold with Aspic Jelly

Many instructions for lining molds say to chill the mold, then pour in a little liquid aspic jelly and turn and tilt the mold until the bottom and sides are coated. This method works and is suitable for some purposes, but it does not produce a smooth, even layer of aspic.

It is not always necessary to line a mold with aspic. Many aspic molds are made by first pouring a thin layer of aspic into the bottom of the mold, chilling it, then adding layers of ingredients (vegetables, meats, mousses, and so on) and covering each layer with a little aspic. The mold is chilled after each layer is added to allow the aspic to set. As long as the layers of solid ingredients are not allowed to touch the sides of the mold, the liquid aspic will fill in these spaces and, in effect, line the mold.

If a mold must be lined with a perfectly even thickness of jelly, the following method is used (see Figure 28.2):

1. Press the mold into a bed of crushed ice so the ice comes all the way to the top edge.

2. Fill the mold with cooled liquid aspic jelly for slicing. Leave the mold in place 10 seconds. Immediately remove the mold from the ice and quickly dump out the jelly that is still liquid. If the layer of jelly remaining inside the mold is too thin, repeat. If it is too thick, remove it, clean the mold, and repeat the procedure, leaving the mold on ice for less time.

3. Decorate the inside of the mold as desired by dipping decorations in liquid aspic and arranging them in place. Chill. Then fill the mold with the selected food product.

FIGURE 28.2 Lining a mold with aspic jelly.

(a) Bury the mold up to the rim in crushed ice.

(b) Fill the mold with liquid aspic.

(c) After 10 seconds, quickly but smoothly pour out the aspic that is still liquid.

(d) An even layer of aspic jelly lines the mold, as can be seen by comparing it with an empty mold.

(e) At this point, you can decorate the mold by dipping vegetable cutouts in liquid aspic and carefully setting them in place in the mold.

KEY POINTS TO REVIEW

- What kind of attention should be paid to sanitation procedures when working with cold foods? Why?

- What are the steps in the procedure for preparing classic aspic jelly?

- What are the steps in the procedures for cooling aspic jelly and then coating foods with it?

- What are the steps in the procedure for lining a mold with aspic jelly?

SPECIAL FORCEMEAT DISHES

This section is concerned with classic meat and poultry dishes called *pâtés*, terrines, and galantines. Some terrines are based on vegetables and other items rather than meats, but these are the subject of a later section. The main ingredients of the items discussed here are a forcemeat and, usually but not always, a garnish.

A **forcemeat** may be defined as a mixture of seasoned, ground meats used as a stuffing or filling. The name comes from the French word *farce*, which means "stuffing."

The **garnish** in a pâté or terrine is not just a decoration but a major ingredient that adds body, flavor, and nutritional value as well as appearance. Garnish usually consists of meats or other foods cut in dice, strips, or other shapes, or left whole if they are small. Classic pâté garnishes include:

Ham	Chicken, duck, or goose livers	Fresh pork fatback
Veal		Tongue
Chicken, duck, or turkey breast	Foie gras	Pistachios
	Game	Truffles

TYPES OF FORCEMEAT

The following are the three basic types of forcemeat:

1. **Straight forcemeat (including country-style forcemeat).**
   This is a mixture of seasoned ground meats. As such, it is basically a form of sausage meat, except the grind is generally, but not always, finer. Consequently, many of the guidelines for making and handling sausage meat, discussed in Chapter 27, apply here as well. It may be helpful to read or to review pages 836–841 in conjunction with this discussion. Straight forcemeats are the basis of most traditional pâtés and terrines and are the major focus of this chapter.

   A *country-style forcemeat* is made the same way, except the grind is coarser. Most country-style forcemeats are made from pork and pork fat and contain some liver.

2. **Gratin forcemeat.**
   This type of forcemeat differs from a straight forcemeat in that a portion of the meat is seared, and thus partially cooked, and cooled before it is ground. Because the partial cooking eliminates some of the binding power of the meat protein, gratin forcemeats usually contain a starch binder called a *panada* (also called *panade*). This type of forcemeat is not used as often as straight or mousseline forcemeats and is not covered in this book.

3. **Mousseline forcemeat**.
   This type of forcemeat consists of white meat (usually poultry or veal) or seafood processed to a purée and combined with heavy cream and egg. See page 865 for a discussion.

STRAIGHT FORCEMEATS

A basic straight forcemeat consists of the following:

50–65 percent lean meat

35–50 percent fat

Seasonings

The many variations on this basic formula depend on the ingredients used and how they are combined.

Meat

Pork is the basic ingredient, but many other meats can be included in addition to or instead of pork, including veal, chicken, turkey, ham, duck, rabbit, and game of all kinds.

Liver

Chicken, goose, duck, or pork liver is often included in forcemeats. Liver gives flavor and also acts as a binder.

Fat

The classic proportion in forcemeats is equal parts fat and meat. Many forcemeats, however, contain less than 50 percent fat, especially in recent years, as people have become more attentive to nutrition and dietary considerations (see the discussion of the fat content of sausages on p. 836). Nevertheless, a certain amount of fat is necessary for both moisture and flavor. A pâté with too little fat tastes dry. Hard fat, such as pork fatback, gives best results. Heavy cream is sometimes used to add fat as well as liquid to a forcemeat.

Note that this discussion of fat content refers only to the solid fat specifically added as a measured ingredient. There is, of course, some fat in the lean meat as well.

Other Ingredients

Eggs or egg whites may be added as a binder. Flour or other starches may be added for the same purpose. Extra binders are not absolutely necessary in a forcemeat made purely of meat and fat because the meat proteins are sufficient to bind the product when cooked. On the other hand, when brandy, cream, and other liquids are added to the forcemeat, extra binders may be needed or at least beneficial.

The Grind

Forcemeat may be ground coarse, medium, or fine. Country-style pâté, or **pâté de campagne** (cawm pah nyuh), is characterized by a coarse texture. Galantines, on the other hand, are usually made from finely ground forcemeats.

Preparing Straight Forcemeats

Many, if not most, pâté and terrine forcemeats contain some liver. Chicken livers or other poultry livers, both economical and widely available, are the most often used. For best results, livers should be soaked in milk and then cleaned according to the following procedure. Pork liver and other larger livers can be cut into pieces and prepared in the same way.

The following recipe can be used with many garnishes to make a great variety of pâtés, terrines, and galantines. It can also be changed according to any of the variations listed by using different meats. Once the basic technique is understood, any kind of pâté can be produced.

The recipe should be viewed as a basic procedure that can be varied in ways other than those indicated following the recipe, just as sausage meat can be varied. The varieties and quantities of spices can be changed. In addition, the fineness of the grind can be varied to make pâtés of varying textures.

The proportion of fat can be increased or reduced, but remember that making the forcemeat too lean will reduce its eating quality. Although at first glance the recipe looks as if it calls for 50 percent fat, this is not the case, as the liver should be included as part of the meat. The proportion of fat is 44 percent. Taking the first column of ingredient quantities as an example, using 1 lb lean pork, 12 oz fat, and 4 oz liver lowers the proportion of fat to 38 percent (not counting, of course, the smaller amount of fat within the meats). Using 12 oz lean pork, 1 lb fat, and 4 oz liver raises the fat proportion to 50 percent.

Just as for sausages (see discussion on p. 839), the meats must be kept well chilled at all times. Chill the grinding equipment before grinding, and return the meats to the refrigerator whenever they lose their chill.

After grinding, the forcemeat should be mixed to develop texture, the same as for sausage forcemeats (p. 840). Mixing binds together the meat proteins, fat, and water so that, after cooking, the forcemeat is smooth and moist rather than dry and crumbly.

Standards of Quality for Forcemeats

1. The main flavor of the forcemeat should be that of the main ingredient. In other words, duck forcemeat should taste like duck, game forcemeat should taste like game, and so on. Additional seasonings and flavorings should enhance the main ingredient, not cover it up or add inappropriate tastes.

2. The texture of the forcemeat should be appropriate to the finished product. For example, forcemeat for country-style terrines should have a coarse texture, while forcemeat for galantines (p. 863) should be smooth. The forcemeat should contain no fragments of bone, cartilage, or connective tissue.

3. The cooked forcemeat should slice easily, and the slices should hold their shape. Forcemeats with a crumbly texture or those that break easily when sliced were incorrectly made.

4. Added garnish should be attractive, correctly cut, and well distributed in the forcemeat. Colors from garnish should not bleed into the forcemeat. Garnish flavors should complement the flavor of the main meat ingredient.

5. The color of the sliced forcemeat should be appropriate to the main meat ingredient, with no gray or discolored areas.

PROCEDURE for Preparing Poultry Livers for Forcemeats

1. Rinse the livers in cold water, drain, then soak 24 hours in enough milk to cover.

2. Drain and rinse thoroughly in cold water. Drain again.

3. Remove all fat and connective tissue. At this point, the livers are ready to be used whole as garnish for pâtés and terrines. If they are to be added to forcemeats, continue with steps 4 and 5.

4. Blend in a blender until liquid.

5. Strain through a chinois or fine strainer to remove all traces of connective tissue.

 # Basic Pork Forcemeat

YIELD: 2 LB (900 G)

U.S.	METRIC	INGREDIENTS	PROCEDURE
14 oz	400 g	Lean pork	**1.** Before beginning, make sure all equipment and all ingredients are well chilled. Forcemeats must be kept cold at all times to prevent the fat from softening or melting.
14 oz	400 g	Pork fat	**2.** Cut the meat and fat into small dice.
		Marinade:	**3.** Sweat the minced shallots in the butter until soft. Add half of the white wine and reduce by half. Cool completely.
1¹/₂ oz	45 g	Shallots, minced	
¹/₂ oz	15 g	Butter	
2 fl oz	60 mL	White wine	**4.** Combine the meat and fat with the shallots, the rest of the wine, the bay leaves, brandy, salt, spice mixture, and pepper. Toss to mix well. Cover and refrigerate overnight.
2	2	Bay leaves	
1 fl oz	30 mL	Brandy	
2¹/₂ tsp	12 mL	Salt	**5.** Remove the bay leaves. Grind the meat and fat twice through the fine blade of a meat grinder.
¹/₂ tsp	2 mL	Pâté spice or quatre épices (see Note)	
¹/₄ tsp	1 mL	White pepper	
4 oz	100 g	Chicken livers, soaked, cleaned, and puréed (see above)	**6.** Combine the ground meats and liver purée.
2	2	Eggs (see step 7)	**7.** Place the ground meats in the chilled bowl of a mixer. Beat the eggs lightly and add to the ground meat. Using a chilled paddle, mix thoroughly, until the forcemeat is well mixed and feels slightly sticky. (*Note:* The eggs are optional and are omitted in many pâtés.)
			8. Make a quenelle (a small ball of forcemeat) and poach it in simmering water. Cool. Taste and correct the seasonings in the forcemeat.
			9. Keep the forcemeat chilled until ready for use.

Per 1 ounce (28.35 g): Calories, 150; Protein, 3 g; Fat, 14 g (89% cal.); Cholesterol, 54 mg; Carbohydrates, 0 g; Fiber, 0 g; Sodium, 200 mg.

Note: Pâté spice may be purchased in various blends, or you may make your own blend to taste. Pâté spice usually contains black and white pepper, cloves, nutmeg, ginger, cayenne, bay leaf, thyme, and marjoram. Grind very fine and sift through a sieve. For quatre épices, see page 843.

VARIATIONS

Omit the pork, fat, livers, and eggs in the basic recipe. Substitute the following ingredients and quantities. Vary seasonings to taste.

Veal Forcemeat

9 oz	250 g	Lean pork
9 oz	250 g	Lean veal
14 oz	400 g	Fresh pork fat
3	3	Eggs

Chicken Forcemeat I

9 oz	250 g	Lean pork
9 oz	250 g	Chicken meat
14 oz	400 g	Fresh pork fat
3	3	Eggs

Chicken Forcemeat II

1 lb 2 oz	500 g	Chicken meat
14 oz	400 g	Fresh pork fat
3	3	Eggs

Duck, Pheasant, or Game Forcemeat

7 oz	200 g	Lean pork, or a mixture of pork and veal
7 oz	200 g	Duck, pheasant, or game meat
14 oz	400 g	Fresh pork fat
4 oz	100 g	Livers, soaked, cleaned, and puréed
1	1	Egg

TERRINES AND PÂTÉS

Terrines and pâtés are baked forcemeats, often but not always containing one or more types of garnish. Strictly speaking, the difference between the two lies in how they are baked. By definition, a **terrine** is baked in an earthenware dish. The dish itself is also called a *terrine*, a word derived from the French *terre*, meaning "earth." Today, other materials besides earthenware, such as glass or metal, may be used for terrines. Terrines may be presented in their baking dish, or they may be unmolded.

Pâtés include products baked in a crust, although chefs disagree about the correct usage of the terms (see sidebar). In this book, we use the term **pâté** to indicate products baked with a crust and *terrine* for products baked without a crust. Popular usage of the terms, however, is much looser.

It should be noted that many kinds of products are called *terrines* because they are prepared in terrine molds. The terrines discussed in this section are based on the straight forcemeats we have just considered. Other kinds of terrines are discussed in a later section.

Preparing Pâtés

The essential difference between a pâté and a terrine, as we are using the terms, is the crust. Although a heavy pastry crust may not be suitable for all kinds of terrine mixtures, the typical baked forcemeat-type terrine under consideration here can usually be made with or without a crust.

This section concentrates on the specific procedures for making the pastry and finishing the assembled pâté. Making the meat filling is the same as for terrines and is not repeated here. To make a pâté en croûte, apply the following procedure to the Veal and Ham Terrine and to any of the variations following the basic recipe (see p. 862).

Pastries used to enclose pâtés are of various types, but the most commonly used are similar to pie pastries, but sturdier. A recipe for this type of **pâte à pâté**, or pâté pastry, is included here. Its advantage over many other types of pâté pastry is that it is relatively good to eat. Some authorities argue about whether the dough around a pâté is meant to be eaten. But because customers are not necessarily aware of this argument, it is best to use a pastry that is reasonably pleasant to eat.

Traditional English pâtés, or raised meat pies, use a hot-water pastry that can be modeled like clay and is sturdy when baked. Pastries used for display—that is, for show platters not intended to be eaten—are also made to be sturdy and easy to handle. These pastries are not considered here.

One type of procedure for assembling a pâté follows the pastry recipe (Figure 28.3).

PÂTÉS AND PASTRY

Many experts argue that a pâté is, by definition, baked in a crust. The word pâté (with an accent on the *e*) is derived from the word *pâte* (without the accent), meaning "pastry." Others insist the root meaning of *pâte* is "paste," so it can apply to any kind of paste, including meat pastes. This agrees with modern usage, in which the word *pâté* is used for products with or without a crust, and even for spreadable meat pastes.

To avoid confusion, perhaps the best advice is to use the term **pâté en croûte** if you want to specify a pâté with a crust.

Pâté Pastry (Pâte à Pâté)

YIELD: 1 LB 12 OZ (900 G)

U.S.	METRIC	INGREDIENTS	PROCEDURE
1 lb	500 g	Flour	1. Place the flour in a large mixing bowl. Add the butter and lard. Rub them in until no lumps of fat remain.
4 oz	125 g	Butter	
3¹/₂ oz	100 g	Lard	
1	1	Eggs	2. Beat the eggs with the water and salt until the salt is dissolved.
3 fl oz	100 mL	Water, cold	3. Add the liquid to the flour mixture. Mix gently until it is completely absorbed.
1¹/₄ tsp	7 mL	Salt	4. Gather the dough into a ball. On a work surface, knead the dough a few minutes, or until it is smooth.
			5. Place the dough in a pan and cover with plastic film. Refrigerate until needed, or at least 4 hours.

Per 1 ounce (28.35 g): Calories, 120; Protein, 2 g; Fat, 7 g (53% cal.); Cholesterol, 20 mg; Carbohydrates, 20 g; Fiber, 0 g; Sodium, 140 mg.

FIGURE 28.3 Making a pâté en croûte.

(a) Collapsible molds are used to make pâtés en croûte because they can be removed from the mold without damaging it. Assemble the mold and grease the inside well.

(b) Lightly roll the pastry into a rectangle, keeping it thick.

(c) Work the dough into a boat shape. Dust heavily with flour and fold the dough lengthwise to make a pocket.

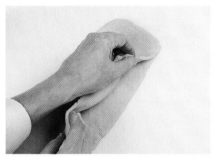

(d) Roll the double thickness of dough into a rectangle the size of the mold. Open the pocket.

(e) Fit the dough into the mold. Carefully work it to fit snugly. A ball of dough dipped in flour helps fit the dough into the corners without tearing it.

(f) Partially fill the mold with forcemeat and arrange the garnish according to the instructions in the specific recipe.

(g) Finish filling the mold, mounding the forcemeat slightly.

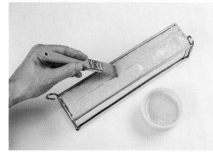

(h) Fold the ends and then sides of the dough over the top of the forcemeat, trimming the dough so it meets in the middle. Egg-wash the dough.

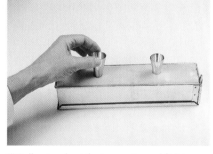

(i) Roll out and cut the top pastry and fit it in place, making sure it is sealed well to the dough below it. Egg-wash. Make holes in the top and fit pastry tubes in them to serve as chimneys to allow the escape of steam and to prevent melted fat from bubbling over the top crust as the pâté bakes.

(j) After the pâté has baked and cooled, pour liquid aspic through the chimneys to fill the spaces left when the forcemeat shrank during baking.

PROCEDURE for Lining and Filling Pâté Molds and Finishing Pâtés

1. Prepare the pastry in advance so it has plenty of time to rest. Remove it from refrigeration long enough ahead of time to allow it to warm up slightly.

2. Prepare the molds by greasing them well on the inside. The directions here pertain to standard rectangular pâté molds. These usually are hinged and collapsible so it is easy to remove the pâté without damaging it. If you are using bottomless molds, then also grease the sheet pans on which they are to set. For molds of other shapes, modify the pastry-molding procedure as necessary to fit the shape.

3. For best results, mold the pastry so it is of even thickness, has no seams, and fits the mold perfectly. First, work the pastry with the hands for a few seconds to make it pliable. Then shape it into a rectangle and roll it slightly with a rolling pin to flatten it, keeping it quite thick.

4. Make an indentation down the center of the dough with the fingers. Gradually make the dough into a sort of boat shape. Dust the inside of the dough shape well with flour (to keep the two layers of dough from sticking together) and fold the dough along the indentation to make a pocket.

5. Gently roll out the dough to make a rectangle the size of the mold. Be careful not to roll the dough too thin, which would make it fragile. Open up the pocket.

6. Fit the opened pocket into the mold. Carefully mold the pastry to the shape of the mold by pushing the dough with your fingers. Make sure there are no air bubbles between the dough and mold. A ball of dough dipped in flour is useful for pushing the dough into the corners of the mold without tearing it.

7. If the pâté is to be made without a top crust, leave a rim of dough about 1/4 inch (5 mm) above the top of the mold. Crimp this rim to make a decorative border. If there is to be a top crust, leave a rim of half the width of the mold and let it hang over the sides. (For an alternative method, see step 10.)

8. The mold is now ready to fill. For display pieces, it is common practice to line the inside of the dough with thin sheets of fatback. For pâtés to be eaten, however, it is more appetizing to omit the fat lining. Fill the mold with the desired forcemeat and garnish as for terrines (see p. 861). Mound the filling slightly so the top crust, if used, will have an attractive domed shape. The dough should hold this shape even as the forcemeat shrinks and settles during baking.

   If the pâté is not to have a top crust, it is now ready for baking. Skip to step 12. Baking without a top crust is easier and allows you to make an attractive aspic glaze with decorations for the top.

9. Fold the rim of the dough from the sides of the mold over the top of the filling. If using a top crust:

- Roll out a sheet of dough.
- Lay the sheet of dough on top of the mold, trim it to size, and remove it.
- Brush the edges of dough from the sides of the mold with egg wash.
- Return the pastry top to the mold and fit it in place, gently sealing it to the egg-washed dough.

10. As an alternative method for fitting the top crust:
- Leave a 1/4-inch (5-mm) rim of dough, as in step 7.
- Brush the inside of this rim with egg wash.
- Roll out and cut a top crust slightly larger than the top of the mold. Place it on top of the filled mold.
- Crimp or pinch the two layers of pastry together with the fingers to seal.

11. Decorate the top crust with pastry cutouts, if desired. Seal the cutouts to the crust with egg wash. Make one or two vent holes in the top crust to allow steam to escape. Fit pastry tubes into these holes to form chimneys in order to keep juices from running over the top crust and spoiling its appearance.

12. To bake:
- Preheat an oven to 400°F (200°C). Place the pâté on a sheet pan (if you are using a bottomless mold, it will, of course, already be on a sheet pan) and put it in the oven.
- After 10 minutes, reduce the heat to 350°F (175°C). The higher initial temperature helps brown the pastry. Bake at this lower temperature until the internal temperature reaches 160°–165°F (72°C).
- For an average rectangular mold, the baking time will be 1–2 hours. Small molds that make 1–4 portions will take 45 minutes or less.
- For very large molds, use a baking temperature of 325°F (160°C) so they cook evenly. Extend the baking time accordingly.

13. Remove the pâté from the oven. Let the pâté cool to room temperature in its mold. For a pâté made without a top crust, first let it cool until it is warm. Then let it finish cooling with a weight on top in order to give the pâté a firmer texture. The weight should be large enough to cover the meat but small enough so it doesn't touch the pastry rim. This can be accomplished by cutting a board to the proper size, laying it in place on the pâté, and placing weights on the board. (Obviously, this cannot be done if there is a top crust.) Refrigerate.

14. When the pâté is cold, prepare an aspic jelly. Melt the aspic and flavor it, if desired, with a little sherry, port, or Madeira. Cool it according to the procedure on page 853. Fill the pâté with the aspic.

- If the pâté has a pastry top, pour the aspic through the vent hole or holes, using a funnel, until the pâté is completely full.
- If the pâté has no top crust, fill it with enough aspic to completely cover the top of the meat.
- Refrigerate until the aspic is set.

15. Remove the pâté carefully from the mold.

16. Pâtés without a top crust may now be decorated and reglazed with aspic if desired. Decorate as desired (using the materials suggested on page 853) by dipping the decorations in liquid aspic and setting them in place. Chill briefly, then apply a little more aspic to glaze the top.

17. For storage, handling, and presentation, see page 850.

Preparing Forcemeat Terrines

Terrines, like pâtés, may be baked in molds of various shapes and sizes. Traditional oval molds, for example, have long been popular. For ease of portion control, however, rectangular molds are the most appropriate.

A terrine may be lined with thin sheets of fatback, although this is optional. The layer of fat does not contribute significantly, as is widely believed, to keeping the meat moist during baking; after all, the terrine mold itself is more moistureproof than the layer of fat. Although such a fat lining is traditional, today's diners are more likely to find a rim of fat unappetizing. Of course, the fat layer can be removed before serving. Alternatively, a sheet of caul fat, which is much thinner than fatback, can be used to line the mold.

PROCEDURE for Preparing Forcemeat Terrines

1. Prepare the desired forcemeat (see p. 857).

2. Prepare the selected garnish. Meat garnishes are usually cut into strips, which are laid lengthwise in the mold.

3. Marinate the garnish as desired. This step is optional but adds to the flavor.

4. Prepare the mold. Do not use a hinged or collapsible mold, which cannot be placed in a water bath. If desired, line the mold with thin sheets of fatback (sliced on a slicing machine) or with a sheet of caul fat, letting the excess hang over the sides. Make the sheets of fat sufficiently large so the amount of fat hanging over the sides can be folded over to cover the top completely. If the mold is not lined with fat, grease it well.

5. Place a layer of forcemeat in the bottom of the mold. If no garnish is used, simply fill the mold. Spread the forcemeat evenly and rap the mold sharply on the workbench to dislodge any air bubbles.

6. Arrange a layer of garnish on top of the forcemeat.

7. Continue adding forcemeat and garnish until they are all used. End with a layer of forcemeat on top. Two or three layers of garnish are usually sufficient.

8. If a fat lining has been used, fold the excess fat over the top of the forcemeat to cover it.

9. Cover the top with a sheet of aluminum foil. Cut a few holes in the foil to allow steam to escape.

10. Place the mold in a water bath for baking. Make sure the bath is deep enough to allow the hot water to come halfway up the sides of the mold. Bake at 350°F (175°C) until the internal temperature registers 165°F (74°C).

11. Remove the terrine from the water bath and place it on a rack to cool. When it has cooled somewhat but is still warm, finish cooling it with a weight, as explained in the procedure for making pâtés. It should not be weighted when it is still hot because it is too fragile and might split or fracture, and the weight might force out too much juice. If a looser texture is desired, cool the terrine without weighting it.

12. When the terrine is completely cool, cover and refrigerate it.

13. The terrine may be sealed with a layer of fat or aspic. This protects the terrine from air and helps preserve it.

- To add a layer of fat, melt lard (or rendered duck fat or other fat appropriate to the terrine), then let stand until cool but still liquid. The terrine should be cool, about 50°F (10°C). Pour in enough fat to cover the meat completely. Let stand until the fat has congealed, then cover and refrigerate. The purpose of this fat is only to extend the keeping quality of the terrine. It should be removed before serving.

- Add aspic to a terrine in the same way as adding a layer of fat; see also the procedure for adding aspic to a pâté, page 860. Unlike melted fat, aspic extends the storage life of a terrine only a few days because the aspic itself dries out. On the other hand, aspic contributes to both flavor and appearance. If desired, apply decorations to the top of the terrine and add another layer of aspic to glaze.

 # Veal and Ham Terrine

YIELD: APPROXIMATELY 2 LB (1 KG)

U.S.	METRIC	INGREDIENTS	PROCEDURE
2 lb	1 kg	Veal Forcemeat (p. 857) Garnish:	1. Prepare the forcemeat according to the recipe on page 857. Refrigerate it until very cold.
4 oz	125 g	Veal, lean, trimmed	2. Cut the veal, ham, and fatback for the garnish into strips about ¹/₄ in. (6 mm) thick. Mix with the brandy and marinate in the refrigerator 1 hour or longer.
4 oz	125 g	Smoked ham	
1 oz	30 g	Fresh pork fatback	
2 fl oz	60 mL	Brandy	
as needed	as needed	Fresh pork fatback or caul fat for lining molds (optional)	3. Have ready a 2-qt (2-L) rectangular terrine mold. 4. If using fatback to line the mold, have the fat very cold. Cut it on a slicer into broad, thin slices less than 1/8 in. (3 mm) thick. Line the mold with the slices, overlapping them by about ¹/₄ in. (5 mm). Let the tops of the slices hang over the edges. If using caul fat, line the mold with a large sheet of caul, letting the edges hang over the side of the mold. If not using caul or fat, grease the mold well. 5. Fill the terrine with alternating layers of forcemeat and garnish, beginning and ending with forcemeat and laying the strips of garnish lengthwise in the terrine. Press the meat firmly into the terrine so there are no air bubbles. 6. If using sheets of fat or caul to line the mold, fold the overhanging fat over the top of the forcemeat to cover. 7. Cover with foil. 8. Set the terrine in a hot-water bath. Bake at 350°F (175°C) until the internal temperature is 165°F (74°C). 9. Remove from the oven and cool until just warm. Weight and continue to cool, following the basic procedure on page 860. Finish, if desired, with a layer of melted fat or aspic, as described in the basic procedure.

Per 1 ounce (28.35 g): Calories, 170; Protein, 5 g; Fat, 15 g (83% cal.); Cholesterol, 50 mg; Carbohydrates, 0 g; Fiber, 0 g; Sodium, 250 mg.

VARIATIONS

Veal and Ham Terrine with Foie Gras

Prepare as in the basic recipe, but place a layer of sliced, cooked foie gras down the center of the terrine. Use slices of foie gras terrine (p. 873) or canned foie gras pâté. A row of sliced truffles may be placed on top of the foie gras layer.

Veal and Tongue Terrine

Use cooked, cured beef tongue in place of the ham.

Rabbit Terrine

Bone out a rabbit, keeping the loin meat in 2 long strips. Make a rabbit forcemeat by following the veal forcemeat recipe but substituting meat from the rabbit legs for all or part of the veal. Soak, clean, and liquefy the rabbit liver according to the procedure on page 857. Add it to the forcemeat. Omit the garnish from the basic recipe, instead using the rabbit loins marinated in the brandy. Fold the thin end of each loin back on itself so it is of uniform thickness. When filling the terrine, put half the forcemeat into the mold, lay the loins end to end down the center of the terrine, then fill with the remaining forcemeat.

Optional step: Make a stock with the rabbit bones. Reduce the stock to a glaze, cool, and mix with the forcemeat.

Optional step: Add a small quantity of nuts, such as skinned pistachios, to the forcemeat.

Country Terrine

Use pork forcemeat, keeping the grind rather coarse. Chop the garnish coarsely and mix with the forcemeat.

Game Terrine

Prepare as in the basic recipe, using Game Forcemeat (p. 857) and strips of game meat instead of the veal and ham for garnish. Optional: Add a small quantity of green peppercorns, rinsed and drained, to the forcemeat.

Duck Terrine

Bone out a duck. Use the leg meat, any trimmings, and the liver for making Duck Forcemeat (p. 857). Flavor the forcemeat lightly with grated orange zest, using the zest of ¹/₂ orange for each 2 lb (1 kg) forcemeat. If desired, flavor the forcemeat with duck stock reduced to a glaze and cooled. Use the breast meat for garnish, omitting the veal and ham from the basic recipe but keeping the fatback. Cut the breast meat into strips and marinate in the brandy with the fatback strips.

GALANTINES

A **galantine** is a ground meat mixture—that is, a forcemeat—wrapped in the skin of the product it is made from, such as chicken or duck. A galantine is almost always poached, although, in some instances, it is roasted.

A galantine is made by rolling up a forcemeat in a large piece of skin, giving it a cylindrical or sausage shape that yields round slices. Consequently, the name *galantine* is also given to forcemeats or other mixtures (such as mousselines) that are rolled into a sausage shape in a piece of parchment, plastic film, or other material.

A finished galantine is often displayed whole, decorated, and glazed with aspic, with a few slices removed to show a cross section. For à la carte service, slices of galantine are served the same way as slices of pâté and terrine.

The following is a representative galantine recipe. The procedure is illustrated in Figure 28.4.

FIGURE 28.4 Making a chicken galantine.

(a) Place the flattened breast meat on the center of the skin so that 1–2 in. (3–5 cm) of skin shows around all sides.

(b) Shape the forcemeat into a cylinder and place it along one edge of the rectangle of chicken as shown.

(c) With the aid of the cheesecloth, roll up the forcemeat in the chicken skin. Do not roll the cheesecloth into the chicken.

(d) Tie the ends of the cheesecloth securely. Proceed as indicated in the recipe.

KEY POINTS TO REVIEW

- What are the three basic types of forcemeat? Describe them. What are their primary ingredients?

- What are the steps in the procedure for preparing poultry livers for use in forcemeats?

- What are the steps in the procedure for preparing forcemeat terrines?

- What are the steps in the procedure for making a pâté en croûte, beginning with lining the mold?

- What is a galantine? How is it made?

Chicken Galantine

YIELD: 3 LB (1.25 KG)

U.S.	METRIC	INGREDIENTS	PROCEDURE
1	1	Roasting chicken, about 5 lb (2.25 kg)	1. One day in advance, prepare the chicken. Cut off the wings at the second joint. Slit the skin of the chicken along the backbone and carefully remove the skin in one piece. Remove the breasts, keeping them whole. Remove the meat from the legs and wings and reserve it for making the forcemeat.
to taste	to taste	Salt	
to taste	to taste	White pepper	
4 fl oz	125 mL	Brandy	2. Lay the skin flat, inside up, and trim it into a neat rectangle. Remove all fat and connective tissue. Place a piece of cheesecloth on a sheet pan and lay the skin in the center of it.
			3. Butterfly the breast meat and pound it flat so the two breasts together make a rectangle. Place the flattened breast meat on the center of the skin. There should be at least 1–2 in. (2–5 cm) of skin showing around all sides of the breast meat rectangle. Sprinkle with salt, white pepper, and half of the brandy. Cover with plastic film and refrigerate overnight.
			4. Trim all fat and connective tissue from the leg and wing meat and measure 9 oz (250 g) for making the forcemeat. Measure another 8 oz (225 g) leg meat for the garnish and mix it with the remaining brandy. Reserve any remaining meat for another use.
			5. Use the carcass and giblets for making stock.
1 lb	450 g	Chicken Forcemeat I (p. 857), made with part of the leg meat (see step 4)	6. Prepare the forcemeat, grinding it very fine by using a food processor or by passing it 3 times through the fine blade of a grinder. Keep it cold at all times.
1	1	Liver from the chicken	7. Soak, clean, and liquefy the livers according to the procedure on page 857. Mix the liver purée with the forcemeat.
		Garnish:	8. Cut the chicken leg meat, ham, tongue, pimiento, and truffle into small dice.
8 oz	225 g	Leg meat from the chicken (from step 4)	9. Mix the diced garnish and the pistachios into the forcemeat until well combined.
2 oz	60 g	Smoked ham	10. Drain the brandy from the chicken skin and pat dry with a clean towel.
2 oz	60 g	Cured beef tongue, cooked	11. Form the forcemeat into a cylinder the length of the breast meat rectangle. Place the forcemeat on the breast meat and roll it up into the skin with the aid of the cheesecloth.
1 oz	30 g	Pimientos, rinsed and dried	
1 oz	30 g	Truffles (optional)	12. Roll the galantine in the cheesecloth and tie the ends. Then roll the galantine in a sheet of parchment, working to get the roll as smooth as possible. Tie the roll loosely at 2-in. (5-cm) intervals. (This method is used when the galantine must be completely smooth, with no tie marks. For a simpler method, tie the cheesecloth roll in 3 or 4 places and at the ends.)
2 oz	60 g	Pistachios, blanched and skinned	
as needed	as needed	Chicken stock	13. Poach the galantine slowly in chicken stock until the internal temperature is 160°F (71°C), 45–60 minutes. Retie the galantine, which will have shrunk, then let it cool completely in the stock.
			14. Remove from the stock, unwrap, and decorate as desired.

Per 1 ounce (28.35 g): Calories, 90; Protein, 5 g; Fat, 7 g (70% cal.); Cholesterol, 30 mg; Carbohydrates, 0 g; Fiber, 0 g; Sodium, 100 mg.

TERRINES BASED ON MOUSSELINES

The body of most fish terrines, as well as some vegetable terrines and other specialty items, consists of a mousseline forcemeat. These are made like traditional terrines, except a mousseline forcemeat takes the place of the straight forcemeat.

A **mousseline forcemeat** consists of raw, puréed fish, poultry, or meat combined with heavy cream and, usually but not always, eggs or egg whites. Because they contain no starch or binder, and because of the large quantity of cream they contain, mousselines are the most delicate of forcemeats. The procedure for making a mousseline forcemeat is detailed in the following basic recipe.

Ingredient proportions in mousseline forcemeat depend on the qualities of the meat or fish being used. The albumin content of the egg white makes the mousseline firm when it is cooked. If the meat or fish you are using has a high albumin content, however, you may be able to reduce the quantity of egg white. In some cases, you may not need any egg. Similarly, the amount of cream depends on the firmness of the fish and on the intended use of the forcemeat. If it is to be the base of a terrine that will be sliced, too much cream will make it too delicate. On the other hand, for small timbale molds and similar items that don't need to be as firm, the quantity of cream can be increased beyond the amount indicated in the following recipe.

As for straight forcemeats, it is important to keep the ingredients cold at all times.

The procedure for assembling and cooking a mousseline terrine is the same as for making a regular forcemeat terrine (see p. 861), except the internal temperature, when done, is slightly lower, 158°–160°F (70°C).

Cooked vegetables, fish fillets, and other appropriate items are used as garnish. Two or more mousselines can be layered in the mold to make multicolored terrines. Alternatively, a mousseline forcemeat can be spread on the bottom and sides of the mold, which is then filled with a different mousseline plus garnish and topped with a layer of the first mousseline.

Basic Mousseline Forcemeat

YIELD: APPROXIMATELY 1 LB 12 OZ (875 G)

U.S.	METRIC	INGREDIENTS
1 lb	500 g	Chicken meat, lean veal, fish, or shellfish (see Note)
2 oz	60 g	Egg whites
12 fl oz	375 mL	Heavy cream (quantity variable)
1 tsp or to taste	5 mL or to taste	Salt
to taste	to taste	White pepper
small pinch	small pinch	Cayenne
to taste	to taste	Nutmeg

PROCEDURE

1. Have all ingredients and equipment very cold. In addition, have ready a bowl set in an ice bath for step 7. It is important to keep all ingredients cold throughout the production process.

2. Purée the meat or fish in a food processor.

3. Add the egg whites and process until they are well blended in and the mixture is smooth.

4. With the machine running, slowly pour in the cream through the feed tube. The mixture should be light, fluffy, and firm enough to hold a shape. If it starts to become softer than this while the cream is being added, stop adding cream.

5. Season. The amount of salt needed depends on the meat or fish and the intended use. For example, freshwater fish may need more salt than ocean fish.

6. Keep the mixture cold while preparing a poach test to check seasonings. Wrap a small quantity of the mousseline tightly in plastic wrap and poach in simmering water. Taste and adjust seasonings.

7. Force the mixture through a fine sieve into a bowl set in ice, in order to remove any bits of sinew, skin, or bone. This step is sometimes omitted, but it much improves the texture.

Per 1 ounce (28.35 g): Calories, 70; Protein, 5 g; Fat, 6 g (73% cal.); Cholesterol, 35 mg; Carbohydrates, 0 g; Fiber, 0 g; Sodium, 120 mg.

Note: Use lean, skinless chicken meat, lean veal, fillets from lean, white fish, or shellfish such as scallops, shrimp, or lobster.

VARIATION

Herbed Mousseline

Add a mixture of chopped fresh herbs to the sieved mousseline mixture.

Terrine of Vegetables with Chicken Mousseline

YIELD: 2 LB (1 KG)

U.S.	METRIC	INGREDIENTS	PROCEDURE
1 lb 8 oz	750 g	Chicken Mousseline Forcemeat (p. 865)	1. Mix the chicken mousseline with the chopped herbs and the glace de volaille, if used.
2 tbsp	7 g	Chopped parsley	
1 tsp	1 g	Chopped fresh tarragon	
$^1/_2$ oz	15 g	Glace de volaille, melted (optional)	
2 oz	60 g	Zucchini, small, trimmed	2. Cut the zucchini into strips $^1/_4$–$^1/_2$ in. (1 cm) wide. Blanch 2 minutes in salted water, drain, and chill.
2 oz	60 g	Red bell pepper, cored and seeded	3. Char and peel the red pepper (see p. 290). Cut it into strips.
2 oz	60 g	Carrots, trimmed and peeled	4. Cut the carrots into strips like the zucchini. Blanch 3 minutes, drain, and chill.
2 oz	60 g	Green beans, trimmed	5. Blanch the green beans 1–2 minutes, depending on their tenderness.
2 oz	60 g	Shiitake mushroom caps	6. Cut the mushroom caps in half. Blanch 30 seconds, drain, cool, and pat dry to remove extra moisture.

7. Butter well the bottom and sides of a 1$^1/_2$-qt (1.5-L) terrine mold.

8. Spread one-third of the mousseline on the bottom of the mold, being sure to eliminate air bubbles.

9. Arrange the carrot strips and beans lengthwise in the mold, pushing them partway into the mousseline. Keep the vegetables at least $^1/_4$ in. (5 mm) from the sides of the mold (see **Figure 28.5**).

10. Spread a thin layer of mousseline over the vegetables. Arrange the mushroom caps down the center of the mold, then cover with another thin layer of mousseline. About one-third of the mousseline should be left.

11. Arrange the pepper and zucchini strips lengthwise in the mold, adding a little more mousseline as necessary.

12. Top with the remaining mousseline, again spreading it carefully to avoid air bubbles. Rap the terrine sharply on the workbench to eliminate any remaining air bubbles. Smooth the top of the mousseline with a spatula.

13. Cover tightly with foil. Set in a hot-water bath and bake in an oven heated to 325°F (165°C) until set firm, about 1 hour and 15 minutes.

14. Cool thoroughly, then chill well in the refrigerator.

15. Unmold. Slice carefully with a knife dipped in hot water. Serve garnished with a few salad greens and an appropriate cold sauce.

Per 1 ounce (28.35 g): Calories, 60; Protein, 4 g; Fat, 5 g (67% cal.); Cholesterol, 25 mg; Carbohydrates, 1 g; Fiber, 0 g; Sodium, 90 mg.

VARIATIONS

Instead of the vegetables indicated, select your choice of seasonal vegetables.

For a more luxurious terrine, include thin slices of truffle with the garnish, or omit the parsley and add minced truffle to the mousseline.

Seafood Terrine with Vegetables

Use a fish or shellfish mousseline instead of the chicken mousseline. Reduce the number and quantity of vegetables. Add to the garnish some strips of smoked salmon.

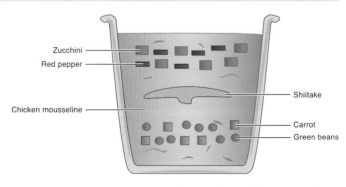

FIGURE 28.5 Cross-section diagram of Terrine of Vegetables with Chicken Mousseline.

TERRINES AND OTHER MOLDS WITH GELATIN

All the terrines we have discussed so far are traditional cooked terrines—that is, the raw force-meat is cooked in the mold. There are many kinds of uncooked terrines as well. These are not cooked after assembly but rather chilled until set. Any ingredients that require cooking are cooked before assembly. The terrines we discuss in this section rely on gelatin for their structure.

Preparing these items in terrine molds allows them to be cut into slices for serving, just as traditional forcemeat terrines are. They can also be made in molds of any other shape, including portion-size molds, which can simply be unmolded and garnished for serving. Cooked terrines, on the other hand, are best made in regularly shaped, symmetrical molds so they cook uniformly.

Most molds bound with gelatin fall into two general categories: those based on aspics and those based on mousses.

ASPIC MOLDS

Aspic-based terrines are simply glorified gelatin molds—that is, they consist of solid ingredients held together by gelatin in the form of aspic jelly.

The proportion of aspic to solids can vary greatly. At one extreme, there may be just enough aspic to hold the solid ingredients together, so the aspic jelly itself is almost not evident. On the other hand, the aspic may predominate, with solid ingredients suspended in it at intervals. For this latter type to succeed, the aspic jelly must be of excellent quality, with good flavor, a firm but not rubbery texture, and sparkling clarity.

The majority of aspic terrines fall between these extremes.

The following procedure is applicable to the production of most aspic terrines and other aspic molds:

1. Either line the mold with aspic, following the procedure on page 854, or pour a layer of aspic into the bottom of the mold. Chill until firm.

2. Arrange a layer of garnish in the mold.

3. Add just enough aspic jelly to cover the solid garnish. Chill until firm.

4. Repeat steps 2 and 3 until the mold is full.

5. For best storage, leave the aspic in the mold, covered tightly with plastic film, until service time.

Terrines made by this method depend on a crystal-clear aspic jelly for their appearance and are often very elegant. Another approach is simply to combine the jelly with a mixture of ingredients and fill the terrine with this mixture. A clarified aspic may not be necessary for this method. Terrines made this way range from coarse, peasant-style dishes to more elaborate constructions.

Headcheese and a number of other commercially made luncheon-meat loaves are examples of this type of terrine. Tripes à la Mode de Caen (p. 509), when properly made, can also be chilled until solid and unmolded because it contains enough natural gelatin from the calves' feet and other ingredients. Jambon Persillé, or Parsleyed Ham (p. 868), is another example of a country-style aspic-based terrine made with unclarified jelly.

MOUSSES

A savory cold **mousse**, as used for the base of a terrine, is a preparation of puréed meat, poultry, fish, vegetable, or other food, bound with gelatin and usually lightened with the addition of partially whipped heavy cream. (It is true that the terms *mousse* and *mousseline* are often used more or less interchangeably, but we use them here in two distinct senses in order to avoid confusion.)

The gelatin used to bind or set the mousse may be added in the form of an aspic jelly or as powdered gelatin softened and dissolved in another liquid ingredient.

Because mousses, like aspics, are not cooked after assembly but merely chilled, they are often prepared not only in terrines but also in decorative, irregularly shaped molds. The production of mousses is relatively simple. The procedure consists of four main steps:

1. Purée the main ingredient.

2. Add the aspic jelly or dissolved gelatin.

3. Fold in the lightly whipped cream and season to taste.

4. Pour into the prepared mold.

Molds are usually lined with aspic jelly and decorated according to the procedures on page 854. As with other kinds of terrine, garnish, if any, is either mixed with the mousse or arranged in the mold as the mousse is added.

Although this method is really little more than mixing together the ingredients in a given order, two precautions must be taken:

1. **Carry out the entire procedure, including the pouring of the mixture into the mold, quickly and in one continuous process.**

   If you stop partway through the procedure, the gelatin is likely to set, and you will have a lumpy, poorly mixed product.

2. **Do not overwhip the cream.**

   Whip it only until it forms soft mounds. When cream is overwhipped, it breaks and becomes grainy. This same effect can be caused by the extra beating the cream gets when it is being folded into the mousse mixture. A mousse made with overwhipped cream tastes dry and grainy, not smooth and creamy.

Mousses can also be made without gelatin or other binders. A soft mousse is simply a puréed or ground food with the addition of lightly whipped cream. Although these soft mousses are too soft to be used in terrines, they can be spooned into neat, oval quenelle shapes onto salad plates, garnished attractively, and served as first courses.

Jambon Persillé (Parsleyed Ham in Aspic)

YIELD: APPROXIMATELY 2 LB (1 KG)

U.S.	METRIC	INGREDIENTS	PROCEDURE
2	2	Pig feet, split	1. Put the pig feet, onion with the clove stuck in it, sachet, and wine in a heavy pot. Add enough white stock to cover. Simmer 2 hours, adding more stock or water as needed.
1	1	Onion	
1	1	Clove	2. Add the ham to the pot. Add more stock as needed. Simmer until the ham is tender. Cooking time will vary greatly, depending on the ham.
		Sachet:	
1	1	Garlic clove	3. Remove the ham. Trim any fat and skin. Cut the ham into large dice. Remove any meat from the pig feet, chop it, and add it to the ham. (Skin from the feet may also be added, if desired.) Chill the meat.
1	1	Bay leaf	
¹/₂ tsp	2 mL	Dried thyme	
¹/₂ tsp	2 mL	Dried tarragon	4. Skim and strain the cooking liquid. The stock may be clarified, but this is not necessary if a traditional rustic look is desired. The pig feet should have yielded sufficient gelatin, but test to make sure and add more gelatin if needed to make a strong aspic; see the procedure on page 852.
12 fl oz	350 mL	White wine	
as needed	as needed	White stock	
1 lb 8 oz	750 g	Mild-cured ham, in 1 or more large pieces	
¹/₂ cup	30 g	Chopped parsley	5. Melt the aspic (if it has congealed), and add the parsley and vinegar. Taste and add salt and pepper if necessary.
¹/₂ fl oz	15 mL	Wine vinegar	6. Select the desired molds; large salad bowls are traditional. Line the bottom of the mold with a thin layer of the parsley aspic. Chill until firm.
if needed	if needed	Salt	7. Combine the ham and aspic and pour into the mold. Chill until set.
if needed	if needed	Pepper	8. To serve, unmold and slice. Serve unadorned or plated with salad greens and vinaigrette.

Per 1 ounce (28.35 g): Calories, 96; Protein, 10 g; Fat, 6 g (49% cal.); Cholesterol, 34 mg; Carbohydrates, 1 g; Fiber, 0 g; Sodium, 385 mg.

Tricolor Vegetable Terrine

YIELD: APPROXIMATELY 1 LB 12 OZ (800 G)

U.S.	METRIC	INGREDIENTS	PROCEDURE
1 lb	450 g	Spinach	1. Have ready a 1-qt (1-L) terrine mold. For ease of unmolding, line with plastic film.
1/4 oz	7 g	Shallot, minced	
1/4 oz	7 g	Butter	2. Trim the stems from the spinach and wash it well in several changes of water. Cook in boiling, salted water until done, about 2 minutes, and drain. Rinse under cold water to cool. Drain. Squeeze dry.
1 tsp	5 mL	Gelatin powder	
1 fl oz	30 mL	Chicken stock, vegetable stock, or water, cold	
to taste	to taste	Salt	3. Chop the spinach into fine pieces by hand or in a food processor.
2 fl oz	60 mL	Heavy cream	
			4. Sweat the shallots in butter until soft. Add the spinach and cook slowly until quite dry. Cool thoroughly, but do not chill.
			5. Soften the gelatin in the stock, then heat until it is dissolved. Cool and stir into the spinach. Add salt to taste.
			6. Quickly whip the cream until it forms soft peaks. Immediately fold it into the spinach mixture. Pour it into the mold and smooth with a spatula. Chill until set.
5 oz	150 g	Cauliflower, trimmed	7. Steam the cauliflower and turnips until they are tender. Purée in a food processor. For the smoothest texture, force the purée through a sieve. Mix the vegetables together. Heat slowly in a large sauté pan to dry the purée slightly. Cool thoroughly, but do not chill.
2 oz	60 g	White turnips, peeled	
1 tsp	5 mL	Gelatin powder	
1 fl oz	30 mL	Chicken stock, vegetable stock, or water, cold	
to taste	to taste	Salt	8. Repeat steps 5 and 6 to make the white mousse. Pour it into the mold on top of the green mousse. Chill.
2 fl oz	60 mL	Heavy cream	
7 oz	200 g	Carrots, trimmed and peeled	9. Trim, cook, and purée the carrots in the same way, and dry the purée as above. Repeat steps 5 and 6 to make the orange mousse, and add it to the terrine (see **Figure 28.6**). Chill until set firm.
1 tsp	5 mL	Gelatin powder	
1 fl oz	30 mL	Chicken stock, vegetable stock, or water, cold	
to taste	to taste	Salt	10. Unmold the terrine and slice to serve. Garnish as desired and serve with an appropriate cold sauce.
2 fl oz	60 mL	Heavy cream	

Per 1 ounce (28.35 g): Calories, 30; Protein, 1 g; Fat, 3 g (65% cal.); Cholesterol, 10 mg; Carbohydrates, 2 g; Fiber, 1 g; Sodium, 20 mg.

VARIATIONS

Other vegetable purées may be substituted for those in the basic recipe.

For a low-fat version, omit the gelatin powder, stock, and heavy cream. In place of the stock and cream, use an equal quantity of a strong aspic. Mix the aspic with the vegetable purées.

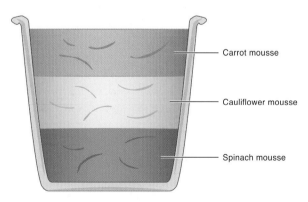

Carrot mousse

Cauliflower mousse

Spinach mousse

FIGURE 28.6 Cross-section diagram of Tricolor Vegetable Terrine.

Ham Mousse

YIELD: 2 LB (1 KG)

U.S.	METRIC	INGREDIENTS
1 lb	500 g	Cooked, lean smoked ham
5 fl oz	150 mL	Chicken velouté
1 tbsp	15 mL	Madeira wine
to taste	to taste	White pepper
to taste	to taste	Dry mustard
to taste	to taste	Salt
1/4 oz	7 g	Gelatin powder
4 fl oz	125 mL	Chicken stock, cold
8 fl oz	250 mL	Heavy cream

PROCEDURE

1. Select a mold or molds totaling about 1 qt (1 L) in capacity. If desired, line the molds with aspic and decorate them according to the procedure on page 854. Keep chilled until needed.
2. Grind the ham until it is very fine.
3. Mix the velouté with the puréed ham. Add the Madeira and season to taste with white pepper, dry mustard, and salt. Salt might not be needed if the ham is very salty.
4. Soften the gelatin in the stock. Heat the stock until the gelatin dissolves, then cool the liquid aspic, but do not let it set.
5. Whip the cream until it forms soft peaks.
6. Add the aspic jelly to the ham mixture and stir until well combined.
7. Quickly and thoroughly fold the cream into the ham mixture. Taste and adjust seasonings if necessary.
8. Fill the prepared molds. Chill several hours or overnight until set firm. Unmold just before serving.

Per 1 ounce (28.35 g): Calories, 50; Protein, 3 g; Fat, 4 g (69% cal.); Cholesterol, 20 mg; Carbohydrates, 1 g; Fiber, 0 g; Sodium, 190 mg.

VARIATIONS

For a denser but less rich mousse, reduce the quantity of cream as desired.

Substitute prosciutto for one-eighth to one-fourth of the cooked ham.

Instead of velouté, substitute mayonnaise thinned with cream to the thickness of velouté.

Mousses of other meats, poultry, and fish may be prepared according to the same procedure, substituting an appropriate stock (such as fish stock for fish mousse) and using appropriate seasonings in place of the mustard and Madeira (for example, salmon mousse flavored with dill, cayenne, and white wine).

Mousse of Foie Gras

YIELD: 1 LB (500 G)

U.S.	METRIC	INGREDIENTS
8 oz	250 g	Foie gras (see Note)
4 fl oz	125 mL	Aspic jelly for coating
4 fl oz	125 mL	Heavy cream
to taste	to taste	Salt
to taste	to taste	White pepper

PROCEDURE

1. Force the foie gras through a sieve to purée it.
2. Melt and cool the aspic according to the procedure on page 853. Add it to the foie gras, mixing it in thoroughly.
3. Whip the cream until it forms soft peaks. Quickly and thoroughly fold it into the foie gras.
4. While folding in the cream, taste and adjust the seasonings with salt and white pepper. It is best to do this while folding in the cream so there is no delay that would allow the gelatin to set too early and to avoid excess mixing, which may overwhip the cream.
5. Pour at once into a terrine or other mold. Cover tightly and chill at least 1 day.
6. This dish is very rich and should be served in small quantities, about 2 oz (60 g) per portion. Serve by dipping a spoon into hot water and drawing it across the surface of the terrine, as though scooping ice cream. Place the spoonful in the center of a plate and serve with Melba toasts and raw vegetable garnish or salad greens.

Per 1 ounce (28.35 g): Calories, 50; Protein, 3 g; Fat, 4 g (66% cal.); Cholesterol, 95 mg; Carbohydrates, 1 g; Fiber, 0 g; Sodium, 85 mg.

Note: See page 871 for a discussion of foie gras. Cooked, not raw, foie gras is called for in this recipe. Terrine of Foie Gras (p. 873) may be used. If fresh foie gras products are not available, canned foie gras may be used.

FOIE GRAS, LIVER TERRINES, AND RILLETTES

The chapter concludes with three traditional terrines that require somewhat different techniques from those already discussed. It should be noted that, although these items are especially high in fat and cholesterol, they are as popular as they have ever been, even in these times of diet consciousness.

FOIE GRAS TERRINES

The most prized and, perhaps, the most famous ingredient for pâtés and terrines in classical cuisine is **foie gras** (fwahgrah). This French term means "fat liver." Foie gras is the fatted liver of specially fed varieties of ducks and geese. Until recently, only canned or processed foie gras products were available in the United States. Now, however, the breed of duck that is raised to produce foie gras (called the *mullard* or *moulard*, a cross between the muscovy and White Pekin ducks) is grown on American farms. Consequently, fresh, raw duck foie gras is now sold in this country. Its availability has created a great deal of enthusiasm among American cooks in spite of its high price.

The special feeding of the ducks makes their livers very large, more than 1 lb (500 g) as a rule, with a high fat content. A good-quality fresh foie gras is a pale yellowish-tan color with a smooth, velvety texture, almost like butter. The liver has two lobes, one large and one small.

It is important to be aware that foie gras consists mostly of fat. Indeed, the rich flavor of the fat is the whole reason foie gras is so highly prized. Any fat that cooks out during preparation is carefully saved and used for another purpose. Those who must avoid fats, especially animal fats, should probably steer clear of this delicacy. For the rest of us, the high price of foie gras helps protect our health by making overindulgence unlikely.

There are usually two grades of domestic duck foie gras. The A grade is larger, usually 1¼ lb (600 g) or more, with relatively few blemishes and blood spots. The B grade is smaller and has more blood spots and veins. (There is also a C grade, but it is not used in food service.)

Preparing Foie Gras for Cooking

No matter how a raw foie gras is to be prepared, it should first be rinsed in cold water and examined closely for green spots. These are caused by bile; they must be cut or scraped away because the bile has a strong, bitter taste. Also, if there are any bits of external fat, remove them.

Next, the liver should be soaked. (This step is not required, but it does improve the product.) Place it in lightly salted ice-cold water or milk to cover. Let stand for up to 2 hours, no longer. Remove from the salted liquid and rinse in fresh, cold water.

For cold preparations such as terrines and mousses, the liver should first be deveined. To devein the foie gras, first let it come to room temperature. Its fat content makes the cold liver too brittle to devein without excessive breakage, which would result in more cooking loss. When the liver is at room temperature, even the heat of the hands melts the fat, so it is important to handle the liver lightly and to work quickly.

Begin by separating the two lobes and laying them, smooth side down, on a clean work surface (Figure 28.7). Carefully trim off any bloody spots. Grasping a lobe with your thumbs at the sides and fingers underneath in the center, very lightly bend the lobe lengthwise. The top, rough surface should open up slightly, revealing a heavy vein that runs lengthwise through the liver. (If it does not open up, help it along with a shallow incision with the point of a paring knife.) Carefully pull out this vein, along with any other heavy veins that are attached, all the while being careful to keep the liver as intact as possible. Repeat with the other lobe. The foie gras is now ready to be made into a terrine.

No matter how a foie gras is cooked, it is essential to avoid even the slightest overcooking. The liver is delicate, and the fat cooks out very quickly. Even a few seconds too long in a sauté pan can reduce a slice of foie gras to a few specks of connective tissue floating in a puddle of very expensive grease.

FIGURE 28.7 Deveining foie gras.

(a) Slowly and gently pull the two lobes apart.

(b) Remove any visible membrane and surface blemishes from each lobe.

(c) Starting at the narrow, top end of each lobe, slit open the lobe about halfway to the other end and about halfway into its depth.

(d) Grasp the thick, top part of the vein network and pull gently while holding back the meat of the liver with the other hand.

LIVER TERRINES

Liver terrines, often called liver pâtés, are popular, inexpensive appetizers—except, of course, for those made with foie gras. The classic liver terrine is a mixture of liquefied livers—that is, cleaned, soaked, blended, and strained according to the procedure on page 857—with eggs and seasonings, baked in a terrine until set. An example of this type of recipe can be found on page 874.

This kind of liver terrine generally contains flour as a stabilizer. Because the liver forcemeat is liquid, the flour improves the texture of the cooked product by helping bind the moisture. Heavy cream is also included in most recipes as a source of fat. Other sources of fat sometimes used in addition to or in place of the cream are ground pork fat, ground bacon, marrow, and rendered foie gras fat left over from making terrines.

Another type of liver terrine consists of a basic pork forcemeat with a liver content high enough for the flavor of the liver to predominate. To make this type of terrine, make the forcemeat on page 857, but use 6 times the quantity of liver. The forcemeat will be quite soft. Follow the basic procedure for making forcemeat terrines (see p. 861), using whole, trimmed chicken livers marinated in brandy as the garnish.

A quick and simple substitute for these more elaborate terrines might be considered a type of rillettes (see the next section), as it consists of a seasoned mixture of cooked meat (liver, in this case) and fat. This is the type of chicken liver pâté found on delicatessen and coffee shop menus. To make this type of pâté, sauté some chicken livers, mash or purée them, and mix the purée with about one-eighth its weight in rendered chicken fat, pork fat, soft butter, or other fat, or else with one-fourth its weight in cream cheese. Season as desired, with salt, pepper, herbs, brandy or sherry, and/or sautéed minced onion.

RILLETTES

In France's Loire Valley, the first thing customers are likely to be served in a typical neighborhood restaurant, whether they order it or not, is a crock of rillettes and some country bread to spread it on. Variations on this unpretentious dish have become widely popular and are served even in elegant North American restaurants.

Rillettes (ree yet) is a dish made of pork cooked slowly until it is very tender, then shredded, mixed with its own fat, seasoned, and packed into crocks or terrines.

Variations of the classic dish can be made by using other meats in addition to or instead of pork. Items rich in fat, such as duck and goose, are especially appropriate. Rillettes of lean meats, such as chicken, turkey, and rabbit, can be made, but some pork fat or other fat must be added to them when they are cooking.

Some chefs even serve rillettes made from fish, such as salmon or cod. The basic procedure is the same, except the cooking time is, of course, much shorter. The cooked fish is shredded, mixed with just enough butter or other fat to give it a pleasant texture, and seasoned well.

A typical recipe for classic pork rillettes is on page 874. Follow the same basic procedure to make duck rillettes and other variations. A recipe for rillettes made with fish is on page 875.

KEY POINTS TO REVIEW

- What are the basic ingredients in a mousseline forcemeat? How is the forcemeat made?

- What is a mousse? How is it made?

- What is foie gras? How is it prepared for cooking?

- What are rillettes? How are they prepared?

- How is a baked liver terrine made?

Terrine of Foie Gras

YIELD: VARIABLE, DEPENDING ON SIZE OF LIVER

U.S.	METRIC	INGREDIENTS
1	1	Fresh, A-grade duck foie gras, about 1^1/$_2$ lb (700 g)
1 tsp	5 mL	Salt
^{1}/$_4$ tsp	1 mL	White pepper
^{1}/$_2$–1 fl oz	15–30 mL	Choice of wine or liquor: port, Madeira, Sauternes, cognac, or Armagnac

PROCEDURE

1. Soak, rinse, and devein the foie gras as described on page 871.

2. Place the liver in a bowl and season with salt and white pepper. Add the selected wine or liquor, using the smaller quantity for cognac or Armagnac; if using a sweet wine, use up to but not more than the larger quantity indicated. Turn the liver gently so that all sides are moistened.

3. Cover and refrigerate. Marinate 24 hours.

4. Remove the livers from refrigeration 1–2 hours before cooking time, and let them come to room temperature. This is essential for the proper cooking of the terrine.

5. Pack the large lobe (or the pieces of the large lobe, if it broke during deveining) into the selected terrine, smooth side down. Top with the small lobe, smooth side up. Press the liver in firmly to eliminate air spaces. Cover with foil.

6. Place several folded kitchen towels on the bottom of a roasting pan or other pan used as a hot-water bath. (This helps insulate the terrine from strong bottom heat.) Place the terrine in the pan and add warm, not hot, water to come halfway up the sides of the terrine.

7. Place the terrine in an oven preheated to 200°–215°F (100°C). Bake until the proper doneness, as determined by an instant-read thermometer (see the following paragraph). This will take from 45 minutes to a little over 1 hour.

   The terrine is done when the thermometer reads 113°–130°F (45°–54°C). At the lower end of this range, the cooled terrine will be rather pink in the center, with a soft, creamy texture. At the higher end, the terrine will be firmer and less pink, but more fat will have cooked out, resulting in a lower yield. The right degree of doneness is a matter of personal preference.

8. Remove the terrine from the hot-water bath and set on a rack to cool. After about 10 minutes, weight it with a board that just fits inside the top of the terrine, or with another terrine, and several pounds of weights. When the terrine is almost cool but the fat is still liquid, pour off all the melted fat and juices. Separate and discard the juices. Reserve the fat. Put the weights back on the terrine and continue to cool.

9. When the terrine is cold, unmold it and remove and discard any bits of blood or juice on the bottom of the foie gras. Clean out the mold and put the foie gras back in it. Heat the reserved fat just until melted and pour it over the terrine. Refrigerate until cold, then cover tightly and refrigerate 3–5 days so the flavors can develop.

10. To unmold, dip the terrine in warm water for a few seconds, then invert on a platter or cutting board. Slice with a sharp knife dipped in hot water before each slice. Serve with a little chopped aspic and toasted brioche, or with salad greens and a mild vinaigrette made with walnut oil.

Per 1 ounce (28.35 g): Calories, 50; Protein, 6 g; Fat, 2 g (32% cal.); Cholesterol, 165 mg; Carbohydrates, 1 g; Fiber, 0 g; Sodium, 160 mg.

Chicken Liver Terrine

YIELD: 2 LB (1 KG)

U.S.	METRIC	INGREDIENTS
1 lb	500 g	Chicken livers
8 fl oz	250 mL	Heavy cream
5	5	Eggs, lightly beaten
2 oz	60 g	Flour
2 tsp	10 mL	Salt
1/2 tsp	2 mL	White pepper
1 1/2 fl oz	50 mL	Brandy
as needed	as needed	Caul fat or thin slices of pork fatback for lining the mold (optional)

PROCEDURE

1. Soak, rinse, liquefy, and strain the livers, following the procedure on page 857.
2. Mix in the remaining ingredients, except caul fat, until smooth. If necessary, strain to eliminate lumps.
3. If possible, cover and refrigerate this mixture overnight. This helps eliminate air bubbles that may have gotten in, and it allows the flour to absorb moisture.
4. Line a terrine mold with the caul or fatback, or grease it very generously with butter or lard.
5. Cover with foil. Bake in a water bath at 300°F (150°C) until set. The water in the water bath should come up to the same level as the liver mixture. Cooking time will depend on the size and shape of the terrine; approximate time is about 2 hours. Check it periodically after about 1 1/2 hours so it does not overbake.
6. Remove from the water bath and cool on a rack. Refrigerate overnight or longer. Unmold and slice, or serve directly from the terrine.

Per 1 ounce (28.35 g): Calories, 60; Protein, 4 g; Fat, 4 g (57% cal.); Cholesterol, 105 mg; Carbohydrates, 2 g; Fiber, 0 g; Sodium, 170 mg.

VARIATIONS

Substitute calf liver or pork liver for the chicken liver.

Rillettes of Pork

YIELD: APPROXIMATELY 1 LB (500 G)

U.S.	METRIC	INGREDIENTS
2 lb	1 kg	Pork butt or shoulder, with fat but without skin and bones
1	1	Onion, small
2	2	Cloves
2	2	Bay leaves
pinch	pinch	Dried thyme
2 fl oz	60 mL	Water
1 1/2 tsp	2 mL	Salt

PROCEDURE

1. Cut the meat, with all the fat, into large dice.
2. Cut off 1–2 oz (30–60 g) of the fat and render it slowly in a large, heavy pot. Add the meat and brown it lightly and gently over moderate heat.
3. Stick the onion with the cloves. Add the onion and the remaining ingredients. Cover and cook slowly in a low oven or on the range over very low heat, until the meat is very tender. This will take several hours. Check periodically to see if the meat has become dry. If it has, add 1–2 oz (30–60 mL) water.
4. Remove the bay leaves, onion, and cloves. Place the contents of the pot in a colander set over a large bowl. Press firmly on the meat and collect the fat and drippings in the bowl. Separate the fat and juices and reserve them separately.
5. Pound the meat with a large pestle or mallet, then shred it with two forks until the meat is a mass of fibers and no lumps. Alternatively, place the meat in the bowl of a mixer and mix with the paddle attachment at lowest speed until the meat is thoroughly shredded.
6. Add as much of the rendered fat as desired and mix it with the shredded meat. Taste and adjust the seasonings. It should be well seasoned because it is to be served cold. If the mixture seems dry, add some of the reserved juices to achieve the desired consistency. The mixture should be thick but spreadable, neither too dry nor too soft.
7. Pack into crocks or terrines and smooth the top. If the rillettes are to be kept for more than 1–2 days, seal the surface from the air by covering with a layer of melted fat. Refrigerate overnight or longer. Remove the layer of fat before serving.
8. Serve with crusty bread and sour pickles.

Per 1 ounce (28.35 g): Calories, 110; Protein, 9 g; Fat, 8 g (67% cal.); Cholesterol, 35 mg; Carbohydrates, 2 g; Fiber, 0 g; Sodium, 240 mg.

VARIATIONS Rillettes of Duck, Goose, Rabbit, Turkey, or Chicken

Substitute any of the above meats for all or part of the pork.
Lean meats should be cooked with additional pork fat to supply enough rendered fat to blend with the shredded meat.

Rillettes of Salmon, Haddock, or Finnan Haddie

YIELD: 1 LB 4 OZ (600 G)

U.S.	METRIC	INGREDIENTS	PROCEDURE
1 lb	500 g	Salmon, haddock, or finnan haddie, skinless and boneless	1. Combine the fish and wine in a saucepan or sauté pan. Poach the fish gently just until it is done. Because there is not enough wine to cover the fish, turn the fish over occasionally during cooking so it cooks evenly.
8 fl oz	250 mL	White wine	2. Drain the fish and cool completely.
4 oz	125 g	Butter, unsalted	3. Break the fish into small pieces, then mash with a fork until there are no lumps.
2 tsp	10 mL	Lemon juice	4. Soften the butter, then mix it with the fish until uniformly blended. Season to taste with lemon juice, salt, white pepper, and hot pepper sauce.
to taste	to taste	Salt	
to taste	to taste	White pepper	5. Pack the mixture into small ramekins or crocks for individual service. Chill.
to taste	to taste	Hot pepper sauce	
as desired	as desired	Caviar, for garnish	6. Top each portion with a small spoonful of caviar just before serving. Use salmon caviar for salmon rillettes, or any desired caviar for haddock or finnan haddie rillettes.
			7. For service, place the ramekin on a small plate lined with a folded napkin or a doily. Arrange slices of toast or bread on the underliner around the ramekin, or serve the toast on the side.

Per 1 ounce (28.35 g): Calories, 70; Protein, 59 g; Fat, 6 g (72% cal.); Cholesterol, 25 mg; Carbohydrates, 0 g; Fiber, 0 g; Sodium, 55 mg.

ADDITIONAL RECIPES

These additional recipes may be found on your CulinarE-Companion recipe management program:

Lentil and Leek Terrine with Smoked Turkey and Prosciutto; Terrine of Vegetables and Chicken in Aspic.

TERMS FOR REVIEW

garde manger	**garnish (pâté)**	**pâte à pâté**	**mousse**
aspic jelly	**pâté de campagne**	**pâté en croûte**	**foie gras**
aspic powder	**terrine**	**galantine**	**rillettes**
forcemeat	**pâté**	**mousseline forcemeat**	

QUESTIONS FOR DISCUSSION

1. Explain how to slice a meat terrine.

2. Why are mustards and vinaigrettes often served with pâtés and terrines?

3. What are three purposes of using aspic as a coating or glaze for cold foods?

4. Describe how to melt and cool aspic jelly.

5. What are the basic ingredients in a typical pork forcemeat?

6. How are raw livers prepared for use in forcemeats?

7. What is the purpose of pouring a layer of melted fat over a terrine after baking?

8. What is the difference between the terms *mousseline* and *mousse*, as used in this chapter?

9. What are the four basic steps in the production of a molded mousse? Why is it important to perform these steps quickly?

10. Why is it important not to overcook foie gras?

FOOD PRESENTATION

Until this point, most of the focus of this book has been on the preparation of food, from the selection of ingredients through mise en place and final cooking. We still have not reached our final goal, however. The point of learning and practicing procedures to prepare food of high quality is to ensure the food is eaten and enjoyed.

In other words, our work isn't done until the food we have prepared is arranged on plates or platters and ready to be presented to the diner.

In traditional classical cuisine, until well after the middle of the twentieth century, the normal practice in fine dining establishments was to send the food on platters and in casseroles and other serving dishes to the dining room, where it would be transferred to dinner plates by the serving staff, sometimes after carving or portioning. The chefs who developed nouvelle cuisine, however, wanted to control the appearance of the food down to the last detail and so began to arrange food on dinner plates in the kitchen. Since that time, many styles of plating have come and gone, as chefs have devoted much attention to the appearance and arrangement of food on dinner plates.

Throughout this book, we stress making food look good as well as taste good. We talk about accurate, neat cutting of vegetables and fruits, about proper trimming of meats, poultry, and fish, about grill-marking steaks, about preserving color in cooked vegetables, and about attractive plating of salads. In this chapter, we continue the discussion of making food attractive.

AFTER READING THIS CHAPTER, YOU SHOULD BE ABLE TO

1. Explain why attractive food presentation is important.

2. Serve food that is attractively arranged on the plate or platter, with proper balance of color, shape, and texture.

3. Plan and arrange attractive food platters for buffets.

HOT FOOD PRESENTATION

We eat for enjoyment as well as for nutrition and sustenance. Cooking is not just a trade but an art that appeals to our senses of taste, smell, and sight.

"The eye eats first" is a well-known saying. Our first impressions of a plate of food set our expectations. The sight of food stimulates our appetite, starts our digestive juices flowing, and makes us eager to dig in. Our meal becomes exciting and stimulating.

On the other hand, if the food looks carelessly served, tossed onto the plate in a sloppy manner, we assume it was cooked with the same lack of care. If the colors are pale and washed out, with no color accent, we expect the flavors to be bland and monotonous. If the size of the plate makes the steak look small (even if it's not), we go away unsatisfied.

Your job as a cook and a chef, then, is to get your customers interested in your food or, better yet, excited about it. You can't afford to turn them off before they even taste it. Your success depends on making your customers happy.

FUNDAMENTALS OF PLATING

When a chef plans a new dish, appearance as well as flavor must be considered. Turn back to page 122 and look once again at the section called "Building Flavor Profiles." Remember that the senses of sight, taste, smell, and touch all come into play when we evaluate and enjoy food. How a dish looks is part of the identity of the dish, just like how it tastes, smells, and feels in the mouth.

In other words, how a dish looks is not something you think about only after you have prepared it. It is something you have in mind from the beginning of preparation. Remembering this helps you create natural-looking presentations, so the food looks like what it is rather than like an artificial construction with a complicated design.

Most of us have had the experience in a restaurant of seeing another dish carried past our table by a server and immediately thinking, "I want whatever that is." Only rarely is this thought prompted by an overelaborate, fussy arrangement. Rather, the dish appeals to us most likely because it simply has the appearance and the aroma of well-prepared food.

THREE ESSENTIALS OF FOOD PRESENTATION

Making food look good requires careful attention to all kitchen tasks. The following three principles should be observed in order to create attractive food. Note that only one of them concerns arranging the food on the plate.

Good Preparation and Cooking Techniques

If vegetables are improperly cut during prep, the plate presentation will look improper. If meat is badly trimmed before cooking, a fancy plating design won't correct it. If a fish is overcooked and dry or a green vegetable is drab and mushy, it won't look good no matter what you do with it. On the other hand, well-prepared and properly cooked food with a good aroma is usually appealing all by itself.

Professional Work Habits

Serving attractive food is largely a matter of being neat and careful and using common sense. This is an aspect of the professionalism we discussed in Chapter 1. Professionals take pride in their work and in the food they serve. They don't send a plate to the dining room with sauce accidentally dribbled across the rim and maybe a thumbprint or two for extra effect—not because their supervisors told them not to or because a rule in a textbook says so, but because pride of workmanship prevents it.

Visual Sense

Beyond just being neat, effective food presentation depends on developing an understanding of techniques involving balance, arrangement, and garniture. These are the subjects of our next sections.

BALANCE

A plate arrangement may consist of the following components: *main item, accompaniments* or *side dishes, sauce* or *sauces,* and *other garnish.* The main item is almost always present. The others may or may not be included. (See Garnish sidebar for more on the meaning of the term.)

In a good plate arrangement, all these components are in balance. *Balance* is a term we used when talking about menu planning in Chapter 4. The rules of good menu balance also apply to plating. Select foods and garnishes that offer variety and contrast while avoiding combinations that are awkward or jarring.

Colors

Two or three colors on a plate are usually more interesting than just one. Visualize this combination: poached chicken breast with suprême sauce, mashed potatoes, and steamed cauliflower. Appetizing? Or how about fried chicken, French fries, and corn? Not quite as bad, but still a little monotonous.

Many hot foods, especially meats, poultry, and fish, have little color other than shades of brown, gold, or white. It helps to select vegetables or accompaniments that add color interest—one reason why green vegetables are so popular.

Shapes

Plan for variety of shape and form as well as of color. For example, you probably do not want to serve meatballs with Brussels sprouts and new potatoes. Too many items of the same shape, in this case round, looks monotonous or even odd. Green beans and whipped potatoes might be better choices to accompany the meatballs. Try for a variety of shapes that work together well.

Cutting vegetables into different shapes gives you great flexibility. Carrots, for example, which can be cut into dice, rounds, or sticks (bâtonnet, julienne, etc.), can be adapted to nearly any plate.

Textures

Textures are not strictly visual considerations, but they are as important in plating as in menu planning (Chapter 4). Good balance requires a variety of textures on the plate. Perhaps the most common error is serving too many soft or puréed foods, such as baked salmon loaf with whipped potatoes and puréed squash.

Flavors

You can't see flavors, either, but this is one more factor you must consider when balancing colors, shapes, and textures on the plate. Consult the menu planning guidelines in Chapter 4.

GARNISH

The word **garnish** is derived from a French word meaning "to adorn" or "to furnish." In English, we use the word to mean "to decorate or embellish a food item by the addition of other items." The word is used also for the decorative items themselves.

In classical cooking, the term *garnish* is used the way we use the term *accompaniments.* In other words, garnishes are any items placed on the platter or plate or in the soup bowl in addition to the main item. It happens that these accompaniments also make the food look more attractive, but that is not the emphasis. The classical French chef had a tremendous repertoire of hundreds of simple and elaborate garnishes, and they all had specific names (see Classical Garnish Terms sidebar).

Eventually the term *garnish* evolved to mean just a sprig of parsley or other small decorative item placed on a plate, often not intended to be eaten. Today, the use of parsley sprigs on every plate, once common, has become rare. We are again using the word in a more traditional way.

CLASSICAL GARNISH TERMS

Many of the classical names for garnishes are still sometimes used in modern kitchens, although they have lost the precise meanings they once had. You will encounter some of these terms in your career, so it is worthwhile learning them. Remember that the following definitions are not the exact classical ones but simply the garnish or accompaniment generally indicated by the terms in today's kitchens.

Bouquetière: bouquet of vegetables

Printanière: assorted spring vegetables

Jardinière: assorted garden vegetables

Primeurs: assorted first spring vegetables

Clamart: peas

Crécy: carrots

Doria: cucumbers (cooked in butter)

Dubarry: cauliflower

Fermière: carrots, turnips, onions, and celery, cut into uniform slices

Florentine: spinach

Forestière: mushrooms

Lyonnaise: onions

Niçoise: tomatoes concassé cooked with garlic

Parmentier: potatoes

Princesse: asparagus

Provençale: tomatoes with garlic, parsley, and, sometimes, mushrooms and/or olives

UNITY

Basically, there is unity when the plate looks like one meal that happens to be made up of several items rather than like several unrelated items that just happen to be on the same plate.

Create a center of attention and relate everything to it. In a main course, the meat or fish is generally the center of attention and is traditionally placed front and center. Other items are placed around and behind it so as to balance it and keep the customer's eyes centered rather than pulled off the edge of the plate.

Visual balance is similar to the balance of flavors discussed on page 122. In that discussion, we introduced the concept of primary flavors and supporting flavors. The primary flavors, you recall, are those of the main ingredients, and the supporting or secondary flavors are those of additional ingredients selected to enhance, harmonize with, or contrast with the primary flavors. Visual design works in a similar way. The main item on the plate is the primary design element. Other items, including side dishes, garnishes, and sauces, are supporting design elements. Each item should enhance, harmonize with, or contrast with the main element and each other in a pleasing way.

Don't add unnecessary elements to the plate, especially unnecessary inedible garnishes. In many or even most cases, the food is attractive and colorful without garnish, and adding it clutters the plate and increases your food cost as well.

In any case, it is usually best to add nothing to the plate that is not intended to be eaten. Before you place the parsley alongside the fish fillet or plant a bushy sprig of rosemary in the mashed potatoes, first consider if the plate needs an extra item. If it does, then consider whether it wouldn't be better to add something edible to enhance the other foods with its taste and texture as well as its appearance.

ARRANGEMENT ON THE PLATE

Until recent years, plated main courses followed a standard pattern: meat or fish item at the front of the plate (closest to the diner), vegetable and starch items at the rear.

This arrangement is still commonly used because it is one of the simplest and most convenient. Nevertheless, many chefs are eager to display their creativity with imaginative plating presentations.

In high-end restaurants in both Europe and North America, for most of the twentieth century, food was not plated in the kitchen. Instead, items were arranged on platters and presented to the diner by the serving staff. After performing various finishing tasks, such as carving small roasts, the dining room staff then plated the main items, side dishes, and sauces and set them before the customers. All this changed in the 1970s, when chefs began plating in the kitchen. Since then, plating styles have been changing constantly.

So far, this chapter has focused on general guidelines applicable to nearly any style of plating, but we haven't yet had much to say about specific styles. Plating styles change greatly over the years. Plate arrangements that were modern and daring during the era of nouvelle cuisine look old-fashioned today.

Because styles change and evolve, almost any modern plating we can describe here is likely to look dated in a few years. In fact, the more unusual and "styled" a plating is, the more quickly it will become out of date.

Not only that, but different styles are suitable to different kinds of restaurants. Guests at a casual neighborhood restaurant are most likely to expect casual or even home-style arrangements on the plate, while at high-end restaurants noted for their creative, modern cuisine, guests will expect more creative and refined presentations.

Using recipes from this book, let's look at some examples of how the same items can be presented in different ways suitable for different venues. Even the simplest dishes, such as soups and salads, can be styled in endless ways.

Creating a Plating Style

We can learn a lot about how creative chefs apply their imaginations to food presentation by looking at an item as simple as the Roasted Beet Salad with Gorgonzola on page 704. This dish consists of only three main elements that need to be arranged on the plate: beets, greens, and gorgonzola cheese. The photo that accompanies the recipe shows how the salad might look when plated exactly according to the recipe instructions.

A chef might decide to keep the recipe essentially the same but arrange the beets on a round plate in a slightly different fashion. While doing so, she adds a few walnut pieces to the salad for an appealing texture contrast and a new flavor element that goes well with the beets and cheese. The result is shown in the Plate 1 photo.

The second plating illustrated here is only slightly more elaborate. It differs from plate 1 in the following ways: (1) Pear slices have been added for a sweet flavor contrast. (2) To keep the presentation simple, the golden beets have been eliminated to compensate for the added complexity of the pear addition. The chef has arranged the red beets in a circle on the plate as in the original recipe. (3) Rather than being crumbled, the gorgonzola is cut into small wedges, and the quantity of the cheese is increased. (4) The walnuts have been candied, so that they harmonize with the sweetness of the pear.

A third chef decides to use the same basic elements—beets, greens, gorgonzola, walnuts, and pear—and transform them into an imaginative and elegant creation for a high-end restaurant. The pear is poached rather than raw. It is sliced lengthwise and arranged in a wave pattern on a long, narrow plate as in the photo of plate 3. Wedges of red beet and slices of golden beet are coated with vinaigrette and arranged at opposite ends of the plate. Instead of plain gorgonzola crumbles, a single Gorgonzola Truffle (p. 799) is placed against the line of pear, next to a spoonful of Pear Chutney (p. 209). The salad is completed with fresh greens and walnut pieces.

Finally, a chef at a new restaurant specializing in modern, cutting-edge cuisine imagines a completely new way to plate this salad of beets, cheese, pear, and walnuts. Instead of plain roast beets, the chef prepares Vanilla Roasted Beets (p. 335). He cuts rectangles of gorgonzola, sprinkles the top with sugar, and caramelizes the top with a torch. Candied walnuts are powdered and placed in a line along the edge of the large, square plate. The photo of Plate 4 shows how these elements are arranged in a modern style. Two new flavor accents are three "swooshes" of Orange Fluid Gel (p. 218) and a few drops of beet-green oil (prepared the same way as Basil Oil, p. 216).

These examples give you an idea of the tremendous range of plate presentation styles, suitable for any kind of restaurant, that can be designed from a single recipe or combination of foods.

Beet salad plate 1

Beet salad plate 2

Beet salad plate 3

Beet salad plate 4

The same ideas and techniques can be applied to nearly any dish, even a simple bowl of soup. Many of the recipes in this book include instructions for plating, but often they don't. In both cases, the cook is free to use creativity to make attractive plate presentations, keeping in mind the general guidelines in the first part of this chapter. Following the pattern that we used with the Roasted Beet Salad, we have selected additional recipes from this text and have designed four different plate presentations for each basic dish, from simple to complex.

Butternut Squash Soup

Base dish: Butternut Squash Soup with Caramelized Apples (p. 249)

Plate 1: Prepare the recipe as directed. Instead of the swirl of cream, finish the soup with a swirl of caramelized juices from the apple.

Plate 2: Prepare the soup as in the basic recipe, but flavor with ground ginger. Replace the swirl of cream or apple juices with a drizzle of sage oil (made using the procedure for Basil Oil, p. 216). Garnish with curried almonds (toss sliced almonds with vegetable oil, salt, and curry powder; toast lightly in the oven).

Plate 3: Instead of cutting the apples into small dice, cut them into thick 1-inch (2.5-cm) disks and caramelize them. Finish the soup with candied squash seeds and Maple Emulsion (p. 220).

Plate 4: Prepare caramelized apple disks as in Plate 3, arrange them in the center of the bowl, and top them with a few slices of smoked chicken breast. Top the chicken with Pickled Apple (p. 703) and a few frisée leaves. Dot the soup with sage oil.

Butternut squash soup plate 1

Butternut squash soup plate 2

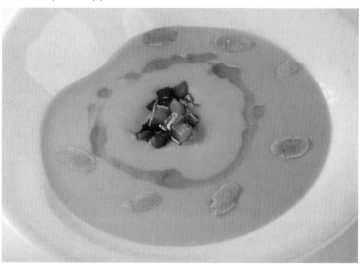

Butternut squash soup plate 3

Butternut squash soup plate 4

Braised Short Ribs

Base dish: Braised Short Ribs (p. 445), plus a vegetable and a starch

Plate 1: Make a simple arrangement, in the center of a round plate, of short ribs, Dauphi-noise Potatoes (p. 358), and steamed green beans sautéed with chopped shallot.

Plate 2: Using the recipe for Grits with Cheddar Cheese (p. 391) as a base, prepare the grits but omit the cheese and, instead, add a little heavy cream. Place grits on center of a round plate and top with Collards with Ham (p. 313). Place short rib on top of the collards, nappé with a little sauce, and top with crispy onion rings.

Plate 3: Prepare Grits with Cheddar Cheese using smoked cheddar. Pour the grits into a hotel pan, smooth, and chill until firm. Cut rectangles and brown lightly in butter. Plate the grits cake on a large square plate with the short ribs, Collards with Ham, and Glazed Tournéed Carrots (p. 315).

Plate 4: Prepare potato purée, flavor with horseradish, and thin with hot cream. Spread a little purée on rectangular plate and top with Pont Neuf potatoes (p. 361). Add short rib and spinach flavored with garlic. Top the short rib with a little salad of frisée and shallot mixed with vinaigrette.

Short rib plate 1

Short rib plate 2

Short rib plate 3

Short rib plate 4

Roast Rack of Lamb

Base dish: Roast Rack of Lamb (p. 490) plus vegetable and a Duchesse Potato (p. 353) preparation

Plate 1: This is a simple arrangement of lamb, lamb jus, duchesse potato, and Brussels Sprouts with Walnuts (p. 326) on a traditional round plate.

Plate 2: Here the lamb and jus are served with Zucchini Sauté Provençale (p. 322) and a few dots of Pesto (p. 399). The duchesse mixture is used to make Potato Croquettes (p. 362).

Plate 3: The idea for this presentation starts with the components of Plate 2, but the rack of lamb is rubbed with pesto before roasting. The duchesse mixture is spread in a pan, topped with a mixture of bread crumbs and parmesan cheese, and browned under a salamander. A rectangle of this mixture takes the place of the croquettes. In addition to the zucchini sauté, the presentation also features Orange and Tarragon Fennel Sous Vide (p. 317) on a large, square plate.

Plate 4: The lamb and jus are arranged on a rectangular plate with a creamy purée of cauliflower. The lamb rests against small mounds of Orange Scented Spinach (p. 310) and Roasted Cauliflower with Almonds and Raisins (p. 334). The duchesse mixture is made into small, round croquettes.

Rack of lamb plate 1

Rack of lamb plate 2

Rack of lamb plate 3

Rack of lamb plate 4

Beef Tenderloin

Base dish: Beef tenderloin, roasted following the basic roasting procedure in Roast Rib of Beef au Jus (p. 486), Bordelaise Sauce (p. 187), potatoes, and broccoli.

Plate 1: A slice of roast tenderloin with Bordelaise sauce is plated in a traditional style with steamed, buttered broccoli and mashed potatoes.

Plate 2: This arrangement of tenderloin with roasted potatoes (p. 356) and Broccoli Mornay (p. 311) is still fairly traditional but looks modern on a large, square plate.

Plate 3: Instead of the simple roast tenderloin, this presentation is lifted to another level by featuring Beef Tenderloin Sous Vide (p. 513), flavored with garlic and herbs. A long, rectangular plate showcases, in the center, the sliced tenderloin resting against sautéed mushrooms and accompanied by broccoli flavored with grated lemon zest and lightly browned, shaved garlic. The potatoes in this plating take the form of Potato Rods Rissolé (p. 363) sprinkled with parsley chiffonade. The long, narrow shape of the potatoes echoes the shape of the plate.

Plate 4: A large square plate holds an arrangement of Beef Tenderloin Sous Vide, garnished with a little special sea salt called fleur de sel and resting on Olive Oil and Garlic Confit Red Potatoes (p. 357), with the same shaved-garlic and lemon-scented broccoli in the rear. The Bordelaise sauce in front is complemented by Caramelized Roasted Onion Purée (p. 330) in the rear. Finally, the beef is topped with crispy fried shallot rings.

Beef tenderloin plate 1

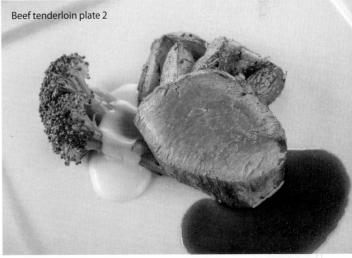

Beef tenderloin plate 2

Beef tenderloin plate 4

Beef tenderloin plate 3

Sautéed Salmon

Base dish: Escalope of Salmon with Red Wine Sauce (p. 630), plus vegetable and starch.

Plate 1: A thicker cut of salmon is used than in the basic recipe and is served on a round plate in a traditional plating with mixed vegetables and parsleyed new potatoes. The salmon is lightly coated with herbs and a fine brunoise of shallot.

Plate 2: The same cut of salmon is now placed on top of side-by-side mounds of Braised Green Cabbage (p. 327) and Potatoes with Bacon and Onions (p. 365). On top of the salmon is a garnish of halved grape tomatoes, tossed with olive oil and herbs and roasted to concentrate flavor.

Plate 3: Complementing the red wine sauce, a swoosh of Root Vegetable Purée (p. 204) gives this presentation a modern feel, even on a traditional round plate. The salmon is served with Roasted Winter Vegetables (p. 332) cut in medium dice, and Green Lentils with Celery Root and Mustard (p. 374). The salmon is topped with a little Pear Salad (p. 702).

Plate 4: A modernist plating on a large, square plate balances the salmon and its sauce on the left against Herb and Garlic Sous Vide Potatoes (p. 350) with Carrot Purée (p. 204) on the right. The salmon is topped with a small mound of delicate salad greens and slivered radish.

Salmon red wine sauce plate 1

Salmon red wine sauce plate 2

Salmon red wine sauce plate 3

Salmon red wine sauce plate 4

Pork tenderloin plate 1

Pork tenderloin plate 2

Pork tenderloin plate 3

Pork tenderloin plate 4

Grilled Pork Tenderloin

Base dish: Grilled Marinated Pork Tenderloin (p. 501) with greens, sweet potato, and apple.

Plate 1: In a traditional arrangement in the center of a square plate, slices of pork tenderloin rest against mashed sweet potatoes and spinach sautéed with diced shallot. Applesauce (p. 213) is a traditional accompaniment for pork.

Plate 2: In this plating, the sliced pork is served on a bed of spinach in the center of a large, square plate, with a spoonful of mashed sweet potatoes alongside. The pork is topped with Apple Relish (p. 208). Serving as a sauce for the arrangement is the juice from the apple relish mixed with a little additional olive oil.

Plate 3: A modern arrangement on a long, rectangular plate features the sliced pork against Roman-Style Spinach (p. 316), next to sticks of Fennel and Coriander Sweet Potatoes (p. 316). For a variation on traditional applesauce, apples are roasted (as for Roasted Winter Vegetables (p. 332) but without the garlic and herbs) before being puréed.

Plate 4: In this most modern of the four platings, the sweet potatoes are puréed until very smooth and enriched with cream and butter. They are spread in a diagonal on a large, square plate to serve as a base for the arrangement. Two thick slices of pork tenderloin, garnished with crushed pink Himalayan salt, are placed at one end of the sweet potato, and Apple and Fennel Salad (p. 702) at the opposite end. Between are 2 small mounds of Roman-Style Spinach and 3 pieces of Fennel and Coriander Sweet Potatoes cut into dice.

KEY POINTS TO REVIEW

- Why is it important to plate food attractively?

- What is meant by balance when applied to plate arrangements? What elements should be balanced?

- What guidelines should be observed when plating food or planning plate arrangements?

COLD FOOD PRESENTATION AND BUFFET SERVICE

The buffet is a popular and profitable form of food presentation found in nearly every kind of food-service operation. There are at least three reasons for this popularity:

1. **Visual appeal.**
   An attractive presentation of foods has the effect of lavishness and ample quantity, and careful arrangement and garnish suggest quality as well.

2. **Efficiency.**
   The buffet allows the restaurant to serve a large number of people in a short time with relatively few service personnel.

3. **Adaptability.**
   Buffet service is adaptable to nearly every kind of food (except items that must be cooked to order, like broiled and deep-fried foods) and to all price ranges, occasions, restaurant styles, and local food customs.

BUFFET ARRANGEMENT AND APPEARANCE

The buffet's visual appeal is, perhaps, its greatest attraction for the customer. Eye appeal of food is always important, but perhaps nowhere more important than on a buffet, because the appearance sells the food. A buffet is not just food service—it is food display.

LAVISHNESS AND ABUNDANCE

Above all else, a buffet should look lavish and plentiful. The appearance of an abundance of food beautifully laid out is exciting and stimulating to the appetite. There are many ways to create this look.

1. **Color.**
   A variety of colors is as vital on a buffet as it is on a single plate. Plan menus and garnish so you have enough color on the table.

2. **Height.**
   Flat foods on flat trays on flat tables are uninteresting to the eye.
   A centerpiece is an important feature, giving height and focus to the buffet. Ice carvings, tallow sculptures, and floral or fruit displays are some possibilities. These should be placed on a separate table behind the food table.
   Centerpieces on individual platters also add height. Large food items such as large cheeses and whole roasts being carved at the table are also effective. Multilevel tables, when available, are used to good effect.

3. **Full platters and bowls.**
   Replenish items as they become depleted. A nearly empty bowl isn't as appetizing as a full one.
   Arrange platters so they still have interest even when portions have been removed (more on this later).

4. **Proper spacing.**
   While you shouldn't crowd the items, don't spread them so far apart that the table looks half empty.

SIMPLICITY

This sounds like a contradiction to the lavishness principle, but it's not. You need to strike a good balance between the two. Lavishness is not the same as clutter.

1. **Overdesigned, overdecorated food scares people away from eating it.**

   How many times have you heard someone say, "Oh, it's so pretty I don't want to touch it!" or, "That looks too good to eat"? Even if they don't say it, they might think it.

   Too much design detracts from the food. Sometimes food is so overdecorated that it no longer looks like food. This completely defeats the purpose. The customer should at least be able to identify the food for what it is.

2. **Excessive garnish is quickly destroyed as customers take portions.**

ORDERLINESS

A buffet should look like it was planned, not like it just happened. Customers prefer food presentations that look carefully done, not just thrown together. Keep three guidelines in mind:

1. Simple arrangements are much easier to keep neat and orderly than complicated designs.

2. Colors and shapes should look lively and varied, but make sure they go together and do not clash.

3. Keep the style consistent. If it's formal, then everything should be formal. If it's casual or rustic, then every part of the presentation should be casual or rustic. If it's a Mexican fiesta, don't include German sauerbraten just because your specialty happens to be sauerbraten.

   This is true not only of the food but of the dishes and serving pieces, too. Don't use ornate silver serving pieces for a country theme, for example.

MENU AND SERVING SEQUENCE

Practical reasons as well as visual appeal determine the order in which foods are arranged on the buffet. As far as possible, it is good to have items in the proper menu order (for example, appetizers first, main course afterward, desserts last) if only to avoid confusing the customers, who might otherwise wonder what the food is and how much they should take. But there are many reasons for changing the order. The following should be taken into account when arranging a buffet:

1. Hot foods are best served last. If served first, hot foods get cool while the guests make other selections from the cold foods. Also, it is more effective, visually, to place the decorative cold platters first and the less attractive chafing dishes last.

2. The more expensive foods are usually placed after the less expensive items. This gives you some control of food cost, as the guests' plates will be nearly full of other attractive foods by the time they get to the costly items.

3. Sauces and dressings should be placed next to the items with which they are to be served. Otherwise, the customer might not match them with the right foods.

4. A separate dessert table is often a good idea. This approach allows guests to make a separate trip for dessert without interfering with the main serving line. It is also possible, if the menu is large, to have a separate appetizer table.

5. Plates, of course, must be the first items on the table. Silverware, napkins, and other items not needed until the guest sits down to eat should be at the end of the buffet table or set in place on the dining tables.

THE COCKTAIL BUFFET

One kind of buffet doesn't conform to this menu order pattern. The cocktail buffet displays appetizers intended to accompany drinks and other refreshments at receptions, cocktail parties, and cocktail hours preceding banquets and dinners. There is no serving line—or, looking at it a different way, there is a separate line for each item. Guidelines for the cocktail buffet are as follows:

1. Only appetizer-type foods are served: tasty, well-seasoned foods in small portions.

2. Stacks of small plates are placed beside each item rather than at the beginning of the table.

3. The table or tables must be easy to get to from all parts of the room and must not block traffic. Do not place them next to the entry because guests gather around them, blocking movement into and out of the room.

COLD PLATTER PRESENTATION

The cold platter is the mainstay of the buffet and offers the most opportunity for visual artistry. It also can be one of the most demanding forms of food presentation, particularly in the case of show platters, which require great precision, patience, and artistic sense.

Cold platters can range from a simple tray of cold cuts to elaborate constructions of pâtés, meats, poultry, or fish decorated with aspic, truffles, and vegetable flowers. In this chapter, we have space only for a discussion of general guidelines you can apply both to formal buffet platters and to simple cold food arrangements. To learn more detailed, complex techniques, you must depend on your instructors, more advanced courses, and on-the-job experience. But this section should help you take the foods available in whatever kitchen you find yourself working and produce an attractive, appetizing buffet.

BASIC PRINCIPLES OF PLATTER PRESENTATION

1. The classic buffet platter has three elements:
   - Centerpiece or **grosse pièce** (gross pyess). This may be an uncut portion of the main food item, such as a pâté or a cold roast, decorated and displayed whole. It may be a separate but related item, such as a molded salmon mousse on a platter of poached slices of salmon in aspic. It may be something as simple as a bowl or **ravier** (rahv yay; an oval relish dish) of sauce or condiment. Or it may be strictly for decoration, such as a butter sculpture or a squash vase filled with vegetable flowers. Whether or not the grosse pièce is intended to be eaten, it should be made of edible materials.
   - The slices or serving portions of the main food item, arranged artistically.
   - The garnish, arranged artistically, in proportion to the cut slices.
2. The food should be easy to handle and serve, so one portion can be removed without ruining the arrangement.
3. A simple design is best. Simple arrangements are easier to serve, more appetizing than overworked food, and more likely to remain attractive when they are half demolished by the guests.

   Simple arrangements may be the hardest to produce. Everything must be perfect because less decoration is available to divide the attention.
4. Attractive platter presentations may be made on silver or other metals, mirrors, china, plastic, wood, or many other materials, as long as they are presentable and suitable for use with food. Metal platters that might cause discoloration or metallic flavors are often covered with a thin layer of aspic before the food is placed on them.
5. Once a piece of food has touched the tray, do not remove it. Shiny silver or mirror trays are easily smudged, and you'll have to wash the tray and start over again. This shows the importance of good planning.

   Following this rule also helps eliminate overhandling of food, which is a bad sanitary practice.
6. Think of the platter as part of the whole buffet. It must look attractive and appropriate not only by itself but among the other presentations on the table. The arrangement should always be planned from the same angle from which it will be seen on the buffet.

DESIGNING THE PLATTER

1. **Plan ahead.**

   Making a sketch is a good idea. Otherwise, you might have half the food on the platter and suddenly realize you have to start over because everything doesn't fit the way you had hoped. The result is wasted time and excessive handling of food.

   One way to start a sketch is to divide the platter into six or eight equal parts, as in Figure 29.1. This helps you avoid lopsided or crooked arrangements by giving you equally spaced markers as guides. It is relatively easy, then, to sketch in a balanced, symmetrical layout, as the examples show.
2. **Plan for movement in your design.**

   This doesn't mean you should mount the food on little wheels. It means that a good design makes the eye move across the platter, following the lines you have set up.

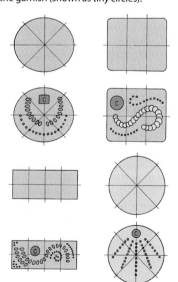

FIGURE 29.1 Begin your planning sketch of a buffet platter by dividing the tray into six or eight sections. This helps you lay out a balanced, symmetrical design. The examples shown here indicate the placement of the main items (usually slices of foods), the centerpiece (labeled **c**), and the garnish (shown as tiny circles).

Most food for platters consists of single small portions arranged in rows or lines. The trick is to put movement into those lines by curving or angling them, as shown in Figure 29.2. In general, curves and angles are said to have movement. Square corners do not.

3. Give the design a focal point.

This is the function of the centerpiece, which emphasizes and strengthens the design by giving it direction and height. This may be done directly, by having the lines point at it, or more subtly, by having the lines angle toward it or sweep around it in graceful curves. Again, see Figures 29.1 and 29.2.

Note that the centerpiece isn't always in the center, in spite of its name. Because of its height, it should be at the back or toward the side so it doesn't hide the food. Remember, you are designing the platter from the customer's point of view.

It's not necessary for every platter on the buffet to have a centerpiece. Some of them should, however, or the buffet will lack height and be less interesting to the eye.

4. Keep items in proportion.

The main items on the platter—the slices of meat, pâté, or whatever—should look like the main items. The centerpiece should not be so large or so tall that it totally dominates the platter. The garnish should enhance, not overwhelm, the main item in size, height, or quantity. The number of portions of garnish should be in proportion to the amount of the main item.

The size of the platter should be in proportion to the amount of food. Don't select one that is so small as to become crowded or so large as to look almost empty even before the first guest has arrived.

Keep enough space between items or between rows so the platter doesn't look jumbled or confused.

Figure 29.1 indicates placement of garnish as well as of the main item. Note how the arrangement of garnish reflects or accents the pattern established by the sliced foods.

5. Let the guest see the best side of everything.

Angle overlapping slices and wedge-shaped pieces toward the customer. Make sure the best side of each slice is face up.

FIGURE 29.2 Arranging rows of foods in curves or angled lines gives movement to the design.

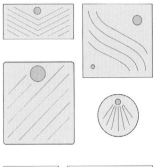

KEY POINTS TO REVIEW

- When planning a buffet, how do you create the impression of lavishness? What guidelines help you achieve simplicity and orderliness?

- In general, in what order should items be arranged on a buffet?

- What guidelines should be observed when planning a cold buffet platter?

TERMS FOR REVIEW

garnish   **classical garnish**   **grosse pièce**   **ravier**

QUESTIONS FOR DISCUSSION

1. Discuss the idea of professionalism and how it applies to the presentation of food.

2. Following are several popular food combinations. Describe what plating problems they present, if any, and how you might efficiently and economically solve them.

> Fish and chips (deep-fried fillets and French fries)
> Prime rib of beef and baked potato
> Meat loaf, mashed potatoes, and gravy
> Open-faced hot turkey sandwich
> Beef stroganoff and egg noodles
> Chicken à la king in a patty shell

3. What is the difference between a cocktail buffet and a luncheon or dinner buffet?

4. Customers like to see a lot of food on a buffet. Is it correct to say, then, that the best way to please customers is to put out as much food as possible? Explain your answer.

5. When you are preparing a cold buffet platter, why is it a good idea to plan ahead by making a sketch? What would you include in the sketch? What do the terms *movement* and *focal point* mean in platter design?

BAKESHOP PRODUCTION: BASIC PRINCIPLES AND INGREDIENTS

At one time, it was common for food-service establishments to produce their own breads, desserts, and other baked goods. Today, many operations find it more economical to buy these products from commercial bakeries. However, many other owners and chefs have discovered that offering fresh, "home-baked" breads, cakes, and pastries attracts customers and increases profits. With little more than an oven and a mixer, many cooks turn out attractive baked items that set their operations apart from competitors.

For this reason, it is important for you to learn the fundamentals of baking, even if you intend to become a cook rather than a baker. These chapters will not make a professional baker out of you. A baker requires far more technical and specialized information than can be presented in this short space. But you will learn the basic methods for producing a wide variety of breads, desserts, and pastries with only the simplest of resources.

In this chapter, we introduce bakeshop production with a discussion of the basic processes and ingredients common to nearly all baked goods. This will give you the understanding necessary to proceed to actual production in the succeeding chapters.

AFTER READING THIS CHAPTER, YOU SHOULD BE ABLE TO

1. Explain why it is important to weigh baking ingredients.

2. Use a baker's balance scale.

3. Calculate formulas based on baker's percentages.

4. Explain the factors that control the development of gluten in baked products.

5. Explain the changes that take place in a dough or batter as it bakes.

6. Prevent or retard the staling of baked items.

7. Describe the major ingredients of baked goods and their functions and characteristics.

BASIC PRINCIPLES OF BAKING

If you consider that most bakery products are made of the same few ingredients—flour, shortening, sugar, eggs, water or milk, and leavening—you should have no difficulty understanding the importance of accuracy in the bakeshop, where slight differences in proportions or procedures can mean great differences in the final product.

If you have begun your food-service studies in a kitchen production laboratory, you surely have been told many times of the importance of measurement, not only for portion control and cost control but also for consistency in the quality of the final product. However, you have, no doubt, also learned there is a great deal of margin for error and that it is possible (if not desirable) to cook many foods without measuring anything. Coming into the bakeshop, where measurement is absolutely essential, may be a bit of a shock to you after your kitchen experiences, but it should reinforce the habits of accuracy you may have let slip.

If, on the other hand, you are beginning your practical studies in the bakeshop, then you will do well to pay particular attention to the principles of measurement presented here. They will be valuable to you throughout your career.

FORMULAS AND MEASUREMENT

Bakers generally talk about **formulas** rather than **recipes**. If this sounds more like the chemistry lab than the kitchen, it is with good reason. The bakeshop is much like a chemistry laboratory both in the scientific accuracy of all the procedures and in the complex reactions that take place during mixing and baking.

MEASUREMENT

All ingredients must be weighed. Accuracy of measurement, as we have already said many times, is critical in the bakeshop. Measurement is by weight rather than by volume because weight is much more accurate. Unlike in recipes for the home baker, you will not see a professional baker's formula calling for 6 cups flour.

To demonstrate to yourself the importance of weighing rather than measuring by volume, measure 1 cup flour in two ways. (1) Sift some flour and lightly spoon it into a dry measure. Level the top and weigh the flour. (2) Scoop some unsifted flour into the same measure and pack it lightly. Level the top and weigh the flour. Note the difference. No wonder home recipes can be so inconsistent!

Digital professional scale
Courtesy of Cardinal Detecto

PROCEDURE for Using a Baker's Balance Scale

Balance scale
Courtesy of Cardinal Detecto

The principle of using a baker's scale is simple: The scale must balance before setting the weights, and it must balance again after scaling. The example cited illustrates using a scale with U.S. units. The same procedure is used for metric scales.

1. Set the scale scoop or other container on the left side of the scale.

2. Balance the scale by placing counterweights on the right side and/or adjusting the ounce weight on the horizontal bar.

3. Set the scale for the desired weight by placing weights on the right side and/or by moving the ounce weight. For example, to set the scale for 1 pound 8 ounces, place a 1-pound weight on the right side and move the ounce weight to the right 8 ounces. If the ounce weight is already over 8 ounces, so that you cannot move it another 8, add 2 pounds to the right side of the scale and subtract 8 ounces by moving the ounce weight 8 places to the left. The result is still 1 pound 8 ounces.

4. Add the ingredient being scaled to the left side until the scale balances.

The baker's term for weighing out ingredients is *scaling*.

The following ingredients may be measured by *volume* because they weigh *1 pound per pint or 1 kilogram per liter*:

    Water   Milk   Eggs

Thus, if a formula calls for 2 pounds eggs, you may measure 2 pints (1 quart). (Liquid flavoring ingredients, such as vanilla extract, normally measured in very small quantities, may also be measured by volume; 1 tablespoon equals $^1/_2$ ounce.) In the metric system, 1 milliliter water weighs 1 gram; 1 liter weighs 1 kilogram. All other liquid ingredients (such as corn syrup and molasses) and all dry ingredients are normally weighed.

BAKER'S PERCENTAGES

Bakers use a simple but versatile system of percentages for expressing their formulas. Bakers' percentages express the amount of each ingredient used as a percentage of the amount of flour used.

To put it differently, the percentage of each ingredient is its total weight divided by the weight of the flour and multiplied by 100 percent, or

$$\frac{\text{Weight of ingredient}}{\text{Weight of flour}} \times 100\% = \% \text{ of ingredient}$$

Thus, flour is always 100 percent. (If two kinds of flour are used, their total is 100 percent.) Any ingredient that weighs the same as the flour is also given as 100 percent. The following ingredients from a cake formula illustrate how these percentages are used. Both U.S. and metric examples are given. (Note that numbers may be rounded off for practical measuring.) Check the figures with the above equation to make sure you understand them.

Ingredient	Weight		Percentage
Cake flour	5 lb		100 %
Sugar	5 lb		100 %
Baking powder		4 oz	5 %
Salt		2 oz	2.5 %
Emulsified shortening	2 lb	8 oz	50 %
Skim milk	3 lb		60 %
Egg whites	3 lb		60 %
	8 lb 14 oz		377.5 %

Ingredient	Weight	Percentage
Cake flour	2500 g	100 %
Sugar	2500 g	100 %
Baking powder	125 g	5 %
Salt	60 g	2.5 %
Emulsified shortening	1250 g	50 %
Skim milk	1500 g	60 %
Egg whites	1500 g	60 %
	9435 g	377.5 %

The advantage of using baker's percentages is that the formula is easily adapted for any yield, and single ingredients may be varied without changing the whole formulation. Please remember that these numbers do not refer to the percentage of the total yield. They are simply a way of expressing *ingredient proportions*. The total of these percentage numbers will always be greater than 100 percent.

PROCEDURE for Calculating the Weight of an Ingredient If the Weight of Flour Is Known

1. Change the ingredient percentage to decimal form by moving the decimal point two places to the left.

2. Multiply the weight of the flour by this decimal to get the weight of the ingredient.

   Example (U.S.): A formula calls for 20 percent sugar and you are using 10 pounds flour. How much sugar do you need?

$$20\% = 0.20$$
$$10 \text{ lb} \times 0.20 = 2 \text{ lb sugar}$$

   Note: In the U.S. system, weights must normally be expressed all in one unit, either ounces or pounds, in order for the calculation to work, as explained in Chapter 4.

   Example (Metric): A formula calls for 20 percent sugar and you are using 5000 grams (5 kg) flour. How much sugar do you need?

$$20\% = 0.20$$
$$5000 \text{ g} \times 0.20 = 1000 \text{ g sugar}$$

PROCEDURE for Converting a Formula to a New Yield

1. Change the total percentage to decimal form by moving the decimal point two places to the left.

2. Divide the desired yield by this decimal figure to get the weight of flour.

3. If necessary, round off this number to the next highest figure. This will allow for losses in mixing, makeup, and panning, and it will make calculations easier.

4. Use the weight of flour and remaining ingredient percentages to calculate the weights of the other ingredients, as in the previous procedure.

   Example: In the previous sample cake formula, how much flour is needed if you require 6 pounds (3000 g) cake batter?

$$6 \text{ lb} = 96 \text{ oz}$$
$$377.5\% = 3.775$$
$$96 \text{ oz} \div 3.775 = 25.43 \text{ oz}$$

or, rounded off, 26 oz (1 lb 10 oz)

$$3000 \text{ g} \div 3.775 = 794.7 \text{ g}$$

or, rounded off, 800 g

Clearly, the percentage system we have been discussing is used only when flour is a major ingredient, as in breads, cakes, and cookies. For these formulas, we use a written format different from our regular recipe format in this book.

In these formulas, the indicated yield is the total weight of the ingredients. This figure indicates the weight of the batter or dough. It is the figure we need to know for the purpose of scaling the dough or batter into loaves or pans. The finished weight of the baked goods will be less because moisture is lost during baking.

Also, please note that all yields, including percentage totals, are rounded off to the next lower whole number. This eliminates unimportant fractions and makes reading and calculating easier.

SELECTION OF INGREDIENTS

In addition to measuring, there is another basic rule of accuracy in the bakeshop: *Use the exact ingredients specified*.

As you will learn in this chapter, different flours, shortenings, and other ingredients do not function alike. Bakers' formulas are balanced for specific ingredients. Do not substitute bread flour for pastry flour or regular shortening for emulsified shortening, for example. They won't work the same way.

Occasionally, a substitution may be made, such as instant yeast for compressed yeast (see p. 906), but not without adjusting the quantities or rebalancing the formula.

KEY POINTS TO REVIEW

- What is the proper way to measure bakeshop ingredients? Why?

- What is the procedure for weighing ingredients using a baker's balance scale?

- How do you use baker's percentages to calculate the weight of ingredients in a baking formula?

MIXING AND GLUTEN DEVELOPMENT

WHAT IS GLUTEN?

Gluten is a substance made up of proteins present in wheat flour; it gives structure and strength to baked goods.

In order for gluten to be developed, the proteins must first absorb water. Then, as the dough or batter is mixed or kneaded, the gluten forms long, elastic strands. As the dough or batter is leavened, these strands capture the gases in tiny pockets or cells, and we say the product rises. When the product is baked, the gluten, like all proteins (see p. 108), coagulates or solidifies and gives structure to the product.

HOW DOES THE BAKER CONTROL GLUTEN?

Flour is mostly starch, but its protein or gluten content, not its starch, concerns the baker most. Without gluten proteins to give structure, baked goods would not hold together.

The baker must be able to control the gluten, however. For example, we want French bread to be firm and chewy, which requires much gluten. On the other hand, we want cakes to be tender, which means we want very little gluten development.

Ingredient proportions and mixing methods are determined, in part, by how they affect the development of gluten. The baker has several methods for adjusting gluten development.

1. **Selection of flours.**

   Wheat flours are classified as *strong* or *weak*, depending on their protein content.

   Strong flours come from *hard wheat* and have a high protein content.

   Weak flours come from *soft wheat* and have a low protein content.

   Thus, we use strong flours for breads and weak flours for cakes.

   Only wheat flour develops gluten. To make bread from rye and other grains, the formula must be balanced with some high-gluten wheat flour, or the bread will be heavy.

2. **Shortening.**

   Any fat used in baking is called a **shortening** because it shortens gluten strands. It does this by surrounding the particles and lubricating them so they do not stick together. Thus, *fats are tenderizers*. A cookie or pastry that is very crumbly due to high fat content is said to be short.

   You can see why French bread has little or no fat, while cakes contain a great deal.

3. **Liquid.**

   Because gluten proteins must absorb water before they can be developed, the amount of water in a formula can affect toughness or tenderness. To keep them tender, pie crusts and crisp cookies are made with very little liquid.

4. **Mixing methods.**

In general, the more a dough or batter is mixed, the more the gluten develops. Thus, bread doughs are mixed or kneaded for a long time to develop the gluten. Cakes, pie crusts, muffins, and other products that must be tender are mixed for a short time.

It is possible to overmix bread dough, however. Gluten strands stretch only so far. They break if the dough is overmixed.

THE BAKING PROCESS

The changes undergone by a dough or batter as it bakes are basically the same for all baked products, from breads to cookies and cakes. You should know what these changes are so you can learn how to control them.

The stages in the baking process take place as follows. Be aware that many of these steps occur at the same time, not one after the other. For example, escape of water vapor and other gases begins almost at once, but it is more rapid later in the baking process.

1. **Melting of fats.**

Different fats melt—and release trapped gases—at different temperatures, so the proper shortening should be selected for each product. Most fats used in baking melt between 90° and 130°F (32° and 55°C).

As the fats melt, they surround the air cells and make the product more tender.

2. **Formation and expansion of gases.**

Some gases are already present in the dough, as in proofed bread dough and in sponge cake batters. As they are heated, the gases expand and leaven the product.

Some gases are not formed until heat is applied. Yeast and baking powder form gases rapidly when first placed in the oven. Steam is also formed as the moisture of the dough is heated.

As the product rises, the cell walls become thinner as they are stretched by the expanding gases. This tenderizes the product.

Leavening and leavening agents are discussed in more detail beginning on page 905.

3. **Killing of yeast and other microorganisms.**

In addition to yeast, doughs may contain other microorganisms, including bacteria and molds. Most of these, including yeast, die when the item reaches about 140°F (60°C). When yeast dies, fermentation stops and no more gas is released.

4. **Coagulation of proteins.**

Like all proteins, gluten and egg proteins coagulate or solidify when they reach high enough temperatures. This is the process that gives structure to baked goods.

Correct baking temperature is important. If the temperature is too high, coagulation will start too soon, before the expansion of gases has reached its peak. The product will have poor volume or a split crust. If the temperature is too low, the proteins will not coagulate soon enough, and the product may collapse.

5. **Gelatinization of starches.**

The starches absorb moisture, expand, and become firmer.

6. **Escape of water and other gases.**

This takes place throughout the baking process, but it is fastest during the later stages of baking. Water evaporation decreases the weight of the product and also enables crust formation.

7. **Crust formation and browning.**

Browning occurs when sugars caramelize and starches and proteins undergo Maillard browning (p. 108). This contributes to flavor. Milk, sugar, and egg increase browning.

A crust is formed as water evaporates from the surface and leaves it dry.

STALING

Staling is the change in texture and aroma of baked goods due to the change in structure and the loss of moisture by the starch granules. Stale baked goods have lost their fresh-baked aroma and are firmer, drier, and more crumbly than fresh products.

Prevention of staling is a major concern of the baker because most baked goods lose quality rapidly.

Staling can be slowed by three techniques:

1. **Protecting the product from air.**

   Wrapping bread in plastic and covering cakes with icing are two examples.

   Unfortunately, hard-crusted breads, which stale rapidly, should not be wrapped, or the crusts will become soft. These bread products should always be served fresh.

2. **Adding moisture retainers to the formula.**

   Fats and sugars are good moisture retainers, and products high in these ingredients keep best.

   Some of the best French bread has no fat at all, and if it is not served within hours of baking, it will begin to stale. For longer keeping, bakers often add a very small amount of fat and/or sugar to the formula.

3. **Freezing.**

   Baked goods frozen *before* they become stale maintain quality for longer periods. They should be served very soon after thawing. Frozen breads may be reheated with excellent results if they are served immediately.

   Refrigerating actually speeds staling rather than slowing it. Only baked goods that could develop health hazards, such as those with cream fillings, are refrigerated.

Loss of crispness is caused by absorption of moisture, so it is, in a sense, the opposite of staling. This is a problem with low-moisture products such as cookies and pie crusts. The problem is usually solved by proper storage in airtight wraps or containers to protect the products from moisture in the air. Prebaked pie shells should be filled as close to service time as possible.

KEY POINTS TO REVIEW

- What is gluten? What are four ways to control gluten development in doughs and batters?

- What are seven changes that take place in doughs and batters as they are baked?

- How can you slow the rate at which baked goods become stale?

 INGREDIENTS

The following introduction to baking ingredients is necessarily simplified. If you decide to pursue a career as a baker, you will need to learn a great deal of technical information. However, the basic information presented here is enough to enable you to produce a full range of baked items in a small bakeshop or restaurant kitchen.

FLOURS, MEALS, AND STARCHES

WHITE WHEAT FLOUR

White wheat flour is milled from wheat kernels after the outer covering, called *bran*, and the germ are removed. Wheat flour normally contains 63 to 73 percent starch and 6 to 15 percent protein. The rest is moisture, fat, sugar, and minerals.

Bread flour

Cake flour

Pastry flour

Wheat flour is the source of the protein called **gluten**, which you remember is one of the essential elements in baking. Bakers select flour on the basis of its gluten content. Flours high in protein are called **strong**, and those low in protein are called **weak**. (**Note**: Rye, barley, oats, and some other grains also contain gluten proteins, but these proteins do not develop into a gluten structure as do the proteins in wheat flour. Thus, for the baker, these other grains in effect do not contain gluten, but people with gluten intolerance may still have to avoid them in their diets.)

For our purposes, in the small bakeshop, we need to know about three kinds of wheat flour:

1. **Bread flour** is a strong flour used for making breads, hard rolls, and any product that requires high gluten. The best bread flours are **patent flours,** which come from the interior portions of the kernel. **Straight** flours are also strong flours.

2. **Cake flour** is a weak or low-gluten flour made from soft wheat. It has a soft, smooth texture and a pure white color. Cake flour is used for cakes and other delicate baked goods that require low gluten content.

3. **Pastry flour** is lower in gluten than bread flour but higher than cake flour. It has the same creamy white color as bread flour, not the pure white of cake flour. Pastry flour is used for cookies, pie pastry, some sweet yeast doughs, biscuits, and muffins.

Being able to identify these three flours by sight and touch is an important skill because, sooner or later, someone will dump a bag of flour into the wrong bin, and you will need to recognize the problem.

Bread flour feels slightly coarse when rubbed between the fingers. If squeezed into a lump, it falls apart as soon as the hand is opened. Its color is creamy white.

Cake flour feels smooth and fine. It stays in a lump when squeezed in the palm of the hand. Its color is pure white.

Pastry flour feels like cake flour but has the creamy color of bread flour.

All-purpose flour, seen in retail markets, is not often found in bakeshops. This flour is formulated to be slightly weaker than bread flour so it can be used for pastries as well. A professional baker, however, prefers to use flours formulated for specific purposes because these give the best results.

Hand test for flour strength (from left to right): bread flour, pastry flour, cake flour

WHOLE WHEAT FLOUR

Whole wheat flour is made by grinding the entire wheat kernel, including the bran and germ. The germ, which is the embryo of a new wheat plant, is high in fat, which can become rancid. This is why whole wheat flour does not keep as well as white flour.

Because it is made from wheat, whole wheat flour contains gluten, so it can be used alone in bread making. However, a bread made with 100 percent whole wheat will be heavy because the gluten strands are cut by the sharp edges of the bran flakes. Also, the fat from the wheat germ contributes slightly to the shortening action. This is why most whole wheat breads are strengthened with white bread flour.

Bran flour is flour to which bran flakes have been added. The bran may be coarse or fine, depending on specifications.

RYE FLOUR

Next to white and whole wheat, rye is the most popular flour in bread making. Because rye flour does not develop much gluten, breads made with it are heavy unless some hard wheat flour is added.

Rye flour is available in three shades, *light*, *medium*, and *dark. Rye meal* or **pumpernickel** is a coarse meal made from the whole rye grain.

Rye blend is a mixture of rye flour and hard wheat flour.

OTHER FLOURS

Products milled from other grains are occasionally used to add variety to baked goods. These include cornmeal, buckwheat flour, soy flour, potato flour, oat flour, and barley flour. The term *meal* is used for products that are not as finely ground as flour.

All these products must normally be used in combination with wheat flour because they do not form gluten.

STARCHES

In addition to flours, other starch products are also used in the bakeshop. Unlike flour, they are used primarily to thicken puddings, pie fillings, and similar products. The principles of thickening with starches are covered in Chapter 8.

The most important starches in dessert production are as follows:

1. *Cornstarch* has a special property that makes it valuable for certain purposes. Products thickened with cornstarch set up almost like gelatin when cooled. For this reason, it is used to thicken cream pies and other products that must hold their shape.

2. *Waxy maize* and other *modified starches* also have valuable properties. They do not break down when frozen, so are used for products that are to be frozen. Also, they are clear when cooked, and give a brilliant, clear appearance to fruit pie fillings.

   Waxy maize does not set up firm like cornstarch but rather makes a soft paste, which has the same consistency hot and cold. Thus, it is not suitable for cream pie fillings.

3. *Instant starches* are precooked or pregelatinized, so they thicken cold liquids without further cooking. They are useful when heat will damage the flavor of the product, as in fresh fruit glazes (such as strawberry).

FATS

We have said that one of the main functions of fats in baking is to shorten gluten strands and tenderize the product. We can summarize the reasons for using fats in baked items as follows:

- To tenderize the product and soften the texture.
- To add moistness and richness.
- To increase keeping quality.
- To add flavor.
- To assist in leavening when used as creaming agents or when used to give flakiness to puff pastry, pie dough, and similar products.

SHORTENINGS

Any fat acts as a shortening in baking because it shortens gluten strands and tenderizes the product. However, we usually use the word *shortening* to mean any of a group of solid fats, usually white and tasteless, that are specially formulated for baking.

Because shortenings are used for many purposes, manufacturers have formulated different kinds of fats with different properties. Following are the three main types of shortening.

Regular Shortenings

These shortenings have a tough, waxy texture, and small particles of the fat tend to hold their shape in a dough or batter. This type of shortening does not melt until a high temperature is reached.

Regular shortening has good creaming ability. This means that a large quantity of air can be mixed into it to give a batter lightness and leavening power. Therefore, it is used in products mixed by the creaming method, such as certain cookies.

Because of its texture, this type of shortening is used for flaky products such as pie crusts and biscuits. It is also used in breads and many pastries. Unless another shortening is specified, regular shortening is generally used.

Emulsified Shortenings

These are soft shortenings that spread easily throughout a batter and quickly coat the particles of sugar and flour. Because of their easy spreading, they give a smoother and finer texture to cakes and make them moister.

Emulsified shortening is often used whenever the weight of sugar in a cake batter is greater than the weight of flour. Because this shortening spreads so well, a simpler mixing method can be used, as explained in Chapter 33. Such cakes are referred to as *high-ratio* cakes, so emulsified shortening is sometimes called *high-ratio shortening*.

In addition, emulsified shortening is used in certain icings because it can hold more sugar and liquid without curdling.

Puff Pastry Shortenings

Puff pastry shortenings are firm like regular shortening. They are especially formulated for puff pastry and other doughs that form layers, such as Danish pastry.

BUTTER AND MARGARINE

Shortenings are manufactured to have certain textures and hardness. Butter, on the other hand, is a natural product that doesn't have these advantages. It is hard and brittle when cold and soft at room temperature, and it melts easily. Consequently, doughs made with butter are hard to handle. Margarine is a little easier to handle, but it has many of the same disadvantages.

On the other hand, butter and margarine have two major advantages:

1. **Flavor.**
   Shortenings are intentionally flavorless, but butter has a highly desirable flavor.
2. **Melting qualities.**
   Butter melts in the mouth. Shortenings do not. After eating pastries or icings made with shortening, one can be left with an unpleasant film of shortening coating the mouth.

For these reasons, many bakers and pastry chefs feel the advantages of butter outweigh its disadvantages for some purposes.

OILS

Oils are liquid fats. They are not often used as shortening in baking because they spread through a batter or dough too thoroughly and shorten too much. Their usefulness in the bakeshop is limited primarily to greasing pans and proofing bowls, to deep-frying doughnuts, and to serving as a wash for some kinds of rolls. A few quick breads and cakes use oil as a shortening.

SUGARS

Sugars or sweetening agents are used for the following purposes in baking:

- To add sweetness and flavor.
- To create tenderness and fineness of texture by weakening the gluten structure.
- To give crust color.
- To increase keeping qualities by retaining moisture.
- To act as creaming agents with fats.

We customarily use the term *sugar* for regular refined sugars derived from sugar cane or beets. The chemical name for these sugars is **sucrose**. However, other sugars of different chemical structure are also used in the bakeshop. The following are the more important sugars.

REGULAR REFINED SUGARS, OR SUCROSE

Refined sugars are classified by the size of grains:

1. **Granulated sugar.**

   Regular granulated, also called *fine granulated* or *table sugar*, is the most familiar and the most commonly used.

   Very fine and *ultrafine* sugars are finer than regular granulated. They are prized for making cakes and cookies because they make a more uniform batter and can support higher quantities of fat.

   Sanding sugars are coarser and are used for coating doughnuts, cakes, and other products.

2. **Confectioners' or powdered sugars.**

   Confectioners' sugars are ground to a fine powder and mixed with a small amount of starch to prevent caking. They are classified by coarseness or fineness.

   10X is the finest sugar. It gives the smoothest texture in icings.

   6X is the standard confectioners' sugar. It is used in icings, toppings, and cream fillings.

   Coarser types (*4X* and *XX*) are used for dusting or for any purposes for which 6X and 10X are too fine.

MOLASSES AND BROWN SUGAR

Molasses is concentrated sugar cane juice. *Sulfured molasses* is a byproduct of sugar refining. It is the product that remains after most of the sugar has been extracted from cane juice. *Unsulfured molasses* is not a byproduct but a specially manufactured sugar product. Its taste is less bitter than that of sulfured molasses.

Molasses contains large amounts of sucrose, plus other sugars, acids, and impurities.

Brown sugar is mostly sucrose, but it also contains varying amounts of molasses and other impurities. The darker grades contain more molasses.

Because molasses and brown sugar contain *acids*, they can be used with baking soda to provide leavening (see p. 906).

Molasses retains moisture in baked goods and so prolongs freshness. However, crisp cookies made with molasses quickly become soft for the same reason.

GLUCOSE CORN SYRUP

Corn syrup is a liquid sweetener manufactured by breaking down the starch molecules of cornstarch into simple sugar molecules called *glucose*. Not all the starch is broken down during the process. Low-conversion syrups are only slightly sweet and are quite thick because of the high starch content. Regular, all-purpose corn syrups are medium-conversion syrups in which half the starch is converted to glucose.

Corn syrup aids in retaining moisture and is used in some icings and in candy making.

HONEY

Honey is a natural sugar syrup consisting largely of glucose and fructose, plus other compounds that give it flavor. Honeys vary considerably in flavor and color, depending on their source. Flavor is the major reason for using honey, especially because it can be expensive.

Honey contains invert sugar, which means it stays smooth and resists crystallizing. Like molasses, it contains *acid*, which enables it to be used with baking soda as a leavening.

MALT SYRUP

Malt syrup is used primarily in yeast breads. It serves as food for the yeast and adds flavor and crust color to the breads.

KEY POINTS TO REVIEW

- What three main types of white wheat flour are used in the bakeshop? Describe them.

- What five functions do fats have in baked goods?

- What is the difference between regular and emulsified shortening? When compared with shortening, what are two advantages and one disadvantage of using butter in baked goods?

- What five functions do sugars have in baked goods? What is the difference between regular granulated sugar and confectioners' sugar?

LIQUIDS

Gluten cannot be developed without moisture, so liquids are essential to the baking process.

Pie crusts provide a good illustration of how liquids function in baking. If too much water is incorporated in a pie dough, a lot of gluten develops and the crust is tough. If no water at all is used, no gluten develops and the crust does not hold together.

Some of the moisture in doughs and batters changes to steam during baking. This contributes to leavening.

WATER

Water is the basic liquid in baking, especially in breads.

Tap water is normally suitable for most baking purposes. However, in some localities, the water may be *hard*, meaning it contains many dissolved minerals. These minerals may strengthen gluten too much, causing the dough to be too elastic and hard to work. In these areas, the water may have to be treated for use in baking.

MILK AND CREAM

Milk products, as described in Chapter 25, are important in baking. These products include liquid whole and skim milk, buttermilk, and dry milk solids.

Milk contributes to the texture, flavor, nutritional value, keeping quality, and crust color of baked goods:

1. *Whole milk* contains fat, which must be calculated as part of the shortening in a dough. For this reason, whole and skim milk are not interchangeable in a formula unless adjustments are made for the fat.

2. *Buttermilk,* which is slightly acid, is often used in conjunction with baking soda as a leavening agent in quick breads.

3. *Cream* is not often used as a liquid in doughs and batters, except in a few specialty products. In these instances, it is used as a shortening as well as a liquid because of its fat content. Cream is more important in the production of fillings and toppings.

4. *Dry milk* is often used because of its convenience and low cost. In some formulas, it is not necessary to reconstitute it. The milk powder is included with the dry ingredients, and water is used as the liquid.

OTHER SOURCES OF LIQUID

Eggs, honey, molasses, and even butter (about 15 percent water) contribute moisture to a dough or batter. In many cookies, for example, eggs are the only liquid in the formula.

EGGS

FORMS

As we discussed in Chapter 24, eggs are purchased in the following forms:

1. Whole shell eggs.
2. Frozen: whites, yolks, whole, and whole with extra yolks.
3. Dried: whole, whites, yolks.

FUNCTIONS

Eggs perform the following functions in baking:

1. **Structure.**
   Like gluten protein, egg protein coagulates to give structure to baked products. This is especially important in high-ratio cakes, where the high sugar and fat content weakens the gluten.
   If used in large quantities, eggs make baked products tough or chewy unless balanced by high fat and sugar, which are tenderizers.

2. **Emulsification of fats.**
   Egg yolks contain natural emulsifiers, which help produce smooth batters. This contributes to both volume and texture.

3. **Leavening.**
   Beaten eggs incorporate air in tiny cells or bubbles. In a batter, this trapped air expands when heated and aids in leavening.

4. **Shortening action.**
   The fat in egg yolks acts as a shortening. This is an important function in products that are low in other fats.

5. **Moisture.**
   Whole eggs are about 70 percent water, egg whites about 86 percent water, and egg yolks about 49 percent water. This moisture must be calculated as part of the total liquid in a formula.

6. **Flavor.**

7. **Nutritional value.**

8. **Color.**
   Yolks impart a yellow color to doughs and batters. Also, eggs brown easily and contribute to crust color.

LEAVENING AGENTS

Leavening is the production or incorporation of gases in a baked product to increase volume and to produce shape and texture. These gases must be retained in the product until the structure is set enough (by the coagulation of gluten and egg protein) to hold its shape.

Exact measurement of leavening agents is important because small changes can produce major defects in baked products.

YEAST

Fermentation is the process by which yeast acts on carbohydrates and changes them into carbon dioxide gas and alcohol. This release of gas produces the leavening action in yeast products. The alcohol evaporates completely during and immediately after baking.

Yeast is a microscopic plant. As a living organism, it is sensitive to temperature.

34°F (1°C)	Inactive; storage temperature.
60°–70°F (15°–20°C)	Slow action.
70°–90°F (20°–32°C)	Best growth; proofing temperature for bread doughs.
Above 100°F (38°C)	Reaction slows.
140°F (60°C)	Yeast is killed.

Yeast is available in three forms:

Fresh yeast

1. **Fresh yeast**, also called *compressed yeast*, is moist and perishable and is preferred by professional bakers. It is usually purchased in 1-lb (450-g) cakes.

2. **Active dry yeast** is a dry, granular form of yeast. Active dry yeast must be rehydrated in 4 times its weight of warm water—about 110°F (43°C)—before use. When using active dry yeast in a bread formula, use part of the water in the formula to dissolve the yeast. Do not add more water.

3. **Instant dry yeast** is also a dry granular form of yeast, but it does not have to be dissolved in water before use. It can be mixed directly with the flour. It also produces more gas than regular dry yeast, so less of it is needed. Instant dry yeast is sometimes called *rapid-rise* or *quick-rise* yeast.

    Instant yeast is the form of yeast used in all the formulas in Chapter 31.

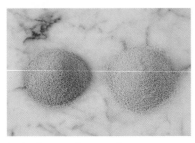

Active dry yeast; instance dry yeast

4. In doughs with a high sugar content, a special type of instant yeast called **osmotolerant yeast** is often indicated in the formula, as this yeast performs better in sweet doughs. If osmotolerant yeast is called for in the formula and you have only regular instant yeast, increase the quantity of yeast by 30 percent.

Until the development of instant yeast, fresh yeast was preferred by professional bakers. Today, however, instant yeast is widely used in bakeshops. In this book, when yeast is required in a formula, instant yeast is specified. To substitute fresh yeast or active dry yeast, use the following guidelines.

To convert instant dry yeast to fresh yeast, multiply the quantity by 3. For example, of the formula calls for 0.5 ounces instant yeast, multiply by 3 to get 1.5 ounces fresh yeast.

To convert instant dry yeast to active dry yeast, multiply the quantity by 1.4. For example, if the formula calls for 30 grams of instant yeast, multiply by 1.4 to get 42 grams active dry yeast.

Yeast contributes flavor in addition to leavening action.

CHEMICAL LEAVENERS

Chemical leaveners are those that release gases produced by chemical reactions.

Baking Soda

Baking soda is the chemical sodium bicarbonate. If *moisture* and an *acid* are present, soda releases carbon dioxide gas, which leavens the product.

Heat is not necessary for the reaction (although the gas is released faster at higher temperatures). For this reason, products leavened with soda must be baked at once, or the gases will escape and leavening power will be lost.

Acids that react with soda in a batter include honey, molasses, buttermilk, fruits, cocoa, and chocolate. Sometimes cream of tartar is used for the acid. The amount of soda used in a formula is generally the amount needed to balance the acid. If more leavening power is needed, baking powder, not more soda, is used.

Baking Powder

Baking powders are mixtures of baking soda plus an acid to react with it.

Because baking powders do not depend on acid ingredients for their leavening power in a formula, they are more versatile.

Single-acting baking powders require only moisture to be able to release gas. Like baking soda, they can be used only if the product is to be baked immediately after mixing.

Double-acting baking powders release some gas when cold, but they require heat for complete reaction. Thus, cake batters made with these can incorporate the leavening agent early in the mixing period and can stand for some time before being baked.

Do not include more baking powder than necessary in a formula because undesirable flavors may be created.

Baking Ammonia

Baking ammonia is the chemical ammonium carbonate. It decomposes during baking to form carbon dioxide gas and ammonia gas. Only heat and moisture are necessary for it to work. No acids are needed.

Because it decomposes completely, it leaves no residue that can affect flavor. However, it can be used only in small products, like cookies, which allow the ammonia gas to be completely driven off.

Baking ammonia releases gases quickly, so it is sometimes used in products like cream puffs where rapid leavening is desired.

AIR

Air is incorporated into a batter primarily by two methods, creaming and foaming. This air expands during baking and leavens the product.

1. **Creaming** is the process of beating fat and sugar together to incorporate air. It is an important technique in cake and cookie making. Some pound cakes and cookies are leavened almost entirely by this method.

2. **Foaming** is the process of beating eggs, with or without sugar, to incorporate air. Foams made with whole eggs are used to leaven sponge cakes, while angel food cakes, meringues, and soufflés are leavened with egg white foams.

STEAM

When water turns to steam, it expands to 1,600 times its original volume. Because all baked products contain some moisture, steam is an important leavening agent.

Puff pastry, cream puffs, popovers, and pie crusts use steam as their major or only leavening agent.

If the starting baking temperature for these products is high, steam is produced rapidly and leavening is greatest.

SALT, FLAVORINGS, AND SPICES

SALT

Salt plays an important role in baking. It is more than just a seasoning or flavor enhancer. It also has these functions:

1. Salt strengthens gluten structure and makes it more stretchable. Thus, it improves the texture of breads.

2. Salt inhibits yeast growth. It is, therefore, important for controlling fermentation in bread doughs and in preventing the growth of undesirable wild yeasts.

For these reasons, the quantity of salt in a formula must be carefully controlled.

CHOCOLATE AND COCOA

Chocolate and cocoa are derived from cocoa or cacao beans. When the beans are roasted and ground, the resulting product is called *chocolate liquor*, which contains a white or yellowish fat called *cocoa butter*.

Cocoa is the dry powder that remains after part of the cocoa butter is removed from chocolate liquor.

Dutch process cocoa is processed with an alkali. It is slightly darker, smoother in flavor, and more easily dissolved in liquids than regular cocoa.

Bitter or *unsweetened chocolate* is straight chocolate liquor. In some less expensive brands, some of the cocoa butter may be replaced by another fat.

Sweetened dark chocolate is bitter chocolate with the addition of sugar in varying amounts. If the percentage of sugar is low, it is sometimes called *semisweet* or *bittersweet*.

Milk chocolate is sweet chocolate with the addition of milk solids. It is used primarily in candy making. (None of the recipes in this book call for milk chocolate.)

Cocoa and chocolate are high in starch. When cocoa is added to a cake formula, it is sometimes considered part of the flour proportion for this reason.

SPICES

Spices are discussed in detail in Chapter 6. The most important spices in the bakeshop are cinnamon, nutmeg, mace, cloves, ginger, caraway, cardamom, allspice, anise, and poppy seed.

Because spices are used in small quantities, it is not much more expensive to use the best quality, and the results are superior.

Spices should be measured by weight unless the quantity is so small that measuring spoons are necessary.

EXTRACTS AND EMULSIONS

Extracts are flavorful oils and other substances dissolved in alcohol. These include vanilla, lemon, and bitter almond.

Emulsions are flavorful oils mixed with water with the aid of emulsifiers such as vegetable gums. Lemon and orange are the most frequently used emulsions.

The flavorings of extracts and emulsions may be natural or artificial. Natural flavorings give the best results, but they are often expensive. Artificial flavorings must be used in moderation to avoid creating strong or undesirable flavors in baked items.

KEY POINTS TO REVIEW

- What are the two main liquids used to make doughs and batters? How do liquids affect gluten development?

- What eight functions do eggs have in baked goods?

- What three types of yeast are used in the bakeshop? How does yeast make bread rise?

- What is the difference between baking soda and baking powder?

TERMS FOR REVIEW

gluten	pastry flour	confectioners' sugar	chemical leavener
strong flours	whole wheat flour	leavening	single- and double-acting baking powders
weak flours	pumpernickel	fermentation	
shortening	rye blend	fresh yeast	creaming
staling	regular shortening	active dry yeast	foaming
bread flour	emulsified shortening	instant dry yeast	extracts
cake flour	sucrose	osmotolerant yeast	emulsions

QUESTIONS FOR DISCUSSION

1. Below are ingredients for a white cake. The weight of the flour is given, and the proportions of other ingredients are indicated by percentages. Calculate the weights required for each.

Cake flour	3 lb or 1500 g (100%)
Baking powder	4%
Shortening	50%
Sugar	100%
Salt	1%
Milk	75%
Egg whites	33%
Vanilla	2%

2. Discuss four factors that affect the development of gluten in doughs and batters.

3. Why do some cakes fall if they are removed from the oven too soon?

4. Which kind of cake would you expect to have better keeping qualities—a sponge cake, which is low in fat, or a high-ratio cake?

5. Why is white wheat flour used in rye breads? in whole wheat breads? Some bakeries in Europe produce a kind of pumpernickel bread with 100 percent rye flour. What would you expect its texture to be like?

6. Describe how to distinguish among bread, pastry, and cake flours by touch and sight.

7. What is the difference between regular and emulsified shortenings?

8. Shortbread is a type of cookie made with flour, butter, and sugar, but no liquid. What would you expect its texture to be like? Why?

YEAST PRODUCTS

In its simplest form, bread is nothing more than a dough of flour and water, leavened by yeast and baked. In fact, some hard-crusted French breads contain only these ingredients, plus salt. Other kinds of bread contain additional ingredients, including sugar, shortening, milk, eggs, and flavorings. But flour, water, and yeast are still the basic building blocks of all breads.

Yet for something that seems so simple, bread can be one of the most exacting and complex products to make. Success in bread making depends largely on your understanding of two basic principles: gluten development, which we discussed in the previous chapter, and yeast fermentation, which we have touched on and which we study in greater detail here.

This chapter focuses on the production of many kinds of yeast products, including breads, dinner rolls, sweet rolls, Danish pastry, and croissants.

AFTER READING THIS CHAPTER, YOU SHOULD BE ABLE TO

1. Prepare breads and dinner rolls.

2. Prepare sweet dough products.

3. Prepare Danish pastry and croissants.

UNDERSTANDING YEAST PRODUCTS

YEAST PRODUCT TYPES

Although all yeast doughs are essentially made according to the same basic principles, it is useful to divide yeast products into categories such as the following.

REGULAR YEAST DOUGH PRODUCTS

Lean Dough Products

A **lean dough** is one that is low in fat and sugar:

- Hard-crusted breads and rolls, including French and Italian breads, kaiser rolls and other hard rolls, and pizza. These are the leanest of all bread products.
- Other white breads and dinner rolls. These have a higher fat and sugar content and, sometimes, also contain eggs and milk solids. Because they are slightly richer, they generally have soft crusts.
- Whole-grain breads. Whole wheat and rye breads are the most common. Many varieties of rye bread are produced with light or dark flours or with pumpernickel flour and various flavorings, especially molasses and caraway seeds.

Rich Dough Products

There is no exact dividing line between rich and lean doughs but, in general, rich doughs contain higher proportions of fat, sugar, and, sometimes, eggs:

- Nonsweet breads and rolls, including rich dinner rolls and brioche. These have a high fat content but low enough sugar to be served as dinner breads. Brioche dough is especially rich, made with a high proportion of butter and eggs.
- Sweet rolls, including coffee cakes and many breakfast and tea rolls. These have high fat and sugar and, often, eggs. They are usually made with a sweet filling or topping.

ROLLED-IN YEAST DOUGH PRODUCTS

Rolled-in doughs, also called **laminated doughs**, are those in which a fat is incorporated into the dough in many layers by means of a rolling and folding procedure. The alternating layers of fat and dough give the baked product a flaky texture.

- Nonsweet rolled-in doughs: croissants.
- Sweet rolled-in doughs: Danish pastry.

MIXING METHODS

Mixing yeast doughs has three main purposes:

1. To combine all ingredients into a uniform, smooth dough.
2. To distribute the yeast evenly throughout the dough.
3. To develop gluten.

Three principal mixing methods are used for yeast doughs: the **straight dough method**, the **modified straight dough method**, and the **sponge method**.

STRAIGHT DOUGH METHOD

There is only one step in this method, as practiced by many bakers.

Some bakers dissolve the compressed yeast in some of the water before adding the remaining ingredients. Others omit this step. Active dry yeast, on the other hand, must be rehydrated before mixing.

The advantage of softening the yeast in water is that it helps ensure that the yeast is evenly distributed in the dough.

PROCEDURE: Straight Dough Mixing Method

If using active dry yeast or fresh yeast, rehydrate yeast in all or part of the water. If using instant dry yeast, mix the dry yeast with the flour. Combine all ingredients in the mixing bowl and mix.

MODIFIED STRAIGHT DOUGH METHOD FOR RICH DOUGHS

For rich sweet doughs, the method is modified to ensure even distribution of the fat and sugar.

PROCEDURE: Modified Straight Dough Method

1. If using fresh or active dry yeast, soften the yeast in part of the water. If using instant dry yeast, mix the dry yeast with the flour.
2. Combine the fat, sugar, salt, milk solids, and flavorings. Mix until well combined, but do not whip until light.
3. Add the eggs gradually, as fast as they are absorbed.
4. Add the liquid and mix briefly.
5. Add the flour and yeast. Mix into a smooth dough.

SPONGE METHOD

Sponge doughs are prepared in two stages.

PROCEDURE: Sponge Method

1. Combine the liquid (or part of the liquid), the yeast, and part of the flour (and, sometimes, part of the sugar). Mix into a thick batter or soft dough. Let ferment until double in bulk.
2. Punch down and add the rest of the flour and remaining ingredients. Mix to a uniform, smooth dough.

KEY POINTS TO REVIEW

- What are the steps in the straight dough method?
- What are the steps in the modified straight dough method?
- What are the steps in the sponge method?

STEPS IN YEAST DOUGH PRODUCTION

The production of yeast breads involves 12 basic steps. These steps are applied to yeast products in general, with variations depending on the particular product.

1. Scaling ingredients
2. Mixing
3. Bulk fermentation
4. Folding or punching
5. Dividing (scaling or portioning of dough)
6. Preshaping or rounding
7. Benching
8. Makeup and panning
9. Proofing
10. Baking
11. Cooling
12. Storing

As you can see, mixing of ingredients into a dough is only one part of a complex procedure.

SCALING INGREDIENTS

All ingredients must be weighed accurately. The only items that may be measured by volume are water, milk, and eggs, which may be scaled at 1 pint per pound (1 L per kg).

MIXING

Use the *dough arm* attachment when using a vertical mixer. Mix for the specified time.

The first two purposes of mixing—combining the ingredients into a dough and distributing the yeast—are accomplished during the first part of mixing. The remaining time is necessary to develop the gluten. Overmixed and undermixed doughs have poor volume and texture. (Review Gluten Development, p. 897.)

It is necessary for you to learn to tell by sight and feel when a dough is thoroughly mixed. This can be done only through experience and with the guidance of your instructor. A properly developed dough feels smooth and elastic. A lean dough should not be sticky.

Sometimes it is necessary to add a little more flour if the dough hasn't lost its stickiness after most of the mixing time has passed.

Rich doughs are generally undermixed slightly because greater tenderness is desired for these products.

Note: Mixing speeds and times given in bread formulas in this book are guidelines only. Small mixers might be damaged if they are run at too high a speed with a stiff dough. In such cases, use a lower speed and extend the mixing time as necessary. Depending on the mixer, developing a dough at first or slow speed requires about twice as much mixing time as at second speed. Follow the manufacturer's recommendations.

BULK FERMENTATION

Fermentation is the process by which yeast acts on the sugars and starches in the dough to produce carbon dioxide gas and alcohol.

Gluten becomes smoother and more elastic during fermentation. An underfermented dough does not develop proper volume, and the texture will be coarse. A dough that ferments too long or at too high a temperature becomes sticky, hard to work, and slightly sour.

An underfermented dough is called a **young dough**. An overfermented dough is called an **old dough**.

Doughs with weak gluten, such as rye doughs and rich doughs, are usually underfermented or "taken to the bench young."

PROCEDURE for Fermenting Yeast Doughs

1. Place the dough in a lightly oiled container and oil the surface to prevent a crust from forming. (This may not be necessary if humidity is high—about 75 percent.)

2. Cover the container lightly and let the dough rise at a temperature of about 80°F (27°C).

3. Fermentation is complete when the dough has doubled in volume. If fermentation is complete, a dent will remain after the hand is pressed into the top of the dough.

FOLDING OR PUNCHING

Punching is *not* hitting the dough with your fist. It is a method of deflating the dough that *expels carbon dioxide, redistributes the yeast* for further growth, *relaxes the gluten*, and *equalizes the temperature* throughout the dough. Professional bakers often prefer the term *folding*, because it more accurately describes the process.

Additional fermentation and folding may or may not take place, depending on the product. In general, doughs with shorter mixing times are folded more times, in order to continue to develop the gluten. On the other hand, if a dough has received a long mix and the gluten has been fully developed, the dough may not be folded at all but immediately divided or scaled.

PROCEDURE for Folding Yeast Doughs

Pull up the dough on one side, fold over the center, and press down. Repeat on the remaining three sides. Then turn the dough upside down in the bowl.

DIVIDING (SCALING OR PORTIONING OF DOUGH)

Using a baker's scale, divide the dough into pieces of uniform weight, according to the product being made.

During scaling, allowance is made for weight loss due to evaporation of moisture in the oven. This weight loss is 10 to 13 percent of the weight of the dough. Allow an extra $1^1/_2$–2 ounces of dough for each 1 pound of baked bread, or 50–65 grams per 500 grams baked bread.

PRESHAPING OR ROUNDING

After scaling, the pieces of dough are shaped into smooth, round balls. This procedure forms a kind of skin by stretching the gluten on the outside of the dough into a smooth layer. Rounding simplifies later shaping of the dough and also helps retain gases produced by the yeast.

Your instructor will demonstrate rounding techniques. Machines are also available that divide and round portions of dough automatically. Figure 31.1 illustrates a piece of dough being rounded by hand.

BENCHING

Rounded portions of dough are allowed to rest on the bench 10–15 minutes. This relaxes the gluten to make shaping the dough easier. Also, fermentation continues during this time.

MAKEUP AND PANNING

The dough is shaped into loaves or rolls and placed in pans or on baking sheets. For all loaves and rolls, the seam must be centered on the bottom to avoid splitting during baking.

Breads and rolls take a great many forms. A variety of shapes and techniques is presented in the next section.

PROOFING

Proofing is a continuation of the process of yeast fermentation, which increases the volume of the shaped dough. Bakers use different terms to distinguish between fermentation of the mixed dough and proofing of the made-up product before baking. Proofing temperatures are generally higher than fermentation temperatures.

If a proof box is not available, come as close to these conditions as you can by covering the products to retain moisture and setting them in a warm place.

Underproofing results in poor volume and dense texture. Overproofing results in coarse texture and some loss of flavor.

Rich doughs are slightly underproofed because their weaker gluten structure will not withstand much stretching.

FIGURE 31.1 To round a piece of dough, roll it on the bench with the palm of your hand. As you rotate the dough, use the edge of your hand to pinch the dough against the bench. This movement stretches the surface of the dough so it is completely smooth except for a seam at the bottom where it was pinched together.

PROCEDURE for Proofing Yeast Dough Products

Place the panned products in a proof box at 80°–85°F (27°–30°C) and 70 to 80 percent humidity, as indicated in the formula. Proof until double in bulk.

BAKING

As you recall from the previous chapter, many changes take place in a dough during baking. The most important changes are these:

1. **Oven spring**, which is the rapid rising in the oven due to production and expansion of trapped gases as a result of the oven heat. The yeast is very active at first but is killed when the temperature inside the dough reaches 140°F (60°C).

2. Coagulation of proteins and gelatinization of starches. In other words, the product becomes firm and holds its shape.

3. Formation and browning of the crust.

Load the ovens carefully, as proofed doughs are fragile until they become set by baking.

Oven temperatures must be adjusted for the product being baked. Rolls spaced apart are baked at a higher temperature than large loaves so they become browned in the short time it takes to bake them. In general, lean breads such as those popular in North America are baked at 400°–425°F (200°–220°C), while some French breads and rolls are baked at 425°–475°F (220°–245°C). Rich doughs and sweet doughs are baked at a lower temperature, 350°–400°F (175°–200°C), because their fat, sugar, and milk content makes the crust brown faster.

Hard-crusted breads are baked with steam injected into the oven during the first part of the baking period. This aids the formation of a thin, crisp crust.

Rye breads also benefit from baking with steam for the first 10 minutes.

A break on the side of the loaf is caused by continued rising after the crust is formed. To allow for this final expansion, hard-crusted breads are cut or scored before baking by making shallow slashes on the top of the loaf with a sharp knife or razor.

Small rolls bake completely without a break, so they are usually not scored.

Baking times vary considerably, depending on the product. A golden-brown crust color is the normal indication of doneness. Loaves that are done sound hollow when thumped.

COOLING

After baking, bread must be removed from pans and cooled rapidly on racks to allow the escape of excess moisture and alcohol created during fermentation.

Rolls baked apart from each other on sheets may be left on them because they will get adequate air circulation.

If soft crusts are desired, breads may be brushed with melted shortening before cooling.

Do not cool in a draft, or crusts may crack.

STORING

Breads to be served within 8 hours may be left on racks. For longer storage, wrap cooled breads in moistureproof bags to retard staling. Bread must be thoroughly cooled before wrapping, or moisture will collect inside the bags.

Wrapping and freezing maintains quality for longer periods. Refrigeration, however, increases staling.

Hard-crusted breads should not be wrapped (unless frozen), or the crusts will soften.

Because of the complexity of bread production, many things can go wrong. To remedy common bread faults, check Table 31.1 for possible causes and correct your procedures.

TABLE 31.1 Bread Faults and Their Causes

Fault	Causes
Shape	
Poor volume	Too much salt
	Too little yeast
	Weak flour
	Under- or overmixing
	Improper fermentation or proofing
	Oven too hot
Too much volume	Too little salt
	Too much yeast
	Too much dough scaled
	Overproofing
Poor shape	Too much liquid
	Improper molding or makeup
	Improper proofing
	Too much steam in oven

Fault	Causes
Shape (*continued*)	
Split or burst crust	Overmixing
	Underfermentation
	Improper molding—seam not on bottom
	Oven too hot
	Not enough steam in oven
Texture and crumb	
Too dense or close-grained	Too little yeast
	Underproofing
	Too much salt
	Too little liquid
Too coarse or open	Too much yeast
	Too much liquid
	Incorrect mixing time
	Improper fermentation
	Overproofing
	Pan too large
Streaked crumb	Improper mixing procedure
	Poor molding or makeup techniques
	Too much flour used for dusting
Poor texture or crumbly	Fermentation time too long or too short
	Overproofing
	Baking temperature too low
	Flour too weak
	Too little salt
Gray crumb	Fermentation time too long or temperature too high
Crust	
Too dark	Too much sugar or milk
	Underfermentation (young dough)
	Oven temperature too high
	Baking time too long
	Insufficient steam at beginning of baking
Too pale	Too little sugar or milk
	Overfermentation (old dough)
	Overproofing
	Oven temperature too low
	Baking time too short
	Too much steam in oven
Too thick	Too little sugar or fat
	Overfermentation (old dough)
	Baked too long and/or at too low a temperature
	Too little steam
Blisters on crust	Too much liquid
	Improper fermentation
	Improper shaping of loaves
Flavor	
Flat taste	Too little salt
Poor flavor	Inferior, spoiled, or rancid ingredients
	Poor bakeshop sanitation
	Under- or overfermentation

DOUGH FORMULAS AND TECHNIQUES

BREAD AND ROLL FORMULAS

The basic yeast dough mixing and baking methods discussed earlier in this chapter apply to the following formulas. Therefore, the methods are not repeated in detail for each formula. The basic procedures are indicated, and you should refer to the first part of this chapter if you need to refresh your memory for details.

Please note that *instant yeast* is used in all the formulas in this chapter.

Makeup techniques for loaves, rolls, and other items are described and illustrated after this recipe section.

Hard Rolls ♥

INGREDIENTS	U.S.			METRIC	PERCENTAGE	
Bread flour	1 lb	6	oz	625 g	100	%
Yeast, instant		0.25	oz	7 g	1.2	%
Water		13	oz	370 g	59	%
Salt		0.5	oz	14 g	2.25	%
Sugar		0.5	oz	14 g	2.25	%
Shortening		0.5	oz	14 g	2.25	%
Egg whites		0.5	oz	14 g	2.25	%
Total weight:	2 lb	5	oz	1058 g	169	%

PROCEDURE

Mixing:
Straight dough method.
4–5 minutes at 1st speed, 10–12 minutes at 2nd speed (see Note, p. 914).

Fermentation:
About 30 minutes at 7°F (25°C).

Scaling and makeup:
Rolls—1 lb (500 g) per dozen.
French-type loaf—18 oz (550 g) per loaf.
See makeup technique after recipe section.
Dock after proofing. Brush with water.

Baking:
425°F (220°C). Steam for first 10 minutes.

Per 1 roll: Calories, 90; Protein, 3 g; Fat, 1 g (10% cal.); Cholesterol, 0 mg; Carbohydrates, 17 g; Fiber, 1 g; Sodium, 200 mg.

Soft Rolls ♥

INGREDIENTS	U.S.			METRIC	PERCENTAGE	
Bread flour	1 lb	5	oz	625 g	100	%
Yeast, instant		0.25	oz	7 g	1.2	%
Water		12.5	oz	375 g	60	%
Salt		0.4	oz	12 g	2	%
Sugar		2	oz	60 g	9.5	%
Nonfat milk powder		1	oz	30 g	4.75	%
Shortening		1	oz	30 g	4.75	%
Butter or margarine		1	oz	30 g	4.75	%
Total weight:	2 lb	7	oz	1165 g	189	%

PROCEDURE

Mixing:
Straight dough method.
4–5 minutes at 1st speed, 10–12 minutes at 2nd speed (see Note, p. 914).

Fermentation:
30 minutes at 77°F (25°C).

Scaling and makeup:
16–20 oz (450–600 g) per dozen rolls.
See makeup techniques after recipe section.

Baking:
400°F (200°C).

Per 1 roll: Calories, 120; Protein, 3 g; Fat, 2.5 g (20% cal.); Cholesterol, 0 mg; Carbohydrates, 20 g; Fiber, 0.5 g; Sodium, 190 mg.

Soft Rolls

French Bread

INGREDIENTS	U.S.		METRIC	PERCENTAGE	
Sponge					
Bread flour	8	oz	250 g	33	%
Water	8	oz	250 g	33	%
Yeast, instant	0.17 oz		5 g	0.7	%
Malt syrup	0.25 oz		8 g	1	%
Dough					
Bread flour	1 lb		500 g	67	%
Water	8	oz	250 g	33	%
Salt	0.44 oz (2 1/4 tsp)		13 g	1.75	%
Total weight:	2 lb 8	oz	1276 g	169	%

Per 1 ounce (28.35 g): Calories, 60; Protein, 2 g; Fat, 0.5 g (7% cal.); Cholesterol, 0 mg; Carbohydrates, 13 g; Fiber, 0 g; Sodium, 140 mg.

PROCEDURE

Mixing:

Sponge method. Ferment sponge for 4 hours at 75°F (24°C) or overnight at 65°F (18°C).

Mix dough 4–5 minutes at 1st speed; 5 minutes more at 2nd speed (see Note, p. 914).

Dough fermentation:

1 1/2 to 2 hours at 77°F (25°C). Fold once or twice.

Scaling and makeup:

French loaves—12 oz (350 g).

Round loaves—18 oz (550 g).

Rolls—16–20 oz (450–600 g) per dozen.

See makeup techniques after recipe section.

Baking:

400°F (200°C). Steam for first 10 minutes.

French breads

White Pan Bread

INGREDIENTS	U.S.		METRIC	PERCENTAGE	
Bread flour	1 lb 4	oz	500 g	100	%
Yeast, instant	0.25 oz		6 g	1.3	%
Water	12	oz	300 g	60	%
Salt	0.5 oz		12 g	2.5	%
Sugar	0.75 oz		18 g	3.75	%
Nonfat milk powder	1	oz	25 g	5	%
Shortening	0.75 oz		18 g	3.75	%
Total weight:	2 lb 3	oz	879 g	176	%

Per 1 ounce (28.35 g): Calories, 70; Protein, 2 g; Fat, 1 g (13% cal.); Cholesterol, 0 mg; Carbohydrates, 13 g; Fiber, 0 g; Sodium, 160 mg.

PROCEDURE

Mixing:

Straight dough method.

4–5 minutes at 1st speed, 10–12 minutes at 2nd speed (see Note, p. 914).

Fermentation:

30 minutes at 77°F (25°C).

Makeup:

Pan loaves. See makeup techniques after recipe section.

Baking:

400°F (200°C).

VARIATION

Whole Wheat Bread

Prepare basic White Pan Bread using

Bread flour	1 lb	500 g	40%
Whole wheat flour	1 lb 8 oz	750 g	60%

Whole Wheat Bread

Rye Bread and Rolls ♥

INGREDIENTS	U.S.		METRIC	PERCENTAGE
Rye flour	8	oz	250 g	40 %
Bread flour	12	oz	375 g	60 %
Yeast, instant	0.25 oz		7 g	1.25 %
Water	12	oz	375 g	60 %
Salt	0.4	oz (2 tsp)	12 g	2.5 %
Shortening	0.5	oz	15 g	2.5 %
Molasses	0.5	oz	15 g	2.5 %
Caraway seeds	0.25 oz		8 g	1.25 %
Total weight:	4 lb 4	oz	1065 g	170 %

PROCEDURE

Mixing:

Straight dough method.

4–5 minutes at 1st speed; 3 minutes more at 2nd speed (see Note, p. 914).

Fermentation:

1¹/₂ to 2 hours at 77°F (25°C). Fold once or twice.

Scaling and makeup:

1 lb (500 g) per pan loaf.

1 lb (500 g) per dozen rolls.

See makeup techniques after recipe section.

Baking:

400°F (200°C). Steam for first 10 minutes.

Per 1 ounce (28.35 g): Calories, 60; Protein, 2 g; Fat, 1 g (14% cal.); Cholesterol, 0 mg; Carbohydrates, 12 g; Fiber, 2 g; Sodium, 160 mg.

Focaccia ♥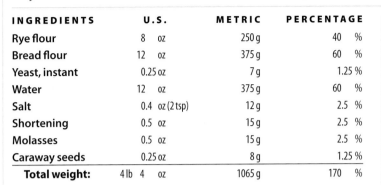

INGREDIENTS	U.S.		METRIC	PERCENTAGE
Bread flour	1 lb 8	oz	750 g	100 %
Yeast, instant	0.12 oz		4 g	0.5 %
Water	15	oz	470 g	62.5 %
Salt	0.5	oz	15 g	2 %
Olive oil	1	oz	25 g	3.5 %
Total weight:	2 lb 8	oz	1264 g	168 %

PROCEDURE

Mixing:

Straight dough method.

8 minutes at 1st speed (see Note, p. 914).

Fermentation:

1¹/₂ hours at 77°F (25°C).

Makeup:

Oil sheet pans with olive oil. Roll out dough to about ³/₄ in. (2 cm) thick and place in pans (see **Figure 31.2**). Brush tops generously with olive oil. After proofing, press dimples into the dough at approximately 3-in. (8-cm) intervals.

Baking:

425°F (220°C). Steam for first 10 minutes.

Per 1 ounce (28.35 g): Calories, 70; Protein, 2 g; Fat, 1 g (13% cal.); Cholesterol, 0 mg; Carbohydrates, 12 g; Fiber, 1 g; Sodium, 115 mg.

VARIATIONS

Rosemary Focaccia

After pressing in the dimples, sprinkle the top with rosemary (preferably fresh) and coarse salt.

Olive Focaccia

Mix 30% (12 oz/375 g) chopped oil-cured black olives into the dough.

FIGURE 31.2 Focaccia.

(a) Roll and stretch the dough into a rectangle large enough to fill the pan.

(b) Place the dough in the pan. Top with olive oil.

(c) With the fingertips, poke holes heavily at regular intervals.

(d) Top with the desired topping, such as fresh herbs and coarse salt.

Brioche

INGREDIENTS	U.S.		METRIC	PERCENTAGE
Milk	2	oz	60 g	20 %
Bread flour	2	oz	60 g	20 %
Yeast, instant osmotolerant	0.2		6 g	2 %
Eggs	5	oz	150 g	50 %
Bread flour	8	oz	240 g	80 %
Sugar	0.5	oz	15 g	5 %
Salt	0.2 oz (1 tsp)		6 g	2 %
Butter, softened	6	oz	180 g	60 %
Total weight:	1 lb 8	oz	717 g	239 %

PROCEDURE

Mixing:

Sponge method:

1. Scald milk and cool to lukewarm. Mix the flour with the yeast and add to the milk. Mix to make a sponge. Let rise until double.

2. Gradually mix in eggs and then dry ingredients (using the paddle attachment) to make a soft dough.

3. Beat in butter, a little at a time, until completely absorbed and dough is smooth. Dough will be very soft and sticky.

Fermentation:

Cover with plastic film and place in retarder overnight.

Makeup:

1¹/₂ oz (50 g) per roll.

See makeup techniques after recipe section. Dough is very soft and is easiest to make up when chilled.

Egg-wash after proofing.

Baking:

400°F (200°C) for small rolls. 375°F (190°C) for larger units.

Per 1 roll: Calories, 160; Protein, 4 g; Fat, 10 g (56% cal.); Cholesterol, 60 mg; Carbohydrates, 14 g; Fiber, 0 g; Sodium, 190 mg.

VARIATION

To make the dough less sticky and less difficult to handle, reduce the butter to 35–50% (14–20 oz/450–625 g). This adjustment also reduces cost. However, the brioche will not be as rich and delicate.

Sweet Roll Dough

INGREDIENTS	U.S.		METRIC	PERCENTAGE
Butter, margarine, or shortening (see note)	4	oz	100 g	20 %
Sugar	4	oz	100 g	20 %
Salt	0.4 oz		10 g	2 %
Nonfat milk solids	1	oz	25 g	5 %
Eggs	3	oz	75 g	15 %
Bread flour	1 lb		400 g	80 %
Cake flour	4	oz	100 g	20 %
Yeast, instant osmotolerant	0.4 oz		10 g	2 %
Water	8	oz	200 g	40 %
Total weight:	2 lb 8	oz	1020 g	204 %

PROCEDURE

Mixing:

Modified straight dough method:

Mix 4 minutes at 1st speed, then 2–3 minutes at second speed.

Fermentation:

45 minutes at 75°F (25°C), then refrigerate.

Makeup:

See makeup techniques after recipe section.

Baking:

375°F (190°C).

Per 1 ounce (28.35 g): Calories, 90; Protein, 2 g; Fat, 3 g (31% cal.): Cholesterol, 15 mg; Carbohydrates, 13 g; Fiber, 0.5 g; Sodium, 90 mg.

Note: Any of the fats listed may be used alone or in combination.

VARIATION

Raised Doughnuts

Prepare basic sweet roll dough, but reduce the fat and sugar by half. Mace, nutmeg, or other spices may be added.

Scaling: 1¹/₂ oz (50 g) each.

Give full proof.

Frying: 360°F (182°C).

Drain. Roll in cinnamon sugar or 6X sugar when cool.

ROLLED-IN DOUGHS: DANISH PASTRY AND CROISSANTS

VIENNOISERIE

Viennoiserie (vee en wahz ree), or Viennese pastry, is the general term given to sweet yeast-raised dough goods, both laminated and nonlaminated. Brioche, Danish, and croissants are classic examples of viennoiserie.

Rolled-in or laminated doughs contain many layers of fat sandwiched between layers of dough. These layers create the flakiness you are familiar with in Danish pastry.

Two basic kinds of rolled-in yeast doughs are made in the bakeshop:

- Danish pastry
- Croissants

Rolled-in doughs are mixed only slightly because the rolling-in procedure continues to develop the gluten.

Butter is the preferred fat for flavor and the melt-in-the-mouth quality of rolled-in doughs. Specially formulated shortenings are available when lower cost and greater ease of handling are more important considerations.

ROLLING-IN Procedure for Danish and Croissant Dough

The rolling-in procedure has two parts.

1. Enclosing the fat in the dough.

In the method illustrated in Figure 31.3, the fat is spotted on two-thirds of the dough and the dough is folded in thirds like a business letter. This results in five layers: three layers of dough and two layers of fat.

2. Rolling out and folding the dough to increase the number of layers.

In these doughs, we use a simple fold, or three-fold, which means we fold the dough in thirds. Each complete rolling and folding step is called a *turn*. We give the dough three turns, creating over 100 layers of dough and fat.

FIGURE 31.3 Rolling-in procedure for Danish and croissant dough.

(a, b) Roll the dough into a rectangle about 3 times as long as it is wide and ¹/₂ to ³/₄ in. (1 to 2 cm) thick. Smear the butter over two-thirds of the length of the dough, leaving a margin at the edges.

(c) Fold the unbuttered third over the center third.

(d) Fold the remaining third on top. Rest the dough in the retarder (under refrigeration) 20 to 30 minutes to allow the gluten to relax.

(e) Place the dough on the bench at right angles to its position in step (d). Take this step before each rolling-out of the dough so the gluten is stretched in all directions, not just lengthwise. Roll the dough into a rectangle.

(f) Fold again into thirds by first folding the top third over the center. Be sure to brush off excess dusting flour from between the folds.

(g) Fold over the remaining third. You have now completed the first turn or fold. Incorporating the butter doesn't count as a turn. Press one finger in the dough near the end to make one indentation. This indicates "1 turn" to anyone who may have to take up where you left off, or to you if you have several batches going. Refrigerate the dough 20–30 minutes to relax the gluten. Repeat the above rolling and folding procedures for a second and third turn, resting the dough between turns. Mark the number of turns in the dough with two or three fingers. After the third turn, rest the dough in the retarder several hours or overnight. Cover it with plastic film to prevent crusting. The dough is then ready for makeup.

In Chapter 35, you will learn an even more complex rolling-in procedure used for puff pastry, which is leavened only by steam, not by yeast. This procedure produces over 1,000 layers!

KEY POINTS TO REVIEW

- What are the 12 steps in yeast dough production?

- What are the steps in the procedure for fermenting yeast doughs?

- What are the steps in the rolling-in procedure for Danish and croissant doughs?

 # Danish Pastry

INGREDIENTS	U.S.		METRIC	PERCENTAGE	
Milk	1 lb		400 g	40	%
Butter	5	oz	125 g	12.5	%
Sugar	6	oz	150 g	15	%
Salt	0.8 oz		20 g	2	%
Cardamom	1	tsp	2 g (5 mL)	0.2	%
Eggs	8	oz	200 g	20	%
Egg yolks	2	oz	50 g	5	%
Bread flour	2 lb		800 g	80	%
Cake flour	8	oz	200 g	20	%
Yeast, instant osmotolerant	0.8 oz		20 g	2	%
Butter	1 lb 4	oz	500 g	50	%
Total weight:	6 lb 4	oz	2467 g	246	%

PROCEDURE

Mixing:

Modified straight dough method:
1. Scald milk. Cool to lukewarm.
2. Mix butter, sugar, salt, and spice until smooth, using paddle. Beat in eggs and yolks.
3. Mix flours with dry yeast. Add liquid (from step 1) and the flour to the bowl with the butter/sugar/egg mixture. With dough arm, mix 3–4 minutes on 2nd speed.
4. Rest in retarder 20–30 minutes.
5. Roll in remaining butter and give 3 three-folds, as shown in **Figure 31.3**.

Makeup:

See makeup techniques after recipe section.

Proofing:

90°F (32°C) with little steam. Egg-wash after proofing.

Baking:

375°F (190°C).

Per 1 ounce (28.35 g): Calories, 110; Protein, 2 g; Fat, 6 g (53% cal.); Cholesterol, 35 mg; Carbohydrates, 10 g; Fiber, 0 g; Sodium, 120 mg.

Croissants

INGREDIENTS	U.S.		METRIC	PERCENTAGE	
Milk	1 lb		450 g	57	%
Sugar	1	oz	30 g	4	%
Salt	0.5 oz		15 g	2	%
Butter, soft	3	oz	80 g	10	%
Bread flour	1 lb 12	oz	800 g	100	%
Yeast, instant	0.4 oz		11 g	1.4	%
Butter	1 lb		450 g	57	%
Total weight:	4 lb 1	oz	1836 g	231	%

PROCEDURE

Mixing:

Straight dough method.

Scald milk, cool to lukewarm. Add remaining ingredients except last 1 lb (450 g) butter. Mix into a smooth dough. Do not overmix.

Fermentation:

1 hour at 75°F (24°C).

Punch down, spread out on flat pan, and rest in retarder 30 minutes.

Roll in last amount of butter and give 3 three-folds (see **Figure 31.3**). Rest in retarder overnight.

Makeup:

See **Figure 31.19** on page 931.

Proofing:

80°F (27°C). Egg-wash after proofing.

Baking:

400°F (200°C).

Per 1 ounce (28.35 g): Calories, 110; Protein, 2 g; Fat, 7 g (57% cal.); Cholesterol, 20 mg; Carbohydrates, 10 g; Fiber, 0 g; Sodium, 160 mg.

Croissants

FILLINGS AND TOPPINGS FOR SWEET DOUGH PRODUCTS AND DANISH

Cinnamon Sugar ♥

YIELD: ABOUT 8 OZ (250 G)

U.S.	METRIC	INGREDIENTS	PROCEDURE
8 oz	250 g	Sugar	Stir together thoroughly.
0.5 oz	15 g	Cinnamon	

Per 1 ounce (28.35 g): Calories, 110; Protein, 0 g; Fat, 0 g (0% cal.); Cholesterol, 0 mg; Carbohydrates, 29 g; Fiber, 0 g; Sodium, 0 mg.

Streusel or Crumb Topping

YIELD: 2 LB (1 KG)

U.S.	METRIC	INGREDIENTS	PROCEDURE
8 oz	250 g	Butter and/or shortening	Rub all ingredients together until crumbly.
5 oz	150 g	Granulated sugar	
4 oz	120 g	Brown sugar	
1/2–1 tsp	2–5 mL	Cinnamon or mace	
1/2 tsp	2 mL	Salt	
1 lb	500 g	Pastry flour	

Per 1 ounce (28.35 g): Calories, 130; Protein, 1 g; Fat, 6 g (40% cal.); Cholesterol, 15 mg; Carbohydrates, 19 g; Fiber, 2 g; Sodium, 85 mg.

VARIATION

Nut Streusel

Add 4 oz (125 g) finely chopped nuts to basic mixture.

Clear Glaze for Coffee Cakes and Danish ♥

YIELD: 2 LB (1 KG)

U.S.	METRIC	INGREDIENTS	PROCEDURE
1 cup	250 mL	Water	1. Mix together and bring to a boil. Stir to ensure the sugar is completely dissolved.
1 lb	500 g	Light corn syrup	2. Brush on while hot.
8 oz	250 g	Granulated sugar	

Per 1 ounce (28.35 g): Calories, 70; Protein, 0 g; Fat, 0 g (0% cal.); Cholesterol, 0 mg; Carbohydrates, 18 g; Fiber, 0 g; Sodium, 15 mg.

Date, Prune, or Apricot Filling

YIELD: 1 LB 8 OZ (750 G)

U.S.	METRIC	INGREDIENTS	PROCEDURE
1 lb	500 g	Dates, prunes (pitted), or dried apricots	1. Chop dried fruit very fine, or pass through a grinder.
3 oz	100 g	Sugar	2. Combine all ingredients in a saucepan. Bring to a boil. Simmer and stir until thick and smooth, about 10 minutes.
8 fl oz	250 mL	Water	3. Cool before using.

Per 1 ounce (28.35 g): Calories, 65; Protein, 0 g; Fat, 0 g (0% cal.); Cholesterol, 0 mg; Carbohydrates, 17 g; Fiber, 1 g; Sodium, 0 mg.

Almond Filling

YIELD: 1 LB 8 OZ (750 G)

U.S.	METRIC	INGREDIENTS	PROCEDURE
8 oz	250 g	Almond paste	1. With paddle attachment, mix almond paste and sugar at low speed until evenly mixed.
8 oz	250 g	Sugar	2. Mix in fat and flour until smooth.
4 oz	125 g	Butter and/or shortening	3. Beat in eggs, a little at a time, until smooth.
2 oz	62 g	Pastry or cake flour	
2 oz	62 g	Eggs	

Per 1 ounce (28.35 g): Calories, 130; Protein, 1 g; Fat, 7 g (48% cal.); Cholesterol, 20 mg; Carbohydrates, 16 g; Fiber, 1 g; Sodium, 45 mg.

Cheese Filling

YIELD: APPROXIMATELY 2 LB 3 OZ (1125 G)

U.S.	METRIC	INGREDIENTS	PROCEDURE
1 lb	500 g	Baker's cheese	1. Using the paddle attachment, cream the cheese, sugar, and salt until smooth.
5 oz	150 g	Sugar	
0.12 oz (5/8 tsp)	3 mL	Salt	
3 oz	100 g	Eggs	2. Add the eggs, butter, and vanilla. Blend in.
3 oz	100 g	Butter and/or shortening, soft	
1 1/2 tsp	7 mL	Vanilla	
1.5 oz	50 g	Cake flour	3. Add the cake flour. Blend until just absorbed.
3–5 oz	100–150 g	Milk	4. Add the milk, a little at a time, adding just enough to bring the mixture to a smooth, spreadable consistency.
4 oz	125 g	Raisins (optional)	5. Stir in the raisins, if desired.

Per 1 ounce (28.35 g): Calories, 60; Protein, 3 g; Fat, 2.5 g (38% cal.); Cholesterol, 20 mg; Carbohydrates, 6 g; Fiber, 0 g; Sodium, 70 mg.

MAKEUP TECHNIQUES

FIGURE 31.4 Rounding small rolls.

(a) Holding the palm of the hand fairly flat, roll the dough in a tight circle on the workbench. Do not use too much flour for dusting, as the dough must stick to the bench a little for the technique to work.

(b) As the ball of dough takes on a round shape, gradually cup your hand.

(c) The finished ball of dough should have a smooth surface, except for a slight pucker on the bottom.

The object of yeast dough makeup techniques is to shape the dough into rolls or loaves that bake properly and have an attractive appearance. When you shape a roll or loaf correctly, you stretch the gluten strands on the surface into a kind of smooth skin. This tight gluten surface holds the item in shape. This is especially important for loaves and rolls that are baked free-standing, not in pans.

Units that are not made up correctly develop irregular shapes and splits and may flatten out on the pan.

Following are a few of the many makeup techniques for yeast doughs.

HARD ROLLS AND BREADS

Round Rolls

1. Scale the dough as indicated in the recipes, usually 1 pound (450 g) per dozen.
2. Round each unit as shown in Figure 31.4.
3. Place rolls 2 inches (5 cm) apart on sheet pans sprinkled with cornmeal.

Round Loaves

1. Flatten the rounded, benched dough into a circle. Fold the four sides over the center, then round again.
2. Place on sheet pans sprinkled with cornmeal.

Club Rolls

1. Make up as shown in Figure 31.5.
2. Place 2 inches (5 cm) apart on sheet pans sprinkled with cornmeal.

Crescent Rolls

1. Scale dough into 20-ounce (600-g) units.
2. After rounding and benching, flatten the dough and roll it out into a circle 12 inches (30 cm) across.
3. With a pastry wheel, cut the dough circle into 12 equal wedges or triangles. (Alternative method: For large quantities of dough, roll it out into a rectangle and cut like croissant dough. See Figure 31.19 on p. 931.)
4. Roll the triangles into crescents using the same technique as for croissants (see Figure 31.19 on p. 931).
   Note: If using soft roll dough, brush the dough with butter before cutting it into triangles. Do not use any fat.

FIGURE 31.5 Making club rolls.

(a) Flatten the piece of dough roughly into a rectangle.

(b) Begin to roll the dough by folding over the back edge of the rectangle. Press the seam firmly with your fingertips.

(c) Continue to roll the dough, always pressing the seam firmly after each turn. As you roll the dough, the front edge will appear to shrink. Stretch the front corners as shown by the arrows to keep the width uniform.

(d) When the roll is finished, seal the seam well so you have a tight roll.

(e) Dock the proofed roll with a single slash to give the baked roll this appearance.

French-Type Loaves

1. Scale the dough into units weighing 12 to 18 ounces (350 to 500 g).

2. Make up as shown in Figure 31.6.

FIGURE 31.6 Making French-type loaves.

(a) Flatten the rounded, relaxed dough with your hands or with a rolling pin.

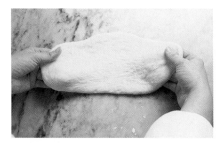

(b) Stretch the oval with the hands to lengthen it.

(c) Roll the dough tightly.

(d) Seal the seam well. If a longer, thinner loaf is required, relax the rolled units again for a few minutes. Flatten them with your palms and stretch the dough lightly to increase its length. Once again, roll tightly and seal the seam. Roll the loaf on the bench under your palms to make it even and to stretch it to the desired shape and length.

(e) Place the dough seam-side down on pans dusted with cornmeal. Proofing the loaves on special trough-shaped pans maintains their shape. Proof. Wash with water. Slash with diagonal cuts or with one lengthwise cut; this can be done before or after proofing.

SOFT ROLL DOUGHS

Tied or Knotted Rolls

1. Scale dough to 16–20 ounces (450–600 g) per dozen.

2. With the palm of the hand, roll each unit on the workbench into a strip or rope of dough.

3. Tie rolls as shown:

   Single-knot rolls: Figure 31.7

   Double-knot rolls: Figure 31.8

   Braided rolls: Figure 31.9

   Figure-eight rolls: Figure 31.10

4. Place 2 inches (5 cm) apart on greased baking sheets.

5. Egg-wash after proofing.

Pan Rolls

1. Scale dough to 16–20 ounces (450–600 g) per dozen.

2. Make up as for round hard rolls.

3. Place on greased pans ½ inch (1 cm) apart.

FIGURE 31.7 Tying a single-knot roll

FIGURE 31.8 Tying a double-knot roll

FIGURE 31.9 Tying a braided roll

FIGURE 31.10 Tying a figure-eight roll

FIGURE 31.11 Parker House rolls.

(a) Round the scaled piece of dough.

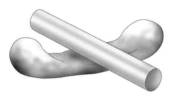

(b) Flatten the center of the dough with a thin rolling pin as shown.

(c) Fold the dough over and press down on the folded edge to make a crease.

(d) The baked roll has this shape.

Parker House Rolls

1. Scale dough to 16–20 ounces (450–600 g) per dozen.

2. Make up as shown in Figure 31.11.

3. Place on greased baking sheet ¹/₂ inch (1 cm) apart.

Cloverleaf Rolls

1. Scale dough to 16–20 ounces (450–600 g) per dozen.

2. Make up and pan as shown in Figure 31.12.

Butterflake Rolls

Make up as shown in Figure 31.13.

PAN LOAVES

Shaping dough into loaves to be baked in loaf pans is illustrated in Figure 31.14.

FIGURE 31.12 Cloverleaf rolls.

(a) Divide each piece of dough into 3 equal parts. Shape into balls. Place 3 balls in the bottom of each greased muffin tin.

(b) The baked roll has this appearance.

FIGURE 31.13 Butterflake rolls.

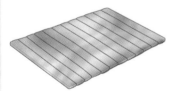

(a) Roll the dough into a thin rectangle. Brush with melted butter. Cut into strips 1 in. (2.5 cm) wide.

(b) Stack 6 strips. Cut into pieces 1¹/₂ in. (3.5 cm) long.

(c) Place the pieces on end in greased muffin tins. Proof.

(d) The baked rolls have this appearance.

FIGURE 31.14 Pan loaves.

(a) Start with the rounded, benched dough. Flatten it with the palms of the hands.

(b) Stretch it into a long rectangle.

(c, d) Fold into thirds.

(e) Roll the dough into a tight roll of the same length as the pan it is to be baked in. Seal the seam well and place the dough seam side down in the greased pan.

BRIOCHE

Brioche dough may be made into many shapes. The traditional shape is shown in Figure 31.15.

FIGURE 31.15 Making brioche.

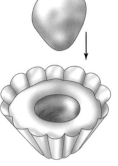

(a) For a small brioche, roll the dough into a round piece.

(b) Using the edge of your hand, pinch off about one-fourth of the dough without detaching it. Roll the dough on the bench so that both parts are round.

(c) Place the dough in the tin, large end first. With your fingertips, press the small ball into the larger one as shown.

(d) For large brioche, separate the two parts of the dough. Place the large ball in the tin and make a hole in the center. Form the smaller ball into a pear shape and fit it into the hole.

(e) A baked large brioche.

SWEET DOUGH PRODUCTS

Note: Many sweet dough products may be glazed with Clear Glaze (p. 924) and/or iced with Flat Icing (p. 959) after baking. Flat icing is drizzled over the cooled products without covering them completely.

Cinnamon Rolls

1. Scale dough into 20-ounce (600-g) units. On a floured board, roll each piece of dough into a rectangle measuring 9 × 12 inches and about $^1/_4$ inch thick (23 × 30 × 0.5 cm).

2. Brush with butter and sprinkle with 2 ounces (60 g) cinnamon sugar.

3. Roll up like a jelly roll 12 inches (30 cm) long, as shown in the illustration.

4. Cut into 1-inch (2.5-cm) rolls.

5. Place cut side down in greased muffin tins or on greased sheet pans. One full-size pan, 18 × 26 inches (46 × 66 cm), holds 48 rolls placed 6 by 8.

For variations on the basic cinnamon roll shape, see Figure 31.16.

Cinnamon Raisin Rolls

Prepare like cinnamon rolls, but add 2 ounces (60 g) raisins to the filling.

FIGURE 31.16 The filled dough roll is the starting point for a variety of sweet dough.

(a) Roll the dough into a rectangle. Brush with butter and sprinkle with cinnamon sugar, or spread with desired filling.

(b) Roll up like a jelly roll.

(c) For cinnamon rolls and similar products, cut off pieces 1 in. (2.5 cm) in length.

(d) For combs or bear claws, make the roll thinner and cut it into longer pieces. Flatten slightly and cut partway through each piece in 3 to 6 places as shown. Leave straight or bend into a curve to open the cuts.

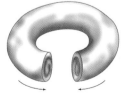

(e) For figure-eight cinnamon rolls, cut the rolls almost through as shown. Open them and lay them flat on the baking sheet.

(f) To make a wreath-shaped coffee cake, join the ends of the dough roll to make a circle.

(g) Cut partway through the dough at 1-in. (2.5-cm) intervals as shown.

(h) Twist each segment outward to open the cuts.

Caramel Rolls

1. Prepare like cinnamon rolls.

2. Before panning, spread the bottoms of the pans or muffin tins with the following mixture. Use about 1 ounce (30 g) of the mixture per roll.

> 2 lb (1 kg) brown sugar
>
> 8 oz (250 g) corn syrup
>
> 10 oz (300 g) butter
>
> 4 oz (125 mL) water

Cream the sugar, corn syrup, and butter. Beat in the water.

Quantities given are enough for 1 sheet pan of 48 rolls.

Caramel Nut Rolls or Pecan Rolls

Prepare like caramel rolls, but sprinkle the sugar-butter mixture in the pans with chopped nuts or pecan halves before placing the rolls in the pans.

Wreath Coffee Cake

1. Make a filled dough roll as for cinnamon rolls, but do not cut it into separate pieces. Other fillings, such as prune or date, may be used instead of butter and cinnamon sugar.

2. Shape the roll into a circle as shown in Figure 31.16 (f–h). Place on a greased baking sheet. Cut and shape as shown in the illustration.

3. Egg-wash after proofing.

Filled Coffee Cake

1. Scale dough into 12-ounce (350-g) units.

2. Roll each unit into a rectangle measuring 9 × 18 inches (23 × 46 cm).

3. Spread half of each rectangle with desired filling, using about 6 ounces (175 g) filling.

4. Fold the unspread half over the spread half to make a 9-inch (23-cm) square.

5. Place in greased 9-inch (23-cm) square pan.

6. Sprinkle with Streusel Topping (p. 924), about 4 ounces (125 g) per pan.

7. Proof and bake.

ROLLED-IN DOUGH PRODUCTS

Danish Rolls and Coffee Cakes

Most of the techniques given in the previous section for sweet dough products may be used for Danish pastry.

Two additional methods are illustrated in Figures 31.17 and 31.18.

Baked Danish dough products are frequently glazed with Clear Glaze (p. 924) and/or iced with Flat Icing (p. 959).

FIGURE 31.17 Spiral Danish rolls.

(a) Roll the dough into a rectangle 16 in. (40 cm) wide and less than $1/4$ in. (0.5 cm) thick. (The length of the rectangle depends on the quantity of dough.) Brush the dough with melted butter. Sprinkle half of it with cinnamon sugar as shown.

(b) Fold the unsugared half over the sugared half. You now have a rectangle 8 in. (20 cm) wide. Roll the dough gently with a rolling pin to press the layers together.

(c) Cut the dough into strips $1/2$ in. (1 cm) wide.

(d) Place a strip crosswise in front of you on the bench.

(e) With the palms of your hands on the ends of the strip, roll one end toward you and the other away from you so the strip twists. Stretch the strip slightly as you twist it.

(f) Curl the strip into a spiral shape on the baking sheet. Tuck the end underneath and pinch it against the roll to seal it in place. If desired, press a hollow in the center of the roll and place a spoonful of filling (such as a fruit filling) in the center.

FIGURE 31.18 Danish pockets.

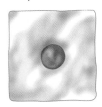

(a) Roll the dough to less than ¼ in. (0.5 cm) thick and cut into 5-in. (13-cm) squares. Place desired filling on the center of each square. Brush the corners lightly with water—this helps them seal when pressed together.

(b) Fold two opposite corners over the center. Press down firmly to seal them. (If desired, rolls may be left in this shape.)

(c) Fold the other two corners over the center and again press them firmly together.

Croissants

The method for making up croissants is illustrated in Figure 31.19.

FIGURE 31.19 Making croissants.

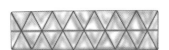

(a) Roll the dough into a rectangle 10 in. (26 cm) wide and about 1/8 in. (3 mm) thick. (The length depends on the amount of dough used.)

(b) Cut into triangles as shown. Special roller cutters are available that do this quickly.

(c) Place a triangle on the bench in front of you. Stretch the back corners outward slightly, as shown by the arrows.

(d) Begin to roll the dough toward the point.

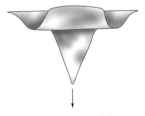

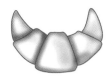

(e) Stretch the point of the triangle slightly as you roll it.

(f) Finish rolling the dough.

(g) Bend the roll into a crescent shape. The point of the triangle must be toward the inside of the crescent and tucked under the roll so it won't pop up during baking.

TERMS FOR REVIEW

lean dough	**straight dough method**	**sponge method**	**punching**
rolled-in or laminated dough	**modified straight dough method**	**fermentation**	**proofing**
		young and old doughs	**oven spring**

QUESTIONS FOR DISCUSSION

1. What are the three major purposes of mixing yeast doughs?
2. Explain the difference in procedure between the straight dough method and the sponge method. How is the straight dough method sometimes modified for sweet doughs, and why is this necessary?
3. What are the 12 steps in the production of yeast products? Explain each briefly.

4. Judging from what you know about fermentation of doughs, do you think it might be necessary for bakers to modify procedures from winter to summer? How?
5. As you know, butter is very hard when cold and melts easily at warm temperatures. What precautions do you think are necessary when using butter as the rolling-in fat for Danish pastry?

QUICK BREADS

Quick breads are the perfect solution for operations that want to offer their patrons fresh, homemade bread products but can't justify the labor cost of making yeast breads. Also, quick breads have the advantage of being easily made in almost unlimited varieties, using such ingredients as whole wheat flour, rye flour, cornmeal, bran, oatmeal, and many kinds of fruits, nuts, and spices.

As their name implies, quick breads are quick to make. Because they are leavened by chemical leaveners and steam, not by yeast, no fermentation time is necessary. And because they are usually tender products with little gluten development, mixing them takes just a few minutes.

Although prepared biscuit and muffin mixes are available, the only extra work required to make these products from scratch is the time to scale a few extra ingredients. With a careful and imaginative selection of ingredients and an understanding of basic mixing methods, you can create superior products.

You may already have studied two kinds of quick bread in the breakfast chapter: pancakes and waffles. In this chapter, we present two basic mixing methods and apply them to biscuits, muffins, quick loaf breads and coffee cakes, and corn breads. In addition, we discuss popovers, which are leavened by steam only.

AFTER READING THIS CHAPTER, YOU SHOULD BE ABLE TO

1. Prepare baking powder biscuits and variations.

2. Prepare muffins, loaf breads, coffee cakes, and corn breads.

3. Prepare popovers.

MIXING AND PRODUCTION METHODS

TYPES OF DOUGH

Dough mixtures for quick breads are generally of two types:

1. *Soft doughs* are used for biscuits. These products are rolled out and cut into desired shapes. They are mixed by the biscuit method.

2. *Batters* may be either **pour batters**, which are liquid enough to pour, or **drop batters**, which are thicker and drop from a spoon in lumps.

Most quick-bread batters are mixed by the **muffin method**, except for drop biscuits, which are mixed by the **biscuit method**, and some rich cakelike muffins and coffee cakes, which are mixed by a cake-mixing method called the *creaming method*. The biscuit and muffin methods are presented in this chapter. The creaming method is presented in Chapter 33, along with other cake methods.

The muffins and loaf breads in this chapter should be thought of as breads rather than as tea cakes. They are lower in fat and sugar than some of the rich, cakelike muffins sometimes seen.

GLUTEN DEVELOPMENT IN QUICK BREADS

Only slight gluten development is desired in most quick breads. Tenderness is the desired quality, in contrast to the chewy quality of yeast breads. In addition, chemical leavening agents do not create the same kind of textures that yeast does and are not strong enough to create a light, tender product if the gluten is too strong.

1. Muffin, loaf bread, and pancake batters are mixed as little as possible, just until the dry ingredients are moistened. This, plus the presence of fat and sugar, keeps gluten development low.

   Overmixing muffins produces not only toughness but also irregular shapes and large, elongated holes inside the product. This latter condition is called **tunneling**.

2. Biscuit dough is often lightly kneaded, enough to develop some flakiness but not enough to toughen the product.

3. Popovers are the exception among quick breads. They are made with a thin batter and leavened by steam only. Large holes develop inside the product during baking, and the structure must be strong enough to hold without collapsing. Thus, bread flour is used, and the batter is mixed well to develop the gluten. The high percentage of egg in popovers also helps build structure.

Tunneling

THE BISCUIT METHOD

PROCEDURE: Biscuit Method

1. Scale all ingredients accurately.

2. Sift the dry ingredients together into a mixing bowl.

3. Cut in the shortening, using the paddle attachment or the pastry knife attachment. If preferred, you may cut in the fat by hand, using a pastry blender or your fingers. Continue until the mixture resembles coarse cornmeal.

4. Combine the liquid ingredients. Biscuits may be prepared in advance up to this point. Portions of each mixture may then be scaled and combined just before baking.

5. Add the liquid to the dry ingredients. Mix just until the ingredients are combined and a soft dough is formed. Do not overmix.

6. Bring the dough to the bench and knead it lightly by pressing it out and folding it in half. Rotate the dough 90 degrees after each fold.

7. Repeat this procedure about 10 to 20 times, or for about 30 seconds. The dough should be soft and slightly elastic but not sticky. Overkneading toughens the biscuits. The dough is now ready for makeup.

Variations on the basic procedure produce different characteristics in the finished product.

1. Using slightly more shortening and cutting it in less—until the pieces are the size of peas—produces a flakier biscuit.

2. Omitting the kneading step produces a tender, crustier biscuit, but one with less volume.

Makeup of Biscuits

1. Roll the biscuit dough into a sheet about ¹/₂ inch (1 cm) thick, being careful to roll it evenly and to a uniform thickness. Biscuits approximately double in height during baking.

2. Cut into desired shapes. When using round hand cutters, cutting straight down produces the best shape after baking. Do not twist the cutter. Space the cuts closely to minimize scraps.

    Cutting into squares or triangles with a pastry cutter or knife eliminates scraps that would have to be rerolled. Roller cutters also eliminate or reduce scraps. Reworked scraps are tougher.

3. Place the biscuits ¹/₂ inch (1 cm) apart on a baking sheet for crisp-crusted biscuits, or touching each other for softer biscuits. Bake as soon as possible.

    If desired, the tops may be brushed with egg wash or milk before baking to aid browning.

THE MUFFIN METHOD

This mixing method is used not only for muffins but also for pancakes, waffles, quick loaf breads, and coffee cakes. Loaf breads and coffee cakes are sometimes higher in fat and sugar than muffins, so they can withstand more mixing without toughening.

The muffin method is not as suitable for formulas high in fat and sugar, unlike the cake-mixing method called the **creaming method**, discussed in Chapter 33. Consequently, quick breads mixed by the muffin method are not as rich and cakelike as many of today's popular muffins, which are closer to cakes than to traditional muffins. Keep this in mind as you practice the muffin method with the recipes in this chapter.

THE RUBBED DOUGH METHOD

The method for mixing biscuit doughs, in its simplest form, consists of two essential steps: rubbing or cutting the fat into the dry ingredients, and mixing in the combined wet ingredients. These are the same essential steps as for mixing pie doughs by a method called the *rubbed dough method*, as detailed on page 983. Although the end product is quite different, many pastry chefs consider these methods to be the same and refer to them by the same name.

PROCEDURE: Muffin Method

1. Sift together the dry ingredients (see Figure 32.1).

2. Combine all liquid ingredients, including melted fat or oil.

3. Add the liquids to the dry ingredients and mix just until all the flour is moistened. The batter will look lumpy. Do not overmix.

4. Pan and bake immediately. The dry and liquid mixtures may be prepared in advance. Once they are combined, the batter should be baked without delay, or loss of volume may result. When portioning batter into muffin tins, be careful not to stir the mix and toughen it. Scoop the batter from the outside edge for best results.

FIGURE 32.1 The muffin method.

(a) Sift together the dry ingredients.

(b) Add the combined liquid ingredients to the dry ingredients and mix just until the flour is moistened.

(c) Pan and bake immediately. Scoop the batter from the outside edge to minimize additional mixing.

THE WELL MIXING METHOD

The muffin method is sometimes called the **well mixing method**. This name comes from the technique, commonly used in European pastry shops, of mixing a dough directly on the workbench rather than in a bowl. In order to keep the liquid ingredients from flowing away, the chef makes a well in the dry ingredients (see Figure 32.2) and pours the liquid ingredients into this well. When you are mixing in a bowl, it is of course not necessary to make a well to confine the liquids. But the habit has stuck, and the name has become attached to this mixing method.

FIGURE 32.2 Make a well in the mound of flour and add the liquids.

SUMMARY: BISCUIT AND MUFFIN METHODS

Biscuit Method
1. Combine dry ingredients and cut in fat.
2. Combine liquid ingredients.
3. Add liquid and dry ingredients mix just until combined.
4. If required, knead very lightly.

Muffin Method
1. Combine dry ingredients.
2. Combine liquid ingredients, including melted fat.
3. Add liquid to dry ingredients and mix just until combined.

KEY POINTS TO REVIEW

- What are the steps in the biscuit method?
- What are the steps in the muffin method?
- Why must popover batter be mixed more thoroughly than muffin and biscuit doughs?

FORMULAS

 Biscuits

INGREDIENTS	U.S.		METRIC	PERCENTAGE	
Bread flour	10	oz	300 g	50	%
Pastry flour	10	oz	300 g	50	%
Salt	0.4	oz	12 g	2	%
Sugar	1	oz	30 g	5	%
Baking powder	1.2	oz	36 g	6	%
Shortening (regular) and/or butter	7	oz	210 g	35	%
Milk	13	oz	390 g	65	%
Total weight:	2 lb 10	oz	1279 g	213	%

PROCEDURE

Mixing and makeup:

Biscuit method.

Scaling:

Approximately 1 lb (500 g) per dozen 2-in. (5-cm) biscuits.

Baking:

425°F (220°C), about 15 minutes.

Per 1 biscuit: Calories, 130; Protein, 2 g; Fat, 7 g (48% cal.); Cholesterol, 0 mg; Carbohydrates, 15 g; Fiber 1 g; Sodium, 260 mg.

VARIATIONS

Buttermilk Biscuits

Use buttermilk instead of regular milk.

Cheese Biscuits

Add 30% (6 oz/180 g) grated cheddar cheese to dry ingredients.

Currant Biscuits

Add 15% (3 oz/90 g) dried currants to dry ingredients. Increase sugar to 10% (2 oz/60 g). Sprinkle tops with cinnamon sugar before baking.

Herb Biscuits

Add 5% (1 oz/30 g) fresh chopped parsley to the dry ingredients.

Biscuits, without and with egg wash

 Plain Muffins

INGREDIENTS	U.S.			METRIC		PERCENTAGE	
Pastry flour	1 lb	4	oz	600	g	100	%
Sugar		10	oz	300	g	50	%
Baking powder		1.2	oz	36	g	6	%
Salt		0.25	oz	7.5	g	1.25	%
Eggs, beaten		6	oz	180	g	30	%
Milk		14	oz	420	g	70	%
Vanilla extract		0.5	oz	15	g	2.5	%
Melted butter or shortening		8	oz	240	g	40	%
Total weight:	3 lb	12	oz	1798	g	299	%

PROCEDURE

Mixing:

Muffin method.

Scaling and panning:

Grease and flour muffin tins, or use paper liners.

Scale batter with a No. 16 scoop, 2 oz (60 g) per unit.

Baking:

400°F (200°C), about 20 minutes.

Per 1 muffin: Calories, 170; Protein, 3 g; Fat, 7 g (36% cal.); Cholesterol, 40 mg; Carbohydrates, 25 g; Fiber, 3 g; Sodium, 230 mg.

VARIATIONS

Raisin Spice Muffins

Add 20% raisins (4 oz/125 g), 1¼ tsp (6 mL) cinnamon, and ½ tsp (2 mL) nutmeg to the dry ingredients.

Date Nut Muffins

Add 15% (3 oz/90 g) each chopped dates and chopped walnuts to the dry ingredients.

Blueberry Muffins

Gently fold 40% (8 oz/240 g) well-drained blueberries into the finished batter.

Whole Wheat Muffins

Use 70% (14 oz/420 g) pastry flour and 30% (6 oz/ 180 g) whole wheat flour. Reduce baking powder to 4% (0.8oz/ 24 g) and add 0.75% (1 tsp/5 mL) baking soda. Add 10% (2 oz/60 g) molasses to the liquid ingredients.

Corn Muffins

Use 65% (13oz/390 g) pastry flour and 35% (7 oz/210 g) yellow cornmeal. (See also Corn Bread formula, p. 939.)

Bran Muffins

Use 30% (6 oz/180 g) bran, 40% (8 oz/240 g) bread flour, and 30% (6 oz/180 g) pastry flour. Add 15% (3 oz/ 90 g) raisins to the dry ingredients. Add 15% (3 oz/90 g) molasses to the liquid ingredients.

Crumb Coffee Cake

Increase fat to 50% (10 oz/300 g). Pour into greased, paper-lined sheet pan and spread smooth. Top with 80% (1 lb/480 g) Streusel Topping (p. 924). Bake at 360°F (180°C), about 30 minutes.

Muffins, clockwise from top: blueberry, corn, bran

Banana Bread

INGREDIENTS	U.S.		METRIC		PERCENTAGE	
Pastry flour	12	oz	350	g	100	%
Sugar	5	oz	140	g	40	%
Baking powder	0.6 oz (3 1/2 tsp)		18	g	5	%
Baking soda	1/2	tsp	2	g	0.5	%
Salt	1	tsp	4.5 g		1.25	%
Chopped walnuts	3	oz	88	g	25	%
Eggs	5	oz	140	g	40	%
Ripe banana pulp, puréed	12	oz	350	g	100	%
Oil, melted shortening, or butter	4	oz	115	g	33	%
Total weight:	2 lb 10	oz	1207	g	344	%

PROCEDURE

Mixing:

Muffin method.

Scaling:

1 lb 10 oz (750 g) per loaf pan measuring 8 1/2 × 4 1/2 in. (22 × 11 cm).

Baking:

375°F (190°C), about 50 minutes.

Per 1 ounce (28.35 g): Calories, 90; Protein, 2 g; Fat, 4.5 g (41% cal.); Cholesterol, 15 mg; Carbohydrates, 12 g; Fiber, 1 g; Sodium, 120 mg.

Popovers

INGREDIENTS	U.S.		METRIC	PERCENTAGE	
Eggs	10	oz	312 g	125	%
Milk	1 lb		500 g (1 L)	200	%
Salt	0.15 oz (3/4 tsp)		4 g (3 mL)	1.5	%
Melted butter or shortening	1	oz	30 g	12.5	%
Bread flour	8	oz	250 g	100	%
Total weight:	2 lb 3	oz	1069 g	439	%

PROCEDURE

Mixing:

1. Beat eggs, milk, and salt with whip attachment until well blended. Add melted fat.
2. Replace whip with paddle. Mix in flour until completely smooth.

Scaling and panning:

Grease every cup of popover tins or every other cup of muffin tins—popovers need room for expansion. Fill cups about two-thirds full, about 1 1/2 oz (50 g) batter per unit.

Baking:

450°F (230°C) for 10 minutes. Reduce heat to 375°F (190°C) for 20–30 minutes.

Before removing them from oven, be sure popovers are dry and firm enough to avoid collapse. Remove from pans immediately.

Per 1 popover: Calories, 70; Protein, 3 g; Fat, 3 g (38% cal.); Cholesterol, 60 mg; Carbohydrates, 8 g; Fiber, 0 g; Sodium, 95 mg.

Popovers

Corn Bread, Muffins, or Sticks

INGREDIENTS	U.S.		METRIC	PERCENTAGE	
Pastry flour	10	oz	300 g	50	%
Cornmeal	10	oz	300 g	50	%
Sugar	3	oz	90 g	15	%
Baking powder	1	oz	30 g	5	%
Salt	0.4	oz	12 g	2	%
Eggs, beaten	4	oz	120 g	20	%
Milk	1 lb 1	oz	500 g (1 L)	85	%
Corn syrup	1	oz	30 g	5	%
Melted butter or shortening	6	oz	180 g	30	%
Total weight:	3 lb 4	oz	1067 g	262	%

PROCEDURE

Mixing:

Muffin method.

Scaling:

60 oz (1700 g) per half-size sheet pan (13 × 18 in./33 × 46 cm).

24 oz (725 g) per 9-in. (23-cm) square pan or per dozen muffins.

10 oz (300 g) per dozen corn sticks.

Grease and flour pans well.

Baking:

400°F (200°C) for corn bread, 25–30 minutes. 425°F (220°C) for muffins or sticks, 15–20 minutes.

Per 1 muffin: Calories, 160; Protein, 3 g; Fat, 7 g (39% cal.); Cholesterol, 35 mg; Carbohydrates, 22 g; Fiber, 2 g; Sodium, 290 mg.

Orange Nut Bread

INGREDIENTS	U.S.		METRIC	PERCENTAGE	
Sugar	16	oz	175 g	50	%
Grated orange zest	0.5	oz	14 g	4	%
Pastry flour	12	oz	350 g	100	%
Nonfat dry milk	1	oz	28 g	8	%
Baking powder	0.5	oz	14 g	4	%
Baking soda	1	tsp	5 g	1.4	%
Salt	1	tsp	5 g (5 mL)	1.4	%
Chopped walnuts	6	oz	175 g	50	%
Eggs	2.55	oz	75 g	20	%
Orange juice	3	oz	85 g	25	%
Water	8	oz	230 g	65	%
Oil, melted butter, or shortening	3	oz	85 g	25	%
Total weight:	2 lb 10	oz	1241 g	344	%

PROCEDURE

Mixing:

Muffin method.

Blend the sugar and orange zest thoroughly before adding remaining dry ingredients to ensure even distribution.

Scaling:

1 lb 10 oz (750 g) per loaf pan measuring 8¹/₂ × 4¹/₂ in. (22 × 11 cm).

Baking:

375°F (190°C), about 50 minutes.

Per 1 ounce (28.35 g): Calories, 80; Protein, 2 g; Fat, 4 g (39% cal.); Cholesterol, 5 mg; Carbohydrates, 12 g; Fiber, 1 g; Sodium, 130 mg.

TERMS FOR REVIEW

pour batter	muffin method	tunneling
drop batter	biscuit method	well mixing method

QUESTIONS FOR DISCUSSION

1. If you made a batch of muffins that came out of the oven with strange, knobby shapes, what would you expect was the reason?
2. What is the most important difference between the biscuit method and the muffin method?
3. How is the mixing method for popovers different from the mixing method for muffins? What is the leavening agent for popovers?

CAKES AND ICINGS

Cakes are the richest and sweetest of all the baked products we have studied so far. From the baker's point of view, producing cakes requires as much precision as producing breads, but for completely opposite reasons. Breads are lean products that require strong gluten development and careful control of yeast action during the long fermentation and proofing periods. Cakes, on the other hand, are high in both fat and sugar. The baker's job is to create a structure that will support these ingredients and yet be as light and delicate as possible. Fortunately, producing cakes in quantity is relatively easy if the baker has good, well-balanced formulas, scales ingredients accurately, and understands basic mixing methods well.

Cakes owe their popularity not only to their richness and sweetness but also to their versatility. They can be presented in many forms, from simple sheet cakes in cafeterias to elaborately decorated works of art for weddings and other important occasions. With only a few basic formulas and a variety of icings, the chef or baker can construct the perfect dessert for any occasion or purpose.

AFTER READING THIS CHAPTER, YOU SHOULD BE ABLE TO

1. Demonstrate the five basic cake mixing methods.

2. Describe the characteristics of high-fat cakes and low-fat cakes.

3. Prepare high-fat, or shortened, cakes and low-fat, or foam-type, cakes.

4. Prepare the six basic types of icing.

5. Assemble and ice layer cakes, small cakes, and sheet cakes.

UNDERSTANDING CAKE MAKING

BASIC MIXING METHODS

The mixing methods presented in this chapter are basic for most cakes prepared in the modern bakeshop. Each method is used for a particular type of formula.

High-fat or shortened cakes	**Low-fat or foam-type cakes**
Creaming method	Foaming or sponge method
Two-stage or blending method	Angel food method
	Chiffon method

We discuss these cake types in detail after you have had a chance to study the actual procedures on pages 943–945.

CREAMING METHOD

The **creaming method**, also called the *conventional method*, was, for a long time, the standard method for mixing butter cakes. The development of emulsified or high-ratio shortenings led to the development of simpler mixing methods for shortened cakes. But the creaming method is still used for many types of butter cakes.

TWO-STAGE METHOD

The **two-stage method**, also called the **blending method**, was developed for use with modern high-ratio shortenings (see Chapter 30). Although it is simpler than the creaming method, it produces a very smooth batter that bakes up into a fine-grained, moist cake. It is called *two-stage* because the liquids are added in two stages.

FOAMING OR SPONGE METHOD

All egg-foam cakes are similar in that they contain little or no shortening and depend for most or all of their leavening on the air trapped in beaten eggs.

One mixing method is usually presented as the basic method for all foam cakes. However, because whole-egg foams and egg-yolk foams are handled differently from egg-white foams, we discuss two separate, although similar, methods: the **foaming method** and the **angel food method**. A third method, the **chiffon method**, is somewhat unusual. It combines an egg-white foam with a high-fat batter made with oil.

ANGEL FOOD METHOD

Angel food cakes are based on egg-white foams and contain no fat. For success in beating egg whites, review the principles of egg foams in Chapter 24.

CHIFFON METHOD

Chiffon cakes and angel food cakes are both based on egg-white foams. But here the similarities in the mixing methods end. In angel food cakes, a dry flour-sugar mixture is folded into the egg whites. In chiffon cakes, a batter containing flour, egg yolks, vegetable oil, and water is folded into the whites.

Whip egg whites for chiffon cakes until they are a little firmer than those for angel food cakes, but not until they are dry. Chiffon cakes contain baking powder, so they do not depend on the egg foam for all their leavening.

PROCEDURE: Creaming Method

1. Scale ingredients accurately. Have all ingredients at room temperature.

2. Place the butter or shortening in the mixing bowl. With the paddle attachment, beat slowly until the fat is smooth and creamy.

3. Add the sugar. Cream the mixture at moderate speed until it is light and fluffy (Figure 33.1).

   Some bakers prefer to add the salt and flavorings with the sugar to ensure uniform distribution.

   If melted chocolate is used, it is added during creaming.

4. Add the eggs, a little at a time. After each addition, beat until the eggs are absorbed before adding more. The mixture should be light and fluffy after the eggs are beaten in.

5. Scrape down the sides of the bowl to ensure even mixing.

6. Add the sifted dry ingredients (including the spices, if they were not added in step 3), alternating with the liquids. This is done as follows:

   • Add one-fourth of the dry ingredients. Mix just until blended in.
   • Add one-third of the liquid. Mix just until blended in.
   • Repeat until all ingredients are used. Scrape down the sides of the bowl occasionally for even mixing.

   The reason for adding dry and liquids alternately is that the batter may not absorb all the liquid unless some of the flour is present.

   Cocoa, if used, is included with the flour.

FIGURE 33.1 Creaming method.

(a) Cream the butter and sugar until the mixture is light and fluffy.

(b) Beat in the eggs a little at a time.

(c) Add one fourth of the dry ingredients and mix in.

(d) Add one third of the liquid ingredients and mix in. Repeat until all the dry and liquid ingredients are incorporated.

PROCEDURE: Two-Stage Method

1. Scale ingredients accurately. Have all ingredients at room temperature.

2. Sift the flour, baking powder, soda, and salt into the mixing bowl and add the shortening. With the paddle attachment, mix at low speed for 2 minutes. Stop the machine, scrape down the bowl and beater, and mix again for 2 minutes.

   If melted chocolate is used, blend it in during this step.

   If cocoa is used, sift it with the flour in this step or with the sugar in step 3.

3. Sift the remaining dry ingredients into the bowl and add part of the water or milk. Blend at low speed 3–5 minutes. Scrape down the sides of the bowl and the beater several times to ensure even mixing.

4. Combine the remaining liquids and lightly beaten eggs. With the mixer running, add this mixture to the batter in three parts. After each part, turn off the machine and scrape down the bowl.

   Continue mixing for a total of 5 minutes in this stage. The finished batter is normally quite liquid.

VARIATIONS

This variation combines steps 2 and 3 above into one step.

1. Scale ingredients as in the basic method.

2. Sift all dry ingredients into the mixing bowl. Add the shortening and part of the liquid. Mix on low speed 7–8 minutes. Scrape down the sides of the bowl and the beater several times.

3. Continue with step 4 in the basic procedure.

PROCEDURE: Foaming or Sponge Method

1. Scale ingredients accurately. Have all ingredients at room temperature. If butter is included, it must be melted.

   If liquid and butter are included, heat them together, just until the butter is melted.

2. Combine the eggs and sugar, and warm to about 110°F (43°C). This may be done in one of two ways:

   • Stir the egg-sugar mixture over a hot-water bath.

   • Warm the sugar on a sheet pan in the oven (do not get it too hot) and gradually beat it into the eggs.

   This step is performed because warm foam attains greater volume.

3. With the whip attachment, beat the eggs at high speed until light and thick. This may take 10–15 minutes.

   This step is important. One of the most frequent causes of failure in the sponge method is not whipping the eggs and sugar enough. The foam must be very thick. When the beater is lifted from the bowl, the foam should fall slowly from it and make a ribbon that slowly sinks into the batter in the bowl (see Figure 33.2).

4. Fold in the sifted flour, being careful not to deflate the foam. Many bakers do this by hand.

   If other dry ingredients are used, such as cornstarch or baking powder, they are first sifted with the flour.

5. If melted butter or a butter-liquid mixture is being used, fold it in at this point. Be careful not to overmix, or the cake will be tough (because of developed gluten).

6. Immediately pan and bake the batter. Delays cause loss of volume.

VARIATIONS

Some formulas contain water or some other liquid, but no butter (so you cannot heat the liquid and butter together, as in the basic procedure). In this case, the liquid is usually added after step 3 and before folding in the flour. Either whip it in in a steady stream or stir it in, as indicated in the recipe.

In some formulas, the egg yolks and whites are separated. Use the yolks and part of the sugar to make the foam in steps 2 and 3. Use the remaining sugar to whip with the whites. Fold the egg-white foam into the batter after step 5.

FIGURE 33.2 Foaming or sponge method.

(a) With a wire whip or the whip attachment of a mixer, beat the eggs and sugar until they are very thick and light.

(b) Fold in the sifted flour in three or four stages, until all the flour is blended in.

(c) If melted butter is used, fold it in after the flour.

PROCEDURE: Angel Food Method

1. Scale ingredients accurately. Have all ingredients at room temperature. You may warm the egg whites slightly for better volume.

2. Sift the flour with half the sugar. This step helps the flour mix more evenly with the foam.

3. Beat the egg whites, using the whip attachment, until they form soft peaks. Add salt and cream of tartar near the beginning of the beating process (Figure 33.3).

4. Gradually beat in the sugar that was not mixed with the flour. Continue to beat until the egg whites form soft, glossy peaks. Do not overbeat.

5. Fold in the flour-sugar mixture just until it is thoroughly absorbed, but no longer.

6. Pan and bake immediately.

(a) Whip the egg whites to soft peaks. Add salt and cream of tartar during the beginning of the beating process.

(b) Gradually beat in the portion of sugar that was not mixed with flour.

(c) Whip to soft, moist peaks.

FIGURE 33.3 Angel food method.

(d) Fold in the flour-sugar mixture until just absorbed.

(e) Deposit in ungreased pans and bake immediately.

PROCEDURE: Chiffon Method

1. Scale all ingredients accurately, and be sure they are all at room temperature. Use a good-quality, flavorless vegetable oil.

2. Sift the dry ingredients, including part of the sugar, into the mixing bowl.

3. Mixing with the paddle attachment at second speed, gradually add the oil (Figure 33.4), then the egg yolks, water, and liquid flavorings, all in a slow, steady stream. While adding the liquids, stop the machine several times and scrape down the bowl and the beater. Mix until smooth, but do not overmix.

4. Whip the egg whites until they form soft peaks. Add the cream of tartar and sugar in a stream and whip to firm, moist peaks.

5. Fold the whipped egg whites into the flour–liquid mixture.

6. Immediately deposit batter in ungreased tube pans (like angel food cakes) or in layer pans that have had the bottoms, but not the sides, greased and dusted (like sponge layers).

FIGURE 33.4 Chiffon method.

(a) Gradually add the oil to the dry ingredients.

(c) Then add the water.

(b) Then add the egg yolks.

(d) Fold the whipped egg whites into the batter.

KEY POINTS TO REVIEW

- What are the steps in the creaming method?

- What are the steps in the two-stage method?

- What are the steps in the foaming or sponge method?

- What are the steps in the angel food method?

- What are the steps in the chiffon method?

PREPARED MIXES

Many cake mixes are available that contain all ingredients except water and, sometimes, egg. These products also contain emulsifiers to ensure even blending of ingredients. To use them, follow the package instructions exactly.

Most mixes produce cakes with excellent volume, texture, and tenderness. Whether or not they also taste good is a matter of opinion. On the other hand, cakes made from scratch are not necessarily better. They are better only if they are carefully mixed and baked and are prepared using good, tested formulas and high-quality ingredients.

CAKE FORMULA TYPES

The proper mixing method for a particular formula depends on the balance of ingredients. A baker can look at the ingredients in a formula and know immediately which mixing method to use.

- If **fat is high**, use the creaming method or the two-stage method.

  The two-stage method may be used if the percentage of sugar is over 100 percent and if the fat is emulsified shortening.

  In other cases, the creaming method is used.

- If **fat is low** and **eggs and sugar are high**, use an egg-foam method.

HIGH-FAT CAKES

The creaming method's major disadvantage is the labor it requires. The two-stage method is quicker, but because the flour is mixed for a long time, two conditions are necessary to prevent the gluten from developing toughness:

1. Increased percentage of sugar (sugar is a tenderizer).

2. Emulsified shortening, which blends thoroughly to prevent toughness.

Cakes made by the two-stage method have good volume and lightness, a fine, velvety texture, and great tenderness. The texture of butter cakes made by the creaming method is coarser, and the tenderness is generally somewhat less.

One factor seems to be neglected when cakes are rated, however—flavor. Shortening contributes no flavor to cakes, only texture. Butter, on the other hand, is highly prized for its flavor. It also influences texture because it melts in the mouth, while shortening does not. Thus, butter cakes are and always will be in demand. Therefore, the creaming method is important for you to know.

LOW-FAT CAKES

High-fat cakes depend on air incorporated by the creaming action of the fat and sugar for some of their leavening and much of their texture. Low-fat or no-fat cakes obviously cannot. They must depend on the foaming action of eggs.

Sponge cakes have a springy texture and are tougher than shortened cakes. This makes them valuable for many kinds of desserts that require much handling to assemble. For example, many European-style cakes or tortes are made by cutting sponge cake layers horizontally into thin layers and stacking them with a variety of rich fillings, creams, icings, and fruits.

Even if a high-ratio cake survived all this without breaking into crumbs, it would probably disintegrate when it absorbed moisture from the fillings. In addition, sponge layers in this kind of cake are usually moistened with a flavored sugar syrup to compensate for their lack of moisture.

The *fruit torte* (illustrated in the photograph) is an example of this type of cake. Genoise layers are split, moistened with dessert syrup (p. 1008), layered and iced with whipped cream, and topped with attractively arranged fruit pieces. The fruit is then coated with glaze (p. 961) to protect it and enhance its appearance.

Sponge sheets for jelly rolls and other rolled cakes are made without shortening, so they do not crack when rolled.

Flour for sponge cakes must be weak to avoid making the cake tough. Cornstarch is often added to cake flour for sponge cakes to weaken the flour further.

Fruit torte.

SCALING AND PANNING

Prepare pans before mixing cake batters so cake batters can be baked without delay:

1. For high-fat cakes, the bottoms of layer pans must be greased, preferably with a commercial pan greasing preparation. If this is not available, dust the greased pan with flour and tap out the excess.

2. For sheet cakes, line the pan with greased parchment.

3. For angel food cakes, do not grease the pan. The batter must be able to cling to the sides in order to rise.

4. For sponge cake layers with a small percentage of fat, grease the bottoms but not the sides.

PROCEDURE: Scaling Creaming-Method Batters

These batters are thick and do not pour easily. Scale cakes as follows:

1. Place the prepared cake pan on the left side of a balance scale. Balance the scale by placing another pan on the right side.

2. Set the scale for the desired weight.

3. Add batter to the left pan until the scale balances.

4. Remove the pan from the scale and spread the batter smooth with a spatula.

5. Repeat with remaining pans.

6. Give the pans several sharp raps on the bench to free large trapped air bubbles. Bake immediately.

ALTERNATIVE PROCEDURE: Scaling Two-Stage Batters

These batters are more liquid than creamed batters. They may be scaled like creamed batters or, for greater speed, they may be scaled as follows:

1. Place an empty volume measure on the left side of a balance scale. Balance the scale to zero.

2. Set the scale for the desired weight.

3. Pour batter into the measure until the scale balances.

4. Note the volume of batter in the measure.

5. Pour the measured batter into a prepared pan, quickly scraping out the measure to get all the batter.

6. Scale the remaining cakes with the volume measure, using the volume noted in step 4.

7. Give the pans several sharp raps on the bench to free large trapped air bubbles. Bake immediately.

PROCEDURE: Scaling Foam Cakes

Foam cake batters should be handled as little as possible and baked immediately in order to avoid deflating the beaten eggs. Although they may be scaled like creamed batters, many bakers prefer to eyeball them in order to minimize handling.

1. Have all prepared pans lined up on the bench.

2. Scale the first pan as for creamed batters.

3. Quickly fill remaining pans to the same level as the first pan, judging the level by eye.

4. Spread the batter smooth and bake immediately.

See Table 33.1 for average scaling weights as well as baking temperatures and times.

TABLE 33.1 Average Cake Scaling Weights, Baking Temperatures, and Times

Pan Type and Size	Scaling Weight		Baking Temperatures		Approximate Baking Time in Minutes
	U.S.	Metric	U.S.	Metric	
High-fat cakes					
Round layers					
6 in. (15 cm)	8–10 oz	230–285 g	375°F	190°C	18
8 in. (20 cm)	14–18 oz	400–510 g	375°F	190°C	25
10 in. (25 cm)	24–28 oz	680–800 g	360°F	180°C	35
12 in. (30 cm)	32–40 oz	900–1100 g	360°F	180°C	35
Sheets and square pans					
18 × 26 in. (46 × 66 cm)	7–8 lb	3.2–3.6 kg	360°F	180°C	35
18 × 13 in. (46 × 33 cm)	3¹/₂–4 lb	1.6–1.8 kg	360°F	180°C	35
9 × 9 in. (23 × 23 cm)	24 oz	680 g	360°F	180°C	30–35
Loaf (pound cake)					
2¹/₄ × 3¹/₂ × 8 in. (6 × 9 × 20 cm)	16–18 oz	450–500 g	350°F	175°C	50–60
2³/₄ × 4¹/₂ × 8¹/₂ in. (7 × 11 × 22 cm)	24–27 oz	680–765 g	350°F	175°C	55–65
Cupcakes per dozen	18 oz	510 g	385°F	195°C	18–20
Foam-type cakes					
Round layers					
6 in. (15 cm)	5–6 oz	140–170 g	375°F	190°C	20
8 in. (20 cm)	10 oz	280 g	375°F	190°C	20
10 in. (25 cm)	16 oz	450 g	360°F	180°C	25–30
12 in. (30 cm)	24 oz	700 g	360°F	180°C	25–30
Sheets (for jelly roll or sponge roll)					
18 × 26 in., ¹/₂ in. thick (46 × 66 cm, 12 mm thick)	2¹/₂ lb	1.2 kg	375°F	190°C	15–20
18 × 26 in., ¹/₄ in. thick (46 × 66 cm, 6 cm thick)	28 oz	800 g	400°F	200°C	7–10
Tube (angel food and chiffon)					
8 in. (20 cm)	12–14 oz	340–400 g	360°F	180°C	30
10 in. (25 cm)	24–32 oz	700–900 g	350°F	175°C	50
Cupcakes per dozen	10 oz	280 g	375°F	190°C	18–20

Note: The weights given are averages. Weights may be increased by 25 percent if thicker layers are desired. Baking times may then need to be increased slightly.

BAKING AND COOLING

BAKING

Cake structure is fragile, so proper baking conditions are essential for high-quality products. The following guidelines will help you avoid cake failures:

1. Preheat the ovens. (To conserve expensive energy, don't preheat longer than necessary.)

2. Make sure ovens and shelves are level.

3. Do not let pans touch each other in the oven. If the pans touch, air circulation is inhibited and the cakes rise unevenly.

4. Bake at the correct temperature.

   Too hot an oven causes the cake to set unevenly or to set before it has fully risen. Crusts will be too dark.

   Too slow an oven causes poor volume and texture because the cake doesn't set fast enough and may fall.

5. Do not open the ovens or disturb the cakes until they have finished rising and are partially browned. Disturbing the cakes before they are set may cause them to fall.

6. If steam in the oven is available, use it for creamed and two-stage batters. These cakes bake with a flatter top if baked with steam because the steam delays the formation of the top crust.

7. Tests for doneness:

   • Shortened cakes shrink away from sides of pan slightly.

   • Cakes are springy. The center of the top springs back when pressed slightly.

   • A cake tester or pick inserted in the center of the cake comes out clean.

COOLING AND REMOVING FROM PANS

1. Cool layer cakes and sheet cakes 15 minutes in pans and then turn out while slightly warm. They are too fragile to turn out when hot, and they may break.

2. Turn out layer cakes onto racks to finish cooling.

3. To turn out sheet cakes:

   • Sprinkle top lightly with granulated sugar.

   • Set an empty sheet pan on top, bottom side down.

   • Invert both pans.

   • Remove top pan.

   • Peel parchment off cake.

4. Cool angel food cakes upside down in pans. Support the edges of the pan so the top of the cake is off the bench. When cool, loosen the cake from the sides of the pan with a knife or spatula and pull out carefully.

COMMON CAKE FAULTS AND THEIR CAUSES

Errors in mixing, scaling, baking, and cooling cakes cause many kinds of defects and failures. For easy reference, these defects and their possible causes are summarized in the trouble-shooting guide in Table 33.2.

TABLE 33.2 Common Cake Faults and Their Causes

Fault	Causes
Volume and Shape	
Poor volume	Too little flour
	Too much liquid
	Too little leavening
	Oven too hot
Uneven shape	Improper mixing
	Batter spread unevenly
	Uneven oven heat
	Oven racks not level
	Cake pans warped
Crust	
Too dark	Too much sugar
	Oven too hot
Too light	Too little sugar
	Oven not hot enough
Burst or cracked	Too much flour or flour too strong
	Too little liquid
	Improper mixing
	Oven too hot
Soggy	Underbaked
	Cooling in pans or with not enough ventilation
	Wrapping before cool
Texture	
Dense or heavy	Too little leavening
	Too much liquid
	Too much sugar
	Too much shortening
	Oven not hot enough
Coarse or irregular	Too much leavening
	Too little egg
	Improper mixing
Crumbly	Too much leavening
	Too much shortening
	Too much sugar
	Wrong kind of flour
	Improper mixing
Tough	Flour too strong
	Too much flour
	Too little sugar or shortening
	Overmixing
Poor flavor	
	Poor-quality ingredients
	Poor storage or sanitation
	Unbalanced formula

KEY POINTS TO REVIEW

- What mixing methods are used for high-fat cakes? for low-fat cakes?
- What are the procedures for scaling cake batters?
- How are cakes tested for doneness?
- How are cakes removed from their baking pans?

ALTITUDE ADJUSTMENTS

At high altitudes, atmospheric pressure is much lower than at sea level. This factor must be taken into account in cake baking. Formulas must be adjusted to suit baking conditions over 2,000 or 3,000 feet (600 or 900 m) above sea level.

Although general guidelines can be given, the exact adjustments required vary for different kinds of cake. Many manufacturers of flour, shortening, and other bakery ingredients supply detailed information and adjusted formulas for any given locality.

In general, the following adjustments must be made above elevations of 2,000 or 3,000 feet (600 or 900 m). See Table 33.3 for more specific adjustments.

LEAVENING

Leavening gases expand more when air pressure is lower, so baking powder and baking soda must be *decreased*.

Creaming and foaming procedures should also be reduced so less air is incorporated.

TOUGHENERS: FLOUR AND EGGS

Cakes require firmer structure at high altitudes. Both eggs and flour must be increased to supply proteins for structure.

TENDERIZERS: SHORTENING AND SUGAR

Shortening and sugar must be decreased so the structure of the cake is firmer.

LIQUIDS

At high altitudes, water boils at a lower temperature and evaporates more easily. Liquids must be *increased* to prevent excess drying both during and after baking. This also helps compensate for the decrease in moisturizers (sugar and fat) and the increase in flour, which absorbs moisture.

BAKING TEMPERATURES

Increase baking temperatures about 25°F (14°C) above 3,500 feet (1050 m).

PAN GREASING

High-fat cakes tend to stick at high altitudes. Grease pans more heavily. Remove baked cakes from pans as soon as possible.

STORING

Wrap or ice cakes as soon as they are cool to prevent drying.

TABLE 33.3 Approximate Formula Adjustment in Shortened Cakes at High Altitudes

		Percentage Adjustment		
Ingredient	Increase or Decrease	2500 Feet (750 m)	5000 Feet (1500 m)	7500 Feet (2300 m)
Baking powder	Decrease	20%	40%	60%
Flour	Increase	—	4%	9%
Eggs	Increase	2.5%	9%	15%
Sugar	Decrease	3%	6%	9%
Fat	Decrease	—	—	9%
Liquid	Increase	9%	15%	22%

To make adjustments, multiply the percentage indicated by the amount of ingredient and add or subtract as indicated.

Example: To adjust 1 lb (16 oz) eggs for 7,500 feet: 0.15 × 16 oz = 2.4 oz

16 oz + 2.4 oz = 18.4 oz

CAKE FORMULAS

CREAMING METHOD

Yellow Butter Cake

INGREDIENTS	U.S.		METRIC	PERCENTAGE	
Butter	12	oz	360 g	80	%
Sugar	13	oz	390 g	87	%
Salt	0.12	oz (²/₃ tsp)	4 g	0.75	%
Eggs	7.5	oz	225 g	50	%
Cake flour	15	oz	450 g	100	%
Baking powder	0.62	oz (3 ³/₄ tsp)	18 g	4	%
Milk	15	oz	450 g	100	%
Vanilla extract	0.25	oz	8 g	1.5	%
Total weight:	3 lb 15	oz	1905 g	423	%

PROCEDURE

Mixing:
Creaming method.
Scaling and baking:
See Table 33.1.

Per 1 ounce (28.35 g): Calories, 90; Protein, 1 g; Fat, 5 g (48% cal.); Cholesterol, 25 mg; Carbohydrates, 11 g; Fiber, 0 g; Sodium, 80 mg.

Chocolate Butter Cake

INGREDIENTS	U.S.		METRIC	PERCENTAGE	
Butter	9	oz	280 g	75	%
Sugar	15	oz	470 g	125	%
Salt	0.2	oz (1 tsp)	6 g	1.5	%
Unsweetened chocolate, melted	6	oz	188 g	50	%
Eggs	8	oz	250 g	67	%
Cake flour	12	oz	250 g	100	%
Baking powder	0.5	oz	15 g	4	%
Milk	14	oz	439 g	115	%
Vanilla	0.25	oz (1 ¹/₂ tsp)	8 mL	2	%
Total weight:	4 lb		1906 g	539	%

PROCEDURE

Mixing:
Creaming method. Blend in the melted chocolate after the fat and sugar are well creamed.
Scaling and baking:
See Table 33.1.

Per 1 ounce (28.35 g): Calories, 90; Protein, 1 g; Fat, 5 g (46% cal.); Cholesterol, 25 mg; Carbohydrates, 12 g; Fiber, 0.5 g; Sodium, 80 mg.

Brown Sugar Spice Cake

INGREDIENTS	U.S.		METRIC		PERCENTAGE	
Butter	12	oz	400	g	80	%
Brown sugar	15	oz	500	g	100	%
Salt	0.25	oz	8	g	1.5	%
Eggs	9	oz	300	g	60	%
Cake flour	15	oz	500	g	100	%
Baking powder	0.5	oz	15	g	3	%
Baking soda	3/8	tsp	1.5	g (1.5 mL)	0.3	%
Cinnamon	1 1/2	tsp	2.5	g (7 mL)	0.5	%
Ground cloves	3/4	tsp	1.5	g (3 mL)	0.3	%
Nutmeg	3/8	tsp	1	g (1.5 mL)	0.2	%
Milk	15	oz	500	g	100	%
Total weight:	4 lb 2	oz	2229	g	445	%

PROCEDURE

Mixing:
Creaming method.
Scaling and baking:
See Table 33.1

Per 1 ounce (28.35 g): Calories, 90; Protein, 1 g; Fat, 5 g (46% cal.); Cholesterol, 30 mg; Carbohydrates, 12 g; Fiber, 0 g; Sodium, 105 mg.

VARIATION

Carrot Nut Cake

Reduce the milk to 90% (13.5 oz/450 g). Add 40% (6 oz/200 g) grated fresh carrots, 20% (3 oz/100 g) finely chopped walnuts, and 0.5% (1 tsp/3 g or 5 mL) grated orange zest after eggs are beaten in. Omit cloves.

Old-Fashioned Pound Cake

INGREDIENTS	U.S.	METRIC	PERCENTAGE
Butter or butter and shortening combined (see Note)	1 lb	500 g	100 %
Sugar	1 lb	500 g	100 %
Vanilla	2 tsp	10 mL	2 %
Eggs	1 lb	500 g	100 %
Cake flour	1 lb	500 g	100 %
Total weight:	4 lb	2000 g	402 %

PROCEDURE

Mixing:
Creaming method. Add the eggs and the cake flour alternately to avoid curdling the mixture.
Scaling and baking:
See Table 33.1.

Per 1 ounce (28.35 g): Calories, 110; Protein, 2 g; Fat, 7 g (51% cal.); Cholesterol, 45 mg; Carbohydrates, 13 g; Fiber, 0 g; Sodium, 70 mg.

Note: If you are using unsalted butter or shortening, add 1.5% (1/4 oz /1 1/4 tsp or 6 mL) salt during the first stage of mixing.

Marble Pound Cake

VARIATIONS

Mace or grated lemon or orange zest may also be used to flavor pound cake.

Raisin Pound Cake

Add 25% (4 oz/125 g) raisins or dried currants that were soaked in boiling water and drained well.

Chocolate Pound Cake

Add 25% (4 oz/125 g) unsweetened chocolate to the butter and sugar after the creaming stage.

Marble Pound Cake

Fill pans one-third full of the basic yellow batter. Add a layer of Chocolate Pound Cake batter, then finish with the yellow batter. Run a spatula blade through the layers to marble them.

TWO-STAGE METHOD

 White Cake

INGREDIENTS	U.S.		METRIC	PERCENTAGE	
Cake flour	12	oz	375 g	100	%
Baking powder	0.75	oz	22 g	6.25	%
Salt	0.25	oz	8 g	2	%
Emulsified shortening	6	oz	188 g	50	%
Sugar	15	oz	470 g	125	%
Skim milk	6	oz	188 g	50	%
Vanilla extract	0.18	oz (1¹⁄₈ tsp)	5 g	1.5	%
Almond extract	0.09	oz (¹⁄₂ tsp)	2 g	0.75	%
Skim milk	6	oz	188 g	50	%
Egg whites	8	oz	250 g	67	%
Total weight:	3 lb 6	oz	1696 g	452	%

PROCEDURE

Mixing:
Two-stage method.
Scaling and baking:
See Table 33.1.

Per 1 ounce (28.35 g): Calories, 90; Protein, 1 g; Fat, 3 g (33% cal.); Cholesterol, 0 mg; Carbohydrates, 13 g; Fiber, 0 g; Sodium, 105 mg.

VARIATIONS

Use water instead of milk and add 10% (1.2 oz/38 g) nonfat dry milk powder to the dry ingredients.

Flavor with lemon extract or emulsion instead of vanilla and almond.

Yellow Cake

Reduce shortening to 45% (5.5 oz/170 g). Substitute whole eggs for egg whites, using the same total weight (67%). Use 2% vanilla (¼ oz/7.5 g) and omit almond extract.

Devil's Food Cake

INGREDIENTS	U.S.		METRIC	PERCENTAGE	
Cake flour	12	oz	375 g	100	%
Cocoa	2	oz	60 g	17	%
Salt	0.25	oz	8 g	2	%
Baking powder	0.375	oz	12 g	3	%
Baking soda	0.25	oz	8 g	2	%
Emulsified shortening	7	oz	220 g	58	%
Sugar	1 lb		500 g	133	%
Skim milk	8	oz	250 g	67	%
Vanilla extract	0.18	oz (1 tsp)	5 g	1.5	%
Skim milk	6	oz	188 g	50	%
Eggs	8	oz	250 g	67	%
Total weight:	3 lb 12	oz	1876 g	500	%

PROCEDURE

Mixing:
Two-stage method.
Scaling and baking:
See Table 33.1.

Per 1 ounce (28.35 g): Calories, 90; Protein, 1 g; Fat, 4 g (39% cal.); Cholesterol, 15 mg; Carbohydrates, 13 g; Fiber, 1 g; Sodium, 105 mg.

FOAMING METHODS

 ## Sponge Cake (Genoise)

INGREDIENTS	U.S.			METRIC	PERCENTAGE
Eggs	1 lb	2	oz	562 g	150 %
Sugar		12	oz	375 g	100 %
Cake flour		12	oz	375 g	100 %
Butter (optional; see p. 944)		4	oz	125 g	33 %
Vanilla extract or lemon flavor		0.25	oz	8 g	2 %
Total weight:	2 lb 14		oz	1445 g	385 %

PROCEDURE

Mixing:
Foaming method.
Scaling and baking:
See Table 33.1.

Per 1 ounce (28.35 g): Calories, 90; Protein, 2 g; Fat, 3 g (31% cal.); Cholesterol, 55 mg; Carbohydrates, 13 g; Fiber, 0 g; Sodium, 35 mg.

VARIATIONS

Chocolate Genoise

Substitute 2 oz (60 g) cocoa powder for 2 oz (60 g) flour. Use the vanilla, not the lemon flavor.

Sponge Roll or Jelly Roll

Prepare the basic formula, but omit the butter. Add the vanilla to the beaten eggs just before folding in the flour. Spread evenly in parchment-lined pans. When baked and cooled, trim the edges. Cut in half or into quarters. Spread each rectangle with desired filling (jelly, buttercream, etc.) and roll up so the long side of the rectangle becomes the length of the roll. Ice or sprinkle with 6X sugar.

Milk and Butter Sponge

INGREDIENTS	U.S.		METRIC	PERCENTAGE
Sugar	10	oz	312 g	125 %
Whole eggs	6	oz	188 g	75 %
Egg yolks	2	oz	60 g	25 %
Salt	0.12	oz (⁵/₈ tsp)	4 g	1.5%
Cake flour	8	oz	250 g	100 %
Baking powder	0.25	oz	8 g	3 %
Skim milk	4	oz	125 g	5 %
Butter	2	oz	60 g	25 %
Vanilla extract	0.25	oz	8 g	3 %
Total weight:	2 lb 4	oz	1015 g	407 %

PROCEDURE

Mixing:
Sponge method. Heat the milk and butter until the butter is melted; fold into batter (step 5 in basic procedure).
Scaling and baking:
Cake layers; see Table 33.1.

Per 1 ounce (28.35 g): Calories, 90; Protein, 2 g; Fat, 2.5 g (25% cal.); Cholesterol, 50 mg; Carbohydrates, 15 g; Fiber, 0 g; Sodium, 90 mg.

Jelly Roll Sponge

INGREDIENTS	U.S.		METRIC	PERCENTAGE
Sugar	11	oz	325 g	100 %
Eggs, whole	10	oz	292 g	90 %
Egg yolks	2	oz	65 g	20 %
Salt	0.25	oz (1 ¼ tsp)	7 g (7 mL)	2 %
Corn syrup	1.5	oz	45 g	14 %
Water	1	oz	30 g	10 %
Vanilla	1	tsp	5 mL	1.5%
Hot water	4	oz	118 g	36 %
Cake flour	11	oz	325 g	100 %
Baking powder	1	tsp	5 g (5 mL)	1.5%
Total weight:	2 lb 8	oz	1217 g	375 %

PROCEDURE

Mixing:
Sponge method. Add the syrup, the first quantity of water, and the vanilla to the sugar and eggs in the first mixing stage. When the foam is completely whipped, stir in the second quantity of water.
Scaling and baking:
See Table 33.1. One recipe makes 1 sheet pan. Line the pans with greased paper. Immediately after baking, turn out of pan onto a sheet of parchment and remove the paper from the bottom of the cake. Spread with jelly and roll up tightly. When cool, dust with confectioners' sugar.

Per 1 ounce (28.35 g): Calories, 80; Protein, 2 g; Fat, 1 g (12% cal.); Cholesterol, 50 mg; Carbohydrates, 15 g; Fiber, 0 g; Sodium, 40 mg.

Yellow Chiffon Cake

INGREDIENTS	U.S.		METRIC	PERCENTAGE	
Cake flour	10	oz	250 g	100	%
Sugar	8	oz	200 g	80	%
Salt	0.25	oz	6 g	2.5	%
Baking powder	0.5	oz	12 g	5	%
Vegetable oil	5	oz	125 g	50	%
Egg yolks	5	oz	125 g	50	%
Water	7.5	oz	188 g	75	%
Vanilla extract	0.25	oz	6 g	2.5	%
Egg whites	10	oz	250 g	100	%
Sugar	5	oz	125 g	50	%
Cream of tartar	0.05 oz ($^5/_8$ tsp)		1 g	0.5	%
Total weight:	3 lb 3	oz	1288 g	515	%

PROCEDURE

Mixing:
Chiffon method.
Scaling and baking:
Set Table 33.1.

Per 1 ounce (28.35 g): Calories, 90; Protein, 1 g; Fat, 3.5 g (38% cal.); Cholesterol, 35 mg; Carbohydrates, 12 g; Fiber, 0 g; Sodium, 95 mg.

VARIATION

Chocolate Chiffon Cake

Add 20% cocoa (2 oz/50 g); sift it with the flour. Increase the egg yolks to 60% (6 oz/150 g).
Increase the water to 90% (9 oz/225 g).

Angel Food Cake

INGREDIENTS	U.S.		METRIC	PERCENTAGE	
Egg whites	2 lb		1000 g	267	%
Cream of tartar	0.25 oz (1 tbsp)		8 g (15 mL)	2	%
Salt	1	tsp	5 g (5 mL)	1.5	%
Sugar	1 lb		500 g	133	%
Vanilla	2	tsp	10 mL	2.5	%
Almond extract	1	tsp	5 mL	1.25	%
Sugar	1 lb		500 g	133	%
Cake flour	12	oz	375 g	100	%
Total weight:	4 lb 12	oz	2403 g	640	%

PROCEDURE

Mixing:
Angel food method.
Scaling and baking:
See Table 33.1.

Per 1 ounce (28.35 g): Calories, 70; Protein, 2 g; Fat, 0 g (0% cal.); Cholesterol, 0 mg; Carbohydrates, 16 g; Fiber, 0 g; Sodium, 50 mg.

VARIATION

Chocolate Angel Food Cake

Substitute 3 oz (90 g) cocoa for 3 oz (90 g) flour.

Angel Food Cake

ICINGS: PRODUCTION AND APPLICATION

PRODUCING AND HANDLING BASIC TYPES

Icings or frostings (the two terms mean the same thing) are sweet coatings for cakes and other baked goods. Icings have three main functions:

1. They improve the keeping qualities of the cake by forming a protective coating around it.

2. They contribute flavor and richness.

3. They improve appearance.

There are six basic kinds of icing:

Fondant	Fudge-type icing
Buttercream	Flat-type icing
Foam-type icing	Royal or decorator's icing

In addition, we consider two other preparations for cakes:

Glazes	Fillings

Use top-quality flavorings for icings so they enhance the cake rather than detract from it. Use moderation when adding flavorings and colors. Flavors should be light and delicate. Colors should be delicate, pastel shades—except chocolate, of course.

GUIDELINES for Using Fondant

1. Heat fondant over a warm-water bath, stirring constantly, to thin the icing and make it pourable. Do not heat over 100°F (38°C), or it will lose its shine.

2. If the fondant is still too thick, thin it with a little simple sugar syrup or water (simple syrup blends in more easily).

3. Add flavorings and colorings as desired.

4. To make chocolate fondant, stir melted bitter chocolate into warm fondant until the desired color and flavor are reached. Chocolate thickens the fondant, so the icing may require more thinning with sugar syrup.

5. Apply fondant by pouring it over the item or by dipping items into it.

FONDANT

Fondant is a sugar syrup that is crystallized to a smooth, creamy white mass. It is familiar as the icing for napoleons, éclairs, petits fours, and some cakes. When applied, it sets up into a shiny, nonsticky coating.

Because it is difficult to make in the bakeshop, fondant is almost always purchased already prepared, either in ready-to-use moist form or in a dry form that requires only the addition of water.

BUTTERCREAM

Buttercream icings are light, smooth mixtures of fat and confectioners' sugar. They may also contain eggs to increase their smoothness or lightness. These popular icings are used for many kinds of cake. They are easily flavored and colored to suit a variety of purposes.

We consider three basic kinds of buttercream:

1. *Simple buttercreams* are made by creaming together fat and sugar to the desired consistency and lightness. A small quantity of egg whites may be whipped in.

   Decorator's buttercream is a simple buttercream used for making flowers and other cake decorations. It is creamed only a little because if too much air is beaten in, it would not be able to hold delicate shapes.

2. *Meringue-type buttercreams* are prepared by first beating egg whites and adding a boiling syrup or just sugar. Soft butter is then mixed into the meringue. This is a very light, smooth icing.

3. *French buttercreams* are similar to the meringue type, but the foam is made with egg yolks (and, sometimes, whole eggs) and boiling syrup. This is a very rich, light icing.

Butter, especially sweet, unsalted butter, is the preferred fat for buttercreams because of its flavor and melt-in-the-mouth quality. Icings made with shortening only can be unpleasant because the fat congeals and coats the inside of the mouth and does not melt. However, butter makes a less stable icing because it melts so easily. There are two ways around this problem:

1. Use buttercreams in cool weather only.
2. Blend a small quantity of emulsified shortening with the butter to stabilize it.

Simple Buttercream

YIELD: 2 LB 2 OZ (1049 G)

U.S.	METRIC	INGREDIENTS
8 oz	250 g	Butter
4 oz	125 g	Shortening
1 lb 4 oz	625 g	Confectioners' sugar (10X)
1.25 oz	40 g	Egg whites, pasteurized
1/2 tsp	2 mL	Lemon juice
3/4 tbsp	7 mL	Vanilla

PROCEDURE

1. Cream together the butter, shortening, and sugar until well blended, using the paddle attachment.
2. Add the egg whites, lemon juice, and vanilla. Blend in at medium speed. Then mix at high speed until light and fluffy.

Per 1 ounce (28.35 g): Calories, 150; Protein, 0 g; Fat, 10 g (58% cal.); Cholesterol, 15 mg; Carbohydrates, 16 g; Fiber, 0 g; Sodium, 65 mg.

VARIATIONS

Decorator's Buttercream

Use 12 oz (750 g) regular shortening, no butter. Omit lemon juice and vanilla. Reduce egg whites to 1 oz (30 g). Blend at low speed until smooth; do not whip.

Cream Cheese Icing

Substitute cream cheese for the butter and shortening. Omit egg whites. If necessary, thin the icing with cream or milk. If desired, flavor with grated lemon or orange zest instead of vanilla.

Meringue-Type Buttercream

YIELD: 1 LB 11 OZ (850 G)

U.S.	METRIC	INGREDIENTS
8 oz	250 g	Sugar
2 fl oz	60 mL	Water
4 oz	125 g	Egg whites
12 oz	375 g	Butter, soft
2 oz	60 g	Emulsified shortening
1/2 tsp	2 mL	Lemon juice
3/4 tsp	4 mL	Vanilla

PROCEDURE

1. Combine the sugar and water in a saucepan. Bring to a boil, stirring to dissolve the sugar.
2. Continue to boil until the syrup reaches a temperature of 240°F (115°C).
3. While the syrup is boiling, beat the egg whites in a clean, grease-free bowl, using the whip attachment, until they form firm, moist peaks. Do not overbeat.
4. As soon as the syrup reaches 240°F (115°C), pour it very slowly into the egg whites while the mixer is running at medium speed.
5. Continue to beat until the meringue is cool and forms stiff peaks. (You have now made an Italian meringue. For more information, see Chapter 35.)
6. With the mixer still running at medium speed, begin adding the butter, a little at a time. Add it just as fast as it can be absorbed by the meringue.
7. When all the butter is beaten in, add the shortening in the same way.
8. Beat in the lemon juice and vanilla.
9. Continue whipping until the buttercream is smooth.

Per 1 ounce (28.35 g): Calories, 130; Protein, 1 g; Fat, 10 g (67% cal.); Cholesterol, 20 mg; Carbohydrates, 10 g; Fiber, 0 g; Sodium, 90 mg.

 French Buttercream

YIELD: 1 LB 6 OZ (688 G)

U.S.	METRIC	INGREDIENTS	PROCEDURE
8 oz	250 g	Sugar	1. Combine the sugar and water in a saucepan. Bring to a boil, stirring to dissolve the sugar.
2 fl oz	60 mL	Water	2. Continue to boil until the syrup reaches a temperature of 240°F (115°C).
3 oz	90 g	Egg yolks	3. While the syrup is boiling, beat the yolks with the whip attachment until they are thick and light.
			4. As soon as the syrup reaches 240°F (115°C), pour it very slowly into the beaten yolks while the mixer is running at second speed.
			5. Continue to beat until the mixture is cool and the yolks are very light and thick.
10 oz	300 g	Butter, soft	6. With the mixer still running, add the butter, a little at a time. Add it just as fast as it can be absorbed by the mixture.
¾ tsp	4 mL	Vanilla	7. Beat in the vanilla. If the icing is too soft, refrigerate until it is firm enough to spread.

Per 1 ounce (28.35 g): Calories, 150; Protein, 1 g; Fat, 12 g (71% cal.); Cholesterol, 80 mg; Carbohydrates, 10 g; Fiber, 0 g; Sodium, 110 mg.

VARIATIONS

Flavored buttercreams are made by adding the desired flavoring to any of the basic buttercream recipes. In addition to the two variations given below, extracts and emulsions such as lemon, orange, and almond may be used.

Chocolate Buttercream

Add 4–5 oz (125–150 g) sweet chocolate, melted and cooled, to each 1 pound (500 g) buttercream.

Coffee Buttercream

For each 1 pound (500 g) buttercream, add 1½ tbsp (22 mL) instant coffee dissolved in 2 tsp (10 mL) hot water.

FOAM-TYPE ICING

Foam icings, sometimes called *boiled icings*, are simply meringues made with a boiling syrup. Some also contain stabilizing ingredients like gelatin. Foam-type icings should be applied thickly to cakes and left in peaks and swirls.

These icings are not stable. They should be used the day they are prepared. Italian meringue, discussed in Chapter 35, is the simplest foam-type icing. Follow the recipes on page 1002 but add 8 ounces (250 g) corn syrup to the sugar and water for the boiled syrup. The meringue is usually flavored with vanilla.

FLAT ICING

Flat icings, also called *water icings*, are simply mixtures of 10X sugar, water, and, sometimes, corn syrup and flavoring. They are used mostly for coffee cakes, Danish pastry, and sweet rolls. Flat icings are warmed to 100°F (38°C) for application and are handled like fondant.

FUDGE-TYPE ICING

Fudge icings are rich cooked icings. Many are made somewhat like candy. Fudge icings are heavy and thick, and they may be flavored with a variety of ingredients. They are used on cupcakes, layer cakes, loaf cakes, and sheet cakes.

Fudge icings are stable and hold up well on cakes and in storage. Stored icings must be covered tightly to prevent drying and crusting.

To use stored fudge icing, warm it in a double boiler until soft enough to spread.

Fudge-type icings do not necessarily contain chocolate. Plain white fudge icings may be flavored with vanilla, almond, maple, coffee, or other desired flavoring.

Flat Icing

YIELD: 1 LB 4 OZ (630 G)

U.S.	METRIC	INGREDIENTS	PROCEDURE
1 lb	500 g	Confectioners' sugar (10X or 6X)	1. Mix all ingredients until smooth.
3 fl oz	90 g	Water, hot	2. To use, place desired amount in a double boiler. Warm
1 fl oz	30 g	Corn syrup	to 100°F (38°C) and apply to the product to be iced.
³/₄ tsp	4 mL	Vanilla	

Per 1 ounce (28.35 g): Calories, 90; Protein, 0 g; Fat, 0 g (0% cal.); Cholesterol, 0 mg; Carbohydrates, 24 g; Fiber, 0 g; Sodium, 0 mg.

Caramel Fudge Icing

YIELD: 2 LB (1 KG)

INGREDIENTS	U.S.		METRIC	SUGAR AT 100% PERCENTAGE
Brown sugar	1 lb 8	oz	750 g	100 %
Milk	12	oz	375 g	50 %
Butter or part butter and part shortening	6	oz	188 g	25 %
Salt	0.1	oz (½ tsp)	2 g	0.4 %
Vanilla extract	0.25	oz	8 mL	1 %

Per 1 ounce (28.35 g): Calories, 130; Protein, 0 g; Fat, 4.5 g (33% cal.); Cholesterol, 15 mg; Carbohydrates, 21 g; Fiber, 0 g; Sodium, 65 mg.

PROCEDURE

1. Combine the sugar and milk in a saucepan. Bring to a boil, stirring to dissolve the sugar. Using a brush dipped in water, wash down the sides of the saucepan to prevent sugar crystals from forming. (See "Sugar Cooking," Chapter 36.)

2. Boil the mixture slowly, without stirring, until it reaches 240°F (115°C).

3. Pour the mixture into the bowl of a mixer. Add the butter and salt. Mix in with the paddle attachment.

4. Turn off the machine. Let the mixture cool to 110°F (43°C).

5. Add the vanilla and turn the machine on low speed. Beat the icing until it is smooth and creamy in texture. If it is too thick, thin it with a little cream or milk.

6. Spread on cooled cake while the icing is warm, or rewarm it in a double boiler.

Quick White Fudge Icing

YIELD: 2 LB 9 OZ (1300 G)

U.S.	METRIC	INGREDIENTS	PROCEDURE
4 fl oz	125 mL	Water	1. Place the water, butter, shortening, syrup, and salt in a saucepan. Bring to a boil.
2 oz	60 g	Butter	
2 oz	60 g	Emulsified shortening	
1.5 oz	45 g	Corn syrup	
½ tsp	2 mL	Salt	
2 lb	1 kg	Confectioners' sugar (10X or 6X)	2. Sift the sugar into the bowl of a mixer.
			3. Using the paddle attachment and with the machine running on low speed, add the boiling water mixture. Blend until smooth. Icing will become lighter the more it is mixed.
1½ tsp	8 mL	Vanilla	4. Blend in the vanilla.
			5. Use while still warm, or rewarm in a double boiler. If necessary, thin with hot water.

Per 1 ounce (28.35 g): Calories, 110; Protein, 0 g; Fat, 2.5 g (20% cal.); Cholesterol, 5 mg; Carbohydrates, 23 g; Fiber, 0 g; Sodium, 25 mg.

VARIATION

Quick Chocolate Fudge Icing

Omit the butter in the basic recipe. Beat in 6 oz (188 g) melted unsweetened chocolate after the boiling water has been added. Thin with more hot water as needed.

Cocoa Fudge Icing

YIELD: 2 LB 6 OZ (1188 G G)

U.S.	METRIC	INGREDIENTS	PROCEDURE
1 lb	500 g	Granulated sugar	1. Combine the granulated sugar, syrup, water, and salt in a saucepan. Bring to a boil, stirring to dissolve the sugar. Boil the mixture until it reaches 240°F (115°C). (See Sugar Cooking, Chapter 36.)
5 oz	150 g	Corn syrup	
4 fl oz	125 mL	Water	
½ tsp	2 mL	Salt	
4 oz	125 g	Butter or part butter and part emulsified shortening	2. While the sugar mixture is cooking, mix the fat, confectioners' sugar, and cocoa until evenly combined, using the paddle attachment of the mixer.
8 oz	250 g	Confectioners' sugar (10X or 6X)	3. With the machine running at low speed, very slowly pour in the hot syrup.
3 oz	90 g	Cocoa	
to taste	to taste	Vanilla	4. Mix in the vanilla. Continue to beat until the icing is smooth, creamy, and spreadable. If necessary, thin with a little hot water.
as needed	as needed	Hot water	
			5. Use while still warm, or rewarm in a double boiler.

Per 1 ounce (28.35 g): Calories, 110; Protein, 0 g; Fat, 2.5 g (20% cal.); Cholesterol, 5 mg; Carbohydrates, 22 g; Fiber, 1 g; Sodium, 60 mg.

VARIATION

Vanilla Fudge Icing

Use evaporated milk or light cream instead of water for the syrup. Omit cocoa.
Adjust consistency with additional confectioners' sugar (to thicken) or water (to thin).
Other flavorings may be used in place of vanilla, such as almond, maple, peppermint, or coffee.

ROYAL ICING

Royal icing, also called *decorating* or *decorator's icing*, is similar to flat icings except that it is much thicker and is made with egg whites, which make it hard and brittle when dry. It is used almost exclusively for decorative work.

To prepare royal icing:

1. Place the desired amount of 10X sugar in a mixing bowl. Add a small quantity of cream of tartar (for whiteness)—about 1/8 teaspoon per pound of sugar (1 g per kg).
2. Beat in egg white, a little at a time, until the sugar forms a smooth paste. You will need 2–3 ounces egg whites per pound of sugar (125 g per k).
3. Keep unused icing covered with a damp cloth at all times to prevent hardening.

GLAZES

Glazes are thin, glossy, transparent coatings that give shine to baked products and help prevent drying.

The simplest glaze is a sugar syrup or diluted corn syrup brushed onto coffee cakes or Danish while the glaze is hot. See Chapter 31 for recipe (p. 924). Syrup glazes may contain gelatin or waxy maize starch. Fruit glazes, the most popular being apricot, are available commercially prepared. They are melted, thinned with a little water, and brushed on while hot.

Fruit glazes may also be made by melting apricot or other preserves and forcing them through a strainer.

One of the most common uses of glazes in cake making is to coat the fruit arranged on the top of fruit tortes (see p. 947).

Ganache is a mixture of heavy cream and melted chocolate. Although it is not transparent like the glazes we have been discussing, it is used like other glazes to give a thin, shiny coating to cakes and other desserts. Making basic ganache is a simple procedure:

1. Heat heavy cream just to the boiling point and remove from the heat.
2. Add chopped chocolate and let stand until the chocolate has melted.
3. Stir until smooth.

Freshly made ganache can be used immediately, or it can be refrigerated and then rewarmed for later use. When it cools, it solidifies and can be used as the base for chocolate truffles.

Chocolate Ganache

YIELD: 1 LB 12 OZ (875 G)

U.S.	METRIC	INGREDIENTS	PROCEDURE
1 lb	500 g	Dark chocolate	1. Chop the chocolate into small pieces.
12 oz	375 g	Heavy cream	2. Bring the cream just to a boil, stirring to prevent scorching. (Use very fresh cream; old cream is more likely to curdle when boiled.)
			3. Remove from the heat and add the chocolate. Stir and let stand for a few minutes. Stir again until the chocolate is completely melted and the mixture is smooth.
			4. Apply to cakes as a glaze by pouring the ganache over the cake, as for fondant.

Per 1 ounce (28.35 g): Calories, 130; Protein, 1 g; Fat, 10 g (67% cal.); Cholesterol, 20 mg; Carbohydrates, 10 g; Fiber, 1 g; Sodium, 5 mg.

VARIATIONS

The proportion of chocolate and cream may be varied. For a firmer product, or if the weather is warm, decrease the cream to as little as half the weight of the chocolate. For a softer ganache, increase the cream to the same weight as the chocolate.

FILLINGS

Fillings are sometimes used instead of icings between cake layers. Fillings are also used in such products as jelly rolls, Danish, and other pastries.

1. **Fruit fillings.**

   Fruit fillings may be cooked or uncooked.

   Cooked fruit fillings are chopped or puréed fruits or fruit juices thickened with starch or eggs. They are prepared somewhat like pie fillings (see Chapter 35).

   Uncooked fruit fillings include jellies and preserves and dried fruits that have been ground and flavored (see recipes in Chapter 31). Fresh fruits, such as the strawberries in strawberry shortcake, are also used.

   Many ready-to-use fruit fillings are on the market.

2. **Cream fillings.**

   Cream fillings include pastry cream (recipes in Chapter 36) and various pudding-type preparations.

   Desserts with cream fillings should be assembled as close to service time as possible and kept refrigerated to avoid health hazards.

3. **Whipped cream.**

   Whipped cream is used as a dessert topping, filling, and frosting. See page 790 for instructions on whipping and handling heavy cream.

   Artificial whipped toppings resemble whipped cream in appearance. They should be used only if your customers actually like them.

ASSEMBLING AND ICING CAKES

SELECTION OF ICING

The flavor, texture, and color of the icing must be compatible with the cake.

1. **In general, use heavy frostings with heavy cakes and light frostings with light cakes.**

   For example, ice angel food cakes with a simple flat icing, fondant, or a light, fluffy boiled icing.

   High-ratio cakes go well with buttercreams and fudge-type icings.

   Shortened sponge layer cakes (genoise) are often combined with fruits or fruit fillings, light French or meringue-type buttercream, whipped cream, or flavored fondant.

2. **Use the best-quality flavorings, and use them sparingly. The flavor of the frosting should not be stronger than that of the cake.**

   Fudge-type icings may be flavored more strongly, as long as the flavor is of good quality.

3. **Use coloring sparingly. Light, pastel shades are more appetizing than loud colors.**

   Paste colors give the best results. Mix a little color with a small portion of the icing, then use this icing to color the rest.

SMALL CAKES

1. Cupcakes are iced by dipping the tops in a soft icing. Twist the cakes slightly and pull them out quickly in one smooth motion. Cupcakes may also be iced by spreading icing on with a spatula. Practice is necessary to develop speed and efficiency.

2. Petits fours are tiny cakes cut from sheet cakes. Select a cake that doesn't crumble easily. Carefully cut it into desired shapes. Remove all crumbs and place the cakes on a rack over a sheet pan. Ice by pouring fondant or flat icing over them to cover completely.

PROCEDURE: Assembling Layer Cakes

1. Cool cake layers completely before assembling and icing.

2. Trim layers, if necessary.
  - Remove any ragged edges.
  - Slightly rounded tops are easily covered by icing, but excessively large bumps may have to be cut off.
  - If desired, layers may be split horizontally. This makes the cake higher and increases the proportion of filling to cake. See Figure 33.5.

3. Brush all crumbs from cakes. Loose crumbs make the icing process difficult.

4. Place the bottom layer upside down (to give a flat surface for the filling) on a cardboard cake circle of the same diameter. Place the cake in the center of a cake turntable.

  If a cake circle or turntable is not available, place the cake on a serving plate and slip sheets of waxed paper or parchment under the edges of the cake to keep the plate clean.

5. Spread filling on the bottom layer out to the edges. If the filling is different from the icing for the outside of the cake, be careful not to spread the filling over the edges.

  Use the proper amount of filling. If applied too heavily, it will ooze out when the top layer is set in place.

6. Place the top layer on the bottom layer, right side up.

7. Ice the cake:
  - If a thin or light icing is used, pour or spread the icing onto the center of the cake. Then spread it to the edges and down the sides with a spatula.
  - If a heavy icing is used, it may be necessary to spread the sides first, then place a good quantity of icing in the center of the top and push it to the edges with the spatula.

  Pushing the icing rather than pulling or dragging it with the spatula prevents pulling up crumbs and getting them mixed with the icing.

  Use enough icing to cover the entire cake generously, but not excessively, with an even layer.

  Smooth the icing with the spatula or leave it textured or swirled, as desired.

  The finished, iced cake should have a perfectly level top and perfectly straight, even sides.

FIGURE 33.5 Cake layers may be split horizontally, using a long-bladed, serrated knife.

SHEET CAKES

Sheet cakes are ideal for volume service because they require little labor to bake, ice, and decorate, and they keep well as long as they are uncut.

  For special occasions, sheet cakes are sometimes decorated as a single unit with a design or picture in colored icing, a "Happy Special Occasion" message, and so on. It is more common, however, to ice them for individual service, as in the following procedure.

PROCEDURE: **Icing Sheet Cakes**

1. Turn out the cake onto the bottom of another sheet pan or tray, as described on page 949. Cool the cake thoroughly.

2. Trim the edges evenly with a serrated knife.

3. Brush all crumbs from the cake.

4. Place a quantity of icing in the center of the cake. With a spatula, push the icing to the edges. Smooth the top with the spatula, giving the entire cake an even layer of icing.

5. With a long knife or spatula, mark the entire cake off into portions, as in Figure 33.6, by pressing the back of the knife lightly into the icing. Do not cut the cake.

6. Using a paper cone or pastry bag fitted with a star tube, pipe a rosette or swirl of icing onto the center of each marked-off portion, or select another decoration, as desired. Whatever decorations you use, keep them simple, and make them the same for every portion. The finished sheet cake will resemble that in Figure 33.7.

7. Hold for service. Cut as close as possible to service time to keep the cake from drying.

FIGURE 33.6 Cake-cutting guides for sheet cakes and round layer cakes. For sheets measuring 13 × 18 in. (33 × 46 cm), simply divide the above diagrams for full-sized sheet cakes in half.

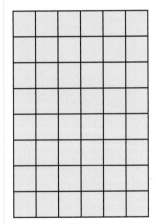

6 x 8 = 48 portions

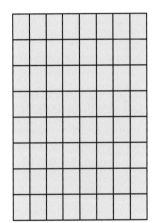

8 x 8 = 64 portions

8 x 12 = 96 portions

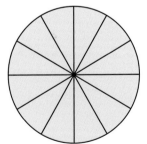

8-10 inch layers
12 portions

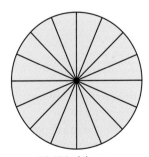

10-12 inch layers
16 portions

FIGURE 33.7 A finished sheet cake marked off into portions and decorated so each portion is identical.

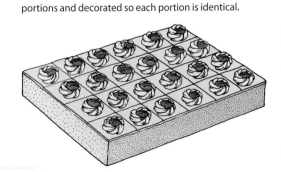

KEY POINTS TO REVIEW

- What are the six basic types of icing?
- What are the steps in the procedure for assembling and icing layer cakes?
- What are the steps in the procedure for icing sheet cakes?

TERMS FOR REVIEW

creaming method

two-stage method

blending method

foaming method

angel food method

chiffon method

fondant

buttercream

foam icing

flat icing

fudge icing

royal icing

glaze

ganache

QUESTIONS FOR DISCUSSION

1. Briefly list the steps in each of the four basic cake-mixing methods presented in this chapter.

2. What are the reasons, in the creaming method, for creaming the butter and sugar until the mixture is light and fluffy?

3. In both the creaming method and the two-stage method, scraping down the sides of the bowl is emphasized. Why is this necessary?

4. What might the finished product be like if you tried to mix a low-fat cake by the two-stage method? Explain.

5. Examine the following cake formulas and indicate which mixing method you would use for each.

6. What is the most important rule to consider when using fondant?

7. Compare the keeping qualities of simple buttercreams and meringue-type buttercreams.

8. List the steps in assembling and icing a three-layer cake.

Cake 1

2 lb			Cake flour	1 kg	
1 lb	2	oz	Emulsified shortening	525 g	
	1	oz	Salt	30 g	
1 lb	1.5 oz		Baking powder	45 g	
2 lb	8	oz	Fine granulated sugar	1250 g	
1 lb			Skim milk	500 g	
1 lb	5	oz	Whole eggs	650 g	
	10	oz	Skim milk	300 g	

Cake 2

3 lb			Whole eggs	1.5 kg	
1 lb			Egg yolks	500	g
2 lb	4 oz		Sugar	1125	g
2 lb			Cake flour	1	kg
	6 oz		Cornstarch	175	g
	6 oz		Melted butter	175	g

Cake 3

1 lb			Butter	500	g
	8	oz	Shortening	250	g
4 lb			Sugar	2	kg
	1	oz	Vanilla	30	g
1 lb	4	oz	Whole eggs	625	g
3 lb	8	oz	Cake flour	1750	g
	10	oz	Cocoa powder	300	g
	1.5 oz		Baking soda	45	g
3 lb			Buttermilk	1.5 kg	
1 lb	8	oz	Water	750	g

34

COOKIES

The word *cookie* means "small cake," and that's exactly what a cookie is. In fact, some cookies are made from cake batter. Some products, such as certain kinds of brownies, are difficult to classify as cakes or cookies.

Most cookie formulas, however, call for less liquid than cake formulas do. Cookie doughs range from soft to very stiff, unlike the thinner batters for cakes. This difference in moisture content means some differences in mixing methods, although the basic procedures are much like those for cakes.

The most apparent differences between cakes and cookies are in the makeup. Because most cookies are individually formed or shaped, a great deal of hand labor is involved. Learning correct methods and practicing diligently are essential for efficiency.

AFTER READING THIS CHAPTER, YOU SHOULD BE ABLE TO

1. List the factors responsible for crispness, softness, chewiness, and spread in cookies.

2. Demonstrate the three basic cookie mixing methods.

3. Prepare the seven basic cookie types: dropped, bagged, rolled, molded, icebox, bar, and sheet.

4. Prepare pans for, bake, and cool cookies.

COOKIE CHARACTERISTICS AND THEIR CAUSES

Cookies come in an infinite variety of shapes, sizes, flavors, and textures. Characteristics that are desirable in some are not desirable in others. For example, we want some cookies to be crisp and others to be soft. We want some to hold their shape and others to spread during baking. In order to produce the characteristics we want and to correct faults, it is useful to know what causes these characteristics.

CRISPNESS

Cookies are crisp if they are very low in moisture. The following factors contribute to crispness:

1. Low proportion of liquid in the mix. Most crisp cookies are made from a stiff dough.
2. High sugar and fat content.
3. Evaporation of moisture during baking due to high temperatures and/or long baking.
4. Small size or thin shape, so the cookies dry quickly during baking.
5. Proper storage. Crisp cookies can become soft if they absorb moisture.

SOFTNESS

Softness is the opposite of crispness, so it has the opposite causes, as follows:

1. High proportion of liquid in mix.
2. Low sugar and fat.
3. Honey, molasses, or corn syrup included in formulas. These sugars are hygroscopic, which means they readily absorb moisture from the air or from their surroundings.
4. Underbaking.
5. Large size or thick shape. The cookies retain moisture.
6. Proper storage. Soft cookies can become stale and dry if not tightly covered or wrapped.

CHEWINESS

Moisture is necessary for chewiness, but other factors are also required. In other words, all chewy cookies are soft, but not all soft cookies are chewy. These factors contribute to chewiness:

1. High sugar and liquid content, but low fat content.
2. High proportion of eggs.
3. Strong flour, or gluten developed during mixing.

SPREAD

Spread is desirable in some cookies, while others must hold their shape. Several factors contribute to spread or lack of spread:

1. **Sugar.**
   High sugar content increases spread. Coarse granulated sugar increases spread, whereas fine sugar or confectioners' sugar reduces spread.
2. **Leavening.**
   High baking soda or baking ammonia content encourages spread. So does long creaming, which incorporates air.
3. **Temperature.**
   Low oven temperature increases spread. High temperature decreases spread because the cookie sets up before it has a chance to spread too much.
4. **Liquid.**
   A slack batter—that is, one with a high liquid content—spreads more than a stiff dough.

5. Flour.

Strong flour or activation of gluten decreases spread.

6. Pan grease.

Cookies spread more if baked on a heavily greased pan.

MIXING METHODS

Cookie-mixing methods are much like cake-mixing methods. The major difference is that less liquid is usually incorporated, so mixing is somewhat easier.

Less liquid means gluten is less developed by the mixing. Also, a smooth, uniform mix is easier to obtain.

The three most important cookie mixing methods are the following:

1. One-stage

2. Creaming

3. Sponge

These methods are subject to many variations due to differences in formulas. The general procedures are as follows. Be sure, however, to follow the exact instructions when a formula indicates a variation in the procedure.

ONE-STAGE METHOD

The **one-stage method** is the counterpart of the blending or two-stage cake-mixing method, discussed in the previous chapter. Cake batters have more liquid, so it must be added in two or more stages in order to blend uniformly. Low-moisture cookies, on the other hand, can be mixed all in one stage.

CREAMING METHOD

The **creaming method** for cookies is nearly identical to the creaming method for cakes. Because cookies require less liquid, it is usually not necessary to add the liquid alternately with the flour. It can be added all at once.

SPONGE METHOD

The **sponge method** for cookies is essentially the same as the egg-foam methods for cakes. The procedure varies considerably, depending on the ingredients. Batches should be kept small because the batter is delicate.

PROCEDURE for One-Stage Method

1. Scale ingredients accurately. Have all ingredients at room temperature.

2. Place all ingredients in mixer. With the paddle attachment, mix at low speed until uniformly blended. Scrape down the sides of the bowl as necessary.

PROCEDURE for Creaming Method

1. Scale ingredients accurately. Have all ingredients at room temperature.

2. Place the fat, sugar, salt, and spices in the mixing bowl. With the paddle attachment, cream these ingredients at low speed.

For light cookies, cream until the mix is light and fluffy, incorporating more air for leavening.

For a dense, chewy cookie, cream only slightly.

3. Add the eggs and liquid, if any, and blend in at low speed.

4. Sift in the flour and leavening. Mix until just combined.

PROCEDURE for Sponge Method

1. Scale all ingredients accurately. Have all ingredients at room temperature, or warm the eggs slightly for greater volume, as for sponge cakes.

2. Following the procedure given in the formula used, whip the eggs (whole, yolks, or whites) and the sugar to the proper stage: soft peaks for whites, thick and light for whole eggs or yolks.

3. Fold in the remaining ingredients as specified in the recipe. Be careful not to overmix or to deflate the eggs.

KEY POINTS TO REVIEW

- What are the factors that determine whether a cookie will be crisp, soft, or chewy?

- What are the factors that determine how much a cookie will spread when baked?

- What are the steps in the one-stage method for mixing cookies?

- What are the steps in the creaming method for mixing cookies?

- What are the steps in the sponge method for mixing cookies?

TYPES AND MAKEUP METHODS

We can classify cookie types by makeup method as well as by mixing method. Grouping by the makeup method is perhaps more useful from the point of view of production because mixing methods are relatively simple, whereas makeup procedures vary considerably.

In this section, we present basic procedures for producing seven cookie types:

1. Dropped
2. Bagged
3. Rolled
4. Molded
5. Icebox
6. Bar
7. Sheet

No matter what makeup method you use, follow one important rule: *Make all cookies of uniform size and thickness*. This is essential for even baking. Because baking times are so short, small cookies may burn before large ones are done.

DROPPED COOKIES

Dropped cookies are made from a soft dough or batter. They are fast and easy to make up. Many sponge or foam-type batters are made up as dropped cookies.

1. Select the proper size scoop for accurate portioning.

   A No. 30 scoop makes a large cookie, about 1 ounce (30 g).

   A No. 40 scoop makes a medium cookie.

   Nos. 50, 60, or smaller scoops make small cookies.

2. Drop the cookies onto the prepared baking sheets. Allow enough space between cookies for spreading.

3. Rich cookies spread by themselves. However, if the formula requires it, flatten the mounds of batter slightly with a weight dipped in sugar.

BAGGED COOKIES

Bagged cookies, or pressed cookies, are also made from soft doughs. The dough must be soft enough to be forced through a pastry bag but stiff enough to hold its shape.

1. Fit a pastry bag with a tip of the desired size and shape. Fill the bag with the cookie dough. Review Figure 12.1 for tips on use of the pastry bag.

2. Press out cookies of desired shape and size directly onto prepared cookie sheets.

ROLLED COOKIES

Rolled cookies, which are cut from a stiff dough, are not often made in commercial food service because they require excessive labor. Also, scraps are always left over after cutting. When rerolled, these scraps make inferior, tough cookies.

1. Chill dough thoroughly.

2. Roll dough to $1/8$ inch (3 mm) thick on a floured canvas or floured workbench. Use as little flour as possible for dusting because the flour can toughen the cookies.

3. Cut out cookies with cookie cutters and place on prepared baking sheets. Cut as close together as possible to reduce the quantity of scraps.

MOLDED COOKIES

The first part of the procedure for **molded cookies** (steps 1 and 2) is simply a fast and fairly accurate way of dividing the dough into equal portions. Each piece is then molded into the desired shape. This usually consists of simply flattening the pieces out with a weight. For some traditional cookies, special molds are used to flatten the dough and, at the same time, stamp it with a design.

The pieces may also be shaped by hand into crescents, fingers, or other shapes:

1. Roll the dough into long cylinders about 1 inch (2.5 cm) thick, or whatever size is required. (Refrigerate the dough if it is too soft to handle.)
2. With a knife or bench scraper, cut the roll into 1-ounce (30-g) pieces, or whatever size is required.
3. Place the pieces on prepared baking sheets, leaving 2 inches (5 cm) of space between them.
4. Flatten cookies with a weight (such as a can) dipped in granulated sugar after pressing each cookie.

    A fork is sometimes used for flattening the dough, as for peanut butter cookies.
5. Alternative method: After step 2, shape the dough by hand into desired shapes.

ICEBOX COOKIES

The icebox method, or refrigerator method, is ideal for operations that wish to have freshly baked cookies on hand at all times. The rolls of dough may be made up in advance and stored. **Icebox cookies** can easily be cut and baked as needed.

1. Scale dough into pieces of uniform size, from $1^1/_2$ pounds (700 g), if you are making small cookies, to 3 pounds (1400 g), for large cookies.
2. Form the dough into cylinders from 1 to 2 inches (2.5 to 5 cm) in diameter, depending on the size cookie desired.

    For accurate portioning, it is important to make all the cylinders of dough the same thickness and length.
3. Wrap the cylinders in parchment or waxed paper, place them on sheet pans, and refrigerate overnight.
4. Unwrap the dough and cut into slices of uniform thickness. The exact thickness required depends on the size of the cookie and how much the dough spreads during baking. The usual range is from $1/_8$ to $1/_2$ inch (3 to 12 mm).

    A slicing machine is recommended for ensuring even thickness. Doughs containing nuts or fruits should be sliced by hand with a knife.
5. Place the slices on prepared baking sheets, allowing 2 inches (5 cm) of space between cookies.

BAR COOKIES

Bar cookies are so called because the dough is shaped into long bars, which are baked and then cut. After cutting, they may be baked again, as in the case of biscotti, which means "twice baked." Do not confuse bar cookies with sheet cookies (see below), which are often called *bars* by consumers.

1. Scale the dough into $1^3/_4$-pound (800-g) units (1-pound units, or 500-g units, may be used for smaller cookies).
2. Shape the pieces of dough into cylinders the length of the sheet pans. Place three strips on each greased pan, spacing them well apart.
3. Flatten the dough with the fingers into strips about 3–4 inches wide and about $1/_4$ inch thick (8–10 cm wide, 6 mm thick).
4. If required, brush with egg wash.
5. Bake as directed in the formula.
6. After baking, while cookies are still warm, cut each strip into bars about $1^3/_4$ inches (4.5 cm) wide.

SHEET COOKIES

Sheet cookies vary so much that it is nearly impossible to give a single procedure for all of them. Some of them are almost like sheet cakes, only denser and richer. They may even be iced like sheet cakes. Others consist of 2 or 3 layers added and baked in separate stages. The following procedure is a general guideline only:

1. Spread the cookie mixture into prepared sheet pans. Make sure the thickness is even.
2. If required, add topping, or brush with an egg wash.
3. Bake as directed. Cool.
4. Apply icing or topping, if any.
5. Cut into individual squares or rectangles.

PANNING, BAKING, AND COOLING

PREPARING THE PANS

1. Use clean, unwarped pans.
2. Lining the sheets with parchment or silicone paper is fast, and it eliminates the necessity of greasing the pans.
3. A heavily greased pan increases the spread of the cookie. A greased and floured pan decreases spread.
4. Some high-fat cookies can be baked on ungreased pans.

BAKING

1. Most cookies are baked at a relatively high temperature for a short time.
2. Too low a temperature increases spreading and may produce hard, dry, pale cookies.
3. Too high a temperature decreases spreading and may burn the edges or bottoms.
4. Even one minute of overbaking can burn cookies, so watch them closely. The heat of the pan continues to bake the cookies even after they are removed from the oven.
5. Doneness is indicated by color. The edges and bottoms should just be turning a light golden color.
6. With some rich doughs, burnt bottoms may be a problem. In this case, double-pan the cookies by placing the sheet pan on a second pan of the same size.

COOLING

1. Remove the cookies from the pans while they are still warm, or they may stick.
2. If the cookies are very soft, do not remove them from the pans until they are cool enough and firm enough to handle. Cookies may be soft when hot but become crisp when cool.
3. Do not cool cookies too rapidly or in cold drafts, or they may crack.
4. Cool completely before storing.

KEY POINTS TO REVIEW

- What are the seven makeup methods for making cookies? Describe the procedure for each.

- How are pans prepared for baking cookies?

- When baking cookies, how do you determine when they are done?

- What is the proper way to cool cookies after baking?

Chocolate Chip Cookies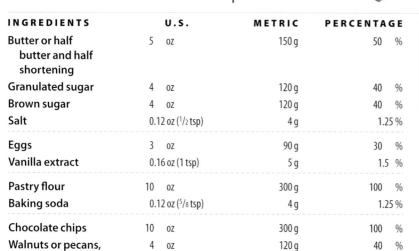

INGREDIENTS	U.S.		METRIC	PERCENTAGE	
Butter or half butter and half shortening	5	oz	150 g	50	%
Granulated sugar	4	oz	120 g	40	%
Brown sugar	4	oz	120 g	40	%
Salt	0.12 oz ($^1/_2$ tsp)		4 g	1.25	%
Eggs	3	oz	90 g	30	%
Vanilla extract	0.16 oz (1 tsp)		5 g	1.5	%
Pastry flour	10	oz	300 g	100	%
Baking soda	0.12 oz ($^5/_8$ tsp)		4 g	1.25	%
Chocolate chips	10	oz	300 g	100	%
Walnuts or pecans, chopped	4	oz	120 g	40	%
Total weight:	2 lb 8	oz	1213 g	404	%

Per 1 ounce (28.35 g): Calories, 130; Protein, 2 g; Fat, 7 g (47% cal.); Cholesterol, 15 mg; Carbohydrates, 16 g; Fiber, 2 g; Sodium, 85 mg.

PROCEDURE

Mixing:

Creaming method. Blend in chocolate chips and nuts last.

Makeup:

Drop method. Use greased or parchment-lined baking sheets.

Baking:

375°F (190°C), 8–12 minutes, depending on size.

VARIATIONS

Brown Sugar Nut Cookies

Omit granulated sugar and use 80% (8 oz/240 g) brown sugar. Omit chocolate chips and increase nuts to 100% (10 oz/300 g).

TOLL HOUSE COOKIES

Chocolate chip cookies, in their many varieties, are the most popular cookies in North America. They owe their origin to the Toll House cookie, said to have been developed in the 1920s or 1930s by Ruth Wakefield, owner of the Toll House Inn in Whitman, Massachusetts. The original Toll House cookies are simple butter cookies with semisweet chocolate morsels mixed into the dough. Today's chocolate chip or chocolate chunk cookies are likely to contain any kind of chocolate plus other ingredients, especially nuts, such as pecans, walnuts, or macadamia nuts.

Oatmeal Raisin Cookies

INGREDIENTS	U.S.		METRIC	PERCENTAGE	
Butter and/or shortening	8	oz	250 g	67	%
Brown sugar	1 lb		500 g	133	%
Salt	1	tsp	5 g (5 mL)	1.5	%
Eggs	4	oz	125 g	33	%
Vanilla	2	tsp	10 mL	3	%
Milk	1	oz	30 g	8	%
Pastry flour	12	oz	375 g	100	%
Baking powder	0.5 oz (1 tbsp)		15 g	4	%
Baking soda	0.25 oz (1$^1/_2$ tsp)		8 g	2	%
Rolled oats (quick cooking)	10	oz	300 g	83	%
Raisins (see Note)	8	oz	250 g	67	%
Total weight:	3 lb 11	oz	1858 g	501	%

Per 1 ounce (28.35 g): Calories, 110; Protein, 2 g; Fat, 4 g (31% cal.); Cholesterol, 15 mg; Carbohydrates, 18 g; Fiber, 1 g; Sodium, 135 mg.

PROCEDURE

Mixing:

Creaming method. Combine oats with other dry ingredients after they are sifted. Mix raisins into dough last.

Makeup:

Drop method. Use greased or parchment-lined baking sheets.

Baking:

375°F (190°C), 10–12 minutes, depending on size.

Note: If raisins are hard and dry, soak them in hot water 30 minutes, drain, and dry well before adding to cookie dough.

Butter Tea Cookies

INGREDIENTS	U.S.		METRIC	PERCENTAGE	
Butter or half butter and half shortening	1 lb		500 g	67	%
Granulated sugar	8	oz	250 g	33	%
Confectioners' sugar	4	oz	125 g	17	%
Eggs	6	oz	175 g	25	%
Vanilla (or almond extract)	1½ tsp		8 mL	1	%
Cake flour	1 lb 8	oz	750 g	100	%
Total weight:	3 lb 10	oz	1823 g	243	%

PROCEDURE

Mixing:

Creaming method.

Makeup:

Bagged method. Make small cookies, about 1 in. (2.5 cm) in diameter, using a star tube or plain tube. Bag out onto ungreased or parchment-lined baking sheets.

Baking:

375°F (190°C), about 10 minutes.

Per 1 ounce (28.35 g): Calories, 130; Protein, 1 g; Fat, 7 g (50% cal.); Cholesterol, 30 mg; Carbohydrates, 15 g; Fiber, 0 g; Sodium, 70 mg.

VARIATIONS

Almond Tea Cookies

Add 17% (4 oz/125 g) almond paste. Blend it thoroughly with the sugar before adding the butter.

Sandwich-Type Cookies

Select cookies with the same size and shape. Turn half of them over and dot the centers of the flat sides with small amount of jam or fudge icing. Sandwich with the remaining cookies.

Chocolate Tea Cookies

Substitute 6 oz (175 g) cocoa for 6 oz (175 g) flour.

Sugar Cookies

INGREDIENTS	U.S.		METRIC	PERCENTAGE	
Butter and/or shortening	8	oz	250 g	40	%
Sugar	10	oz	310 g	50	%
Salt	0.16	oz (¾ tsp)	5 g	0.8	%
Eggs	2	oz	60 g	10	%
Milk	2	oz	60 g	10	%
Vanilla extract	0.25	oz	8 g	1.25	%
Cake flour	1 lb 4	oz	625 g	100	%
Baking powder	0.625	oz	18 g	3	%
Total weight:	2 lb 11	oz	1336 g	215	%

PROCEDURE

Mixing:

Creaming method.

Makeup:

Rolled method. Before cutting the rolled-out dough, wash with milk and sprinkle with sugar. Use greased or parchment-lined baking sheets.

Baking:

375°F (190°C), 8–10 minutes.

Per 1 ounce (28.35 g): Calories, 120; Protein, 1 g; Fat, 4.5 g (36% cal.); Cholesterol, 15 mg; Carbohydrates, 17 g; Fiber, 0 g; Sodium, 150 mg.

VARIATIONS

Lemon rind, extract, or emulsion may be used in place of vanilla.

Rolled Brown Sugar Cookies

Increase butter to 50% (10 oz/310 g). Omit granulated sugar and use 60% (12 oz/375 g) brown sugar.

Rolled Chocolate Cookies

Substitute 2 oz (60 g) cocoa for 2 oz (60 g) flour.

Shortbread Cookies

INGREDIENTS	U.S.		METRIC	PERCENTAGE	
Butter	12	oz	375 g	75	%
Sugar	8	oz	250 g	50	%
Salt	0.12 oz (3/4 tsp)		4 g (3 mL)	0.75	%
Egg yolks (see Note)	4	oz	125 g	25	%
Flavoring (optional; see Note)					
Pastry flour	1 lb		500 g	100	%
Total weight:	2 lb 8	oz	2508 g	250	%

PROCEDURE

Mixing:

Creaming method.

Makeup:

Rolled method. Roll dough ¼ inch (0.5 cm) thick (this is thicker that most rolled cookies). Use greased or parchment-lined baking sheets.

Baking:

350°F (175°C), about 15 minutes.

Per 1 ounce (28.35 g): Calories, 130; Protein, 2 g; Fat, 8 g (53% cal.); Cholesterol, 55 mg; Carbohydrates, 14 g; Fiber, 2 g; Sodium, 105 mg.

Note: Traditional Scottish shortbread is made with butter, flour, and sugar—no eggs, flavoring, or liquid. Because the dough is crumbly, it is not rolled out but, rather, pressed into pans or molds and baked, and then cut while still hot. For the recipe given here, you may make the cookies without added flavoring or flavor to taste with vanilla, almond, or lemon.

Molasses Cookies

INGREDIENTS	U.S.		METRIC	PERCENTAGE	
Butter	13	oz	405 g	54	%
Brown sugar	1 lb 3	oz	590 g	79	%
Eggs	4.5	oz	140 g	19	%
Molasses	10	oz	315 g	42	%
Pastry flour	1 lb 8	oz	750 g	100	%
Baking soda	0.5	oz	15 g	2	%
Salt	0.33 oz		10 g	1.4	%
Ginger	0.2	oz	6 g	0.8	%
Cinnamon	0.2		6 g	0.8	%
Ground cloves	0.1		3 g	0.4	%
Total weight:	4 lb 7	oz	2240 g	299	%

PROCEDURE

Mixing:

Creaming method

Makeup:

Molded method. For large cookies, roll dough into cylinders 1½ inches (4 cm) thick, and cut into 2-oz (60-g) portions. Roll each piece into a ball and roll in sugar. Pan on parchment-lined sheets. Flatten slightly with weight.

Baking:

375°F (175°C) for about 10–12 minutes.

Per cookie: Calories, 240; Protein, 2 g; Fat, 9 g (34% cal.); Cholesterol, 35 mg; Carbohydrates, 38 g; Fiber, <1 g; Sodium, 290 mg.

Molasses Cookies

Snickerdoodles

INGREDIENTS	U.S.		METRIC		PERCENTAGE	
Butter		15 oz	450	g	75	%
Sugar	1 lb		480	g	80	%
Eggs		5 oz	150	g	25	%
Vanilla extract		0.5 oz	15	g	2.5	%
Pastry flour	1 lb	4 oz	600	g	100	%
Baking powder		0.2 oz	6	g	1	%
Salt		0.16 oz	4.8	g	0.8	%
Total dough weight:	3 lb	8 oz	1705	g	283	%
For coating:						
Cinnamon Sugar (p. 924)	as needed		as needed			

PROCEDURE

Mixing:

Creaming method

Makeup:

Molded method. For large cookies, roll dough into cylinders 1¹/₂ inches (4 cm) thick, and cut into 2-oz (60-g) portions. Roll each piece into a ball and roll in cinnamon sugar. Pan on parchment-lined sheets.

Baking:

375°F (175°C) for about 10–12 minutes. Do not allow to brown.

Per cookie: Calories, 270; Protein, 3 g; Fat, 13 g (43% cal.); Cholesterol, 50 mg; Carbohydrates, 35 g; Fiber, <1 g; Sodium, 200 mg.

Raisin Spice Bars

INGREDIENTS	U.S.		METRIC		PERCENTAGE	
Sugar	1 lb	4 oz	580 g		83	%
Butter and/or shortening		8 oz	230 g		33	%
Eggs		8 oz	230 g		33	%
Molasses		4 oz	115 g		17	%
Pastry flour	1 lb	8 oz	700 g		100	%
Cinnamon		2 tsp	3 g (10 mL)		0.5	%
Ground cloves		¹/₂ tsp	1 g (2 mL)		0.16	%
Ground ginger		1 tsp	2 g (5 mL)		0.3	%
Baking soda		³/₄ tsp	3 g (3 mL)		0.5	%
Salt		1 tsp	5 g (5 mL)		0.75	%
Raisins (see note)	1 lb		470 g		67	%
Total weight:	5 lb		2339 g		335	%

PROCEDURE

Mixing:

One-stage method.

Makeup:

Bar method. Eggwash with whole eggs or egg whites.

Baking:

350°F (175°C), about 15 minutes.

Per 1 ounce (28.35 g): Calories, 100; Protein, 1 g; Fat, 2.5 g (21% cal.); Cholesterol, 15 mg; Carbohydrates, 20 g; Fiber, 1 g; Sodium, 65 mg.

Note: If raisins are hard and dry, soak them in hot water 30 minutes, drain, and dry well before adding to the mix.

Raisin Spice Bars

Peanut Butter Cookies

INGREDIENTS	U.S.	METRIC	PERCENTAGE
Butter and/or shortening	12 oz	375 g	75 %
Brown sugar	8 oz	250 g	50 %
Granulated sugar	8 oz	250 g	50 %
Salt	1 tsp	5 g (5 mL)	1 %
Peanut butter	12 oz	375 g	75 %
Eggs	4 oz	125 g	25 %
Vanilla extract	2 tsp	10 g	2 %
Pastry flour	1 lb	500 g	100 %
Baking soda	1 tsp	5 g (5 mL)	1 %
Total weight:	3 lb 12 oz	1895 g	379 %

PROCEDURE

Mixing:

Creaming method. Cream peanut butter with the fat and sugar.

Makeup:

Molded method. Use a fork instead of a weight to flatten the cookies. Use greased or parchment-lined baking sheets.

Baking:

375°F (190°C), 8–12 minutes, depending on size.

Per 1 ounce (28.35 g): Calories, 130; Protein, 2 g; Fat, 8 g (53% cal.); Cholesterol, 20 mg; Carbohydrates, 14 g; Fiber, 1 g; Sodium, 135 mg.

Icebox Cookies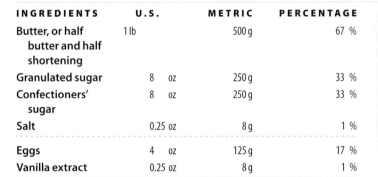

INGREDIENTS	U.S.			METRIC	PERCENTAGE
Butter, or half butter and half shortening	1 lb			500 g	67 %
Granulated sugar		8	oz	250 g	33 %
Confectioners' sugar		8	oz	250 g	33 %
Salt		0.25	oz	8 g	1 %
Eggs		4	oz	125 g	17 %
Vanilla extract		0.25	oz	8 g	1 %
Pastry flour	1 lb	8	oz	750 g	100 %
Total weight:	3 lb	12	oz	1891 g	252 %

PROCEDURE

Mixing:

Creaming method.

Makeup:

Icebox method. Scale dough strips to 1 1/2 lb (750 g) each. Slice cookies 1/4 in. (1/2 cm) thick. Bake on ungreased pans.

Baking:

375°F (190°C), about 12 minutes.

Per 1 ounce (28.35 g): Calories, 120; Protein, 1 g; Fat, 6 g (44% cal.); Cholesterol, 25 mg; Carbohydrates, 16 g; Fiber, 1 g; Sodium, 110 mg.

VARIATIONS

To reduce spread, use all confectioners' sugar.

Butterscotch Icebox Cookies

In place of sugars in basic recipe, use 67% (1 lb/500g) brown sugar and use only butter. Increase eggs to 20% (5 oz/150 g). Add 1/2 tsp (2 g or 2 mL) baking soda to the flour.

Chocolate Icebox Cookies

Add 17% (4 oz/125 g) melted unsweetened chocolate to the creamed butter and sugar.

Nut Icebox Cookies

Add 25% (6 oz/188 g) finely chopped nuts to the sifted flour in the basic recipe or the Butterscotch or Chocolate Cookie recipes.

Brownies 🌹

INGREDIENTS	U.S.		METRIC	PERCENTAGE
Unsweetened chocolate	1 lb		450 g	100 %
Butter	1 lb 8	oz	675 g	150 %
Eggs	1 lb 8	oz	675 g	150 %
Sugar	3 lb		1350 g	300 %
Salt	0.25 oz (1½ tsp)		7 g (7 mL)	1.5 %
Vanilla	1	oz	30 mL	6 %
Cake flour	1 lb		450 g	100 %
Chopped walnuts or pecans	1 lb		450 g	100 %
Total weight:	9 lb 1	oz	4087 g	907 %

PROCEDURE

Mixing:

Sponge method.

1. Melt chocolate and butter together in a double boiler. Stir so that the mixture is smooth. Let it cool to room temperature.
2. Blend the eggs, sugar, and salt until well mixed, but do not whip. Add the vanilla.
3. Blend in the chocolate mixture.
4. Sift the flour and fold it in.
5. Fold in the nuts.

Makeup:

Sheet method. Grease and flour the pans or line them with parchment. Quantity of basic recipe is enough for 1 full sheet pan, 18 × 26 in. (46 × 66 cm), 2 half-size sheet pans, 4 pans measuring 9 × 13 in. (23 × 33 cm), or 6 square pans measuring 9 in. (23 cm) per side.

If desired, batter may be sprinkled with an additional 50% (8 oz/255 g) chopped nuts after panning.

Baking:

325°F (165°C), about 60 minutes. For 2-in. (5-cm) square brownies, cut sheet pan 8 × 12 to yield 96 pieces.

Per 1 brownie: Calories, 190; Protein, 3 g; Fat, 12 g (54% cal.); Cholesterol, 45 mg; Carbohydrates, 20 g; Fiber, 1 g; Sodium, 95 mg.

VARIATION

Butterscotch Brownies or Blondies

Omit chocolate. Use brown sugar instead of granulated sugar. Increase flour to 1 lb 6 oz (600 g).

Almond Biscotti

INGREDIENTS	U.S.		METRIC	PERCENTAGE	
Eggs	10 oz		300 g	35	%
Sugar	1 lb	2 oz	550 g	65	%
Salt		0.4 oz (2 tsp)	12 g (10 mL)	2	%
Vanilla		0.3 oz (2 tsp)	8 mL	1	%
Grated orange zest		0.1 oz (1¼ tsp)	3 g	0.5	%
Pastry flour	1 lb	12 oz	850 g	100	%
Baking powder		0.7 oz	20 g	2.5	%
Blanched whole almonds		10 oz	300 g	35	%
Total weight:	4 lb	3 oz	2047 g	241	%

PROCEDURE

Mixing:

Sponge method.

1. Combine the eggs, sugar, and salt. Stir over hot water to warm the mixture. Whip until thick and light.
2. Fold in the vanilla and orange zest.
3. Sift together the flour and baking powder. Fold in the egg mixture.
4. Mix in the almonds.

Makeup:

Bar method. Shape into logs about 2–2½ in. (6 cm) thick. Dust your hands and the workbench with flour. The dough will be soft, sticky, and difficult to handle, but the logs do not have to be perfectly shaped. Egg wash.

Baking:

325°F (160°C) about 30–40 minutes, until light golden.

Finishing:

Let cool slightly. Slice diagonally about ½ inch (12 mm) thick. Place slices cut side down on sheet pans. Bake at 275°F (135°C) until toasted and golden brown, about 30 minutes.

Per 1 ounce (28.35 g): Calories, 110; Protein, 3 g; Fat, 4.5 g (36% cal.); Cholesterol, 15 mg; Carbohydrates, 15 g; Fiber, 2 g; Sodium, 110 mg.

ADDITIONAL RECIPES

These additional recipes may be found on your CulinarE-Companion recipe management program:

Cinnamon Cookies; Coconut Macaroons (Meringue Type); Ladyfingers.

TERMS FOR REVIEW

one-stage method	dropped cookies	molded cookies	bar cookies
creaming method	bagged cookies	icebox cookies	sheet cookies
sponge method	rolled cookies		

QUESTIONS FOR DISCUSSION

1. What makes cookies crisp, and how can you keep them crisp after they are baked?
2. If you baked some cookies that were unintentionally chewy, how would you correct them in the next batch?
3. Describe briefly the difference between the creaming method and the one-stage method.
4. Besides cost control, why are accurate scaling and uniform sizing important when making up cookies?

PIES AND PASTRIES

On the North American frontier, it was not uncommon for the pioneer house-wife to bake 21 pies a week—one for every meal. Pies were so important to the settlers that in winter, when fruits were unavailable, cooks would bake pies for dessert out of whatever materials were available, such as potatoes, vinegar, and soda crackers.

Few of us today eat pie at every meal. Nevertheless, pies are still a favorite dessert. Most customers will order and pay a higher price for a piece of chocolate cream pie than for chocolate pudding, even if the pie filling is the same as the pudding and even if they leave the crust uneaten.

In this chapter, we present the preparation of pie crusts and fillings. In addition, we discuss how to make puff pastry, éclair paste, meringues, and fruit desserts.

AFTER READING THIS CHAPTER, YOU SHOULD BE ABLE TO

1. Prepare flaky pie dough and mealy pie dough.
2. Prepare crumb crusts and short, or cookie, crusts.
3. Assemble and bake pies.
4. Prepare the following pie fillings: fruit fillings using the cooked juice method, the cooked fruit method, and the old-fashioned method; custard or soft fillings; cream pie fillings; and chiffon fillings.
5. Prepare puff pastry dough and puff dough products.
6. Prepare éclair paste and éclair paste products.
7. Prepare standard meringues and meringue desserts.
8. Prepare fruit desserts.

PIES

PIE DOUGHS

Before you begin studying this section, it would be a good idea for you to review the section on gluten development in Chapter 30. Pie pastry is a simple product in terms of its ingredients: flour, shortening, water, and salt. Yet success or failure depends on how the shortening and flour are mixed and how the gluten is developed. The key to making pie dough is proper technique, and you will remember the techniques better if you understand why they work.

INGREDIENTS

Flour

Pastry flour is the best choice for pie doughs. It has enough gluten to produce the desired structure and flakiness, yet is low enough in gluten to yield a tender product, if handled properly.

If stronger flours are used, the percentage of shortening should be increased to provide more tenderness.

Fat

Regular hydrogenated shortening is the most popular fat for pie crusts because it has the right plastic consistency to produce a flaky crust. It is firm and moldable enough to make an easily worked dough. Emulsified shortening should not be used because it blends too quickly with the flour, making a flaky pastry difficult to achieve.

Butter contributes excellent flavor to pie pastry, but it is not frequently used in volume production for two reasons: It is expensive, and it melts very easily, making the dough difficult to work.

It is desirable, if costs permit, to blend a quantity of butter into the shortening used for pie crusts to improve flavor. The quantity of pie crust dumped in the garbage after customers have eaten out the filling is evidence that many people are not satisfied with the taste of pie crusts made with shortening.

If all butter is used in place of shortening, the percentage of fat in the formula should be increased by about one-fourth. (If 1 pound shortening is called for, use 1 pound 4 ounces butter. If 500 grams shortening are called for, use 625 grams butter.) The liquid should be reduced slightly, as butter contains moisture.

Lard is an excellent shortening for pies because it is firm and plastic. Some people dislike its flavor, however, so it is not widely used in food service.

Liquid

Water is necessary to develop some gluten in the flour and to give structure and flakiness to the dough. If too much water is used, the crust will become tough because of too much gluten development. If not enough water is used, the crust will fall apart.

Milk makes a rich dough that browns quickly. However, the crust is less crisp, and the production cost is higher.

Whether water or milk is used, it must be added cold (40°F/4°C or colder) to maintain proper dough temperature.

Salt

Salt has some tenderizing and conditioning effect on the gluten. However, its main contribution is to flavor.

Salt must be dissolved in the liquid before being added to the mix to ensure even distribution.

TEMPERATURE

Pie dough should be kept cool, about 60°F (15°C), during mixing and makeup for two reasons:

1. Shortening has the best consistency when cool. If it is warm, it blends too quickly with the flour. If it is very cold, it is too firm to be easily workable.

2. Gluten develops more slowly at cool temperatures than at warm temperatures.

PIE DOUGH TYPES

There are two basic types of pie dough: flaky and mealy.

The difference between the two is how the fat is blended with the flour. Complete mixing procedures are given later. First, it is important to understand the basic distinction between the two types.

Flaky Pie Dough

For **flaky pie dough**, the fat is cut or rubbed into the flour until the particles of shortening are about the size of peas or hazelnuts. That is, the flour is not completely blended with the fat, and the fat is left in pieces. (Many bakers distinguish between this crust, which they call *short-flake*, and *long-flake* crusts, in which the fat is left in pieces the size of walnuts and the flour is even less coated with shortening.)

When water is added, the flour absorbs water and develops some gluten. When the dough is rolled out, the lumps of fat and moistened flour are flattened and become flakes of dough separated by layers of fat.

Mealy Pie Dough

For **mealy pie dough**, the fat is blended into the flour more thoroughly, until the mixture looks like coarse cornmeal. Because the flour is more completely coated with fat:

- The crust is very short and tender because less gluten can develop.
- Less water is needed in the mix because the flour won't absorb as much as in flaky dough.
- The baked dough is less likely to absorb moisture from the filling and become soggy.

Mealy dough is used for bottom crusts in baked fruit pies and soft or custard-type pies because it resists sogginess. Flaky doughs are used for top crusts and for prebaked pie shells.

Fat-and-flour mixtures for flaky pie dough and mealy pie dough.

Trimmings

Reworked scraps or trimmings are tougher than freshly made dough. They may be combined with mealy dough and used for bottom crusts only.

MIXING PIE DOUGHS

Hand mixing is best for small quantities of dough, especially flaky dough, because it gives more control over the mixing. Quantities up to 10 pounds (5 kg) can be mixed almost as quickly by hand as by machine.

For machine mixing, use a pastry knife or paddle attachment and blend at low speed.

The mixing method for pie doughs is called the rubbed dough method. Although the procedure has several steps, the two main steps are characteristic of the method:

1. Rub the fat into the sifted dry ingredients.
2. Carefully mix the combined liquid ingredients into the dry ingredients.

The seven steps in the pie dough recipe that follows explain the rubbed dough method in more detail. Study this procedure well, because most pie doughs and several other basic pastries are mixed with this procedure or a variation. Also, compare this procedure with the biscuit method outlined on page 934. Although biscuit dough is softer and contains leavening, it is mixed with a similar procedure.

Procedures for rolling pie doughs and lining pie pans are discussed in the next section, Assembly and Baking.

> ## KEY POINTS TO REVIEW
>
> - What are the steps in the rubbed dough method for mixing pie doughs?
> - How do you mix pie doughs in order to get flaky dough? mealy dough?
> - Flaky dough is best for what kind of pie crusts? Mealy dough is best for what kind of pie crusts?

 Flaky Pie Dough   #  Mealy Pie Dough

INGREDIENTS	U.S.		METRIC	PERCENTAGE	U.S.		MEALY METRIC	PERCENTAGE
Pastry flour	1 lb 4	oz	500 g	100 %	1 lb 4	oz	500 g	100 %
Shortening	14	oz	350 g	70 %	13	oz	325 g	65 %
Salt	0.4 oz (2 tsp)		10 g	2 %	0.4 oz (2 tsp)		10 g	2 %
Water	6	oz	150 g	30 %	5	oz	125 g	25 %
Total weight:	2 lb 9	oz	1010 g	202 %	2 lb 6	oz	960 g	192 %

Per 1 ounce (28.35 g): Calories, 130; Protein, 1 g; Fat, 10 g (65% cal.); Cholesterol, 0 mg;
Carbohydrates, 11 g; Fiber, 20 g; Sodium, 105 mg.

PROCEDURE

Mixing:

1. Collect all equipment.
2. Collect and scale ingredients.
3. Dissolve salt in water. Set aside.
4. Place flour and shortening in mixing bowl.
5. Rub or cut shortening into flour to the proper degree:
   For mealy dough—until it resembles coarse cornmeal.
   For flaky dough—until fat particles are the size of peas or hazelnuts.
6. Add salt and water. Mix very gently, just until water is absorbed. Do not overwork the dough.
7. Place the dough in pans, cover with plastic film, and place in refrigerator or retarder for several hours.

OTHER PIE CRUSTS

Crumb Crusts

Graham cracker crusts are popular because they have an appealing flavor and are much easier to make than pastry crusts. For variation, vanilla or chocolate wafer crumbs or gingersnap crumbs may be used instead of graham cracker crumbs. Ground nuts may be added for special desserts.

Crumb crusts are used only for unbaked pies, such as cream pies and chiffon pies. Be sure the flavor of the crust is compatible with the filling. Lime chiffon filling in a chocolate crumb crust is not an appealing combination. Some cream fillings are so delicate that they would be overwhelmed by a crust that is too flavorful.

Baking the crust makes a firmer, less crumbly crust and increases flavor.

Short-Dough Crusts

Short pastry is actually a kind of cookie dough. It is richer than regular pie pastry and contains butter, sugar, and eggs. Because **short dough** is difficult to handle, it is used primarily for small fruit tarts.

KEY POINTS TO REVIEW

- What are the ingredients and procedure for making a graham cracker crust?

- How is mixing a short dough crust different from the rubbed-dough method for mixing pie crust?

Graham Cracker Crust

YIELD: 2 LB (900G)   CRUSTS FOR: FIVE 8-IN. (20-CM) PIES
FOUR 9-IN. (23-CM) PIES
THREE 10-IN. (25-CM) PIES

U.S.	METRIC	INGREDIENTS
1 lb	450 g	Graham cracker crumbs
8 oz	225 g	Sugar
8 oz	225 g	Butter, melted

PROCEDURE

1. Mix crumbs and sugar in mixing bowl.
2. Add butter and mix until evenly blended and crumbs are all moistened by the melted butter.
3. Scale the mixture into pie pans: 8 oz (225 g) for 9-in. (23-cm) pans; 6 oz (175g) for 8-in. (20-cm) pans.
4. Spread mixture evenly on bottom and sides of pan. Press another pan on top to pack crumbs evenly.
5. Bake at 350°F (175°C) for 10 minutes.
6. Cool thoroughly before filling.

Per 1 ounce (28.35 g): Calories, 140; Protein, 1 g: Fat, 7 g (45% cal); Cholesterol, 15 mg; Carbohydrates, 18 g; Fiber, 0 g; Sodium, 140 mg.

VARIATIONS

Substitute chocolate or vanilla wafer crumbs or gingersnap crumbs for the cracker crumbs.

Short Dough

INGREDIENTS	U.S.		METRIC	PERCENTAGE	
Butter or butter and shortening	8	oz	250 g	67	%
Sugar	3	oz	90 g	25	%
Salt	0.06 oz (¹/₄ tsp)		2 g	0.5	%
Eggs	2.25 oz		70 g	19	%
Pastry flour	12	oz	375 g	100	%
Total weight:	1 lb 9	oz	787 g	211	%

PROCEDURE

1. Using the paddle attachment, mix the butter, sugar, and salt at low speed until smooth and evenly blended.
2. Add the eggs and mix just until absorbed.
3. Sift the flour and add it to the mixture. Mix just until evenly blended.
4. Chill several hours before using.

Per 1 ounce (28.35 g): Calories, 130; Protein, 2 g; Fat, 8 g (53% cal); Cholesterol, 30 mg; Carbohydrates, 14 g; Fiber, 2 g; Sodium, 100 mg.

PROCEDURE for Making Small Fruit Tarts

1. Roll out chilled short dough on a floured surface until it is slightly less than ¹/₄ inch (5 mm) thick.
2. With a round cutter about ¹/₂ inch (1 cm) larger than the top diameter of your individual tart shells, cut the dough into circles.
3. For each shell, fit a circle of dough into a tin and press it well against the bottom and sides. If you are using fluted tins, make sure the dough is thick enough on the sides so it won't break apart at the ridges.
4. Fit paper liners inside the shells and fill with dried beans to keep the dough from blistering or puffing while baking.
5. Bake at 400°F (200°C) about 15 minutes, or until the shells are fully baked. Remove the paper liners and the beans.
6. Cool the shells completely and remove them from the tins.
7. Fill the shells half full of vanilla pastry cream (see Chapter 36).
8. Arrange well-drained fresh, cooked, or canned fruits over the pastry cream.
9. Brush the top with apricot glaze, melted currant jelly, or other desired glaze (see Chapter 33).
10. Keep refrigerated until service.

ASSEMBLY AND BAKING

TYPES OF PIES

Pies may be classified into two groups based on method of assembling and baking.

1. **Baked pies.**

   Raw pie shells are filled and then baked. **Fruit pies** contain fruit fillings and usually have a top crust. **Soft pies** are those with custard-type fillings or, in other words, liquid fillings that become firm when their egg content coagulates. They are usually baked as single-crust pies.

2. **Unbaked pies.**

   Baked pie shells are filled with a prepared filling, chilled, and served when the filling is firm enough to slice. **Cream pies** are made with pudding or boiled custard-type fillings. **Chiffon pies** are made with fillings that are lightened by the addition of beaten egg white.

   Baking a pie or tart shell without a filling is called **baking blind**.

PROCEDURE for Rolling Pie Dough and Lining Pans

1. **Scale the dough.**

8 ounces (225 g) for 9-inch (23-cm) bottom crusts	6 ounces (175 g) for 8-inch (20-cm) bottom crusts
6 ounces (175 g) for 9-inch (23-cm) top crusts	5 ounces (150 g) for 8-inch (20-cm) top crusts

   Experienced bakers are able to roll crusts using less dough because less must be trimmed when the dough is rolled to a perfect circle of the exact size needed.

2. **Dust the bench and rolling pin lightly with flour.**

   Too much dusting flour toughens the dough. Use no more than needed to prevent sticking.

3. **Roll out the dough.**

   Flatten the dough lightly and roll it out to a uniform ⅛-inch (3-mm) thickness. Use even strokes and roll from the center outward in all directions. Lift the dough frequently to make sure it is not sticking. The finished dough should form a perfect circle.

4. **Place the dough in a pan.**

   To lift the dough without breaking it, roll it lightly around the rolling pin. Allow the dough to drop into the pan and press it into the corners without stretching it. Stretched dough shrinks during baking. There should be no air bubbles between the dough and the pan.

5. **For single-crust pies, flute the edges, if desired, and trim off excess dough.**

   Some bakers feel that fluted edges add to the appearance of the product. Others feel that fluting takes too much time and produces only a rim of heavy dough that customers leave on their plates.

6. **For two-crust pies.**

   Fill with cold filling, place the second crust on top, and seal the top and bottom crusts together at edges. Flute, if desired, and trim excess dough. Apply desired wash or glaze to top.

7. **Bake as directed in the recipe.**

THE SOGGY BOTTOM

A common pie fault is an underbaked bottom crust or a crust that soaks up moisture from the filling. Soggy bottoms can be avoided in several ways:

1. Use mealy dough for bottom crusts. Mealy dough absorbs less liquid than flaky dough.

2. Use high bottom heat, at least at the beginning of baking, to set the crust quickly. Bake the pies at the bottom of the oven.

3. Do not add hot fillings to unbaked crusts.

4. Use dark metal pie tins, which absorb heat. (If you use disposable aluminum pans, choose pans with the bottoms colored black.)

PROCEDURE for Preparing Baked Pies

Note: For pies without a top crust, omit steps 3 through 7.

1. Line the pie pan with pie dough as in the basic procedure (Figure 35.1).

2. Fill with cooled filling. See Table 35.1 for scaling instructions. Do not drop filling on the rim of the pie shell; this makes it hard to seal the rim to the top crust. To avoid spilling custard filling, place the empty shell on the rack in the oven and then pour in the filling.

3. Roll out the dough for the top crust.

4. Perforate the top crust to allow steam to escape during baking.

5. Moisten the rim of the bottom crust to help seal it to the top crust.

6. Fit the top crust in place. Seal the edges together firmly and trim excess dough. An easy way to do this is to press the rim with the tines of a fork. Alternatively, the rim may be fluted. An efficient way to trim excess dough is to rotate the pie tin while pressing on the edges with the palms of the hands.

7. Brush the top with the desired wash: milk, cream, eggs and milk, or water. Sprinkle with granulated sugar if desired.

8. Place the pie on the lower level of an oven preheated to 425°–450°F (220°–230°C). The high initial heat helps set the bottom crust to avoid soaking. Fruit pies are usually baked at this high heat until done. For custard pies, reduce the heat to 325°–350°F (165°–175°C) after 10 minutes to avoid overcooking and curdling the custard. Custard pies include those containing large quantities of egg, such as pumpkin pie and pecan pie.

TABLE 35.1 Scaling Guidelines for Baked Pies

Pie Size		Weight of Filling	
U.S.	Metric	U.S.	Metric
8 in.	20 cm	26–30 oz	750–850 g
9 in.	23 cm	32–40 oz	900–1150 g
10 in.	25 cm	40–50 oz	1150–1400 g

Note: Weights are guidelines only. Exact weights may vary, depending on the filling and the depth of the pans.

FIGURE 35.1 Preparing baked pies.

(a) Line the pie pans with the pie dough.

(b) Fill with cooled fillings.

(c) Fit the top crusts in place.

(d) Seal the top crust to the rim and trim the dough from the edges.

PROCEDURE for Preparing Unbaked Pies

1. Line a pie pan with pie dough as in the basic procedure.

2. Dock the crust well with a fork to prevent blistering.

3. Place another pan inside the first one so the dough is between 2 pans. This is called double-panning.

4. Place the pans upside down in an oven preheated at 450°F (230°C). Baking upside down helps keep the dough from shrinking down into the pan.

   Some bakers like to chill the crusts before baking to relax the gluten and help reduce shrinkage.

5. Bake at 450°F (230°C) for 10–15 minutes. The top pan may be removed during the last part of baking so the crust can brown.

6. Cool the baked crust completely.

7. Fill with cream or chiffon filling. Fill as close as possible to service time to prevent soaking the crust.

8. Chill the pie until it is set enough to slice.

KEY POINTS TO REVIEW

- What are the steps in the procedure for rolling pie dough and lining pans?

- What are the steps in the procedure for preparing baked pies?

- What are the steps in the procedure for preparing unbaked pies?

FILLINGS

STARCHES FOR FILLINGS

Many kinds of pie filling, especially fruit fillings and cream fillings, depend on starch for their thick texture.

Types

Cornstarch is used for cream pies because it sets up into a firm gel that holds its shape when sliced. Cornstarch may also be used for fruit pies.

Waxy maize and other *modified starches* are best for fruit pies because they are clear when set and make a soft paste rather than a firm gel. Waxy maize should be used for pies that are to be frozen because it is not broken down by freezing.

Flour, tapioca, and other starches are used less frequently. Flour has less thickening power than other starches and makes the product cloudy.

Instant starch or *pregelatinized starch* needs no cooking because it has already been cooked. When used with certain fruit fillings, it eliminates the need to cook the filling before making up the pie. It has no advantage, however, if the filling contains such ingredients as raw fruit or eggs and so must be cooked.

Starches differ in thickening power, so follow the formulas exactly.

Cooking Starches

To avoid lumping, starches must be mixed with a cold liquid or sugar before being added to a hot liquid.

Sugar and *strong acids* reduce the thickening power of starch. When possible, all or part of the sugar and strong acids like lemon juice should be added *after the starch has thickened*.

FRUIT FILLINGS

Fruit pie fillings consist of fruits and fruit juices, sugar, spices, and a starch thickener.

Fruits for Pie Fillings

Fresh fruits make excellent pies if they are at their seasonal peak. Fresh apples are used extensively for high-quality pies. But the quality of fresh fruits can vary considerably, and fresh fruit pies require a lot of labor.

Frozen fruits are widely used for pies because they are consistent in quality and readily available.

Canned fruits can also be of high quality. Solid pack (with little juice) gives a higher yield of fruit per can than syrup or water pack.

Dried fruits must be rehydrated by soaking and, usually, simmering before they are made into pie fillings.

Fruits must have sufficient acid (tartness) to make flavorful fillings. If they lack natural acid, you may need to add lemon, orange, or pineapple juice to supply the acid.

Cooked Juice Method

The advantage of the **cooked juice method** is that only the juice is cooked. The fruit retains better shape and flavor because it is subjected to less heat and handling. This method is used when the fruit requires little or no cooking before filling the pie. Examples: cherry, peach, most frozen or canned fruits. Fresh berries can also be prepared by this method. Some of the berries are cooked or puréed to provide juice. The remaining berries are mixed with the finished gel.

Cooked Fruit Method

The **cooked fruit method** is used when the fruit requires cooking or there is not enough liquid for the cooked juice method. Examples: fresh apple, raisin, rhubarb.

PROCEDURE: **Cooked Juice Method**

1. Drain the juice from the fruit.

2. Measure the juice and, if necessary, add water or other fruit juice to bring it to the desired volume.

3. Bring the juice to a boil.

4. Dissolve the starch in cold water and stir it into the boiling juice. Return the juice to the boil and cook until it is clear and thickened.

5. Add sugar, salt, and flavorings, and stir until dissolved.

6. Pour the thickened juice over the drained fruit and mix gently. Be careful not to break or mash the fruit.

7. Cool.

PROCEDURE: **Cooked Fruit Method**

1. Bring the fruit and its juice or water to a boil. Some sugar may be added to the fruit to draw out juices.

2. Dissolve the starch in cold water and stir it into the fruit. Return the fruit mixture to a boil and cook until it is clear and thickened. Stir while cooking.

3. Add sugar, salt, flavorings, and other ingredients and stir until dissolved.

4. Cool as quickly as possible.

VARIATION

Some fruits, such as fresh apples, may be cooked in butter rather than boiled in water for better flavor.

Old-Fashioned Method

This method is best suited to pies made with fresh apples or peaches. It is not as widely used in food service as the other methods because it is more difficult to control the thickening of the juices.

PROCEDURE: **Old-Fashioned Method**

1. Mix the starch and spices with the sugar until uniformly blended.

2. Mix the fruit with the sugar mixture.

3. Fill the unbaked pie shell with the fruit.

4. Place lumps of butter on top of the filling.

5. Cover with a top crust or with Streusel (p. 924) and bake.

KEY POINTS TO REVIEW

- What are the steps in the cooked juice method for making fruit pie fillings?

- What are the steps in the cooked fruit method for making fruit pie fillings?

- What are the steps in the old-fashioned method for making fruit pie fillings?

COOKED JUICE METHOD

 # Apple Pie Filling (Canned Fruit)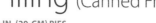

YIELD: ABOUT 9¹/₂ LB (4.5 KG) FIVE 8-IN. (20-CM) PIES
FOUR 9-IN. (23-CM) PIES
THREE 10-IN. (25-CM) PIES

U.S.	METRIC	INGREDIENTS	PROCEDURE
6 lb 8 oz	3 kg	Canned apples (1 No. 10 can)	1. Drain the apples, reserving the juice.
as needed	as needed	Water	2. Add enough water to the juice to measure 1¹/₂ pt (750 mL).
8 fl oz	250 mL	Water, cold	3. Mix the cold water and starch.
3 oz	90 g	Cornstarch or modified starch	4. Bring the juice mixture to a boil.
			5. Stir in the starch mixture and return to a boil.
1 lb 4 oz	575 g	Sugar	6. Add the sugar, salt, cinnamon, nutmeg, and butter. Simmer until the sugar is dissolved.
¹/₄ oz (1¹/₄ tsp)	7 g (6 mL)	Salt	7. Pour the syrup over the apples and mix gently. Cool completely.
¹/₄ oz (3¹/₂ tsp)	7 g (17 mL)	Cinnamon	8. Fill pie shells. Bake at 425°F (220°C) for 30–40 minutes.
1 tsp	2 g (5 mL)	Nutmeg	
3 oz	90 g	Butter	

Per 1 ounce (28.35 g): Calories, 30; Protein, 0 g; Fat, 0.5 g (14% cal.); Cholesterol, 0 mg; Carbohydrates, 7 g; Fiber, 0 g; Sodium, 25 mg.

VARIATIONS

Dutch Apple Pie Filling

Simmer 8 oz (250 g) raisins in water. Drain and add to Apple Pie Filling.

Cherry Pie Filling

Use 1 No. 10 can sour cherries instead of apples. Increase starch to 4 oz (125 g). Add 1¹/₂ fl oz (45 mL) lemon juice in step 6. Increase the sugar to 1 lb 12 oz (800 g). Omit cinnamon and nutmeg. Add almond extract to taste (optional). If desired, color with 2–3 drops red coloring.

Peach Pie Filling

Use 1 No. 10 can sliced peaches, preferably solid or heavy pack, instead of apples. Omit cinnamon and nutmeg.

Pineapple Pie Filling

Use 1 No. 10 can crushed pineapple instead of apples. Increase liquid in step 1 to 1 qt (1 L). Increase starch to 4 oz (125 g). Use 1 lb 8 oz (700 g) sugar and 8 oz (250 g) corn syrup. Omit cinnamon and nutmeg.

COOKED FRUIT METHOD

Rhubarb Pie Filling ♥

YIELD: ABOUT 5 LB 8 OZ (2.5 KG) THREE 8-IN. (20-CM) PIES
TWO 10-IN. (25-CM) PIES

U.S.	METRIC	INGREDIENTS
3 lb 8 oz	1.6 kg	Fresh rhubarb
8 fl oz	250 mL	Water
8 oz	225 g	Sugar
4 fl oz	125 mL	Water
2 oz	60 g	Cornstarch
8 oz	225 g	Sugar
1 tsp	5 mL	Salt
1 oz	30 g	Butter

PROCEDURE

1. Cut the rhubarb into 1-in. (2.5-cm) pieces.
2. Combine the rhubarb, first quantity of water, and sugar in a saucepan. Bring to a boil and simmer 2 minutes.
3. Mix the second quantity of water and starch. Stir into the rhubarb and boil until thick and clear.
4. Add the remaining ingredients. Stir gently until the sugar is dissolved and the butter is melted.
5. Cool completely.
6. Fill pie shells. Bake at 425°F (220°C), 30–40 minutes.

Per 1 ounce (28.35 g): Calories, 30; Protein, 0 g; Fat, 5 g (14% cal.); Cholesterol, 0 mg; Carbohydrates, 7 g; Fiber, 0 g; Sodium, 30 mg.

VARIATION

Fresh Apple Pie Filling

Use 5 lb (2.25 kg) fresh peeled and sliced apples instead of rhubarb. Flavor with $1\frac{1}{2}$ tsp (7 mL) cinnamon, $\frac{1}{2}$ tsp (2 mL) nutmeg, and $\frac{1}{2}$–1 fl oz (15–30 mL) lemon juice during step 4.

Raisin Pie Filling ♥ 🌸

YIELD: ABOUT 2 LB (1 KG) ONE 9-IN. (23-CM) PIE

U.S.	METRIC	INGREDIENTS
13 oz	360 g	Raisins
13 fl oz	400 mL	Water
2 fl oz	50 mL	Water, cold
$\frac{1}{2}$ oz	15 g	Cornstarch or modified starch
4 oz	114 g	Sugar
$\frac{1}{2}$ tsp	2 mL	Salt
$3\frac{1}{2}$ tsp	18 mL	Lemon juice
$\frac{1}{2}$ tsp	2 mL	Grated lemon zest
$\frac{1}{5}$ tsp	1 mL	Cinnamon
0.6 oz	18 g	Butter

PROCEDURE

1. Combine the raisins and water in a saucepan. Simmer 5 minutes.
2. Mix the cold water and starch. Stir into the raisins and simmer until thickened.
3. Add the remaining ingredients. Stir until sugar is dissolved and mixture is uniform.
4. Cool thoroughly.
5. Fill pie shells. Bake at 425°F (220°C) for 30–40 minutes.

Per 1 ounce (28.35 g): Calories 50, Protein, 0 g; Fat, 0 g; (0% cal.); Cholesterol, 0 mg. Carbohydrates, 12 g; Fiber, 5 g; Sodium, 35 mg.

OLD-FASHIONED METHOD

Old-Fashioned Apple Pie Filling ♥

YIELD: ABOUT 11 LB (5 KG) SIX 8-IN. (20-CM) PIES
FIVE 9-IN. (23-CM) PIES
FOUR 10-IN. (25-CM) PIES

U.S.	METRIC	INGREDIENTS	PROCEDURE
9 lb EP	4.1 kg EP	Fresh peeled, sliced apples	1. Select firm, tart apples.
2 fl oz	60 mL	Lemon juice	2. Combine apple slices and lemon juice in a large mixing bowl and toss to coat apples with the juice.
2 lb	900 g	Sugar	3. Mix the sugar, starch, salt, and spices.
3 oz	90 g	Cornstarch	4. Add to the apples and toss gently until well mixed.
1/4 oz (1 1/4 tsp)	7 g (6 mL)	Salt	
1/4 oz (3 1/2 tsp)	7 g (17 mL)	Cinnamon	
1 tsp	5 mL	Nutmeg	
3 oz	90 g	Butter	5. Fill pie shells. Dot the filling with butter before setting the top crusts in place. Bake at 400°F (200°C) about 45 minutes.

Per 1 ounce (28.35 g): Calories, 40; Protein, 0 g; Fat, 0.5 g (11% cal.); Cholesterol, 0 mg;
Carbohydrates, 9 g; Fiber, 0 g; Sodium, 20 mg.

APPLES FOR PIES

What varieties of apples are best for making pies? Two criteria are important: taste and texture. First, the apples should have a good flavor and a good level of acidity. Apples that are very mild make pies with little flavor. The sugar content, or sweetness, of the apple is less important, as the sugar in the recipe can be adjusted.

Second, the apples should hold their shape when cooked. Apples that turn to mush, such as McIntosh, are better for applesauce than for pies.

Popular apple varieties that have good taste and texture for use in pie fillings include Granny Smith, Jonathan, Jonagold, Newton Pippin, Rome, Macoun, Pink Lady, Stayman-Winesap, and Golden Delicious.

CUSTARD OR SOFT FILLINGS

Custard, pumpkin, pecan, and similar pies are made with an uncooked liquid filling containing eggs. The eggs coagulate when the pie is baked, setting the filling.

The greatest difficulty in cooking soft pies is cooking the crust completely yet not over-cooking the filling. Start the pie at the bottom of a hot oven (425°–450°F/220°–230°C) for first 10 minutes to set the crust. Then reduce the heat to 325°–350°F (165°–175°C) to cook the filling slowly.

To test for doneness:

1. Shake the pie very gently. If it is no longer liquid, it is done. The center will still be slightly soft but will continue cooking in its own heat after the pie is removed from the oven.

2. Insert a thin knife 1 inch (2.5 cm) from the center. It will come out clean if the pie is done.

Custard Pie Filling ♥ 🌹

YIELD: 2 LB (0.9 KG)   ONE 9-IN. (23-CM) PIE

U.S.	METRIC	INGREDIENTS
8 oz	225 g	Eggs
4 oz	112 g	Sugar
1/4 tsp	1 mL	Salt
1 1/2 tsp	7 mL	Vanilla
1.25 pt	600 mL	Milk (see Note)
1/4–1/2 tsp	1-2 mL	Nutmeg

PROCEDURE

1. Beat the eggs lightly. Add sugar, salt, and vanilla. Blend until smooth. Do not whip air into the mixture.

2. Stir in the milk. Skim off any foam.

3. Pour into the unbaked pie shells.

4. Sprinkle tops with nutmeg.

5. Bake at 450°F (230°C) for 15 minutes. Reduce heat to 325°F (165°C) and bake until set, about 20–30 minutes more.

Per 1 ounce (28.35 g): Calories, 35; Protein, 2 g; Fat, 1.5 g (33% cal.); Cholesterol, 35 mg; Carbohydrates, 5 g; Fiber, 0 g; Sodium, 35 mg.

Note: For a richer custard, use part milk and part cream.

VARIATION

Coconut Custard Pie Filling

Use 2.5 oz (70 g) unsweetened, flaked coconut. Sprinkle coconut into pie shells before adding custard mixture. Coconut may be toasted lightly in oven before adding to pies, if desired. Omit nutmeg.

Pecan Pie Filling ♥ 🌹

YIELD: 1 LB 12 OZ (1640 G) FILLING PLUS 5 OZ (142 G) PECANS   ONE 9-IN. (23-CM) PIE

U.S.	METRIC	INGREDIENTS
7 oz	200 g	Sugar (see Note)
2 oz	160 g	Butter
1/4 tsp	1 mL	Salt
7 oz	200 g	Eggs
12 oz (about 8 1/2 fl oz)	350 g	Dark corn syrup
1 1/2 tsp	8 mL	Vanilla
5 oz	142 g	Pecans

PROCEDURE

1. Using the paddle attachment at low speed, blend the sugar, butter, and salt until evenly blended.

2. With the machine running, add the eggs, a little at a time, until they are all absorbed.

3. Add the syrup and vanilla. Mix until well blended.

4. To assemble pies, distribute pecans evenly in pie shells. Fill with syrup mixture.

5. Bake at 450°F (230°C) for 10 minutes. Reduce heat to 325°F (165°C). Bake about 40 minutes more, or until set.

Per 1 ounce (28.35 g): Calories, 120; Protein, 1 g; Fat, 5 g (38% cal.); Cholesterol, 35 mg; Carbohydrates, 17 g; Fiber, 0 g; Sodium, 70 mg.

Note: Brown sugar may be used if darker color and stronger flavor are desired.

Pumpkin Pie Filling ♥ 🥧

YIELD: ABOUT 4¼ LB (2 KG)   TWO 9-IN. (23-CM) PIES

U.S.	METRIC	INGREDIENTS	PROCEDURE
1 lb 10½ oz	750 g	Pumpkin purée, 1 No. 2½ can	1. Place pumpkin purée in the bowl of a mixer fitted with a whip attachment.
1 oz	30 g	Pastry flour	2. Sift together the flour, spices, and salt.
2 tsp	10 mL	Cinnamon	3. Add the flour mixture and sugar to the pumpkin. Mix at 2nd speed until smooth and well blended.
¼ tsp	1 mL	Nutmeg	
¼ tsp	1 mL	Ground ginger	
⅛ tsp	0.5 mL	Ground cloves	
⅝ tsp	3 mL	Salt	
10 oz	290 g	Brown sugar	4. Add the eggs and mix in. Scrape down the sides of the bowl.
10 oz	300 g	Eggs	5. Turn the machine to low speed. Gradually pour in the syrup-molasses mixture, then the milk. Mix until evenly blended.
2 oz	60 g	Corn syrup or half corn syrup and half molasses	6. Fill the pie shells. Bake at 450°F (230°C) for 15 minutes. Lower heat to 350°F (175°C) and bake until set, about 30–40 minutes more.
1 pt 4 oz	600 mL	Milk	

Per 1 ounce (28.35 g): Calories, 30; Protein, 1 g; Fat, 0.5 g (14% cal.); Cholesterol, 10 mg; Carbohydrates, 6 g; Fiber, 0 g; Sodium, 45 mg.

VARIATIONS

Sweet Potato Pie Filling

Substitute canned sweet potatoes, drained and puréed, for the pumpkin.

Squash Pie Filling

Substitute puréed squash for the pumpkin.

CREAM PIE FILLINGS

Cream pie fillings are the same as puddings, which, in turn, are the same as basic pastry creams with added flavorings such as vanilla, chocolate, or coconut. Lemon filling is made by the same method, using water and lemon juice instead of milk.

The one difference between puddings and pie fillings you should note is that *cream pie fillings are made with cornstarch* so slices hold their shape when cut. Puddings may be made with flour, cornstarch, or other starches.

Techniques and recipes for these fillings are included in Chapter 36, along with other basic creams and puddings.

CHIFFON PIES

Chiffon fillings are made by adding gelatin to a cream filling or to a thickened fruit and juice mixture and then folding in egg whites and/or whipped cream. The mixture is then poured into baked pie shells and allowed to set.

These preparations are the same as chiffon desserts, bavarians, and some mousses and cold soufflés. To avoid unnecessary repetition, techniques and recipes for these products are included in Chapter 36 with other puddings and creams.

PASTRIES, MERINGUES, AND FRUIT DESSERTS

In addition to pie dough, two other pastries have great importance in bakeshops and kitchens: *puff pastry*, used for such products as napoleons and turnovers, and *éclair* or *choux paste*, used for éclairs and cream puffs. These products are also used in the hot food kitchen and the pantry in the preparation of a number of hors d'oeuvres, entrées, and side dishes.

Meringues and fruit desserts are also covered in this section. Meringues are not only important as pie toppings but also can be formed, baked until crisp, and used in many of the same ways as pastry shells for desserts.

PUFF PASTRY

Puff pastry is one of the most remarkable products of the bakeshop. Although it includes no added leavening agent, it can rise to 8 times its original thickness when baked.

Puff pastry is a rolled-in dough, like Danish and croissant dough. This means it is made up of many layers of fat sandwiched between layers of dough. Unlike Danish dough, however, puff pastry contains no yeast. Steam, created when the moisture in the dough layers is heated, is responsible for the spectacular rising power of puff pastry.

Puff pastry or puff dough is one of the most difficult of all bakery products to prepare. Because it consists of over 1000 layers, many more than Danish dough, the rolling-in procedure requires a great deal of time and care.

As for so many other products, there are nearly as many versions of puff pastry as there are bakers. Both formulas and rolling-in techniques vary. The formula provided here contains no eggs, for example, although some bakers add them.

The folding-in technique used here differs somewhat from that used by European pastry chefs, although it is widely used by American bakers. (See Figure 35.2.)

Butter is the preferred fat for rolling in because of its flavor and melt-in-the-mouth quality. Special puff pastry shortening is also available. This shortening is much easier to work with than butter because it is not as hard when refrigerated and doesn't soften and melt as easily as butter at warm temperatures. It is also less expensive than butter. However, puff pastry shortening can be unpleasant to eat because it tends to congeal and coat the inside of the mouth.

Skill at producing puff pastry requires careful attention to your instructor and diligent practice. Take special note of alternative methods your instructor may present.

Puff Pastry

INGREDIENTS	U.S.		METRIC	PERCENTAGE	
Bread flour	12	oz	375 g	75	%
Cake flour	4	oz	125 g	25	%
Butter, softened	2	oz	60 g	12.5	%
Salt	0.25 oz		8 g	1.5	%
Water, cold	9	oz	282 g	56	%
Butter	1 lb		500 g	100	%
Bread flour (see Note)	2	oz	60 g	12.5	%
Total weight:	2 lb 13	oz	1410 g	282	%

PROCEDURE

Mixing:

1. Place the first quantities of flour and butter in a mixing bowl. With a paddle attachment, mix at low speed until well blended.
2. Dissolve the salt in the cold water.
3. Add the salted water to the flour and mix at low speed until a soft dough forms. Do not overmix.
4. Remove the dough from the mixer and let rest in the refrigerator 20 minutes.
5. Cream the last quantities of butter and flour at low speed in the mixer until the mixture is about the same consistency as the dough, neither too hard nor too soft.
6. Roll the butter into the dough following the procedure shown in **Figure 35.2**. Give the dough 4 four-folds or 5 three-folds.

Per 1 ounce (28.35 g): Calories, 120; Protein, 1 g; Fat 9 g (69% cal.); Cholesterol, 25 mg; Carbohydrates, 8 g; Fiber, 0 g; Sodium, 150 mg.

Note: The purpose of the 4 oz (125 g) bread flour is to absorb some of the moisture of the butter. Omit if puff paste shortening is used instead of butter.

FIGURE 35.2 Rolling-in procedure for puff pastry.

(a) Dust the bench lightly with flour. Roll dough to a rectangle about 3 times as long as it is wide and about ½ in. (1–1.5 cm) thick. Make the corners as square as possible. Form the butter into a rectangle two-thirds the size of the dough, leaving room around the edges, and place on the dough as shown.

(b) Fold the third without fat over the center third.

(c) Fold the remaining third on top. Fold all ends and corners evenly and squarely. This procedure, enclosing the butter in the dough, does not count as one of the folds. The folding procedure starts with the next step.

(d) Turn the dough 90 degrees on the bench so the length becomes the width. This step must be taken before each rolling-out so the gluten is stretched in all directions, not just lengthwise. Failure to do this results in products that deform or shrink unevenly when they bake. Before rolling, beat the dough lightly as shown so the butter is evenly distributed. Roll the dough into a rectangle. Make sure the corners are square. Roll smoothly and evenly. Do not press down when rolling, or the layers may stick together and the product not rise properly.

(e) Brush excess flour from the top of the dough.

(f) Fold the top edge of the dough to the center. Make sure the corners are square and even. Again brush off excess flour.

(g) Fold the bottom edge to the center.

(h) Fold the dough in half like closing a book. You have now given the dough 1 four-fold. Refrigerate the dough 15–20 minutes to relax the gluten. Do not refrigerate it too long, or the butter will become too hard. (If it does, let it soften a few minutes at room temperature before proceeding.)

(i) Give the dough another 3 four-folds, as in steps (f)–(h). After another rest, the dough is ready to be rolled out and made up into the desired products. (Alternative method: Instead of giving the dough 4 four-folds, you may give it 5 three-folds.) See Figure 31.3 for the three-fold method.

BLITZ PUFF PASTRY

Blitz puff pastry is much easier and quicker to make than classic puff dough. (**Blitz** is German for "lightning.") It does not rise nearly as high as true puff pastry, so it is not suitable for patty shells and other products where a high, light pastry is desirable. However, it bakes up crisp and flaky and is perfectly suitable for napoleons and similar desserts that are layered with cream fillings.

Blitz puff paste, as you will see, is actually a flaky pie dough that is rolled and folded like regular puff dough.

Blitz Puff Pastry 🌹

INGREDIENTS	U.S.		METRIC	PERCENTAGE	
Bread flour	8	oz	250 g	50	%
Pastry flour	8	oz	250 g	50	%
Butter, slightly softened	1 lb		500 g	100	%
Salt	0.25 oz		8 g	1.5	%
Water, cold	8	oz	250 g	50	%
Total weight:	2 lb	8 oz	1258 g	251	%

Per 1 ounce (28.35 g): Calories, 120; Protein, 1 g; Fat 9 g (69% cal.); Cholesterol, 25 mg; Carbohydrates, 8 g; Fiber, 1 g; Sodium, 160 mg.

PROCEDURE

Mixing:

1. Sift the two flours together into a mixing bowl.
2. Cut the butter into the flour as for pie dough, but leave the fat in very large lumps, 1 in. (2.5 cm) across.
3. Dissolve the salt in the water.
4. Add the water to the flour-butter mixture. Mix until the water is absorbed.
5. Let the dough rest for 15 minutes. Refrigerate if the bakeshop is warm.
6. Dust the bench with flour and roll out the dough into a rectangle. Give the dough 3 four-folds.

VARIATION

Reduce the butter to 75% (1 lb 8 oz/750 g).

GENERAL Guidelines for Makeup of Puff Dough Products

1. The dough should be cool and firm when it is rolled and cut. If it is too soft, the layers may stick together at the cuts, preventing proper rising.

2. Cut with straight, firm, even cuts. Use a sharp cutting tool.

3. Avoid touching the cut edges with your fingers; this can make the layers stick together.

4. For best rising, place units upside down on baking sheets. Even sharp cutting tools may press the top layers of dough together. Baking upside down puts the stuck-together layers at the bottom.

5. Avoid letting egg wash run down the edges. Egg wash can cause the layers to stick together at the edges.

6. Rest made-up products for 30 minutes in a cool place or in the refrigerator before baking. This relaxes the gluten and reduces shrinkage.

7. Press trimmings together, keeping the layers in the same direction. After being rolled out and given another three-fold, they may be used again, although they will not rise as high.

8. Baking temperatures of 400°–425°F (200°–220°C) are best for most puff dough products. Cooler temperatures do not create enough steam in the products to leaven them well. Higher temperatures set the crust too quickly.

PROCEDURE for Making Turnovers

See Figure 35.3.

1. Roll out puff pastry dough to 1/8 inch (3 mm) thick.
2. Cut the dough into 4-inch (10-cm) squares. Wash the edges of each with water.
3. Portion the desired filling into the center of each square.
4. Fold diagonally and press the edges together.
5. Puncture the tops with a knife in two or three places to allow steam to escape. Let rest 30 minutes.
6. Brush the tops with egg wash, if desired, or brush with milk or water and sprinkle with sugar.
7. Bake at 400°F (200°C) until crisp and brown.

FIGURE 35.3 Makeup of turnovers.

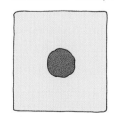

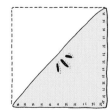

(a) Cut the dough into 4-in. (10-cm) squares. Wash the edges with water and place filling in the center of each square.

(b) Fold over diagonally and press the edges together. Puncture 2 or 3 steam holes in top.

PROCEDURE **for Making Pinwheels**

See Figure 35.4.

1. Roll out puff dough to $\frac{1}{8}$ inch (3 mm) thick.

2. Cut the dough into 5-inch (12-cm) squares.

3. Wash the centers with water.

4. Cut diagonally from the corners to about 1 inch (2.5 cm) from the centers.

5. Fold every other corner into the centers and press in place.

6. Bake at 400°F (200°C).

7. Let cool. Spoon desired fruit filling into the centers. Dust lightly with confectioners' sugar. (Pinwheels may also be filled before baking if the filling is thick and not likely to burn.)

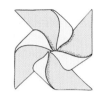

FIGURE 35.4 Makeup of pinwheels.

(a) Cut the dough into 5-in. (12-cm) squares. Wash the centers with water. Cut diagonally from the corners to 1 in. (2.5 cm) from center. **(b, c, d)** Fold every other corner to the center and press down. Fill the center with fruit filling before or after baking.

PROCEDURE **for Making Patty Shells**

See Figure 35.5.

1. Roll out puff dough to $\frac{1}{8}$ inch (3 mm) thick.

2. Roll a second piece of dough to $\frac{1}{4}$ inch (6 mm) thick.

3. Cut out the same number of circles from each piece of dough with a round 3-inch (7.5-cm) cutter.

4. Using a 2-inch (5-cm) cutter, cut out the centers of the thick circles.

5. Wash the thin circles with water or egg wash and place a ring on top of each. Wash the top carefully with egg wash (do not drip wash down the edges). Let rest 30 minutes.

6. Place a sheet of greased parchment over the tops of the shells to prevent their toppling over while baking.

7. Bake at 400°F (200°C) until brown and crisp.

FIGURE 35.5 Makeup of patty shells.

(a) Roll 1 sheet of puff dough $\frac{1}{8}$ in. (3 mm) thick and another sheet $\frac{1}{4}$ in. (6 mm) thick. Cut an equal number of 3-in. (7.5-cm) circles from each. Cut out the centers of the thick circles with a 2-in. (5-cm) cutter.

(b) Wash the thin circles with water or egg wash and place the thick circles on top.

PROCEDURE **for Making Cream Horns**

See Figure 35.6.

1. Roll out puff dough into a sheet $\frac{1}{8}$ inch (3 mm) thick and about 15 inches (38 cm) wide.

2. Cut out strips $1\frac{1}{4}$ inches (3 cm) wide by 15 inches (38 cm) long.

3. Wash the strips with water.

4. With the washed side out, roll the strips diagonally onto cream horn tubes, making a spiral. Overlap the edges by about $\frac{3}{8}$ inch (1 cm). If you are using conical tubes, start at the small end.

5. Roll in granulated sugar and lay on baking sheets. The end of the dough strip should be on the bottom so it does not pop up during baking. Let rest 30 minutes.

6. Bake at 400°F (200°C) until brown and crisp.

7. Slip out the tubes while still warm.

8. Just before service, fill the horns from both ends with whipped cream or pastry cream, using a pastry bag with a star tip. Dust with confectioners' sugar.

FIGURE 35.6 Makeup of cream horns.

(a) Roll puff dough to $\frac{1}{8}$ in. (3 mm) thick and cut it into strips $1\frac{1}{4}$ in. (3 cm) wide and 15 in. (38 cm) long. Wash the strips with water and press one end (washed side out) onto one end of a cream horn tube as shown.

(b) Roll the dough strip in a spiral by turning the tube. Overlap the edges by about $\frac{3}{8}$ in. (1 cm). Do not stretch the dough.

(c) Roll completely and press the end in place to seal.

PROCEDURE for Making Napoleons

1. Roll puff dough into a very thin sheet about the size of a sheet pan. Blitz puff paste or rerolled trimmings may be used.

2. Place on sheet pan and let rest 30 minutes.

3. Dock with a fork to prevent blistering.

4. Bake at 400°F (200°C) until brown and crisp.

5. Trim the edges of the pastry sheet and cut with a serrated knife into equal strips 4 inches (10 cm) wide. Set the best one aside for the top layer. (If one of the strips breaks, don't be upset. It can be used as the middle layer.)

6. Spread one rectangle with Vanilla Pastry Cream (p. 1011) or with a mixture of pastry cream and whipped cream.

7. Top with a second sheet of pastry.

8. Spread with another layer of pastry cream.

9. Place a third pastry rectangle on top, flattest side up.

10. Ice top with fondant (p. 956).

11. To decorate, pipe 4 strips of chocolate fondant lengthwise on the white fondant. Draw a spatula or the back of a knife across the top in opposite directions, 1 inch (2.5 cm) apart, as shown in Figure 35.7.

12. Cut into strips 2 inches (5 cm) wide.

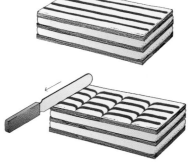

FIGURE 35.7 Decorating napoleons.

(a) Spread the top of the assembled napoleon with white fondant. With a paper cone, pipe 4 strips of chocolate fondant.

(b) Draw a spatula or the back of a knife across the icing at 2-in. (5-cm) intervals.

(c) Draw the spatula in the opposite direction in the center of these 2-in. (5-cm) intervals as shown.

(d) Cut the napoleon into strips 2 in. (5 cm) wide.

ÉCLAIR PASTE

Éclairs and cream puffs are made from a dough called **éclair paste** or *choux paste*. The French name **pâte à choux** (pot a shoo) means "cabbage paste," referring to the fact that cream puffs look like little cabbages.

Unlike puff pastry, éclair paste is extremely easy to make. The dough itself can be prepared in just a few minutes. This is fortunate because for best baking results, the dough should not be prepared ahead of time.

In principle, éclair paste is similar to popover batter, even though one is a thick dough and the other a thin batter. Both products are leavened by steam, which expands the product rapidly and forms large holes in the center. The heat of the oven then coagulates the gluten and egg proteins to set the structure and make a firm product. A strong flour is necessary for sufficient structure.

Éclair paste must be firm enough to hold its shape when piped from a pastry bag. Occasionally, you may find a formula that produces too slack a dough. Correct such a formula by reducing the water or milk slightly.

Proper baking temperatures are important. Start at a high temperature (425°–475°F/ 215°–245°C) for the first 10 minutes to develop steam. Then reduce the heat to 375°–425°F (190°–215°C) to finish baking and set the structure. The products must be firm and dry before being removed from the oven. If they are removed too soon or cooled too quickly, they may collapse. Some bakers like to leave them in a turned-off oven with the door ajar. However, if the oven must be heated again for other products, this may not be the best idea, especially in these times of high energy costs. It may be better to bake the products thoroughly, remove them carefully from the oven, and let them cool slowly in a warm place.

Éclair Paste or Pâte à Choux

INGREDIENTS	U.S.		METRIC	PERCENTAGE
Water, milk, or half water, half milk	1 lb 2 oz		560 g	150 %
Butter or regular shortening	9 oz		280 g	75 %
Salt	0.18 oz (1 tsp)		5 g	1.5 %
Bread flour	12 oz		375 g	100 %
Eggs	1 lb 4 oz		625 g	167 %
Total weight:	3 lb 11 oz		1845 g	493 %

PROCEDURE

Mixing:

1. Combine liquid, butter, and salt in a heavy saucepan and bring to a boil.
2. Remove pan from heat and add the flour all at once. Stir quickly.
3. Return the pan to moderate heat and stir vigorously until the dough forms a ball and pulls away from the sides of the pan (**Figure 35.8**).
4. Transfer the dough to the bowl of a mixer. If you wish to mix by hand, leave it in the saucepan.
5. With the paddle attachment, mix at low speed until the dough has cooled slightly. It should be about 140°F (60°C)—still very warm, but not too hot to touch.
6. At medium speed, beat in the eggs, a little at a time. Add no more than one-fourth of the eggs at once, and wait until each quantity is completely absorbed before adding the next. When all the eggs are absorbed, the paste is ready to use.

Per 1 ounce (28.35 g): Calories, 70; Protein, 2 g; Fat, 4 g (56% cal.); Cholesterol, 35 mg; Carbohydrates, 5 g; Fiber, 0 g; Sodium, 90 mg.

FIGURE 35.8. Making eclair paste

(a) Cook the mixture of water, butter, and flour until the paste forms a ball and pulls away from the sides of the pan.

(b) Cool the paste to 140°F (60°C).

(c) Beat in the eggs a little at a time.

(d) The finished paste should be smooth and moist but thick enough to hold its shape.

PROCEDURE for Making Cream Puffs and Profiteroles

1. Line sheet pans with silicone paper, or butter them lightly.

2. Fit a large pastry bag with a plain tube. Fill the bag with the choux paste.

3. For cream puffs, pipe round mounds of dough about 1¹/₂ inches (4 cm) in diameter onto the lined baking sheets (see Figure 35.9).

   For profiteroles, pipe round mounds of dough about ³/₄ inch (2 cm) in diameter.

   If you prefer, you may drop the dough from a spoon.

4. Bake at 425°F (215°C) for 10 minutes. Lower the heat to 375°F (190°C) until the puffs are well browned and very crisp.

5. Remove from the oven and cool slowly in a warm place.

6. When cool, cut a slice from the top of each puff. Fill with whipped cream, Vanilla Pastry Cream (p. 1011), or desired filling, using a pastry bag with a star tube.

7. Replace the tops and dust with confectioners' sugar.

8. Fill the puffs as close to service as possible. If cream-filled puffs must be held, refrigerate them.

9. Unfilled and uncut puffs, if they are thoroughly dry, may be held in plastic bags in the refrigerator for 1 week. Recrisp in the oven for a few minutes before use.

FIGURE 35.9 For cream puffs or profiteroles, pipe choux paste into bulbs of desired size onto greased sheet pans, or onto pans that have been lined with parchment.

PROCEDURE for Making Éclairs

1. Proceed as for cream puffs, except pipe the dough into strips about ³/₄ inch (2 cm) wide and 3–4 inches (8–10 cm) long (see Figure 35.10). Bake as for cream puffs.

2. Fill baked, cooled éclair shells with pastry cream. Two methods may be used:
   - Make a small hole in one end of the shell and fill using a pastry bag or a doughnut-filling pump.
   - Cut a slice lengthwise from the top and fill using a pastry bag.

3. Dip the tops of the éclairs in chocolate fondant (p. 956).

4. For service and holding, see cream puffs procedure.

VARIATION: FROZEN ÉCLAIRS OR PROFITEROLES

1. Fill éclairs or profiteroles with ice cream. Keep frozen until service.

2. At service time, top with chocolate syrup.

FIGURE 35.10 For éclairs, pipe choux paste into fingers of desired size onto greased sheet pans, or onto pans that have been lined with parchment.

PROCEDURE for Making French Crullers or French Doughnuts

1. Cut sheets of parchment paper to the same width as your deep fryer.

2. Using a pastry bag with a star tube, pipe choux paste onto the parchment in circles (doughnut shapes) about 2 inches (5 cm) across.

3. Slide the paper with the paste into a deep fryer heated to 375°F (190°C). Remove the paper as the doughnuts release and float free.

4. Fry the doughnuts on both sides until golden brown. French doughnuts must be completely fried, or they may collapse when cooling. Remove and drain on absorbent paper.

5. When cooled, drizzle fondant icing over the tops.

MERINGUES

Meringues are beaten egg whites sweetened with sugar. Their most frequent use in North America is for pie toppings and cake icings (known as boiled icing). They are also used to give volume and lightness to buttercream icings and to such preparations as dessert soufflés.

Another excellent use for meringues is to bake them in a slow oven until crisp. In this form, they can be used in place of cake layers or pastry shells to make light, elegant desserts. Chopped nuts may be folded into meringue before forming and baking to make these desserts more flavorful.

Basic rules for beating egg whites are discussed in Chapter 24, page 767. Please review this section before attempting to make any of the following preparations. We repeat one rule here because it is so important:

Make sure that all equipment is free of any trace of fat or grease, and that the egg whites have no trace of yolk in them. *Even a small trace of fat will prevent the whites from foaming properly*.

Soft meringues, used for pie topping, may be made with as little as 1 pound (500 g) sugar per pound (500 g) of egg whites. **Hard meringues**, baked until crisp, are made with up to twice as much sugar as egg whites.

BASIC MERINGUES

The stiffness to which meringues are beaten may vary, as long as they are not beaten until they are too stiff and dry. For most purposes, they are beaten until they form stiff, or nearly stiff, moist peaks.

1. **Common meringue** is made from egg whites at room temperature, beaten with sugar. It is the easiest to make, and it is reasonably stable due to the high percentage of sugar.

2. **Swiss meringue** is made from egg whites and sugar warmed over a double boiler while beating. Warming gives this meringue better volume and stability.

3. **Italian meringue** is made by beating a hot sugar syrup into the egg whites. This meringue is the most stable of the three because the egg whites are actually cooked by the heat of the syrup. When flavored with vanilla, this meringue is also known as *boiled icing*. It is also used in meringue-type buttercream icings.

 Meringue ♥ 🌹

INGREDIENTS	COMMON MERINGUE	SWISS MERINGUE	ITALIAN MERINGUE
Egg whites	8 oz/250 g	8 oz/250 g	8 oz/250 g
Sugar	1 lb/500 g	1 lb/500 g	1 lb/500 g
Water	—	—	4 fl oz/125 mL

Per 1 ounce (28.35 g): Calories, 80; Protein, 1 g; Fat, 0 g (0% cal.); Cholesterol, 0 mg; Carbohydrates, 19 g; Fiber, 0 g; Sodium, 15 mg.

Note: For soft meringues to top pies, use half the amount of sugar.

PROCEDURE

Common Meringue:

1. With the whip attachment, beat the egg whites at high speed until they form soft peaks.
2. Gradually add the sugar with the machine running.
3. Continue to beat until the meringue forms stiff but moist peaks.

Swiss Meringue:

1. Place the egg whites and sugar in a stainless-steel bowl or the top of a double boiler. Beat with a wire whip over hot water until the mixture is warm (about 120°F/50°C).
2. Transfer the mixture to the bowl of a mixing machine and whip at high speed until stiff peaks form.

Italian Meringue:

1. Heat the sugar and water in a saucepan until the sugar dissolves and the mixture boils. Boil until a candy thermometer placed in the mixture registers 240°F (115°C).
2. While the syrup is cooking, beat the egg whites in a mixing machine until they form soft peaks.
3. With the machine running, very slowly beat the hot syrup into the egg whites.
4. Continue beating until the meringue forms firm peaks.

MERINGUE DESSERTS

PROCEDURE for Making Pie Topping

1. Make common meringue or Swiss meringue using equal parts sugar and egg whites. Beat until just stiff.

2. Spread a generous amount (2–3 cups/500–700 mL) of meringue on still-warm pies. Mound it slightly, and be sure to attach it to the edge of the crust all around. If this is not done, the meringue may slide around on the finished pie. Leave the meringue in ripples or peaks.

3. Bake at 400°F (200°C) until the surface is attractively browned.

4. Remove from oven and cool.

PROCEDURE for Making Baked Meringue Shells

1. Beat common or Swiss meringue until stiff.

2. Using a pastry bag or a spoon, form the meringue into small nest shapes on a parchment-lined baking sheet.

3. Bake at 200°–225°F (about 100°C) until crisp but not browned. This will take 1–3 hours.

4. Cool the shells and remove from the parchment. Be careful, as they may be fragile.

5. Use in place of pastry shells for fruit tarts, fill with whipped cream and fresh strawberries or raspberries, or fill with a scoop of ice cream and garnish with chocolate or raspberry sauce. Crisp meringues with ice cream make a dessert called **meringue glacée** (glah say).

PROCEDURE for Making Japonaise Meringues

Japonaise (zhah po nez) meringues are used like cake layers. They may be filled and iced with light buttercream, chocolate mousse, whipped cream, or similar light icings and creams.

1. Prepare 1 recipe (1 lb/500 g egg whites plus 2 lb/1 kg sugar) Swiss meringue.

2. Quickly but carefully fold in 1 pound (500 g) finely chopped hazelnuts.

3. With a pastry bag, form circles of the desired diameter and about ¹/₂ inch (1.5 cm) thick on parchment-lined sheet pans, as shown in Figure 35.11.

4. Bake as for meringue shells.

FIGURE 35.11 To make meringue layers, mark a circle on a sheet of parchment and pipe the meringue in a spiral to fill the circle.

PROCEDURE for Making Baked Alaska

1. Pack softened ice cream into a dome-shaped mold of the desired size. Freeze solid.

2. Prepare a layer of sponge cake the same size as the flat side of the mold and about ¹/₂ inch (1.5 cm) thick.

3. Unmold the frozen ice cream onto the cake layer so the cake forms a base for the ice cream.

4. With a spatula, cover the entire dessert with a thick layer of meringue. If desired, decorate with more meringue forced from a pastry bag.

5. Bake at 450°F (230°C) until the meringue is golden brown.

6. Serve immediately.

FRUIT DESSERTS

Fruit desserts are included here because many are similar to pies or pie fillings. Special favorites include cobblers, which are much like fruit pies made in large baking pans without a bottom crust; crisps, which are like cobblers, but with brown-sugar streusel topping instead of a pastry crust; and betties, which have alternate layers of rich cake crumbs and fruit. Also, don't overlook fresh fruits for dessert, served plain, lightly sweetened, or with cream.

KEY POINTS TO REVIEW

- What are the steps in the rolling-in procedure for making puff pastry dough?

- What are the steps in mixing éclair paste or pâte à choux?

- What are the three basic types of meringue? Describe how to prepare each.

Apple Crisp

YIELD: 1 PAN, 12 × 20 IN. (30 × 50 CM)   PORTION: 48   PORTION SIZE: 4 OZ (125 G)

U.S.	METRIC	INGREDIENTS	PROCEDURE
8 lb EP	4 kg EP	Apples, peeled and sliced	1. Toss the apple gently with the granulated sugar and lemon juice.
4 oz	125 g	Granulated sugar	
2 fl oz	60 mL	Lemon juice	2. Spread the apples evenly in a baking pan measuring 12 × 20 in. (30 × 50 cm) (see Note).
1 lb	500 g	Butter	3. Rub the butter, brown sugar, cinnamon, and flour together until well blended and crumbly.
1 lb 8 oz	750 g	Brown sugar	
2 tsp	10 mL	Cinnamon	4. Sprinkle evenly over the apples.
1 lb 8 oz	750 g	Pastry flour	5. Bake at 350°F (175°C) about 45 minutes, or until the top is browned and the apples are tender.

Per serving: Calories, 180; Protein, 2 g; Fat, 8 g (37% cal.); Cholesterol, 20 mg; Carbohydrates, 28 g; Fiber, 2 g; Sodium, 85 mg.

Note: If possible, use stainless-steel pans instead of aluminum pans. The acid of the fruit will react with aluminum and create an undesirable flavor.

VARIATION

Peach, Cherry, or Rhubarb Crisp

Substitute the indicated fruit for the apples. If rhubarb is used, increase the sugar in step 1 to 12 oz (350 g).

Apple Crisp

Poached Pears ♥

PORTIONS: 24   PORTION SIZE: 2 PEAR HALVES

U.S.	METRIC	INGREDIENTS	PROCEDURE
2 qt	2 L	Water	1. Combine the water and sugar in a trunnion kettle or large saucepot. Bring to a boil, stirring until the sugar is dissolved.
3 lb	1.5 kg	Sugar	
4 tsp	20 mL	Vanilla	2. Remove from heat and add the vanilla.
24	24	Pears	3. Peel the pears. Cut them in half and remove the cores with a melon ball cutter.
			4. Add the pears to the syrup and simmer very slowly until just tender.
			5. Let the pears cool in the syrup. When cool, refrigerate in the syrup until needed for service.

Per serving: Calories, 320; Protein, 1 g; Fat, 0.5 g (1% cal.); Cholesterol, 0 mg; Carbohydrates, 82 g; Fiber, 4 g; Sodium, 5 mg.

VARIATIONS

Pears in Wine

Substitute red or white table wine for the water. Omit vanilla. Add 1 sliced lemon to the syrup. Peel the pears, but leave them whole.

Poached Peaches

Substitute peaches for pears in basic recipe.

Peaches in Wine

Substitute peaches for pears in recipe for Pears in Wine.

Raspberry or Cherry Gratin

PORTIONS: 1

U.S.	METRIC	INGREDIENTS	PROCEDURE
3 oz	90 g	Raspberries or sweet, pitted cherries	1. Select a shallow gratin dish or other heatproof dish large enough to hold the fruit in a shallow layer.
1	1	Genoise layer (see step 2)	2. Cut a thin slice of genoise, about ³/₈ in. (1 cm) thick, to cover the bottom of the dish.
2 oz	60 g	Pastry cream	3. Arrange the fruit on top of the genoise.
1 oz	30 g	Whipped cream	4. Combine the pastry cream, whipped cream, and flavoring. Spread the mixture over the fruit to cover completely.
to taste	to taste	Optional flavoring: kirsch, orange liqueur, or raspberry or cherry brandy	
¹/₄ oz	7 g	Sliced almonds	5. Mix the almonds and butter and sprinkle over the pastry cream. Dredge the top heavily with confectioners' sugar.
¹/₄ oz	7 g	Melted butter	
as needed	as needed	Confectioners' sugar	6. Place under a broiler or in the top of a hot oven for a few minutes to brown the top. Serve hot.

Per serving: Calories, 480; Protein, 7 g; Fat, 27 g (50% cal.); Cholesterol, 165 mg; Carbohydrates, 53 g; Fiber, 7 g; Sodium, 145 mg.

ADDITIONAL RECIPES

These additional recipes may be found on your CulinarE-Companion recipe management program:

Apple Betty; Blueberry Pie Filling; Fresh Strawberry Pie Filling; Fruit Cobbler.

TERMS FOR REVIEW

flaky pie dough	soft pie	cooked fruit method	hard meringue
mealy pie dough	cream pie	puff pastry	common meringue
rubbed dough method	chiffon pie	blitz puff pastry	Swiss meringue
crumb crust	baking blind	éclair paste	Italian meringue
short dough	instant starch	pâte à choux	meringue glacée
fruit pie	cooked juice method	soft meringue	

QUESTIONS FOR DISCUSSION

1. Discuss the factors that affect tenderness, toughness, and flakiness in pie dough. Why should emulsified shortening not be used for pie dough?

2. What kind of crust or crusts would you use for a pumpkin pie? an apple pie? a banana cream pie?

3. What would happen to a flaky pie dough if you mixed it too long before adding the water? after adding the water?

4. How can you prevent shrinkage when baking pie shells?

5. What are the remedies for soggy or undercooked bottom pie crusts?

6. What starch would you use to thicken apple pie filling? chocolate pie filling? lemon pie filling? peach pie filling?

7. Why is lemon juice added to lemon pie filling after the starch has thickened the water? Wouldn't this thin the filling?

8. Why is it important to bake cream puffs and éclairs thoroughly and to cool them slowly?

9. Briefly describe the difference between common, Swiss, and Italian meringues.

36

CREAMS, CUSTARDS, PUDDINGS, FROZEN DESSERTS, AND SAUCES

A quick glance at this last chapter may give you the impression that you will be overwhelmed with a great many recipes and techniques within a few pages. Among the subjects covered are custard sauces, pastry cream, puddings, custards, mousses, bavarians, soufflés, ice cream, and dessert sauces.

It's all much simpler than it seems. Once you have learned three basic preparations—vanilla custard sauce, pastry cream, and baked custard—you will have learned most of the rest. Vanilla custard sauce, also called crème anglaise or English cream, is the basis for bavarians, ice cream, and some dessert sauces. Pastry cream, with a variety of flavorings, is also used for pie fillings and puddings and is the basis for some soufflés. Many baked puddings are baked custard with added starch or fruit ingredients.

There seems little point in giving you recipes for cream pie fillings in the pie section, a recipe for pastry cream filling for napoleons in the puff pastry section, and recipes for boiled puddings in the pudding section, and never telling you they are all basically the same preparation. You are not just learning a collection of unrelated recipes; you are learning to cook and to understand what you are cooking.

AFTER READING THIS CHAPTER, YOU SHOULD BE ABLE TO

1. Cook sugar syrups to the seven stages of hardness.

2. Prepare crème anglaise, pastry cream, and baked custard.

3. Prepare starch-thickened puddings and baked puddings.

4. Prepare bavarians, chiffons, mousses, and dessert soufflés.

5. Assemble frozen desserts.

6. Prepare dessert sauces.

SUGAR COOKING

Understanding sugar cooking is important in the preparation of desserts and confections because sugar syrups of various strengths are often required (see, for example, Italian Meringue, p. 1002).

BASIC PRINCIPLES

The principle of sugar cooking is fairly simple. A solution or syrup of sugar and water is boiled to evaporate part of the water. As the water is boiled off, the temperature of the syrup gradually rises. When all the water has evaporated, what you have left is melted sugar. The sugar then begins to caramelize or turn brown and to change flavor. If heating continues, the sugar continues to darken and then burn.

A syrup cooked to a high temperature is harder when it is cooled than a syrup cooked to a lower temperature. For example, a syrup cooked to 240°F (115°C) forms a soft ball when cooled. A syrup cooked to 300°F (150°C) is hard and brittle when cooled.

One part water (by weight) is enough to dissolve and cook 3–4 parts sugar. There is no point in adding more water than is necessary because you just have to boil it off.

SIMPLE SYRUP

Simple syrup is a solution of equal weights of sugar and water. Combine equal weights of water and granulated sugar in a saucepan, stir, and bring to a boil to dissolve the sugar. Cool the syrup.

Dessert syrup is a flavored simple syrup used to moisten and flavor some cakes (see p. 947). (Many chefs use 2 or 3 parts water to 1 part sugar for a less sweet syrup.) Flavorings may be extracts, such as vanilla, or liquors, such as rum or kirsch. Add flavorings after the syrup has cooled because flavor may be lost if they are added to hot syrup. Syrups may also be flavored by boiling them with lemon or orange rind.

CRYSTALLIZATION

Graininess is a common fault in many candies and desserts. Graininess results when cooked sugar **crystallizes**, or turns to tiny sugar crystals, rather than staying dissolved in the syrup. If even one sugar crystal comes in contact with a cooked syrup, it can start a chain reaction that turns the whole thing into a mass of sugar crystals.

To avoid crystallization during the first stages of boiling, use one of the following techniques:

1. Wash down the sides of the saucepan with a brush dipped in water. This removes crystals that may seed the whole batch.

2. When first bringing the syrup to a boil, cover the pan and boil for several minutes. Condensed steam will wash down the sides of the pan. Uncover and finish cooking without stirring.

Sometimes an acid such as cream of tartar is added to a syrup before cooking. Acids change some of the sugar to *invert sugar*, which resists crystallizing. Corn syrup is sometimes added for the same reason.

Washing down the sides of a pan

STAGES OF SUGAR COOKING

Testing the temperature with a candy thermometer is the most accurate way to determine the desired doneness of a syrup.

In the old days, syrups were tested by dropping a little bit into a bowl of cold water and checking the hardness of the cooled sugar. The stages of doneness were given names that described their hardness. Table 36.1 lists these stages of sugar cooking.

BASIC CUSTARDS AND CREAMS

The three preparations presented in this section are among the most basic and useful preparations in the bakeshop. All three can be classified as custards because they consist of a liquid thickened by the coagulation of eggs.

TABLE 36.1 Stages of Doneness in Sugar Cooking

Stage	Temperature °F	°C
Thread	230	110
Soft ball	240	115
Firm ball	245	118
Hard ball	250–260	122–127
Small crack	265–270	130–132
Crack	275–280	135–138
Hard crack	290–310	143–155
Caramel	320–340	160–170

Crème anglaise (krem awn glezz), or vanilla custard sauce, is a stirred custard. It consists of milk, sugar, and egg yolks (and vanilla) stirred over very low heat until lightly thickened.

Pastry cream contains starch thickeners as well as eggs, resulting in a much thicker and more stable product. It is used as a cake and pastry filling, as a filling for cream pies, and as a pudding. With additional liquid, it is used as a custard sauce.

Baked custard, like vanilla custard sauce, consists of milk, sugar, eggs, and flavoring (usually whole eggs are used for greater thickening power). But, unlike the sauce, it is baked rather than stirred over heat, so it sets and becomes firm. Baked custard is used as a pie filling, as a dessert by itself, and as a basis for many baked puddings.

All of these preparations are subject to a wide range of variation. Because they are based on eggs, it would be helpful for you to review the basic egg cooking principles discussed in Chapter 24.

CRÈME ANGLAISE

The following recipe gives the method for preparing vanilla custard sauce, or crème anglaise. Special care is necessary in preparing this sauce because the eggs can curdle easily if overcooked. The following guidelines will help you succeed:

1. Use clean, sanitized equipment, and follow strict sanitation procedures. Egg mixtures are good breeding grounds for bacteria that cause food poisoning.

2. When combining the egg yolks and sugar, whip the mixture as soon as the sugar is added. Letting sugar and egg yolks stand together without mixing creates lumps that cannot be beaten out (Figure 36.1). Using a stainless-steel bowl for this step makes the cooking and stirring easier in step 5.

3. Heat the milk to scalding (just below simmering) before combining with the egg yolks. This makes the final cooking much shorter. To avoid scorching, you can set the pan of milk in a pan of boiling water. This takes longer than using direct heat, but the pan can be left unattended for a few minutes while you perform other tasks.

4. *Slowly* beat the hot milk into the beaten eggs and sugar. This raises the temperature of the eggs gradually and helps prevent curdling.

5. Set the bowl containing the egg mixture in a pan of simmering water and stir constantly to prevent curdling.

6. To test for doneness, two methods are available. Keep in mind that this is a very light sauce, so you can't expect a lot of thickening.

   • Check the temperature with a thermometer. When it reaches 185°F (85°C), the sauce is cooked. Never let the temperature go above 190°F (87°C), or the sauce is likely to curdle.

   • When the mixture lightly coats the back of a spoon instead of running off like milk, the sauce is cooked (Figure 36.2).

FIGURE 36.1 Combining sugar and egg yolks without immediately beating them creates hard lumps.

FIGURE 36.2 Crème anglaise coating the back of a spoon.

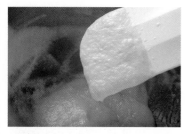

FIGURE 36.3 Crème anglaise that has curdled because it was overheated.

7. *Immediately* cool the sauce by setting the pan or bowl in ice water. Stir occasionally to cool it evenly.

8. If the sauce curdles (Figure 36.3), it is sometimes possible to save it. Immediately stir in 1–2 ounces (30–60 mL) cold milk, transfer the sauce to a blender, and blend at high speed.

Crème Anglaise (Vanilla Custard Sauce)

YIELD: ABOUT 2¹/₂ PT (1.25 L)

U.S.	METRIC	INGREDIENTS	PROCEDURE
12	12	Egg yolks	**1.** Review the guidelines for preparing vanilla custard sauce preceding this recipe.
8 oz	250 g	Sugar	**2.** Combine the egg yolks and sugar in a stainless-steel bowl. Whip until thick and light.
1 qt	1 L	Milk	**3.** Scald the milk in a boiling-water bath or over direct heat.
			4. Very gradually, pour the scalded milk into the egg yolk mixture while stirring constantly with the whip.
			5. Set the bowl over simmering water. Heat it slowly, stirring constantly, until it thickens enough to coat the back of a spoon (or until it reaches 185°F/85°C).
1 tbsp	15 mL	Vanilla	**6.** Immediately remove the bowl from the heat and set it in a pan of cool water. Stir in the vanilla. Stir the sauce occasionally as it cools.

Per 1 fl oz (29.57 mL): Calories, 60; Protein, 2 g; Fat, 2.5 g (38 % cal.); Cholesterol, 65 mg; Carbohydrates, 7 g; Fiber, 0 g; Sodium, 15 mg.

VARIATIONS

For a richer crème anglaise, substitute heavy cream for up to half the milk. To flavor with a vanilla bean instead of vanilla extract, first split the bean in half lengthwise. Scrape the pulp from inside the bean with a paring knife (Figure 36.4). Add the pulp and the split bean to the milk before heating in step 3.

Chocolate Crème Anglaise

Melt 6 oz (175 g) sweetened chocolate. Stir into the crème anglaise while it is still warm (not hot).

FIGURE 36.4 Vanilla bean technique.

(a) Split the bean in half lengthwise.

(b) Scrape the pulp from inside the bean with a paring knife.

PASTRY CREAM

Although it requires more ingredients and steps, pastry cream is easier to make than custard because it is less likely to curdle. Pastry cream contains a starch thickening agent that stabilizes the eggs. It can actually be boiled without curdling. In fact, it must be brought to a boil, or the starch will not cook completely and the cream will have a raw, starchy taste.

Strict observance of all sanitation rules is essential when preparing pastry cream because of the danger of bacterial contamination. Use clean, sanitized equipment. Do not put your fingers in the cream, and do not taste it except with a clean spoon. Chill the finished cream rapidly in shallow pans. Keep the cream and all cream-filled products refrigerated at all times.

The procedure for preparing pastry cream is given in the following recipe. Note that the basic steps are similar to those for custard sauce. In this case, however, a starch is mixed with the eggs and half the sugar to make a smooth paste. (In some recipes with lower egg content, it is necessary to add a little cold milk to provide enough liquid to make a paste.) Meanwhile, the milk is scalded with the other half of the sugar (the sugar helps protect the milk from scorching on the bottom of the pan).

The egg mixture is then tempered with some of the hot milk, returned to the kettle, and brought to a boil. Some chefs prefer to add the cold paste gradually to the hot milk, but the tempering procedure given here seems to give better protection against lumping and curdling.

🎩 Vanilla Pastry Cream

YIELD: ABOUT 2¼ PT (1.12 L)

U.S.	METRIC	INGREDIENTS	PROCEDURE
4 oz	125 g	Sugar	1. In a heavy saucepan or trunnion kettle, dissolve the sugar in the milk and bring just to a boil.
1 qt	1 L	Milk	
3 oz	90 g	Egg yolks	2. With a whip, beat the egg yolks and whole eggs in a stainless-steel bowl (see **Fig. 36.5**).
4 oz	125 g	Whole eggs	3. Sift the starch and sugar into the eggs. Beat with the whip until perfectly smooth.
2.5 oz	75 g	Cornstarch	4. Temper the egg mixture by slowly beating in the hot milk in a thin stream.
4 oz	125 g	Sugar	5. Return the mixture to the heat and bring to a boil, stirring constantly.
2 oz	60 g	Butter	6. When the mixture comes to a boil and thickens, remove from the heat.
1 tbsp	15 mL	Vanilla	7. Stir in the butter and vanilla. Mix until the butter is melted and completely blended in.
			8. Pour into a clean, sanitized hotel pan or other shallow pan. Dust lightly with sugar and cover with waxed paper to keep a crust from forming. Cool and chill as quickly as possible.
			9. For filling pastries such as éclairs and napoleons, whip the chilled pastry cream until smooth before using.

Per 1 fl oz (29.57 mL): Calories, 70; Protein, 2 g; Fat, 3 g (38% cal.); Cholesterol, 45 mg; Carbohydrates, 9 g; Fiber, 0 g; Sodium 30 mg.

VARIATIONS

For a lighter pastry cream filling, fold whipped heavy cream into the chilled pastry cream. Quantities may be varied to taste. For every 1 qt (1 L) pastry cream, whip ½–1 cups (125-250 mL) heavy cream.

Chocolate Pastry Cream

For each 12 oz (375 g) pastry cream, melt together 4 oz (125 g) semisweet chocolate. Stir into the pastry cream while it is still warm.

Coffee Pastry Cream

Add 2 tbsp (30 mL) instant coffee powder to the milk in step 1.

CREAM PIE FILLINGS

Quantities for two 8-inch (20-cm) pies

Vanilla Cream Pie Filling

This is the same as Vanilla Pastry Cream. Fill prebaked pie shells with cooled but not chilled filling.

Coconut Cream Pie Filling

Add 4 oz (125 g) toasted unsweetened coconut to Vanilla Pastry Cream.

Banana Cream Pie Filling

Using Vanilla Cream Pie Filling, pour half of the filling into the pie shells, cover with sliced bananas, and fill with remaining filling. (Bananas may be dipped in lemon juice to help prevent browning.)

Chocolate Cream Pie Filling I

This is the same as Chocolate Pastry Cream, above.

Chocolate Cream Pie Filling II

In step 1, use only 1¾ pt (875 mL) milk. Add 1½ oz (45 g) cocoa to dry ingredients (sugar and cornstarch) in basic Vanilla Pastry Cream recipe. Add 4 fl oz (125 mL) cold milk to the eggs.

Butterscotch Cream Pie Filling

Combine 1 lb (450 g) brown sugar and 5 oz (150 g) butter in a saucepan over low heat. Heat and stir until butter is melted and ingredients are blended. Omit all the sugar from the basic Vanilla Pastry Cream recipe (steps 1 and 3). Increase the starch to 3 oz (90 g). As the mixture is nearing a boil in step 5, gradually stir in the brown sugar mixture. Finish as in basic recipe.

Lemon Pie Filling

Follow the procedure for Vanilla Pastry Cream, but make the following ingredient adjustments:

1. Use water instead of milk.
2. Increase sugar in step 1 to 8 oz (250 g).
3. Increase the cornstarch to 3 oz (90 g).
4. Add the grated zest of 1 lemon to the egg mixture.
5. Add 4 fl oz (125 mL) lemon juice to the finished, hot cream in place of the vanilla.

CREAM PUDDINGS

Vanilla Pudding

Coconut Pudding

Banana Cream Pudding

Chocolate Pudding I and II

Butterscotch Pudding

For each of these puddings, prepare the corresponding pie filling but use only half the cornstarch.

FIGURE 36.5 Preparing pastry cream.

(a) Beat the eggs with the sugar and cornstarch mixture until smooth.

(b) Temper by slowly beating in the hot milk.

(c) Return the mixture to heat and bring to a boil, stirring constantly.

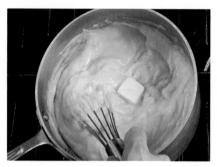

(d) Remove from heat and stir in the butter and vanilla.

(e) Cool in a shallow pan, covered with plastic film directly on the surface.

Pastry Cream Variations

Cream pie fillings and puddings are actually pastry cream flavored with various ingredients.

Cornstarch should be used as the thickening agent when the cream is to be used as a pie filling so the cut slices hold their shape. For other uses, either cornstarch or flour may be used. Remember that twice as much flour as cornstarch is required for the same thickening power. Other variations are possible, as you will see in the recipes. Sometimes whipped cream is folded into cold pastry cream to lighten it and make it creamier.

Lemon pie filling is also a variation of pastry cream. It is made with water instead of milk, and it is flavored with lemon juice and grated lemon rind.

BAKED CUSTARD

Baked custard is a mixture of eggs, milk, sugar, and flavorings that is baked until the eggs coagulate and the custard is set. A good custard holds a clean, sharp edge when cut.

The following recipe gives the procedure for making baked custard. Note these points in particular:

FIGURE 36.6 Testing a baked pudding for doneness.

1. Scald the milk before beating it slowly into the eggs. This reduces cooking time and helps the product cook more evenly.

2. Remove any foam, which would mar the appearance of the finished product.

3. Bake at 325°F (165°C). High temperatures increase the risk of overcooking and curdling.

4. Bake in a water bath so the outside edges are not overcooked before the inside is set.

5. To test for doneness, insert a thin-bladed knife 1–2 inches (3–5 cm) from the center (Figure 36.6). If it comes out clean, the custard is done. The center may not be completely set, but the custard will continue to cook in its own heat after removal from the oven.

KEY POINTS TO REVIEW

- What is the proper procedure to cook a sugar syrup so it doesn't crystallize? How do you tell when a syrup is cooked to the proper stage?

- What is the procedure for cooking crème anglaise?

- What is the procedure for cooking pastry cream?

- What is the procedure for cooking a plain baked custard?

Baked Custard

PORTIONS: 12   **PORTION SIZE: 5 OZ (150 G)**

U.S.	METRIC	INGREDIENTS
1 lb	500 g	Eggs
8 oz	250 g	Sugar
½ tsp	2 mL	Salt
½ fl oz	15 mL	Vanilla
2½ pt	1.25 L	Milk

PROCEDURE

1. Combine the eggs, sugar, salt, and vanilla in a mixing bowl. Mix until thoroughly blended, but do not whip.

2. Scald the milk in a double boiler or in a saucepan over low heat.
3. Gradually pour the milk into the egg mixture, stirring constantly.
4. Skim off all foam from the surface of the liquid.
5. Arrange custard cups in a shallow baking pan. (Butter the insides of the cups if the custards are to be unmolded.)
6. Carefully pour the custard mixture into the cups. If any bubbles form during this step, skim them off.
7. Set the baking pan on the oven shelf. Pour enough hot water into the pan around the cups so the level of the water is about as high as the level of the custard mixture.
8. Bake at 325°F (165°C) until set, about 45 minutes.
9. Carefully remove from the oven and cool. Store, covered, in refrigerator.

Per serving: Calories, 190; Protein, 8 g; Fat, 7 g (33% cal.); Cholesterol, 175 mg; Carbohydrates, 24 g; Fiber, 0 g; Sodium, 190 mg.

VARIATION

Crème Caramel

Cook 12 oz (375 g) sugar and 2 fl oz (60 mL) water until it caramelizes (see the section on sugar cooking at the beginning of this chapter). Line the bottoms of the custard cups with the hot caramel. (Be sure the cups are clean and dry.) Fill with custard and bake as in basic recipe. When done, cool and refrigerate overnight. Unmold onto a plate for service.

PUDDINGS

It is difficult to give a definition of **pudding** that includes everything called by that name. The term is used for such different dishes as chocolate pudding, blood sausages (blood puddings), and steak-and-kidney pudding. In this chapter, however, we consider only the more popular dessert puddings.

Two kinds of pudding, starch-thickened and baked, are the most frequently prepared in food-service kitchens. These are the types we discuss here. A third type, steamed pudding, is less often served, and then mainly in cold weather, because steamed puddings are usually rather heavy and filling.

STARCH-THICKENED PUDDINGS

These are also called *boiled puddings* because they are boiled in order to cook the starch that thickens them.

1. **Cornstarch pudding or blancmange.**
   Cornstarch pudding consists of milk, sugar, and flavorings and is thickened with cornstarch (or, sometimes, another starch). If enough cornstarch is used, the hot mixture may be poured into molds, chilled, and unmolded for service.

2. **Cream puddings.**
   Cream puddings, as you learned in the previous section, are the same as pastry cream. Puddings are usually made with less starch, however, and may contain any of several flavoring ingredients, such as coconut or chocolate. Butterscotch pudding is given its flavor by using brown sugar instead of white sugar.

   If you look again at the recipe for Vanilla Pastry Cream (p. 1011), you will see the only difference between cornstarch puddings and cream puddings is that the latter contain eggs. In fact, cream puddings may be made by stirring hot cornstarch pudding into beaten eggs, then heating the entire mixture to just below the simmer. Care must be taken to avoid curdling the eggs if this method is used.

   A basic recipe for cornstarch pudding follows. Recipes for cream puddings are included among the variations following the recipe for Vanilla Pastry Cream.

Blancmange, English Style

PORTIONS: 12   **PORTION SIZE: 4 OZ (125 G)**

U.S.	METRIC	INGREDIENTS	PROCEDURE
1 qt	1 L	Milk	1. Combine the milk, sugar, and salt in a heavy saucepan and bring to a simmer.
6 oz	190 g	Sugar	
1/4 tsp	1 mL	Salt	
4 oz	125 g	Cornstarch	2. Mix the cornstarch and cold milk until perfectly smooth.
8 fl oz	250 mL	Milk, cold	3. Pouring in a thin stream, add about 1 cup (250 mL) hot milk to the cornstarch mixture.
			4. Stir this mixture back into the hot milk.
			5. Stir over low heat until the mixture thickens and comes to a boil.
1 1/2 tsp	7 mL	Almond or vanilla extract	6. Remove from heat and add desired flavoring.
			7. Pour into 1/2-cup (125-mL) molds. Cool, then chill. Unmold for service.

Per serving: Calories, 150; Protein, 3 g; Fat, 3.5 g (20% cal.); Cholesterol, 15 mg; Carbohydrates, 28 g; Fiber, 0 g; Sodium, 100 mg.

Note: French blancmange is very different from English. The French style is made with almond milk and gelatin.

VARIATIONS

Blancmange or cornstarch pudding may be flavored in any way cream puddings are. See the variations following the Vanilla Pastry Cream recipe (p. 1011).

BAKED PUDDINGS

Baked puddings are custards that contain additional ingredients, usually in large quantities. Bread pudding, for example, is made by pouring a custard mixture over pieces of bread in a pan and baking it in the oven. Rice pudding, another popular item, is made of cooked rice and custard.

The procedure for making baked puddings is the same as for making baked custard. A water bath may not be necessary if the starch content of the pudding is high.

Soft pie fillings, such as pumpkin, could also be considered baked puddings.

Rice Pudding

PORTIONS: 12   PORTION SIZE: 5 OZ (150 G)

U.S.	METRIC	INGREDIENTS	PROCEDURE
8 oz	225 g	Rice (medium- or long-grain)	1. Wash the rice well. Drain.
1 1/2 qt	1.5 L	Milk	2. Combine the rice, milk, vanilla, and salt in a heavy saucepan. Cover and simmer over very low heat until the rice is tender, about 30 minutes. Stir occasionally to be sure the mixture doesn't scorch on the bottom. Remove from heat when cooked.
1 tsp	5 mL	Vanilla	
1/4 tsp	1 mL	Salt	
1	1	Whole egg	3. Combine the egg, yolks, sugar, and cream in a mixing bowl. Mix until evenly combined.
2	2	Egg yolks	
8 oz	250 g	Sugar	4. Ladle some of the hot milk from the cooked rice into this mixture and mix well. Then very slowly stir the egg mixture into the hot rice.
8 fl oz	250 mL	Light cream	
as needed	as needed	Cinnamon	5. Pour into a buttered baking pan, 12 × 20 in. (30 × 50 cm). Sprinkle the top with cinnamon.
			6. Bake in a water bath at 350°F (175°C) for 30–40 minutes, or until set. Serve warm or chilled.

Per serving: Calories, 270; Protein, 7 g; Fat, 9 g (30 % cal.); Cholesterol, 80 mg; Carbohydrates, 40 g; Fiber, 0 g; Sodium, 120 mg.

VARIATION

Raisin Rice Pudding

Add 8 oz (250 g) raisins to the cooked rice and milk mixture.

Bread and Butter Pudding

PORTIONS: 12   PORTION SIZE: 6 1/2 OZ (200 G)

U.S.	METRIC	INGREDIENTS	PROCEDURE
1 lb	500 g	White bread, in thin slices	1. Cut each slice of bread in half. Brush both sides of each piece with melted butter.
4 oz	125 g	Melted butter	2. Arrange the bread overlapping in a buttered baking pan, 12 × 20 in. (30 × 50 cm).
1 lb	500 g	Eggs	3. Mix the eggs, sugar, salt, and vanilla until thoroughly combined.
8 oz	250 g	Sugar	4. Gradually stir in the milk.
1/2 tsp	2 mL	Salt	
1/2 fl oz	15 mL	Vanilla	
2 1/2 pt	1.25 L	Milk	
as needed	as needed	Cinnamon	5. Pour the custard mixture over the bread slices in the pan. Let stand, refrigerated, 1 hour or longer, so the bread absorbs the custard mixture.
as needed	as needed	Nutmeg	6. Sprinkle the top lightly with cinnamon and nutmeg.
			7. Set the pan in a larger pan containing about 1 in. (3 cm) hot water.
			8. Place in an oven preheated to 350°F (175°C). Bake about 1 hour, or until set.
			9. Serve warm or cold with whipped cream or light custard sauce, or dusted with confectioners' sugar.

Per serving: Calories, 350; Protein, 11 g; Fat, 16 g (41% cal.); Cholesterol, 190 mg; Carbohydrates, 41 g; Fiber, 1 g; Sodium, 440 mg.

BAVARIANS, CHIFFONS, MOUSSES, AND SOUFFLÉS

All the preparations in this section have one thing in common: They all have a light, fluffy, or puffed texture created by the addition of whipped cream, beaten egg whites, or both.

Although these particular products may be new to you, you should have little trouble learning to prepare them if you have already studied the previous chapter and the first part of this chapter. Once you have learned to prepare crème anglaise, pastry cream, starch-thickened fruit fillings, meringues, and whipped cream and have learned to work with gelatin (read p. 710 if you have not yet studied gelatin), all you have to do is combine these products in different ways to make bavarians, chiffons, mousses, and soufflés.

Let's look at these four items separately to see what they are made of. Afterward, we examine the procedures for assembling them.

BAVARIANS

A **bavarian**, also known as *bavarian cream* or *bavarois*, is made of three basic elements: crème anglaise (flavored as desired), gelatin, and whipped cream.

That's all there is to it. Gelatin is softened in cold liquid, stirred into the hot crème anglaise until dissolved, and chilled until almost set. Whipped cream is then folded in, and the mixture is poured into a mold until set. It is unmolded for service.

Accurate measuring of the gelatin is important. If not enough gelatin is used, the dessert will be too soft to hold its shape. If too much is used, it will be too firm and rubbery.

CHIFFONS

Chiffons are most popular as fillings for chiffon pies, but they may also be served more simply as puddings and chilled desserts.

The major difference between chiffons and bavarians is that beaten egg whites are used in place of or in addition to whipped cream. In other words, chiffons are made of a base plus gelatin plus beaten egg whites. (Some chiffons also contain whipped cream.)

Bases for chiffons include the following three main types:

1. **Thickened with starch.**
   The procedure is the same as for fruit pie fillings made by the cooked juice or cooked fruit method, except the fruit is finely chopped or puréed. Most fruit chiffons are made this way.

2. **Thickened with egg.**
   The procedure is the same as for custard sauce or crème anglaise. Many chocolate chiffons are made this way, as is pumpkin chiffon.

3. **Thickened with egg and starch.**
   The procedure is the same as for pastry cream. Lemon chiffon is usually made this way.

MOUSSES

There are so many varieties of mousse that it is impossible to give a rule for all of them. In general, we define a **mousse** as any soft or creamy dessert made light and fluffy by the addition of whipped cream, beaten egg whites, or both. Note that bavarians and chiffons both fit this description. In fact, they are often served as mousses, but with the gelatin reduced or left out so the mousse is softer.

Many kinds of base are used for mousses. The base may be nothing more than melted chocolate or puréed fresh fruit, or it may be more complex, like the bases for chiffons.

Some mousses contain both beaten egg whites and whipped cream. When this is the case, most chefs prefer to fold in the egg whites first, even though they may lose some volume. The reason is that if the cream is added first, there is more danger it will be overbeaten and turn to butter during the folding and mixing procedure.

If egg whites are folded into a hot base, they will cook or coagulate, and the mousse will be firmer and more stable. Whipped cream should never be folded into hot mixtures, or it will melt and deflate.

DESSERT SOUFFLÉS

Soufflés are lightened with beaten egg whites, then baked. Baking causes the soufflé to rise like a cake because the air in the egg foam expands when heated.

To understand the structure of dessert soufflés, we can divide their preparation into four stages:

1. **Base.**
   Many kinds are used for dessert soufflés. Most are heavy, starch-thickened preparations, such as pastry cream or sweetened white sauce.

2. **Egg yolks.**
   When used, these are added to the base.

3. **Egg whites.**
   Whenever possible, egg whites should be whipped with some of the sugar. This makes dessert soufflés more stable than entrée soufflés.

4. **Baking.**
   Review the section on entrée soufflés (pp. 777–778) so you understand the general principles of baking soufflés.

SUMMARY AND COMPARISON

1. **Bavarian**
   Base: Custard sauce
   Gelatin
   Whipped cream

2. **Chiffon**
   Base: Starch-thickened (fruit filling type)
        Egg-thickened (custard type)
        Egg- and starch-thickened (pastry
          cream type)
   Gelatin
   Egg whites
   (Optional whipped cream)

3. **Mousse**
   Base: Many varieties
   Little or no gelatin
   Egg whites and/or whipped cream

4. **Soufflé**
   Base: Many varieties, usually containing
       egg yolk
   Egg whites
   Baked

GENERAL PROCEDURE for Making Bavarians, Chiffons, Mousses, and Soufflés

The following is a general procedure only. It is not a detailed method for one specific dessert, but it will give you a basic understanding that will help you tackle many recipes. These basic steps apply to most bavarians, chiffons, mousses, and soufflés.

1. Prepare the base.

2. If gelatin is used, soften it in cold liquid and stir it into the hot base until dissolved. Chill until almost set.

3. Fold in the beaten egg whites and/or whipped cream.

4. Chill (bavarians, chiffons, and mousses) or bake (soufflés).

FIGURE 36.7. Preparing Bavarian and Bavarian-type creams.

(a) Dissolve the gelatin in the hot base.

(b) Mix in flavorings.

(c) Cool until thick but not set. Then fold in whipped cream.

KEY POINTS TO REVIEW

- What is the difference between a cornstarch pudding and a cream pudding?

- Rice pudding and bread pudding are made the same way as which of the three basic custard preparations discussed in the previous section?

- In what ways are bavarians, chiffons, and mousses similar? In what ways are they different?

- What are the four basic components of a baked soufflé?

Bavarian Cream

PORTIONS: 12   **PORTION SIZE: 3 OZ (90 G)**

U.S.	METRIC	INGREDIENTS	PROCEDURE
0.75 oz	22 g	Gelatin (unflavored)	1. Soak the gelatin in cold water.
5 oz	150 mL	Cold water	
		Crème anglaise:	2. Prepare the crème anglaise; Whip the egg yolks and sugar until thick and light. Scald the milk and slowly stir it into the egg-yolk mixture, beating constantly. Add vanilla. Cook in a hot-water bath, stirring constantly, until it just thickens slightly. (Review pp. 1009–1010 for details on making crème anglaise.)
6	6	Egg yolks	
4 oz	125 g	Sugar	
1 pt	500 mL	Milk	
1½ tsp	8 mL	Vanilla	
			3. While the sauce is still hot, add the softened gelatin. Stir until the gelatin is dissolved.
			4. Cool in the refrigerator or over crushed ice, stirring occasionally to keep the mixture smooth.
1 pt	500 mL	Heavy cream	5. While the custard sauce is cooling, whip the cream until it forms soft, not stiff, peaks. Do not overwhip.
			6. When the custard sauce is very thick but not yet set, fold in the whipped cream.
			7. Pour into molds or into serving dishes.
			8. Chill until completely set. If prepared in molds, unmold for service.

Per serving: Calories, 240; Protein, 5 g; Fat, 19 g (70 % cal.); Cholesterol, 165 mg; Carbohydrates, 13 g; Fiber, 0 g; Sodium, 45 mg.

VARIATIONS

Chocolate Bavarian Cream

Add 6 oz (175 g) sweetened chocolate, chopped or grated, to the hot crème anglaise. Stir until completely melted and blended in.

Coffee Bavarian Cream

Add 1½ tbsp (22 mL) instant coffee powder to the hot crème anglaise.

Strawberry Bavarian Cream

Reduce the milk to 8 fl oz (250 mL) and the sugar to 3 oz (90 g) when making the crème anglaise. Mash 8 oz (250 g) strawberries with 3 oz (90 g) sugar, or use 12 oz (350 g g) frozen, sweetened strawberries. Stir this purée into the custard sauce before adding the whipped cream.

Raspberry Bavarian Cream

Prepare like Strawberry Bavarian Cream, using raspberries.

Strawberry Chiffon Dessert or Pie Filling ♥

YIELD: 1 LB 5 OZ; ONE 9-INCH (23-CM) PIE

U.S.	METRIC	INGREDIENTS	PROCEDURE
13 oz	360 g	Frozen sweetened strawberries (see Note)	1. Thaw and drain the strawberries, reserving the juice. Chop the strawberries coarsely.
¹/₅ tsp	1 ml	Salt	2. Place the reserved juice and salt in a saucepan. Bring to a boil.
0.2 oz	6 g	Cornstarch	3. Dissolve the cornstarch in the water and stir into the strawberry juice. Cook until thick. Remove from heat.
1 fl oz	24 mL	Water	
0.2 oz	6 g	Gelatin	4. Soften the gelatin in the water. Add to the hot, thickened fruit juice and stir until completely dissolved.
1.5 fl oz (3 tbsp)	45 mL	Water, cold	5. Stir in the lemon juice and the drained strawberries.
0.2 fl oz (1¼ tsp)	6 mL	Lemon juice	6. Chill the mixture until thickened but not set.
3 oz	90 g	Egg whites, pasteurized	7. Beat the egg whites until they form soft peaks.
2.5 oz	70 g	Sugar	8. Gradually add the sugar and continue to beat until a thick, glossy meringue is formed.
			9. Fold the meringue into the fruit mixture.
			10. Portion into individual serving dishes or fill baked pie shells.
			11. Chill until set.

Per 1 ounce (28.35 g): Calories, 40; Protein, 1 g; Fat, 0 g (0% cal.); Cholesterol, 0 mg; Carbohydrates, 9 g; Fiber, 0 g; Sodium, 30 mg.

Note: To use fresh strawberries, slice or dice 3 lb (1.4 kg) fresh, hulled strawberries and mix with 1 lb (450 g) sugar. Let stand 2 hours in refrigerator. Drain and reserve juice. Proceed as in basic recipe.

VARIATIONS

For a creamier chiffon, reduce egg whites to 2.4 oz (70 g). Whip 3.25 fl oz (100 mL) heavy cream and fold it in after the meringue.

Raspberry Chiffon Dessert or Pie Filling

Substitute raspberries for strawberries in basic recipe.

Pineapple Chiffon Dessert or Pie Filling

Use 10 oz (285) crushed pineapple. Mix the drained juice with an additional 3.25 fl oz (100 mL) pineapple juice and add 1.6 oz (50 g) sugar.

Frozen Strawberry or Raspberry Mousse

Omit gelatin and second quantity of water from basic recipe or from raspberry variation. Reduce egg whites to 1.6 oz (50 g). Whip 5 fl oz (150 mL) heavy cream and fold it in after the meringue. Pour into molds or other containers and freeze.

Strawberry Chiffon Pie

Vanilla Souffle 🌸

PORTIONS: 10   PORTION SIZE: 4¹/₂ OZ (125 G)

U.S.	METRIC	INGREDIENTS	PROCEDURE
3 oz	90 g	Flour	1. Work the flour and butter together to form a smooth paste.
3 oz	90 g	Butter	
4 oz	125 g	Sugar	2. Dissolve the sugar in the milk and bring to a boil. Remove from the heat.
1 pt	500 mL	Milk	3. With a wire whip, beat in the flour paste. Beat vigorously to make sure there are no lumps.
			4. Return the mixture to the heat and bring to a boil, beating constantly. Simmer for several minutes, until the mixture is very thick and no starchy taste remains.
			5. Transfer the mixture to a mixing bowl. Cover and let cool 5–10 minutes.
as needed	as needed	Butter	6. While the mixture is cooling, butter the soufflé dishes well and coat with sugar. For 1 recipe, use 1 dish, 10 in. (25 cm) in diameter; 2 dishes, 7 in. (18 cm) in diameter; or 10 single-portion dishes.
as needed	as needed	Sugar	
8	8	Egg yolks	7. Quickly beat the egg yolks and vanilla into the milk mixture.
2 tsp	10 mL	Vanilla	
8–10	8–10	Egg whites	8. Beat the egg whites until they form soft peaks. Add the sugar and beat until the mixture forms firm, moist peaks.
2 oz	60 g	Sugar	9. Fold the egg whites into the soufflé base.
			10. Pour the mixture into the prepared baking dishes and smooth the tops.
			11. Bake at 375°F (190°C). Approximate baking times are 45–50 minutes for a 10-in. (25-cm) dish, 30–40 minutes for a 7-in. (18-cm) dish, and 15 minutes for single-portion dishes.

Per serving: Calories, 270; Protein, 8 g; Fat, 13 g (44% cal.); Cholesterol, 195 mg; Carbohydrates, 29 g; Fiber, 0 g; Sodium, 150 mg.

FIGURE 36.8. Preparing vanilla soufflé.

(a) Cook the roux-thickened base until thick and no starchy taste remains.

(b) Beat in the egg yolks.

(c) Butter and sugar soufflé dishes.

(d) Fold whipped egg whites into the base.

VARIATIONS

Chocolate Soufflé

Add 3 oz (90 g) melted unsweetened chocolate and 1 oz (30 g) melted sweet chocolate to the base after step 5.

Lemon Soufflé

Instead of vanilla, use the grated zest of 2 lemons for flavoring.

Liqueur Soufflé

Flavor with 2–3 fl oz (60–90 mL) liqueur, such as kirsch or Grand Marnier, added after step 5.

Coffee Soufflé

Flavor with 2 tbsp (30 mL) instant coffee powder, added to the milk in step 2.

Vanilla Soufflé

Chocolate Mousse

YIELD: ABOUT 2³/₄ LB (1.25 KG) OR 2¹/₂ PT (1.25 L)   PORTIONS: 12   PORTION SIZE: 4¹/₂ FL OZ (145 ML)

U.S.	METRIC	INGREDIENTS	PROCEDURE
1 lb	500 g	Bittersweet chocolate	1. Melt the chocolate over hot water.
4 oz	125 g	Butter	2. Remove from the heat and add the butter. Stir until the butter is melted and completely mixed in.
6 oz	180 g	Egg yolks, pasteurized	3. Add the egg yolks, one at a time. Mix in each yolk completely before adding the next.
8 oz	250 g	Egg whites, pasteurized	4. Beat the egg whites until they form soft peaks. Add the sugar and beat until the egg whites form stiff but moist peaks. Do not overbeat.
2¹/₂ oz	75 g	Sugar	5. Fold the egg whites into the chocolate.
8 fl oz	250 mL	Heavy cream	6. Whip the heavy cream until it forms soft peaks. Fold it into the chocolate mixture.
			7. Spoon the mousse into serving dishes, or use a pastry bag fitted with a star tube.
			8. Chill the mousse well before serving.

Per serving: Calories, 380; Protein, 6 g; Fat, 29 g (64% cal.); Cholesterol, 220 mg; Carbohydrates, 30 g; Fiber, 2 g; Sodium, 120 mg.

Note: This mixture may also be used as a filling for cakes, pastries, and baked meringues.

FROZEN DESSERTS

The popularity of ice cream needs no explanation. Whether a plain scoop of vanilla ice cream in a dish or an elaborate assemblage of fruits, syrups, toppings, and numerous flavors of ice cream and sherbet, frozen desserts appeal to just about everyone.

CLASSIFICATION

1. **Ice cream.**

   Ice cream is a smooth, frozen mixture of milk, cream, sugar, flavorings, and, sometimes, eggs. **Philadelphia-style ice cream** contains no eggs, while **French-style ice cream** contains egg yolks. The eggs add richness and help make a smoother product because of the emulsifying properties of the yolks.

   Ice milk is like ice cream, but with a lower butterfat content.

   Frozen yogurt contains yogurt in addition to the normal ingredients for ice cream or ice milk.

2. **Sherbet.**

   Sherbets and **ices** are made from fruit juices, water, and sugar. American sherbets usually contain milk or cream and, sometimes, egg whites. The egg whites increase smoothness and volume. Ices, also called **water ices**, contain only fruit juice, water, sugar, and, sometimes, egg whites. They do not contain milk products. The French word **sorbet** (sor bay) is sometimes used for these products. **Granité** (grah nee tay) is a coarse, crystalline ice made without egg white.

3. **Still-frozen dessert.**

   Ice cream and sherbet are churn-frozen, meaning they are mixed constantly while being frozen. If they were not churned, they would freeze into solid blocks of ice. The churning keeps the ice crystals small and incorporates air into the ice cream.

   Frozen soufflés and *frozen mousses* are made like chilled mousses and bavarians—that is, whipped cream, beaten egg whites, or both are folded in to give them lightness. This allows them to be still-frozen in an ordinary freezer.

PRODUCTION AND QUALITY

Until recently, few establishments made their own ice cream because of the labor involved, the equipment required, and the convenience of commercially made products. Also, in some areas, strict health codes made it difficult for all but large producers to make ice cream. Today, more and more restaurants are making their own ice creams and sorbets.

A basic ice cream mix is simply a crème anglaise or custard sauce mixed with 1 or 2 parts heavy cream for every 4 parts milk used in the sauce. This base is flavored as desired with vanilla, melted chocolate, instant coffee, crushed strawberries, and so on. It is then chilled thoroughly and then frozen according to the instructions for the equipment being used.

When the mix has frozen, it is transferred to containers and placed in a deep-freeze at below 0°F (–18°C) to harden. (Soft-frozen or soft-serve ice creams are served directly as they come from the churn freezer without being hardened.)

Whether you make ice cream or buy it, you should be aware of the following quality factors:

1. *Smoothness* is related to the size of the ice crystals in the product. Ice cream should be frozen rapidly and churned well during freezing so large crystals don't have a chance to form.

   Rapid hardening helps keep crystals small. So do eggs and emulsifiers or stabilizers added to the mix.

   Large crystals may form if the ice cream is not stored at a low enough temperature (below 0°F/–18°C).

2. **Overrun** is the increase in volume due to the incorporation of air when freezing ice cream. It is expressed as a percentage of the original volume of the mix. For example, if the ice cream doubles in volume, the amount of increase is equal to the original volume and the overrun is 100 percent.

   Some overrun is necessary to give a smooth, light texture. Ice cream with too much overrun is airy and foamy and lacks flavor. It was once thought ice cream should have from 80 to 100 percent overrun and that less would make it heavy and pasty. This may be true for ice creams containing gums and other stabilizers. However, some high-quality manufacturers produce rich (and expensive) ice cream with as little as 20 percent overrun.

3. *Mouthfeel* or *body* depends, in part, on smoothness and overrun as well as on other qualities. Good ice cream melts in the mouth to a smooth, not too heavy liquid. Some ice creams have so many stabilizers they never melt to a liquid. Unfortunately, many people have become so accustomed to these products that an ice cream that actually does melt strikes them as not rich enough.

STORAGE AND SERVICE

1. Store ice creams and sherbets at 0°F (–18°C) or lower. This low temperature helps prevent the formation of large ice crystals.

2. For service, temper frozen desserts at 8° to 15°F (–13° to –9°C) for 24 hours so they are soft enough to serve.

3. When serving, avoid packing the ice cream. The best method is to draw the scoop across the surface of the product so the product rolls into a ball in the scoop.

4. Use standard scoops for portioning ice cream. Normal portions for popular desserts are as follows:

Parfait	3 No. 30 scoops
Banana split	3 No. 30 scoops
À la mode topping for pie or cake	1 No. 20 scoop
Sundae	2 No. 20 scoops
Plain dish of ice cream	1 No. 10, 12, or 16 scoop

5. Measure syrups, toppings, and garnishes for portion control. For syrups, use pumps that dispense measured quantities, or use standard ladles.

POPULAR ICE CREAM DESSERTS

1. **Parfaits** are made by alternating layers of ice cream and fruit or syrup in a tall, narrow glass. They are usually named after the syrup or topping. For example: a **chocolate parfait** comprises three scoops of vanilla or chocolate ice cream alternating with layers of chocolate syrup and topped with whipped cream and shaved chocolate.

2. **Sundaes** or **coupes** consist of one or two scoops of ice cream or sherbet in a dish or glass and topped with syrups, fruits, toppings, and garnishes. They are quick to prepare, unlimited in variety, and as simple or as elegant as you could wish—served in an ordinary soda fountain glass, a silver cup, or a crystal champagne glass.

   Two sundaes have become classics:

   Peach Melba. Vanilla ice cream topped with a fresh, poached, or canned peach half, napped with sweetened raspberry purée (Melba sauce), and garnished with slivered almonds.

   Pear Belle Hélène. Vanilla ice cream topped with a poached or canned pear half, napped with chocolate sauce, and garnished with toasted sliced almonds.

3. **Bombes** are ice cream molds made by lining a chilled mold with softened ice cream, freezing it hard, and then filling the center with another flavor of ice cream or sherbet and freezing it again. (More than two flavors may be used.) The dessert is unmolded onto a cold platter for service and decorated as desired with whipped cream, fruits, and/or confections.

4. *Meringue glacée*. See discussion of meringues, page 1002.

5. *Baked Alaska*. See discussion of meringues, page 1003.

6. *Frozen éclairs* and *profiteroles*. See discussion of éclair pastries, page 1000.

DESSERT SAUCES

Most dessert sauces fall into one of three categories.

1. **Custard sauces.**
   Vanilla custard sauce, or crème anglaise, is presented early in this chapter. It is among the most basic preparations in dessert cookery. Chocolate or other flavors may be added to create varieties. See recipe on page 1010.

2. **Fruit purées.**
   These are simply purées of fresh or cooked fruits, sweetened with sugar. Other flavorings and spices are sometimes added. Some fruit sauces are thickened with cornstarch or other starch.

   Raspberry sauce and strawberry sauce, two popular items, can be made by simply puréeing frozen sweetened berries or by puréeing fresh berries and adding sugar to taste. See also the recipe for Applesauce (p. 213).

3. **Syrups.**
   This is a broad category that includes such products as chocolate sauce and caramel sauce. An understanding of sugar cooking is necessary to produce many of these sauces.

   The following recipes are popular examples of this category of sauce.

KEY POINTS TO REVIEW

- What are the key temperatures for storing ice cream and for holding it for service?

- How do you control portion size when preparing standard ice cream desserts?

- What are the three main categories of dessert sauce?

Chocolate Sauce

YIELD: 1 QT (1 L)

U.S.	METRIC	INGREDIENTS
1 pt	500 mL	Water
2 lb	1 kg	Sugar
6 oz	190 g	Corn syrup
8 oz	250 g	Unsweetened chocolate, melted
2 oz	60 g	Butter

PROCEDURE

1. Combine the water, sugar, and corn syrup and bring to a boil, stirring to dissolve the sugar.
2. Boil 1 minute and remove from heat. Let cool 1–2 minutes.
3. Melt the chocolate and butter together over low heat. Stir until smooth.
4. Very slowly stir the hot syrup into the chocolate.
5. Place over moderate heat and bring to a boil. Boil 4 minutes.
6. Remove from heat and cool.

Per 1 fl oz (29.57 mL): Calories, 170; Protein, 1 g; Fat, 5 g (24% cal.); Cholesterol, 5 mg; Carbohydrates, 34 g; Fiber, 1 g; Sodium, 25 mg.

Caramel Sauce

YIELD: 1¹/₂ PT (750 ML)

U.S.	METRIC	INGREDIENTS
8 oz	250	Sugar
2 fl oz	60 mL	Water
³/₄ tsp	4 mL	Lemon juice
6 fl oz	190 mL	Heavy cream
4 fl oz	125 mL	Milk

PROCEDURE

1. Combine the sugar, water, and lemon juice in a heavy saucepan. Bring to a boil, stirring to dissolve the sugar.
2. When the sugar is dissolved, cover the pan and boil 2 minutes.
3. Uncover and cook to the caramel stage (see pp. 1008–1009). Toward the end of the cooking time, turn the heat very low to avoid burning the sugar or getting it too dark. It should be a golden color.
4. Remove from heat and cool 5 minutes.
5. Bring the heavy cream to a boil. Add a few ounces (50–100 mL) of the heavy cream to the caramel.
6. Stir and continue to add the cream slowly.
7. Return to the heat and stir until all the caramel is dissolved.
8. Let cool completely.
9. Stir the milk into the cooled caramel to thin it.

Per 1 fl oz (29.57 g): Calories, 130; Protein, 1 g; Fat, 6 g (39% cal.); Cholesterol, 20 mg; Carbohydrates, 20 g; Fiber, 0 g; Sodium, 10 mg.

VARIATIONS

Hot Caramel Sauce

Prepare as directed through step 7. Omit the milk.

Clear Caramel Sauce

Substitute 2¹/₂–3 fl oz (75-90 mL) water for the heavy cream and omit the milk. If the sauce is too thick when cool, add more water.

Clear caramel sauce and caramel sauce with cream.

ADDITIONAL RECIPES

These additional recipes may be found on your CulinarE-Companion recipe management program:

Chocolate Chiffon Dessert or Pie Filling; Lemon Chiffon Dessert or Pie Filling; Pumpkin Chiffon Dessert or Pie Filling.

TERMS FOR REVIEW

simple syrup	**chiffon**	**sorbet**
dessert syrup	**mousse**	**granité**
crystallize	**ice cream**	**overrun**
crème anglaise	**Philadelphia-style ice cream**	**parfait**
pastry cream	**French-style ice cream**	**sundae**
baked custard	**ice milk**	**coupe**
blancmange	**frozen yogurt**	**Peach Melba**
cream pudding	**sherbet**	**Pear Belle Hélène**
baked pudding	**ices**	**bombe**
bavarian	**water ices**	

QUESTIONS FOR DISCUSSION

1. How can you avoid unwanted crystallization when cooking sugar syrups?

2. Crème anglaise and pastry cream both contain eggs. Why is it possible to boil pastry cream but not custard sauce?

3. Explain the importance of sanitation in the production of pastry cream. What specific steps should you take to ensure a safe product?

4. Crème anglaise, pastry cream, and baked custard are made with basic techniques that are also used for the following preparations. Identify which of the three techniques is used for each.

Coconut cream pie	French vanilla ice cream
Baked rice pudding	Pumpkin pie
Butterscotch pudding	Custard pie
Chocolate bavarian	Lemon meringue pie

5. Briefly describe the differences among bavarians, chiffons, mousses, and soufflés.

6. When making dessert soufflés, what is the advantage of beating the egg whites with part of the sugar?

7. When making a bavarian or a chiffon, what difficulty would you encounter if you chilled the gelatin mixture too long before folding in the whipped cream or egg whites?

APPENDIX 1

METRIC CONVERSION FACTORS

Weight

1 ounce = 28.35 grams

1 gram = 0.035 ounce

1 pound = 454 grams

1 kilogram = 2.2 pounds

Volume

1 fluid ounce = 29.57 milliliters

1 milliliter = 0.034 ounce

1 cup = 237 milliliters

1 quart = 946 milliliters

1 liter = 33.8 fluid ounces

Length

1 inch = 25.4 millimeters

1 centimeter = 0.39 inch

1 meter = 39.4 inches

Temperature

To convert Fahrenheit to Celsius:

Subtract 32. Then multiply by $5/9$.

Example: Convert 140°F to Celsius.

$140 - 32 = 108$

$108 \times 5/9 = 60°C$

To convert Celsius to Fahrenheit:

Multiply by $9/5$. Then add 32.

Example: Convert 150°C to Fahrenheit.

$150 \times 9/5 = 270$

$270 + 32 = 302°F$

Note: The metric equivalents in the recipes in this book are rounded off. See pages 73–74 for complete explanation.

APPENDIX 2

STANDARD CAN SIZES

Can Name	Volume		Approximate Weight[a]	
	U.S	Metric	U.S.	Metric
6 oz	5.75 fl oz	170 mL	6 oz	170 g
8 oz	8.3 fl oz	245 mL	8 oz	227 g
No. 1 picnic	10.5 fl oz	311 mL	10.5 oz	298 g
No. 211 cylinder	12 fl oz	355 mL	12 oz	340 g
No. 300	13.5 fl oz	399 mL	14 oz	397 g
No. 303	15.6 fl oz	461 mL	16–17 oz	454–482 g
No. 2	20 fl oz	591 mL	1 lb 4 oz	567 g
No. 2½	28.5 fl oz	843 mL	1 lb 13 oz	822 g
No. 3 cylinder	46 fl oz	1360 mL	3 lb	1360 g
No. 5	56 fl oz	1656 mL	3 lb 8 oz	1588 g
No. 10	103.7 fl oz	3067 mL	6½–7 lb	2722–2948 g

Because the density of food varies, so does the net weight for any given can size.

APPENDIX 3

APPROXIMATE WEIGHT-VOLUME EQUIVALENTS OF DRY FOODS

The following equivalents are rough averages only. Actual weight per volume varies considerably. For accurate measurement, all ingredients should be weighed.

Bread flour, sifted

1 pound = 4 cups

1 cup = 4 ounces

Bread flour, unsifted

1 pound = $3^1/3$ cups

1 cup = 4.75 ounces

Cake flour, sifted

1 pound = $4^1/4$ cups

1 cup = 3.75 ounces

Cake flour, unsifted

1 pound = $3^1/2$ cups

1 cup = 4.5 ounces

Granulated sugar

1 pound = $2^1/4$ cups

1 cup = 7 ounces

Confectioners' sugar, sifted

1 pound = 4 cups

1 cup = 4 ounces

Confectioners' sugar, unsifted

1 pound = $3^1/2$ cups

1 cup = 4.5 ounces

Cornstarch, sifted

1 pound = 4 cups

1 cup = 4 ounces

1 ounce = 4 tablespoons = $^1/4$ cup

1 tablespoon = 0.25 ounce

Cornstarch, unsifted

1 pound = $3^1/2$ cups

1 cup = 4.5 ounces

1 ounce = $3^1/2$ tablespoons

1 tablespoon = 0.29 ounce

Cocoa, unsifted

1 pound = 5 cups

1 cup = 3.2 ounces

1 ounce = 5 tablespoons

1 tablespoon = 0.2 ounce

Gelatin, unflavored

1 ounce = 3 tablespoons

$^1/4$ ounce = $2^1/4$ teaspoons

1 tablespoon = 0.33 ounce

1 teaspoon = 0.11 ounce

Baking soda

1 ounce = 1 tablespoon + $2^1/4$ teaspoons

0.25 ounce = $1^1/3$ teaspoons

1 tablespoon = 0.57 ounce

1 teaspoon = 0.19 ounce

Baking powder (phosphate type and sodium aluminum sulfate type)

1 ounce = 2 tablespoons

0.25 ounce = $1^1/2$ teaspoons

1 tablespoon = 0.5 ounce

1 teaspoon = 0.17 ounce

Cream of tartar

1 ounce = 4 tablespoons

0.25 ounce = 1 tablespoon

1 teaspoon = 0.08 ounce

Salt

1 ounce = $4^1/4$ teaspoons

0.25 ounce = $1^1/8$ teaspoons

1 teaspoon = 0.22 ounce

Paprika and ground chiles

1 ounce = 17 teaspoons

0.25 ounce = $4^1/4$ teaspoons

1 teaspoon = 0.06 ounce

Ground spices (except paprika and ground chiles)

1 ounce = 14 teaspoons

0.25 ounce = $3^1/2$ teaspoons

1 teaspoon = 0.07 ounce

Grated lemon zest

1 ounce = 4 tablespoons

1 teaspoon = 0.08 ounce

Dried beans

1 cup = 6.5 ounces

1 pound = $2^1/2$ cups
(yields 6 cups cooked)

Rice, long-grain

1 cup = 7 ounces

1 pound = $2^1/4$ cups
(yields 8 cups cooked)

APPENDIX 4

KITCHEN MATH EXERCISES: METRIC VERSIONS

This appendix includes metric sample calculations corresponding to the calculations in the text that use U.S. measures. Refer to the appropriate pages in the text for explanations.

RECIPE CONVERSION, PAGES 75–81

Beef Tenderloin Tips and Mushrooms à la Crème

PORTIONS: 8   PORTION SIZE: 250 G

Butter	60 g
Onions	125 g
Flour	15 mL
Mushrooms	250 g
Beef tenderloin	1250 g
White wine	125 mL
Prepared mustard	10 mL
Brown sauce	750 mL
Heavy cream	250 mL
Salt	to taste
Pepper	to taste

To determine quantities for 18 portions, divide the new yield by the old yield to find the conversion factor:

$$\frac{\text{new yield}}{\text{old yield}} = \frac{18}{8} = 2.25$$

EXAMPLE 1

Ingredient	Quantity	Times	Conversion Factor	Equals	New Quantity (rounded off)
Butter	60 g	×	2.25	=	135 g
Onions	125 g	×	2.25	=	275 g
Flour	15 mL	×	2.25	=	35 mL
Mushrooms	250 g	×	2.25	=	575 g
Beef tenderloin	1250 g	×	2.25	=	2800 g
White wine	125 mL	×	2.25	=	275 mL
Prepared mustard	10 mL	×	2.25	=	23 mL
Brown sauce	750 mL	×	2.25	=	1700 mL
Heavy cream	250 mL	×	2.25	=	575 mL

To determine quantities for 40 portions at 175 grams each, first find the total yield of the old recipe. Multiply the portions by the portion size:

$$8 \text{ (portions)} \times 250 \text{ g} = 2000 \text{ g}$$

Do the same calculation for the desired yield:

$$40 \text{ (portions)} \times 175 \text{ g} = 7000 \text{ g}$$

Divide the desired yield by the old yield to find the conversion factor:

$$7000 \div 2000 = 3.5$$

EXAMPLE 2

Ingredient	Quantity	Times	Conversion Factor	Equals	New Quantity (rounded off)
Butter	60 g	×	3.5	=	200 g
Onions	125 g	×	3.5	=	450 g
Flour	15 mL	×	3.5	=	50 mL
Mushrooms	250 g	×	3.5	=	875 g
Beef tenderloin	1250 g	×	3.5	=	4375 g
White wine	125 mL	×	3.5	=	450 mL
Prepared mustard	10 mL	×	3.5	=	35 mL
Brown sauce	750 mL	×	3.5	=	2625 mL
Heavy cream	250 mL	×	3.5	=	875 mL

COMPLETED RAW YIELD TEST FORM (METRIC), PAGE 83

Item: veal leg to scaloppine   Test number: 3   Date: 3/1/2014
Purveyor: ABC Meats   Price per kilogram: $11.00   Total cost: $148.50
AP weight (1): 13.5 kg   Kg price (2): $11.00   Total cost (3): $148.50

Trim, salvage, and waste:

	Item	Weight	Value/kg	Total Value (kg × value)
(4)	fat	1.14 kg	$0.25	$0.29
(5)	bone	1.5 kg	$0.88	$1.32
(6)	ground veal	0.95 kg	$9.75	$9.26
(7)	stew meat	1.4 kg	$10.95	$15.33
(8)	unusable trim	0.4 kg	0	0
(9)	cutting loss	0.09 kg	0	0
(10)				

Total weight (4 − 10) (11): 5.48 kg   Total value (4 thru 10) (12): $26.20
Total yield of item (13): 8.02 kg
Net cost (3 − 12) (14): $122.30
Cost per kg (14 ÷ 13) (15): $15.25
Percentage of increase (15 ÷ 2) (16): 1.39 (139%)

COMPLETED COOKED YIELD TEST FORM (METRIC), PAGE 85

Item: roast fresh ham   Test number: 2   Date: 3/1/2014
AP price per kg: $7.75
Cooking temperature: 165°C
Net raw weight (1): 5.5 kg   Net cost per kg (2): $8.73
   Total net cost (3): $48.02
Weight as served (4): 3.75 kg
Cooked cost per kg (3 ÷ 4) (5): $12.81
Shrinkage (1 − 4) (6): 1.75 kg
Percentage of shrinkage (6 ÷ 1) (7): 32%
Total percentage of cost increase (5 ÷ AP price per kg) (8): 165%

METRIC EXAMPLE: COSTING A RECIPE, PAGE 86
ITEM: BAKED RICE

Ingredient	Recipe Quantity	AP Quantity	Price	Total
Rice, long-grain	2 kg	2 kg	$1.59/kg	$3.18
Butter	375 g	0.375 kg	$4.25/kg	$1.59
Onions	500 g	0.5 kg	$0.79/kg	$0.40
Chicken stock	4 L	4 L	$0.30/L	$1.20
Salt	30 g	0.03 kg	$0.35/kg	$0.01
			Total cost	$6.38
			Number of portions	50
			Cost per portion	$0.13

APPENDIX 5

EGGS AND SAFETY

EGGS AND EGG-BASED PRODUCTS

Outbreaks of salmonellosis have been traced to clean, whole, uncracked-shell eggs contaminated with *Salmonella enteritidis*. Whole-shell eggs are now classified as potentially hazardous foods by the U.S. Food and Drug Administration (FDA).

The following guidelines, which treat eggs as potentially hazardous foods during storage, handling, preparation, and service, must be followed to prevent the possibility of food-borne illness:

- Store eggs at refrigerated temperatures of 45°F (7°C) or lower until time of use. Do not freeze eggs in shells.

- Cook eggs thoroughly until both the yolk and white are firm, not runny. Generally, eggs should be cooked to 145°F (63°C) or higher for at least 15 seconds.

- To hold cooked eggs for later service, cook them to 155°F (68°C) or higher for 15 seconds, then hold them at 135°F (57°C) or higher.

- Avoid pooling raw eggs for holding. Eggs may be pooled in small quantities for immediate cooking and serving.

- For lightly cooked egg items, such as custards, French toast, mousses, and meringues, use pasteurized eggs.

- Avoid raw egg menu items. Review menus, recipes, and preparation procedures using raw eggs. Pasteurized eggs may be substituted in Caesar salad, hollandaise and béarnaise sauces, eggnog, ice cream, and egg-fortified beverages.

- Pasteurized eggs require the same time and temperature handling as other potentially hazardous foods.

- Wash hands with hot, soapy water before and after handling eggs and egg products.

- Wash and sanitize utensils, equipment, and the work area after handling eggs and egg products.

- Do not reuse a container that has held a raw egg mixture. Use a clean, sanitized container for each batch.

Source: National Restaurant Association Educational Foundation, *ServSafe Coursebook*, 5th ed.

BIBLIOGRAPHY

A

Achatz, Grant. *Alinea.* Berkeley, California: Ten Speed Press, 2008.

Amendola, Joseph. *The Baker's Manual for Quantity Baking and Pastry Making,* 5th ed. Hoboken, New Jersey: John Wiley & Sons, 2002.

American Culinary Federation. *Culinary Fundamentals.* Upper Saddle River, New Jersey: Prentice Hall, 2006.

Anderson, Jean. *The Food of Portugal.* New York: Morrow, 1986.

Anderson, Jean, and Hedy Wurz. *The New German Cookbook.* New York: HarperCollins, 1993.

Andoh, Elizabeth. *At Home with Japanese Cooking.* New York: Knopf, 1980.

B

Bayless, Rick. *Authentic Mexican.* New York: Morrow, 1987.

Bertolli, Paul, and Alice Waters. *Chez Panisse Cooking.* New York: Random House, 1988.

Bickel, Walter, ed. *Hering's Dictionary of Classical and Modern Cookery.* London: Virtue, 1991.

Bissel, Frances. *The Book of Food.* New York: Henry Holt, 1994.

Blocker, Linda, and Julia Hill. *Culinary Math,* 3rd ed. Hoboken, New Jersey: John Wiley & Sons, 2007.

Bocuse, Paul. *Paul Bocuse's French Cooking.* New York: Pantheon, 1977.

Boni, Ada. *Italian Regional Cooking.* New York: Bonanza, 1969.

Bugialli, Giuliano. *Classic Techniques of Italian Cooking.* New York: Simon & Schuster, 1982.

————. *The Fine Art of Italian Cooking.* New York: Times Books, 1977.

C

Casas, Penelope. *The Foods and Wines of Spain.* New York: Knopf, 1987.

Claiborne, Craig, and Virginia Lee. *The Chinese Cookbook.* Philadelphia: Lippincott, 1972.

Cordon Bleu, Le. *Kitchen Essentials.* Hoboken, New Jersey: John Wiley & Sons, 2001.

Cox, Beverly. *Cooking Techniques.* Boston: Little, Brown, 1981.

Culinary Institute of America. *Garde Manger: The Art and Craft of the Cold Kitchen,* 4th ed. Hoboken, New Jersey: John Wiley & Sons, 2012.

————. *The Professional Chef,* 9th ed. Hoboken, New Jersey: John Wiley & Sons, 2011.

————. *The Professional Chef's Knife Kit.* New York: John Wiley & Sons, 2000.

————. *Techniques of Healthy Cooking,* 3rd ed. Hoboken, New Jersey: John Wiley & Sons, 2008.

D

David, Elizabeth. *French Provincial Cooking.* Harmondsworth, England: Penguin, 1960.

————. *Italian Food.* Harmondsworth, England: Penguin, 1954.

Davidson, Alan. *The Oxford Companion to Food,* 2nd ed. Oxford: Oxford University Press, 2006.

Dornenberg, Andrew, and Karen Page. *Culinary Artistry.* New York: John Wiley & Sons, 1996.

E

Egan, Maureen, and Susan Davis Allen. *Healthful Quantity Baking.* New York: John Wiley & Sons, 1992.

Escoffier, A. *The Escoffier Cook Book.* New York: Crown, 1969.

F

Feinstein, Andrew Hale, and John M. Stefanelli. *Purchasing: Selection and Procurement for the Hospitality Industry,* 7th ed. Hoboken, New Jersey: John Wiley & Sons, 2008.

Friberg, Bo. *The Professional Pastry Chef,* 4th ed. Hoboken, New Jersey: John Wiley & Sons, 2002.

G

Gisslen, Wayne. *Advanced Professional Cooking.* New York: John Wiley & Sons, 1992.

————. *Essentials of Professional Cooking.* Hoboken, New Jersey: John Wiley & Sons, 2003.

————. *Professional Baking,* 6th ed. Hoboken, New Jersey: John Wiley & Sons, 2013.

Graham, Kevin. *Grains, Rice, and Beans.* New York: Artisan, 1995.

H

Hazan, Marcella. *The Classic Italian Cookbook.* New York: Knopf, 1976.

————. *More Classic Italian Cooking.* New York: Knopf, 1978.

Hom, Ken. *Chinese Technique.* New York: Simon & Schuster, 1981.

K

Kapoor, Sandy. *Professional Healthy Cooking.* New York: John Wiley & Sons, 1995.

Katsigris, Costas, and Chris Thomas. *Design and Equipment for Restaurants and Foodservice,* 3rd ed. Hoboken, New Jersey: John Wiley & Sons, 2009.

Keller, Thomas. *Under Pressure: Cooking Sous Vide.* New York: Artisan, 2008.

Kennedy, Diana. *The Cuisines of Mexico,* 2nd ed. New York: Harper & Row, 1986.

————. *Mexican Regional Cooking.* New York: Harper & Row, 1978.

Kinsella, John, and David T. Harvey. *Professional Charcuterie.* New York: John Wiley & Sons, 1996.

Knight, John B., and Lendel H. Kotschevar. *Quantity Food Production, Planning, and Management,* 3rd ed. New York: John Wiley & Sons, 2000.

L

Labensky, Sarah, and Alan M. Hause. *On Cooking,* 5th ed. Upper Saddle River, New Jersey: Prentice Hall, 2012.

Lang, George. *The Cuisine of Hungary*. New York: Bonanza, 1971.

Larousse, David Paul. *The Professional Garde Manger*. New York: John Wiley & Sons, 1996.

———. *The Sauce Bible*. New York: John Wiley & Sons, 1993.

Librairie Larousse. *Larousse Gastronomique*. New York: Clarkson Potter, 2001.

Loken, Joan K. *The HACCP Food Safety Manual*. New York: John Wiley & Sons, 1995.

M

McClane, A.J. *The Encyclopedia of Fish*. New York: Holt, Rinehart & Winston, 1977.

McGee, Harold. *The Curious Cook*. San Francisco: North Point, 1990.

———. *On Food and Cooking*: The Science and Lore of the Kitchen, rev. ed. New York: Scribner, 2004.

McSwane, David, Nancy Roberts Rue, and Richard Linton, *Essentials of Food Safety and Sanitation,* 4th ed. Upper Saddle River, New Jersey: Prentice Hall, 2004.

Madison, Deborah. *The Greens Cookbook*. New York: Broadway Books, 1987.

———. *Vegetarian Cooking for Everyone*. New York: Broadway Books, 1997.

Miller, Gloria Bley. *The Thousand Recipe Chinese Cookbook*. New York: Grosset & Dunlap, 1970.

Mizer, David A., Mary Porter, Beth Sonnier, and Karen Eich Drummond. *Food Preparation for the Professional*, 3rd ed. New York: John Wiley & Sons, 2000.

Molt, Mary K. *Food for Fifty*, 13th ed. Upper Saddle River, New Jersey: Prentice Hall, 2010.

Myhrvold, Nathan, Chris Young, and Maxime Bilet. *Modernist Cuisine: The Art and Science of Cooking*. Bellevue, Washington: The Cooking Lab, 2011.

N

National Restaurant Association Educational Foundation. *ServSafe Coursebook*, 6th ed. Chicago: National Restaurant Association Educational Foundation, 2012.

North American Meat Processors. *The Meat Buyer's Guide,* 6th ed. NAMP, 2010.

P

Pauli, Eugen. *Classical Cooking the Modern Way: Recipes*, 3rd ed. Arno Schmidt, trans., and Margaret Schmidt, ed. New York: John Wiley & Sons, 1997.

———. *Classical Cooking the Modern Way: Methods and Techniques*, 3rd ed. Arno Schmidt, trans., and Margaret Schmidt, ed. New York: John Wiley & Sons, 1999.

Pepin, Jacques. *The Art of Cooking*. New York: Knopf, 1987.

———. *La Technique: The Fundamental Techniques of Cooking: An Illustrated Guide*. New York: Quadrangle/Times Books, 1976.

Peterson, James. *Fish and Shellfish*. New York: Morrow, 1996.

———. *Sauces*, 3rd ed. Hoboken, New Jersey: John Wiley & Sons, 2008.

———. *Splendid Soups: Recipes and Master Techniques for Making the World's Best Soups*. New York: John Wiley & Sons, 2001.

R

Roca, Juan, and Salvador Brugués. *Sous Vide Cuisine*. Barcelona: Montagud Editores, 2005.

S

Sackett, Lou, Jaclyn Pestka, and Wayne Gisslen. *Professional Garde Manger: A Comprehensive Guide to Cold Food Preparation*. Hoboken, New Jersey, 2010.

Saulnier, L. *La Répertoire de la Cuisine*. Woodbury, New York: Barron's, 1976.

Schmidt, Arno, and Inja Nam. *The Book of Hors d'Oeuvres and Canapés*. New York: John Wiley & Sons, 1996.

Schneider, Elizabeth. *Uncommon Fruits and Vegetables: A Commonsense Guide*. New York: Harper & Row, 1986.

———. *Vegetables from Amaranth to Zucchini*. New York: William Morrow, 2001.

Sheraton, Mimi. *The German Cookbook*. New York: Random House, 1965.

Somerville, Annie. *Field of Greens*. New York: Bantam, 1993.

Sonnenschmidt, Frederic H., and Jean F. Nicolas. *The Professional Chef's Art of Garde Manger*, 5th ed. New York: John Wiley & Sons, 1993.

Styler, Christopher. *Working the Plate: The Art of Food Presentation*. Hoboken, New Jersey: John Wiley & Sons, 2006.

Sultan, William J. *Practical Baking*, 5th ed. New York: John Wiley & Sons, 1990.

T

Torres, Marimar. *The Spanish Table*. Garden City, New York: Doubleday, 1986.

Tsuji, Shizuo. *Japanese Cooking: A Simple Art*. Tokyo: Kodansha, 1980.

W

Waters, Alice. *Chez Panisse Vegetables*. New York: HarperCollins, 1996.

Willan, Anne. *La Varenne Pratique*. New York: Crown, 1989.

GLOSSARY AND COOKING VOCABULARY

Note: Phonetic guides are included for difficult French words, giving the approximate pronunciation using English sounds. Exact rendering is impossible in many cases because French has a number of sounds that don't exist in English.

A

Abaisser (ah bess say) To roll a dough to the desired thickness with the aid of a rolling pin.

Abats (ah bah) Offal; internal organs of butchered animals. Also called variety meats.

Aboyeur (ah bwah yer) Kitchen worker who accepts and transmits orders from waiters, calls for orders to be finished, inspects finished dishes, and passes them to the dining room staff.

Accompaniment Salad A salad served as a side dish— that is, at the same time as a main course.

Acidifier (ah si di fee ay) To add lemon juice or vinegar to fruits, vegetables, and fish to prevent oxidation.

Aciduler (ah see dyoo lay) To make a preparation slightly acidic, tart, or tangy by adding a little lemon juice or vinegar.

Active Dry Yeast A dry, granular form of yeast that must be rehydrated in 4 times its weight of warm water before use.

Adductor Muscle The muscle with which a mollusk closes its shell. In the case of American and Canadian scallops, this is usually the only part that is eaten.

Aerobic Requiring oxygen to live and grow; said of bacteria.

Aging Holding meats in coolers under controlled conditions to allow natural tenderizing to take place.

Aiguillette (ay gwee yet) A long, narrow slice of meat cut from the breast of poultry (especially duck) and game birds.

À la Carte (1) Referring to a menu on which each individual item is listed with a separate price. (2) Referring to cooking to order, as opposed to cooking ahead in large batches.

Al Dente Firm, not soft or mushy, to the bite. Said of vegetables and pasta.

Allemande (1) German style. (2) A sauce made of velouté (usually veal), a liaison, and lemon juice.

Allergen A substance that causes an allergic reaction.

All-Purpose Potato An irregularly shaped potato suitable for most purposes, though not usually for baking due to its shape; not as dry, starchy, or expensive as a russet.

Allumette Cut into matchstick shapes; usually refers to potatoes.

Amaranth A tiny, yellow-brown seed high in good-quality protein and having a somewhat spicy, nutty flavor when cooked; often used in vegetarian diets.

Americano Espresso diluted with hot water.

Amino Acids Long chains of smaller compounds that, when joined in various combinations, make up over 100,000 proteins in the human body.

Amuse Bouche A tiny appetizer or hors d'oeuvre offered to guests seated at their tables, either before or after they have ordered from the menu, in order to welcome guests and showcase an aspect of the chef's cooking style and talent.

Anadromous Referring to fish that live in salt water but spawn in fresh water.

Anaerobic Requiring an absence of oxygen to live and grow; said of bacteria.

Angel Food Method Mixing method for angel food cakes, involving folding dry ingredients into an egg-white foam.

Anthocyanins Red or purple pigments in vegetables and fruits.

Anthoxanthin (an tho zan thin) A type of white or pale yellow pigment in vegetables.

Antipasto Italian hors d'oeuvre.

AP Weight As purchased; the weight of an item before trimming.

Appareil (ah pa ray) A mixture of the principal elements of a final recipe (usually egg-based).

Appetizer A generally small-portioned first course of a multicourse meal, used to whet the appetite, often to the accompaniment of drinks.

Appetizer Salad Salad used to stimulate appetite with fresh, crisp ingredients; tangy, flavorful dressing; and attractive appearance.

Arborio Rice A variety of short-grain rice from Italy.

Argenteuil (ar zhawn toy) Garnished with asparagus.

Aromate (ah row mat) A condiment or vegetable with a characteristic smell or taste. Often used in reference to a combination of flavoring vegetables, such as carrot, onion, leek, and celery.

Artisan(al) Cheese A cheese produced primarily by hand, in small batches, with particular attention to the tradition of the cheese maker's art and using as little mechanization as possible.

Aspic Jelly A clarified stock that contains enough gelatin to solidify when cold.

Aspic Powder Unflavored gelatin mixed with a powdered stock base.

As Purchased (AP) Term for the untrimmed quantity of a food item, in the form in which it is purchased.

AS Weight As served; the weight of an item as sold or served, after processing and/or cooking.

Au Gratin (oh gra tan) Having a browned or crusted top, often made by topping with bread crumbs, cheese, and/or a rich sauce and passing under the broiler or salamander.

Au Jus (oh zhoo) Served with its natural juices, usually unthickened pan drippings.

Au Sec (oh seck) Until dry.

Avgolemono Greek soup made of chicken stock, egg, and lemon juice.

B

Bacteria Microscopic organisms, some of which cause disease, including food-borne disease.

Bagged Cookies Cookies made from dough that is forced through a pastry bag into various shapes.

Bain-Marie A container of hot water used for keeping foods hot.

Bake To cook foods by surrounding them with hot, dry air. Similar to roast, but the term bake usually applies to breads, pastries, vegetables, and fish.

Baked Alaska A dessert consisting of ice cream on a sponge cake base, covered with meringue and browned in the oven.

Baked Egg Egg baked in an individual serving dish. Also called shirred egg and egg en cocotte.

Baked Pudding Custard that contains additional ingredients, usually starchy ingredients in large quantities, and is baked in the oven.

Baking Blind Baking a pie or tart shell without a filling.

Baking Powder, Single- and Double-Acting A mixture of baking soda and a reacting acid used to leaven baked goods. While single-acting baking powder requires only moisture to release gas, double-acting requires heat for a complete reaction.

Ballotine (ball oh teen) A piece of meat or poultry that is boned and stuffed.

Barbecue To cook with dry heat created by the burning of hardwood or by the hot coals of this wood.

Bar Cookies Cookies for which the dough is shaped into long bars, then baked and cut. They may be baked again after cutting.

Barding Tying thin slices of fat, such as pork fatback, over meats with no natural fat cover to protect them while roasting.

Barley Type of grain, usually purchased as pearled barley.

Barquette (bar ket) A small, long oval pastry mold, or an item baked in such a mold.

Basic Grind Referring to sausages made simply by grinding meats to various stages of coarseness or fineness.

Basmati Rice A variety of long-grain rice from India.

Basquaise (bas kez) In the Basque style; usually indicates the presence of red peppers in the dish.

Basted Style of fried egg cooked covered so the top is cooked by retained steam.

Batch Cooking Cooking method that involves dividing food into batches and cooking them one a time, as needed, in order to cook them as close as possible to serving time.

Bâtonnet (bah toh nay) Cut into sticks, $1/4 \times 1/4 \times 2$–$2^1/2$ inches (6 mm $\times$ 6 mm $\times$ 5–6 cm).

Batter Semiliquid mixture containing flour or other starch, used for the production of such products as cakes and breads and for coating products to be deep-fried.

Batterie (bat tree) Set; complete set of kitchen utensils.

Bavarian Cream A dessert made of custard sauce, gelatin, and whipped cream.

Bavarois (ba var wah) Bavarian cream.

Bean Curd See Tofu.

Bean Paste See Miso.

Bean Thread Thin noodle made with mung bean starch.

Béarnaise (bare nez) A sauce made of butter and egg yolks and flavored with a reduction of vinegar, shallots, tarragon, and peppercorns.

Béchamel A sauce made by thickening milk with a roux.

Beignet Fritter.

Bercy (bare see) A white or brown sauce flavored with a reduction of white wine and shallots.

Beurre (burr) Butter.

Beurre Blanc (burr blahn) Butter-based sauce consisting of an emulsion of butter with a reduction of dry white wine, vinegar, and shallots.

Beurre Clarifié (burr cla ri fee ay) Clarified butter; butter that is gently melted in order to remove the impurities that float to the top and the whey that sinks to the bottom.

Beurre Composé (burr com po zay) Butter mixed with one or more flavoring ingredients.

Beurre Manié (burr mahn yay) Equal parts raw butter and flour mixed into a smooth paste.

Beurre Noir (burr nwahr) Butter heated until it is dark brown, then flavored with vinegar.

Beurre Noisette (burr nwah zett) Whole butter heated until it is light brown.

Beurrer (burr ray) (1) To lightly coat a container with butter in order to prevent sticking. (2) To add butter to a sauce or dough.

Biscuit Method Mixing method involving rubbing solid fat into dry ingredients and then mixing with combined wet ingredients. Similar to *rubbed method*, but using more liquid.

Bisque A cream soup made from shellfish.

Bivalve A mollusk with a pair of hinged shells, such as clam and oyster.

Blanc (blahn) (1) White. (2) A mixture of flour and acidulated water (usually with lemon juice), used to prevent certain foods from discoloring during cooking.

Blanch To cook an item partially and briefly in boiling water or hot fat. Usually a pre-preparation technique, as to loosen peels from vegetables, fruits, and nuts, to partially cook French fries or other foods before service, to prepare for freezing, or to remove undesirable flavors.

Blanch-and-Chill To partially cook, chill, and finish-cook foods as needed in order to reduce the amount of time required to cook completely to order.

Blancmange (1) An English pudding thickened with cornstarch. (2) A French almond-flavored pudding containing gelatin and milk.

Blanquette A white stew made of white meat or poultry simmered without preliminary browning and served with a white sauce.

Blending Method See *two-stage method*.

Blitz Puff Pastry A pastry similar to puff pastry but quickly made by a variation of the Rubbed Dough Method and then rolled and folded to increase flakiness.

Blond Roux Roux cooked until it begins to change to a slightly darker color; used for veloutés, or sauces based on white stocks.

Blue Corn Corn with a blue or purplish color, derived from early varieties of corn grown by Native Americans.

Boar Wild pig, or the meat from this animal.

Boeuf à la Mode A classic French dish of braised beef.

Boil To cook in water or other liquid that is bubbling rapidly, about 212°F (100°C) at sea level and at normal pressure.

Bolster A raised ridge of metal at the heel end of a knife blade.

Bombe A molded ice cream or sherbet dessert.

Bordelaise A brown sauce flavored with a reduction of red wine, shallots, pepper, and herbs and garnished with marrow.

Botulism A deadly food-borne intoxication usually associated with improperly canned foods.

Bouchée (boo shay) A small round of puff pastry that can be filled with different mixtures.

Bouillir (boo year) To boil; to bring a liquid to the boiling point.

Boulangère (boo lawn zhare) Cooked with stock, onions, and potatoes; a style of meat preparation.

Bound Salad A salad mixed with a heavy dressing in order to bind the ingredients together.

Bouquet Garni A combination of fresh herbs tied together, used for flavoring.

Bouquetière (book tyair) Garnished with an assortment or bouquet of fresh vegetables, such as artichokes, carrots, turnips, green beans, peas, cauliflower, and potatoes.

Braise (1) To cook covered in a small amount of liquid, usually after preliminary browning. (2) To cook (certain vegetables) slowly in a small amount of liquid without preliminary browning.

Bran The tough but edible layer that covers the endosperm of a grain.

Bread Flour A strong flour used for making breads, hard rolls, and any product requiring high gluten.

Breakfast Cook A type of short-order cook who is skilled in quickly and efficiently cooking egg dishes and other breakfast items to order.

Breakfast Sausage Fresh pork that has been ground and seasoned; in patty, link, or bulk form.

Breve (bray vay) A mixture of espresso and steamed half-and-half.

Brine A water-based solution of salt and other ingredients, used to cure meats and other foods.

Brine Cure A curing method in which the food is immersed in a solution (brine) made of the curing ingredients dissolved in water.

Brioche Rich yeast dough containing large amounts of eggs and butter, or the product made from this dough.

Brochette (broe shet) (1) A skewer made of wood or bamboo. (2) Small pieces of food stuck on a long piece of metal or wood and grilled.

Broil To cook with radiant heat from above.

Broth A flavorful liquid obtained from the simmering of meats and/or vegetables.

Brown Rice Rice with the bran layer left on, which gives it a light brown color, a slightly coarse, crunchy texture, and a nutty flavor.

Brown Roux Roux cooked until it takes on a light brown color and a nutty aroma. When heavily browned, it contributes flavor and color to brown sauces.

Brunoise (broon wahz) (1) Cut into very small (1/8 in./3 mm) dice. (2) Garnished with vegetables cut in this manner.

Bruschetta (broo sket ta) A slice of toasted Italian bread served as an appetizer, usually rubbed with garlic and moistened with olive oil, often served with additional toppings.

Buckwheat A type of seed used as a grain, either whole or ground into flour. Technically not a grain, it is the seed of a plant with branched stems and broad, arrow-shaped leaves.

Buffet-Style Service Offering hors d'oeuvres arranged attractively on one or more tables for guests to help themselves.

Bulgur A type of cracked wheat that has been partially cooked.

Butcher To kill and dress a meat animal.

Butler-Style Hors d'Oeuvre Service Offering hors d'oeuvres to guests by service staff carrying small trays as they pass among the assembled party.

Butter Fat derived from milk, usually consisting of about 80 percent fat, with the remainder being water and milk solids.

Buttercream An icing made of butter and/or shortening blended with confectioners' sugar or sugar syrup and, sometimes, other ingredients.

Butterfat Milk fat.

Butterflied Cut partially through and spread open to increase the surface area.

Buttermilk (1) Fresh, liquid milk, usually skim milk, that has been cultured or soured by bacteria. (2) The liquid that remains after butter is removed from churned cream.

C

Cacao (ka ka oh) Cocoa.

Café au Lait French term meaning "coffee with milk." It is similar to a latte, but may be made with strong regular dark roast coffee rather than espresso.

Cake Flour A weak or low-gluten flour made from soft wheat.

Calamari Italian for "squid" (plural).

Calorie The amount of heat needed to raise the temperature of 1 kg water by 1°C. Used as a measure of food energy. More correctly called a *kilocalorie*.

Calvados (cal vah dose) An alcoholic beverage distilled from cider, made exclusively in the Normandy region of France.

Canapé (can ah pay) Tiny, open-faced sandwich, served as an hors d'oeuvre.

Canard (can arr) Duck.

Caneton (can e tone) Male duckling.

Canette (can net) Female duckling.

Capon A castrated male chicken.

Cappuccino Mixture of equal parts espresso and frothy, steamed milk.

Caraméliser (care a mel ee zay) To caramelize; to coat a mold with cooked sugar; to cook sugar until dark for use in other preparations (to coat or to make a sauce).

Caramelization The browning of sugars caused by heat.

Carbohydrate Any of a group of compounds, including starches and sugars, that supply energy to the body.

Carême, Marie-Antoine Famous nineteenth-century French chef, often considered the founder of classical cuisine.

Carotenoids Yellow or orange pigments in vegetables and fruits.

Carpaccio Very thin slices of meat or fish, served raw.

Carryover Cooking The rise in temperature inside roast meat after it is removed from the oven.

Carve To cut cooked meat for serving.

Catadromous Referring to fish that live in fresh water but spawn in the ocean.

Caul A fatty membrane that covers the stomach of a pig; used for wrapping meats for cooking and for lining terrines.

Caviar (1) The salted roe or eggs of sturgeon. (2) The salted roe of another fish, such as salmon or whitefish, if that fish is designated in the name, such as whitefish caviar.

Cellophane Noodle Another term for bean thread noodle.

Celsius Scale The metric system of temperature measurement, with 0°C set at the freezing point of water and 100°C set at the boiling point of water.

Centi- Prefix in the metric system meaning "one-hundredth."

Cêpe (sepp) Bolete or porcini mushroom.

Cephalopod A member of the class of mollusks that includes octopus and squid.

Certified Pork Pork that is guaranteed or certified to be free of trichinosis.

Chai A sweetened blend of spiced milk and tea.

Chamber Vacuum Packer Packaging equipment specifically designed to seal food in a plastic vacuum bag by pulling air from the bag at various pressures.

Champignon (shamp in yon) Mushroom.

Chantilly (shawn tee yee) Whipped cream to which sugar and vanilla have been added.

Chapelure (shap a lure) Dried bread crumbs made from both the crust and center of dried bread. Used for breading.

Charcuterie (shar koo tree) The art of preparing fresh and cured pork products, including sausages and pâtés.

Charcutier (shar koo tyay) One who prepares and sells pork products, including sausages and pâtés.

Charlotte (shar lott) (1) A dessert, often containing Bavarian cream, made in a special mold. (2) A savory preparation made in this mold.

Chasoba Buckwheat noodles made with powdered green tea in addition to the buckwheat.

Chasseur (sha sur) "Hunter style," usually referring to items served with a brown sauce containing mushrooms, tomato, and white wine.

Château Potato Potato tournéed to about 2 inches (5 cm) long.

Chaud-Froid Sauce An opaque sauce containing gelatin, used to coat certain cold foods.

Chef Person in charge of a kitchen or of a department of a kitchen.

Chef de Cuisine French term meaning "head of the kitchen." The cook who runs the stove department of the kitchen and to whom the meat chef and pastry chef report. Also known as the *cuisinier*.

Chemical Leavener Leavener that releases gases produced by chemical reactions.

Chèvre (1) Goat. (2) Goat cheese (short for *fromage de chèvre*).

Chiffon (1) A light, fluffy dessert or pie filling containing gelatin and beaten egg whites. (2) A type of cake made with an egg-white foam and with oil as a shortening.

Chiffonade Cut into fine shreds; usually said of leafy vegetables and herbs.

Chiffon Method A cake-mixing method involving folding an egg-white foam into a mixture of flour, egg yolks, flour, sugar, and liquids.

Chiffon Pie Pie made with filling lightened by the addition of beaten egg white.

China Cap A cone-shaped strainer.

Chinois (shee nwah) A fine conical strainer.

Chitterlings Pork intestines.

Chlorophyll Green pigment in vegetables and fruits.

Cholesterol A fatty substance found in foods derived from animal products and in the human body; it has been linked to heart disease.

Chop To cut into irregularly shaped pieces.

Choucroute (shoo kroot) Sauerkraut.

Choucroute Garni Sauerkraut cooked with sausage, pork, and, sometimes, poultry products. A specialty of Alsace, France.

Chowder A hearty American soup made from fish, shellfish, and/or vegetables, usually containing milk and potatoes.

Chukasoba Wheat noodles made with flour and water with an alkali additive to the water; typically used in ramen dishes.

Chutney Any of several types of spicy condiment or relish.

Cilantro The fresh coriander plant, used as an herb.

Clamart Garnished with or containing peas.

Clarification The mixture of ingredients used to clarify a stock.

Clarified Butter Purified butterfat, with water and milk solid removed.

Clarifier (clare re fee ay) (1) To clarify; to clear a cloudy liquid by straining, heating, and gently simmering with egg whites. (2) To separate butterfat from the milk solids and water in whole butter.

Class A, B, C, and K Fires The four classes of fire identified by the type of fuel involved. Class A: ordinary combustibles, such as wood, paper, and cloth; Class B: burning liquids, such as grease, oil, gasoline, and solvents; Class C: electrical equipment, such as switches and motors; Class K: combustible cooking appliances and products, such as vegetable or animal oils and fats.

Classical Garnish In classical cuisine, any food or combination of foods placed on a plate or platter to accompany the main item; each garnish combination is assigned a standard name.

Clearmeat A mixture of ground meat, egg whites, and flavoring ingredients, used to clarify consommés.

Clear Soup A soup based on a clear, unthickened broth or stock that may be served plain or garnished with a variety of vegetables and meats.

Club Sandwich A sandwich consisting of three slices of toast and filled with such ingredients as sliced chicken or turkey, lettuce, tomato, and bacon.

Coagulation The process by which proteins become firm, usually when heated.

Cockle A type of small bivalve mollusk from a different family than clam. It may be cooked like a clam, however, and is almost always served in the shell.

Cocktail A type of appetizer generally made of seafood or fruit and often served with a tart or tangy sauce.

Cocotte Potato Potato tournéed to about 1½ inches (4 cm) long.

Cold Smoking A smoking method in which foods are smoked at a low temperature, usually at or below 85°F (30°C), so they are not cooked in the process.

Collagen A type of connective tissue in meats that dissolves when cooked with moisture.

Collagen Casing An edible artificial sausage casing molded from animal materials.

Coller (cole lay) To thicken or set using gelatin, as in making jelly or fruit mousse.

Colloid A mixture in which one substance (the dispersed phase) is evenly mixed throughout another substance (the continuous phase).

Combi (Combination) Oven An oven that can operate in conventional, convection, and steamer modes.

Commercial Dried Pasta Pasta dough that has been shaped and dried; usually made of semolina flour and water.

Common Meringue Meringue made from egg whites at room temperature, beaten with sugar.

Complementary Proteins Proteins supplied by foods that, if eaten together, supply all the amino acids necessary in the human diet.

Complete Protein A protein that supplies all the amino acids necessary in the human diet.

Composed Salad A salad made by arranging two or more ingredients attractively on the plate rather than by mixing them.

Compound Butter A mixture of raw butter and various flavoring ingredients.

Concasser (cone cas say) To break up coarsely with a knife or mortar. To chop coarsely.

Concassé (de tomates) (cone cas say duh to maht) Peeled, seeded, and diced tomatoes.

Condensed Milk Whole milk heavily sweetened with sugar, with about 60 percent of the water removed.

Condiment Any edible substance that can be added to a food to contribute flavoring.

Conduction The transfer of heat from one item to something touching it or to a cooler part of the first item.

Confectioners' Sugar Fine powdered sugar mixed with a small amount of starch to prevent caking.

Confit (cone fee) A food saturated with one of the following: vinegar (for vegetables); sugar (for fruits); alcohol (for fruits); fat (for poultry and meat). Literally, "preserved."

Connective Tissue Certain proteins in meat, poultry, and fish that contribute to toughness; some are broken down by heat and some are not.

Consommé A rich, flavorful seasoned stock or broth clarified to make it perfectly clear and transparent.

Contaminated Containing harmful substances not originally present in food.

Convection The transfer of heat by the movement of a liquid or gas.

Convection Oven An oven in which hot air is circulated by a fan.

Convenience Food Any food product that has been partially or completely prepared or processed by the manufacturer.

Conversion Factor The number used to increase or decrease the amount of each ingredient when converting a recipe to a different yield.

Cooked Juice Method Method for making fruit pie fillings, used when only the juice requires cooking.

Cooked Fruit Method Method for making pie fillings, used when the fruit requires cooking, or when there is not enough liquid for the cooked juice method.

Cooking The art or practice of preparing edible food by applying heat and/or combining select measured ingredients in an ordered process.

Coq au Vin (coke oh van) A French dish of chicken braised in wine.

Coquille (coe kee) Shell.

Coral The roe or eggs of certain shellfish.

Corn A grain with a set of husks covering the entire seed head, or ear. Unlike other grains, it may be eaten as a fresh vegetable.

Corne (corn) Plastic tool used for scraping the contents out of containers.

Corrective Action A procedure that must be followed whenever a critical limit is not met. Corrective actions should be identified in written procedures that clearly communicate to the worker what must be done in a particular situation.

Coucher (koo shay) (1) To lay; to place a rolled piece of dough on a baking sheet. (2) To spread; to spread a layer of cream or other garnish. (3) To pipe; to cover with a layer using a piping bag.

Coulis (koo lee) A vegetable or fruit purée, used as a sauce.

Coupe (koop) (1) Cup. (2) A dessert consisting of one or two scoops of ice cream or sherbet in a dish or glass, topped with syrups, fruits, toppings, and/or garnishes; a sundae.

Couper (koo pay) To cut.

Course A food or group of foods served at one time or intended to be eaten at the same time.

Court Bouillon (koor bwee yohn) Water containing seasonings, herbs, and, usually, an acid; used for cooking fish.

Couscous A type of granular pasta from North Africa that resembles a grain.

Cracked Wheat Whole wheat grain that has been cut into pieces.

Crayfish A freshwater crustacean resembling a small lobster; also called *crawfish*.

Creaming The process of beating fat and sugar together to incorporate air.

Creaming Method Cake-mixing method involving mixing flour and eggs into creamed sugar and fat; also used for some muffins and coffee cakes.

Cream Pie Pie made with pudding or boiled custard-type filling.

Cream Pudding Milk-based pudding thickened with starch and eggs; essentially pastry cream with various flavorings.

Cream Soup A soup thickened with roux or another thickening agent and containing milk and/or cream.

Crécy (kray see) Garnished with or containing carrots. Also, the name of an area known for its carrot production.

Crème Anglaise (krem awng lezz) A light vanilla-flavored custard sauce made of milk, sugar, and egg yolks.

Crème Fouettée (krem foo eh tay) Whipped cream; cream that has been whisked in order to incorporate air.

Crème Fraîche A thick, slightly aged heavy cream.

Crème Pâtissière (krem pa tis see air) Pastry cream; sweetened milk thickened with starch and eggs, used for pastry making.

Crémer (kray may) (1) To cream together sugar and butter. (2) To add cream.

Crêpe (krep) Very thin pancake.

Crépinette A sausage patty wrapped in caul.

Critical Control Point (CCP) An action that can be taken to eliminate or minimize a food safety hazard.

Croissant A crescent-shaped roll made from a rich, rolled-in yeast dough.

Croquette (crow kett) Food that has been puréed or bound with a thick sauce, made into small shapes, breaded, and fried.

Cross-Contamination The transfer of bacteria to food from another food or from equipment or work surfaces.

Croustade (krew stahd) (1) A crisp crust that is fried. (2) An empty pastry case.

Croûte (kroot) Crust; the brown outer covering of bread. A meat or fish en croûte is one that is wrapped in a crust.

Croûton (kroo tohn) A slice or piece of toasted bread.

Crudité (croo dee tay) A raw vegetable served as a relish.

Crumb Crust Pie crust made of cookie crumbs, butter, and sugar.

Crustacean A sea animal with a segmented shell and jointed legs, such as lobster and shrimp.

Cryovac® Brand name of a machine used to vacuum-pack meats and other foods in plastic. The air- and moisture-proof packaging protects the contents from bacteria and mold and prevents weight loss due to drying.

Crystallize To form sugar crystals. May occur when sugar is cooked.

Cuisson (kwees sohn) (1) The liquid used for cooking a food. (2) The cooking; the action and manner of cooking a food.

Curdle; Curdling The separation of protein solids, such as egg solids from liquids due to coagulation.

Custard A liquid that is thickened or set firm by the coagulation of egg protein.

Custard, baked A combination of beaten eggs and liquid, usually milk or cream, baked until it sets to a solid, due to the coagulation of egg protein.

Cutting Loss Loss of weight of meat due to drying or to particles of meat and fat sticking to the cutting board.

Cuttlefish A cephalopod similar to squid, but with a chalky interior bone and a squatter body shape.

Cycle Menu A menu that changes every day for a certain period, then repeats the same daily items in the same order.

D

Dal The generic term in India for dried legume.

Danish A rich, sweet, flaky yeast dough containing layers of rolled-in fat.

Dariole (dahr ree ole) A small, thimble-shaped mold.

Dark Meat Poultry meat with darker color and more fat and connective tissue, found in drumsticks and thighs.

Darne (darn) Thick slice, containing the central bone, cut from round fish.

Dashi (dah shee) Japanese soup stock.

Daube (dobe) Stew of meat braised in red wine.

Decaffeinated Coffee Coffee from which the caffeine has been removed by solvents.

Deci- Prefix in the metric system meaning "one-tenth."

Découper (day koo pay) To cut; to cut using scissors, a knife, or pastry cutter.

Deep-Fried Sandwich A sandwich dipped in beaten egg mixture and, sometimes, in bread crumbs, and then deep-fried.

Deep-Fry To cook submerged in hot fat.

Déglacer (day gla say) To deglaze.

Deglaze To swirl a liquid in a sauté pan or other pan to dissolve cooked particles or food remaining on the bottom.

Dégraisser (day gray say) To degrease; to remove excess fat from the surface of a food or sauce.

Degree Celsius Metric unit for measuring temperature; also called *degree centigrade*.

Demi-Glace A rich brown sauce that has been reduced by half.

Demitasse Literally, "half-cup." Strong, black coffee served in small cups after dinner.

Denature To change the structure of protein molecules by means of heat or chemicals.

Dessert Salad A salad containing sweet ingredients such as fruits, sweetened gelatin, nuts, and cream. It is best served as a dessert or as part of a buffet or party menu.

Dessert Syrup A flavored simple syrup, used to moisten and flavor some cakes.

Détrempe (day trompe) Dough made of flour and water; used for puff pastry.

Dice To cut into small cubes.

Dip Accompaniment to potato chips, crackers, and raw vegetables that is thick enough to stick to items used as dippers.

Doneness The degree of a meat's protein coagulation, as determined by internal temperature, or the degree to which its connective tissues have broken down, as determined by tenderness.

Doria Garnished with cucumbers cooked in butter.

Double-Crème Cheese A rich cheese containing at least 60 percent fat.

Dough Arm Agitator attachment for electric mixers, used for mixing and kneading yeast doughs.

Doughs, Young and Old Dough is considered young when underfermented and old when overfermented.

Drawn With entrails removed.

Dressed (1) Poultry market form: killed, bled, and plucked. (2) Fish market form: viscera, scales, head, tail, and fins removed.

Dresser (dres say) To arrange prepared food on a plate or platter before serving.

Dried Whole Milk Whole milk that has been dried to a powder.

Drop Batter A batter that is too thick to pour but that drops from a spoon in lumps.

Dropped Cookies Cookies made from soft dough or batter dropped on baking sheets with a spoon, scoop, or other measuring implement.

Dry Aging The process of storing meats, typically large cuts, under carefully controlled conditions of temperature, humidity, and air circulation in order to develop flavor and tenderness.

Dry Cure A curing method in which the curing ingredients are packed or rubbed over the food.

Dry-Heat Methods Cooking methods in which heat is conducted to foods without the use of moisture.

Dubarry Garnished with or containing cauliflower.

Duchesse Potatoes (doo shess) Potato purée mixed with butter and egg yolks.

Duck A bird with a thick layer of fat under the skin and low yield, compared with chicken or turkey. It is usually roasted whole, though its parts are sometimes cooked separately.

Dugléré (dew glay ray) A classic preparation for fish incorporating white wine and tomato; named for a nineteenth-century chef.

Dumpling Any of a variety of small starch products made from soft dough or batter and cooked by simmering or steaming.

Duxelles (duke sell) A coarse paste or hash made of finely chopped mushrooms sautéed with minced shallots; used as a garnish or filling.

E

Éclair Paste Dough used to make éclairs and cream puffs.

E. coli A bacterium (*Escherichia coli*) that causes severe illness, either as an intoxication or an infection, typically due to consumption of raw or undercooked red meats, unpasteurized dairy products, or fish from contaminated water.

Edible Portion (EP) The quantity of a raw, uncooked food item after it is trimmed.

Egg Pasta Pasta containing at least 5 percent egg solids in addition to flour and water; usually in the form of flat noodles of various widths.

Elastin A type of connective tissue in meats that does not dissolve when cooked.

émincer (eh man say) To cut into very thin slices.

Empty Calorie A food that supplies few nutrients per calorie.

Emulsified Grind Referring to sausages made by processing meat and fat to a purée, usually with the addition of water or another liquid.

Emulsified Shortening An easy-spread shortening used when the weight of sugar in a cake batter is greater than the weight of flour.

Emulsion A uniform mixture of two unmixable substances—in the kitchen, usually liquids.

Emulsion Grind See *emulsified grind.*

En Croûte (on kroot) Wrapped in pastry.

Endosperm The starchy mass that forms most of a grain kernel.

Enriched Rice Rice that has received a coating of vitamins to compensate for some of the nutrients lost during milling.

Enrober (on robe bay) To coat; to completely cover with various ingredients, such as chocolate or dough.

Entremet (on tre may) Literally, "between courses"; originally a course served between the roast and the dessert. Today the term is used for various desserts, usually containing cream.

Entremetier (awn truh met yay) The cook who prepares vegetables, starches, soups, and eggs.

Epazote (ep ah so tay) A pungent herb used in Mexican cooking.

EP Weight Edible portion; the weight of an item after all trimming and preparation is done.

Escaloper (eh scal oh pay) To cut scallops; to cut meat or fish on a bias.

Escoffier, Georges-Auguste Great chef of the early twentieth century and the father of modern cookery.

Essence (ess sahns) Essence; concentrated extract, used as a flavoring (e.g., coffee essence).

Essential Amino Acid Any of 9 of the 20 amino acids, excluding those that can be made in the body, that must be included in the diet in order for the body to produce all the proteins it needs.

Essential Fatty Acid A fatty acid that must be consumed in the diet because it can't be made by the body.

Espagnole A sauce made of brown stock and flavoring ingredients and thickened with a brown roux.

Espresso, Expresso Strong, dark coffee made from beans roasted until almost black, ground very fine, and brewed under steam pressure.

Étuver (eh too vay) To stew or cook slowly; to gently cook a food covered with fat and a little water without changing the color of the ingredients.

Evaporated Milk Sterilized and canned whole or skim milk with about 60 percent of the water removed.

Evaporation The process by which water turns from a liquid to a gas at any temperature.

Executive Chef The manager of a large kitchen or food production department.

Extended Meal Service Service of a meal at which customers eat at different times.

Expediter Kitchen worker who accepts and transmits orders from waiters, calls for orders to be finished, inspects finished dishes, and passes them to the dining room staff.

Extract Flavorful oil or other substance dissolved in alcohol; used for flavoring.

F

Fabricate To cut raw meat into pieces.

Fabricated Cuts Raw meat that is cut up into pieces and trimmed.

Facultative Able to live and grow with or without the presence of oxygen; said of bacteria.

Farce Forcemeat stuffing; a mixture of ground ingredients (meat, herbs, vegetables) used to fill poultry, fish, vegetables, etc.

Farci (far see) Stuffed.

Farmstead Cheese Cheese made entirely with milk from a farmer's own herd or flock on the farm where the animals are raised.

Farro A grain that is the ancestor of modern wheat.

Fat Fish Fish with high fat content.

Fermentation The process by which yeast acts on carbohydrates to change them into carbon dioxide gas and alcohol.

Fermière (fair myair) (1) Garnished with carrots, turnips, onions, and celery cut into uniform slices. (2) Farm-made or farm-raised.

Fettuccine Flat egg noodles.

Fiber A group of indigestible carbohydrates found in grains, fruits, and vegetables.

Filet Boneless tenderloin.

Fillet Boneless side of fish; to remove the fillet from the fish bone.

Fines Herbes (feen zairb) Mixture of edible aromatic plants used as seasoning (parsley, tarragon, chives).

Fin Fish Fish with fins and internal skeletons.

Fish Carpaccio Very thin slices of firm, meaty fish served with garnishes and typically with a piquant sauce.

Fish Tartare A mixture of chopped raw fish, condiments, and seasonings.

Flaking An indication of doneness of cooked fish, when the flesh breaks apart into its natural separations.

Flaky Pie Dough Dough made by cutting or rubbing but not entirely blending fat into flour, leaving pieces of fat that contribute to a flaky texture.

Flamber (flahm bay) To light alcohol in a preparation (e.g., crêpes Suzette); to flame.

Flan (flahn) (1) Open pastry case or shell. (2) A custard tart.

Flatfish A flat type of fish, such as flounder and sole, with both eyes on one side of the head.

Flat Icing A mixture of 10X sugar, water, and, sometimes, corn syrup and flavoring. Also called *water icing*.

Flavonoids White pigments in vegetables and fruits.

Flavored Milk Milk containing flavoring ingredients.

Flavoring Adding a new flavor to a food, therefore changing or modifying the original flavor.

Flavor Profile The combination of flavors and aromas that make up the total taste impression of a dish.

Flaxseed A type of seed containing beneficial fiber and omega-3 fatty acids, used mostly in small quantities in breads and commercial breakfast cereal preparations.

Fleuron (flur rohn) Puff pastry piece cut into a crescent shape; served as decoration with fish dishes.

Florentine Garnished with or containing spinach.

Flow of Food The path that food travels in a food-service operation from receiving to serving.

Foam Icing Meringue made with boiling syrup that may contain stabilizing ingredients like gelatin. Also called *boiled icing*.

Foaming The process of beating eggs, with or without sugar, to incorporate air.

Foaming Method A cake-mixing method that incorporates whipped eggs into a batter.

Foie Gras (fwah grah) Liver of specially fattened geese and ducks.

Fond (fohn) Stock.

Fondant A smooth, creamy white icing or candy consisting of very finely crystallized sugar syrup.

Fond Lié A sauce made by thickening brown stock with cornstarch or a similar starch.

Fondu (fone dew) Melted.

Fondue, Swiss A dish consisting of melted Gruyère and Emmentaler cheeses and white wine into which cubes of bread are dipped and eaten. From the French word meaning "melted."

Food Cost Percentage The raw food cost, or portion cost, divided by the menu price.

Food Danger Zone The temperature range of 41°–135°F (5°–57°C) in which bacteria grow rapidly.

Forcemeat A seasoned mixture of ground meats and other foods, used as a filling or stuffing or as a base for terrines and pâtés.

Forestière Garnished with mushrooms.

Four-Hour Rule The sanitary practice of permitting foods to remain in the food danger zone for a cumulative total of no more than four hours between receiving and serving.

Four Parts of a Salad The base or underliner, body, garnish, and dressing.

Free-Range Referring to animals, usually poultry, that are allowed to move relatively freely outdoors as they are raised for market.

French Dressing Salad dressing made of oil, vinegar, and seasonings.

French-Style Ice Cream Ice cream containing egg yolks.

Fresh Not frozen, canned, or dried.

Fresh Yeast Form of yeast that is moist and perishable; also called *compressed yeast*.

Fricassée A white stew in which the meat is cooked in fat without browning before liquid is added.

Frire (freer) To deep-fry.

Frisée A variety of curly endive or chicory that is more tender and lighter in color than curly endive.

Frittata A flat, unfolded omelet.

Friture (free tur) (1) Deep fryer. (2) Deep-fried foods.

Frozen Yogurt Ice cream that contains yogurt in addition to the usual ingredients.

Fruit Pie Pie that contains fruit filling; usually has a top crust.

Fruit Salad A salad containing fruits as its main ingredients.

Fry To cook in hot fat.

Fudge Icing Rich cooked icing that is heavy and thick; it may be flavored with a variety of ingredients, and it is often somewhat like candy.

Full Slip Describes melons picked ripe, with no portion of the stems attached.

Fumet (foo may) A flavorful stock, usually fish stock.

Fusion Cuisine The use of ingredients and techniques from more than one regional or international cuisine in a single dish.

G

Galantine (ga lawn teen) A forcemeat wrapped in the skin of the animal from which it is made, such as a chicken or duck, or rolled into a cylinder without the skin.

Game Meat from animals and birds normally found in the wild; many game animals are now farm-raised.

Ganache A rich cream made of heavy cream and chocolate.

Garde Manger (gard mawn zhay) (1) The cook in charge of cold food production, including salads and buffet items. (2) The department of a kitchen in which these foods are prepared.

Garni Garnished; having had garnish added to it.

Garnish (1) Decorative edible item used to ornament or enhance the eye appeal of another food item. (2) To add such a decorative item to food.

Garniture (1) Garnish. (2) The act or process of garnishing.

Gastrique A mixture of caramelized sugar and vinegar, used to flavor sauces.

Gaufrette (go frett) Waffle.

Gazpacho A cold Spanish soup made of puréed raw vegetables.

Gelatinization The process by which starch granules absorb water and swell in size.

Gelatin Salad A salad made from flavored or unflavored gelatin, fruits, and, sometimes, vegetables.

Gelée Aspic jelly.

Genoise (zhen wahz) A French sponge cake.

Germ The portion of a whole grain consisting of a tiny embryo that forms the new plant once the seed sprouts.

Glaçage (glah sahj) Glaze; mixture of ingredients with a syrupy consistency, sweet or savory, used to coat pastries, candies, and certain savory foods.

Glace (glahss) (1) Ice cream. (2) Glaze; stock reduced until thick and syrupy.

Glacé (glah say) (1) Glazed (usually refers to vegetables). (2) Frozen. (3) Served with ice cream (e.g., meringue glacée).

Glace de Poisson Fish glaze; a reduction of fish stock.

Glace de Viande (glahss duh vee awnd) Meat glaze; a reduction of brown stock.

Glace de Volaille Chicken glaze; a reduction of chicken stock.

Glacer (glah say) To glaze; to cover or coat pastries with a glaze.

Glaze (1) A stock reduced until it coats the back of a spoon. (2) A shiny coating, such as a syrup, applied to a food. (3) To make a food shiny or glossy by coating it with a glaze or by browning it under a broiler or in a hot oven.

Gluten A substance made of proteins present in wheat flour that gives structure and strength to baked goods.

Glutinous or Sticky Rice A type of short-grain rice that becomes sticky and chewy when cooked.

Goujonnettes (goo zhone nett) Strips of fish, breaded and deep-fried.

Goulash A Hungarian stew flavored with paprika.

Grading Designation of the quality of a meat based on its texture, firmness, color, marbling, and the age or maturity of the animal.

Gram The basic unit of weight in the metric system; equal to about one-thirtieth of an ounce.

Grandmère (grahn mare) A classic garniture made from bacon, sautéed button mushrooms, and glazed pearl onions.

Granité (grah nee tay) A coarse, crystalline frozen dessert made of water, sugar, and fruit juice or other flavoring.

Gras-Double (grah doo bl') A type of beef tripe that is smooth rather than honeycombed.

Gratiner (gra tee nay) (1) To brown under the grill or salamander. (2) To glaze.

à la Grecque (ah la grek) Refers to a preparation of vegetables cooked in white wine.

Green Meat Meat that has not had enough time after slaughter to develop tenderness and flavor.

Green Wheat Wheat that is harvested while immature, then dried.

Griddle (1) To cook on a flat, solid cooking surface. (2) The surface itself.

Grill To cook on an open grid over a heat source.

Grillardin (gree ar dan) Broiler cook.

Grilled Sandwich A simple sandwich that is buttered on the outside and browned on the griddle, in a hot oven, or in a panini grill.

Griller (gree yay) To grill; to cook on a grill.

Grosse Pièce (gross pyess) Centerpiece of a buffet platter.

Guinea A domestically raised relative of the pheasant.

H

HACCP Hazard Analysis Critical Control Point; a food safety system of self-inspection designed to highlight hazardous foods and to control food handling with the goal of avoiding hazards.

Hacher (ah shay) To chop; to reduce to small pieces with a knife.

Half-and-Half Fresh cream that contains 10 to 18 percent fat, which is too low a proportion for the liquid to be called cream.

Hard Meringue Meringue made with up to twice as much sugar as egg whites.

Hare A game animal similar to rabbit, with dark red, lean meat.

Haricot (ahr ree co) Bean.

Haricot Bean A variety of green bean that is allowed to ripen until the seed is mature and dry.

Haricot Blanc (ahr ree co blahn) White bean.

Haricot Vert (ahr ree co vare) Green bean.

Hash (1) To chop. (2) A dish made of chopped foods.

Hazard A potentially dangerous food condition due to contamination, growth of pathogens, survival of pathogens, or presence of toxins.

HDL High-density lipoprotein, a compound that helps remove cholesterol from the bloodstream and eliminate it from the body, preventing heart disease.

Herbal Tea Beverage that is brewed like tea but made with herbs, spices, dried fruits, and other plant ingredients in place of or in addition to tea leaves.

Herb The leaves of certain plants, used in flavoring. These plants usually grow in temperate climates.

Hidden (Food) Cost The cost of supplementary ingredients, such as garnishes and condiments.

High-Fat Cakes Cakes with a relatively high fat content, best made by means of the creaming method or the two-stage method.

Holding Temperatures Temperatures at which certain products are kept for service or for storage.

Hollandaise A sauce made of butter, egg yolks, and flavorings (especially lemon juice).

Homard (oh mahr) Lobster.

Homemade Made on the premises.

Hominy Corn that has been treated with lye.

Homogenized Milk Milk that has been processed so the cream doesn't separate out.

Hongroise (ong grwahz) Hungarian style.

Hors d'Oeuvre A small food item usually served before or separately from a meal; see *appetizer.* French for "outside the work."

Hot Smoking A smoking method in which foods are smoked at a temperature high enough to cook or partially cook them.

Huile (weel) Oil.

Huile d'Olive (weel doe leave) Olive oil.

Huile d'Arachide (weel da rah sheed) Peanut oil.

Huile de Noix (weel de nwah) Walnut oil.

Hydrocolloid A colloid in which the continuous phase is water.

Hygroscopic Readily absorbing moisture.

I

Ice Frozen dessert made from fruit juices, water, sugar, and, sometimes, egg whites.

Icebox Cookies Cookies for which rolls of dough are made in advance, refrigerated, and then sliced and baked as needed.

Ice Cream Smooth, frozen mixture of milk, cream, sugar, flavorings, and, sometimes, eggs.

Iced Coffee Cold coffee made from double-strength brewed coffee to compensate for dilution by melting ice.

Ice Milk Frozen dessert similar to ice cream, but with a lower butterfat content.

Immersion Circulator A device for maintaining circulating water in a water bath at a precisely steady temperature.

Imported Coming from outside a country.

Incomplete Proteins Proteins that are missing or do not contain a high enough concentration of one or more of the essential amino acids.

Induction Cooktop A type of cooktop that works by using magnetic energy to make pots hot without getting hot itself.

Infection Disease, including much food-borne disease, caused by bacteria in the body.

Infrared A type of radiation used to cook food, such as when broiling.

Insoluble Fiber Fiber that absorbs less water than soluble fiber and forms bulk in the intestines. It is found in cell walls and other structural parts of plants.

Inspection Government-mandated examination of meat and other foods, intended to guarantee their wholesomeness and fitness for human consumption.

Instant Coffee A powdered, soluble extract from coffee beans made by brewing regular coffee and drying it.

Instant Dry Yeast A dry, granular form of yeast that does not have to be dissolved in water before use, as it absorbs water more quickly than regular dry yeast.

Instant Rice Rice that has been precooked and dried so it can be prepared quickly.

Instant Starch Starch that has already been cooked and dried. Also called *pregelatinized starch.*

Institution Meat Purchase Specifications (IMPS) A set of specifications followed by food-service suppliers that lists all meat and poultry cuts by number and describes them in detail for purchasers.

Integral Sauce A sauce based on the juices released during the cooking of a meat, poultry, fish, or vegetable.

Intoxication Disease caused by poisons that bacteria produce while they are growing in food.

Irradiation A process of exposing foods to radiation in order to kill bacteria, parasites, and other potentially harmful organisms.

Italian Meringue Meringue made by beating a hot sugar syrup into egg whites. It is the most stable meringue because the egg whites are cooked by the heat of the syrup.

J

Jambon (zhom bohn) Ham.

Jambonnette (zhom bo nett) Stuffed poultry leg made to resemble a small ham.

Jardinière (zhar din yair) Garnished with fresh garden vegetables, such as carrots, turnips, green beans, peas, and cauliflower.

Jasmine Rice A type of aromatic rice from Southeast Asia.

Jerk A traditional Jamaican cooking style in which meats are marinated in a special spice mixture containing allspice and Scotch bonnet peppers before grilling.

Joue (zhoo) Cheek (beef, pork, or veal).

Judic Garnished with braised lettuce.

Julienne (zhoo lee yen) (1) Cut into small, thin strips, about $\frac{1}{8} \times \frac{1}{8} \times 1$–2 inches (3 mm × 3 mm × 25–50 mm). (2) Garnished with foods cut in this manner.

Jus (zhoo) (1) Juice. (2) Unthickened juices from a roast. (3) Liquid made from pressing a fruit or vegetable.

Jus Lié Thickened juices from a roast.

K

Kamut An ancient relative of wheat, similar to spelt in composition and flavor.

Kasha Whole buckwheat groats.

Katsuobushi (kaht soo oh boo shee) Japanese shaved, dried bonito. Used to make stock.

Ketone Body A toxic compound that can form in the blood if fats are burned with no carbohydrates present.

Ketosis A condition in which the blood becomes unable to carry oxygen, sometimes the result of consuming insufficient carbohydrates.

Kilo- Prefix in the metric system meaning "one thousand."

Kombu A type of seaweed used to make Japanese soup stock.

L

Lacto-Ovo-Vegetarian Referring to a vegetarian diet that includes dairy products and eggs.

Lacto-Vegetarian Referring to a vegetarian diet that includes milk and other dairy products.

Lag Phase The time needed for bacteria to adjust to a new environment before they start to multiply.

Lait (lay) Milk.

Langoustine (lawn goo steen)/Langostino (1) Prawn; a type of crustacean. (2) A smaller relative of the rock lobster, marketed as *rock shrimp*.

Lard (1) The rendered fat of hogs. (2) To insert strips of pork fat into lean meats, using a larding needle, to prevent the meat from drying out during cooking.

Larder (lahr day) To lard.

Lardon (lahr doan) A small piece or strip of slab bacon.

Lasagna Broad, flat egg noodle.

Lasagne A baked, layered casserole made with lasagna noodles. ("Lasagne" is the plural of "lasagna.")

Latte Short for *caffe latte*. A mixture of 1 part espresso and 2 or more parts steamed milk.

LDL Low-density lipoprotein, the most important carrier of cholesterol. If too much is present in the bloodstream, it may deposit excess cholesterol inside the arteries and block blood flow.

Leading Sauce A basic sauce used in the production of other sauces. The five leading hot sauces are béchamel, velouté, espagnole, tomato, and hollandaise. Mayonnaise and vinaigrette are often considered leading cold sauces. Also called a *mother sauce*.

Lean Dough Dough that is low in fat and sugar.

Lean Fish Fish with low fat content.

Leavening The production or incorporation of gases in a baked product to increase volume and to produce shape and texture.

Legume (leh gyoom) (1) A plant that bears seed pods that split along two opposite sides when ripe. (2) French word meaning "vegetable."

Lemongrass A tropical grass with the aroma of lemon, used for flavoring.

Lentil A small, lens-shaped legume.

Lentille (lawn teey) Lentil.

Levain (le vanh) Starter dough; a dough made from live yeast and flour, used to make breads.

Levure (le vure) Yeast.

Liaison A binding agent, usually made of cream and egg yolks, used to thicken sauces and soups.

Lier (lee ay) To thicken; to change the consistency of a liquid by adding a roux, starch, egg, flour, or beurre manié.

Light Cream Fresh cream that contains 18 to 30 percent fat. Also called *table cream* or *coffee cream*.

Light Meat Poultry meat with lighter color and less fat and connective tissue than dark meat; breast meat.

Limiting Amino Acid An amino acid that is in short supply, therefore limiting the usefulness of other amino acids in forming complete proteins.

Line Cook A cook responsible for preparing or finishing hot à la carte items during service in a restaurant.

Lipid Any of a group of compounds that includes fats and cholesterol.

Lipoprotein Combinations of protein and fat that carry cholesterol and fat through the bloodstream. There are two important types: low-density lipoprotein (LDL) and high-density lipoprotein (HDL).

Liter The basic unit of volume in the metric system; equal to slightly more than 1 quart.

London Broil Flank steak or other cut of beef broiled rare and cut in thin slices.

Long-Grain Rice Rice with long, slender grains that stay separate and fluffy when properly cooked.

Lotte (lot) Monkfish.

Low-Fat Cakes Cakes with low fat but high egg and sugar content, best mixed using the egg-foam method.

Low-Fat Milk Milk with a fat content of 0.5 to 2 percent.

Lozenge Diamond-shape cut.

Lyonnaise (lee oh nez) Containing or garnished with onions.

M

Macaroni Dried noodle product made of flour (usually semolina) and water.

Macchiato (mah kee ah toe) Espresso topped with frothed milk.

Mâche A small, tender leafy green with a delicate taste.

Macédoine (mass e dwan) A mixture of vegetables or fruit, cut into small cubes.

Macérer (mass e ray) To macerate; to soak an element in alcohol in order to flavor it (usually done for pastry).

Magret (mah gray) The boneless breast of the moulard duck.

Maillard Reaction A complex chemical reaction that occurs when heated proteins react with carbohydrate molecules, resulting in browning and flavor changes.

Main-Course Salad A large salad, served as a full meal, containing a substantial portion of protein.

Maître d'Hôtel Butter (may truh doh tell) Compound butter containing parsley and lemon juice.

Major Mineral Mineral that must be consumed in relatively large amounts—greater than 100 milligrams daily—such as calcium chloride, magnesium, phosphorus, sulfur, sodium, and potassium.

Mandoline (man do leen) A slicer with several blades that allow for various cuts and thicknesses of fruits and vegetables.

Marbling The fat deposited within muscle tissue.

Margarine A manufactured product made of vegetable or animal fats and intended to resemble butter in taste, texture, and appearance.

Marinate To soak a food in a seasoned liquid.

Marsala A flavorful sweet to semidry wine from Sicily.

Matignon (mah teen yohn) A type of mirepoix containing ham in addition the vegetables and moistened with Madeira. A matignon is usually cooked in butter before being used.

Maturity The age of an animal, a major consideration when selecting among classes of poultry.

Mayonnaise A semisolid cold sauce or dressing consisting of oil and vinegar emulsified with egg yolks.

Mayonnaise Chaud-Froid A mixture of aspic jelly and mayonnaise, used like regular chaud-froid.

Mealy Pie Dough Dough for which the fat is blended into the flour until the mixture looks like coarse cornmeal.

Médaillon (may die yohn) Medallion; round slice of meat, fowl, fish, or crustacean, served hot or cold.

Medium-Grain Rice Rice with small, short kernels that become sticky when cooked.

Meringue A foam made of beaten egg whites and sugar.

Meringue Glacée Baked meringue shells served with ice cream.

Mesclun A mixture of tender baby lettuces.

Meter The basic unit of length in the metric system; slightly longer than 1 yard.

Metric System An international system of measurement used in most countries outside of the United States.

Meunière Referring to fish prepared by dredging in flour and sautéing, served with brown butter, lemon juice, and parsley.

Microorganism A tiny, usually single-celled organism visible only through a microscope. Some types can contaminate food and cause disease.

Microwave Radiation generated in special ovens and used to cook or heat foods.

Mie de Pain (mee de pan) Fresh bread crumbs.

Milk Fat Fat content of milk. Also known as *butterfat*.

Millet A small, round, yellow grain high in protein that is an important food source in much of Africa and Asia.

Milli- Prefix in the metric system meaning "one-thousandth."

Mince To chop into very fine pieces.

Minestrone Italian vegetable soup.

Minimum Internal Cooking Temperature The lowest temperature to which a food item must be heated and at which it must be held for a given time in order to be considered safe.

Minimum-Use Ingredient Ingredient used in very small quantities in the preparation of an operation's menu.

Mirepoix (meer pwah) A mixture of rough-cut or diced vegetables, herbs, and spices, used for flavoring.

Mise en Place (meez on plahss) French term meaning "put in place" or "everything in place." The setup for food production. All the preparation and organization that must be achieved before actual production can begin.

Miso A paste made of fermented soybeans, sometimes with the addition of other grains. Also called *bean paste*.

Mocha (moh kah) (1) A variety of Arabian coffee. (2) A mixture of espresso and hot chocolate or cocoa. (3) A flavoring made of coffee and chocolate.

Modified Straight Dough Method Mixing method used for rich sweet doughs to ensure even distribution of fat and sugar in the dough.

Moist-Heat Methods Cooking methods in which heat is conducted to foods by water or other liquid or by steam.

Molded Cookies Cookies for which the dough is divided into equal portions and then molded into the desired shape.

Molecular Gastronomy The study of the chemical and physical processes that occur in cooking and the application of this science to food preparation. In general, a range of techniques used by avant-garde chefs.

Mollusks Soft-bodied sea animals, usually inside a pair of hinged shells, such as clam and oyster.

Monounsaturated Fat A type of fat, liquid at room temperature, typically found in olive oil and canola oil.

Monter (mohn tay) (1) To whisk (egg whites, cream) in order to incorporate air and increase volume. (2) To add butter to a sauce in small pieces.

Monter au Beurre (mohn tay oh burr) To finish a sauce or soup by swirling in raw butter until it is melted.

Morille (moh reey) Morel mushroom.

Mornay A sauce made of béchamel and Gruyère cheese.

Mother Sauce See *leading sauce*.

Moulard A breed of duck with a thick, meaty breast, raised for its large, fatty liver.

Mousse A soft, creamy food, either sweet or savory, made light by the addition of whipped cream, beaten egg whites, or both.

Mousseline Forcemeat A forcemeat made of puréed fish, poultry, or meat, heavy cream, and, usually, egg whites.

Mozzarella A mild unripened cheese, used in pizzas and many other Italian-style dishes.

Muffin Method Mixing method in which combined liquid ingredients are mixed with combined dry ingredients; used for many muffins, other quick breads, and pancakes.

Multidecker Sandwich Sandwich made with more than two slices of bread.

N

à la Nage Literally, "swimming." A style of cooking and serving poached seafood and other items in their poaching liquids.

Nappé Having the proper texture in a sauce to lightly coat foods.

Napper (nap pay) To coat; to cover a food, savory, or sweet with a light layer of sauce, aspic, or jelly.

Natural Casing A sausage casing made from the intestines of meat animals.

Navarin A brown lamb stew.

New England Boiled Dinner A dish consisting of simmered corned beef and simmered vegetables, served together.

New Potato Immature potato with tender, thin skin, harvested while the plant top is still green.

Niçoise (nee swahz) (1) Prepared in the style of Nice, France. (2) Garnished with or containing tomato concassé cooked with garlic.

Nitrosamine A cancer-causing compound formed when meats containing sodium nitrate are subjected to high heat.

Noisette (nwah set) (1) Hazelnut. (2) Small, nut-size cuts made with a ball cutter. See also Beurre Noisette.

Nonfat Dry Milk Skim milk that has been dried to a powder.

Nouvelle Cuisine A modern style of cooking that emphasizes lightness of sauces and seasonings, shortened cooking times, and new and sometimes startling combinations of foods.

Nutrient Density The amount of nutrients per calorie in a food.

O

Oats Type of cereal grain that is most familiar in North America as a breakfast food.

Oblique Cut Diagonal cut used for long, cylindrical vegetables. Also called the *roll cut*.

Oeuf (euf) Egg.

Offal Variety meats.

Oie (wah) Goose.

Graisse d'Oie (gress dwah) Goose fat.

Oignon Brûlé (awn yohn broo lay) French for "burnt onion." A halved onion whose cut surface has been caramelized. Used to color stocks.

Oignon Piqué (awn yohn pee kay) An onion to which a bay leaf is fastened by piercing with a whole clove. Used to flavor simmering liquids.

Oil A fat that is normally liquid at room temperature.

Omega-3 Fatty Acid One of a family of essential unsaturated fatty acids that play vital roles in growth, in the immune system, in proper eyesight, and in cell structure. The family consists of three major nutrients: alpha-linolenic acid (ALA), docosahexaenoic acid (DHA), and eicosapentaenoic acid (EPA).

One-Stage Cooling Method A method for cooling hot foods to a safe, cold temperature within a limited time (no more than 4 hours).

One-Stage Method Method of mixing for low-moisture cookies, combining all ingredients in one step.

Oolong A greenish-brown, partially fermented tea.

Open-Faced Sandwich (Hot or Cold) Sandwich made with a single slice of bread and either hot or cold toppings.

Organic Grown or raised without chemical growth enhancers or medications or, for plants, without artificial fertilizers or pesticides.

Os (ohss) Bone.

Osmotolerant Yeast A special type of instant yeast that is often used in doughs with high sugar content, as this yeast performs better in sweet doughs.

Oven Spring The rapid rise of yeast goods in the oven due to the production and expansion of trapped gases as a result of the oven heat.

Over Easy Style of fried egg that is fried, flipped over, and cooked just until the white is set but the yolk is still liquid.

Over Hard Style of fried egg that is fried, flipped over, and cooked until the yolk is completely set.

Overhead Broiler A broiler that generates heat from above, with food items placed on a grate beneath the heat source.

Over Medium Style of fried egg that is fried, flipped over, and cooked until the yolk is partially set.

Overrun The increase in volume of ice cream or frozen desserts due to the incorporation of air while freezing.

Ovo-Vegetarian Referring to a vegarian diet that includes eggs in addition to plant products.

P

Paddle A flat-bladed mixer attachment used for general mixing.

Paner (pan ay) To coat a food with fresh or dry bread crumbs after dipping in an egg wash.

Pan Gravy A type of sauce made with the pan drippings of the meat or poultry it is served with.

Pan-Broil To cook uncovered in a sauté pan or skillet without fat.

Pan-Fry To cook in a moderate amount of fat in an uncovered pan.

Panini Plural of *panino*.

Panino (1) Originally, a small Italian sandwich made with a dinner roll. (2) A grilled sandwich usually made in a device that grills both sides at once while compressing the sandwich.

Panko (pahn ko) Coarse Japanese-style bread crumbs.

Pan-Smoking See Rangetop Smoke-Roasting.

Pan-Steaming Cooking (vegetables) in a small amount of water in a covered pan.

Papillote (pa pee yote) (1) Buttered or oiled paper, used to wrap fruits, meats, fish, etc., for cooking. (2) Paper frill used to decorate the ends of bones of certain poultry and meats.

en Papillote (on poppy yote) Wrapped in paper or foil for cooking so the food steams in its own moisture.

Parasite An organism that survives by living on or inside another organism, the host. It passes from one host organism to another as it completes its life cycle.

Parboil To cook partially in a boiling or simmering liquid.

Parboiled or Converted Rice Long-grain rice that is specially processed to provide higher vitamin and mineral content than regular milled white rice.

Parcook To partially cook by any method.

Parfait (1) A dessert consisting of alternating layers of ice cream and fruit or syrup in a tall, narrow glass. (2)

A frozen dessert made of a sweet egg-yolk foam and whipped cream.

Parisienne Large, spherical cuts made with a ball cutter.

Parmentier (par mawn tyay) Garnished with or containing potatoes.

Par Stock The inventory of goods an operation must have on hand to continue operating between deliveries.

Pasta General term for any shape of macaroni product or egg noodle.

Pasteurized Heat-treated to kill bacteria that might cause disease or spoilage.

Pastry Cream A thick custard sauce containing eggs and starch.

Pastry Flour Flour that is lower in gluten than bread flour but higher than cake flour; used for making pie and pastry doughs, quick breads, and other tender items.

Partridge A game bird usually weighing about 1 pound (500 g).

Pâte (paht) (1) Dough or batter. (2) Pasta. (French for "paste.")

Pâté (pah tay) A dish made of a baked forcemeat, usually in a crust.

Pâte à Choux (pot a shoo) A soft dough used for making éclairs and cream puffs. Also called *éclair paste*.

Pâte à Pâté Dough or pastry used to make a crust for pâté.

Pâté de Campagne A pâté or terrine characterized by a coarse texture.

Pâté en Croûte A pâté in a pastry crust.

Pathogen A bacteria microorganism that causes disease.

Pâtissier (pat tee see ay) Pastry chef.

Paysanne (pie zahn) In thin slices, roughly 1/2 × 1/2 × 1/8 inches (12 mm × 12 mm × 3 mm), round, square, or rectangular.

Peach Melba A sundae consisting of vanilla ice cream, a peach half, and Melba (raspberry) sauce.

Pear Belle Hélène Vanilla ice cream dessert topped with a poached or canned pear half, napped with chocolate sauce, and garnished with toasted almond slices.

Pearled Barley Type of barley that has been milled to remove the outer bran layers.

Persillade (pear see yahd) A mixture of bread crumbs, parsley, and garlic, used to coat roast meat items, usually lamb.

Pesco-Vegetarian Referring to a vegetarian diet that includes fish.

Pheasant Game bird with delicate, light-colored meat and a subtle flavor similar to that of chicken.

Philadelphia-Style Ice Cream Ice cream containing no eggs.

Physical Contamination The contamination of food with objects that might be nontoxic but might cause injury or discomfort.

Pigment Any substance that gives color to an item.

Pilaf Rice or other grain product that is first cooked in fat, then simmered in a stock or other liquid, usually with onions, seasonings, or other ingredients.

Pintade (pan todd) Guinea fowl.

Piquer (pee kay) (1) To pick; to lard a piece of meat, using a larding needle, in order to keep the meat from drying out during cooking. (2) To make small holes in dough, using a fork, to prevent it from rising too much.

Pizza A thin sheet of lean bread dough baked with a topping.

Pluche (ploosh) Sprig; small leaves picked off their stems (e.g., sprig of chervil).

Poach To cook gently in water or another liquid that is hot but not actually bubbling, 160°–180°F (71°–82°C).

Poêlé (pwah lay) (1) In classical cuisine, a type of roast or pot roast in which the item, usually poultry or white meat, is baked on a bed of matignon and basted with butter during cooking. (1) In modern French terminology, pan-fried.

Poissonier (pwah so nyay) Fish cook.

Polenta Italian-style cornmeal.

Polyunsaturated Fat Fat typically found in vegetable oils, such as corn oil, safflower oil, sunflower oil, and cottonseed oil; liquid at room temperature.

Portion Control The measurement of portions to ensure the correct amount of an item is served.

Portion-Controlled Cuts Ready-to-cook meats cut according to a customer's specifications.

Portion Cost Total cost of all the ingredients in a recipe divided by the number of portions served. Also called *raw food cost*.

Potage A general term for soup, though it is sometimes associated specifically with thick, hearty soups.

Potentially Hazardous Food A food that provides a good environment for the growth of hazardous bacteria.

Pot Roast A large cut of meat cooked by braising.

Poulet (poo lay) Chicken.

Poulette Allemande sauce flavored with mushrooms, parsley, and lemon juice.

Pour Batter A batter that is liquid enough to pour.

Poussin A young chicken weighing 1 pound (500 g) or less.

Pozole (poh soh leh) Whole-grain hominy.

Prague Powder #1 A blend of 6 percent sodium nitrite and 94 percent sodium chloride (table salt), used to cure meats; also called *curing salt* and *tinted curing mix*.

Prague Powder #2 A curing mixture similar to Prague Powder #1 but containing sodium nitrate in addition to sodium nitrite.

Praline (prah leen) Caramelized sugar with almonds or hazelnuts, ground to a smooth paste; used to flavor and decorate pastries.

Prawn Large shrimp or langoustine.

Pressure-Fry To deep-fry in a special covered fryer that traps the steam given off by the foods being cooked, increasing the pressure inside the kettle.

Primal Cuts One of the primary divisions of meat quarters, foresaddles, hindsaddles, and carcasses as they are broken down into smaller cuts.

Primary Flavor The basic flavor of a main ingredient of a dish; one of the components of a flavor profile.

Primeur (pree mur) Garnished with fresh spring vegetables such as carrots, turnips, green beans, peas, cauliflower, and small potatoes.

Princesse Garnished with asparagus.

Printanière (pran tawn yair) Garnished with fresh spring vegetables such as carrots, turnips, pearl onions, peas, green beans, and asparagus.

Prix Fixe (pree feex) French term meaning "fixed price"; referring to a menu offering a complete meal, with a choice of courses, for one given price.

Process Cheese A product made by grinding and melting one or more cheeses, blending them with other ingredients, and pouring the mixture into molds to solidify.

Professionalism An unwritten code of behavior and set of attitudes followed by the successful food-service employee.

Profiterole A tiny round pastry made from éclair paste, filled with savory fillings and served as an hors d'oeuvre or filled with ice cream and served as a dessert.

Progressive Grinding Process of grinding meat, starting with the largest die and continuing with the next smaller die, until the desired texture is achieved.

Proofing The process during which the volume of shaped dough increases due to fermentation.

Protein Large molecules consisting of amino acids, which our bodies and the cells in our bodies need to function properly. Proteins are essential for growth, for building body tissues, and for basic body functions.

Provençale (pro vawn sal) Garnished with or containing tomatoes, garlic, parsley, and, sometimes, mushrooms and olives.

Puff Pastry A light, flaky pastry made from a rolled-in dough and leavened by steam.

Pullman Loaf A long, rectangular loaf of bread.

Pumpernickel (1) Coarsely ground rye flour. (2) Bread made with this flour.

Punching A method of deflating dough by pulling up the dough on all sides, folding over the center, pressing down, and turning the dough upside down in the bowl. Also called *folding*.

Purée (1) A food product that has been mashed or strained to a smooth pulp. (2) To make such a pulp by mashing or straining a food.

Purée soup A soup naturally thickened by puréeing one or more of its ingredients.

Q

Quail A small game bird, now domestically raised, usually weighing 6 ounces (175 g) or less.

Quatre épices (kaht ray peace) A spice mixture commonly used to season sausages and forcemeats; French for "four spices."

Quenelle (kuh nel) (1) Dumpling made of meat, poultry, or fish purée, mixed with egg white and cream. Usually molded with two spoons into an egg shape. (2) Oval three-sided shape made using two large spoons.

Quiche A savory tart or pie consisting of a custard baked in a pastry shell.

Quick Bread A bread leavened by chemical leaveners or steam rather than yeast.

Quinoa (keen wah) A tiny, ivory-colored, round grain native to the South American Andes, high in good-quality protein and lower in carbohydrates than other grains.

R

Radiation The transfer of energy by waves, such as infrared or light waves.

Rafraîchir (rah fray sheer) To refresh, to cool, to chill; to quickly cool in cold water a food that has been blanched.

Raft The coagulated clearmeat that forms when stock is clarified.

Rangetop Smoke-Roasting Procedure using wood chips in a closed container to create smoke and cook small, tender, quick-cooking items. Also called *pan-smoking*.

Râpé (rah pay) Grated.

Ratatouille (ra ta twee) A Southern French vegetable stew of onions, tomatoes, zucchini, eggplant, and green bell peppers.

Ratite A category of farm-raised birds including ostrich and emu.

Ravier (rahv yay) Oval relish dish.

Ravioli Dumplings consisting of filled egg noodles.

Recipe A set of instructions for producing a certain dish.

Reduce To cook by simmering or boiling until quantity is decreased; often done to concentrate flavors.

Reduction (1) A liquid concentrated by cooking it to evaporate part of the water. (2) The process of making such a liquid.

Regular Shortening Shortening with good creaming ability, which allows a large quantity of air to be mixed into the batter to give it lightness and leavening power.

Relish (1) A type of appetizer consisting of raw or pickled vegetables. (2) A mixture of chopped vegetables and, sometimes, fruits, at least one of which has been pickled in vinegar or a salt solution.

Remonter (re moan tay) To remount; to repair a sauce or a cream that has separated by returning it to its proper appearance and texture.

Remouillage (ray mwee yahzh) A stock made from bones that were already used once to make stock.

Rice Noodle Noodle made from rice flour or starch.

Rice Stick Another term for rice noodle.

Rice Vermicelli Fine, hairlike rice noodles often cooked by deep-frying them dry.

Ricotta An Italian-style cheese similar to cottage cheese but smoother, moister, and sweeter in flavor.

Rillettes (ree yet) A seasoned mixture of meat, such as pork, and fat, mashed to a paste; served as an appetizer.

Ring-Top Range A type of flattop that has removable rings, allowing access to more intense heat from the flames underneath.

Ripening Process that converts freshly made curds into distinctive, flavorful cheeses.

Risotto A moist Italian dish of rice cooked in butter and stock.

Rissolé (riss oh lay) Browned. Often referring to potatoes cut in small shapes, parboiled, and browned in hot fat.

Roast To cook foods by surrounding them with hot, dry air in an oven or on a spit over an open fire.

Roaster A young chicken of either sex, usually 3–5 pounds (1.6–2.3 kg).

Rock Shrimp Small crustacean resembling shrimp, also referred to as *langoustine* or *langostino*.

Roe Fish eggs.

Rösti Potatoes Boiled potatoes that are grated, formed into small cakes, and pan-fried until crisp.

Rolled Cookies Cookies cut from a stiff rolled-out dough.

Rolled-in or Laminated Dough Dough into which a fat was incorporated in many layers by means of a rolling and folding procedure.

Rolled Oats Whole grains that have been steamed until soft and then flattened between rollers, which reduces their cooking time considerably.

Rondeau A round, broad, shallow, heavy-duty pot with straight sides, used for browning, braising, and stewing meats. Also called a *brazier*.

Rondelle (ron dell) A round or bias-round cut that varies in diameter and thickness.

Roquefort A blue-veined cheese made in Roquefort, France, from sheep's milk.

Rotisserie An item of cooking equipment that slowly rotates meat or other foods in front of a heating element.

Rôtisseur (ro tee sur) Cook who prepares roasted, braised, and broiled meats.

Rough Prep The preliminary processing of ingredients to the point at which they can be used in cooking.

Round Fish A round-shaped fish, such as cod and trout; distinct from flatfish, such as flounder.

Roux A cooked mixture of equal parts flour and fat.

Royal Icing An icing made of confectioners' sugar and egg whites, used for decorating.

Rubbed Dough Method Mixing method for pie doughs that requires fat to be rubbed into the sifted dry ingredients.

Russet Starchy potato often used for baking and deep-frying.

Rye Blend A mixture of rye flour and hard wheat flour.

S

Sabayon (sa ba yohn) A thick, frothy sauce, either sweet or savory, made by whisking egg yolks and liquid over low heat.

Sachet d'épices (sa shay day peace) A mixture of herbs and spices tied in a cheesecloth bag. Often called simply *sachet*.

Salamander Small broiler used primarily for browning or glazing the tops of certain items.

Salmonella A food-borne disease spread by improper food handling and inadequate sanitation.

Salsa The Spanish and Italian term for "sauce." A mixture of raw or cooked chopped vegetables, herbs, and, occasionally, fruits.

Sanitize To kill disease-causing bacteria, usually by means of heat or chemical disinfectants.

Saturated Fat A fat that is normally solid at room temperature.

Sauce A flavorful liquid, usually thickened, used to season, flavor, and enhance other foods.

Saucier (so see ay) Sauce cook; prepares sauces and stews and sautés foods to order.

Sauerbraten A German dish consisting of beef marinated in and then cooked with vinegar and other ingredients.

Sausage A mixture of ground meat, usually pork, and seasonings, usually stuffed into casings.

Sausage, Cured A sausage that contains nitrites or nitrates.

Sausage, Fresh A sausage that contains no nitrites or nitrates.

Sausage, Smoked A cured sausage that is hot smoked or cold smoked.

Sauté To cook quickly in a small amount of fat.

Sauteuse A slope-sided sauté pan.

Sautoir A straight-sided sauté pan.

Scampi A kind of shellfish similar to large shrimp. In the United States, the term is often used for large shrimp, especially if broiled with garlic butter.

Scorching The burning of milk due to the coagulation of proteins on the bottom of the pan.

Sear To brown the surface of a food quickly at high temperature.

Seasoning (1) Enhancing the natural flavor of a food without significantly changing its flavor. (2) Any ingredient, such as salt, used for this purpose.

Semolina A hard, high-protein flour often used for the best-quality macaroni products.

Separate-Course Salad A light salad, usually with delicate greens and light dressing, served after the main course.

Set Meal Service Service of a meal at which all customers eat at one time.

Seviche A preparation of raw seafood marinated in an acid mixture, which coagulates the protein so the texture of the raw fish resembles that of cooked fish. Also called *ceviche*, it is native to Latin America.

Shallow Poaching Poaching an item only partially submerged in the liquid.

Sheet Cookies Cookies for which the dough is spread out and baked in sheet pans and then cut into individual shapes.

Shell (Egg) The hard but fragile outer covering of an egg. Because it is porous, odors and flavors can be absorbed by the egg, which also loses moisture over time even when intact.

Shellfish Fish with external shells but without internal bone structure.

Sherbet Frozen dessert made from fruit juice, water, and sugar. It also may contain milk, cream, or egg whites.

Shirred Egg Egg baked in a shallow buttered dish.

Short Having a high fat content, which makes the product (such as a cookie or pastry) crumbly and tender.

Short Dough A kind of pastry or cookie dough that is richer than regular pie pastry and contains butter, sugar, and eggs.

Shortening (1) Any fat used in baking to tenderize the product by shortening gluten strands. (2) A white, tasteless, solid fat formulated for baking or deep-frying.

Short-Grain Rice Rice with small, round kernels that become sticky when cooked. It is used for rice pudding and rice molds as well as for making sushi and daily eating in Japanese cuisine.

Short-Order Cook The cook responsible for the preparation of foods that are quickly prepared to order during service time. May handle the broiler, deep fryer, griddle, sandwich production, and sautéed items.

Shred To cut into thin but irregular strips, either with the coarse blade of a grater or with a knife.

Sieve Size Size of individual pieces, usually of canned vegetables.

Silverskin A thin layer or sheet of connective tissue covering parts of some muscles.

Simmer To cook in water or other liquid that is bubbling gently, 185°–200°F (85°–93°C).

Simple Sandwich (Hot or Cold) Sandwich made with two slices of bread and either hot or cold fillings.

Simple Syrup A solution of equal weights of sugar and water.

Singer (san jay) To sprinkle with flour at the start of cooking in order to eventually give a certain consistency to the sauce.

Sirniki Russian pan-fried cheesecakes.

Skim (Nonfat) Milk Milk that has had all or nearly all of the fat removed.

Slurry A mixture of raw starch and cold liquid, used for thickening.

Small Sauce A sauce made by adding one or more ingredients to a leading sauce.

Smoke Point The temperature at which fats begin to deteriorate rapidly and smoke when heated.

Smoker An enclosure used for making hot-smoked and cold-smoked foods.

Smoke-Roast To cook with dry heat in the presence of wood smoke.

Soba Thin Japanese noodles made with buckwheat and wheat flour.

Sodium Nitrate A compound, $NaNO_3$, used to cure certain meats, especially air-dried meats.

Sodium Nitrite A compound, $NaNO_2$, used to cure meats.

Soft Meringue Meringue with a lower sugar content than hard meringue, typically used for pie topping.

Soft Pie Pie with custard-type filling, usually baked with a single crust.

Soft-Shell Crab A just-molted crab whose new shell has not yet hardened.

Solanine A poisonous substance found in potatoes that have turned green.

Soluble Fiber Fiber that absorbs water and forms a kind of gel. It is found inside and between plant cells.

Somen Thin, white Japanese noodles made from wheat flour and water, usually packed in small bundles.

Sorbet (sor bay) Sherbet, usually made without milk products.

Soufflé A light, fluffy baked egg dish consisting of a base (such as a heavy white sauce) mixed with egg yolks and flavoring ingredients into which beaten egg whites are folded just before baking. May be sweet or savory.

Sour Cream Cream that has been fermented or cultured by added lactic acid bacteria, making it thick and slightly tangy.

Sous Chef (soo shef) Cook who supervises food production and who reports to the executive chef.

Sous Vide (soo veed) French for "under vacuum"; a technique for cooking vacuum-packed foods at precise temperatures.

Soy Milk A liquid milk substitute made by soaking dried soybeans, grinding them with water, and finally straining the mixture.

Spaetzle Small dumplings or noodles made from a thin egg and flour batter.

Specialty Cheese Cheese in limited production, made with particular attention to natural flavor and texture profiles.

Spelt A type of wheat grain similar to farro.

Spice Any part of a plant, other than the leaves, used in flavoring foods.

Sponge Method (1) Bread-mixing method involving fermenting part of the flour and water before adding the remaining ingredients. (2) Cake-mixing method involving folding flour and other ingredients into an egg foam.

Spread The capacity of cookie dough to expand during baking.

Squab Young domestically raised pigeon.

Staling The change in texture and aroma of baked goods due to the loss of moisture by its starch granules.

Standard Breading Procedure The procedure for coating a food product with bread crumbs (or other crumbs or meal) by passing it through flour, then egg wash, then crumbs.

Standardized Recipe A set of instructions describing the way a particular establishment prepares a particular dish.

Staphylococcus, Staph A bacterium that causes food-borne disease by producing a toxin or poison in improperly stored foods.

Static Menu A menu that offers the same dishes every day.

Station Chef Cook in charge of a particular department in a kitchen or food production facility.

Steam To cook by direct contact with steam.

Steel-Cut Oats Whole-grain oats cut into small pieces; usually cooked as porridge.

Stew (1) To simmer a food or foods in a small amount of liquid that is usually served with the food as a sauce. (2) A dish cooked by stewing, usually one whose main ingredients are cut in small pieces.

Stewing Cooking small pieces of meat by simmering or braising.

Still-Frozen Frozen without being churned in an ice-cream freezer.

Stock A clear, thin—that is, unthickened—liquid flavored with soluble substances extracted from meat, poultry, or fish, their bones, and vegetables and seasonings.

Straight Dough Method Bread-mixing method involving mixing all ingredients, including presoaked yeast, in one step.

Strength of Acidity The acetic acid content of vinegar, which determines the tartness of vinegar and of dressings made from it.

Streusel (stroy zel) A crumbly topping for baked goods, consisting of fat, sugar, and flour rubbed together.

Strong Flour Flour with a high protein or gluten content.

Submersion Method Method by which fish and other items are cooked completely submerged in court bouillon.

Sucrose Chemical name for the sugar that constitutes table sugar.

Suer (soo ay) To sweat; to gently cook vegetables in a little fat, without coloring them, in order to bring out their flavor.

Sundae Dessert consisting of one or more scoops of ice cream or sherbet in a dish or glass, topped with syrups, fruits, toppings, and garnishes.

Sunny Side Up Style of fried egg that is cooked slowly without flipping until the white is completely set but the yolk is still soft and yellow.

Supporting Flavor Flavors of seasoning and flavoring ingredients and other secondary ingredients of a dish; flavors that support and enhance the primary flavors.

Suprême (soo prem) (1) The breast part of the fowl. (2) A filet of fish.

Suprême Sauce A sauce made of chicken velouté and heavy cream.

Surimi A processed seafood product manufactured to resemble shellfish such as crab.

Sushi A seasoned Japanese short-grain rice that is often, but not always, garnished with raw fish.

Sweat To cook in a small amount of fat over low heat, sometimes covered.

Sustainable Agriculture Methods of raising healthful food in ways that are profitable to farms and farming communities and that provide living wages and benefits to workers while at the same time preserving and enhancing the soil, water, and air.

Sweetbreads The thymus glands of calves and young animals, used as food.

Sweet or Unsalted Butter Butter that contains no salt. It has a fresher, sweeter taste than salted butter and is more perishable as well.

Swiss Meringue Meringue made from egg whites and sugar warmed over a double boiler while beating.

Swiss Steak Beef round steaks braised in brown sauce.

T

Table d'Hôte (tobble dote) (1) Referring to a fixed-price menu with no choices. (2) Referring to a menu on which prices are listed for complete meals rather than for each separate item.

Tagine (tah zheen) A spiced stew from North Africa, originally made in an earthenware dish with a conical lid; the dish is also called a *tagine*.

Tamis A drum sieve, consisting of a screen mesh in a round metal or wood frame.

Tamiser (tah mee zay) To sift; to press through a fine drum sieve.

Tang The portion of a metal knife blade that is inside the handle.

Tapa Any of a variety of Spanish-style appetizers, intended to be served with wine or other drinks.

Tapenade (ta peh nahd) Purée of black olives, anchovies, and olive oil.

Tasting Menu A type of fixed-price menu designed to showcase the chef's art by presenting a series of small courses.

Tatsoi A leafy vegetable or salad green related to mustard and watercress.

Tea Sandwich Any small, fancy sandwich generally made from light, delicate ingredients and bread trimmed of the crust.

Tempeh A fermented soy product with a dense, meaty texture.

Temper To raise the temperature of a cold liquid gradually by slowly stirring in a hot liquid.

Temporary Emulsion An unstable emulsion whose components will eventually separate.

Teriyaki A grilled or pan-broiled food finished with a soy sauce–based glaze.

Terrine (tare reen) (1) A deep rectangular mold, traditionally made of white porcelain, used to cook seasoned ground meats, fish, or poultry. (2) The food cooked in such a mold.

Textured Vegetable Protein (TVP) A high-protein product made from defatted soy flour, processed to give it a spongelike texture; it is available unflavored or flavored to resemble various meats.

Timbale (tam ball) (1) A mold in the shape of a large thimble. (2) Type of dough shaped as a container, baked, and filled with various foods.

Tinted Curing Mix (TCM) See Prague Powder #1.

Tofu A bland, white food made by coagulating soy milk. Also called *bean curd*.

Tomalley The liver of lobsters and some other shellfish.

Tournant (toor nawn) Cook who replaces other station cooks; relief cook or swing cook.

Tourné To cut a vegetable into a neat seven-sided oval shape.

Tournedos (toor nuh doe) A small beef steak cut from the tenderloin.

Tourner (tour nay) Literally, "to turn." (1) To give certain vegetables a regular long shape, using a knife (see Tourné). (2) To mix ingredients using a circular motion.

Toxin-Mediated Infection An infection that occurs when pathogens enter the body and multiply, producing toxins in the body.

Trace Mineral An essential mineral nutrient that must be consumed in small amounts—less than 100 milligrams daily—such as chromium, copper, fluoride, iodine, iron, manganese, molybdenum, selenium, and zinc.

Trans Fat A solid fat, usually manufactured by hydrogenation, that limits the body's ability to rid itself of cholesterol.

Treviso A red-leafed relative of radicchio and Belgian endive with elongated leaves.

Trichinosis A food-borne disease caused by a parasite sometimes found in undercooked pork.

Tripe The muscular stomach lining of beef or other meat animals.

Triple-Crème Cheese A very rich cheese with at least 75 percent fat, dry weight.

Triticale A high-protein hybrid of wheat and rye with a nutty, sweet flavor.

Truffer (troo fay) (1) To add chopped truffles to a dish, stuffing, or foie gras. (2) To slide a thin slice of truffle under the skin of poultry.

Truite au Bleu Poached trout that was alive until cooking time and that turns blue when cooked in court bouillon.

Trunnion Kettle Steam-jacketed kettle that can be tilted for emptying.

Trussing Tying poultry into a compact shape for cooking.

Tunneling A condition of muffin products characterized by large, elongated holes; caused by overmixing.

Turban (toor bahn) (1) A type of ring mold. (2) Food prepared in such a mold.

TVP See Textured Vegetable Protein.

Two-Stage Cooling Method A method for cooling hot foods to a safe, cold temperature within two consecutive periods totaling no more than 6 hours.

Two-Stage Method Mixing method for high-fat, high-ratio cakes using emulsified shortening; it requires the addition of liquids in two stages.

U

Udon Thick white Japanese noodles made from wheat flour.

Ultra-High-Temperature (UHT) Pasteurization Process by which a product is heated at a temperature higher than for ultra-pasteurized products. The resulting product is packaged in shelf-stable, sterile cartons.

Ultra-Pasteurized Describes a product whose shelf life has been expanded through heating at a much higher temperature than for normal pasteurization. This process destroys nearly all organisms that cause spoilage.

Umami Known as the *fifth taste*, often described as the perception of meatiness.

Univalve A mollusk with a single shell, such as abalone.

Unsalted Butter See Sweet or Unsalted Butter.

Unsaturated Fat A fat that is normally liquid at room temperature.

V

Vapeur (va purr) Steam.

Variety Meats Organs, glands, and other meats that don't form a part of the dressed carcass.

Vegan Referring to a vegetarian diet that omits all animal products, including dairy products and eggs.

Vegetable Salad A salad containing vegetables other than lettuce or other leafy greens as its main ingredients.

Vegetable Soup A clear, seasoned stock or broth with the addition of one or more vegetables. Meat or poultry products and starches are sometimes also added.

Vegetarian Referring to a diet consisting entirely or mostly of foods derived from plants.

Velouté A sauce made by thickening white stock with a roux.

Venison The meat of wild or farm-raised deer.

Venting Allowing circulation or escape of a liquid or gas, such as by setting a pot of hot stock on blocks in a cold-water bath so the cold water can circulate around the pot.

Viande (vee awnd) French for "meat."

Vichyssoise (vee she swahz) Cold purée of leek and potato soup with cream.

Vin Wine.

Vin Blanc White wine.

Vin Rouge Red wine.

Vinaigre (vee negre) Vinegar.

Vinaigrette Dressing or sauce made of oil, vinegar, and flavoring ingredients.

Vitamin Any of a group of compounds that are present in foods in very small quantities and that are necessary for regulating body functions.

Volaille (voe lye) Poultry.

Blanc de Volaille (blahn duh voe lye) Chicken breast.

Cuisse de Volaille (kweese duh voe lye) Chicken leg.

Volatile Evaporating quickly when heated.

W

Wash (1) To brush or coat a food item with a liquid such as egg wash or milk. (2) The liquid used in this procedure.

Water Activity (a$_w$) Measured by the availability of water to bacteria. The scale runs for 0 (no water available) to 1.0.

Water Ice See Ice.

Waxy Potato A young potato high in sugar and low in starch.

Weak Flour Flour with a low protein or gluten content.

Well Mixing Method Mixing method wherein a well is made in the dry ingredients so the liquid ingredients, when poured into it, are confined.

Welsh Rabbit A dish made of melted cheddar cheese and, usually, ale or beer. Sometimes called *Welsh rarebit*.

Wheatberry The whole wheat grain without the hull.

Whipping Cream Fresh cream with a fat content of 30 percent or more that can be whipped into a foam.

White The part of an egg that is clear and soluble when raw but white and firm when coagulated. It consists primarily of albumin protein and also contains sulfur.

White Pekin The most common breed of domestic duck in the United States.

White Roux Roux cooked for a just few minutes, just enough to cook out the raw taste. It is made from butter and flour and is used for béchamel and other white sauces based on milk.

Whitewash A thin mixture or slurry of flour and cold water.

Whole Milk Fresh milk as it comes from the cow with nothing removed or added, usually containing about 3$\frac{1}{2}$ percent milk fat.

Whole Wheat Flour Flour made by grinding the entire wheat kernel, including the bran and germ.

Wild Rice Not a rice but rather the seed of an unrelated grass native to the northern United States and Canada. Scarce and expensive, it has long, slender, hard, dark brown grains with a distinctive nutty flavor.

Winterized Oil Vegetable oil that stays clear and liquid when refrigerated.

Wire Whip A hand tool or mixer attachment used for whipping tasks, such as beating cream and eggs and making mayonnaise.

Working Chef The cook in charge of operations of a kitchen not large enough to have an executive chef. May work one or more production stations.

Wrap A sandwich in which the filling is wrapped, like a Mexican burrito, in a large flour tortilla or similar flatbread.

Y

Yield (1) The amount of usable meat in proportion to fat. (2) The edible amount remaining after trimming any food. (3) The total quantity of food produced using a given recipe.

Yield Grade The designation of the quality of meat according to the amount of usable meat in proportion to fat.

Yield Test A test to determine the cost per unit of weight of meat after trimming and boning.

Yogurt Milk cultured by special bacteria and having a custardlike consistency.

Yolk The yellow part of the egg; it has a high fat and protein content and contains iron and several vitamins.

Z

Zest The colored part of the peel of citrus fruits.

Zester (zes tay) To zest; to remove the colored part of the skin of a citrus fruit (e.g., orange, lemon).

SUBJECT INDEX

Note: Please see Recipe Index that follows for specific recipes.

RECIPE INDEX